Although aging is not without problems, older adults can age successfully by adopting strategies and behaviors for optimal aging.

Development in Late Adulthood

Physical (Chapter 14)

Average life expectancy has increased steadily over the last few decades. Genes and environment affect longevity.

Endurance, vision, and hearing decline in old age. Sense of smell declines after age 70 in many people.

The risk of cardiovascular disease (e.g., heart attack, stroke, hypertension) increases with age. However, risks are affected by lifestyle, and overall death rates from these diseases have declined in recent decades.

Older adults suffer shortness of breath and have increased risk of chronic obstructive pulmonary disorder (COPD). Emphysema and asthma are two common forms of COPD.

Sleep disturbances as well as cancer risk increase with age.

Biological Theories of Aging	
Programmed cell death	Aging is biologically or genetically programmed.
Wear-and-tear	Aging is caused by the body's systems wearing out.
Cellular	Processes within cells cause buildup of harmful substances or cell deterioration and damage over time.

Cognitive (Chapter 14)

Age differences in attention tasks depend on the level of difficulty. On easy tasks, older and younger adults have few differences; on difficult tasks, younger adults do better.

Older adults' psychomotor speed is slower than that of young adults. Practice, task expertise, and being physically fit lessen the amount of slowing.

Working memory typically declines with age. Older adults usually do worse on tests of episodic recall; recognition tasks are less affected by age. Semantic memory and implicit memory are affected little by aging. Memory aids can help older adults compensate.

Contrary to popular belief, wisdom is correlated with life experience, not age. After early middle age, creativity declines with increasing age, but an individual's creative peak varies across disciplines and occupations.

© Lynn Morales/Alamy

Socioemotional (Chapter 14)

Rates of depression are reduced from young adulthood to old age, and symptoms vary by age. Treatment for older adults includes medication, behavior therapy, and cognitive therapy.

Anxiety disorders increase in older adulthood due to loss of health, relocation stress, isolation, and fear of losing independence. Medication and psychotherapy are effective treatments.

© ALAN ODDIE/Photo Edit

Dementia causes severe cognitive impairment. Alzheimer's disease is the most common form of irreversible dementia and is fatal. Various interventions can improve the patient's quality of life.

D1464987

Three Factors That Contribute to Wisdom

Factor	Trait
General personal condition	Mental ability
Specific expertise condition	Practice or mentoring
Facilitative life context	Education or leadership experience

Older adults' life satisfaction is strongly related to the number and quality of their friendships. Sibling relationships are important in old age.

Socioemotional (Chapters 15, 16)

Erikson proposed that older adults struggle between *integrity versus despair,* primarily through a life review.

Whether older adults achieve subjective well-being depends on hardiness, chronic illness, marital status, social network, and stress.

Older adults use religion and spiritual support more often than any other strategy to cope with problems. Older adults who are committed to their faith have better physical and mental health.

Two Theories of Psychosocial Aging

Continuity Theory	Older adults cope with daily life by applying familiar strategies based on past experience to maintain internal and external structures.
Competence and Environmental Press Theory	Older adults adapt optimally when there is a balance between their ability to cope and the level of environmental demands placed on them.

Most people retire by choice, although some are forced to because of health problems or job loss. Financial security, good health, and friends are correlated with satisfaction with retirement. Most retirees maintain their health, friendships, and activity levels.

Long-term marriages tend to be happy until one partner develops serious health problems. Caring for a spouse considerably strains the relationship.

Abuse and neglect of older adults is an increasing problem. Most perpetrators are family members.

Older adults are the most politically active age group.

Older adults are less anxious about death and deal with it better than any other age group.

Dealing with grief can take 1 to 2 years; unexpected death is usually more difficult to handle. Normal grief reactions include sorrow, sadness, denial, disbelief, guilt, and anniversary reactions.

The death of one's partner is a deeply personal loss.

Human Development

A Life-Span View SIXTH EDITION

Robert V. Kail
Purdue University

John C. Cavanaugh
Pennsylvania State System of Higher Education

WADSWORTH
CENGAGE Learning

Australia • Brazil • Japan • Korea • Mexico • Singapore • Spain • United Kingdom • United States

Human Development: A Life-Span View,
Sixth Edition, International Edition
Robert V. Kail and John C. Cavanaugh

Publisher: Jon-David Hague

Executive Editor: Jaime Perkins

Senior Development Group Manager:
  Jeremy Judson

Senior Developmental Editor:
  Kristin Makarewycz

Developmental Editor: Barbara Armentrout

Consulting Development Editor: Trina Tom

Assistant Editor: Paige Leeds

Editorial Assistant: Jessica Alderman

Senior Media Editor: Mary Noel

Executive Marketing Manager: Kimberly Russell

Marketing Manager: Christine Sosa

Marketing Communications Manager:
  Laura Localio

Senior Content Project Manager: Pat Waldo

Design Director: Rob Hugel

Senior Art Director: Vernon T. Boes

Manufacturing Planner: Judy Inouye

Rights Acquisitions Specialist: Dean Dauphinais

Production Service: Cassie Carey, Graphic
  World Inc.

Text Designer: Jeanne Calabrese

Photo Researcher: Josh Garvin,
  Bill Smith Group

Text Researcher: Isabel Saraiva

Copy Editor: Graphic World Inc.

Cover Designer: Denise Davidson

Cover photos: Sam Edwards/OJO images/
  GettyImages; Digital Vision

Compositor: Graphic World Inc.

Library of Congress Control Number: 2011935668

International Edition:

ISBN-13: 978-1-111-83539-2

ISBN-10: 1-111-83539-X

Cengage Learning International Offices

Asia
www.cengageasia.com
tel: (65) 6410 1200

Australia/New Zealand
www.cengage.com.au
tel: (61) 3 9685 4111

Brazil
www.cengage.com.br
tel: (55) 11 3665 9900

India
www.cengage.co.in
tel: (91) 11 4364 1111

Latin America
www.cengage.com.mx
tel: (52) 55 1500 6000

UK/Europe/Middle East/Africa
www.cengage.co.uk
tel: (44) 0 1264 332 424

**Represented in Canada by
Nelson Education, Ltd.**
www.nelson.com
tel: (416) 752 9100 / (800) 668 0671

Cengage Learning is a leading provider of customized learning solutions with office locations around the globe, including Singapore, the United Kingdom, Australia, Mexico, Brazil, and Japan. Locate your local office at: **www.cengage.com/global**

For product information and free companion resources:
www.cengage.com/international
Visit your local office: **www.cengage.com/global**
Visit our corporate website: **www.cengage.com**

Printed in Canada
2 3 4 5 6 7 15 14 13

To Dea and Chris

About the Authors

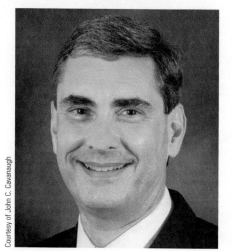

ROBERT V. KAIL

ROBERT V. KAIL is Professor of Psychological Sciences at Purdue University. His undergraduate degree is from Ohio Wesleyan University and his Ph.D. is from the University of Michigan. Kail is editor of *Psychological Science*, the flagship journal of the Association for Psychological Science, and is the incoming editor of *Child Development Perspectives*. He received the McCandless Young Scientist Award from the American Psychological Association, was named the Distinguished Sesquicentennial Alumnus in Psychology by Ohio Wesleyan University, and is a fellow of the Association for Psychological Science. Kail has also written *Children and Their Development*. His research focuses on cognitive development during childhood and adolescence. Away from the office, he enjoys photography and working out.

JOHN C. CAVANAUGH

JOHN C. CAVANAUGH is Chancellor of the Pennsylvania State System of Higher Education. He received his undergraduate degree from the University of Delaware and his Ph.D. from the University of Notre Dame. Cavanaugh is a fellow of the American Psychological Association, the Association for Psychological Science, and the Gerontological Society of America, and has served as president of the Adult Development and Aging Division (Division 20) of the APA. Cavanaugh has also written (with the late Fredda Blanchard-Fields) *Adult Development and Aging*. His research interests in gerontology concern family caregiving as well as the role of beliefs in older adults' cognitive performance. For enjoyment, he backpacks, writes poetry, and, while eating chocolate, ponders the relative administrative abilities of James T. Kirk, Jean-Luc Picard, Kathryn Janeway, Benjamin Sisko, and Jonathan Archer.

Brief Contents

Contents

© Masterfile

3   Tools for Exploring
the World: Physical,
Perceptual, and Motor
Development 81

© Elyse Lewin/Jupiterimages Corporation

© 2011 David Oldfield/Jupiter Images

iStockphoto.com/track5

13 Making It in Midlife: The Biopsychosocial Challenges of Middle Adulthood 455

© Michael Krasowitz/Getty Images

Preface

"What do you want to be when you grow up?" "Where do you see yourself in the next 5 or 10 years?" "What kind of person do you want to become?" These and other questions about "becoming" confront us across our lives. Answering them requires us to understand ourselves in very thorough ways. It requires us to understand how we develop.

Human development is both the most fascinating and the most complex science there is. *Human Development: A Life-Span View, Sixth Edition,* introduces you to the issues, forces, and outcomes that make us who we are.

Contemporary research and theory on human development consistently emphasize the multidisciplinary approach needed to describe and explain how people change (and how they stay the same) over time. Moreover, the great diversity of people requires an appreciation for individual differences in the course of development. *Human Development: A Life-Span View, Sixth Edition,* incorporates both and aims to address three specific goals:

- To provide a comprehensive, yet highly readable, account of human development across the life span.
- To provide theoretical and empirical foundations that enable students to become educated and critical interpreters of developmental information.
- To provide a blend of basic and applied research, as well as controversial topics and emergent trends, to demonstrate connections between the laboratory and life and the dynamic science of human development.

Organization

A Modified Chronological Approach.

The great debate among authors and instructors in the field of human development is whether to take a *chronological approach* (focusing on functioning at specific stages of the life span, such as infancy, adolescence, and middle adulthood) or a *topical approach* (following a specific aspect of development, such as personality, throughout the life span). Both approaches have their merits. We have chosen a modified chronological approach that we believe combines the best aspects of both. The overall organization of the text is chronological: We trace development from conception through late life in sequential order and dedicate several chapters to topical issues pertaining to particular points in the life span (such as infancy and early childhood, adolescence, young adulthood, middle adulthood, and late life).

But because the developmental continuity of such topics as social and cognitive development gets lost with narrowly defined, artificial age-stage divisions, we dedicate some chapters to tracing their development over larger segments of the life span. These chapters provide a much more coherent description of important developmental changes, emphasize the fact that development is not easily divided into "slices," and provide students with understandable explications of developmental theories.

Balanced Coverage of the Entire Life Span.

A primary difference between *Human Development: A Life-Span View, Sixth Edition,* and similar texts is that this book provides a much richer and more complete description of adult development and aging. Following the introductory chapter, the remaining 15 chapters of the text are evenly divided between childhood, adolescence, adulthood, and aging. This balanced treatment reflects not only the rapid emergence of adult development and aging as a major emphasis in the science of human development but also a recognition that roughly three fourths of a person's life occurs beyond adolescence.

As a reflection of our modified chronological approach, *Human Development: A Life-Span View, Sixth Edition,* is divided into four main parts. After an introduction to the science of human development (Chapter 1), Part One includes a discussion of the biological foundations of life (Chapter 2) and development during infancy and early childhood (Chapters 3–5). Part Two focuses on development during middle childhood and adolescence (Chapters 6–9). Part Three (Chapters 10–13) focuses on young and middle adulthood. Part Four examines late adulthood (Chapters 14 and 15) and concludes with a consideration of dying and bereavement (Chapter 16).

Content and Approach

The Biopsychosocial Emphasis.

Our text provides comprehensive, up-to-date coverage of research and theory from conception to old age and death. We explicitly adopt the biopsychosocial framework as an organizing theme, describing it in depth in Chapter 1, then integrating it throughout the text—often in combination with other developmental theories.

An Engaging Personal Style.

On several occasions, we communicate our personal involvement with the issues being discussed by providing examples from our own experiences as illustrations of how human development plays itself out in people's lives. Additionally, every major section of a chapter opens with a short vignette, helping to personalize a concept just before it is discussed. Other rich examples are integrated throughout the text narrative and showcased in the *Real People* feature in nearly every chapter.

Emphasis on Inclusiveness.

In content coverage, in the personalized examples used, and in the photo program, we emphasize diversity—within the United States and around the world—in ethnicity, gender, race, age, ability, and sexual orientation.

Learn More About It.

At the end of each chapter are descriptions of the book-specific website Psychology CourseMate (www.cengagebrain.com), the personalized online study tool CengageNOW (www.cengagebrain.com). Additionally, to reinforce students' understanding of current research data, these online resources include questions about the *Spotlight on Research* boxes (indicated with an icon at the end of each box).

Changes in the Sixth Edition.

Besides updating the sixth edition with new graphics and several hundred new reference citations to works from the past 3 years, the major change in this edition is a neuroscience theme. Neuroscience is a rapidly growing field that combines research on how the brain develops and works with the behavior that we exhibit. It provides insights into how and why we are able to think about things differently as we grow up and older, as well as how certain diseases impair those abilities. A Neuroscience Index has been added to the Diversity Index (pp. xxxii-xxxv).

The boxes about current trends and controversies have been recast as "What Do You Think?" discussions to encourage students to think critically about the developmental aspects of issues such as stem cell research, marriage education, and Social Security. From Kristin Jaymes Stewart to Nelson Mandela to many ordinary people, many new *Real People* boxes illustrate life-span topics.

Of particular note are these content additions, updates, and revisions:

CHAPTER 1: THE STUDY OF HUMAN DEVELOPMENT

- Neuroscience added as a subsection, introducing this new theme in the book.
- Stem cell discussion updated in *What Do You Think?*
- "Recurring Issues" now discussed as "and" instead of "versus" to make it clear that both sides are important.
- In the discussion of forces on development, interaction now follows the four forces.
- New *Real People* box on Lena Horne.

CHAPTER 2: BIOLOGICAL FOUNDATIONS

- New coverage of research into prenatal memory of sensory experiences, and revision of the *Spotlight on Research* feature to focus on the affect of low birth weight on memory.
- New material about pregnant women's cell-phone use and prenatal development.
- Updated discussion of the delayed effects of the teratogen DES.
- Expanded section about combined effects of prenatal risk factors.
- Updated section on fetal surgery and gene replacement.

CHAPTER 3: TOOLS FOR EXPLORING THE WORLD

- Updated and reorganized section on brain specialization.
- New section on theory of mind in children with autism.
- New face perception research discussed in *Spotlight on Research*.
- Expanded coverage of the Neonatal Behavioral Assessment Scale (NBAS).
- Expanded coverage of the roles of heredity and environment in temperament and of stability of temperament.
- Updated discussion of how infants process sensory information.

CHAPTER 4: THE EMERGENCE OF THOUGHT AND LANGUAGE

- Expanded coverage of the impact of video on young children's word learning, including a new *Spotlight on Research* box about learning from infant-oriented media.
- New *Real People* box illustrating accommodation and assimilation in action.
- Updated guidelines for interviewing preschoolers for eyewitness testimony.
- Updated coverage of how infants perceive speech and identify words.
- Expanded discussion of bilingualism.

CHAPTER 5: ENTERING THE SOCIAL WORLD

- New section on functional view of emotions.
- New table summarizing emotional development in infancy.
- Expanded discussion of acquisition of negative emotions.
- New section on the impact of genetics on prosocial behavior.

- New table summarizing factors that promote children's prosocial behavior.
- Updated discussion of consequences of type of infant–parent attachment.
- Updated coverage of gender schemas and gender stereotypes, including mathematical ability.

CHAPTER 6: OFF TO SCHOOL
- Revised and updated coverage of intellectual disability, based on guidelines of the American Association of Intellectual and Developmental Disabilities (AAIDD).
- Updated and expanded coverage of learning disabilities.
- Updated findings from the Multimodal Treatment Study of Children with ADHD (MTA) concluding that ADHD is a chronic condition.
- Updated discussions of reading and writing skills.
- Expanded coverage of physical fitness in elementary-school children.

CHAPTER 7: EXPANDING SOCIAL HORIZONS
- Section on effects of television updated and now includes new subsection about use of computers to play games and to stay in touch with friends.
- New *Spotlight on Research* box about helping parents and children adjust to life after a divorce.
- New material on preventing child maltreatment in high-risk families.
- Updated coverage of consequences of physical punishment, including new material on countries that ban it.
- Updated discussion of ways that children influence parents.
- Expanded discussion of susceptibility to peer pressure.
- Updated and expanded coverage of young children's descriptions of other people.

CHAPTER 8: RITES OF PASSAGE
- Expanded coverage of neuroscience, including more material on:
  a. links between adolescent brain development and risk-proneness, and
  b. links between adolescent brain development and information-processing efficiency.

CHAPTER 9: MOVING INTO THE ADULT SOCIAL WORLD
- Expanded coverage of identity development.
- Expanded and updated discussion of parent–child relationships in adolescence.
- Updated coverage of romantic relationships in adolescence and of dating violence.
- Expanded and updated section on depression in adolescence.
- Much-revised discussion of factors leading to antisocial behavior, including new material on cascading effects of risk factors.

CHAPTER 10: BECOMING AN ADULT
- Section on emerging adulthood rewritten, with new material on neuroscience work on brain maturation.
- New *Real People* box about Kristen Jaymes Stewart (of *Twilight Saga*).
- Revised discussion of binge drinking for *What Do You Think?* box.
- Revised section on primary and secondary mental abilities.
- New subsection on fluid and crystallized intelligence.

- New integrated figure showing connections among primary and secondary mental abilities and fluid and crystallized intelligence.
- New subsection on neuroscience research on intelligence, including the parieto-frontal integration theory (P-FIT).
- Inclusion of research using reflective judgment in teacher education in the section on postformal thinking.

CHAPTER 11: BEING WITH OTHERS
- Additional material on online friendships.
- Added material on online dating and speed dating.
- New international information about marriage and singlehood.
- Information about military marriages added to discussion of early years of marriage.
- New *What Do You Think?* box on marriage education.
- Expanded coverage on the context of remarriage.

CHAPTER 12: WORK AND LEISURE
- Social cognitive career theory added to occupation choice section.
- Occupational expectations section includes more on the relationship between adolescent expectations and adult reality as well as research on Millennials' expectations.
- New *Spotlight on Research* box on job satisfaction in a cross-cultural study of teachers.
- Vallerand's Passion Model of employee burnout added.
- Concept of "glass cliff" added.
- New *Real People* box on coping with layoffs.

CHAPTER 13: MAKING IT IN MIDLIFE
- Updated discussion of diagnosis and treatment of osteoporosis, including bone mineral density assessment through a DXA test.
- Updated coverage of rheumatoid arthritis.
- Thorough updating of menopause symptoms, outcomes, and hormone replacement therapy.
- Updated discussion of prostate cancer screening.
- Post-traumatic stress disorder added to consequences of stress.
- Expanded coverage of cross-cultural aspects of personality traits.
- Expanded discussion of the empty nest, with more on "boomerang kids" and international comparisons.

CHAPTER 14: THE PERSONAL CONTEXT OF LATER LIFE
- New *What Do You Think?* box on personal decisions about optimal length of life.
- Discussion of biological theories of aging reorganized.
- New neuroscience material added to sections on physiological change and cognition, including creativity.
- New *Real People* box on creativity in late life.
- Updating of genetics and testing research in Alzheimer's disease.

CHAPTER 15: SOCIAL ASPECTS OF LATER LIFE
- New section on housing options, including aging in place, assisted living, Eden Alternative, Green House Project, and cohousing.
- New *Real People* box about Nelson Mandela.
- New discussion about emotion regulation.

- New *Spotlight on Research* about neuroimaging and emotion regulation.
- Added discussion of online friendships.
- Major revision of section on sibling relationships.

CHAPTER 16: THE FINAL PASSAGE
- New section on complicated or prolonged grief disorder, proposed for DSM-5.
- Reorganized euthanasia section.
- Neuroimaging research cited regarding death anxiety.

Special Features

Three special features are a significant reason why this textbook is unique. These features are woven seamlessly into the narrative, signaled by a distinct icon for each—not boxed off from the flow of the chapter. Each box appears in nearly every chapter. The three features are:

Spotlight on research These features emphasize a fuller understanding of the science and scope of life-span development.

What do you think? These features ask students to think critically about social and developmental issues.

Real People Applying Human Development These features illustrate the everyday applications of life-span development issues.

Pedagogical Features

Among the most important aspects of *Human Development: A Life-Span View, Sixth Edition,* is its exceptional integration of pedagogical features, designed to help students maximize their learning.

- *Integration of Features.* One of the first things you may notice in paging through this text is that the three special features described earlier, which are normally set apart in boxes in other texts (boxes that students often skip!), are integrated directly into the narrative. This *unrivaled* integration is meant to help the student stay focused on a seamless presentation of human development across the life span.
- *Section-by-Section Pedagogy.* Each major section (every chapter has four or five) has been carefully crafted: It opens with a set of learning objectives, a vignette, typically includes one or more *Think About It* questions in the margin encouraging critical thinking, and ends with a set of questions called *Test Yourself* that reinforces key elements of the section. For easy assignment and to help readers visually organize the material, major units within each chapter are numbered.
- *Chapter-by-Chapter Pedagogy.* Each chapter opens with a table of contents and concludes with a bulleted, detailed *Summary* (broken down by learning objective within each major section), followed by a list of *Key Terms* (with page references), and *Learn More About It* (which lists websites with online resources to support this textbook).

In sum, we believe that our integrated pedagogical system will give the student all the tools she or he needs to comprehend the material and study for tests.

Supplementary Materials

INSTRUCTOR'S RESOURCE MANUAL
ISBN: 978-1-111-83521-7

The Instructor's Resource Manual by Ashley Lewis Presser contains resources designed to streamline and maximize the effectiveness of course preparation. The contents include chapter overviews and outlines, learning objectives, critical thinking discussion questions, instructional goals, lecture expanders, suggested websites and Internet activities, video recommendations, and handouts.

TEST BANK
ISBN: 978-1-111-83522-4

The Test Bank by Ashley Lewis Presser contains more than 3,000 text-specific questions with multiple-choice, fill-in-the-blank, essay, and true/false questions for each chapter.

POWERLECTURE™
ISBN: 978-1-111-83524-8

The fastest, easiest way to build powerful, customized media-rich lectures, PowerLecture provides a collection of book-specific PowerPoint® lecture slides (written by Tamara Ferguson) and class tools to enhance the educational experience.

PSYCHOLOGY COURSEMATE

Psychology CourseMate includes:

- an interactive eBook
- the following interactive teaching and learning tools:
  - quizzes
  - flashcards
  - videos
  - tools formerly found in the study guide
  - and more
- Engagement Tracker, a first-of-its-kind tool that monitors student engagement in the course

Go to login.cengage.com to access these resources.

CENGAGENOW™

CengageNOW offers all of your teaching and learning resources in one intuitive program organized around the essential activities you perform for class—lecturing, creating assignments, grading, quizzing, and tracking student progress and performance. CengageNOW's intuitive "tabbed" design allows you to navigate to all key functions with a single click, and a unique homepage tells you just what needs to be done and when. CengageNOW provides students access to an integrated eBook, and interactive tutorials, videos, and animations that help them get the most out of your course.

WEBTUTOR ON BLACKBOARD AND WEBCT

Jump-start your course with customizable, rich, text-specific content within your Course Management System.

- Jump-start—Simply load a WebTutor cartridge into your Course Management System.
- Customizable—Easily blend, add, edit, reorganize, or delete content.
- Content—Rich, text-specific content, media assets, quizzes, web links, discussion topics, interactive games and exercises, and more.

Whether you want to web-enable your class or put an entire course online, WebTutor delivers. Visit webtutor.cengage.com to learn more.

Acknowledgments

Textbook authors do not produce books on their own. We owe a debt of thanks to many people who helped take this project from a first draft to a bound book. Thanks to Jim Brace-Thompson, for his enthusiasm, good humor, and sage advice at the beginning of this project; to Jaime Perkins for taking the reins and guiding the sixth edition; to Barbara Armentrout and Kristin Makarewycz, for providing helpful feedback as we revised; Mary Noel for her work on the book's media; Trina Tom, Paige Leeds, and Jessica Alderman for their attention to a host of details; and to Pat Waldo, for shepherding the book through production. We also thank Vernon Boes, art director; Kim Russell, senior marketing director; and Chris Sosa, marketing manager, for their contributions to the book. We are grateful to Cassie Carey at Graphic World for keeping everything on track. We also thank Josh Garvin, Dean Dauphinais, Isabel Saraiva, and Jeanne Calabrese.

We would also like to thank the many reviewers who generously gave their time and effort to help us sharpen our thinking about human development and, in so doing, shape the development of this text.

Reviewers

Sixth Edition Reviewers

MAIDA BERENBLATT
Suffolk County Community College

KAREN IHNEN
St. Cloud Technical and Community College

AMY LANDERS
Old Dominion University

MICHAEL JASON MCCOY
Cape Fear Community College

JULIE ANN MCINTYRE
Russell Sage College

MARIBETH PALMER-KING
Broome Community College

STACIE SHAW
Presentation College

LINDA SPERRY
Indiana State University

SUSAN D. TALLEY
Utah State University

VIRGINIA TOMPKINS
Ohio State University

Fifth Edition Reviewers

CYNTHIA B. CALHOUN
Southwest Tennessee Community College

PAUL ANDERER CASTILLO
State University of New York, Canton

LISA DAVIDSON
Northern Illinois University

DOUG FRIEDRICH
University of West Florida

ALYCIA M. HUND
Illinois State University

RICHARD KANDUS
Mt. San Jacinto College

JOHN W. OTEY
Southern Arkansas University

SHANA PACK
Western Kentucky University

LISA ROUTH
Pikes Peak Community College

CARRIE SWITZER
University of Illinois, Springfield

CAITLIN WILLIAMS
San Jose State University

Fourth Edition Reviewers

L. RENÉ BERGERON
University of New Hampshire

JANINE P. BUCKNER
Seton Hall University

CHARLES TIMOTHY DICKEL
Creighton University

DOUGLAS FRIEDRICH
University of West Florida

LANA-LEE HARDACRE
Conestoga College

JULIE A. HASELEU
Kirkwood Community College

BRETT HEINTZ
Delgado Community College

HEATHER M. HILL
University of Texas, San Antonio

MARY ANNE O'NEILL
Rollins College Hamilton Holt School

SHANA PACK
Western Kentucky University

IAN PAYTON
Bethune-Cookman College

Third Edition Reviewers

GARY ALLEN
University of South Carolina

KENNETH E. BELL
University of New Hampshire

BELINDA BEVINS-KNABE
University of Arkansas at Little Rock

CATHERINE DEERING
Clayton College and State University

JUDITH DIETERLE
Daytona Beach Community College

SANDY EGGERS
University of Memphis

WILLIAM FABRICIUS
Arizona State University

DOUGLAS FRIEDRICH
University of West Florida

TRESMAINE R. GRIMES
Iona College

SUSAN HORTON
Mesa Community College

JENEFER HUSMAN
University of Alabama

ERWIN J. JANEK
Henderson State University

WAYNE JOOSE
Calvin College

MARGARET D. KASIMATIS
Carroll College

MICHELLE L. KELLEY
Old Dominion University

KIRSTEN D. LINNEY
University of Northern Iowa

BLAKE TE-NEIL LLOYD
University of South Carolina

SUSAN MAGUN-JACKSON
University of Memphis

MARION G. MASON
Bloomsburg University of Pennsylvania

JULIE ANN MCINTYRE
Russell Sage College

EDWARD J. MORRIS
Owensboro Community College

JANET D. MURRAY
University of Central Florida

ELLEN E. PASTORINO
Valencia Community College

ROBERT F. RYCEK
University of Nebraska at Kearney

JEFF SANDOZ
University of Louisiana at Lafayette

BRIAN SCHRADER
Emporia State University

CAROLYN A. SHANTZ
Wayne State University

CYNTHIA K. SHINABARGER REED
Tarrant County College

TRACY L. SPINRAD
Arizona State University

KELLI W. TAYLOR
Virginia Commonwealth University

LORRAINE C. TAYLOR
University of South Carolina

BARBARA TURNAGE
University of Central Florida

YOLANDA VAN ECKE
Mission College

CAROL G. WEATHERFORD
Clemson University

SANDY WURTELE
*University of Colorado at Colorado
   Springs*

Second Edition Reviewers

GARY L. ALLEN
University of South Carolina

ANN M.B. AUSTIN
Utah State University

DAVID BISHOP
Luther College

ELIZABETH M. BLUNK
Southwest Texas State University

JOSETTE BONEWITZ
Vincennes University

LANTHAN D. CAMBLIN, JR.
University of Cincinnati

SHELLEY M. DRAZEN
SUNY, Binghamton

KENNETH ELLIOTT
University of Maine, Augusta

NOLEN EMBRY
Lexington Community College

JAMES GARBARINO
Cornell University

CATHERINE HACKETT RENNER
West Chester University

SANDRA HELLYER
Indiana University-Purdue University at Indianapolis

JOHN KLEIN
Castleton State College

WENDY KLIEWER
Virginia Commonwealth University

NANCY MACDONALD
University of South Carolina, Sumter

LISA MCGUIRE
Allegheny College

MARTIN D. MURPHY
University of Akron

JOHN PFISTER
Dartmouth College

BRADFORD PILLOW
Northern Illinois University

GARY POPOLI
Hartford Community College

ROBERT PORESKY
Kansas State University

JOSEPH M. PRICE
San Diego State University

ROSEMARY ROSSER
University of Arizona

TIMOTHY O. SHEARON
Albertson College of Idaho

MARCIA SOMER
University of Hawaii-Kapiolani Community College

NANCI STEWART WOODS
Austin Peay State University

ANNE WATSON
West Virginia University

FRED A. WILSON
Appalachian State University

KAREN YANOWITZ
Arkansas State University

CHRISTINE ZIEGLER
Kennesaw State University

First Edition Reviewers

POLLY APPLEFIELD
University of North Carolina at Wilmington

DANIEL R. BELLACK
Trident Technical College

DAVID BISHOP
Luther College

LANTHAN CAMBLIN, JR.
University of Cincinnati

KENNETH ELLIOTT
University of Maine at Augusta

MARTHA ELLIS
Collin County Community College

STEVE FINKS
University of Tennessee

LINDA FLICKINGER
St. Clair County Community College

REBECCA GLOVER
University of North Texas

J. A. GREAVES
Jefferson State Community College

PATRICIA GUTH
Westmoreland County Community College

PHYLLIS HEATH
Central Michigan University

MYRA HEINRICH
Mesa State College

SANDRA HELLYER
Indiana University-Purdue University at Indianapolis

SHIRLEY-ANNE HENSCH
University of Wisconsin Center

THOMAS HESS
North Carolina State University

KATHLEEN HURLBURT
University of Massachusetts–Lowell

HEIDI INDERBITZEN
University of Nebraska at Lincoln

SANFORD LOPATER
Christopher Newport University

BILL MEREDITH
University of Nebraska at Omaha

MARIBETH PALMER-KING
Broome Community College

HARVE RAWSON
Franklin College

VIRGINIA WYLY
State University of New York College at Buffalo

To the Student

Human Development is written with you, the student, in mind. In the next few pages, we describe several features of the book that will make it easier for you to learn. Please don't skip this material; it will save you time in the long run.

Learning and Study Aids

Each chapter includes several distinctive features to help you learn the material and organize your studying.

- Each chapter opens with an overview of the main topics and a detailed outline.
- Each major section within a chapter begins with a set of learning objectives. There is also a brief vignette introducing one of the topics to be covered in that section and providing an example of the developmental issues people face.
- When key terms are introduced in the text, they appear in bold, blue type and are defined in the margin. This should make key terms easy to find and learn.
- Key developmental theories are introduced in Chapter 1 and are referred to throughout the text.
- Critical thinking questions appear in the margins. These *Think About It* questions are designed to help you make connections across sections within a chapter or across chapters.
- The end of each section includes a feature called *Test Yourself,* which will help you check your knowledge of major ideas you just read about. The Test Yourself questions serve two purposes. First, they give you a chance to spot-check your understanding of the material. Second, the questions will relate the material you have just read to other facts, theories, or the biopsychosocial framework you read about earlier.
- Text features that expand or highlight a specific topic are integrated with the rest of the material. This book includes the following three features, each identified by a distinctive icon:
  - *Spotlight on Research* elaborates a specific research study discussed in the text and provides more details on the design and methods used.
  - *What Do You Think?* offers thought-provoking discussions about current issues affecting development.
  - *Real People: Applying Human Development* is a case study that illustrates how an issue in human development discussed in the chapter is manifested in the life of a real person.
- The end of each chapter includes several special study tools. A *Summary* organized by learning objective within major section headings provides a review of the key ideas in the chapter. Next is a list of *Key Terms* that appear in the chapter. Drawing the chapter to a close is *Learn More About It,* which contains a list of online resources with interactive learning tools designed for this book.

We strongly encourage you to take advantage of these learning and study aids as you read the book. We have also left room in the margins for you to make notes to yourself on the material, so you can more easily integrate the text with your class and lecture material.

Your instructor will probably assign about one chapter per week. Don't try to read an entire chapter in one sitting. Instead, on the first day, preview the chapter. Read the introduction and notice how the chapter fits into the entire book; then page through the chapter, reading the learning objectives, vignettes, and major headings. Also read the italicized sentences and the boldfaced terms. Your goal is to get a general overview of the entire chapter—a sense of what it's all about.

Now you're ready to begin reading. Go to the first major section and preview it again, reminding yourself of the topics covered. Then start to read. As you read, think about what you're reading. Every few paragraphs, stop briefly. Try to summarize the main ideas in your own words; ask yourself if the ideas describe your own experience or that of others you know; tell a friend about something interesting in the material.

In other words, read actively—get involved in what you're reading. Don't just stare glassy-eyed at the page!

Continue this pattern—reading, summarizing, thinking—until you finish the section. Then answer the Test Yourself questions to determine how well you've learned what you've read. If you've followed the read-summarize-think cycle as you worked your way through the section, you should be able to answer most of the questions.

The next time you sit down to read (preferably the next day), start by reviewing the second major section. Then complete it with the read-summarize-think cycle. Repeat this procedure for all the major sections.

When you've finished the last major section, wait a day or two and then review each major section. Pay careful attention to the italicized sentences, the boldfaced terms, and the Test Yourself questions. Also, use the study aids at the end of the chapter to help you integrate the ideas in the chapters.

With this approach, it should take several 30- to 45-minute study sessions to complete each chapter. Don't be tempted to rush through an entire chapter in a single session. Research consistently shows that you learn more effectively by having daily (or nearly daily) study sessions devoted to both reviewing familiar material *and* taking on a relatively small amount of new material.

Psychology CourseMate

Human Development: A Life-Span View, Sixth Edition, includes Psychology Course-Mate, which helps you make the grade. Psychology CourseMate includes:

- an interactive eBook with highlighting, note-taking, and search capabilities
- the following interactive learning tools:
  o quizzes
  o flashcards
  o videos
  o animations
  o and more!

Go to login.cengage.com to access these resources, and look for this icon *CourseMate to find resources related to your text in Psychology CourseMate.

Terminology

A few words about terminology before we embark. Certain terms will be used to refer to different periods of the life span. Although you may already be familiar with the terms, we would like to clarify how they will be used in this text. The following terms will refer to a specific range of ages:

Newborn: birth to 1 month

Infant: 1 month to 1 year

Toddler: 1 year to 2 years

Preschooler: 2 years to 6 years

School-age child: 6 years to 12 years

Adolescent: 12 years to 20 years

Young adult: 20 years to 40 years

Middle-age adult: 40 years to 60 years

Young-old adult: 60 years to 80 years

Old-old adult: 80 years and beyond

Sometimes, for the sake of variety, we will use other terms that are less tied to specific ages, such as babies, youngsters, and older adults. However, you will be able to determine the specific ages from the context.

Organization

Authors of textbooks on human development always face the problem of deciding how to organize the material into meaningful segments across the life span. This book is organized into four parts: Prenatal Development, Infancy, and Early Childhood; School-Age Children and Adolescents; Young and Middle Adulthood; and Late Adulthood. We believe this organization achieves two major goals. First, it divides the life span in ways that relate to the divisions encountered in everyday life. Second, it enables us to provide a more complete account of adulthood than other books do.

Because some developmental issues pertain only to a specific point in the life span, some chapters are organized around specific ages. Overall, the text begins with conception and proceeds through childhood, adolescence, adulthood, and old age to death. But because some developmental processes unfold over longer periods of time, some of the chapters are organized around specific topics.

Part One covers prenatal development, infancy, and early childhood. Here we will see how genetic inheritance operates and how the prenatal environment affects a person's future development. During the first two years of life, the rate of change in both motor and perceptual arenas is amazing. How young children acquire language and begin to think about their world is as intriguing as it is rapid. Early childhood also marks the emergence of social relationships, as well as an understanding of gender roles and identity. By the end of this period, a child is reasonably proficient as a thinker, uses language in sophisticated ways, and is ready for the major transition into formal education.

Part Two covers the years from elementary school through high school. In middle childhood and adolescence, the cognitive skills formed earlier in life evolve to adult-like levels in many areas. Family and peer relationships expand. During adolescence, there is increased attention to work, and sexuality emerges. The young person begins to learn how to face difficult issues in life. By the end of this period, a person is on the verge of legal adulthood. The typical individual uses logic and has been introduced to most of the issues that adults face.

Part Three covers young adulthood and middle age. During this period, most people achieve their most advanced modes of thinking, achieve peak physical performance, form intimate relationships, start families of their own, begin and advance within their occupations, manage to balance many conflicting roles, and begin to confront aging. Over these years, many people go from breaking away from their families to having their children break away from them. Relationships with parents are redefined, and the pressures of being caught between the younger and older generations are felt. By the end of this period, most people have shifted focus from time since birth to time until death.

Part Four covers the last decades of life. The biological, physical, cognitive, and social changes associated with aging become apparent. Although many changes reflect decline, many other aspects of old age represent positive elements: wisdom, retirement, friendships, and family relationships. We conclude this section, and the text, with a discussion of the end of life. Through our consideration of death, we will gain additional insights into the meaning of life and human development.

We hope the organization and learning features of the text are helpful to you—making it easier for you to learn about human development. After all, this book tells the story of people's lives, and understanding the story is what it's all about.

Human Development

A Life-Span View

Neuroscience Index

Diversity Index

The Study of Human Development

JEANNE CALMENT WAS ONE of the most important people to have ever lived. Her achievement, though notable, was not made in sports, government, or any other profession. When she died in 1996 at age 122 years and 164 days, she set the world record for the longest verified human life span. Jeanne lived her whole life in Arles, France. During her lifetime, she met Vincent Van Gogh, experienced the invention of the lightbulb, automobiles, airplanes, space travel, computers, and all sorts of everyday conveniences. Longevity ran in her family: her older brother, François, lived to the age of 97, her father to 93, and her mother to 86. Jeanne was extraordinarily healthy her whole life, hardly ever being ill. She was also active; she learned fencing when she was 85, and she was still riding a bicycle at age 100. She lived on her own until she was 110, when she moved to a nursing home. Her life was documented in the 1995 film *Beyond 120 Years with Jeanne Calment.* Shortly before her 121st birthday, Musicdisc released *Time's Mistress,* a CD of Jeanne speaking over a background of rap and hip-hop music.

Did you ever wonder what your life span will be? The people you will meet and the experiences you will have? Did you ever think about how you managed to go from being a young child to the more experienced person you are now? Or what might lie ahead in your future over the next few years or decades? Take a moment and think about your life to this point. Make a note to yourself about—or share with someone else—your fondest memories from childhood or the events and people who have most influenced you. And also make a note about what you think you might experience across the rest of

Jeanne Calment experienced many changes in society during her 122-year life span.

© Masterfile

Georges GOBET / Newscom

your life. (Then, many years from now, retrieve it and see if you were right.)

Thinking about your past experiences is the beginning of an exciting personal journey. Think about major moments or experiences you've had. What happened? Why do you think things happened the way they did? What major forces shaped that event, and have shaped your life?

Likewise, looking ahead to what your future may hold is also exciting. Will you be able to create your own destiny? What forces are out there to shape you years from now? How will personal changes affect your future?

In this course, you will have the opportunity to ask some of life's most basic questions: How did your life begin? How did you go from a single cell—about the size of the period at the end of a sentence in this text—to the fully grown, complex adult person you are today? Will you be the same or different by the time you reach late life? How do you influence other people's lives? How do they influence yours? How do the various roles you have throughout life—child, teenager, partner, spouse, parent, worker, grandparent—shape your development? How do we deal with our own and others' deaths?

human development
the multidisciplinary study of how people change and how they remain the same over time

These are examples of the questions that create the scientific foundation of **human development,** *the multidisciplinary study of how people change and how they remain the same over time.* Answering these questions requires us to draw on theories and research in the physical and social sciences, including biology, genetics, neuroscience, chemistry, allied health and medicine, psychology, sociology, demography, ethnography, economics, and anthropology. The science of human development reflects the complexity and uniqueness of each person and each person's experiences as well as commonalities and patterns across people. As a science, human development is firmly grounded in theory and research as it seeks to understand human behavior.

Before our journey begins, we need to collect some things to make the trip more rewarding. In this chapter, we pick up the necessary road maps that point us in the proper direction: a framework to organize theories and research, common issues and influences on development, and the methods developmentalists use to make discoveries. Throughout the book, we will point out how the various theories and research connect to your own experience. Pack well, and bon voyage.

Thinking About Development

LEARNING OBJECTIVES

- What fundamental issues of development have scholars addressed throughout history?

- What are the basic forces in the biopsychosocial framework? How does the timing of these forces affect their impact?

- How does neuroscience enhance our understanding of human development?

Javier Suarez smiled broadly as he held his newborn grandson for the first time. So many thoughts rushed into his mind—What would Ricardo experience growing up? Would the poor neighborhood they live in prevent him from reaching his potential? Would the family genes for good health be passed on? How would Ricardo's life growing up as a Latino in the United States be different from Javier's own experiences in Mexico?

LIKE MANY GRANDPARENTS, Javier wonders what the future holds for his grandson. The questions he asks are interesting in their own right, but they are important for another reason: They bear on general issues of human development that have intrigued philosophers and scientists for centuries. In the next few pages we introduce some of these issues, which surface when any aspect of development is being investigated.

Recurring Issues in Human Development

Think about your life up until now. What factors have shaped it? You might suspect such things as your genetic heritage, your family or neighborhood, the suddenness of some changes in your life and the gradualness of others, and the culture(s) in which you grew up or now live. You might have also noticed that you are like some people you know, and very much unlike others (and they from you). So you might suspect that everyone's life is shaped by a complex set of factors.

Your speculations capture three fundamental characteristics of human development: nature and nurture, continuity and discontinuity, and universal and context-specific development. A person's development is a blend of these characteristics; for example, some of your characteristics remain the same through life (continuity) and others change (discontinuity). Because these characteristics apply to all the topics in this book, let's examine each one.

Nature and Nurture

Think for a minute about a particular feature that you and several people in your family have, such as intelligence, good looks, or a friendly and outgoing personality. Why is this feature so prevalent? Did you inherit it from your parents, and they from your grandparents? Or is it mainly because of where and how you and your parents were brought up? *Answers to these questions illustrate different positions on the* **nature–nurture issue,** *which involves the degree to which genetic or hereditary influences (nature) and experiential or environmental influences (nurture) determine the kind of person you are.* The key point is that development is always shaped by both: Nature and nurture are mutually interactive influences.

For example, in Chapter 2 you will see that some individuals inherit a disease that leads to mental retardation if they eat dairy products. However, if their environment contains no dairy products, they develop normal intelligence. Similarly, in Chapter 10 you will learn that one risk factor for cardiovascular disease is heredity but that lifestyle factors such as diet and smoking play important roles in determining who has heart attacks.

As these examples illustrate, a major aim of human development research is to understand how heredity and environment jointly determine development. For Javier, it means his grandson's development will surely be shaped both by the genes he inherited and by the experiences he will have.

nature–nurture issue
the degree to which genetic or hereditary influences (nature) and experiential or environmental influences (nurture) determine the kind of person you are

THINK ABOUT IT

Think of some common, everyday behaviors, such as dancing or playing basketball with your friends. How do nature and nurture influence these behaviors?

Hi and Lois

Continuity and Discontinuity

Think of some ways in which you remain similar to how you were as a 5-year-old. Maybe you were outgoing and friendly at that age and remain outgoing and friendly today. Examples like these suggest a great deal of continuity in development. Once a person begins down a particular developmental path—for example, toward friendliness or intelligence—he or she tends to stay on that path throughout life, other things being equal. From a continuity perspective, if Ricardo is a friendly and smart 5-year-old then he should be friendly and smart as a 25-year-old and as a 75-year-old.

The other view—that development is not always continuous—is illustrated in the Hi and Lois cartoon. Sweet and cooperative Trixie has become assertive and demanding. In this view, people can change from one developmental path to another and perhaps several times in their lives. Consequently, Ricardo might be smart and friendly at age 5, smart but obnoxious at 25, and wise but aloof at 75!

continuity–discontinuity issue
whether a particular developmental phenomenon represents a smooth progression throughout the life span (continuity) or a series of abrupt shifts (discontinuity)

The **continuity–discontinuity issue** *concerns whether a particular developmental phenomenon represents a smooth progression throughout the life span (continuity) or a series of abrupt shifts (discontinuity).* Of course, on a day-to-day basis, behaviors often look nearly identical, or continuous. But when viewed over the course of many months or years, the same behaviors may have changed dramatically, reflecting discontinuous change. Throughout this book, you will find examples of developmental changes that represent continuities and others that are discontinuities. For example, in Chapter 5 you will see evidence of continuity: Infants who have satisfying emotional relationships with their parents typically become children with satisfying peer relationships. But in Chapter 15 you will see an instance of discontinuity: After spending most of adulthood trying to ensure the success of the next generation and to leave a legacy, older adults turn to evaluating their own lives in search of closure and a sense that what they have done has been worthwhile.

Universal and Context-Specific Development

In some cities in Brazil, 10- to 12-year-olds sell fruit and candy to pedestrians and passengers on buses. Although they have little formal education and often cannot identify the numbers on the money, they handle money proficiently (Saxe, 1988). In contrast, 10- to 12-year-olds in the United States are formally taught at home or school to identify numbers and to perform the kinds of arithmetic needed to handle money. Can one theory explain development in both groups of children? *The* **universal and context-specific development issue** *concerns whether there is just one path of development or several.* Some theorists argue that, despite what look like differences in development, there is really only one fundamental developmental process for everyone. According to this view, differences in development are simply variations on a fundamental developmental process in much the same way that cars as different as a Chevrolet, a Honda, and a Porsche are all products of fundamentally the same manufacturing process.

universal versus context-specific development issue
whether there is just one path of development or several paths

The alternative view is that differences among people are not simply variations on a theme. Advocates of this view argue that human development is inextricably intertwined with the context within which it occurs. A person's development is a product of complex interaction with the environment, and that interaction is *not* fundamentally the same in all environments. Each environment has its own set of unique procedures that shape development, just as the "recipes" for different cars yield vehicles as different as a MINI Cooper and a stretch limousine.

As is the case for the nature–nurture and continuity–discontinuity issues, the end result is a blend; individual development reflects both universal and context-specific influences. For example, the basic order of development of physical skills in infancy is essentially the same in all cultures. But how those skills are focused or encouraged in daily life differs across cultures.

Putting all three issues together and using personality to illustrate, we can ask how heredity and environment interact to influence the development of personality, whether the development of personality is continuous or discontinuous, and whether personality develops in much the same way around the world. To answer these kinds of questions, we need to look at the forces that combine to shape human development.

Basic Forces in Human Development: The Biopsychosocial Framework

When trying to explain why people develop as they do, scientists usually consider four interactive forces:

- *Biological forces that include all genetic and health-related factors that affect development.*
- *Psychological forces that include all internal perceptual, cognitive, emotional, and personality factors that affect development.*
- *Sociocultural forces that include interpersonal, societal, cultural, and ethnic factors that affect development.*
- *Life-cycle forces that reflect differences in how the same event affects people of different ages.*

Each person is a unique combination of these forces. To see why each force is important, think about whether a mother decides to breast-feed her infant. Her decision will be based on biological variables (e.g., the quality and amount of milk she produces), her attitudes about the virtues of breast-feeding, the influences of other people (e.g., the father), and her cultural traditions about appropriate ways to feed infants. Additionally, her decision will reflect her age and stage of life. Only by focusing on all of these forces can we have a complete view of the mother's decision.

One useful way to organize the biological, psychological, and sociocultural forces on human development is with the **biopsychosocial framework**. As you can see in ■ Figure 1.1, the biopsychosocial framework emphasizes that each of the forces in-

Even with little formal education, this Brazilian boy has well-developed mathematical skills, an example of cultural contextual forces on development.

Mathias Oppersdorff / Photo Researchers, Inc.

biopsychosocial framework
a useful way to organize the biological, psychological, and sociocultural forces on human development

■ **Figure 1.1**
The biopsychosocial framework shows that human development results from interacting forces.

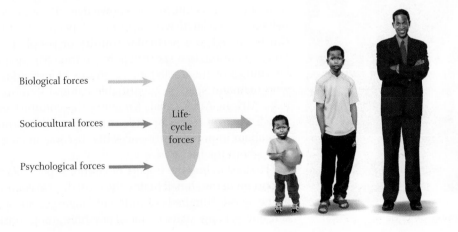

Biological forces

Sociocultural forces

Psychological forces

Life-cycle forces

Copyright © Cengage Learning 2010

Biological influences on development help explain why relatives tend to look alike.

teracts with the others to make up development. Let's look at the different elements of the biopsychosocial model in more detail.

Biological Forces: Genetics and Health

Prenatal development, brain maturation, puberty, and menopause may occur to you as outcomes of biological forces. Indeed, major aspects of each process are determined by our genetic code. For example, many children resemble their parents, which shows biological influences on development. But biological forces also include the effects of lifestyle factors, such as diet and exercise. Collectively, biological forces can be viewed as providing the raw material necessary and as setting the boundary conditions (in the case of genetics) for development.

Psychological Forces: Known by Our Behavior

Psychological forces seem familiar because they are the ones used most often to describe the characteristics of a person. For example, think about how you describe yourself to others. Most of us say that we have a nice personality and are intelligent, honest, self-confident, or something similar. Concepts like these reflect psychological forces.

In general, psychological forces are all the internal cognitive, emotional, personality, perceptual, and related factors that influence behavior. Psychological forces have received the most attention of the three main developmental forces. Much of what we discuss throughout this text reflects psychological forces. For example, we will see how the development of intelligence enables individuals to experience and think about their world in different ways. We'll also see how the emergence of self-esteem is related to the beliefs people have about their abilities, which in turn influence what they do.

Sociocultural Forces: Race, Ethnicity, and Culture

People develop in the world, not in a vacuum. To understand human development, we need to know how people and their environments interact and relate to each other. That is, we need to view an individual's development as part of a much larger system in which no individual part can act without influencing all other aspects of the system. This larger system includes one's parents, children, and siblings as well as important individuals outside the family, such as friends, teachers, and co-workers. The system also includes institutions that influence development, such as schools, television, and the workplace. At a broader level, the society in which a person grows up plays a key role.

All of these people and institutions fit together to form a person's culture: the knowledge, attitudes, and behavior associated with a group of people. Culture can be linked to a particular country or people (e.g., French culture), to a specific point in time (e.g., popular culture of the 2000s), or to groups of individuals who maintain specific, identifiable cultural traditions (e.g., African Americans). Knowing the culture from which a person comes provides some general information about important influences that become manifest throughout the life span.

Understanding the impact of culture is particularly important in the United States, the most diverse country in the world. Hundreds of different languages are spoken, and in many states no racial or ethnic group consti-

The culture in which you grow up influences how you experience life.

tutes a majority. The many customs that people from different cultures bring offer insights into the broad spectrum of human experience and attest to the diversity of the U.S. population.

Although the U.S. population is changing rapidly, much of the research we describe in this text was conducted on middle-class European Americans. Accordingly, we must be careful *not* to assume that findings from this group necessarily apply to people in other groups. Indeed, there is a great need for research on different cultural groups. Perhaps, as a result of taking this course, you will help fill this need by becoming a developmental researcher yourself.

Another practical problem that we face is how to describe racial and ethnic groups. Terminology changes over time. For example, the terms *colored people, Negroes, black Americans,* and *African Americans* have all been used to describe Americans of African ancestry. In this book, we use the term *African American* because it emphasizes their unique cultural heritage. Following the same reasoning, we use *European American* (instead of *Caucasian* or *white*), *Native American* (instead of *Indian* or *American Indian*), *Asian American,* and *Latino American* (rather than *Hispanic*).

These labels are not perfect. In some cases, they blur distinctions among ethnic groups. For example, people from both Puerto Rico and Mexico may be described as Latinos. However, their cultural backgrounds vary on several important dimensions, so we should not view them as being from a homogeneous group. Similarly, the term *Asian American* blurs variations among people whose heritage is, for example, Japanese, Chinese, or Korean. Throughout this text, whenever researchers have identified the subgroups in their research sample, we will use the more specific terms in describing results. When we use the more general terms, remember that conclusions may not apply to all subgroups within the group described by the more general term.

Life-Cycle Forces: Timing Is Everything

Consider the following two females. Jacqui, a 32-year-old, has been happily married for 6 years. She and her husband have a steady income. They decide to start a family, and a month later Jacqui learns she is pregnant. Jenny, a 17-year-old, lives in the same neighborhood as Jacqui. She has been sexually active for about 6 months but is not in a stable relationship. After missing her period, Jenny took a pregnancy test and discovered that she is pregnant.

Although both Jacqui and Jenny became pregnant, the outcome of each pregnancy will certainly be affected by factors in each woman's situation such as her age, her financial situation, and the extent of her social support systems. The example illustrates life-cycle forces: The same event can have different effects depending on when it happens in a person's life. In the scenarios with Jacqui and Jenny, the same event—pregnancy—produces happiness and eager anticipation for one woman but anxiety and concern for the other.

Due to life-cycle forces, experiencing pregnancy in high school is different from experiencing it later in adulthood.

Due to life cycle forces, pregnancy in one's 30s is experienced differently than pregnancy in one's teens.

The Forces Interact

So far, we've described the four forces in the biopsychosocial framework as if they were independent. But as we pointed out, each force shapes the others. Consider eating habits. When the authors of this text were growing up, a "red meat and potatoes" diet was common and was thought to be healthy. Scientists later discovered that high-fat diets may lead to cardiovascular disease and some forms of cancer. Consequently, social pressures changed what people eat; advertising campaigns were begun; and restaurants began to indicate which menu items were low in fat. Thus, the biological forces of fat in the diet were influenced by the social forces of the times, whether in support of or in opposition to eating beef every evening. As your authors became more educated about diets and their effects on health, the psychological forces of thinking and reasoning also influenced their choice of diets. (We confess, however, that dark chocolate remains a passion for one of us!) However, research on the effects of fat sometimes collided with centuries of cultural food traditions and with the ability to afford healthier alternatives, meaning that some folks changed their diets more easily than others. Finally, the age of the person when this research became widely known mattered, too. Young children eat what their parents provide, whereas adults may have the ability to make choices.

This example illustrates that no aspect of human development can be fully understood by examining the forces in isolation. All four must be considered in interaction. In fact, we'll see later in this chapter that integration across the major forces of the biopsychosocial framework is one criterion by which the adequacy of a developmental theory can be judged.

Combining the four developmental forces gives a view of human development that encompasses the life span yet appreciates the unique aspects of each phase of life. From this perspective we can view each life story as a complex interplay among the four forces. One way to see this is to look back on life from the perspective of old age. Lena Horne, discussed in the Real People feature, is a good example of this.

Real People
Applying Human Development — Lena Horne

The developmental forces are valuable in helping us understand how people's lives unfold. Lena Horne's (1917–2010) life is a great example. At the age of 2, she was on the cover of the magazine of the National Association for the Advancement of Colored People (NAACP); her paternal grandmother, Cora Calhoun Horne, was an early member of the NAACP. When she went to Hollywood in 1941 to sing in a night-club, blacks were not allowed to live in that city. When her neighbors found out she was going to live in an apartment that a white man had initially rented and turned over to her, it took Humphrey Bogart to defend her rights to the community.

Lena broke many barriers in her career by signing a long-term contract with MGM and through her subsequent fame and great success as a jazz singer. In the 1960s she became very active in the civil rights movement. In 1981 she won a Tony Award for her one-woman show *Lena Horne: The Lady and Her Music*.

The combined influence of the developmental forces is evident in Lena's life. Her mother had a career on the stage in Harlem, and her paternal grandparents were active in the NAACP. Growing up in New York in the 1920s and 1930s exposed her to opportunities, such as at the famous Cotton Club, that would not have been available in a small town. The psychological impact of discrimination, due to the sociocultural environment of the times, influenced her to fight for civil rights for all African Americans. And the fact that she was still singing and recording into the 1990s gave her the chance to influence several generations of people.

Looking back on her life when she was 80, Lena said: "My identity is very clear to me now. I am a black woman. I'm free. I no longer have to be a 'credit.' I don't have to be a symbol to anybody; I don't have to be a first to anybody. I don't have to be an imitation of a white woman that Hollywood sort of hoped I'd become. I'm me, and I'm like nobody else."

Lena Horne

Neuroscience: A Window Into Human Development

Understanding that the four developmental forces interact is one thing. But what if, like the comic book hero Superman, you could use your X-ray vision and actually *see* these forces interact? That's what is possible in the field of neuroscience. *Applied to human development,* **neuroscience** *is the study of the brain and the nervous system, especially in terms of brain–behavior relationships.* Neuroscientists use a variety of methods to do this, from molecular analyses of individual brain cells to sophisticated techniques that yield images of patterns of activity in the brain.

Neuroscientific approaches are being applied to a wide range of issues in human development, especially those involving memory, reasoning, and emotion (Blanchard-Fields, 2010). For example, neuroscientists are beginning to unlock relations between developmental changes in specific regions of the brain to explain well-known developmental phenomena such as adolescents' tendency to engage in risky behavior and older adults' short-term memory problems.

Neuroscience brings an important perspective to human development. Identifying patterns of brain activity helps to reveal interactions between biological, psychological, sociocultural, and life-cycle forces, which allows a better understanding of how each person is a unique expression of these forces.

neuroscience
the study of the brain and nervous system, especially in terms of brain-behavior relationships

Test Yourself

RECALL

1. The nature–nurture issue involves the degree to which _____ and the environment influence human development.
2. Azar remarked that her 14-year-old son is incredibly shy and has been ever since he was a little baby. This illustrates the _____ of development.
3. _____ forces include genetic and health factors.
4. Neuroscience examines _____ relations.

INTERPRET

How does the biopsychosocial framework provide insight into the recurring issues of development (nature–nurture, continuity–discontinuity, universal–context-specific)?

APPLY

How does your life experience reflect the four developmental forces?

Recall answers: (1) genetics, (2) continuity, (3) Biological, (4) brain–behavior

1.2 Developmental Theories

LEARNING OBJECTIVES

- What is a developmental theory?
- How do psychodynamic theories account for development?
- What is the focus of learning theories of development?
- How do cognitive-developmental theories explain changes in thinking?
- What are the main points in the ecological and systems approach?
- What are the major tenets of life-span and life-course theories?

Marcus has just graduated from high school, first in his class. For his proud mother, Betty, this is a time to reflect on her son's past and to ponder his future. Marcus has always been a happy, easygoing child—a joy to rear. And he's constantly been interested in learning. Betty wonders why he is so perpetually good-natured and so curious. If she knew the secret, she laughed, she could write a best-selling book and be a guest on *The Colbert Report*!

The Eight Stages of Psychosocial Development in Erikson's Theory

Psychosocial Stage	Age	Challenge
Basic trust vs. mistrust	Birth to 1 year	To develop a sense that the world is safe, a "good place"
Autonomy vs. shame	1 to 3 years	To realize that one is an independent person who can make decisions and doubt
Initiative vs. guilt	3 to 6 years	To develop the ability to try new things and to handle failure
Industry vs. inferiority	6 years to adolescence	To learn basic skills and to work with others
Identity vs. identity confusion	Adolescence	To develop a lasting, integrated sense of self
Intimacy vs. isolation	Young adulthood	To commit to another in a loving relationship
Generativity vs. stagnation	Middle adulthood	To contribute to younger people through child rearing, child care, or other productive work
Integrity vs. despair	Late life	To view one's life as satisfactory and worth living

theory
an organized set of ideas that is designed to explain development

TO ANSWER BETTY'S QUESTIONS ABOUT HER SON'S GROWTH, developmental researchers need to provide a theory of his development. Theories are essential because they provide the "why's" for development. What is a theory? *In human development, a* **theory** *is an organized set of ideas that is designed to explain development.* For example, suppose friends of yours have a baby who cries often. You could imagine several explanations for her crying. Maybe the baby cries because she's hungry; maybe she cries to get her parents to hold her; maybe she cries because she's simply a cranky, unhappy baby. Each of these explanations is a very simple theory: It tries to explain why the baby cries so much. Of course, actual theories in human development are much more complicated, but the purpose is the same—to explain behavior and development.

There are no truly comprehensive theories of human development to guide research (Newman & Newman, 2007). Instead, five general perspectives influence current research: psychodynamic theory; learning theory; cognitive theory; ecological and systems theory; and theories involving the life-span perspective, selective optimization with compensation, and the life-course perspective. Let's consider each approach briefly.

Psychodynamic Theory

psychodynamic theories
theories proposing that development is largely determined by how well people resolve conflicts they face at different ages

Psychodynamic theories *hold that development is largely determined by how well people resolve conflicts they face at different ages.* This perspective traces its roots to Sigmund Freud's theory that personality emerges from conflicts that children experience between what they want to do and what society wants them to do.

Building on Freud's idea, Erik Erikson (1902–1994) proposed the first comprehensive life-span view, his psychosocial theory, which remains an important theoretical framework today.

Erikson's Theory

psychosocial theory
Erikson's proposal that personality development is determined by the interaction of an internal maturational plan and external societal demands

In his **psychosocial theory**, *Erikson proposed that personality development is determined by the interaction of an internal maturational plan and external societal demands.* He proposed that the life cycle is composed of eight stages and that the order of the stages is biologically fixed (the eight stages shown in ● Table 1.1). You can see that the name of each stage reflects the challenge people face at a particular age. For example, the challenge for young adults is to become involved in a loving relationship. Challenges are met through a combination of inner psychological influences and outer social influences. When challenges are met successfully, people are well prepared to meet the challenge of the next stage.

epigenetic principle
in Erikson's theory, the idea that each psychosocial strength has its own special period of particular importance

The sequence of stages in Erikson's theory is based on the **epigenetic principle**, *which means that each psychosocial strength has its own special period of particular impor-*

tance. The eight stages represent the order of this ascendancy. Because the stages extend across the whole life span, it takes a lifetime to acquire all of the psychosocial strengths. Moreover, Erikson realizes that present and future behavior must have its roots in the past, because later stages are built on the foundation laid in previous stages.

The psychodynamic perspective emphasizes that the trek to adulthood is difficult because the path is strewn with challenges. Outcomes of development reflect the manner and ease with which children surmount life's barriers. When children overcome early obstacles easily, they are better able to handle the later ones. A psychodynamic theorist would tell Betty that her son's cheerful disposition and his academic record suggest that he has handled life's early obstacles well, which is a good sign for his future development.

Erik Erikson

Learning Theory

In contrast to psychodynamic theory, learning theory concentrates on how learning influences a person's behavior. This perspective emphasizes the role of experience, examining whether a person's behavior is rewarded or punished. This perspective also emphasizes that people learn from watching others around them. Two influential theories in this perspective are behaviorism and social learning theory.

Behaviorism

Early in the 20th century, John Watson (1878–1958) believed that infants' minds were essentially "blank slates" and argued that learning determines what people will become. He assumed that, with the correct techniques, anything could be learned by almost anyone. In Watson's view, then, experience was just about all that mattered in determining the course of development.

Watson did little research to support his claims, but B. F. Skinner (1904–1990) filled this gap. *Skinner studied* **operant conditioning**, *in which the consequences of a behavior determine whether a behavior is repeated in the future.* Skinner showed that two kinds of consequences were especially influential. *A* **reinforcement** *is a consequence that increases the future likelihood of the behavior that it follows.* Positive reinforcement consists of giving a reward such as chocolate, gold stars, or paychecks to increase the likelihood of a previous behavior. A father who wants to encourage his daughter to help with chores may reinforce her with praise, food treats, or money whenever she cleans her room. Negative reinforcement consists of rewarding people by taking away unpleasant things. The same father could use negative reinforcement by saying that whenever his daughter cleans her room she doesn't have to wash the dishes or fold laundry.

A **punishment** *is a consequence that decreases the future likelihood of the behavior that it follows.* Punishment suppresses a behavior either by adding something aversive or by withholding a pleasant event. Should the daughter fail to clean her room, the father may punish her by nagging (adding something aversive) or by not allowing her to watch television (withholding a pleasant event).

Skinner's research was done primarily with animals, but human development researchers showed that the principles of operant conditioning could be extended readily to people, too (Baer & Wolf, 1968). Applied properly, reinforcement and punishment are indeed powerful influences on children, adolescents, and adults.

B. F. Skinner

operant conditioning
learning paradigm in which the consequences of a behavior determine whether a behavior is repeated in the future

reinforcement
a consequence that increase the future likelihood of the behavior that it follows

punishment
a consequence that decreases the future likelihood of the behavior that it follows

THINK ABOUT IT

Try to use the basic ideas of operant conditioning to explain how children create theories of the physical and social world.

Social Learning Theory

Researchers discovered that people sometimes learn without reinforcement or punishment. *People learn much by simply watching those around them, which is known as* **imitation or observational learning**. Imitation is occurring when one toddler throws a toy after seeing a peer do so or when a school-age child offers to help an older adult carry groceries because she's seen her parents do the same.

Perhaps imitation makes you think of "monkey-see, monkey-do," in which people simply mimic what they see. Early investigators had this view too, but research

imitation or observational learning
learning that occurs by simply watching how others behave

Albert Bandura

self-efficacy
people's beliefs about their own abilities and talents

quickly showed that this was wrong. People do not always imitate what they see around them. People are more likely to imitate if the person they see is popular, smart, or talented. They're also more likely to imitate when the behavior they see is rewarded than when it is punished. Findings like these imply that imitation is more complex than sheer mimicry. People are not mechanically copying what they see and hear; instead, they look to others for information about appropriate behavior. When peers are reinforced for behaving in a particular way, this encourages imitation.

Albert Bandura (1925–) based his social cognitive theory on this more complex view of reward, punishment, and imitation. Bandura's theory is "cognitive" because he believes people actively try to understand what goes on in their world; the theory is "social" because, along with reinforcement and punishment, what other people do is an important source of information about the world.

Bandura also argues that experience gives people a sense of **self-efficacy**, *which refers to people's beliefs about their own abilities and talents*. Self-efficacy beliefs help to determine when people will imitate others. A child who sees himself as not athletically talented, for example, will not try to imitate LeBron James dunking a basketball despite the fact that LeBron is obviously talented and popular. Thus, whether an individual imitates others depends on who the other person is, on whether that person's behavior is rewarded, and on the individual's beliefs about his or her own abilities.

Bandura's social cognitive theory is a far cry from Skinner's operant conditioning. The operant conditioned person who responds mechanically to reinforcement and punishment has been replaced by the social cognitive person who actively interprets these and other events. Nevertheless, Skinner, Bandura, and all learning theorists share the view that experience propels people along their developmental journeys. These theorists would tell Betty that she can thank experience for making her son Marcus both happy and successful academically.

Cognitive-Developmental Theory

Another way to approach development is to focus on thought processes and the construction of knowledge. In cognitive-developmental theory, the key is how people think and how thinking changes over time. Three distinct approaches have developed.

One approach postulates that thinking develops in a universal sequence of stages; Piaget's theory of cognitive development (and its recent extensions) is the best-known example. The second approach proposes that people process information as computers do, becoming more efficient over much of the life span; information-processing theory is an example of this view. The third approach emphasizes the contributions of culture on cognitive growth.

Piaget's Theory

The cognitive-developmental perspective focuses on how children construct knowledge and how their constructions change over time. Jean Piaget (1896–1980), who was the most influential developmental psychologist of the 20th century, proposed the best-known of these theories. Piaget believed that children naturally try to make sense of their world. Throughout infancy, childhood, and adolescence, youngsters want to understand the workings of both the physical and the social world. For example, infants want to know about objects: "What happens when I push this toy off the table?" And they want to know about people: "Who is this person who feeds and cares for me?" In their efforts to comprehend their world, Piaget argued that children act like scientists, creating theories about the physical and social worlds. Children try to weave all that they know about objects and people into a complete theory, which is tested daily by experience because their theories lead children to expect certain things to happen. As with real scientific theories, when the predicted events do occur, a child's belief in her theory grows stronger. When the predicted events do not occur, the child must revise her theory.

Piaget's Four Stages of Cognitive Development

Stage	Approximate Age	Characteristics
Sensorimotor	Birth to 2 years	Infant's knowledge of the world is based on senses and motor skills; by the end of the period, uses mental representation
Preoperational thought	2 to 6 years	Child learns how to use symbols such as words and numbers to represent aspects of the world but relates to the world only through his or her perspective
Concrete operational thought	7 years to early adolescence	Child understands and applies logical operations to experiences provided they are focused on the here and now
Formal operational thought	Adolescence and beyond	Adolescent or adult thinks abstractly, deals with hypothetical situations, and speculates about what may be possible

Let's take the perspective of an infant whose theory of objects includes the idea that "Toys pushed off the table fall to the floor." If the infant pushes some other object—a plate or an article of clothing—she will find that it, too, falls to the floor, and she can then make the theory more general: "Objects pushed off the table fall to the floor." Piaget also believed children begin to construct knowledge in new ways at a few critical points in development. When this happens, they revise their theories radically. These changes are so fundamental that the revised theory is, in many respects, a brand-new theory. Piaget claimed that these changes occur three times in development: once at about age 2 years, a second time at about age 7, and a third time just before adolescence. These changes mean that children go through four distinct stages in cognitive development. Each stage represents a fundamental change in how children understand and organize their environment, and each stage is characterized by more sophisticated types of reasoning. For example, the first or *sensorimotor* stage begins at birth and lasts until about 2 years of age. As the name implies, sensorimotor thinking refers to an infant's constructing knowledge through sensory and motor skills. This stage and the three later stages are shown in ●Table 1.2.

Jean Piaget

Piaget's theory has had an enormous influence on how developmentalists and practitioners think about cognitive development. The theory has been applied in many ways—from the creation of discovery learning toys for children to the ways teachers plan lessons. In Chapter 4 we'll see how Piaget explained thinking during infancy and the preschool years; in Chapters 6 and 8 we'll learn about his description of thinking in school-age children and adolescents.

Piaget also had his critics. In Chapter 10 we will see how some have argued that cognitive development does not stop with adolescence but continues well into adulthood.

Information-Processing Theory

Information-processing theorists draw heavily on how computers work to explain thinking and how it develops through childhood and adolescence. *Just as computers consist of both hardware (disk drives, random-access memory, and central processing unit) and software (the programs it runs),* **information-processing theory** *proposes that human cognition consists of mental hardware and mental software.* Mental hardware refers to cognitive structures, including different memories where information is stored. Mental software includes organized sets of cognitive processes that enable people to complete specific tasks, such as reading a sentence, playing a video game, or hitting a baseball. For example, an information-processing psychologist would say that, for a student to do well on an exam, she must encode the informa-

information-processing theory
theory proposing that human cognition consists of mental hardware and mental software

Information-processing theory helps explain how this girl learns, stores, and retrieves information so she can pass the exam she is studying for.

tion as she studies, store it in memory, and then retrieve the necessary information during the test.

How do information-processing psychologists explain developmental changes in thinking? To answer this question, think about improvements in personal computers. Today's personal computers can accomplish much more than computers built just a decade ago. Why? Today's computers have better hardware (e.g., more memory and a faster processor) as well as more sophisticated software that takes advantage of the better hardware. Like modern computers, older children and adolescents have better hardware and better software than younger children, who are more like last year's out-of-date model. For example, older children typically solve math word problems better than younger children because they have greater memory capacity to store the facts in the problem and because their methods for performing arithmetic operations are more efficient.

Some researchers also point to deterioration of the mental hardware—along with declines in the mental software—as explanations of cognitive aging. In Chapter 14 we will see, for example, that normal aging brings with it significant changes in people's ability to process information.

Vygotsky's Theory

Lev Vygotsky (1896–1934) was one of the first theorists to emphasize that children's thinking does not develop in a vacuum but rather is influenced by the sociocultural context in which children grow up. A Russian psychologist, Vygotsky believed that, because a fundamental aim of all societies is to enable children to acquire essential cultural values and skills, every aspect of a child's development must be considered against this backdrop. For example, most parents in the United States want their children to work hard in school and be admitted to college because earning a degree is one of the keys to finding a good job. However, in Mali (an African country), Bambara parents want their children to learn to farm, herd animals such as cattle and goats, gather food such as honey, and hunt because these skills are key to survival in their environment. Vygotsky viewed development as an apprenticeship in which children develop as they work with skilled adults, including teachers and parents.

For Piaget, information-processing theorists, and Vygotsky, children's thinking becomes more sophisticated as they develop. Piaget explained this change as resulting from the more sophisticated knowledge that children construct from more sophisticated thinking; information-processing psychologists attribute it to more sophisticated mental hardware and mental software. What would these theorists say to Betty about Marcus's good nature? Vygotsky would point out that Betty communicated key aspects of the culture to Marcus, which influenced his good nature. Neither Piaget nor information-processing theorists would have much to say, because their theories do not handle personality issues very well. What about Marcus's academic success? That's a different story. Piaget would explain that all children naturally want to understand their world; Marcus is simply unusually skilled in this regard. An information-processing psychologist would point to superior hardware and superior software as the keys to his academic success. Vygotsky would again emphasize Betty's influence in cultural transmission.

Lev Vygotsky

The Ecological and Systems Approach

ecological theory
theory based on idea that human development is inseparable from the environmental contexts in which a person develops

Most developmentalists agree that the environment is an important force in many aspects of development. However, only ecological theories (which get their name from the branch of biology dealing with the relation of living things to their environment and to one another) have focused on the complexities of environments and their links to development. *In **ecological theory**, human development is inseparable from the environmental contexts in which a person develops.* The ecological approach proposes that all aspects of development are interconnected, much like the threads of a spider's web, so that no aspect of development can be isolated from others and understood independently. An ecological theorist would emphasize that, to understand why adolescents behave as they do, we need to consider the many different systems that influence them, including parents, peers, teachers, television, the neighborhood, and social policy.

We will consider two examples of the ecological and systems approach: Bronfenbrenner's theory and the competence–environmental press framework.

Urie Bronfenbrenner

Bronfenbrenner's Theory

The best-known proponent of the ecological approach was Urie Bronfenbrenner (1917–2005), who proposed that the developing person is embedded in a series of complex and interactive systems (Bronfenbrenner, 1979, 1989, 1995). Bronfenbrenner divided the environment into the four levels shown in ■ Figure 1.2: the microsystem, the mesosystem, the exosystem, and the macrosystem.

At any point in life, the **microsystem** *consists of the people and objects in an individual's immediate environment.* These are the people closest to a child, such as parents or siblings. Some children may have more than one microsystem; for example, a young child might have the microsystems of the family and of the day-care setting. As you can imagine, microsystems strongly influence development.

Microsystems themselves are connected to create the mesosystem. *The* **mesosystem** *provides connections across microsystems, because what happens in one microsystem is likely to influence others.* Perhaps you've found that if you have a stressful day at work or school then you're often grouchy at home. This indicates that your mesosystem is alive and well; your microsystems of home and work are interconnected emotionally for you.

The **exosystem** *refers to social settings that a person may not experience firsthand but that still influence development.* For example, changes in government policy regarding welfare may mean that poor children have less opportunity for enriched preschool experiences. Although the influence of the exosystem is no more than secondhand, its effects on human development can be quite strong.

microsystem
the people and objects in an individual's immediate environment

mesosystem
provides connections across microsystems

exosystem
social settings that a person may not experience firsthand but that still influence development

■ **Figure 1.2**
Bronfenbrenner's ecological approach emphasizing the interaction across different systems in which people operate.

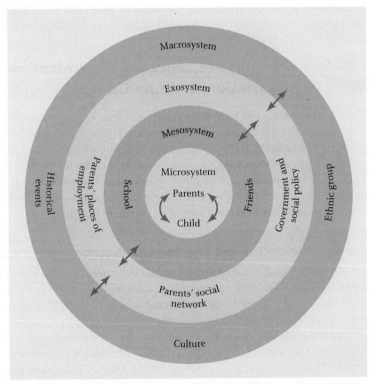

Source: Kopp, Clarie B. Kopp/Krakow, Joanne B., *The Child: Development in Social Context*, © 1982, p. 648. Reprinted by permission of Pearson Education, Inc., Upper Saddle River, New Jersey.

macrosystem
the cultures and subcultures in which the microsystem, mesosystem, and exosystem are embedded

The broadest environmental context is the **macrosystem**, *the cultures and subcultures in which the microsystem, mesosystem, and exosystem are embedded.* A mother, her workplace, her child, and the child's school are part of a larger cultural setting, such as Asian Americans living in Southern California or Italian Americans living in large cities on the East Coast. Members of these cultural groups share a common identity, a common heritage, and common values. The macrosystem evolves over time; each successive generation may develop in a unique macrosystem.

Bronfenbrenner's ecological theory emphasizes the many levels of influence on human development. People are affected directly by family members and friends and indirectly by social systems such as neighborhoods and religious institutions—which, in turn, are affected by the beliefs and heritage of one's culture.

Competence–Environmental Press Theory

Another view of the influence of environments on human development comes from Lawton and Nahemow's (1973) competence–environmental press theory. *According to this theory, people adapt most effectively when there is a good match between their* **competence** *or abilities and the* **environmental press,** *or the demands put on them by the environment.* This theory was originally proposed to account for the ways in which older adults function in their environment, but it applies as well throughout the life span. For example, the match between a child's social skills and her peer group's demands can determine whether she is accepted by the peer group. As with Bronfenbrenner's theory, competence–environmental press theory emphasizes that, in order to understand people's functioning, it is essential to understand the systems in which they live.

competence
a person's abilities

environmental press
demands put on people by the environment

Ecological theorists would agree with learning theorists in telling Betty that the environment has been pivotal in her son's amiable disposition and his academic achievements. However, the ecological theorist would insist that environment means much more than the reinforcements, punishments, and observations that are central to learning theory; such a theorist would emphasize the different levels of environmental influence on Marcus. Betty's ability to balance home (microsystem) and work (mesosystem) so skillfully (which meant that she was usually in a good mood herself) contributed positively to Marcus's development, as did Betty's membership in a cultural group (exosystem) that emphasized the value of doing well in school.

Life-Span Perspective, Selective Optimization With Compensation, and Life-Course Perspective

Most of the theories of human development that we have considered so far pay little attention to the adult years of the life span. Historically, adulthood was downplayed owing to the belief that it was a time when abilities had reached a plateau (rather than continuing to develop) and that adulthood was followed by inevitable decline in old age. However, modern perspectives emphasize the importance of viewing human development as a lifelong process. These perspectives view development in terms of where a person has been and where he or she is heading.

Life-Span Perspective and Selective Optimization With Compensation

What would it be like to try to understand your best friend without knowing anything about his or her life? We cannot understand adults' experiences without appreciating their childhood and adolescence. Placing adults' lives in this broader context is what the life-span perspective does.

According to the **life-span perspective,** *human development is multiply determined and cannot be understood within the scope of a single framework.* The basic premise of the life-span perspective is that aging is a lifelong process of growing up and growing old, beginning with conception and ending with death. No single period of a person's life (such as childhood, adolescence, or middle age) can be understood apart from its origins and its consequences. To understand a specific period, we must know

life-span perspective
view that human development is multiply determined and cannot be understood within the scope of a single framework

what came before and what is likely to come afterward (Riley, 1979). In addition, how one's life is played out is affected by social, environmental, and historical change. Thus, the experiences of one generation may not be the same as those of another.

Paul Baltes (1939–2006) and colleagues provide many of the main approaches to human development from a life-span perspective (Baltes & Smith, 2003) in a model that has influenced a wide range of research, especially on adult development and aging. A key point in their model is that human development is complex and cannot be understood from a single discipline. They identify four key features of the life-span perspective as follows.

Paul Baltes

- *Multidirectionality.* Development involves both growth and decline; as people grow in one area they may lose in another and at different rates. For example, people's vocabulary ability tends to increase throughout life, but reaction time tends to slow down.

- *Plasticity.* One's capacity is not predetermined or carved in stone. Many skills can be learned or improved with practice, even in late life. For example, people can learn better ways to remember information, which may help them deal with the declines in memory ability that accompany aging. There are limits to the degree of potential improvement, though, as described in later chapters.

- *Historical context.* Each of us develops within a particular set of circumstances determined by the historical time in which we are born and the culture in which we grow up. For example, living in a middle-class suburb in 1950s Indianapolis has little in common with living in a poor Latino neighborhood in 1990s Texas.

- *Multiple causation.* How we develop results from the biological, psychological, sociocultural, and life-cycle forces that we mentioned previously. For example, two children growing up in the same family will have different experiences if one has a developmental disability and the other does not.

Taken together, the principles of the life-span perspective create a way to describe and explain the successful adaptation of people to the changes that occur with aging by proposing an interaction between three processes: selection, compensation, and optimization (Baltes, 1997; Baltes et al., 2006; B. Baltes & Heydens-Gahir, 2003; M. Baltes & Carstensen, 1999). Selection processes serve to choose goals, life domains, and life tasks, whereas optimization and compensation concern maintaining or enhancing chosen goals. *The basic assumption of the* **selective optimization with compensation (SOC) model** *is that the three processes form a system of behavioral action that generates and regulates development and aging.*

As people mature and grow old, they select from a range of possibilities or opportunities. This selection occurs for two main reasons. *Elective selection* occurs when one chooses to reduce one's involvement to fewer domains as a result of new demands or tasks, such as when a college student drops out of some organizations because of the amount of work required in the courses she is taking that term. *Loss-based selection* occurs when this reduced involvement happens as a result of anticipated losses in personal or environmental resources, such as when an older person stops going to church because he can no longer drive. In either case, selection can involve the continuation of previous goals on a lesser scale or the substitution of new goals, and it may be either proactive or reactive.

Compensation occurs when a person's skills have decreased so that they no longer function well in a particular domain. When a person compensates, she searches for an alternative way to accomplish the goal; for example, if an injury reduces one's ability to drive then one might compensate by taking the bus. Sometimes, compensation requires learning a new skill; for example, an older adult who is experiencing short-term memory problems might compensate by learning to use a personal digital assistant. Thus, compensation differs from selection in that the task or goal is maintained—although other means are used to achieve it.

Optimization involves minimizing losses and maximizing gains. The main idea is to find the best match possible between one's resources (biological, psychological, and

selective optimization with compensation (SOC) model
model in which three processes (selection, optimization, and compensation) form a system of behavioral action that generates and regulates development and aging

At age 93, cellist Pablo Casals said he still practiced 3 hours a day because he was "beginning to notice some improvement."

life-course perspective
description of how various generations experience the biological, psychological, and sociocultural forces of development in their respective historical contexts

sociocultural) and one's desired goals. Because people cannot achieve optimal outcomes in everything, development becomes a dynamic process of selecting the right goals and compensating when possible to help maximize the odds of achieving them.

One can see the SOC model at work in many situations. For example, aging musicians may reduce the number of pieces they play (selection), rehearse them more often (optimization), and sing them in a lower key (compensation). This way, they can continue playing concerts later in life. A college athlete who excels at ice hockey and baseball may decide to concentrate on hockey (selection), work on training all year (optimization), and develop a wicked wrist shot to make up for a mediocre slap shot (compensation).

The life-span perspective and the SOC model have provided important approaches to the contemporary study of human development. The emphasis on the need for a multidisciplinary approach and for recognizing many interactive forces will be developed throughout this text.

Life-Course Perspective

Adults often describe their lives as a story that includes several key life transitions (e.g., going to school, getting a first job, getting married, having children). Such stories show how people move through their lives and experience unique interactions of the four forces of development.

The **life-course perspective** *describes the ways in which various generations experience the biological, psychological, and sociocultural forces of development in their respective historical contexts.* Specifically, it lets researchers examine the effects of historical time on how people create their lives (Dannefer & Miklowski, 2006; Hagestad & Dannefer, 2001; Hareven, 1995, 2001; Mayer, 2009). A key feature of the life-course perspective is the dynamic interplay between the individual and society. This interplay creates three major dimensions, which all involve timing and underlie the life-course perspective:

- *The individual timing of life events in relation to external historical events.* This dimension addresses the question: How do people time and sequence their lives (e.g., getting a first job) in the context of changing historical conditions (e.g., economic good times or recession)?

- *The synchronization of individual transitions with collective familial ones.* This dimension addresses the question: How do people balance their own lives (e.g., work obligations) with those of their family (e.g., children's soccer games)?

- *The impact of earlier life events, as shaped by historical events, on subsequent ones.* This dimension addresses the question: How does experiencing an event earlier in life (e.g., a male turning 18 years old) at a particular point in history (e.g., when there is a military draft) affect one's subsequent life (e.g., choosing a particular career)?

Research from the life-course perspective has clearly shown that major life transitions such as marriage, childbearing, starting and ending a career, and completing one's education occur at many different ages across people and generations. These differences first appear after adolescence, when people begin to have much more control over the course of their lives. Research has also shown that life transitions are more continuous and multidirectional than previously thought. For example, completing an education was relegated to early adulthood in traditional models, yet current trends toward lifelong learning make this view obsolete. Finally, research shows that the various domains of people's lives are highly interdependent; for example, the decision to have a child is often made in the context of where one is in one's career and education.

The emphasis in the life course perspective on interrelations between the individual and society with reference to historical time has made it a dominant view in the social sciences. In particular, this approach is useful in helping researchers understand how the various aspects of people's experiences (work, family, education) interact to create unique lives.

Overall, life-span and life-cycle theories have greatly enhanced the general body of developmental theory by drawing attention to the role of aging in the broader context

Theoretical Perspectives on Human Development

Perspective	Examples	Main Idea	Emphases in Biopsychosocial Framework	Positions on Developmental Issues
Psychodynamic	Erikson's psychosocial theory	Personality develops through sequence of stages	Psychological, social, and life-cycle forces crucial; less emphasis on biological	Nature–nurture interaction, discontinuity, universal sequence but individual differences in rate
Learning	Behaviorism (Watson, Skinner)	Environment controls behavior	In all theories, some emphasis on biological and psychological, major focus on social, little recognition of life cycle	In all theories, strongly nurture, continuity, and universal principles of learning
Cognitive	Social learning theory (Bandura)	People learn through modelling and observing		
	Piaget's theory (and extensions)	For Piaget, thinking develops in a sequence of stages	For Piaget main emphasis on biological and social forces, less on psychological, little on life cycle	For Piaget, strongly nature, discontinuity, and universal sequence of stages
	Information-processing theory	Thought develops by increases in efficiency at handling information	Emphasis on biological and psychological, less on social and life cycle	Nature–nurture interaction, continuity, individual differences in universal structures
	Vygotsky's theory	Development influenced by culture	Emphasis on psychological and social forces	Nature–nurture interaction, continuity, individual differences
Ecological and Systems	Bronfenbrenner's theory	Developing person embedded in a series of interacting systems	Low emphasis on biological, moderate on psychological and life cycle, heavy on social	Nature–nurture interaction, continuity, context-specific
	Competence–environmental press (Lawton and Nahemow)	Adaptation is optimal when ability and demands are in balance	Strong emphasis on biological, psychological, and social, moderate on life cycle	Nature–nurture interaction, continuity, context-specific
Life-Span Perspective/ SOC and	Baltes's life-span perspective and selective optimization with compensation (SOC)	Development is multiply determined; optimization of goals	Strong emphasis on the interactions of all four forces; cannot consider any in isolation	Nature–nurture interaction, continuity and discontinuity, context-specific
Life-Course Perspective	Life-course theory	Life course transitions decreasingly tied to age; increased continuity over time; specific life paths across domains are interdependent	Strong emphasis on psychological, sociocultural, life cycle; less on biological	Nature–nurture interaction, continuity and discontinuity, context-specific

of human development, even though they still have yet to reach their full explanatory potential (Mayer, 2009). These theories have played a major role in conceptualizing adulthood and have greatly influenced the research we consider in Chapters 10 through 15. Theorists espousing a life-span or life-course perspective would tell Betty that Marcus will continue to develop throughout his adult years and that this developmental journey will be influenced by biopsychosocial forces, including his own family.

The Big Picture

As summarized in ● Table 1.3, each of the theories provides ways of explaining how the biological, psychological, sociocultural, and life-cycle forces create human development. But because no single theory provides a complete explanation of all aspects of development, we must rely on the biopsychosocial framework to help piece together an account based on many different theories. Throughout the remainder of this text, you will read about many theories that differ in focus and in scope. To help you understand them better, each theory will be introduced in the context of the issues that it addresses.

Because one of the criteria for a theory is that it be testable, developmentalists have adopted certain methods to help accomplish this. The next section describes the methods by which developmentalists conduct research and test their theories.

Test Yourself

RECALL

1. _____organize knowledge in order to provide testable explanations of human behaviors and the ways in which they change over time.

2. The _____ perspective proposes that development is determined by the interaction of an internal maturational plan and external societal demands.

3. According to social cognitive theory, people learn from reinforcements, from punishments, and through _____.

4. Piaget's theory and Vygotsky's theory are examples of the _____ perspective.

5. According to Bronfenbrenner, development occurs in the context of the _____, mesosystem, exosystem, and macrosystem.

6. A belief that human development is characterized by multidirectionality and plasticity is fundamental to the _____ perspective.

INTERPRET

How are the information-processing perspective and Piaget's theory similar? How are they different?

APPLY

Using three different developmental theories, explain how LeBron James or Lady Gaga achieved their success.

Recall answers: (1) Theories, (2) psychosocial, (3) observing others, (4) cognitive-developmental, (5) microsystem, (6) life-span

---

1.3 Doing Developmental Research

LEARNING OBJECTIVES

- How do scientists measure topics of interest in studying human development?

- What research designs are used to study human development?

- How do researchers integrate results from multiple studies?

- What ethical procedures must researchers follow?

- How do investigators communicate results from research studies?

- How does research affect public policy?

Leah and Joan are both mothers of 10-year-old boys. Their sons have many friends, but the basis for the friendships is not obvious to the mothers. Leah believes "opposites attract"—children form friendships with peers who have complementary interests and abilities. Joan doubts this; her son seems to seek out other boys who are near clones of himself in terms of interests and abilities.

SUPPOSE LEAH AND JOAN KNOW that you're taking a course in human development, so they ask you to settle their argument. Leah believes complementary children are more often friends, whereas Joan believes similar children are more often friends. You know that research could show whose ideas are supported under which circumstances, but how? In fact, human development researchers must make several important decisions as they prepare to study a topic. They need to decide how to measure the topic of interest; they must design their study; they must choose a method for studying development; and they must decide whether their plan respects the rights of individuals participating in the research.

Human development researchers do not always stick to this sequence of steps. For example, often researchers consider the research participants while making other decisions—perhaps rejecting a measurement procedure because it violates the rights of participants. Nevertheless, for simplicity we will adhere to this sequence as we describe each of the steps in doing developmental research.

Measurement in Human Development Research

The first step in doing developmental research is deciding how to measure the topic or behavior of interest. So, the first step toward answering Leah and Joan's question about friendships would be to decide how to measure friendships.

Human development researchers use one of four approaches: observing systematically, using tasks to sample behavior, asking people for self-reports, and taking physiological measures.

Systematic Observation

As the name implies, **systematic observation** *involves watching people and carefully recording what they do or say.* Two forms of systematic observation are common. *In* **naturalistic observation**, *people are observed as they behave spontaneously in a real-life situation.* There's a catch with observation, though. Researchers can't keep track of everything that someone does, so beforehand they must decide what variables to record. For example, researchers studying friendship might decide to observe children as they start their first year in middle school (chosen because many children will be making new friends at this time). They could decide to focus on and record where children sit in the lunchroom and who talks to whom.

Structured observations *differ from naturalistic observations in that the researcher creates a setting that is likely to bring out the behavior of interest.* Structured observations are particularly useful for studying behaviors that are difficult to observe naturally, such as how people respond to emergencies. An investigator relying on natural observations to study people's responses to emergencies wouldn't make much progress with naturalistic observation because emergencies don't occur at predetermined times and locations. However, using a structured observation, an investigator might stage an emergency—perhaps cooperating with authorities to simulate an accident—in order to observe other people's responses.

Other behaviors are difficult for researchers to observe because they occur in private settings, not public ones. For example, much interaction between friends takes place at home, where it would be difficult for investigators to observe unobtrusively. However, friends could be asked to come to the researcher's laboratory, which might be furnished to resemble a room in a typical house. Friends would then be asked to perform some activity typical of friends, such as playing a game. The researchers would then observe their activity from another room, through a one-way mirror, or by videotaping them.

Structured observations are valuable in enabling researchers to observe behavior(s) that would otherwise be difficult to study. But there are limits as well. For example, observing friends as they discuss a problem in a mock family room has many artificial aspects to it: The friends are not in their own homes, they were told in general terms what to do, and they know they're being observed. Any of these factors may cause friends to behave differently than they would in the real world. Researchers must be careful that their method does not make the behavior they are observing unnatural or unrealistic.

systematic observation
watching people and carefully recording what they do or say

naturalistic observation
technique in which people are observed as they behave spontaneously in some real-life situation

structured observations
technique in which a researcher creates a setting that is likely to elicit the behavior of interest

Copyright © Cengage Learning 2010

Sampling Behavior With Tasks

When investigators can't observe a behavior directly, another popular alternative is to create tasks that are thought to sample the behavior of interest. One task often used to measure older adults' memory is "digit span": Adults listen as a sequence of digits is presented aloud. After the last digit is presented, they try to repeat the digits in order.

Another example is shown in ■ Figure 1.3. To study the ability to recognize emotions, the child has been asked to look at the photographs and point to the face that looks happy. A child's answers on this sort of task are useful in determining his or her ability to recognize emotions. This approach is popular and convenient; a potential problem with it, however, is that the task may not provide a realistic sample of the behavior of interest. For example, asking children to judge emotions from photographs may not be valid because it underestimates what they do in real life. Can you think of reasons why this might be the case? We mention several reasons on page 35, just after Test Yourself.

Self-Reports

Self-reports actually represent a special case of using tasks to measure people's behavior. **Self-reports** *are people's answers to questions about the topic of interest.*

When questions are posed in written form, the self-report is a questionnaire; when questions are posed orally, the self-report is an interview. Either way, questions are created that probe different aspects of the topic of interest. For example, if you believe that adults are more often friends when they have interests in common, then you might tell your research participants the following:

> Tom and Dave just met each other at work. Tom likes to read, plays the clarinet, and is not interested in sports; Dave likes to Tweet his friends, enjoys tinkering with his car, and watches sports all weekend. Do you think Tom and Dave will become friends?

The participants would then decide, perhaps using a rating scale, whether the men are likely to become friends.

Self-reports are useful because they can lead directly to information on the topic of interest. They are also relatively convenient, particularly when they can be administered to groups of participants or online.

However, self-reports are not always a good measure of people's behavior because not all answers are accurate. Why? When asked about past events, for example, people may not remember them accurately. For example, an older adult asked about ado-

self-reports
people's answers to questions about the topic of interest

THINK ABOUT IT

If you were studying middle-aged adults caring for their aging parents, what would be the relative advantages of systematic observation, sampling behavior with tasks, and self-reports?

Measuring Behaviors of Interest in Human Development Research

Method	Strength	Weakness
Systematic observation		
Naturalistic observation	Captures people's behavior in its natural setting	Difficult to use with behaviors that are rare or that typically occur in private settings
Structured observation	Can be used to study behaviors that are rare or that typically occur in private settings	May be invalid if structured setting distorts the behavior
Sampling behavior with tasks	Convenient—can be used to study most behaviors	May be invalid if the task does not sample behavior as it occurs naturally
Self-reports	Convenient—can be used to study most behaviors	May be invalid because people answer incorrectly (due to either forgetting or response bias)
Physiological measures	Provide a more direct measure of underlying behavior	Highly specific in what they measure and thus cannot be applied broadly

lescent friends may not remember those friendships well. Sometimes people answer incorrectly as a result of "response bias": For many questions, some responses are more socially acceptable than others, and people are more likely to select socially acceptable answers than socially unacceptable ones. For example, many would be reluctant to admit that they have no friends at all. As long as investigators remain aware of these potential biases, though, self-report can be a valuable tool for research in human development.

Physiological Measures

One less common but potentially powerful form of measurement is measuring people's physiological responses. Earlier we saw that brain activity is used in neuroscience research to track certain behaviors, such as memory. Another measure is heart rate, which often slows down when people are paying close attention to something interesting. Consequently, researchers often measure heart rate to determine a person's degree of attention. As another example, the hormone cortisol is often secreted in response to stress. By measuring cortisol levels in people's saliva, scientists can determine when they are experiencing stress.

As these examples suggest, physiological measures are usually specialized—they focus on a particular aspect of a person's behavior (memory, attention, and stress in the examples). What's more, they're often used with other behaviorally oriented methods. A researcher studying stress might observe several people for overt signs of stress, ask parents/partners/friends to rate the target people's stress, and also measure cortisol in the target people's saliva. If all three measures lead to the same conclusions about stress, then the researcher can be much more confident about those conclusions.

The strengths and weaknesses of the four approaches to measurement are summarized in ●Table 1.4.

Reliability and Validity

After researchers choose a method, they must show that it is both reliable and valid. *The **reliability** of a measure is the extent to which it provides a consistent index of a characteristic.* A measure of friendship, for example, is reliable if it gives a consistent estimate of a person's friendship network each time you administer it. But reliability isn't enough. A measure must also be valid to be useful in research.

*The **validity** of a measure refers to whether it really measures what researchers think it measures.* For example, a measure of friendship is valid if it actually measures friendship and not, for example, popularity. Validity is often established by showing that the measure in question is closely related to another measure that is known to be valid.

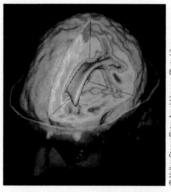

■ **Figure 1.4**
Brain imaging techniques provide a physiological measure that helps researchers understand brain–behavior relations.

reliability
extent to which a measure provides a consistent index of a characteristic

validity
extent to which a measure actually assesses what researchers think it does

Representative Sampling

Valid measures also depend on the people who are tested. *Researchers are usually interested in broad groups of people called* **populations**. Examples of populations would be all American 7-year-olds or all African American grandparents. *Virtually all studies include only a* **sample** *of people, which is a subset of the population.* Researchers must take care that their sample really is representative of the population of interest, because an unrepresentative sample can lead to invalid conclusions. For example, if a study of friendship in older adults tested only people who had no siblings, you would probably decide that this sample is not representative of the population of older adults and question the validity of its results.

As you read on, you'll soon discover that much of the research we describe was conducted with samples of middle-class European American people. Are these samples representative of all people in the United States? Of all people in the world? Sometimes, but not always. To make samples more representative, some U.S. federal agencies now require researchers to include certain groups (e.g., ethnic minorities, women, children) unless there is a compelling reason not to do so. This policy may make it possible to obtain a broader view of developmental processes. But until we have representative samples in all developmental research, we cannot know whether a particular phenomenon applies only to the group studied or to people more generally.

General Designs for Research

Having selected a way to measure the topic or behavior of interest, researchers must next embed this measure in a research design that yields useful, relevant results. Human development researchers rely on two primary designs in their work: correlational studies and experimental studies.

Correlational Studies

In a **correlational study**, *investigators look at relations between variables as they exist naturally in the world.* Imagine a researcher who wants to test the idea that smarter people have more friends. To find out, the researcher would measure two variables for each person, the number of friends that the person has and the person's intelligence, then see whether the two variables are related.

The results of a correlational study are usually measured by calculating a **correlation coefficient**, *which expresses the strength and direction of a relation between two variables. Correlations can range from −1.0 to 1.0.* The correlation coefficient reflects one of three possible relations between intelligence and the number of friends:

- People's intelligence is unrelated to the number of friends they have, reflected in a correlation of 0.
- People who are smart tend to have more friends than people who are not as smart. That is, *more* intelligence is associated with having *more* friends. In this case, the variables are positively related and the correlation is between 0 and 1.
- People who are smart tend to have fewer friends than people who are not as smart. That is, *more* intelligence is associated with having *fewer* friends. In this case, the variables are negatively related so the correlation is between −1 and 0.

In interpreting a correlation coefficient, you need to consider both the sign *and* the size of the correlation. The sign indicates the *direction* of the relation between variables: a positive sign means that larger values of one variable are associated with larger values of the second variable, whereas a negative sign means that larger values of one variable are associated with smaller values of the second variable.

The *size* or *strength* of a relation is measured by how much the correlation differs from 0, either positively or negatively. A correlation of .9 between intelligence and

populations
broad groups of people that are of interest to researchers

sample
a subset of the population

correlational study
investigation looking at relations between variables as they exist naturally in the world

correlation coefficient
an expression of the strength and direction of a relation between two variables

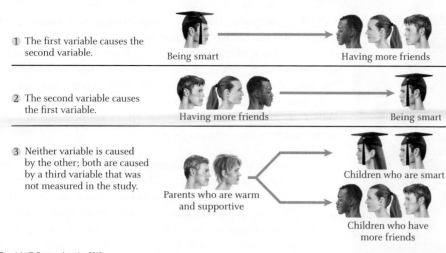

1 The first variable causes the second variable.

Being smart → Having more friends

2 The second variable causes the first variable.

Having more friends → Being smart

3 Neither variable is caused by the other; both are caused by a third variable that was not measured in the study.

Parents who are warm and supportive → Children who are smart / Children who have more friends

■ **Figure 1.5**
There are three basic interpretations of a correlation coefficient because there is no direct way to assess cause and effect.

number of friends would indicate a very strong relation: Knowing a person's intelligence, you could accurately predict how many friends the person has. If instead the correlation were only .3, then the link between intelligence and number of friends would be relatively weak: Although more intelligent people would have more friends on average, there would be many exceptions to this rule. Similarly, a correlation of −.9 would indicate a strong negative relation between intelligence and number of friends whereas a correlation of −.3 would indicate a weak negative relation.

Although a correlational study can determine whether variables are related, it doesn't address the question of cause and effect between the variables. For example, suppose a researcher finds that the correlation between intelligence and number of friends is .7, which indicates that people who are smarter have more friends than people who are not as smart. This correlation has three possible interpretation, shown in ■ Figure 1.5: (1) perhaps being smart causes people to have more friends; (2) having more friends causes people to be smarter; or (3) neither variable causes the other, instead, both intelligence and number of friends are caused by a third variable (such as parents who are supportive) that was not measured in the study. Any of these interpretations could be true, but they cannot be distinguished in a correlational study. Investigators who wish to track down causes must resort to a different design, called an experimental study.

Experimental Studies

An **experiment** *is a systematic way of manipulating the key factor(s) that the investigator thinks causes a particular behavior. The factor being manipulated is called the **independent variable**; the behavior being observed is called the **dependent variable**.* In human development, an experiment requires that the investigator begin with one or more independent variables (usually treatments, interventions, experiences, events, and so forth) that are thought to affect the behavior of interest. People are then assigned randomly to conditions that differ in the amount of the independent variable they are given. Finally, an appropriate measure (the dependent variable) is taken of all participants to see whether the treatment or treatments had the expected effect. Because each person has an equal chance of being assigned to each treatment condition (the definition of random assignment), any differences between the groups can be attributed to the differential treatment people received in the experiment rather than to other factors.

Suppose, for example, that an investigator believes older adults perform better in a driving simulator when they are not talking on a mobile phone. The photo shows how we might test this hypothesis. Older adults come to the laboratory, where they are randomly assigned to a driving simulation during which they either do or do not talk on a mobile phone.

experiment
a systematic way of manipulating the key factor(s) that the investigator thinks causes a particular behavior

independent variable
the factor being manipulated

dependent variable
the behavior being observed

An experiment using a driving simulator to examine skills of older drivers. In this "no phone" condition, the driver is not talking on a mobile phone.

All participants take the same driving test under circumstances that are held as constant as possible—except for the presence or absence of the mobile phone. If scores on the driving test are, on average, greater in the "no phone" condition than in the "mobile phone" condition, then the investigator may say with confidence that the mobile phone has a harmful effect on driving skill. Conclusions about cause and effect are possible in this example because the direct manipulation occurred under controlled conditions.

Human development researchers usually conduct experiments in laboratory-like settings because this allows better control over the variables that may influence the outcome of the research. Thus, a shortcoming of laboratory experiments is that the behavior of interest is not studied in its natural setting. For example, performance on a driving simulator may not be the same as actual driving performance. Consequently, there is always the potential problem that the results may be invalid because they are artificial—specific to the laboratory setting and not representative of the behavior in the "real world."

Each research design used by developmentalists has both strengths and weaknesses. There is no one best method. Consequently, no single investigation can definitively answer a question. Researchers rarely rely on one study or even one method to reach conclusions. Instead, they prefer to find converging evidence from as many different kinds of studies as possible.

Qualitative Studies

Suppose you live near a children's playground. Each day, you watch the children play various games with each other and on the swings and sliding board. Because you are interested in learning more about how children go about playing, you decide to watch more carefully. With the parents' permission, you video the children's play each day for several weeks. You then watch the videos and notice whether there are specific patterns that emerge.

qualitative research
method that involves gaining in-depth understanding of human behavior and what governs it

What you have done is to conduct one type of **qualitative research**, *which involves gaining in-depth understanding of human behavior and what governs it.* Unlike quantitative research, qualitative research seeks to uncover reasons underlying various aspects of behavior. Because qualitative research typically involves intensive observation of behavior over extended periods of time, the need is for smaller but focused samples rather than large random samples. Frequently used techniques include video recording and detailed interviews, from which qualitative researchers categorize the data into patterns as the primary basis for organizing and reporting results. In contrast, quantitative research relies on numerical data and statistical tests as the bases for reporting results.

Qualitative research can be conducted for its own sake, as a preliminary step, or as a complement to quantitative research. Research reports of qualitative research are usually richer and provide more details about the behavior being observed.

Designs for Studying Development

Research in human development usually concerns differences or changes that occur over time. In these cases—in addition to deciding how to measure the behavior of interest and whether the study will be correlational or experimental—investigators must also choose one of three designs that allow them to examine development: longitudinal, cross-sectional, or sequential.

Who were the investigators and what was the aim of the study? Frank Fujita and Ed Diener (2005) were interested in learning whether life satisfaction stays the same or changes across adulthood. So they compared measures of life satisfaction to physiological and demographic data over a 17-year period.

How did the investigators measure the topic of interest? Life satisfaction was measured by self-ratings to the question "How happy are you at present with your life as a whole?" from 0 (totally unhappy) to 10 (totally happy). Monthly household income was also obtained through self-report. Additionally, measures of height, weight, body mass index (BMI), systolic and diastolic blood pressure, and personality traits were obtained.

Who were the participants in the study? The participants were 3,608 Germans (1,709 males and 1,899 females) who had answered a question about life satisfaction every year from 1984 to 2000 as part of the German Socio-Economic Panel study. The sample is a nationally representative sample of German adults born between 1902 and 1968.

What was the design of the study? The study used a longitudinal design, with assessments conducted annually.

Were there ethical concerns with the study? All of the participants were provided information about the purpose of the study and the tests they would take. Each participant provided informed consent.

What were the results? ■ Figure 1.6 shows correlations between successive measurements. Variables such as height, weight, and BMI are very stable—they are consistent from one measurement to the next, which yields very strong positive correlations (nearly 1.0). In contrast, blood pressure, personality traits, income, and—most important—life satisfaction are less stable. And

these factors show greater change as the time between measurements increases. One of the most critical results showed that roughly 25% of the sample changed significantly in their life satisfaction ratings over the course of the study.

What did the investigators conclude? Fujita and Diener argued that life satisfaction is somewhat stable but can vary over time. This is important, particularly for people whose life satisfaction may be low; there is the chance that things could improve.

What converging evidence would strengthen these conclusions? Because the sample included only German adults, it would be necessary to study people from other cultures to find out whether the results generalize cross-culturally.

Go to Psychology CourseMate at www.cengagebrain.com to enhance your understanding of this research.

■ **Figure 1.6**
Note that overall life satisfaction showed some decrease over a 17-year period compared to height, weight, body mass index, and personality, but was less stable over time than personality.

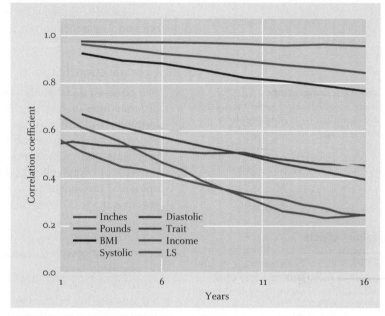

From Fujita, F., & Diener, E. (2005). Life satisfaction set point stability and change. *Journal of Personality and Social Psychology, 88*(1), p. 161. Copyright © 2005 by the American Psychological Association.

Longitudinal Studies

In a **longitudinal study,** *the same individuals are observed or tested repeatedly at different points in their lives.* As the name implies, the longitudinal approach examines development over time. It is the most direct way to identify change. More important, it is the only way to answer certain questions about the stability or instability of behavior: Will characteristics such as aggression, dependency, or mistrust observed in infancy or early childhood persist into adulthood? Will a regular exercise program begun in middle age have benefits in later life? Does people's satisfaction with their lives remain the same or change across adulthood? Such questions can be explored only by testing people at one point in their development and then retesting them later.

The Spotlight on Research feature focuses on the last question about life satisfaction. As you read it, pay close attention to the questions and how the researchers ap-

longitudinal study
longitudinal study research design in which the same individuals are observed or tested repeatedly at different points in their lives

proached them. Doing so gives you insight into the creative process of research and into the potential strengths and weaknesses of a longitudinal research study.

As powerful as it is in identifying what does—or does not—change over time, though, the longitudinal approach has disadvantages that frequently offset its strengths. One is cost: The expense of keeping up with a large sample of individuals can be staggering. A related problem is the need to keep the sample together over the course of the research. Maintaining contact with people over years or decades can be a challenge. Even among those who do not move away, some lose interest and choose not to continue; others may die. These "dropouts" are often significantly different from their peers, and this fact may also distort the study's outcome. For example, a study may seem to show that a group of older adults shows intellectual stability late in life. What may have happened, however, is that those who found earlier testing most difficult quit the study and thereby raised the group average on the next round. Even when the sample remains constant, taking the same test many times may make people "test-wise." Improvement (or even the lack of change) over time may be attributed to development when it actually stems from practice with a particular test. Changing the test from testing session to testing session solves the practice problem but raises a new question: how to compare responses to different tests.

Because of these and other problems with the longitudinal method, human development researchers often use cross-sectional studies instead.

Cross-Sectional Studies

cross-sectional study
study in which developmental differences are identified by testing people of different ages

In a **cross-sectional study**, *developmental differences are identified by testing people of different ages in the study.* Development is charted by noting the differences between individuals of different ages at the same point in calendar time. The cross-sectional approach avoids the problems of repeated testing, and the costs of tracking a sample over time. But cross-sectional research has its own weaknesses. Because people are tested at only one point in their lives, we learn nothing about the continuity of development. Consequently, we cannot tell whether an aggressive 14-year-old remains aggressive at age 30 because the person would be tested either at age 14 or age 30 but not at both.

cohort effects
problem with cross-sectional designs in which differences between age groups (cohorts) may result as easily from environmental events as from developmental processes

Cross-sectional studies are also affected by **cohort effects**, *meaning that differences between age groups (cohorts) may result as much from environmental events as from developmental processes.* In a simple cross-sectional study, we typically compare people from two age groups. Differences that we find are attributed to the difference in age, but this needn't be the case. Why? The cross-sectional study assumes that, when the older people were younger, they resembled the people in the younger age group. This isn't always true, and this fact—rather than difference in age—may be responsible for differences between the groups. Suppose that young adults were found to be more imaginative than middle-aged adults. Should we conclude that imagination declines between these ages? Not necessarily. Perhaps a new curriculum to nourish creativity was introduced after the middle-aged adults completed school. Because the younger adults experienced the curriculum but the middle-aged adults did not, the difference between them is difficult to interpret.

Sequential Studies

sequential design
developmental research design based on cross-sectional and longitudinal designs

Some researchers use another, more complex research approach, called a **sequential design**, *that is based on both cross-sectional and longitudinal designs.* Basically, a sequential design begins with a simple cross-sectional or longitudinal design. At some regular interval, the researcher then adds more cross-sectional or longitudinal studies, resulting in a sequence of these studies. For example, suppose a researcher wants to learn whether adults' memory ability changes with age. One way to do this would be to follow several groups of people of different ages over time, creating a sequence of longitudinal studies. The start would be a typical cross-sectional study in which 60-

Designs Used in Human Development Research

Type of Design	Definition	Strengths	Weaknesses
General Designs			
Correlational	Observe variables as they exist in the world and determine their relations	Behavior is measured as it occurs naturally	Cannot determine cause and effect
Experimental	Manipulate independent variable and determine effect on dependent variable	Control of variables allows conclusions about cause and effect	Work is often laboratory-based, which can be artificial
Developmental Designs			
Longitudinal	One group of people is tested repeatedly as they develop	Only way to chart an individual's development and look at the stability of behavior over time	Expensive, participants drop out, and repeated testing can distort performance
Cross-sectional	People of different ages are tested at the same time	Convenient—solves all problems associated with longitudinal studies	Cannot study stability of behavior; cohort effects complicate interpretation of differences between groups
Sequential	Multiple groups of people are tested over time, based on either multiple longitudinal or cross-sectional designs	Best way to address limitation of single longitudinal and cross-sectional designs	Very expensive and time consuming; may not completely solve limitations of longitudinal and cross-sectional designs

and 75-year-olds are tested. Then the two groups would be retested every 3 years, creating two separate longitudinal studies.

Although sequential designs are relatively rare because they are so expensive, they have several advantages. Most important, they help address most of the limitations described earlier concerning single cross-sectional and longitudinal studies. For example, sequential designs help isolate cohort effects and also help determine whether age-related changes are due to participant dropout or to some other cause. We will encounter examples of sequential designs when we consider some of the large studies examining the normal processes of aging in Chapter 14.

●Table 1.5 summarizes the strengths and weaknesses of the general research designs and the developmental designs. You'll read about each of these designs in this book, although the two cross-sectional types (cross-sectional experimental and cross-sectional correlational) occur more frequently than the other combinations of general and developmental designs. Why? The relative ease of conducting cross-sectional studies more than compensates for their limitations.

Integrating Findings From Different Studies

Several times in the past few pages, we've emphasized the value of using different methods to study the same phenomenon. The advantage of this approach is that conclusions are most convincing when the results are consistent regardless of method.

In reality, though, findings are often inconsistent. For example, suppose many researchers find that people often share highly personal information with friends (on Facebook, for instance), some researchers find that people share occasionally with friends, and a few researchers find that people never share with friends. What should we conclude? **Meta-analysis** *allows researchers to synthesize the results of many quantitative studies to estimate actual relations between variables* (Wilson, 2010). In conducting a meta-analysis, investigators find all studies published on a topic over a substantial period of time (e.g., 10 to 20 years) and then record and analyze the results and important methodological variables.

meta-analysis
a tool that enables researchers to synthesize the results of many studies to estimate relations between variables

The usefulness of meta-analysis is illustrated in a study by McClure (2000), who asked whether boys and girls differ in their ability to recognize emotions in facial expressions. She found 60 studies, published between 1931 and 1999, that included nearly 10,000 children and adults. In each study, participants were administered some sort of task (such as the one shown in Figure 1.3 on page 24), in which the aim is to select a face expressing a particular emotion. Analyzing across the results of all 60 studies, McClure found that, overall, girls recognized emotions in facial expressions more accurately than boys. The gender difference was constant from 4 to 16 years of age and was the same when the faces were shown as photos and as drawings. However, the gender difference was larger when participants judged emotions in children's faces than when they judged emotions in adults' faces.

Thus, meta-analysis is a particularly powerful tool because it allows scientists to determine whether a finding generalizes across many studies that used different methods. In addition, meta-analysis can reveal the impact of those different methods on results.

Conducting Research Ethically

Choosing a good research design involves more than just selecting a particular method. Researchers must determine whether the methods they plan on using are ethical. That is, when designing a research study, investigators must do so in a way that does not violate the rights of people who participate. To verify that every research

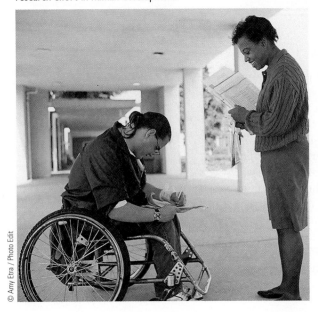

Informed consent is a necessary aspect of any research effort in human development.

© Amy Etra / Photo Edit

project incorporates these protections, local panels of experts and community representatives review proposed studies before any data are collected. Only with the approval of this panel can scientists begin their study. If the review panel objects to some aspects of the proposed study, then the researcher must revise those aspects and present them anew for the panel's approval. Likewise, each time a component of a study is changed, the review panel must be informed and give its approval.

To guide review panels, professional organizations (e.g., the American Psychological Association; see www.apa.org/research/responsible/human/index.aspx) and government agencies (e.g., the National Institutes of Health; see grants.nih.gov/grants/policy/hs/index.htm) have codes of ethical conduct that specify the rights of research participants as well as procedures to protect these participants. The following essential guidelines are included in all of these codes.

■ *Minimize risks to research participants.* Use methods that have the least potential for causing harm or stress for research participants. During the research, monitor the procedures to ensure avoidance of any unforeseen stress or harm.

■ *Describe the research to potential participants so they can determine whether they wish to participate.* Prospective participants must be told the purpose of the project, what they will be asked to do, whether there are any risks or potential harm or any benefits they may receive, that they are free to discontinue participation at any time without penalty, that they are entitled to a complete debriefing at the end of the project, and any other relevant information the review panel deems appropriate. After the study has been explained, each participant signs a document that states he or she understands what they will do in the study. Special caution must be exercised in obtaining consent for the participation of children and adolescents, as well as people who have conditions that affect intellectual functioning (e.g., Alzheimer's disease, severe head injury). In these cases, consent from a parent, legal guardian, or other responsible person—in addition to consent of the participant—is necessary.

■ *Avoid deception; if participants must be deceived, provide a thorough explanation of the true nature of the experiment as soon as possible.* Providing complete information about a study in advance sometimes biases or distorts a person's responses. Consequently, investigators may provide participants with partial information about the study or even mislead them about its true purpose. As soon as it is feasible—typically, just after the experiment—any false information that was given to research participants must be corrected, and the reasons for the deception must be provided.

- *Results should be anonymous or confidential.* Research results should be anonymous, which means that people's data cannot be linked to their name. When anonymity is not possible, research results should be confidential, which means that the identity of participants is known only to the investigator(s) conducting the study.

Conducting research ethically is an obligation of every investigator. If you conduct a project, even in connection with a course, you should submit your procedures for review. If you are a participant in someone else's project, make sure you are given appropriate and complete information and read it thoroughly.

Communicating Research Results

When the study is complete and the data analyzed, researchers write a report of their work that describes what the researchers did and why, their results, and the meaning(s) behind their results. The researchers will submit the report to one of several scientific journals that specialize in human development research. Some of these are *Child Development, Developmental Psychology, Psychology and Aging,* and the *Journals of Gerontology.* If the journal editor accepts the report then it will appear in the journal, where other human development researchers can learn of the results.

These reports of research are the basis for virtually all the information we present in this book. You have already encountered many citations of research in the format of names in parentheses, followed by a date, like this:

(Smith & Jones, 2012)

This indicates the last names of the person(s) who did the research and the year in which the research was published. By looking in the Reference section at the end of the book, which is organized alphabetically by the first author's last name, you can find the title of the article and the journal where it was published.

All of these different steps in research may seem tedious and involved to you. For a human development researcher, however, much of the fun of doing research is planning a study that no one has done before and that will provide useful information to other specialists. This is one of the most creative and challenging parts of human development research.

Applying Research Results: Social Policy

One question many people have about research is whether any of it really matters. Actually, research on human development has a strong influence on policy makers and politicians. For example, every state in the United States and many countries around the world have laws against child abuse and laws that govern child labor practices. Many countries have laws setting minimum ages for certain activities, such as consuming alcohol. Some states in the United States are changing the way older drivers are screened when they renew their driver's licenses. Human development research played a role in the establishing of all of these laws and regulations.

Other examples of how developmental research affects social policy include the elimination of mandatory retirement, the Americans with Disabilities Act, many educational reform laws, and indices courts use to decide whether an adolescent offender should be tried as a juvenile or as an adult. Clearly, the research done by developmentalists influences many aspects of daily life that are governed by laws and societal rules.

At several points in the text, we will describe some important connections between human development research and social policy. As you will see, these connections are broad ranging and include areas that you may even take for granted. For example, lead-based paint can no longer be used in the United States, mainly because research by developmentalists showed that infants and young children who were exposed to lead-based paint (and who sometimes ate paint chips as they flaked off) suffered brain damage and learning problems. Research on human development not only provides many insights into what makes people tick but can also provide ways to improve the quality of life.

However, the views of scientists, ethicists, public citizens, and government sometimes collide in ways that result in significant debate concerning research. Such is the case with stem cell research, a hotly debated topic explored in the What Do *You* Think? feature.

What do you think? — Stem Cell Research

Imagine you were unable to walk because of a severe spinal cord injury received in an accident. Imagine further that there was a possible cure: inducing the damaged nerve cells in your spine to regenerate, thereby enabling you to walk again. Would you support every effort possible to ensure that this research was pursued, especially if you knew that there were positive outcomes in research (see Nistor et al., 2005) on other animal species? But what if you knew also that applying this research to humans would require using stem cells from human embryos? Would that affect your opinion about the research?

Research on regeneration of nerve cells and for treatment of other diseases—such as Alzheimer's disease, leukemia, and Parkinson's disease—could be revolutionized through the use of stem cells. Researchers such as Dr. Hans Keirstead, co-director of the Sue and Bill Gross Stem Cell Research Center at the University of California, Irvine, believe that stem cell research represents the future for medical treatment breakthroughs. His lab developed the first treatment for spinal cord injuries based on human stem cells approved by the U.S. Food and Drug Administration (FDA) in 2009.

Stem cells are unspecialized human or animal cells that can produce mature specialized body cells and at the same time replicate themselves. There are two basic types of stem cells: embryonic and "adult." Embryonic stem cells are derived from a blastocyst, which is a very young embryo that contains 200 to 250 cells and is shaped like a hollow sphere (see Chapter 2 for more information). The stem cells themselves are the cells in the blastocyst that ultimately would develop into a person or animal. "Adult" stem cells are derived from the umbilical cord and placenta or from blood, bone marrow, skin, and other tissues.

Medical researchers are interested in using stem cells to repair or replace damaged body tissues because stem cells are less likely than other foreign cells to be rejected by the immune system when they are implanted in the body. (Tissue and organ rejection is a major problem following transplant surgery, for example.) Embryonic stem cells have the capacity to develop into every type of tissue found in your body. Stem cells have been used experimentally to form the blood-making cells of the bone marrow as well as heart, blood vessel, muscle, and insulin-producing tissue. Embryonic germ line cells have been used to help paralyzed mice regain some of their ability to move. Since the 1990s, umbilical cord blood stem cells have been used to treat heart and other defects in children who have rare metabolic diseases and to treat children with certain anemias and leukemias. Stem cell research has also produced several positive results regarding Parkinson's disease.

Even though stem cell research holds much promise and has already been shown to be effective in treating some conditions, it is extremely controversial. The Bush administration restricted federally funded research to certain existing colonies of embryonic stem cells, pointing out that creating additional embryonic stem cells involves the destruction of human embryos, which they considered wrong. But proponents of the research argue that most embryonic stem cells used in research are left over from fertility clinics and are destined to be destroyed anyway. The Obama administration moved to ease restrictions through an executive order in 2009, which many researchers thought settled the issue. However, a federal judge overturned that in 2010, ruling that the previous restrictions regarding federal funding of research were still in effect. Clearly, the legal and political aspects of stem cell research remain far from settled.

The relation between research and social policy regarding stem cells is also informed by individuals such as Dr. Rev. Stephen Bellamy, a scientist and Anglican theologian/ethicist (Lako, Trounson, & Daher, 2010). Bellamy distinguishes among various approaches to both reproduction and research, arguing that there is no one stance that applies equally to all situations.

For more information about stem cell research, the related controversies, and the U.S. government's policy, check the websites at the National Institutes of Health (stemcells.nih.gov) and the American Association for the Advancement of Science (www.aaas.org/spp/cstc/briefs/stemcells). What do *you* think should be done?

Test Yourself

Problems With Using Photographs to Measure Understanding of Emotions

On page 24, we invited you to consider why asking children to judge emotions from photos may not be valid. Children's judgments of the emotions depicted in photographs may be less accurate than they would be in real life because, in real life: (1) facial features are usually moving—not still as in the photographs—and movement may be one of the clues children naturally use to judge emotions; (2) facial expressions are often accompanied by sounds, and children use both sight and sound to judge emotions; and (3) children most often judge facial expressions of people they know (parents, siblings, peers, teachers), and knowing the "usual" appearance of a face may help children judge emotions accurately.

Copyright © Cengage Learning 2010

SUMMARY

1.1 THINKING ABOUT DEVELOPMENT

What fundamental issues of development have scholars addressed throughout history?

■ Three main issues are prominent in the study of human development. The nature–nurture issue involves the degree to which genetics and the environment influence human development. In general, theorists and researchers view nature and nurture as mutually interactive influences; development is always shaped by both. The continuity–discontinuity issue concerns whether the same explanations (continuity) or different explanations (discontinuity) must be used to explain changes in people over time. Continuity approaches emphasize quantitative change; discontinuity approaches emphasize qualitative change. In the issue of universal versus context-specific development, the question is whether development follows the same general path in all people or is fundamentally dependent on the sociocultural context.

What are the basic forces in the biopsychosocial framework? How does the timing of these forces make a difference in their impact?

■ Development is based on the combined impact of four primary forces. Biological forces include all genetic and health-related factors that affect development. Many of these biological forces are determined by our genetic code.

■ Psychological forces include all internal cognitive, emotional, perceptual, and personality factors that influence development. Collectively, psychological forces explain the most noticeable differences in people.

■ Sociocultural forces include interpersonal, societal, cultural, and ethnic factors that affect development. Culture consists of the knowledge, attitudes, and behavior associated with a group of people. Overall, sociocultural forces provide the context or backdrop for development.

- Life-cycle forces provide a context for understanding how people perceive their current situation and its effects on them.

- The biopsychosocial framework emphasizes that the four forces are mutually interactive; development cannot be understood by examining the forces in isolation. Furthermore, the same event can have different effects, depending on when it happens.

How does neuroscience enhance our understanding of human development?

- Neuroscience is the study of the brain and the nervous system, especially in terms of brain–behavior relationships. Identifying patterns of brain activity helps demonstrate how developmental forces interact.

1.2 DEVELOPMENTAL THEORIES

What is a developmental theory?

- Developmental theories organize knowledge so as to provide testable explanations of human behaviors and the ways in which they change over time. Current approaches to developmental theory focus on specific aspects of behavior. At present, there is no single unified theory of human development.

How do psychodynamic theories account for development?

- Psychodynamic theories propose that behavior is determined by the way people deal with conflicts they face at different ages. Erikson proposed a life-span theory of psychosocial development, consisting of eight universal stages, each characterized by a particular struggle.

What is the focus of learning theories of development?

- Learning theory focuses on the development of observable behavior. Operant conditioning is based on the notions of reinforcement, punishment, and environmental control of behavior. Social learning theory proposes that people learn by observing others.

How do cognitive-developmental theories explain changes in thinking?

- Cognitive-developmental theory focuses on thought processes. Piaget proposed a four-stage universal sequence based on the notion that, throughout development, people create their own theories to explain how the world works. According to information-processing theory, people deal with information like a computer does; development consists of increased efficiency in handling information. Vygotsky emphasized the influence of culture on development.

What are the main points in the ecological and systems approach?

- Bronfenbrenner proposed that development occurs in the context of several interconnected systems of increasing complexity. The competence–environmental press theory postulates that there is a "best fit" between a person's abilities and the demands placed on that person by the environment.

What are the major tenets of life-span and life course theories?

- According to the life-span perspective, human development is characterized by multidirectionality, plasticity, historical context, and multiple causation. All four developmental forces are critical.

- Selective optimization with compensation refers to the developmental trends to focus one's efforts and abilities in successively fewer domains as one ages and to acquire ways to compensate for normative losses.

- The life course perspective refers to understanding human development within the context of the historical time period in which a generation develops, which creates unique sets of experiences.

1.3 DOING DEVELOPMENTAL RESEARCH

How do scientists measure topics of interest in studying human development?

- Research typically begins by determining how to measure the topic of interest. Systematic observation involves recording people's behavior as it takes place, in either a natural environment (naturalistic observation) or a structured setting (structured observation). Researchers sometimes create tasks to obtain samples of behavior. In self-reports, people answer questions posed by the experimenter. Physiological measures provide a way to examine body–behavior relationships.

- Researchers must determine that their measures are reliable and valid; they must also obtain a sample representative of some larger population.

What research designs are used to study human development?

- In correlational studies, investigators examine relations among variables as they occur naturally. This relation is often measured by a correlation coefficient, which can vary from −1 (strong inverse relation) to 0 (no relation) to +1 (strong positive relation). Correlational studies cannot determine cause and effect, so researchers do experimental studies in which an independent variable is manipulated and the impact of this manipulation on a dependent variable is recorded. Experimental studies allow conclusions about cause and effect, but the required strict control of other variables often makes the situation artificial. The best approach is to use both experimental and correlational studies to provide converging evidence. Qualitative research permits more in-depth analysis of behavior and is often used as a preliminary step for, or in conjunction with, quantitative research.

- To study development, some researchers use a longitudinal design in which the same people are observed repeatedly as they grow. This approach provides evidence concerning actual patterns of individual growth but has several shortcomings as well: It is time consuming, some people drop out of the project, and repeated testing can affect performance.

- An alternative, the cross-sectional design, involves testing people of different ages. This design avoids the problems of the longitudinal design but provides no information about individual growth. Also, what appear to be age differences may be cohort effects. Because neither design is problem-free, the best approach involves using both to provide converging evidence.

- Sequential designs are based on multiple longitudinal or cross-sectional designs.

How do researchers integrate results from multiple studies?

- Meta-analysis provides a way for researchers to look for trends across multiple studies to estimate the relations among variables.

What ethical procedures must researchers follow?

- Planning research also involves selecting methods that preserve the rights of research participants. Experiment-ers must minimize the risks to participants, describe the research so that candidates can make an informed decision about participating, avoid deception, and keep results anonymous or confidential.

How do investigators communicate results from research studies?

- Once research data are collected and analyzed, investigators publish the results in scientific outlets such as journals and books. Such results form the foundation of knowledge about human development.

How does research affect public policy?

- Research results are sometimes used to inform and shape public policy. Controversial topics such as stem cell research also form the basis for public policy in terms of what types of research are permitted.

KEY TERMS

human development (5)

nature–nurture issue (5)

continuity–discontinuity issue (6)

universal and context-specific development issue (6)

biological forces (7)

psychological forces (7)

sociocultural forces (7)

life-cycle forces (7)

biopsychosocial framework (7)

neuroscience (11)

theory (12)

psychodynamic theories (12)

psychosocial theory (12)

epigenetic principle (12)

operant conditioning (13)

reinforcement (13)

punishment (13)

imitation or observational learning (13)

self-efficacy (14)

information-processing theory (15)

ecological theory (16)

microsystem (17)

mesosystem (17)

exosystem (17)

macrosystem (18)

competence (18)

environmental press (18)

life-span perspective (18)

selective optimization with compensation (SOC) model (19)

life-course perspective (20)

systematic observation (23)

naturalistic observation (23)

structured observations (23)

self-reports (24)

reliability (25)

validity (25)

populations (26)

sample (26)

correlational study (26)

correlation coefficient (26)

experiment (27)

independent variable (27)

dependent variable (27)

qualitative research (28)

longitudinal study (29)

cross-sectional study (30)

cohort effects (30)

sequential design (30)

meta-analysis (31)

stem cells (34)

Log in to **www.cengagebrain.com** to access the resources your instructor requires. For this book, you can access:

 Psychology CourseMate

- CourseMate brings course concepts to life with interactive learning, study, and exam preparation tools that support the printed textbook. A textbook-specific website, Psychology CourseMate includes an integrated interactive eBook and other interactive learning tools including quizzes, flashcards, videos, and more.

CENGAGENOW™

- CengageNOW Personalized Study is a diagnostic study tool containing valuable text-specific resources—and because you focus on just what you don't know, you learn more in less time to get a better grade.

WebTUTOR™

- More than just an interactive study guide, WebTutor is an anytime, anywhere customized learning solution with an eBook, keeping you connected to your textbook, instructor, and classmates.

PART ONE

Prenatal Development, Infancy, and Early Childhood

Petit Format/Photo Researchers, Inc.

© Elyse Lewin/Jupiterimages Corporation

Masterfile

© 2011 Ariel Skelley/Jupiterimages Corporation

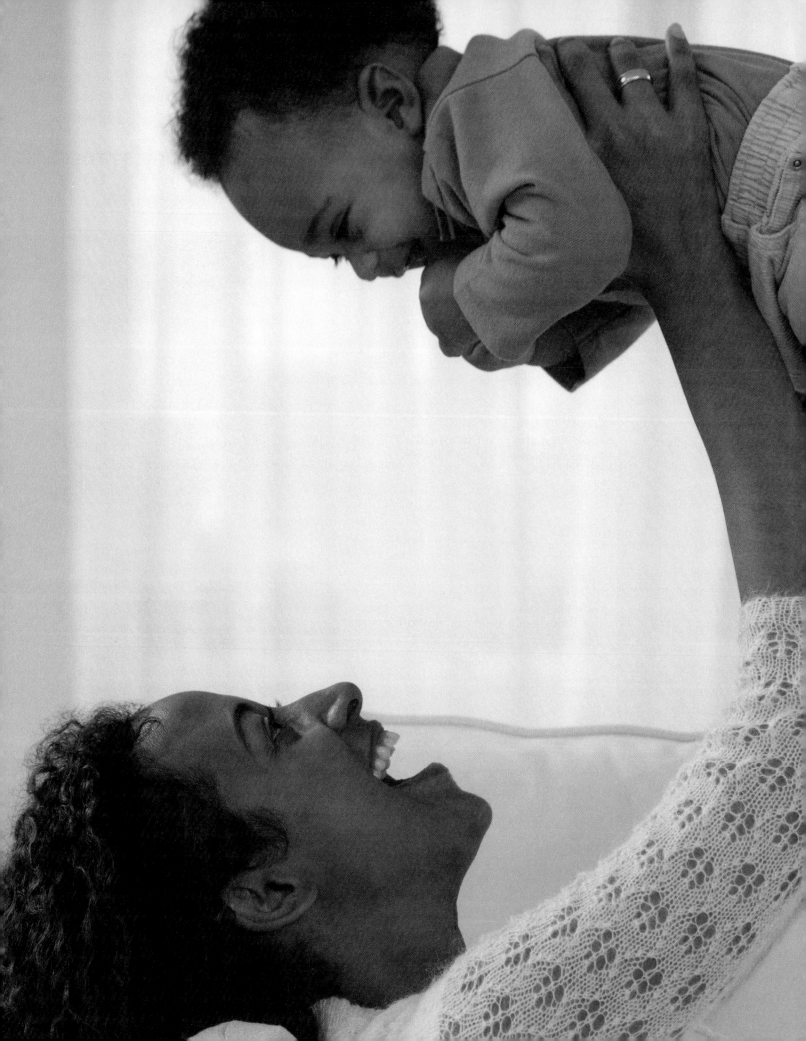

Biological Foundations
Heredity, Prenatal Development, and Birth

2

IF YOU ASK PARENTS to name the most memorable experiences of their lives, many immediately mention the events associated with the birth of their children. From the initial exciting news that a woman is pregnant through birth 9 months later, the entire experience of pregnancy and birth evokes awe and wonder.

The period before birth is the foundation for all human development and the focus of this chapter. Pregnancy begins when egg and sperm cells unite and exchange hereditary material. In the first section, you'll see how this exchange takes place and, in the process, learn about inherited factors that affect development. The second section of the chapter traces the events that transform sperm and egg into a living, breathing human being. You'll learn about the timetable that governs development before birth and, along the way, get answers to common questions about pregnancy. We talk about some of the problems that can occur during development before birth in the third section of the chapter. The last section focuses on birth and the newborn baby. You'll find out how an expectant mother can prepare for birth and what labor and delivery are like.

2.1 | In the Beginning: 23 Pairs of Chromosomes

LEARNING OBJECTIVES

- What are chromosomes and genes? How do they carry hereditary information from one generation to the next?

- What are common problems involving chromosomes and what are their consequences?

- How is children's heredity influenced by the environment in which they grow up?

Leslie and Glenn are excited at the thought of starting their own family. At the same time, they're nervous because Leslie's grandfather had sickle-cell disease and died when he was just 20 years old. Leslie is terrified that her baby may inherit the disease that killed her grandfather. She and Glenn wish that someone could reassure them that their baby will be okay.

HOW CAN WE REASSURE LESLIE AND GLENN? For starters, we need to know more about sickle-cell disease. Red blood cells carry oxygen and carbon dioxide to and from the body. When a person has sickle-cell disease, the red blood cells are long and curved like a sickle. These stiff, misshapen cells cannot pass through small capillaries, so oxygen cannot reach all parts of the body. The trapped sickle cells also block the way of white blood cells that are the body's natural defense against bacteria. As a result, many people with sickle-cell disease—including Leslie's grandfather and many other African Americans, who are more prone to this painful disease than other groups—die from infections before the age of 20.

Sickle-cell disease is inherited and, because Leslie's grandfather had the disorder, it runs in her family. Will Leslie's baby inherit the disease? To answer this question, we need to examine the mechanisms of heredity.

Mechanisms of Heredity

At conception, egg and sperm unite to create a new organism that incorporates some characteristics of each parent. *Each egg and sperm cell has 23* **chromosomes**, *threadlike structures in the nucleus that contain genetic material.* When a sperm penetrates an egg, their chromosomes combine to produce 23 pairs of chromosomes. *The first 22*

chromosomes
threadlike structures in the nuclei of cells that contain genetic material

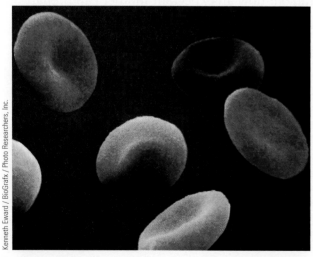

Red blood cells carry oxygen throughout the body.

Kenneth Eward / BioGrafx / Photo Researchers, Inc.

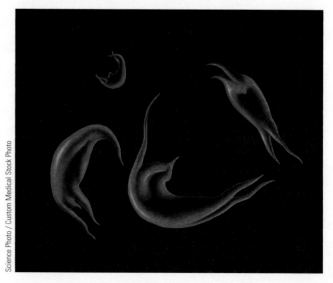

Sickle-shaped blood cells associated with sickle-cell disease cannot pass through the body's smallest blood vessels.

Science Photo / Custom Medical Stock Photo

pairs of chromosomes are called **autosomes**. *The 23rd pair determines the sex of the child, so these are known as the* **sex chromosomes**. When the 23rd pair consists of an X and a Y chromosome, the result is a boy; two X chromosomes produce a girl.

Each chromosome actually consists of one molecule of **deoxyribonucleic acid**—*DNA for short.* To understand the structure of DNA, imagine four different colors of beads placed on two strings. The strings complement each other precisely: Wherever a red bead appears on one string, a blue bead appears on the other; wherever a green bead appears on one string, a yellow one appears on the other. DNA is organized this way, except that the four colors of beads are actually four different chemical compounds: adenine, thymine, guanine, and cytosine. The strings, which are made up of phosphates and sugars, wrap around each other and so create the double helix shown in

■ Figure 2.1.

The order in which the chemical compound "beads" appear is really a code that causes the cell to create specific amino acids, proteins, and enzymes—important biological building blocks. For example, three consecutive thymine "beads" make up the instruction to create the amino acid phenylalanine. *Each group of compounds that provides a specific set of biochemical instructions is a* **gene**. Thus, genes are the functional units of heredity because they determine the production of chemical substances that are, ultimately, the basis for all human characteristics and abilities.

autosomes
first 22 pairs of chromosomes

sex chromosomes
23rd pair of chromosomes; these determine the sex of the child

deoxyribonucleic acid (DNA)
molecule composed of four nucleotide bases that is the biochemical basis of heredity

gene
group of nucleotide bases that provides a specific set of biochemical instructions

■ **Figure 2.1**
DNA is organized in a double helix, with strands of phosphates and sugars linked by nucleotide bases.

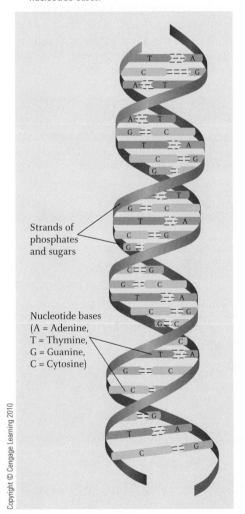

Strands of phosphates and sugars

Nucleotide bases
(A = Adenine,
T = Thymine,
G = Guanine,
C = Cytosine)

Copyright © Cengage Learning 2010

Humans have 23 pairs of chromosomes, including 22 pairs of autosomes and 1 pair of sex chromosomes.

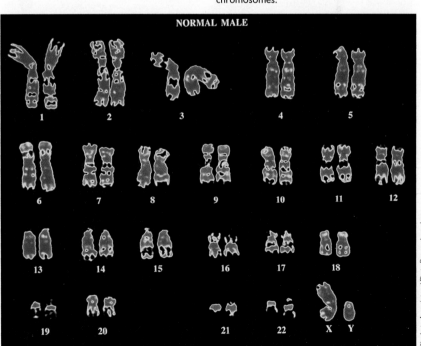

NORMAL MALE

Biophoto Associates / Photo Researchers, Inc.

Altogether, a person's 46 chromosomes include roughly 25,000 genes (Pennisi, 2005). Chromosome 1 has the most genes (nearly 3,000) and the Y chromosome has the fewest (just over 200). Most of these genes are the same for all people—fewer than 1% of genes cause differences between people (Human Genome Project, 2003). Through biochemical instructions that are coded in DNA, genes regulate the development of all human characteristics and abilities. *The complete set of genes makes up a person's heredity and is known as the person's* **genotype**. *Genetic instructions, in conjunction with environmental influences, produce a* **phenotype**, *an individual's physical, behavioral, and psychological features.*

How do genetic instructions produce the misshapen red blood cells of sickle-cell disease? *Genes come in different forms that are known as* **alleles**. In the case of red blood cells, for example, two alleles can be present on chromosome 11. One allele has instructions for normal red blood cells; another allele has instructions for sickle-shaped red blood cells. *The alleles in the pair of chromosomes are sometimes the same, which is known as being* **homozygous**. *The alleles sometimes differ, which is known as being* **heterozygous**. Leslie's baby would be homozygous if it had two alleles for normal cells *or* two alleles for sickle-shaped cells. The baby would be heterozygous if it had one allele of each type.

How does a genotype produce a phenotype? With sickle-cell disease, for example, how do genotypes lead to specific kinds of blood cells? The answer is simple if a person is homozygous. When both alleles are the same—and therefore have chemical instructions for the same phenotype—that phenotype results. If Leslie's baby had an allele for normal red blood cells on both of its 11th chromosomes, then the baby would be almost guaranteed to have normal cells. If, instead, the baby had two alleles for sickle-shaped cells, then it would almost certainly suffer from the disease.

When a person is heterozygous, the process is more complex. *Often one allele is* **dominant**, *which means that its chemical instructions are followed while those of the other,* **recessive** *allele are ignored.* In sickle-cell disease, the allele for normal cells is dominant and the allele for sickle-shaped cells is recessive. This is good news for

genotype
person's hereditary makeup

phenotype
physical, behavioral, and psychological features that result from the interaction between one's genes and the environment

alleles
variations of genes

homozygous
when the alleles in a pair of chromosomes are the same

heterozygous
when the alleles in a pair of chromosomes differ from each other

dominant
form of an allele whose chemical instructions are followed

recessive
allele whose instructions are ignored in the presence of a dominant allele

■ **Figure 2.2**
In single-gene inheritance, a heterozygous father and a heterozygous mother can have a healthy child, a child with sickle-cell trait, or a child with sickle-cell disease.

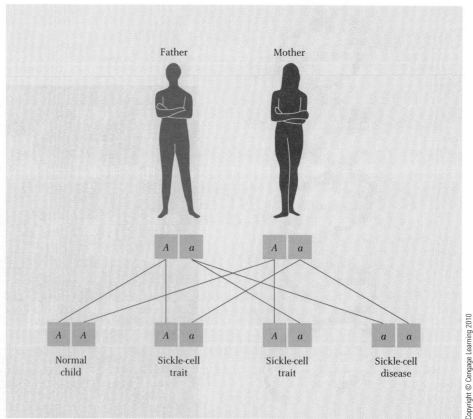

Some Common Phenotypes Associated with Single Pairs of Genes

Dominant Phenotype	Recessive Phenotype
Curly hair	Straight hair
Normal hair	Pattern baldness (men)
Dark hair	Blond hair
Thick lips	Thin lips
Cheek dimples	No dimples
Normal hearing	Some types of deafness
Normal vision	Nearsightedness
Farsightedness	Normal vision
Normal color vision	Red–green color blindness
Type A blood	Type O blood
Type B blood	Type O blood
Rh-positive blood	Rh-negative blood

SOURCE: McKusick, 1995.

Leslie: As long as either she or Glenn contributes the allele for normal red blood cells, their baby will not develop sickle-cell disease.

■ Figure 2.2 summarizes what we've learned about sickle-cell disease: *A* denotes the allele for normal blood cells, and *a* denotes the allele for sickle-shaped cells. Depending on the alleles in Leslie's egg and in the sperm that fertilizes that egg, three outcomes are possible. Only if the baby inherits two recessive alleles for sickle-shaped cells is it likely to develop sickle-cell disease. But this is unlikely in Glenn's case: He is positive that no one in his family has had sickle-cell disease, so he almost certainly has the allele for normal blood cells on both of the chromosomes in his 11th pair.

Even though Glenn's sperm will carry the gene for normal red blood cells, this doesn't guarantee that their baby will be healthy. Why? *Sometimes one allele does not dominate another completely, a situation known as* **incomplete dominance**. In incomplete dominance, the phenotype that results often falls between the phenotype associated with either allele. This is the case for the genes that control red blood cells. *Individuals with one dominant and one recessive allele have* **sickle-cell trait**: *In most situations they have no problems, but when seriously short of oxygen they suffer a temporary, relatively mild form of the disease.* Sickle-cell trait is likely to appear when the person exercises vigorously or is at high altitudes (Sullivan, 1987). Leslie and Glenn's baby would have sickle-cell trait if it inherits a recessive gene from Leslie and a dominant gene from Glenn.

The simple genetic mechanism responsible for sickle-cell disease—which involves a single gene pair with one dominant allele and one recessive allele—is also responsible for numerous other common traits, as shown in ● Table 2.1. In each of these instances, individuals with the recessive phenotype have two recessive alleles, one from each parent. Individuals with the dominant phenotype have at least one dominant allele.

Most of the traits listed in Table 2.1 are biological and medical phenotypes. This same genetic mechanism can cause serious disorders, as we'll see in the next section.

incomplete dominance
situation in which one allele does not dominate another completely

sickle-cell trait
disorder in which individuals show signs of mild anemia only when they are seriously deprived of oxygen; occurs in individuals who have one dominant allele for normal blood cells and one recessive sickle-cell allele

Genetic Disorders

Some people are affected by heredity in a special way: They have genetic disorders that disrupt the usual pattern of development. Genetics can derail development in two ways. First, some disorders are inherited. Sickle-cell disease is one example of an inherited disorder. Second, sometimes eggs or sperm do not include the usual 23 chromosomes but have more or fewer chromosomes instead. In the next few pages, we'll see how inherited disorders and abnormal numbers of chromosomes can alter a person's development.

Inherited Disorders

phenylketonuria (PKU)
inherited disorder in which the infant lacks a liver enzyme

You know that sickle-cell disease is a disorder that affects people who inherit two recessive alleles. *Another disorder that involves recessive alleles is **phenylketonuria** (PKU), a disorder in which babies are born lacking an important liver enzyme.* This enzyme converts phenylalanine—an amino acid found in dairy products, bread, diet soda, and fish—into tyrosine (another amino acid). Without this enzyme, phenylalanine accumulates and produces poisons that harm the nervous system, resulting in mental retardation (Diamond et al., 1997; Mange & Mange, 1990).

Most inherited disorders are like sickle-cell disease and PKU in that they are carried by recessive alleles. Relatively few serious disorders are caused by dominant alleles. Why? If the allele for the disorder is dominant, every person with at least one of these alleles would have the disorder. Individuals affected with these disorders typically do not live long enough to reproduce, so dominant alleles that produce fatal disorders soon vanish from the species. *An exception is **Huntington's disease**, a fatal disease characterized by progressive degeneration of the nervous system.* Huntington's disease is caused by a dominant allele found on chromosome 4. Individuals who inherit this disorder develop normally through childhood, adolescence, and young adulthood. During middle age, however, nerve cells begin to deteriorate, which produces symptoms such as muscle spasms, depression, and significant changes in personality (Shiwach, 1994). By this age, many adults with Huntington's disease have already reproduced, creating children who may well later display the disease themselves.

Huntington's disease
progressive and fatal type of dementia caused by dominant alleles

Abnormal Chromosomes

Sometimes individuals do not receive the normal complement of 46 chromosomes. If they are born with extra, missing, or damaged chromosomes, development is always disturbed. The best example is Down syndrome. People with Down syndrome have almond-shaped eyes and a fold over the eyelid. Their head, neck, and nose are usually smaller than normal. During the first several months of life, development of babies with Down syndrome seems to be normal. Thereafter, their mental and behavioral development begins to lag behind the average child's. For example, a child with Down syndrome might first sit up without help at about 1 year, walk at 2, and talk at 3, reaching each of these developmental milestones months or even years behind children without Down syndrome. By childhood, most aspects of cognitive and social development are seriously retarded. Rearing a child with Down syndrome presents special challenges. During the preschool years, children with Down syndrome need special programs to prepare them for school. Educational achievements of children with Down syndrome are likely to be limited, and their life expectancy ranges from 25 to 60 years (Yang, Rasmussen, & Friedman, 2002). Nevertheless, as you'll see in Chapter 6, many individuals with Down syndrome lead full, satisfying lives.

Children with Down syndrome typically have upward slanting eyes with a fold over the eyelid, a flattened facial profile, and a smaller than average nose and mouth.

What causes Down syndrome? Individuals with Down syndrome typically have an extra 21st chromosome that is usually provided by the egg (Machatkova et al., 2005). Why the mother provides two 21st chromosomes is unknown. However, the odds that a woman will bear a child with Down syndrome increase markedly as she gets older. For a woman in her late 20s, the risk of giving birth to a baby with Down syndrome is about 1 in 1,000; for a woman in her early 40s, the risk is about 1 in 50. Why? A woman's eggs have been in her ovaries since her own prenatal development. Eggs may deteriorate over time as

Common Disorders Associated with the Sex Chromosomes

Disorder	Sex Chromosomes	Frequency	Characteristics
Klinefelter's syndrome	XXY	1 in 500 male births	Tall, small testicles, sterile, below-normal intelligence, passive
XYY complement	XYY	1 in 1,000 male births	Tall, some cases apparently have below-normal intelligence
Turner's syndrome	X	1 in 2,500–5,000 female births	Short, limited development of secondary sex characteristics, problems perceiving spatial relations
XXX syndrome	XXX	1 in 500–1,200 female births	Normal stature but delayed motor and language development

part of aging or because an older woman has a longer history of exposure to hazards in the environment, such as X-rays, that may damage her eggs.

An extra autosome (as in Down syndrome), a missing autosome, or a damaged autosome always has far-reaching consequences for development because the autosomes contain huge amounts of genetic material. In fact, nearly half of all fertilized eggs abort spontaneously within 2 weeks—primarily because of abnormal autosomes. Thus, most eggs that cannot develop normally are removed naturally (Moore & Persaud, 1993).

Abnormal sex chromosomes can also disrupt development. ● Table 2.2 lists four of the more frequent disorders associated with atypical numbers of X and Y chromosomes. Keep in mind that "frequent" is a relative term; although these disorders are more frequent than PKU or Huntington's disease, most are uncommon. Notice that there are no disorders consisting solely of Y chromosomes. The presence of an X chromosome appears to be necessary for life.

Fortunately, most of us receive the correct number of chromosomes and do not inherit life-threatening illnesses. For most people, heredity reveals its power in creating a unique individual—a person unlike any other.

Now that you understand the basic mechanisms of heredity, we can learn how heredity and environment work together to produce behavioral and psychological development.

Heredity, Environment, and Development

Many people mistakenly view heredity as a set of phenotypes unfolding automatically from the genotypes that are set at conception. Nothing could be further from the truth. Although genotypes are fixed when the sperm fertilizes the egg, phenotypes are not. Instead, phenotypes depend both on genotypes and on the environment in which individuals develop.

To begin our study of heredity and environment, we need to look first at the methods that developmental scientists use.

Behavioral Genetics: Mechanisms and Methods

Behavioral genetics *is the branch of genetics that deals with inheritance of behavioral and psychological traits.* Behavioral genetics is complex, in part, because behavioral and psychological phenotypes are complex. Traits controlled by single genes are usually "either–or" phenotypes. A person either has dimpled cheeks or not; a person either has normal color vision or red–green color blindness; a person's blood either clots normally or it does not. In contrast, most important behavioral and psychological characteristics are *not* of an "either–or" nature; rather, a range of different outcomes is possible. Take extraversion as an example. Imagine trying to classify 10 people that you know well as either extroverts or introverts. This would be easy for a few extremely outgoing individuals (extroverts) and a few intensely shy persons (introverts). Most people are neither extroverts nor introverts but "in between." The result is a distribution of individuals ranging from extreme introversion at one end to extreme extroversion at the other.

behavioral genetics
the branch of genetics that studies the inheritance of behavioral and psychological traits

Introversion–extroversion is an example of a psychological characteristic that defines a continuum. Think of other psychological characteristics like this, in which outcomes are not "either–or" but are distributed across a range.

polygenic inheritance

when phenotypes are the result of the combined activity of many separate genes

Many behavioral and psychological characteristics are distributed in this fashion, including intelligence and many aspects of personality. *When phenotypes reflect the combined activity of many separate genes, the pattern is known as* **polygenic inheritance**. Because so many genes are involved in polygenic inheritance, we usually cannot trace the effects of each gene. But we can use a hypothetical example to show how many genes work together to produce a behavioral phenotype that spans a continuum. Let's suppose that four pairs of genes contribute to extroversion, that the allele for extroversion is dominant, and that the total amount of extroversion is simply the sum of the dominant alleles. If we continue to use uppercase letters to represent dominant alleles and lowercase letters to represent the recessive allele, then the four gene pairs would be Aa, Bb, Cc, and Dd.

These four pairs of genes produce 81 different genotypes and 9 distinct phenotypes. For example, a person with the genotype AABBCCDD has 8 alleles for extroversion (the proverbial party animal). A person with the genotype aabbccdd has no alleles for extroversion (the proverbial wallflower). All other genotypes involve some combination of dominant and recessive alleles, so these are associated with phenotypes representing intermediate levels of extroversion. In fact, ■ Figure 2.3 shows that the most common outcome is for people to inherit exactly 4 dominant and 4 recessive alleles, and 19 of the 81 genotypes (e.g., AABbccDd, aaBbcCDd) produce this pattern. A few extreme cases (very outgoing or very shy), when coupled with many intermediate cases, produce the familiar bell-shaped distribution that characterizes many behavioral and psychological traits.

Remember, this example is completely hypothetical. Extroversion is *not* based on the combined influence of eight pairs of genes. However, the sample shows how several genes working together *could* produce a continuum of phenotypes. Something like our example is probably involved in the inheritance of many human behavioral traits, except that many more pairs of genes are involved. What's more, the environment also influences the phenotype.

If many behavioral phenotypes involve countless genes, how can we hope to unravel the influence of heredity? Twins and adopted children provide some important clues to the role of heredity. In twin studies, researchers compare identical and frater-

■ **Figure 2.3**

Many behavioral phenotypes represent a continuum (with many people falling at the middle of the continuum), an outcome that can be caused by many genes working together.

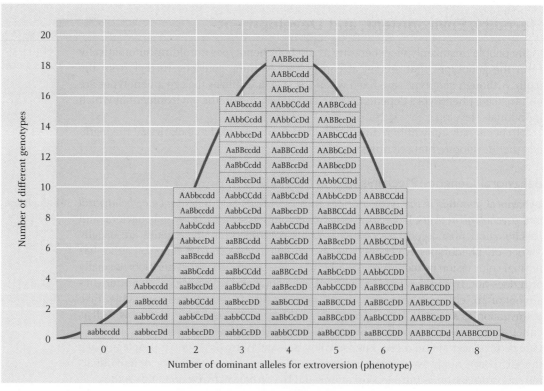

nal twins. *Identical twins are called* **monozygotic twins** *because they come from a single fertilized egg that splits in two*. Because identical twins come from the same fertilized egg, the same genes control their body structure, height, and facial features, which explains why identical twins look alike. *In contrast, fraternal or* **dizygotic twins** *come from two separate eggs fertilized by two separate sperm*. Genetically, fraternal twins are just like any other siblings—on average, about half their genes are the same. In twin studies, scientists compare identical and fraternal twins to measure the influence of heredity. When identical twins are more alike than are fraternal twins, this implicates heredity (Phelps, Davis, & Schartz, 1997).

A similar logic is used in adoption studies, in which adopted children are compared with their biological parents and their adoptive parents. The idea here is that biological parents provide their child's genes whereas adoptive parents provide the child's environment. Consequently, if a behavior has important genetic roots, then adopted children should behave more like their biological parents than like their adoptive parents.

These and other methods are not foolproof. Perhaps you recognized a potential flaw in twin studies: Parents and other people may treat monozygotic twins more similarly than they treat dizygotic twins. This would make monozygotic twins more similar than dizygotic twins in their experiences as well as in their genes. Each method of study has its unique pitfalls, but if different methods converge on the same conclusion about the influence of heredity then we can be confident of that result. Throughout this book, you'll see many instances where twin studies and adoption studies have pointed to genetic influences on human development.

Behavioral geneticists are now moving beyond the traditional methods of twin and adoption studies (Dick & Rose, 2002; Plomin & Crabbe, 2000). Today it is possible to isolate particular segments of DNA in human chromosomes. These segments then serve as markers for identifying specific alleles. The procedure is complicated, but the basic approach often begins by identifying people who differ in the behavioral or psychological trait of interest. For example, researchers might identify children who are outgoing and children who are shy. Or they might identify children who read well and children who read poorly. The children rub the inside of their mouths with a cotton swab, which yields cheek cells that contain DNA. The cells are analyzed in a lab, and the DNA markers for the two groups are compared. If the markers differ consistently, then the alleles near the marker probably contribute to the differences between the groups.

Techniques of this sort have the potential to identify the many different genes that contribute to complex behavioral and psychological traits. Of course, these new methods have limits. Some require large samples of people, which can be hard to obtain when studying a rare disorder. Also, some studies require that an investigator have a specific idea at the outset about which chromosomes to search for and where. These can be major hurdles. However, when used with traditional methods of behavioral genetics (e.g., adoption studies), the new methods promise a much greater understanding of how genes influence behavior and development.

Throughout the rest of this book, you'll encounter many instances that show the interactive influences of heredity and environment on human development. In the next few pages, however, we want to mention some general principles of heredity–environment interactions.

Paths From Genes to Behavior

How do genes work together—for example, to make some children brighter than others and some more outgoing than others? That is, how does the information in strands of DNA influence a person's behavioral and psychological development? The

monozygotic twins
the result of a single fertilized egg splitting to form two new individuals; also called identical twins

dizygotic twins
the result of two separate eggs fertilized by two sperm; also called fraternal twins

iStockphoto.com / Noriko Cooper

Identical twins are called monozygotic twins because they came from a single fertilized egg that split in two; consequently, they have identical genes.

specific paths from genes to behavior are largely uncharted (Meaney, 2010), but in the next few pages we'll discover some of their general properties.

1. *Heredity and environment interact dynamically throughout development.* A traditional but simple-minded view of heredity and environment is that heredity provides the clay of life and experience does the sculpting. In fact, genes and environments constantly interact to produce phenotypes throughout a person's development (Meaney, 2010; Rutter, 2007). Although we often think there is a direct link between a genotype and a phenotype—given a certain genotype, a specific phenotype occurs, necessarily and automatically—in fact, the path from genotype to phenotype is massively more complicated and less direct than this. A more accurate description would be that a genotype leads to a phenotype but only if the environment "cooperates" in the usual manner.

A good example of this genotype–phenotype link is seen in the disease phenylketonuria (PKU for short). As we saw on page 46, PKU is a homozygous recessive trait in which phenylalanine accumulates in the child's body, damaging the nervous system and leading to retarded mental development. Phenylalanine is abundant in many foods that most children eat regularly—meat, chicken, eggs, cheese—so that the environment usually provides the input (phenylalanine) necessary for the phenotype (PKU) to emerge. However, in the middle of the 20th century, the biochemical basis for PKU was discovered and now newborns are tested for the disorder. Infants who have the genotype for the disease are immediately placed on a diet that limits phenylalanine and the disease does not appear; their nervous system develops normally. In more general terms, a genotype is expressed differently (no disease) when it is exposed to a different environment (one lacking phenylalanine).

The effect can work in the other direction, too, with the environment triggering genetic expression. That is, people's experiences can help to determine how and when genes are activated. For instance, teenage girls begin to menstruate at a younger age if they've had a stressful childhood (Belsky, Houts, & Fearon, 2010). The exact pathway of influence is unknown (though it probably involves the hormones that are triggered by stress and those that initiate ovulation), but this is a clear case in which the environment advances the genes that regulate the developmental clock (Ellis, 2004).

We've used a rare disease (PKU) and a once-in-a-lifetime event (onset of menstruation) to show intimate connections between nature and nurture in human development. These examples may make it seem as if such connections are relatively rare but nothing could be further from the truth. At a biological level, genes always operate in a cellular environment. There is constant interaction between genetic instructions and the nature of the immediate cellular environment, which can be influenced by a host of much broader environmental factors (e.g., hormones triggered by a child's experiences). This continuous interplay between genes and multiple levels of the environment (from cells to culture) that drives development is known as epigenesis.

Returning to the analogy of sculpting clay, an epigenetic view of molding would be that new and different forms of genetic clay are constantly being added to the sculpture, leading to resculpting by the environment, which causes more clay to be added, and the cycle continues. Hereditary clay and environmental sculpting are continuously interweaving and influencing each other.

Because of the epigenetic principle, you need to be wary when you read statements like "X percent of a trait is due to heredity." In fact, *behavioral geneticists often use correlations from twin and adoption studies to calculate a* **heritability coefficient,** *which estimates the extent to which differences between people reflect heredity.* For example, intelligence has a heritability coefficient of about .5, which means that about 50% of the differences in intelligence between people is due to heredity (Bouchard, 2004).

Why be cautious? One reason is that many people mistakenly interpret heritability coefficients to mean that 50% of *an individual's* intelligence is due

heritability coefficient
a measure (derived from a correlation coefficient) of the extent to which a trait or characteristic is inherited

to heredity; this is incorrect because heritability coefficients apply to groups of people, not to a single person.

A second reason for caution is that heritability coefficients apply only to a specific group of people living in a specific environment. They cannot be applied to other groups of people living in the same environment or to the same people living elsewhere. For example, a child's height is certainly influenced by heredity, but the value of a heritability coefficient depends on the environment. When children grow in an environment that has ample nutrition—allowing all children to grow to their full genetic potential—heritability coefficients will be large. But when some children receive inadequate nutrition, this aspect of their environment will limit their height and, in the process, reduce the heritability coefficient.

Similarly, the heritability coefficient for reading disability is larger for parents who are well educated than for parents who are not (Friend, DeFries, & Olson, 2008). Why? Well-educated parents more often provide the academically stimulating environment that fosters a child's reading; consequently, reading disability in this group usually reflects heredity. In contrast, less educated parents less often provide the needed stimulation; hence their children's reading disability reflects a mixture of genetic and environmental influences.

This brings us back to the principle that began this section: "*Heredity and environment interact dynamically throughout development.*" Both genes and environments are powerful influences on development, but we can understand one only by considering the other, too. This is why it is essential to expand research beyond the middle-class, European American participants who have dominated the samples of scientists studying human development. Only by studying diverse groups of people can we truly understand the many ways in which genes and environments propel children along their developmental journeys.

2. *Genes can influence the kind of environment to which a person is exposed.* In other words, "nature" can help to determine the kind of "nurturing" that a child receives (Scarr, 1992; Scarr & McCartney, 1983). A person's genotype can lead others to respond in a specific way. For example, imagine someone who is bright and outgoing as a result, in part, of her genes. As a child, she may receive plenty of attention and encouragement from teachers. In contrast, someone who is not as bright and is more withdrawn (again, due in part to heredity) may be easily overlooked by teachers. In addition, as children grow and become more independent, they actively seek environments that fit their genetic makeup. Children who are bright may actively seek peers, adults, and activities that strengthen their intellectual development. Similarly, people who are outgoing may seek the company of other people, particularly extroverts like themselves. *This process of deliberately seeking environments that fit one's heredity is called* **niche-picking**. Niche-picking is first seen in childhood and becomes more common as children grow older and can control their environments. The Real People feature shows niche-picking in action.

3. *Environmental influences typically make children within a family different.* One of the fruits of behavioral genetic research is greater understanding of the manner in which environments influence people. Traditionally, scientists considered some environments beneficial for people and others detrimental. This view has been especially strong in regard to family environments. Some parenting practices are thought to be more effective than others, and parents who use these effective practices are believed to have children who are, on average, better off than children of parents who don't use these practices. This view leads to a simple prediction: Children within a family should be similar because they all receive the same type of effective (or ineffective) parenting. However, dozens of behavioral genetic studies show that, in

niche-picking
process of deliberately seeking environments that are compatible with one's genetic makeup

Children who are outgoing often like to be with other people and deliberately seek them out; this is an example of niche-picking.

Ben and Matt Kail were born 25 months apart. Even as a young baby, Ben was always a "people person." He relished contact with other people and preferred play that involved others. From the beginning, Matt was different. He was more withdrawn and was quite happy to play alone. The first separation from parents was harder for Ben than for Matt, because Ben relished parental contact more. When they entered school, Ben enjoyed increasing the scope of his friendships; Matt liked all the different activities that were available and barely noticed the new faces. Though brothers, Ben and Matt are quite dissimilar in terms of their sociability, a characteristic known to have important genetic components (Braungart et al., 1992).

As Ben and Matt have grown up (they're now adults), they have consistently sought environments that fit their differing needs for social stimulation. Ben was involved in team sports and now enjoys teaching. Matt took art and photography classes and now is happy working at his computer. Ben and Matt have chosen very different niches, and their choices have been driven in part by the genes that regulate sociability.

---

reality, siblings are not very much alike in their cognitive and social development (Plomin & Spinath, 2004).

Does this mean that family environment is not important? No. *These findings point to the importance of* **nonshared environmental influences**, *the forces within a family that make children different from one another.* Although the family environment is important, it usually affects each child in a unique way, which makes siblings differ. Each child is likely to have different experiences in daily family life. For example, parents may be more affectionate with one child than another, they may use more physical punishment with one child than another, or they may have higher expectations for school achievement by one child than another. All these contrasting parental influences tend to make siblings different, not alike (Liang & Eley, 2005). Family environments are important, but—as we describe their influence throughout this book—you should remember that families actually create multiple unique environments, one for each person in the family.

nonshared environmental influences
forces within a family that make siblings different from one another

Children's experiences within a family typically make them different from one another, not more alike.

Much of what we have said about genes, environment, and development is summarized in ■ Figure 2.4 (Lytton, 2000). Parents are the source of children's genes and, at least for young children, the primary source of children's experiences. Children's genes also influence the experiences they have and the impact of those experiences on them. However, to capture the idea of nonshared environmental influences we would need a separate diagram for each child, reflecting the fact that parents provide unique genes and a unique family environment for each of their offspring.

Most of this book explains the links between nature, nurture, and development. We can first see the interaction of nature and nurture during prenatal development, which we examine in the next section of this chapter.

■ **Figure 2.4**
Parents influence their children by providing genes and by providing experiences; children's genes and their environments work together to shape development.

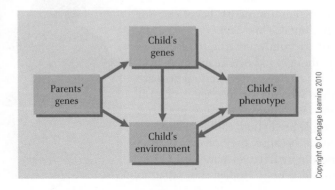

Test Yourself

RECALL

1. The first 22 pairs of chromosomes are called
   _____ .

2. _____ reflects the combined activity of a number of distinct genes.

3. Individuals with _____ have an extra 21st chromosome, usually inherited from the mother.

4. When a fertilized egg has defective autosomes, the usual result is that _____ .

5. Nonshared environmental influences tend to make siblings _____ .

INTERPRET

Explain how niche-picking shows the interaction between heredity and environment.

APPLY

Leslie and Glenn, the couple concerned that their baby could have sickle-cell disease, are already charting their baby's life course. Leslie, who has always loved to sing, is confident that her baby will be a fantastic musician and imagines a routine of music lessons, rehearsals, and concerts. Glenn, a pilot, is just as confident that his child will share his love of flying; he is already planning trips the two of them can take together. What advice might you give to Leslie and Glenn about factors they are ignoring?

Recall answers: (1) autosomes, (2) Polygenic inheritance, (3) Down syndrome, (4) the fertilized egg is aborted spontaneously, (5) different from each other

---

2.2 From Conception to Birth

LEARNING OBJECTIVES

- What happens to a fertilized egg in the first 2 weeks after conception?

- When do body structures and internal organs emerge in prenatal development?

- When do body systems begin to function well enough to support life?

Eun Jung has just learned that she is pregnant with her first child. Like many other parents-to-be, she and her husband, Kinam, are ecstatic. But they also soon realize how little they know about "what happens when" during pregnancy. Eun Jung is eager to visit her obstetrician to learn more about the normal timetable of events during pregnancy.

PRENATAL DEVELOPMENT BEGINS WHEN A SPERM SUCCESSFULLY FERTILIZES AN EGG. *The many changes that transform the fertilized egg into a newborn human constitute* **prenatal development**. Prenatal development takes an average of 38 weeks, which are divided into three periods: the period of the zygote, the period of the embryo, and the period of the fetus.* Each period gets its name from the scientific term used to describe the baby-to-be at that point in its prenatal development.

In this section, we'll trace the major developments of each of these periods. As we do, you'll learn the answers to the "what happens when" question that intrigues Eun Jung.

prenatal development
the many changes that turn a fertilized egg into a newborn human

Period of the Zygote (Weeks 1–2)

The teaspoon or so of seminal fluid produced during a fertile male's ejaculation contains from 200 to 500 million sperm. Of the sperm released into the vagina, only a few hundred will actually complete the 6- or 7-inch journey to the Fallopian tubes. Here,

---

*Perhaps you've heard that pregnancy lasts 40 weeks and wonder why we say that prenatal development lasts 38 weeks. The reason is that the 40 weeks of pregnancy are measured from the start of a woman's last menstrual period, which typically is about 2 weeks before conception.

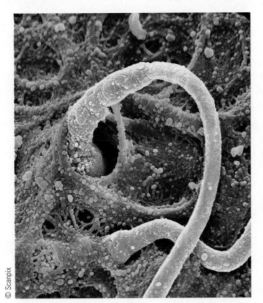
© Scanpix

Fertilization begins when sperm cells burrow their way through the outer layers of an egg cell. In this photo, the tails of the sperm can be seen clearly but one sperm has burrowed so deeply that the head is barely visible.

an egg arrives monthly, hours after it is released by an ovary. If an egg is present, many sperm will simultaneously begin to burrow their way through the cluster of nurturing cells that surround the egg. When one sperm finally penetrates the cellular wall of the egg, chemical changes occur in the wall immediately, blocking out all other sperm. Then the nuclei of the egg and sperm fuse, and the two independent sets of 23 chromosomes are interchanged. The development of a new human being is under way!

For nearly all of history, sexual intercourse was the only way for egg and sperm to unite and begin the development that results in a human being. This is no longer the only way, as we see in the What Do *You* Think? feature.

Whether by artificial means as just described or by natural means, fertilization begins the period of the **zygote**, *the technical term for the fertilized egg.* This period ends when the zygote implants itself in the wall of the uterus. During these 2 weeks, the zygote grows rapidly through cell division. ■ Figure 2.5 traces the egg cell from the time it is released from the ovary until the zygote becomes implanted in the wall of the uterus. The zygote travels down the Fallopian tube toward the uterus. Within hours, the zygote divides for the first time; it then continues to do so every 12 hours. Occasionally, the zygote separates into two clusters that develop into identical twins. Fraternal twins, which are more common, are created when two eggs are released and each is fertilized by a different sperm cell.

What do you think? | Conception in the 21st Century

More than 30 years ago, Louise Brown captured the world's attention as the first test-tube baby—conceived in a petri dish instead of in her mother's body. Today, this reproductive technology is no longer experimental; it is used more than 140,000 times annually by American women and produces more than 55,000 babies each year (Centers for Disease Control and Prevention, 2007d). Many new techniques are available to couples who cannot conceive a child through sexual intercourse. *The best-known technique,* **in vitro fertilization**, *involves mixing sperm and egg together in a petri dish and then placing several fertilized eggs in the mother's uterus, with the hope that they will become implanted in the uterine wall.* Other methods include injecting many sperm directly into the Fallopian tubes or a single sperm directly into an egg.

The sperm and egg usually come from the prospective parents, but sometimes they are provided by donors. Typically, the fertilized eggs are placed in the uterus of the prospective mother, but sometimes they are placed in the uterus of a surrogate mother who carries the baby to term. This means that a baby could have as many as five "parents": the man and woman who provided the sperm and egg; the surrogate mother who carried the baby; and the mother and father who will rear the baby.

New reproductive techniques offer hope for couples who have long wanted a child, and stud-

ies of the first generation of children conceived via these techniques indicate that their social and emotional development is perfectly normal (Golombok et al., 2006; MacCallum, Golombok, & Brinsden, 2007). But there are difficulties as well. Only about one third of attempts at in vitro fertilization succeed. What's more, when a woman becomes pregnant, she is more likely to have twins or triplets because multiple eggs are transferred to increase the odds that at least one fertilized egg will implant in the mother's uterus. (An extreme example of this would be "Octomom," a woman who had octuplets following in vitro fertilization.) She is also at greater risk for giving birth to a baby with low birth weight or birth defects. Finally, the procedure is expensive—the average cost in the United States of a single cycle of treatment is between $10,000 and $15,000—and often is not covered by health insurance.

These problems emphasize that, although technology has increased the alternatives for infertile couples, pregnancy on demand is still in the realm of science fiction. At the same time, the new technologies have led to much controversy because of some complex ethical issues associated with their use. One concerns the prospective parents' right to select particular egg and sperm cells; another involves who should be able to use this technology.

Pick your egg and sperm cells from a catalog? Until recently, prospective parents have known nothing about egg and sperm donors. Today, however, they are sometimes able to select egg and sperm based on physical and psychological characteristics of the donors, including appearance and race. Some claim that such prospective parents have a right to be fully informed about the person who provides the genetic material for their baby. *Others argue that this amounts to* **eugenics**, *which is the effort to improve the human species by allowing only certain people to mate and pass along their genes to subsequent generations.*

Available to all? Most couples who use in vitro fertilization are in their 30s and 40s, but a number of older women have begun to use the technology. Many of these women cannot conceive naturally because they have gone through menopause and no longer ovulate. Some argue that it is unfair to a child to have parents who may not live until the child reaches adulthood. Others point out that people are living longer and that middle-aged (or older) adults make better parents. (We discuss this issue in more depth in Chapter 13.)

What do you think? Should prospective parents be allowed to browse a catalog with photos and biographies of prospective donors? Should new reproductive technologies be available to all, regardless of age?

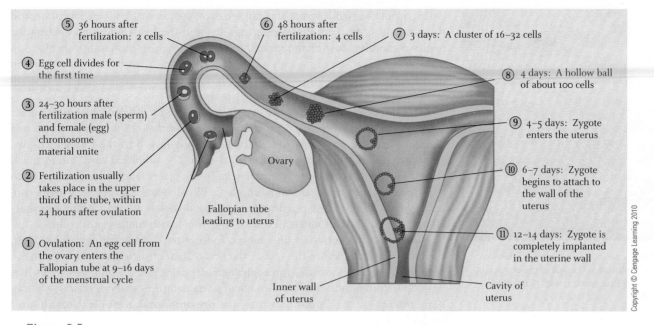

① Ovulation: An egg cell from the ovary enters the Fallopian tube at 9–16 days of the menstrual cycle

② Fertilization usually takes place in the upper third of the tube, within 24 hours after ovulation

③ 24–30 hours after fertilization male (sperm) and female (egg) chromosome material unite

④ Egg cell divides for the first time

⑤ 36 hours after fertilization: 2 cells

⑥ 48 hours after fertilization: 4 cells

⑦ 3 days: A cluster of 16–32 cells

⑧ 4 days: A hollow ball of about 100 cells

⑨ 4–5 days: Zygote enters the uterus

⑩ 6–7 days: Zygote begins to attach to the wall of the uterus

⑪ 12–14 days: Zygote is completely implanted in the uterine wall

Ovary

Fallopian tube leading to uterus

Inner wall of uterus

Cavity of uterus

■ **Figure 2.5**
The period of the zygote spans 14 days, beginning with fertilization of the egg in the Fallopian tube and ending with implantation of the fertilized egg in the wall of the uterus.

After about 4 days, the zygote includes about 100 cells and resembles a hollow ball. The inner part of the ball is destined to become the baby. The outer layer of cells will form a number of structures that provide a life-support system throughout prenatal development.

By the end of the first week, the zygote reaches the uterus. *The next step is* **implantation**, *in which the zygote burrows into the uterine wall and establishes connections with a woman's blood vessels.* Implantation takes about a week to complete and triggers hormonal changes that prevent menstruation, letting the woman know that she has conceived.

The implanted zygote is less than a millimeter in diameter, yet its cells have already begun to differentiate. *A small cluster of cells near the center of the zygote, the* **germ disc**, *will eventually develop into the baby.* The other cells are destined to become structures that support, nourish, and protect the developing organism. *For example, the layer of cells closest to the uterus will become the* **placenta**, *a structure through which nutrients and wastes are exchanged between the mother and the developing organism.*

Implantation and differentiation of cells mark the end of the period of the zygote. Comfortably settled in the shelter of the uterus, the zygote is well prepared for the remaining 36 weeks of the marvelous trek leading up to birth.

Period of the Embryo (Weeks 3–8)

Once the zygote is completely embedded in the uterine wall, it is called an **embryo**. This new period typically begins the third week after conception and lasts until the end of the eighth week. During the period of the embryo, body structures and internal organs develop. At the beginning of this period, three layers begin to form in the embryo. *The outer layer or* **ectoderm** *becomes hair, the outer layer of skin, and the nervous system; the middle layer or* **mesoderm** *forms muscles, bones, and the circulatory system; the inner layer or* **endoderm** *forms the digestive system and the lungs.*

One dramatic way to see these changes is to compare a 3-week-old embryo with an 8-week-old embryo. The 3-week-old embryo is about 2 millimeters long. Specialization of cells is under way, but the organism looks more like a salamander than a human being. However, growth and specialization proceed so rapidly that an 8-week-old

in vitro fertilization
process by which sperm and an egg are mixed in a petri dish to create a zygote, which is then placed in a woman's uterus

eugenics
effort to improve the human species by letting only people whose characteristics are valued by a society mate and pass along their genes

zygote
fertilized egg

implantation
step in which the zygote burrows into the uterine wall and establishes connections with a woman's blood vessels

germ disc
small cluster of cells near the center of the zygote that will eventually develop into a baby

placenta
structure through which nutrients and wastes are exchanged between the mother and the developing child

embryo
term given to the zygote once it is completely embedded in the uterine wall

ectoderm
outer layer of the embryo, which will become the hair, the outer layer of skin, and the nervous system

mesoderm
middle layer of the embryo, which becomes the muscles, bones, and circulatory system

endoderm
inner layer of the embryo, which becomes the lungs and the digestive system

By the end of the period of the zygote, the fertilized egg has been implanted in the wall of the uterus and has begun to make connections with the mother's blood vessels.

At 3 weeks after conception, the fertilized egg is about 2 millimeters long and resembles a salamander.

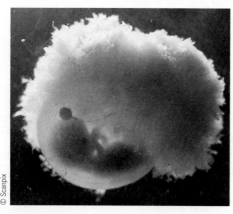

At 8 weeks after conception, near the end of the period of the embryo, the fertilized egg is obviously recognizable as a baby-to-be.

embryo looks very different: You can see eyes, jaw, arms, and legs. The brain and the nervous system are developing rapidly, and the heart has been beating for nearly a month. Most of the organs found in a mature human are in place, in some form. (The sex organs are a notable exception.) Yet because it is only an inch long and weighs but a fraction of an ounce, the embryo is much too small for the mother to feel its presence.

The embryo's environment is shown in ■ Figure 2.6. *The embryo rests in a sac called the **amnion**, which is filled with **amniotic fluid** that cushions the embryo and maintains a constant temperature.* The embryo is linked to the mother via two structures, the placenta and the umbilical cord. *The **umbilical cord** houses blood vessels that join the embryo to the placenta.* In the placenta, the blood vessels from the umbilical cord run close to the mother's blood vessels but aren't actually connected to them. The close proximity of the blood vessels allows nutrients, oxygen, vitamins, and waste products to be exchanged between mother and embryo.

Growth in the period of the embryo follows two important principles. First, the head develops before the rest of the body. *Such growth, from the head to the base of the spine, illustrates the **cephalocaudal principle**.* Second, arms and legs develop before hands and feet. *Growth of parts near the center of the body before those that are more distant illustrates the **proximodistal principle**.* Growth after birth also follows these principles.

With body structures and internal organs in place, the embryo has passed another major milestone in prenatal development. What's left is for these structures and organs to begin working properly. This is accomplished in the final period of prenatal development, as we'll see next.

Period of the Fetus (Weeks 9–38)

*The final and longest phase of prenatal development, the **period of the fetus**, begins at the ninth week (when cartilage begins to turn to bone) and ends at birth.* During this period, the baby-to-be becomes much larger and its bodily systems begin to work. The increase in size is remarkable. At the beginning of this period, the fetus weighs less than an ounce. At about 4 months, the fetus weighs roughly 4 to 8 ounces, which is large enough for the mother to feel its movements. In the last 5 months of pregnancy, the fetus will gain an additional 7 or 8 pounds before birth. ■ Figure 2.7, which depicts the fetus at one eighth of its actual size, shows these incredible increases in size.

During the fetal period, the finishing touches are placed on the many systems that are essential to human life, such as respiration, digestion, and vision. Some highlights of this period include the following.

■ At 4 weeks after conception, a flat set of cells curls to form a tube. One end of the tube swells to form the brain; the rest forms the spinal cord. By the start of the fetal period, the brain has distinct structures and has begun to regulate body functions. *During the period of the fetus, all regions of the brain grow— particularly the **cerebral cortex**, the wrinkled surface of the brain that regulates many important human behaviors.* (On pages 94–95 in Chapter 3, we describe fetal brain development in detail.)

■ Near the end of the embryonic period, male embryos develop testes and female embryos develop ovaries. In the 3rd month, the testes in a male fetus secrete a hormone that causes a set of cells to become a penis and scrotum; in a female fetus this hormone is absent, so the same cells become a vagina and labia.

■ During the 5th and 6th months after conception, eyebrows, eyelashes, and scalp hair emerge. *The skin thickens and is covered with a thick greasy substance, or **vernix**, that protects the fetus during its long bath in amniotic fluid.*

■ By about 6 months after conception, fetuses differ in their usual heart rates and in how much their heart rate changes in response to physiological stress. In one

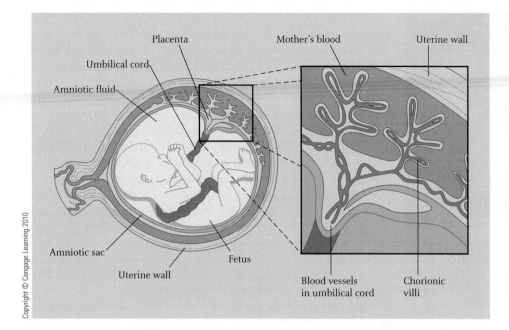

Placenta
Umbilical cord
Amniotic fluid

Mother's blood
Uterine wall

Amniotic sac
Uterine wall
Fetus

Blood vessels
in umbilical cord
Chorionic
villi

Copyright © Cengage Learning 2010

■ **Figure 2.6**
The fetus is wrapped in the amniotic sac and connected to the mother through the umbilical cord.

amnion
inner sac in which the developing child rests

amniotic fluid
fluid that surrounds the fetus

umbilical cord
structure containing veins and arteries that connects the developing child to the placenta

cephalocaudal principle
a principle of physical growth that states that structures nearest the head develop first

proximodistal principle
principle of physical growth that states that structures nearest the center of the body develop first

■ **Figure 2.7**
The baby-to-be becomes much larger during the period of the fetus, and its bodily systems start to work.

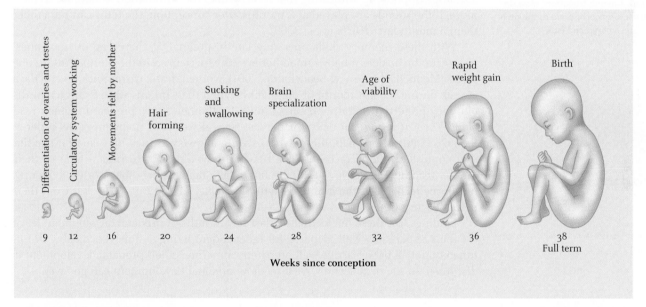

Differentiation of ovaries and testes
Circulatory system working
Movements felt by mother
Hair forming
Sucking and swallowing
Brain specialization
Age of viability
Rapid weight gain
Birth

9   12   16   20   24   28   32   36   38
Full term

Weeks since conception

From *Before We Are Born,* Fourth Edition, by K.L. Moore and T.V.N. Persaud, p. 130. Copyright © 1993 W.B. Saunders. Reprinted with permission.

study (DiPietro et al., 2007), fetuses with greater heart-rate variability were, as 2-month-olds, more advanced in their motor, mental, and language development. Greater heart-rate variability may be a sign that the nervous system is responding efficiently to environmental change (as long as the variability is not extreme).

With these and other rapid changes, by 22 to 28 weeks most systems function well enough that a fetus born at this time has a chance to survive, which is why this age range is called the **age of viability**. By this age, the fetus has a distinctly baby-like look, but babies born this early have trouble breathing because their lungs are not yet mature. Also, they cannot regulate their body temperature very well because they lack the insulating layer of fat that appears in the 8th month after conception. With modern neonatal intensive care, infants born this early can survive; but they face other challenges, as we'll see later in this chapter.

period of the fetus
longest period of prenatal development, extending from the 9th until the 38th week after conception

cerebral cortex
wrinkled surface of the brain that regulates many functions that are distinctly human

vernix
substance that protects the fetus's skin during development

age of viability
age at which a fetus can survive because most of its bodily systems function adequately; typically at 7 months after conception

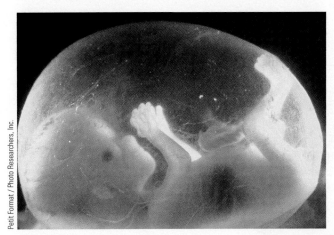

At 22–28 weeks after conception, the fetus has achieved the age of viability, meaning that it has a chance of surviving if born prematurely.

THINK ABOUT IT

Health care professionals often divide pregnancy into three 3-month trimesters. How do these three trimesters correspond to the periods of the zygote, embryo, and fetus?

During the fetal period, the fetus actually starts to behave (Joseph, 2000). The delicate movements that were barely noticeable at 4 months are now obvious. In fact, the fetus is a budding gymnast and kick-boxer rolled into one. It will punch or kick and turn somersaults. When active, the fetus moves about once a minute (DiPietro et al., 2004). These bursts of activity are followed by times when the fetus is still, as regular activity cycles emerge. Although movement is common in a healthy pregnancy, some fetuses are more active than others and these differences predict infants' behavior: An active fetus is more likely than an inactive fetus to be an unhappy, difficult baby (DiPietro et al., 1996).

Another sign of growing behavioral maturity is that the senses work. There's not much to see in the uterus (imagine being in a cave with a flashlight that has a weak battery), but there are sounds galore. The fetus can hear the mother's heart beating and can hear her food digesting. More important, the fetus can hear her speak and can hear others speak to her (Lecanuet, Granier-Deferre, & Busnel, 1995). And there are tastes: As the fetus swallows amniotic fluid, it responds to different flavors in the fluid.

Not only can the fetus detect sounds and flavors, it can remember these sensory experiences later. For example, when sounds are played through a loudspeaker placed on a pregnant woman's abdomen, the fetus usually responds to the sound and vibrations by moving. However, if the sound is repeated every 30 seconds, the fetus gradually stops responding, indicating that it recognizes the stimulation as familiar. What's more, if the sounds are played at 8 months after conception, the fetus can remember them a month later (Dirix et al., 2009).

With these memory skills operating late in pregnancy, there's an obvious question: After birth, does a baby remember events experienced in the uterus? Yes. In one study (Mennella, Jagnow, & Beauchamp, 2001), women drank carrot juice several days a week during the last month of pregnancy. When their infants were 5 and 6 months old, they preferred cereal with carrot juice. In another study, pregnant women read aloud *The Cat in the Hat* daily for the last several weeks of pregnancy (DeCasper & Spence, 1986). After birth, the newborns were allowed to suck on a special pacifier that controlled a tape recorder. The newborns would suck to hear a tape of their mother reading *The Cat in the Hat* but not to hear her reading other stories. Evidently, newborns recognized the familiar, rhythmic quality of *The Cat in the Hat* from their prenatal story times.

Findings like these tell us that the last few months of prenatal development leave the fetus remarkably well prepared for independent living as a newborn baby. Unfortunately, not all babies arrive well prepared. Sometimes their prenatal development is disrupted. In the next section, we'll see how prenatal development can go awry.

Test Yourself

RECALL

1. The period of the zygote ends _____ .

2. Body structures and internal organs are created during the period of the _____ .

3. _____ is called the age of viability because this is when most body systems function well enough to support life.

4. In the last few months of prenatal development, the fetus has regular periods of activity and _____ , which are the first signs of fetal behavior.

INTERPRET

Compare the events of prenatal development that precede the age of viability with those that follow it.

APPLY

In the last few months before birth, the fetus has some basic perceptual and motor skills; a fetus can hear, see, taste, and move. What are the advantages of having these skills in place months before they're really needed?

Recall answers: (1) at 2 weeks after conception (when the zygote is completely implanted in the wall of the uterus), (2) embryo, (3) Between 22 and 28 weeks, (4) the eyes and ears respond to stimulation

- How is prenatal development influenced by a pregnant woman's age, her nutrition, and the stress she experiences while pregnant?

- How do diseases, drugs, and environmental hazards sometimes affect prenatal development?

- What general principles affect the ways that prenatal development can be harmed?

- How can prenatal development be monitored? Can abnormal prenatal development be corrected?

Chloe was 2 months pregnant at her first prenatal checkup. As her appointment drew near, she began a list of questions to ask her obstetrician. "I use my cell phone a lot. Is radiation from the phone harmful to my baby?" "When my husband and I get home from work, we'll have a glass of wine to help unwind from the stress of the day. Is moderate drinking like this okay?" "I'm 38. I know older women give birth to babies with mental retardation more often. Can I know if my baby will be mentally retarded?"

EACH OF CHLOE'S QUESTIONS CONCERNS HARM TO HER BABY-TO-BE. She worries about the safety of her cell phone, about her nightly glass of wine, and about her age. Chloe's concerns are well founded. Many factors influence the course of prenatal development, and they are the focus of this section. If you're sure you can answer *all* of Chloe's questions, then skip this section and go directly to page 69. Otherwise, read on to learn about problems that sometimes arise in pregnancy.

General Risk Factors

As the name implies, general risk factors can have widespread effects on prenatal development. Scientists have identified three general risk factors: nutrition, stress, and a mother's age.

Nutrition

The mother is the developing child's sole source of nutrition, so a balanced diet that includes foods from each of the five major food groups is vital. Most pregnant women need to increase their intake of calories by about 10 to 20% to meet the needs of prenatal development. A woman should expect to gain between 25 and 35 pounds during pregnancy, assuming that her weight was normal before pregnancy. A woman who was underweight before becoming pregnant may gain as much as 40 pounds; a woman who was overweight should gain at least 15 pounds (Institute of Medicine, 1990). Of this gain, about one third reflects the weight of the baby, the placenta, and the fluid in the amniotic sac; another third comes from increases in a woman's fat stores; and yet another third comes from the increased volume of blood and increases in the size of the woman's breasts and uterus (Whitney & Hamilton, 1987).

Sheer amount of food is only part of the equation for a healthy pregnancy. *What a pregnant woman eats is also very important.* Proteins, vitamins, and minerals are essential for normal prenatal development. For example, folic acid (one of the B vitamins) is important for the baby's nervous system to develop properly (Shaw et al., 1995). *When mothers do not consume adequate amounts of folic acid, their babies are at risk for* **spina bifida**, *a disorder in which the embryo's neural tube does not close properly during the first month of pregnancy.* Since the neural tube develops into the brain and spinal cord, the result when it does not close properly is permanent damage to the spinal cord and the nervous system. Many children with spina bifida need crutches, braces, or wheelchairs. Other prenatal problems have also been traced to inadequate proteins, vitamins, or minerals, so health care providers typically recom-

spina bifida
disorder in which the embryo's neural tube does not close properly

mend that pregnant women supplement their diet with additional proteins, vitamins, and minerals.

When a pregnant woman does not provide adequate nourishment, the infant is likely to be born prematurely and to be underweight. Inadequate nourishment during the last few months of pregnancy can particularly affect the nervous system, because this is a time of rapid brain growth. Finally, babies who do not receive adequate nourishment are vulnerable to illness (Morgane et al., 1993).

Stress

Does a pregnant woman's mood affect the zygote, embryo, or fetus in her uterus? Is a woman who is happy during pregnancy more likely to give birth to a happy baby? Is a harried office worker more likely to give birth to an irritable baby? *These questions address the impact on prenatal development of chronic* **stress**, *which refers to a person's physical and psychological responses to threatening or challenging situations.* We can answer these questions with some certainty for nonhumans. When pregnant female animals experience constant stress—such as repeated electric shocks or intense overcrowding—their offspring are often smaller than average and prone to other physical and behavioral problems (DiPietro, 2004). In addition, stress seems to cause greater harm when experienced early in pregnancy (Schneider et al., 1999).

stress
physical and psychological responses to threatening or challenging conditions

When pregnant women experience chronic stress, they're more likely to give birth early or have smaller babies; this may be because women who are stressed are more likely to smoke or drink and less likely to rest, exercise, and eat properly.

Determining the impact of stress on human pregnancy is more difficult because we must rely solely on correlational studies. (It would be unethical to do an experiment that assigned some pregnant women to a condition of extreme stress.) Studies typically show that women who report greater anxiety during pregnancy more often give birth early or have babies who weigh less than average (Copper et al., 1996; Paarlberg et al., 1995). What's more, when women are anxious throughout pregnancy, their children are less able to pay attention as infants and more prone to behavioral problems as preschoolers (Huizink et al., 2002; O'Connor et al., 2002). Similar results emerge in studies of pregnant women exposed to disasters, such as the September 11 attacks on the World Trade Center: their children's physical and behavioral development is affected (Engel et al., 2005; Laplante et al., 2004). Finally, the harmful effects of stress are not linked to anxiety in general but are specific to worries about pregnancy, particularly worries in the first few months (Davis & Sandman, 2010; DiPietro et al., 2006).

Increased stress can harm prenatal development in several ways. First, when a pregnant woman experiences stress, her body secretes hormones that reduce the flow of oxygen to the fetus while increasing its heart rate and activity level (Monk et al., 2000). Second, stress can weaken a pregnant woman's immune system, making her more susceptible to illness (Cohen & Williamson, 1991) that can, in turn, damage fetal development. Third, pregnant women under stress are more likely to smoke or drink alcohol and are less likely to rest, exercise, and eat properly (DiPietro et al., 2004). All these behaviors endanger prenatal development.

We want to emphasize that the results described here apply to women who experience prolonged, extreme stress. Virtually all women sometimes become anxious or upset while pregnant. Occasional, relatively mild anxiety is not thought to have any harmful consequences for prenatal development.

Mother's Age

Traditionally, the 20s were thought to be the prime childbearing years. Teenage women as well as women who were 30 or older were considered less fit for the rigors of pregnancy. Is being a 20-something really important for a successful pregnancy? Let's answer this question separately for teenage and older women. Compared to women in their 20s, teenage women are more likely to have problems during pregnancy, labor, and delivery. This is largely because pregnant teenagers are more likely to be economically disadvantaged and to lack good prenatal care—either because they are unaware of the need for it or because they cannot afford it. For example, in one study (Turley, 2003) children of teenage moms were compared with their cousins, whose mothers were the older sisters of the teenage moms but had given birth when they were in their 20s. The two groups of children were similar in academic skills and behavioral problems, indicating that it's not the age but rather the typical family background of teenage moms that creates problems. Similarly, research done on African American adolescents indicates that, when differences in prenatal care are taken into account, teenagers are just as likely as women in their 20s to have problem-free pregnancies and to give birth to healthy babies (Goldenberg & Klerman, 1995).

For teenage mothers and their babies, life is often a struggle because the mothers are unable to complete their education and often live in poverty.

Nevertheless, even when a teenager receives adequate prenatal care and gives birth to a healthy baby, all is not rosy. Children of teenage mothers generally do less well in school and more often have behavioral problems (D'Onofrio et al., 2009; Fergusson & Woodward, 2000). The problems of teenage motherhood—incomplete education, poverty, and marital difficulties—affect the child's later development (Moore & Brooks-Gunn, 2002).

Of course, not all teenage mothers and their infants follow this dismal life course. Some teenage mothers finish school, find good jobs, and have happy marriages; their children do well in school, academically and socially. These "success stories" are more likely when teenage moms live with a relative—typically, the child's grandmother (Gordon, Chase-Lansdale, & Brooks-Gunn, 2004). However, teenage pregnancies with "happy endings" are definitely the exception; for most teenage mothers and their children, life is a struggle. Educating teenagers about the true consequences of teen pregnancy is crucial.

Are older women better suited for pregnancy? This is an important question because today's American woman is waiting longer than ever to have her first child. Completing an education and beginning a career often delay childbearing. In fact, the birth rate in the early 2000s among 40- to 44-year-olds was at its highest rate since the 1960s (Hamilton et al., 2010).

Traditionally, older women were thought to have more difficult pregnancies and more complicated labor and deliveries. Today, we know that women in their 20s are twice as fertile as women in their 30s (Dunson, Colombo, & Baird, 2002). For women 35 years of age and older, the risks of miscarriage and stillbirth increase rapidly. Among 40- to 45-year-olds, for example, nearly half of all pregnancies result in miscarriage (Andersen et al., 2000). Also, women in their 40s are more liable to give birth to babies with Down syndrome. As mothers, however, older women are quite effective. For example, they are just as able to provide the sort of sensitive, responsive caregiving that promotes a child's development (Bornstein et al., 2006).

In general, then, prenatal development is most likely to proceed normally when women are between 20 and 35 years of age, are healthy and eat right, get good health care, and lead lives that are free of chronic stress. But even in these optimal cases, prenatal development can be disrupted, as we'll see in the next section.

Teratogens: Drugs, Diseases, and Environmental Hazards

In the late 1950s, many pregnant women in Germany took thalidomide, a drug that helped them sleep. Soon, however, came reports that many of these women were giving birth to babies with deformed arms, legs, hands, or fingers. *Thalidomide is a powerful* **teratogen**, *an agent that causes abnormal prenatal development.* Ultimately, more than 7,000 babies worldwide were harmed before thalidomide was withdrawn from the market (Kolberg, 1999).

teratogen
an agent that causes abnormal prenatal development

● TABLE 2.3

Teratogenic Drugs and Their Consequences

Drug	Potential Consequences
Alcohol	Fetal alcohol syndrome, cognitive deficits, heart damage, retarded growth
Aspirin	Deficits in intelligence, attention, and motor skills
Caffeine	Lower birth weight, decreased muscle tone
Cocaine and heroin	Retarded growth, irritability in newborns
Marijuana	Lower birth weight, less motor control
Nicotine	Retarded growth, possible cognitive impairments

fetal alcohol spectrum disorder (FASD) disorder affecting babies whose mothers consumed large amounts of alcohol while they were pregnant

When pregnant women drink large amounts of alcohol, their children often have fetal alcohol syndrome; they tend to have a small head and a thin upper lip as well as retarded mental development.

Prompted by the thalidomide disaster, scientists began to study teratogens extensively. Today, we know a great deal about many teratogens that affect prenatal development. Most teratogens fall into one of three categories: drugs, diseases, or environmental hazards. Let's look at each.

Drugs

Thalidomide illustrates the harm that drugs can cause during prenatal development. ● Table 2.3 lists several other drugs that are known teratogens. Most of the drugs in the list are substances you may use routinely—alcohol, aspirin, caffeine, nicotine. Nevertheless, when consumed by pregnant women, they do present special dangers (Behnke & Eyler, 1993).

Cigarette smoking is typical of the potential harm from teratogenic drugs (Cornelius et al., 1995; Fried, O'Connell, & Watkinson, 1992). The nicotine in cigarette smoke constricts blood vessels and thus reduces the oxygen and nutrients that can reach the fetus over the placenta. Therefore, pregnant women who smoke are more likely to miscarry (abort the fetus spontaneously) and to bear children who are smaller than average at birth (Cnattingius, 2004; Ernst, Moolchan, & Robinson, 2001). And, as children develop, they are more likely to show signs of impaired attention, language, and cognitive skills as well as behavioral problems (Brennan et al., 2002; Wakschlag et al., 2006). Finally, even secondhand smoke harms the fetus: When pregnant women don't smoke but fathers do, babies tend to be smaller at birth (Friedman & Polifka, 1996). The message is clear and simple: Pregnant women shouldn't smoke, and they should avoid others who do smoke.

Alcohol also carries serious risk. *Pregnant women who regularly consume quantities of alcoholic beverages may give birth to babies with* **fetal alcohol spectrum disorder** (FASD). The most extreme form, fetal alcohol syndrome (FAS), is most likely among pregnant women who are heavy recreational drinkers— that is, women who drink 5 or more ounces of alcohol a few times each week (Jacobson & Jacobson, 2000; Lee, Mattson, & Riley, 2004). Children with FAS usually grow more slowly than normal and have heart problems and misshapen faces. Like the children in the photo, youngsters with FAS often have a small head, a thin upper lip, a short nose, and widely spaced eyes. FAS is the leading cause of developmental disabilities in the United States, and children with FAS have serious

© David Young-Wolff / Photo Edit

Teratogenic Diseases and Their Consequences

Disease	Potential Consequences
AIDS	Frequent infections, neurological disorders, death
Chlamydia	Premature birth, low birth weight, eye inflammation
Chicken pox	Spontaneous abortion, developmental delays, mental retardation
Cytomegalovirus	Deafness, blindness, abnormally small head, mental retardation
Genital herpes	Encephalitis, enlarged spleen, improper blood clotting
Rubella (German measles)	Mental retardation; damage to eyes, ears, and heart
Syphilis	Damage to the central nervous system, teeth, and bones
Toxoplasmosis	Damage to eye and brain; learning disabilities

attentional, cognitive, and behavioral problems (e.g., Howell et al., 2006; Sokol, Delaney-Black, & Nordstrom, 2003).

Is there any amount of drinking that's safe during pregnancy? Maybe, but scientists have yet to determine one. This inconclusiveness stems from two factors. First, drinking is often estimated from women's responses to interviews or questionnaires. These replies may be incorrect, leading to inaccurate estimates of the harm associated with drinking. Second, any safe level of consumption is probably not the same for all women. Based on their health and heredity, some women may be able to consume more alcohol safely than others.

These factors make it impossible to offer guaranteed statements about safe levels of alcohol or any of the other drugs listed in Table 2.3. For this reason, the best policy is for women to avoid all drugs throughout pregnancy.

Diseases

Sometimes women become ill while pregnant. Most diseases, such as colds and many strains of the flu, do not affect the fetus. However, several bacterial and viral infections can be quite harmful; several are listed in ●Table 2.4.

Some diseases pass from the mother through the placenta to attack the embryo or fetus directly. AIDS, cytomegalovirus, rubella, and syphilis are examples of diseases that are transmitted through the placenta. Other diseases attack during birth: The virus is present in the lining of the birth canal, and babies are infected as they pass through the canal. AIDS and genital herpes are two such diseases.

The only way to guarantee that these diseases will not harm prenatal development is for a woman to be sure that she does not contract them either before or during her pregnancy. Medicines that may help to treat a woman after she has become ill do not prevent the disease from damaging the fetus.

Environmental Hazards

As a by-product of life in an industrialized world, people are often exposed to toxins in food they eat, fluids they drink, and air they breathe. Chemicals associated with industrial waste are the most common form of environmental teratogens. The quantity involved is usually minute; however, as with drugs, amounts that go unnoticed in an adult can cause serious damage to the fetus.

Polychlorinated biphenyls (PCBs) illustrate the danger of environmental teratogens. These were used in electrical transformers and paints until the U.S. government banned them in the 1970s. However, PCBs (like many industrial by-products) seeped into the waterways, where they contaminated fish and wildlife. The amount of PCBs in a typical contaminated fish does not affect adults, but when pregnant women ate

THINK ABOUT IT

A pregnant woman reluctant to give up her morning cup of coffee and nightly glass of wine says, "I drink so little coffee and wine that it couldn't possibly hurt my baby." What do you think?

TABLE 2.5

Environmental Teratogens and Their Consequences

Hazard	Potential Consequences
Lead	Mental retardation
Mercury	Retarded growth, mental retardation, cerebral palsy
PCBs	Impaired memory and verbal skills
X-rays	Retarded growth, leukemia, mental retardation

Copyright © Cengage Learning 2010

large numbers of PCB-contaminated fish, their children's cognitive skills and reading achievement were impaired (Jacobson & Jacobson, 1996).

Several environmental hazards that are known teratogens are listed in ● Table 2.5. You may be wondering about one ubiquitous feature of modern environments that doesn't appear in Table 2.5: cell phones. Is a pregnant woman's cell phone usage hazardous to the health of her fetus? At this point, there's no definitive answer to that question. The radiofrequency radiation that cell phones generate has sometimes been linked to health risks in adults (e.g., cancer) but the findings are very inconsistent (National Radiological Protection Board, 2004; Verschaeve, 2009). There are few scientific studies of the impact of cell phones on prenatal development. In a study conducted in Denmark, cell phone use during *and after* pregnancy was associated with increased risk for behavior problems in childhood (Divan et al., 2008). In another study, conducted in Spain, cell phone use late in pregnancy was associated with lower motor development but greater mental development in 14-month-olds (Vrijheid et al., 2010). At this point, more research is needed to know if radiofrequency radiation from a pregnant woman's cell phone is a health risk. We do know, of course, another way in which cell phones represent a huge health risk for pregnant women: Talking while driving is incredibly distracting and reduces a driver's performance to the level seen by people driving under the influence of alcohol (Strayer, Drews & Crouch, 2006). So, while we wait for research to provide more information, the best advice for a pregnant woman would be to keep a cell phone at a distance when it's not being used and never use it while driving.

Environmental teratogens are treacherous because people are unaware of their presence in the environment. For example, the women studied by Jacobson, Jacobson, and Humphrey (1990) did not realize they were eating PCB-laden fish. Thus, the invisibility of some environmental teratogens makes it more difficult for a pregnant woman to protect herself from them. The best advice is for a pregnant woman to be particularly careful about the foods she eats and the air she breathes. Be sure that all foods are cleaned thoroughly to rid them of insecticides. Try to avoid convenience foods, which often contain many chemical additives. Stay away from air that's been contaminated by household products such as cleansers, paint strippers, and fertilizers. Women in jobs (such as housecleaning or hairdressing) that require contact with potential teratogens should switch to less potent chemicals, if possible. For example, they should use baking soda instead of more chemically laden cleansers; and they should wear protective gloves, aprons, and masks to reduce their contact with potential teratogens. Finally, because environmental teratogens continue to increase, check with a health care provider to learn whether other materials should be avoided.

How Teratogens Influence Prenatal Development

By assembling all the evidence on the harm caused by drugs, diseases, and environmental hazards, scientists have identified five important general principles about how teratogens usually work (Hogge, 1990; Jacobson & Jacobson, 2000; Vorhees & Mollnow, 1987).

1. *The impact of a teratogen depends on the genotype of the organism.* A substance may be harmful to one species but not to another. To determine its safety, thalidomide was tested on pregnant rats and rabbits, and their offspring had normal limbs. Yet when pregnant women took the same drug in comparable doses, many had children with deformed limbs. Moreover, some women who took thalidomide gave birth to babies with normal limbs whereas others, taking comparable doses of thalidomide at the same time in their pregnancies, gave birth to babies with deformed arms and legs. Apparently, heredity makes some individuals more susceptible than others to a teratogen.

2. *The impact of teratogens changes over the course of prenatal development.* The timing of exposure to a teratogen is very important. Teratogens typically have different effects in the three periods of prenatal development. ■ Figure 2.8 shows how the consequences of teratogens differ for the periods of the zygote, embryo, and fetus. During the period of the zygote, exposure to teratogens usually results in spontaneous abortion of the fertilized egg. During the period of the embryo, exposure to teratogens produces major defects in bodily structure. For instance, women who took thalidomide during the period of the embryo had babies with ill-formed or missing limbs, and women who contract rubella during the period of the embryo have babies with heart defects. During the period of the fetus, exposure to teratogens either produces minor defects in bodily structure or causes body systems to function improperly. For example, when women drink large quantities of alcohol during this period, the fetus develops fewer brain cells.

Even within the different periods of prenatal development, developing body parts and systems are more vulnerable at some times than others. The red shading in the chart indicates a time of maximum vulnerability; orange shading indicates a time when the developing organism is less vulnerable. The

■ **Figure 2.8**

The effects of a teratogen on an unborn child depend on the stage of prenatal development.

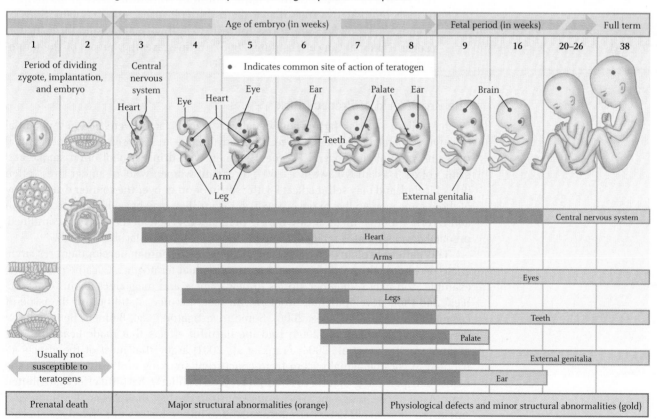

From *Before We Are Born,* Fourth Edition, by K.L. Moore and T.V.N. Persaud, p. 130. Copyright © 1993 W.B. Saunders. Reprinted with permission.

heart, for example, is most sensitive to teratogens during the first half of the embryonic period. Exposure to teratogens before this time rarely produces heart damage, and exposure after this time results in relatively mild damage.

3. *Each teratogen affects a specific aspect (or aspects) of prenatal development.* Said another way, teratogens do not harm all body systems; instead, damage is selective. When women contract rubella, their babies often have problems with their eyes, ears, and heart but have normal limbs. When mothers consume PCB-contaminated fish, their babies typically have normal body parts and normal motor skill but below-average verbal and memory skills.

4. *The impact of teratogens depends on the dose.* Just as a single drop of oil won't pollute a lake, small doses of teratogens may not harm the fetus. In research on PCBs, for example, cognitive skills were affected only among children who had the greatest prenatal exposure to these by-products. In general, the greater the exposure, the greater the risk for damage (Adams, 1999). An implication of this principle is that researchers should be able to determine safe levels for a teratogen. In reality, this is extremely difficult because sensitivity to teratogens is not the same for all people (and it's not practical to establish separate safe amounts for each person). Hence, the safest rule is zero exposure to teratogens.

5. *Damage from teratogens is not always evident at birth but may appear later in life.* In the case of malformed limbs or babies born addicted to cocaine, the effects of a teratogen are obvious immediately. Sometimes, however, the damage from a teratogen becomes evident only as the child develops. For example, when women ate PCB-contaminated fish, their babies were normal at birth. Their below-average cognitive skills were not evident until several months later.

An even more dramatic example of the delayed impact of a teratogen involves the drug diethylstilbestrol (DES). Between 1947 and 1971, many pregnant women took DES, a synthetic version of the female hormone estrogen, to prevent miscarriages. Their babies were apparently normal at birth. As adults, however, daughters of women who took DES are more likely to have breast cancer or a rare cancer of the vagina. And they sometimes have abnormalities in their reproductive tract that make it difficult to become pregnant. Sons of women who took DES are at risk for testicular abnormalities and for testicular cancer (National Cancer Institute, 2006). In this case, the impact of the teratogen is not evident until decades after birth.

The Real World of Prenatal Risk

We have discussed risk factors individually as if each factor were the only potential threat to prenatal development. In reality, many infants are exposed to multiple general risks and multiple teratogens. Pregnant women who drink alcohol often smoke and drink coffee (Haslam & Lawrence, 2004). Pregnant women who are under stress often drink alcohol, and may self-medicate with aspirin or other over-the-counter drugs. Many of these same women live in poverty, which means they may have inadequate nutrition and receive minimal medical care during pregnancy. When all the risks are combined, prenatal development is rarely optimal (Yumoto, Jacobson, & Jacobson, 2008).

This pattern explains why it's often challenging for human development researchers to determine the harm associated with individual teratogens. Cocaine is a perfect example. You may remember stories in newspapers and magazines about "crack babies" and their developmental problems. In fact, the jury is still out on the issue of cocaine as a teratogen (Jones, 2006). Some investigators (e.g., Bennett, Bendersky, & Lewis, 2008; Dennis et al., 2006) find the harmful effects that made headlines, but others (e.g., Brown et al., 2004; Frank et al., 2001) argue that most of the effects attributed to cocaine actually stem from concurrent smoking and drinking and to the inadequate parenting that these children receive. Similarly, harmful effects attributed to smoking during pregnancy may also stem from the fact that pregnant women who smoke are more likely to be less educated and to have a history of psychological problems, including antisocial behavior (D'Onofrio et al., 2010).

Of course, findings like these don't mean that pregnant women should feel free to light up (or, for that matter, to shoot up). Instead, they highlight the difficulties involved in determining the harm associated with a single risk factor (e.g., smoking) when it usually occurs alongside many other risk factors (e.g., inadequate parenting, continued exposure to smoke after birth).

From what we've said so far in this section, you may think that the developing child has little chance of escaping harm. But most babies *are* born in good health. Of course, a good policy for pregnant women is to avoid diseases, drugs, and environmental hazards that are known teratogens. This, coupled with thorough prenatal medical care and adequate nutrition, is the best recipe for normal prenatal development.

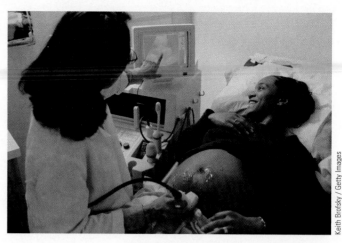

A standard part of prenatal care is ultrasound, in which sound waves are used to generate an image of the fetus that can be used to determine its position in the uterus.

Keith Brofsky / Getty Images

Prenatal Diagnosis and Treatment

"I really don't care whether I have a boy or girl, just as long as it's healthy." Legions of parents worldwide have felt this way, but until recently, all they could do was hope for the best. Today, however, advances in technology mean that parents can have a much better idea of whether their baby is developing normally.

Genetic Counseling

Often the first step in deciding whether a couple's baby is likely to be at risk is genetic counseling. A counselor asks about family medical history and constructs a family tree for each parent to assess the odds that their child would inherit a disorder. If the family tree suggests that a parent is likely to be a carrier of the disorder, blood tests can determine the parent's genotype. With this information, a genetic counselor then advises prospective parents about their choices. A couple might simply go ahead and attempt to conceive a child "naturally." Alternatively, they may decide to use sperm or eggs from other people. Yet another choice might be adoption.

Prenatal Diagnosis

After a woman is pregnant, how can we know whether prenatal development is progressing normally? Traditionally, obstetricians tracked the progress of prenatal development by feeling the size and position of the fetus through a woman's abdomen. This technique was not very precise and, of course, couldn't be done at all until the fetus was large enough to feel. Today, however, several new techniques have revolutionized our ability to monitor prenatal growth and development. *A standard part of prenatal care in the United States is* **ultrasound**, *in which sound waves are used to generate a picture of the fetus.* In this procedure, a tool about the size of a hair dryer is rubbed over the woman's abdomen, and the image appears on a nearby computer monitor. The pictures generated are hardly portrait quality; they are grainy, and it takes an expert's eye to distinguish what's what. Nevertheless, parents are often thrilled to see their baby and to watch it move.

Ultrasound typically can be used as early as 4 or 5 weeks after conception; prior to this time, the embryo is not large enough to generate an interpretable image. Ultrasound pictures are quite useful for determining the position of the fetus within the uterus and, at 16–20 weeks after conception, its sex. Ultrasound is also helpful in detecting twins or triplets. Finally, ultrasound is used to identify gross physical deformities, such as abnormal growth of the head.

In pregnancies where a genetic disorder is suspected, two other techniques are particularly valuable because they provide a sample of fetal cells that can be analyzed. *In* **amniocentesis**, *a needle is inserted through the mother's abdomen to obtain a sample of the amniotic fluid that surrounds the fetus.* As you can see in ■ Figure 2.9, ultrasound is

ultrasound
prenatal diagnostic technique that uses sound waves to generate an image of the fetus

amniocentesis
prenatal diagnostic technique that uses a syringe to withdraw a sample of amniotic fluid through the mother's abdomen

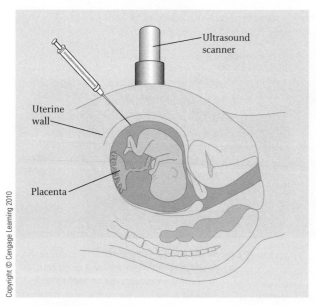

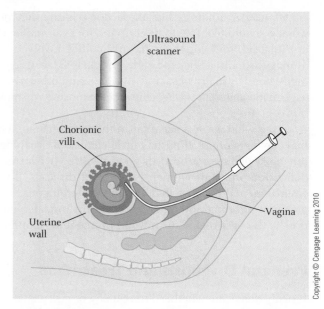

■ **Figure 2.9**

In amniocentesis, a sample of fetal cells is extracted from the fluid in the amniotic sac.

■ **Figure 2.10**

In chorionic villus sampling, fetal cells are extracted from the placenta.

chorionic villus sampling (CVS)
prenatal diagnostic technique that involves taking a sample of tissue from the chorion

THINK ABOUT IT

Imagine that you are 42 years old and pregnant. Would you want to have amniocentesis or chorionic villus sampling to determine the genotype of the fetus? Why or why not?

fetal medicine
field of medicine concerned with treating prenatal problems before birth

used to guide the needle into the uterus. The fluid contains skin cells that can be grown in a laboratory dish and then analyzed to determine the genotype of the fetus.

A procedure that can be used much earlier in pregnancy is **chorionic villus sampling (CVS)** *in which a sample of tissue is obtained from part of the placenta.* ■ Figure 2.10 shows that a small tube—typically inserted through the vagina and into the uterus but sometimes through the abdomen—is used to collect a small plug of cells from the placenta. This procedure can be used within 9–12 weeks after conception, much earlier than amniocentesis.

Results are returned from the lab in about two weeks following amniocentesis and in 7–10 days following CVS. (The wait is longer for amniocentesis because genetic material can't be evaluated until enough cells have reproduced for analysis.) With the samples obtained from either technique, roughly 200 different genetic disorders, including Down syndrome, can be detected. These procedures are virtually free of errors but come at a price: Miscarriages are slightly—1 or 2%—more likely after amniocentesis or chorionic villus sampling (Wilson, 2000). A woman must decide whether the information gained from amniocentesis or chorionic villus sampling justifies the slightly increased risks of a possible miscarriage.

Fetal Medicine

Ultrasound, amniocentesis, and chorionic villus sampling have made it much easier to determine whether prenatal development is progressing normally. But what happens when it is not? Traditionally, a woman's options have been limited: She could continue the pregnancy or end it. Today the list of options is expanding. *A whole new field called* **fetal medicine** *is concerned with treating prenatal problems before birth.* Many tools are now available to solve problems that are detected during pregnancy (Rodeck & Whittle, 2009). One approach is to treat disorders medically by administering drugs or hormones to the fetus. For example, in fetal hypothyroidism, the fetal thyroid gland does not produce enough hormones. This can lead to retarded physical and mental development, but the disorder can be treated by injecting the necessary hormones directly into the amniotic cavity, resulting in normal growth. Another example is congenital adrenal hyperplasia, an inherited disorder in which the fetal adrenal glands produce too much androgen; this causes early maturation of boys or masculinization of girls. Treatment consists of injecting hormones into the mother that reduce the amount of androgen secreted by the fetal adrenal glands (Evans, Platt, & De La Cruz, 2001).

Another way to correct prenatal problems is fetal surgery (Warner, Altimier, & Crombleholme, 2007). For instance, spina bifida can be corrected with fetal surgery in the seventh or eighth month of pregnancy. Surgeons cut through the mother's abdominal wall to expose the fetus and then cut through the fetal abdominal wall; the spinal cord is repaired, and the fetus is returned to the uterus. However, the procedure is far from foolproof: The best techniques and the ideal times to use them are still unknown (Adzick, 2010).

Fetal surgery has also been used to treat a disorder affecting identical twins in which one twin—the "donor"—pumps blood through its own and the other twin's circulatory system. The donor twin usually fails to grow, but surgery can correct the problem by sealing off the unnecessary blood vessels between the twins (Baschat, 2007). Fetal surgery holds great promise, but it is still highly experimental and thus is viewed as a last resort.

Yet another approach is genetic engineering, in which defective genes are replaced by synthetic normal genes. Take sickle-cell disease as an example. Recall from page 44 that, when a baby inherits the recessive allele for sickle-cell disease from both parents, the child has misshaped red blood cells that can't pass through capillaries. In theory, it should be possible to take a sample of cells from the fetus, remove the recessive genes from the 11th pair of chromosomes, and replace them with the dominant genes. These "repaired" cells could then be injected into the fetus, where they would multiply and cause normal red blood cells to be produced (David & Rodeck, 2009). As with fetal surgery, however, translating idea into practice is challenging. Researchers are still studying these techniques with mice and sheep and there have been some successful applications with older children (Coutelle et al., 2005; Maguire et al., 2009). However, routine use of this method in fetal medicine is still years away.

Answers to Chloe's Questions. Now you can return to Chloe's questions in the section-opening vignette (page 59) and answer them for her. If you're not certain, here are the pages in this chapter where the answers appear:

- About her cell phone—page 64
- About her nightly glass of wine—page 63
- About giving birth to a baby with mental retardation—page 61

Test Yourself

RECALL

1. General risk factors in pregnancy include a woman's nutrition, _____ , and her age.

2. _____ are some of the most dangerous teratogens because a pregnant woman is often unaware of their presence.

3. During the period of the zygote, exposure to a teratogen typically results in _____ .

4. Two techniques used to determine whether a fetus has a hereditary disorder are amniocentesis and _____ .

INTERPRET

Explain how the impact of a teratogen changes over the course of prenatal development.

APPLY

What would you say to a 45-year-old woman who is eager to become pregnant but is unsure about the possible risks associated with pregnancy at this age?

Recall answers: (1) prolonged stress, (2) Environmental hazards, (3) spontaneous abortion of the fertilized egg, (4) chorionic villus sampling

LEARNING OBJECTIVES

- What are the different phases of labor and delivery?
- What are "natural" ways of coping with the pain of childbirth? Is childbirth at home safe?
- What adjustments do parents face after a baby's birth?

- What are some complications that can occur during birth?
- What contributes to infant mortality in developed and least developed countries?

Dominique is 6 months pregnant; soon she and her partner will begin childbirth classes at the local hospital. She is relieved that the classes are finally starting because this means that pregnancy is nearly over. But all the talk she has heard about "breathing exercises" and "coaching" sounds mysterious to her. Dominique wonders what's involved and how the classes will help her during labor and delivery.

AS WOMEN LIKE DOMINIQUE NEAR THE END OF PREGNANCY, they find that sleeping and breathing become more difficult, that they tire more rapidly, that they become constipated, and that their legs and feet swell. Women look forward to birth, both to relieve their discomfort and, of course, to see their baby. In this section, you'll see the different steps involved in birth, review different approaches to childbirth, and look at problems that can arise. Along the way, we'll look at classes like those Dominique will take and the exercises that she'll learn.

Stages of Labor

Labor is an appropriate name for childbirth, which is the most intense, prolonged physical effort that humans experience. Labor is usually divided into the three stages shown in ■ Figure 2.11.

- In stage 1, which may last from 12 to 24 hours for a first birth, the uterus starts to contract. The first contractions are weak and irregular. Gradually, they become stronger and more rhythmic, enlarging the cervix (the opening from the uterus to the vagina) to approximately 10 centimeters.
- In stage 2, the baby passes through the cervix and enters the vagina. The mother helps push the baby along by contracting muscles in her abdomen. *Soon the top of the baby's head appears, an event known as* **crowning**. Within about an hour, the baby is delivered.

crowning
appearance of the top of the baby's head during labor

■ **Figure 2.11**
Labor includes three stages, beginning when the uterus contracts and ending when the placenta is expelled.

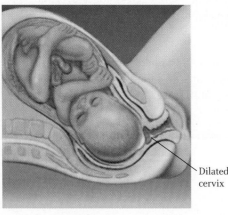

Stage 1 — Dilated cervix

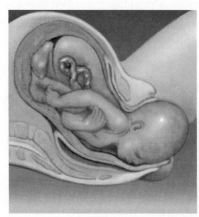

Stage 2

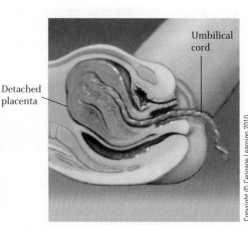

Detached placenta — Umbilical cord

Stage 3

Copyright © Cengage Learning 2010

- In stage 3, which lasts only minutes, the mother pushes a few more times to expel the placenta (also called, appropriately, the *afterbirth*).

The times given for each of the stages are only approximations; the actual times vary greatly among women. For most women, labor with their second and subsequent children is much more rapid; stage 1 may last 4 to 6 hours, and stage 2 may be as brief as 20 minutes.

Approaches to Childbirth

When your authors were born in the 1950s, women in labor were admitted to a hospital and administered a general anesthetic. Fathers waited anxiously in a nearby room for news of the baby. These were standard hospital procedures and virtually all American babies were born this way.

But no longer is this true. In the middle of the 20th century, two European physicians—Grantly Dick-Read (1959) and Ferdinand Lamaze (1958)—criticized the traditional view in which labor and delivery had come to involve elaborate medical procedures that were often unnecessary and that often left women afraid of giving birth. This fear led them to be tense, thereby increasing the pain they experienced during labor. These physicians argued for a more "natural" or prepared approach to childbirth, viewing labor and delivery as life events to be celebrated rather than medical procedures to be endured.

Today many varieties of prepared childbirth are available to pregnant women. However, most share some fundamental beliefs. One is that birth is more likely to be problem-free and rewarding when mothers and fathers understand what's happening during pregnancy, labor, and delivery. Consequently, prepared childbirth means going to classes to learn basic facts about pregnancy and childbirth (like the material presented in this chapter).

A second common element is that natural methods of dealing with pain are emphasized over medication. Why? When a woman is anesthetized with either general anesthesia or regional anesthesia (in which only the lower body is numbed), she can't use her abdominal muscles to help push the baby through the birth canal. Without this pushing, the obstetrician may have to use mechanical devices to pull the baby through the birth canal, which involves some risk to the baby (Johanson et al., 1993). Also, drugs that reduce the pain of childbirth cross the placenta and can affect the baby. Consequently, when a woman receives large doses of pain-relieving medication, her baby is often withdrawn or irritable for days or even weeks (Brazelton, Nugent, & Lester, 1987; Ransjoe-Arvidson et al., 2001). These effects are temporary, but they may give the new mother the impression that she has a difficult baby. It is best, therefore, to minimize the use of pain-relieving drugs during birth.

Relaxation is the key to reducing birth pain without drugs. Because pain often feels greater when a person is tense, pregnant women learn to relax during labor, through deep breathing or by visualizing a reassuring, pleasant scene or experience. Whenever they begin to experience pain during labor, they use these methods to relax.

A third common element of prepared childbirth is to involve a supportive "coach." The father-to-be, a relative, or close friend attends childbirth classes with the mother-to-be. The coach learns the techniques for coping with pain and, like the man in the photo, practices them with the pregnant woman. During labor and delivery, the coach is present to help the woman use the techniques she has learned and to offer support and encouragement. This preparation and support is effective in reducing the amount of medication that women like Dominique from the vignette take during labor (Maimburg et al., 2010).

Another element of the trend to natural childbirth is the idea that birth need not always take place in a hospital. Virtually all babies in the United States are born in hospitals, with only 1% born at home (Studelska, 2006). Yet around the

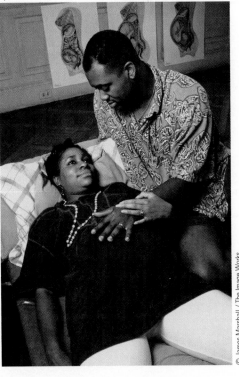

During childbirth preparation classes, pregnant women learn exercises that help them relax and reduce the pain associated with childbirth.

© James Marshall / The Image Works

In many countries around the world, such as Uzbekistan, a midwife delivers the baby.

world—in Europe, South America, and Asia—many children are born at home, reflecting a cultural view that the best place to welcome a new family member is at home, surrounded by family members.

For Americans accustomed to hospital delivery, home delivery can seem like a risky proposition. In fact, in the least developed countries of the world, where hospital delivery is far less common, the neonatal mortality rate (number of infants who live less than a month) is nine times higher than in the United States. In India alone, more than a million babies die before they are a month old; many parents do not name their newborns so that they will not become attached to a child who is likely to die (UNICEF, 2008).

The statistics are shocking, but you should not take them as an argument for the necessity of hospital births. In many of the least developed countries of the world, traditionally no trained health care professionals have been present at birth. When such professionals (typically a midwife) are present, labor and delivery become much safer for mother and infant alike, even when delivery takes place at home. Of course, sometimes problems emerge during pregnancy and labor; in these instances, ready access to a medical facility is essential. Combining these two elements—a health care professional present at every birth and specialized facilities available for problems—reduces neonatal mortality substantially (WHO, 2005).

This combination also works well in developed countries. Birth at home is safe if a woman is healthy, her pregnancy has been problem free, labor and delivery are expected to be problem free, a trained health care professional is there to assist, and comprehensive medical care is available should the need arise (Wax, Pinette & Cartin, 2010). Most women are more relaxed during labor in their homes and enjoy the greater control they have over labor and birth in a home delivery. But if there is any reason to believe that problems requiring medical assistance might occur, labor and delivery should take place in the hospital.

Adjusting to Parenthood

For parents, the time immediately after a trouble-free birth is full of excitement, pride, and joy—the much-anticipated baby is finally here! But it is also a time of adjustments for parents. A woman experiences many physical changes after birth. Her breasts begin to produce milk and her uterus gradually becomes smaller, returning to its normal size in 5 or 6 weeks. And levels of female hormones (e.g., estrogen) drop.

Parents must also adjust psychologically. They reorganize old routines, particularly for first-born children, to fit the young baby's sleep–wake cycle. In the process, fathers sometimes feel left out when mothers devote most of their attention to the baby.

Becoming a parent can be a huge adjustment, so it's not surprising that roughly half of all new mothers find that their initial excitement gives way to irritation, resentment, and crying spells—the so-called "baby blues." These feelings usually last a week or two and probably reflect both the stress of caring for a new baby and the physiological changes that take place as a woman's body returns to a nonpregnant state (Brockington, 1996).

For 10 to 15% of new mothers, however, irritability continues for months and is often accompanied by feelings of low self-worth, disturbed sleep, poor appetite, and apathy—a condition known as postpartum depression. Postpartum depression does not strike randomly. Biology contributes: Particularly high levels of hormones during the later phases of pregnancy place women at risk for postpartum depression (Harris et al., 1994). Experience also contributes: Women are more likely to experience postpartum depression when they were depressed before pregnancy, are coping with other life stresses (e.g., death of a loved one or moving to a new residence), did not plan to become pregnant, and lack other adults (e.g., the father) to support their adjustment to motherhood (O'Hara, 2009).

Women who are lethargic and emotionless do not mother warmly and enthusiastically. They don't touch and cuddle their new babies much or talk to them. And depressed moms are less effective in the common but essential tasks of feeding and sleep routines (Field, 2010). When postpartum depression persists over years, children's development is affected (Wachs, Black, & Engle, 2009). For example, anti-social behavior is more common (Kim-Cohen et al., 2005) and such effects are stronger when children have few opportunities to interact with nondepressed adults.

Thus, postpartum depression is a serious condition that can harm moms and babies alike; if a mom's depression doesn't lift after a few weeks, she should seek help. Home visits by trained health care professionals can be valuable. During these visits, these visitors show mom better ways to cope with the many changes that accompany her new baby. They also provide emotional support by being a caring, sensitive listener, and they can refer the mother to other resources in the community if needed. Finally, one simple way to reduce the risk of postpartum depression is worth mentioning—breast-feeding. Moms who breast-feed are less likely to become depressed, perhaps because breast-feeding releases hormones that act as antidepressants (Gagliardi, 2005).

Birth Complications

Women who are healthy when they become pregnant usually have a normal pregnancy, labor, and delivery. When women are not healthy or don't receive adequate prenatal care, problems can surface during labor and delivery. (Of course, even healthy women can have problems, but not as often.) The more common birth complications are listed in ●Table 2.6.

Some of these complications, such as a prolapsed umbilical cord, are dangerous because they can disrupt the flow of blood through the umbilical cord. *If this flow of blood is disrupted then infants do not receive adequate oxygen, a condition known as* **hypoxia**. Hypoxia sometimes occurs during labor and delivery because the umbilical cord is pinched or squeezed shut, cutting off the flow of blood. Hypoxia is serious because it can lead to mental retardation or death (Hogan et al., 2006).

hypoxia
a birth complication in which umbilical blood flow is disrupted and the infant does not receive adequate oxygen

To guard against hypoxia, fetal heart rate is monitored during labor, either by ultrasound or with a tiny electrode that is passed through the vagina and attached to the scalp of the fetus. An abrupt change in heart rate can be a sign that the fetus is not receiving enough oxygen. If the heart rate does change suddenly, a health care professional will try to confirm that the fetus is in distress, perhaps by measuring fetal heart rate with a stethoscope on the mother's abdomen.

When a fetus is in distress or when the fetus is in an irregular position or is too large to pass through the birth canal, a physician may decide to remove it from the mother's uterus surgically (Guillemin, 1993). *In a* **cesarean section** (*or* **C-section**) *an incision is made in the abdomen to remove the baby from the uterus.* A C-section is riskier for mothers than a vaginal delivery because of increased bleeding and greater danger of infection. A C-section poses little risk for babies, although they are often briefly lethargic from the anesthesia that the mother receives before the operation. And mother–infant interactions are much the same for babies delivered vaginally or by planned or unplanned C-sections (Durik, Hyde, & Clark, 2000).

cesarean section (C-section)
surgical removal of infant from the uterus through an incision made in the mother's abdomen

Birth complications are hazardous not just for a newborn's health; they have long-term effects, too. When babies experience many birth complications, they are at risk for becoming aggressive or violent and for developing schizophrenia (Cannon et al., 2000; de Haan et al., 2006). This is particularly true for newborns with birth complications who later experience family adversity, such as living in poverty (Arseneault et al., 2002). These outcomes underscore the importance of excellent health care through pregnancy and labor and the need for a supportive environment throughout childhood.

THINK ABOUT IT

A friend of yours has just given birth 6 weeks prematurely. The baby is average size for a baby born prematurely and seems to be faring well, but your friend is concerned nonetheless. What could you say to reassure your friend?

TABLE 2.6

Common Birth Complications

Complication	Features
Cephalopelvic disproportion	When the infant's head is larger than the pelvis, making it impossible for the baby to pass through the birth canal.
Irregular position	In shoulder presentation, the baby is lying crosswise in the uterus and the shoulder appears first; in breech presentation, the buttocks appear first.
Preeclampsia	A pregnant woman has high blood pressure, protein in her urine, and swelling in her extremities (due to fluid retention).
Prolapsed umbilical cord	The umbilical cord precedes the baby through the birth canal and is squeezed shut, cutting off oxygen to the baby.

Copyright © Cengage Learning 2010

preterm (premature)
babies born before the 36th week after conception

low birth weight
newborns who weigh less than 2,500 grams (5 pounds)

very low birth weight
newborns who weigh less than 1,500 grams (3 pounds)

extremely low birth weight
newborns who weigh less than 1,000 grams (2 pounds)

Problems also arise when babies are born too early or too small. Normally, a baby spends about 38 weeks developing before being born. *Babies born before the 36th week are called* **preterm** *or* **premature**. In the first year or so, premature infants often lag behind full-term infants in many facets of development. However, by 2 or 3 years of age, such differences have vanished, and most premature infants develop normally (Greenberg & Crnic, 1988).

Prospects are usually not as bright for babies who are "small for date." These infants are most often born to women who smoke or drink alcohol frequently during pregnancy or who do not eat enough nutritious food (Chomitz, Cheung, & Lieberman, 1995). *Newborns who weigh 2,500 grams (5.5 pounds) or less are said to have* **low birth weight***; newborns weighing less than 1,500 grams (3.3 pounds) are said to have* **very low birth weight***; and those weighing less than 1,000 grams (2.2 pounds) are said to have* **extremely low birth weight***.

Babies with very or extremely low birth weight do not fare well. Many do not survive, and those who live often lag behind in the development of intellectual and motor skills (Kavsek & Bornstein, 2010). These impaired cognitive processes are shown in the Spotlight on Research feature.

The odds are better for newborns who weigh more than 1,500 grams. Most survive, and their prospects are better if they receive appropriate care. Small-for-date babies are typically placed in special, sealed beds where temperature and air quality are regulated carefully. These beds effectively isolate infants, depriving them of environmental stimulation. You might think that stimulation is the last thing that these fragile creatures need, but sensory stimulation actually helps small-for-date babies to develop. Consequently, they often receive auditory stimulation, such as a tape recording of soothing music or their mother's voice, or visual stimulation provided from a mobile placed over the bed. Infants also receive tactile stimulation—they are "massaged" several times daily. These forms of stimulation foster physical and cognitive development in small-for-date babies (Field & Diego, 2010).

This special care should continue when infants leave the hospital for home. Consequently, intervention programs for small-for-date babies typically include training programs designed for parents of infants and young children. In these programs, parents learn how to respond appropriately to their child's behaviors. For example, they are taught the signs that a baby is in distress, overstimulated, or ready to interact. Parents also learn how to use games and activities to foster their child's development. In addition, children are enrolled in high-quality child care centers where the curriculum is coordinated with parent training. This sensitive care promotes devel-

Small-for-date babies often survive, but their cognitive and motor development usually is delayed.

© Abid Katib / Getty Images

Who were the investigators, and what was the aim of the study? Cognitive development is often delayed in low birth weight babies. Susan Rose and her colleagues (2009) hoped to understand whether memory developed normally in low birth weight babies.

How did the investigators measure the topic of interest? Memory is an essential skill for it allows us to benefit from past experiences. Without memory, everything would be experienced as brand new, à la Drew Barrymore's character in *50 First Dates.* Psychologists have devised many tasks to study different facets of memory; we'll see some of these in Chapter 4. Rose and her colleagues used several memory tasks, including an imitation task in which an experimenter demonstrated a brief sequence of novel events, such as making a gong from two posts, a base, and a metal plate. After a brief delay, children were given the parts and encouraged to reproduce what they'd seen.

Who were the children in the study? The sample included 144 full-term babies who weighed at least 2,500 grams at birth and 59 babies born prematurely who weighed, on average, about 1,100 grams at birth. The two groups of babies were matched by gender (about even numbers of boys and girls), by race (about 90% of the infants were African American or Latino American), and by mother's education (an average of just over 13 years of education). The memory tasks were administered when children were 2-year-olds and again when they were 3-year-olds.

What was the design of the study? The study was correlational because the investigators were interested in the relation that existed naturally between two variables: birth weight and memory

skill. The study was longitudinal because children were tested twice: at 2 and 3 years of age. (Actually, they were also tested at younger and older ages but for simplicity we're focusing on these two ages.)

Were there ethical concerns with the study? No. The tasks were ones commonly used with toddlers; they posed no known risks to children. The investigators obtained permission from the parents for the children to participate.

What were the results? The graph in ■ Figure 2.12 shows the percentage of actions in each event that children successfully imitated. Memory obviously improves substantially between 2 and 3 years. Nevertheless, at each age, children who had been born prematurely imitated a smaller percentage of events than did children born after a full term.

What did the investigators conclude? Low birth weight impairs basic cognitive processes—in this case, memory—and during the toddler years there's no evidence that children born prematurely "catch up": They're just as far behind at age 3 as at age 2.

What converging evidence would strengthen these conclusions? The results show that low birth weight affects children's memory. More convincing would be additional longitudinal results showing that low birth weight children with impaired basic skills are more likely to be diagnosed with a learning disability, more likely to repeat a grade, or less likely to graduate from high school.

 Go to Psychology CourseMate at www.cengagebrain.com to enhance your understanding of this research.

■ **Figure 2.12**
At both 2 years and 3 years, preterm children remember less than full-term children do.

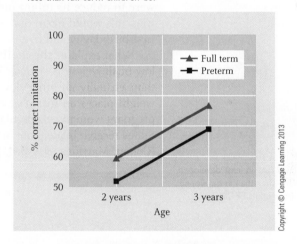

Copyright © Cengage Learning 2013

opment in low birth weight babies; for example, sometimes they catch up to full-term infants in terms of cognitive development (Hill, Brooks-Gunn, & Waldfogel, 2003).

Long-term positive outcomes for these infants depend critically on providing a supportive and stimulating home environment. Unfortunately, not all at-risk babies have optimal experiences. Many receive inadequate medical care because their families live in poverty. Others experience stress and disorder in their family life. For these low birth weight babies, development is usually delayed and sometimes permanently diminished.

The importance of a supportive environment for low birth weight babies is underscored by the results of a 30-year longitudinal study by Werner (1989, 1995) covering all children born on the Hawaiian island of Kauai in 1955. When low birth weight children grew up in stable homes—defined as having two mentally healthy parents throughout childhood—they were indistinguishable from children born without birth complications. However, when low birth weight children experienced an unstable family environment—defined as including divorce, parental alcoholism, or parental mental illness—they lagged behind their peers in intellectual and social development.

Thus, when biological and sociocultural forces are both harmful—low birth weight *plus* inadequate medical care or family stress—the prognosis for babies is grim. The message to parents of low birth weight newborns is clear: Do not despair, because excellent caregiving can compensate for all but the most severe birth problems (Werner, 1994; Werner & Smith, 1992).

Infant Mortality

infant mortality
the number of infants out of 1,000 births who die before their first birthday

If you were the proud parent of a newborn and a citizen of Afghanistan, the odds are 1 in 6 that your baby would die before his or her first birthday—worldwide, *Afghanistan has the highest* **infant mortality** *rate, defined as the percentage of infants who die before their first birthday*. In contrast, if you were a parent and a citizen of the Czech Republic, Iceland, Finland, or Japan, the odds are less than 1 in 300 that your baby would die within a year, because these countries have among the lowest infant mortality rates.

The graph shown as ■ Figure 2.13 puts these numbers in a broader, global context, depicting infant mortality rates for 15 developed nations as well as for 15 least developed countries. Not surprisingly, risks to infants are far greater—about 20 times, on average—in the least developed nations compared to developed nations (UNICEF, 2007). In fact, the differences are so great that the graphs for the two groups of nations must be drawn on different scales.

If you're an American, you may be surprised to see that the United States ranks near the bottom of the list of developed nations. The difference is small, but if the U.S. were to reduce its infant mortality rate to the 4% that's common in European countries, this would mean that 8,000 American babies who now die annually before their first birthday would live.

What explains these differences in infant mortality rates? For American infants, low birth weight is critical. The United States has more babies with low birth weight than virtually all other developed countries, and we've already seen that low birth weight places an infant at risk. Low birth weight can usually be prevented when a pregnant woman gets regular prenatal care, but many pregnant women in the United States receive inadequate or no prenatal care because they have no health insurance (Cohen, Martinez, & Ward, 2010). Virtually all the countries that rank ahead of the

■ **Figure 2.13**
The infant mortality rate in least developed countries is much higher than in developed countries. Data from UNICEF 2007.

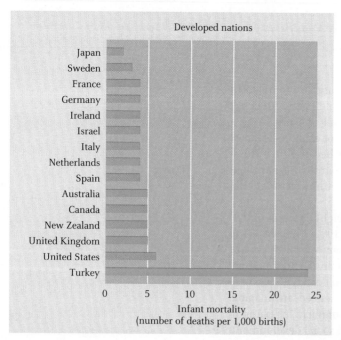

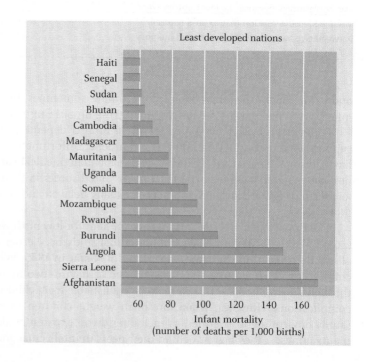

United States provide complete prenatal care at little or no cost. Many of these countries also provide for paid leaves of absence for pregnant women (OECD, 2006).

In least developed countries, inadequate prenatal care is common and mothers often have inadequate nutrition. After birth, infants in these countries face the twin challenges of receiving adequate nutrition and avoiding disease. However, with improved prenatal care and improved health care and nutrition for infants, the global infant mortality has been cut in half since 1990 (UNICEF, 2007). With continued improvements in such care, the main challenges for infants worldwide will be walking, talking, and bonding with parents—not sheer survival.

Test Yourself

RECALL

1. In the third stage of labor, the _____ is delivered.

2. Two problems with using anesthesia during labor are that a woman can't use her abdominal muscles to help push the baby down the birth canal and _____.

3. Home delivery is safe when a pregnant woman is healthy, has had a problem-free pregnancy, expects to have a problem-free delivery, and _____.

4. When the supply of oxygen to the fetus is disrupted because the umbilical cord is squeezed shut, _____ results.

INTERPRET

Explain why some at-risk newborns develop normally but others do not.

APPLY

Lynn is pregnant with her first child and would like to give birth at home. Her husband is totally against the idea and claims that it's much too risky. What advice would you give them?

Recall answers: (1) placenta, (2) the pain-relieving medication crosses the placenta and affects the baby, (3) when trained health care professionals are present to deliver the baby, (4) hypoxia

SUMMARY

2.1   IN THE BEGINNING: 23 PAIRS OF CHROMOSOMES

What are chromosomes and genes? How do they carry hereditary information from one generation to the next?

- At conception, the 23 chromosomes in the sperm merge with the 23 chromosomes in the egg. Each chromosome is one molecule of DNA; a section of DNA that provides specific biochemical instructions is called a gene.

- All of a person's genes make up a genotype; the phenotype refers to the physical, behavioral, and psychological characteristics that develop when the genotype is exposed to a specific environment.

- Different forms of the same gene are called alleles. A person who inherits the same allele on a pair of chromosomes is homozygous; in this case, the biochemical instructions on the allele are followed. A person who inherits different alleles is heterozygous; in this case, the instructions of the dominant allele are followed and those of the recessive allele ignored.

What are common problems involving chromosomes and what are their consequences?

- Most inherited disorders are carried by recessive alleles. Examples include sickle-cell disease and phenylketonuria, in which toxins accumulate and cause mental retardation. Sometimes fertilized eggs do not have 46 chromosomes. Usually they are aborted spontaneously soon after conception. An exception is Down syndrome, in which individuals usually have an extra 21st chromosome. Down syndrome individuals have a distinctive appearance and are mentally retarded. Disorders of the sex chromosomes are more common because these chromosomes contain less genetic material than do autosomes.

How is children's heredity influenced by the environment in which they grow up?

- Behavioral and psychological phenotypes that reflect an underlying continuum (such as intelligence) often involve polygenic inheritance. In polygenic inheritance, the phenotype reflects the combined activity of many distinct genes. Polygenic inheritance has been examined traditionally by studying twins and adopted children and, more recently, by identifying DNA markers.

- The impact of heredity on a child's development depends on the environment in which the genetic instructions are carried out, and these heredity–environment interactions occur throughout a child's life. A child's genotype can af-

2.2 FROM CONCEPTION TO BIRTH

What happens to a fertilized egg in the first 2 weeks after conception?

■ The first period of prenatal development lasts 2 weeks. It begins when the egg is fertilized by the sperm in the Fallopian tube and ends when the fertilized egg has implanted itself in the wall of the uterus. By the end of this period, cells have begun to differentiate.

When do body structures and internal organs emerge in prenatal development?

■ The second period of prenatal development begins 2 weeks after conception and ends 8 weeks after. This is a period of rapid growth in which most major body structures are created. Growth in this period is cephalocaudal (the head develops first) and proximodistal (parts near the center of the body develop first).

When do body systems begin to function well enough to support life?

■ The third period of prenatal development begins 9 weeks after conception and lasts until birth. The highlights of this period are a remarkable increase in the size of the fetus and changes in body systems that are necessary for life. By 7 months, most body systems function well enough to support life.

2.3 INFLUENCES ON PRENATAL DEVELOPMENT

How is prenatal development influenced by a pregnant woman's age, her nutrition, and the stress she experiences while pregnant?

■ Parents' age can affect prenatal development. Teenagers often have problem pregnancies, mainly because they rarely receive adequate prenatal care. After age 35, pregnant women are more likely to have a miscarriage or to give birth to a child with mental retardation. Prenatal development can also be harmed if a pregnant mother has inadequate nutrition or experiences considerable stress.

How do diseases, drugs, and environmental hazards sometimes affect prenatal development?

■ Teratogens are agents that can cause abnormal prenatal development. Many drugs that adults take are teratogens. For most drugs, scientists have not established amounts that can be consumed safely.

■ Several diseases are teratogens. Only by avoiding these diseases entirely can a pregnant woman escape their harmful consequences.

■ Environmental teratogens are particularly dangerous because a pregnant woman may not know that these substances are present in the environment.

What general principles affect the ways that prenatal development can be harmed?

■ The impact of teratogens depends on the genotype of the organism, the period of prenatal development when the organism is exposed to the teratogen, and the amount of exposure. Sometimes the effect of a teratogen is not evident until later in life.

How can prenatal development be monitored? Can abnormal prenatal development be corrected?

■ Many techniques are used to track the progress of prenatal development. A common component of prenatal care is ultrasound, which uses sound waves to generate a picture of the fetus. This picture can be used to determine the position of the fetus, its sex, and whether there are gross physical deformities.

■ When genetic disorders are suspected, amniocentesis and chorionic villus sampling are used to determine the genotype of the fetus.

■ Fetal medicine is a new field in which problems of prenatal development are corrected medically via surgery or genetic engineering.

2.4 LABOR AND DELIVERY

What are the different phases of labor and delivery?

■ Labor consists of three stages. In stage 1, the muscles of the uterus contract. The contractions, which are weak at first and gradually become stronger, cause the cervix to enlarge. In stage 2, the baby moves through the birth canal. In stage 3, the placenta is delivered.

What are "natural" ways of coping with the pain of childbirth? Is childbirth at home safe?

■ Natural or prepared childbirth is based on the assumption that parents should understand what takes place during pregnancy and birth. In natural childbirth, pain-relieving medications are avoided because this medication prevents women from pushing during labor and because it affects the fetus. Instead, women learn to cope with pain through relaxation, imagery, and the help of a supportive coach.

■ Most American babies are born in hospitals, but many European babies are born at home. Home delivery is safe when the mother is healthy, when pregnancy and birth are trouble-free, and when a health care professional is present to deliver the baby.

What adjustments do parents face after a baby's birth?

■ Following the birth of a child, a woman's body undergoes several changes: her breasts fill with milk, her uterus becomes smaller, and hormone levels drop. Both parents also adjust psychologically, and sometimes fathers feel left out. After giving birth, some women experience postpartum depression: they are irritable, have poor appetite and disturbed sleep, and are apathetic.

What are some complications that can occur during birth?

■ During labor and delivery, the flow of blood to the fetus can be disrupted because the umbilical cord is squeezed shut. This causes hypoxia, a lack of oxygen to the fetus. Some babies are born prematurely and others are "small for date." Premature babies develop more slowly at first but catch up by 2 or 3 years of age. Small-for-date babies often do not fare well, particularly if they weigh less than 1,500 grams at birth and if their environment is stressful.

What contributes to infant mortality in developed and least developed countries?

■ Infant mortality is relatively high in many countries around the world, primarily because of inadequate care before birth and disease and inadequate nutrition after birth.

KEY TERMS

chromosomes (42)
autosomes (43)
sex chromosomes (43)
deoxyribonucleic acid (DNA) (43)
gene (43)
genotype (44)
phenotype (44)
alleles (44)
homozygous (44)
heterozygous (44)
dominant (44)
recessive (44)
incomplete dominance (44)
sickle-cell trait (44)
phenylketonuria (PKU) (46)
Huntington's disease (46)
behavioral genetics (47)
polygenic inheritance (48)
monozygotic twins (49)
dizygotic twins (49)

heritability coefficient (50)
niche-picking (51)
nonshared environmental influences (52)
prenatal development (53)
in vitro fertilization (54)
eugenics (54)
zygote (54)
implantation (55)
germ disc (55)
placenta (55)
embryo (55)
ectoderm (55)
mesoderm (55)
endoderm (55)
amnion (56)
amniotic fluid (56)
umbilical cord (56)
cephalocaudal principle (56)
proximodistal principle (56)

period of the fetus (56)
cerebral cortex (56)
vernix (56)
age of viability (57)
spina bifida (59)
stress (60)
teratogen (61)
fetal alcohol spectrum disorder (62)
ultrasound (67)
amniocentesis (67)
chorionic villus sampling (68)
fetal medicine (68)
crowning (70)
hypoxia (73)
cesarean section (C-section) (73)
preterm (premature) (74)
low birth weight (74)
very low birth weight (74)
extremely low birth weight (74)
infant mortality (76)

LEARN MORE ABOUT IT

Log in to **www.cengagebrain.com** to access the resources your instructor requires. For this book, you can access:

Psychology CourseMate

■ CourseMate brings course concepts to life with interactive learning, study, and exam preparation tools that support the printed textbook. A textbook-specific website, Psychology CourseMate includes an integrated interactive eBook and other interactive learning tools including quizzes, flashcards, videos, and more.

CENGAGENOW™

■ CengageNOW Personalized Study is a diagnostic study tool containing valuable text-specific resources—and because you focus on just what you don't know, you learn more in less time to get a better grade.

WebTUTOR™

■ More than just an interactive study guide, WebTutor is an anytime, anywhere customized learning solution with an eBook, keeping you connected to your textbook, instructor, and classmates.

Tools for Exploring the World

Physical, Perceptual, and Motor Development

3

THINK ABOUT WHAT you were like 2 years ago. Whatever you were doing, you probably look, act, think, and feel in much the same way today as you did then. Two years in an adult's life usually doesn't result in profound changes, but 2 years makes a big difference early in life. The changes that occur in the first few years after birth are incredible. In less than 2 years, an infant is transformed from a seemingly helpless newborn into a talking, walking, havoc-wreaking toddler. No changes at any other point in the life span come close to the drama and excitement of these early years.

In this chapter, our tour of these 2 years begins with the newborn and then moves to physical growth—changes in the body and the brain. The third section of the chapter examines motor skills. You'll discover how babies learn to walk and how they learn to use their hands to hold and then manipulate objects. In the fourth section, we'll examine changes in infants' sensory abilities that allow them to comprehend their world.

As children begin to explore their world and learn more about it, they also learn more about themselves. They learn to recognize themselves and begin to understand more about their thoughts and others' thoughts. We'll explore these changes in the last section of the chapter.

LEARNING OBJECTIVES

- How do reflexes help newborns interact with the world?
- How do we determine whether a baby is healthy and adjusting to life outside the uterus?

- What behavioral states are common among newborns?
- What are the different features of temperament? Do they change as children grow?

Lisa and Steve, proud but exhausted parents, are astonished at how their lives revolve around 10-day-old Dan's eating and sleeping. Lisa feels as if she is feeding Dan around the clock. When Dan naps, Lisa thinks of many things she should do but usually naps herself because she is so tired. Steve wonders when Dan will start sleeping through the night so that he and Lisa can get a good night's sleep themselves.

THE NEWBORN BABY THAT THRILLS PARENTS LIKE LISA AND STEVE IS ACTUALLY RATHER HOMELY. Newborns arrive covered with blood and vernix, a white-colored "wax" that protected the skin during the many months of prenatal development. In addition, the baby's head is temporarily distorted from its journey through the birth canal, and the newborn has a beer belly and is bowlegged.

What can newborns like Dan do? We'll answer that question in this section and, as we do, you'll learn when Lisa and Steve can expect to resume getting a full night's sleep.

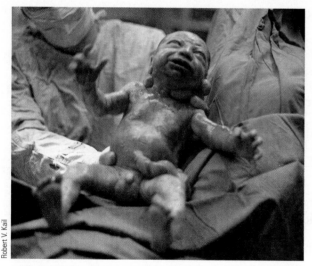

Robert V. Kail

This newborn baby (Ben Kail at 20 seconds old) is covered with vernix and is bow-legged; his head is distorted from the journey down the birth canal.

reflexes
unlearned responses triggered by specific stimulation

The Newborn's Reflexes

Most newborns are well prepared to begin interacting with their world. The newborn is endowed with a rich set of **reflexes**, *unlearned responses that are triggered by a specific form of stimulation.* ●Table 3.1 shows the variety of reflexes commonly found in newborn babies.

You can see that some reflexes are designed to pave the way for newborns to get the nutrients they need to grow: The rooting and sucking reflexes ensure that the newborn is well prepared to begin a new diet of life-sustaining milk. Other reflexes seem designed to protect the newborn from danger in the environment. The eye blink, for example, helps newborns avoid unpleasant stimulation.

Still other reflexes serve as the foundation for larger, voluntary patterns of motor activity. For example, the stepping reflex motions look like precursors to walking, so it probably won't surprise you to learn that babies who practice the stepping reflex often learn to walk earlier than those who don't practice this (Zelazo, 1993).

Reflexes are also important because they can be a useful way to determine whether the newborn's nervous system is working properly. For example, infants with damage to the sciatic nerve, which is found in the spinal cord, do not show the withdrawal reflex. Infants who have problems with the lower part of the spine do not show the Babinski reflex. If these or other reflexes are weak or missing altogether, a thorough physical and behavioral assessment is called for. Similarly, many of these reflexes normally vanish during infancy; if they linger then this, too, indicates the need for a thorough physical examination.

Assessing the Newborn

Imagine that a mother has just asked you if her newborn baby is healthy. How would you decide? You would probably check to see whether the baby seems to be breathing and if her heart seems to be beating. In fact, breathing and heartbeat are two vital

Some Major Reflexes Found in Newborns

Name	Response	Age When Reflex Disappears	Significance
Babinski	A baby's toes fan out when the sole of the foot is stroked from heel to toe	8–12 months	Perhaps a remnant of evolution
Blink	A baby's eyes close in response to bright light or loud noise	Permanent	Protects the eyes
Moro	A baby throws its arms out and then inward (as if embracing) in response to loud noise or when its head falls	6 months	May help a baby cling to its mother
Palmar	A baby grasps an object placed in the palm of its hand	3–4 months	Precursor to voluntary walking
Rooting	When a baby's cheek is stroked, it turns its head toward the stroking and opens its mouth	3–4 weeks (replaced by voluntary head turning)	Helps a baby find the nipple
Stepping	A baby who is held upright by an adult and is then moved forward begins to step rhythmically	2–3 months	Precursor to voluntary walking
Sucking	A baby sucks when an object is placed in its mouth	4 months (replaced by voluntary sucking)	Permits feeding

signs included in the Apgar score, which provides a quick, approximate assessment of the newborn's status by focusing on the body systems needed to sustain life. The other vital signs are muscle tone, presence of reflexes such as coughing, and skin tone. Each of the five vital signs receives a score of 0, 1, or 2, where 2 is the optimal score. For example, a newborn whose muscles are completely limp receives a 0; a baby who shows strong movements of arms and legs receives a 2. The five scores are added together, with a total score of 7 or more indicating a baby who is in good physical condition. A score of 4–6 means that the newborn needs special attention and care. A score of 3 or less signals a life-threatening situation that requires emergency medical care (Apgar, 1953).

For a comprehensive evaluation of the newborn's well-being, pediatricians and other child-development specialists sometimes administer the Neonatal Behavioral Assessment Scale or NBAS for short (Brazelton & Nugent, 1995). The NBAS is used with newborns to 2-month-olds to provide a detailed portrait of the baby's behavioral repertoire. The scale includes 28 behavioral items (e.g., the baby's response to light, sound, and touch) along with 18 items that test reflexes like those that we've just described. The baby's performance is used to evaluate the functioning of these four systems:

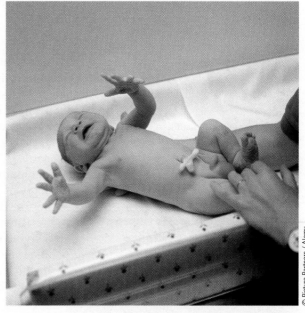

Newborns exhibit the Moro reflex, opening their arms and then bringing them inward in response to loud noise or when their head falls.

- *Autonomic:* the newborn's ability to control body functions such as breathing and temperature regulation
- *Motor:* the newborn's ability to control body movements and activity level
- *State:* the newborn's ability to maintain a state (e.g., staying alert or staying asleep)
- *Social:* the newborn's ability to interact with people

The NBAS is based on the view that newborns are remarkably competent individuals who are well prepared to interact with the environment. Reflecting this view, examiners go to great lengths to bring out a baby's best performance. They do everything possible to make a baby feel comfortable and secure during testing. And if the infant does not at first succeed on an item, the examiner provides some assistance (Alberts, 2005).

Not only is the NBAS useful to clinicians in evaluating the well-being of individual babies, researchers have found it a valuable tool. Sometimes performance on the

THINK ABOUT IT

Newborns seem to be extremely well prepared to begin to interact with their environment. Which of the theories described in Chapter 1 predict such preparedness? Which do not?

NBAS is used as a dependent variable. For example, harm associated with teratogens has been shown to predict lower scores on the NBAS (e.g., Engel et al., 2009). Researchers also use scores on the NBAS to predict later development (e.g., Stjernqvist, 2009).

The Newborn's States

Newborns spend most of each day alternating among four different states (St. James-Roberts & Plewis, 1996; Wolff, 1987):

- **Alert inactivity**—*The baby is calm with eyes open and attentive; the baby seems to be deliberately inspecting the environment.*
- **Waking activity**—*The baby's eyes are open but they seem unfocused; the arms or legs move in bursts of uncoordinated motion.*
- **Crying**—*The baby cries vigorously, usually accompanied by agitated but uncoordinated motion.*
- **Sleeping**—*The baby alternates from being still and breathing regularly to moving gently and breathing irregularly; eyes are closed throughout.*

Of these states, crying and sleeping have captured the attention of parents and researchers alike.

Crying

Newborns spend 2–3 hours each day crying or on the verge of crying. If you've not spent much time around newborns, you might think that all crying is pretty much alike. In fact, scientists and parents can identify three distinctive types of cries (Snow, 1998). *A* **basic cry** *starts softly and then gradually becomes more intense; it usually occurs when a baby is hungry or tired. A* **mad cry** *is a more intense version of a basic cry; and a* **pain cry** *begins with a sudden, long burst of crying followed by a long pause and gasping.* Thus, crying represents the newborn's first venture into interpersonal communication. By crying, babies tell their parents that they are hungry or tired, angry or hurt. By responding to these cries, parents are encouraging their newborn's efforts to communicate.

Parents are naturally concerned when their baby cries, and if they can't quiet a crying baby, their concern mounts and can easily give way to frustration and annoyance. It's no surprise, then, that parents develop little tricks for soothing their babies. Many Western parents lift a baby to the shoulder and walk or gently rock the baby. Sometimes they will also sing lullabies, pat the baby's back, or give the baby a pacifier. Yet another method is to put a newborn into a car seat and go for a drive; this technique was used once, as a last resort, at 2 a.m. with Ben Kail when he was 10 days old. After about the 12th time around the block, he finally stopped crying and fell asleep!

Another useful technique is swaddling, in which an infant is wrapped tightly in a blanket. Swaddling is used in many cultures around the world, including Turkey and Peru, as well as by Native Americans. Swaddling provides warmth and tactile stimulation that usually works well to soothe a baby (Delaney, 2000).

Parents are sometimes reluctant to respond to their crying infant for fear of producing a baby who cries constantly. Yet they hear their baby's cry as a call for help that they shouldn't ignore. What to do? Should parents respond? "Yes, usually" is probably the best answer (Hubbard & van IJzendoorn, 1991). If parents respond *immediately, every time* their infant cries, the result may well be a fussy, whiny baby. Instead, parents need to consider why their infant is crying and the intensity of the crying. When a baby wakes during the night and cries quietly, a parent might wait before responding, giving the baby a chance to calm itself. However, when parents hear a loud noise from an infant's bedroom followed by a mad cry, they should respond immediately. Parents need to remember that crying is

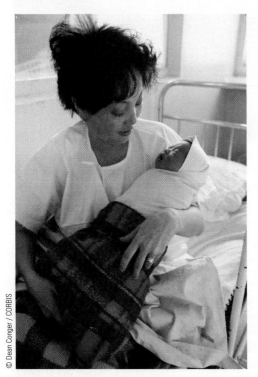

In many countries worldwide, infants are wrapped tightly in blankets as a way to keep them soothed.

actually the newborn's first attempt to communicate with others. They need to decide what the infant is trying to tell them and whether that warrants a quick response or whether they should let the baby soothe itself.

THINK ABOUT IT

When Mary's 4-month-old son cries, she rushes to him immediately and does everything possible to console him. Is this a good idea?

Sleeping

Crying may get parents' attention, but sleep is what newborns do more than anything else. They sleep 16–18 hours daily. The problem for tired parents is that newborns sleep in naps taken round-the-clock. Newborns typically go through a cycle of wakefulness and sleep about every 4 hours. That is, they will be awake for about an hour, sleep for 3 hours, and then start the cycle anew. During the hour when newborns are awake, they regularly move between the different waking states several times. Cycles of alert inactivity, waking activity, and crying are common.

As babies grow older, the sleep–wake cycle gradually begins to correspond to the day–night cycle (St. James-Roberts & Plewis, 1996). By 3 or 4 months, many babies sleep for 5–6 hours straight, and by 6 months many are sleeping for 10–12 hours at night, a major milestone for bleary-eyed parents like Lisa and Steve.

Co-sleeping, in which infants and young children sleep with their parents, is common in many countries around the world.

By 6 months, most North American infants are sleeping in a crib in their own room. Although this practice seems "natural" to North American parents, in much of the rest of the world children sleep with their parents throughout infancy and the preschool years. Such parent–child "co-sleeping" is commonly found in cultures where people define themselves less as independent individuals and more as part of a group. For parents in cultures that value such interdependence—including Egypt, Italy, Japan, and Korea, as well as the Maya in Guatemala and the Inuit in Canada—co-sleeping is an important step in forging parent–child bonds, just as sleeping alone is an important step toward independence in cultures that value self-reliance (Nelson, Schiefenhoevel, & Haimerl, 2000; Tan, 2009; Worthman & Brown, 2007).

How does co-sleeping work? Infants may sleep in a cradle placed next to their parents' bed or in a basket that's in their parents' bed. When they outgrow this arrangement, they sleep in the bed with their mother; depending on the culture, the father may sleep in the same bed, in another bed in the same room, in another room, or in another house altogether!

You might think that co-sleeping would make children more dependent on their parents, but research provides no evidence of this (Cortesi et al., 2004; Okami, Weisner, & Olmstead, 2002). Plus, co-sleeping has the benefit of avoiding the lengthy and elaborate rituals that are often required to get youngsters to sleep in their own room alone. With co-sleeping, children and parents simply go to bed together with few struggles.

Roughly half of newborns' sleep is **irregular** *or* **rapid-eye-movement (REM) sleep,** *a time when the body is quite active.* During REM sleep, newborns move their arms and legs; they may grimace and their eyes may dart beneath their eyelids. Brain waves register fast activity, the heart beats more rapidly, and breathing is more rapid. *In* **regular** *or* **nonREM sleep,** *breathing, heart rate, and brain activity are steady and newborns lie quietly without the twitching associated with REM sleep.* REM sleep becomes less frequent as infants grow. By 4 months, only 40% of sleep is REM sleep. By the first birthday, REM sleep will drop to 25%—not far from the adult average of 20% (Halpern, MacLean, & Baumeister, 1995).

The function of REM sleep is still debated. Older children and adults dream during REM sleep, and brain waves during REM sleep resemble those of an alert, awake person. Consequently, many scientists believe that REM sleep provides stimulation for the brain that fosters growth in the nervous system (Halpern et al., 1995; Roffwarg, Muzio, & Dement, 1966).

irregular or rapid-eye-movement (REM) sleep
irregular sleep in which an infant's eyes dart rapidly beneath the eyelids while the body is quite active

regular (nonREM) sleep
sleep in which heart rate, breathing, and brain activity are steady

By the toddler and preschool years, sleep routines are well established. Most 2-year-olds spend about 13 hours sleeping, compared to just under 11 hours for 6-year-olds. By age 4, most youngsters give up their afternoon nap and sleep longer at nighttime to compensate. This can be a challenging time for parents and caregivers who use naptime as an opportunity to complete some work or to relax.

Following an active day, most preschool children drift off to sleep easily. However, most children will have an occasional night when bedtime is a struggle. Furthermore, for approximately 20 to 30% of preschool children, bedtime struggles occur nightly (Lozoff, Wolf, & Davis, 1985). More often than not, these bedtime problems reflect the absence of a regular bedtime routine that's followed consistently. The key to a pleasant bedtime is to establish a nighttime routine that helps children to "wind down" from busy daytime activities. This routine should start at about the same time every night ("It's time to get ready for bed . . .") and end at about the same time (when the parent leaves the child and the child tries to fall asleep). This nighttime routine may be anywhere from 15 to 45 minutes long, depending on the child. Also, as children get older, parents can expect them to perform more of these tasks independently. A 2-year-old will need help all along the way, but a 5-year-old can do many of these tasks alone. But remember to follow the routine consistently; this way, children know that each step is getting them closer to bedtime and falling asleep.

sudden infant death syndrome (SIDS)
when a healthy baby dies suddenly for no apparent reason

■ **Figure 3.1**

This poster is one part of an effective campaign to reduce SIDS by encouraging parents to have their babies sleep on their backs.

National Institute of Child Health and Development.

Sudden Infant Death Syndrome

For many parents of young babies, however, sleep is a cause of concern. *In sudden infant death syndrome (SIDS)*, *a healthy baby dies suddenly for no apparent reason.* Approximately 1–3 of every 1,000 American babies dies from SIDS. Most of them are between 2 and 4 months of age (Wegman, 1994).

Scientists don't know the exact causes of SIDS, but one idea is that 2- to 4-month-old infants are particularly vulnerable to SIDS because many newborn reflexes are waning during these months and thus infants may not respond effectively when breathing becomes difficult. They may not reflexively move their head away from a blanket or pillow that is smothering them (Lipsitt, 2003).

Researchers have also identified several risk factors associated with SIDS (Sahni, Fifer, & Myers, 2007). Babies are more vulnerable if they were born prematurely or with low birth weight. They are also more vulnerable when their parents smoke. SIDS is more likely when a baby sleeps on its stomach (face down) than when it sleeps on its back (face up). Finally, SIDS is more likely during winter, when babies sometimes become overheated from too many blankets and sleepwear that is too heavy (Carroll & Loughlin, 1994). Evidently, SIDS infants, many of whom were born prematurely or with low birth weight, are less able to withstand physiological stresses and imbalances that are brought on by cigarette smoke, breathing that is temporarily interrupted, or overheating (Simpson, 2001).

In 1992, based on mounting evidence that SIDS occurred more often when infants slept on their stomachs, the American Academy of Pediatrics (AAP) began advising parents to put babies to sleep on their backs or sides. In 1994 the AAP joined forces with the U.S. Public Health Service to launch a national program to educate parents about the dangers of SIDS and the importance of putting babies to sleep on their backs. The "Back to Sleep" campaign was widely publicized through brochures, posters like the one shown in ■ Figure 3.1, and videos. Since the "Back to Sleep" campaign began, research shows that far more infants are now sleeping on their backs and that the incidence of SIDS has dropped (Dwyer & Ponsonby, 2009). However, it became clear that African American infants were still twice as likely to die from SIDS, apparently because they were much more likely to be placed on their stomachs to sleep. Consequently, in the 21st century the National Institutes of Health has partnered with groups such as the Women in the NAACP and the National Council of 100 Black Women to train thousands of people to convey the Back to Sleep message in a culturally appropriate manner to African American communities (NICHD, 2004). The goal is for African American infants to benefit from the lifesaving benefits of the Back to Sleep program. The message for all parents—particularly if their babies were premature or small-for-date—is to keep their babies away from smoke, to put them on their backs to sleep, and to not overdress them or wrap them too tightly in blankets (Willinger, 1995).

Temperament

So far, we've talked as if all babies are alike. But if you've seen a number of babies together, you know this isn't true. Perhaps you've seen some babies who are quiet most of the time alongside others who cried often and impatiently? Maybe you've known infants who responded warmly to strangers next to others who seemed shy? *These characteristics of infants indicate a consistent style or pattern to an infant's behavior, and collectively they define an infant's* **temperament**.

Alexander Thomas and Stella Chess (Thomas & Chess, 1977; Thomas, Chess, & Birch, 1968) pioneered the study of temperament with the New York Longitudinal Study, in which they traced the lives of 141 individuals from infancy through adulthood. Thomas and Chess interviewed parents about their babies and had individuals unfamiliar with the children observe them at home. From these interviews and observations, Thomas and Chess suggested that infants' behavior varied along nine temperamental dimensions. One dimension was activity, which referred to an infant's typical level of motor activity. A second was persistence, which referred to the amount of time that an infant devoted to an activity, particularly when obstacles were present.

The New York Longitudinal Study launched research on infant temperament, but today we know that Thomas and Chess overestimated the number of temperamental dimensions. Instead of nine dimensions, scientists now propose from two to six dimensions. For example, Mary K. Rothbart (2007) has devised an influential theory of temperament that includes three different dimensions:

- *Surgency/extroversion* refers to the extent to which a child is generally happy, active, vocal, and regularly seeks interesting stimulation.

- *Negative affect* refers to the extent to which a child is angry, fearful, frustrated, shy, and not easily soothed.

- *Effortful control* refers to the extent to which a child can focus attention, is not readily distracted, and can inhibit responses.

These dimensions of temperament emerge in infancy, continue into childhood, and are related to dimensions of personality that are found in adolescence and adulthood (Gartstein, Knyazev, & Slobodskaya, 2005). However, the dimensions are not independent: infants who are high on effortful control tend to be high on surgency/extroversion and low on negative affect. In other words, babies who can control their attention and inhibit responses tend to be happy and active but not angry or fearful.

Hereditary and Environmental Contributions to Temperament

Most theories agree that temperament reflects both heredity and experience (Caspi, Roberts, & Shiner, 2005). The influence of heredity is shown in twin studies: Identical twins are more alike in most aspects of temperament than fraternal twins (Goldsmith, Pollak, & Davidson, 2008). If, for example, one identical twin is temperamentally active then the other usually is, too. However, the impact of heredity also depends on the temperamental dimension and the child's age. For example, negative affect is more influenced by heredity than the other dimensions, and temperament in childhood is more influenced by heredity than is temperament in infancy (Wachs & Bates, 2001).

The environment also contributes to children's temperament. Positive emotionality—youngsters who laugh often, seem

temperament
consistent style or pattern of behavior

Twin studies show the impact of heredity on temperament: If one identical twin is active, the other one usually is.

to be generally happy, and express pleasure often—seems to reflect environmental influences (Goldsmith et al., 1997). What's more, infants are less emotional when parents are responsive (Hane & Fox, 2006; Leerkes, Blankson, & O'Brien, 2009). Conversely, infants become increasingly fearful when their mothers are depressed (Gartstein et al., 2010). And some temperamental characteristics are more common in some cultures than in others. Asian babies tend to be less emotional than European American babies. For instance, Asian babies cry less often and less intensely than European American babies, but Russian infants are more fearful and emotionally negative (Gartstein, Slobodskaya, & Kinsht, 2003; Kagan et al., 1994).

Of course, in trying to determine the contributions of heredity and environment to temperament, the most likely explanation is that both contribute (Henderson & Wachs, 2007). In fact, one view is that temperament may make some children particularly susceptible to environmental influences—either beneficial or harmful (Belsky, Bakermans-Kranenburg, & van IJzendoorn, 2007). For example, in one study (Kochanska, Aksan, & Joy, 2007), emotionally fearful children were more likely to cheat in a game when their parents' discipline emphasized asserting power (e.g., "Do this now and don't argue!"). Yet these children were the least likely to cheat when parents were nurturing and supportive. In other words, temperamentally fearful preschoolers can become dishonest or scrupulously honest, depending on their parents' disciplinary style.

There's no question that heredity and experience cause babies' temperaments to differ, but how stable is temperament? We'll find out in the next section.

Stability of Temperament

Do calm, easygoing babies grow up to be calm, easygoing children, adolescents, and adults? Are difficult, irritable infants destined to grow up to be cranky, whiny children? In fact, temperament is moderately stable throughout infancy, childhood, and adolescence (Janson & Mathiesen, 2008; Wachs & Bates, 2001). For example, newborns who cry under moderate stress tend, as 5-month-olds, to cry when they are placed in stressful situations (Stifter & Fox, 1990). In addition, when inhibited toddlers are adults, they respond more strongly to unfamiliar stimuli. Schwartz et al. (2003) had adults who were either inhibited or uninhibited as toddlers view novel and familiar faces. Records of brain activity showed that when adults who had been inhibited as toddlers viewed novel faces, they had significantly more activity in the amygdala, a brain region that regulates perception of fearful stimuli. Thus, the same individuals who avoided strangers as 2-year-olds had, as adults, the strongest response to novel faces.

Children's temperament influences the way that adults treat them; for example, parents engage in more vigorous play when their children are temperamentally active.

Thus, evidence suggests that temperament is at least somewhat stable throughout infancy and the toddler years (Lemery et al., 1999). Of course, the links are not perfect. Sam, an emotional 1-year-old, is more likely to be emotional as a 12-year-old than Dave, an unemotional 1-year-old. However, it's not a "sure thing" that Sam will still be emotional as a 12-year-old. Instead, think of temperament as a predisposition. Some infants are naturally predisposed to be sociable, emotional, or active; others *can* act in these ways, too, but only if the behaviors are nurtured by parents and others.

Though temperament is only moderately stable during infancy and toddlerhood, it can still shape development in important ways. For example, an infant's temperament may determine the experiences that parents provide. Parents may read more to quiet babies but play more physical games with their active babies. These different experiences, driven by the infants' temperament, contribute to each infant's development despite the fact that the infants' temperament may change over the years. Thus, although infants have many features in common, temperament characteristics remind us that each baby also seems to have its own unique personality from the very start.

THINK ABOUT IT

How would a learning theorist explain why children have different temperaments?

Test Yourself

RECALL

1. Some reflexes help infants get necessary nutrients, other reflexes protect infants from danger, and still other reflexes _____.

2. The _____ is based on five vital functions and provides a quick indication of a newborn's physical health.

3. A baby lying calmly with its eyes open and focused is in a state of _____.

4. Newborns spend more time asleep than awake, and about half this time asleep is spent in _____, a time thought to foster growth in the central nervous system.

5. The campaign to reduce SIDS emphasizes that infants should _____.

6. Research on the stability of temperament in infants and young children typically finds that _____.

INTERPRET

Compare the Apgar and the NBAS as measures of a newborn baby's well-being.

APPLY

Based on what you know about the stability of temperament, what would you say to a parent who's worried that her 15-month-old seems shy and inhibited?

Recall answers: (1) serve as the basis for later motor behaviors, (2) Apgar score, (3) alert inactivity, (4) REM sleep, (5) sleep on their backs, (6) temperament is moderately stable in these years.

3.2 Physical Development

LEARNING OBJECTIVES

- How do height and weight change from birth to 2 years of age?

- What nutrients do young children need? How are they best provided?

- What are the consequences of malnutrition? How can it be treated?

- What are nerve cells, and how are they organized in the brain?

- How does the brain develop? When does it begin to function?

While crossing the street, 4-year-old Martin was struck by a passing car. He was in a coma for a week but then gradually became more alert. Now he seems to be aware of his surroundings. Needless to say, Martin's mother is grateful that he survived the accident, but she wonders what the future holds for her son.

FOR PARENTS AND CHILDREN ALIKE, physical growth is a topic of great interest and a source of pride. Parents marvel at the speed with which babies add pounds and inches, and 2-year-olds proudly proclaim, "I bigger now!" In this section, we examine some of the basic features of physical growth, see how the brain develops, and discover how the accident affected Martin's development.

Growth of the Body

Growth is more rapid in infancy than during any other period after birth. Typically, infants double their birth weight by 3 months of age and triple it by their first birthday. This rate of growth is so rapid that, if continued throughout childhood, a typical 10-year-old boy would be nearly as long as a jumbo jet and weigh almost as much (McCall, 1979).

Average heights and weights for young children are represented by the lines marked 50th percentile in ▪ Figure 3.2. An average girl weighs about 7 pounds at birth, about 21 pounds at 12 months, and about 26 pounds at 24 months. If perfectly average, she would be 19–20 inches long at birth, grow to 29–30 inches at 12 months, and 34–35 inches at 24 months. Figures for an average boy are similar, but weights are slightly greater at ages 12 and 24 months.

Boys and girls grow taller and heavier from birth to 3 years of age, but the range of normal heights and weights is quite wide.

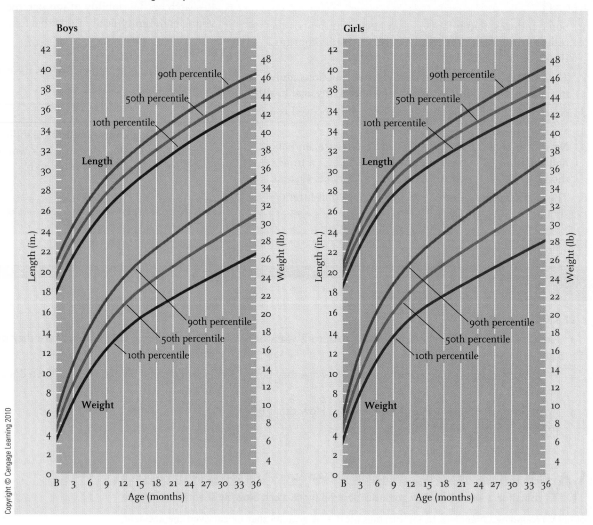

These charts also highlight how much children of the same age vary in weight and height. The lines marked 90th percentile in Figure 3.2 represent heights and weights for children who are larger than 90% of their peers; the lines marked 10th percentile represent heights and weights for children who are smaller than 90% of their peers. Any heights and weights between these lines are considered normal. At age 1, for example, normal weights for boys range from about 19 to 27 pounds. This means that an extremely light but normal boy weighs only two thirds as much as his extremely heavy but normal peer!

The important message here is that average height and normal height are not one and the same. Many children are much taller or shorter than average but are still perfectly normal. This applies to all of the age norms that we mention in this book. Whenever we provide a typical or average age for a developmental milestone, remember that the normal range for passing the milestone is much wider.

Whether an infant is short or tall depends largely on heredity. Both parents contribute to their children's height. In fact, the correlation between the average of the two parents' heights and their child's height at 2 years of age is about .7 (Plomin, 1990). As a general rule, two tall parents will have tall offspring; two short parents will have short offspring; and one tall parent and one short parent will have offspring of medium height.

So far we have emphasized the quantitative aspects of growth, such as height. This ignores an important fact: Infants are not simply scaled-down versions of adults.

THINK ABOUT IT

In Chapter 2 we explained how polygenic inheritance is often involved when phenotypes form a continuum. Height is such a phenotype. Propose a simple polygenic model to explain how height might be inherited.

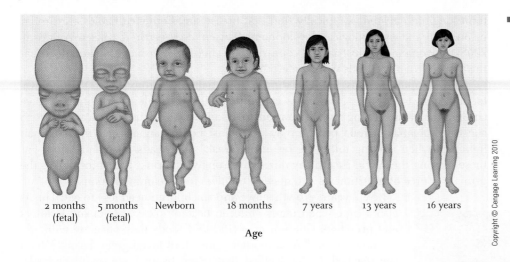

2 months (fetal)   5 months (fetal)   Newborn   18 months   7 years   13 years   16 years

Age

■ Figure 3.3 shows that, compared to adolescents and adults, infants and young children look top-heavy because their heads and trunks are disproportionately large. As growth of the hips, legs, and feet catches up later in childhood, their bodies take on more adult proportions. This pattern of growth, in which the head and trunk develop first, follows the cephalocaudal principle introduced in Chapter 2 (page 56).

Growth of this sort requires energy. Let's see how food and drink provide the fuel to grow.

"You Are What You Eat": Nutrition and Growth

In a typical 2-month-old, roughly 40% of the body's energy is devoted to growth. Most of the remaining energy is used for basic bodily functions such as digestion and respiration. A much smaller portion is consumed in physical activity.

Because growth requires so much high energy, young babies must consume an enormous number of calories relative to their body weight. A typical 12-pound 3-month-old, for example, should ingest about 600 calories daily, or about 50 calories per pound of body weight. An adult, by contrast, needs to consume only about 15–20 calories per pound, depending on the person's level of activity.

Breast-feeding is the best way to ensure that babies get the nourishment they need. Human milk contains the proper amounts of carbohydrates, fats, protein, vitamins, and minerals for babies. Breast-feeding also has several other advantages compared to bottle-feeding (Dewey, 2001). First, breast-fed babies are ill less often because breast milk contains the mother's antibodies. Second, breast-fed babies are less prone to diarrhea and constipation. Third, breast-fed babies typically make the transition to solid foods more easily, apparently because they are accustomed to changes in the taste of breast milk that reflect a mother's diet. Fourth, breast milk cannot be contaminated, which is a significant problem in developing countries when formula is used to bottle-feed babies.

Because of these many advantages, the American Academy of Pediatrics recommends that children be breast-fed for the first year, with iron-enriched solid foods introduced gradually. Cereal is a good first semi-solid food, followed by vegetables, fruits, and then meats. A good rule is to introduce only one food at a time. A 7-month-old having cheese for the first time, for instance, should have no other new foods for a few days. In this way, allergies that may develop—skin rash or diarrhea—can be linked to a particular food, making it easier to prevent recurrences.

The many benefits of breast-feeding do not mean that bottle-feeding is harmful. Formula, when prepared in sanitary conditions, provides generally the same nutrients as human milk. But infants are more prone to develop allergies from formula, and formula does not protect infants from disease. Even so, bottle-feeding has advantages of its own. A mother who cannot readily breast-feed can still enjoy the intimacy of feeding her baby, and other family members can participate in feeding. In fact, long-term longitudinal studies typically find that breast- and bottle-fed babies are similar in physical and psychological development (Jansen, de Weerth, & Riksen-Walraven, 2008), so women in industrialized countries can choose either method and know that their babies' dietary needs will be met.

In developing nations, bottle-feeding is potentially disastrous. Often the only water available to prepare formula is contaminated; the result is that infants have chronic diarrhea, leading to dehydration and sometimes death. Or, in an effort to conserve valuable formula, parents may ignore instructions and use less formula than indicated when making milk; the resulting "weak" milk leads to malnutrition. For these reasons, the World Health Organization strongly advocates breast-feeding as the primary source of nutrition for infants and toddlers in developing nations.

Toddlers and preschool children often become picky eaters. This can be annoying but should not concern parents.

By 2 years, growth slows and so children need less to eat. This is also a time when many children become picky eaters, and toddlers and preschool children may find that foods they once ate willingly are now "yucky." As a toddler, Laura Kail loved green beans. When she reached 2, she decided that green beans were awful and adamantly refused to eat them. Though such finickiness can be annoying, it may actually be adaptive for increasingly independent preschoolers. Because toddlers don't know what is safe to eat and what isn't, eating only familiar foods protects them from potential harm (Aldridge, Dovey, & Halford, 2009).

Parents should not be overly concerned about this finicky period. Although some children do eat less than before (in terms of calories per pound), virtually all picky eaters get adequate food for growth. Nevertheless, picky-eating children can make mealtime miserable for all. What's a parent to do? Experts recommend several guidelines for encouraging children to be more open-minded about foods and to deal with them when they aren't (Aldridge et al., 2009; American Academy of Pediatrics, 2008).

- When possible, allow children to choose among different healthy foods (e.g., milk versus yogurt).
- Allow children to eat foods in any order they want.
- Offer children new foods one at a time and in small amounts; encourage but don't force children to eat new foods.
- Don't force children to "clean their plates."
- Don't spend mealtimes talking about what the child is or is not eating; instead, talk about other topics that interest the child.
- Never use food to reward or punish children.

By following these guidelines, mealtimes can be pleasant and children can receive the nutrition they need to grow.

Malnutrition

malnourished
being small for one's age because of inadequate nutrition

An adequate diet is only a dream to many of the world's children. *Worldwide, about one in four children under age 5 is* **malnourished**, *as indicated by being small for their age* (UNICEF, 2006). Many are from third-world countries. In fact, nearly half of the world's undernourished children live in India, Bangladesh, and Pakistan (UNICEF, 2006). But malnutrition is regrettably common in industrialized countries, too. Many American children growing up homeless and in poverty are malnourished. Approximately 10% of American households do not have adequate food (Nord, Andrews, & Carlson, 2007).

Malnourishment is especially damaging during infancy because growth is so rapid during these years. By the school-age years, children with a history of infant malnutrition often have difficulty maintaining attention in school; they are easily distracted. Malnutrition during rapid periods of growth apparently damages the brain, affecting a child's abilities to pay attention and learn (Benton, 2010; Morgane et al., 1993).

Malnutrition would seem to have a simple cure—an adequate diet. But the solution is more complex than you might expect. Malnourished children are often listless and inactive (Ricciuti, 1993). They are unusually quiet and express little interest in what goes on around them. These behaviors are useful to children whose diet is inadequate because they conserve limited energy. Unfortunately, these behaviors may also deprive youngsters of experiences that would further their development. For example, when children are routinely unresponsive and lethargic, parents often come to believe that their actions have little impact on the children. That is, when children do not respond to parents' efforts to stimulate their development, this discourages parents from providing additional stimulation in the future. Over time, parents tend to provide fewer experiences that foster their children's development. The result is a self-perpetuating cycle in which malnourished children are forsaken by parents who feel as if they can do little to contribute to their children's growth. Thus, a biological influence (lethargy stemming from insufficient nourishment) causes a profound change in the experiences (parental teaching) that shape a child's development (Worobey, 2005).

To break the vicious cycle, these children need more than an improved diet. Their parents must be taught how to foster their children's development and must be encouraged to do so. Programs that combine dietary supplements with parent training offer promise in treating malnutrition (Engle & Huffman, 2010). Children in these programs often catch up with their peers in physical and intellectual growth, showing that the best way to reduce the effect of malnutrition on psychological forces is by addressing both biological and sociocultural forces (Super, Herrera, & Mora, 1990).

Many children around the world are malnourished, such as these two in a Somali refugee camp.

The Emerging Nervous System

The physical changes we see as infants grow are impressive. Even more awe-inspiring are the changes we cannot see—those involving the brain and the nervous system. An infant's feelings of hunger or pain, its smiles or laughs, and its efforts to sit upright or to hold a rattle all reflect the functioning of the brain and the rest of the emerging nervous system.

How does the brain accomplish these many tasks? To begin to answer this question, we need to look at the organization of the brain. *The basic unit in the brain and the rest of the nervous system is the* **neuron**, *a cell that specializes in receiving and transmitting information.* Neurons have the basic elements shown in ■ Figure 3.4. *The* **cell body**, *in the center of the cell, contains the basic biological machinery that keeps the neuron alive. The receiving end of the neuron, the* **dendrite**, *looks like a tree with its many branches.* This structure allows one neuron to receive input from thousands

neuron
basic cellular unit of the brain and nervous system that specializes in receiving and transmitting information

cell body
center of the neuron that keeps the neuron alive

dendrite
end of the neuron that receives information; it looks like a tree with many branches

■ **Figure 3.4**
A nerve cell includes dendrites that receive information, a cell body that has life-sustaining machinery, and, for sending information, an axon that ends in terminal buttons.

Dendrites

Terminal buttons

Axon

Cell body

Direction of information flow

axon
tubelike structure that emerges from the cell body and transmits information to other neurons

terminal buttons
small knobs at the end of the axon that release neurotransmitters

neurotransmitters
chemicals released by the terminal buttons that allow neurons to communicate with each other

cerebral cortex
wrinkled surface of the brain that regulates many functions that are distinctly human

hemispheres
right and left halves of the cortex

corpus callosum
thick bundle of neurons that connects the two hemispheres

frontal cortex
brain region that regulates personality and goal-directed behavior

of other neurons (Morgan & Gibson, 1991). *The tubelike structure that emerges from the other side of the cell body, the* **axon**, *transmits information to other neurons. At the end of the axon are small knobs called* **terminal buttons**, *which release chemicals called* **neurotransmitters**. These neurotransmitters are the messengers that carry information to nearby neurons.

Take 50–100 billion neurons like these, and you have the beginnings of a human brain. An adult's brain weighs a little less than 3 pounds and would easily fit into your hands. *The wrinkled surface of the brain is the* **cerebral cortex**; *made up of 10 billion neurons, the cortex regulates many of the functions that we think of as distinctly human. The cortex consists of left and right halves, called* **hemispheres**, *linked by a thick bundle of neurons called the* **corpus callosum**. The characteristics you value the most—your engaging personality, your "way with words," or your uncanny knack for "reading" others' emotions—are all controlled by specific regions in the cortex (■ Figure 3.5). *For example, your personality and your ability to make and carry out plans are largely centered in an area in the front of the cortex called (appropriately enough) the* **frontal cortex**. For most people, the ability to produce and understand language is mainly housed in neurons in the left hemisphere of the cortex. When you recognize that others are happy or sad, neurons in your right hemisphere are usually at work.

Now that we know a bit about the organization of the mature brain, let's look at how the brain grows and begins to function.

The Making of the Working Brain

The brain weighs only three quarters of a pound at birth, which is roughly 25% of the weight of an adult brain. But the brain grows rapidly during infancy and the preschool years. At 3 years of age, for example, the brain has achieved 80% of its ultimate weight. Brain weight doesn't tell us much, however, about the fascinating sequence of changes that take place to create a working brain. Instead, we need to move back to prenatal development.

Emerging Brain Structures

The beginnings of the brain can be traced to the period of the zygote. *At roughly 3 weeks after conception, a group of cells form a flat structure known as the* **neural plate**. At 4 weeks, the neural plate folds to form a tube that ultimately becomes the brain and spinal cord. When the ends of the tube fuse shut, neurons are produced in one small region of the neural tube. Production of neurons begins about 10 weeks after conception, and by 28 weeks the developing brain has virtually all the neurons

neural plate
flat group of cells present in prenatal development that becomes the brain and spinal cord

■ **Figure 3.5**
The brain on the left, viewed from above, shows the left and right hemispheres. The brain on the right, viewed from the side, shows the major regions of the cortex and their primary functions.

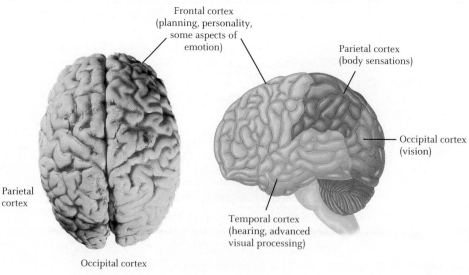

Frontal cortex
(planning, personality, some aspects of emotion)

Parietal cortex
(body sensations)

Occipital cortex
(vision)

Parietal cortex

Temporal cortex
(hearing, advanced visual processing)

Occipital cortex

Courtesy of Dr. Dana Copeland.

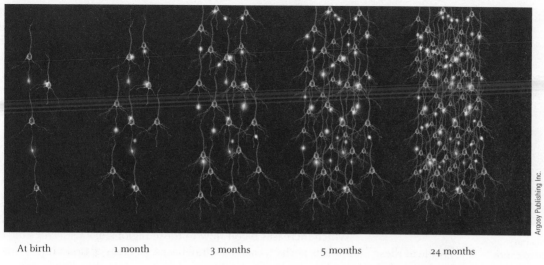

At birth   1 month   3 months   5 months   24 months

From birth to 2 years, neurons grow and create many new synapses with other neurons

it will ever have. During these weeks, neurons form at the incredible rate of more than 4,000 per second (Kolb, 1989).

From the neuron-manufacturing site in the neural tube, neurons migrate to their final positions in the brain. The brain is built in stages, beginning with the innermost layers. Neurons in the deepest layer are positioned first, followed by neurons in the second layer, and so on. This layering process continues until all six layers of the mature brain are in place, which occurs about 7 months after conception (Rakic, 1995).

In the fourth month of prenatal development, axons begin to acquire **myelin**—*the fatty wrap that speeds neural transmission.* This process continues through infancy and into childhood and adolescence (Casaer, 1993). Neurons that carry sensory information are the first to acquire myelin; neurons in the cortex are among the last. You can see the effect of more myelin in improved coordination and reaction times. The older the infant and (later) the child, the more rapid and coordinated his or her reactions.

In the months after birth, the brain grows rapidly. Axons and dendrites grow longer, and, like a maturing tree, dendrites quickly sprout new limbs. As the number of dendrites increases, so does the number of synapses, reaching a peak at about the first birthday. *Soon after, synapses begin to disappear gradually, a phenomenon known as* **synaptic pruning**. Thus, beginning in infancy and continuing into early adolescence, the brain goes through its own version of "downsizing," weeding out unnecessary connections between neurons. This pruning depends on the activity of the neural circuits: synapses that are active are preserved, but those that aren't active are eliminated (Webb, Monk, & Nelson, 2001). Pruning is completed first for brain regions associated with sensory and motor functions. Regions associated with basic language and spatial skills are completed next, followed by regions associated with attention and planning (Casey et al., 2005).

Growth of a Specialized Brain

Because the mature brain is specialized, with different psychological functions localized in particular regions, developmental researchers have had a keen interest in determining the origins and time course of the brain's specialization. For many years, the only clues to specialization came from children who had suffered brain injury. The logic here was to link the location of the injury to the impairment that results: If a region of the brain regulates a particular function (e.g., understanding speech), then damage to that region should impair the function.

myelin
fatty sheath that wraps around neurons and enables them to transmit information more rapidly

synaptic pruning
gradual reduction in the number of synapses, beginning in infancy and continuing until early adolescence

Electrodes placed on an infant's scalp can detect electrical activity that is used to create an electroencephalogram, a pattern of the brain's response to stimulation.

electroencephalography
the study of brain waves recorded from electrodes that are placed on the scalp

functional magnetic resonance imaging (fMRI)
method of studying brain activity by using magnetic fields to track blood flow in the brain

Fortunately, relatively few children suffer brain injury. But this meant that scientists needed other methods to study brain development. *One of them,* **electroencephalography,** *involves measuring the brain's electrical activity from electrodes placed on the scalp, as shown in the left photo.* If a region of the brain regulates a function, then the region should show distinctive patterns of electrical activity while a child is using that function. *A newer technique,* **functional magnetic resonance imaging (fMRI)**, *uses magnetic fields to track the flow of blood in the brain.* With this method, shown in the bottom photo, the research participant's brain is literally wrapped in an incredibly powerful magnet that can track blood flow in the brain as participants perform different cognitive tasks (Casey et al., 2005). The logic here is that active brain regions need more oxygen, which increases blood flow to those regions.

None of these methods is perfect; each has drawbacks. In cases of brain injury, for example, multiple areas of the brain may be damaged, making it hard to link impaired functioning to a particular brain region. fMRI is used sparingly because it's very expensive and participants must lie still for several minutes at a time.

Despite these limitations, the combined outcome of research using these different approaches has identified some general principles that describe the brain's specialization as children develop.

1. *Specialization is early in development.* Maybe you expect the brain to be completely unspecialized? In fact, many regions are already specialized very early in infancy. For example, early specialization of the frontal cortex is shown by the finding that damage to this region in infancy results in impaired decision making and abnormal emotional responses (Anderson et al., 2001). Similarly, studies using electroencephalography show that a newborn infant's left hemisphere generates more electrical activity in response to speech than the right hemisphere (Molfese & Burger-Judisch, 1991). Thus, by birth, the cortex of the left hemisphere is already specialized for language processing. As we'll see in Chapter 4, this specialization allows language to develop rapidly during infancy. Finally, studies of children with prenatal brain damage indicate that by infancy the right hemisphere is specialized for understanding certain kinds of spatial relations (Stiles et al., 2005).

2. *Specialization takes two specific forms.* First, with development the brain regions active during processing become more focused and less diffuse—an analogy would be to a thunderstorm that covers a huge region versus one that packs the same power in a much smaller region (Durston et al., 2006). Second, the kinds of stimuli that trigger brain activity shift from being general to specific

In fMRI, a magnet is used to track the flow of blood to different regions of the brain as children and adults perform cognitive tasks.

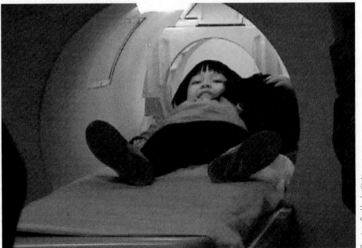

(Johnson, Grossman, & Cohen Kadosh, 2009). For example, studies of how the brain processes faces (Scherf et al., 2007) show both forms of specificity: face processing becomes focused in a particular area and becomes tuned narrowly to faces (i.e., more brain activity in response to faces compared to other stimuli).

3. *Different brain systems specialize at different rates.* Think of a new housing development involving construction of many multistory homes. In each house, the first floor is completed before higher floors but some houses are finished before others are even started. In this same way, brain regions involving basic sensory and perceptual processes specialize well before those regions necessary for higher order processes (Fox, Levitt, & Nelson, 2010). Similarly, some brain systems that are sensitive to reward reach maturity in adolescence but the systems responsible for self-control aren't fully specialized until adulthood (Somerville & Casey, 2010).

4. *Successful specialization requires stimulation from the environment.* To return to the analogy of the brain as a house, the newborn's brain is perhaps best conceived as a partially finished, partially furnished house: A general organizational framework is there, with preliminary neural pathways designed to perform certain functions. The left hemisphere no doubt has some language pathways and the frontal cortex has some emotion-related pathways. However, completing the typical organization of the mature brain requires input from the environment (Greenough & Black, 1992). *In this case, environmental input influences* **experience-expectant growth**—*over the course of evolution, human infants have typically been exposed to some forms of stimulation that are used to adjust brain wiring, strengthening some circuits and eliminating others.* For example, under normal conditions, healthy human infants experience moving visual patterns (e.g., faces) and varied sounds (e.g., voices). Just as a newly planted seed depends on a water-filled environment for growth, a developing brain depends upon environmental stimulation to fine-tune circuits for vision, hearing, and other systems (Black, 2003).

Of course, experiences later in life also sculpt the brain (and we'll see this in several chapters later in this book). **Experience-dependent growth** *denotes changes in the brain that are not linked to specific points in development and that vary across individuals and across cultures.* Experience-dependent growth is illustrated by a preschool child's learning of a classmate's name, an elementary-school child's discovery of a shortcut home from school, and an adolescent's mastery of the functions of a new cell phone. In each case, brain circuits are modified in response to an individual's experiences. With today's technology, we can't see these daily changes in the brain. But when they accumulate over many years—as when individuals acquire expertise in a skill—brain changes can be detected. For example, skilled cellists have extensive brain regions devoted to controlling the fingers of the left hand as they are positioned on the strings (Elbert et al., 1995). And years of driving a taxicab produces changes in the hippocampus, a region of the brain implicated in navigation and way-finding (Maguire, Woollett, & Spiers, 2006).

5. *The immature brain's lack of specialization confers a benefit—greater plasticity.* Just as the structures in a housing development follow a plan that specifies the location of each house and its design, brain development usually follows a predictable course that reflects epigenetic interactions (page 50) between the genetic code and required environmental input. Sometimes, however, the normal course is disrupted. A person may experience events harmful to the brain (e.g., injured in an accident) or may be deprived of some essential ingredients of successful "brain building" (e.g., necessary experiences).

Research that examines the consequences of these atypical experiences shows that the brain has some flexibility: it is plastic. Remember Martin, the child in the vignette whose brain was damaged when he was struck by a car? His language skills were impaired after the accident. This was not surprising, because the left hemisphere of Martin's brain had absorbed most of the force of

experience-expectant growth
process by which the wiring of the brain is organized by experiences that are common to most humans

experience-dependent growth
process by which an individual's unique experiences over a lifetime affect brain structures and organization

THINK ABOUT IT
What's an example of experience-expectant growth for U.S. 3-year-olds? What's a common type of experience-dependent growth at this age?

the collision. But within several months, Martin had completely recovered his language skills. Apparently other neurons took over language-related processing from the damaged neurons. This recovery of function is not uncommon—particularly for young children—and shows that the brain is plastic. In other words, young children often recover more skills after brain injury than older children and adults, apparently because functions are more easily reassigned in the young brain (Demir, Levine, & Goldin-Meadow, 2010; Stiles et al., 2005).

Test Yourself

RECALL

1. Compared to older children and adults, an infant's head and trunk are _____.

2. Because of the high demands of growth, infants need _____ calories per pound than adults.

3. The most effective treatment for malnutrition is improved diet and _____.

4. The _____ is the part of the neuron that contains the basic machinery to keep the cell alive.

5. The frontal cortex is the seat of personality and regulates _____.

6. Human speech typically elicits the greatest electrical activity from the _____ of an infant's brain.

7. A good example of brain plasticity is that, although children with brain damage often have impaired cognitive processes, _____.

INTERPRET

Compare growth of the brain before birth with growth of the brain after birth.

APPLY

How does malnutrition illustrate the influence on development of life-cycle forces in the biopsychosocial framework?

Recall answers: (1) disproportionately large, (2) more, (3) parent training, (4) cell body, (5) planning, (6) left hemisphere, (7) they often regain their earlier skills over time

---

Moving and Grasping: Early Motor Skills

LEARNING OBJECTIVES

- What are the component skills involved in learning to walk? At what age do infants master them?

- How do infants learn to coordinate the use of their hands?

Nancy is 14 months old and a world-class crawler. Using hands and knees, she can go just about anywhere she wants to. Nancy does not walk and seems uninterested in learning how. Nancy's dad wonders whether he should be doing something to help Nancy progress beyond crawling. Deep down, he worries that perhaps he was negligent in not providing more exercise for Nancy when she was younger.

DO YOU REMEMBER WHAT IT WAS LIKE TO LEARN TO TYPE, to drive a car with a stick shift, to play a musical instrument, or to play a sport? *Each of these activities involves* **motor skills**: *coordinated movements of the muscles and limbs.* Success demands that each movement be done in a precise way, in exactly the right sequence, and at exactly the right time. For example, in the few seconds that it takes you to type "human development," if you don't move your fingers in exactly the correct sequence to the precise location on the keyboard, you might get "jinsj drveo;nrwnt."

These activities are demanding for adults, but think about similar challenges for infants. *Infants must learn to move about in the world, to* **locomote**. At first unable to move independently, infants soon learn to crawl, to stand, and to walk. Once the child can move through the environment upright, the arms and hands are free. To take advantage of this arrangement, the human hand has fully independent fingers (instead of a paw), with the thumb opposing the remaining four fingers. *Infants must*

motor skills
coordinated movements of the muscles and limbs

locomote
ability to move around in the world

*learn the **fine motor skills** associated with grasping, holding, and manipulating objects.* In the case of feeding, for example, infants progress from being fed by others, to holding a bottle, to feeding themselves with their fingers, to eating with utensils.

Together, locomotion and fine motor skills give children access to an enormous variety of information about shapes, textures, and features in their environment. In this section, we'll see how locomotion and fine motor skills develop and, as we do, we'll see whether Nancy's dad should worry about her lack of interest in walking.

fine motor skills
motor skills associated with grasping, holding, and manipulating objects

Locomotion

Advances in posture and locomotion transform the infant in little more than a year. ■ Figure 3.6 shows some of the important milestones in motor development and the age by which most infants have achieved them. By about 5 months of age, most babies will have rolled from back to front and will be able to sit upright with support. By 7 months infants can sit alone, and by 10 months they can creep. A typical 14-month-old is able to stand alone briefly and walk with assistance. *This early, unsteady form of walking is called* **toddling** *(hence the term* **toddler***)*. Of course, not all children walk at exactly the same age. Some walk before their first birthday;

toddling
early, unsteady form of walking done by infants

toddlers
young children who have just learned to walk

■ **Figure 3.6**
Locomotor skills improve rapidly in the 15 months after birth, and progress can be measured by many developmental milestones.

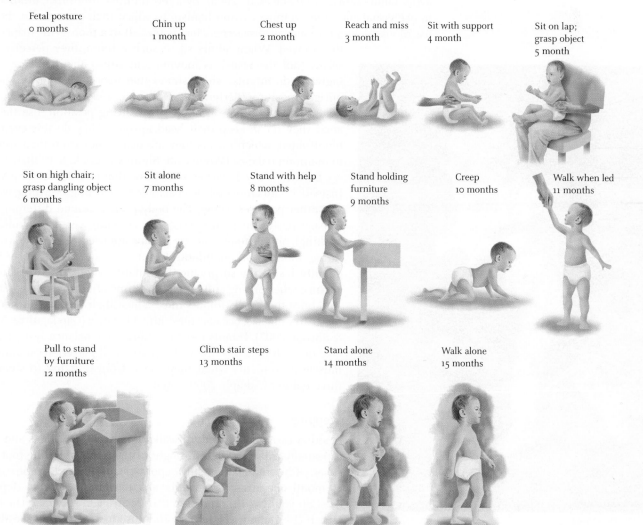

Fetal posture
0 months

Chin up
1 month

Chest up
2 month

Reach and miss
3 month

Sit with support
4 month

Sit on lap;
grasp object
5 month

Sit on high chair;
grasp dangling object
6 months

Sit alone
7 months

Stand with help
8 months

Stand holding
furniture
9 months

Creep
10 months

Walk when led
11 months

Pull to stand
by furniture
12 months

Climb stair steps
13 months

Stand alone
14 months

Walk alone
15 months

Based on Shirley, 1931, and Bayley, 1969.

others—like Nancy, the world-class crawler in the vignette—take their first steps as late as 18 or 19 months of age. By 24 months, most children can climb steps, walk backward, and kick a ball.

Researchers once thought these developmental milestones reflected maturation (e.g., McGraw, 1935). Walking, for example, emerged naturally when the necessary muscles and neural circuits matured. Today, however, locomotion—and, in fact, all of motor development—is viewed from a new perspective. *According to* **dynamic systems theory**, *motor development involves many distinct skills that are organized and reorganized over time to meet the demands of specific tasks.* For example, walking includes maintaining balance, moving limbs, perceiving the environment, and having a reason to move. Only by understanding each of these skills and how they are combined to allow movement in a specific situation can we understand walking (Thelen & Smith, 1998).

dynamic systems theory
theory that views motor development as involving many distinct skills that are organized and reorganized over time to meet specific needs

Posture and Balance

The ability to maintain an upright posture is fundamental to walking. But upright posture is virtually impossible for newborns and young infants because of the shape of their body. Cephalocaudal growth means that an infant is top-heavy. Consequently, as soon as a young infant starts to lose her balance, she tumbles over. Only with growth of the legs and muscles can infants maintain an upright posture (Thelen, Ulrich, & Jensen, 1989).

Once infants can stand upright, they must continuously adjust their posture to avoid falling down (Metcalfe et al., 2005). By a few months after birth, infants begin to use visual cues and an inner-ear mechanism to adjust their posture. To show the use of visual cues for balance, researchers had babies sit in a room with striped walls that moved. When adults sit in such a room, they perceive themselves (not the walls) as moving and adjust their posture accordingly; so do infants, which shows that they use vision to maintain upright posture (Bertenthal & Clifton, 1998). In addition, when 4-month-olds who are propped in a sitting position lose their balance, they try to keep their head upright. They do this even when blindfolded, which means they are using cues from their inner ear to maintain balance (Woollacott, Shumway-Cook, & Williams, 1989).

Balance is not, however, something that infants master just once. Instead, infants must relearn balancing for sitting, crawling, walking, and other postures. Why? The body rotates around different points in each posture (e.g., the wrists for crawling versus the ankles for walking), and different muscle groups are used to generate compensating motions when infants begin to lose their balance. Consequently, it's hardly surprising that infants who easily maintain their balance when sitting still topple over time after time when crawling. And once they walk, infants must adjust their posture further when they carry objects, because these affect balance (Garciaguirre, Adolph, & Shrout, 2007). Infants must recalibrate the balance system as they take on each new posture, just as basketball players recalibrate their muscle movements when they move from dunking to shooting a three-pointer (Adolph, 2000, 2003).

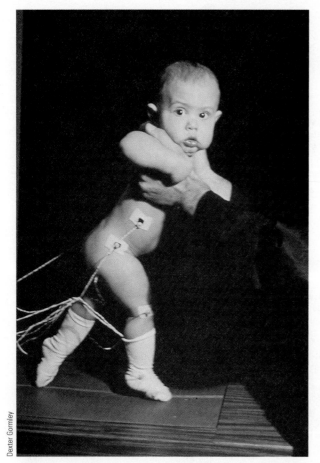

Infants are capable of stepping—moving the legs alternately—long before they can walk alone.

Stepping

Another essential element of walking is moving the legs alternately, repeatedly transferring the weight of the body from one foot to the other. Children don't step spontaneously until approximately 10 months because they must be able to stand in order to step.

Can younger children step if they are held upright? Thelen and Ulrich (1991) devised a clever procedure to answer this question. Infants were placed on a treadmill and held upright by an adult. When the belt on the treadmill started to move, infants could respond in one of several ways. They might simply let both legs be dragged rearward by the belt. Or they might let their legs be dragged briefly, then move them forward together in a hopping motion. Many 6- and 7-month-olds demonstrated the mature pattern of alternating steps on each leg. Even more amazing is that—when the treadmill was equipped with separate belts for each leg that moved at different speeds—babies adjusted, stepping more rapidly on the faster belt.

Apparently, the alternate stepping motion that is essential for walking is evident long before infants walk alone. Walking unassisted is not possible, though, until other component skills are mastered.

Environmental Cues

Many infants learn to walk in the relative security of flat, uncluttered floors at home. But they soon discover that the environment offers a variety of surfaces, some more conducive to walking than others. Infants use cues in the environment to judge whether a surface is suitable for walking. For example, they are more likely to cross a bridge when it's wide and has a rigid handrail than when it is narrow and has a wobbly handrail (Berger, Adoph, & Lobo, 2005). And if they can't decide whether a surface is safe, they depend on an adult's advice (Tamis-LeMonda et al., 2008). Research results like these show that infants use perceptual cues to decide whether a surface is safe for walking.

Coordinating Skills

Dynamic systems theory emphasizes that learning to walk demands orchestration of many individual skills. Each component skill must first be mastered alone and then integrated with the other skills (Werner, 1948). That is, *mastery of intricate motions requires both* **differentiation** *(mastery of component skills) and* **integration**—*combining the motions in proper sequence into a coherent, working whole.* In the case of walking, not until 12 to 15 months of age have children mastered the component skills to be coordinated and so allow independent, unsupported walking.

differentiation
distinguishing and mastering individual motions

integration
linking individual motions into a coherent, coordinated whole

Mastering individual skills and coordinating them well does not happen overnight. Instead, each takes time and repeated practice. For example, with concentrated practice, toddlers learn to change their stride to walk more slowly down steep slopes (Gill, Adolph, & Vereijken, 2009). However, improvements are limited to the movements that were trained. For example, when infants practice crawling on steep slopes, there is no transfer to walking on steep slopes, because the motions differ (Adolph, 1997). Thus, just as daily practice kicking a soccer ball won't improve your golf game, infants who receive much practice in one motor skill don't usually improve in others.

In many African cultures, infants are routinely carried piggyback style; this strengthens the infant's legs, which allows them to walk at a younger age.

These findings from laboratory research are not the only evidence that practice promotes motor development; cross-cultural research points to the same conclusion. In Europe and North America, most infants typically walk alone near their first birthday. But infants in other cultures often begin to walk (and reach the other milestones listed on page 99) at an earlier age because child-care practices allow children to practice their emerging motor skills. For example, in some traditional African cultures, infants sit and walk at younger ages. Why? Infants are commonly carried by their parents "piggyback" style, which helps develop muscles in the infants' trunk and legs.

Some cultures even take a further step. They believe that practice is essential for motor skills to develop normally and so they (or siblings) provide daily training sessions. For example, the Kipsigis of Kenya help children learn to sit by having them sit while propped up (Super, 1981). And among the West Indians of Jamaica, mothers have an elaborate exercise routine that allows babies to practice walking (Hopkins & Westra, 1988). This training provides additional opportunities for children to learn the elements of different motor skills; not surprisingly, infants with these opportunities learn to sit and walk earlier.

You may be surprised that some cultures do just the opposite—they have practices that *discourage* motor development. The Ache, an indigenous group in

© Vic Bider / PhotoEdit

Locomotor skills develop rapidly in preschool children, making it possible for them to play vigorously.

Paraguay, protect infants and toddlers from harm by carrying them constantly (Kaplan & Dove, 1987). In Chinese cities, parents often allow their children to crawl only on a bed surrounded by pillows, in part because they don't want their children crawling on a dirty floor (Campos et al., 2000). In both cases, infants reach motor milestones a few months later than the ages listed in the chart on page 99.

Even European and North American infants are crawling at older ages today than they did in previous generations (Dewey et al., 1998). This generational difference reflects the effectiveness of the Back to Sleep campaign described on page 86. Because today's babies spend less time on their tummies, they have fewer opportunities to discover that they can propel themselves by creeping, which would otherwise prepare them for crawling.

Thus, cultural practices can accelerate or delay the early stages of motor development, depending on the nature of practice that infants and toddlers receive. In the long run, however, the age of mastering various motor milestones is not critical for children's development. All healthy children learn to walk, and whether this happens a few months before or after the "typical" ages shown on page 99 has no bearing on children's later development.

Beyond Walking

If you can recall the feeling of freedom that accompanied your first driver's license, you can imagine how the world expands for infants and toddlers as they learn to move independently. The first tentative steps are soon followed by others that are more skilled. With more experience, infants take longer, straighter steps. And, like adults, they begin to swing their arms, rotating the left arm forward as the right leg moves then repeating with the right arm and left leg (Ledebt, 2000; Ledebt, van Wieringen, & Savelsbergh, 2004).

Most children learn to run a few months after they walk alone. Most 2-year-olds have a "hurried walk" instead of a true run; they move their legs stiffly (rather than bending them at the knees) and are not "airborne" as is the case with true running. By 5 or 6 years, children run easily, quickly changing directions or speed. Hopping also shows young children's growing skill: A typical 2- or 3-year-old will hop a few times on one foot, typically keeping the upper body very stiff; by 5 or 6, children can hop long distances on one foot or alternate hopping first on one foot a few times, then on the other.

With their advanced motor skills, older preschoolers delight in unstructured play. They enjoy swinging, climbing over jungle gyms, and balancing on a beam. Some learn to ride a tricycle or to swim.

THINK ABOUT IT

How does learning to hop on one foot demonstrate differentiation and integration of motor skills?

Fine Motor Skills

A major accomplishment of infancy is skilled use of the hands (Bertenthal & Clifton, 1998). Newborns have little apparent control of their hands, but 1-year-olds are extraordinarily talented.

Reaching and Grasping

At about 4 months, infants can successfully reach for objects (Bertenthal & Clifton, 1998). These early reaches often look clumsy—and for good reason. When infants reach, they don't move their arm and hand directly and smoothly to the desired object (as older children and adults do). Instead, the infant's hand moves like a ship under the direction of an unskilled navigator: It moves a short distance, slows, then moves again in a slightly different direction—a process that's repeated until the hand finally contacts the object (McCarty & Ashmead, 1999). As infants grow, their reaches have fewer movements, though they are still not as continuous and smooth as reaches by older children and adults (Berthier, 1996).

Reaching requires that an infant move the hand to the location of a desired object. Grasping poses a different challenge: Now the infant must coordinate movements of individual fingers to grab an object. Grasping, too, becomes more efficient during infancy. Most 4-month-olds just use their fingers to hold objects, wrapping the object tightly with their fingers alone. Not until 7 or 8 months do most infants use their thumbs to hold objects (Siddiqui, 1995). At about this same age, infants begin to position their hands to make it easier to grasp an object. If trying to grasp a long thin rod, for example, infants place their fingers perpendicular to the rod, which is the best position for grasping (Wentworth, Benson, & Haith, 2000). Infants need not see their hand to position it correctly: They position the hand just as accurately in reaching for a lighted object in a darkened room as when reaching in a lighted room (McCarty et al., 2001).

A typical 4-month-old grasps an object with fingers alone.

Infants' growing control of each hand is accompanied by greater coordination of the two hands. Although 4-month-olds use both hands, their motions are not coordinated; rather, each hand seems to have a mind of its own. Infants may hold a toy motionless in one hand while shaking a rattle in the other. At roughly 5 to 6 months of age, infants can coordinate the motions of their hands so that each hand performs different actions that serve a common goal. So a child might, for example, hold a toy animal in one hand and pet it with the other (Karniol, 1989). These skills continue to improve after the child's first birthday: 1-year-olds reach for most objects with one hand; by 2 years, they reach with one or two hands, as appropriate, depending on the size of the object (van Hof, van der Kamp, & Savelsbergh, 2002).

These gradual changes in fine motor coordination are well illustrated by the ways children feed themselves. Beginning at roughly 6 months of age, many infants experiment with "finger foods" such as sliced bananas and green beans. Infants can easily pick up such foods, but getting them into their mouths is another story. The hand grasping the food may be raised to the cheek, then moved to the edge of the lips, and finally shoved into the mouth. Mission accomplished, but only after many detours along the way! However, infants' eye–hand coordination improves rapidly, and foods varying in size, shape, and texture are soon placed directly in the mouth.

At about the first birthday, many parents allow their children to try eating with a spoon. Youngsters first simply play with the spoon, dipping it in and out of a dish filled with food or sucking on an empty spoon. Soon they learn to fill the spoon with food and place it in their mouth, but the motions are awkward. For example, most 1-year-olds fill a spoon by first placing it directly over a dish. Then, they lower it until the bowl of the spoon is full. In contrast, 2-year-olds typically scoop food from a dish by rotating their wrist, which is the same motion adults use.

By age 5, fine-motor skills are developed to the point that most youngsters can dress themselves.

As preschoolers, children become much more dextrous and are able to make many precise and delicate movements with their hands and fingers. Greater fine motor skill means that preschool children can begin to care for themselves. No longer must they rely primarily on parents to feed and clothe them; instead, they become increasingly skilled at feeding and dressing themselves. A 2- or 3-year-old, for example, can put on some simple clothing and use zippers but not buttons; by 3 or 4 years, children can fasten buttons and take off their clothes when going to the bathroom; and most 5-year-olds can dress and undress themselves—except for tying shoes, which children typically master at about age 6.

Greater fine motor coordination also leads to improvements in preschool children's printing and drawing. Given a crayon or marker, 2-year-olds will scribble, expressing delight in the simple lines that are created just by moving a crayon or marker across paper. By 4 or 5 years of age, children use their drawings to depict recognizable objects.

All of these actions illustrate the principles of differentiation and integration that were introduced in our discussion of locomotion. Complex acts involve many simple movements. Each must be performed correctly and in the proper sequence. Development involves first mastering the separate elements and then assembling them into a smoothly functioning whole.

Handedness

When young babies reach for objects, they don't seem to prefer one hand over the other; they use their left and right hands interchangeably. They may shake a rattle with their left hand and moments later pick up blocks with their right. By the first birthday, most youngsters are emergent right-handers. They use their left hand to steady the toy while the right hand manipulates the object. This early preference for one hand becomes stronger and more consistent during the preschool years and is well established by kindergarten (Marschik et al., 2008; Rönnqvist & Domellöff, 2006).

What determines whether children become left- or right-handed? Some scientists believe that a gene biases children toward right-handedness (Annett, 2008). Consistent with this idea, identical twins are more likely than fraternal twins to have the same handedness—both are right-handed or both are left-handed (Sicotte, Woods, & Mazziotta, 1999). But experience also contributes to handedness. Modern industrial cultures favor right-handedness. School desks, scissors, and can openers, for example, are designed for right-handed people and can be used by left-handers only with difficulty. In the United States, elementary school teachers used to urge left-handed children to use their right hands. As this practice has diminished in the last 50 years, the percentage of left-handed children has risen steadily (Levy, 1976). Thus, handedness seems to have both hereditary and environmental influences.

Test Yourself

RECALL

1. According to _____, motor development involves many distinct skills that are organized and reorganized over time, depending on task demands.

2. When 4-month-olds tumble from a sitting position, they usually try to keep their head upright. This happens even when they are blindfolded, which means that the important cues to balance come from _____.

3. Skills important in learning to walk include maintaining upright posture and balance, stepping, and _____.

4. Akira uses both hands simultaneously, but not in a coordinated manner; each hand seems to be "doing its own thing." Akira is probably _____ months old.

5. Before the age of _____, children show no signs of handedness; they use their left and right hands interchangeably.

INTERPRET

Compare and contrast the milestones of locomotor development in the first year with the fine motor milestones.

APPLY

Describe how the mastery of a fine motor skill—such as learning to use a spoon or a crayon—illustrate the integration of biological, psychological, and sociocultural forces in the biopsychosocial framework.

Recall answers: (1) dynamic systems theory, (2) the inner ear, (3) using perceptual information, (4) 4, (5) 1 year

Coming to Know the World: Perception

LEARNING OBJECTIVES

- Are infants able to smell, to taste, and to experience pain?
- Can infants hear? How do they use sound to locate objects?
- How well can infants see? Can they see color and depth?
- How do infants coordinate information between different sensory modalities, such as between vision and hearing?

Darla is mesmerized by her newborn daughter, Olivia. Darla loves holding Olivia, talking to her, and simply watching her. Darla is certain that Olivia is already getting to know her and is coming to recognize her face and the sound of her voice. Darla's husband, Steve, thinks she is crazy: "Everyone knows that babies are born blind, and they probably can't hear much either." Darla doubts Steve and wishes someone could tell her the truth about Olivia's vision and hearing.

TO ANSWER DARLA'S QUESTIONS, we need to define what it means for an infant to experience or sense the world. Humans have several kinds of sense organs, each of which is receptive to a different kind of physical energy. For example, the retina at the back of the eye is sensitive to some types of electromagnetic energy, and sight is the result. The eardrum detects changes in air pressure, and hearing is the result. Cells at the top of the nasal passage detect the passage of airborne molecules, and smell is the result. In each case, the sense organ translates the physical stimulus into nerve impulses that are sent to the brain. *The processes by which the brain receives, selects, modifies, and organizes these impulses is known as* **perception**. This is simply the first step in the complex process of accumulating information that eventually results in "knowing."

Darla's questions are really about her newborn daughter's perceptual skills. By the end of this section, you'll be able to answer her questions because we're going to look at how infants use different senses to experience the world.

perception
processes by which the brain receives, selects, modifies, and organizes incoming nerve impulses that are the result of physical stimulation

Smell, Taste, and Touch

Newborns have a keen sense of smell. Infants respond positively to pleasant smells and negatively to unpleasant smells (Mennella & Beauchamp, 1997). They have a relaxed and contented facial expression when they smell honey or chocolate, but they frown or turn away when they smell rotten eggs or ammonia. Young babies can also recognize familiar odors. Newborns will look in the direction of a pad that is saturated with their own amniotic fluid (Schaal, Marlier, & Soussignan, 1998). They will also turn toward a pad saturated with the odor of their mother's breast or her perfume (Porter & Winberg, 1999).

Newborns also have a highly developed sense of taste. They readily differentiate salty, sour, bitter, and sweet tastes (Rostenstein & Oster, 1997). Most infants seem to have a "sweet tooth." They react to sweet substances by smiling, sucking, and licking their lips (Steiner et al., 2001) but grimace when fed bitter or sour substances (Kaijura, Cowart, & Beauchamp, 1992). Infants are also sensitive to changes in the taste of breast milk that reflect a mother's diet and will nurse more after their mother has consumed a sweet-tasting substance such as vanilla (Mennella & Beauchamp, 1996).

Newborns are sensitive to touch. As we saw earlier in this chapter, many areas of the newborn's body respond reflexively when touched. Touching an infant's cheek, mouth, hand, or foot produces reflexive movements, documenting that infants perceive touch. What's more, babies' behavior in response to apparent pain-provoking stimuli suggests that they experience pain (Warnock & Sandrin,

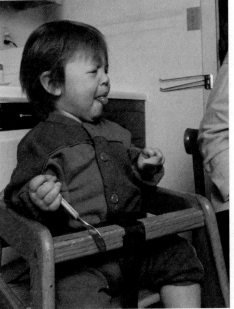

Infants and toddlers do not like bitter and sour tastes!

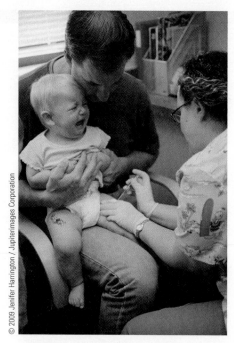

An infant's response to an inoculation—a distinctive facial expression coupled with a distinctive cry—clearly suggests that the baby feels pain.

2004). Look, for example, at the baby in the photo who is receiving an inoculation. He's opened his mouth to cry and, although we can't hear him, the sound of his cry is probably the unique pattern associated with pain. The pain cry begins suddenly, is high-pitched, and is not easily soothed. This baby is agitated, his heart rate has jumped, and he's trying to move his hands, arms, and legs (Craig et al., 1993; Goubet, Clifton, & Shah, 2001). All together, these signs strongly suggest that babies experience pain.

Perceptual skills are extraordinarily useful to newborns and young babies. Smell and touch help them recognize their mothers. Smell and taste make it much easier for them to learn to eat. Early development of smell, taste, and touch prepares newborns and young babies to learn about the world.

Hearing

Do you remember, from Chapter 2, the study in which mothers read aloud *The Cat in the Hat* late in pregnancy? This research showed that the fetus can hear at 7 or 8 months after conception. As you would expect from these results, newborns typically respond to sounds in their surroundings. If a parent is quiet but then coughs, an infant may startle, blink his eyes, and move his arms or legs. These responses may seem natural, but they do indeed indicate that infants are sensitive to sound.

Overall, adults can hear better than infants (Saffran, Werker, & Werner, 2006). Adults can hear some very quiet sounds that infants can't. More interestingly, infants best hear sounds that have pitches in the range of human speech: neither very high-pitched nor very low-pitched. Infants can differentiate speech sounds, such as vowels from consonant sounds, and by 4 or 5 months they can recognize their own names (Jusczyk, 1995; Mandel, Jusczyk, & Pisoni, 1995).

Infants also can distinguish different musical sounds and can remember lullabies and songs that parents sing to them (Trainor, Wu, & Tsang, 2004). They can distinguish different melodies and prefer melodies that are pleasant sounding over those that are unpleasant sounding or dissonant (Trainor & Heinmiller, 1998). And infants are sensitive to the rhythmic structure of music: After hearing a simple sequence of notes, they can tell the difference between another sequence that matches the original versus one that doesn't (Hannon & Trehub, 2005). This early sensitivity to music is remarkable but perhaps not so surprising when you consider that music is (and has been) central in all cultures.

Thus, by the middle of the first year, infants respond to much of the information that is provided by sound. In Chapter 4, we will reach the same conclusion when we examine the perception of language-related sounds.

Seeing

If you've ever watched infants, you've probably noticed that they spend much of their waking time looking around. Sometimes they seem to be generally scanning their environment, and sometimes they seem to be focusing on nearby objects. What do they see as a result? Perhaps their visual world is a sea of confusing gray blobs. Or maybe they see the world essentially as adults do. Actually, neither of these descriptions is entirely accurate, but the second is closer to the truth.

The various elements of the visual system—the eye, the optic nerve, and the brain—are relatively well developed at birth. Newborns respond to light and can track moving objects with their eyes. How well do infants see? *The clarity of vision, called* **visual acuity**, *is defined as the smallest pattern that can be distinguished dependably.* You've undoubtedly had your acuity measured, probably by being asked to read rows of progressively smaller letters or numbers from a chart. To assess newborns' acuity we use the same approach, adjusted somewhat because we can't use words to explain to infants what we'd like them to do. Most infants will look at patterned stimuli instead of plain, patternless stimuli. For example, if we were to show the two stimuli in

visual acuity
smallest pattern that one can distinguish reliably

■ Figure 3.7 to an infant, most babies would look longer at the striped pattern than at the gray pattern. As we make the lines narrower (along with the spaces between them), there comes a point at which the black and white stripes become so fine that they simply blend together and appear gray—just like the other pattern.

To estimate an infant's acuity, we pair the gray square with squares in which the widths of the stripes differ, like the ones in ■ Figure 3.8: When infants look at the two stimuli equally, this indicates that they are no longer able to distinguish the stripes of the patterned stimulus. By measuring the width of the stripes and their distance from an infant's eye, we can estimate acuity, with detection of thinner stripes indicating better acuity. Measurements of this sort indicate that newborns and 1-month-olds see at 20 feet what normal adults would see at 200–400 feet. But by the first birthday, an infant's acuity is essentially the same as that of an adult with normal vision (Kellman & Arterberry, 2006).

Color

Not only do infants begin to see the world with greater acuity during the first year, they also begin to see it in color! How do we perceive color? The wavelength of light is the basis of color perception. In ■ Figure 3.9, light that we see as red has a relatively long wavelength, whereas violet (at the other end of the color spectrum) has a much shorter wavelength. *Concentrated in the back of the eye, along the retina, are specialized neurons called* **cones**. Some cones are particularly sensitive to short-wavelength light (blues and violets). Others are sensitive to medium-wavelength light (greens and yellows), and still others are sensitive to long-wavelength light (reds and oranges). These different kinds of cones are linked by complex circuits of neurons, and this circuitry is responsible for our ability to see the world in colors.

These circuits gradually begin to function in the first few months after birth. Newborns and young babies can perceive few colors, but by 3 months the three kinds of cones and their associated circuits are working and infants are able to see the full range of colors (Kellman & Arterberry, 2006). In fact, by 3 to 4 months, infants' color perception seems similar to that of adults (Adams & Courage, 1995; Franklin, Pilling, & Davies, 2005). In particular, infants, like adults, tend to see categories of color: they see the spectrum as a group of reds, a group of yellows, a group of greens, and the like (Dannemiller, 1998).

Depth

People see objects as having three dimensions: height, width, and depth. The retina of the eye is flat, so height and width can be represented directly on its two-dimensional surface. But the third dimension, depth, cannot be represented directly

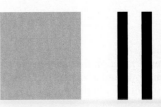

■ **Figure 3.7**
Infants usually prefer looking at striped patterns to plain ones, a preference that can be used to measure an infant's visual acuity.

■ **Figure 3.8**
Visual acuity can be measured by determining the thinnest stripes that the infant prefers to view.

cones
specialized neurons in the back of the eye that sense color

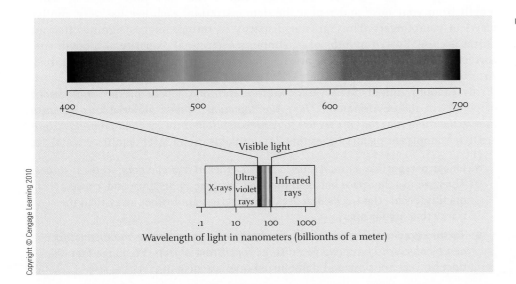

Visible light

| X-rays | Ultra-violet rays | | Infrared rays |

.1   10   100   1000

Wavelength of light in nanometers (billionths of a meter)

■ **Figure 3.9**
The visible portion of light ranges from a wavelength of about 400 nanometers (which looks violet) to nearly 700 nanometers (which looks red).

on this flat surface, so how do we perceive depth? We use perceptual processing to *infer* depth.

Depth perception tells us whether objects are near or far, which was the basis for some classic research by Eleanor Gibson and Richard Walk (1960) on the origins of depth perception. In their work, *babies were placed on a glass-covered platform, a device known as the* **visual cliff**. On one side of the platform, a checkerboard pattern appeared directly under the glass; on the other side, the pattern appeared several feet below the glass. The result was that the first side looked shallow but the other looked deep, like a cliff.

Mothers stood on each side of the visual cliff and tried to coax their infants across the deep or the shallow side. Most babies willingly crawled to their mothers when they stood on the shallow side. In contrast, almost every baby refused to cross the deep side, even when the mothers called them by name and tried to lure them with an attractive toy. Clearly, infants can perceive depth by the time they are old enough to crawl.

What about younger babies who cannot yet crawl? When babies as young as 6 weeks are simply placed on the visual cliff, their hearts beat more slowly when they are placed on the deep side of the cliff. Heart rate often decelerates when people notice something interesting, so this would suggest that 6-week-olds notice that the deep side is different. At 7 months, infants' heart rate accelerates, a sign of fear. Thus, although young babies can detect a difference between the shallow and the deep sides of the visual cliff, only older, crawling babies are actually afraid of the deep side (Campos et al., 1978).

How do infants infer depth, on the visual cliff or anywhere? They use several kinds of cues. *Among the first are* **kinetic cues**, *in which motion is used to estimate depth.* **Visual expansion** *refers to the fact that as an object moves closer, it fills an ever-greater proportion of the retina.* Visual expansion is why we flinch when someone unexpectedly tosses a soda can toward us, and it's what allows a batter to estimate when a baseball will arrive over the plate. *Another cue,* **motion parallax**, *refers to the fact that nearby moving objects move across our visual field faster than those at a distance.* Motion parallax is in action when you look out the side window in a moving car: Trees next to the road move rapidly across the visual field, but mountains in the distance move much more slowly. Babies use these cues in the first weeks after birth; for example, a 1-month-old baby will blink if a moving object looks as if it's going to hit him in the face (Nánez & Yonas, 1994).

Another cue becomes important at about 4 months. **Retinal disparity** *is based on the fact that the left and right eyes often see slightly different versions of the same scene.* You can demonstrate retinal disparity by touching your nose with your finger. If you look at your finger with one eye (closing the other eye), each eye has a very different view of your finger. But if you hold your finger at arm's length from your nose and repeat the demonstration, each eye has very similar views of your fingers. Thus, greater disparity in retinal images signifies that an object is close. By 4–6 months of age, infants use retinal disparity as a depth cue, correctly inferring that objects are nearby when disparity is great (Kellman & Arterberry, 2006).

By 7 months, infants use several cues for depth that depend on the arrangement of objects in the environment. *These are sometimes called* **pictorial cues** *because they're the same cues that artists use to convey depth in drawings and paintings.* Here are two examples of pictorial cues that 7-month-olds use to infer depth.

- **Linear perspective**: *Parallel lines come together at a single point in the distance.* Thus, we use the space between the lines as a cue to distance and, consequently, decide the tracks that are close together are farther away than the tracks that are far apart.

- **Texture gradient**: *The texture of objects changes from coarse and distinct for nearby objects to finer and less distinct for distant objects.* We judge that distinct flowers are close and that blurred ones are distant.

Infants avoid the "deep side" of the visual cliff, indicating that they perceive depth.

visual cliff
glass-covered platform that appears to have a "shallow" and a "deep" side; used to study infants' depth perception

kinetic cues
cues to depth perception in which motion is used to estimate depth

visual expansion
kinetic cue to depth perception that is based on the fact that an object fills an ever-greater proportion of the retina as it moves closer

motion parallax
kinetic cue to depth perception based on the fact that nearby moving objects move across our visual field faster than do distant objects

retinal disparity
way of inferring depth based on differences in the retinal images in the left and right eyes

pictorial cues
cues to depth perception that are used to convey depth in drawings and paintings

linear perspective
a cue to depth perception based on the fact that parallel lines come together at a single point in the distance

texture gradient
perceptual cue to depth based on the fact that the texture of objects changes from coarse and distinct for nearby objects to finer and less distinct for distant objects

Texture gradient is used to infer depth: We interpret the distinct flowers as being closer than the flowers with the coarse texture.

Linear perspective is one cue to depth: We interpret the railroad tracks that are close together as being more distant than the tracks that are far apart.

Not only do infants use visual cues to judge depth, they also use sound. Remember that infants correctly judge quieter objects to be more distant than louder objects. Given such an assortment of cues, it is not surprising that infants gauge depth so accurately.

Perceiving Objects

Perceptual processes enable us to interpret patterns of lines, textures, and colors as objects. That is, our perception actually creates an object from sensory stimulation. This is particularly challenging because we often see only parts of objects—nearby objects often obscure parts of more distant objects. Nevertheless, we recognize these objects despite this complexity in our visual environment.

Perception of objects is limited in newborns, but it develops rapidly in the first few months after birth (Johnson, 2001). By 4 months, infants use a number of cues to determine which elements go together to form objects. One important cue is motion: Elements that move together are usually part of the same object (Kellman & Arterberry, 2006). For example, at the left of ■ Figure 3.10, a pencil appears to be moving back and forth behind a colored square. If the square were removed, you would be surprised to see a pair of pencil stubs, as shown on the right side of the diagram. The common movement of the pencil's eraser and point leads us to believe that they're part of the same pencil.

Young infants, too, are surprised by demonstrations like this. If they see a display like the moving pencils, they will then look very briefly at a whole pencil, apparently because they expected it. In contrast, if after seeing the moving pencil they're shown the two pencil stubs, they look much longer—as if trying to figure out what happened (Amso & Johnson, 2006; Eizenman & Bertenthal, 1998). Evidently, even very young babies use common motion to create objects from different parts.

THINK ABOUT IT

Psychologists often refer to "perceptual-motor skills," which implies that the two are closely related. Based on what you've learned in this chapter, how might motor skills influence perception? How could perception influence motor skills?

Good Shoot / Super Stock

Many cues tell us that these are two objects, not one unusually shaped object: The two objects differ slightly in color, the glass of juice has a different texture than the orange, and the glass has a well-defined edge.

THINK ABOUT IT

When 6-month-old Sebastian watches his mother type on a keyboard, how does he know that her fingers and the keyboard are not simply one big unusual object?

Motion is one clue to object unity, but infants use others including color, texture, and aligned edges. As you can see in ■ Figure 3.11, infants more often group features together (i.e., believe they're part of the same object) when they're the same color, have the same texture, and their edges are aligned (Johnson, 2001).

Perceiving Faces

One object that's particularly important for infants is the human face. Some scientists argue that babies are innately attracted to stimuli that are facelike. The claim here is that some aspect of the face—perhaps three high-contrast blobs close together—constitutes a distinctive stimulus that is readily recognized, even by newborns. For example, newborns turn their eyes to follow a moving face more than they turn their eyes for nonface stimuli (Johnson, Grossman, & Farroni, 2008). This preference for faces and facelike stimuli supports the view that infants are innately attracted to faces. However, preference for tracking a moving face changes abruptly at about 4 weeks of age—infants now track all moving stimuli. One idea is that newborns' face tracking is a reflex, based on primitive circuits in the brain, that is designed to enhance attention to facelike stimuli. Starting at about 4 weeks, circuits in the brain's cortex begin to control infants' looking at faces and other stimuli (Morton & Johnson, 1991).

By 7 or 8 months, infants process faces in much the same way that adults do: as a configuration in which the internal elements (e.g., eyes, nose, mouth) are arranged and spaced in a unique way. Younger infants, in contrast, often perceive faces as an independent collection of facial features, as if they have not yet learned that the arrangement and spacing of features is critical (Bhatt et al., 2005; Schwarzer, Zauner, & Jovanovic, 2007).

Through the first 6 months after birth, infants have a very general prototype for a face—one that includes human and nonhuman faces (Pascalis, de Haan, & Nelson, 2002). However, between 6 and 12 months of age, infants fine-tune their prototype of a face so that it reflects those kinds of faces that are familiar in their environments. In the Spotlight on Research feature, we'll see that this age-related refinement of facial configurations results in a highly unusual outcome—3-month-olds outperform 9-month-olds!

The findings from the Spotlight on Research study suggest a crucial role for experience: Older infants' greater familiarity with faces leads to a more precise configura-

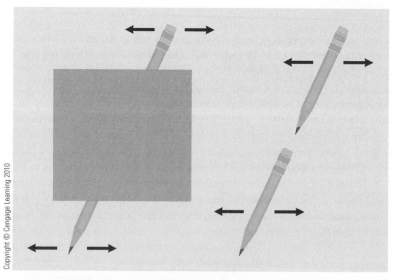

Copyright © Cengage Learning 2010

■ **Figure 3.10**

After infants have seen the pencil ends moving behind the square, they are surprised to see two pencils when the square is removed; this shows that babies use common motion as a way to determine what makes up an object.

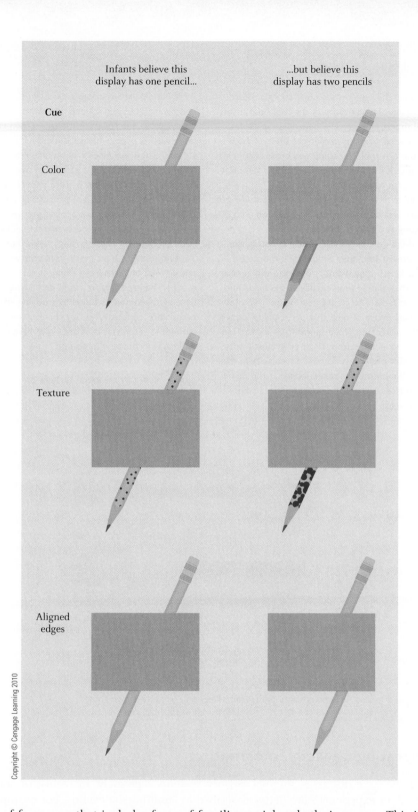

Infants believe this display has one pencil...

...but believe this display has two pencils

Cue

Color

Texture

Aligned edges

tion of faces, one that includes faces of familiar racial and ethnic groups. This interpretation is supported by the finding that individuals born in Asia but adopted as infants by European parents recognize European faces better than Asian faces (Sangrigoli et al., 2005).

These changes in face-recognition skill show the role of experience in fine-tuning infants' perception, a theme that will emerge again in the early phases of language learning (Chapter 4). And these improved face-recognition skills are adaptive, for they provide the basis for social relationships that infants form during the rest of the first year, which we'll examine in Chapter 5.

Who were the investigators, and what was the aim of the study? All human faces have the same basic features—eyes, nose, and mouth in the familiar configuration. But faces of different groups differ in their details. For example, people of African descent often have a relatively broad nose and people of Asian descent often have a fold of skin in the upper eyelid that covers the inner corner of the eye. As infants are exposed to faces in their environments and fine-tune their face-recognition processes, they might *lose* the ability to recognize some kinds of faces. For example, a young infant's broadly tuned face-recognition processes might work well for faces of Asian, African, and European individuals but an older infant's more finely tuned processes might only recognize faces from familiar groups. Testing this hypothesis was the aim of a study by David Kelly, Shaoying Liu, Kang Lee, Paul Quinn, Olivier Pascalis, Alan Slater, and Liezhong Ge (2009).

How did the investigators measure the topic of interest? Kelly and colleagues wanted to determine whether infants could recognize faces from different groups equally well. Consequently, they had infants view a photo of an adult's face (e.g., an Asian man). Then that face was paired with a novel face of the same group (e.g., a different Asian man). Experimenters recorded participants' looking at the two faces—the expectation was that, if the participants recognized the familiar face, they would look longer at the novel face.

Who were the participants in the study? The study included forty-six 3-month-olds and forty-one 9-month-olds from Hangzhou, China. (The researchers also tested 6-month-olds but for simplicity we are not describing their results.) One third of the infants at each age saw Asian faces; another third saw African faces; and another third saw European faces.

What was the design of the study? This study was experimental. The independent variables

included the type of face (African, Asian, European) and the familiarity of the face on the test trial (novel, familiar). The dependent variable was the participants' looking at the two faces on the test trial. The study was cross-sectional because it included 3-month-olds and 9-month-olds, each tested once.

Were there ethical concerns with the study? No. There was no obvious harm associated with looking at pictures of faces.

What were the results? If infants recognized the familiar face, they should look more at the novel face; if they did not recognize the familiar face, they should look equally at the novel and familiar faces. The graphs in ■ Figure 3.12 show the percentage of time that participants looked at the novel face. The 3-month-olds looked longer at novel faces from all three groups (more than 50% preference for the novel face). In contrast, 9-month-olds looked longer at the novel Asian faces but not the novel African or European faces.

What did the investigators conclude? Kelly and colleagues (2009) concluded that during the first year "...the ability to recognize own-race faces was retained, whereas the capacity to individuate other-race faces was simultaneously reduced, demonstrating a pattern of perceptual narrowing" (p. 111). That is, from experience infants finely tune their face-processing systems to include only faces from familiar groups.

What converging evidence would strengthen these conclusions? These findings show that 3-month-olds' face processing systems work equally well on faces from different racial groups. The investigators could determine how broadly the system is tuned by studying infants' recognition of faces of young children and comparing the responses of infants who have older siblings with those who do not.

 Go to Psychology CourseMate at www.cengagebrain.com to enhance your understanding of this research.

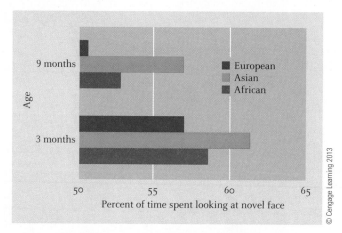

■ **Figure 3.12**
At 9 months, infants recognize faces only from their own group.

© Cengage Learning 2013

Integrating Sensory Information

So far, we have discussed infants' sensory systems separately. In reality, of course, most infant experiences are better described as "multimedia events." For example, a nursing mother provides visual and taste cues to her baby. A rattle stimulates vision, hearing, and touch. In fact, much stimulation is not specific to one sense but spans multiple senses. Temporal information, such as duration or tempo, can be seen or heard. For example, you can detect the rhythm of a person clapping by seeing the hands meet or by hearing the sound of hands striking. Similarly, the texture of a surface—whether it's rough or smooth—can be detected by sight or by feel.

Infants readily perceive many of these relations. For example, infants can recognize visually an object that they have only touched previously (Sann & Streri, 2007). Similarly, they can detect relations between information presented visually and auditorily. For example, babies look longer when an object's motion matches its sound (it makes higher-pitched sounds while rising but lower-pitched sounds while falling) than when it doesn't (Walker et al., 2010). And they can link the temporal properties of visual and auditory stimulation, such as duration and rhythm (Lewkowicz, 2000). Finally, they link their own body movement to their perceptions of musical rhythm, giving new meaning to the phrase "feel the beat, baby!" (Gerry, Faux, & Trainor, 2010).

Traditionally, coordinating information from different senses (e.g., vision with hearing, vision with touch) was thought to be a challenging task for infants. But a new view is that cross-modal perception is actually *easier* for infants because regions in the brain devoted to sensory processing are not yet specialized in infancy. For example, regions in an adult's brain that respond only to visual stimuli respond to visual *and* auditory input in the infant's brain (Spector & Maurer, 2009). *And some researchers have argued that the infant's sensory systems are particularly attuned to* **intersensory redundancy**, *that is, to information that is presented simultaneously to different sensory modes* (Bahrick & Lickliter, 2002; Bahrick, Lickliter, and Flom, 2004). Perception is best when information is presented redundantly to multiple senses. When an infant sees and hears the mother clapping (visual, auditory information), he focuses on the information conveyed to both senses and pays less attention to information that's only available in one sense, such as the color of the mother's nail polish or the sounds of her humming along with the tune. Or the infant can learn that the mom's lips are chapped from seeing the flaking skin and by feeling the roughness as the mother kisses him. According to intersensory redundancy theory, it's as if infants follow the rule "Any information that's presented in multiple senses must be important, so pay attention to it!" (Flom & Bahrick, 2007).

Integrating information from different senses is yet another variation on the theme that has dominated this chapter: Infants' sensory and perceptual skills are impressive. Darla's newborn daughter, from the opening vignette, can definitely smell, taste, and feel pain. She can distinguish sounds. Her vision is a little blurry now but will improve rapidly; in a few months, she'll see the full range of colors and perceive depth. In short, Darla's daughter, like most infants, is exceptionally well prepared to begin to make sense out of her environment.

A mother who breast-feeds provides her baby with a multimedia event: the baby sees, smells, hears, feels, and tastes her!

intersensory redundancy
infants' sensory systems are attuned to information presented simultaneously to different sensory modes

Test Yourself

RECALL

1. Infants respond negatively to substances that taste sour or _____.

2. Infants respond to _____ with a high-pitched cry that is hard to soothe.

3. Infants' hearing is best for sounds that have the pitch of _____.

4. At age _____, infants' acuity is like that of an adult with normal vision.

5. _____ are specialized neurons in the retina that are sensitive to color.

6. The term _____ refers to the fact that images of an object in the left and right eyes differ for nearby objects.

7. When elements consistently move together, infants decide that they are _____.

8. Infants readily integrate information from different senses, and their sensory systems seem to be particularly attuned to _____.

INTERPRET

Compare the impact of nature and nurture on the development of infants' sensory and perceptual skills.

APPLY

Perceptual skills are quite refined at birth and mature rapidly. What evolutionary purposes are served by this rapid development?

Recall answers: (1) bitter, (2) pain, (3) human speech, (4) 1 year, (5) Cones, (6) retinal disparity, (7) part of the same object, (8) information presented redundantly to multiple senses

LEARNING OBJECTIVES

- When do children begin to realize that they exist?
- What are toddlers' and preschoolers' self-concepts like?

- When do preschool children begin to acquire a theory of mind?

When Ximena brushes her teeth, she puts her 20-month-old son, Christof, in an infant seat facing the bathroom mirror. She's been doing this for months, and Christof always seems to enjoy looking at the images in the mirror. Lately, he seems to pay special attention to his own reflection. Ximena thinks that sometimes Christof deliberately frowns or laughs just to see what he looks like. Is this possible, Ximena wonders, or is her imagination simply running wild?

AS INFANTS' PHYSICAL, MOTOR, AND PERCEPTUAL SKILLS GROW, they learn more and more about the world around them. As part of this learning, infants and toddlers begin to realize that they exist independently of other people and objects in the environment and that their existence continues over time. In this last section, you'll see how children become self-aware and learn what Christof knows about himself.

Origins of Self-Concept

When do children begin to understand that they exist? Measuring the onset of this awareness is not easy. Obviously, we can't simply ask a 3-year-old, "So, tell me, when did you first realize you existed and weren't just part of the furniture?" Investigators need a less direct approach, and a mirror offers one route. Babies sometimes touch the face in the mirror or wave at it, but none of their behaviors indicate that they recognize themselves in the mirror. Instead, babies act as if the face in the mirror is simply a very interesting stimulus.

How would we know that infants recognize themselves in a mirror? One clever approach is to have mothers place a red mark on their infant's nose; they do this surreptitiously, while wiping the baby's face. Then the infant is returned to the mirror. Many 1-year-olds touch the red mark on the mirror, showing that they notice the mark on the face in the mirror. By 15 months, however, an important change occurs: Babies see the red mark in the mirror, then reach up and touch their own noses. By age 2, virtually all children do this (Bullock & Lütkenhaus, 1990; Lewis & Brooks-Gunn, 1979). When these older children notice the red mark in the mirror, they understand that the funny looking nose in the mirror is their own!

We don't need to rely solely on the mirror task to know that self-awareness emerges between 18 and 24 months. During this same period, toddlers look more at photographs of themselves than at photos of other children. They also refer to themselves by name or with a personal pronoun, such as "I" or "me," and sometimes they know their age and their gender. These changes, which often occur together, suggest that self-awareness is well established in most children by age 2 (Lewis & Ramsay, 2004; Moore, 2007).

Soon toddlers and young children begin to recognize continuity in the self over time; the "I" in the present is linked to the "I" in the past (Nelson, 2001). Awareness of a self that is extended in time is fostered by conversations with parents about the past and the future. Through such

Not until 15 to 18 months of age do babies recognize themselves in the mirror, which is an important step in becoming self-aware.

conversations, a 3-year-old celebrating a birthday understands that she's an older version of the same person who had a birthday a year previously.

Children's growing awareness of a self extended in time is also revealed by their understanding of ownership (Fasig, 2000). When a toddler sees his favorite toy and says "mine," this implies awareness of continuity of the self over time: "In the past, I played with that." And when toddlers say "Mine!" they are often not being aggressive or selfish; instead, "mine" is a way of indicating ownership in the process defining themselves. They are not trying to deny the toy to another but simply saying that playing with this toy is part of who they are (Levine, 1983).

Once children fully understand that they exist, they begin to wonder who they are. They want to define themselves. Throughout the preschool years, possessions continue to be one of the ways in which children define themselves. Preschoolers are also likely to mention physical characteristics ("I have blue eyes"), their preferences ("I like spaghetti"), and their competencies ("I can count to 50"). What these features have in common is a focus on a child's characteristics that are observable and concrete (Harter, 2006).

As children enter school, their self-concepts become even more elaborate (Harter, 1994), changes that we'll explore in Chapter 9.

Preschool children use their toys to define themselves; saying "Mine" is shorthand for "This is my doll and I like to play with dolls a lot!"

Theory of Mind

As youngsters gain more insights into themselves as thinking beings, they begin to realize that people have thoughts, beliefs, and intentions. They also understand that thoughts, beliefs, and intentions often cause people to behave as they do. Amazingly, even infants understand that people's behavior is often intentional—designed to achieve a goal. Imagine a father who says, "Where are the crackers?" in front of his 1-year-old daughter and then begins opening kitchen cabinets, moving some objects to look behind them. Finding the box of crackers, he says, "There they are!" An infant who understands intentionality would realize how her father's actions (searching, moving objects) were related to the goal of finding the crackers.

Many clever experiments have revealed that 1-year-olds do indeed have this understanding of intentionality. For example, in one study infants observed an adult reaching over a barrier for a ball, but failing because the ball was just out of reach. Then the barrier was removed and infants saw an adult using the same "over the barrier" reaching motion or reaching directly for the ball; in both cases, the adult grasped the ball. By 10 months, infants were surprised to see the adult relying on the "over the barrier" reach when it was no longer needed. In other words, with the barrier removed, infants expected to see the adult reach directly because that was the best way to achieve the goal of getting the ball; they were surprised when the actor relied on the familiar but no longer necessary method of reaching (Brandone & Wellman, 2009).

From this early understanding of intentionality, young children's naïve psychology expands rapidly. *Between 2 and 5, children develop a* **theory of mind**, *a naïve understanding of the relations between mind and behavior*. One of the leading researchers on theory of mind, Henry Wellman (1993, 2002), believes that children's theory of mind moves through three phases during the preschool years. In the earliest phase, 2-year-olds are aware of desires and often speak of their wants and likes, as in "Lemme see" or "I wanna sit." And they often link their desires to their behavior, such as "I happy there more cookies" (Wellman, 1993). Thus, by age 2, children understand that people have desires and that desires can cause behavior.

By about age 3, children clearly distinguish the mental world from the physical world. For example, if told about a girl who has a cookie and another who is thinking

theory of mind
ideas about connections between thoughts, beliefs, intentions, and behavior that create an intuitive understanding of the link between mind and behaviour

THINK ABOUT IT

Suppose you believe that a theory of mind develops faster when preschoolers spend much time with other children. What sort of correlational study would you devise to test this hypothesis? How could you do an experimental study to test the same hypothesis?

about a cookie, 3-year-olds know that only the first girl can see, touch, and eat her cookie (Harris et al., 1991). Most 3-year-olds also use "mental verbs" like "think," "believe," "remember," and "forget," which suggests that they have some understanding of different mental states (Bartsch & Wellman, 1995). Although 3-year-olds talk about thoughts and beliefs, they nevertheless emphasize desires when trying to explain why people act as they do.

Not until 4 years of age do mental states really take center stage in children's understanding of their own actions and the actions of others. That is, by age 4, children understand that behavior is often based on a person's beliefs about events and situations, even when those beliefs are wrong. This developmental transformation is particularly evident when children are tested on false-belief tasks such as the one shown in ■ Figure 3.13. In all false-belief tasks, a situation is set up so that the child being tested has accurate information but someone else does not. For example, in the story in Figure 3.13, the child being tested knows the ball is really in the box, but Sally, the girl in the story, believes that the ball is still in the basket. Remarkably, although 4-year-olds correctly say that Sally will look for the ball in the basket (i.e., will act on her false belief), most 3-year-olds claim that she will look for the ball in the box. The 4-year-olds understand that Sally's behavior is based on her beliefs even though her beliefs are incorrect (Frye, 1993).

This basic developmental progression is remarkably robust. Wellman, Cross, and Watson (2001) conducted a meta-analysis of approximately 175 studies in which more than 4,000 young children were tested on false-belief tasks. Before 3½ years, children typically make the false-belief error: Attributing their own knowledge of the ball's location to Sally, they say she will search in the correct location. Yet a mere 6 months later, children now understand that Sally's false belief will cause her to look for the ball in the box. And this general developmental pattern is evident in many different cultures around the world (Callaghan et al., 2005; Liu et al., 2008).

This pattern therefore signifies a fundamental change in children's understanding of the centrality of beliefs in a person's thinking about the world. By age 4 children "realize that people not only have thoughts and beliefs, but also that thoughts and beliefs are crucial to explaining why people do things; that is, actors' pursuits of their desires are inevitably shaped by their beliefs about the world" (Bartsch & Wellman, 1995, p. 144).

You can see preschool children's growing understanding of false belief in the next Real People feature.

Real People
Applying Human Development

"Seeing Is Believing . . ." for 3-Year-Olds

Preschoolers gradually recognize that people's behavior is sometimes guided by mistaken beliefs. We once witnessed an episode at a day-care center that documented this growing understanding. After lunch, Karen, a 2-year-old, saw ketchup on the floor and squealed, "blood, blood!" Lonna, a 3-year-old, said in a disgusted tone, "It's not blood—it's ketchup." Then, Shenan, a 4-year-old, interjected, "Yeah, but Karen *thought* it was blood." A similar incident took place a few weeks later, on the day after Halloween. This time Lonna put on a monster mask and scared Karen. When Karen began to cry, Lonna said, "Oh stop. It's just a mask." Shenan broke in again, saying, "You know it's just a mask. But she *thinks* it's a monster." In both cases, only Shenan understood that Karen's behavior was based on her beliefs (that the ketchup is blood and that the monster is real), even though her beliefs were false.

The early stages of children's theory of mind seem clear. But just *how* this happens is very much a matter of debate. One view emphasizes the contribution of language, which develops rapidly during the same years that theory of mind emerges (as we'll see in Chapter 4). Some scientists believe that children's language skills contribute to growth of theory of mind, perhaps reflecting the benefit of an expanding vocabulary that includes verbs describing mental states, such as *think, know, believe* (Pascual et al., 2008). Or the benefits may reflect children's mastery of

Copyright © Cengage Learning 2010

This is Sally. Sally has a basket.

This is Anne. Anne has a box.

Sally puts her ball in her basket.

Sally goes out for a walk.

Anne takes the ball from the basket and puts it into the box.

Now Sally comes back. She wants to play with her ball. Where will she look for her ball?

■ **Figure 3.13**
In a false-belief task, most 3-year-olds say that Sally will look for the ball in the box, showing that they do not understand how people can act on their beliefs (where the ball is) even when those beliefs are wrong.

grammatical forms that can be used to describe a setting where a person knows that another person has a false belief (Low, 2010).

A different view is that a child's theory of mind emerges from interactions with other people, interactions that provide children with insights into different mental states (Dunn & Brophy, 2005; Peterson & Slaughter, 2003). Through conversations with parents and older siblings that focus on other people's mental states, children learn facts of mental life, and this helps children to see that others often have different perspectives than they do. In other words, when children frequently participate in conversations that focus on other people's moods, their feelings, and their intentions, they learn that people's behavior is based on their beliefs, regardless of the accuracy of those beliefs (Taumoepeau & Ruffman, 2008).

These different views are also used to explain why theory of mind develops very slowly in children with autism, which is our last topic in this chapter.

Theory of Mind in Children with Autism

Autism is the most serious of a family of disorders known as autism spectrum disorders (ASD). Individuals with ASD acquire language later than usual and their speech often echoes what others say to them; they sometimes become intensely interested in objects (e.g., making the same actions with a toy over and over); and they often seem uninterested in other people and when they do interact, those exchanges are often

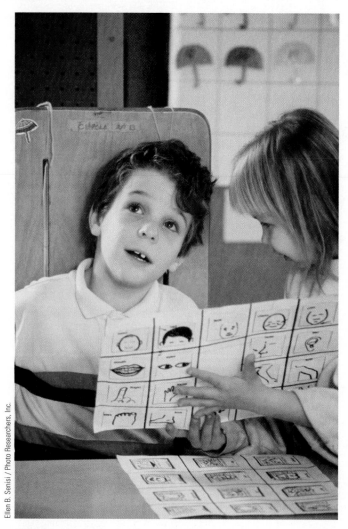

Ellen B. Senisi / Photo Researchers, Inc.

Children with autism master language later than usual and are often much more interested in objects than in people.

awkward as if individuals with ASD aren't following the rules that govern social interactions. Symptoms usually emerge early in life, typically by 18–24 months of age. Roughly one out of every 200–300 U.S. children is diagnosed with ASD; about 80% of them are boys (Mash & Wolfe, 2010). ASD is heritable and many studies point to atypical brain functioning, perhaps due to abnormal levels of neurotransmitters (NINDS, 2009).

Children with ASD grasp false belief very slowly and this performance leads some researchers to conclude that the absence of a theory of mind—sometimes called "mindblindness" (Baron-Cohen, 1995)—is the defining characteristic of ASD (Tager-Flusberg, 2007). Other scientists aren't convinced. Although no one doubts that autistic children find false-belief tasks puzzling, some scientists say that mindblindness is a by-product of other deficits and not the cause of the symptoms associated with ASD. One idea is that ASD reflects problems in inhibiting irrelevant actions and in shifting smoothly between actions (Pellicano, 2010). Another idea emphasizes a focused processing style that is common in ASD. For example, children with ASD find hidden objects faster than typically developing children do (Joseph et al., 2009). But this emphasis on perceptual details usually comes at the expense of maintaining a coherent overall picture. Consequently, in social interactions, children with ASD may focus on one facet of another person's behavior (e.g., her gestures) but ignore other verbal and nonverbal cues (e.g., speech, facial expressions, body language) that collectively promote fluid interactions. Research to evaluate these claims is still ongoing; it's likely that the answers will indicate that multiple factors contribute to ASD.

ASD can't be cured. However, therapy can be used to improve language and social skills in children with autism. In addition, medications can be used to reduce some of the symptoms, such as repetitive behavior (NINDS, 2009). When ASD is diagnosed early and autistic children grow up in supportive, responsive environments and receive appropriate treatments, they can lead satisfying and productive lives.

Test Yourself

RECALL

1. Apparently children are first self-aware at age 2 because this is when they first recognize themselves in a mirror and in photographs and when they first use _____.

2. During the preschool years, children's self-concepts emphasize _____, physical characteristics, preferences, and competencies.

3. Unlike 4-year-olds, most 3-year-olds don't understand that other people's behavior is sometimes based on _____.

INTERPRET

Compare and contrast different explanations of the growth of theory of mind during the preschool years.

APPLY

Self-concept emerges over the same months that toddlers show rapid gains in locomotor skills. How might changes in locomotor skill contribute to a toddler's emerging sense of self?

Recall answers: (1) personal pronouns such as "I" and "me," (2) possessions, (3) false beliefs

3.1 THE NEWBORN

How do reflexes help newborns interact with the world?

- Babies are born with a number of different reflexes. Some help them adjust to life outside of the uterus, some help protect them from danger, and some serve as the basis for later voluntary motor behavior.

How do we determine whether a baby is healthy and adjusting to life outside the uterus?

- The Apgar scale measures five vital signs to determine a newborn baby's physical well-being. The Neonatal Behavioral Assessment Scale provides a comprehensive evaluation of a baby's behavioral and physical status.

What behavioral states are common among newborns?

- Newborns spend their day in one of four states: alert inactivity, waking activity, crying, and sleeping. A newborn's crying includes a basic cry, a mad cry, and a pain cry. The best way to calm a crying baby is by putting it on the shoulder and rocking.

- Newborns spend approximately two thirds of every day asleep and go through a complete sleep–wake cycle once every 4 hours. By 3 or 4 months, babies sleep through the night. Newborns spend about half of their time asleep in REM sleep, an active form of sleep that may stimulate growth in the nervous system.

- Some healthy babies die from sudden infant death syndrome. Factors that contribute to SIDS are prematurity, low birth weight, and smoking. Also, babies are vulnerable to SIDS when they sleep on their stomach and when they are overheated. The goal of the Back to Sleep campaign is to prevent SIDS by encouraging parents to have infants sleep on their backs.

What are the different features of temperament? Do they change as children grow?

- Temperament refers to a consistent style or pattern to an infant's behavior. Modern theories list two to six dimensions of temperament, including (for example) extroversion and negative affect. Temperament is influenced both by heredity and by environment and is a reasonably stable characteristic of infants and young children.

3.2 PHYSICAL DEVELOPMENT

How do height and weight change from birth to 2 years of age?

- Physical growth is particularly rapid during infancy, but babies of the same age differ considerably in their height and weight. Size at maturity is largely determined by heredity.

- Growth follows the cephalocaudal principle, in which the head and trunk develop before the legs. Consequently, infants and young children have disproportionately large heads and trunks.

What nutrients do young children need? How are they best provided?

- Infants must consume a large number of calories relative to their body weight, primarily because of the energy required for growth. Breast-feeding and bottle-feeding both provide babies with adequate nutrition.

- Malnutrition is a worldwide problem that is particularly harmful during infancy, when growth is so rapid. Treating malnutrition adequately requires improving children's diets and training their parents to provide stimulating environments.

What are nerve cells, and how are they organized in the brain?

- A nerve cell, called a neuron, includes a cell body, a dendrite, and an axon. The mature brain consists of billions of neurons, organized into nearly identical left and right hemispheres connected by the corpus callosum. The cerebral cortex regulates most of the functions we think of as distinctively human. The frontal cortex is associated with personality and goal-directed behavior; the left hemisphere of the cortex with language; and the right hemisphere of the cortex with nonverbal processes such as perceiving emotions.

How does the brain develop? When does it begin to function?

- Brain specialization is evident in infancy; further specialization involves more focused brain areas and narrowing of stimuli that trigger brain activity. Different systems specialize at different rates. Specialization depends upon stimulation from the environment. The relative lack of specialization in the immature brain makes it better able to recover from injury

3.3 MOVING AND GRASPING: EARLY MOTOR SKILLS

What are the component skills involved in learning to walk? At what age do infants master them?

- Infants acquire a series of locomotor skills during their first year, culminating in walking a few months after the first birthday. Like most motor skills, learning to walk involves differentiation of individual skills, such as maintaining balance and using the legs alternately, and then integrating these skills into a coherent whole.

How do infants learn to coordinate the use of their hands?

- Infants first use only one hand at a time, then both hands independently, then both hands in common actions, and finally, at about 5 months of age, both hands in different actions with a common purpose.

- Most people are right-handed, a preference that emerges after the first birthday and becomes well established during the preschool years. Handedness is determined by heredity but can also be influenced by cultural values.

3.4 COMING TO KNOW THE WORLD: PERCEPTION

Are infants able to smell, to taste, and to experience pain?

- Newborns are able to smell, and some can recognize their mother's odor; they also taste, preferring sweet substances and responding negatively to bitter and sour tastes.

- Infants respond to touch. They probably experience pain because their responses to painful stimuli are similar to those of older children.

How well can infants hear?

- Babies can hear. More important, they can distinguish different sounds.

How well can infants see? Can they see color and depth?

- A newborn's visual acuity is relatively poor, but 1-year-olds can see as well as an adult with normal vision.

- Color vision develops as different sets of cones begin to function, a process that seems to be complete by 3 or 4 months of age. Infants perceive depth based on kinetic cues, retinal disparity, and pictorial cues. They also use motion to recognize objects.

- Infants are attracted to faces and experience leads infants to form a face template based on faces they see often.

How do infants process and combine information from different sensory modalities, such as between vision and hearing?

- Infants coordinate information from different senses. They can recognize, by sight, an object they've felt previously. Infants are often particularly attentive to information presented redundantly to multiple senses.

3.5 BECOMING SELF-AWARE

When do children begin to realize that they exist?

- Beginning at about 15 months, infants begin to recognize themselves in the mirror, which is one of the first signs of self-recognition. They also begin to prefer looking at pictures of themselves, begin referring to themselves by name (or using personal pronouns), and sometimes know their age and gender. Evidently, by 2 years of age, most children are self-aware.

What are toddlers' and preschoolers' self-concepts like?

- Preschoolers often define themselves in terms of observable characteristics, such as possessions, physical characteristics, preferences, and competencies.

When do preschool children begin to acquire a theory of mind?

- Theory of mind—which refers to a person's ideas about connections between thoughts, beliefs, intentions, and behavior—develops rapidly during the preschool years. Most 2-year-olds know that people have desires and that desires can cause behavior. By age 3, children distinguish the mental world from the physical world but still emphasize desire in explaining others' actions. By age 4, however, children understand that behavior is based on beliefs about the world, even when those beliefs are wrong. Children with autism have a limited theory of mind.

KEY TERMS

reflexes (82)
alert inactivity (84)
waking activity (84)
crying (84)
sleeping (84)
basic cry (84)
mad cry (84)
pain cry (84)
irregular or rapid-eye-movement (REM) sleep (85)
regular (nonREM) sleep (86)
sudden infant death syndrome (SIDS) (86)
temperament (87)
malnourished (92)
neuron (93)
cell body (93)
dendrite (93)
axon (94)

terminal buttons (94)
neurotransmitters (94)
cerebral cortex (94)
hemispheres (94)
corpus callosum (94)
frontal cortex (94)
neural plate (94)
myelin (95)
synaptic pruning (95)
electroencephalography (96)
functional magnetic resonance imaging (fMRI) (96)
experience-expectant growth (97)
experience-dependent growth (97)
motor skills (98)
locomote (98)
fine motor skills (99)
toddling (99)

toddlers (99)
dynamic systems theory (100)
differentiation (101)
integration (101)
perception (105)
visual acuity (106)
cones (107)
visual cliff (108)
kinetic cues (108)
visual expansion (108)
motion parallax (108)
retinal disparity (108)
pictorial cues (108)
linear perspective (108)
texture gradient (108)
intersensory redundancy (113)
theory of mind (115)

Log in to **www.cengagebrain.com** to access the resources your instructor requires. For this book, you can access:

Psychology CourseMate

■ CourseMate brings course concepts to life with interactive learning, study, and exam preparation tools that support the printed textbook. A textbook-specific website, Psychology CourseMate includes an integrated interactive eBook and other interactive learning tools including quizzes, flashcards, videos, and more.

CENGAGENOW™

■ CengageNOW Personalized Study is a diagnostic study tool containing valuable text-specific resources—and because you focus on just what you don't know, you learn more in less time to get a better grade.

WebTUTOR™

■ More than just an interactive study guide, WebTutor is an anytime, anywhere customized learning solution with an eBook, keeping you connected to your textbook, instructor, and classmates.

The Emergence of Thought and Language:

Cognitive Development in Infancy and Early Childhood

4

ON THE TV SHOW *Family Guy,* Stewie is a 1-year-old who can't stand his mother (Stewie: "Hey, mother, I come bearing a gift. I'll give you a hint. It's in my diaper and it's not a toaster.") and hopes to dominate the world. Much of the humor, of course, turns on the idea that babies are capable of sophisticated thinking and just can't express it. But what thoughts do lurk in the mind of an infant who is not yet speaking? How does cognition develop during infancy and early childhood? What makes these changes possible?

These questions provide the focus of this chapter. We begin with what has long been considered the definitive account of cognitive development, Jean Piaget's theory, as well as a later variant, core knowledge theories. The next two sections of the chapter concern alternative accounts of cognitive development. One account, the information-processing perspective, traces children's emerging cognitive skills in many specific domains, including memory skills. The other, Lev Vygotsky's theory, emphasizes the cultural origins of cognitive development and explains why children sometimes talk to themselves as they play or work.

Throughout development, children express their thoughts in oral and written language. In the last section of this chapter, you'll see how children master the sounds, words, and grammar of their native language.

LEARNING OBJECTIVES

- According to Piaget, how do schemes, assimilation, and accommodation provide the foundation for cognitive development throughout the life span?

- How does thinking become more advanced as infants progress through the sensorimotor stage?

- What are the distinguishing characteristics of thinking during the preoperational stage?

- What are the strengths and weaknesses of Piaget's theory?

- How have contemporary researchers extended Piaget's theory?

Three-year-old Jamila loves talking to her grandmother ("Gram") on the telephone. Sometimes these conversations are not very successful because Gram asks questions and Jamila replies by nodding her head "yes" or "no." Jamila's dad has explained that Gram (and others on the phone) can't see her nodding—that she needs to say "yes" or "no." But Jamila invariably returns to head-nodding. Her dad can't see why such a bright and talkative child doesn't realize that nodding is meaningless over the phone.

WHY DOES JAMILA INSIST ON NODDING HER HEAD WHEN SHE'S TALKING ON THE PHONE? This behavior is quite typical, according to the famous Swiss psychologist Jean Piaget (1896–1980). In Piaget's theory, children's thinking progresses through four qualitatively different stages. In this section, we'll begin by describing some of the general features of Piaget's theory, then examine Piaget's account of thinking during infancy and during the preschool years, and finally consider some of the strengths and weaknesses of the theory.

Basic Principles of Cognitive Development

Piaget believed that children are naturally curious. They constantly want to make sense of their experience and, in the process, construct their understanding of the world. For Piaget, children at all ages are like scientists in that they create theories about how the world works. Of course, children's theories are often incomplete. Nevertheless, children's theories are valuable to them because they make the world seem more predictable.

According to Piaget, children understand the world with **schemes**, *psychological structures that organize experience.* Schemes are mental categories of related events, objects, and knowledge. During infancy, most schemes are based on actions. That is, infants group objects based on the actions they can perform on them. For example, infants suck and grasp, and they use these actions to create categories of objects that can be sucked and objects that can be grasped.

Schemes are just as important after infancy, but they are now based primarily on functional or conceptual relationships, not action. For example, preschoolers learn that forks, knives, and spoons form a functional category of "things I use to eat." Or they learn that dogs, cats, and goldfish form a conceptual category of "pets."

Like preschoolers, older children and adolescents have schemes based on functional and conceptual schemes. But they also have schemes that are based on increasingly abstract properties. For example, an adolescent might put fascism, racism, and sexism in a category of "ideologies I despise."

Thus, schemes of related objects, events, and ideas are present throughout development. But as children develop, their rules for creating schemes shift from physical activity to functional, conceptual, and, later, abstract properties of objects, events, and ideas.

Assimilation and Accommodation

Schemes change constantly, adapting to children's experiences. In fact, intellectual adaptation involves two processes working together: assimilation and accommodation. **Assimilation** *occurs when new experiences are readily incorporated into existing schemes.* Imagine a baby who has the familiar grasping scheme. She will soon dis-

scheme
according to Piaget, a mental structure that organizes information and regulates behavior

assimilation
according to Piaget, taking in information that is compatible with what one already knows

cover that the grasping scheme also works well on blocks, toy cars, and other small objects. Extending the existing grasping scheme to new objects illustrates assimilation. **Accommodation** *occurs when schemes are modified based on experience.* Soon the infant learns that some objects can only be lifted with two hands and that some can't be lifted at all. Changing the scheme so that it works for new objects (e.g., using two hands to grasp heavy objects) illustrates accommodation.

Assimilation and accommodation are often easier to understand when you remember Piaget's belief that infants, children, and adolescents create theories to try to understand events and objects around them. The infant whose theory is that objects can be lifted with one hand finds that her theory is confirmed when she tries to pick up small objects, but she's in for a surprise when she tries to pick up a heavy book. The unexpected result forces the infant, like a good scientist, to revise her theory to include this new finding.

The Real People feature shows how accommodation and assimilation allow young children to understand their worlds.

accommodation
according to Piaget, changing existing knowledge based on new knowledge

equilibration
according to Piaget, a process by which children reorganize their schemes to return to a state of equilibrium when disequilibrium occurs

Real People
Applying Human Development
Learning About Butterflies: Accommodation and Assimilation in Action

When Ethan, an energetic 2½-year-old, saw a monarch butterfly for the first time, his mother, Kat, told him, "Butterfly, butterfly; that's a butterfly, Ethan." A few minutes later, a zebra swallowtail butterfly landed on a nearby bush and Ethan shouted in excitement, "Butterfly, Mama, butterfly!" A bit later, a moth flew out of another bush; with even greater excitement in his voice, Ethan shouted, "Butterfly, Mama, more butterfly!" As Kat was telling Ethan, "No, honey, that's a moth, not a butterfly," she marveled at how rapidly

Ethan seemed to grasp new concepts with so little direction from her. How was this possible?

Piaget's explanation would be that when Kat named the monarch butterfly for Ethan, he formed a simple theory, something like "butterflies are bugs with big wings." The second butterfly differed in color but was still a bug with big wings, so it was readily *assimilated* into Ethan's new theory of butterflies. However, when Ethan referred to the moth as a butterfly, Kat corrected him. Presumably, Ethan was then

forced to *accommodate* to this new experience. The result was that he changed his theory of butterflies to make it more precise; the new theory might be something like "butterflies are bugs with thin bodies and big, colorful wings." He also created a new theory, something like "a moth is a bug with a bigger body and plain wings." Accommodation and assimilation work together to help Ethan make sense of his experiences.

Equilibration and Stages of Cognitive Development

Assimilation and accommodation are usually in balance, or equilibrium. Children find that many experiences are readily assimilated into their existing schemes but that they sometimes need to accommodate their schemes to adjust to new experiences. This balance between assimilation and accommodation was illustrated by our infant with a theory about lifting objects. Periodically, however, this balance is upset, and a state of disequilibrium results. That is, children discover that their current schemes are not adequate because they are spending too much time accommodating and much less time assimilating. *When disequilibrium occurs, children reorganize their schemes to return to a state of equilibrium, a process that Piaget called* **equilibration**. To restore the balance, current but outmoded ways of thinking are replaced by a qualitatively different, more advanced set of schemes.

One way to understand equilibration is to return to the metaphor of the child as a scientist. As we discussed in Chapter 1, good scientific theories readily explain some phenomena but usually must be revised to explain others. Children's theories allow them to understand many experiences by predicting, for example, what will happen ("It's morning, so it's time for breakfast") or who will do what ("Mom's gone to work, so Dad will take me to school"), but the theories must be modified when predictions go awry ("Dad thinks I'm old enough to walk to school, so he won't take me").

Sometimes scientists find that their theories contain critical flaws that can't be fixed simply by revising; instead, they must

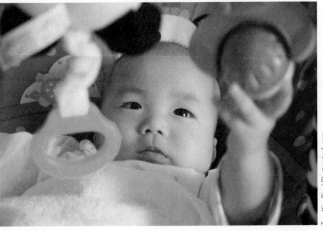

This baby will learn that many objects can be grasped easily with one hand—illustrating assimilation—but will also discover that bigger, heavier objects can be grasped only with two hands—illustrating accommodation.

create a new theory that draws upon the older theory but is fundamentally different. For example, when the astronomer Copernicus realized that the earth-centered theory of the solar system was fundamentally wrong, his new theory built on the assumption that the sun is the center of the solar system. In much the same way, periodically children reach states in which their current theories seem to be wrong much of the time, so they abandon these theories in favor of more advanced ways of thinking about their physical and social worlds.

According to Piaget, these revolutionary changes in thought occur three times over the life span, at approximately 2, 7, and 11 years of age. This divides cognitive development into the following four stages:

Period of Development	Age Range
Sensorimotor period	Infancy (0–2 years)
Preoperational period	Preschool and early elementary school years (2–7 years)
Concrete operational period	Middle and late elementary school years (7–11 years)
Formal operational period	Adolescence and adulthood (11 years and up)

The ages listed are only approximate. Some youngsters move through the periods more rapidly than others, depending on their ability and their experience. However, the only route to formal operations—the most sophisticated type of thought—is through the first three periods, in sequence. Sensorimotor thinking always gives rise to preoperational thinking; a child cannot "skip" preoperational thinking and move directly from the sensorimotor to the concrete operational period.

In the next few pages of this chapter, we consider Piaget's account of sensorimotor and preoperational thinking, the periods from birth to approximately 7 years of age. In Chapter 6, we will return to Piaget's theory to examine his account of concrete and formal operational thinking in older children and adolescents.

Sensorimotor Thinking

Piaget (1951, 1952, 1954) believed that the first 2 years of life form a distinct phase in human development. *The* **sensorimotor period**, *from birth to roughly 2 years of age, is the first of Piaget's four periods of cognitive development.* In the 24 months of this stage, infants' thinking progresses remarkably along three important fronts.

Adapting to and Exploring the Environment

Newborns respond reflexively to many stimuli, but between 1 and 4 months reflexes are first modified by experience. An infant may inadvertently touch his lips with his thumb, thereby initiating sucking and the pleasing sensations associated with sucking. Later, the infant tries to re-create these sensations by guiding his thumb to his mouth. Sucking no longer occurs only reflexively when a mother places a nipple at the infant's mouth; instead, the infant has found a way to initiate sucking himself.

At about 8 months, infants reach a watershed: the onset of deliberate, intentional behavior. For the first time, the "means" and "end" of activities are distinct. If, for example, a father places his hand in front of a toy, an infant will move his father's hand to be able to play with the toy. "The moving the hand" scheme is the means to achieve the goal of "grasping the toy." Using one action as a means to achieve another end is the first indication of purposeful, goal-directed behavior during infancy.

Beginning at about 12 months, infants become active experimenters. An infant may deliberately shake a number of different objects trying to discover which produce sounds and which do not. Or an infant may decide to drop different objects to see what happens. An infant will discover that stuffed animals land quietly whereas bigger toys often make a more satisfying "clunk" when they hit the ground. These actions represent a significant extension of intentional behavior; now babies repeat actions with different objects solely for the purpose of seeing what will happen.

sensorimotor period
first of Piaget's four stages of cognitive development, which lasts from birth to approximately 2 years

Understanding Objects

Objects fill the world. Some, including dogs, spiders, and college students, are animate; others, including cheeseburgers, socks, and this textbook, are inanimate. But they all share a fundamental property—they exist independently of our actions and thoughts toward them. Much as we may dislike spiders, they still exist when we close our eyes or wish they would go away. *Piaget's term for this understanding that objects exist independently is* **object permanence**. And Piaget made the astonishing claim that infants lacked this understanding for much of the first year. That is, he proposed that an infant's understanding of objects could be summarized as "out of sight, out of mind." For infants, objects exist when in sight and no longer exist when out of sight.

Piaget concluded that infants have little understanding of objects. If a tempting object such as an attractive toy is placed in front of a 4- to 8-month-old, the infant will probably reach and grasp the object. If, however, the object is subsequently hidden by a barrier or covered with a cloth, the infant will neither reach nor search. Instead, the infant seems to lose all interest in the object, as if the now hidden object no longer exists. Paraphrasing the familiar phrase, "out of sight, out of existence!"

Beginning at about 8 months, infants search for an object that an experimenter has covered with a cloth. In fact, many 8- to 12-month-olds love to play this game: an adult covers the object and the infant sweeps away the cover, laughing and smiling all the while! But despite this accomplishment, their understanding of object permanence remains incomplete, according to Piaget. If 8- to 10-month-olds see an object hidden under one container several times and then see it hidden under a second container, they usually reach for the toy under the first container. Piaget claimed that this behavior shows only a fragmentary understanding of objects because infants do not distinguish the object from the actions they use to locate it, such as reaching for a particular container.

Piaget argued that not until approximately 18 months do infants have full understanding of object permanence. However, in a few pages, we'll see that infants know more about objects than Piaget claimed.

object permanence
understanding, acquired in infancy, that objects exist independently of oneself

When interesting toys are covered so that they can't be seen, young babies lose interest, as if "out of sight" means "out of existence."

Using Symbols

By 18 months, most infants have begun to talk and gesture, evidence of their emerging capacity to use symbols. Words and gestures are symbols that stand for something else. When a baby waves, it's a symbol that's just as effective as saying "good-bye" to bid farewell. Children also begin to engage in pretend play, another use of symbols. A 20-month-old may move her hand back and forth in front of her mouth, pretending to brush her teeth.

Once infants can use symbols, they can begin to anticipate the consequences of actions mentally instead of having to perform them. Imagine that an infant and parent construct a tower of blocks next to an open door. Leaving the room, a 12- to 18-month-old might close the door, knocking over the tower because he cannot foresee this outcome of closing the door. But an 18- to 24-month-old can anticipate the consequence of closing the door and move the tower beforehand.

In just 2 years, the infant progresses from reflexive responding to actively exploring the world, understanding objects, and using symbols. These achievements are remarkable and set the stage for preoperational thinking, which we'll examine next.

Preoperational Thinking

Once they have crossed into preoperational thinking, the magical power of symbols is available to young children. Of course, mastering this power is a lifelong process; the preschool child's efforts are tentative and sometimes incorrect

Toddlers frequently gesture, a sign of their growing competence at using symbols.

(DeLoache, 1995). Piaget identified a number of characteristic shortcomings in preschoolers' fledgling symbolic skills. Let's look at three.

Egocentrism

Preoperational children typically believe that others see the world—both literally and figuratively—exactly as they do. **Egocentrism** *is difficulty in seeing the world from another's outlook.* When youngsters stubbornly cling to their own way, they are not simply being contrary. Preoperational children simply do not comprehend that other people differ in their ideas, convictions, and emotions.

One of Piaget's famous experiments, the three-mountains problem, demonstrates preoperational children's egocentrism (Piaget & Inhelder, 1956, chap. 8). Youngsters were seated at a table like the one shown in ▪ Figure 4.1. When preoperational children were asked to choose the photograph that corresponded to another person's view of the mountains, they usually picked the photograph that showed their own view of the mountains, not the other person's. Preoperational youngsters evidently suppose that the mountains are seen the same way by all; they presume that theirs is the only view, not one of many conceivable views. According to Piaget, only concrete operational children fully understand that all people do not experience an event in exactly the same way.

Recall that, in the vignette, 3-year-old Jamila nods her head during phone conversations with her grandmother. This, too, reflects preoperational egocentrism. Jamila assumes that, because she is aware that her head is moving up and down (or side to side), her grandmother must be aware of it, too. Because of this egocentrism, preoperational youngsters often attribute their own thoughts and feelings to others. *They may even credit inanimate objects with life and lifelike properties, a phenomenon known as* **animism** (Piaget, 1929). A preschool child may think that the sun is unhappy on a cloudy day or that a car hurts when it's in an accident. Caught up in their egocentrism, preoperational youngsters believe that inanimate objects have feelings just as they do.

egocentrism
difficulty in seeing the world from another's point of view; typical of children in the preoperational period

animism
crediting inanimate objects with life and lifelike properties such as feelings

▪ **Figure 4.1**
Egocentrism: When asked to select the photograph that shows the mountains as the adult sees them, preschool children often select the photograph that shows how the mountain looks to them.

Copyright © Cengage Learning 2010

Centration

A second characteristic of preoperational thinking is that children seem to have the psychological equivalent of tunnel vision: They often concentrate on one aspect of a problem but totally ignore other, equally relevant aspects. **Centration** *is Piaget's term for this narrowly focused thought that characterizes preoperational youngsters.*

Piaget demonstrated centration in his experiments involving conservation. In the conservation experiments, Piaget wanted to determine when children realize that important characteristics of objects (or sets of objects) stay the same despite changes in their physical appearance. Some tasks that Piaget used to study conservation are shown in ■ Figure 4.2. Each begins with identical objects (or sets of objects). Then one

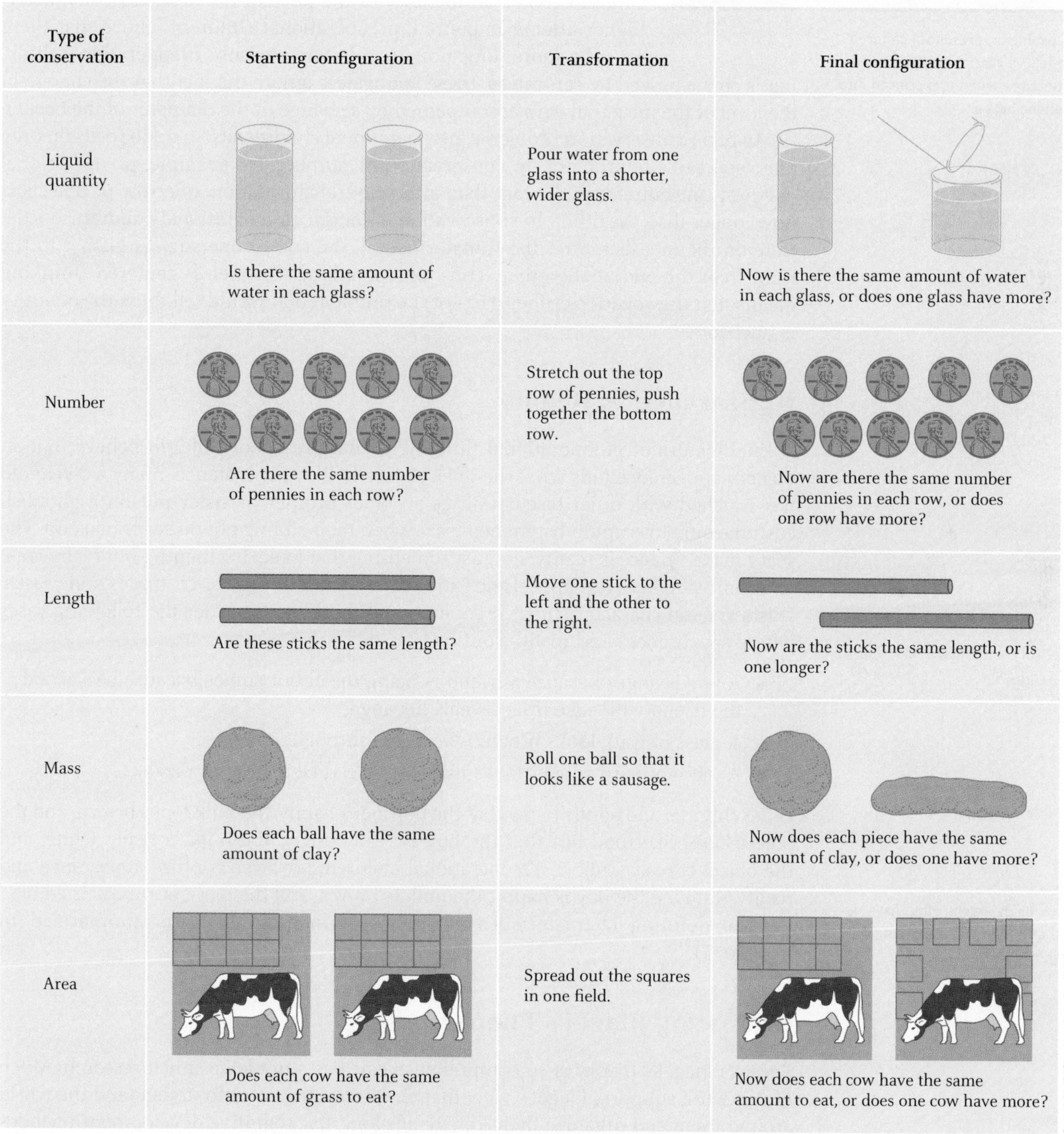

centration
according to Piaget, narrowly focused type of thought characteristic of preoperational children

Type of conservation	Starting configuration	Transformation	Final configuration
Liquid quantity	Is there the same amount of water in each glass?	Pour water from one glass into a shorter, wider glass.	Now is there the same amount of water in each glass, or does one glass have more?
Number	Are there the same number of pennies in each row?	Stretch out the top row of pennies, push together the bottom row.	Now are there the same number of pennies in each row, or does one row have more?
Length	Are these sticks the same length?	Move one stick to the left and the other to the right.	Now are the sticks the same length, or is one longer?
Mass	Does each ball have the same amount of clay?	Roll one ball so that it looks like a sausage.	Now does each piece have the same amount of clay, or does one have more?
Area	Does each cow have the same amount of grass to eat?	Spread out the squares in one field.	Now does each cow have the same amount to eat, or does one cow have more?

■ **Figure 4.2**
Children in the preoperational stage of development typically have difficulty solving conservation problems, in which important features of an object (or objects) stay the same despite changes in physical appearance.

In conservation problems, preschool children typically do not believe that the quantity of a liquid remains the same when it is poured into a taller, more slender beaker.

of the objects (or sets) is transformed, and children are asked if the objects are the same in terms of some important feature.

A typical conservation problem involves conservation of liquid quantity. Children are shown identical beakers filled with the same amount of juice. After children agree that the two beakers have the same amount of juice, the juice is poured from one beaker into a taller, thinner beaker. The juice looks different in the tall, thin beaker—it rises higher—but of course the amount is unchanged. Nevertheless, preoperational children claim that the tall, thin beaker has more juice than the original beaker. (And, if the juice is poured into a wider beaker, they believe it has less.)

What is happening here? According to Piaget, preoperational children center on the level of the juice in the beaker. If the juice is higher after it is poured, preoperational children believe that there must be more juice now than before. Because preoperational thinking is characterized by centration, these youngsters ignore the fact that the change in the level of the juice is always accompanied by a change in the diameter of the beaker.

In other conservation problems, preoperational children also tend to focus on only one aspect of the problem. In conservation of number, for example, preoperational children concentrate on the fact that, after the transformation, one row of objects is now longer than the other. In conservation of length, preoperational children concentrate on the fact that, after the transformation, the end of one stick is farther to the right than the end of the other. Thus, preoperational children's "centered" thinking means that they overlook other parts of the problem that would tell them the quantity is unchanged.

Appearance as Reality

A final feature of preoperational thinking is that preschool children believe that an object's appearance tells what the object is really like. For instance, many a 3-year-old has watched with quiet fascination as an older brother or sister put on a ghoulish costume only to erupt in frightened tears when their sibling put on scary makeup. The scary made-up face is reality, not just something that looks frightening but really isn't.

Confusion between appearance and reality is not limited to costumes and masks. It is a general characteristic of preoperational thinking. Consider the following cases where appearances and reality conflict:

- A boy is angry because a friend is being mean but smiles because he's afraid the friend will leave if he reveals his anger.
- A glass of milk looks brown when seen through sunglasses.
- A piece of hard rubber looks like food (e.g., like a piece of pizza).

Older children and adults know that the boy looks happy, the milk looks brown, and the object looks like food but that the boy is really angry, the milk is really white, and the object is really rubber. Preoperational children, however, confuse appearance and reality, thinking the boy is happy, the milk is brown, and the piece of rubber is edible.

The defining characteristics of preoperational thought are summarized in ●Table 4.1.

THINK ABOUT IT

Children with low birth weight often have delayed intellectual development. According to Piaget, what form might the delay take?

Evaluating Piaget's Theory

Because Piaget's theory is so comprehensive, it has stimulated much research. Much of this work supports Piaget's view that children actively try to understand the world around them and organize their knowledge and that cognitive development includes major qualitative changes (Brainerd, 1996; Flavell, 1996). One important contribution of Piaget's theory is that many teachers and parents have found it a rich source of ideas about ways to foster children's development.

Characteristics of Preoperational Thinking

Characteristic	Definition	Example
Egocentrism	Child believes that all people see the world as he or she does	A child gestures during a telephone conversation, not realizing that the listener cannot see the gestures
Centration	Child focuses on one aspect of a problem or situation but ignores other relevant aspects	In conservation of liquid quantity, child pays attention to the height of the liquid in the beaker but ignores the diameter of the beaker
Appearance as reality	Child assumes that an object really is what it appears to be	Child believes that a person smiling at another person is really happy even though the other person is being mean

Guidelines for Fostering Cognitive Development

Piaget's theory has several straightforward implications for the conditions that promote cognitive growth.

- Cognitive growth occurs as children construct their own understanding of the world, so the teacher's role is to create environments where children can discover for themselves how the world works. A teacher shouldn't simply try to tell children how addition and subtraction are complementary but instead should provide children with materials that allow them to discover the complementarity themselves.

- Children profit from experience only when they can interpret this experience with their current cognitive structures. It follows, then, that the best teaching experiences are slightly ahead of the children's current level of thinking. As youngsters begin to master basic addition, don't jump right to subtraction but go to slightly more difficult addition problems.

- Cognitive growth can be particularly rapid when children discover inconsistencies and errors in their own thinking. Teachers should therefore encourage children to look at the consistency of their thinking but then let children take the lead in sorting out the inconsistencies. If a child is making mistakes in borrowing on subtraction problems, a teacher shouldn't correct the error directly but should encourage the child to look at a large number of these errors to discover what he or she is doing wrong.

According to Piaget's theory of cognitive development, children need to learn by doing.

Criticisms of Piaget's Theory

Although Piaget's contributions to child development are legendary, some elements of his theory have held up better than others (Siegler & Alibali, 2005).

- *Piaget's theory underestimates cognitive competence in infants and young children and overestimates cognitive competence in adolescents.* In Piaget's theory, cognitive development is steady in early childhood but not particularly rapid. In contrast, a main theme of modern child-development science is that of the extraordinarily competent infant and toddler. By using more sensitive tasks than Piaget's, modern investigators have shown that infants and toddlers are vastly more capable than expected based on Piaget's theory. For example, in a few pages we'll see that infants have much greater understanding of objects than Piaget believed. Paradoxically, however, Piaget overestimated cognitive skill in adolescents, who often fail to reason according to formal operational principles and revert to less sophisticated reasoning.

- *Piaget's theory is vague with respect to processes and mechanisms of change.* One important shortcoming is that many of the key components of the theory, such as accommodation and assimilation, are too vague to test scientifically. Consequently, scientists abandoned them in favor of other cognitive processes that could be evaluated more readily and hence could provide more convincing accounts of children's thinking.

■ *Piaget's stage model does not account for variability in children's performance.* An even more important criticism is that cognitive development is nowhere near as stagelike as Piaget believed. In Piaget's view, each stage of intellectual development has unique characteristics that leave their mark on everything a child does. Preoperational thinking is defined by egocentrism and centration; formal operational thinking is defined by abstract and hypothetical reasoning. Consequently, children's performance on different tasks should be very consistent. On the conservation and the three-mountains tasks, for instance, according to Piaget a 4-year-old should always respond in a preoperational way: He should say the water is not the same after pouring, and that another person sees the mountains the same way he does. In fact, children's thinking falls far short of this consistency. A child's thinking may be sophisticated in some domains but naive in others (Siegler, 1981). This inconsistency does not support Piaget's view that children's thinking should always reflect the distinctive imprint of their current stage of cognitive development.

■ *Piaget's theory undervalues the influence of the sociocultural environment on cognitive development.* Returning to the metaphor of the child as scientist, Piaget describes the child as a lone scientist, constantly trying to figure out by herself how her theory coordinates with data. In reality, a child's effort to understand her world is a far more social enterprise than Piaget described. Her growing understanding of the world is profoundly influenced by interactions with family members, peers, and teachers, and it takes place against the backdrop of cultural values. Piaget did not ignore these social and cultural forces entirely, but they are not prominent in his theory.

These criticisms do not mean that Piaget's theory is invalid or should be abandoned. As noted previously, it remains the most complete account of cognitive development. However, in recent years researchers have attempted to round out our understanding of cognitive development using other theoretical perspectives, such as the information-processing approach examined later in this chapter.

Extending Piaget's Account: Children's Naive Theories

Piaget believed that children, like scientists, formulate theories about how the world works. Children's theories are usually called "naive theories" because, unlike real scientific theories, they are not created by specialists and are rarely evaluated by formal experimentation. Naive theories are nevertheless valuable because they allow children (and adults) to understand new experiences and predict future events.

In Piaget's view, children formulate a grand, comprehensive theory that attempts to explain an enormous variety of phenomena—including reasoning about objects, people, and morals, for example—within a common framework. More recent views retain the idea of children as theorists but propose that children, like real scientists, develop specialized theories about much narrower areas. For example, *according to the* **core knowledge hypothesis**, *infants are born with rudimentary knowledge of the world; this knowledge is elaborated based on children's experiences* (Spelke & Kinzler, 2007; Wellman & Gelman, 1998). Some of the theories young children first develop concern physics, psychology, and biology. That is, infants and toddlers rapidly develop theories that organize their knowledge about properties of objects, people, and living things (Wellman & Gelman, 1998).

We examined children's developing theory of mind in Chapter 3; in the next few pages, we'll look at children's naive theories of physics and biology.

core knowledge hypothesis
infants are born with rudimentary knowledge of the world, which is elaborated based on experiences

Naive Physics

As adults, we know much about objects and their properties. For example, we know that if we place a coffee cup on a table, it will remain there unless moved by another person; it will not move by itself or simply disappear. And we don't release a coffee cup in midair because we know that an unsupported object will fall. Child-development researchers have long been interested in young children's understanding of objects, in part because Piaget claimed that understanding of objects develops slowly and takes many months to become complete. However, by devising some clever procedures, other investigators have shown that babies understand objects much earlier than Piaget claimed. Renée Baillargeon (1987, 1994), for example, assessed object permanence by using a procedure in which infants first saw a silver screen that appeared to be rotating back and forth. After an infant became familiar with this display, one of two new displays was shown. In the "realistic" event, a red box appeared in a position behind the screen, making it impossible for the screen to rotate as far back as it had previously. Instead, the screen moved away from the infant until it made contact with the box, then moved back toward the infant. In the "unrealistic" event, shown in ■ Figure 4.3, the red box appeared but the screen continued to move as before. The screen moved away from the infant until it was flat, then moved forward, again revealing the red box. The illusion was possible because the box was mounted on a movable platform that allowed it to drop out of the way of the moving screen. However, from the infant's perspective, it appeared as if the box vanished behind the screen, only to reappear.

The disappearance and reappearance of the box violates the idea that objects exist permanently. Consequently, an infant who understands the permanence of objects

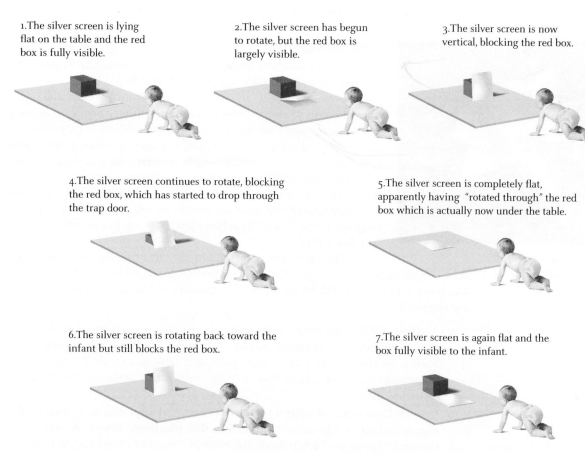

1. The silver screen is lying flat on the table and the red box is fully visible.

2. The silver screen has begun to rotate, but the red box is largely visible.

3. The silver screen is now vertical, blocking the red box.

4. The silver screen continues to rotate, blocking the red box, which has started to drop through the trap door.

5. The silver screen is completely flat, apparently having "rotated through" the red box which is actually now under the table.

6. The silver screen is rotating back toward the infant but still blocks the red box.

7. The silver screen is again flat and the box fully visible to the infant.

■ **Figure 4.3**
Infants are surprised to see the silver screen rotate flat, which suggests that they understand the "permanence" of the red box. Based on Baillargeon (1987).

should find the unrealistic event a truly novel stimulus and look at it longer than the realistic event. Baillargeon found that 4½-month-olds consistently looked longer at the unrealistic event than at the realistic event. Infants apparently thought that the unrealistic event was novel, just as we are surprised when an object vanishes from a magician's scarf. Evidently, then, infants have some understanding of object permanence early in the first year of life.

Of course, understanding that objects exist independently is just a start; objects have numerous other important properties and infants know many of them. By about 6 months, infants are surprised when an object that's released in midair doesn't fall, when an object remains stationary after being hit, or when an object passes through another solid object (Luo, Kaufman, & Baillargeon, 2009). At this age, infants are surprised when a tall object is completely hidden when placed behind a shorter object, apparently because it violates their expectations about concealment (Walden et al., 2007; Wang & Baillargeon, 2005). Finally, by 5 months, infants expect liquids but not solids to change their shape as they're moved from one container to another (Hespos, Ferry, & Rips, 2009).

These amazing demonstrations attest to the fact that the infant is indeed an accomplished naive physicist (Baillargeon, 2004). Of course, the infant's theories are far from complete, since physical properties can be understood at many different levels (Hood, Carey, & Prosada, 2000). Using gravity as an example, infants can expect that unsupported objects will fall, elementary-school children know that such objects fall due to gravity, and physics students know that the force of gravity equals the mass of an object times the acceleration caused by gravity. Obviously, infants do not understand objects at the level of physics students. However, the important point is that infants rapidly create a reasonably accurate theory of some basic properties of objects, a theory that helps them to expect that objects such as toys will act in predictable ways.

Naive Biology

Fundamental to adults' naive theories is the distinction between living and nonliving things. Adults know that living things, for example, are made of cells, inherit properties from parents, and move spontaneously. Adults' theories of living things begin in infancy, when youngsters first distinguish animate objects (e.g., people, insects, other animals) from inanimate objects (e.g., rocks, plants, furniture, tools). Motion is critical in early understanding of the difference between animate and inanimate objects: That is, infants and toddlers use motion to identify animate objects, and by 12–15 months they have determined that animate objects are self-propelled, can move in irregular paths, and act to achieve goals (Biro & Leslie, 2007; Rakison & Hahn, 2004).

By the preschool years, children's naive theories of biology have come to include many of the specific properties associated with living things (Wellman & Gelman, 1998). Many 4-year-olds' theories of biology include the following elements.

© iStockphoto.com/BanksPhotos

Toddlers distinguish animate objects, such as goats, from inanimate objects, such as furniture and tools.

- *Movement*: Children understand that animals can move themselves but that inanimate objects can be moved only by other objects or by people. Shown an animal and a toy hopping across a table in exactly the same manner, preschoolers claim that only the animal can move itself (Gelman & Gottfried, 1996).
- *Growth*: Children understand that, from their first appearance, animals get bigger and physically more complex but that inanimate objects do not change in this way. They believe, for example, that sea otters and termites become larger as time goes by but that teakettles and teddy bears do not (Rosengren et al., 1991).

- *Internal parts*: Children know that the insides of animate objects contain different materials than the insides of inanimate objects. Preschool children judge that blood and bones are more likely to be inside an animate object but that cotton and metal are more likely to be inside an inanimate object (Simons & Keil, 1995).

- *Inheritance*: Children realize that only living things have offspring that resemble their parents. Asked to explain why a dog is pink, preschoolers believe that some biological characteristic of the parents probably made the dog pink; asked to explain why a can is pink, preschoolers rely on mechanical causes (e.g., a worker used a machine), not biological ones (Springer & Keil, 1991; Weissman & Kalish, 1999). And both U.S. and Brazilian children believe that a baby pig that is adopted by a cow will grow up to look like and behave like a pig (Sousa, Altran, & Medin, 2002).

- *Illness*: Preschoolers believe that relatively permanent conditions such as color blindness or food allergies are more likely to be inherited from parents but that temporary illnesses such as a sore throat or a runny nose are more likely to be transmitted through contact with other people (Raman & Gelman, 2005). And they understand that people can become ill when they eat contaminated food (Legare, Wellman, & Gelman, 2009).

- *Healing*: Children understand that, when damaged, animate things heal by regrowth whereas inanimate things must be fixed by humans. Preschoolers know that hair will grow back when cut from a child's head but must be repaired by a person when cut from a doll's head (Backscheider, Shatz, & Gelman, 1993).

By 4 years, children's understanding of living things is so sophisticated that children aren't fooled by lifelike robots: 4-year-olds know that robots are machines that (a) do not eat or grow and (b) are made by people and can break (Jipson & Gelman, 2007).

A fundamental part of young children's theory of living things is a commitment to **teleological explanations**: *children believe that livings things and parts of living things exist for a purpose.* Lions exist so that people can see them in a zoo. Fish have smooth skin so that they won't cut other fish that swim alongside them (Kelemen, 2003). One view is that teleological explanations are based on children's knowledge that objects such as tools and machines are usually made with a purpose in mind. Children may follow a similar logic in thinking that living things (and their parts) were designed with a specific purpose in mind (Kelemen & DiYanni, 2005). This teleological thinking echoes the animistic thinking described on page 128: children attribute their own intentions and goals to other living objects.

Young children's theories of living things are also rooted in **essentialism**: *children believe that all living things have an essence that can't be seen but gives a living thing its identity.* All birds share an underlying "bird-ness" that distinguishes them from dogs, which of course share an underlying "dog-ness." And bird-ness is what allows birds to fly and sing (Gelman, 2003). Young children's essentialism explains why 4-year-olds believe that a baby kangaroo adopted by goats will still hop and have a pouch and why they believe that a watermelon seed planted in a cornfield will produce watermelons (Gelman & Wellman, 1991). The baby kangaroo and the watermelon seed have kangaroo-ness and watermelon-ness that cause properties of kangaroos and watermelons to emerge in maturity.

Where do children get this knowledge of living things? Some of it comes just by watching animals, which children love to do. But parents also contribute: When reading books about animals to preschoolers, mothers frequently mention the properties that distinguish animals, including self-initiated motion (e.g., "the seal is jumping in the water") and psychological properties (e.g., "the bear is really mad!"). Such talk helps to highlight important characteristics of animals for youngsters (Gelman et al., 1998).

teleological explanations
children's belief that living things and parts of living things exist for a purpose

essentialism
children's belief that all living things have an essence that can't be seen but gives a living thing its identity

Of course, although preschoolers' naive theories of biology are complex, their theories aren't complete. Preschoolers don't know, for instance, that genes are the biological basis for inheritance (Springer & Keil, 1991). Preschoolers' theories include some misconceptions: They believe that body parts have intentions—that the heart "wants" to pump blood and bones "want" to grow (Morris, Taplin, & Gelman, 2000). And, although preschoolers know that plants grow and heal, they nevertheless don't consider plants to be living things. It's not until 7 or 8 years of age that children routinely decide that plants are alive. Preschoolers' reluctance to call plants living things may stem from their belief in goal-directed motion as a key property of living things. This is not easy to see in plants, but when 5-year-olds are told that plants move in goal-directed ways—for example, tree roots turn toward a source of water or a venus fly-trap closes its leaves to trap an insect—they decide that plants are alive after all (Opfer & Siegler, 2004).

Despite these limits, children's naive theories of biology, when joined with their naive theory of physics, provide powerful tools for making sense of their world and for understanding new experiences.

Test Yourself

RECALL

1. The term _____ refers to modification of schemes based on experience.

2. According to Piaget, _____ are psychological structures that organize experience.

3. Piaget believed that infants' understanding of objects could be summarized as _____.

4. By 18 months, most infants talk and gesture, which shows that they have the capacity _____.

5. Preschoolers are often _____, meaning that they are unable to take another person's viewpoint.

6. Preoperational children sometimes attribute thoughts and feelings to inanimate objects; this is called _____.

7. One criticism of Piaget's theory is that it underestimates cognitive competence in _____.

8. Most 4-year-olds know that living things move, _____, have internal parts, resemble their parents, and heal when injured.

INTERPRET

Piaget championed the view that children participate actively in their own development. How do the sensorimotor child's contributions differ from the formal-operational child's contributions?

APPLY

Based on what you know about Piaget's theory, what would his position have been on the continuity–discontinuity issue discussed in Chapter 1?

Recall answers: (1) accommodation, (2) schemes, (3) "out of sight, out of mind," (4) to use symbols, (5) egocentric, (6) animism, (7) infants and young children, (8) grow

---

LEARNING OBJECTIVES

- What is the basis of the information-processing approach?
- How well do young children pay attention?
- What kinds of learning take place during infancy?
- Do infants and preschool children remember?
- What do infants and preschoolers know about numbers?

When Claire, a bubbly 3-year-old, is asked how old she'll be on her next birthday, she proudly says, "Four!" while holding up five fingers. Asked to count four objects, whether they're candies, toys, or socks, Claire almost always says, "1, 2, 6, 7 ... SEVEN!" Claire's older brothers find all this very funny, but her mother thinks that, the obvious mistakes notwithstanding, Claire's behavior shows that she knows a lot about numbers and counting. But what, exactly, does Claire understand? That question has her mother stumped!

TODAY, MANY DEVELOPMENTALISTS BORROW FROM COMPUTER SCIENCE TO FORMULATE their ideas about human thinking and how it develops (Kail & Bisanz, 1992; Plunkett, 1996). As you recall from Chapter 1, this approach is called information processing. In this section, we'll see what information processing has revealed about young children's thinking and, along the way, see what to make of Claire's counting.

General Principles of Information Processing

In the information-processing view, human thinking is based on both mental hardware and mental software. **Mental hardware** *refers to mental and neural structures that are built in and that allow the mind to operate.* **Mental software** *refers to mental programs that are the basis for performing particular tasks.* According to information-processing psychologists, it is the combination of mental hardware and mental software that allows children to accomplish a specific task. Information-processing psychologists claim that, as children develop, their mental software becomes more complex, more powerful, and more efficient.

In the next few pages, we'll look at the development of many important cognitive processes in infants, toddlers, and preschoolers, beginning with attention.

Attention

Hannah was only 3 days old and was often startled by the sounds of traffic outside her family's apartment. Hannah's parents worried that she might not get enough sleep. Yet, within a few days, traffic sounds no longer disturbed Hannah; she slept blissfully. Why was a noise that had been so troubling no longer a problem? *The key is* **attention**, *a process that determines which sensory information receives additional cognitive processing.*

Hannah's response was normal not only for infants but also for children and adolescents. *When presented with a strong or unfamiliar stimulus, an* **orienting response** *usually occurs: A person startles, fixes the eyes on the stimulus, and shows changes in heart rate and brain-wave activity.* Collectively, these responses indicate that the infant has noticed the stimulus. Remember, too, that Hannah soon ignored the sounds of traffic. After repeated presentations of a stimulus, people recognize it as familiar and the orienting response gradually disappears. **Habituation** *is the diminished response to a stimulus as it becomes more familiar.*

The orienting response and habituation are both useful to infants. On the one hand, orienting makes the infant aware of potentially important or dangerous events in the environment. On the other hand, constantly responding to insignificant stimuli is wasteful, so habituation keeps infants from devoting too much energy to biologically nonsignificant events (Rovee-Collier, 1987).

Preschool children gradually learn how to focus their attention, but as compared to older children and adults they are often not very attentive (Hatania & Smith, 2010). Preschoolers are easily distracted by extraneous information. However, we can help children to pay attention better. One straightforward approach is to make relevant information stand out. For example, closing a classroom door may not eliminate competing sounds and smells entirely, but it does make them less noticeable. When preschoolers are working at a table or desk, we can remove other objects that are not necessary for the task. Another useful tactic, particularly for young children, is to remind them to pay attention to relevant information and to ignore the rest.

Learning

An infant is always learning. For example, a 5-month-old learns that a new toy makes a noise every time she shakes it. Infants are born with many mechanisms that enable them to learn from experience. This learning can take several forms, including habituation, classical conditioning, operant conditioning, and imitation.

mental hardware
mental and neural structures that are built in and that allow the mind to operate

mental software
mental "programs" that are the basis for performing particular tasks

attention
processes that determine which information will be processed further by an individual

orienting response
an individual views a strong or unfamiliar stimulus, and changes in heart rate and brain-wave activity occur

habituation
becoming unresponsive to a stimulus that is presented repeatedly

Infants (and older children) pay attention to loud stimuli at first but then ignore them if they aren't interesting or dangerous.

© LIU JIN / AFP / Getty Images

Classical Conditioning

Some of the most famous experiments in psychology were conducted with dogs by the Russian physiologist Ivan Pavlov. Dogs salivate when fed. Pavlov discovered that, if something always happened just before feeding—for example, if a bell sounded— then dogs would begin to salivate to that event. *In **classical conditioning**, a neutral stimulus elicits a response that was originally produced by another stimulus.* In Pavlov's experiments, the bell was a neutral stimulus that did not naturally cause dogs to salivate. However, by repeatedly pairing the bell with food, the bell began to elicit salivation. Similarly, infants will suck reflexively when sugar water is placed in their mouth with a dropper; if a tone precedes the drops of sugar water, infants will suck when they hear the tone (Lipsitt, 1990).

classical conditioning
a form of learning that involves pairing a neutral stimulus and a response originally produced by another stimulus

operant conditioning
view of learning, proposed by B. F. Skinner, that emphasizes reward and punishment

Classical conditioning is important because it gives infants a sense of order in their environment. That is, through classical conditioning, infants learn that a stimulus is a signal for what will happen next. A youngster may smile when she hears the family dog's collar because she knows the dog is coming to play with her. Or a toddler may frown when he hears water running in the bathroom because he realizes this means it's time for a bath.

Infants and toddlers are definitely capable of classical conditioning when the stimuli are associated with feeding or other pleasant events. It is much more difficult to demonstrate classical conditioning in infants and toddlers when the stimuli are aversive, such as loud noises or shock (Fitzgerald & Brackbill, 1976). Yet because adults care for and protect very young children, learning about potentially dangerous stimulation is not a common biological problem for infants and toddlers (Rovee-Collier, 1987).

Toddlers who have fun playing in the water will welcome the sound of the bathtub being filled.

Operant Conditioning

In classical conditioning, infants form expectations about what will happen in their environment. **Operant conditioning** *focuses on the relation between the consequences of behavior and the likelihood that the behavior will recur.* When a child's behavior leads to pleasant consequences, the child will probably behave similarly in the future; when the child's behavior leads to unpleasant consequences, the child will probably not repeat the behavior. When a baby smiles, an adult may hug the baby in return; this pleasing consequence makes the baby more likely to smile in the future. When a baby grabs a family heirloom, an adult may become angry and shout at the baby; these unpleasant consequences make the baby less likely to grab the heirloom in the future.

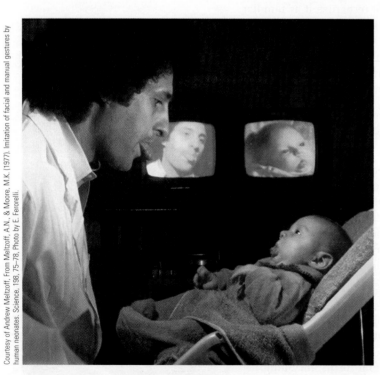

Newborns imitate an adult's facial expressions.

Imitation

Older children, adolescents, and young adults learn much simply by watching others behave. For example, children learn new sports moves by watching pro athletes, they learn how to pursue romantic relationships by watching TV, and they learn how to play new computer games by watching peers. Infants, too, are capable of imitation (Barr & Hayne, 1999). A 10-month-old may imitate an adult waving her finger back and forth or imitate another infant who knocks down a tower of blocks.

More startling is the claim that even newborns imitate. Meltzoff and Moore (1989, 1994) found that 2- to

3-week-olds would stick out their tongue or open and close their mouth to match an adult's acts. This work is controversial because other researchers do not consistently obtain these results. In addition, because the newborns' behavior is not novel—newborns are already capable of sticking out their tongues as well as opening and closing their mouths—some researchers do not consider this to be a "true" form of imitation (Anisfeld, 1991, 1996). This work may well be describing an early, limited form of imitation; over the course of the first year of life, infants are able to imitate a rapidly expanding range of behaviors.

Memory

Young babies remember events for days or even weeks at a time. Some of the studies that opened our eyes to the infant's ability to remember used the following method devised by Rovee-Collier (1997, 1999). A ribbon from a mobile is attached to a 2- or 3-month-old's leg; within a few minutes, the babies learn to kick to make the mobile move. When Rovee-Collier brought the mobile to the infants' homes several days or a couple of weeks later, babies would still kick to make the mobile move. If Rovee-Collier waited several weeks to return, most babies forgot that kicking moved the mobile. When that happened, Rovee-Collier gave them a reminder—she moved the mobile herself without attaching the ribbon to their foot. Then she would return the next day, hook up the apparatus, and the babies would kick to move the mobile.

Rovee-Collier's experiments show that three important features of memory exist as early as 2 and 3 months of age: (1) an event from the past is remembered, (2) over time, the event can no longer be recalled, and (3) a cue can serve to dredge up a memory that seems to have been forgotten.

From these humble origins, memory improves rapidly in older infants and toddlers. Youngsters can recall more of what they experience and can remember it longer (Bauer & Lukowski, 2010; Pelphrey et al., 2004). When youngsters are shown novel actions with toys and later are asked to imitate what they saw, toddlers can remember more than infants and remember the actions for longer periods (Bauer, 2007). For example, if shown how to make a rattle by first placing a wooden block inside a container and then putting a lid on the container, toddlers are more likely than infants to remember the necessary sequence of steps.

These improvements in memory can be traced, in part, to growth in the brain regions that support memory (Bauer, 2007; Richmond & Nelson, 2007). On the one hand, the brain structures primarily responsible for the initial storage of information, including the hippocampus and amygdala, seem to develop very early—by age 6 months. On the other hand, the structure responsible for retrieving these stored memories, the prefrontal cortex, develops much later—into the 2nd year. What's more, part of the hippocampus is not mature until about 20–24 months. Development of memory during the first 2 years therefore reflects growth in these two different brain regions, shown in ■ Figure 4.4. In other words, as the hippocampus and prefrontal cortex mature over the first 24 months, children's memory skills gradually improve.

Autobiographical Memory

A novel feature of memory emerges in the preschool years. **Autobiographical memory** *refers to people's memory of the significant events and experiences of*

Several days after infants have learned that kicking moves a mobile, they will kick when they see the mobile, showing that they remember the connection between kicking and the mobile's movements.

autobiographical memory
memories of the significant events and experiences of one's own life

■ **Figure 4.4**
The amygdala, hippocampus, and prefrontal cortex are brain structures that support memory.

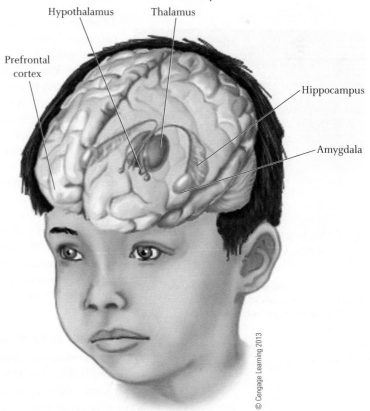

Hypothalamus Thalamus Prefrontal cortex Hippocampus Amygdala

their own lives. You can access your own autobiographical memory by answering these questions:

> Who was your teacher in fourth grade?
> Where (and with whom!) was your first kiss?
> Was your high-school graduation indoors or outdoors?

In answering these questions you searched your memory, just as you would search your memory to answer such questions as "What is the capital of Ohio?" and "Who invented the sewing machine?" However, answers to questions about Ohio and sewing machines are based on general knowledge that you have not experienced personally; answers to questions about *your* fourth-grade teacher, *your* first kiss, and *your* high-school graduation are based on knowledge unique to your own life. Autobiographical memory is important because it helps people construct a personal life history. In addition, autobiographical memory allows people to relate their experiences to others, creating socially shared memories (Bauer, 2006).

Autobiographical memory originates in the preschool years. According to one influential theory (Nelson & Fivush, 2004), autobiographical memory emerges gradually as children acquire the component skills. Infants and toddlers have the basic memory skills that enable them to remember past events. Layered on top of these memory skills during the preschool years are language skills and a child's sense of self. Language allows children to become conversational partners. After infants begin to talk, parents often converse with them about past and future events, particularly about personal experiences in the child's past and future. Parents may talk about what the child did today at day care or remind the child about what the child will be doing this weekend. In conversations like these, parents teach their children the important features of events and how events are organized (Fivush, Reese, & Haden, 2006). Children's autobiographical memories are richer when parents talk about past events in detail and, specifically, when they encourage children to expand their description of past events by, for example, using open-ended questions (e.g., "Where did Mommy go last night?"). When parents use this conversational style with their preschool children, as young adolescents they have earlier memories of childhood (Jack et al., 2009).

The richness of parent–child conversations also helps to explain a cultural difference in autobiographical memory. Compared to adults living in China, Japan, and Korea, Europeans and North Americans typically remember more events from their early years and remember those events in more detail (Peterson, Wang, & Hou, 2009; Wang, 2006). This difference in early memories can be traced to cultural differences in parent–child conversational styles: the elaborative style is less common among Asian parents, which means that Asian youngsters have fewer opportunities for the conversations about past events that foster autobiographical memory (Kulkofsky, Wang, & Koh, 2009; Wang, 2007).

How does an emergent sense of self contribute to autobiographical memory? During the first 2 years, infants rapidly acquire a sense that they exist independently in space and time. An emerging sense of self thus provides coherence and continuity to children's experience. Children realize that the self who went to the park a few days ago is the same self who is now at a birthday party and is the same self who will read a book with dad before bedtime. The self provides a personal timeline and anchors a child's recall of the past (and anticipation of the future). In sum, a sense of self, language skills that enable children to converse with parents about past and future, and basic memory skills all contribute to the emergence of autobiographical memory in preschool children.

Asian parents are less likely than Europeans or North Americans to talk to their children about events that foster autobiographical memory.

Preschoolers as Eyewitnesses

Research on children's autobiographical memory has played a central role in cases of suspected child abuse. When abuse is suspected, the victim is usually the sole witness. To prosecute the alleged abuser, the child's testimony is needed. But can preschoolers accurately recall these events? Answering this question is not as easy as it might seem. One obstacle to accurate testimony is that young children are often interviewed repeatedly during legal proceedings, which can cause them to confuse what actually happened with what others suggest may have happened. When the questioner is an adult in a position of authority, children often believe that what is suggested by the adult actually happened (Candel et al., 2009; Ceci & Bruck, 1998). They will tell a convincing tale about "what really happened" simply because adults have led them to believe things must have happened that way. Young children's storytelling can be so convincing that—even though enforcement officials and child protection workers believe they can usually tell if children are telling the truth—professionals often cannot distinguish true and false reports (Gordon, Baker-Ward, & Ornstein, 2001).

Preschool children are particularly suggestible. Why? One idea is that preschool children are more suggestible because of limited source-monitoring skills (Poole & Lindsay, 1995). Older children, adolescents, and adults often know the source of information that they remember. For example, a father recalling his daughter's piano recitals will know the source of many of his memories: Some are from personal experience (he attended the recital), some he saw on videotape, and some are based on his daughter's descriptions. Preschool children are not particularly skilled at such source monitoring. When recalling past events, preschoolers are often confused about who did or said what; when confused in this manner, they frequently assume that they must have experienced something personally. Consequently, when preschool children are asked leading questions (e.g., "When the man touched you, did it hurt?"), this information is also stored in memory but without the source. Because preschool children are not skilled at monitoring sources, they have trouble distinguishing what they actually experienced from what interviewers imply that they experienced (Ghetti, 2008).

Although preschoolers are easily misled, they can provide reliable testimony if interviewers follow several guidelines derived from research on children's memory. Specifically, interviewers should:

- Interview children as soon as possible after the event in question.
- Encourage children to tell the truth, to feel free to say "I don't know" to questions, and to correct interviewers when they say something that's incorrect.
- Start by asking children to describe the event in their own words ("Tell me what happened after school . . ."), follow up with open-ended questions ("Can you tell me more about what happened while you were walking home?"), and minimize the use of specific questions (because they may suggest to children events that did not happen).
- Allow children to understand and feel comfortable in the interview format by beginning with a neutral event (e.g., a birthday part or holiday celebration) before moving to the event of interest.
- Ask questions that consider alternate explanations of the event (i.e., explanations that don't involve abuse).

Following guidelines like these can foster the conditions under which children are likely to recall past events more accurately and thereby be better witnesses (Lamb et al., 2007).

Learning Number Skills

Powerful learning and memory skills allow infants and preschoolers to learn much about their worlds. This rapid growth is well illustrated by research on children's understanding of the concept of number. Basic number skills originate in infancy, long before babies learn the names of numbers. Many babies experience daily variation in quantity. They play with two blocks and see that another baby has three; they watch as a father sorts laundry and finds two black socks but only one blue sock, and they eat one hot dog for lunch while an older brother eats three.

From these experiences, babies apparently come to appreciate that quantity or amount is one of the ways in which objects in the world can differ. That is, research

Infants are surprised when they see objects added or removed but the original number of objects are still present when the screen is removed; this pattern suggests some basic understanding of addition and subtraction.

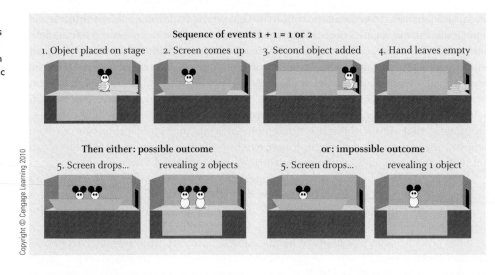

Sequence of events 1 + 1 = 1 or 2

1. Object placed on stage 2. Screen comes up 3. Second object added 4. Hand leaves empty

Then either: possible outcome

5. Screen drops... revealing 2 objects

or: impossible outcome

5. Screen drops... revealing 1 object

Copyright © Cengage Learning 2010

suggests that 5-month-olds can distinguish two objects from three and (though less often) three objects from four (Cordes & Brannon, 2009; Wynn, 1996). Apparently, infants' perceptual processes enable them to distinguish differences in quantity. That is, just as colors (reds, blues) and shapes (triangles, squares) are basic perceptual properties, small quantities ("twoness" and "threeness") are as well.

What's more, young babies can perform simple addition and subtraction. In experiments using the method shown in ■ Figure 4.5, infants view a stage with one mouse. A screen hides the mouse and then a hand appears with a second mouse, which is placed behind the screen. When the screen is removed and reveals one mouse, 5-month-olds look longer than when two mice appear. Apparently, 5-month-olds expect that one mouse plus another mouse should equal two mice, and they look longer when this expectancy is violated (Wynn, 1992). Likewise, when the stage first has two mice and one of them is removed, infants are surprised when the screen is removed and two mice are still on the stage. These experiments only work with very small numbers, indicating that the means by which infants add and subtract are quite simple and probably unlike the processes that older children use (Mix, Huttenlocher, & Levine, 2002).

Finally, scientists have shown that infants can compare quantities. One way to relate two quantities is by their ratio; amazingly, 6-month-olds are sensitive to ratio (McCrink & Wynn, 2007). Once they are shown stimuli that feature *two* blue circles for every yellow circle (e.g., 8 blue and 4 yellow or 30 blue and 15 yellow) infants look longer when they're next shown stimuli that have a ratio of *four* blue circles to every yellow circle (e.g., 36 blue and 9 yellow). Infants can also detect the larger of two quantities. If 10-month-olds watch an adult place two crackers in one container but three crackers in a second container, the infants usually reach for the container with more crackers (Feigenson, Carey, & Hauser, 2002).

Learning to Count

By 2 years of age, most youngsters know some number words and have begun to count. This counting, however, is usually full of mistakes. They might count "1, 2, 6, 7"—skipping 3, 4, and 5. Gelman and Meck (1986) charted preschoolers' understanding of counting. They simply placed several objects in front of a child and asked, "How many?" By analyzing children's answers to many of these questions, Gelman and Meck discovered that, by age 3, most children have mastered three basic principles of counting—at least when it comes to counting up to five objects.

one-to-one principle
counting principle that states that there must be one and only one number name for each object counted

■ **One-to-one principle**: *There must be one and only one number name for each object that is counted.* A child who counts three objects as "1, 2, a" understands this principle because the number of names matches the number of objects to be counted, even though the third name is a letter.

- **Stable-order principle**: *Number names must be counted in the same order.* A child who counts in the same sequence—for example, consistently counting four objects as "1, 2, 4, 5"—shows understanding of this principle.
- **Cardinality principle**: *The last number name differs from the previous ones in a counting sequence by denoting the number of objects.* Typically, 3-year-olds reveal their understanding of this principle by repeating the last number name, often with emphasis: "1, 2, 4, 8 . . . EIGHT!"

stable-order principle
counting principle that states that number names must always be counted in the same order

cardinality principle
counting principle that the last number name denotes the number of objects being counted

During the preschool years, children master these basic principles and apply them to increasingly larger sets of objects. By age 5, most youngsters can apply these counting principles to as many as nine objects. Of course, children's understanding of these principles does not mean that they always count accurately. To the contrary, children can apply all these principles consistently while counting incorrectly. They must master the conventional sequence of number names and the counting principles to learn to count accurately.

This turns out to be easier when infants are frequently exposed to number words at home (Levine et al., 2008). It's also easier when youngsters learn languages that use plural nouns. English, for example, usually indicates plural by adding "s" to a noun. But in some languages (e.g., Japanese), the noun is the same regardless of the number of objects; toddlers speaking these languages learn number words more slowly (Sarnecka et al., 2007).

Learning the number names beyond 9 is easier because the counting words can be generated based on rules for combining decade number names (20, 30, 40) with unit names (1, 2, 3, 4). Later, similar rules are used for hundreds, thousands, and so on. By age 4, most youngsters know the numbers to 20, and some can count to 99. Usually, they stop counting at a number ending in 9 (29, 59), apparently because they don't know the next decade name (Siegler & Robinson, 1982).

Thus far we have not considered the impact of social context on children's thinking. In the next section we'll examine a theory developed by Vygotsky, who believed that cognitive development has its roots in social interactions.

By age 5, children have mastered the three counting principles and can apply them to large sets of objects.

Test Yourself

RECALL

1. One way to improve preschool children's attention is to make irrelevant stimuli _____.

2. Four-month-old Tanya has forgotten that kicking moves a mobile. To remind her of the link between kicking and the mobile's movement, we could _____.

3. Preschoolers may be particularly suggestible because they are less skilled at _____.

4. When a child who is counting a set of objects repeats the last number, usually with emphasis, this indicates the child's understanding of the _____ principle of counting.

INTERPRET

Do the developmental mechanisms in the information-processing perspective emphasize nature, nurture, or both? How?

APPLY

Describe how research on children's eyewitness testimony illustrates connections among emotional, cognitive, and social development.

Recall answers: (1) less noticeable, (2) let her view a moving mobile, (3) monitoring the sources of their memories, (4) cardinality

- What is the zone of proximal development? How does it help explain how children accomplish more when they collaborate with others?

- Why is scaffolding a particularly effective way of teaching youngsters new concepts and skills?

- When and why do children talk to themselves as they solve problems?

Victoria, a 4-year-old, enjoys solving jigsaw puzzles, coloring, and building towers with blocks. While busy with these activities, she often talks to herself. For example, once as she was coloring a picture, she said, "Where's the red crayon? Stay inside the lines. Color the blocks blue." These remarks were not directed at anyone else; after all, Victoria was alone. Why did she say these things? What purpose did they serve?

HUMAN DEVELOPMENT IS OFTEN REFERRED TO AS A JOURNEY THAT TAKES PEOPLE ALONG MANY DIFFERENT PATHS. For Piaget and for information-processing psychologists, children make the journey alone. Other people (and culture in general) certainly influence the direction that children take, but fundamentally the child is a solitary adventurer-explorer, boldly forging ahead. Lev Vygotsky (1896–1934), a Russian psychologist, proposed a very different account: Development is an apprenticeship in which children advance when they collaborate with others who are more skilled. According to Vygotsky (1934/1986), children rarely make much headway on the developmental path when they walk alone; they progress when they walk hand in hand with an expert partner.

For Vygotsky and other sociocultural theorists, the social nature of cognitive development is captured in the concept of **intersubjectivity**, *which refers to mutual, shared understanding among participants in an activity.* For example, when parents and children play board games together, they share an understanding of the goals of their activity and of their roles in playing the games. Such shared understanding allows parents and children to work together in complementary fashion on the puzzles. *Such interactions typify* **guided participation**, *in which cognitive growth results from children's involvement in structured activities with others who are more skilled than they.* Through guided participation, children learn from others how to connect new experiences and new skills with what they already know (Rogoff, 2003). Guided participation is shown when a child learns a new video game from a peer or an adolescent learns a new karate move from a partner.

Vygotsky died of tuberculosis at the age of 37, so he never had the opportunity to develop his theory fully. He did not provide a complete theory of cognitive development throughout childhood and adolescence (as Piaget did), nor did he give definitive accounts of cognitive change in specific domains (as information-processing theorists do). However, many of his ideas are influential, largely because they fill in some gaps in the Piagetian and information-processing accounts. In the next few pages, we'll look at three of Vygotsky's most important contributions—the zone of proximal development, scaffolding, and private speech—and learn more about why Victoria talks to herself.

The Zone of Proximal Development

Four-year-old Ian and his father often solve puzzles together. Although Ian does most of the work, his father encourages him, sometimes finds a piece that he needs, or shows Ian how to put parts together. When Ian tries to assemble the

intersubjectivity
mutual, shared understanding among participants in an activity

guided participation
children's involvement in structured activities with others who are more skilled, typically producing cognitive growth

same puzzles by himself, he can rarely complete them. *The difference between what Ian can do with assistance and what he does alone defines his* **zone of proximal development**. That is, the zone is the area between the level of performance a child can achieve when working independently and a higher level of performance that is possible when working under the guidance or direction of more skilled adults or peers (Wertsch & Tulviste, 1992). For example, elementary-school children are often asked to solve arithmetic story problems. Many youngsters have trouble with these problems, often because they simply don't know where to begin. By structuring the task for them—"first decide what you're supposed to figure out, then decide what information you're told in the problem"—teachers can help children accomplish what they cannot do by themselves. Thus, just as training wheels help children learn to ride a bike by allowing them to concentrate on certain aspects of bicycling, collaborators help children perform more effectively by providing structure, hints, and reminders.

The idea of a zone of proximal development follows naturally from Vygotsky's basic premise: Cognition develops first in a social setting and only gradually comes under the child's independent control. What factors aid this shift? This leads us to the second of Vygotsky's key contributions.

Scaffolding

Have you ever had the good fortune to work with a master teacher, one who seemed to know exactly when to say something to help you over an obstacle but otherwise let you work uninterrupted? **Scaffolding** *is a style in which teachers gauge the amount of assistance they offer to match the learner's needs.* Early in learning a new task, children know little, so teachers give much direct instruction about how to do all the different elements of a task. As the children catch on, teachers need to provide much less direct instruction; they are more likely to be giving reminders.

Worldwide, parents attempt to scaffold their children's learning, but not always using the same methods. Rogoff and her colleagues (1993) observed mothers in four countries—Guatemala, India, Turkey, and the United States—as they showed their toddlers how to operate a novel toy. In all cultures, most mothers attempted to scaffold their children's learning, either by dividing a difficult task into easier subtasks or by doing parts of the task themselves, particularly the more complicated parts. However, mothers in different cultures accomplish scaffolding in different ways. Mothers in Turkey and the United States relied primarily on verbal instruction. Mothers in India and Guatemala used verbal instruction but also used touches (e.g., nudging a child's elbow) or gaze (e.g., winking or staring) to guide their youngsters. Evidently, parents worldwide try to simplify learning tasks for their children, but they use different methods.

The defining characteristic of scaffolding—giving help but not more than is needed—clearly promotes learning (Cole, 2006). Youngsters do not learn readily when they are constantly told what to do or when they are simply left to struggle through a problem unaided. However, when teachers collaborate with them, allowing children to take on more and more of a task as they master its different elements, they learn more effectively (Murphy & Messer, 2000). Scaffolding is an important technique for transferring skills from others to the child, both in formal settings such as schools and in informal settings such as the home or playground.

Young children can often accomplish far more with some adult guidance than they can accomplish alone; Vygotsky referred to this difference as the zone of proximal development.

zone of proximal development
difference between what children can do with assistance and what they can do alone

scaffolding
a style in which teachers gauge the amount of assistance they offer to match the learner's needs

THINK ABOUT IT

Vygotsky emphasized cognitive development as collaboration. How could such collaboration be included in Piaget's theory? In information processing?

Private Speech

private speech
a child's comments that are not intended for others but are designed instead to help regulate the child's own behavior

Remember Victoria, the 4-year-old in the vignette who talked to herself as she colored? *Her behavior demonstrates* **private speech**: *comments that are not intended for others but are designed to help children regulate their own behavior* (Vygotsky, 1934/1986). Thus, Victoria's remarks are simply an effort to help herself color the picture.

Vygotsky viewed private speech as an intermediate step toward self-regulation of cognitive skills (Fernyhough, 2010). At first, children's behavior is regulated by speech from other people that is directed toward them. When youngsters first try to control their own behavior and thoughts without others present, they instruct themselves by speaking aloud. Private speech seems to be children's way of guiding themselves, of making sure that they do all the required steps in solving a problem. Finally, as children gain ever greater skill, private speech becomes *inner speech*, which was Vygotsky's term for thought.

If private speech functions in this way, then children should use private speech more often on difficult tasks than on easy tasks, because children are most likely to need extra guidance on harder tasks. Also, children should be more likely to use private speech after a mistake than after a correct response. These predictions are generally supported by research (Berk, 2003), which suggests the power of language in helping children learn to control their own behavior and thinking.

Young children often regulate their own behavior by talking to themselves, particularly while performing difficult tasks.

Thus, Vygotsky's work has characterized cognitive development not as a solitary undertaking but as a collaboration between expert and novice. His work reminds us of the importance of language, which we'll examine in detail in the last section of this chapter.

Test Yourself

RECALL

1. The _____ is the difference between the level of performance that youngsters can achieve with assistance and the level they can achieve alone.

2. The term _____ refers to a style in which teachers adjust their assistance to match a child's needs.

3. According to Vygotsky, _____ is an intermediate step between speech from others and inner speech.

INTERPRET

How would scaffolding that's appropriate for infants differ from the scaffolding that's appropriate for preschool children?

APPLY

Review Piaget's description of the conditions that foster cognitive development (page 131). How would a comparable list derived from Vygotsky's theory compare?

Recall answers: (1) zone of proximal development, (2) scaffolding, (3) private speech

---

4.4 LANGUAGE

LEARNING OBJECTIVES

■ When do infants first hear and make speech sounds?

■ When do children start to talk? How do they learn word meanings?

■ How do young children learn grammar?

■ How well do youngsters communicate?

Nabina is just a few weeks away from her first birthday. For the past month, she has seemed to understand much of her mother's speech. If her mom asks, "Where's Garfield?" (the family cat), Nabina scans the room and points toward Garfield. Yet Nabina's own speech is still gibberish: She "talks" constantly, but her mom can't understand a word of it. If Nabina apparently understands others' speech, why can't she speak herself?

AN EXTRAORDINARY HUMAN ACHIEVEMENT OCCURS SOON AFTER THE FIRST BIRTHDAY: Most children speak their first word, which is followed in the ensuing months by several hundred more. This marks the beginning of a child's ability to communicate orally with others. Through speech, youngsters impart their ideas, beliefs, and feelings to family, friends, and others.

Actually, the first spoken words represent the climax of a year's worth of language growth. To tell the story of language acquisition properly and explain Nabina's seemingly strange behavior, we must begin with the months preceding the first words.

A baby's first form of communication—crying—is soon joined by other, language-based ways of communicating.

The Road to Speech

When a baby is upset, a concerned mother tries to console it. This familiar situation is rich in language-related information. The infant, not yet able to talk, is conveying its displeasure by one of the few means of communication available to it—crying. The mother, for her part, is using both verbal and nonverbal measures to cheer her baby, to send the message that the world is really not as bad as it may seem now.

The situation also raises two questions about infants as nonspeaking creatures. First, can babies who are unable to speak understand any of the speech that is directed at them? Second, how do infants progress from crying to more effective methods of oral communication, such as speech? Let's start by answering the first question.

Perceiving Speech

Even newborn infants hear remarkably well (page 106); the left hemisphere of a newborn's brain is already particularly sensitive to language (page 96), and they prefer to listen to speech over comparably complex nonspeech sounds (Vouloumanos et al., 2010). But can babies distinguish speech sounds? To answer this question, we first need to know more about the elements of speech. *The basic building blocks of language are* **phonemes**, *which are unique sounds that can be joined to create words.* Phonemes include consonant sounds, such as the sound of *t* in *toe* and *tap*, along with vowel sounds such as the sound of *e* in *get* and *bed*. Infants can distinguish many of these sounds, some of them as early as 1 month after birth (Aslin, Jusczyk, & Pisoni, 1998).

phonemes
unique sounds used to create words; the basic building blocks of language

How do we know that infants can distinguish between different vowels and consonants? Researchers have devised a number of clever techniques to determine whether babies respond differently to distinct sounds. In one approach, a rubber nipple is connected to a computer so that sucking causes the computer to play a sound out of a loudspeaker. In just a few minutes, 1-month-olds learn the relation between their sucking and the sound: They suck rapidly to hear a tape that consists of nothing more than the sound of *p* as in *pin, pet*, and *pat* (pronounced "puh").

After a few more minutes, infants seemingly tire of this repetitive sound and suck less often, which represents the habituation phenomenon described on page 137. But if the computer presents a different sound—such as the sound of *b* in *bed, bat*, or *bird* (pronounced "buh")—babies begin sucking rapidly again. Evidently, they recognize that the sound of *b* is different from *p* because they suck more often to hear the new sound (Jusczyk, 1995).

Of course, the same sound is not pronounced exactly the same way by all people. For example, two native speakers of English may say *baby* differently and a nonnative speaker's pronunciation could differ even more. Only older infants consistently recognize the same words across variations in pronunciation (Schmale & Seidl, 2009).

THE IMPACT OF LANGUAGE EXPOSURE Not all languages use the same set of phonemes, so a distinction that is important in one language may be ignored in another. For example, French and Polish (unlike English) differentiate between nasal and nonnasal vowels.

To hear the difference, say the word *rod*. Now repeat it, but holding your nose. The subtle difference between the two sounds illustrates a nonnasal vowel (the first version of *rod*) and a nasal one (the second).

Because an infant might be exposed to any of the world's languages, it would be adaptive for young infants to be able to perceive a wide range of phonemes. In fact, research shows that infants can distinguish phonemes that are not used in their native language. For example, Japanese does not distinguish the consonant sound of *r* in *rip* from the sound of *l* in *lip*, and Japanese adults trying to learn English have great difficulty distinguishing these sounds. At about 6–8 months, Japanese and American infants can distinguish these sounds equally well. However, by 10–12 months, perception of *r* and *l* improves for American infants—presumably because they hear these sounds frequently—but declines for Japanese babies (Kuhl et al., 2006).

Newborns apparently are biologically capable of hearing the entire range of phonemes in all languages worldwide. But as babies grow and are more exposed to a particular language, they begin to notice only the linguistic distinctions that are meaningful in their own language (Maye, Weiss, & Aslin, 2008). Thus, specializing in one language apparently comes at the cost of making it more difficult to hear sounds in other languages (Best, 1995). This pattern of greater specialization in speech perception is very reminiscent of the profile for face perception (pages 110–112). With greater exposure to human faces, babies develop a more refined notion of a human face, just as they develop a more refined notion of the sounds that are important in their native language.

IDENTIFYING WORDS Of course, hearing individual phonemes is only the first step in perceiving speech. One of the biggest challenges for infants is identifying recurring patterns of sounds—words. Imagine, for example, an infant overhearing this conversation between a parent and an older sibling:

SIBLING: Jerry got a new *bike*.
PARENT: Was his old *bike* broken?
SIBLING: No. He'd saved his allowance to buy a new mountain *bike*.

An infant listening to this conversation hears *bike* three times. Can the infant learn from this experience? Yes. When 7- to 8-month-olds hear a word repeatedly in different sentences, they later pay more attention to this word than to words they haven't heard previously. Evidently, 7- and 8-month-olds can listen to sentences and recognize the sound patterns that they hear repeatedly (Houston & Jusczyk, 2003; Saffran, Aslin, & Newport, 1996). By 6 months, infants pay more attention to content words (e.g., nouns, verbs) than to function words (e.g., articles, prepositions), and they look at the correct parent when they hear "mommy" or "daddy" (Shi & Werker, 2001; Tincoff & Jusczyk, 1999).

In normal conversation, there are no silent gaps between words, so how do infants pick out words? Stress is one important clue. English contains many one-syllable words that are stressed and many two-syllable words that have a stressed syllable followed by an unstressed syllable (e.g., *dough'-nut, tooth'-paste, bas'-ket*). Infants pay more attention to stressed syllables than unstressed syllables, which is a good strategy for identifying the beginnings of words (Mattys et al., 1999; Thiessen & Saffran, 2003). And infants learn words more readily when they appear at the beginning and ends of sentences, probably because the brief pause between sentences makes it easier to identify first and last words (Seidl & Johnson, 2006).

Of course, stress is not a foolproof sign. Many two-syllable words have stress on the second syllable (e.g., *gui-tar', sur-prise'*), so infants need other methods to identify words in speech. One method is statistical. Infants notice syllables that go together frequently (Jusczyk, 2002). For example, in many studies, 8-month-olds heard the following sounds, which consisted of four three-syllable artificial words, said over and over in random order:

pa bi ku go la tu da ro pi ti bu do da ro pi go la tu pa bi ku da ro pi . . . .

We've underlined the words and inserted gaps between them so you can see them more easily, but in the studies there were no breaks at all—just a steady flow of syllables for 3 minutes. Later, infants listened to these words less than to new words that were novel combinations of the same syllables. They had detected *pa bi ku, go la tu, da ro pi,* and *ti bu do* as familiar patterns and hence listened to them less than to new "words" like *tu da ro,* even though the latter were made up from syllables they'd already heard (Aslin, Saffran, & Newport, 1998; Pelucci, Hay, & Saffran, 2009).

Yet another way that infants identify words is through their emerging knowledge of how sounds are used in their native language. For example, think about these two pairs of sounds: *s* followed by *t* and *s* followed by *d.* Both pairs of sounds are quite common at the end of one word and the beginning of the next: bu*s t*akes, ki*ss t*ook; thi*s d*og, pa*ss d*irectly. However, *s* and *t* occur frequently within a word (*st*op, li*st,* pe*st, st*ink) but *s* and *d* do not. Consequently, when *d* follows an *s,* it probably starts a new word. In fact, 9-month-olds follow rules like this one because— when they hear novel words embedded in continuous speech—they're more likely to identify the novel word when the final sound in the preceding word occurs infrequently with the first sound of the novel word (Mattys & Jusczyk, 2001). Thus, infants use many powerful tools to identify words in speech. Of course, they don't yet understand the meanings of these words; they just recognize a word as a distinct configuration of sounds.

Another strategy that infants use is to rely on familiar function words, such as the articles *a* and *the,* to break up the speech stream. These words are very common in adults' speech; by 6 months most infants recognize these words and use them to determine the onset of a new word (Shi & Lepage, 2008). For example, for infants familiar with *a* the sequence like *aballabataglove* becomes *a ball a bat a glove.* The new words are isolated by the familiar ones.

Parents (and other adults) often help infants master language sounds by talking in a distinctive style. *In **infant-directed speech**, adults speak slowly and with exaggerated changes in pitch and loudness.* If you listen to a mother talking to her baby, you will notice that she alternates between speaking softly and loudly and between high and low pitches and that her speech seems emotionally expressive (Liu, Tsao, & Kuhl, 2007; Trainor, Austin, & Desjardins, 2000). Infant-directed speech is also known as *motherese* because this form of speaking was first noted in mothers, although it's now known that most caregivers talk this way to infants.

Infant-directed speech may attract infants' attention more than adult-directed speech because its slower pace and accentuated changes provide infants with more (and more salient) language clues (Cristia, 2010). For example, infants can segment words more effectively when they hear them in infant-directed speech (Thiessen, Hill, & Saffran, 2005). In addition, infant-directed speech includes especially good examples of vowels (Kuhl et al., 1997), which may help infants learn to distinguish these sounds. When talking to infants, speaking clearly is a good idea. In one study (Liu, Kuhl, & Tsao, 2003), infants who could best distinguish speech sounds had mothers who spoke most clearly.

Infant-directed speech, then, helps infants perceive the sounds that are fundamental to their language. But how do infants accomplish the next step, producing speech? We answer this question next.

infant-directed speech
speech that adults use with infants that is slow and has exaggerated changes in pitch and volume; it is thought to aid language acquisition

When mothers and other adults talk to young children, they often use infant-directed speech in which they speak slowly and with exaggerated changes in pitch and loudness.

Steps to Speech

As any new parent can testify, newborns and young babies make many sounds—they cry, burp, and sneeze. Language-based sounds don't appear immediately. *At 2 months, infants begin to produce vowel-like sounds, such as "ooooooo" or "ahhhhhh," a phenomenon known as **cooing**.* Sometimes infants become quite excited as they coo, perhaps reflecting the joy of simply playing with sounds.

cooing
early vowel-like sounds that babies produce

babbling
speechlike sounds that consist of vowel–consonant combinations; common at about 6 months

After cooing comes **babbling**, *speechlike sound that has no meaning*. A typical 6-month-old might say "dah" or "bah," utterances that sound like a single syllable consisting of a consonant and a vowel. Over the next few months, babbling becomes more elaborate as babies apparently experiment with more complex speech sounds. Older infants sometimes repeat a sound, as in "bahbahbah," and begin to combine different sounds, such as "dahmahbah" (Hoff, 2009).

Babbling is not just mindless playing with sounds; it is a precursor to real speech. We know this, in part, from video records of people's mouths while speaking. When adults speak, their mouth is open somewhat wider on the right side than on the left side, reflecting the left hemisphere's control of language and muscle movements on the body's right side (Graves & Landis, 1990). Infants do the same when they babble but not when making other nonbabbling sounds, which suggests that babbling is fundamentally linguistic (Holowka & Petitto, 2002).

Other evidence for the linguistic nature of babbling comes from studies of developmental change in babbling: At roughly 8 to 11 months, infants' babbling sounds more like real speech because infants stress some syllables and vary the pitch of their speech (Snow, 2006). In English declarative sentences, for example, pitch first rises and then falls toward the end of the sentence. In questions, however, the pitch is level and then rises toward the end of the question. Older babies' babbling reflects these patterns: Babies who are brought up by English-speaking parents have both the declarative and question patterns of intonation in their babbling. Babies exposed to a language with different patterns of intonation, such as Japanese or French, reflect their language's intonation in their babbling (Levitt & Utman, 1992).

The appearance of intonation in babbling indicates a strong link between perception and production of speech: Infants' babbling is influenced by the characteristics of the speech that they hear (Goldstein & Schwade, 2008). Beginning in the middle of the first year, infants try to reproduce the sounds of language that others use in trying to communicate with them (or, in the case of deaf infants with deaf parents, the signs that others use). Hearing *dog*, an infant may first say "dod" and then "gog" before finally saying "dog" correctly. In the same way that beginning typists gradually link movements of their fingers with particular keys, through babbling infants learn to use their lips, tongue, and teeth to produce specific sounds, gradually making sounds that approximate real words (Poulson et al., 1991). Fortunately, learning to produce language sounds is easier for most babies than the cartoon suggests!

These developments in production of sound, coupled with the 1-year-old's advanced ability to perceive speech sounds, clearly set the stage for the infant's first true words.

First Words and Many More

Recall that Nabina, the 1-year-old in the vignette, looks at the family cat when she hears its name. This phenomenon is common in 10- to 14-month-olds. They appear to understand what others say despite the fact that they have yet to speak. In response to "Where is the book?" children will go find the book. They grasp the question, even

THINK ABOUT IT
Compare and contrast the steps in learning to make speech sounds with Piaget's account of the sensorimotor period.

though their own speech is limited to advanced babbling (Fenson et al., 1994; Hoff-Ginsberg, 1997). Evidently, children have made the link between speech sounds and particular objects, even though they cannot yet manufacture the sounds themselves. As fluent adult speakers, we forget that speech is a motor skill requiring perfect timing and tremendous coordination.

A few months later, most youngsters utter their first words. In many languages, those words are similar (Nelson, 1973; Tardif et al., 2008) and include terms for mother and father, greetings (*Hi, bye-bye*) as well as foods and toys (*juice, ball*). By age 2, most youngsters have a vocabulary of a few hundred words, and by age 6, a typical child's vocabulary includes more than 10,000 words (Bloom, 1998).

The Grand Insight: Words as Symbols

To make the transition from babbling to real speech, infants need to learn that speech is more than just entertaining sound. They need to know that particular sounds form words that can refer to objects, actions, and properties. Put another way, infants must recognize that words are symbols—entities that stand for other entities. Piaget believed that this insight occurs at roughly 18 months of age and marks the beginning of transition from the sensorimotor to the preoperational stage. However, a glimmer of understanding of symbols occurs earlier, soon after the first birthday. By this age, children have already formed concepts such as "round, bouncy things" or "furry things that bark," based on their own experiences. With the insight that speech sounds can denote these concepts, infants begin to identify a word that goes with each concept (Reich, 1986).

If this argument is correct then we should find that children use symbols in other areas, not just in language. They do. Gestures are symbols, and infants begin to gesture shortly before their first birthday (Goodwyn & Acredolo, 1993). Young children may smack their lips to indicate hunger or wave "bye-bye" when leaving. In these cases, gestures and words convey a message equally well.

What's more, gestures sometimes pave the way for language. Before knowing an object's name, infants often point to it or pick it up for a listener, as if saying, "I want this!" or "What's this?" In one study, 50% of all objects were first referred to by gesture and, about 3 months later, by word (Iverson & Goldin-Meadow, 2005). Given this connection between early gestures and first spoken words, it's not surprising that toddlers who are more advanced in their use of gesture tend to have, as preschoolers, more complex spoken language (Rowe & Goldin-Meadow, 2009).

What's What? Fast Mapping of Words

After children develop the insight that a word can symbolize an object or action, their vocabularies grow, but slowly at first. A typical 15-month-old, for example, may learn two to three new words each week. However, at about 18 months, many children experience a naming explosion during which they learn new words—particularly names of objects—much more rapidly than before. Children now learn ten or more new words each week (Fenson et al., 1994; McMurray, 2007).

This rapid rate of word learning is astonishing when we realize that most words have many plausible but incorrect referents. To illustrate, imagine what's going through the mind of a child when her mother points to a flower and says, "Flower. This is a flower. See the flower." This all seems crystal clear to you and incredibly straightforward. But what might the child learn from this episode? Perhaps the correct referent for "flower." But a youngster could, just as reasonably, conclude that "flower" refers to a petal, to the color of the flower, or to the mother's actions in pointing at the flower.

Surprisingly, though, most youngsters learn the proper meanings of simple words in just a few presentations. *Children's ability to connect new words to referents so rapidly that they cannot be considering all possible meanings for the new word is termed* **fast mapping**. How can young children learn new words so rapidly? Researchers believe that many distinct factors contribute to young children's rapid word learning (Hollich, Hirsh-Pasek, & Golinkoff, 2000).

fast mapping
a child's connections between words and referents that are made so quickly that he or she cannot consider all possible meanings of the word

One of the challenges for theories of language learning is to explain how children figure out that the parent's words refer to the object, not to its color or texture and not to the parent.

JOINT ATTENTION Parents encourage word learning by carefully watching what interests their children. When toddlers touch or look at an object, parents often label it for them. When a youngster points to a banana, a parent may say, "Banana, that's a banana." And parents usually simplify the task for children by using just one label for an object (Callanan & Sabbagh, 2004).

Of course, to take advantage of this help, infants must be able to tell when parents are labeling instead of just conversing. In fact, when adults label an unfamiliar object, 18- to 20-month-olds assume that the label is the object's name *only* when adults show signs that they are referring to the object. For example, toddlers are more likely to learn the name of an object or action when adults look at the object or action while saying its name than when adults look elsewhere while labeling (Liebal et al., 2009; Nurmsoo & Bloom, 2008). Young children also consider an adult's credibility as a source: If an adult seems uncertain or has given incorrect names for words in the past, preschoolers are less likely to learn words from them (Birch, Akmal, & Frampton, 2010; Koenig & Woodward, 2010). Thus, beginning in the toddler years, parents and children work together to create conditions that foster word learning: Parents label objects and youngsters rely on adults' behavior to interpret the words they hear. Finally, although joint attention helps children to learn words, it is not required: Children learn new words when they are used in ongoing conversation and when they overhear others use novel words (Akhtar, Jipson, & Callanan, 2001).

CONSTRAINTS ON WORD NAMES Joint attention simplifies word learning for children, but the problem still remains: How does a toddler know that banana refers to the object that she's touching, as opposed to her activity (touching) or to the object's color? Many researchers believe that young children follow several simple rules that limit their conclusions about what labels mean.

A study by Au and Glusman (1990) shows how researchers have identified rules that young children use. Au and Glusman presented preschoolers with a stuffed animal with pink horns that otherwise resembled a monkey and called it a *mido. Mido* was then repeated several times, always referring to the monkeylike stuffed animal with pink horns. Later, these youngsters were asked to find a *theri* in a set of stuffed animals that included several *mido*. Never having heard of a *theri*, what did the children do? They never picked a *mido*; instead, they selected other stuffed animals. Knowing that *mido* referred to monkeylike animals with pink horns, evidently they decided that *theri* must refer to one of the other stuffed animals.

Apparently children were following this simple but effective rule for learning new words:

- If an unfamiliar word is heard in the presence of objects that already have names and objects that don't, the word refers to one of the objects that doesn't have a name.

Researchers have discovered several other simple rules that help children match words with the correct referent (Hoff, 2009; Woodward & Markman, 1998):

- A name refers to a whole object, not its parts or its relation to other objects, and refers not just to this particular object but to all objects of the same type (Hollich, Golinkoff, & Hirsh-Pasek, 2007). For example, when a grandparent points to a stuffed animal on a shelf and says "dinosaur," children conclude that *dinosaur* refers to the entire dinosaur, not just its ears or nose, not to the fact that the dinosaur is on a shelf, and not to this specific dinosaur but to all dinosaurlike objects.

- If an object already has a name and another name is presented, the new name denotes a subcategory of the original name. If the child who knows the meaning of *dinosaur* sees a brother point to another dinosaur and hears the brother say "T-rex," the child will conclude that *T-rex* is a special type of dinosaur.

■ Given many similar category members, a word applied consistently to only one of them is a proper noun. If a child who knows *dinosaur* sees that one of a group of dinosaurs is always called "Dino," the child will conclude that *Dino* is the name of that dinosaur.

Rules like these make it possible for children like Nabina, the child in the vignette, to learn words rapidly because they reduce the number of possible referents. The child being shown a flower follows these rules to decide that *flower* refers to the entire object, not its parts or the action of pointing to it.

SENTENCE CUES Children hear many unfamiliar words embedded in sentences containing words they already know. The other words and the overall sentence structure can be helpful clues to a word's meaning (Yuan & Fisher, 2009). For example, when a parent describes an event using familiar words but an unfamiliar verb, children often infer that the verb refers to the action performed by the subject of the sentence (Fisher, 1996; Woodward & Markman, 1998). When youngsters hear, "The man is juggling," they will infer that *juggling* refers to the man's actions with the bowling pins because they already know *man* and because *-ing* refers to ongoing actions.

As another example of how sentence context aids word learning, look at the blocks in ■ Figure 4.6 and point to the "boz block." You probably pointed to the middle block. Why? In English, adjectives usually precede the nouns they modify, so you inferred that *boz* is an adjective describing *block*. Since *the* before *boz* implies that only one block is *boz*, you picked the middle one, having decided that *boz* means "winged." Toddlers, too, use sentence cues like these to judge word meanings. Hearing "This is a zav," 2-year-olds will interpret *zav* as a category name; but hearing "This is Zav" (without the article *a*), they interpret *zav* as a proper name (Hall, Lee, & Belanger, 2001).

COGNITIVE FACTORS The naming explosion coincides with a time of rapid cognitive growth, and children's increased cognitive skill helps them to learn new words. As children's thinking becomes more sophisticated and, in particular, as they start to have goals and intentions, language becomes a means to express those goals and to achieve them. Thus, intention provides children with an important motive to learn language—to help achieve their goals (Bloom & Tinker, 2001).

In addition, young children's improving attentional and perceptual skills also promote word learning. Smith (2000, 2009), for example, argues that shape plays a central role in learning words. Infants and young children spontaneously pay attention to an object's shape, and they use this bias to learn new words. In Smith's theory, children first associate names with a single object: "ball" is associated with a specific tennis ball, and "cup" is associated with a favorite sippy cup. As children encounter new balls and new cups, however, they hear the same words applied to similarly shaped objects and reach the conclusion that balls are round and cups are cylinders with handles. With further experience, children derive an even more general rule: Objects that have the same shape have the same name. From this, children realize that paying attention to shape is an easy way to learn names. Consistent with this theory, the shape bias and the naming explosion typically occur at about the same time (Gershkoff-Stowe & Smith, 2004).

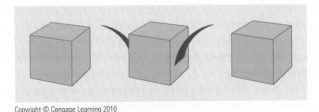

Copyright © Cengage Learning 2010

■ Figure 4.6
"The boz block" probably refers to the middle block because "the" implies that only one block is "boz" and the middle block is the only one with wings.

DEVELOPMENTAL CHANGE IN WORD LEARNING Some of the word-learning tools described in the past few pages are particularly important at different ages (Hirsh-Pasek & Golinkoff, 2008). Before 18 months, infants learn words relatively slowly—often just one new word each day. At this age, children rely heavily on simple attentional processes (e.g., the shape bias) to learn new words. But by 24 months, most children are learning many new words daily. This faster learning reflects children's greater use of language cues (e.g., constraints on names) and a speaker's social cues. At any age, infants and toddlers rely on a mixture of word-learning tools, but with age they gradually move away from attentional cues to language and social cues.

NAMING ERRORS Of course, these rules for learning new words are not perfect; initial mappings of words onto meanings are often only partially correct (Hoff & Naigles, 2002). *A common mistake is* **underextension**, *defining a word too narrowly*. Using *car* to refer only to the family car and *ball* to a favorite toy ball are examples of underextension. *Between 1 and 3 years, children sometimes make the opposite error,* **overextension**, *defining a word too broadly*. Children may use *car* to also refer to buses and trucks or use *doggie* to refer to all four-legged animals.

The overextension error occurs more frequently when children are producing words than when they are comprehending words. Two-year-old Jason may say "doggie" to refer to a goat but nevertheless correctly point to a picture of a goat when asked. Because overextension is more common in word production, it may actually reflect another fast-mapping rule that children follow: "If you can't remember the name for an object, say the name of a related object" (Naigles & Gelman, 1995). Both underextension and overextension disappear gradually as youngsters refine meanings for words after increased exposure to language.

Individual Differences in Word Learning

The naming explosion typically occurs at about 18 months, but like many developmental milestones, the timing of this event varies widely for individual children. Some youngsters have a naming explosion as early as 14 months, but for others it may be as late as 22 months (Goldfield & Reznick, 1990). Another way to make this point is to look at variation in the size of children's vocabulary at a specific age. At 18 months, for example, an average child's vocabulary would have about 75 words, but a child in the 90th percentile would know nearly 250 words and a child in the 10th percentile fewer than 25 words (Fenson et al., 1994).

This range in vocabulary size for typical 18-month-olds is huge—from 25 to 250 words! What can account for this difference? Heredity contributes: Twin studies find that vocabulary size is more similar in identical twins than in fraternal twins (Dionne et al., 2003). But the difference is fairly small, indicating a relatively minor role for genetics.

More important are two other factors. *One is* **phonological memory**, *the ability to remember speech sounds briefly*. This is often measured by saying a nonsense word to children—"ballop" or "glistering"—and asking them to repeat it immediately. Children's skill in recalling such words is strongly related to the size of their vocabulary (Gathercole et al., 1992; Leclercq & Majerus, 2010). Children who have difficulty remembering speech sounds accurately find word learning particularly challenging, which is not surprising because word learning involves associating meaning with an unfamiliar sequence of speech sounds.

However, the single most important factor in growth of vocabulary is the child's language environment. Children have larger vocabularies when they are exposed to much high-quality language. The more words that children hear, the better (Hurtado, Marchman, & Fernald, 2008). Specifically, children learn more words when their parents' speech is rich in different words and is grammatically sophisticated (Hoff, 2003; Huttenlocher et al., 2010), and when parents respond promptly and appropriately to their children's talk (Tamis-Lemonda & Bornstein, 2002).

underextension
when children define words more narrowly than adults do

overextension
when children define words more broadly than adults do

phonological memory
ability to remember speech sounds briefly; an important skill in acquiring vocabulary

THINK ABOUT IT

Gavin and Mitch are both 16-month-olds. Gavin's vocabulary includes about 14 words, but Mitch's has about 150 words, more than 10 times as many as Gavin. What factors contribute to this difference?

BILINGUALISM Millions of American children grow up in bilingual households; these youngsters usually speak English and another language. When infants learn two languages simultaneously, they often progress somewhat slowly at first. They mix words from the two languages and are less skilled at using language-specific sounds to guide word learning (Fennell, Byers-Heinlein, & Werker, 2007). Soon, however, they separate the languages, and bilingual children reach most language milestones at about the same age as monolingual children (Pettito et al., 2001). When each language is considered separately, bilingual children often have somewhat smaller vocabularies than monolingual children (Umbel et al., 1992). However, because bilingual youngsters often know words in one language but not the other, their total vocabulary (i.e., words known in both languages plus words known in either language but not both) is greater than that of monolingual children.

Being bilingual also has some important language and cognitive advantages. Bilingual children better understand that words are simply arbitrary symbols. Bilingual youngsters, for instance, are more likely than monolingual children to understand that, as long as all English speakers agreed, *dog* could refer to cats and *cat* could refer to dogs (Bialystok, 1988; Campbell & Sais, 1995). And they are more skilled at switching back and forth between tasks and often are better able to inhibit inappropriate responses (Bialystok, 2010; Carlson & Meltzoff, 2008), perhaps reflecting their experience of switching between languages and inhibiting relevant words from the "other" language (e.g., when shown a photo of a dog and asked, "What's this?" preschoolers bilingual in French and English must respond "dog" while suppressing "chien").

referential style
language-learning style of children whose vocabularies are dominated by names of objects, persons, or actions

expressive style
language-learning style of children whose vocabularies include many social phrases that are used like one word

WORD LEARNING STYLES As youngsters expand their vocabulary, they often adopt a distinctive style of learning language (Bates, Bretherton, & Snyder, 1988; Nelson, 1973). *Some children have a* **referential style***; their vocabularies mainly consist of words that name objects, persons, or actions.* For example, Rachel, a referential child, had 41 name words in her 50-word vocabulary but only two words for social interaction or questions. *Other children have an* **expressive style***; their vocabularies include some names but also many social phrases that are used like a single word, such as "go away," "what'd you want?" and "I want it."* Elizabeth, an expressive child, had a more balanced vocabulary, with 14 words for social interactions and questions and 24 name words.

Referential and expressive styles represent end points on a continuum; most children are somewhere in between. For children with referential emphasis, language is primarily an intellectual tool: a means of learning and talking about objects (Masur, 1995). In contrast, for children with expressive emphasis, language is more of a social tool: a way of enhancing interactions with others. Of course, both of these functions—intellectual and social—are important functions of language, which explains why most children blend the referential and expressive styles of learning language.

Encouraging Language Growth

How can parents and other adults help children learn words? For children to expand their vocabularies, they need to hear others speak. Not surprisingly, then, children learn words more rapidly if their parents speak to them frequently (Huttenlocher et al., 1991; Roberts, Burchinal, & Durham, 1999). Of course, sheer quantity of parental speech is not all that matters. Parents can foster word learning by naming objects that are the focus of a child's attention (Dunham, Dunham, & Curwin, 1993). Parents can name different products on store shelves as they point to them. During a walk, parents can label the objects—birds, plants, vehicles—that the child sees.

Parents can also help children learn words by reading books with them. Reading together is fun for parents and children alike and pro-

Researchers have found that parents are more effective than videos in teaching new words to their children.

© Leila Cutler / Alamy

vides opportunities for children to learn new words. However, the way that parents read makes a difference. When parents carefully describe pictures as they read, preschoolers' vocabularies increase (Reese & Cox, 1999). Asking children questions during reading also helps (Sénéchal, Thomas, & Monker, 1995). When an adult reads a sentence (e.g., "Arthur is angling"), then asks a question (e.g., "What is Arthur doing?"), a child must match the new word (*angling*) with the pictured activity (*fishing*) and say the word aloud. When parents read without questioning, children can ignore words they don't understand. Questioning forces children to identify meanings of new words and practice saying them.

Of course, video is an essential part of the lives of infants and young children. Does it help them to learn new words? For preschool children the answer is "yes," at least under some circumstances. For example, preschool children who regularly watch *Sesame Street* usually have larger vocabularies than preschoolers who watch *Sesame Street* only occasionally (Wright et al., 2001). Other programs that promote word learning are those that tell a story (e.g., *Thomas the Tank Engine*) as well as programs like *Blue's Clues* and *Dora the Explorer*, which directly ask questions of the viewer. And the benefits of these programs are greatest when preschoolers watch them with adults, in part because the video contents become the focus of joint attention, as described on page 152. In contrast, most cartoons have no benefit for language learning (Linebarger & Vaala, 2010).

Spotlight on research — Do Infants Learn Words From Watching Infant-Oriented Media?

Who were the investigators, and what was the aim of the study? Although marketing and some testimonials suggest that infants expand their vocabulary from watching infant-oriented video, there's very little experimental work on the issue. Because correlational studies (e.g., Zimmerman et al., 2007) suggested a negative relation between exposure to infant videos and the size of infants' vocabularies (i.e., more exposure was associated with smaller vocabularies), Judy DeLoache and her colleagues (2010) conducted an experiment to determine the impact of exposure to infant-oriented videos on word learning.

How did the investigators measure the topic of interest? DeLoache et al. created four different conditions. In two of them, parents were given a commercially available DVD that's designed to teach new words to young children. The video includes 25 common objects and each object is labeled three times (e.g., This is a *clock*). In both conditions, infants watched the video at home five times a week for four weeks. However, in one condition they watched it together with a parent; in another condition, they watched it alone (although the parent was usually in the same room). In a third condition, parents were given a list of the 25 words presented in the video and encouraged to teach the words to their infant ". . . in whatever way seems natural to you" (p. 1571) over the same four-week period. Finally, in a control condition, infants were not exposed to the 25 words in any way;

instead, they were simply tested at the beginning and the end of the four-week period to determine which words they understood. Infants in the other three conditions were also tested in this manner: Infants were shown a replica of one of the objects shown in the video (e.g., a clock) along with a replica of an object not shown in the video (e.g., a fan) and the experimenter asked infants to show the target object (e.g., "Can you show me the clock?").

Who were the children in the study? DeLoache and her colleagues tested seventy-two 12- to 18-month-olds.

What was the design of the study? This study was experimental: The independent variable was the nature of the infants' exposure to the 25 words in the video (video with parental interaction, video only, parental teaching, no systematic exposure). The dependent variable was the percentage of times that the infants selected the correct replica. The study was not developmental (12- to 18-month-olds were tested just once), so it was neither cross-sectional nor longitudinal. However, the study nicely illustrates a field experiment (i.e., an experiment conducted in infants' homes).

Were there ethical concerns with the study? No. The task posed no danger to the infants, who were seated on a caregiver's lap throughout testing.

What were the results? During the initial testing, most infants knew 6–7 words; these

words were not presented during the final testing. During final testing, infants learned the most words when parents taught them directly. These infants learned about half of the words. In contrast, the infants in the remaining conditions (the two video conditions and the control condition) only learned about one third of the words. In other words, regular daily exposure to the 25 words through the video produced no greater word learning than incidental, casual exposure that took place in the control condition.

What did the investigators conclude? The findings indicate that the video was ineffective in promoting word learning. In the words of DeLoache et al., ". . . the degree to which babies actually learn from baby videos is negligible" (p. 1573).

What converging evidence would strengthen these conclusions? DeLoache and colleagues tested only a single video; extending the work to other DVDs would be useful. In addition, most of the infants came from middle-class homes; it would be important to determine whether these videos have any effectiveness in promoting language for infants from families with lower socioeconomic status.

 Go to Psychology CourseMate at www.cengagebrain.com to enhance your understanding of this research.

What about videos claiming that they promote word learning in infants? Most of the evidence suggests that before 18 months of age, infant-oriented videos (e.g., *Baby Einstein, Brainy Baby*) are not effective in promoting infants' word learning (Linebarger & Vaala, 2010). The Spotlight on Research feature describes a study reporting this sort of negative evidence.

Why do baby videos seem to have no benefit for infants' word learning, particularly when videos for preschool children are effective? According to Linebarger and Vaala (2010), one reason is that these videos are ". . . poorly designed, insufficient to support language processing, and developmentally inappropriate" (p. 184). Another reason is that 12- to 18-month olds have limited understanding of relationships between real objects and their depictions in photographs and video. In other words, they have difficulty relating what they see in the video to those objects and actions as experienced in their own lives (Troseth, Pierroutsakos, & DeLoache, 2004).

Research on video and on parents' influence points to a simple but powerful conclusion: Children are most likely to learn new words when they participate in activities that force them to understand the meanings of new words and use those new words.

Speaking in Sentences: Grammatical Development

Within months after children say their first words, they begin to form simple two-word sentences. Such sentences are based on "formulas" that children figure out from their own experiences (Braine, 1976; Radford, 1995). Armed with a few formulas, children can express an enormous variety of ideas:

Formula	Example
actor + action	Mommy sleep, Timmy run
action + object	Gimme cookie, throw ball
possessor + possession	Kimmy pail, Maya shovel

Each child develops a unique repertoire of formulas, reflecting his or her own experiences. However, the formulas listed here are commonly used by many children growing up in different countries around the world.

From Two Words to Complex Sentences

Children rapidly move beyond two-word sentences, first doing so by linking two-word statements together: "Rachel kick" and "Kick ball" become "Rachel kick ball." Even longer sentences soon follow; sentences with 10 or more words are common in 3-year-olds' speech. For example, at 1½ years, Laura Kail would say "Gimme juice" or "Bye-bye Ben." As a 2½-year-old, she had progressed to "When I finish my ice cream, I'll take a shower, okay?" and "Don't turn the light out—I can't see better!"

Children's two- and three-word sentences often fall short of adults' standards of grammar. Youngsters will say "He eating" rather than "He is eating," or "two cat" rather than "two cats." *This sort of speech is called* **telegraphic** *because, like telegrams of days gone by, children's speech includes only words directly relevant to meaning, and nothing more.* Before cell phones and e-mail, people sent urgent messages by telegraph, and the cost was based on the number of words. Consequently, telegrams were brief and to the point, containing only the important nouns, verbs, adjectives, and adverbs—much like children's two-word speech. *The missing elements,* **grammatical morphemes,** *are words or endings of words (such as* -ing, -ed, *or* -s) *that make a sentence grammatical.* During the preschool years, children gradually acquire the grammatical morphemes, first mastering those that express simple relations like *-ing*, which is used to denote that the action expressed by the verb is ongoing. More complex forms, such as appropriate use of the various forms of the verb *to be*, are mastered later (Peters, 1995).

Children's use of grammatical morphemes is based on their growing knowledge of grammatical rules, not simply memory for individual words. This was first demon-

telegraphic speech
speech used by young children that contains only the words necessary to convey a message

grammatical morphemes
words or endings of words that make a sentence grammatical

This is a wug.

Now there is another one.
There are two of them.
There are two _____.

Berko, 1958.

■ **Figure 4.7**
When shown the two birds, young children usually refer to them as two "wugs," spontaneously adding an s to "wug" to make it plural.

overregularization
grammatical usage that results from applying rules to words that are exceptions to the rule

strated in a landmark study by Berko (1958) in which preschoolers were shown pictures of nonsense objects like the one in ■ Figure 4.7. The experimenter labeled it, saying, "This is a wug." Then youngsters were shown pictures of two of the objects, and the experimenter said, "These are two." Most children spontaneously said, "wugs." Because both the singular and plural forms of this word were novel for these youngsters, they could have generated the correct plural form only by applying the familiar rule of adding -s.

Children growing up in homes where English is spoken face the problem that their native tongue is highly irregular, with many exceptions to the rules. *Sometimes children apply rules to words that are exceptions to the rule, errors called* **overregularizations**. With plurals, for example, youngsters may incorrectly add an -s instead of using an irregular plural—two "mans" instead of two "men." With the past tense, children may add -ed instead of using an irregular past tense: "I goed home" instead of "I went home" (Marcus et al., 1992; Mervis & Johnson, 1991).

These examples give some insight into the complexities of mastering the grammatical rules of one's language. Not only must children learn an extensive set of specific rules, they must also absorb—on a case-by-case basis—all of the exceptions. Despite the enormity of this task, most children have mastered the basics of their native tongue by the time they enter school. How do they do it? Biological, psychological, and sociocultural forces all contribute.

How Do Children Acquire Grammar?

Most youngsters can neither read nor do arithmetic when they enter kindergarten, but virtually all have mastered the fundamentals of grammar of their native tongue. How do they do it? Theorists have proposed several different answers to this question.

THE BEHAVIORIST ANSWER Probably the simplest explanation for learning grammar is that children imitate the grammatical forms they hear. In fact, B. F. Skinner (1957) and other learning theorists once claimed that all aspects of language—sounds, words, grammar, and communication—are learned through imitation and reinforcement (Moerk, 2000; Whitehurst & Vasta, 1975).

Critics were quick to point to some flaws in this theory. One problem is that most of children's sentences are novel, which is difficult to explain in terms of simple imitation of adults' speech. For example, when young children create questions by inserting a *wh* word at the beginning of a sentence ("What she doing?"), who are they imitating? Also troublesome is that, even when children imitate adult sentences, they do not imitate adult grammar. In simply trying to repeat "I am drawing a picture," young children will say "I draw picture." Finally, linguists (see, e.g., Chomsky, 1957, 1995) have argued that grammatical rules are far too complex for toddlers and preschoolers to infer them solely on the basis of speech that they hear.

THE LINGUISTIC ANSWER Many scientists believe that children are born with mechanisms that simplify the task of learning grammar (Slobin, 1985). According to this view, children are born with neural circuits in the brain that allow them to infer the grammar of the language that they hear. That is, grammar itself is not built into the child's nervous system, but processes that guide the learning of grammar are. Many findings indirectly support this view:

1. If children are born with a "grammar learning processor," then specific regions of the brain should be involved in learning grammar. As we discussed on page 96, the left hemisphere of the brain plays a critical role in understanding language. And by 2 years of age, specific regions of the left hemisphere are activated when sentences break simple grammatical rules, such as a noun appearing when a verb would be expected (Bernal et al., 2010).
2. If learning grammar depends on specialized neural mechanisms that are unique to humans, then efforts to teach grammar to nonhumans should fail. This prediction has been tested by trying to teach grammar to chimpanzees, the species closest to humans on the evolutionary ladder. The result: Chimps mas-

ter a handful of grammatical rules governing two-word speech, but only with massive effort that is completely unlike the preschool child's learning of grammar (Savage-Rumbaugh, 2001; Seyfarth & Cheney, 1996).

3. The period from birth to about 12 years is a critical period for acquiring language generally and mastering grammar particularly. If children do not acquire language in this period, they never truly master language later (Newport, 1991; Rymer, 1993).

4. The mastery of grammar is closely related to vocabulary growth in a way that suggests both are part of a common, emerging language system (Dixon & Marchman, 2007). For example, one idea is that, as children learn words, they learn not only a word's meaning but also about the kinds of sentences in which a word appears and its position in those sentences. They learn the meaning of "teacher" and that "teacher" can appear as the actor and object in transitive sentences. Grammar then emerges naturally as children learn more and more words.

Although these findings are consistent with the idea that children have innate grammar-learning mechanisms, they do not prove the existence of such mechanisms. Consequently, scientists have continued to look for other explanations.

THE COGNITIVE ANSWER Some theorists (Braine, 1992) believe that children learn grammar through powerful cognitive skills that help them rapidly detect regularities in their environment, including patterns in the speech they hear. According to this approach, it's as if children establish a huge spreadsheet that has the speech they've heard in one column and the context in which they heard it in a second column; periodically infants scan the columns looking for recurring patterns (Maratsos, 1998). For example, children might be confused the first time they hear -s added to the end of a familiar noun. However, as the database expands to include many instances of familiar nouns with an added -s, children discover that -s is always added to a noun when there are multiple instances of the object. Thus, they create the rule: noun + -s = plural. With this view, children learn language by searching for regularities across many examples that are stored in memory, not through an inborn grammar-learning device (Bannard & Matthews, 2008).

THE SOCIAL-INTERACTION ANSWER This approach is eclectic, drawing on each of the views we've considered. From the behaviorist approach, it takes an emphasis on the environment; from the linguistic approach, that language learning is distinct; and from the cognitive view, that children have powerful cognitive skills they can use to master language. The unique contribution of this perspective is emphasizing that much language learning takes place in the context of interactions between children and adults, with both parties eager for better communication (Bloom & Tinker, 2001). Children have an ever-expanding repertoire of ideas and intentions that they wish to convey to others, and caring adults want to understand their children, so both parties work to improve language skills as a means toward better communication. Thus, improved communication provides an incentive for children to master language and for adults to help them.

None of these accounts provides a comprehensive explanation of how grammar is mastered. But many scientists believe the final explanation will include contributions from the linguistic, cognitive, and social-interaction accounts. That is, children's learning of grammar will be explained in terms of some mechanisms specific to learning grammar, children actively seeking to identify regularities in their environment, and linguistically rich interactions between children and adults (MacWhinney, 1998).

Communicating With Others

Imagining two preschoolers arguing is an excellent way to learn what is needed for effective communication. Both youngsters probably try to speak at the same time; their remarks may be rambling or incoherent; and they neglect to listen to each other altogether. These actions reveal three key elements in effective oral communication with others (Grice, 1975):

- People should take turns, alternating as speaker and listener.
- When speaking, your remarks should be clear from the listener's perspective.
- When listening, pay attention and let the speaker know if his or her remarks don't make sense.

Complete mastery of these elements is a lifelong pursuit. After all, even adults often miscommunicate with one another, violating each of these prescriptions in the process. However, youngsters grasp many of the basics of communication early in life.

Taking Turns

Many parents begin to encourage turn-taking long before infants have said their first words (Field & Widmayer, 1982):

PARENT: Can you see the bird?
INFANT: (cooing) ooooh.
PARENT: It *is* a pretty bird.
INFANT: ooooh.
PARENT: You're right, it's a cardinal.

Soon after 1-year-olds begin to speak, parents encourage their youngsters to participate in conversational turn-taking. To help their children along, parents often carry both sides of the conversation to show how the roles of speaker and listener are alternated (Hoff, 2009):

PARENT: (initiating conversation) What's Kendra eating?
PARENT: (illustrating reply for child) She's eating a cookie.

Help of this sort is needed less often by age 2, when spontaneous turn-taking is common in conversations between youngsters and adults (Barton & Tomasello, 1991). By 3 years of age, children have progressed to the point that, when a listener fails to reply promptly, the child will often repeat his or her remarks to elicit a response and keep the conversation moving (Garvey & Berninger, 1981).

Arguments can often be traced to people's failure to follow the fundamental conversational rules of taking turns, speaking clearly, and listening carefully.

Speaking Effectively

When do children first try to initiate communications with others? In fact, what appear to be the first deliberate attempts to communicate typically emerge at 10 months (Golinkoff, 1993). Infants at this age may touch or point to an object while simultaneously looking at another person. They continue this behavior until the person acknowledges them. It's as if the child is saying, "This is a neat toy! I want you to see it, too."

Beginning at 10 months, an infant may point, touch, or make noises to get an adult to do something. An infant in a playpen who wants a toy that is out of reach may make noises while pointing to the toy. The noises capture an adult's attention, and the pointing indicates what the baby wants (Tomasello, Carpenter, & Liszkowski, 2007). The communication may be a bit primitive by adult standards, but it works for babies! And mothers typically translate their baby's pointing into words, so that gesture paves the wave for learning words (Goldin-Meadow, Mylander, & Franklin, 2007). After the first birthday, children begin to use speech to communicate and often initiate conversations with adults (Bloom et al., 1996). Toddlers' first conversations are about themselves, but their conversational scope expands rapidly to include objects in the environment (e.g., toys, food). Later, conversations begin to include more abstract notions, such as hypothetical objects and past or future events (Foster, 1986).

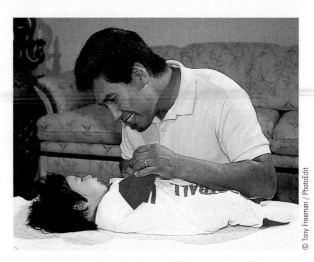

In early parent–child "conversations," parents usually carry both sides of the conversation, alternating as speaker and listener.

Of course, young children are not always skilled conversational partners. At times their communications are confusing, leaving a listener to wonder, "What was that all about?" Every message—whether an informal conversation or a formal lecture—should have a clear meaning. But saying something clearly is often difficult because clarity can only be judged by considering the listener's age, experience, and knowledge of the topic, along with the context of the conversation. For example, think about the simple request, "Please hand me the Phillips screwdriver." This message may be clear to older listeners who are familiar with variants of screwdrivers, but it is vague to younger listeners, to whom all screwdrivers come from the same mold. Of course, if the toolbox is filled with Phillips screwdrivers of assorted sizes, the message is ambiguous even to a knowledgeable listener.

Consistently constructing clear messages is a fine art, which we would hardly expect young children to have mastered. By the preschool years, however, youngsters have made their initial attempts to calibrate messages, adjusting them to match the listener and the context. For example, preschool children give more elaborate messages to listeners who lack access to critical information than to listeners who have this information (Nadig & Sedivy, 2002; O'Neill, 1996). For example, a child describing where to find a toy will give more detailed directions to a listener whose eyes were covered when the toy was hidden. And if a word's meaning might be ambiguous in the context of the conversation (e.g., bat as an animal versus a piece of sporting equipment), young children sometimes gesture to indicate the meaning (Kidd & Holler, 2009).

Listening Well

Sometimes messages are vague or confusing; in such situations, a listener needs to ask the speaker to clarify the message. Preschoolers do not always realize when a message is ambiguous. Told to find "the red toy," they may promptly select the red ball from a pile that includes a red toy car, a red block, and a red toy hammer. Instead of asking the speaker to refer to a specific red toy, preschool listeners often assume they know which toy the speaker had in mind (Beal & Belgrad, 1990). During the elementary-school years, youngsters gradually master the many elements involved in determining whether another person's message is consistent and clear (Ackerman, 1993).

THINK ABOUT IT

Compare Piaget's theory, Vygotsky's theory, and the information-processing approach in their emphasis on the role of language in cognitive development.

Improvement in communication skill is yet another astonishing accomplishment in language during the first 5 years of life; changes are summarized in ●Table 4.2. By the time children are ready to enter kindergarten, they use language with remarkable proficiency and are able to communicate with growing skill.

● TABLE 4.2

Major Milestones of Language Development

Age	Milestones
Birth to 1 year	Babies hear phonemes from birth. They begin to coo between 2 and 4 months, then begin to babble at about 6 months.
About the 1st birthday	Babies begin to talk and to gesture, showing they have begun to use symbols.
1–3 years	Vocabulary expands rapidly (due to fast mapping); particularly at about 18 months. Two-word sentences emerge in telegraphic speech at about 18 months and more complex sentences are evident by 3 years. Turn-taking is evident in communication by 2 years.
3–5 years	Vocabulary continues to expand; grammatical morphemes are added; and children begin to adjust their speech to listeners but, as listeners, often ignore problems in messages they receive.

Copyright © Cengage Learning 2010

Test Yourself

RECALL

1. _____ are fundamental sounds used to create words.

2. Infants' mastery of language sounds may be fostered by _____, in which adults speak slowly and exaggerate changes in pitch and loudness.

3. Older infants' babbling often includes _____, a pattern of rising and falling pitch that distinguishes statements from questions.

4. Youngsters with a(n) _____ style have early vocabularies dominated by words that are names and use language primarily as an intellectual tool.

5. In _____, a young child's meaning of a word is broader than an adult's meaning.

6. Answers to the question, "How do children acquire grammar?" include linguistic, cognitive, and _____ influences.

7. When talking to listeners who lack critical information, preschoolers _____.

INTERPRET

How do the various explanations of grammatical development differ in their view of the child's role in mastering grammar?

APPLY

According to Piaget's theory, preschoolers are egocentric. How should this egocentrism influence their ability to communicate? Are the findings we have described on children's communication skills consistent with Piaget's view?

Recall answers: (1) Phonemes, (2) infant-directed speech, (3) intonation, (4) referential, (5) overextension, (6) social-interaction, (7) provide more elaborate messages

SUMMARY

4.1 THE ONSET OF THINKING: PIAGET'S ACCOUNT

According to Piaget, how do schemes, assimilation, and accommodation provide the foundation for cognitive development throughout the life span?

■ In Piaget's view, children construct their own understanding of the world by creating schemes, categories of related events, objects, and knowledge. Infants' schemes are based on actions, but older children's and adolescents' schemes are based on functional, conceptual, and abstract properties.

■ Schemes change constantly. In assimilation, experiences are readily incorporated into existing schemes. In accommodation, experiences cause schemes to be modified.

■ When accommodation becomes much more common than assimilation, this signals that schemes are inadequate and so children reorganize them. This reorganization produces four different phases of mental development from infancy through adulthood.

How does thinking become more advanced as infants progress through the sensorimotor stage?

- The first 2 years of life constitute Piaget's sensorimotor period. Over these 2 years, infants begin to adapt to and explore their environment, understand objects, and learn to use symbols.

What are the distinguishing characteristics of thinking during the preoperational stage?

- From 2 to 7 years of age, children are in Piaget's preoperational period. Although now capable of using symbols, their thinking is limited by egocentrism—the inability to see the world from another's point of view. Preoperational children are also centered in their thinking and sometimes confuse appearance with reality.

What are the strengths and weaknesses of Piaget's theory?

- One important contribution of Piaget's theory is the view that children actively try to understand their world. Another contribution is specifying conditions that foster cognitive development.

- However, the theory has been criticized because it underestimates infants' and preschoolers' competence, is vague regarding processes of change, does not account for variability in performance, and undervalues the influence of the sociocultural environment.

How have contemporary researchers extended Piaget's theory?

- In contrast to Piaget's idea that children create a comprehensive theory that integrates all their knowledge, the modern view is that children are specialists who generate naive theories in particular domains, including physics and biology. Infants understand many properties of objects; they know how objects move, what happens when objects collide, and that objects fall when not supported.

- Infants understand the difference between animate and inanimate objects. As preschoolers, children know that—unlike inanimate objects—animate objects move themselves, grow, have distinct internal parts, resemble their parents, and repair through healing.

4.2 INFORMATION PROCESSING DURING INFANCY AND EARLY CHILDHOOD

What is the basis of the information-processing approach?

- According to the information-processing view, cognitive development involves changes in mental hardware and in mental software.

How well do young children pay attention?

- Infants use habituation to filter unimportant stimuli. Compared to older children, preschoolers are less able to pay attention to task-relevant information. Their attention can be improved by making irrelevant stimuli less noticeable.

What kinds of learning take place during infancy?

- Infants are capable of many forms of learning, including classical conditioning, operant conditioning, and imitation.

Do infants and preschool children remember?

- Infants can remember and can be reminded of events they seem to have forgotten. Memory improves during infancy, reflecting growth of the brain. Autobiographical memory emerges in the preschool years, reflecting children's growing language skills and their sense of self.

- Preschoolers sometimes testify in cases of child abuse. When questioned repeatedly, they often have difficulty distinguishing what they experienced from what others may suggest they have experienced. Inaccuracies of this sort can be minimized by following certain guidelines when interviewing children, such as interviewing them as soon as possible after the event.

What do infants know about numbers?

- Infants are able to distinguish small quantities, such as "twoness" from "threeness." By 3 years of age, children can count small sets of objects and in so doing adhere to the one-to-one, stable-order, and cardinality principles.

- Learning to count to larger numbers involves learning rules about unit and decade names.

4.3 MIND AND CULTURE: VYGOTSKY'S THEORY

What is the zone of proximal development? How does it help explain how children accomplish more when they collaborate with others?

- Vygotsky believed that cognition develops first in a social setting and only gradually comes under the child's independent control. The difference between what children can do with assistance and what they can do alone constitutes the zone of proximal development.

Why is scaffolding a particularly effective way of teaching youngsters new concepts and skills?

- Control of cognitive skills is most readily transferred to the child through scaffolding, a teaching style in which teachers let children take on more and more of a task as they master its different components. Scaffolding is common worldwide, but the specific techniques for scaffolding children's learning vary from one cultural setting to the next.

When and why do children talk to themselves as they solve problems?

- Children often talk to themselves, particularly when the task is difficult or after they have made a mistake. Such private speech is one way that children regulate their behavior, and it represents an intermediate step in the transfer of control of thinking from others to the self.

4.4 LANGUAGE

When do infants first hear and make speech sounds?

■ Phonemes are the basic units of sound from which words are constructed. Infants can hear phonemes soon after birth. They can even hear phonemes that are not used in their native language, but this ability diminishes after the first birthday.

■ Infant-directed speech is adults' speech to infants that is slower and has greater variation in pitch and loudness. Infants prefer infant-directed speech, perhaps because it gives them additional language clues.

■ Newborns' communication is limited to crying, but babies coo at about 3 months of age. Babbling soon follows, consisting of a single syllable; over several months, infants' babbling comes to include longer syllables as well as intonation.

When do children start to talk? How do they learn word meanings?

■ After a brief period in which children appear to understand others' speech but do not speak themselves, most infants begin to speak around the first birthday. The first use of words is triggered by the realization that words are symbols. Soon after, the child's vocabulary expands rapidly.

■ Most children learn the meanings of words much too rapidly for them to consider all plausible meanings systematically. Instead, children use certain rules to determine the probable meanings of new words. The rules do not always yield the correct meaning. An underextension is a child's meaning that is narrower than an adult's meaning; an overextension is a child's meaning that is broader.

■ Individual children differ in vocabulary size, differences that are due to phonological memory and the quality of the child's language environment. Bilingual children learn language readily and better understand the arbitrary nature of words. Some youngsters use a referential word-learning style that emphasizes words as names and that views language as an intellectual tool. Other children use an expressive style that emphasizes phrases and views language as a social tool.

■ Children's vocabulary is stimulated by experience. Parents can foster the growth of vocabulary by speaking with children and reading to them. Video helps preschoolers to learn new words but is ineffective with infants.

How do young children learn grammar?

■ Soon after children begin to speak, they create two-word sentences that are derived from their own experiences. Moving from two-word to more complex sentences involves adding grammatical morphemes. Children first master grammatical morphemes that express simple relations and later those that denote complex relations. Mastery of grammatical morphemes involves learning not only rules but also exceptions to the rules.

■ Behaviorists proposed that children acquire grammar through imitation, but that explanation is incorrect. Today's explanations come from three perspectives: the linguistic perspective emphasizes inborn mechanisms that allow children to infer the grammatical rules of their native language, the cognitive perspective emphasizes cognitive processes that allow children to find recurring patterns in the speech they hear, and the social-interaction perspective emphasizes social interactions with adults in which both parties want improved communication.

How well do young children communicate?

■ Parents encourage turn-taking even before infants begin to talk, and later they demonstrate both the speaker and listener roles for their children. By 3 years of age, children spontaneously take turns and prompt one another to take their turn.

■ Preschool children adjust their speech in a rudimentary fashion to fit the listener's needs. However, preschoolers are unlikely to identify ambiguities in another's speech; instead, they are likely to assume they knew what the speaker meant.

KEY TERMS

scheme (124)
assimilation (124)
accommodation (125)
equilibration (125)
sensorimotor period (126)
object permanence (127)
egocentrism (128)
animism (128)
centration (129)
core knowledge hypothesis (132)
teleological explanations (135)

essentialism (135)
mental hardware (137)
mental software (137)
attention (137)
orienting response (137)
habituation (137)
classical conditioning (138)
operant conditioning (138)
autobiographical memory (139)
one-to-one principle (142)
stable-order principle (143)

cardinality principle (143)
intersubjectivity (144)
guided participation (144)
zone of proximal development (145)
scaffolding (145)
private speech (146)
phonemes (147)
infant-directed speech (149)
cooing (149)
babbling (150)
fast mapping (151)

underextension (154)	referential style (155)	grammatical morphemes (157)
overextension (154)	expressive style (155)	overregularization (158)
phonological memory (154)	telegraphic speech (157)	

LEARN MORE ABOUT IT

Log in to **www.cengagebrain.com** to access the resources your instructor requires. For this book, you can access:

 Psychology CourseMate

■ CourseMate brings course concepts to life with interactive learning, study, and exam preparation tools that support the printed textbook. A textbook-specific website, Psychology CourseMate includes an integrated interactive eBook and other interactive learning tools including quizzes, flashcards, videos, and more.

CENGAGENOW™

■ CengageNOW Personalized Study is a diagnostic study tool containing valuable text-specific resources—and because you focus on just what you don't know, you learn more in less time to get a better grade.

WebTUTOR™

■ More than just an interactive study guide, WebTutor is an anytime, anywhere customized learning solution with an eBook, keeping you connected to your textbook, instructor, and classmates.

Entering the Social World

Socioemotional Development in Infancy and Early Childhood

HUMANS ENJOY ONE ANOTHER'S COMPANY. Social relationships of all sorts—friends, lovers, spouses, parents and children, co-workers, and teammates—make our lives both interesting and satisfying.

In this chapter, we trace the origins of these social relationships. We begin with the first social relationship—between an infant and a parent. You will see how this relationship emerges over the first year and how it is affected by the separation that comes when parents work full-time. Interactions with parents and others are often full of emotions—happiness, satisfaction, anger, and guilt, to name just a few. In the second section, you'll see how children express different emotions and how they recognize others' emotions.

In the third section, you'll learn how children's social horizons expand beyond parents to include peers. Then you'll discover how children play and how they help others in distress.

As children's interactions with others become more wide-ranging, they begin to learn about the social roles they are expected to play. Among the first social roles children learn are those associated with gender—how society expects boys and girls to behave. We'll explore children's awareness of gender roles in the last section of the chapter.

LEARNING OBJECTIVES

- What are Erikson's first three stages of psychosocial development?

- How do infants form emotional attachments to mother, father, and other significant people in their lives?

- What are the different varieties of attachment relationships, how do they arise, and what are their consequences?

- Is attachment jeopardized when parents of infants and young children are employed outside of the home?

Kendra's son Roosevelt is a happy, affectionate 18-month-old. Kendra so loves spending time with him that she is avoiding an important decision. She wants to return to her job as a loan officer at the local bank. Kendra knows a woman in the neighborhood who has cared for some of her friends' children, and they all think she is a fantastic babysitter. But Kendra still has a nagging feeling that going back to work isn't a "motherly" thing to do—that being away during the day may hamper Roosevelt's development.

THE SOCIOEMOTIONAL RELATIONSHIP THAT DEVELOPS BETWEEN AN INFANT AND A PARENT (USUALLY, BUT NOT NECESSARILY, THE MOTHER) IS SPECIAL. This is a baby's first relationship, and scientists and parents believe it should be satisfying and trouble free to set the stage for later relationships. In this section, we'll look at the steps involved in creating the baby's first emotional relationship. Along the way, you'll see how this relationship is affected by the separation that sometimes comes when a parent like Kendra works full-time.

Erikson's Stages of Early Psychosocial Development

Some of our keenest insights into the nature of psychosocial development come from a theory proposed by Erik Erikson (1982). We first encountered Erikson's theory in Chapter 1; recall that he describes development as a series of eight stages, each with a unique crisis for psychosocial growth. When a crisis is resolved successfully, an area of psychosocial strength is established. When the crisis is not resolved, that aspect of psychosocial development is stunted, which may limit the individual's ability to resolve future crises.

In Erikson's theory, infancy and the preschool years are represented by three stages, shown in ●Table 5.1. Let's take a closer look at each stage.

Basic Trust Versus Mistrust

Erikson argues that a sense of trust in oneself and others is the foundation of human development. Newborns leave the warmth and security of the uterus for an unfamiliar world. If parents respond to their infant's needs consistently, the infant comes to trust and feel secure in the world. Of course, the world is not always pleasant and can sometimes be dangerous. Parents may not always reach a falling baby in time, or they may accidentally feed an infant food that is too hot. Erikson sees value in these experiences, because infants learn mistrust. *With a proper balance of trust and mistrust, infants can acquire* **hope**, *which is an openness to new experience tempered by wariness that discomfort or danger may arise.*

hope
according to Erikson, an openness to new experience tempered by wariness that occurs when trust and mistrust are in balance

Autonomy Versus Shame and Doubt

Between 1 and 3 years of age, children gradually come to understand that they can control their own actions. With this understanding, children strive for autonomy, for independence from others. However, autonomy is counteracted by doubt that the child can handle demanding situations and by shame that may result from failure. *A*

TABLE 5.1		
Erikson's First Three Stages		
Age	**Crisis**	**Strength**
Infancy	Basic trust vs. mistrust	Hope
1–3 years	Autonomy vs. shame and doubt	Will
3–5 years	Initiative vs. guilt	Purpose

Copyright © Cengage Learning 2010

*blend of autonomy, shame, and doubt gives rise to **will**, the knowledge that, within limits, youngsters can act on their world intentionally.*

Initiative Versus Guilt

Most parents have their 3- and 4-year-olds take some responsibility for themselves (by dressing themselves, for example). Youngsters also begin to identify with adults and their parents; they begin to understand the opportunities that are available in their culture. Play begins to have purpose as children explore adult roles, such as mother, father, teacher, athlete, or writer. Youngsters start to explore the environment on their own, ask innumerable questions about the world, and imagine possibilities for themselves.

This initiative is moderated by guilt as children realize that their initiative may place them in conflict with others; they cannot pursue their ambitions with abandon. **Purpose** *is achieved with a balance between individual initiative and a willingness to cooperate with others.*

One of the strengths of Erikson's theory is its ability to tie together important psychosocial developments across the entire life span. We will return to the remaining stages in later chapters. For now, let's concentrate on the first of Erikson's crises—the establishment of trust in the world—and look at the formation of bonds between infants and parents.

The Growth of Attachment

In explaining the essential ingredients of these early social relationships, most modern accounts take an evolutionary perspective. *According to* **evolutionary psychology,** *many human behaviors represent successful adaptation to the environment.* That is, over human history, some behaviors have made it more likely that people will reproduce and pass on their genes to following generations. For example, we take it for granted that most people enjoy being with other people. But evolutionary psychologists argue that our "social nature" is a product of evolution: For early humans, being in a group offered protection from predators and made it easier to locate food. Thus, early humans who were social were more likely than their asocial peers to live long enough to reproduce, passing on their social orientation to their offspring (Gaulin & McBurney, 2001). Over many, many generations, "being social" had such a survival advantage that nearly all people are socially oriented (though in varying amounts, as we know from the research on temperament discussed in Chapter 3).

Applied to child development, evolutionary psychology highlights the adaptive value of children's behavior at different points in development (Bjorklund & Pellegrini, 2000). For example, think about the time and energy parents invest in child rearing. Without such effort, infants and young children would die before they were sexually mature, which means that a parent's genes could not be passed along to grandchildren (Geary, 2002). Here, too, parenting just seems "natural" but really represents an adaptation to the problem of guaranteeing that one's helpless offspring can survive until they're sexually mature.

will
according to Erikson, a young child's understanding that he or she can act on the world intentionally; this occurs when autonomy, shame, and doubt are in balance

purpose
according to Erikson, balance between individual initiative and the willingness to cooperate with others

evolutionary psychology
theoretical view that many human behaviors represent successful adaptations to the environment

Evolutionary psychology emphasizes the adaptive value of parents nurturing their offspring.

attachment
enduring socioemotional relationship
between infants and their caregivers

Steps Toward Attachment

An evolutionary perspective of early human relationships comes from John Bowlby (1969, 1991). According to Bowlby, *children who form an* **attachment** *to an adult—that is, an enduring socioemotional relationship—are more likely to survive.* This person is usually the mother but need not be; the key is a strong emotional relationship with a responsive, caring person. Attachments can form with fathers, grandparents, or someone else. Bowlby described four phases in the growth of attachment:

- *Preattachment* (birth to 6–8 weeks). During prenatal development and soon after birth, infants rapidly learn to recognize their mothers by smell and sound, which sets the stage for forging an attachment relationship (Hofer, 2006). What's more, evolution has endowed infants with many behaviors that elicit caregiving from an adult. When babies cry, smile, or gaze intently at a parent's face, the parent usually smiles back or holds the baby. The infant's behaviors and the responses they evoke in adults create an interactive system that is the first step in the formation of attachment relationships.

- *Attachment in the making* (6–8 weeks to 6–8 months). During these months, babies begin to behave differently in the presence of familiar caregivers and unfamiliar adults. Babies now smile and laugh more often with the primary caregiver. And when babies are upset, they're more easily consoled by the primary caregiver. Babies are gradually identifying the primary caregiver as the person they can depend on when they're anxious or distressed.

- *True attachment* (6–8 months to 18 months). By approximately 7 or 8 months, most infants have singled out the attachment figure—usually the mother—as a special individual. The attachment figure is now the infant's stable socioemotional base. For example, a 7-month-old will explore a novel environment but periodically look toward his mother, as if seeking reassurance that all is well. The behavior suggests that the infant trusts his mother and indicates that the attachment relationship has been established. In addition, this behavior reflects important cognitive growth: It means that the infant has a mental representation of the mother, an understanding that she will be there to meet the infant's needs (Lewis et al., 1997).

- *Reciprocal relationships* (18 months on). Infants' growing cognitive and language skills and their accumulated experience with their primary caregivers make infants better able to act as true partners in the attachment relationship. They often take the initiative in interactions and negotiate with parents ("Please read me another story!"). They begin to understand parents' feelings and goals and sometimes use this knowledge to guide their own behavior (e.g., social referencing, which we describe on pages 180–181). In addition, they cope with separation more effectively because they can anticipate that parents will return.

THINK ABOUT IT

Based on Piaget's description of infancy (pages 126–127), what cognitive skills might be important prerequisites for the formation of an attachment relationship?

Fathers and mothers differ in how they play with children. Fathers are much more likely to engage in vigorous physical play.

Father–Infant Relationships

Attachment typically first develops between infants and their mothers because mothers are usually the primary caregivers of American infants. Babies soon become attached to fathers, too, even though fathers spend less time in caregiving tasks (e.g., feeding or bathing a child) than mothers do (Pleck & Masciadrelli, 2004). Instead, fathers spend spend more time playing with their babies than taking care of them. And even their style of play differs. Physical play like that shown in the photo is the norm for fathers, whereas mothers spend more time reading and talking to babies, showing them toys, and playing games like patty-cake (Paquette, 2004). Given the opportunity to play with mothers or fathers, infants more often choose their fathers. However, when infants are distressed, mothers are preferred (Field, 1990). Thus, although most infants become attached to both parents, mothers and fathers typically have distinctive roles in their children's early social development.

Sequence of Events in the Strange Situation

1. An observer shows the experimental room to the mother and infant, then leaves the room.

2. The infant is allowed to explore the playroom for 3 minutes; the mother watches but does not participate.

3. A stranger enters the room and remains silent for 1 minute, then talks to the baby for a minute, then approaches the baby. The mother leaves unobtrusively.

4. The stranger does not play with the baby but attempts to comfort the baby if necessary.

5. After 3 minutes, the mother returns, greets, and consoles the baby.

6. When the baby has returned to play, the mother leaves again, this time saying "bye-bye" as she leaves.

7. The stranger attempts to calm and play with the baby.

8. After 3 minutes, the mother returns, and the stranger leaves.

Forms of Attachment

Thanks to biology, virtually all infants behave in ways that elicit caregiving from adults, and because of this behavior attachment almost always develops between infant and caregiver by 8 or 9 months of age. However, attachment can take different forms, and environmental factors help determine the quality of attachment between infants and caregivers. Mary Ainsworth (1978, 1993) pioneered the study of attachment relationships using a procedure that has come to be known as the Strange Situation. You can see in ●Table 5.2 that the Strange Situation involves a series of episodes, each about 3 minutes long. The mother and infant enter an unfamiliar room filled with interesting toys. The mother leaves briefly, then mother and baby are reunited. Meanwhile, the experimenter observes the baby and records its response to both separation and reunion.

Copyright © Mary Kate Denny / PhotoEdit

When infants have an attachment relationship with the mother, they use her as a secure base from which to explore the environment.

Based on how the infant reacts to separation from—and reunion with—the mother, Ainsworth and other researchers have discovered four primary types of attachment relationships (Ainsworth, 1993; Thompson, 2006). One is a secure attachment and three are different types of insecure attachment (avoidant, resistant, and disorganized).

■ **Secure attachment**: *The baby may or may not cry when the mother leaves, but when she returns the baby wants to be with her, and if the baby is crying it stops.* Babies in this group seem to be saying, "I missed you terribly, I'm delighted to see you, but now that all is well, I'll get back to what I was doing." Approximately 60–65% of American babies have secure attachment relationships.

■ **Avoidant attachment**: *The baby is not upset when the mother leaves and, when she returns, may ignore her by looking or turning away.* Infants with an avoidant attachment look as if they're saying, "You left me *again*. I always have to take care of myself!" About 20% of American infants have avoidant attachment relationships, which is one of the three forms of insecure attachment.

■ **Resistant attachment**: *The baby is upset when the mother leaves, and it remains upset or even angry when she returns and is difficult to console.* These babies seem to be telling the mother, "Why do you do this? I need you desperately and yet you just leave me without warning. I get so angry when you're like this." About 10–15% of American babies have this resistant attachment relationship, which is another form of insecure attachment.

secure attachment
relationship in which infants have come to trust and depend on their mothers

avoidant attachment
relationship in which infants turn away from their mothers when they are reunited following a brief separation

resistant attachment
relationship in which, after a brief separation, infants want to be held but are difficult to console

disorganized (disoriented) attachment

relationship in which infants don't seem to understand what's happening when they are separated and later reunited with their mothers

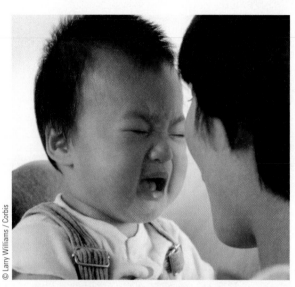

When infants who have a resistant attachment relationship are reunited with the mother, they're typically tearful, angry, and difficult to console.

■ **Disorganized (disoriented) attachment**: *The baby seems confused when the mother leaves and when she returns, as if not really understanding what's happening.* The baby often behaves in contradictory ways, such as nearing the mother when she returns but not looking at her, as if wondering, "What's happening? I want you to be here, but you left and now you're back. I don't get what's going on!" About 5–10% of American babies have this disorganized attachment relationship, the last of the three kinds of insecure attachment.

The Strange Situation is an important tool for studying attachment, but some scientists have criticized its emphasis on separation and reunion as the primary means for assessing quality of attachment. They suggest that what is considered an appropriate response to separation may not be the same in all cultures (Rothbaum et al., 2000). Consequently, investigators now use other methods to complement the Strange Situation. One of them, the Attachment Q-Set, can be used with young children as well as infants and toddlers. In this method, trained observers watch mothers and children interact at home; then the observer rates the interaction on many attachment-related behaviors (e.g., "Child greets mother with a big smile when she enters the room"). The ratings are totaled to provide a measure of the security of the child's attachment. Scores obtained with the Q-Set converge with assessments derived from the Strange Situation (van IJzendoorn et al., 2004).

Whether measured with the Strange Situation or the Attachment Q-Set, the quality of attachment during infancy predicts parent–child relations during childhood, adolescence, and young adulthood. Infants with secure attachment relationships tend to report, as adolescents and young adults, that they depend on their parents for care and support. In contrast, infants with insecure attachment relationships often, as adolescents and young adults, report being angry with their parents or deny being close to them. However, consistency is far from perfect. Stressful life events—death of a parent, divorce, life-threatening illness, poverty—help to determine stability and change in attachment. Stressful life events are associated with insecure attachments during adolescence and young adulthood. Consequently, when infants with insecure attachments experience stressful life events, their attachment tends to remain insecure; when infants with secure attachment experience these same events, their attachment often becomes insecure, perhaps because stress makes parents less available and less responsive to their children (Hamilton, 2000; Moss et al., 2005; Waters et al., 2000).

Consequences of Attachment

Erikson and other theorists (e.g., Waters & Cummings, 2000) believe that infant–parent attachment, the first social relationship, lays the foundation for all of the infant's later social relationships. In this view, infants who experience the trust and compassion of a secure attachment should develop into preschool children who interact confidently and successfully with their peers. In contrast, infants who do not experience a successful, satisfying first relationship should be more prone to problems in their social interactions as preschoolers.

In fact, children with secure attachment relationships have higher-quality friendships and fewer conflicts in their friendships than children with insecure attachment relationships (McElwain et al., 2008). What's more, secure attachment in infancy is associated with more stable and higher-quality romantic relationships in adolescence (Collins, Welsh, & Furman, 2009). Finally, research consistently points to links between disorganized attachment and behavior problems involving anxiety, anger, and aggressive behavior (Fearon et al., 2010; Moss et al., 2006).

The conclusion seems inescapable: As they grow, infants who have secure attachment relationships tend to have satisfying social interactions but infants with disorganized attachment do not. Why? One explanation focuses on the lasting impact of this first social relationship. Secure attachment evidently leads infants to trust and confide in other humans, which leads to more skilled social interactions later in childhood. Another view does not discount the impact of this early relationship but adds another wrinkle: Theorists who emphasize continuity of caregiving argue that parents who establish secure attachments with infants tend to be warm and supportive throughout their child's development (Lamb et al., 1985; Thompson, 2006). Thus, it is continuous exposure to high-quality parenting that promotes secure attachment in infancy and positive social relationships in childhood and adolescence. These accounts *are not* mutually exclusive: A successful first relationship and continued warm parenting likely work together to foster children's development.

Of course, attachment is only the first of many steps along the long road of social development. Infants with insecure attachments are not forever damned, but this initial misstep *can* interfere with their social development. Consequently, we need to look at the conditions that determine the quality of attachment.

What Determines Quality of Attachment?

Because secure attachment is so important to a child's later development, researchers have tried to identify the factors involved. Undoubtedly the most important is the interaction between parents and their babies (De Wolff & van IJzendoorn, 1997; Tomlinson, Cooper, & Murray, 2005). A secure attachment is most likely when parents respond to infants predictably and appropriately. For example, when the mother promptly responds to her baby's crying and reassures the baby, the mother's behavior evidently conveys that social interactions are predictable and satisfying. This behavior seems to instill in infants the trust and confidence that are the hallmarks of secure attachment.

Why does predictable and responsive parenting promote secure attachment relationships? To answer this question, think about your own friendships and romantic relationships. These relationships are usually most satisfying when we believe we can trust the other people and depend on them in times of need. The same formula seems to hold for infants. *Infants develop an* **internal working model**, *a set of expectations about parents' availability and responsiveness, generally and in times of stress.* When parents are dependable and caring, babies come to trust them, knowing they can be relied on for comfort. That is, babies develop an internal working model in which they believe their parents are concerned about their needs and will try to meet them (Huth-Bocks et al., 2004; Thompson, 2000).

In a particularly clever demonstration of infants' working models of attachment (Johnson, Dweck, & Chen, 2007), 13-month-olds were shown animated videos depicting a large ellipse (mother) paired with a small ellipse (child). The video began with the mother and child ellipses together, then the mother moved away from the child, who began to cry. On some trials, the mother ellipse returned to the child ellipse; on other trials, she continued to move away. Securely attached 13-month-olds looked longer at the trials depicting an unresponsive mother but insecurely attached infants looked longer at the trials when the mother returned. Evidently, each group has a working model of how parents respond—securely attached infants expect parents to respond but insecurely attached infants do not—and they look longer at the trials that violated their expectations of maternal behavior.

Many research findings attest to the importance of a caregiver's sensitivity for developing secure attachment.

- In a study conducted at a kibbutz in Israel, infants were less likely to develop secure attachment when they slept in dormitories with other children under 12 and where they received inconsistent (if any) attention when they became upset overnight (Sagi et al., 1994).
- In a study conducted in the Netherlands, infants were more likely to form a secure attachment when their mother had 3 months of training that emphasized monitoring an infant's signals and responding appropriately and promptly (van den Boom, 1994, 1995).

internal working model
infant's understanding of how responsive and dependable the mother is; thought to influence close relationships throughout the child's life

Perhaps the most important ingredient in fostering a secure attachment relationship is responding predictably and appropriately to the infant's needs.

■ Infants living in orphanages in Romania were more likely to be attached to their institutional caregiver when the caregivers were emotionally involved and responsive to them (Zeanah et al., 2005).

Thus, secure attachment is most likely when parents are sensitive and responsive. Of course, not all caregivers react to babies in a reliable and reassuring manner. Some respond intermittently or only after the infant has cried long and hard. And when these caregivers finally respond, they are sometimes annoyed by the infant's demands and may misinterpret the infant's intent. Over time, these babies tend to see social relationships as inconsistent and often frustrating, conditions that do little to foster trust and confidence.

Another factor contributing to the quality of attachment is temperament. Babies with difficult temperaments are somewhat less likely to form secure attachment relationships (Goldsmith & Harman, 1994; Seifer et al., 1996). That is, babies who fuss often and are difficult to console are more prone to insecure attachment. This may be particularly likely when a difficult, emotional infant has a mother whose personality is rigid and traditional than when the mother is accepting and flexible (Mangelsdorf et al., 1990). Rigid mothers do not adjust well to the often erratic demands of their difficult babies; instead, they want the baby to adjust to them. This means that rigid mothers less often provide the responsive, sensitive care that leads to secure attachment.

Fortunately, even brief training for mothers of newborns can help them respond to their babies more effectively (Bakermans-Kranenburg, Van IJzendoorn, & Juffer, 2003). Mothers can be taught how to interact more sensitively, affectionately, and responsively, paving the way for secure attachment and the lifelong benefits associated with a positive internal working model of interpersonal relationships.

The formation of attachment illustrates well the combined influence of the different components of the biopsychosocial framework. Many infant behaviors that elicit caregiving in adults—smiling and crying, for example—are biological in origin. When the caregiver is responsive to the infant (a sociocultural force), a secure attachment forms in which the infant trusts caregivers and knows that they can be relied on in stressful situations (a psychological force).

Millions of American infants and preschoolers attend day-care or nursery-school programs.

Attachment, Work, and Alternative Caregiving

Since the 1970s, more women in the workforce and more single-parent households have made arranging for child care a reality of parenting for many American families. Today, millions of infants and toddlers are cared for by someone other than their mother. Some are cared for in their home by their father, a grandparent, or another relative. Others receive care in the provider's home; the provider is often but not always a relative. Still others attend day-care or nursery-school programs. The patterns are very similar for European American, African American, and Latino American youngsters (Singer et al., 1998; U.S. Census Bureau, 2008).

Parents and policy makers alike have been concerned about the impact of such care on children generally and, specifically, its impact on attachment. Is there, for example, a maximum amount of time per week that infants should spend in care outside the home? Is there a minimum age below which infants should not be placed in care outside the home? To answer these questions, a comprehensive longitudinal study of early child care was initiated by the National Institute of Child Health and Human Development (NICHD). Planning for the early child-care study began in 1989, and by 1991 the study was under way. Researchers recruited 1,364 mothers and their newborns from 12 U.S. cities. Both mothers and children have been tested repeatedly (and the testing continues because the study is ongoing).

From the outset, one of the concerns was the impact of early child care on mother–infant attachment. In fact, the results so far show no overall effects of child care experience on mother–infant attachment for either 15- or 36-month-olds (NICHD Early Child Care Research Network, 1997, 2001). In other words, a secure mother–infant attachment was just as likely regardless of the quality of child care, the amount of time the child spent in care, the age when the child began care, how frequently the parents changed child-care arrangements, or the type of child care (e.g., a child-care center or in the home with a nonrelative).

However, when the effects of child care were considered along with characteristics of the mothers, an important pattern was detected: At 15 and 36 months, insecure attachments were more common when less sensitive mothering was combined with low-quality or large amounts of child care (NICHD Early Child Care Research Network, 1997, 2001). As the investigators put it, "poor quality, unstable, or more than minimal amounts of child care apparently added to the risks already inherent in poor mothering, so that the combined effects were worse than those of low maternal sensitivity and responsiveness alone" (1997, p. 877). These conclusions are particularly convincing because the same pattern of results was found in Israel in a large-scale study of child care and attachment that was modeled after the NICHD early child-care study (Sagi et al., 2002). Finally, although children in child care often become attached to their nonparental caregivers, this is not at the expense of secure attachment to parents (Ahnert, Pinquart, & Lamb, 2006).

These results provide clear guidelines for parents like Kendra, the mother in the vignette. They can enroll their infants and toddlers in high-quality day-care programs with no fear of harmful consequences. When children are enrolled in high-quality day care, other factors (e.g., the type of child care or the amount of time the child spends in child care) typically do not affect the mother–child attachment relationship.

The results of the early child-care study are reassuring for parents, who often have misgivings about their infants and toddlers spending so much time in the care of others. Nevertheless, they raise another, equally important question: What are the features of high-quality child care? That is, what should parents look for when trying to find care for their children? In general, high-quality child care has the following features (Burchinal et al., 2000; Lamb, 1999):

- A low ratio of children to caregivers
- Well-trained, experienced staff
- Low staff turnover
- Ample opportunities for educational and social stimulation
- Effective communication between parents and day-care workers concerning the general aims and routine functioning of the program

Collectively, these variables do not guarantee that a child will receive high-quality care. Sensitive, responsive caregiving—the same behavior that promotes secure attachment relationships—is the real key to high-quality child care. Centers that have well-trained, experienced staff caring for a relatively small number of children are more likely to provide good care, but the only way to know the quality of care with certainty is to see for yourself (Lamb, 1999).

Fortunately, today many employers know that convenient, high-quality child care makes for a better employee. Many cities have modified their zoning codes so that new shopping complexes and office buildings include child-care facilities. Many businesses realize that the availability of excellent child care helps attract and retain a skilled labor force. With effort, organization, and help from the community and business, full-time employment and high-quality caregiving *can* be compatible. We will return to this issue in Chapter 12 from the perspective of the parents. For now, the Real People feature provides one example of a father who stays home to care for his daughter while her mother works full-time.

Responding to the needs of working families, many employers now provide child care on site.

Lois, 46, and Bill, 61, had been married nearly 4 years when Lois gave birth to Sarah. Lois, a kindergarten teacher, returned to work full-time 4 months after Sarah was born. Bill, who had been halfheartedly pursuing a Ph.D. in education, became a full-time househusband. Bill does the cooking and takes care of Sarah during the day. Lois comes home from school at noon so that the family can eat lunch together, and she is home from work by 4 in the afternoon. Once a week, Bill takes Sarah to a parent–infant play program. The other parents, all mothers in their 20s or 30s, first assumed that Bill was Sarah's grandfather and had trouble relating to him as an older father. Soon, however, he was an accepted member of the group. On weekends, Lois's and Bill's grown children from previous marriages often visit and enjoy caring for and playing with Sarah. By all accounts, Sarah looks to be a healthy, happy, outgoing 9-month-old. Is this arrangement nontraditional? Clearly. Is it effective for Sarah, Lois, and Bill? Definitely. Sarah receives the nurturing care she needs, Lois goes to work assured that Sarah is in Bill's knowing and caring hands, and Bill relishes being the primary caregiver.

Test Yourself

RECALL

1. _____ proposed that maturational and social factors come together to pose eight unique challenges for psychosocial growth during the life span.

2. Infants must balance trust and mistrust to achieve _____, an openness to new experience that is coupled with awareness of possible danger.

3. By approximately _____ months of age, most infants have identified a special individual—usually but not always the mother—as the attachment figure.

4. Joan, a 12-month-old, was separated from her mother for about 15 minutes. When they were reunited, Joan would not let her mother pick her up. When her mother approached, Joan would look the other way or toddle to another part of the room. This behavior suggests that Joan has a(n) _____ attachment relationship.

5. The single most important factor in fostering a secure attachment relationship is _____.

6. Tim and Douglas, both 3-year-olds, rarely argue; when they disagree, one goes along with the other's ideas. The odds are good that both boys have _____ attachment relationships with their parents.

7. An insecure attachment relationship is likely when an infant receives poor-quality child care and _____.

INTERPRET

Compare the infant's contributions to the formation of mother–infant attachment with the mother's contributions.

APPLY

Based on what you know about the normal developmental timetable for the formation of mother–infant attachment, what would seem to be the optimal age range for children to be adopted?

Recall answers: (1) Erik Erikson, (2) hope, (3) 6 or 7, (4) avoidant insecure, (5) responding consistently and appropriately, (6) secure, (7) insensitive, unresponsive mothering

---

5.2 Emerging Emotions

LEARNING OBJECTIVES

- At what ages do children begin to express basic emotions?
- What are complex emotions, and when do they develop?
- When do children begin to understand other people's emotions? How do they use this information to guide their own behavior?

Nicole is ecstatic that she is finally going to see her 7-month-old nephew, Claude. She rushes into the house and, seeing Claude playing on the floor with blocks, sweeps him up in a big hug. After a brief, puzzled look, Claude bursts into angry tears and begins thrashing his arms and legs, as if saying to Nicole, "Who are you? What do you want? Put me down! Now!" Nicole quickly hands Claude to his mother, who is surprised by her baby's outburst and even more surprised that he continues to sob while she rocks him.

THIS VIGNETTE ILLUSTRATES THREE COMMON EMOTIONS. Nicole's initial joy, Claude's anger, and his mother's surprise are familiar to all of us. In this section, we look at when children first express emotions, how children come to understand emotions in others, and, finally, how children regulate their emotions. As we do, we'll learn why Claude reacted to Nicole as he did and how Nicole could have prevented Claude's outburst.

The Function of Emotions

Why do people feel emotions? Wouldn't life be simpler if humans were emotionless like computers or residents of Mr. Spock's planet Vulcan? Probably not. Think, for example, about activities that most adults find pleasurable: a good meal, sex, holding one's children, and accomplishing a difficult but important task. These activities were and remain essential to the continuity of humans as a species, so it's not surprising that they elicit emotions (Gaulin & McBurney, 2001).

Modern theories emphasize the functional value of emotion. That is, according to the functional approach, emotions are useful because they help people adapt to their environment (Izard & Ackerman, 2000; Saarni et al., 2006). Take fear as an example. Most of us would rather not be afraid, but there are instances in which feeling fearful is very adaptive. Imagine you are walking alone, late at night, in a poorly lighted section of campus. You become frightened and, as a consequence, are particularly attentive to sounds that might signal the presence of threat, and you probably walk quickly to a safer location. Thus, fear is adaptive because it organizes your behavior around an important goal—avoiding danger (Cosmides & Tooby, 2000).

Similarly, other emotions are adaptive. Happiness, for example, is adaptive in contributing to stronger interpersonal relationships: When people are happy with another person, they smile, and this often causes the other person to feel happy too, strengthening their relationship (Izard & Ackerman, 2000). Disgust is adaptive in keeping people away from substances that might make them ill: As we discover that the milk in a glass is sour, we experience disgust and push the glass away (Oaten, Stevenson, & Case, 2009). Thus, according to the functional approach to human emotion, most emotions developed over the course of human history to meet unique life challenges and help humans to survive.

Experiencing and Expressing Emotions

The three emotions from the vignette—joy, anger, and fear—are considered "basic emotions," as are interest, disgust, distress, sadness, and surprise (Draghi-Lorenz, Reddy, & Costall, 2001). **Basic emotions** *are experienced by people worldwide, and each consists of three elements: a subjective feeling, a physiological change, and an overt behavior* (Izard, 2007). For example, suppose you wake to the sound of a thunderstorm and then discover your roommate has left for class with your umbrella. Subjectively, you might feel ready to explode with anger; physiologically, your heart would beat faster; and behaviorally, you would probably be scowling.

basic emotions
emotions experienced by humankind and that consist of three elements: a subjective feeling, a physiological change, and an overt behavior

Development of Basic Emotions

Using facial expressions and other overt behaviors, scientists have traced the growth of basic emotions in infants. According to one influential theory (Lewis, 2000), newborns experience only two general emotions: pleasure and distress. Rapidly, though, more discrete emotions emerge, and by 8 or 9 months of age infants are thought to experience all basic emotions. For example, joy emerges at about 2 or 3 months. *At this age* **social smiles** *first appear; infants smile when they see another human face.* Sometimes social smiling is accompanied by cooing, the early form of vocalization described in Chapter 4 (Sroufe & Waters, 1976). Smiling and cooing seem to be the infant's way of expressing pleasure at seeing another person. Sadness is also observed at about this age: Infants look sad, for example, when their mothers stop playing with them (Lewis, 2000).

social smiles
smile that infants produce when they see a human face

Social smiles emerge at about 2-3 months, when infants smile in response to a human face.

stranger wariness

first distinct signs of fear that emerge around 6 months of age when infants become wary in the presence of unfamiliar adults

THINK ABOUT IT

How might an infant's ability to express emotions relate to the formation of attachment? To the temperamental characteristics described on pages 87–88?

Young babies are often wary of strangers; consequently, they're unhappy when strangers hold them before giving them a chance to "warm up."

Anger is one of the first negative emotions to emerge from generalized distress, typically between 4 and 6 months. Infants will become angry, for example, if a favorite food or toy is taken away (Sullivan & Lewis, 2003). Reflecting their growing understanding of goal-directed behavior (see Section 4.1), older infants become increasingly angry when their attempts to achieve a goal are frustrated (Braungart-Rieker, Hill-Soderlund, & Karrass, 2010). For example, if a parent restrains an infant trying to pick up a toy, the guaranteed result is a very angry baby.

Like anger, fear emerges later in the first year. *At about 6 months, infants become wary in the presence of an unfamiliar adult, a reaction known as* **stranger wariness**. When a stranger approaches, a 6-month-old typically looks away and begins to fuss (Mangelsdorf, Shapiro, & Marzolf, 1995). If a grandmother picks up her grandchild without giving the infant a chance to warm up to her, the outcome is as predictable as it was with Claude, the baby boy in the vignette who was frightened by his aunt: He cries, looks frightened, and reaches with arms outstretched in the direction of someone familiar.

How wary an infant feels around strangers depends on a number of factors (Thompson & Limber, 1991). First, infants tend to be less fearful of strangers when the environment is familiar and more fearful when it is not. Many parents know this firsthand from traveling with their infants: Enter a friend's house for the first time and the baby clings tightly to its mother. Second, the amount of anxiety depends on the stranger's behavior. Instead of rushing to greet or pick up the baby, as Nicole did in the vignette, a stranger should talk with other adults and, in a while, perhaps offer the baby a toy (Mangelsdorf, 1992). Handled this way, many infants will soon be curious about the stranger instead of afraid.

Wariness of strangers is adaptive because it emerges at the same time that children begin to master creeping and crawling (described on pages 99–102). Like Curious George, the monkey in a famous series of children's books, babies are inquisitive and want to use their new locomotor skills to explore their world. Being wary of strangers provides a natural restraint against the tendency to wander away from familiar caregivers. However, as youngsters learn to interpret facial expressions and recognize when a person is friendly, their wariness of strangers declines.

Of the negative emotions, we know the least about disgust. Preschool children may respond with disgust at the odor of feces or at being asked to touch a maggot or being asked to eat a piece of candy that's resting on the bottom of a brand-new potty seat. Parents likely play an important role in helping children to identify disgusting stimuli: Mothers respond quite vigorously to disgust-eliciting stimuli when in presence of their children. They might say "That's revolting!" while moving away from the stimulus (Stevenson et al., 2010). A child's early sensitivity to disgust is useful because many of the cues that elicit disgust are also signals of potential illness: disgusting stimuli such as feces, vomit, and maggots can all transmit disease.

Emergence of Complex Emotions

In addition to basic emotions such as joy and anger, people feel complex emotions such as pride, guilt, and embarrassment. Most scientists (e.g., Lewis, 2000; Mascolo, Fischer, & Li, 2003) believe that complex emotions don't surface until 18 to 24 months of age because they depend on the child having some understanding of the self, which typically occurs between 15 and 18 months. Children feel guilty or embarrassed, for example, when they've done something they know they shouldn't have done (Kochanska et al., 2002): A child who breaks a toy is thinking, "You told me to be careful. But I wasn't!" Similarly, children feel pride when they accomplish a challenging

Infants' Expression of Emotions

	Defined	Emerge	Examples
Basic	Experienced by people world-wide and include a subjective feeling, a physiological response, and an overt behavior	Birth to 9 months	Happiness, anger, fear
Self-conscious	Responses to meeting or failing to meet expectations or standards	18 to 24 months	Pride, guilt, embarrassment

task for the first time. Thus, children's growing understanding of themselves enables them to experience complex emotions like pride and guilt (Lewis, 2000).

The features of basic and self-conscious emotions are summarized in ● Table 5.3.

Later Developments

As children grow, they continue to experience basic and complex emotions, but different situations or events elicit these emotions. In the case of complex emotions, cognitive growth means that elementary-school children experience shame and guilt in situations they would not have when they were younger (Reimer, 1996). For example, unlike preschool children, many school-age children would be ashamed if they neglected to defend a classmate who had been wrongly accused of a theft.

Fear is another emotion that can be elicited in different ways, depending on a child's age. Many preschool children are afraid of the dark and of imaginary creatures. These fears typically diminish during the elementary-school years as children grow cognitively and better understand the difference between appearance and reality. Replacing these fears are concerns about school, health, and personal harm (Silverman, La Greca, & Wasserstein, 1995). Such worries are common and not a cause for concern in most children. In some youngsters, however, they become so extreme that they overwhelm the child (Chorpita & Barlow, 1998). For example, a 7-year-old's worries about school would not be unusual unless her concern grew to the point that she refused to go to school.

Cultural Differences in Emotional Expression

Children worldwide express many of the same basic and complex emotions. However, cultures differ in the extent to which emotional expression is encouraged (Hess & Kirouac, 2000). In many Asian countries, for example, outward displays of emotion are discouraged in favor of emotional restraint. Consistent with these differences, in one study (Camras et al., 1998), European American 11-month-olds cried and smiled more often than Chinese 11-month-olds. In another study (Camras et al., 2006), U.S. preschoolers were more likely than Chinese preschoolers to smile at funny pictures and to express disgust after smelling a cotton swab dipped in vinegar.

Cultures also differ in the events that trigger emotions, particularly complex emotions. Situations that evoke pride in one culture may evoke embarrassment or shame in another. For example, American elementary-school children often show pride at personal achievement, such as getting the highest grade on a test or coming in first place in a county fair. In contrast, Asian elementary-school children are embarrassed by a public display of individual achievement but show great pride when their entire class is honored for an achievement (Lewis et al., 2010; Stevenson & Stigler, 1992).

THINK ABOUT IT

Explain how the different forces in the biopsychosocial framework contribute to the development of basic and complex emotions.

American children are often quite proud of personal achievement, but Asian children would be embarrassed by such a public display of individual accomplishments.

Copyright © David Young-Wolff / PhotoEdit

Expression of anger also varies around the world. Suppose a child has just completed a detailed drawing when a classmate spills a drink, ruining the drawing. Most American children would respond with anger. In contrast, children growing up in East Asian countries that practice Buddhism (e.g., Mongolia, Thailand, Nepal) rarely respond with anger because this goes against the Buddhist tenet to extend loving kindness to all people, even those whose actions hurt others. Instead, they would probably remain quiet and experience shame that they had left the drawing in a vulnerable position (Cole, Tamang, & Shrestha, 2006).

Thus, culture can influence when and how much children express emotion. Of course, expressing emotion is only part of the developmental story. Children must also learn to recognize others' emotions, which is our next topic.

Recognizing and Using Others' Emotions

Imagine you are broke (only temporarily, of course) and plan to borrow $20 from your roommate when she returns from class. Shortly, she storms into your apartment, slams the door, and throws her backpack on the floor. Immediately, you change your plans, realizing that now is hardly a good time to ask for a loan. This example reminds us that we often need to recognize others' emotions and sometimes change our behavior as a consequence.

When can infants first identify emotions in others? Perhaps as early as 4 months and definitely by 6 months, infants begin to distinguish facial expressions associated with different emotions. They can, for example, distinguish a happy, smiling face from a sad, frowning face (Bornstein & Arterberry, 2003; Montague & Walker-Andrews, 2001), and fearful, happy, and neutral faces elicit different patterns of electrical activity in the infant's brain (Leppanen et al., 2007), which also shows the ability to differentiate facial expressions of emotion. What's more, like adults, infants are biased toward negative emotions (Vaish, Woodward, & Grossmann, 2008). They attend more rapidly to faces depicting negative emotions (e.g., anger) and pay attention to them longer than emotionless or happy faces (LoBue & DeLoache, 2010; Peltola et al., 2008).

Of course, infants might be able to distinguish an angry face from a happy one but not know the emotional significance of the two faces. How can we tell whether infants understand the emotions expressed in a face? The best evidence is that infants often match their own emotions to other people's emotions. When happy mothers smile and talk in a pleasant voice, infants express happiness themselves. If mothers are angry or sad, infants become distressed, too (Haviland & Lelwica, 1987; Montague & Walker-Andrews, 2001).

Also like adults, infants use others' emotions to direct their behavior. *Infants in an unfamiliar or ambiguous environment often look at their mother or father as if searching for cues to help them interpret the situation, a phenomenon known as* **social referencing**. If a parent looks afraid when shown a novel object, 12-month-olds are less likely to play with the toy than if a parent looks happy (Repacholi, 1998). Furthermore, an infant can use parents' facial expressions or their vocal expressions alone to decide whether they want to explore an unfamiliar object (Mumme, Fernald, & Herrera, 1996). And infants' use of parents' cues is precise (see ■ Figure 5.1). If two unfamiliar toys are shown to a parent who expresses disgust at one toy but not the other, 12-month-olds will avoid the toy that elicited the disgust but not the other toy (Moses et al., 2001). By 14 months of age, infants even remember this information: They avoid a toy that elicited disgust an hour earlier (Hertenstein & Campos, 2004). By 18 months, they're more sophisticated still: If one adult demonstrates an unfamiliar toy and a second adult comments, in an angry tone, "That's really annoying! That's so irritating!" then 18-month-olds play less with the toy compared to when the second adult makes neutral remarks in a mild manner. These youngsters apparently decided that it wasn't such a good idea to play with the toy if it might upset the second adult again (Repacholi & Meltzoff, 2007; Repacholi, Meltzoff, & Olsen, 2008). Thus, social refer-

social referencing
behavior in which infants in unfamiliar or ambiguous environments look at an adult for cues to help them interpret the situation

■ **Figure 5.1**

If parents seem frightened by an unfamiliar object, then babies are also wary or even afraid of it.

encing shows that infants are remarkably skilled in using the emotions of adults to help them direct their own behavior.

Although infants and toddlers are remarkably adept at recognizing others' emotions, their skills are far from mature. Adults are much more skilled than infants—and school-aged children, for that matter—in recognizing the subtle signals of an emotion (Thomas et al., 2007). Adults are also better able to tell when others are "faking" emotions; they can distinguish the face of a person who's really happy from the face of a person who's faking happiness (Del Giudice & Colle, 2007). Thus, facial expressions of emotion are recognized with steadily increasing skill throughout childhood and into adolescence.

What experiences contribute to children's understanding of emotions? Parents and children frequently talk about past emotions and why people felt as they did, and this is particularly true for negative emotions such as fear and anger (Lagattuta & Wellman, 2002). Not surprisingly, children learn about emotions when parents talk about feelings, explaining how they differ and the situations that elicit them (Brown & Dunn, 1996; Cervantes & Callanan, 1998). Also, a positive and rewarding relationship with parents and siblings is related to children's understanding of emotions (Brown & Dunn, 1996; Thompson, Laible, & Ontai, 2003). The nature of this connection is still a mystery. One possibility is that, within positive parent–child and sibling relationships, people express a fuller range of emotions (and do so more often) and are more willing to talk about why they feel as they do, providing children with more opportunities to learn about emotions.

Regulating Emotions

Think back to a time when you were *really* angry at a good friend. Did you shout at the friend? Did you try to discuss matters calmly? Or did you simply ignore the situation altogether? Shouting is a direct expression of anger, but calm conversation and overlooking a situation are purposeful attempts to regulate emotion. People often regulate emotions; for example, we routinely try to suppress fear (because we know there's no real need to be afraid of the dark), anger (because we don't want to let a friend know just how upset we are), and joy (because we don't want to seem like we're gloating over our good fortune).

As these examples illustrate, skillfully regulating emotions depends on cognitive processes like those described in Chapter 4 (see Zelazo & Cunningham, 2007). Attention is an important part of emotion regulation: We can control emotions such as fear by diverting attention to other less emotional stimuli, thoughts, or feelings (Rothbart & Sheese, 2007; Watts, 2007). We can also use strategies to reappraise the meaning of an event (or of feelings or thoughts) so that it provokes less emotion (John & Gross, 2007). For example, a soccer player nervous about taking a penalty kick can reinterpret her state of physiological arousal as being "pumped up" instead of being "scared to death."

Because cognitive processes are essential for emotional regulation, the research described in Chapters 4 and 6 would lead us to expect that successful regulation develops gradually through childhood and adolescence and that at any age some children will be more skilled than others are at regulating emotions (Thompson, Lewis, & Calkins, 2008). In fact, child-development researchers have documented both patterns. Emotion regulation clearly begins in infancy. By 4 to 6 months, infants use simple strategies to regulate their emotions (Buss & Goldsmith, 1998; Rothbart & Rueda, 2005). When something frightens or confuses an infant—for example, a stranger or a mother who suddenly stops responding—he or she often looks away (just as older children and even adults often turn away or close their eyes to block out disturbing stimuli). Frightened infants also move closer to a parent, another effective way of helping to control their fear (Parritz, 1996). And by 24 months, a distressed toddler's face typically expresses sadness instead of fear or anger; apparently by this age toddlers have learned that a sad facial expression is the best way to get a mother's attention and support (Buss & Kiel, 2004).

Older children encounter a wider range of emotional situations, so it's fortunate they develop a number of related new ways to regulate emotion (Eisenberg & Morris, 2002) as follows.

- Children begin to regulate their own emotions and rely less on others to do this for them. A fearful child no longer runs to a parent but instead devises her own methods for dealing with fear (e.g., "I know the thunderstorm won't last long, and I'm safe inside the house").

- Children more often rely on mental strategies to regulate emotions. For example, a child might reduce his disappointment at not receiving a much-expected gift by telling himself that he didn't really want the gift in the first place.

- Children more accurately match the strategies for regulating emotion with the particular setting. For example, when faced with emotional situations that are unavoidable (e.g., a child must go to the dentist to have a cavity filled), children adjust to the situation (e.g., thinking of the positive consequences of treating the tooth) instead of trying to avoid it.

Collectively, these age-related trends give children tools for regulating emotions.

Unfortunately, not all children regulate their emotions well, and those who don't will tend to have adjustment problems and problems interacting with peers (Eisenberg & Morris, 2002; Eisenberg et al., 2005). When children can't control their anger, worry, or sadness, they often have difficulty resolving the conflicts that inevitably surface in peer relationships (Fabes et al., 1999). For example, when children are faced with a dispute over who gets to play with a toy, their unregulated anger can interfere with finding a mutually satisfying solution. Thus, ineffective regulation of emotions leads to more frequent conflicts with peers and, as a result, less satisfying peer relationships and less adaptive adjustment to school (Eisenberg et al., 2001; Olson et al., 2005).

Test Yourself

RECALL

1. Basic emotions include a subjective feeling, a physiological change, and _____.

2. The first detectable form of fear is _____, which emerges at about 6 months.

3. Wariness of strangers is adaptive because it emerges at about the same time that _____.

4. Complex emotions, such as guilt and shame, emerge later than basic emotions because _____.

5. In social referencing, infants use a parent's facial expression _____.

6. Infants often control fear by looking away from a frightening event or by _____.

INTERPRET

Distinguish basic emotions from complex emotions.

APPLY

Cite similarities between developmental change in infants' expression and regulation of emotion and developmental change in infants' comprehension and expression of speech (described on pages 147–150 of Chapter 4).

Recall answers: (1) an overt behavior, (2) wariness of strangers, (3) infants master creeping and crawling, (4) complex emotions require more advanced cognitive skills, (5) to direct their own behavior (e.g., deciding if an unfamiliar situation is safe or frightening), (6) moving closer to a parent

LEARNING OBJECTIVES

- When do youngsters first begin to play with each other? How does play change during infancy and the preschool years?

- What determines whether children help one another?

Six-year-old Juan got his finger trapped in the DVD player when he tried to remove a disk. While he cried and cried, his 3-year-old brother, Antonio, and his 2-year-old sister, Carla, watched but did not help. Later, when their mother had soothed Juan and concluded that his finger was not injured, she worried about her younger children's reactions. In the face of their brother's obvious distress, why did Antonio and Carla do nothing?

INFANTS' INITIAL INTERACTIONS ARE WITH PARENTS, but soon they begin to interact with other people, notably their peers. In this section, we'll trace the development of these interactions and learn why children like Antonio and Carla don't always help others.

The Joys of Play

Peer interactions begin surprisingly early in infancy. Two 6-month-olds together will look, smile, and point at one another. Over the next few months, infants laugh and babble when with other infants (Rubin, Bukowski, & Parker, 2006).

In parallel play, children play independently but actively watch what other children are doing.

Soon after the first birthday, children begin **parallel play**, *in which each youngster plays alone but maintains a keen interest in what another is doing.* Two toddlers may each have his or her own toys, but each will watch the other's play, too. Exchanges between youngsters also become more common. When one toddler talks or smiles, the other usually responds (Howes, Unger, & Seidner, 1990).

Beginning at roughly 15 to 18 months, toddlers no longer simply watch one another at play. *Instead, they engage in similar activities and talk or smile at one another, illustrating* **simple social play**. Play has now become truly interactive (Howes & Matheson, 1992). An example of simple social play would be two 20-month-olds pushing toy cars along the floor, making "car sounds," and periodically trading cars.

Toward the second birthday, **cooperative play** emerges: Now a distinct theme organizes children's play, and they take on special roles based on the theme. They may play "hide-and-seek" and alternate the roles of hider and seeker, or they may have a tea party and take turns being the host and the guest (Parten, 1932).

The nature of young children's play changes dramatically in a few years (Howes & Matheson, 1992). In a typical day-care center, 1-year-olds spend most of their time in parallel play, and other forms of play are relatively rare. In contrast, cooperative play is the norm among 3-year-olds.

parallel play
when children play alone but are aware of and interested in what another child is doing

simple social play
play that begins at about 15 to 18 months; toddlers engage in similar activities as well as talk and smile at each other

cooperative play
play that is organized around a theme, with each child taking on a different role; begins at about 2 years of age

Make-Believe

During the preschool years, cooperative play often takes the form of make-believe. Preschoolers have telephone conversations with imaginary partners or pretend to drink imaginary juice. In the early phases of make-believe, children rely on realistic

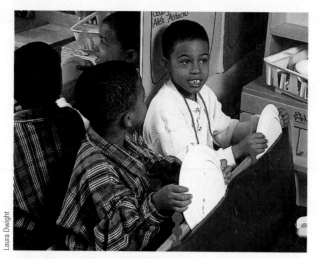

Make-believe—a favorite of preschoolers everywhere—fosters their cognitive development.

props to support their play. While pretending to drink, younger preschoolers use a real cup; while pretending to drive a car, they use a toy steering wheel. In the later phases of make-believe, children no longer need realistic props; instead, they can imagine that a block is the cup or that a paper plate is the steering wheel. Of course, this gradual movement toward more abstract make-believe is possible because of cognitive growth that occurs during the preschool years (Striano, Tomasello, & Rochat, 2001).

Although make-believe is a particularly striking feature of preschoolers' play, this capacity emerges in even younger children. By age 16–18 months, toddlers have an inkling of the difference between pretend play and reality. If toddlers see an adult who pretends to fill two glasses with water and then drinks from one of the glasses, they will pretend to drink from the other glass (Bosco, Friedman, & Leslie, 2006).

Of course, the first time that a parent pretends, this must be puzzling for toddlers. They probably wonder why mom is drinking from an empty glass or eating cereal from an empty bowl. But mothers help toddlers make sense out of this behavior: When mothers pretend, they typically look directly at the child and grin, as if to say, "This is just for fun—it's not real!" And toddlers return the smile, as if responding, "I get it! We're playing!" (Nishida & Lillard, 2007). When children are older, they usually tell play partners that they want to pretend ("Let's pretend"), then describe those aspects of reality that are being changed ("I'll be the pilot and this is my plane," referring to the couch). It's as if children mutually agree to enter a parallel universe that's governed by its own set of rules (Rakoczy, 2008; Weisberg & Bloom, 2009).

As you might suspect, make-believe reflects the values important in a child's culture (Gosso, Morais, & Otta, 2007). For example, adventure and fantasy are favorite themes for European American youngsters, but family roles and everyday activities are favorites of Korean American children. In addition, European American children are more assertive in their make-believe and more likely to disagree with their play partner's ideas about pretending ("*I* want to be the king; *you* be the mom!") but Korean American children are more polite ("Could I *please* be king?"). Thus, cultural values influence both the content and the form of make-believe (Farver & Shin, 1997).

Make-believe play is not only entertaining for children but also promotes cognitive development. Children who spend much time in make-believe play tend to be more advanced in language, memory, and reasoning (Bergen & Mauer, 2000). They also tend to have a more sophisticated understanding of other people's thoughts, beliefs, and feelings (Lindsey & Colwell, 2003).

Another benefit of make-believe is that it allows children to explore topics that frighten them. Children who are afraid of the dark may reassure a doll who is also afraid of the dark. By explaining to the doll why she need not be afraid, children come to understand and regulate their own fear of darkness. Or children may pretend that a doll has misbehaved and must be punished, which allows them to experience the parent's anger and the doll's guilt. With make-believe, children explore other emotions as well, including joy and affection (Gottman, 1986).

For many preschool children, make-believe play involves imaginary companions. Children can usually describe their imaginary playmates in some detail, mentioning sex and age as well as hair and eye color. Imaginary companions were once thought to be fairly rare but many preschoolers, particularly firstborn and only children, report imaginary companions (Taylor et al., 2004). What's more, an imaginary companion is associated with many *positive* social characteristics (Gleason & Hohmann, 2006; Roby & Kidd, 2008): Preschoolers with imaginary friends tend to be more sociable and have more real friends than other preschoolers. And among older children who are at risk for developing behavior problems, an imaginary companion promotes better adjustment during adolescence (Taylor, Hulette, & Dishion, 2010).

THINK ABOUT IT

How might Jean Piaget have explained the emergence of make-believe during the preschool years? How would Erik Erikson explain it?

Solitary Play

At times throughout the preschool years, many children prefer to play alone. Should parents be worried? Usually, no. Solitary play comes in many forms and most are normal—even healthy. Spending playtime alone coloring, solving puzzles, or assembling LEGOs is not a sign of maladjustment. Many youngsters enjoy solitary activities and, at other times, choose very social play.

However, some forms of solitary play *are* signs that children are uneasy interacting with others (Coplan et al., 2001; Harrist et al., 1997). One type of unhealthy solitary play is wandering aimlessly. Sometimes children go from one preschool activity center to the next, as if trying to decide what to do. But really they just keep wandering, never settling into play with others or into constructive solitary play. Another unhealthy type of solitary play is hovering: A child stands nearby peers who are playing, watching them play but not participating. Over time, these behaviors do not bode well for youngsters (Coplan & Armer, 2007), so it's best for these youngsters to see a professional who can help them overcome their reticence in social situations.

Gender Differences in Play

Between 2 and 3 years of age, children begin to prefer playing with same-sex peers (Martin & Fabes, 2001). Little boys play together with cars, and little girls play together with dolls. This preference increases during childhood, reaching a peak in preadolescence. By age 10 or 11, the vast majority of peer activity is with same-sex children, and most of this involves sex-typed play: Boys are playing sports or playing with cars or action figures; girls are doing artwork or playing with pets or dolls (McHale et al., 2004). This tendency for boys to play with boys and girls with girls has several distinctive features (Maccoby, 1998).

- Given the opportunity, children spontaneously select same-sex playmates. Adult pressure (e.g., "James, why don't you play with John, not Amy") is not necessary.
- Children resist parents' efforts to get them to play with members of the opposite sex. Girls are often unhappy when parents encourage them to play with boys, and boys are unhappy when parents urge them to play with girls.
- Children's reluctance to play with members of the opposite sex is not restricted to gender-typed games, such as playing house or playing with cars. Boys and girls prefer same-sex playmates even in gender-neutral activities such as playing tag or doing puzzles.

Why do boys and girls seem so attracted to same-sex play partners? One reason is that their styles of play are very different. Boys specifically prefer rough-and-tumble play and generally are more competitive and dominating in their interactions. In contrast, when girls play, they are much more cooperative, prosocial, and conversation-oriented (Rose & Rudolph, 2006). Boys don't enjoy the way that girls play and girls are aversive to boys' play (Maccoby, 1990, 1998).

Second, when girls and boys play together, girls do not readily influence boys. *Girls' interactions with one another are typically* **enabling**—*their actions and remarks tend to support others and sustain the interaction.* When drawing together, one girl might say to another, "Cool picture" or "What do you want to do now?" *In contrast, boy's interactions are often* **constricting**—*one partner tries to emerge as the victor by threatening or contradicting the other, by exaggerating, and so on.* In the same drawing task, one boy might say to another, "My picture's better" or "Drawing is stupid—let's watch TV." When these styles are brought together, girls find their enabling style is ineffective with boys. The same subtle overtures that work with other girls have no impact on boys. Boys ignore girls' polite suggestions

enabling actions
individuals' actions and remarks that tend to support others and sustain the interaction

constricting actions
interaction in which one partner tries to emerge as the victor by threatening or contradicting the other

At about 2 or 3 years of age, boys and girls start to prefer playing with members of their own sex.

about what to do and ignore girls' efforts to resolve conflicts with discussion (Rose & Rudolph, 2006).

Some theorists believe that these contrasting styles may have an evolutionary basis (Geary et al., 2003). Boys' concerns about dominating others may stem from a concern with establishing one's rank among a group of males because those males at the upper ranks have better access to mates and better access to resources needed for offspring. Girls' concerns about affiliation may be a by-product of the fact that women traditionally left their own communities (and relatives) to live in a husband's community. Having no relatives nearby enhanced the value of a close friend, which placed a premium on the affiliative behaviors that lead to and maintain friendships.

Regardless of the exact cause, early segregation of playmates by style of play means that boys learn primarily from boys and girls from girls. Over time, such social segregation by sex reinforces gender differences in play. When young boys spend most of their time playing with other boys, their play becomes more active and more aggressive. In contrast, when young girls spend most of their time playing with other girls their play becomes less active and less aggressive (Martin & Fabes, 2001).

Parental Influence

Parents become involved in their preschool children's play in several ways (Parke & O'Neill, 2000):

- *Playmate.* Many parents enjoy the role of playmate (and many parents deserve an Oscar for their performances). They use the opportunity to scaffold their children's play (see page 145), often raising it to more sophisticated levels (Tamis-LeMonda & Bornstein, 1996). For example, if a toddler is stacking toy plates, a parent might help the child stack the plates (play at the same level) or might pretend to wash each plate (play at a more advanced level). When parents demonstrate more advanced forms of play, their children often play at the more advanced levels later (Lindsey & Mize, 2000).

- *Social director.* It takes two to interact, and young children rely on parents to create opportunities for social interactions. Many parents of young children arrange visits with peers, enroll children in activities (e.g., preschool programs), and take children to settings that attract young children (e.g., parks, swimming pools). All this effort is worth it: Children whose parents provide them with frequent opportunities for peer interaction tend to get along better with their peers (Ladd & Pettit, 2002).

- *Coach.* Successful interactions are based on a host of skills, including how to initiate an interaction, make joint decisions, and resolve conflicts. When parents help their children acquire these skills, children tend to be more competent socially and to be more accepted by their peers (Parke et al., 2004). But there's a catch: The coaching needs to be constructive for children to benefit. Parent-coaches sometimes make suggestions that aren't very clear or are actually misguided. Bad coaching is worse than none at all because it harms children's peer relations (Russell & Finnie, 1990).

- *Mediator.* When young children play, they often disagree, argue, and sometimes fight. However, children play more cooperatively and longer when parents are present to help iron out conflicts (Mize, Pettit, & Brown, 1995). When young children can't agree on what to play, a parent can negotiate a mutually acceptable activity. When both youngsters want to play with the same toy, a parent can arrange for them to share. Here, too, parents scaffold their preschoolers' play, smoothing the interaction by providing some of the social skills that preschoolers lack.

In addition to these direct influences on children's play, parents influence children's play indirectly via the quality of the

THINK ABOUT IT

Suppose friends of yours ask you how their preschool daughter could get along well with peers. What advice would you give them?

Parents influence their children's play in many ways, perhaps none more important than by mediating the disputes that arise when preschoolers play.

© David Young-Wolff / PhotoEdit

parent–child attachment relationship. Recall that children's relationships with peers are most successful when, as infants, they had a secure attachment relationship with their mother (Bascoe et al., 2009; Wood, Emmerson, & Cowan, 2004). A child's relationship with his or her parents is the internal working model for all future social relationships. When the parent–child relationship is of high quality and emotionally satisfying, children are encouraged to form relationships with other people. Another possibility is that a secure attachment relationship with the mother makes an infant feel more confident about exploring the environment, which in turn provides more opportunities to interact with peers. These two views are not mutually exclusive; both may contribute to the relative ease with which securely attached children interact with their peers (Hartup, 1992).

Helping Others

Prosocial behavior *is any behavior that benefits another person.* Cooperation—that is, working together toward a common goal—is one form of prosocial behavior. Of course, cooperation often "works" because individuals gain more than they would by not cooperating. *In contrast,* **altruism** *is behavior that is driven by feelings of responsibility toward other people, such as helping and sharing, in which individuals do not benefit directly from their actions.* If two youngsters pool their funds to buy a candy bar to share, this is cooperative behavior. If one youngster gives half of her lunch to a peer who forgot his own, this is altruism.

Many scientists believe that humans are biologically predisposed to be helpful, to share, to cooperate, and to be concerned for others (Hastings, Zahn-Waxler, & McShane, 2006). Why has prosocial behavior evolved over time? The best explanation has nothing to do with lofty moral principles; it's much more pragmatic: People who frequently help others are more likely to receive help themselves, which increases the odds that they'll pass along their genes to future generations. In fact, basic acts of altruism can be seen by 18 months of age. When toddlers and preschoolers see other people who are obviously hurt or upset, they appear concerned, like the child in the photo. They try to comfort the person by hugging him or patting him (Zahn-Waxler et al., 1992). Apparently, at this early age, children recognize signs of distress. And if an adult is in obvious need of help—a teacher accidentally drops markers on a floor—most 18-month-olds spontaneously help get the markers (Warneken & Tomasello, 2006).

During the toddler and preschool years, children gradually begin to understand others' needs and learn appropriate altruistic responses (van der Mark, van IJzendoorn, & Bakermans-Kranenburg, 2002). When 3-year-old Alexis sees that her infant brother is crying because he's dropped his favorite bear, she retrieves it for him; when 4-year-old Darren sees his mom crying while watching a TV show, he may turn off the TV. These early attempts at altruistic behavior are limited because young children's knowledge of what they can do to help is modest. As youngsters acquire more strategies to help others, their preferred strategies become more adultlike (Eisenberg, Fabes, & Spinrad, 2006).

Let's look at some specific skills that set the stage for altruistic behaviors.

prosocial behavior
any behavior that benefits another person

altruism
prosocial behavior such as helping and sharing in which the individual does not benefit directly from his or her behavior

By 18 months of age, toddlers try to comfort others who are hurt or upset.

Skills Underlying Altruistic Behavior

Remember from Chapter 4 that preschool children are often egocentric, so they may not see the need for altruistic behavior. For example, young children might not share candy with a younger sibling because they cannot imagine how unhappy the sibling is without the candy. In contrast, school-age children, who can more easily take another person's perspective, would perceive the unhappiness and would be more inclined to share. In fact, research consistently indicates that altruistic behavior is related to perspective-taking skill. Youngsters who understand others' thoughts and feelings share better with others and help them more often (Strayer & Roberts, 2004; Vaish, Carpenter, & Tomasello, 2009).

empathy
experiencing another person's feelings

Related to perspective taking is **empathy**, *which is the actual experiencing of another's feelings.* Children who deeply feel another individual's fear, disappointment, sorrow, or loneliness are more inclined to help that person than children who do not feel those emotions (Eisenberg et al., 2006; Malti et al., 2009). In other words, youngsters who are obviously distressed by what they are seeing are most likely to help if they can.

Of course, perspective taking, empathy, and moral reasoning skills do not guarantee that children always act altruistically. Even though children have the skills needed to act altruistically, they may not because of the particular situation, as we'll see next.

When children can empathize with others who are sad or upset, they're more likely to offer to help.

Situational Influences

Kind children occasionally disappoint us by being cruel, and children who are usually stingy sometimes surprise us by their generosity. Why? The setting helps determine whether children act altruistically or not.

- *Feelings of responsibility.* Children act altruistically when they feel responsible for the person in need. For example, children may help siblings and friends more often than strangers simply because they feel a direct responsibility for people they know well (Costin & Jones, 1992). And they're more likely to help when prompted with photos showing two people who look to be friends (Over & Carpenter, 2009). In other words, a simple reminder of the importance of friendship (or affiliation with others) can be enough to elicit helping.

- *Feelings of competence.* Children act altruistically when they feel they have the skills to help the person in need. Suppose, for example, that a preschooler is growing more and more upset because she can't figure out how to work a computer game; a peer who knows little about computer games is not likely to come to the young girl's aid because the peer doesn't know what to do to help. If the peer tries to help, he or she could end up looking foolish (Peterson, 1983).

- *Mood.* Children act altruistically when they are happy or feeling successful but not when they are sad or feeling as if they have failed (Wentzel, Filisetti, & Looney, 2007). A preschool child who has just spent an exciting morning as the "leader" in nursery school is more inclined to share treats with siblings than is a preschooler who was punished by the teacher (Eisenberg, 2000).

- *Costs of altruism.* Children act altruistically when such actions entail few or modest sacrifices. A preschool child who was given a snack that she doesn't particularly like is more inclined to share it with others than one who was given her very favorite food (Eisenberg & Shell, 1986).

So when are children most likely to help? They help when they feel responsible for the person in need, have the needed skills, are happy, and believe they will give up little by helping. When are children least likely to help? They are least likely to help when they feel neither responsible nor capable of helping, are in a bad mood, and believe that helping will entail a large personal sacrifice.

With these guidelines in mind, can you explain why Antonio and Carla, the children in the opening vignette, watched idly as their older brother cried? The last two factors—mood and costs—are not likely to be involved. However, the first two factors may explain the failure of Antonio and Carla to help their older brother. Our explanation appears on page 190, just before Test Yourself.

So far, we've seen that altruistic behavior is determined by children's skills (such as perspective taking) and by characteristics of situations (such as whether children feel competent to help in a particular situation). Whether children are altruistic is also determined by genetics and by socialization, the topic of the remaining two sections in this module.

THINK ABOUT IT

Suppose some kindergarten children want to raise money for a gift for one of their classmates who is ill. Based on the information presented here, what advice can you give the children as they plan their fund-raising?

© Catchlight Visual Services / Alamy

The Contribution of Heredity

As we mentioned on page 187, many scientists believe that prosocial behavior represents an evolutionary adaptation: people who help others are more likely to be helped themselves and thus are more likely to survive and have offspring. According to this argument, we should expect to find evidence for heritability of prosocial behavior and in fact that's the case: twin studies consistently find that identical twins are more similar in their prosocial behavior than are fraternal twins (Gregory et al., 2009).

Genes probably affect prosocial behavior indirectly, by their influence on temperament. For example, children who are temperamentally less able to regulate their emotions (in part due to heredity) may help less often because they're so upset by another's distress that taking action is impossible (Eisenberg et al., 2007). Another temperamental influence may be via inhibition (shyness). Children who are temperamentally shy are often reluctant to help others, particularly people they don't know well (Young, Fox, & Zahn-Waxler, 1999). Even though shy children realize that others need help and are upset by another person's apparent distress, their reticence keeps these feelings from being translated into action. Thus, in both cases, children are aware that others need help. But in the first instance they're too upset themselves to figure out how to help and in the second instance they know how to help but are too inhibited to follow through.

Socialization of Altruism

Dr. Martin Luther King Jr. said that his pursuit of civil rights for African Americans was particularly influenced by three people: Henry David Thoreau (a 19th-century American philosopher), Mohandas Gandhi (leader of the Indian movement for independence from England), and his father, Dr. Martin Luther King Sr. As is true of many humanitarians, Dr. King's prosocial behavior started in childhood, at home. But how do parents foster altruism in their children? Several factors contribute.

- *Modeling.* When children see adults helping and caring for others, they often imitate such prosocial behavior (Eisenberg et al., 2006). Of course, parents are the models to whom children are most continuously exposed, so they exert a powerful influence. Parents who report frequent feelings of warmth and concern for others tend to have children who experience stronger feelings of empathy. When a mother is helpful and responsive, her children often imitate her by being cooperative, helpful, sharing, and less critical of others. In a particularly powerful demonstration of the impact of parental modeling, people who had risked their lives during World War II to protect Jews from the Nazis often reported their parents' emphasis on caring for all people (Oliner & Oliner, 1988).

- *Disciplinary practices.* Children behave prosocially more often when their parents are warm and supportive, set guidelines, and provide feedback; in contrast, prosocial behavior is less common when parenting is harsh, threatening, and includes frequent physical punishment (Asbury et al., 2003; Eisenberg & Fabes, 1998). Particularly important is parents' use of reasoning as a disciplinary tactic with the goal of helping children see how their actions affect others. For example, after 4-year-old Annie grabbed some crayons from a playmate, her father told Annie, "You shouldn't just grab things away from people. It makes them angry and unhappy. Ask first, and if they say 'no' then you mustn't take them."

- *Opportunities to behave prosocially.* You need to practice to improve motor skills and the same is true of prosocial behaviors—children and adolescents are more likely to act prosocially when they're routinely given the opportunity to help and cooperate with others. At home, children can help with household tasks, such as cleaning and setting the table. Adolescents can be encouraged to participate in community service, such as working at a food bank or tutoring younger children. Experiences like these help sensitize children and adolescents to the needs of others and allow them to enjoy the satisfaction of helping (Grusec, Goodnow, & Cohen, 1996; McLellan & Youniss, 2003).

Thus, parents can foster altruism in their youngsters by behaving altruistically them-selves, using reasoning to discipline their children, and encouraging their children to help at home and elsewhere. Situational factors also play a role, and altruism requires perspective taking and empathy. Combining these ingredients, we can give a general account of children's altruistic behavior. As children get older, their perspective-taking and empathic skills develop, which enables them to see and feel another's needs. Nonetheless, children are never invariably altruistic (or, fortunately, invariably nonal-truistic) because particular contexts affect altruistic behavior, too. These factors are summarized in ●Table 5.4.

Postscript: Why Didn't Antonio and Carla Help?

Here are our explanations. First, neither Antonio nor Carla may have felt sufficiently responsible to help, because (a) with two children who could help, each child's feeling of individual responsibility is reduced, and (b) younger children are less likely to feel responsible for an older sibling. Second, it's our guess that neither child has had many opportunities to use the DVD player. In fact, it's likely that they both have been strongly discouraged from touching it. Consequently, they don't feel competent to help because neither knows how it works or what they should do to help Juan remove his finger.

THINK ABOUT IT

Paula worries that her son Elliot is too selfish and wishes that he were more caring and com-passionate. As a parent, what could Paula do to encourage Elliot to be more concerned about the welfare of others?

Test Yourself

RECALL

1. Toddlers who are 12–15 months old often engage in _____ play, in which they play separately but look at one another and sometimes communicate verbally.

2. One of the advantages of _____ play is that chil-dren can explore topics that frighten them.

3. When girls interact, conflicts are typically resolved through _____; boys more often resort to intimidation.

4. _____ is the ability to understand and feel another person's emotions.

5. Contextual influences on prosocial behavior include feelings of responsibility, feelings of competence, _____, and the costs associated with behaving prosocially.

6. Parents can foster altruism in their youngsters by behav-ing altruistically themselves, using reasoning to discipline their children, and _____.

INTERPRET

Why must a full account of children's prosocial behavior in-clude an emphasis on skills (e.g., empathy) as well as situa-tions (e.g., whether a child feels responsible)?

APPLY

How might children's temperament, which we discussed in Chapter 3, influence the development of their play with peers?

Recall answers: (1) parallel, (2) make-believe, (3) discussion and compromise, (4) Empathy, (5) mood, (6) providing children with opportunities to practice being altruistic

LEARNING OBJECTIVES

- What are our stereotypes about males and females? How well do they correspond to actual differences between boys and girls?

- How do young children learn gender roles?
- How are gender roles changing? What further changes might the future hold?

Meda and Perry want their 6-year-old daughter, Hope, to pick activities, friends, and ultimately a career based on her interests and abilities rather than on her gender. They have done their best to encourage gender-neutral values and behavior. Both are therefore astonished that Hope seems to be totally indistinguishable from other 6-year-olds reared by conventional parents. Hope's close friends are all girls. When Hope is with her friends, they play house or play with dolls. What seems to be going wrong with Meda and Perry's plans for a gender-neutral girl?

FAMILY AND WELL-WISHERS ARE ALWAYS EAGER TO KNOW THE SEX OF A NEWBORN. Why are people so interested in a baby's sex? The answer is that being a "boy" or "girl" is not simply a biological distinction. Instead, these terms are associated with distinct social roles. *Like a role in a play, a* **social role** *is a set of cultural guidelines as to how a person should behave, particularly with other people.* The roles associated with gender are among the first that children learn, starting in infancy. Youngsters rapidly learn about the behaviors that are assigned to males and females in their culture. At the same time, they begin to identify with one of these groups. As they do, they take on an identity as a boy or a girl.

In this section, you'll learn about the "female role" and the "male role" in North America today, and you'll also discover why Meda and Perry are having so much trouble rearing a gender-neutral girl.

social role
set of cultural guidelines about how one should behave, especially with other people

Images of Men and Women: Facts and Fantasy

All cultures have **gender stereotypes**—*beliefs and images about males and females that may or may not be true.* For example, many men and women believe that males are rational, active, independent, competitive, and aggressive. At the same time, many men and women claim that females are emotional, passive, dependent, sensitive, and gentle (Ruble, Martin, & Berenbaum, 2006).

Based on gender stereotypes, we expect males and females to act and feel in particular ways, and we respond to their behavior differently depending on their gender (Smith & Mackie, 2000). For example, if you saw a toddler playing with a doll then you would probably assume that she is a girl, based on her taste in toys. What's more, your assumption would lead you to believe both that she plays more quietly and that she is more readily frightened than a boy (Karraker, Vogel, & Lake, 1995). Once we assume the child is a girl, our gender stereotypes lead to a host of inferences about behavior and personality.

gender stereotypes
beliefs and images about males and females that are not necessarily true

If you assume the child is a girl, this leads to a host of other inferences about her personality and behavior.

Learning Gender Stereotypes

Children don't live in a gender-neutral world for long. Although 12-month-old boys and girls look equally at gender-stereotyped toys, 18-month-olds do not: Girls look longer at pictures of dolls than pictures of trucks but boys look longer at pictures of trucks (Serbin et al., 2001). By 4 years of age, children's knowledge of gender-stereotyped activities is extensive: They believe girls play hopscotch but boys play football; girls help bake cookies but boys take out the trash; and women feed babies but men chop wood (Gelman, Taylor, & Nguyen, 2004). They've also begun to learn about behaviors and traits that are stereotypically masculine or feminine. Preschoolers believe that boys are more often aggressive physically but that girls tend to be aggressive verbally (Giles & Heyman, 2005).

By the time children are ready to enter elementary school, they are well on their way to learning gender stereotypes. For example, 5-year-olds believe boys are strong and dominant and girls are emotional and gentle (Heyman & Legare, 2004).

Beyond the preschool years, children also learn that the traits and occupations associated with males tend to earn more money and have greater power than those associated with females (Weisgram, Bigler, & Liben, 2010). At some point children apparently internalize a general belief about male versus female occupations—something like "Jobs for men are more prestigious than jobs for women." And although older children are more familiar with gender stereotypes, they see these stereotypes as general guidelines for behavior that are not necessarily binding for all boys and girls (Banse et al., 2010).

Gender-Related Differences

So far we've only considered people's *beliefs* about differences between males and females, and many of them are false. Research reveals that males and females often do not differ in the ways specified by cultural stereotypes. What are the bona fide differences between males and females? Of course, in addition to the obvious anatomical differences, males are typically larger and stronger than females throughout most of the life span. As infants, boys are more active than girls, and this difference increases during childhood (Else-Quest et al., 2006; Saudino, 2009). In contrast, girls have a lower mortality rate and are less susceptible to stress and disease (Zaslow & Hayes, 1986).

When it comes to social roles, activities for males tend to be more strenuous, involve more cooperation with others, and often require travel. Activities for females are usually less demanding physically, more solitary, and take place closer to home. This division of roles is much the same worldwide (Whiting & Edwards, 1988).

The extent of gender differences in the intellectual and psychosocial arenas remains uncertain. Research suggests differences between males and females in the following areas.

- *Verbal ability*. During the toddler years, girls have larger vocabularies than boys (Leaper & Smith, 2004). During elementary school and high school, girls read, write, and spell better than boys, and more boys have reading and other language-related problems such as stuttering (Halpern et al., 2007; Wicks-Nelson & Israel, 2006).

- *Mathematics*. On math achievement tests, boys used to get higher scores but that difference has diminished substantially over the past 25 years; now boys have a negligible advantage (Lindberg et al., 2010). This change apparently reflects efforts to encourage girls to pursue mathematics generally and to take more math courses specifically. Around the world, gender differences in math are negligible in countries where females have similar access to education, occupations, and political power as males, but gender differences remain where females are limited to traditionally feminine-stereotyped occupations that do not require math skills (Else-Quest, Hyde, & Linn, 2010).

- *Spatial ability*. On problems like those in ■ Figure 5.2, which measure the ability to manipulate visual information mentally, you must decide which figures are rotated variants of the standard shown at the left. Males typically respond more rapidly and accurately than females, and these differences are observed in infancy (Halpern et al., 2007; Quinn & Liben, 2008).

- *Social influence*. Girls are more likely than boys to comply with the directions of adults (Maccoby & Jacklin, 1974). Girls and women are also more readily influenced by others in a variety of situations, particularly when they are under

■ **Figure 5.2**

On spatial ability tasks, which involve visualizing information in different orientations, males tend to respond more rapidly and more accurately than females.

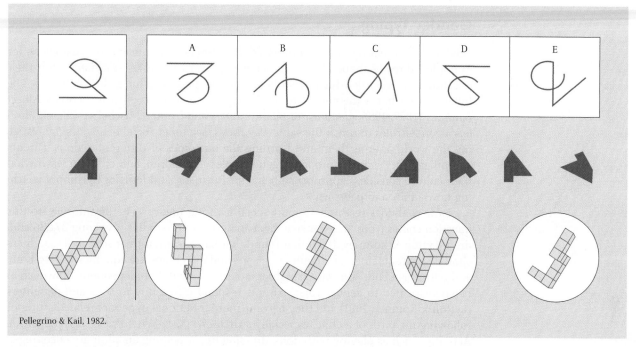

Pellegrino & Kail, 1982.

From "Process Analysis of Spatial Aptitude," by J. W. Pellegrino and R. V. Kail, in R. J. Sternberg (Ed.), *Intelligence,* Vol. 1, p. 316. Copyright © 1982 Lawrence Erlbaum Associates, Inc. Reprinted with permission.

group pressure (Becker, 1986; Eagly, Karau, & Makhijani, 1995). However, these gender differences may simply reflect that females value group harmony more than males do and thus seem to give in to others (Miller, Danaher, & Forbes, 1986; Strough & Berg, 2000). For instance, at a meeting to plan a school function, girls are just as likely as boys to recognize the flaws in a bad idea, but girls are more willing to go along simply because they don't want the group to start arguing.

■ *Aggression.* In virtually all cultures that have been studied boys are more physically aggressive than girls, and this is true as early as 17 months of age (Baillargeon et al., 2007; Card et al., 2008). This difference continues throughout the life span (Sanson et al., 1993). In contrast, *girls are more likely to resort to* **relational aggression** *in which they try to hurt others by damaging their relationships with peers.* They may call children names, make fun of them, spread rumors about them, or pointedly ignore them (Ostrov & Godleski, 2010).

relational aggression
aggression used to hurt others by undermining their social relationships

■ *Emotional sensitivity.* Girls are better able to express their emotions and interpret others' emotions (Hall & Halberstadt, 1981; Weinberg et al., 1999). For example, throughout infancy and childhood, girls identify facial expressions (e.g., a happy face versus a sad face) more accurately than boys do (McClure, 2000). And in their interactions with peers, girls are more empathic—they're better able to feel how other children and adolescents are feeling (Rose & Rudolph, 2006).

In most other intellectual and social domains, boys and girls are similar. When thinking about areas in which sex differences have been found, keep in mind that gender differences often depend on a person's experiences (Casey, 1996; Serbin, Powlishta, & Gulko, 1993). Also, gender differences may fluctuate over time, reflecting historical change in the contexts of childhood for boys and girls. Finally, each result just described refers to a difference in the *average* performance of boys and girls. These differences tend to be small, which means that they do not apply to all boys and girls (Hyde, 2007). Many girls have greater spatial ability than some boys, and many boys are more susceptible to social influence than are some girls.

Gender Typing

Folklore holds that parents and other adults—teachers and television characters, for example—directly shape children's behavior toward the roles associated with their sex. Boys are rewarded for boyish behavior and punished for girlish behavior. The folklore even has a theoretical basis. According to social cognitive theorists Albert Bandura (1977, 1986; Bandura & Bussey, 2004) and Walter Mischel (1970), children learn gender roles in much the same way they learn other social behaviors: by watching the world around them and learning the outcomes of different actions. Parents and others thus shape appropriate gender roles in children, and children learn what their culture considers appropriate behavior for males and females by simply watching how adults and peers act.

How well does research support social learning theory? The best answer to this question comes from an extensive meta-analysis of 172 studies involving 27,836 children (Lytton & Romney, 1991) that found that parents often treat sons and daughters similarly. Parents interact equally with sons and daughters, are equally warm to both, and encourage both sons and daughters to achieve and be independent. However, in behavior related to gender roles, parents respond differently to sons and daughters (Lytton & Romney, 1991) and they become more traditional in gender-related attitudes following the birth of a child, especially a firstborn (Katz-Wise, Priess, & Hyde, 2010). Activities such as playing with dolls, dressing up, or helping an adult are encouraged more often in daughters than in sons; rough-and-tumble play and playing with blocks are encouraged more in sons than in daughters. Parents tolerate mild aggression in sons to a greater degree than in daughters (Martin & Ross, 2005). As we'll see in the Spotlight on Research feature, when young children make stereotyped comments, their mothers usually go along with them.

THINK ABOUT IT

The women's liberation movement became a powerful social force in North America during the 1960s. Describe how you might do research to determine whether the movement has changed the gender roles that children learn.

Spotlight on research — How Mothers Talk to Children About Gender

Who were the investigators, and what was the aim of the study? Imagine a mother and her preschool son reading a picture book together. Seeing a picture of a girl catching a frog, he says, "Girls hate frogs!" Seeing a picture of a boy playing football, he exclaims, "Yes, Daniel and I like playing football!" When mothers hear the children make gender-stereotyped statements like these, what do they do? Susan Gelman, Marianne Taylor, and Simone Nguyen (2004) conducted a study to answer this question.

How did the investigators measure the topic of interest? Gelman and colleagues created books that included 16 pictures. Half of the pictures showed a child or adult in a gender-stereotyped activity (e.g., a girl sewing, a man driving a truck); half showed a child or adult in an activity that was counter to gender stereotypes (e.g., a boy baking, a woman firefighter). Mothers were simply asked to go through the picture book with the children, as they might do at home. They

were not told about the investigators' interest in gender. Mothers and children were videotaped as they looked at the books.

Who were the participants in the study? The study included 72 pairs of mothers and children: of the children, 24 were 2-year-olds, 24 were 4-year-olds, and 24 were 6-year-olds. At each age, half of the children were girls.

What was the design of the study? This study was correlational because Gelman and colleagues were interested in the relations that existed naturally between children's speech and a mother's reply to children's speech, the child's age, and the child's sex. The study was cross-sectional because it included 2-year-olds, 4-year-olds, and 6-year-olds, each tested once.

Were there ethical concerns with the study? No; most children enjoy reading picture books with their parents.

What were the results? First the investigators determined the number of stereotyped state-

ments that children made, including those that endorsed a stereotype—"Jackie and Sherry love to play with dolls!"—as well as those that deny a counterstereotype—"Boys aren't ballet dancers!" The 2-year-olds averaged about 24 of the statements; the 4- and 6-year-olds averaged about 30. Then the mother's response to stereotyped comments was classified in one of eight categories; we'll describe just three of the categories for simplicity.

- The mother *affirmed* her child's remark: "Yes, girls do like playing with dolls!"
- The mother *repeated* the child's remark as a question: "Are you sure boys aren't dancers?"
- The mother *negated* the child's remark: "Oh yes, boys *can* be dancers."

The percentage of times that mothers used each of these responses is shown in ■ Figure 5.3. You can see that mothers agreed with their child's

■ **Figure 5.3**

Mothers often affirm their children's stereotyped comments or repeat them as questions, but they rarely negate them. Data from Gelman et al. (2004).

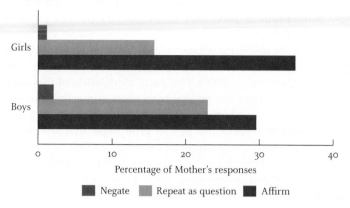

Percentage of Mother's responses

■ Negate   ■ Repeat as question   ■ Affirm

remarks about one third of the time. They almost never disagreed directly with their children: Fewer than 2% of mothers' comments fell in this category. But they did rephrase their children's statements as questions about 20% of the time, which is a subtle way for a mother to dispute her child. Overall, mothers' responses were very similar for sons and daughters; they also were much the same for 2-, 4-, and 6-year-olds, so the data in the graphs are averaged across the three age groups.

What did the investigators conclude? Gelman and colleagues (2004) concluded that "mothers are surprisingly accepting of children's stereotyping statements . . . Mothers rarely directly contradicted a child's gender stereotype statement, and in fact more often affirmed the child's stereotype than questioned it."

What converging evidence would strengthen these conclusions? The mothers in this sample were very well educated: Most were college graduates. It would be important to see whether mothers with less education respond in a similar fashion. In addition, it would be valuable to know the impact of a mother's reply on her child's gender stereotyping. When mothers question children's stereotyped statements, do children rethink their concepts? Or is this form of feedback too subtle to affect preschool children, particularly younger ones?

 Go to Psychology CourseMate at www.cengagebrain.com to enhance your understanding of this research.

Fathers are more likely than mothers to treat sons and daughters differently. More than mothers, fathers often encourage gender-related play. Fathers punish their sons more, but they accept dependence in their daughters (Snow, Jacklin, & Maccoby, 1983). A father, for example, may urge his frightened young son to jump off the diving board ("Be a man!") but not insist that his daughter do so ("That's okay, honey"). Apparently mothers are more likely to respond based on their knowledge of the needs of individual children but fathers respond based on gender stereotypes. A mother responds to her son knowing that he's smart but unsure of himself; a father may respond based on what he thinks boys generally should be like.

Of course, adults differ in their views on the relative rights and roles of males and females. Some have very traditional views and believe that men should be hired preferentially for some jobs and that it's more important for sons than daughters to attend college; others have more gender-neutral views and believe that women should have the same business and professional opportunities as men and that daughters should have the same educational opportunities as sons. It would be surprising if parents did not convey these attitudes to their children, and indeed they do (Crouter et al., 2007). A meta-analysis of 48 studies including more than 10,000 pairs of parents and children showed that children's gender-related interests, attitudes, and self-concepts are more traditional when their parents have traditional views but are more gender-neutral when their parents have nontraditional views (Tenenbaum & Leaper, 2002).

Peers are also influential. By 3 years, most children's play shows the impact of gender stereotypes—boys prefer blocks and trucks, whereas girls prefer tea sets and dolls—and youngsters are critical of peers who engage in gender-inappropriate play (Aspenlieder, 2009). This is particularly true of boys who like feminine toys or who choose feminine

Fathers are more likely than mothers to encourage their children's gender-related play.

Preschoolers discourage peers from cross-gender play.

activities. Boys who play with dolls and girls (like the one in the photo) who play with trucks will both be ignored, teased, or ridiculed by their peers, but a boy will receive harsher treatment than a girl (Levy, Taylor, & Gelman, 1995). Once children learn rules about gender-typical play, they often harshly punish peers who violate those rules.

Peers influence gender roles in another way, too. We've seen that, by 2 or 3 years of age, children most often play with same-sex peers (Martin & Fabes, 2001). This early segregation of playmates based on a child's gender means that boys learn primarily from boys and girls from girls. This helps solidify a youngster's emerging sense of membership in a particular gender group and sharpens the contrast between their own gender and the other gender.

Thus, through encouraging words, critical looks, and other forms of praise and punishment, other people influence boys and girls to behave differently (Jacobs & Eccles, 1992). However, children learn more than simply the specific behaviors associated with their gender. *A child gradually begins to identify with one group and to develop a* **gender identity**—*a sense of the self as a male or a female.*

gender identity
sense of oneself as male or female

Gender Identity

If you were to listen to a typical conversation between two preschoolers, you might hear something like this:

MARIA: When I grow up, I'm going to be a singer.
JUANITA: When I grow up, I'm going to be a papa.
MARIA: No, you can't be a papa—you'll be a mama.
JUANITA: No, I wanna be a papa.
MARIA: You can't be a papa. Only boys can be papas, and you're a girl!

Obviously, Maria's understanding of gender is more developed than Juanita's. How can we explain these differences? According to Lawrence Kohlberg (1966; Kohlberg & Ullian, 1974), children gradually develop a basic understanding that they are of either the female or the male sex. Gender then serves to organize many perceptions, attitudes, values, and behaviors. Full understanding of gender is said to develop gradually in three steps.

gender labeling
young children's understanding that they are either boys or girls and naming themselves accordingly

gender stability
understanding in preschool children that boys become men and girls become women

gender constancy
understanding that maleness and femaleness do not change over situations or personal wishes

■ **Gender labeling**: *By age 2 or 3, children understand that they are either boys or girls and label themselves accordingly.*

■ **Gender stability**: *During the preschool years, children begin to understand that gender is stable: Boys become men and girls become women.* However, children in this stage may believe that a girl who wears her hair like a boy will become a boy and a boy who plays with dolls will become a girl (Fagot, 1985).

■ **Gender constancy**: *Between 4 and 7 years, most children understand that maleness and femaleness do not change over situations or according to personal wishes.* They understand that a child's sex is unaffected by the clothing a child wears or the toys a child likes. Juanita and Maria both know that they're girls, but Maria has developed a greater sense of gender stability and gender constancy.

According to Kohlberg's theory, children begin learning about gender roles after they have mastered gender constancy—that is, after they know that gender is fixed across time and situation. Research generally supports this prediction (Martin, Ruble, & Szkrybalo, 2002). However, some work suggests that children begin learning about gender-typical behavior as soon as they master gender stability (invariance across time) and that, as they understand gender consistency (invariance across situation), their understanding of gender roles becomes more flexible. They agree that it's okay for boys to play with dolls or for girls to play with trucks (Ruble et al., 2007).

Kohlberg's theory specifies *when* children begin learning about gender-appropriate behavior and activities but not *how* such learning takes place. A theory proposed by Carol Martin (Martin et al., 1999; Martin & Ruble, 2004) addresses how children learn about gender (see ▪ Figure 5.4). *In **gender-schema theory**, children first decide if an object, activity, or behavior is associated with females or males; then they use this information to decide whether they should learn more about the object, activity, or behavior.* That is, once children know their gender, they pay attention primarily to experiences and events that are gender appropriate (Martin & Halverson, 1987). According to gender-schema theory, a preschool boy watching a group of girls playing in sand will decide that playing in sand is for girls and that, because he is a boy, playing in sand is not for him. Seeing a group of older boys playing football, he will decide that football is for boys and that, because he is a boy, football is acceptable and he should learn more about it.

According to gender-schema theory, after children understand gender, it's as if they see the world through special glasses that allow only gender-typical activities to be in focus (Liben & Bigler, 2002). This pattern was evident in a study by Zosuls and colleagues (2009), who recorded children's language development from 10 to 21 months, looking for occasions when children referred to themselves as a boy or as a girl. In addition, at 17 and 21 months, children were observed as they played with several gender stereotypic toys (truck, doll) and gender-neutral toys (telephone, miniature people). The investigators found that children who referred to themselves by gender played more often with gender-stereotyped toys. In other words, Beth, who has referred to herself as a girl, plays with dolls but not with trucks. In contrast, James, who has never referred to himself as a boy, plays with dolls and trucks equally.

After children understand gender, their tastes in TV programs begin to shift along gender-specific lines (Luecke-Aleksa et al., 1995). In addition, they begin to use gender labels to evaluate toys and activities. Shown an unfamiliar toy and told that children of a specific sex *really* like this toy, children like the toy much more if others of their sex do, too (Shutts, Banaji, & Spelke, 2010). As Martin and Ruble (2004) put it: "Children are gender detectives who search for cues about gender—who should or should not engage in a particular activity, who can play with whom, and why girls and boys are different" (p. 67).

This selective viewing of the world explains a great deal about children's learning of gender roles, but one final important element needs to be considered: biology.

gender-schema theory
theory that states that children want to learn more about an activity only after first deciding whether it is masculine or feminine

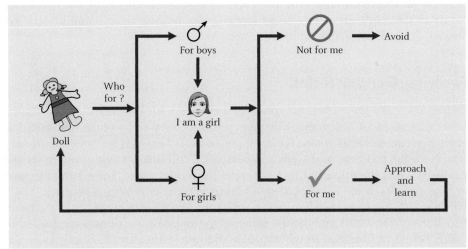

▪ **Figure 5.4**
According to gender-schema theory, children first decide if an object, activity, or behavior is for females or males and then learn more about the objects, activities, or behaviors that are appropriate for their own gender.

Copyright © Cengage Learning 2013

Biological Influences

Most child-development researchers agree that biology contributes to gender roles and gender identity. Evolutionary developmental psychology, for example, reminds us that men and women performed vastly different roles for much of human history. Women were more invested in child rearing, and men were more invested in providing important resources (e.g., food and protection) for their offspring (Geary, 2002). In adapting to these roles, different traits and behaviors evolved for men and women. For example, men became more aggressive because that was adaptive in helping them ward off predators.

If gender roles are based in part on our evolutionary heritage, then behavior genetic research should show the impact of heredity on gender-role learning. Indeed, twin studies show a substantial hereditary impact on gender-role learning (Iervolino et al., 2005). For identical twins, if one strongly prefers sex-typical toys and activities the other one usually does, too. Fraternal twins are also similar in their preference for sex-typical toys and activities, but not to the extent of identical twins.

Twin studies point to a biological basis for gender-role learning but the studies don't tell us what factors are responsible. Some scientists believe that the sex hormones are key players, and consistent with this idea, the amount of testosterone in amniotic fluid predicts a child's preference for masculine sex-typed activities: Auyeung and colleagues (2009) found that male and female fetuses who were exposed to greater amounts of testosterone at 4 months after conception had greater interest in masculine sex-typed activities during the elementary school years. This same conclusion was reached in studies of children with congenital adrenal hyperplasia (CAH), a genetic disorder in which, beginning in prenatal development, the adrenal glands secrete large amounts of androgen. The extra androgen doesn't affect a baby boy's physical development, but in baby girls it can enlarge the clitoris so that it resembles a penis. Girls affected by CAH have surgery during infancy to correct their physical appearance, and they receive hormone therapy to correct the imbalance of androgen. Nevertheless, during childhood and adolescence, girls with CAH prefer masculine activities (such as playing with cars instead of dolls) and male playmates to a much greater extent than girls not exposed to these amounts of androgen (Berenbaum & Snyder, 1995; Pasterski et al., 2005). These effects are largest for girls who have the greatest exposure to androgen during prenatal development (Berenbaum, Duck, & Bryk, 2000; Servin et al., 2003). Apparently, the androgen not only masculinizes the genitals in baby girls but also affects the prenatal development of brain regions critical for masculine and feminine gender-role behavior.

Perhaps the most accurate conclusion to draw is that biology, the socializing influence of people and media, and the child's own efforts to understand gender-typical behavior all contribute to gender roles and differences. Recognizing the interactive nature of these influences on gender learning also enables us to better understand how gender roles are changing today, which we consider next.

Evolving Gender Roles

Gender roles are not etched in stone; they change with the times. In the United States, the range of acceptable roles for girls and boys and women and men has never been greater than today. For example, some fathers stay home to be primary caregivers for children, and some mothers work full-time as sole support for the family. What is the impact of these changes on gender roles? Some insights come from the results of the Family Lifestyles Project (Weisner & Wilson-Mitchell, 1990). This research has examined families in which the adults were members of the counterculture of the 1960s and 1970s. Some of the families are deeply committed to living their own lives and to rearing their children without traditional gender stereotypes. In these families, men and women share the household, financial, and child-care tasks.

The results of this project show that parents like Meda and Perry, from the opening vignette, can influence some aspects of gender stereotyping more readily than others. On the one hand, children in these families tend to have same-sex friends and to like sex-typed activities: The boys enjoy physical play, and the girls enjoy drawing and reading. On the other hand, the children have few stereotypes concerning occupations: They agree that girls can be president of the United States and drive trucks and that boys can be nurses and secretaries. They also have fewer sex-typed attitudes about the use of objects. They claim that boys and girls are equally likely to use an iron, a shovel, a hammer and nails, and a needle and thread.

Apparently, some features of gender roles and identities are more readily influenced by experience than others. This is as it should be. For most of our history as a species, Homo sapiens have existed in small groups of families, hunting animals and gathering vegetation. Women have given birth to the children and cared for them. Over the course of human history, it has been adaptive for women to be caring and nurturing because this increases the odds of a secure attachment and, ultimately, the survival of the infant. Men's responsibilities have included protecting the family unit from predators and hunting with other males, roles for which physical strength and aggressiveness are crucial.

Circumstances of life in the 21st century are, of course, substantially different: Often, both men and women are employed outside the home and both men and women care for children. Nevertheless, the cultural changes of the past few decades cannot erase hundreds of thousands of years of evolutionary history (Geary, 2002). We should not be surprised that boys and girls play differently, that girls tend to be more supportive in their interactions with others, and that boys are usually more aggressive physically.

Despite the wide range of roles acceptable today for men and women, boys and girls still tend to have same-sex friends and to enjoy many sex-typed activities.

© Tetra Images / Alamy

Test Yourself

RECALL

1. _____ are beliefs and images about males and females that may or may not be true.

2. Research on intellectual functioning and social behavior has revealed sex differences in verbal ability, _____, social influence, and aggression.

3. _____ may be particularly influential in teaching gender roles, because they more often treat sons and daughters differently.

4. According to Kohlberg's theory, understanding of gender includes gender labeling, gender stability, and

   _____.

5. Children studied in the Family Lifestyles Project, whose parents were members of the counterculture of the 1960s and 1970s, had traditional gender-related views toward friends and _____.

INTERPRET

How do the different forces in the biopsychosocial framework contribute to the development of gender roles?

APPLY

What advice would you give to a mother who wants her daughter to grow to be gender-free in her attitudes, beliefs, and aspirations?

Recall answers: (1) Gender stereotypes, (2) spatial ability (3) Fathers, (4) gender constancy, (5) preferred activities

SUMMARY

5.1   BEGINNINGS: TRUST AND ATTACHMENT

What are Erikson's first three stages of psychosocial development?

■ In Erikson's theory of psychosocial development, individuals face certain psychosocial crises at different phases in development. The crisis of infancy is to establish a balance between trust and mistrust of the world, producing hope; between 1 and 3 years of age, youngsters must blend autonomy and shame to produce will; and between 3 and 5 years, initiative and guilt must be balanced to achieve purpose.

How do infants form emotional attachments to mother, father, and other significant people in their lives?

■ Attachment is an enduring socioemotional relationship between infant and parent. For both adults and infants, many of the behaviors that contribute to the formation of attachment are biologically programmed. Bowlby's theory of attachment is rooted in evolutionary psychology and describes four stages in the development of attachment: preattachment, attachment in the making, true attachment, and reciprocal relationships.

What are the different varieties of attachment relationships, how do they arise, and what are their consequences?

■ Research with the Strange Situation, in which infant and mother are separated briefly, reveals four primary forms of attachment. Most common is a secure attachment in which infants have complete trust in the mother. Less common are three types of attachment relationships in which this trust is lacking. In avoidant relationships, infants deal with the lack of trust by ignoring the mother; in resistant relationships, infants often seem angry with her; in disorganized relationships, infants do not appear to understand the mother's absence.

■ Children who have had secure attachment relationships during infancy often interact with their peers more readily and more skillfully. Secure attachment is most likely to occur when mothers respond sensitively and consistently to their infants' needs.

■ Responsive caregiving results in infants developing an internal working model that parents will try to meet their needs. Secure attachment can be harder to achieve when infants are temperamentally difficult.

Is attachment jeopardized when parents of young children are employed outside of the home?

■ Many U.S. children are cared for at home by a father or other relative, in a day-care provider's home, or in a day-care center. Infants and young children are not harmed by such care as long as it is of high quality and parents remain responsive to their children.

5.2   EMERGING EMOTIONS

At what age do children begin to express basic emotions?

■ Basic emotions—which include joy, anger, and fear—emerge in the first year. Fear first appears in infancy as stranger wariness.

What are complex emotions and when do they develop?

■ Complex emotions have an evaluative component and include guilt, embarrassment, and pride. They appear between 18 and 24 months and require more sophisticated cognitive skills than the basic emotions of happiness and fear. Cultures differ in the rules for expressing emotions and in the situations that elicit particular emotions.

When do children begin to understand other people's emotions? How do they use this information to guide their own behavior?

■ By 6 months, infants have begun to recognize the emotions associated with different facial expressions. They use this information to help them evaluate unfamiliar situations. Beyond infancy, children learn more about the causes of different emotions.

■ Infants use simple strategies to regulate emotions such as fear. As children grow, they become better skilled at regulating their emotions. Children who do not regulate emotions well tend to have problems interacting with others.

5.3   INTERACTING WITH OTHERS

When do youngsters first begin to play with each other? How does play change during infancy and the preschool years?

■ Even infants notice and respond to one another, but the first real interactions (at about 12 to 15 months) take the form of parallel play in which toddlers play alone while watching each other. A few months later, simple social play emerges in which toddlers engage in similar activities and interact with one another. At about 2 years of age, cooperative play organized around a theme becomes common. Make-believe play is also common; in addition to being fun, it allows children to examine frightening topics. Most forms of solitary play are harmless.

What determines whether children help one another?

■ Prosocial behaviors, such as helping or sharing, are more common in children who understand (by perspective taking) and experience (by empathy) another's feelings.

■ Prosocial behavior is more likely when children feel responsible for the person in distress. Also, children help more often when they believe they have the skills needed, when they are feeling happy or successful, and when they perceive that the costs of helping are small.

- Twin studies show that prosocial behavior is influenced by heredity, probably through its impact on behavioral control (inhibition) and emotion regulation.
- Parents can foster altruism in their youngsters by behaving altruistically themselves, using reasoning to discipline their children, and by encouraging their children to help at home and elsewhere.

5.4 GENDER ROLES AND GENDER IDENTITY

What are our stereotypes about males and females? How well do they correspond to actual differences between boys and girls?

- Gender stereotypes are beliefs about males and females that are often used to make inferences about a person that are based solely on his or her gender; by 4 years of age, children know these stereotypes well.
- Studies of gender differences reveal that girls have greater verbal skill but that boys have greater spatial skill. Differences in math are negligible when females have access to education and occupations where math is valuable. Girls are better able to interpret emotions and are more prone to social influence, but boys are more aggressive. These differences vary based on a number of factors, including the historical period.

How do young children learn gender roles?

- Parents treat sons and daughters similarly, except in sex-typed activities. Fathers may be particularly important in sex typing because they are more likely to treat sons and daughters differently.
- In Kohlberg's theory, children gradually learn that gender is stable over time and cannot be changed according to personal wishes. After children understand gender stability, they begin to learn gender-typical behavior. According to gender-schema theory, children learn about gender by paying attention to behaviors of members of their own sex and ignoring behaviors of members of the other sex.
- Evolutionary developmental psychology reminds us that different roles for males and females caused different traits and behaviors to evolve for men and women. The idea that biology influences some aspects of gender roles is also supported by research on females exposed to male hormones during prenatal development.

How are gender roles changing?

- Gender roles have changed considerably in the past 50 years. However, studies of nontraditional families indicate that some components of gender stereotypes are more readily changed than others.

KEY TERMS

hope (168)
will (169)
purpose (169)
evolutionary psychology (169)
attachment (170)
secure attachment (171)
avoidant attachment (171)
resistant attachment (171)
disorganized (disoriented) attachment (172)
internal working model (173)

basic emotions (177)
social smiles (177)
stranger wariness (178)
social referencing (180)
parallel play (183)
simple social play (183)
cooperative play (183)
enabling actions (185)
constricting actions (185)
prosocial behavior (187)

altruism (187)
empathy (188)
social role (191)
gender stereotypes (191)
relational aggression (193)
gender identity (196)
gender labeling (196)
gender stability (196)
gender constancy (196)
gender-schema theory (197)

Log in to **www.cengagebrain.com** to access the resources your instructor requires. For this book, you can access:

Psychology CourseMate

- CourseMate brings course concepts to life with interactive learning, study, and exam preparation tools that support the printed textbook. A textbook-specific website, Psychology CourseMate includes an integrated interactive eBook and other interactive learning tools including quizzes, flashcards, videos, and more.

CENGAGENOW™

- CengageNOW Personalized Study is a diagnostic study tool containing valuable text-specific resources—and because you focus on just what you don't know, you learn more in less time to get a better grade.

WebTUTOR™

- More than just an interactive study guide, WebTutor is an anytime, anywhere customized learning solution with an eBook, keeping you connected to your textbook, instructor, and classmates.

Off to School:
Cognitive and Physical Development
in Middle Childhood

EVERY FALL, AMERICAN 5- AND 6-YEAR-OLDS trot off to kindergarten, starting an educational journey that lasts 13 or more years. As the journey begins, many children can read only a few words and know little math; by the end, most can read complete books and many have learned algebra and geometry. This mastery of complex academic skills is possible because of profound changes in children's thinking, changes described in the first section of this chapter.

For most American schoolchildren, intelligence and aptitude tests are a common part of their educational travels. In the second section of this chapter, you'll see what tests measure and why some children get lower scores on tests. In the third section, you'll discover how tests are often used to identify schoolchildren with atypical or special needs.

Next we look at the way that students learn to read, write, and do math. In this section, you'll discover some of the educational practices that seem to foster students' learning.

Finally, children's growing cognitive skills, when coupled with improved motor coordination, enable them to participate in sports. In the last section, we'll look at such participation and the physical changes that make it possible.

LEARNING OBJECTIVES

- What are the distinguishing characteristics of thought during Piaget's concrete-operational and formal-operational stages?

- How do children use strategies and monitoring to improve learning and remembering?

Adrian, a sixth grader in middle school, just took his first social studies test— and failed. He is shocked because he'd always received A's and B's in elementary school. Adrian realizes that glancing through the textbook chapter once before a test is probably not going to work in middle school, but he's not sure what else he should be doing.

ADRIAN'S COGNITIVE SKILLS FAR SURPASS THOSE OF THE INFANTS AND TODDLERS that we examined in Chapter 4. Yet the vignette shows that his skills aren't flawless. In this section, we'll learn more about cognitive growth in childhood, first from the perspective of Piaget's theory and then by considering the information-processing account.

More Sophisticated Thinking: Piaget's Version

You probably remember Jean Piaget from Chapters 1 and 4. Piaget believed that thought develops in a sequence of stages. The first two stages, sensorimotor and preoperational thinking, characterize infancy and the preschool years. In the next few pages we describe the remaining two stages, the concrete-operational and formal-operational stages, which apply to school-age children and adolescents.

The Concrete-Operational Period

Let's start by reviewing three important limits of preoperational thinking described in Chapter 4:

- Preschoolers are egocentric, believing that others see the world as they do.
- Preschoolers sometimes confuse appearances with reality.
- Preschoolers are unable to reverse their thinking.

None of these limits applies to children in the concrete-operational stage, which extends from approximately 7 to 11 years. Egocentrism wanes gradually. Why? As youngsters have more experiences with friends and siblings who assert their own perspectives on the world, children realize that theirs is not the only view. The understanding that events can be interpreted in different ways leads to the realization that appearances can be deceiving. *Also, thought can be reversed, because school-age children have acquired* **mental operations**, *which are actions that can be performed on objects or ideas and that consistently yield a result.* Recall from Chapter 4 that, on the conservation task, concrete-operational children realize that the amount of liquid is the same after it has been poured into a different beaker—pointing out that the pouring can always be reversed.

In discussing the concrete-operational period, we have emphasized how acquiring mental operations is advantageous to children. At the same time, as the name implies, concrete-operational thinking is limited to the tangible and real, to the here and now. The concrete-operational youngster takes "an earthbound, concrete, practical-minded sort of problem-solving approach" (Flavell, 1985, p. 98). Thinking abstractly and hypothetically is beyond the ability of concrete-operational children; these skills are acquired in the formal-operational period, as you'll see in the next section.

The Formal-Operational Period

With the onset of the formal-operational period, which extends from roughly age 11 into adulthood, children and adolescents expand beyond thinking about only the concrete and the real. Instead, they apply psychological operations to abstract enti-

mental operations
cognitive actions that can be performed on objects or ideas

THINK ABOUT IT

Piaget, Freud, and Erikson each propose unique stages for ages 7 to 11 years. How similar are the stages they propose? How do they differ?

ties too; they are able to think hypothetically and reason abstractly (Siegler & Alibali, 2004).

To illustrate these differences, let's look at problem solving, where formal-operational adolescents often take a very different approach from concrete-operational children. In one of Piaget's experiments (Inhelder & Piaget, 1958), children and adolescents were presented with several flasks, each containing what appeared to be the same clear liquid. They were told that one combination of the clear liquids would produce a blue liquid, and they were asked to determine the necessary combination. A typical concrete-operational youngster plunges right in, mixing liquids from different flasks in a haphazard way. In contrast, formal-operational adolescents understand that setting up the problem in abstract terms is the key. The problem is not really about pouring liquids but about combining different elements until all possible combinations have been tested. So a teenager might mix liquid from the first flask with liquids from each of the other flasks. If none of those combinations produces a blue liquid, the teenager would conclude that the liquid in the first flask is not an essential part of the mixture. The next step would be to mix the liquid in the second flask with each of the remaining liquids. A formal-operational thinker would continue in this manner until he or she finds the critical pair that produces the blue liquid. For adolescents, the problem does not involve the concrete acts of pouring and mixing. Instead, they understand that it involves identifying possible combinations and then evaluating each one. This sort of adolescent combinatorial reasoning is illustrated in the Real People feature.

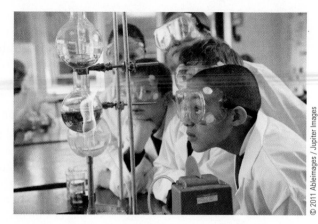

Concrete-operational thinkers often solve problems haphazardly, but formal-operational thinkers more often set up problems in abstract terms.

Real People
Applying Human Development
Combinatorial Reasoning Goes to the Races

As a 15-year-old, Robert Kail delivered the *Indianapolis Star*. In the spring of 1965, the newspaper announced a contest for all newspaper carriers. The task was to list the most words that could be created from the letters contained in the words "SAFE RACE." Whoever listed the most words would win two tickets to the Indianapolis 500 auto race.

Kail realized that this was a problem in combinatorial reasoning. All he needed to do was create all possible combinations of letters, then look them up. Following this procedure, he had to win (or, at worst, tie). So he created exhaustive lists of possible words, beginning with each of the letters individually, then all possible combinations of two letters, and working his way up to all possible combinations of all eight letters

(e.g., SCAREEFA, SCAREEAF). This was monotonous enough, but no more so than the next step: looking up all those possible words in a dictionary. (Remember, this was in the days before computerized spell-checkers.) Weeks later, he had generated a list of 126 words. As predicted, a few months later, he learned that he had won the contest. Combinatorial reasoning has its payoffs!

Adolescents' more sophisticated thinking is also shown in their ability to make appropriate conclusions from facts, which is known as **deductive reasoning**. Suppose we tell a person the following statement:

If you hit a glass with a hammer, the glass will break.

If you then tell the person, "You hit the glass with a hammer" he or she would conclude, of course, that "The glass will break," a conclusion that formal-operational adolescents do reach.

Concrete-operational youngsters sometimes reach this conclusion too—but based on their experience, not because the conclusion is logically necessary. To see the difference, imagine that the statement now is:

If you hit a glass with a feather, the glass will break.

Told "You hit the glass with a feather," the conclusion "the glass will break" follows just as logically as it did in the first example. In this instance, however, the conclusion is contrary to fact: it goes against what experience tells us is really true. Concrete-operational 10-year-olds resist reaching conclusions that are contrary to

deductive reasoning
drawing conclusions from facts; characteristic of formal-operational thought

known facts, whereas formal-operational 15-year-olds often reach such conclusions (De Neys & Everaerts, 2008). Formal-operational teenagers understand that these problems are about abstractions that need not correspond to real-world relations. In contrast, concrete-operational youngsters reach conclusions based on their knowledge of the world.

Comments on Piaget's View

We mentioned in Chapter 4 that, although Piaget's theory provides our single most comprehensive theory of cognitive development, it has some shortcomings. Specifically, it overestimates cognitive competence in adolescents, is vague concerning processes of change, does not account for variability in children's performance, and undervalues the influence of the sociocultural environment. Because of these limits to Piaget's theory, we need to look at other approaches to complete our account of mental development during childhood and adolescence. In the next few pages, we'll focus on the information-processing approach that we first examined in Chapter 4.

Information-Processing Strategies for Learning and Remembering

You'll remember from Chapter 4, information-processing psychologists believe that cognitive development proceeds by increases in the efficiency with which children process information. In other words, just as personal computers have become progressively more sophisticated in their hardware and software, information-processing psychologists believe that cognitive development reflects change in mental computing power, including mental hardware and mental software.

One of the key issues in this approach concerns the means by which children store information in permanent memory and retrieve it when needed later. *According to information-processing psychologists, most human thought takes place in* **working memory***, where a relatively small number of thoughts and ideas can be stored briefly.* As you read these sentences, for example, the information is stored in working memory. However, as you read additional sentences, they displace the contents of sentences you read earlier. *For you to learn this information, it must be transferred to* **long-term memory***, a permanent storehouse of knowledge that has unlimited capacity.* If information you read is not transferred to long-term memory, it is lost, just as information vanishes from a computer's memory when the power fails.

Memory Strategies

How do you try to learn the information in this book or your other textbooks? If you're like many college students, you will probably use some combination of highlighting key sentences, outlining chapters, taking notes, writing summaries, and testing yourself. These are all effective learning strategies that make it easier for you to store text information in long-term memory.

Children begin to use simple strategies fairly early. For example, 7- or 8-year-olds use rehearsal, a strategy of repetitively naming information that is to be remembered. As children grow older, they learn other memory strategies. *One memory strategy is* **organization***—structuring information to be remembered so that related information is placed together.* For example, a sixth grader trying to remember major battles of the American Civil War could organize them geographically (e.g., Shiloh and Fort Donelson in Tennessee, Antietam and Monocacy in Maryland) or chronologically (e.g., Fort Sumter and First Manassas in 1861, Gettysburg and Vicksburg in 1863).

Another memory strategy is **elaboration***—embellishing information to be remembered to make it more memorable.* To see elaboration in action, imagine a child who can never remember if the second syllable of *rehearsal* is spelled *her* (as it sounds) or *hear*. The child could remember the correct spelling by reminding herself that *rehearsal* is like *re-hear-ing*. Thus, thinking about the derivation of *rehearsal* makes it

working memory
type of memory in which a small number of items can be stored briefly

long-term memory
permanent storehouse for memories that has unlimited capacity

organization
as applied to children's memory, a strategy in which information to be remembered is structured so that related information is placed together

elaboration
memory strategy in which information is embellished to make it more memorable

easier to remember how to spell it. Finally, as children grow, they're also more likely to use external aids to memory, such as making notes and writing down information on calendars so they won't forget future events (Eskritt & McLeod, 2008).

As children grow, they make more use of memory aids, such as taking notes.

Metacognition

Just as there's not much value to a filled toolbox if you don't know how to use the tools, memory strategies aren't much good unless children know when to use them. For example, rehearsal is a great strategy for remembering phone numbers but lousy for remembering amendments to the U.S. Constitution or the plot of *Hamlet*. During the elementary-school years and adolescence, children gradually learn to identify different kinds of memory problems and the memory strategies most appropriate to each. For example, when reading a textbook or watching a television newscast, outlining or writing a summary are good strategies because they identify the main points and organize them. Children gradually become more skilled at selecting appropriate strategies, but even high-school students do not always use effective learning strategies when they should (Pressley & Hilden, 2006).

After children choose a memory strategy, they need to monitor its effectiveness. That is, they need to decide if the strategy is working. If it's not, they need to begin anew, reanalyzing the memory task to select a better approach. If the strategy is working, they should determine the portion of the material they have not yet mastered and concentrate their efforts there. Monitoring improves gradually with age. For example, elementary-school children can accurately identify which material they have not yet learned, but they do not consistently focus their study efforts on this material (Bjorklund, 2005).

Diagnosing memory problems accurately and monitoring the effectiveness of memory strategies are two important elements of **metamemory**, *which refers to a child's intuitive understanding of memory.* That is, as children develop, they learn more about how memory operates and devise naive theories of memory that represent an extension of the theory of mind described on pages 115–118 (Lockl & Schneider, 2007). For example, children learn that memory is fallible (i.e., they sometimes forget!) and that some types of memory tasks are easier than others (e.g., remembering the main idea of the Gettysburg address is simpler than remembering it word for word). This growing knowledge of memory helps children to use memory strategies more effectively, just as an experienced carpenter's accumulated knowledge of wood tells her when to use nails, screws, or glue to join two boards.

Of course, children's growing understanding of memory is paralleled by their increased understanding of all cognitive processes. *Such knowledge and awareness of cognitive processes is called* **metacognitive knowledge**. Metacognitive knowledge grows rapidly during the elementary-school years: Children come to know much about perception, attention, intentions, knowledge, and thinking (Flavell, 2000; McCormick, 2003). For example, school-age children know that sometimes they deliberately direct their attention—as in searching for a parent's face in a crowd—but that sometimes events capture attention—as with an unexpected clap of thunder (Parault & Schwanenflugel, 2000).

One of the most important features of children's metacognitive knowledge is their understanding of the connections among goals, strategies, monitoring, and outcomes, shown in ■ Figure 6.1. Children come to realize that for a broad spectrum of tasks—ranging from learning words in a spelling list to learning to spike a volleyball to learning to get along with an overly talkative classmate seated nearby—they need to regulate their learning by understanding the goal and selecting a means to achieve that goal. Then they determine whether the chosen method is working. *Effective* **cognitive self-regulation**—*that is, skill at identifying goals, selecting effective strategies, and monitoring accurately—is a characteristic of successful students* (Usher & Pajares, 2008; Zimmerman, 2001). A student may decide that writing each spelling word twice before the test is a good way to get all the words right. When the student gets only 70% correct on the first test, he switches to a new strategy (e.g., writing each word

metamemory
person's informal understanding of memory; includes the ability to diagnose memory problems accurately and to monitor the effectiveness of memory strategies

metacognitive knowledge
a person's knowledge and awareness of cognitive processes

THINK ABOUT IT

Which elements of the biopsychosocial framework are emphasized in the information-processing approach to cognitive development?

cognitive self-regulation
skill at identifying goals, selecting effective strategies, and accurate monitoring; a characteristic of successful students

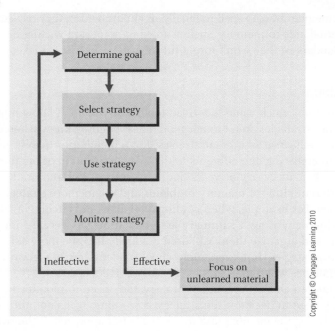

■ **Figure 6.1**

Effective learning involves understanding the goals of the task, selecting an appropriate strategy, and monitoring the effectiveness of the chosen strategy.

Determine goal

Select strategy

Use strategy

Monitor strategy

Ineffective   Effective

Focus on unlearned material

Copyright © Cengage Learning 2010

four times, plus writing its definition), showing the adaptive nature of cognitive processes in self-regulated learners.

Perhaps this has a familiar ring to it. It should, for the diagram simply summarizes an important set of study skills. Analyzing, strategizing, and monitoring are key elements of productive studying. The study goals change when you move from this book to your math text to a novel that you are reading for English, but the basic sequence still holds. Studying should always begin with a clear understanding of what goal you are trying to achieve, because this sets the stage for all the events that follow. Too often we see students like Adrian—the student in the vignette—who simply read text material without any clear idea of what they should be getting out of it. Instead, students should be active readers (Adams, Treiman, & Pressley, 1998). Always study with a plan. Start by skimming the text to become familiar with the material. Before you read more carefully, try to anticipate some of the topics that the author will cover in detail. When you reach natural breaks in the material, try to summarize what you've read and think of questions that a teacher might ask about the material. Finally, when you don't understand something in the text, stop and determine the source of your confusion. Perhaps you don't know a word's meaning. Maybe you skipped or misunderstood an earlier section of the material. By reading actively and using strategies like these, you'll be much more likely to understand and remember what you've read (Pressley & Hilden, 2006).

Test Yourself

RECALL

1. During Piaget's _____ stage, children are first able to represent objects mentally in different ways and to perform mental operations.

2. Hypothetical and deductive reasoning are characteristic of children in Piaget's _____ stage.

3. Children and adolescents often select a memory strategy after they have _____.

4. The term _____ refers to periodic evaluation of a strategy to determine whether it is working.

INTERPRET

Do developmental improvements in memory strategies and metacognition emphasize nature, nurture, or both? How?

APPLY

Formal-operational adolescents are able to reason abstractly. How might this ability help them use the study skills shown in ■ Figure 6.1 more effectively?

Recall answers: (1) concrete-operational, (2) formal-operational, (3) determined the goal of the memory task, (4) monitoring

LEARNING OBJECTIVES

- What is the nature of intelligence?
- Why were intelligence tests first developed? What are their features?
- How well do intelligence tests work?

- How do heredity and environment influence intelligence?
- How and why do test scores vary for different racial and ethnic groups?

Diana is an eager fourth-grade teacher who loves history. Consequently, every year she's frustrated when she teaches a unit on the American Civil War. Although she's passionate about the subject, her enthusiasm is *not* contagious. Instead, her students' eyes glaze over and she can see young minds drifting off and, of course, they never seem to grasp the war's historical significance. Diana wishes there was a different way to teach this unit, one that would engage her students more effectively.

BEFORE YOU READ FURTHER, how would you define intelligence? If you're typical of most Americans, your definition probably includes the ability to reason logically, connect ideas, and solve real problems. You might mention verbal ability, meaning the ability to speak clearly and articulately. You might also mention social competence: for example, an interest in the world at large and an ability to admit when you make a mistake (Sternberg & Kaufman, 1998).

As you'll see in this section, many of these ideas about intelligence are included in psychological theories of intelligence. We'll begin by considering the theories of intelligence, where we'll get some insights into ways that Diana could make the Civil War come alive for her class. Next, you'll see how intelligence tests were devised initially to assess individual differences in intellectual ability. Then we'll look at a simple question: "How well do modern tests work?" Finally, we'll examine how race, ethnicity, social class, gender, environment, and heredity influence intelligence.

Theories of Intelligence

Psychometricians *are psychologists who specialize in measuring psychological characteristics such as intelligence and personality.* When psychometricians want to research a particular question, they usually begin by administering a large number of tests to many individuals. Then they look for patterns in performance across the different tests. The basic logic underlying this technique is similar to the logic a jungle hunter uses to decide whether some dark blobs in a river are three separate rotting logs or a single alligator (Cattell, 1965). If the blobs move together then the hunter decides they are part of the same structure, an alligator. If they do not move together then they are three different structures, three logs. Similarly, if changes in performance on one test are accompanied by changes in performance on a second test—that is, if they move together—then one could assume that the tests are measuring the same attribute or factor.

Suppose, for example, that you believe there is such a thing as general intelligence. That is, you believe that some people are smart regardless of the situation, task, or problem, whereas others are not so smart. According to this view, children's performance should be very consistent across tasks. Smart children should always receive high scores, and the less smart youngsters should always receive lower scores. In fact, more than 100 years ago, Charles Spearman (1904) reported findings supporting the idea that a general factor for intelligence, or *g*, is responsible for performance on all mental tests.

Other researchers, however, have found that intelligence consists of distinct abilities. For example, Thurstone and Thurstone (1941) analyzed performance on a wide

psychometricians
psychologists who specialize in measuring psychological traits such as intelligence and personality

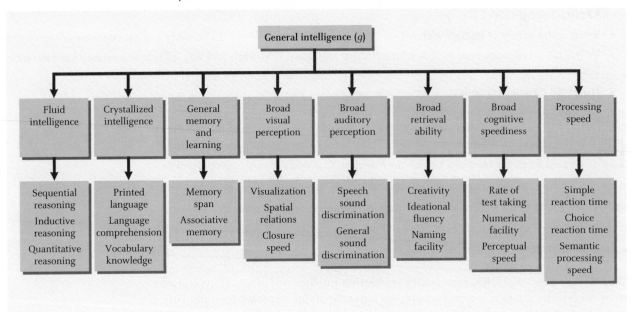

From Carroll, 1993.

range of tasks and identified seven distinct patterns, each reflecting a unique ability: perceptual speed, word comprehension, word fluency, spatial relations, number proficiency, memory, and inductive reasoning. Thurstone and Thurstone also acknowledged a general factor that operated in all tasks, but they emphasized that the specific factors were more useful in assessing and understanding intellectual ability.

The Hierarchical View of Intelligence

These conflicting findings have led many psychometric theorists to propose hierarchical theories of intelligence that include both general and specific components. John Carroll (1993, 1996), for example, proposed the hierarchical theory with three levels shown in ■ Figure 6.2. At the top of the hierarchy is *g*, general intelligence. In the level underneath *g* are eight broad categories of intellectual skill, ranging from fluid intelligence to processing speed. Each of the abilities in the second level is further divided into the skills listed in the third and most specific level. Crystallized intelligence, for example, includes understanding printed language, comprehending language, and knowing vocabulary.

Carroll's hierarchical theory is, in essence, a compromise between the two views of intelligence—general versus distinct abilities. But some critics find it unsatisfactory because it ignores the research and theory on cognitive development. They believe we need to look beyond the psychometric approach to understand intelligence. In the remainder of this section, then, we'll look at two newer theories that have gained a following.

Gardner's Theory of Multiple Intelligences

Only recently have psychologists viewed intelligence from the perspective of Piaget's theory and information-processing psychology. These new theories present a much broader theory of intelligence and how it develops. Among the most ambitious is Howard Gardner's (1983, 2002, 2006) theory of multiple intelligences. Rather than using test scores as the basis for his theory, Gardner draws on research in child development, studies of brain-damaged persons, and studies of exceptionally talented people.

Nine Intelligences in Gardner's Theory of Multiple Intelligences

Type of Intelligence	Definition
Linguistic	Knowing the meanings of words, having the ability to use words to understand new ideas, and using language to convey ideas to others
Logical-mathematical	Understanding relations that exist among objects, actions, and ideas as well as the logical or mathematical operations that can be performed on them
Spatial	Perceiving objects accurately and imagining in the "mind's eye" the appearance of an object before and after it has been transformed
Musical	Comprehending and producing sounds varying in pitch, rhythm, and emotional tone
Bodily-kinesthetic	Using one's body in highly differentiated ways as dancers, craftspeople, and athletes do
Interpersonal	Identifying different feelings, moods, motivations, and intentions in others
Intrapersonal	Understanding one's emotions and knowing one's strengths and weaknesses
Naturalistic	Understanding the natural world, distinguishing natural objects from artifacts, grouping and labeling natural phenomena
Existential	Considering "ultimate" issues, such as the purpose of life and the nature of death

SOURCE: Gardner, 1983, 1999, 2002.

Using these criteria, Gardner identified seven distinct intelligences when he first proposed the theory in 1983. In subsequent work, Gardner (1999, 2002) has identified two additional intelligences; the complete list is shown in ●Table 6.1.

The first three intelligences in this list—linguistic intelligence, logical-mathematical intelligence, and spatial intelligence—are included in traditional theories of intelligence. The last six intelligences are not: Musical, bodily-kinesthetic, interpersonal, intrapersonal, naturalistic, and existential intelligences are unique to Gardner's theory. According to Gardner, Carlos Santana's wizardry on the guitar, Roger Federer's remarkable shots on the tennis court, and Oprah Winfrey's grace and charm in dealing with people are all features of intelligence that are totally ignored in traditional theories.

How did Gardner arrive at these nine distinct intelligences? First, each has a unique developmental history. Linguistic intelligence, for example, develops much earlier than the others. Second, each intelligence is regulated by distinct regions of the brain, as shown in studies of brain-damaged persons. Spatial intelligence, for example, is regulated by particular regions in the right hemisphere of the brain. Third, each intelligence has special cases of talented individuals. There are well-known instances of musically intelligent people, for example, who exhibit incredible talent at an early age. Claudio Arrau, one of the 20th century's greatest pianists, could read musical notes before he could read words. And Yo-Yo Ma, the famed cellist, performed in concert at 7 years of age for President John F. Kennedy.

Prompted by Gardner's theory, researchers have begun to look at other nontraditional aspects of intelligence. *Probably the best known is* **emotional intelligence**, *which is the ability to use one's own and others' emotions effectively for solving problems and living happily.* Emotional intelligence made headlines in 1995 as a result of a best-selling book, *Emotional Intelligence*, in which Daniel Goleman (1995) argued that "emotions [are] at the center of aptitudes for living" (p. xiii). One major model of emotional intelligence (Salovey & Grewal, 2005; Mayer, Salovey, & Caruso, 2008) includes several distinct facets, including perceiving emotions accurately (e.g., recognizing a happy face), understanding emotions (e.g., distinguishing happiness from ecstasy),

emotional intelligence
ability to use one's own and others' emotions effectively for solving problems and living happily

and regulating one's emotions (e.g., hiding one's disappointment). People who are emotionally intelligent tend to have more satisfying interpersonal relationships, to have greater self-esteem, and to be more effective in the workplace (Joseph & Newman, 2010; Mayer, Roberts, & Barsade, 2008).

The theory of multiple intelligences has important implications for education. Gardner (1993, 1995) believes that schools should foster all intelligences, not just the traditional linguistic and logical-mathematical intelligences. Teachers should capitalize on the strongest intelligences of individual children. That is, teachers need to know a child's profile of intelligence—the child's strengths and weaknesses—and gear instruction to the strengths (Chen & Gardner, 2005). For example, Diana, the fourth-grade teacher in the opening vignette, could help some of her students understand the Civil War by studying music of that period (musical intelligence). Other students might benefit by emphasizing maps that show the movement of armies in battle (spatial intelligence). Still others might profit from focusing on the experiences of African Americans living in the North and the South (interpersonal intelligence).

These guidelines do not mean that teachers should gear instruction solely to a child's strongest intelligence, pigeonholing youngsters as "numerical learners" or "spatial learners." Instead, whether the topic is the signing of the Declaration of Independence or Shakespeare's *Hamlet*, instruction should try to engage as many different intelligences as possible (Gardner, 1999, 2002). The typical result is a much richer understanding of the topic by all students.

Some American schools have enthusiastically embraced Gardner's ideas (Gardner, 1993). Are these schools better than those that have not? Educators in schools using the theory think so; they cite evidence that their students benefit in many ways (Kornhaber, Fierros, & Veenema, 2004) but some critics are not yet convinced (Waterhouse, 2006). Nevertheless, there is no doubt that Gardner's work has helped liberate researchers from narrow, psychometric-based views of intelligence.

Sternberg's Theory of Successful Intelligence

Robert Sternberg has studied intelligence for more than 35 years. He began by asking how adults solve problems on intelligence tests. Over the years, this work led to a comprehensive theory of intelligence in which "intelligence" is defined as using one's abilities skillfully to achieve one's personal goals (Sternberg, 1999, 2008). Goals can be short-term—such as getting an A on a test, making a snack in the microwave, or winning the 100-meter hurdles—or longer term, such as having a successful career and a happy family life. Achieving these goals by using one's skills defines successful intelligence.

In achieving personal goals, people use three different kinds of abilities. **Analytic ability** *involves analyzing problems and generating different solutions*. Suppose a 12-year-old wants to download songs to her iPod, but something isn't working. Analytic intelligence is shown in considering different causes of the problem—maybe the iPod is broken or maybe the software to download songs wasn't installed correctly. Analytic intelligence also involves thinking of different solutions: She could surf the Internet for clues about what's wrong or ask a sibling for help.

Creative ability *involves dealing adaptively with novel situations and problems*. Returning to our 12-year-old, suppose that she discovers her iPod is broken just as she's ready to leave on a daylong car trip. Lacking the time (and money) to buy a new player, creative intelligence is shown in dealing successfully with a novel goal: finding an enjoyable activity to pass the time on a long drive.

Finally, **practical ability** *involves knowing what solution or plan will actually work*. Problems can be solved in different ways in principle, but in reality only one solution may be practical. Our 12-year-old may realize that the only way to figure out why her iPod isn't working is to surf the Internet: She doesn't want to ask for help because her parents wouldn't approve of many of the songs, and she doesn't want a sibling to know that she's downloading them anyway.

analytic ability
in Sternberg's theory of intelligence, the ability to analyze problems and generate different solutions

creative ability
in Sternberg's theory of intelligence, the ability to deal adaptively with novel situations and problems

practical ability
in Sternberg's theory of intelligence, the ability to know which problem solutions are likely to work

Like Gardner, Sternberg (1999) argues that instruction is most effective when it is geared to a child's strength. A child with strong analytic ability, for example, may find algebra simpler when the course emphasizes analyses and evaluation; a child with strong practical ability may be at his best when the material is organized around practical applications. Thus, the theory of successful intelligence shows how instruction can be matched to students' strongest abilities, enhancing students' prospects for mastering the material (Grigorenko, Jarvin, & Sternberg, 2002).

Sternberg emphasizes that successful intelligence is revealed in people's pursuit of goals. Of course, these goals vary from one person to the next and, just as important, often vary even more in different cultural or ethnic groups. That is, intelligence is always partly defined by the demands of an environment or cultural context. What is intelligent for children growing up in cities in North America may not be intelligent for children growing up in the Sahara desert, the Australian outback, or on a remote island in the Pacific Ocean. For example, in Brazil, many school-age boys sell candy and fruit to bus passengers and pedestrians. These children often cannot identify the numbers on paper money, yet they know how to purchase their goods from wholesale stores, make change for customers, and keep track of their sales (Saxe, 1988).

If the Brazilian vendors were given the tests that measure intelligence in American students, they would fare poorly. Does this mean they are less intelligent than American children? Of course not. The skills important to American conceptions of intelligence and that are assessed on our intelligence tests are less valued in these other cultures and so are not cultivated in the young. Each culture defines what it means to be intelligent, and the specialized computing skills of vendors are just as intelligent in their cultural settings as verbal skills are in American culture (Sternberg & Kaufman, 1998).

As with Gardner's theory, researchers are still evaluating Sternberg's theory and are still debating the question of what intelligence is. However it is defined, the facts are that individuals differ substantially in intellectual ability and that numerous tests have been devised to measure these differences. We'll examine the construction, properties, and limitations of these tests in the next section.

For street vendors in Brazil, successful intelligence involves sophisticated arithmetic operations for buying products, making change, and keeping track of sales.

Binet and the Development of Intelligence Testing

American schools faced a crisis at the beginning of the 20th century. Between 1890 and 1915, school enrollment nearly doubled nationally because of an influx of immigrants and because reforms restricted child labor and emphasized education (Giordano, 2005). With the increased enrollment, teachers were confronted with increasing numbers of students who did not learn as readily as the "select few" who had populated their classes previously. How to deal with "feebleminded" children was one of the pressing issues of the day for U.S. educators.

These problems were not unique to the United States. In 1904, the Minister of Public Instruction in France asked two noted psychologists of the day, Alfred Binet and Theophile Simon, to formulate a way of recognizing children who would be unable to learn in school without special instruction. Binet and Simon's approach was to select simple tasks that French children of different ages ought to be able to do, such as naming colors, counting backward, and remembering numbers in order. Based on preliminary testing, Binet and Simon identified problems that typical 3-year-olds could solve, that typical 4-year-olds could solve, and so on. *Children's* **mental age,** *or* **MA**, *referred to the difficulty of the problems they could solve correctly.* A child who solved problems that the average 7-year-old could solve would have an MA of 7.

Binet and Simon used mental age to distinguish "bright" from "dull" children. A bright child would have the MA of an older child—for example, a 6-year-old with an MA of 9. A dull child would have the MA of a younger child—for example, a 6-year-old with an MA of 4. Binet and Simon confirmed that bright children identified using

mental age (MA)
in intelligence testing, a measure of children's performance corresponding to the chronological age of those whose performance equals the child's

THINK ABOUT IT

If Jean Piaget were to create an intelligence test, how would it differ from the type of test Binet created?

intelligence quotient (IQ)
mathematical representation of how a person scores on an intelligence test in relation to how other people of the same age score

their test did better in school than dull children. Voilà—the first objective measure of intelligence!

The Stanford-Binet

Lewis Terman, of Stanford University, revised Binet and Simon's test substantially and published a version known as the Stanford-Binet in 1916. *Terman described performance as an* **intelligence quotient,** *or IQ, which was simply the ratio of mental age to chronological age (CA), multiplied by 100*:

$$IQ = MA/CA \times 100$$

At any age, children who are perfectly average have an IQ of 100, because their mental age equals their chronological age. Furthermore, roughly two thirds of children taking a test will have IQ scores between 85 and 115. The IQ score can also be used to compare intelligence in children of different ages. A 4-year-old girl with an MA of 5 has an IQ of 125 (5/4 × 100), just like that of an 8-year-old boy with an MA of 10 (10/8 × 100).

IQ scores are no longer computed as the ratio of MA to CA. Instead, children's IQ scores are determined by comparing their test performance to the average IQ score of others their age. When children perform at the average for their age, their IQ is 100. Children who perform above the average have IQs greater than 100; children who perform below the average have IQs less than 100. Nevertheless, the concept of IQ as the ratio of MA to CA helped to popularize the Stanford-Binet test.

By the 1920s the Stanford-Binet had been joined by many other intelligence tests. Educators greeted these new devices enthusiastically because they seemed to offer an efficient and objective way to assess a student's chances of succeeding in school (Chapman, 1988). Today, nearly 100 years later, the Stanford-Binet remains a popular test; the latest version was revised in 2003. Like the earlier versions, the modern Stanford-Binet consists of various cognitive and motor tasks ranging from the extremely easy to the extremely difficult. The Stanford-Binet, the Wechsler Intelligence Scale for Children-IV (WISC-IV), and the Kaufman Assessment Battery for Children-II are the primary individualized tests of intelligence in use today.

Do Tests Work?

If tests work, they should predict important outcomes in children's lives. That is, children who receive higher IQ scores should be more successful in school and after they leave school. In fact, IQ scores are remarkably powerful predictors of developmental outcomes. One expert argued that "IQ is the most important predictor of an individual's ultimate position within American society" (Brody, 1992). Of course, since IQ tests were devised to predict school success, it's not surprising that they do this quite well. IQ scores predict school grades, scores on achievement tests, and number of years of education with correlations that are usually between .5 and .7 (Brody, 1992; Geary, 2005).

These correlations are far from perfect, which reminds us that some youngsters with high test scores do not excel in school and others with low test scores manage to get good grades. In fact, some researchers find that self-discipline predicts grades in school even better than IQ scores do (Duckworth & Seligman, 2005). In general, however, tests do a reasonable job of predicting school success.

Not only do intelligence scores predict success in school, they predict occupational success (Deary, Batty, & Gale, 2008; Strenze, 2007). Individuals with higher IQ scores are more likely to hold high-paying, high-prestige positions within medicine, law, and engineering (Schmidt & Hunter, 1998) and, among scientists with equal education, those with higher IQ scores have more patents and more articles published in scientific journals (Park, Lubinski, & Benbow, 2008). Some of the link between IQ and occupational success occurs because these professions require more education and we've already seen that IQ scores predict educational success. But even within a profession—where all individuals have the same amount of education—IQ scores

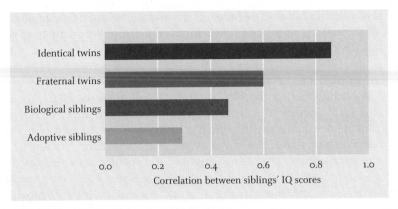

Figure 6.3
Siblings' IQ scores become more similar as the siblings become more similar genetically, which is evidence for a hereditary basis for intelligence.

Based on the data in Bouchard and McGue, 1981.

predict job performance and earnings, particularly for more complex jobs (Henderson, 2010; Schmidt & Hunter, 2004). If, for example, two teenagers have summer jobs running tests in a biology lab, the smarter of the two will probably learn the procedures more rapidly and, once learned, conduct them more accurately.

Hereditary and Environmental Factors

In a typical U.S. elementary school, several first graders will have IQs greater than 120 and a similar number will have IQ scores in the low 80s. What accounts for the 20-point difference in these youngsters' scores? Heredity plays an important role (Bouchard, 2009), as does experience (Bronfenbrenner & Morris, 2006).

Some of the evidence for hereditary factors is shown in ■ Figure 6.3. If genes influence intelligence, then siblings' test scores should become more alike as siblings become more similar genetically (Plomin & Petrill, 1997). In other words, because identical twins are identical genetically, they should have virtually identical test scores, which would be a correlation of 1. Fraternal twins have about 50% of their genes in common, just like other siblings of the same biological parents. Consequently, their test scores should be (a) less similar than scores for identical twins, (b) similar to other siblings who have the same biological parents, and (c) more similar than the scores of children and their adopted siblings. You can see in ■ Figure 6.3 that each of these predictions is supported.

Studies of adopted children support this conclusion: If heredity helps determine IQ, then children's IQs should be more like their biological parents' IQs than their adoptive parents' IQs. In fact, the correlation between children's IQ and their biological parents' IQ is usually greater than the correlation between children's IQ and their adoptive parents' IQ (Plomin et al., 1997). What's more, the relation between children's IQs and their biological parents' IQ actually gets *stronger* as children grow older. In other words, as adopted children become older, their test scores increasingly resemble their biological parents' scores.

Do these results mean that heredity is the sole determiner of intelligence? No. Three areas of research show the importance of environment on intelligence: characteristics of the home environment, changes in IQ scores, and intervention programs. Let's start with research on the characteristics of families and homes. If intelligence were due solely to heredity, then environment should have little or no impact on children's intelligence. But we know that many characteristics of

Following adoption, children's scores on intelligence tests tend to resemble the scores of their biological parents, not their adoptive parents, which shows the impact of heredity on intelligence.

parents' behavior and home environments are related to children's intelligence. For example, children with high test scores tend to come from homes that are well organized and have plenty of appropriate play materials (Bradley et al., 2001; Tamis-LeMonda et al., 2004).

The impact of the environment on intelligence is also implicated by a dramatic rise in IQ test scores during the 20th century (Flynn & Weiss, 2007). For example, scores on the WISC increased by nearly 10 points over a 25-year period (Flynn, 1999). Heredity cannot account for such a rapid increase over a few decades (a mere fraction of a second in genetic time). Consequently, the rise must reflect the impact of some aspect of the environment. The change might reflect smaller, better-educated families with more leisure time (Daley et al., 2003; Dickens & Flynn, 2001). Or it might be due to movies, television, and (more recently) computers and the Internet providing children with an incredible wealth of virtual experience (Greenfield, 1998). Yet another possibility is suggested by the fact that improvements in IQ scores are particularly striking at the lower end of the distribution: Fewer children are receiving very low IQ scores, which may show the benefits of improved health care, nutrition, and education for children who had limited access to these resources in previous generations (Geary, 2005). Although the exact cause remains a mystery, the increase does show the effect of changing environmental conditions on intelligence.

The importance of a stimulating environment for intelligence is also demonstrated by intervention programs that prepare economically disadvantaged children for school. When children grow up in never-ending poverty, the cycle is predictable and tragic: Youngsters have few of the intellectual skills to succeed in school, so they fail; lacking an education, they find minimal jobs (if they can work at all); and this practically guarantees that their children, too, will grow up in poverty.

Since Project Head Start began in 1965, massive educational intervention has been an important tool in the effort to break this repeated cycle of poverty. Head Start and other intervention programs teach preschool youngsters basic school readiness skills and social skills and also offer guidance to parents (Administration for Children and Families, 2010; Campbell et al., 2001). When children participate in these enrichment programs, their test scores go up and school achievement improves, particularly when intervention programs are extended beyond preschool and into the elementary-school years (Ludwig & Phillips, 2007). And, as adults, children who participated in enrichment programs are much more likely to attend college and have skilled employment (Pungello et al., 2010). Of course, massive intervention over many years is expensive. But so are the economic consequences of poverty, unemployment, and their by-

High-quality preschool programs provide stimulating environments that can increase children's scores on intelligence tests and improve their school performance.

products. Intervention programs show that the repetitive cycle of school failure and education can be broken. And, in the process, they show that intelligence is fostered by a stimulating and responsive environment.

The Impact of Ethnicity and Socioeconomic Status

On many intelligence tests, ethnic groups differ in their average scores: Asian Americans tend to have the highest scores, followed by European Americans, Latino Americans, and African Americans (Hunt & Carlson, 2007). To a certain extent, these differences in test scores reflect group differences in socioeconomic status. Children from economically advantaged homes tend to have higher test scores than children from economically disadvantaged homes, and European American and Asian American families are more likely to be economically advantaged whereas Latino American and African American families are more likely to be economically disadvantaged. Nevertheless, when children from comparable socioeconomic status are compared, group differences in IQ test scores are reduced but not eliminated (Magnuson & Duncan, 2006). Let's look at four explanations for these differences.

A Role for Genetics?

On page 217, you learned that heredity helps determine a child's intelligence: Smart parents tend to beget smart children. Does this also mean that group differences in IQ scores reflect genetic differences between groups? No. Most researchers agree that there is no evidence that some ethnic groups have more "smart genes" than others. Instead, they believe that the environment is largely responsible for these differences (Bronfenbrenner & Morris, 2006; Neisser et al., 1996).

A popular analogy (Lewontin, 1976) demonstrates the thinking here. Imagine two kinds of corn: Each kind produces both short and tall plants, and height is known to be due to heredity. If one kind of corn grows in a good soil—plenty of water and nutrients—the mature plants will reach their genetically determined heights: some short, some tall. If the other kind of corn grows in poor soil, few of the plants will reach their full height and overall the plants of this kind will be much shorter. Even though height is quite heritable for each type of corn, the difference in height between the two groups is due solely to the quality of the environment. Similarly, though IQ scores may be quite heritable for different groups, limited exposure to stimulating environments may mean that one group ends up with lower IQ scores overall (just as the group of plants growing up in poor soil do not reach their full height).

The same conclusion applies to ethnic groups. Differences within ethnic groups are partly due to heredity, but differences between groups apparently reflect environmental influences. Three potential influences have been studied, and we'll look at these next.

Experience With Test Contents

Some critics contend that differences in test scores reflect bias in the tests themselves. They argue that test items reflect the cultural heritage of the test creators, most of whom are economically advantaged European Americans, and so tests are biased against economically disadvantaged children from other groups (Champion, 2003). Such critics point to test items like this one:

A conductor is to an orchestra as a teacher is to what?
book   school   class   eraser

Children whose background includes exposure to orchestras are more likely to answer this question correctly than children who lack this exposure.

The problem of bias has led to the development of **culture-fair intelligence tests**, *which include test items based on experiences common to many cultures.* An exam-

culture-fair intelligence tests
intelligence tests devised using items common to many cultures

■ **Figure 6.4**
Culture-fair intelligence tests are designed
to minimize the impact of experiences that
are unique to some cultures or to some
children within a culture.

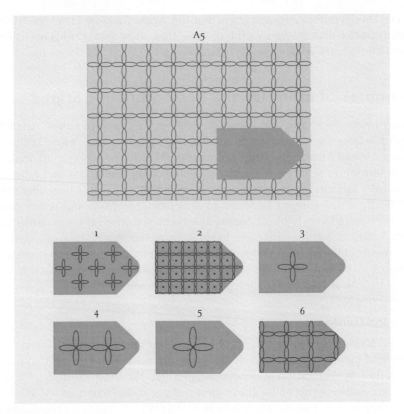

Based on Raven's Progressive Matrices.

ple is Raven's Progressive Matrices, which consists solely of items like the one shown in ■ Figure 6.4. Examinees are asked to select the piece that would complete the design correctly (piece 6, in this case). Although items like this are thought to reduce the impact of specific experience on test performance, ethnic group differences remain on so-called culture-fair intelligence tests (Herrnstein & Murray, 1994). Apparently, familiarity with test-related items is not the key factor responsible for group differences.

Test-Taking Skills

The impact of experience and cultural values can extend beyond particular items to a child's familiarity with the entire testing situation. Tests underestimate a child's intelligence if, for example, the child's culture encourages children to solve problems in collaboration with others and discourages them from excelling as individuals. Moreover, because they are wary of questions posed by unfamiliar adults, many economically disadvantaged children often answer test questions by saying, "I don't know." Obviously, this strategy guarantees an artificially low test score. When these children are given extra time to feel at ease with the examiner, they respond less often with "I don't know" and their test scores improve considerably (Zigler & Finn-Stevenson, 1992).

Stereotype Threat

When people know that they belong to a group that is said to lack skill in a domain, this makes them anxious when performing in that domain for fear of confirming the stereotype; as a result, they often do poorly. *This self-fulfilling prophecy, in which knowledge of stereotypes leads to anxiety and reduced performance consistent with the original stereotype, is called* **stereotype threat**. Applied to intelligence, the argument is that African American children experience stereotype threat when they take intelligence tests, and this contributes to their lower scores (Steele, 1997; Steele &

stereotype threat
an evoked fear of being judged in
accordance with a negative stereotype
about a group to which you belong

Aronson, 1995). For example, imagine two 10-year-olds taking an intelligence test for admission to a special program for gifted children. The European American child worries that if he fails the test he won't be admitted to the program. The African American child has the same fears but also worries that if he does poorly it will confirm the stereotype that African American children don't get good scores on IQ tests (Suzuki & Aronson, 2005). Consistent with this idea, when African American students experience self-affirmation—they remind themselves of values that are important to them and why—the threat is reduced and their performance improves (Cohen et al., 2006).

Interpreting Test Scores

If all tests reflect cultural influences to at least some degree, how should we interpret test scores? Remember that tests assess successful adaptation to a particular cultural context. Most intelligence tests predict success in a school environment, which usually espouses middle-class values. Regardless of ethnic group, a child with a high test score has the intellectual skills needed for academic work based on middle-class values (Hunt & Carlson, 2007). A child with a low test score apparently lacks those skills. Does a low score mean that a child is destined to fail in school? No. It simply means that, based on the child's current skills, he or she is unlikely to do well. We know from intervention projects that improving children's skills improves their school performance.

By focusing on groups of people, it's easy to overlook the fact that individuals within these groups differ in intelligence. The average difference in IQ scores between various ethnic groups is relatively small compared to the entire range of scores for these groups (Sternberg, Grigorenko, & Kidd, 2005). You can easily find youngsters with high IQ scores from all ethnic groups, just as you can find youngsters with low IQ scores from all groups. In the next section, we'll look at children at the extremes of ability.

Test Yourself

RECALL

1. If some children consistently have high scores on different intelligence tests while other children consistently have lower scores on the same tests, this would support the view that intelligence _____.

2. According to _____ theories, intelligence includes both general intelligence and more specific abilities, such as verbal and spatial skill.

3. Gardner's theory of multiple intelligences includes linguistic, logical-mathematical, and spatial intelligences, which are included in psychometric theories, as well as musical, _____, interpersonal, and intrapersonal intelligences, which are ignored in psychometric theories.

4. Based on Gardner's view of intelligence, teachers should _____.

5. According to Sternberg, successful intelligence depends on _____, creative, and practical abilities.

6. As adopted children get older, their IQ scores increasingly resemble the IQ scores of their _____ parents.

7. Evidence for the impact of environment on children's intelligence comes from studies of children's homes, from historical change in IQ scores, and from _____.

8. The problem of cultural bias on intelligence tests led to the development of _____.

INTERPRET

Compare and contrast the major perspectives on intelligence in terms of the extent to which they make connections between different aspects of development. That is, to what extent does each perspective emphasize cognitive processes versus integrating physical, cognitive, social, and emotional development?

APPLY

Suppose that a local government official proposes to end all funding for preschool programs for disadvantaged children. Write a letter to this official in which you describe the value of these programs.

Recall answers: (1) consists of a general factor, (2) hierarchical, (3) bodily-kinesthetic, (4) teach in a manner that engages as many different intelligences as possible, (5) analytic, (6) biological, (7) intervention studies, (8) culture-fair intelligence tests

LEARNING OBJECTIVES

- What are the characteristics of gifted and creative children?
- What are different forms of disability?
- What are the distinguishing features of attention-deficit hyperactivity disorder?

Sanjit, a second grader, has taken two separate intelligence tests, and both times he had above-average scores. His parents took him to an ophthalmologist, who determined that his vision is 20/30—nothing wrong with his eyes. Nevertheless, Sanjit absolutely cannot read. Letters and words are as mysterious to him as Kanye West's music would be to Mozart. What is wrong?

THROUGHOUT HISTORY, SOCIETIES HAVE RECOGNIZED CHILDREN WITH UNUSUAL ABILITIES AND TALENTS. Today, we know much about the extremes of human skill. Let's begin with a glimpse at gifted and creative children.

Gifted and Creative Children

Traditionally, giftedness was defined by scores on intelligence tests: a score of 130 or greater was the criterion for being gifted. Today, however, definitions of giftedness are broader and include exceptional talent in an assortment of areas, such as art, music, creative writing, and dance (Robinson & Clinkenbeard, 1998; Winner, 2000).

Exceptional talent—whether defined solely by IQ scores or more broadly—seems to have several prerequisites (Rathunde & Csikszentmihalyi, 1993):

- The child's love for the subject and overwhelming desire to master it
- Instruction, beginning at an early age, with inspiring and talented teachers
- Support and help from parents, who are committed to promoting their child's talent

The message here is that exceptional talent must be nurtured. Without encouragement and support from stimulating and challenging mentors, a youngster's talents will wither, not flourish. Talented children need a curriculum that is challenging and complex; they need teachers who know how to foster talent; and they need like-minded peers who stimulate their interests (Feldhusen, 1996). With this support, gifted children's achievement can be remarkable. In a 20-year longitudinal study, gifted teens were, as adults, extraordinarily successful in school and in their careers (Lubinski et al., 2006).

The stereotype is that gifted children are often thought to be emotionally troubled and unable to get along with their peers. In reality, gifted children and adults tend to be more mature than their peers and have fewer emotional problems (Luthar, Zigler, & Goldstein, 1992; Simonton & Song, 2009), and as adults, they report being highly satisfied with their careers, relationships with others, and life in general (Lubinski et al., 2006).

Creativity

Mozart and Salieri were rival composers in Europe during the 18th century. Both were talented, ambitious musicians. Yet more than 200 years later, Mozart's work is revered but Salieri's is all but forgotten. Why? Then and now, Mozart's work was recognized as creative, but Salieri's was not. What is creativity, and how does it differ from intelligence? *Intelligence is often associated with* **convergent thinking,** *which means using the information provided to determine a standard, correct answer. In contrast, creativity is often linked to* **divergent thinking,** *in which the aim is not a single*

convergent thinking
using information to arrive at one standard and correct answer

divergent thinking
thinking in novel and unusual directions

correct answer (often there isn't one) but instead to think in novel and unusual directions (Callahan, 2000).

Divergent thinking is often measured by asking children to produce a large number of ideas in response to some specific stimulus (Kogan, 1983). Children might be asked to name different uses for a common object, such as a coat hanger. Or they might be given a page filled with circles and be asked to draw as many different pictures as they can, as shown in ■ Figure 6.5. Both the number of responses and their originality are used to measure creativity.

Creativity, like giftedness, must be cultivated. Youngsters are more likely to be creative when their home and school environments value nonconformity and encourage children to be curious. When schools, for example, emphasize mastery of factual material and discourage self-expression and exploration, creativity usually suffers (Thomas & Berk, 1981). In contrast, creativity can be enhanced by experiences that stimulate children to be flexible in their thinking and to explore alternatives (Starko, 1988).

Gifted and creative children represent one extreme of human ability. At the other extreme are youngsters with disability, the topic of the next section.

Children With Disability

"Little David," so named because his father was also named David, was the oldest of four children. He learned to sit only days before his first birthday, he began to walk at 2, and he said his first words as a 3-year-old. By age 5, David was far behind his age-mates developmentally. David had Down syndrome, a disorder described in Section 2.1 that is caused by an extra 21st chromosome.

Children With Intellectual Disability

Down syndrome is an example of a condition that leads to **intellectual disability**, *which refers to substantial limitations in intellectual ability as well as problems adapting to an environment, with both emerging before 18 years of age.* Limited intellectual skill is often defined as a score of 70 or less on an intelligence test such as the Stanford-Binet. Adaptive behavior includes conceptual skills important for successful adaptation (e.g., literacy, understanding money and time), social skills (e.g., interpersonal skill), and practical skills (e.g., personal grooming, occupational skills). It is usually evaluated from interviews with a parent or other caregiver. Only individuals who are under 18, have problems adapting in these areas, *and* IQ scores of 70 or less are considered to have an intellectual disability (AAIDD Ad Hoc Committee on Terminology and Classification, 2010).*

Modern explanations pinpoint four factors that place individuals at risk for intellectual disability:

- Biomedical factors, including chromosomal disorders, malnutrition, traumatic brain injury
- Social factors, such as poverty and impaired parent–child interactions
- Behavioral factors, such as child neglect or domestic violence
- Educational factors, including impaired parenting and inadequate special education services

No individual factor in this list necessarily leads to intellectual disability. Instead, the risk for intellectual disability grows as more of these factors are present (AAIDD Ad

■ **Figure 6.5**
One way to measure creativity is to determine how many original responses children can make to a specific stimulus.

intellectual disability
substantially below-average intelligence and problems adapting to an environment that emerge before the age of 18

---

*What we now call intellectual disability was long known as mental retardation and much federal and state law in the United States still uses the latter term. However, intellectual disability is the preferred term because it better reflects the condition not as a deficit in the person but as a poor ". . . fit between the person's capacities and the context in which the person is to function" (AAIDD Ad Hoc Committee on Terminology and Classification, 2010, p. 13).

Hoc Committee on Terminology and Classification, 2010). For example, the risk is great for a child with Down syndrome whose parents live in poverty and cannot take advantage of special education services.

As you can imagine, the many factors that can lead to intellectual disability means that the term encompasses an enormous variety of individuals. One way to describe this variation is in terms of the kind and amount of support that they need. At one extreme, some people have so few skills that they must be supervised constantly. Consequently, they usually live in institutions for persons with intellectual disability, where they can sometimes be taught self-help skills such as dressing, feeding, and toileting (Reid, Wilson, & Faw, 1991). At the other extreme are individuals who go to school and master many academic skills, but not as quickly as a typical child. They often work and many marry. With comprehensive training programs that focus on vocational and social skills, they're often productive citizens and satisfied human beings (Ellis & Rusch, 1991).

learning disability
when a child with normal intelligence has difficulty mastering at least one academic subject

THINK ABOUT IT

How might our definitions of giftedness and intellectual disability differ if they were based on Gardner's theory of multiple intelligences?

Children with reading disabilities often have trouble associating sounds with letters.

© Bubbles Photolibrary / Alamy

Children With Learning Disability

A key element of the definition of intellectual disability is substantially below-average intelligence. In contrast, by definition children with learning disability have normal intelligence. That is, *children with* **learning disability** *(a) have difficulty mastering an academic subject, (b) have normal intelligence, and (c) are not suffering from other conditions that could explain poor performance, such as sensory impairment or inadequate instruction.*

In the United States, about 5% of school-age children are classified as learning disabled, which translates into nearly 3 million youngsters. The number of distinct disabilities and the degree of overlap among them are still debated (Torgesen, 2004). However, most scientists would agree that three are particularly common (Hulme & Snowling, 2009): difficulties in reading individual words, sometimes known as developmental dyslexia; difficulties understanding words that have been read successfully, which is called impaired reading comprehension; and, finally, difficulties in mathematics, which is termed mathematical learning disability or developmental dyscalculia.

Understanding learning disabilities is complicated because each type has its own causes (Landerl et al., 2009) and thus requires its own treatment. For example, developmental dyslexia is the most common type of learning disability. (It's so common that sometimes it's just referred to as reading disability.) Many children with this disorder have problems in distinguishing sounds in written and oral language. For children with developmental dyslexia—like Sanjit (in the opening vignette) or the boy in the photograph—distinguishing *bis* from *bep* or *bis* from *dis* is very difficult; apparently the words all sound very similar (Ziegler et al., 2010). The Spotlight on Research feature illustrates research that has examined this problem in detail.

Spotlight on research — Phonological Representations in Children With Reading Disability

Who were the investigators, and what was the aim of the study? Most reading experts agree that, compared to children who read normally, children with reading disability have difficulty with phonological processing, that is, with translating print into sound. Where experts disagree is on the nature of this problem. One idea is that phonological representations—information in long-term memory about the sounds of words—may be less detailed or less precise in children who have reading disability. For example, think about pairs of similar-sounding words such as *bit* and *bet* or *but* and *bed*. In each pair, only the vowels distinguish the two words and the vowels themselves sound similar. If phonological representations in children with reading disability have less precise information about vowel sounds, this could cause children to read more slowly and less accurately.

According to this hypothesis, reading disability is really a language-related disability and should be apparent when children use language sounds in nonreading tasks. Jennifer Bruno and her colleagues—Frank Manis, Patricia Keating, Anne Sperling, Jonathan Nakamoto, and Mark Seidenberg (2007)—tested this hypothesis by determining how well children with reading disability recognized familiar words that were presented auditorily.

How did the investigators measure the topic of interest? The task was simple: Familiar one-syllable words (e.g., *bone, boat*) were presented on audiotape, and children were asked to say what they were. What made the task difficult for children is that only a portion of the word was presented at a time, beginning with just the initial consonant and a small portion of the vowel. If children could not recognize the word on this initial presentation (most couldn't), the word was repeated with a bit more of the vowel presented. This process was repeated, adding more of the vowel and, later, the final consonant, until the child recognized the word. (All of this was

possible because the experimenters recorded an adult saying each of the words, then used specially-designed software that allowed them to edit each word so that a precise amount of vowel was presented.)

Who were the children in the study? Bruno and her colleagues tested twenty-three 8- to 14-year-olds with reading disability along with twenty-three 8- to 14-year-olds with normal reading skills.

What was the design of the study? The study was both experimental and correlational. In the experimental part of the study, the independent variable was the type of consonant sound that ended the word. Some words ended in stop consonants (*dot, seat*), some ended in lateral consonants (*coal, feel*), and some in nasal consonants (*cone, pan*). The dependent variable was how much of the word had to be presented before children recognized it. The study was also correlational because the investigators were interested in the relation between reading skill (reading disabled versus normal reading skill) and ease of recognizing words. The investigators did not

look at age differences, so the study was neither longitudinal nor cross-sectional.

Were there ethical concerns with the study? No. The tasks are frequently used in research, with no known risks.

What were the results? The graph shown in ■ Figure 6.6 shows what proportion of a word needed to be presented until children recognized it. Words ending in stop consonants were easiest for both groups of readers—they recognized these words based on hearing just less than half of the word. Words ending in lateral and nasal consonants were more difficult—children needed to hear more of the word in order to recognize it—and this was particularly true for children with reading disability.

What did the investigators conclude? For words that end with lateral and nasal consonants, children with reading disability need to hear more of a word to recognize it. Bruno and colleagues argued that this result reflects subtle differences in the phonological representations of these simple words in long-term memory of children with reading disability. That is, because phonological representations are less precise for children with reading disability, they must hear more of a word before they can definitely recognize it. Of course, the differences in the graph are small, but these small differences add up quickly when children repeatedly access the sounds of words during reading.

What converging evidence would strengthen these conclusions? This study focused on stop, lateral, and nasal consonants; it would be useful to extend this work to a broader range of vowel and consonant sounds. Doing so would allow researchers to generate a more complete profile of the phonological representations of children with reading disability.

 Go to Psychology CourseMate at www.cengagebrain.com to enhance your understanding of this research.

■ **Figure 6.6**

Children with reading disability need to hear more of a word before they recognize it.

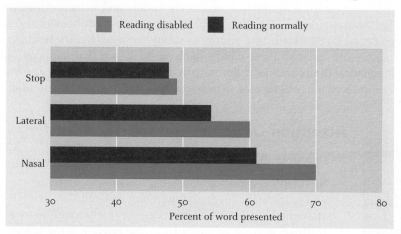

Data from Bruno et al. (2007).

Children with developmental dyslexia typically benefit from two kinds of instruction: training in phonological awareness—experiences that help them to identify subtle but important differences in language sounds—along with explicit instruction on the connections between letters and their sounds. With intensive instruction of this sort, youngsters with developmental dyslexia can read much more effectively (Hulme & Snowling, 2009).

Children with impaired reading comprehension, another common learning disability, have no trouble reading individual words. But they understand far less of what they read. Asked to read sentences such as *The man rode the bus to go to work* or *The dog chased the cat through the woods,* they do so easily but find it difficult to answer questions about what they've read (e.g., *What did the man ride? Where did the man go?*). These problems seem to reflect a limited spoken vocabulary (they simply know fewer words) as well as problems linking words in a sentence together to create coherent meaning (Hulme & Snowling, 2009). Told to select the picture showing children sitting on a table, they may point to a picture of children sitting on a rug or to a picture of children playing a game on a table, but not sitting on it (Nation et al., 2004). In other words, for these youngsters, impaired reading comprehension seems to be a by-product of impaired oral (spoken) language. Consistent with this view, these children understand much more of what they read following extensive instruction in vocabulary and other language skills that are not specific to reading (Clarke et al., 2010).

A third common form of learning disability is mathematical disability. Roughly 5 to 10% of young children struggle with arithmetic instruction from the very beginning. These youngsters progress slowly in their efforts to learn to count, to add, and to subtract; many are also diagnosed with reading disability. As they move into second and third grade (and beyond), these children often use inefficient methods for computing solutions, for example, still using their fingers as third graders to solve problems such as 9 + 7 (Geary, 2005; Jordan, 2007).

We know far less about mathematical learning disability, largely because mathematics engages a broader set of skills than reading (which really involves just two broad classes: decoding and comprehension). Some scientists propose that the heart of the problem is a poorly developed number sense, which includes such skills as understanding and comparing quantities (e.g., 9 > 6) and representing quantity on a number line (Berch, 2005; Jordan, 2007). Another possibility is that youngsters with mathematical disability are impaired in counting and retrieving arithmetic facts from memory (Hulme & Snowling, 2009). Still others suggest that mathematical disability reflects problems in the basic cognitive processes that are used in doing arithmetic, such as working memory and processing speed (Geary et al., 2007).

Because mathematical disability is so poorly understood, effective interventions are not yet available. When the core problems that define mathematical disability have been defined, researchers and educators should be able to craft instruction specifically tailored to improve these child's math skills. When that happens, children with mathematical disability, like children with developmental dyslexia and impaired reading comprehension, will be able to develop their full intellectual potential.

Children with ADHD are typically hyperactive as well as being inattentive and impulsive.

© Catherine Ledner / Getty Images

Attention-Deficit Hyperactivity Disorder

Let's begin with a case study of Stuart, an 8-year-old:

> [His] mother reported that Stuart was overly active as an infant and toddler. His teachers found him difficult to control once he started school. He is described as extremely impulsive and distractible, moving tirelessly from one activity to the next. . . . His teacher reports that he is immature and restless, responds best in a structured, one-on-one situation, but is considered the class pest because he is continually annoying the other children and is disobedient. (Rapaport & Ismond, 1990, p. 120)

For many years, children who were restless and impulsive (like Stuart) were said to have "hyperactive child syndrome" (Barkley, 1996). In the 1960s and 1970s, researchers realized that these children often also had difficulty paying attention. By the 1980s, the disorder had been renamed as attention-deficit hyperactivity disorder (ADHD).

About 3 to 5% of all school-age children are diagnosed with ADHD (Rapport, 1995); boys outnumber girls by a 3:1 ratio (Wicks-Nelson & Israel, 2006). Three symptoms are at the heart of ADHD (American Psychiatric Association, 1994):

- *Hyperactivity.* Children with ADHD are unusually energetic, fidgety, and unable to keep still—especially in situations like the one in the photo where they need to limit their activity.

- *Inattention.* Youngsters with ADHD do not pay attention in class and seem unable to concentrate on schoolwork; instead, they skip from one task to another.
- *Impulsivity.* Children with ADHD often act before thinking; they may run into a street before looking for traffic or interrupt others who are already speaking.

Not all children with ADHD show all these symptoms to the same degree. Most children with ADHD are hyperactive and either impulsive or inattentive (Barkley, 2003). Children with ADHD often have problems with academic performance, conduct, and getting along with their peers (Murray-Close et al., 2010; Stevens & Ward-Estes, 2006).

Many myths surround ADHD. Some concern causes. At one time or another, TV, food allergies, sugar, and poor home life have all been proposed as causes of ADHD, but research does not consistently support any of these theories (e.g., Wolraich et al., 1994). Instead, heredity is an important factor (Saudino & Plomin, 2007). Twin studies show that identical twins are often both diagnosed with ADHD, but this is uncommon for fraternal twins (Pennington, Willcutt, & Rhee, 2005). In addition, prenatal exposure to alcohol and other drugs can place children at risk for ADHD (Milberger et al., 1997).

Another myth is that most children "grow out of" ADHD in adolescence or young adulthood. More than half of the children diagnosed with ADHD will have problems related to overactivity, inattention, and impulsivity as adolescents and young adults (Biederman et al., 2010). Few of these young adults complete college, and some will have work- and family-related problems (Biederman et al., 2006; Murphy, Barkley, & Bush, 2002). One final myth is that many healthy children are wrongly diagnosed with ADHD. The number of children diagnosed with ADHD increased substantially during the 1990s, but not because children were being routinely misdiagnosed; the increased numbers reflected growing awareness of ADHD and more frequent diagnoses of ADHD in girls and adolescents (Goldman et al., 1998).

Because ADHD affects academic and social success throughout childhood and adolescence, researchers have worked hard to find effective treatments. By the mid 1980s, it was clear that ADHD could be treated. For example, children with ADHD often respond well to stimulant drugs such as Ritalin. It may seem odd that stimulants are given to children who are already overactive, but these drugs stimulate the parts of the brain that normally inhibit hyperactive and impulsive behavior. Thus, stimulants actually have a calming influence for many youngsters with ADHD, allowing them to focus their attention (Barkley, 2004).

Drug therapy was not the only approach used: Psychosocial treatments also worked and were designed to improve children's cognitive and social skills; treatments often included home-based intervention and intensive summer programs (Richters et al., 1995). For example, children can be taught to remind themselves to read instructions before starting assignments. And they can be reinforced by others for inhibiting impulsive and hyperactive behavior (Barkley, 2004).

This variety of treatments led scientists to wonder which was the most effective. Consequently, in the mid 1990s, the Multimodal Treatment Study of Children with ADHD (for short, the MTA) was begun. Nearly 600 elementary-school children with ADHD participated; they were assigned to different treatment modes and received treatment for 14 months. The impact of treatment has been measured every few years for several different domains of children's development.

The initial results—obtained at the end of the 14 months of treatment—showed that medication alone was the best way to treat hyperactivity per se. However, for a variety of other measures, including academic and social skills as well as parent–child relations, medication plus psychosocial treatment was somewhat more effective than medication alone (The MTA Cooperative Group, 1999). In contrast, in follow-up studies conducted 6 and 8 years after the 14-month treatment period ended, the treatment groups no longer differed and all groups fared worse than children without ADHD: Children with ADHD were more likely to be inattentive, hyperactive, and impulsive; they were more aggressive; and they were less likely to succeed in school (Molina et al., 2009).

For researchers, parents, and children with ADHD, these are disappointing results. Yet they point to an important conclusion, one with implications for policy: Several months of intensive treatment will not "cure" ADHD; instead, ADHD is perhaps better considered as a chronic condition, like diabetes or asthma, that requires ongoing monitoring and treatment (Hazell, 2009).

Tragically, many children who need these treatments do not receive them. African American and Hispanic American children are far less likely than European American youngsters to be diagnosed with and treated for ADHD, even when they have the same symptoms (Miller, Nigg, & Miller, 2009; Stevens, Harman, & Kelleher, 2005). Why? Income plays a role. African American and Hispanic American families are more often economically disadvantaged and consequently are less able to pay for diagnosis and treatment. Racial bias also contributes: Parents and professionals often attribute the symptoms of ADHD in European American children to a biological problem that can be treated medically, but they more often attribute these symptoms in African American or Hispanic American children to poor parenting, life stresses, or other sources that can't be treated (Kendall & Hatton, 2002).

Obviously, all children with ADHD deserve appropriate treatment. Teachers and other professionals dealing with children need to be sure that poverty and racial bias do not prevent children from receiving the care they need.

Test Yourself

RECALL

1. A problem with defining giftedness solely in terms of IQ score is that _____.

2. Creativity is associated with _____ thinking, in which the goal is to think in novel and unusual directions.

3. Intellectual disability involves substantial limits in intellectual ability and _____ that emerge before 18 years of age.

4. Biomedical, social, _____, and educational factors place some children at risk for intellectual disability.

5. In developmental dyslexia, children have difficulty with

_____ .

6. Key symptoms of attention-deficit hyperactivity disorder are overactivity, _____, and impulsivity.

7. The results of the MTA show that, in the short run, the best way to treat the full spectrum of symptoms of ADHD is through stimulant drugs combined with _____ .

INTERPRET

Compare and contrast traditional and modern definitions of giftedness.

APPLY

How might Jean Piaget have explained differences in intellectual functioning between children with intellectual disability and children without intellectual disability? How might an information-processing psychologist explain these differences?

Recall answers: (1) it excludes talents in areas such as art, music, and dance, (2) divergent, (3) problems adapting to the environment, (4) behavioral, (5) phonological awareness (distinguishing language sounds), (6) inattentiveness, (7) psychosocial treatment that improves children's cognitive and social skills

LEARNING OBJECTIVES

■ What are the components of skilled reading?

■ As children develop, how does their writing improve?

■ How do arithmetic skills change during the elementary-school years? How do U.S. students compare to students from other countries?

■ What are the hallmarks of effective schools and effective teachers?

Angelique is a fifth grader who absolutely loves to read. As a preschooler, Angelique's parents read Dr. Seuss stories to her, and now she has progressed to the point where she can read (and understand!) 400-page novels intended for teens. Her parents marvel at this accomplishment and wish they better understood the skills that were involved so they could help Angelique's younger brother learn to read as well as his sister does.

READING IS INDEED A COMPLEX TASK AND LEARNING TO READ WELL IS A WONDERFUL ACCOMPLISHMENT. Much the same can be said for writing and math. We'll examine each of these academic skills in this section. As we do, you'll learn about the skills that underlie Angelique's mastery of reading. We'll end the section by looking at characteristics that make some schools and some teachers better than others.

Reading

Try reading the following sentence:

Sumisu-san wa nawa o naifu de kirimashita.

You probably didn't make much headway, did you? (Unless you know Japanese.) Now try this one:

Snore secretary green plastic sleep trucks.

These are English words, and you probably read them quite easily, but did you get anything more out of this sentence than the one in Japanese?

These examples show two important processes involved in skilled reading. **Word recognition** *is the process of identifying a unique pattern of letters*. Unless you know Japanese, your word recognition was not successful in the first sentence. You did not know that *nawa* means *rope* or that *kirimashita* is the past tense of the English verb *cut*. Furthermore, because you could not recognize individual words, you had no idea of the meaning of the sentence.

Comprehension *is the process of extracting meaning from a sequence of words*. In the second sentence your word recognition was perfect, but comprehension was still impossible because the words were presented in a random sequence. These examples remind us just how difficult learning to read can be.

In the next few pages, we'll look at some of the skills children must acquire if they are to learn to read and to read well. We'll start with prereading skills and then move to word recognition and comprehension.

word recognition
the process of identifying a unique pattern of letters

comprehension
the process of extracting meaning from a sequence of words

Foundations of Reading Skill

English words are made up of individual letters, so children need to know their letters before they can learn to read. Children learn more about letters and word forms when they're frequently involved in literacy-related activities such as reading with an adult, playing with magnetic letters, or trying to print simple words. And, not surprisingly, children who know more about letters and word forms learn to read more easily than their peers who know less (Levy et al., 2006; Treiman & Kessler, 2003).

A second essential skill is sensitivity to language sounds. *The ability to distinguish the sounds in spoken words is known as* **phonological awareness**. English words consist of syllables and a syllable is made up of a vowel that's usually but not always accompanied by consonants. For example, *dust* is a one-syllable word that includes the initial consonant *d*, the vowel *u*, and the final consonant cluster *st*. Phonological awareness is shown when children can decompose words in this manner by, for example, correctly answering "What's the first sound in *dust*?" or "*Dust* without the *d* sounds like what?" Phonological awareness is strongly related to success in learning to read: Children who can readily identify different sounds in spoken words learn to read more readily than children who do not (Muter et al., 2004).

Learning to read in English is particularly challenging because English is incredibly inconsistent in the way that letters are pronounced (e.g., compare the sound of "a" in *bat, far, rake,* and *was*) and the way that sounds are spelled (e.g., the long "e" sound is the same in each of these spellings: team, feet, piece, lady, receive, magazine). In contrast, many other languages—Greek, Finnish, German, Italian, Spanish, Dutch—are far more consistent, which simplifies the mapping of sounds to letters. In Italian, for example, most letters are pronounced in the same way; reading a word like "domani" (tomorrow) is simple because beginning readers just move from left to right, converting each letter to sound, using simple rules: *d, m,* and *n* are pronounced as in English, *o* as in *cold, a* as in *car,* and *i* as in *see* (Barca, Ellis, & Burani, 2007).). In fact, children learn to read more rapidly in languages where letter-sound rules are more consistent, but phonological awareness remains the single best predictor of reading success in many languages (Lervåg, Bråten & Hulme, 2009; Ziegler et al., 2010).

If phonological skills are so essential, how can we help children master them? Reading to children is one approach that's fun for children and parents alike. When parents read stories, their children learn many language-related skills that prepare them for reading (Justice, Pullen & Pence, 2008; Raikes et al., 2006). And the benefits are not limited to the first steps in learning to read but persist into the middle elementary-school years and are just as useful for children learning to read other languages, such as Chinese (Chow et al., 2008; Sénéchal & LeFevre, 2002).

Recognizing Words

At the very beginning of reading, children sometimes learn to read a few words "by sight" but they have no understanding of the links between printed letters and the word's sound. However, the first step in true reading is learning to decode printed words by sounding out the letters in them: Beginning readers like the boy in the photo often say the sounds associated with each letter and then blend the sounds to produce a recognizable word. After a word has been sounded out a few times, it becomes a known word that can be read by retrieval directly from long-term memory: As the individual letters in a word are identified, long-term memory is searched to see if there is a matching sequence of letters. After the child knows that the letters are, in sequence, *c-a-t,* long-term memory is searched for a match and the child recognizes the word as *cat* (Rayner et al., 2001).

Thus, from their very first efforts to read, most children use retrieval for some words. From that point on, the general strategy is to try retrieval first and, if that fails, to sound out the word or ask a more skilled reader for help. With more experience, the child sounds out fewer words and retrieves more (Siegler, 1986). That is, by sounding out novel words, children increase their store of information about words in long-term memory that is required for direct retrieval (Cunningham et al., 2002; Share, 2008).

So far, word recognition may seem like a one-way street where readers first recognize letters and then recog-

Beginning readers usually rely heavily on "sounding out" a word.

nize words. In reality, we know that information flows both ways: Readers constantly use context to help them recognize letters and words (Rayner et al., 2001). For example, read these two sentences:

Readers also use the sentence context to speed word recognition. Read these two sentences:

The last word in this sentence is cat.
The little girl's pet dog chased the cat.

Most readers recognize *cat* more rapidly in the second sentence because the first seven words put severe limits on the last word: It must be something "chaseable" and, because the "chaser" is a *dog, cat* is a likely candidate. In contrast, the first seven words in the first sentence put no limits on the last word; virtually any word could end the sentence. Beginning and skilled readers both use sentence context like this to help them recognize words (Archer & Bryant, 2001; Kim & Goetz, 1994).

Comprehension

Once individual words are recognized, reading begins to have a lot in common with understanding speech. That is, the means by which people understand a sequence of words is much the same whether the source of words is printed text or speech or, for that matter, Braille or sign language (Oakhill & Cain, 2004).

As children gain more reading experience, they better comprehend what they read. Several factors contribute to this improved comprehension (Siegler & Alibali, 2005).

- *Children become more skilled at recognizing words, allowing more working memory capacity to be devoted to comprehension* (Zinar, 2000). When children struggle to recognize individual words, they often cannot link them to derive the meaning of a passage. In contrast, when children recognize words effortlessly, they can focus their efforts on deriving meaning from the whole sentence.

- *Working memory capacity increases.* Older and better readers can store more of a sentence in memory as they try to identify the propositions it contains (De Beni & Palladino, 2000; Nation et al., 1999). This extra capacity is handy when readers move from sentences like "Kevin hit the ball" to "In the bottom of the ninth, with the bases loaded and the Cardinals down 7–4, Kevin put a line drive into the left-field bleachers, his fourth home run of the series."

- *Children acquire more general knowledge of their physical, social, and psychological worlds.* This allows them to understand more of what they read (Ferreol-Barbey, Piolat, & Roussey, 2000). For example, even if a 6-year-old could recognize all of the words in the longer sentence about Kevin's home run, the child would not fully comprehend the meaning of the passage because he or she lacks the necessary knowledge of baseball.

- *With experience, children use more appropriate reading strategies.* The goal of reading and the nature of the text dictate how you read. When reading a novel, for example, do you often skip sentences (or perhaps paragraphs or entire pages) to get to "the good parts"? This approach makes sense for pleasure reading but not for reading textbooks or recipes or how-to manuals. Reading a textbook requires attention to both the overall organization and the relation of details to that organization. Older, more experienced readers are better able to select a reading strategy that suits the material being read (Brown et al., 1996; Cain, 1999).

- *With experience, children better monitor their comprehension.* When readers don't grasp the meaning of a passage because it is difficult or confusing, they read it again (Baker, 1994). Try this sentence (adapted from Carpenter & Daneman, 1981): "The Midwest State Fishing Contest would draw fishermen from all around the region, including some of the best bass guitarists in Michigan." When you first encountered "bass guitarists" you probably interpreted "bass" as a fish. This didn't make much sense, so you reread the phrase to determine that "bass" refers to a type of guitar. Older readers are better able to realize that their understanding is not complete and take corrective action.

Thus, several factors contribute to improved comprehension as children get older.

One way to summarize what we've learned about reading comprehension is with the Simple View of Reading model proposed by Gough and Tunmer (1986). In the model, reading comprehension is the product of two general processes: word decoding and language comprehension. Children can't comprehend what they read when either a word can't be decoded or it's decoded but not recognized as a familiar word. Thus the best way to ensure that children understand what they read is to help them master fast, accurate decoding and language comprehension (e.g., increasing their vocabulary, improving their mastery of grammar). And reading instruction is most likely to succeed when it is tailored to a child's weaknesses, emphasizing letter-sound skills for children whose decoding skills are limited and working on vocabulary and grammar when children need stronger language comprehension skills (Connor et al., 2007).

THINK ABOUT IT
Reading and speaking are both important elements of literacy. How is learning to read like learning to speak? How do they differ?

Writing

Though few of us end up being a Maya Angelou, a Sandra Cisneros, or a J. K. Rowling, most adults do write, both at home and at work. The basics of good writing are remarkably straightforward, but writing skill develops only gradually during childhood, adolescence, and young adulthood. Research indicates that a number of factors contribute to improved writing as children develop (Adams et al., 1998; Siegler & Alibali, 2004).

Knowledge About Topics

Writing is about telling "something" to others. With age, children have more to tell as they gain more knowledge about the world and incorporate this knowledge into their writing (Benton et al., 1995). For example, asked to write about a mayoral election, 8-year-olds are apt to describe it as much like a popularity contest but 12-year-olds more often describe it in terms of political issues that are both subtle and complex. Of course, students are sometimes asked to write about topics quite unfamiliar to them. In this case, older children's and adolescents' writing is usually better because they are more adept at finding useful reference material and incorporating it into their writing.

Organizing Writing

One difficult aspect of writing is organization, arranging all the necessary information in a manner that readers find clear and interesting. In fact, children and young adolescents organize their writing differently than do older adolescents and adults (Bereiter & Scardamalia, 1987). *Young writers often use a* **knowledge-telling strategy**, *writing down information on the topic as they retrieve it from memory.* For example, asked to write about the day's events at school, a second grader wrote:

> It is a rainy day. We hope the sun will shine. We got new spelling books. We had our pictures taken. We sang Happy Birthday to Barbara. (Waters, 1980, p. 155)

The story has no obvious structure. The first two sentences are about the weather, but the last three deal with completely independent topics. Apparently, the writer simply described each event as it came to mind.

Toward the end of the elementary-school years, children begin to use a **knowledge-transforming strategy**, *deciding what information to include and how best to organize it for the point they wish to convey to the reader.* This approach involves considering the purpose of writing (e.g., to inform, to persuade, to entertain) and the information

knowledge-telling strategy
writing down information as it is retrieved from memory, a common practice for young writers

knowledge-transforming strategy
deciding what information to include and how best to organize it to convey a point

needed to achieve this purpose. It also involves considering the needs, interests, and knowledge of the anticipated audience.

Asked to describe the day's events, older children's writing can take many forms, depending on the purpose and audience. An essay written to entertain peers about humorous events at school would differ from one written to convince parents about problems in schoolwork (Midgette, Haria, & MacArthur, 2008). And both of these essays would differ from one written to inform an exchange student about a typical day in a U.S. middle school. In other words, although children's knowledge-telling strategy gets words on paper, the more mature knowledge-transforming strategy produces a more cohesive text for the reader.

Writing can be particularly hard for young children who are still learning how to print or write cursive letters.

The Mechanical Requirements of Writing

Compared to speaking, writing is more difficult because we need to worry about spelling, punctuation, and actually forming the letters. These many mechanical aspects of writing can be a burden for all writers, but particularly for young writers. For example, when youngsters like the child in the photo are absorbed by the task of printing letters correctly, the quality of their writing usually suffers (Graham, Harris, & Fink, 2000; Olinghouse, 2008). As children master printed and cursive letters, they can pay more attention to other aspects of writing. Similarly, correct spelling and good sentence structure are particularly hard for younger writers; as they learn to spell and to generate clear sentences, they write more easily and more effectively (Graham et al., 1997; McCutchen et al., 1994).

Skill in Revising

Few authors get it down right the first time. Instead, they revise and revise, then revise some more. Unfortunately, young writers often don't revise at all—the first draft is usually the final draft. To make matters worse, when young writers revise, the changes do not necessarily improve their writing (Fitzgerald, 1987). Effective revising requires being able to detect problems and knowing how to correct them (Baker & Brown, 1984; Beal, 1996). As children develop, they're better able to find problems and to know how to correct them, particularly when the topic is familiar (Chanquoy, 2001; McCutchen, Francis, & Kerr, 1997).

If you look over these past few paragraphs, it's quite clear why good writing is so gradual in developing. Many different skills are involved, and each is complicated in its own right. Word processing software makes writing easier by handling some of these skills (e.g., checking spelling, simplifying revising), and research indicates that writing improves when people use word processing software (Clements, 1995; Rogers & Graham, 2008).

Fortunately, students *can* be taught to write better. When instruction focuses on the building blocks of effective writing—strategies for planning, drafting, and revising text—students' writing improves substantially (Graham & Perin, 2007; Tracy, Reid, & Graham, 2009). For example, one successful program for teaching writing—the Self-Regulated Strategy Development in Writing program—tells students that POW + TREE is a trick that good writers use. As you can see in ■ Figure 6.7, POW provides young writers with a general plan for writing; TREE tells them how to organize their writing in a nicely structured paragraph (Harris et al., 2008).

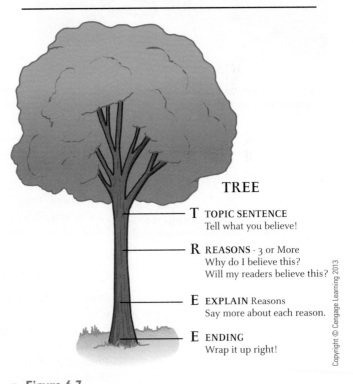

POW

P   Pick my idea

O   Organize my notes

W   Write and say more

TREE

T   TOPIC SENTENCE
Tell what you believe!

R   REASONS - 3 or More
Why do I believe this?
Will my readers believe this?

E   EXPLAIN Reasons
Say more about each reason.

E   ENDING
Wrap it up right!

■ **Figure 6.7**
The POW + TREE strategy for good writing provides young writers with a general plan for writing (POW) and a structure for a paragraph (TREE).

Of course, mastering the full set of writing skills is a huge challenge, one that spans all of childhood, adolescence, and adulthood. Much the same can be said for mastering math skills, as we'll see in the next section.

Math Skills

In Chapter 4 we saw that preschoolers understand many of the principles underlying counting, even if they sometimes stumble over the mechanics of counting. By kindergarten, children have mastered counting and use this skill as the starting point for learning to add. For instance, suppose you ask a kindergartner to solve the following problem: "John had four oranges. Then Mary gave him two more oranges. How many oranges does John have now?" Many 6-year-old children solve the problem by counting. They first count out four fingers on one hand, then count out two more on the other. Finally, they count all six fingers on both hands. To subtract, they do the same procedure in reverse (Siegler & Jenkins, 1989; Siegler & Shrager, 1984). After children begin to receive formal arithmetic instruction in first grade, addition problems are less often solved by counting aloud or by counting fingers (Jordan et al., 2008). Instead, children add and subtract by counting mentally. That is, children act as if they are counting silently, beginning with the larger number and then adding on. By age 8 or 9, children have learned the addition tables so well that sums of the single-digit integers (from 0 to 9) are facts that can be simply retrieved from memory (Ashcraft, 1982).

These counting strategies do not occur in a rigid developmental sequence. Individual children use many or all of these strategies, depending upon the problem. Children usually begin by trying to retrieve an answer from memory; if they are not reasonably confident that the retrieved answer is correct, they resort to counting aloud or on fingers (Siegler, 1988). Retrieval is most likely for problems with small numbers (e.g., 1 + 2, 2 + 4) because these problems are presented frequently in textbooks and by teachers. Consequently, the sum is highly associated with the problem, which makes the child confident that the retrieved answer is correct. In contrast, problems with larger addends, such as 9 + 8, are presented less often. The result is a weaker link between the addends and the sum and consequently a greater chance that children will need to determine an answer by counting.

Of course, arithmetic skills continue to improve as children move through elementary school. They become more proficient in addition and subtraction, learn multiplication and division, and move on to the more sophisticated mathematical concepts involved in algebra, geometry, trigonometry, and calculus.

Young children often solve addition problems by counting, either on their fingers or in their head.

THINK ABOUT IT

What information-processing skills may contribute to growth in children's arithmetic skills?

Comparing U.S. Students With Students in Other Countries

When compared to students worldwide in terms of math skills, American students don't fare well. For example, ■ Figure 6.8 shows the results of a major international comparison involving students in 20 countries (Gonzales et al., 2008). Students in the United States have substantially lower scores than students in the leading nations. Phrased another way, the very best U.S. students perform only at the level of average students in Asian countries like Singapore and Korea. Furthermore, the cultural differences in math achievement hold for both math operations and math problem solving (Stevenson & Lee, 1990).

Why do American students rate so poorly? The Real People feature has some answers.

Shin-ying is an 11-year-old attending school in Taipei, the largest city in Taiwan. Like most fifth graders, Shin-ying is in school from 8 a.m. until 4 p.m. daily. Most evenings, she spends 2 to 3 hours doing homework. This academic routine is grueling by U.S. standards, where fifth graders typically spend 6 to 7 hours in school each day and less than an hour doing homework. We asked Shin-ying what she thought of school and schoolwork. Her answers surprised us.

US: Why do you go to school?

SHIN-YING: I like what we study.

US: Any other reasons?

SHIN-YING: The things that I learn in school are useful.

US: What about homework? Why do you do it?

SHIN-YING: My teacher and my parents think it's important. And I like doing it.

US: Do you think you would do nearly as well in school if you didn't work so hard?

SHIN-YING: Oh no. The best students are always the ones who work the hardest.

Schoolwork is the focal point of Shin-ying's life. Although many American schoolchildren are unhappy when schoolwork intrudes on time for play and television, Shin-ying is enthusiastic about school and school-related activities.

Shin-ying is not unusual among Asian elementary-school students. Many of her comments are typical of students from a comprehensive comparison of students in Japan, Taiwan, and the United States (Perry, 2000; Stevenson & Lee, 1990; Stigler, Gallimore, & Hiebert, 2000).

- *Time in school and how it is used.* By fifth grade, students in Japan and Taiwan spend 50% more time than American students in school, and more of this time is devoted to academic activities than in the United States.

- *Time spent on homework and attitudes toward it.* Students in Taiwan and Japan spend more time on homework and value homework more than American students.

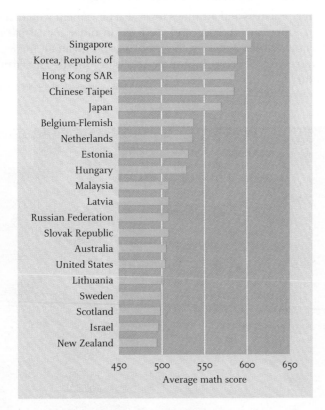

- **Figure 6.8**
Compared to students in other developed countries, U.S. students fare poorly on tests of math skills.

From International Association for the Evaluation of Educational Achievement (IEA), Trends in International Mathematics and Science Study (TIMSS), 2003, p. 5.

- *Parents' attitudes.* American parents are more often satisfied with their children's performance in school; in contrast, Japanese and Taiwanese parents set much higher standards for their children.

- *Parents' beliefs about effort and ability.* Japanese and Taiwanese parents believe more strongly than American parents that effort, not native ability, is the key factor in school success.

Thus, students in Japan and Taiwan excel because they spend more time both in and out of school on academic tasks. What's more, their parents (and teachers) set loftier scholastic goals and believe that students can attain these goals with hard work. Japanese classrooms even post a motto describing ideal students: *gambaru kodomo*—they who strive the hardest.

Parents underscore the importance of schoolwork in many ways to their children. For example, even though homes and apartments in Japan and China are very small by U.S. standards, Asian youngsters typically have a desk in a quiet area where they can study undisturbed (Stevenson & Lee, 1990). For Japanese and Taiwanese teachers and parents, academic excellence is paramount, and it shows in their children's success.

What can Americans learn from Japanese and Taiwanese educational systems? From their experiences with Asian students, teachers, and schools, Stevenson and Stigler (1992) suggest several ways U.S. schools could be improved:

- Give teachers more free time to prepare lessons and correct students' work.

- Improve teachers' training by allowing them to work closely with older, more experienced teachers.

- Organize instruction around sound principles of learning, such as providing multiple examples of concepts and giving students adequate opportunities to practice newly acquired skills.

- Set higher standards for children, who need to spend more time and effort in school-related activities to achieve those standards.

Changing teaching practices and attitudes toward achievement would begin to reduce the gap between American students and students in other industrialized countries, particularly Asian countries. Ignoring the problem will mean an increasingly undereducated workforce and citizenry in a more complex world—an alarming prospect for the 21st century.

Schoolchildren in Asian countries usually have a quiet place at home where they can study.

Effective Schools, Effective Teachers

Because education is run locally in the United States, American education is a smorgasbord. Schools differ along many dimensions, including their emphasis on academic goals and the involvement of parents. Teachers, too, differ in many ways, such as how they run their classrooms and how they teach. These and other variables do affect student achievement, as you'll see in the next few pages. Let's begin with school-based influences.

School-Based Influences on Student Achievement

Roosevelt High School, in the center of Detroit, has an enrollment of 3,500 students in grades 9–12. Opened in 1936, the building shows its age. The rooms are drafty, the desks are decorated with generations of graffiti, and new technology means an overhead projector. Nevertheless, attendance at Roosevelt is good. Most students graduate, and many continue their education at community colleges and state universities. Southport High School, in Newark, has about the same enrollment as Roosevelt High, and the building is about the same age. Yet truancy is commonplace at Southport, where fewer than half the students graduate and almost none go to college.

Although these schools are hypothetical, they accurately depict a common outcome in the United States. Some schools are much more successful than others, whether success is defined in terms of the percentage of students who are literate, graduate, or go to college. Why? Researchers (El Nokali, Bachman, & Votruba-Drzal, 2010; Good & Brophy, 2008; Hill & Taylor, 2004) have identified a number of characteristics of schools where students typically succeed rather than fail.

- *Staff and students alike understand that academic excellence is the primary goal of the school and of every student in the school.* The school day emphasizes instruction (not simply filling time with nonacademic activities), and students are recognized publicly for their academic accomplishments.
- *The school climate is safe and nurturant.* Students know that they can devote their energy to learning (instead of worrying about being harmed in school) and that the staff truly cares that they succeed.
- *Parents are involved.* In some cases, this may be through formal arrangements such as parent–teacher organizations. Or involvement may be informal: Parents may spend some time each week in school grading papers or tutoring a child. Such involvement signals both teachers and students that parents are committed to students' success.

In successful schools, the child's parents are involved—often as tutors.

- *Progress of students, teachers, and programs is monitored.* The only way to know whether schools are succeeding is by measuring performance. Students, teachers, and programs need to be evaluated regularly, using objective measures that reflect academic goals.

In schools where these guidelines are followed regularly, students usually succeed. In schools where the guidelines are ignored, students more often fail.

Of course, on a daily basis, individual teachers have the most potential for impact. Let's see how teachers can influence their students' achievement.

Teacher-Based Influences on Student Achievement

In most schools, some teachers are highly sought after because their classes are very successful: students learn and the classroom climate is usually positive. What are the keys to the success of these master teachers? Research reveals that several factors are critical for students' achievement (Good & Brophy, 2008; Stevenson & Stigler, 1992; Walberg, 1995). Students tend to learn the most when teachers:

Peer tutoring can be very effective; both the tutored student and the tutor usually learn.

- *Manage the classroom effectively so they can devote most of their time to instruction.* When teachers spend a lot of time disciplining students or when students do not move smoothly from one class activity to the next, instructional time is wasted and students are apt to learn less.

- *Believe they are responsible for their students' learning and that their students will learn when taught well.* When students don't understand a new topic, these teachers may repeat the original instruction (in case the student missed something) or create new instructions (in case the student heard everything but just didn't "get it"). These teachers keep plugging away because they feel at fault if students don't learn.

- *Emphasize mastery of topics.* Teachers should introduce a topic, then give students many opportunities to understand, practice, and apply the topic. Just as you'd find it hard to go directly from driver's ed to driving a race car, students more often achieve when they grasp a new topic thoroughly, then gradually move to other, more advanced topics.

- *Teach actively.* Effective teachers don't just talk or give students an endless stream of worksheets. Instead, they demonstrate topics concretely or have hands-on demonstrations for students. They also have students participate in class activities and encourage students to interact, generating ideas and solving problems together.

- *Pay careful attention to pacing.* Teachers present material slowly enough so that students can understand a new concept but not so slowly that students get bored.

- *Value tutoring.* Teachers work with students individually or in small groups so they can gear their instruction to each student's level and check each student's understanding. They also encourage peer tutoring, in which more capable students tutor less capable students. Children who are tutored by peers *do* learn, and so do the tutors because teaching helps tutors to organize their knowledge.

- *Teach students techniques for monitoring and managing their own learning.* Students are more likely to achieve when they are taught how to recognize the aims of school tasks as well as effective strategies (such as those described on pages 208–210 for achieving those aims.

THINK ABOUT IT

Would some of these ways to promote students' learning be more appropriate for students in Piaget's concrete-operational stage? Would some be better for students in the formal-operational stage?

When teachers rely on most of these guidelines for effective teaching most of the time, their students generally learn the material and enjoy doing so (Good & Brophy, 1994; Stevenson & Stigler, 1992; Walberg, 1995).

Test Yourself

RECALL

1. Important prereading skills include knowing letters and _____.

2. Beginning readers typically recognize words by sounding them out; with greater experience, readers are more likely able to _____.

3. Older and more experienced readers understand more of what they read because the capacity of working memory increases, they have more general knowledge of the world, _____, and they are more likely to use appropriate reading strategies.

4. Children typically use a _____ to organize their writing.

5. Children write best when _____.

6. The simplest way of solving addition problems is to _____; the most advanced way is to retrieve sums from long-term memory.

7. Compared to students in U.S. elementary schools, students in Japan and Taiwan spend more time in school, and a greater proportion of that time is _____.

8. In schools where students usually succeed, academic excellence is a priority, the school is safe and nurturant, progress of students and teachers is monitored, and _____.

9. Effective teachers manage classrooms well, believe they are responsible for their students' learning, _____, teach actively, pay attention to pacing, value tutoring, and show children how to monitor their own learning.

INTERPRET

Review the research on pages 231–232 regarding factors associated with skilled reading comprehension. Which of these factors—if any—might also contribute to skilled writing?

APPLY

Imagine two children, both entering first grade. One has mastered prereading skills, can sound out many words, and recognizes a rapidly growing set of words. The second child knows most of the letters of the alphabet but knows only a handful of letter–sound correspondences. How are these differences in reading skills likely to lead to different experiences in first grade?

6.5 PHYSICAL DEVELOPMENT

LEARNING OBJECTIVES

- How much do school-age children grow?
- How do motor skills improve during the elementary-school years?
- Are American children physically fit?
- What are the consequences of participating in sports?

Miguel and Dan are 9-year-olds playing organized baseball for the first time. Miguel's coach is always upbeat. He constantly emphasizes the positive. When they lost a game 12 to 2, the coach complimented all the players on their play in the field and at bat. In contrast, Dan's coach was livid when the team lost, and he was extremely critical of three players who made errors that contributed to the loss. Miguel thinks that baseball is great, but Dan can hardly wait for the season to be over.

DURING THE ELEMENTARY-SCHOOL YEARS, children steadily grow and their motor skills continue to improve. We'll trace these changes in the first two parts of this section. Then we'll see whether U.S. children are physically fit. We'll end the section by examining children's participation in sports and see how coaches like those in the vignette influence children in organized sports.

Growth

Physical growth during the elementary-school years continues at the steady pace established during the preschool years. From ■ Figure 6.9 you can see that a typical 6-year-old weighs about 45 pounds and is 45 inches tall but grows to about 90 pounds and 60 inches by age 12. In other words, most children gain about 8 pounds and 2 to 3 inches per year. Many parents notice that their elementary-school children outgrow shoes and pants more rapidly than they outgrow sweaters, shirts, or jackets; this is because most of the increase in height comes from the legs, not the trunk.

Boys and girls are about the same size for most of these years (which is why they are combined in the figure), but girls are much more likely than boys to enter puberty toward the end of the elementary-school years. Once girls enter puberty, they

During the elementary- and middle-school years, some children are much taller than average and others much shorter.

© 2010 David Leahy / Jupiter Images Corporation

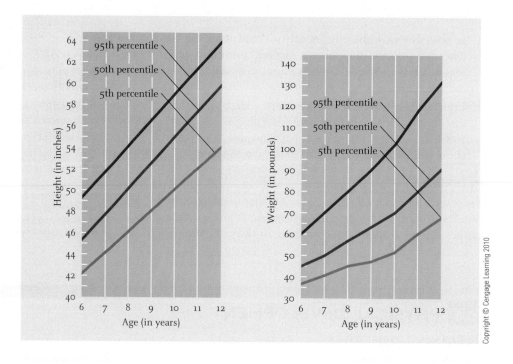

grow rapidly and become much bigger than the boys their age. (We have more to say about this in Chapter 8.) Thus, at ages 11 and 12, the average girl is about half an inch taller than the average boy.

As was true in infancy and early childhood, individuals of the same age often differ markedly in their height and weight. Ethnic differences are also evident in children's growth. In these years, African American children tend to be taller than European American children, who in turn are taller than Asian American children (Webber et al., 1995).

To support this growth and to provide energy for their busy lives, school-age children need to eat more. Although preschool children need only consume about 1,500 to 1,700 calories per day, the average 7- to 10-year-old needs about 2,400 calories each day. Of course, the exact figure depends on the child's age and size and can range anywhere from roughly 1,700 to 3,300 calories daily.

As was true for preschool children, elementary-school children need a well-balanced diet. They should eat regularly from each of the major food groups: grains, vegetables, fruits, milk, meat, and beans. Too often children consume "empty" calories from sweets that have very little nutritional value.

It's also important that school-age children eat breakfast. At this age, many children skip breakfast—often because they're too rushed in the morning. In fact, breakfast should provide about one fourth of a child's daily calories. When children don't eat breakfast, they often have difficulty paying attention or remembering in school (Pollitt, 1995). Therefore, parents should organize their mornings so that their children have enough time for breakfast.

Development of Motor Skills

Elementary-school children's greater size and strength contribute to improved motor skills. During these years, children steadily run faster and jump farther. For example, ■ Figure 6.10 shows how far a typical boy and girl can throw a ball and how far they can jump (in the standing long jump). By the time children are 11 years old, they can throw a ball three times farther than they could at age 6 and can jump nearly twice as far.

Fine motor skills also improve as children move through the elementary-school years. Children's greater dexterity is evident in a host of activities ranging from typing, writing, and drawing to working on puzzles, playing the piano, and building model cars. Children gain much greater control over their fingers and hands, making them much more nimble. This greater fine motor coordination is obvious in children's handwriting.

Gender Differences in Motor Skills

In both gross and fine motor skills, there are gender differences in performance levels. Girls tend to excel in fine motor skills; their handwriting tends to be better than that of boys, for example. Girls also excel in gross motor skills that require flexibility and balance, such as tumbling. On gross motor skills that emphasize strength, boys usually have the advantage. ▪ Figure 6.10 shows that boys throw and jump farther than girls.

Some of the gender differences in gross motor skills that require strength reflect the fact that, as children approach and enter puberty, girls' bodies have proportionately more fat and less muscle than do boys' bodies. This difference explains why, for example, boys can hang by their hands or arms from a bar much longer than girls can. However, for other gross motor skills such as running, throwing, and catching, body composition is much less important (Duff, Ericsson, & Baluch, 2007; Smoll & Schutz, 1990). In these cases, children's experience is crucial. During recess, elementary-school girls are more often found on a swing set, jumping rope, or perhaps talking quietly in a group; in contrast, boys are playing football or shooting baskets. Many girls and their parents believe that sports and physical fitness are less valuable for girls than boys. Consequently, girls spend less time in these sports and fitness-related activities than boys, depriving them of the opportunities to practice that are essential for developing motor skills (Fredricks & Eccles, 2005).

Physical Fitness

Being active physically has many benefits for children: it helps to promote growth of muscles and bone, promotes cardiovascular health (Best, in press; Hillman et al., 2009; National High Blood Pressure Education Program Working Group, 1996), and can help

▪ **Figure 6.10**
Between 6 and 11 years, children's motor skills improve considerably.

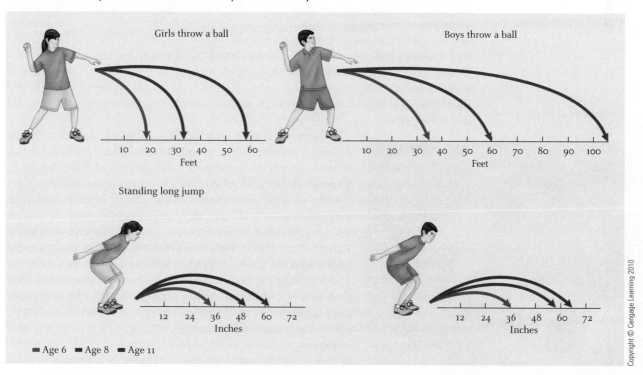

to establish a lifelong pattern of exercise (Perkins et al., 2004). During the elementary-school years, most U.S. school-age children meet the current guidelines of being physically active at least 60 minutes daily (President's Council on Physical Fitness and Sports, 2004).

Unfortunately, when children are tested with a full battery of fitness tests, such as the mile run, pull-ups, and sit-ups, fewer than half usually meet standards for fitness on all tasks (Morrow, 2005). And, as we'll discuss in Chapter 8, the U.S. Surgeon General pronounced that obesity has reached epidemic proportions among American children and adolescents (U.S. Department of Health and Human Services, 2001).

Many factors contribute to low levels of fitness. In most schools, physical education classes meet only once or twice a week and are usually not required of high-school students (Johnston, Delva, & O'Malley, 2007). And, even when students are in these classes, they spend a surprisingly large proportion of time—nearly half—standing around instead of exercising (Lowry et al., 2001; Parcel et al., 1989). Television and other sedentary leisure-time activities may contribute, too. Youth who spend much time online or watching TV often tend to be less fit physically (Lobelo et al., 2009), but the nature of this relation remains poorly understood: Children glued to a TV or computer screen likely have fewer opportunities to exercise, but it might be that children in poor physical condition chose sedentary activities over exercise.

Many experts believe that U.S. schools should offer physical education more frequently each week. And many suggest that physical education classes should offer a range of activities in which all children can participate and that can be the foundation for a lifelong program of fitness (National Association for Sport and Physical Fitness, 2004). Thus, instead of emphasizing team sports such as touch football, physical education classes should emphasize activities like running, walking, racket sports, and swimming; these can be done throughout adolescence and adulthood—either alone or with another person. Families can encourage fitness, too. Instead of spending an afternoon watching TV and eating popcorn, they can go biking, hiking, or swimming together.

Participating in Sports

Children's greater motor skill means they are able to participate in many team sports, including baseball, softball, basketball, and soccer. Obviously, when children play sports, they get exercise and improve their motor skills. But there are other benefits as well. Sports can enhance participants' self-esteem and can help them to learn initiative (Bowker, 2006; Donaldson & Ronan, 2006). Sports can provide children with a chance to learn important social skills, such as how to work effectively (often in complementary roles) as part of a group. And playing sports allows children to use their emerging cognitive skills as they devise new playing strategies or modify the rules of a game.

When adult coaches encourage their team instead of criticizing mistakes, children are likely to enjoy playing sports.

These benefits of participating in sports are balanced by potential hazards. Several studies have linked youth participation in sports to delinquent and antisocial behavior (e.g., Gardner, Roth & Brooks-Gunn, 2009). However, outcomes are usually positive when sports participation is combined with participation in activities that involve adults, such as school, religious, or youth groups (Linver, Roth, & Brooks-Gunn, 2009; Zarrett et al., 2009). But these potential benefits hinge on the adults who are involved. When adult coaches like the one in the opening vignette encourage their players and emphasize skill development, children usually enjoy playing, often improve their skills, and increase their self-esteem (Coatsworth & Conroy, 2009). In contrast, when coaches emphasize winning over skill development and criticize or punish players for bad plays, children lose interest and stop playing (Bailey & Rasmussen, 1996;

Smith & Smoll, 1996). And when adolescents find sports too stressful, they often get "burned out"—they lose interest and quit (Raedeke & Smith, 2004).

To encourage youth to participate, adults (and parents) need to have realistic expectations for children and coach positively, praising children instead of criticizing them. And they need to remember that children play games for recreation, which means they should have fun!

THINK ABOUT IT

What skills of concrete-operational thinking make it possible for children to participate in organized sports?

Test Yourself

RECALL

1. Boys and girls grow at about the same rate during elementary-school years, but at the end of this period, girls _____.

2. When children skip breakfast, _____.

3. Boys typically have the advantage of gross motor skills that emphasize strength, but girls tend to have the advantage of _____.

4. Children may lose interest in sports and quit playing if coaches _____ and criticize or punish players for mistakes.

INTERPRET

What are the pros and cons of children and adolescents participating in organized sports?

APPLY

Describe how participation in sports illustrates connections between motor, cognitive, and social development.

Recall answers: (1) are more likely to enter puberty and grow rapidly, (2) they often have difficulty paying attention and remembering in school, (3) fine motor skills that emphasize dexterity, (4) emphasize winning over skill development

SUMMARY

6.1 COGNITIVE DEVELOPMENT

What are the distinguishing characteristics of thought during Piaget's concrete-operational and formal-operational stages?

- In progressing to Piaget's stage of concrete operations, children become less egocentric, rarely confuse appearances with reality, and are able to reverse their thinking. They now solve perspective-taking and conservation problems correctly. Thinking at this stage is limited to the concrete and the real.

- With the onset of formal-operational thinking, adolescents can think hypothetically and reason abstractly. In deductive reasoning, they understand that conclusions are based on logic, not on experience.

How do children use strategies and monitoring to improve learning and remembering?

- Rehearsal and other memory strategies are used to transfer information from working memory, a temporary store of information, to long-term memory, a permanent store of knowledge. Children begin to rehearse at about age 7 or 8 and take up other strategies as they grow older.

- Effective use of strategies for learning and remembering begins with an analysis of the goals of a learning task. It also includes monitoring one's performance to determine whether the strategy is working. Collectively, these processes make up an important group of study skills.

6.2 APTITUDES FOR SCHOOL

What is the nature of intelligence?

- Traditional approaches to intelligence include theories that describe intelligence as a general factor as well as theories that include specific factors. Hierarchical theories include both general intelligence and various specific skills, such as verbal and spatial ability.

- Gardner's theory of multiple intelligences proposes nine distinct intelligences. Three are found in psychometric theories (linguistic, logical-mathematical, and spatial intelligence), but six are new (musical, bodily-kinesthetic, interpersonal, intrapersonal, naturalistic, and existential intelligence). Gardner's theory has stimulated research on nontraditional forms of intelligence, such as emotional intelligence. The theory also has implications for education—suggesting, for example, that schools should adjust teaching to each child's unique intellectual strengths.

- According to Robert Sternberg, intelligence is defined as using abilities to achieve short- and long-term goals and depends on three abilities: analytic ability to analyze a problem and generate a solution, creative ability to deal adaptively with novel situations, and practical ability to know what solutions will work.

were intelligence tests first developed? What are their .cures?

◢ Binet created the first intelligence test in order to identify students who would have difficulty in school. Using this work, Terman created the Stanford-Binet in 1916; it remains an important intelligence test. The Stanford-Binet introduced the concept of the intelligence quotient (IQ): MA/CA × 100.

How well do intelligence tests work?

■ Intelligence tests are reasonably valid measures of achievement in school. They also predict people's performance in the workplace.

How do heredity and environment influence intelligence?

■ Evidence for the impact of heredity on IQ comes from the findings that (a) siblings' IQ scores become more alike as siblings become more similar genetically, and (b) adopted children's IQ scores are more like their biological parents' test scores than their adoptive parents' scores. Evidence for the impact of the environment comes from the finding that children who live in responsive, well-organized home environments tend to have higher IQ scores, as do children who participate in intervention programs.

How and why do test scores vary for different racial and ethnic groups?

■ There are substantial differences between ethnic groups in their average scores on IQ tests. This difference is attributed to the greater likelihood of Latino American and African American youth being economically disadvantaged and the fact that tests assess knowledge based on middle-class experiences. Stereotype threat and test-taking skills also contribute to group differences. IQ scores remain valid predictors of school success because middle-class experience is often a prerequisite for school success.

6.3 SPECIAL CHILDREN, SPECIAL NEEDS

What are the characteristics of gifted and creative children?

■ Traditionally, gifted children are those with high scores on IQ tests. Modern definitions of giftedness have been broadened to include exceptional talent in the arts. However defined, giftedness must be nurtured by parents and teachers alike. Contrary to folklore, gifted children usually are socially mature and emotionally stable.

■ Creativity is associated with divergent thinking, in which the aim is to think in novel and unusual directions. Tests of divergent thinking can predict which children are most likely to be creative. Creativity can be fostered by experiences that encourage children to think flexibly and to explore alternatives.

What are the different forms of disability?

■ Individuals with intellectual disability have IQ scores of 70 or lower and deficits in adaptive behavior. Biomedical, social, behavioral, and educational factors place children at risk for intellectual disability. Children with learning

disability have normal intelligence but have difficulty mastering specific academic subjects. The most common is developmental dyslexia, which involves difficulty reading individual words because children haven't mastered language sounds.

What are the distinguishing features of attention-deficit hyperactivity disorder?

■ Children with ADHD are distinguished by being hyperactive, inattentive, and impulsive. According to the Multimodal Treatment Study of Children with ADHD, treating ADHD with medication and psychosocial treatment is effective in the short run but does not "cure" children of the disorder.

6.4 ACADEMIC SKILLS

What are the components of skilled reading?

■ Reading includes a number of component skills. Prereading skills include knowing letters and the sounds associated with them. Word recognition is the process of identifying a word. Beginning readers more often accomplish this by sounding out words; advanced readers more often retrieve a word from long-term memory. Comprehension, the act of extracting meaning from text, improves with age as a result of several factors: working memory capacity increases, readers gain more world knowledge, and readers are better able to monitor what they read and to match their reading strategies to the goals of the reading task.

As children develop, how does their writing improve?

■ As children develop, their writing improves, which reflects several factors: They know more about the world and so have more to say; they use more effective ways of organizing their writing; they master the mechanics of writing (e.g., handwriting, spelling); and they become more skilled at revising their writing.

How do arithmetic skills change during the elementary-school years? How do U.S. students compare to students in other countries?

■ Children first add and subtract by counting, but soon they use more effective strategies such as retrieving addition facts directly from memory. In mathematics, U.S. students lag behind students in most other industrialized countries, chiefly because of cultural differences in time spent on schoolwork and homework and in parents' attitudes toward school, effort, and ability.

What are the hallmarks of effective schools and effective teachers?

■ Schools influence students' achievement in many ways. Students are most likely to achieve when their school emphasizes academic excellence, has a safe and nurturing environment, monitors pupils' and teachers' progress, and encourages parents to be involved.

■ Students achieve at higher levels when their teachers manage classrooms effectively, take responsibility for their students' learning, teach mastery of material, pace material well, value tutoring, and show children how to monitor their own learning.

6.5 PHYSICAL DEVELOPMENT

How much do school-age children grow?

- Elementary-school children grow at a steady pace, but more so in their legs than in the trunk. Boys and girls tend to be about the same size for most of these years, but there are large individual and ethnic differences.

- School-age children need approximately 2,400 calories daily, preferably drawn from each of the basic food groups. Children need to eat breakfast, a meal that should provide approximately one fourth of their calories. Without breakfast, children often have trouble concentrating in school.

How do motor skills develop during the elementary-school years?

- Fine and gross motor skills improve substantially over the elementary-school years, reflecting children's greater

size and strength. Girls tend to excel in fine motor skills that emphasize dexterity as well as in gross motor skills that require flexibility and balance; boys tend to excel in gross motor skills that emphasize strength. Although some of these differences reflect differences in body makeup, they also reflect differing cultural expectations regarding motor skills for boys and girls.

Are American children physically fit?

- Many American schoolchildren don't meet today's standards for being physically fit and childhood obesity is a growing concern.

What are the consequences of participating in sports?

- Many school-age children participate in team sports. Benefits of participation include exercise, enhanced self-esteem, and improved social skills. But participation sometimes leads to antisocial behavior, and when adults are involved they sometimes overemphasize competition, which can turn "play" into "work."

KEY TERMS

mental operations (206)

deductive reasoning (207)

working memory (208)

long-term memory (208)

organization (208)

elaboration (208)

metamemory (209)

metacognitive knowledge (209)

cognitive self-regulation (209)

psychometricians (211)

emotional intelligence (213)

analytic ability (214)

creative ability (214)

practical ability (214)

mental age (MA) (215)

intelligence quotient (IQ) (216)

culture-fair intelligence tests (219)

stereotype threat (220)

convergent thinking (222)

divergent thinking (222)

intellectual disability (223)

learning disability (224)

word recognition (229)

comprehension (229)

phonological awareness (230)

knowledge-telling strategy (232)

knowledge-transforming strategy (232)

LEARN MORE ABOUT IT

Log in to **www.cengagebrain.com** to access the resources your instructor requires. For this book, you can access:

 Psychology CourseMate

- CourseMate brings course concepts to life with interactive learning, study, and exam preparation tools that support the printed textbook. A textbook-specific website, Psychology CourseMate includes an integrated interactive eBook and other interactive learning tools including quizzes, flashcards, videos, and more.

CENGAGENOW™

- CengageNOW Personalized Study is a diagnostic study tool containing valuable text-specific resources—and because you focus on just what you don't know, you learn more in less time to get a better grade.

WebTUTOR™

- More than just an interactive study guide, WebTutor is an anytime, anywhere customized learning solution with an eBook, keeping you connected to your textbook, instructor, and classmates.

Expanding Social Horizons

Socioemotional Development in Middle Childhood

ALTHOUGH YOU'VE NEVER had a course called "Culture 101," your knowledge of your culture is deep. Like all human beings, you have been learning, since birth, to live in your culture. *Teaching children the values, roles, and behaviors of their culture—**socialization**—is a major goal of all peoples.* In most cultures, the task of socialization falls initially to parents. In the first section of this chapter, we see how parents set and try to enforce standards of behavior for their children.

Soon other powerful forces contribute to socialization. In the second section, you'll discover how peers become influential, through both individual friendships and social groups. Next, you'll learn how the media—particularly television—contribute to socialization as well.

As children become socialized, they begin to understand more about other people. We'll examine this growing understanding in the last section of the chapter.

Family Relationships

LEARNING OBJECTIVES

- What is a systems approach to parenting?
- What are the primary dimensions of parenting? How do they affect children's development?
- What determines how siblings get along? How do first-born, later-born, and only children differ?

- How do divorce and remarriage affect children?
- What factors lead children to be maltreated?

socialization
teaching children the values, roles, and behaviors of their culture

Tanya and Sheila, both sixth graders, wanted to go to a Miley Cyrus concert with two boys from their school. When Tanya asked if she could go, her mom said, "No way!" Tanya replied defiantly, "Why not?" Her mother blew up: "Because I say so. That's why. Stop bugging me." Sheila wasn't allowed to go either. When she asked why, her mom said, "I just think that you're still too young to be dating. I don't mind your going to the concert. If you want to go just with Tanya, that would be fine. What do you think of that?"

THE VIGNETTE ILLUSTRATES WHAT WE ALL KNOW WELL FROM PERSONAL EXPERIENCE— PARENTS GO ABOUT CHILD REARING IN MANY DIFFERENT WAYS. We'll study these different approaches in this chapter and learn how Tanya and Sheila are likely to be affected by their mothers' styles of parenting. But first we'll start by considering parents as key components of a broader family system.

The Family as a System

Families are rare in the animal kingdom. Only human beings and a handful of other species form familylike units. Why? Compared to the young in other species, children develop slowly. And because children are immature—unable to care for themselves— for many years, the family structure evolved as a way to protect and nurture young children during their development (Bjorklund, Yunger, & Pellegrini, 2002). Of course, modern families serve many other functions as well—they're economic units, and they provide emotional support—but child rearing remains the most salient and probably the most important family function.

As we think about how families function, it's tempting to believe that parents' actions are all that really matter. That is, through their behavior, parents directly and indirectly determine their children's development. This view of parents as "all powerful" was part of early psychological theories (e.g., Watson, 1925) and is held even today by some first-time parents. But most theorists now view families from a contextual perspective (described in Chapter 1). That is, families form a system of interacting elements: parents and children influence one another (Bronfenbrenner & Morris, 2006; Schermerhorn & Cummings, 2008), and families are part of a much larger system that includes extended family, friends, and teachers as well as institutions that influence development (e.g., schools).

In the systems view, parents still influence their children, both directly (e.g., by encouraging them to study hard) and indirectly (e.g., by being generous and kind to others). However, the influence is no longer exclusively from parents to children but is mutual: Children influence their parents, too. By their behaviors, attitudes, and interests, children affect how their parents behave toward them. When children resist discipline, for example, parents may become less willing to reason and more inclined to use force (Ritchie, 1999).

Even more subtle influences become apparent when families are viewed as systems of interacting elements. For example, fathers' behaviors can affect mother–child relationships; a demanding husband may leave his wife with little time, energy, or interest in helping her daughter with her homework. Or, when siblings argue constantly, parents may become preoccupied with avoiding problems rather than encouraging their children's development.

These many examples show that narrowly focusing on parents' impact on children misses the complexities of family life. But there is even more to the systems view. The family itself is embedded in other social systems, such as neighborhoods and religious institutions (Parke & Buriel, 1998). These other institutions can affect family dynamics. Sometimes they simplify child rearing, as when neighbors are trusted friends and can help care for each others' children. Sometimes, however, they complicate child rearing. Grandparents who live nearby and visit constantly can create friction within the family. At times, the impact of the larger systems is indirect, as when work schedules cause a parent to be away from home or when schools must eliminate programs that benefit children.

■ Figure 7.1 summarizes the numerous interactive influences that exist in a systems view of families. In the remainder of this section, we'll describe parents' influences on children and then how children affect their parents' behavior.

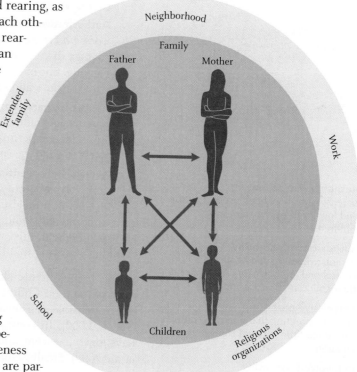

■ **Figure 7.1**

In a systems view of families, parents and children influence each other; this interacting family unit is also influenced by other forces outside of the family.

Copyright © Cengage Learning 2010

Dimensions and Styles of Parenting

Parenting can be described in terms of general dimensions that are like personality traits in that they represent stable aspects of parental behavior—aspects that hold across different situations (Holden & Miller, 1999). When parenting is viewed in this way, two general dimensions of parental behavior emerge. One is the degree of warmth and responsiveness that parents show their children. At one end of the spectrum are parents who are openly warm and affectionate with their children. They are involved with them, respond to their emotional needs, and spend considerable time with them. At the other end of the spectrum are parents who are relatively uninvolved with their children and sometimes even hostile toward them. These parents often seem more focused on their own needs and interests than on those of their children. Warm parents enjoy hearing their children describe the day's activities; uninvolved or hostile parents aren't interested, considering it a waste of their time. Warm parents see when their children are upset and try to comfort them; uninvolved or hostile parents pay little attention to their children's emotional states and invest little effort in comforting them when they're upset. As you might expect, children benefit from warm and responsive parenting (Pettit, Bates, & Dodge, 1997; Zhou et al., 2002).

A second general dimension of parental behavior involves control. Some parents are dictatorial: They try to regulate every facet of their children's lives, like a puppeteer controlling a marionette. At the other extreme are parents who exert little or no control over their children: These children do whatever they want without asking their parents first or worrying about their parents' response. What's best for children is an intermediate amount of control, when parents set reasonable standards for their children's behavior, expect their children to meet those standards, and monitor their children's behavior (i.e., they also usually know where their children are, what they're doing, and with whom). When parents have reasonable expectations for their children and keep tabs on their activities—for example, a mother knows that her 12-year-old is staying after school for choir practice, then going to the library—their children tend to be better adjusted (Kilgore, Snyder, & Lentz, 2000).

Parenting Styles

Combining the dimensions of warmth and control produces four prototypic styles of parenting, as shown in ■ Figure 7.2 (Baumrind, 1975, 1991).

Copyright © Cengage Learning 2010

Figure 7.2

Combining the two dimensions of parental behavior (warmth and control) creates four prototypic styles of parenting.

	Parental control	
Parental involvement	**High**	**Low**
High	Authoritative	Permissive
Low	Authoritarian	Uninvolved

authoritarian parenting

parents who show high levels of control and low levels of warmth toward their children

authoritative parenting

parents who use a moderate amount of control and are warm and responsive to their children

permissive parenting

style of parenting that offers warmth and caring but little parental control over children

uninvolved parenting

style of parenting that provides neither warmth nor control and that minimizes the amount of time parents spend with children

■ **Authoritarian parenting** *combines high control with little warmth.* These parents lay down the rules and expect them to be followed without discussion. Hard work, respect, and obedience are what authoritarian parents wish to cultivate in their children. There is little give-and-take between parent and child because authoritarian parents do not consider children's needs or wishes. This style is illustrated by Tanya's mother in the opening vignette. She feels no obligation whatsoever to explain her decision.

■ **Authoritative parenting** *combines a fair degree of parental control with being warm and responsive to children.* Authoritative parents explain rules and encourage discussion. This style is exemplified by Sheila's mother in the opening vignette. She explained why she did not want the girls going to the concert with the boys and encouraged her daughter to discuss the issue with her.

■ **Permissive parenting** *offers warmth and caring but little parental control.* These parents generally accept their children's behavior and punish them infrequently. An indulgent-permissive parent would readily agree to Tanya's or Sheila's request to go to the concert simply because it is something the child wants to do.

■ **Uninvolved parenting** provides neither warmth nor control. Indifferent-uninvolved parents provide for their children's basic physical and emotional needs but little else. They try to minimize the amount of time spent with their children and avoid becoming emotionally involved with them. If Tanya had parents with this style, she might have simply gone to the concert without asking, knowing that her parents wouldn't care and would rather not be bothered.

Research consistently shows that authoritative parenting is best for most children most of the time. Children with authoritative parents tend to have higher grades and are responsible, self-reliant, and friendly (Amato & Fowler, 2002; Aunola, Stattin, & Nurmi, 2000). In contrast, children with authoritarian parents are often unhappy, have low self-esteem, and frequently are overly aggressive (e.g., Silk et al., 2003; Zhou et al., 2008). Finally, children with permissive parents are often impulsive and have little self-control, whereas children with uninvolved parents often do poorly in school and are aggressive (Aunola et al., 2000; Barber & Olsen, 1997; Driscoll, Russell, & Crockett, 2008). Thus, children typically thrive on a parental style that combines control, warmth, and affection.

VARIATIONS ASSOCIATED WITH CULTURE AND SOCIOECONOMIC STATUS. The general aim of child rearing—helping children become contributing members of their culture—is much the same worldwide (Whiting & Child, 1953), and warmth and control are universal aspects of parents' behavior. But views about the "proper" amount of warmth and the "proper" amount of control vary with particular cultures. European Americans want their children to be happy and self-reliant individuals, and they believe these goals are best achieved when parents are warm and exert moderate control (Goodnow, 1992). In many Asian and Latin American countries, however, individualism is less important than cooperation and collaboration (Okagaki & Sternberg, 1993; Wang, Pomerantz, & Chen, 2007). In China, for example, Confucian principles dictate that emotional restraint is the key to family harmony (Chao, 2001) and, consistent with their cultural values, mothers and fathers in China are less likely to express affection than are mothers and fathers in the United States (Lin & Fu, 1990). In much the same vein, Latino culture typically places greater emphasis on strong family ties and respecting the roles of all family members, particularly adults; these values lead parents to be more protective of their children and to set more rules for them (Halgunseth, Ispa, & Rudy, 2006). Thus, cultural values help specify appropriate ways for parents to interact with their offspring.

Parental styles vary not only across cultures but also within cultures, depending on parents' socioeconomic status. Within the United States, parents of lower socioeconomic status tend to be

Authoritative parents are warm and responsive with children and encourage discussion.

Rob Marmion / Shutterstock.com

more controlling and more punitive—characteristics associated with the authoritarian parenting style—than are parents of higher socioeconomic status (Hoff-Ginsberg & Tardif, 1995). This difference may reflect educational differences that help to define socioeconomic status. Parents of higher socioeconomic status are, by definition, more educated and consequently often see development as a more complex process requiring the more nuanced and child-friendly approach that marks authoritative parenting (Skinner, 1985). Parents who are relatively uneducated often find themselves employed in positions where they're used to taking orders from others; when they're at home, these parents reverse roles and order their children around (Greenberger, O'Neil, & Nagel, 1994).

Another contributing factor derives from another variable that defines socioeconomic status: income (Melby et al., 2008). Because of their limited financial resources, parents of lower socioeconomic status often lead more stressful lives (e.g., they wonder whether they'll have enough money at the end of the month for groceries) and are far more likely to live in neighborhoods where violence, drugs, and crime are commonplace. Thus, parents of lower socioeconomic status may be too stressed to invest the energy needed for authoritative parenting, and the authoritarian approach—with its emphasis on the child's immediate compliance—may actually protect children growing up in dangerous neighborhoods (Parke & Buriel, 1998).

As important as these different dimensions and styles are for understanding parenting, there is more to effective child rearing, as we'll see in the next section.

Parental Behavior

Dimensions and styles are general characterizations of how parents typically behave. If, for example, we describe a parent as warm or controlling, you immediately have a sense of that parent's usual style in dealing with his or her children. Nevertheless, the price for such a broad description is that it tells us little about how parents behave in specific situations and how these parental behaviors influence children's development. Put another way, what specific behaviors can parents use to influence their children? Researchers who study parents name three: direct instruction, modeling, and feedback.

DIRECT INSTRUCTION. Parents often tell their children what to do. But simply playing the role of drill sergeant and ordering children around—"Clean your room!" "Turn off the TV!"—is not very effective. *A better approach is* **direct instruction**, *which involves telling a child what to do, when, and why*. Instead of just shouting, "Share your candy with your brother!" a parent should explain when and why it's important to share with a sibling.

direct instruction
telling a child what to do, when, and why

In addition, just as coaches help athletes master sports skills, parents can help their youngsters master social and emotional skills. Parents can explain links between emotions and behavior: "Catlin is sad because you broke her crayon" (Gottman, Katz, & Hooven, 1996). They can also teach children how to deal with difficult social situations: "When you ask Lindsey if she can sleep over, do it privately so you won't hurt Kaycee's or Hannah's feelings" (Mize & Pettit, 1997). In general, children who get this sort of parental "coaching" tend to be more socially skilled and, not surprisingly, get along better with their peers.

Direct instruction and coaching are particularly powerful when paired with modeling. Urging children to act in a particular way, such as sharing with others, is more compelling when children also see others sharing. In the next section, we'll see how children learn by observing others.

MODELING. Children learn a great deal from parents simply by watching them. The parents' modeling and the youngsters' observational learning leads to imitation, so children's behavior resembles the behavior they observe. *Observational learning can also produce* **counterimitation**, *learning what should not be done*. If an older sibling kicks a friend and parents punish the older sibling, the younger child may learn not to kick others.

counterimitation
learning what should not be done by observing the behavior

Parents can use reinforcement to encourage their children to complete tasks that they don't enjoy, such as household chores

reinforcement
consequence that increases the likelihood that a behavior will be repeated in the future

punishment
applying an aversive stimulus (e.g., a spanking) or removing an attractive stimulus (e.g., TV viewing)

negative reinforcement trap
unwittingly reinforcing a behavior you want to discourage

THINK ABOUT IT

When 10-year-old Dylan's family got a puppy, he agreed to walk it every day after school. But when his mom asks him to do this, he gets angry because he'd rather watch TV. They argue for about 15 minutes, then Dylan's mom gives up and walks the dog herself. And Dylan goes back to watching TV. Analyze this situation. What could Dylan's mom do to prevent these regular arguments?

Observational learning likely contributes to intergenerational continuity of parenting behavior. Parental behavior is often consistent from one generation to the next. When, for example, parents often use harsh physical punishment to discipline their children, these children will, when they are parents, follow suit (Bailey et al., 2009).

FEEDBACK. By giving feedback to their children, parents indicate whether a behavior is appropriate and should continue or is inappropriate and should stop. Feedback comes in two general forms. **Reinforcement** *is any action that increases the likelihood of the response that it follows.* Parents may use praise to reinforce a child's studying or give a reward for completing household chores. **Punishment** *is any action that discourages the recurrence of the response that it follows.* Parents may forbid children to watch television when they get poor grades in school or make children go to bed early for neglecting household chores.

Of course, parents have been rewarding and punishing their children for centuries, so what do psychologists know that parents don't know already? In fact, researchers have made some surprising discoveries concerning the nature of reward and punishment. *Parents often unwittingly reinforce the very behaviors they want to discourage, a situation called the* **negative reinforcement trap** (Patterson, 1980). The negative reinforcement trap occurs in three steps, most often between a mother and her son. In the first step, the mother tells her son to do something he doesn't want to do. She might tell him to clean up his room, to come inside while he's outdoors playing with friends, or to study instead of watching television. In the next step, the son responds with some behavior that most parents find intolerable: He argues, complains, or whines for an extended period of time. In the last step, the mother gives in— saying that the son needn't do as she told him initially—simply to get the son to stop the behavior that is so intolerable. The feedback to the son is that arguing (or complaining or whining) works; the mother rewards that behavior by withdrawing the request that the son did not like.

As for punishment, research shows that it works best when

- administered directly after the undesired behavior occurs, not hours later.
- an undesired behavior *always* leads to punishment, not usually or occasionally.
- accompanied by an explanation of why the child was punished and how punishment can be avoided in the future.
- the child has a warm, affectionate relationship with the person administering the punishment.

At the same time, research reveals some serious drawbacks to punishment. One is that punishment is primarily suppressive: Punished responses are stopped, but only temporarily if children do not learn new behaviors to replace those that were punished. For example, denying TV to brothers who are fighting stops the undesirable behavior, but fighting is likely to recur unless the boys learn new ways of solving their disputes.

A second drawback is that punishment can have undesirable side effects. Children become upset when they are being punished, which means they often miss the feedback that punishment is meant to convey. A child denied TV for misbehaving may become angry over the punishment itself and ignore why he's being punished. What's more, when children are punished physically, this often leads them to behave aggressively and this is true in a range of countries that differ in their general approval of physical punishment (Gershoff et al., 2010). And the impact of harsh punishment is not limited to aggression: it's also associated with a range of negative outcomes including mental health problems, impaired parent–child relationships, and delayed cognitive development (Berlin et al., 2009; Gershoff & Bitensky, 2007). Because physical punishment is so harmful to children many countries around the world (e.g., Costa Rica, the Netherlands, New Zealand, Spain) have banned it altogether (Global Initiative to End All Corporal Punishment of Children, 2011).

One method combines the best features of punishment while avoiding its shortcomings. *In* **time-out,** *a child who misbehaves must briefly sit alone in a quiet, unstimulating location.* Some parents have children sit alone in a bathroom; others have children sit in a corner of a room. Time-out is punishing because it interrupts the child's ongoing activity and isolates the child from other family members, toys, books, and, generally, all forms of rewarding stimulation.

A time-out period usually lasts just a few minutes, which helps parents use the method consistently. During time-out, both parent and child typically calm down. Then, when time-out is over, a parent can talk with the child and explain why the punished behavior is objectionable and what the child should do instead. "Reasoning" like this—even with preschool children—is effective because it emphasizes why a parent punished initially and how punishment can be avoided in the future.

Thus, parents can influence children by direct instruction, by modeling behavior that they value and not modeling what they don't want their children to learn, by giving feedback, and through the parenting styles that we examined in the beginning of this section. In the next section, we'll explore less direct ways that parents influence their children's development.

An effective form of punishment is time-out, in which children sit alone briefly.

time-out
punishment that involves removing children who are misbehaving from a situation to a quiet, unstimulating environment

Influences of the Marital System

When Derek returned from 7-Eleven with a six-pack of beer and chips instead of diapers and baby food, Anita exploded in anger. "How could you! I used the last diaper an hour ago!" Huddled in the corner of the kitchen, their son Randy watched yet another episode in the daily soap opera that featured Derek and Anita.

Although Derek and Anita aren't arguing about Randy—in fact, they're so wrapped up in their conflict that they forget he's in the room—it's hard to conceive that a child would emerge unscathed from such constant parental conflict. Indeed, research shows that chronic parental conflict is harmful for children. When parents are constantly in conflict, children and adolescents often become anxious, withdrawn, and aggressive, and they're more prone to chronic diseases (Miller & Chen, 2010; Rhoades, 2008). Parental conflict affects children's development through three distinct mechanisms. First, seeing parents fight jeopardizes a child's feeling that the family is stable and secure, making a child feel anxious, frightened, and sad (Sturge-Apple et al., 2008). Second, chronic conflict between parents often spills over into the parent–child relationship. A wife who finds herself frequently arguing with and confronting her husband may adopt a similar ineffective style when interacting with her children (Cox, Paley, & Harter, 2001). Third, when parents invest time and energy fighting with each other, they're often too tired or too preoccupied to invest themselves in high-quality parenting (Katz & Woodin, 2002).

Of course, all long-term relationships experience conflict at some point. Does this mean that all children bear at least some scars? Not necessarily. Many parents resolve conflicts in a manner that's constructive instead of destructive. To see this, suppose one parent believes their child should attend a summer camp but the other parent believes it's too expensive and not worth it because the child attended the previous summer. Instead of shouting and name calling (e.g., "You're always so cheap!"), some parents seek mutually acceptable solutions: The child could attend the camp if she earns money to cover part of the cost, or the child could attend a different, less expensive camp. When disagreements are routinely resolved in this way, children actually respond positively to conflict, apparently because it shows that their family is cohesive and able to withstand life's problems (Goeke-Moray et al., 2003).

The extent and resolution of conflict is an obvious way in which the parental system affects children, but it's not the only way. Many mothers and fathers form an effective parental team, working together in a coordinated and complementary fashion toward goals that they share for their child's development. For example, parents may agree that their daughter is smart and athletically skilled and that she should excel in both domains. Consequently, they're quite happy to help her achieve these goals. One parent gives her basketball tips and the other edits her writing.

But not all parents work well together. Sometimes they don't agree on goals: One parent values sports over schoolwork and the other reverses these priorities. Sometimes parents actively compete for their child's attention: One parent may want to take the child shopping but the other wants to take her to a ball game. Finally, parents sometimes act as gatekeepers, limiting one another's participation in parenting. A mother may feel that infant care is solely her turf and not allow the father to participate. Or the father may claim all school-related tasks and discourage the mother from getting involved.

Just as a doubles tennis team won't win many matches if each player ignores the other, parenting is far less effective when each parent tries to "go it alone" instead of working together to achieve shared goals using methods that they both accept. When parents don't work together, when they compete, or when they limit each other's access to their children, problems can result; for example, children can become withdrawn (McHale et al., 2002).

So far we've seen that, to understand parents' impact on children's development, we need to consider the nature of the marital relationship as well as parenting style and specific parenting behaviors (e.g., use of feedback). In addition, Figure 7.1 reminds us that forces outside the family can influence parenting and children's development. To illustrate, let's consider work-related influences. One such influence is a parent's job security: Children and adolescents lose self-esteem and find it difficult to concentrate in school when their parents become unemployed or, for that matter, when they simply worry that their parents may become unemployed (Barling, Zacharatos, & Hepburn, 1999; Kalil & Ziol-Guest, 2005).

Another well-known factor is work-related stress. Not surprisingly, when men and women lead stressful lives at work, they parent less effectively. Sometimes frazzled parents withdraw from family interactions. Over time, this gives the appearance that the parent is detached and uninterested, which makes children anxious and upset. And sometimes work-stressed parents are less accepting and less tolerant, leading to conflicts with their children (Crouter & Bumpus, 2001; Maggi et al., 2008).

Thus, a person's work life can profoundly affect children and adolescents by changing the parenting they experience. For now, another way to view family systems in action is by switching perspectives to see how children affect parenting behavior.

Children's Contributions: Reciprocal Influence

At the beginning of this chapter, we emphasized that the family is a dynamic, interactive system with parents and children influencing each other. In fact, children begin at birth to influence the way their parents treat them. Let's look at two characteristics of children that contribute to this influence.

■ *Age.* Parenting changes as children grow. The same parenting that is marvelously effective with infants and toddlers is inappropriate for adolescents. These age-related changes in parenting are evident in the two basic dimensions of parental behavior, warmth and control. Warmth is beneficial throughout development, because toddlers and teens alike enjoy knowing that others care about them. But the manifestation of parental affection changes, becoming more reserved as children develop. The enthusiastic hugging and kissing that delights toddlers embarrasses adolescents. Parental control also changes as children develop (Shanahan et al., 2007). As children develop cognitively and are better able to make their own decisions, parents gradually relinquish control and expect children to be responsible for themselves. As children enter adolescence, they believe that parents have less authority to make decisions for them, especially in the personal domain (Darling, Cumsille, & Martínez, 2008). In fact, parents gradually relinquish control—though sometimes not as rapidly as adolescents want them to—and increases in decision-making autonomy are associated with greater adolescent well-being (Qin, Pomerantz, & Wang, 2009; Wray-Lake, Crouter, & McHale, 2010).

- *Temperament and behavior.* A child's temperament can have a powerful effect on parental behavior (Brody & Ge, 2001). To illustrate the reciprocal influence of parents and children, imagine two children with different temperaments as they respond to a parent's authoritative style. The first child has an "easy" temperament; she readily complies with parental requests and responds well to family discussions about parental expectations. These parent–child relations are a textbook example of successful authoritative parenting. But suppose the second child has a "difficult temperament" and complies reluctantly or sometimes not at all. Over time, the parent becomes more controlling and less affectionate. The child in turn complies even less in the future, leading the parent to adopt an authoritarian parenting style (Bates et al., 1998; Paulussen-Hoogeboom et al., 2007).

When children respond defiantly to discipline, parents often resort to harsher discipline in the future.

As this example illustrates, parenting behaviors and styles often evolve as a consequence of the child's behavior. With a moderately active young child who is eager to please adults, a parent may discover that a modest amount of control is adequate. But for a very active child who is not as eager to please, a parent may need to be more controlling and directive (Brody & Ge, 2001; Hastings & Rubin, 1999). Influence is reciprocal: Children's behavior helps determine how parents treat them, and the resulting parental behavior influences children's behavior, which in turn causes parents to again change their behavior (Schermerhorn, Chow, & Cummings, 2010).

As time goes by, these reciprocal influences lead many families to adopt routine ways of interacting with each other. Some families end up functioning smoothly: Parents and children cooperate, anticipate each other's needs, and are generally happy. Unfortunately, other families end up troubled: Disagreements are common, parents spend much time trying unsuccessfully to control their defiant children, and everyone is often angry and upset. Still others are characterized by disengagement: Parents withdraw from each other and are not available to their children (Sturge-Apple, Davies, & Cummings, 2010). Over the long term, such troubled families do not fare well, so it's important that these negative reciprocal influences are nipped in the bud (Carrere & Gottman, 1999; Christensen & Heavey, 1999).

Although parent–child relationships are central to human development, other relationships within the family are also influential. For many children, relationships with siblings are very important, as we'll see in the next few pages.

Siblings

For most of a year, all first-born children are only children. Some children remain "onlies" forever but most get brothers and sisters, particularly when first-borns are outgoing and smart (Jokela, 2010). Some first-borns are joined by many siblings in rapid succession; others are joined by a single brother or sister. As the family acquires these new members, parent–child relationships become more complex. Parents can no longer focus on a single child but must adjust to the needs of multiple children. Just as important, siblings influence each other's development.

From the very beginning, sibling relationships are complicated. On the one hand, most expectant parents are excited by the prospect of another child, and their enthusiasm is contagious: Their children, too, eagerly await the arrival of the newest family member. On the other hand, the birth of a sibling is often distressing for older children, who may become withdrawn or return to more childish behavior because of the changes that occur in their lives, particularly the need to share parental attention and affection (Gottlieb & Mendelson, 1990). However, distress can be

Charlie Edward / Shutterstock.com

In many cultures, older siblings regularly provide care for younger siblings.

avoided if parents remain responsive to their older children's needs (Howe & Ross, 1990). In fact, one of the benefits of a sibling's birth is that fathers become more involved with their older children because mothers must devote more time to a newborn (Stewart et al., 1987).

Many older siblings enjoy helping their parents take care of newborns. Older children play with the baby, console it, feed it, or change its diapers. In middle-class Western families, such caregiving often occurs in the context of play, with parents nearby. But in many developing nations, children—particularly girls—play an important role in providing care for their younger siblings (Zukow-Goldring, 2002). As the infant grows, interactions between siblings become more frequent and more complicated. For example, toddlers tend to talk more to parents than to older siblings. But by the time the younger sibling is 4 years old, the situation is reversed: Now young siblings talk more to older siblings than to their mother (Brown & Dunn, 1992). Older siblings become a source of care and comfort for younger siblings when they are distressed or upset (Gass, Jenkins, & Dunn, 2007; Kim et al., 2007). Older siblings also serve as teachers for their younger siblings, teaching them to play games or how to cook simple foods (Maynard, 2002). Finally, when older children do well in school and are popular with peers, younger siblings often follow suit (Brody et al., 2003).

As time goes by, some siblings grow close, becoming best friends in ways that nonsiblings can never be. Other siblings constantly argue, compete, and, overall, simply do not get along with each other. The basic pattern of sibling interaction seems to be established early in development and remains fairly stable (Kramer, 2010). In general, siblings who get along as preschoolers continue to get along as young adolescents, and siblings who quarrel as preschoolers often quarrel as young adolescents (Dunn, Slomkowski, & Beardsall, 1994).

Why are some sibling relationships so filled with love and respect while others are dominated by jealousy and resentment? Put more simply, what factors contribute to the quality of sibling relationships? Biological, psychological, and sociocultural forces all help determine how well siblings get along. Among the biological forces are the child's sex and temperament. Sibling relations are more likely to be warm and harmonious between siblings of the same sex than between siblings of the opposite sex (Dunn & Kendrick, 1981) and when neither sibling is too emotional (Brody, Stoneman, & McCoy, 1994). Age is also important: Sibling relationships generally improve as the younger child approaches adolescence because siblings begin to perceive one another as equals (Buhrmester & Furman, 1990; Kim et al., 2006).

Parents contribute to the quality of sibling relationships, both directly and indirectly (Brody, 1998). The direct influence stems from parents' treatment. Siblings more often get along when they believe that parents have no "favorites" but treat all siblings fairly (McGuire & Shanahan, 2010). When parents lavishly praise one child's accomplishments while ignoring another's, children notice the difference, and their sibling relationship suffers (Updegraff, Thayer, et al., 2005).

This doesn't mean that parents must treat all their children the same. Children understand that parents should treat their kids differently—based on their age or personal needs. Only when differential treatment is not justified do sibling relationships deteriorate (Kowal & Kramer, 1997). In fact, during adolescence, siblings get along better when each has a unique, well-defined relationship with parents (Feinberg et al., 2003).

The indirect influence of parents on sibling relationships stems from the quality of the parents' relationship with each other: A warm, harmonious relationship between parents fosters positive sibling relationships, and conflict between parents is associated with conflict between siblings (Erel, Margolin, & John, 1998; Volling & Belsky, 1992). When parents don't get along, they no longer treat their children comparably, leading to conflict among siblings (Brody et al., 1994).

Many of the features associated with high-quality sibling relationships, such as the sex of the siblings, are common across different ethnic groups. But some unique features also emerge. For example, in a study of African American families, sibling relations were more positive when children had a stronger ethnic identity (McHale et al., 2007). And a study of Mexican American families found that siblings feel closer and spend more time together when siblings have a strong commitment to their family—that is, they felt obligated to their family and viewed it as an important source of support (Updegraff, McHale, et al., 2005).

A biopsychosocial perspective on sibling relationships makes it clear that, in their pursuit of family harmony (what many parents call "peace and quiet"), parents can influence some of the factors affecting sibling relationships but not others. Parents *can* help reduce friction between siblings by being equally affectionate, responsive, and caring to all of their children and by caring for one another. At the same time, parents (and prospective parents!) must realize that some dissension is natural in families, especially those with young boys and girls. Children's different interests lead to conflicts that youngsters cannot resolve because their social skills are limited.

Adopted Children

The U.S. government doesn't keep official statistics on the number of adopted children, but the best estimate is that 2 to 4% of U.S. children are adopted. The majority of adoptive parents are middle-class European Americans and, until the 1960s, so were most adopted children. However, improved birth control and legalized abortion meant that few European American infants were available for adoption. Consequently, parents began to adopt children from other races and other countries. Also, adoption of children with special needs, such as those with chronic medical problems or exposure to maltreatment, became more common (Brodzinsky & Pinderhughes, 2002; Gunnar, Bruce, & Grotevant, 2000).

With these increases in adoptions came reports that adopted children are more prone to behavioral problems, substance use, and criminal activity. In fact, when compared to children living with biological parents, adopted children are quite similar in terms of temperament, mother–infant attachment, self-esteem, and cognitive development (Brodzinsky & Pinderhughes, 2002; Juffer & van IJzendoorn, 2007). However, adopted children *are* more prone to having problems adjusting to school and to conduct disorders, such as being overly aggressive (Miller et al., 2000). To a certain extent, this finding reflects that adoptive parents are more likely to seek help for their adoptive children because they are more affluent and can afford it (Hellerstedt et al., 2008). Also, the extent of these problems hinges on the age when the child was adopted and the quality of care prior to adoption (Brodzinsky & Pinderhughes, 2002; Lee et al., 2010). Problems are much more common when children are adopted at an older age (and thus probably are separated from an attachment figure) and when their care before adoption was poor (e.g., they were institutionalized or lived in a series of foster homes).

Perhaps the best way to summarize this research is that adoption per se is not a fundamental developmental challenge for most children. Quality of life before adoption certainly places some adopted children at risk, but most adopted children fare quite well.

Impact of Birth Order

First-born children are often "guinea pigs" for most parents, who have lots of enthusiasm but little practical experience rearing children. Parents typically have high expectations for their first-borns and are both more affectionate and more punitive toward them. As more children arrive, most parents become more adept at their roles, having learned "the tricks of the trade" from earlier children. With later-born children, parents have more realistic expectations and are more relaxed in their discipline (Baskett, 1985).

The different approaches that parents use with their first- and later-born children help explain differences that are commonly observed between these children. First-

THINK ABOUT IT

Calvin, age 8, and his younger sister, Hope, argue over just about everything and constantly compete for their parents' attention. Teenage sisters Melissa and Caroline love doing everything together and enjoy sharing clothes and secrets about their teen romances. Why might Calvin and Hope get along so poorly while Melissa and Caroline get along so well?

© Alex Gore/Alamy

Beginning in the 1960s, many European American parents adopted children from other racial and ethnic groups.

Contrary to folklore, only children are often smarter and more mature.

born children generally have higher scores on intelligence tests and are more likely to go to college. They are also more willing to conform to parents' and adults' requests. In contrast, perhaps because later-born children are less concerned about pleasing parents and adults, they are more popular with their peers and are more innovative (Beck, Burnet, & Vosper, 2006; Bjerkedal et al., 2007).

What about only children? According to conventional wisdom, parents dote on onlies, who therefore become selfish and egotistical. Is the folklore correct? In a comprehensive analysis of more than 100 studies, only children were not worse off than other children on any measure. In fact, only children were found to succeed more in school and to have higher levels of intelligence, leadership, autonomy, and maturity (Falbo & Polit, 1986).

The possibility that only children are selfish has been a concern in China, where only children are common because of governmental efforts to limit population growth. However, comparisons in China between only children and children with siblings often find no differences; when differences are found, the advantage usually goes to the only child (Jiao, Ji, & Jing, 1996; Liu, Lin, & Chen, 2010). Thus, contrary to the popular stereotype, only children are not "spoiled brats" (nor, in China, are they "little emperors" who boss around parents, peers, and teachers). Instead, only children are, for the most part, much like children who grow up with siblings.

Whether U.S. children grow up with siblings or as onlies, they are more likely than children in other countries to have their family relationships disrupted by divorce. What is the impact of divorce on children and adolescents?

Divorce and Remarriage

Today, many North American children experience their parents' divorce. According to all theories of child development, divorce is distressing for children because it involves conflict between parents and, usually, separation from one of them. (Of course, divorce is also distressing to parents, as we describe in Chapter 11.) Do the disruptions, conflict, and stress associated with divorce affect children? Of course they do. Having answered this easy question, however, many more difficult questions remain: Are all aspects of children's lives affected equally by divorce? How does divorce influence development? Why is divorce more stressful for some children than for others?

What Aspects of Children's Lives Are Affected by Divorce?

Hundreds of studies of divorce have been conducted that involve tens of thousands of preschool- through college-age children. Comprehensive meta-analyses of this research reveal that in school achievement, conduct, adjustment, self-concept, and parent–child relations, children whose parents had divorced fared poorly compared to children from intact families (Amato, 2001; Amato & Keith, 1991; Lansford, 2009). However, the effects of divorce dropped from the 1970s to the 1980s, perhaps because as divorce became more frequent in the 1980s it became more familiar and less frightening. The effects of divorce increased again in the 1990s, perhaps reflecting a widening gap in income between single- and two-parent families (Amato, 2001).

When children of divorced parents become adults, the effects of divorce persist. As adults, children of divorce are more likely to experience conflict in their own marriages, to have negative attitudes toward marriage, and to become divorced themselves. Also, they report less satisfaction with life and are more likely to become depressed (Hetherington & Kelly, 2002; Segrin, Taylor, & Altman, 2005). These findings don't mean that children of divorce are destined to have unhappy, conflict-ridden marriages that inevitably lead to divorce, but children of divorce are at greater risk for such an outcome.

The first year following a divorce is often rocky for parents and children alike. But beginning in the second year, most children begin to adjust to their new circumstances (Hetherington & Kelly, 2002). Children adjust to divorce more readily if their divorced parents cooperate with each other, especially on disciplinary matters (Buchanan & Heiges, 2001). *In* **joint custody**, *both parents retain legal custody of the children.* Children benefit from joint custody if their parents get along (Bauserman, 2002).

Of course, many parents do not get along after a divorce, which eliminates joint custody as an option. Traditionally, mothers have been awarded custody; when this happens, children benefit when fathers remain involved in parenting (Fabricius & Luecken, 2007). In recent years fathers have increasingly often been given custody, especially of sons. This practice coincides with findings that children often adjust better when they live with same-sex parents: Boys often fare better with fathers and girls fare better with mothers (McLanahan, 1999). One reason boys are often better off with their fathers is that boys are likely to become involved in negative reinforcement traps (described on page 252) with their mothers. Another explanation is that both boys and girls may forge stronger emotional relationships with same-sex parents than with opposite-sex parents (Zimiles & Lee, 1991).

joint custody
when both parents retain legal custody of their children following divorce

How Does Divorce Influence Development?

Divorce usually results in several changes to family life that affect children (Amato & Keith, 1991). First, the absence of one parent means that children lose a role model, a source of parental help and emotional support, and a supervisor. For instance, a single parent may have to choose between helping one child complete an important paper or watching another child perform in a school play. She can't do both, and one child will miss out.

Second, single-parent families experience economic hardship, which creates stress and often means that activities once taken for granted are no longer affordable (Lansford, 2009). A family may no longer be able to pay for books for pleasure reading, music lessons, or other activities that promote child development. Moreover, when a single parent worries about having enough money for food and rent, she has less energy and effort to devote to parenting.

Third, conflict between parents is extremely distressing to children and adolescents (Leon, 2003), particularly for children who are emotionally insecure (Davies & Cummings, 1998). In fact, many of the problems ascribed to divorce are actually caused by marital conflict occurring before the divorce (Erel & Burman, 1995; Shaw, Winslow, & Flanagan, 1999). Children whose parents are married but fight constantly often show many of the same effects associated with divorce (Katz & Woodin, 2002).

WHICH CHILDREN ARE MOST AFFECTED BY DIVORCE? Why are some children more affected by divorce than others? Amato and Keith's (1991) analysis, for example, showed that—although the overall impact of divorce is the same for boys and girls— divorce is more harmful when it occurs during childhood and adolescence than during the preschool or college years. Also, children who are temperamentally more emotional tend to be more affected by divorce (Lengua et al., 1999).

Some children suffer more from divorce because of their tendency to interpret events negatively. Suppose, for example, that a father forgets to take a child on a promised outing. One child might believe that an emergency prevented the father from taking the child. A second child might believe that the father hadn't really wanted to spend time with the child in the first place and will never make similar plans again. Children like the second child—who tend to interpret life events negatively—are more likely to have behavioral problems following divorce (Mazur et al., 1999).

Finally, when children actively cope with problems brought on by divorce, either by trying to solve them or by trying to make them feel less threatening, they gain confidence in their ability to control future events in their lives. This confidence acts as a buffer against anxiety or depression, which can be triggered when children feel that problems brought on by divorce are insurmountable (Sandler et al., 2000).

Just as children can reduce the harm of divorce by being active problem solvers, parents can reduce divorce-related stress and help children adjust to their new life circumstances. Parents should explain together to children why they are divorcing and what their children can expect to happen to them. They should reassure children that they will always love them and always be their parents; parents must back up these words with actions by remaining involved in their children's lives despite the increased difficulty of doing so. Finally, parents must expect that their children will sometimes be angry or sad about the divorce, and they should encourage children to discuss these feelings with them.

To help children deal with divorce, parents should *not* compete with each other for their children's love and attention; children adjust to divorce best when they maintain good relationships with both parents. Parents should neither take out their anger with each other on their children nor criticize their ex-spouse in front of the children. Finally, parents should not ask children to mediate disputes; parents should work out problems without putting children in the middle.

Following all these rules all the time is not easy. After all, divorce is stressful and painful for adults, too. Fortunately, there are effective programs available that can help parents and children adjust to life following divorce. The Spotlight on Research feature describes one such program.

Spotlight on research — Evaluation of a Program to Help Parents and Children Adjust to Life After Divorce

Who were the investigators, and what was the aim of the study? We've seen that divorce puts children at risk for reduced school achievement, behavior problems, and other less desirable outcomes. Clorinda Vélez, Sharlene Wolchik, Jenn-Yun Tein, and Irwin Sandler (2011) wanted to determine the benefits for children of an intervention program for mothers that focused primarily on the quality of the mother–child relationship and effective disciplinary methods.

How did the investigators measure the topic of interest? Vélez and her colleagues assigned mothers to one of two conditions: in the intervention condition, mothers participated in five group sessions that discussed ways that a mother can foster quality relationships with her children and three sessions that were devoted to discipline. In the control condition, mothers were simply provided books that described how to adjust to divorce and a reading guide. Before mothers were assigned to conditions and at four points later, mothers and children completed several questionnaires designed to measure parenting quality (defined here as being warm and communicating effectively). In addition, children completed questionnaires designed to measure whether they were coping effectively with divorce-related adjustment (e.g., being proactive in making changes, being optimistic).

Who were the participants in the study? The study included 240 mothers who had been

divorced within the previous two years and who had at least one child between 9 and 12 years of age. The mothers had not remarried and had no plans for doing so in the near future.

What was the design of the study? This study was experimental because Vélez and her colleagues assigned mothers randomly to either an intervention condition or a control condition. The study was longitudinal because mothers and children were tested five times: prior to the experimental treatment, immediately after the treatment, and at 3-month, 6-month, and 6-year intervals after the treatment.

Were there ethical concerns with the study? No; the questionnaires that parents and children completed were ones commonly used to study parent–child relationships and family interactions.

What were the results? Correlations were computed between experimental conditions, relationship quality, and children's active coping. The results revealed that parent–child relationships (as reported by both mothers and children) were higher quality when mothers participated in the intervention condition and that higher quality relationships were associated with more active coping on the part of children. In other words, the intervention condition improved mother–child relationships and this improvement, in turn, resulted in children's use of more active coping to deal with their problems.

What did the investigators conclude? Vélez and her colleagues (2011) concluded that "... by increasing one of children's most important interpersonal resources, mother–child relationship quality, the [intervention program] improved youth's coping efficacy and active coping" (p. 255). In other words, when children have a high-quality relationship with their mother— she is warm to them and communicates well with them—they are empowered to deal with the unique challenges they face as they adjust to life after their parents' divorce.

What converging evidence would strengthen these conclusions? There are two limits to these findings. First, the children were all in middle childhood; it is unclear if intervention would be equally effective with preschool children or with adolescents. Second, mothers and children were mainly middle-class. Would intervention work as well with divorced women living in poverty, who face additional stresses and obstacles to effective parenting? Answering these questions would provide more convincing evidence of the effectiveness of intervention programs designed to help children and mothers adjust to life following divorce.

 Go to Psychology CourseMate at www.cengagebrain.com to enhance your understanding of this research.

Blended Families

Following divorce, most children live in a single-parent household for about 5 years. However, more than two thirds of men and women eventually remarry (Sweeney, 2010). *The resulting unit, consisting of a biological parent, stepparent, and children, is known as a* **blended family**. (Other terms for this family configuration are "remarried family" and "reconstituted family.")

Because mothers are more often granted custody of children, the most common form of blended family is a mother, her children, and a stepfather. Most stepfathers do not participate actively in child rearing and often seem reluctant to become involved (Clarke-Stewart & Brentano, 2005). Nevertheless, boys typically benefit from the presence of a stepfather, particularly when he is warm and involved. Preadolescent girls, however, do not adjust readily to their mother's remarriage, apparently because it disrupts the intimate relationship they have established with her (Visher, Visher, & Pasley, 2003).

These adjustments are more difficult when mothers of adolescents remarry. Adolescents do not adapt to the new family circumstances as easily as children do; they're more likely to challenge a stepfather's authority. And adjustment is more difficult when a stepfather brings his own biological children. In such families, parents sometimes favor their biological children over their stepchildren—they're more involved with and warmer toward their biological children. Such preferential treatment almost always leads to conflict and unhappiness (Dunn & Davies, 2001; Sweeney, 2010). Similarly, when the mother and stepfather argue, children usually side with their biological parent (Dunn, O'Connor, & Cheng, 2005).

The best strategy for stepfathers is to be interested in their new stepchildren but to avoid encroaching on established relationships. Newly remarried mothers must be careful that their enthusiasm for their new spouse does not come at the expense of time and affection for their children. Both parents and children need to have realistic expectations. The blended family can be successful, but it takes effort because of the complicated relationships, conflicting loyalties, and jealousies that usually exist (Sweeney, 2010; White & Gilbreth, 2001).

Over time, children adjust to the blended family. If the marriage is happy, most children profit from the presence of two caring adults. Nevertheless, when compared to children from intact families, children in blended families do less well in school and experience more symptoms of depression (Halpern-Meekin & Tach, 2008). Unfortunately, second marriages are slightly more likely than first marriages to end in divorce, particularly when stepchildren are involved (Teachman, 2008). This means that many children relive the trauma of divorce. Fortunately, programs are available to help members of blended families adjust to their new roles (Bullard et al., 2010). Such programs emphasize effective co-parenting (described on pages 253–254) and, in particular, ways of dealing with behavior problems that children often display with stepparents. These programs result in fewer behavior problems and greater marital satisfaction.

blended family
family consisting of a biological parent, a stepparent, and children

Following divorce, most men and women remarry, creating a blended family.

Parent–Child Relationships Gone Awry: Child Maltreatment

The first time that 7-year-old Max came to school with bruises on his face, he explained to his teacher that he had fallen down the basement steps. When Max had similar bruises a few weeks later, his teacher spoke with the school principal, who contacted local authorities. It turned out that Max's mother thrashed him with a paddle for even minor misconduct; for serious transgressions, she beat Max and made him sleep alone in a dark, unheated basement.

Unfortunately, cases like Max's occur far too often in modern America. Maltreatment comes in many forms (Cicchetti & Toth, 2006). The two that often first come to mind are physical abuse involving assault that leads to injuries, and sexual abuse involving fondling, intercourse, or other sexual behaviors. Another form of maltreatment is neglect, not giving children adequate food, clothing, or medical care. And children can also be harmed by psychological abuse—ridicule, rejection, and humiliation (Wicks-Nelson & Israel, 2006).

The frequency of these various forms of child maltreatment is difficult to estimate because so many cases go unreported. According to the U.S. Department of Health and Human Services (2010), nearly three-quarters of a million children annually suffer maltreatment or neglect. About 75% are neglected, about 15% are abused physically, about 10% are abused sexually, and 5% are maltreated psychologically.

Who Are the Abusing Parents?

Why would a parent abuse a child? You might think parents would have to be severely disturbed to harm their own flesh and blood, but most abusing parents are not suffering from any specific mental or psychological disorder (Wolfe, 1985). Instead, a host of factors put some children at risk for abuse and protect others; the number and combination of factors determine if the child is a likely target for abuse (Cicchetti & Toth, 2006). Let's look at three of the most important factors: those associated with the cultural context, those associated with parents, and those associated with children themselves.

The most general category of contributing factors are those dealing with cultural values and the social conditions in which parents rear their children. Many countries in Europe and Asia have strong cultural prohibitions against physical punishment. It simply isn't done and would be viewed in much the same way we would view an American parent who punished by not feeding the child for a few days. In Sweden, for example, spanking is against the law, and children can report their parents to the police. But spanking is common in the United States. Countries that do not condone physical punishment tend to have lower rates of child maltreatment than the United States.

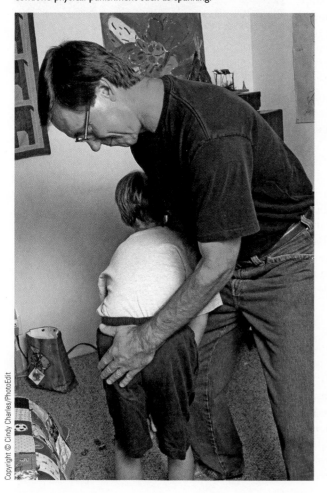

Child abuse is more common in societies that condone physical punishment such as spanking.

What social conditions seem to foster maltreatment? Poverty is one. Maltreatment is more common among children living in poverty, in part because lack of money increases the stress of daily life (Duncan & Brooks-Gunn, 2000). When parents are worrying about whether they can buy groceries or pay rent, they are more likely to punish their children physically instead of making the extra effort to reason with them. Similarly, abuse is more common among military families when a soldier is deployed in a combat zone (Gibbs et al., 2007). In this case, maltreatment may be rooted in stress stemming from concern over the absent parent and temporary single parenthood.

Social isolation is a second force. Abuse is more likely when families are socially isolated from other relatives or neighbors. When a family lives in relative isolation, it deprives children of adults who could protect them and deprives parents of social support that would help them better deal with life's stresses (Coulton et al., 2007).

Cultural factors clearly contribute to child abuse, but they are only part of the puzzle. Although maltreatment is more common among families living in poverty, it does not occur in a majority of these families and it does occur in middle- and upper-class families, too. Consequently, we need to look for additional factors to explain why abuse occurs in some families but not in others.

Child development researchers have identified several important factors (Berlin, Appleyard, & Dodge, 2011; Bugental & Happaney,

2004). First, parents who maltreat their children often were maltreated themselves, which may lead them to believe that abuse is simply a normal part of childhood. This does not mean that abused children inevitably become abusing parents—only about one third do. But a history of child abuse clearly places adults at risk for mistreating their own children (Cicchetti & Toth, 2006; Serbin & Karp, 2003). Second, parents who mistreat their children often use ineffective parenting techniques (e.g., inconsistent discipline), have such unrealistic expectations that their children can never meet them, and often believe that they are powerless to control their children. For example, when abusive parents do not get along with their children, they often chalk this up to factors out of their control, such as children having a difficult temperament or being tired that day; they're less likely to think that their own behavior contributed to unpleasant interactions. Third, in families where abuse occurs, the couple's interactions are often unpredictable, unsupportive, and unsatisfying for both husbands and wives. In other words, mistreatment of children is simply one symptom of family dysfunction. This marital discord makes life more stressful and makes it more difficult for parents to invest effort in child rearing.

To place the last few pieces in the puzzle, we must look at the abused children themselves. Our earlier discussion of the reciprocal influence between parents and children should remind you that children may inadvertently, through their behavior, contribute to their own abuse. In fact, infants and preschoolers are more often abused than older children, probably because they are less able to regulate aversive behaviors that may elicit abuse (Sidebotham et al., 2003). You've probably heard stories about a parent who shakes a baby to death because the baby wouldn't stop crying. Because younger children are more likely to cry or whine excessively—behaviors that irritate all parents sooner or later—they are more likely to be the targets of abuse.

For much the same reason, children who are frequently ill are more often abused. When children are sick, they're more likely to cry and whine, annoying parents. Also, when children are sick, they need medical care (which means additional expense) and can't go to school (which often means that working parents must arrange alternative child care). Because sick children increase the level of stress in a family, they can inadvertently become the targets of abuse. By behaving immaturely or being ill, children unintentionally place themselves at risk for maltreatment (Rogosch et al., 1995).

Stepchildren form another group at risk for abuse (Daly & Wilson, 1996). Just as Cinderella's stepmother doted on her biological children but abused Cinderella, so stepchildren are more prone to abuse and neglect than biological children. Adults are less invested emotionally in their stepchildren, and this lack of emotional investment leaves stepchildren more vulnerable.

Thus cultural, parental, and child factors all contribute to child maltreatment. Any single factor will usually not result in abuse. For instance, a sick infant who cries constantly would not be maltreated in countries where physical punishment is not tolerated. Maltreatment becomes a possibility only when cultures condone physical punishment, parents lack effective skills for dealing with children, and a child's behavior is frequently aversive.

Effects of Abuse on Children

You probably aren't surprised to learn that the prognosis for youngsters like Max is not very good. Some, of course, suffer permanent physical damage. Even when there is no lasting physical damage, children's social and emotional development is often disrupted. They tend to have poor relationships with peers, often because they are too aggressive (Appleyard, Yang, & Runyan, 2010; Cullerton-Sen et al., 2008). Their cognitive development and academic performance are also disturbed. Abused youngsters typically get lower grades in school, score lower on standardized achievement tests, and are more frequently retained in a grade rather than promoted. Also, school-related behavior problems (e.g., being disruptive in class) are common, in part because maltreated children are often socially unskilled and they don't regulate their emotions well (Burack et al., 2006; Maughan & Cicchetti, 2002). Abuse often leads children and

THINK ABOUT IT

Kevin has never physically abused his 10-year-old son, Alex, but he constantly torments him emotionally. For example, when Alex got an F on a spelling test, Kevin screamed, "I skipped Monday Night Football just to help you but you still flunked. You're such a dummy." When Alex began to cry, Kevin taunted, "Look at Alex, crying like a baby." These interactions occur nearly every day. What are the likely effects of such repeated episodes of emotional abuse?

adolescents to become depressed (Appleyard et al., 2010; Harkness, Lumley, & Truss, 2008). Finally, adults who were abused as children are more prone to think about or attempt suicide and are also more likely to abuse spouses and their own children (Malinosky-Rummell & Hansen, 1993). In short, when children are maltreated, the effects are usually widespread and long-lasting.

RESILIENCE. Although the overall picture is bleak, some children are remarkably resilient to the impact of abuse. One factor that protects children is their **ego resilience**, *which denotes children's ability to respond adaptively and resourcefully to new situations.* For example, Flores, Cicchetti, and Rogosch (2005) found that, among a sample of Latino children who had been maltreated, the effects of abuse were smallest for children who were rated by observers as being high in ego resilience—they were flexible in responding to novel and challenging social situations. Another preventive factor is a positive mother–child relationship: When children have a positive representation of their mother—they describe her as "kind" and "loving," for example—they suffer relatively few symptoms of maltreatment (Valentino et al., 2008). However, the buffering value of such a positive view only holds for children who have been neglected. Children who had been abused physically suffer the typical maltreatment-related symptoms even when they have a positive representation of their mother.

Preventing Abuse and Maltreatment

The complexity of child abuse dashes any hopes for a simple solution (Kelly, 2011). Because maltreatment is more apt to occur when several contributing factors are present, eradicating child maltreatment requires many different approaches.

American attitudes toward "acceptable" levels of punishment and poverty would have to change. American children will be abused as long as physical punishment is considered acceptable and effective and as long as poverty-stricken families live in chronic stress from simply trying to provide food and shelter. Parents also need counseling and training in parenting skills. Abuse will continue as long as parents remain ignorant of effective methods of parenting and discipline.

It would be naive to expect all of these changes to occur overnight. However, by focusing on some of the more manageable factors, the risk of maltreatment can be reduced. Social supports help. When parents know they can turn to other helpful adults for advice and reassurance, they better manage the stresses of child rearing that might otherwise lead to abuse. Families can also be taught more effective ways of coping with situations that might otherwise trigger abuse (Wicks-Nelson & Israel, 2006). Through role-playing sessions, parents can learn the benefits of authoritative parenting and effective ways of using feedback and modeling to regulate children's behavior.

Providing social supports and teaching effective parenting are typically done when maltreatment and abuse have already occurred. Of course, preventing maltreatment in the first place is more desirable and more cost-effective. For prevention, one useful tool is familiar: early childhood intervention programs. That is, maltreatment and abuse can be cut in half when families participate for 2 or more years in intervention programs that include preschool education along with family support activities aimed at encouraging parents to become more involved in their children's education (Reynolds & Robertson, 2003). When parents participate in these programs, they become more committed to their children's education. This leads their children to be more successful in school, reducing a source of stress and enhancing parents' confidence in their child-rearing skills, thereby reducing the risks of maltreatment.

Another successful approach focuses specifically on parenting skills in families where children are at risk for maltreatment. In one program (Bugental & Schwartz, 2009), mothers of infants at risk for abuse (due to medical problems at birth) participated in an extensive training program in which they learned to identify likely

ego resilience
a person's ability to respond adaptively and resourcefully to new situations

causes of problems associated with recurring problems encountered caring for their babies (e.g., problems associated with feeding, sleeping, crying). Then they were given help in devising methods to deal with those problems and in monitoring the effectiveness of the methods. When mothers participated in the program, they were less likely to use harsh punishment (a known risk factor for child maltreatment) and their children were less likely to suffer injuries at home (a common measure of parental neglect).

There are also effective programs targeted for parents of older children who are at risk for maltreatment. One program, Parent-Child Interaction Therapy, focuses on helping parents improve in the following two sets of skills: (1) building warm and positive relationships with their children, and (2) developing reasonable expectations for their children and using more effective disciplinary practices. When parents of at-risk children participate in this program, they report less stress, their behavior with their children becomes more positive (more praise and fewer commands), and, critically, suspected abuse is reduced (Thomas & Zimmer-Gimbeck, 2011).

Finally, we need to remember that most parents who have mistreated their children need our help. Although we must not tolerate child maltreatment, most of these parents and children are attached to each other; maltreatment is typically a consequence of ignorance and burden, not malice.

Test Yourself

RECALL

1. According to the systems approach, the family consists of interacting elements that influence each other, and the family itself is _____.

2. A(n) _____ parental style combines high control with low involvement.

3. Most children seem to benefit when parents rely on a(n) _____ style.

4. Parental behaviors that influence children include direct instruction, modeling (learning through observation), and _____.

5. Some parents do not make a good team: they don't work together, they compete with each other for their children's attention, and _____.

6. With later-born children, parents often have more realistic expectations and are _____.

7. Among the effects of divorce on children are inadequate supervision of children, conflict between parents, and _____.

8. When mothers remarry, daughters do not adjust as readily as sons because _____.

9. Children are more likely to be abused when they are younger and when they are _____.

INTERPRET

How can child abuse be explained in terms of the biological, psychological, and sociocultural forces in the biopsychosocial framework?

APPLY

Imagine a family in which Mom and Dad both work full-time outside the home. Mom's employer wants her to take a new position in a distant small town. Mom is tempted because the position represents a promotion with much more responsibility and much higher pay. However, because the town is so small, Dad couldn't get a job comparable to the one he has now, which he loves. Based on the family systems theory shown in Figure 7.1, how might the move affect the couple's 10-year-old daughter and 4-year-old son?

Recall answers: (1) embedded in other social systems, such as neighborhoods, (2) authoritarian, (3) authoritative, (4) feedback (reward and punishment), (5) they limit each other's access to the child, (6) more relaxed in their discipline, (7) economic hardship, (8) the remarriage disrupts an intimate mother–daughter relationship, (9) often ill

LEARNING OBJECTIVES

- What are the benefits of friendship?

- What are the important features of groups of children and adolescents? How do these groups influence individuals?

- Why are some children more popular than others? What are the causes and consequences of being rejected?

- What are some effects of childhood aggression? Why are some children chronic victims of aggression?

Only 36 hours had passed since the campers arrived at Crab Orchard Summer Camp. Nevertheless, groups had already formed spontaneously based on the campers' main interests: arts and crafts, hiking, and swimming. Within each group, leaders and followers had already emerged. This happens every year, but the staff is always astonished at how quickly a "social network" emerges at camp.

THE GROUPS THAT FORM AT SUMMER CAMPS—as well as in schools and neighborhoods—represent one of the more complex forms of peer relationships: Many children are involved, and there are multiple relationships. We'll examine these kinds of interactions later in this section. Let's start by looking at a simpler social relationship, friendship.

Friendships

friendship
voluntary relationship between two people involving mutual liking

Over time, children develop special relationships with certain peers. **Friendship** *is a voluntary relationship between two people involving mutual liking.* By age 4 or 5, most children claim to have a "best friend." If you ask them how they can tell a child is their best friend, their response will probably resemble 5-year-old Katelyn's:

INTERVIEWER:	Why is Heidi your best friend?
KATELYN:	Because she plays with me. And she's nice to me.
INTERVIEWER:	Are there any other reasons?
KATELYN:	Yeah, Heidi lets me play with her dolls.

Thus, the key elements of friendship for younger children are that children like each other and enjoy playing together.

As children develop, their friendships become more complex. For older elementary-school children (ages 8 to 11), mutual liking and shared activities are joined by features that are more psychological in nature: trust and assistance. At this age, children expect that they can depend on their friends—their friends will be nice to them, will keep their promises, and won't say mean things about them to others. Children also expect friends to step forward in times of need: A friend should willingly help with homework or willingly share a snack.

Adolescence adds another layer of complexity to friendships. Mutual liking, common interests, and trust remain. In fact, trust becomes even more important in adolescent friendships. New to adolescence is intimacy—friends now confide in one another, sharing personal thoughts and feelings. Teenagers will reveal their excitement over a new romance or disappointment at not being cast in a school musical. Intimacy is more common in friendships among girls, who are more likely than boys to have one exclusive "best friend" (Markovits, Benenson, & Dolenszky, 2001). Because intimacy is at the core of their friendships, girls are also more likely to be concerned about the faithfulness of their friends and worry about being rejected (Benenson & Christakos, 2003; Poulin & Chan, 2010).

The emergence of intimacy in adolescent friendships means that friends also come to be seen as sources of social and emotional support. Elementary-school children generally rely on close family members—parents, siblings, and grandparents—

as primary sources of support when they need help or are bothered by something. But adolescents turn to close friends instead. Because adolescent friends share intimate thoughts and feelings, they can provide support during emotional or stressful periods (del Valle, Bravo, & Lopez, 2010; Levitt, Guacci-Franco, & Levitt, 1993).

Hand in hand with the emphasis on intimacy is loyalty. Having confided in friends, adolescents expect friends to stick with them through good and bad times. If a friend is disloyal, adolescents are afraid that they may be humiliated because their intimate thoughts and feelings will become known to a much broader circle of people (Berndt & Perry, 1990).

Who Are Friends?

Most friends are alike in age, gender, and race (Hamm, 2000; Mehta & Strough, 2009). Because friends are supposed to treat each other as equals, friendships are rare between an older, more experienced child and a younger, less experienced child. Because children typically play with same-sex peers, boys and girls rarely become friends.

Friendships are more common between children and adolescents from the same race or ethnic group than between those from different groups, reflecting racial segregation in American society. Friendships among children of different groups are more common in schools when a child's school and neighborhood are ethnically diverse (Quillian & Campbell, 2003). Although cross-group friendships are uncommon, they are valuable: Children from majority groups typically form more positive attitudes toward a minority group following a friendship with a youth from that group (Feddes, Noack, & Rutland, 2009).

Friends tend to be alike in age, race, and sex.

Of course, friends are usually alike not only in age, sex, and race, but also in attitudes toward school, recreation, drug use, and plans for the future (Hamm, 2000; Newcomb & Bagwell, 1995). Children and adolescents befriend others who are similar to themselves and, as time passes, friends become more similar in their attitudes and values (Popp et al., 2008; Van Zalk et al., 2010). Nevertheless, friends are not photocopies of each other; friends are less similar, for example, than spouses or dizygotic twins (Rushton & Bons, 2005).

Although children's friendships are overwhelmingly with members of their own sex, a few children have friendships with opposite-sex children. Who are these children, and why do they have opposite-sex friendships? Boys and girls are equally likely to have opposite-sex friendships. The important factor in understanding these children is whether they have same- *and* opposite-sex friends or *only* opposite-sex friends. Children with same- and opposite-sex friendships tend to be very well adjusted, whereas children with only opposite-sex friendships tend to be unpopular, to be less competent academically and socially, and to have lower self-esteem. Apparently, children with both same- and opposite-sex friends are so socially skilled and popular that both boys and girls are eager to be their friends. In contrast, children with only opposite-sex friendships are socially unskilled, unpopular youngsters who are rejected by their same-sex peers and form friendships with opposite-sex children as a last resort (Bukowski, Sippola, & Hoza, 1999).

Quality and Consequences of Friendships

If you think back to your childhood friendships, you probably remember some that were long-lasting and satisfying as well as others that rapidly wore thin and soon dissolved. What accounts for these differences in the quality and longevity of friendships? Sometimes friendships are brief because children lack the skills to sustain them (Jiao, 1999; Parker & Seal, 1996). As friends they can't keep secrets or they're too bossy. Sometimes friendships end because, when conflicts arise, children are more concerned about their own interests and are unwilling to compromise or negotiate (Fonzi et al., 1997; Rose & Asher, 1999). And sometimes friendships end when children discover that their needs and interests aren't as similar as they initially thought (Gavin & Furman, 1996).

When youth have good friends, they're better able to cope with life's stresses.

Considering that friendships disintegrate for many reasons, you're probably reminded that truly good friends are to be treasured. In fact, researchers consistently find that children benefit from having good friends (Berndt & Murphy, 2002). Compared to children who lack friends, children with good friends have higher self-esteem, are less likely to be lonely and depressed, and more often act prosocially by sharing and cooperating with others (Burk & Laursen, 2005; Hartup & Stevens, 1999). Children with good friends cope better with life stresses, such as the transition from elementary school to middle school or junior high (Berndt & Keefe, 1995) or being rejected by peers (McDonald et al., 2010); they're also less likely to be victimized by peers (Schwartz et al., 2000). The benefits of friendship are also long-lasting: Children who have friends have greater self-worth as young adults (Bagwell, Newcomb, & Bukowski, 1998).

Although children and adolescents benefit from their friends' support, there can be costs as well. *Sometimes friends spend much of their time together discussing each other's personal problems, which is known as* **co-rumination**. Girls do this more than boys (consistent with the fact that intimacy is more important to girls' friendships). Such co-rumination strengthens girls' friendships but also puts them at risk for greater depression and anxiety. In other words, when Avanti and Shruti spend day after day talking about problems with their parents and their schoolwork, they grow closer but more troubled (Brendgen et al., 2010; Rose, Carlson, & Waller, 2007).

There are other ways in which friendships can be hazardous (Bagwell, 2004). For example, when aggressive children are friends, they often encourage each other's aggressive behavior (Dishion, Poulin, & Burraston, 2001; Piehler & Dishion, 2007). Similarly, when teens engage in risky behavior (e.g., when they drink, smoke, or have sex), they often reinforce each other's risky behavior (Bot et al., 2005; Henry et al., 2007).

Thus, friends are one important way in which peers influence children's development. Peers also influence development through groups, the topic of the next section.

co-rumination
conversations about one's personal problems, common among adolescent girls

clique
small group of friends who are similar in age, sex, and race

crowd
large group including many cliques that have similar attitudes and values

Groups

At the summer camp in the vignette, new campers form groups based on common interests. Groups are just as prevalent in American schools. "Jocks," "preps," "burnouts," "druggies," "nerds," and "brains"—you may remember these or similar terms referring to groups of older children and adolescents. During late childhood and early adolescence, the peer group becomes the focal point of social relationships for youth (Rubin, Bukowski, & Parker, 1998). *The starting point is often a* **clique**—*a small group of children or adolescents who are friends and tend to be similar in age, sex, race, and attitudes.* Members of a clique spend time together and often dress, talk, and act alike. *A* **crowd** *is a larger mixed-sex group of older children or adolescents who have similar values and attitudes and are known by a common label such as "jocks" or "nerds"* (Brown & Klute, 2003).

Some crowds have more status than others. For example, students in many junior and senior high schools claim that the jocks are the most prestigious crowd, whereas the burnouts are among the least prestigious. Self-esteem in older children and adolescents often reflects the status of their crowd. During the school years, youths from high-status crowds tend to have greater self-esteem than those from low-status crowds (Sussman et al., 2007). Some crowds typically dislike others. Crowds that support adult values, such as jocks and preppies, usually dislike crowds that don't support these values (Laursen et al., 2010).

Why do some students become nerds while others join the burnouts? Adolescents' interests and abilities matter, obviously. Brighter students who enjoy school gravitate to the brain or nerd crowds while athletically talented teens become part of the jock crowd (Prinstein & LaGreca, 2004). Adolescents' crowds also reveal parents' influence. When parents practice authoritative parenting—they are warm but controlling—their children become involved with crowds that endorse adult standards of behavior (for example, normals, jocks, brains). But, when parents' style is neglectful or permissive, their children are less likely to identify with adult standards of behavior and, instead, join crowds like druggies that disavow adult standards. And this seems to be true of African American, Asian American, European American, and Hispanic American children and their parents (Brown et al., 1993).

Group Structure

Groups—be they in school, at a summer camp as in the vignette, or elsewhere—typically have a well-defined structure. *Often groups have a **dominance hierarchy**, headed by a leader to whom all other members of the group defer.* Other members know their position in the hierarchy. They yield to members who are above them in the hierarchy and assert themselves over members who are below them. A dominance hierarchy is useful in reducing conflict within groups because every member knows his or her place.

What determines where members stand in the hierarchy? In children, especially boys, physical power is often the basis for the dominance hierarchy. The leader is usually the member who is the most intimidating physically (Hawley, 1999). Among girls and older boys, hierarchies are often based on individual traits that relate to the group's main function. At Crab Orchard Summer Camp, for example, the leaders are most often the children with the greatest camping experience. Among Girl Scouts, girls chosen to be patrol leaders tend to be bright and goal oriented and to have new ideas (Edwards, 1994). These characteristics are appropriate because the primary function of patrol leaders is to help plan activities for the entire troop of Girl Scouts. Similarly, in a study of classroom discussion groups, the children who became leaders had good ideas and were outgoing (Li et al., 2007). Thus, leadership based on key skills is effective because it gives the greatest influence to those with the skills most important to group functioning.

Peer Pressure

Groups establish norms—standards of behavior that apply to all group members—and may pressure members to conform to these norms. Such "peer pressure" is often characterized as an irresistible, harmful force. The stereotype is that teenagers exert enormous pressure on each other to behave antisocially. In reality, peer pressure is neither always powerful nor always evil. For example, most junior and senior high students *resist* peer pressure to behave in ways that are clearly antisocial, such as stealing (Brown, Lohr, & McClenahan, 1986) and such resistance increases from mid- to late adolescence (Steinberg & Monahan, 2007). And peer pressure can be positive, too; peers often urge one another to work hard in school, to participate in school activities, such as trying out for a play or working on the yearbook, or to become involved in community action projects, such as Habitat for Humanity (Kindermann, 2007; Molloy, Gest, & Rulison, 2011).

Peer pressure is *not* all powerful. Instead, peer influence is stronger when one or more of the following conditions are present: (1) youth are younger and more socially anxious; (2) peers have high status; (3) peers are friends; and (4) standards for appropriate behavior are not clear-cut, as in the case of tastes in music or clothing, or standards for smoking and drinking (Anderson et al., 2011; Brechwald & Prinstein, 2011). Thus, when 14-year-old Doug's best friend (who's one of the most popular kids in school) gets his hair cut like Justin Bieber, Doug may go along because he's young, the peer is popular and his friend, and there are no fixed standards for hairstyle. But when an unpopular kid that 18-year-old Kelly barely knows suggests to her that they go to the mall and shoplift some earrings, Kelly will resist because she's older, the peer is unpopular and not a friend, and norms for shoplifting are clear.

dominance hierarchy
ordering of individuals within a group in which group members with lower status defer to those with greater status

THINK ABOUT IT

Chapter 5 described important differences in the ways that boys and girls interact with same-sex peers. How might these differences help explain why boys' and girls' dominance hierarchies differ?

Group leaders tend to be those who have skills that are valuable to the group: Girl Scout patrol leaders, for example, tend to be goal oriented and to have good ideas.

© Mary Kate Denny/PhotoEdit

Popularity and Rejection

Eileen is, without question, the most popular child in her fourth-grade class. Most of the other youngsters like to play with her and want to sit near her at lunch or on the school bus. Whenever the class must vote to pick a child for something special—to be class representative to the student council, to recite the class poem on Martin Luther King Day, or to lead the classroom to the lunchroom—Eileen invariably wins.

Jay is not as fortunate as Eileen. In fact, he is the least popular child in the class. His presence is obviously unwanted in any situation. When he sits down at the lunch table, other kids move away. When he tries to join a game of four-square, the others quit. Students in the class detest Jay as much as they like Eileen.

Popular and rejected children like Eileen and Jay are common. In fact, studies of popularity (Hymel et al., 2004) reveal that most children in elementary-school classrooms can be placed, fairly consistently, in one of these five categories:

- **Popular children** *are liked by many classmates.*
- **Rejected children** *are disliked by many classmates.*
- **Controversial children** *are both liked and disliked by classmates.*
- **Average children** *are liked and disliked by some classmates, but without the intensity found for popular, rejected, or controversial children.*
- **Neglected children** *are ignored by classmates.*

Of these categories, we know most about popular and rejected children. Each of these categories actually includes two subtypes. Most popular children are skilled academically and socially. They are good students who are usually friendly, cooperative, and helpful. They are more skillful at communicating and better at integrating themselves into an ongoing conversation or play session—they "fit in" instead of "barging in" (Graziano, Keane, & Calkins, 2007; Rubin, Bukowski, & Parker, 2006; Véronneau et al., 2010). A smaller group of popular children includes physically aggressive boys who pick fights with peers and relationally aggressive girls who, like the "Plastics" in the film *Mean Girls*, thrive on manipulating social relationships. Although these youth are not particularly friendly, their antisocial behavior nevertheless apparently has a certain appeal to peers (Cillessen & Rose, 2005; Xie et al., 2006).

Being well liked seems straightforward: Be pleasant and friendly, not obnoxious. Share, cooperate, and help instead of being disruptive. These results don't apply just to American children; they hold for children in many areas of the world, including

popular children
children who are liked by many classmates

rejected children
as applied to children's popularity, children who are disliked by many classmates

controversial children
as applied to children's popularity, children who are intensely liked or disliked by classmates

average children
as applied to children's popularity, children who are liked and disliked by different classmates, but with relatively little intensity

neglected children
as applied to children's popularity, children who are ignored—neither liked nor disliked—by their classmates

Most popular children are good students and they're socially skilled: they tend to be friendly and helpful with peers.

© MBI / Alamy

Canada, Europe, Israel, and China. Sometimes, however, popular children have other characteristics unique to their cultural setting. In Israel, for example, popular children are more likely to be assertive and direct than in other countries (Krispin, Sternberg, & Lamb, 1992). In China, popular children are more likely to be shy than in other countries (Chen et al., 2009). Evidently, good social skills are at the core of popularity in most countries, but other features are important, reflecting culturally specific values.

As for rejected children, many are overly aggressive, hyperactive, socially unskilled, and unable to regulate their emotions. These children are usually much more hostile than popular aggressive children and seem to be aggressive for the sheer fun of it, which peers dislike, instead of using aggression as a means toward other ends, which peers may not actually like but grudgingly respect (Prinstein & Cillessen, 2003). Other rejected children are shy, withdrawn, timid, and, not surprisingly, lonely (Asher & Paquette, 2003; Rubin, Coplan, & Bowker, 2009).

Causes and Consequences of Rejection

No one enjoys being rejected. For children, repeated peer rejection in childhood can have serious long-term consequences less often seen in other groups, including dropping out of school, committing juvenile offenses, and suffering from psychopathology (Ladd, 2006; Rubin et al., 1998).

Peer rejection can be traced, at least in part, to the influences of parents (Ladd, 1998). As expected from Bandura's social cognitive theory, children see how their parents respond in different social situations and often imitate these responses later. Parents who are friendly and cooperative with others demonstrate effective social skills for their youngsters. Parents who are belligerent and combative demonstrate tactics that are much less effective. In particular, when parents typically respond to interpersonal conflict with intimidation or aggression, their children may imitate them, hampering the development of their social skills and making them less popular in the long run (Keane, Brown, & Crenshaw, 1990).

Parents also contribute to their children's social skills and popularity through their disciplinary practices. Inconsistent discipline—punishing a child for misbehaving one day and ignoring the same behavior the next—is associated with antisocial, aggressive behavior, paving the way for rejection. Consistent punishment that does not rely on power assertion but is tied to parental love and affection is more likely to promote social skills and, in the process, popularity (Dekovic & Janssens, 1992; Rubin, Stewart, & Chen, 1995).

Thus, the origins of rejection are clear: Socially awkward, aggressive children are often rejected because they rely on an aggressive interpersonal style, which can be traced to parenting. The implication is that, by teaching youngsters (and their parents) more effective ways of interacting with others, we can make rejection less likely. With improved social skills, rejected children would not need to resort to antisocial behaviors that peers deplore. Training of this sort does work. Rejected children can learn skills that lead to peer acceptance and thereby avoid the long-term harm associated with being rejected (LaGreca, 1993; Mize & Ladd, 1990).

Aggressive Children and Their Victims

By the time toddlers are old enough to play with one another, they show aggression. For example, 1- and 2-year-olds sometimes use physical aggression to resolve their conflicts (Dodge, Coie, & Lynam, 2006). *In* **instrumental aggression,** *a child uses aggression to achieve an explicit goal.* By the start of the elementary-school years, another form of aggression emerges (Coie et al., 1991). **Hostile aggression** *is unprovoked and seems to have as its sole goal to intimidate, harass, or humiliate another child.* Hostile aggression is illustrated by a child who spontaneously says, "You're stupid!" and then kicks a classmate. A third form of aggression is relational aggression, in which children try to hurt others by undermining their social relationships (see Chap-

THINK ABOUT IT

Effective parents and popular children have many characteristics in common. What are they?

instrumental aggression
aggression used to achieve an explicit goal

hostile aggression
unprovoked aggression that seems to have the sole goal of intimidating, harassing, or humiliating another child

Jennie Woodcock; Reflections Photolibrary/Corbis

Children are likely to become chronic victims of aggression if they refuse to defend themselves.

ter 5). Examples would include telling friends to avoid a particular classmate or spreading malicious gossip (Crick et al., 2004).

Children's tendencies to behave aggressively are stable over time, particularly among those children who are highly aggressive at a young age. For example, in one study among 7-year-old boys who were highly aggressive, more than half had committed serious acts of delinquency (e.g., stealing a car, attacking others) by age 17 (Raine et al., 2005). Similarly, in a study involving more than 200 German preschool children (Asendorpf, Denissen, & van Anken, 2008), those children who were judged by teachers to be most aggressive were, as young adults, 12 times more likely than the least aggressive children to have been charged for criminal activity. And violent behavior in adulthood is not the only long-term outcome of childhood aggression; poor adjustment to high school (e.g., dropping out, failing a grade) and unemployment are others (Asendorpf et al., 2008; Ladd, 2003). Clearly, aggression is not simply a case of playful pushing and shoving that children always outgrow. To the contrary, a small minority of children who are highly aggressive develop into young adults who create havoc in society.

Most schoolchildren are the targets of an occasional aggressive act—a shove or kick to gain a desired toy, or a stinging insult by someone trying to save face. However, some youngsters are chronic targets of bullying, either through physical aggression (e.g., a child who is beat up daily on the playground) or through relational aggression (e.g., a child who is constantly the subject of rumors spread by their classmates). In recent years, some youth are victims of electronic bullying in which they are harassed via cell phones or the Internet (Raskauskas & Stoltz, 2007).

As you can imagine, being tormented daily by their peers is hard on children. When children are chronic victims of aggression, they're often lonely, anxious, and depressed; they dislike school and have low self-esteem (Ladd & Ladd, 1998; Rudolph, Troop-Gordon, & Flynn, 2009). Ironically, the impact of bullying is reduced when children see others being bullied, apparently because they feel that they're not being singled out for harassment (Nishina & Juvonen, 2005). Although most children are happier when no longer victimized, the harmful effects linger for some children: They are still lonely and sad despite not having been victims for 1 or 2 years (Kochenderfer-Ladd, & Wardrop, 2001).

Why do some children suffer the sad fate of being victims? Some victims are actually aggressive themselves (Veenstra et al., 2005). These youngsters often overreact, are restless, and are easily irritated. Their aggressive peers soon learn that these children are easily baited. A group of children will, for example, insult or ridicule them, knowing that they will probably start a fight even though they are outnumbered. Other victims tend to be withdrawn and submissive. They are unwilling or unable to defend themselves from their peers' aggression, and so they are usually referred to as passive victims (Guerra, Williams, & Sadek, 2011; Ladd & Ladd, 1998; Salmivalli & Isaacs, 2005). When attacked, they show obvious signs of distress and usually give in to their attackers, thereby rewarding the aggressive behavior. Thus, both aggressive and withdrawn submissive children end up as victims.

Victimized children can be taught ways of dealing with aggression: they can be encouraged to not respond in kind when insulted and to not show fear when threatened. In addition, increasing self-esteem can help. When attacked, children with low self-esteem may think, "I'm a loser and have to put up with this because I have no choice." Increasing children's self-esteem makes them less tolerant of personal attacks (Egan, Monson, & Perry, 1998). Another useful way to help victims is to foster their friendships with peers. When children have friends, they're not as likely to be victimized (Veenstra et al., 2010). Finally, the best solution is to prevent bullying and victimization altogether; an effective way to do this is to create a school climate in which bullying is not condoned and victims are supported by their peers (Kärnä et al., 2011).

Test Yourself

RECALL

1. Friends are usually similar in age, sex, race, and _____.

2. Children with friends have higher self-esteem, are less likely to be lonely, and _____ than children without friends.

3. As a group forms, a _____ typically emerges, with the leader at the top.

4. Peer pressure is most powerful when _____.

5. Popular children often share, cooperate, and are _____.

6. Rejected youngsters are more likely to drop out of school, to commit juvenile offenses, and _____.

7. Some children who are chronic victims of aggression overreact and are easily irritated; other chronic victims are _____.

INTERPRET

How could developmental change in the nature of friendship be explained in terms of Piaget's stages of intellectual development, discussed in Chapters 4 and 6?

APPLY

On page 270 you met Jay, who is the least popular child in his class. Jay's mom is worried about her son's lack of popularity and wants to know what she can do to help her son. Jay's dad thinks that Jay's mom is upset over nothing—he argues that, like fame, popularity is fleeting and that Jay will turn out okay in the end. What advice would you give to Jay's parents?

Recall answers: (1) interests, (2) more often act prosocially (sharing and cooperating), (3) dominance hierarchy, (4) standards for appropriate behavior are vague, (5) socially skilled, (6) to suffer from psychopathology, (7) unwilling or unable to defend themselves

---

LEARNING OBJECTIVES

- What is the impact of watching television on children's attitudes, behavior, and cognitive development?

- How do children use computers at home?

Every day, 7-year-old Roberto follows the same routine when he gets home from school: He watches one action-adventure cartoon after another until it's time for dinner. Roberto's mother is disturbed by her son's constant TV viewing, particularly because of the amount of violence in the shows he likes. Her husband tells her to stop worrying: "Let him watch what he wants to. It won't hurt him and, besides, it keeps him out of your hair."

IN GENERATIONS PAST, CHILDREN LEARNED THEIR CULTURE'S VALUES from parents, teachers, religious leaders, and print media. These sources of cultural knowledge are still with us, but they coexist with new technologies that do not always portray parents' values. Think about some of the technological developments that contemporary children take for granted that were completely foreign to youngsters growing up just a few decades ago. Your list would probably include satellite TV, CD and DVD players, video game players, personal computers, cell phones that take pictures and play music, pagers, and the Internet. More forces than ever before can potentially influence children's development. Two of these technologies—television and computers—are the focus of this section. As we look at their influence, we'll see if Roberto's mother should be worried.

Television

The cartoon exaggerates TV's impact on North American children, but only somewhat. After all, think about how much time you spent in front of a TV while you were growing up. The typical U.S. high school graduate has watched 20,000 hours of TV—the equivalent of 2 full years of watching TV 24/7! No wonder social scien-

"MRS. HORTON, COULD YOU STOP BY SCHOOL TODAY?"

tists and laypeople alike have come to see TV as an important contributor to the socialization of North American children.

It is hard to imagine that such massive viewing of TV would have no effect on children's behavior. After all, 30-second TV ads are designed to influence children's preferences in toys, cereals, and hamburgers, so the programs themselves ought to have even more impact. And research consistently shows that TV is indeed a powerful influence (Browne & Hamilton-Giachritsis, 2005; Huesmann, 2007). For example, Roberto's mother should be concerned: Children become more aggressive after viewing violence on television (Konijn, Bijvank, & Bushman, 2007) and children learn about gender stereotypes from TV (Huesmann, 2007). Watching can also help children learn to be more generous and cooperative and have greater self-control. Youngsters who watch TV shows that emphasize prosocial behavior, such as *Mister Rogers' Neighborhood,* are more likely to behave prosocially (Wilson, 2008). However, prosocial behaviors are portrayed on TV far less frequently than aggressive behaviors, so opportunities to learn the former from television are limited; we are far from harnessing the power of television for prosocial uses.

The biggest positive influence of TV on American children has been *Sesame Street.* Big Bird, Bert, Ernie, and their friends have been helping to educate preschool children for more than 40 years. Mothers and fathers who watched *Sesame Street* as preschoolers are now watching with their own youngsters. Remarkably, the time preschool children spend watching *Sesame Street* predicts their grades in high school and the amount of time they spend reading as adolescents (Anderson et al., 2001).

Today *Sesame Street* is joined by programs designed to teach young children about language and reading skills (*Arthur, Martha Speaks, Super Why!*) and programs that teach basic science and math concepts (*Curious George, Sid the Science Kid, The Dinosaur Train*). Programs like these (and older programs, such as *Electric Company, 3-2-1 Contact,* and *Square One TV*) show that the power of TV can be harnessed to help children learn important academic skills and useful social skills.

Television has its critics who argue that the medium itself—independent of the content of programs—has harmful effects on viewers, particularly children (Huston & Wright, 1998). One common criticism is that, because TV programs consist of many brief segments presented in rapid succession, children who watch a lot of TV develop short attention spans and have difficulty concentrating in school. Another concern heard frequently is that because TV provides ready-made, simple-to-interpret images, children who watch a lot of TV become passive, lazy thinkers and become less creative.

In fact, as stated, neither of these criticisms is consistently supported by research (Huston & Wright, 1998). The first criticism—TV watching reduces attention—is the easiest to dismiss. Research repeatedly shows that increased TV viewing does not lead to reduced attention, greater impulsivity, reduced task persistence, or increased activity levels (Foster & Watkins, 2010). The contents of TV programs can influence these dimensions of children's behavior—children who watch impulsive models behave more impulsively themselves—but TV per se does not harm children's ability to pay attention.

As for the criticism that TV viewing fosters lazy thinking and stifles creativity, the evidence is mixed. Many studies find no link between amount of TV viewing and creativity (e.g., Anderson et al., 2001). Some find a negative relation in which, as children watch more TV, they tend to get lower scores on tests of creativity (Valkenburg & van der Voort, 1994, 1995). Researchers don't know why the negative effects aren't found more consistently, although one idea is that the effects depend on what programs children watch, not simply the amount of TV watched.

In general, then, the sheer amount of TV that children watch is not a powerful influence on development (although a large amount of TV watching does put children at risk for obesity; Hancox & Poulton, 2006). Most of the impact of TV—for good or bad—comes through the contents of TV programs that children watch.

Computers

Some observers believe that computers are creating a "digital childhood"—an era in which new media are transforming the lives of American children. In this section, we'll look at children's use of computers at home. Most American children and adolescents have computers at home and they use them mainly to access the Internet. For example, they search the web for information for school assignments. However, a far more common use of computers, particularly for boys, is to play games online. Research reveals that computer games affect youth in much the same way that TV does—content matters. In other words, just as children and adolescents are influenced by the TV programs they watch, they're influenced by the contents of the computer games they play. On the one hand, many games, including *Tetris* and *Star Fox*, emphasize perceptual–spatial skills, such as estimating the trajectory of a moving object, and responding rapidly. When children play such games frequently, their spatial skills often improve (Subrahmanyam et al., 2001), as does their processing speed (Mackey et al., 2011). On the other hand, many popular games, such as *Manhunt* and *Grand Theft Auto*, are violent, with players killing game characters in extraordinarily gruesome ways. Just as exposure to televised violence can make children behave more aggressively, playing violent video games can make children more aggressive (Anderson et al., 2010).

When children play violent video games, they often become more aggressive.

What's more, a minority—roughly 10%—of youth get "hooked" on video games (Gentile, 2009). They show many of the same symptoms associated with pathological gambling: Playing video games comes to dominate their lives, it provides a "high," and it leads to conflict with others. Not surprisingly, extreme video-game playing is associated with less success in school, apparently because youth spend time playing games instead of studying (Weis & Ceranskosky, 2010).

The other main use of home computers (and cell phones) is to communicate with peers, often through social networking sites such as Facebook. Many children and adolescents use the Internet (and other communication technologies) to maintain existing "real" social connections. In fact, online communication seems to promote self-disclosure, which produces high-quality friendships and, in turn, well-being (Valkenburg & Jochen, 2009). Boys, in particular, benefit from online communication because self-disclosure is much easier for them online than face-to-face (Schouten, Valkenburg, & Peter, 2007).

In many respects, new technologies have changed the *how* of childhood and adolescence but not the *what*. As with previous generations, children and adolescents still play games, connect with peers, and do homework. Technology like a home computer simply provides a different means for accomplishing these tasks.

Test Yourself

RECALL

1. When children watch a lot of TV violence, they often become _____.

2. Contrary to popular criticisms, frequent TV viewing is not consistently related to reduced attention or to a lack of _____.

3. Children use computers at home to do schoolwork, play games, and _____.

4. When children frequently play violent video games, they _____.

INTERPRET

Compare and contrast the ways in which TV viewing and web surfing might affect children's development.

APPLY

What if you had the authority to write new regulations for children's TV programs. What shows would you encourage? What shows would you want to limit?

Recall answers: (1) more aggressive, (2) creativity, (3) communicate with friends, (4) often behave more aggressively

LEARNING OBJECTIVES

- As children develop, how do their descriptions of others change?
- How does understanding of others' thinking change as children develop?

- When and why do children develop prejudice toward others?

When 12-year-old Ian agreed to babysit his 5-year-old brother, Kyle, their mother reminded Ian to keep Kyle out of the basement because Kyle's birthday presents were there, unwrapped. But as soon as their mother left, Kyle wanted to go to the basement to ride his tricycle. When Ian told him no, Kyle burst into angry tears and shouted, "I'm gonna tell Mom that you were mean to me!" Ian wished he could explain to Kyle, but he knew that would just cause more trouble!

AS CHILDREN SPEND MORE TIME WITH OTHER PEOPLE (EITHER DIRECTLY OR VICARIOUSLY, through television), they begin to understand other people better. In this vignette, for example, Ian realizes why Kyle is angry, and he knows that if he gives in to Kyle now, his mother will be angry when she returns. Children's growing understanding of others is the focus of this section. We begin by looking at how children describe others and then examine their understanding of how others think. Finally, we'll also see how children's recognition of different social groups can lead to prejudices.

Describing Others

As children develop, more sophisticated cognitive processes cause self-descriptions to become richer, more abstract, and more psychological. These same changes also occur in children's descriptions of others. Children begin by describing other people in terms of concrete features, such as behavior, and progress to describing them in terms of abstract traits (Barenboim, 1981; Livesley & Bromley, 1973). The Real People feature shows this progression in one child.

Real People
Applying Human Development
Tell Me About a Girl That You Like a Lot

Every few years, Tamsen was asked to describe a girl that she liked a lot. Each time, she described a different girl. More important, the contents of her descriptions changed, focusing less on behavior and emphasizing psychological properties. Let's start with the description she gave as a 7-year-old:

Vanessa is short. She has black hair and brown eyes. She uses a wheelchair because she can't walk. She's in my class. She has dolls just like mine. She likes to sing and read.

Tamsen's description of Vanessa is probably not too different from the way she would have described herself: The emphasis is on concrete characteristics, such as Vanessa's appearance, possessions, and preferences. Contrast this with the following description, which Tamsen gave as a 10-year-old:

Kate lives in my apartment. She is a very good reader and is also good at math and science. She's nice to everyone in our class. And she's very funny. Sometimes her jokes make me laugh so-o-o hard! She takes piano lessons and likes to play soccer.

Tamsen's account still includes concrete features, such as where Kate lives and what she likes to do. However, psychological traits are also evident: Tamsen describes Kate as nice and funny. By age 10, children move beyond the purely concrete and observable in describing others. During adolescence, descriptions become even more complex, as you can see in the following, from Tamsen as a 16-year-old:

Jeannie is very understanding. Whenever anyone at school is upset, she's there to give a helping hand. Yet, in private, Jeannie can be so sarcastic. She can say some really nasty things about people. But I know she'd never say that stuff if she thought people would hear it because she wouldn't want to hurt their feelings.

This description is more abstract: Tamsen now focuses on psychological traits like understanding and concern for others' feelings. It's also more integrated: Tamsen tries to explain how Jeannie can be both understanding and sarcastic.

Each of Tamsen's three descriptions is very typical. As a 7-year-old, she emphasized concrete characteristics; as a 10-year-old, she began to include psychological traits; and as a 16-year-old, she tried to integrate traits to form a cohesive account.

The progression in how children perceive others was illustrated vividly in a classic study by Livesley and Bromley (1973). They interviewed 320 students, 7 to 15 years old, who attended school in Merseyside, England (near Liverpool, home of the Beatles). All participants were asked to describe eight people they knew: two boys, two girls, two men, and two women. The examiner told the participants, "I want you to describe what sort of person they are. I want you to tell me what you think about them and what they are like" (p. 97).

The participants at different ages typically produced descriptions much like Tamsen's at different ages. Livesley and Bromley then categorized the contents of the descriptions. Some of their results appear in Figure 7.3. Descriptions referring to appearances or possessions become less common as children grow older, as do descriptions giving general information, such as the person's age, gender, religion, or school. In contrast, descriptions of personality traits (e.g., "friendly" or "conceited") increase between 8 and 14 years of age. Thus, children's descriptions of others begin with the concrete and later become more conceptual.

More recent research also supports the trend to more abstract and richer psychological descriptions of others but indicates that young children's understanding of other people is more sophisticated than is suggested by their verbal descriptions of people they know (Heyman, 2009). Indeed, modern work indicates that 4- and 5-year-olds have begun to think about other people in terms of psychological traits such as being smart, friendly, helpful, and shy. They can use behavioral examples to infer an underlying trait: Told about a child who won't share cookies or won't allow another child to play with a toy, 4- and 5-year-olds accurately describe the child as selfish. In addition, given information about a trait, they correctly predict future behavior: Told about a child who is shy, they believe that the child will not volunteer to help a puppeteer and will be quiet at a meal with many relatives (Liu, Gelman, & Wellman, 2007).

One idiosyncrasy of young children's descriptions of others is they see others "through rose-colored glasses"—that is, until about 10 years of age, children have a bias to look for positive traits, not negative traits, in others. Young children are willing to believe that someone is smart (or friendly or helpful) based on relatively little evidence (and based on inconsistent evidence), but they require much more evidence (and more consistent evidence) to decide that someone is mean or stupid.

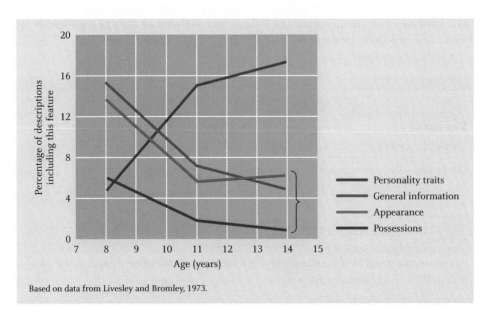

■ **Figure 7.3**
 In describing other people, personality becomes more important with development while general information, appearance, and physical possessions become less important.

Based on data from Livesley and Bromley, 1973.

Understanding What Others Think

One trademark of the preschooler's thinking is difficulty in seeing the world from another's view. Piaget's term for this is *egocentrism,* and it is a defining characteristic of his preoperational stage of development. As children move beyond the preschool years, though, they realize that others see the world differently, both literally and figuratively. For example, in the vignette, Ian knows why his little brother Kyle is angry: Kyle thinks that Ian is being bossy and mean. Ian understands that Kyle doesn't know there is a good reason why he can't go to the basement.

Sophisticated understanding of how others think is achieved gradually throughout childhood and adolescence. According to a theory proposed by Robert Selman (1980, 1981), understanding other people begins with the egocentric thinking characteristic of preoperational children—they think that others think as they do. As children develop, they become able to take the perspective of other people. In Selman's theory, this perspective-taking skill progresses through five stages, which are shown in ● Table 7.1.

To see the progression from stage to stage, imagine two boys arguing about what to do after school. One wants to go to a playground and the other wants to watch TV. If the boys were 5-year-olds (undifferentiated stage), neither would really understand why the other wants to do something different. Their reasoning is stone simple: "If I want to go to the playground, you should too!"

During the early elementary-school years (social-informational stage), each child understands that the other wants to do something different and they explain their differing views in terms of the other person lacking essential information. Their thinking is along the lines, "I know that you want to watch TV, but if you knew what I knew, you'd want to go to the playground." By the late elementary-school years (self-reflective stage), the boys would understand that each wants to do something different and they could "step into the other's shoes" to understand why: "I know you want to go to the playground because you haven't been there all week."

In early adolescence (third-person stage), the boys could step even farther apart and imagine how another person (e.g., a parent or teacher) could view the disagreement. Finally, in late adolescence (societal stage), the boys (now young men, really) can remove themselves even further and appreciate, for example, that many people would think it's silly to watch TV on a beautiful sunny day.

As predicted by Selman's theory, children's reasoning moves through each stage, in sequence, as they grow older. In addition, regardless of age, children at more advanced cognitive levels tend to be at more advanced stages in perspective taking (Gurucharri & Selman, 1982; Krebs & Gillmore, 1982). However, many scientists are not convinced that more sophisticated perspective taking occurs in such a stagelike fashion; they believe that it improves steadily throughout childhood and adolescence (just as cognitive development is now seen to be more continuous than Piaget's theory predicted).

THINK ABOUT IT

How do Selman's stages of perspective taking correspond to Piaget's and Erikson's stages?

● TABLE 7.1

Selman's Stages of Perspective Taking

Stage	Approximate Ages	Description
Undifferentiated	3–6 years	Children know that self and others can have different thoughts and feelings but often confuse the two.
Social-informational	4–9 years	Children know that perspectives differ because people have access to different information.
Self-reflective	7–12 years	Children can step into another's shoes and view themselves as others do; they know that others can do the same.
Third-person	10–15 years	Children can step outside of the immediate situation to see how they and another person are viewed by a third person.
Societal	14 years to adult	Adolescents realize that a third-person perspective is influenced by broader personal, social, and cultural contexts.

Some investigators have linked improved perspective taking to the developing theory of mind, described in Chapter 3 (Chandler & Carpendale, 1998). The traditional false-belief task, for example, reveals children's understanding that another person's actions are often based on their beliefs even when those beliefs are wrong. As an illustration, suppose children hear the following story:

Lindsay and Angela are in the park and see some kids playing softball. Lindsay wants to play, so she runs home for her glove. Angela waits at the park for her, but while Lindsay's away, the kids decided it's too hot for softball and leave to get some ice cream.

Children understand false belief if they say that Lindsay will return to the ball field (acting on her false belief that the kids are still playing ball). But we can add a new wrinkle to the story.

As the kids are leaving the park, one of them thinks that Lindsay might like to join them for ice cream, so she phones Lindsay and tells her the plan.

Now children are asked: "Where does Angela think Lindsay thinks the kids are?" Children understand second-order belief if they say that Angela thinks that Lindsay will go to the ball field. *This sort of "he thinks that she thinks . . ." reasoning is known as* **recursive thinking**. It emerges at about 5 or 6 years of age and improves steadily during the elementary-school years, due to the combined effects of increased language skill and greater executive functioning (Miller, 2009).

One of the benefits of a developing appreciation of others' thoughts and viewpoints is that it allows children to get along better with their peers. That is, children who can readily take another's perspective are typically well-liked by their peers (FitzGerald & White, 2003; LeMare & Rubin, 1987). In the photo, for example, the children with the soccer ball evidently recognize that the girl on the sideline wants to play, so they're inviting her to join them.

Of course, mere understanding does not guarantee good social behavior; sometimes children who understand what another child is thinking take advantage of that child. In general, however, greater understanding of others seems to promote positive interactions.

Socially skilled youth understand what others are thinking; in this case, they invite the girl on the sideline to join them.

Prejudice

As children learn more about others, they discover that people belong to different social groups that are based on variables such as gender, ethnicity, and social class. By the preschool years, most children can distinguish males from females and can identify people from different ethnic groups (Aboud, 1993). *Once children learn their membership in a specific group, they typically show* **prejudice**, *a negative view of others based on their membership in a different group.* Actually, in young children prejudice is not so much a negative view of others as it is an enhanced view of one's own group. That is, preschool and kindergarten children attribute to their own group many positive traits such as being friendly and smart and few negative traits such as being mean (Bigler, Jones, & Lobliner, 1997; Patterson & Bigler, 2006).

Negative views of other groups form more slowly. In young children, negative views typically don't involve overt hostility; it's simply that other groups "come up short" when compared to one's own group (Aboud, 2003). However, when children believe that children from other groups dislike them or think they're better, children's views of other groups become more negative (Nesdale et al., 2005).

As children move into the elementary-school years, their knowledge of racial stereotypes and prejudices increases steadily; by 10 or 11 years of age, most children are aware of broadly held racial stereotypes (Pauker, Ambady, & Apfelbaum, 2010). During these years, prejudice declines some, in part because children learn norms that discourage openly favoring their own group over others (Apfelbaum et al., 2008).

During early adolescence, prejudice sometimes increases again. This resurgence apparently reflects two different processes (Black-Gutman & Hickson, 1996; Teichman, 2001). One is experiential: Exposed to prejudices of those around them, children and adolescents internalize some of these views (Castelli, Zogmaister, & Tomelleri, 2009). A second process concerns adolescents' identity. In the search for identity (described on pages 316–320), adolescents' preferences for their own groups often intensify (Rutland, Killen, & Abrams, 2010). Thus, greater prejudice in older children and adolescents reflects a more positive view of their own group as well as a more negative view of other groups. Bob, a 14-year-old European American growing up in Arizona, becomes more prejudiced because he views his own European American heritage more positively and acquires prejudicial attitudes toward Native Americans from his parents and peers.

Identifying *how* children form actual prejudices is challenging because ethical concerns limit us to correlational studies. (Obviously, we could not do an experiment in which some children are deliberately exposed to biased information about actual groups of children.) Consequently, to study the processes underlying prejudice, researchers sometimes conduct experiments in which children are temporarily assigned to different groups.

Some scientists believe that bias and prejudice emerge naturally out of children's efforts to understand their social world (Bigler & Liben, 2007). Young children actively categorize animate and inanimate objects as they try to understand the world around them. As children's social horizons expand beyond their parents to include peers, they continue to categorize and try to decide how different groups of people "go together." They use perceptually salient features (e.g., race, gender, age) as well as verbal labels that adults may apply to different groups (e.g., "Girls go to lunch first, then the boys"). After children have identified the salient features that define peers in their environment, they begin to classify people that they encounter along these dimensions. Jacob is now seen as a white boy; Kalika is now seen as a black girl (Patterson & Bigler, 2006).

What can parents, teachers, and other adults do to rid children of prejudice? One way is to encourage friendly and constructive contacts between children from different groups. However, contact alone usually accomplishes little. Intergroup contact reduces prejudice only when the participating groups of children are equal in status, when the contact between groups involves pursuing common goals (not competing), and when parents and teachers support the goal of reducing prejudice (Cameron et al., 2006; Killen & McGlothlin, 2005). For example, adults might have children from different groups work together toward common goals. In school, this might be a class

One effective way to reduce prejudice is for children from different races to work together toward a common goal, such as completing a class project.

project. In sports, it might be mastering a new skill. By working together, Gary starts to realize that Vic acts, thinks, and feels as he does simply because he's Vic, not because he's an Italian American.

Another useful approach is to ask children to play different roles (Davidson & Davidson, 1994; Tynes, 2007). They can be asked to imagine that—because of their race, ethnic background, or gender—they have been insulted verbally or not allowed to participate in special activities. A child might be asked to imagine that she can't go to a private swimming club because she's African American or that she wasn't invited to a party because she's Hispanic American. Afterward, children reflect on how they felt when prejudice and discrimination was directed at them. They're also asked to think about what would be fair: What should be done in situations like these?

A final strategy against prejudice involves education. In one study (Hughes, Bigler, & Levy, 2007), European American elementary-school children learned about the racism that famous African Americans experienced. For example, they learned that Jackie Robinson played for a team in the old Negro Leagues because the white people in charge of major league baseball wouldn't allow any African Americans to play. The study also included a control group in which the biographies omitted the experiences of racism. When children learned about racism directed at African Americans, they had much more positive attitudes toward African Americans.

From experiences like these, children and adolescents discover for themselves that a person's membership in a social group tells us very little about that person. They learn, instead, that all children are different and that each person is a unique mix of experiences, skills, and values.

Test Yourself

RECALL

1. When adolescents describe others, they usually _____.

2. In the most advanced stage of Selman's theory, adolescents _____.

3. Young adolescents often become more prejudiced, reflecting the views of those around them and _____.

INTERPRET

How might an information-processing theorist describe the stages of Selman's perspective-taking theory?

APPLY

Based on what you've learned in this section, what can parents and teachers do to discourage prejudice in children?

Recall answers: (1) try to provide a cohesive, integrated account, (2) provide a third-person perspective on situations and recognize the influence of context on this perspective, (3) greater affiliation with their own group

SUMMARY

7.1   FAMILY RELATIONSHIPS

What is a systems approach to parenting?

■ According to the systems approach, the family consists of interacting elements; that is, parents and children influence each other. The family itself is influenced by other social systems, such as neighborhoods and religious organizations.

What are the primary dimensions of parenting? How do they affect children's development?

■ One key factor in parent–child relationships is the degree of warmth that parents express: Children clearly benefit from warm, caring parents. A second factor is control, which is complicated because neither too much nor too little control is desirable. Effective parental control involves setting appropriate standards, enforcing them, and trying to anticipate conflicts.

■ Taking into account both warmth and control, four prototypic parental styles emerge: Authoritarian parents are controlling but uninvolved; authoritative parents are fairly controlling but are also responsive to their children; permissive parents are loving but exert little control; and uninvolved parents are neither warm nor controlling. Authoritative parenting seems best for children in terms of both cognitive and social development, but there are important exceptions associated with culture and socioeconomic status.

■ Parents influence development by direct instruction and coaching. In addition, parents serve as models for their children, who sometimes imitate parents' behavior directly. Sometimes children behave in ways that are similar to what they have seen, and sometimes in ways that are the opposite of what they've seen (counterimitation).

■ Parents also use feedback to influence children's behavior. Sometimes parents fall into the negative reinforcement trap, inadvertently reinforcing behaviors that they want to discourage.

■ Punishment is effective when it is prompt, consistent, accompanied by an explanation, and delivered by a person with whom the child has a warm relationship. Punishment has limited value because it suppresses behaviors but does not eliminate them, and it often has side effects. Time-out is one useful form of punishment.

■ Chronic conflict is harmful to children, but children can actually benefit when their parents solve problems constructively. Parenting is a team sport, but not all parents play well together because they may disagree in child-rearing goals or parenting methods.

- Parenting is influenced by characteristics of children themselves. A child's age and temperament will influence how a parent tries to exert control over the child.

What determines how siblings get along? How do first-born, later-born, and only children differ?

- The birth of a sibling can be stressful for children, particularly when they are still young and when parents ignore their needs. Siblings get along better when they are of the same sex, believe that parents treat them similarly, enter adolescence, and have parents who get along well.

- As adoption became more common in the United States, a myth grew that adopted children are more prone to problems. Research shows that adopted children are similar to children living with biological parents in many respects, although they are more prone to some problems such as adjusting to school and conduct disorders. However, these results depend strongly on the child's age when adopted and the quality of care prior to adoption, which suggests that adoption per se is not a problem for children's development.

- Parents have higher expectations for first-born children, which explains why such children are more intelligent and more likely to go to college. Later-born children are more popular and more innovative. Contradicting the folklore, only children are almost never worse off than children with siblings; in some respects (such as intelligence, achievement, and autonomy), they are often better off.

How do divorce and remarriage affect children?

- Divorce can harm children in a number of areas, ranging from school achievement to adjustment. The impact of divorce stems from less supervision of children following divorce, economic hardship, and conflict between parents. Children often benefit when parents have joint custody following divorce or when they live with the same-sex parent.

- When a mother remarries, daughters sometimes have difficulty adjusting because the new stepfather encroaches on an intimate mother–daughter relationship.

What factors lead children to be maltreated?

- Factors that contribute to child abuse include poverty, social isolation, and a culture's views on violence. Parents who abuse their children were often neglected or abused themselves and tend to be unhappy, socially unskilled individuals. Younger or unhealthy children are more likely to be targets of abuse. Children who are abused often lag behind in cognitive and social development.

7.2 PEERS

What are the benefits of friendship?

- Friendships among preschoolers are based on common interests and getting along well. As children grow, loyalty, trust, and intimacy become more important features in their friendships. Friends are usually similar in age, sex, race, and attitudes. Children with friends are more skilled socially and are better adjusted.

What are the important features of groups of children and adolescents? How do these groups influence individuals?

- Older children and adolescents often form cliques—small groups of like-minded individuals—that become part of a crowd. Some crowds have higher status than others, and members of higher-status crowds often have higher self-esteem than members of lower-status crowds.

- Common to most groups is a dominance hierarchy, a well-defined structure with a leader at the top. Physical power often determines the dominance hierarchy, particularly among boys. However, with older children and adolescents, dominance hierarchies are more often based on skills that are important to the group.

- Peer pressure is neither totally powerful nor totally evil. In fact, groups influence individuals primarily in areas where standards of behavior are unclear, such as tastes in music or clothing or concerning drinking, drug use, and sex.

Why are some children more popular than others? What are the causes and consequences of being rejected?

- Most popular children are socially skilled. They often share, cooperate, and help others. A far smaller number of popular children use aggression to achieve their social goals.

- Some children are rejected by their peers because they are too aggressive. Others are rejected for being too timid or withdrawn. Repeated peer rejection often leads to school failure and behavioral problems.

What are some effects of childhood aggression? Why are some children chronic victims of aggression?

- Many highly aggressive children end up being violent and poorly adjusted as adults. Children who are chronic victims of aggression typically either overreact or refuse to defend themselves.

7.3 ELECTRONIC MEDIA

What is the impact of watching television on children's attitudes, behavior, and cognitive development?

- TV programs can cause children to become more aggressive, to adopt gender stereotypes, and to act prosocially. Programs designed to foster children's cognitive skills, such as *Sesame Street*, are effective. Many criticisms about TV as a medium (e.g., it shortens children's attention span) are not supported by research.

How do children use computers at home?

- At home, children use computers for schoolwork, to play video games (and are influenced by the contents of the games they play), and to communicate with friends via the Internet.

7.4 UNDERSTANDING OTHERS

As children develop, how do their descriptions of others change?

- Children's descriptions of others change in much the same way that children's descriptions of themselves change. During the early elementary-school years, descriptions emphasize concrete characteristics. In the late elementary-school years, they emphasize personality traits. In adolescence, they emphasize providing an integrated picture of others.

How does understanding of others' thinking change as children develop?

- According to Selman's theory, children's understanding of how others think progresses through five stages. In the first (undifferentiated) stage, children often confuse their own and another's view. In the last (societal) stage, adolescents can take a third-person perspective and know that this perspective is influenced by context.

When and why do children develop prejudice toward others?

- Prejudice emerges in the preschool years, soon after children recognize different social groups. Prejudice often increases in early adolescence, reflecting exposure to prejudiced views of others and adolescents' greater affiliation with their own group as they seek an identity. Prejudice can be reduced with positive contact between groups, role playing, and greater knowledge of racism directed at minority groups.

KEY TERMS

socialization (248)
authoritarian parenting (250)
authoritative parenting (250)
permissive parenting (250)
uninvolved parenting (250)
direct instruction (251)
counterimitation (251)
reinforcement (252)
punishment (252)
negative reinforcement trap (252)

time-out (253)
joint custody (259)
blended family (261)
ego resilience (264)
friendship (266)
co-rumination (268)
clique (268)
crowd (268)
dominance hierarchy (269)
popular children (269)

rejected children (269)
controversial children (269)
average children (269)
neglected children (269)
instrumental aggression (271)
hostile aggression (271)
recursive thinking (279)
prejudice (279)

LEARN MORE ABOUT IT

Log in to **www.cengagebrain.com** to access the resources your instructor requires. For this book, you can access:

Psychology CourseMate

- CourseMate brings course concepts to life with interactive learning, study, and exam preparation tools that support the printed textbook. A textbook-specific website, Psychology CourseMate includes an integrated interactive eBook and other interactive learning tools including quizzes, flashcards, videos, and more.

CENGAGENOW™

- CengageNOW Personalized Study is a diagnostic study tool containing valuable text-specific resources—and because you focus on just what you don't know, you learn more in less time to get a better grade.

WebTUTOR™

- More than just an interactive study guide, WebTutor is an anytime, anywhere customized learning solution with an eBook, keeping you connected to your textbook, instructor, and classmates.

Rites of Passage

Physical and Cognitive Development in Adolescence

8

AT AGE 10, MICHELLE WIE BECAME THE YOUNGEST PLAYER ever to qualify for a golf tournament for adult players sponsored by the U.S. Golf Association; at age 13, she won the U.S. Women's Amateur Public Links Championship; and at age 16, she became the youngest female golfer to make the cut in a professional golf tournament for men. Michelle's steady march to the top of her sport over her adolescent years is a remarkable feat. Yet, in a less dramatic and less public way, these years are times of profound changes for all adolescents. In this chapter, we'll examine the physical and cognitive developments in adolescence. We'll begin by describing the important features of physical growth in the teenage years. Then we'll consider some of the necessary ingredients for healthy growth in adolescence. Next, we'll examine the nature of information processing during adolescence. Finally, we'll end the chapter by examining how adolescents reason about moral issues.

Robert V. Kail

LEARNING OBJECTIVES

■ What physical changes occur in adolescence that mark the transition to a mature young adult?

■ What factors cause the physical changes associated with puberty?

■ How do physical changes affect adolescents' psychological development?

Pete just celebrated his 15th birthday, but as far as he is concerned, there is no reason to celebrate. Although most of his friends have grown about 6 inches in the past year or so, have a much larger penis and larger testicles, and mounds of pubic hair, Pete looks just as he did when he was 10 years old. He is embarrassed by his appearance, particularly in the locker room, where he looks like a little boy among men. "Won't I ever change?" he wonders.

THE APPEARANCE OF BODY HAIR, the emergence of breasts, and the enlargement of the penis and testicles are all signs that the child is gone and the adolescent has arrived. Many adolescents take great satisfaction in these signs of maturity. Others, like Pete, worry through their teenage years as they wait for the physical signs of adolescence.

In this section, we'll begin by describing the normal pattern of physical changes that take place in adolescence and look at the mechanisms responsible for them. Then we'll discover the impact of these physical changes on adolescents' psychological functioning. As we do, we'll learn about the possible effects of Pete's maturing later than his peers.

Signs of Physical Maturation

puberty
collection of physical changes that marks the onset of adolescence, including a growth spurt and the growth of breasts or testes

Puberty *denotes two general types of physical changes that mark the transition from childhood to young adulthood.* The first are bodily changes, including a dramatic increase in height and weight, as well as changes in the body's fat and muscle content. The second concern sexual maturation, including change in the reproductive organs and the appearance of secondary sexual characteristics, such as facial and body hair and growth of the breasts.

Physical Growth

When it comes to physical growth, the elementary-school years represent the calm before the adolescent storm. ■ Figure 8.1 shows that, in an average year, a typical 6- to 10-year-old gains about 5 to 7 pounds and grows 2 to 3 inches. In contrast, during the

■ **Figure 8.1**
Children grow steadily taller and heavier until puberty, when they exerience a rapid increase known as the adolescent growth spurt.

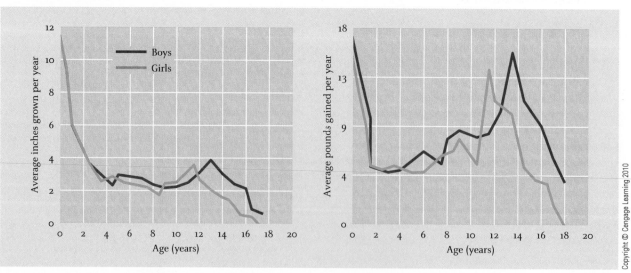

During the growth spurt, girls are often much taller than boys of the same age.

peak of the adolescent growth spurt, a girl may gain as much as 14–15 pounds in a year and a boy 16–17 pounds (Tanner, 1970).

Figure 8.1 also shows that girls typically begin their growth spurt about 2 years before boys do. That is, girls typically start the growth spurt at about age 11, reach their peak rate of growth at about 12, and achieve their mature stature at about 15. In contrast, boys start the growth spurt at age 13, hit peak growth at 14, and reach mature stature at 17. This 2-year difference in the growth spurt can lead to awkward social interactions between 11- and 12-year-old boys and girls because the girls are often taller and look much more mature than the boys.

Body parts don't all mature at the same rate. Instead, the head, hands, and feet usually begin to grow first, followed by growth in the arms and legs. The trunk and shoulders are the last to grow (Tanner, 1990). The result of these differing growth rates is that an adolescent's body sometimes seems to be out of proportion—teens have a head and hands that are too big for the rest of their body. Fortunately, these imbalances don't last long as the later developing parts catch up.

During the growth spurt, bones become longer (which, of course, is why adolescents grow taller) and denser. Bone growth is accompanied by several other changes that differ for boys and girls. Muscle fibers become thicker and denser during adolescence, producing substantial increases in strength. However, muscle growth is much more pronounced in boys than in girls (Smoll & Schutz, 1990). Body fat also increases during adolescence, but much more rapidly in girls than in boys. Finally, heart and lung capacity increases more in adolescent boys than in adolescent girls. Together, these changes help to explain why the typical adolescent boy is stronger, quicker, and has greater endurance than the typical adolescent girl.

Brain Growth in Adolescence

At the beginning of adolescence, the brain is nearly full size—it's about 95% of the size and weight of an adult's brain. Nevertheless, adolescence is important for fine-tuning the brain's functioning. Two features of brain development that begin early in life (and that we discussed in Chapter 3) are nearly complete in adolescence: myelination, which is the acquisition of fatty insulation that makes neurons transmit information faster; and synaptic pruning, which is the weeding out of unnecessary connections between neurons (Ben Bashat et al., 2005; Toga et al., 2006; Wozniak & Lim, 2006). These changes mean that different regions in the adolescent brain are well connected and information is rapidly conveyed between them, which means that adolescents can process information much more efficiently than the child, a theme that we will explore more on pages 301–302.

THINK ABOUT IT

Compare and contrast the events of puberty for boys and girls.

■ **Figure 8.2**
Adolescence is a vulnerable time because the reward- and pleasure-seeking centers of the brain (limbic system) mature more rapidly than the behavioral control systems (frontal cortex); the gap between the two systems is particularly great in adolescence.

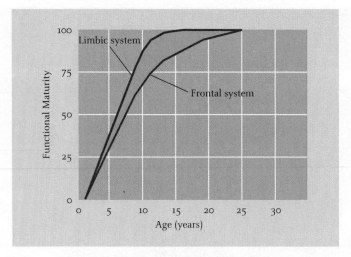

From Casey et al., 2008, "The Adolescent Brain," *Annals of the New York Academy of Sciences,* 1124, Fig. 3, p. 116. Reprinted by permission of John Wiley & Sons.

Another distinguishing feature of the adolescent brain is that some, but not all, brain regions reach maturity. Notably, the limbic system, which helps to regulate the experience of reward, pleasure, and emotion, reaches maturity in early adolescence. For example, when judging the emotion depicted in a facial expression, the amygdala, a key element of the limbic system, is more active in the adolescent brain than in the child or adult brain (Hare et al., 2008). In contrast, systems in the frontal cortex that are associated with deliberate control of behavior are still developing during adolescence (Bava & Tapert, 2010). As shown in Figure 8.2, this makes adolescents vulnerable: the reward- and pleasure-seeking systems are more mature than the systems for controlling behavior. Consequently, even though adolescents may know that behaviors involve risk, the anticipated rewards and pleasure of risky behavior sometimes swamp the adolescent's ability to suppress the desire to engage in such activities (Somerville & Casey, 2010). As we'll see on pages 299–300, this makes adolescents vulnerable to high-risk activities (e.g., drinking, unprotected sex).

Sexual Maturation

Not only do adolescents become taller and heavier, they also become mature sexually. *Sexual maturation includes change in* **primary sex characteristics**, *which refer to organs that are directly involved in reproduction.* These include the ovaries, uterus, and vagina in girls and the scrotum, testes, and penis in boys. *Sexual maturation also includes changes in* **secondary sex characteristics**, *which are physical signs of maturity not directly linked to the reproductive organs.* These include the growth of breasts and the widening of the pelvis in girls, the appearance of facial hair and the broadening of shoulders in boys, and the appearance of body hair and changes in voice and skin in both boys and girls.

Changes in primary and secondary sexual characteristics occur in a predictable sequence for boys and for girls. ■ Figure 8.3 shows these changes and the ages when they typically occur for boys and girls. For girls, puberty begins with growth of the breasts and the growth spurt, followed by the appearance of pubic hair. **Menarche**, *the onset of menstruation, typically occurs at about age 13.* Early menstrual cycles are usually irregular and without ovulation.

For boys, puberty usually commences with the growth of the testes and scrotum, followed by the appearance of pubic hair, the start of the growth spurt, and growth of the penis. *At about age 13, most boys reach* **spermarche**, *the first spontaneous ejaculation of sperm-laden fluid.* Initial ejaculations often contain relatively few sperm; only months or sometimes years later are there sufficient sperm to fertilize an egg (Chilman, 1983).

primary sex characteristics
physical signs of maturity that are directly linked to the reproductive organs

secondary sex characteristics
physical signs of maturity that are not directly linked to reproductive organs

menarche
onset of menstruation

spermarche
first spontaneous ejaculation of sperm

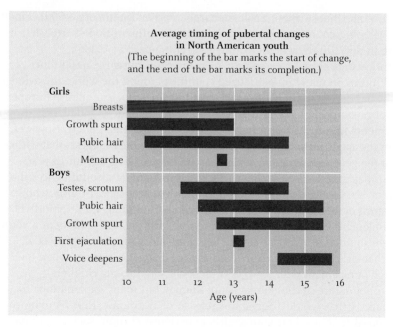

From Malinda, R. M., & Bouchard, C. (1991). Growth, maturation, and physical activity. Champaign IL: Human Kinetics Books.

■ **Figure 8.3**
The events that make up puberty typically start about 2 years earlier in girls than in boys.

Mechanisms of Maturation

What causes the many physical changes that occur during puberty? The pituitary gland is the key player: It helps to regulate physical development by releasing growth hormone. In addition, the pituitary regulates pubertal changes by signaling other glands to secrete hormones. During the early elementary-school years—long before there are any outward signs of puberty—the pituitary signals the adrenal glands to release androgens, initiating the biochemical changes that will produce body hair. A few years later, in girls the pituitary signals the ovaries to release estrogen, which causes the breasts to enlarge, the female genitals to mature, and fat to accumulate. In boys the pituitary signals the testes to release the androgen hormone testosterone, which causes the male genitals to mature and muscle mass to increase.

Although estrogen is often described as a "female hormone" and androgen as a "male hormone," estrogen and androgen are present in both boys and girls. As we've seen, in girls the adrenal glands secrete androgens. The amount is very small compared to that secreted by boys' testes but is enough to influence the emergence of body hair. In boys, the testes secrete very small amounts of estrogen, which explains why some boys' breasts temporarily enlarge early in adolescence.

The timing of pubertal events is regulated, in part, by genetics. This is shown by the closer synchrony of pubertal events in identical twins than in fraternal twins: If one identical twin has body hair, the odds are that the other twin will too (Mustanski et al., 2004). Genetic influence is also shown by the fact that a mother's age at menarche is related to her daughter's age at menarche (Belsky et al., 2007). However, these genetic forces are strongly influenced by the environment, particularly an adolescent's nutrition and health. In general, puberty occurs earlier in adolescents who are well nourished and healthy than in adolescents who are not. For example, puberty occurs earlier in girls who are heavier and taller but later in girls who are afflicted with chronic illnesses or who receive inadequate nutrition (St. George, Williams, & Silva, 1994).

Three other findings underscore the importance of nutrition and health in the onset of puberty. Cross-cultural comparisons reveal that menarche occurs earlier in areas of the world where nutrition and health care are adequate. For example, menarche occurs an average of 2 to 3 years earlier in Western European and North American countries than in African countries. And, within regions, socioeconomic status matters: Girls from affluent homes are more likely to receive adequate nutrition and health care and hence they reach menarche earlier (Steinberg, 1999). Finally, girls from developing countries who are adopted into affluent homes experience puberty earlier than peers in their home countries (Teilmann et al., 2006).

Historical data point to the same conclusion concerning the importance of nutrition and health care. In many industrialized countries around the world, the average age of menarche has declined steadily over the past 150 years. For example, in Europe the average age of menarche was 17 in 1840 compared with about 13 today. This drop reflects improvements in nutrition and better health care over this period. However, in these countries the age of menarche is no longer dropping, which suggests that with adequate nutrition the genetic lower limit for menarche is, on average, about 13 years.

What may surprise you is that the social environment also influences the onset of puberty, at least for girls. Menarche occurs at younger ages in girls who experience chronic stress or who are depressed (Belsky, Steinberg, & Draper, 1991; Moffit et al., 1992). For example, Ellis and Garber (2000) found that girls entered puberty at a younger age when their mothers' romantic relationships were stressful. Also, Belsky et al. (2007) discovered that girls have their first menstrual period at a younger age when their mothers used harsh punishment with them as preschoolers and young children.

The exact nature of these links is not known, but many explanations focus on the circumstances that would trigger the release of hormones that regulate menarche. One proposal is that when young girls experience chronic socioemotional stress—their family life is harsh and they lack warm, supportive parents—the hormones elicited by this stress may help to activate the hormones that trigger menarche. This mechanism would even have an evolutionary advantage: If events of a girl's life suggest that her future reproductive success is uncertain—as indicated by chronic socioemotional stress—then it may be adaptive to reproduce as soon as possible instead of waiting until later when she would be more mature and better able to care for her offspring. That is, the evolutionary gamble in this case might favor "lower quality" offspring early over "higher quality" offspring later (Ellis, 2004).

One variant of this explanation is *paternal investment* theory (Ellis & Essex, 2007; Ellis et al., 2003), which emphasizes the role of fathers in determining the timing of puberty. In this theory, when girls' childhood experiences indicate that fathers are invested in child rearing, this may delay the timing of maturation. But when those experiences indicate that fathers are uninvolved, this may trigger early maturation. Delaying puberty is adaptive when high-quality fathers are plentiful, because it allows the girl to mature herself; but accelerating puberty is adaptive when high-quality fathers are rare, because it allows a girl to be mature sexually should a high-quality father become available and because it means that her mother is likely to be young enough to help with child care. Consistent with this argument, girls experience menarche earlier when their father is absent due to divorce and the effect is compounded when the father is psychologically distant or has mental health problems (Tither & Ellis, 2008).

Psychological Impact of Puberty

Of course, teenagers are well aware of the changes taking places in their bodies. Not surprisingly, some of these changes affect adolescents' psychological development.

Body Image

Compared to children and adults, adolescents are much more concerned about their overall appearance. Many teenagers look in the mirror regularly, checking for signs of additional physical change. Generally, girls worry more than boys about appearance and are more likely to be dissatisfied with their appearance (Vander Wal & Thelen, 2000). Girls are particularly likely to be unhappy with their looks when appearance is a frequent topic of conversation with friends, leading girls to spend more time comparing their own appearance with that of their peers. Peers have relatively

Young adolescents are often quite concerned about their appearance.

The Apache celebrate menarche with a special ceremony in which a girl is said to become a legendary hero.

THINK ABOUT IT

The Apache have an elaborate celebration for menarche. Can you think of other similar ceremonies—perhaps not as elaborate—that take place to celebrate other milestones of adolescent development?

little influence on boys' satisfaction with their looks; instead, boys are unhappy with their appearance when they expect to have an idealized strong, muscular body but do not (Carlson Jones, 2004).

Response to Menarche and Spermarche

Carrie was horror writer Stephen King's first novel (and later a movie starring Sissy Spacek); it opens with a riveting scene in which the title character has her first menstrual period in the shower at school and, not knowing what is happening, fears that she will bleed to death. Fortunately, most adolescent girls today know about menstruation beforehand—usually from discussions with their mothers. Being prepared, their responses are usually fairly mild. Most girls are moderately pleased at this new sign of maturity but moderately irritated by the inconvenience and messiness of menstruation (Brooks-Gunn & Ruble, 1982). Girls usually tell their moms about menarche right away, and after two or three menstrual periods they tell their friends, too (Brooks-Gunn & Ruble, 1982).

Menarche is usually a private occasion for adolescents living in industrialized countries, but in traditional cultures it is often celebrated. For example, the Western Apache, who live in the southwest portion of the United States, traditionally have a spectacular ceremony to celebrate a girl's menarche (Basso, 1970). After a girl's first menstrual period, a group of older adults decide when the ceremony will be held and select a sponsor—a woman of good character and wealth (she helps to pay for the ceremony) who is unrelated to the initiate. On the day before the ceremony, the sponsor serves a large feast for the girl and her family; at the end of the ceremony, the family reciprocates, symbolizing that the sponsor is now a member of their family.

The ceremony itself begins at sunrise and lasts a few hours; it includes eight distinct phases in which the initiate, dressed in ceremonial attire, dances or chants, sometimes accompanied by her sponsor or a medicine man. The intent of these actions is to transform the girl into "Changing Woman," a heroic figure in Apache myth. With this transformation comes longevity and perpetual strength. The ceremony is a signal to all in the community that the initiate is now an adult, and it tells the initiate herself that her community now has adultlike expectations for her.

In contrast to menarche, much less is known about boys' reactions to spermarche. Most boys know about spontaneous ejaculations beforehand, and they get their information by reading, not by asking parents (Gaddis & Brooks-Gunn, 1985). When boys are prepared for spermarche, they feel more positively about it. Nevertheless, boys rarely tell parents or friends about this new development (Stein & Reiser, 1994).

Moodiness

Adolescents are often thought to be extraordinarily moody, moving from joy to sadness to irritation to anger over the course of a morning or afternoon. And the source of teenage moodiness is often presumed to be the influx of hormones associated with puberty—"hormones running wild." In fact, evidence indicates that adolescents are moodier than children or adults, but this is not primarily due to hormones (Steinberg, 1999). Scientists often find that rapid increases in hormone levels are associated with greater irritability and greater impulsivity, but the correlations tend to be small and are found primarily in early adolescence (Buchanan, Eccles, & Becker, 1992).

If hormones are not responsible, then what causes teenage moodiness? Some insights come from an elaborate study in which teenagers carried electronic pagers for a week (Csikszentmihalyi & Larson, 1984). When paged by researchers, the adolescents briefly described what they were doing and how they felt. The record of a typical adolescent is shown in ■ Figure 8.4. His mood shifts frequently from positive to negative, sometimes several times in a single day. For this boy, like most of the adolescents in the study, mood shifts were associated with changes in activities and social settings. Teens are more likely to report being in a good mood when with friends or when recreating; they tend to report being in a bad mood when in adult-regulated settings such as school classrooms or at a part-time job. Because adolescents often change activities and social settings many times in a single day, they appear to be moodier than adults.

Rate of Maturation

Although puberty begins at age 10 in the average girl and age 12 in the average boy, for many children puberty begins months or even years before or after these norms. An early-maturing boy might begin puberty at age 11, whereas a late-maturing boy might start at 15 or 16. An early-maturing girl might start puberty at age 9, a late-maturing girl at 14 or 15.

Maturing early or late has psychological consequences that differ for boys and girls. Several longitudinal studies show that early maturation can be harmful for girls. Girls who mature early often lack self-confidence, are less popular, are more likely to be depressed and have behavior problems, and are more likely to smoke and drink (Ge, Conger, & Elder, 2001; Mendle, Turkheimer, & Emery, 2007). Early maturation can also have life-changing effects on early-maturing girls who are pressured into sex and become mothers while still teenagers: as adults, they typically have less prestigious, lower-paying jobs (Mendle et al., 2007). These harmful outcomes are more likely when girls enter puberty early and their family life is marked by poverty or conflict with parents (Lynne-Landsman, Graber, & Andrews, 2010; Rudolph & Troop-Gordon, 2010).

These ill effects of early maturation are not necessarily the same for all groups of U.S. adolescents. In one study that included a nationally representative sample of American adolescents (Cavanagh, 2004), European American and Latina girls who matured early were twice as likely to be sexually active, but maturing early had no impact on sexual activity in African American girls. What's more, although the peer group influenced whether early-maturing girls were sexually active, the nature of that peer-group influence differed for European American and Latina girls. For early maturing European American girls, sexual activity was associated with having friends who did poorly in school and who engaged in problem behavior (e.g., drinking, fighting, skipping school). In contrast, for early maturing Latinas, sexual activity was associated with having older boys in the peer group, who apparently encourage them to engage in activities (e.g., drinking, smoking, sex) for which they are ill prepared.

The good news here is that the harmful effects of early maturation can be offset by other factors: For example, an early-maturing girl who has warm and supportive parents is less likely to suffer the consequences of early maturation (Ge et al., 2002).

The findings for boys are much more confusing. Some early studies suggested that early maturation benefits boys. For example, in an extensive longitudinal study of adoles-

Because children enter puberty at different ages, early-maturing teens tower over their later-maturing age-mates.

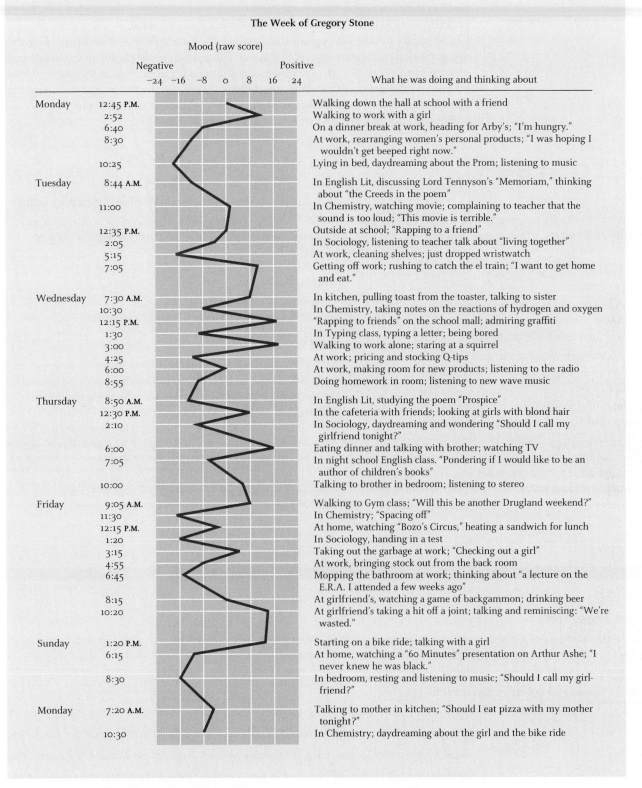

■ Figure 8.4

For most adolescents, mood shifts many times daily (from positive to negative and back), reflecting the many different activities and settings (some positive, some negative) that adolescents experience daily.

The Week of Gregory Stone

Mood (raw score)

Negative — Positive

−24 −16 −8 0 8 16 24

What he was doing and thinking about

Day	Time	What he was doing and thinking about
Monday	12:45 P.M.	Walking down the hall at school with a friend
	2:52	Walking to work with a girl
	6:40	On a dinner break at work, heading for Arby's; "I'm hungry."
	8:30	At work, rearranging women's personal products; "I was hoping I wouldn't get beeped right now."
	10:25	Lying in bed, daydreaming about the Prom; listening to music
Tuesday	8:44 A.M.	In English Lit, discussing Lord Tennyson's "Memoriam," thinking about "the Creeds in the poem"
	11:00	In Chemistry, watching movie; complaining to teacher that the sound is too loud; "This movie is terrible."
	12:35 P.M.	Outside at school; "Rapping to a friend"
	2:05	In Sociology, listening to teacher talk about "living together"
	5:15	At work, cleaning shelves; just dropped wristwatch
	7:05	Getting off work; rushing to catch the el train; "I want to get home and eat."
Wednesday	7:30 A.M.	In kitchen, pulling toast from the toaster, talking to sister
	10:30	In Chemistry, taking notes on the reactions of hydrogen and oxygen
	12:15 P.M.	"Rapping to friends" on the school mall; admiring graffiti
	1:30	In Typing class, typing a letter; being bored
	3:00	Walking to work alone; staring at a squirrel
	4:25	At work; pricing and stocking Q-tips
	6:00	At work, making room for new products; listening to the radio
	8:55	Doing homework in room; listening to new wave music
Thursday	8:50 A.M.	In English Lit, studying the poem "Prospice"
	12:30 P.M.	In the cafeteria with friends; looking at girls with blond hair
	2:10	In Sociology, daydreaming and wondering "Should I call my girlfriend tonight?"
	6:00	Eating dinner and talking with brother; watching TV
	7:05	In night school English class. "Pondering if I would like to be an author of children's books"
	10:00	Talking to brother in bedroom; listening to stereo
Friday	9:05 A.M.	Walking to Gym class; "Will this be another Drugland weekend?"
	11:30	In Chemistry; "Spacing off"
	12:15 P.M.	At home, watching "Bozo's Circus," heating a sandwich for lunch
	1:20	In Sociology, handing in a test
	3:15	Taking out the garbage at work; "Checking out a girl"
	4:55	At work, bringing stock out from the back room
	6:45	Mopping the bathroom at work; thinking about "a lecture on the E.R.A. I attended a few weeks ago"
	8:15	At girlfriend's, watching a game of backgammon; drinking beer
	10:20	At girlfriend's taking a hit off a joint; talking and reminiscing: "We're wasted."
Sunday	1:20 P.M.	Starting on a bike ride; talking with a girl
	6:15	At home, watching a "60 Minutes" presentation on Arthur Ashe; "I never knew he was black."
	8:30	In bedroom, resting and listening to music; "Should I call my girlfriend?"
Monday	7:20 A.M.	Talking to mother in kitchen; "Should I eat pizza with my mother tonight?"
	10:30	In Chemistry; daydreaming about the girl and the bike ride

cents growing up in Milwaukee during the 1970s (Simmons & Blyth, 1987), the early-maturing boys dated more often and had more positive feelings about their physical development and their athletic abilities. But other studies have supported the "off-time hypothesis" for boys. In this view, being early or late is stressful for boys, who strongly prefer to be "on time" in their physical development (Natsuaki, Biehl, & Ge, 2009). Yet another view is that puberty per se is stressful for boys but the timing is not (Ge et al., 2003).

Scientists cannot yet explain this bewildering pattern of results, but it's clear that the transition to puberty seems to have few long-lasting effects for boys. In contrast to what happens with girls, the effects associated with puberty and its timing vanish by young adulthood. When Pete, the late-maturing boy in the vignette, finally matures, others will treat him like an adult, and the few extra years of being treated like a child will not be harmful (Weichold & Silbereisen, 2005).

Test Yourself

RECALL

1. Puberty refers to changes in height and weight, to changes in the body's fat and muscle contents, and to _____.

2. Girls tend to have their growth spurts about _____ earlier than boys.

3. During adolescent physical growth, boys have greater muscle growth than girls, acquire less _____, and have greater increases in heart and lung capacity.

4. Primary sex characteristics are organs directly related to reproduction, whereas secondary sex characteristics are _____.

5. During puberty, the ovaries secrete estrogen, which causes the breasts to enlarge, the genitals to mature, and _____.

6. We know that nutrition and health determine the timing of puberty because puberty is earlier in girls who are taller and heavier, in regions of the world where nutrition and health care are adequate, and _____

7. Adolescents are moodier than children and adults primarily because _____.

8. Early maturation tends to be harmful to girls because _____.

INTERPRET

Compare and contrast the impact of rate of maturation—that is, maturing early versus late—on boys and girls.

APPLY

At first blush, the onset of puberty would seem to be due entirely to biology. In fact, the child's environment influences the onset of puberty. Summarize the ways in which biology and experience interact to trigger the onset of puberty.

Recall answers: (1) sexual maturation, (2) 2 years, (3) fat, (4) physical signs of maturity that are not linked directly to reproductive organs, such as the appearance of body hair, (5) fat to accumulate, (6) today, compared to earlier in history, (7) they change activities and social settings frequently, and their moods track these changes, (8) it leads them to associate with older adolescents and so they may become involved in activities for which they are ill prepared, such as drinking and sex.

8.2 Health

LEARNING OBJECTIVES

- What are the elements of a healthy diet for adolescents? Why do some adolescents suffer from eating disorders?

- Do adolescents get enough exercise? What are the pros and cons of participating in sports in high school?

- What are common obstacles to healthy growth in adolescence?

Dana had just started the seventh grade and was overjoyed that he could try out for the middle-school football team. He'd always excelled in sports and was usually the star when he played football on the playground or in gym class. But this was Dana's first opportunity to play on an actual team—with a real helmet, jersey, pads, and everything—and he was jazzed! Dana's dad played football in high school and thought Dana could benefit from the experience. His mom wasn't so sure—she was afraid that he'd be hurt and have to deal with the injury for the rest of his life.

ADOLESCENCE IS A TIME OF TRANSITION WHEN IT COMES TO HEALTH. On the one hand, teens are much less affected by the minor illnesses that would have kept them at home, in bed, as children. On the other hand, teens are at much greater risk for harm because of their unhealthy and risky behaviors. In this section, we'll look at some of the factors essential to adolescent health and see whether Dana's mother should be worried about sports-related injuries. We'll start with nutrition.

Many American teenagers eat far too many fast food meals, which are notoriously high in calories.

Nutrition

The physical growth associated with puberty means that the body has special nutritional needs. A typical teenage girl should consume about 2,200 calories per day; a typical boy should consume about 2,700 calories. (The exact levels depend on a number of factors, including body composition, growth rate, and activity level.) Teenagers also need calcium for bone growth and iron to make extra hemoglobin, the matter in red blood cells that carries oxygen. Boys need additional hemoglobin because of their increased muscle mass; girls need hemoglobin to replace that lost during menstruation.

Unfortunately, although many U.S. teenagers consume enough calories each day, too much of their intake consists of fast food rather than well-balanced meals. The result of too many meals of burgers, french fries, and a shake is that teens may get inadequate iron or calcium and far too much sodium and fat. With inadequate iron, teens are often listless and moody; with inadequate calcium, bones may not develop fully, placing the person at risk later in life for osteoporosis.

body mass index (BMI)
an adjusted ratio of weight to height; used to define "overweight"

basal metabolic rate
the speed at which the body consumes calories

Obesity

In part because of a diet high in fast foods, many American children and adolescents are overweight. *The technical definition of "overweight" is based on the* **body mass index** *(BMI), which is an adjusted ratio of weight to height.* Children and adolescents who are in the upper 5% (very heavy for their height) are defined as being overweight. Using these standards, in 2001 the U.S. Surgeon General announced that childhood obesity had reached epidemic proportions. In the past 25 to 30 years, the number of overweight children has doubled and the number of overweight adolescents has tripled, so that today roughly one child or adolescent out of six is overweight (U.S. Department of Health and Human Services, 2010).

Overweight youngsters are often unpopular and have low self-esteem (Puhl & Latner, 2007). They are also at risk for many medical problems throughout life, including high blood pressure and diabetes, because the vast majority of overweight children and adolescents become overweight adults (U.S. Department of Health and Human Services, 2010).

Heredity plays an important role in juvenile obesity. In adoption studies, children's and adolescents' weight is related to the weight of their biological parents, not to the weight of their adoptive parents (Stunkard et al., 1986). Genes may influence obesity by influencing a person's activity level. In other words, being genetically more prone to inactivity makes it more difficult to burn off calories and easier to gain weight. *Heredity may also help set the* **basal metabolic rate**, *the speed at which the body consumes calories.* Children and adolescents with a slower basal metabolic rate burn off calories less rapidly, making it easier for them to gain weight (Epstein & Cluss, 1986).

One's environment is also influential. Television advertising, for example, encourages youth to eat tasty but fattening foods. Parents play a role too. They may inadvertently encourage obesity by emphasizing external eating signals—"finish what's on your plate!"—rather than internal cues such as feelings of hunger. Thus, obese children and adolescents may overeat in part because they rely on external cues and disregard internal cues to stop (Coelho et al., 2009; Wansink & Sobal, 2007).

Childhood obesity has reached epidemic proportions in the United States.

Obese youth *can* lose weight. The most effective programs have the following features in common (Epstein et al., 2007; Foreyt & Goodrick, 1995; Israel et al., 1994).

- The focus of the program is to change obese children's eating habits, encourage them to become more active, and discourage sedentary behavior.

- As part of the treatment, children learn to monitor their eating, exercise, and sedentary behavior. Goals are established in each area, and rewards are earned when the goals are met.

- Parents are trained to help children set realistic goals and to use behavioral principles to help children meet these goals. Parents also monitor their own lifestyles to be sure they aren't accidentally fostering their child's obesity.

When programs incorporate these features, obese children do lose weight. However, even after losing weight, many of these children remain overweight. Consequently, it is best to avoid overweight and obesity in the first place; the *Surgeon General's Call for Action* emphasizes the role of increased physical activity and good eating habits in warding off overweight and obesity (U.S. Department of Health and Human Services, 2001). For example, children and adolescents can be encouraged to eat healthier foods by making them more available and by reducing their price (Faith et al., 2007). Frankly, however, we know relatively little about how to prevent obesity: a recent meta-analysis of obesity prevention programs (Stice, Shaw, & Marti, 2006) found that only 20% of these programs work—the remaining 80% were ineffective. The programs that did limit obesity were targeted at a broad range of healthy behaviors (e.g., not smoking, encouraging physical activity) and did not focus on obesity per se.

Fast food is not the only risky diet common among adolescents. Many teenage girls worry about their weight and are attracted to the "lose 10 pounds in 2 weeks!" diets advertised on TV and in teen magazines. Many of these diets are flatly unhealthy—they deprive youth of the many substances necessary for growth. Similarly, for philosophical or health reasons, many adolescents decide to eliminate meat from their diets. Vegetarian diets can be healthy for teens, but only when adolescents do more than eliminate meat. That is, vegetarians need to adjust the rest of their diet to assure that they have adequate sources of protein, calcium, and iron.

Yet another food-related problem common in adolescence are two similar eating disorders, anorexia and bulimia.

Anorexia and Bulimia

In 2006, Brazilian supermodel Ana Carolina Reston died of kidney failure just months after turning 21. At her death she weighed less than 90 pounds and had a body mass index of about 13—much lower than the 16 that is the benchmark for starvation. *Reston suffered from* **anorexia nervosa**, *a disorder marked by a persistent refusal to eat and an irrational fear of being overweight*. Individuals with anorexia nervosa have a grossly distorted image of their own body and claim to be overweight despite being painfully thin (Wilson, Heffernan, & Black, 1996). Anorexia is a very serious disorder, often leading to heart damage. Without treatment, as many as 15% of adolescents with anorexia die (Wang & Brownell, 2005).

A related eating disorder is bulimia nervosa. *Individuals with* **bulimia nervosa** *alternate between binge eating periods, when they eat uncontrollably, and purging through self-induced vomiting or with laxatives*. The frequency of binge eating varies remarkably among people with bulimia nervosa, from a few times a week to more than 30 times a week. What's common to all is the feeling that they cannot stop eating (Mizes & Palermo, 1997).

Adolescent girls with anorexia nervosa believe that they are overweight and they refuse to eat.

© Bubbles Photolibrary / Alamy

anorexia nervosa
persistent refusal to eat accompanied by an irrational fear of being overweight

bulimia nervosa
disease in which people alternate between binge eating—periods when they eat uncontrollably—and purging with laxatives or self-induced vomiting

Anorexia and bulimia are alike in many respects. Both disorders primarily affect females and emerge in adolescence (Wang & Brownell, 2005), and many of the same factors put teenage girls at risk for both eating disorders. Corinna Jacobi and her colleagues (2004) conducted a meta-analysis of more than 300 longitudinal and cross-sectional studies of individuals with eating disorders. They concluded that heredity puts some girls at risk, and molecular genetic studies have implicated genes that regulate both anxiety and food intake (Klump & Culbert, 2007). Several psychosocial factors also put people at risk for eating disorders. When children have a history of eating problems, such as being a picky eater or being diagnosed with pica (i.e., eating nonfood objects such as chalk, paper, or dirt), they're at greater risk for anorexia and bulimia during adolescence. Teenagers who experience negative self-esteem or mood or anxiety disorders are at risk (Hutchinson, Rapee, & Taylor, 2010). However, the most important risk factor for adolescents is being overly concerned about one's body and weight and having a history of dieting (George & Franko, 2010). And why do some teens become concerned about being thin? Peers and the media are the main factors. Teen-age girls worry about being overweight when they have friends who diet to stay thin and when they frequently watch TV shows that emphasize attractive, thin characters (Dohnt & Tiggemann, 2006; Paxton, Eisenberg, & Neumark-Sztainer, 2006).

The meta-analysis also identified some risk factors that are unique to anorexia and bulimia. For example, overprotective parenting is associated with adolescents becoming anorexic but not bulimic. In contrast, obesity in childhood is associated with adolescent bulimia but not anorexia.

Although eating disorders are far more common in girls, boys make up about 10% of diagnosed cases of eating disorders. Because boys with eating disorders are far less common, researchers have conducted much less research. However, some of the known risk factors are childhood obesity, low self-esteem, pressure from parents and peers to lose weight, and participating in sports that emphasize being lean (Ricciardelli & McCabe, 2004; Shoemaker & Furman, 2009).

Fortunately, there are programs that can help protect teens from eating disorders (Stice & Shaw, 2004). The most effective programs are designed for at-risk youth, such as those who already say they are unhappy with their body. The best programs are interactive and enable youth to become involved and to learn new skills, such as how to resist social pressure to be thin. They also work to change critical attitudes (such as ideals regarding thinness) and critical behaviors (such as dieting and overeating). At-risk adolescents who participate in these programs are helped; they are more satisfied with their appearance and less likely to diet or overeat. For those teens affected by eating disorders, treatment is available: Like prevention programs, treatment typically focuses on modifying key attitudes and behaviors (Puhl & Brownell, 2005).

THINK ABOUT IT

Describe how obesity, anorexia, and bulimia represent the different forces in the biopsychosocial network.

Physical Fitness

Being physically active promotes mental and physical health, both during adolescence and throughout adulthood. Individuals who regularly engage in physical activity reduce their risk for obesity, cancer, heart disease, diabetes, and psychological disorders, including depression and anxiety. "Regular activity" typically means exercising for 30 minutes, at least three times a week, at a pace that keeps an adolescent's heart rate at about 140 beats per minute (President's Council on Physical Fitness and Sports, 2004). Running, vigorous walking, swimming, aerobic dancing, biking, and cross-country skiing are all examples of activities that can provide this level of intensity.

Unfortunately, all the evidence indicates that most adolescents rarely get enough exercise. For example, in one study the researchers (Kann et al., 1995) asked high-school students whether they had exercised at least three times for 20 minutes during the past week at a level that made them sweat and breathe hard. In ninth grade, about 75% of boys and 65% of girls said they had; by twelfth grade, these figures had dropped to 65% for boys and 40% for girls. Part of the problem here is that, for many high-school students, physical education classes provide the only regular opportunity

THINK ABOUT IT

Many teenagers do not eat well-balanced meals, and many do not get enough exercise. What would you do to improve teenagers' dietary and exercise habits?

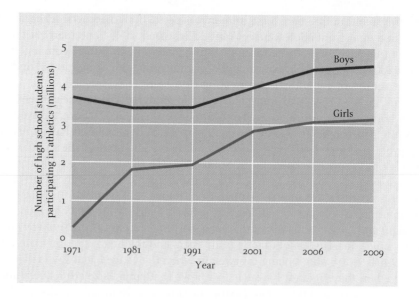

Girls' participation in sports has increased steadily since 1972, when the U.S. government required equal athletic opportunities for boys and girls. Data from National Federation of State High School Associations (2011).

for exercise, yet a minority of high school students are enrolled in physical education and most who are enrolled do not attend daily.

Many teenagers get exercise by participating in organized sports. Today, approximately 4.3 million boys and 3 million girls participate in sports. Although about 1.1 million more boys than girls participate, ■ Figure 8.5 shows that the difference is smaller than it once was. In 1971, about 3.7 million boys participated compared with only about 300,000 girls. However, in 1972 the U.S. government required that schools receiving public funds provide equal educational and athletic opportunities for boys and girls. Since that time, girls' participation in sports has grown steadily (National Federation of State High School Associations, 2008, 2011).

The most popular sport for boys is football; more than a million boys play high-school football. The next most popular sports are track and field, basketball, and baseball. For girls, the most popular sport is track and field; approximately half a million girls participate. The next most popular sports are basketball, volleyball, and fast-pitch softball (National Federation of State High School Associations, 2011).

Participating in sports has many benefits for youth. Sports can enhance participants' self-esteem and can help them to learn initiative (Bowker, 2006; Donaldson & Ronan, 2006). Sports can also provide adolescents a chance to learn important social skills, such as how to work effectively as part of a group, often in complementary roles. At the same time, there are some potential costs. About 15% of high school athletes will be injured and require some medical treatment. Boys are most likely to be injured while playing football or wrestling; girls are injured while participating in cross-country or soccer (Rice, 1993). Fortunately, most of these injuries are not serious and are more likely to involve mere bruises or strained muscles (Nelson, 1996). Dana's mom can rest easy; the odds are that he won't be injured, and if he is, it won't be serious.

A more serious problem is the use of illegal drugs to improve performance (American Academy of Pediatrics, 2005). Some athletes use anabolic steroids—drugs that are chemically similar to the male hormone testosterone—to increase muscle size and strength and to promote more rapid recovery from injury. Approximately 2% of high school students report having used anabolic steroids, with use more frequent in boys and in younger students (Dunn & White, 2011; vandenBerg, Neumark-Sztainer, & Wall, 2007). This is disturbing because steroid use can damage the liver, reproductive

Track is now the most popular sport for adolescent girls.

The Plain Dealer / Landov

system, skeleton, and cardiovascular system (increasing blood pressure and cholesterol levels); in addition, use of anabolic steroids is associated with mood swings, aggression, and depression (Kanayama, Hudson, & Pope, 2008). Parents, coaches, and
health professionals need to be sure that high school athletes are aware of the dangers
of steroids and should encourage youth to meet their athletic goals through methods
that do not involve drug use (American Academy of Pediatrics, 2005).

Threats to Adolescent Well-Being

Every year, approximately 1 U.S. adolescent out of 1,000 dies. Relatively few die from
disease; instead, they are killed in accidents that typically involve automobiles or
firearms. ■ Figure 8.6 shows that the pattern of adolescent death depends, to a large
extent, on gender and ethnicity. Among boys, most deaths are due to accidents involving motor vehicles or firearms. For European American, Latino American, and Asian
American boys, motor vehicles are more deadly than guns, but the reverse is true for
African American boys. Among girls, most deaths are due to natural causes or accidents involving motor vehicles. For European American girls, motor vehicle accidents
account for nearly half of all deaths; for African American girls, natural causes account for nearly half of all deaths; and for Latina American and Asian American girls,
natural causes and motor vehicles account for about the same number of deaths and
together account for about two thirds of all deaths (Federal Interagency Forum on
Child and Family Statistics, 2005).

Sadly, many of these deaths are completely preventable. Deaths in automobile accidents are often linked to driving too fast, drinking alcohol, and not wearing seatbelts
(U.S. Department of Health and Human Services, 1997). And deaths due to guns are
often linked to "all too easy" access to firearms in the home: In far too many homes,
firearms are stored loaded, unlocked, or both (Johnson et al., 2006).

■ **Figure 8.6**
Adolescent boys are much more likely than
adolescent girls to die from accidents or
use of firearms, and this is particularly true
for African American and Latino American
teenage boys.

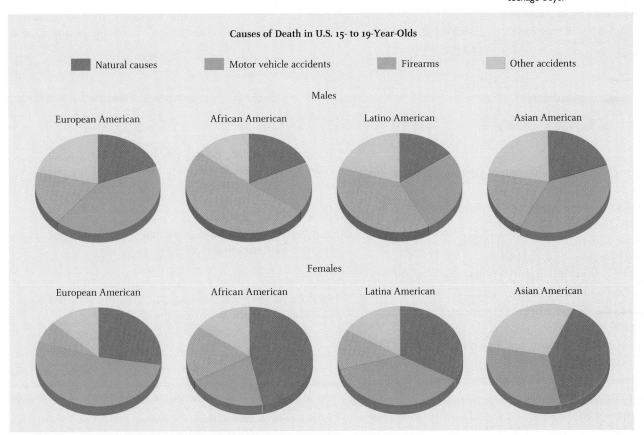

From Federal Interagency Forum on Child and Family Statistics, 2005.

Adolescents are far more accident prone than children or adults, in part because they believe that the rewards of risky behavior far outweigh the potential harm.

Adolescent deaths from accidents can be explained, in part, because adolescents take risks that adults often find unacceptable (Nell, 2002). Teens take unnecessary risks while riding skateboards, scooters, or bicycles. They drive cars recklessly, engage in unprotected sex, and sometimes use illegal and dangerous drugs (we'll discuss this more in Chapter 9). Although it is tempting to call such behavior "stupid" or "irrational," research suggests that adolescent risk taking is quite complex. In fact, adolescents vastly *overestimate* the actual likelihood of harm associated with many risky behaviors. They exaggerate the overall dangers of drunk driving and unprotected sex (Millstein & Halpern-Felsher, 2002). But at the same time they believe that, as individuals, they are much less likely than their peers to experience the harmful consequences of risky behaviors. In other words, driving drunk and having unprotected sex are viewed as dangerous, but only to others (Reyna & Farley, 2006).

The illusion of invulnerability is a piece of the puzzle, but it can't be a complete account of adolescent risk taking: Young adults have the same illusion of invulnerability, yet they are far less likely to engage in risky behaviors. If both adolescents and young adults believe they are invulnerable, why are adolescents more often engaged in high-risk behaviors? They find the rewards associated with risky behavior far more appealing than adults do—so much so that they're willing to ignore the risks. For many adolescents, the pleasure, excitement, and intimacy of sex far outweigh the risks of disease and pregnancy, just as the relaxation associated with smoking outweighs the threat of lung cancer (Halpern-Felsher & Cauffman, 2001; Reyna & Farley, 2006). And, as we saw earlier in this chapter (pages 287–288), the appeal of high-risk behaviors reflects the maturity of the pleasure-seeking brain regions relative to those regions that control behavior (Somerville & Casey, 2010).

Test Yourself

RECALL

1. An adolescent's diet should contain adequate calories, _____, and iron.

2. A vegetarian diet can be healthy for teens, but only when adolescents _____.

3. Individuals with _____ alternate between binge eating and purging.

4. During adolescence, the most important risk factors for anorexia and bulimia are _____.

5. Regular physical activity helps to promote _____ and physical health.

6. Girls' participation in sports has grown steadily since 1972 when _____.

7. Some teenage athletes use anabolic steroids to increase muscular strength and to _____.

8. More teenage girls die from _____ than any other single cause.

9. Because they place greater emphasis on the _____ actions, adolescents make what adults think are risky decisions.

INTERPRET

Distinguish the biological factors that contribute to obesity from the environmental factors.

APPLY

How does adolescent risk taking illustrate the idea that individuals help to shape their own development?

Recall answers: (1) calcium, (2) adjust the rest of their diet so they consume adequate protein, calcium, and iron, (3) bulimia nervosa, (4) being overly concerned about one's body and a history of dieting, (5) mental health, (6) the U.S. government required that schools receiving public funds provide equal athletic opportunities for boys and girls, (7) promote more rapid recovery from an injury, (8) automobile accidents, (9) rewards associated with their

LEARNING OBJECTIVES

■ How do working memory and processing speed change in adolescence?

■ How do increases in content knowledge, strategies, and metacognitive skill influence adolescent cognition?

■ What changes in problem-solving and reasoning take place in adolescence?

Calvin, a 14-year-old boy, was an enigma to his mother, Crystal. On one hand, Calvin's growing reasoning skills impressed and sometimes even surprised her: He not only readily grasped technical discussions of her medical work, he also was becoming adept at finding loopholes in her explanations of why he wasn't allowed to do some things with his friends. On the other hand, sometimes Calvin was a real teenage "space cadet." Simple problem solving stumped him, or he made silly mistakes and got the wrong answer. Calvin didn't correspond to Crystal's image of the formal-operational thinker that she remembered from her college human development class.

FOR INFORMATION-PROCESSING THEORISTS, adolescence does not represent a distinct, qualitatively different stage of cognitive development. Instead, adolescence is considered to be a transitional period between the rapidly changing cognitive processes of childhood and the mature cognitive processes of young adulthood. Cognitive changes do take place in adolescence, but they are small compared to those seen in childhood. Adolescence is a time when cognitive processes are "tweaked" to adult levels. We'll describe these changes in this section and, as we do, we'll see why adolescents like Crystal's son don't always think as effectively as they might.

Working Memory and Processing Speed

Working memory is the site of ongoing cognitive processing, and processing speed is the speed with which individuals complete basic cognitive processes. Both of these capacities achieve adultlike levels during adolescence. Adolescents' working memory has about the same capacity as adults' working memory, which means that teenagers are better able than children to store information needed for ongoing cognitive processes. In addition, ■ Figure 8.7 illustrates changes in processing speed, exemplified in this case by performance on a simple response-time task in which individuals press

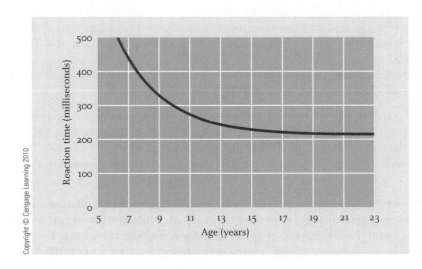

Copyright © Cengage Learning 2010

■ **Figure 8.7**
Response time declines steadily during childhood and reaches adultlike levels during middle adolescence. Data from Kail (2004).

a button as rapidly as possible in response to a visual stimulus. The time needed to respond drops steadily during childhood—from about one third of a second at age 8 to one quarter of a second at age 12—but changes little thereafter. This pattern of change is not specific to simple response time but is, instead, found for a wide range of cognitive tasks: Adolescents generally process information just about as quickly as young adults (Kail, 2004). Change in working memory and processing speed means that, compared to children, adolescents process information very efficiently.

These changes in efficiency reflect the maturational changes to the brain that were described earlier in this chapter (pages 287–288). In particular, increases in myelination during adolescence allow nerve impulses to travel more rapidly, which contributes to more rapid and more efficient information processing during this period (Mabbott et al., 2006; Schmithorst & Yuan, 2010).

Content Knowledge, Strategies, and Metacognitive Skill

As children move into adolescence, they acquire adultlike levels of knowledge and understanding in many domains. Children, for example, may enjoy baseball or computers, but as adolescents they acquire true expertise. For example, many parents turn to their teens for help in learning how to use fancy features on their cell phone. This increased knowledge is useful for its own sake, but it also has the indirect effect of enabling adolescents to learn, understand, and remember more of new experiences (Schneider & Bjorklund, 1998; Schneider & Pressley, 1997). Imagine two middle school students—one a baseball expert, the other not—watching a baseball game. Compared to the novice, the adolescent expert would understand many of the nuances of the game and, later, remember many more features of the game.

As their content knowledge increases, adolescents also become much better skilled at identifying strategies appropriate for a specific task, then monitoring the chosen strategy to verify that it is working (Schneider & Pressley, 1997). For example, adolescents are more likely to outline and highlight information in a text. They are more likely to make lists of material they don't know well and should study more, and they more often embed these activities in a master study plan (e.g., a list of assignments, quizzes, and tests for a 2-week period). All these activities help adolescents learn more effectively and remember more accurately (Schneider & Pressley, 1997; Thomas et al., 1993).

Problem Solving and Reasoning

Adolescents typically solve problems more readily than children, in part because their approach is more sophisticated. Often children rely upon heuristics, rules of thumb that do not guarantee a solution but are useful in solving a range of problems. Heuristics tend to be fast and require little effort. In contrast, adolescents are more likely to solve problems analytically—determining an answer mathematically or logically, depending upon the nature of the problem (Klaczynski, 2004; Stanovich, Toplak, & West, 2008).

To see the difference between heuristic and analytic solutions, think about the following problem:

> Erica wants to go to a baseball game to try to catch a fly ball. She calls the main office and learns that almost all fly balls have been caught in section 43. Just before she chooses her seats, she learns that her friend Jimmy caught two fly balls last week sitting in section 10. Which section is most likely to give Erica the best chance to catch a fly ball? (Kokis et al., 2002, p. 34).

One solution to this problem—more common among children—involves a heuristic that relies on personal experience: When in doubt, imitate other people who have been successful. In this case, that means sitting where the friend sat. The analytic solution—more common among adolescents—involves relying upon the statistical

Adolescents often have adultlike skills in some domains, such as using cell phones, which allows them to teach adults.

THINK ABOUT IT

Students typically are introduced to the study of complex topics such as philosophy and experimental science during adolescence. Explain how their maturing cognitive skills contribute to the study of these and other subject areas.

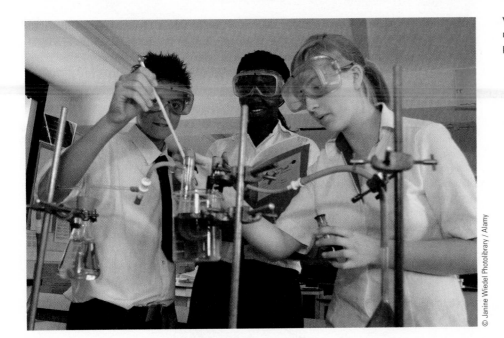

When children reach adolescence, they are more likely to use reasoning skills and solve problems analytically.

© Janine Wiedel Photolibrary / Alamy

information that, historically, the odds of catching a fly ball are greatest in section 43 (Kokis et al., 2002).

Not only do adolescents have more sophisticated approaches to reasoning and problem solving, they are better skilled at finding weaknesses in arguments. In scientific reasoning, for example, adolescents recognize the hazards in making generalizations from extremely small samples. They would be wary of concluding that people from another country are particularly friendly based on meeting just two people from that country (Klaczynski & Lavallee, 2005). And, as we see in the Spotlight on Research feature, they can pinpoint certain kinds of flaws in logical arguments.

Spotlight on research

Adolescents Can Identify Fallacies in Arguments

Who were the investigators, and what was the aim of the study? In formal debates as well as informal conversations, people sometimes rely upon arguments that are irrelevant to the discussion. One of the most common is an *ad hominem* argument, in which one attacks the person making a claim, not the claim itself. If one adolescent is arguing that *Grey's Anatomy* is the best TV show ever, a peer's *ad hominem* argument would be to say, "You only think that because your sister is a surgical intern." This statement may be true (the sister is an intern and that's why the peer likes *Grey's Anatomy*), but it's irrelevant to the debate about the quality of the TV show. Michael Weinstock and his colleagues at Ben Gurion University of the Negev—Yair Neuman and Amnon Glassner (2006)—wanted to determine how well adolescents could identify these reasoning fallacies like the *ad hominem* argument.

How did the investigators measure the topic of interest? The researchers created eight brief scenarios involving a discussion between two indi-

viduals. In two of the scenarios, one person made an *ad hominem* argument. For example, in a scenario involving an argument about the presence of living creatures elsewhere in the universe, an *ad hominem* argument involved saying that a person does not believe in life elsewhere because the person is unimaginative. Other scenarios involved *ad populum* arguments, which assert that a claim must be true because most people believe it to be true. In other words, this is "truth by popular vote." In the scenario about life elsewhere in the universe, an *ad populum* argument would say that "most people in Europe and the United States think there are living creatures on other planets in the universe." Finally, some scenarios involved *ad ignorantiam* arguments, which assert that a claim must be true because no one has shown that it's false: "There must be living creatures on other planets because nobody has proven that Earth is the only planet with living creatures." There were also scenarios that had no false

arguments; participants read each scenario and were asked to identify any problems in the arguments.

Who were the children in the study? Weinstock and his colleagues tested fifty-three 13-year-olds, fifty-eight 15-year-olds, and eighty-two 17-year-olds. The sample included approximately 40% boys and 60% girls. These adolescents attended a combined junior and senior high school in rural Israel. For simplicity, we'll just discuss the results for the youngest and oldest students.

What was the design of the study? This study was experimental because Weinstock and colleagues included two independent variables: the age of the participant and the nature of the false argument in the scenarios (*ad hominem*, *ad populum*, *ad ignorantiam*). The dependent variable was the percentage of times that the students detected the flawed argument. The study was cross-sectional because 13-, 15-, and 17-year-olds were all tested at approximately the same time.

[continued]

Were there ethical concerns with the study? No. The scenarios involved topics that students were likely to encounter and discuss in daily life.

What were the results? The results are illustrated in ■ Figure 8.8 and show the percentage of false arguments that were detected by each of the two age groups. Two patterns are evident in these results. First, older students had greater scores for each type of flawed argument; that is, older adolescents were more likely to detect each kind of flawed argument. Second, at both ages the *ad populum* arguments ("truth by popular vote") were easiest to detect while *ad ignorantiam* arguments were the most difficult. It's also important to notice that most of the aver-

ages shown in the graph are below 50%, which means that adolescents miss more than half of the flawed arguments.

What did the investigators conclude? Weinstock and colleagues concluded that the ability to detect flawed arguments improves substantially during adolescence, although even the oldest adolescents in the sample were far from perfect in identifying the *ad hominem* and *ad ignorantiam* fallacies. The researchers suggested several reasons for the improvement, including greater experience with argumentation and improved metacognitive skills.

What converging evidence would strengthen these conclusions? An obvious way to bolster

these results would be to conduct a longitudinal study that showed how these skills unfold over time for an individual. For example, the averages shown in Figure 8.8 indicate that *ad populum* fallacies were the easiest to detect and *ad ignorantiam* flaws were the hardest. A longitudinal study could confirm this sequence by showing that, as individuals develop, they first master the *ad populum* fallacy, then the *ad hominem* fallacy, and finally the *ad ignorantiam* fallacy.

 Go to Psychology CourseMate at www.cengagebrain.com to enhance your understanding of this research.

■ **Figure 8.8**

Older adolescents detect more flawed arguments; *ad populum* flaws are the easiest to detect and *ad ignorantiam* flaws are the most difficult. Data from Weinstock et al. (2006).

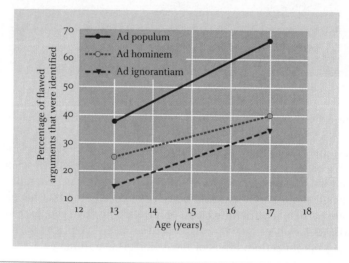

The ability to detect flawed arguments is yet another demonstration of improved information processing during adolescence. Of course, adolescents may not always use their skills effectively. Sometimes they resort to heuristics because they take less effort and are "good enough" for the problem. Also, sometimes adolescents' beliefs interfere with effective thinking: When evidence is inconsistent with adolescents' beliefs, they may dismiss the evidence as being irrelevant or try to reinterpret the evidence to make it consistent with their beliefs. For example, a Baptist teenager may be quick to find flaws in a study showing that Baptists are poor leaders but overlook similar flaws in a study showing that Baptists are smarter than most people (Klaczynski & Lavallee, 2005; Klaczynski & Narasimham, 1998). Thus, adolescents use their reasoning skills selectively, raising their standards to dismiss findings that threaten their beliefs and lowering them to admit findings compatible with their beliefs.

Findings like these tell us that Crystal, the mother in the opening vignette, should not be so perplexed by her son's seemingly erratic thinking: Adolescents (and adults, for that matter) do not always use the most powerful levels of thinking that they possess. The information-processing account of intellectual functioning in adolescence is really a description of how children and adolescents *can* think, not how they always or even usually think.

These changing features of information processing are summarized in ● Table 8.1. Change in each of these elements of information processing occurs gradually. When combined, they contribute to the steady progress to mature thinking that is the destination of adolescent cognitive development.

Information Processing During Adolescence

Feature	State in Adolescence
Working memory and processing speed	Adolescents have adultlike working memory capacity and processing speed, enabling them to process information efficiently.
Content knowledge	Adolescents' greater knowledge of the world facilitates understanding and memory of new experiences.
Strategies and metacognition	Adolescents are better able to identify task-appropriate strategies and to monitor the effectiveness of those strategies.
Problem-solving and reasoning	Adolescents often solve problems analytically by relying on mathematics or logic, and they are able to detect weaknesses in scientific evidence and logical arguments.

Test Yourself

RECALL

1. According to information-processing theorists, adolescence is a time of important changes in working memory, processing speed, _____, strategies, and metacognition.

2. Information-processing theorists view adolescence as a time of _____.

3. When solving problems, children often rely upon heuristics but adolescents are more likely to solve problems _____.

4. When evidence is inconsistent with their beliefs, adolescents often _____.

INTERPRET

The information-processing account of cognitive change in adolescence emphasizes working memory, knowledge, and strategies. How might each of these factors be influenced by nature? By nurture?

APPLY

How can the information-processing skills described here—and, in particular, the limits in adolescent thinking—help to explain adolescent risk taking described on pages 299–300?

Recall answers: (1) content knowledge, (2) gradual cognitive change, (3) analytically, using mathematics or logic, depending upon the problem, (4) ignore or dismiss the evidence

---

8.4 Reasoning About Moral Issues

LEARNING OBJECTIVES

- How do adolescents reason about moral issues?
- Is moral reasoning similar in all cultures?
- How do concern for justice and caring for other people contribute to moral reasoning?
- What factors help promote more sophisticated reasoning about moral issues?

Howard, the least popular boy in the entire eighth grade, had been wrongly accused of stealing a sixth grader's iPod. Min-shen, another eighth grader, knew that Howard was innocent but said nothing to the school principal for fear of what his friends would say about siding with Howard. A few days later, when Min-shen's father heard about the incident, he was upset that his son apparently had so little "moral fiber." Why hadn't Min-shen acted in the face of an injustice?

ONE DAY THE LOCAL PAPER HAD TWO ARTICLES ABOUT YOUTH FROM THE AREA. One article was about a 15-year-old girl who was badly burned while saving her younger brothers from a fire in their apartment. Her mother said she wasn't surprised by her daughter's actions because she had always been an extraordinarily caring person. The other article was about two 17-year-old boys who had beaten an elderly man to death.

They had only planned to steal his wallet, but when he insulted them and tried to punch them, they became enraged.

Reading articles like these, you can't help but question why some teenagers (and adults, as well) act in ways that earn our deepest respect and admiration, whereas others earn our utter contempt as well as our pity. And, at a more mundane level, we wonder why Min-shen didn't tell the truth about the stolen iPod to the principal. In this section, we'll start our exploration of moral reasoning with an influential theory proposed by Lawrence Kohlberg.

Kohlberg's Theory

Some of the world's great novels are based on moral dilemmas. Victor Hugo's *Les Misérables*, for example, begins with the protagonist, Jean Valjean, stealing a loaf of bread to feed his sister's starving child. You could probably think of many reasons Valjean should have stolen the bread as well as arguments why he shouldn't have stolen the bread. Lawrence Kohlberg created stories like this one in which decisions were difficult because every alternative involved some undesirable consequences. In fact, there is no "correct" answer—that's why the stories are referred to as moral "dilemmas." Kohlberg was more interested in the reasoning used to justify a decision— Why should Jean Valjean steal the bread? Why should he not steal the bread?—than in the decision itself.

Kohlberg's (1969) best-known moral dilemma is this story about Heinz, whose wife is dying:

> In Europe, a woman was near death from cancer. One drug might save her, a form of radium that a druggist in the same town had recently discovered. The druggist was charging $2,000, ten times what the drug cost him to make. The sick woman's husband, Heinz, went to everyone he knew to borrow the money, but he could only get together about half of what it cost. He told the druggist that his wife was dying and asked him to sell it cheaper or let him pay later. But the druggist said, "No." The husband got desperate and broke into the man's store to steal the drug for his wife. (p. 379)

Thus, Heinz and Jean Valjean both face moral dilemmas in which the various alternative courses of action have desirable and undesirable features.

Kohlberg analyzed children's, adolescents', and adults' responses to a large number of dilemmas and identified three levels of moral reasoning, each divided into two stages. Across the six stages, the basis for moral reasoning shifts. In the earliest stages, moral reasoning is based on external forces, such as the promise of reward or the threat of punishment. At the most advanced levels, moral reasoning is based on a personal, internal moral code and is unaffected by others' views or society's expectations. Let's take a closer look.

Kohlberg identified three levels of moral reasoning: preconventional, conventional, and postconventional. Each level is further subdivided into two substages. *At the* **preconventional level**, *moral reasoning is based on external forces.* For most children, many adolescents, and some adults, moral reasoning is controlled almost exclusively by rewards and punishments. *Individuals in Stage 1 moral reasoning assume an* **obedience orientation**, *which means believing that authority figures know what is right and wrong.* Consequently, Stage 1 individuals do what authorities say is right to avoid being punished. At this stage, one might argue that Heinz shouldn't steal the drug because an authority figure (e.g., parent or police officer) said he shouldn't do it. Alternatively, one might argue that he should steal the drug because he would get into trouble if he let his wife die.

In Stage 2 of the preconventional level, people adopt an **instrumental orientation**, *in which they look out for their own needs.* Stage 2 individuals are nice to others because they expect the favor to be returned in the future. Someone at this stage could

preconventional level
first level of reasoning in Kohlberg's theory, where moral reasoning is based on external forces

obedience orientation
characteristic of Kohlberg's Stage 1, in which moral reasoning is based on the belief that adults know what is right and wrong

instrumental orientation
characteristic of Kohlberg's Stage 2, in which moral reasoning is based on the aim of looking out for one's own needs

justify stealing the drug because Heinz's wife might do something nice for Heinz in return. Or, they might argue that Heinz shouldn't steal the drug because it will create more problems for him if his wife remains bedridden and he is burdened with caring for her.

At the **conventional level**, *adolescents and adults look to society's norms for moral guidance.* In other words, people's moral reasoning is largely determined by others' expectations of them. *In Stage 3, adolescents' and adults' moral reasoning is based on* **interpersonal norms**. The aim is to win the approval of other people by behaving as "good boys" and "good girls" would. Stage 3 individuals might argue that Heinz shouldn't steal the drug because he must keep his reputation as an honest man, or that he should steal the drug because no one would think negatively of him for trying to save his wife's life.

Stage 4 of the conventional level focuses on **social system morality**. Here, adolescents and adults believe that social roles, expectations, and laws exist to maintain order within society and to promote the good of all people. Stage 4 individuals might reason that Heinz shouldn't steal the drug, even though his wife might die, because it is illegal and no one is above the law. Alternatively, they might claim that he should steal it to live up to his marriage vow of protecting his wife, even though he will face negative consequences for his theft.

At the **postconventional level**, *moral reasoning is based on a personal moral code.* The emphasis is no longer on external forces like punishment, reward, or social roles. *In Stage 5, people base their moral reasoning on a* **social contract**. Adults agree that members of social groups adhere to a social contract because a common set of expectations and laws benefits all group members. However, if these expectations and laws no longer promote the welfare of individuals, they become invalid. Consequently, Stage 5 individuals might reason that Heinz should steal the drug because social rules about property rights no longer benefit individuals' welfare. (Indeed, the Declaration of Independence, written by Thomas Jefferson in 1776, made a similar argument about the laws of England.) They could alternatively argue that he shouldn't steal it because it would create social anarchy.

Finally, in Stage 6 of the postconventional level, **universal ethical principles** *dominate moral reasoning.* Abstract principles such as justice, compassion, and equality form the basis of a personal code that may sometimes conflict with society's expectations and laws. Stage 6 individuals might argue that Heinz should steal the drug because saving a life takes precedence over everything, including the law. Or they might claim that Heinz's wife has a right to die and that he should not force his views on her by stealing and administering the drug.

Putting the stages together, the entire sequence of moral development looks like this:

Preconventional Level: Punishment and Reward
    Stage 1: Obedience to authority
    Stage 2: Nice behavior in exchange for future favors

Conventional Level: Social Norms
    Stage 3: Live up to others' expectations
    Stage 4: Follow rules to maintain social order

Postconventional Level: Moral Codes
    Stage 5: Adhere to a social contract when it is valid
    Stage 6: Personal moral system based on abstract principles

The developmental sequence described by Kohlberg usually takes many years to unfold. But on occasion, we may see the process occur much more dramatically, such as when individuals undergo a major transformation in their moral motivation. One noteworthy example of such a transformation was depicted in Steven Spielberg's Oscar-winning movie *Schindler's List*, as described in the Real People feature.

conventional level
second level of reasoning in Kohlberg's theory, where moral reasoning is based on society's norms

interpersonal norms
characteristic of Kohlberg's Stage 3, in which moral reasoning is based on winning the approval of others

social system morality
characteristic of Kohlberg's Stage 4, in which moral reasoning is based on maintenance of order in society

postconventional level
third level of reasoning in Kohlberg's theory, in which morality is based on a personal moral code

social contract
characteristic of Kohlberg's Stage 5, in which moral reasoning is based on the belief that laws are for the good of all members of society

universal ethical principles
characteristic of Kohlberg's Stage 6, in which moral reasoning is based on moral principles that apply to all

The outbreak of war typically provides numerous opportunities for shrewd businesspeople to profit from the increased demand for manufactured goods. The outbreak of World War II in Europe in 1939 was no exception. Oskar Schindler was one such entrepreneur who made a great deal of money working for the Germans after they conquered Poland. His flamboyant demeanor brought him to the attention of the local German commanders, for whom Schindler did favors. Motivated at first strictly by the potential for personal profit, he opened—with few, if any, qualms—a factory in which he employed Jews as slave labor.

Schindler's company was quite successful. But as the war continued, official German policy toward Jews changed to one of extermination. Jewish citizens in Poland and other countries were rounded up and shipped to concentration camps or summarily executed. Schindler was deeply disturbed by this, and his attitudes began

to change. His employees suggested that he give the Germans a list of workers essential to the factory's continued operation. The list provided protection because the plant's products were used in the war effort. This, of course, also kept the profits rolling in. But Schindler's motivation gradually underwent a profound transformation as well. No longer driven by profit, he went to great lengths to preserve life, at no small danger to himself. He created cover stories to support his claims that certain employees were essential, and he went to Auschwitz to rescue employees who were sent there by mistake.

Oskar Schindler's list saved many lives. Profits were made (and helped provide the perfect cover), but he employed Jews in his factory primarily to save them from the gas chamber. Schindler may have begun the war at Kohlberg's preconventional level—where he was motivated solely by personal profit—but he ultimately moved to the postconventional level—where he

was motivated by the higher principle of saving lives. It is at the postconventional level that heroes are made.

During World War II, Oskar Schindler saved the lives of many Jews by adding their names to lists of employees who were essential for his factory's operation.

Support for Kohlberg's Theory

Kohlberg proposed individuals move through the six stages only in the order listed and in only that order. Consequently, older and more advanced thinkers should be more advanced in their moral development, and indeed they usually are (Stewart & Pascual-Leone, 1992). In addition, longitudinal studies show that individuals progress through each stage in sequence, and virtually no individuals skip stages (Colby et al., 1983).

Further support for Kohlberg's theory comes from research on the link between moral reasoning and moral behavior. Less advanced moral reasoning reflects the influence of external forces (e.g., rewards), but more advanced reasoning is based on a personal moral code. Therefore, individuals at the preconventional and conventional levels would act morally when external forces demand, but otherwise they might not. In contrast, individuals at the postconventional level, where reasoning is based on personal principles, should be compelled to moral action even when external forces may not favor it.

Teenagers who engage in moral behavior, such as participating in protest marches, often reason at high levels in Kohlberg's theory.

Consistent with this claim, adolescents who would defend their principles in difficult situations tend to be more advanced in Kohlberg's stages (Gibbs et al., 1986). For example, students like those in the photograph who protest social conditions tend to have higher moral reasoning scores. This explains why Min-shen, the boy in the vignette, said nothing. Speaking out on behalf of the unpopular student is unlikely to lead to reward and violates social norms against "squealing" on friends. Consequently, an eighth grader—who is probably in the preconventional or conventional level of moral reasoning—would probably let the unpopular student be punished unfairly.

On some other features, Kohlberg's theory does not fare as well. One is that moral reasoning is not as consistent as would be expected from the theory. Teenagers reasoning at the conventional level should always base their moral decisions based on others' expectations; in reality, however, such consistency is not the norm. Moral reasoning may be advanced on some problems but much less sophisticated on others (Krebs & Denton, 2005).

Another concern is Kohlberg's claim that his sequence of stages is universal: All people in all cultures should progress through the six-stage sequence. Indeed, children and adolescents in cultures worldwide reason about moral dilemmas at Stages 2 or 3, just like North American children and adolescents (Krebs & Denton, 2005). But we'll see in the next section that, in other cultures, moral reasoning beyond the earliest stages is often not described well by Kohlberg's theory (Turiel, 2006).

Cultural Differences in Moral Reasoning

Many critics note that Kohlberg's emphasis on individual rights and justice reflects traditional American culture and Judeo-Christian theology. Not all cultures and religions share this emphasis; consequently, moral reasoning might be based on different values in other cultures (Turiel, 2006).

The Hindu religion, for example, emphasizes duty and responsibility to others, not individual rights and justice (Simpson, 1974). Accordingly, children and adults reared with traditional Hindu beliefs might emphasize caring for others in their moral reasoning more than individuals brought up in the Judeo-Christian tradition.

Miller and Bersoff (1992) tested the hypothesis that cultural differences affect moral reasoning by constructing dilemmas with both justice- and care-based solutions. For example:

> Ben planned to travel to San Francisco in order to attend the wedding of his best friend. He needed to catch the very next train if he was to be on time for the ceremony, as he had to deliver the wedding rings. However, Ben's wallet was stolen in the train station. He lost all of his money as well as his ticket to San Francisco.
>
> Ben approached several officials as well as passengers . . . and asked them to loan him money to buy a new ticket. But, because he was a stranger, no one was willing to lend him the money he needed.
>
> While Ben . . . was trying to decide what to do next, a well-dressed man sitting next to him walked away . . . Ben noticed that the man had left his coat unattended. Sticking out of the man's coat pocket was a train ticket to San Francisco . . . He also saw that the man had more than enough money in his coat pocket to buy another train ticket. (p. 545)

One solution emphasized individual rights and justice:

> Ben should not take the ticket from the man's coat pocket even though it means not getting to San Francisco in time to deliver the wedding rings to his best friend. (p. 545)

The other solution placed a priority on caring for others:

> Ben should go to San Francisco to deliver the wedding rings to his best friend even if it means taking the train ticket from the other man's coat pocket. (p. 545)

When children and adults living in the United States responded to dilemmas like this one about Ben, a slight majority selected the justice-based alternative. In contrast, when Hindu children and adults living in India responded to the same dilemmas, the overwhelming majority selected the care-based alternative.

Clearly, moral reasoning reflects the culture in which a person is reared. Consistent with Kohlberg's theory, judgments by American children and adults reflect their culture's emphasis on individual rights and justice. But judgments by Indian children and adults reflect their culture's emphasis on caring for other people. The bases of moral reasoning are not universal as Kohlberg claimed; instead, they reflect cultural values.

Beyond Kohlberg's Theory

Findings like those described in the previous section indicate that Kohlberg's theory is most useful in understanding moral reasoning in cultures with Western philosophical and religious traditions. But researcher Carol Gilligan (1982; Gilligan & Attanucci, 1988)

THINK ABOUT IT

Research shows that people sometimes do not reason at the most advanced levels of which they are capable; instead, they revert to simpler, less mature levels. Might this happen in the realm of moral reasoning too? What factors might make it more likely for a person's moral reasoning to revert to a less sophisticated level?

According to Gilligan, moral reasoning is driven by the need to care for others.

Copyright © Jeff Greenberg / PhotoEdit

argued that Kohlberg's emphasis on justice applies more to males than to females, whose reasoning about moral issues is often rooted in concern for others. According to Gilligan, this "ethic of care" leads females to put a priority on fulfilling obligations to other people and those obligations guide their moral decision-making.

What does research tell us about the importance of justice and care in moral reasoning? Do females and males differ in the bases of their moral reasoning? In a comprehensive meta-analysis (Jaffee & Hyde, 2000), males tended to get slightly greater scores on problems that emphasized justice, whereas females tended to get slightly greater scores on problems that emphasized caring. But the differences were small and do not indicate that moral reasoning by females is predominated by a concern with care and that moral reasoning by males is predominated by a concern with justice. Instead, girls and boys as well as men and women reason about moral issues similarly. Most people think about moral issues in terms of *both* justice and caring, depending upon the nature of the moral dilemma and the context (Turiel, 2006).

Promoting Moral Reasoning

Whether it is based on justice or care, most cultures and most parents want to encourage adolescents to think carefully about moral issues. What can be done to help adolescents develop more mature forms of moral reasoning? Sometimes simply being exposed to more advanced moral reasoning is sufficient to promote developmental change (Walker, 1980). Adolescents may notice, for example, that older friends do not wait to be rewarded to help others. Or a teenager may notice that respected peers take courageous positions regardless of the social consequences. Such experiences apparently cause adolescents to reevaluate their reasoning on moral issues and propel them toward more sophisticated thinking.

Discussion can be particularly effective in revealing shortcomings in moral reasoning (Berkowitz et al., 2006). When people reason about moral issues with others whose reasoning is at a higher level, the usual result is that individuals reasoning at lower levels improve. This is particularly true when the conversational partner with the more sophisticated reasoning makes an effort to understand the other's view, by requesting clarification or paraphrasing what the other child is saying (Walker, Hennig, & Krettenauer, 2000).

Adolescents' moral reasoning (and moral behavior) is also influenced by their involvement in religion. Adolescents who are more involved in religion have greater

When adolescents discuss moral issues together, the thinking of those at lower stages in Kohlberg's theory is often influenced by those whose thinking is at the higher stages; that is, individuals who reason at the lower levels typically move their thinking to a more sophisticated stage.

concern for others and place more emphasis on helping them (Youniss, McLellan, & Yates, 1999). An obvious explanation for this link is that religion provides moral beliefs and guidelines for adolescents. But participation in religion can promote moral reasoning in a second, less direct way. Involvement in a religious community—typically through youth groups associated with a church, synagogue, or mosque—connects teens to an extended network of caring peers and adults. From interacting with individuals in this network, earning their trust, and sharing their values, adolescents gain a sense of responsibility to and concern for others (King & Furrow, 2004).

Research findings such as these send an important message to parents: Discussion is probably the best way for parents to help their children think about moral issues in more mature terms (Walker & Taylor, 1991). Research consistently shows that mature moral reasoning comes about when adolescents are free to express their opinions on moral issues to their parents, who in turn express their own opinions and thus expose their adolescent children to more mature moral reasoning (Hoffman, 1988, 1994).

Test Yourself

RECALL

1. Kohlberg's theory includes the preconventional, conventional, and _____ levels.

2. For children and adolescents in the preconventional level, moral reasoning is strongly influenced by _____.

3. Supporting Kohlberg's theory are findings that level of moral reasoning is associated with age, that people progress through the stages in the predicted sequence, and that _____.

4. Gilligan's view of morality emphasizes _____ instead of justice.

5. When boys' and girls' moral reasoning is compared, the typical result is that _____.

6. If parents wish to foster their children's moral development, they should _____ with them.

INTERPRET

How similar is Piaget's stage of formal operational thought to Kohlberg's stage of conventional moral reasoning?

APPLY

Imagine that you were the father of Min-shen, the boy in the vignette who did not stand up for the boy who was wrongly accused of stealing the iPod. Based on the research described in this section, what might you do to try to advance Min-shen's level of moral reasoning?

Recall answers: (1) postconventional, (2) reward or punishment, (3) more advanced moral reasoning is associated with moral action, (4) caring for others, (5) they do not differ. (6) discuss moral issues

8.1 PUBERTAL CHANGES

What physical changes occur in adolescence that mark the transition to a mature young adult?

■ Puberty includes bodily changes in height and weight as well as sexual maturation. Girls typically begin the growth spurt earlier than boys, who acquire more muscle, less fat, and greater heart and lung capacity. The brain communicates more effectively and the frontal cortex continues to mature. Sexual maturation, which includes primary and secondary sex characteristics, occurs in predictable sequences for boys and girls.

What factors cause the physical changes associated with puberty?

■ Pubertal changes take place when the pituitary gland signals the adrenal gland, ovaries, and testes to secrete hormones that initiate physical changes. The timing of puberty is influenced strongly by health and nutrition. Its timing is also influenced by the social environment; for example, puberty occurs earlier for girls who experience family conflict or depression.

How do physical changes affect adolescents' psychological development?

■ Pubertal changes affect adolescents' psychological functioning. Teens, particularly girls, become concerned about their appearance. When forewarned, adolescents respond positively to menarche and spermarche. Adolescents are moodier than children or adults primarily because their moods shift in response to frequent changes in activities and social setting. Early maturation tends to be harmful to girls.

8.2 HEALTH

What are the elements of a healthy diet for adolescents? Why do some adolescents suffer from eating disorders?

■ For proper growth, teenagers need to consume adequate calories, calcium, and iron. Unfortunately, many teenagers do not eat properly and do not receive adequate nutrition.

■ Anorexia and bulimia, eating disorders that typically affect adolescent girls, are characterized by an irrational fear of being overweight. Several factors contribute to these disorders, including heredity, a childhood history of eating problems, and (during adolescence) negative self-esteem and a preoccupation with one's body and weight. Treatment and prevention programs emphasize changing adolescents' views toward thinness and their eating-related behaviors.

Do adolescents get enough exercise? What are the pros and cons of participating in sports in high school?

■ Individuals who work out at least three times weekly often have improved physical and mental health. Unfortunately, many high-school students do not get enough exercise.

■ Millions of American boys and girls participate in sports. Football and track are the most popular sports for boys and girls, respectively. The benefits of participating in sports include improved physical fitness, enhanced self-esteem, and understanding about teamwork. The potential costs include injury and abuse of performance-enhancing drugs.

What are common obstacles to healthy growth in adolescence?

■ Accidents involving automobiles or firearms are the most common cause of death in American teenagers. Many of these deaths could be prevented if, for example, adolescents did not drive recklessly (e.g., too fast and without wearing seatbelts). Adolescents often overestimate the harm of risky behavior in general, but they don't see themselves as being personally at risk and often place greater value on the rewards associated with risky behavior.

8.3 INFORMATION PROCESSING DURING ADOLESCENCE

How do working memory and processing speed change in adolescence?

■ Working memory increases in capacity and processing speed becomes faster. Both achieve adultlike levels during adolescence.

How do increases in content knowledge, strategies, and metacognitive skill influence adolescent cognition?

■ Content knowledge increases, to expertlike levels in some domains; and strategies and metacognitive skills become much more sophisticated.

What changes in problem solving and reasoning take place in adolescence?

■ Adolescents often solve problems analytically, using mathematics or logic. They also acquire skill in detecting weaknesses in scientific evidence and in logical arguments.

8.4 REASONING ABOUT MORAL ISSUES

How do adolescents reason about moral issues?

■ Kohlberg proposed that moral reasoning includes preconventional, conventional, and postconventional levels. Moral reasoning is first based on rewards and punishments and, much later, on personal moral codes. As predicted by Kohlberg's theory, people progress though the stages in sequence and do not regress, and morally advanced reasoning is associated with more frequent moral behavior. However, few people attain the most advanced levels.

Is moral reasoning similar in all cultures?

■ Not all cultures emphasize justice in moral reasoning. For instance, the Hindu religion emphasizes duty and responsibility to others and, consistent with these beliefs, Hindu children and Indians emphasize caring for other people in their moral reasoning.

How do concern for justice and caring for other people contribute to moral reasoning?

■ Gilligan proposed that females' moral reasoning is based on caring and responsibility for others, not justice. Research does not support consistent sex differences in moral reasoning but has found that males and females both consider caring as well as justice in their moral judgments, depending on the situation.

What factors help promote more sophisticated reasoning about moral issues?

■ Many factors can promote more sophisticated moral reasoning, including (a) observing others reasoning at more advanced levels, (b) discussing moral issues with peers, teachers, and parents, and (c) involvement in a religious community that connects adolescents to a network of caring peers and adults.

KEY TERMS

puberty (286)

primary sex characteristics (288)

secondary sex characteristics (288)

menarche (288)

spermarche (289)

body mass index (BMI) (295)

basal metabolic rate (295)

anorexia nervosa (296)

bulimia nervosa (296)

preconventional level (306)

obedience orientation (306)

instrumental orientation (306)

conventional level (307)

interpersonal norms (307)

social system morality (307)

postconventional level (307)

social contract (307)

universal ethical principles (307)

LEARN MORE ABOUT IT

Log in to **www.cengagebrain.com** to access the resources your instructor requires. For this book, you can access:

Psychology CourseMate

■ CourseMate brings course concepts to life with interactive learning, study, and exam preparation tools that support the printed textbook. A textbook-specific website, Psychology CourseMate includes an integrated interactive eBook and other interactive learning tools including quizzes, flashcards, videos, and more.

CENGAGENOW™

■ CengageNOW Personalized Study is a diagnostic study tool containing valuable text-specific resources—and because you focus on just what you don't know, you learn more in less time to get a better grade.

WebTUTOR™

■ More than just an interactive study guide, WebTutor is an anytime, anywhere customized learning solution with an eBook, keeping you connected to your textbook, instructor, and classmates.

Moving Into the Adult Social World

Socioemotional Development in Adolescence

YOU PROBABLY HAVE VIVID MEMORIES of your teenage years. Remember the exhilarating moments—high school graduation, your first paycheck from a part-time job, and your first feelings of love and sexuality? There were, of course, also painful times—your first day on the job when you couldn't do anything right, not knowing what to say on a date with a person you desperately wanted to impress, and countless arguments with your parents. Feelings of pride and accomplishment accompanied by feelings of embarrassment and bewilderment are common to individuals who are on the threshold of adulthood.

Adolescence represents the transition from childhood to adulthood and is a time when individuals grapple with their identity; many have their first experiences with love and sex, and some enter the world of work. In the first three sections of this chapter, we investigate these challenging developmental issues. Then we look at the special obstacles that sometimes make adolescence difficult to handle.

LEARNING OBJECTIVES

- How do adolescents achieve an identity?
- What are the stages and results of acquiring an ethnic identity?

- How does self-esteem change in adolescence?

Dea was born in Seoul of Korean parents but was adopted by a Dutch couple in Michigan when she was 3 months old. Growing up, she considered herself a red-blooded American. In college, however, Dea realized that others saw her as an Asian American, an identity about which she had never given much thought. She began to wonder, "Who am I really? American? Dutch American? Asian American?"

LIKE DEA, do you sometimes wonder who you are? Self-concept refers to the attitudes, behaviors, and values that make a person unique. In adolescence, self-concept takes on special significance as individuals struggle to achieve an identity that will allow them to participate in the adult world. Through self-reflection, youth search for an identity to integrate the many different and sometimes conflicting elements of the self. In this section we'll learn more about the adolescent search for an identity. Along the way, we'll learn more about Dea's struggle to learn who she is.

The Search for Identity

Erik Erikson's (1968) account of identity formation has been particularly influential in our understanding of adolescence. Erikson argued that adolescents face a crisis between identity and role confusion. This crisis involves balancing the desire to try out many possible selves and the need to select a single self. Adolescents who achieve a sense of identity are well prepared to face the next developmental challenge: establishing intimate, sharing relationships with others. However, Erikson believed that teenagers who are confused about their identity can never experience identity in any human relationship. Instead, throughout their lives they remain isolated and respond to others stereotypically.

How do adolescents achieve an identity? They use the hypothetical reasoning skills of the formal-operational stage to experiment with different selves to learn more about possible identities. Adolescents' advanced cognitive skills enable them to imagine themselves in different roles.

Much of the testing and experimentation is career oriented. Some adolescents may envision themselves as rock stars; others may imagine being professional athletes, Peace Corps workers, or best-selling novelists. Other testing is romantically oriented. Teens may fall in love and imagine living with the loved one. Still other exploration involves religious and political beliefs (Harre, 2007; King, Elder, & Whitbeck, 1997). Teens give different identities a trial run just as you might test-drive different cars before selecting one. By fantasizing about their future, adolescents begin to discover who they will be.

As adolescents strive to achieve an identity, they often progress through different phases or statuses as shown in ●Table 9.1. Unlike Piaget's stages, these four phases do not necessarily occur in sequence. Most young adolescents are in a state of diffusion or foreclosure. The common element in these phases is that teens are not exploring alternative identities. They are avoiding the crisis altogether or have resolved it by taking on an identity suggested by parents or other adults. However, as individuals move beyond adolescence and into young adulthood and have more opportunity to explore alternative identities, diffusion and foreclosure become less common, and achievement and moratorium become more common (Meeus et al., 2010).

Marcia's Four Identity Statuses

Status	Definition	Example
Diffusion	The individual is overwhelmed by the task of achieving an identity and does little to accomplish the task.	Larry hates the idea of deciding what to do with his future so he spends most of his free time playing video games.
Foreclosure	The individual has a status determined by adults rather than by personal exploration.	For as long as she can remember, Sakura's parents have told her that she should be an attorney and join the family law firm. She plans to study prelaw in college, though she's never given the matter much thought.
Moratorium	The individual is examining different alternatives but has yet to find one that's satisfactory.	Brad enjoys almost all of his high-school classes. Some days he thinks it would be fun to be a chemist, some days he wants to be a novelist, and some days he'd like to be an elementary-school teacher. He thinks it's a little weird to change his mind so often, but he also enjoys thinking about different jobs.
Achievement	The individual has explored alternatives and has deliberately chosen a specific identity.	Throughout middle school, Efrat wanted to play in the WNBA. During 9th and 10th grades, she thought it would be cool to be a physician. In 11th grade, she took a computing course and everything finally "clicked"—she'd found her niche. She knew that she wanted to study computer science in college.

© Juice Images / Alamy

As part of their search for an identity, adolescents often try on different roles, for example, imagining what life might be like as a rock star.

Typically, young people do not reach the achievement status for all aspects of identity at the same time (Goossens, 2001; Kroger & Green, 1996). Some adolescents may reach the achievement status for occupations before achieving it for religion and politics. Others reach the achievement status for religion before other domains. Evidently, few youth achieve a sense of identity all at once; instead, the crisis of identity is first resolved in some areas and then in others.

During the search for identity, adolescents reveal a number of characteristic ways of thinking. They are often very self-oriented. *The self-absorption that marks the teenage search for identity is referred to as* **adolescent egocentrism** (Elkind, 1978). Unlike preschoolers, adolescents know that others have different perspectives on the world. Adolescents are simply *much* more interested in their own feelings and experiences than in anyone else's experiences. In addition, as they search for an identity, many adolescents wrongly believe that they are the focus of others' thinking. A teen who spills food on herself may imagine that all her friends are

adolescent egocentrism
self-absorption that is characteristic of teenagers as they search for identity

Adolescents often believe that others are constantly watching them, a phenomenon known as "imaginary audience"; as a result, they're often upset or embarrassed when they make obvious mistakes or blunders, such as spilling food or drink.

imaginary audience
adolescents' feeling that their behavior is constantly being watched by their peers

personal fable
attitude of many adolescents that their feelings and experiences are unique and have never been experienced by anyone else before

illusion of invulnerability
adolescents' belief that misfortunes cannot happen to them

ethnic identity
feeling of belonging to a specific ethnic group

THINK ABOUT IT

Although Piaget's theory of cognitive development was not concerned with identity formation, how might his theory explain why identity is a central issue in adolescence?

thinking only about the stain on her blouse and how sloppy she is. *Many adolescents feel that they are, in effect, actors whose performance is watched constantly by their peers, a phenomenon known as the* **imaginary audience.**

Adolescent self-absorption is also demonstrated by the **personal fable***, teenagers' tendency to believe that their experiences and feelings are unique and that no one has ever felt or thought as they do.* Whether it is the excitement of first love, the despair of a broken relationship, or the confusion of planning for the future, adolescents often believe they are the first to experience these feelings and that no one else could possibly understand the power of their emotions (Elkind & Bowen, 1979). *Adolescents' belief in their uniqueness also contributes to an* **illusion of invulnerability***: the belief that misfortune happens only to others.* They think they can have sex without becoming pregnant or contracting a sexually transmitted disease or that they can drive recklessly without being in an auto accident. These characteristics of adolescents' thinking are summarized in ●Table 9.2.

As adolescents make progress toward achieving an identity, adolescent egocentrism, imaginary audiences, personal fables, and the illusion of invulnerability become less common. What circumstances help adolescents achieve identity? Parents are influential (Marcia, 1980). When parents encourage discussion and recognize children's autonomy, their children are more likely to reach the achievement status. Apparently these youth feel encouraged to undertake the personal experimentation that leads to identity. In contrast, when parents set rules with little justification and enforce them without explanation, children are more likely to remain in the foreclosure status. These teens are discouraged from experimenting personally; instead, their parents simply tell them what identity to adopt. Overall, adolescents are most likely to establish a well-defined identity in a family atmosphere where parents encourage children to explore alternatives on their own but do not pressure or provide explicit direction (Luyckx et al., 2007; Smits et al., 2010).

Beyond parents, peers are also influential. When adolescents have close friends that they trust, they feel more secure exploring alternatives (Meeus, Oosterwegel, & Vollebergh, 2002). The broader social context also contributes (Bosma & Kunnen, 2001). Exploration takes time and access to resources; neither may be readily available to adolescents living in poverty (e.g., they can't explore because they drop out of school to support themselves and their family). Finally, through their personality, adolescents themselves may affect the ease with which they achieve an identity. Individuals who are more open to experience and are more agreeable (friendly, generous, helpful) are more likely to achieve an identity (Crocetti et al., 2008).

Ethnic Identity

Roughly one third of the adolescents and young adults living in the United States are members of ethnic minority groups, including African Americans, Asian Americans, Latino Americans, and Native Americans. *These individuals typically develop an* **ethnic identity***: They feel a part of their ethnic group and learn the special customs and traditions of their group's culture and heritage* (Phinney, 2005).

Achieving an ethnic identity seems to occur in three phases. Initially, adolescents have not examined their ethnic roots. A teenage African American girl in this phase remarked, "Why do I need to learn about who was the first Black woman to do this or that? I'm just not too interested" (Phinney, 1989, p. 44). For this girl, ethnic identity is not yet an important personal issue.

In the second phase, adolescents begin to explore the personal impact of their ethnic heritage. The curiosity and questioning that is characteristic of this stage is captured in the comments of a teenage Mexican American girl who said, "I want to know what we do and how our culture is different from others. Going to festivals and

science), and these beliefs contribute to their overall academic self-concept. A teen who believes she is skilled at English and math but not so skilled in science will probably have a positive academic self-concept overall. But a teen who believes he is untalented in most academic areas will have a negative academic self-concept.

During adolescence, the social component of self-esteem becomes particularly well differentiated. Adolescents distinguish self-worth in many different social relationships. A teenager may, for example, feel very positive about her relationships with her parents but believe that she's a loser in romantic relationships. Another teen may feel loved and valued by his parents but believe the co-workers at his part-time job can't stand him (Harter, Waters, & Whitesell, 1998).

Growth of self-worth among U.S. children and adolescents also varies depending on their ethnicity. Compared to European American children, African Americans and Hispanic Americans have lower self-esteem during most of the elementary-school years. However, in adolescence the gap narrows for Hispanic Americans and actually reverses for African American adolescents, who have greater self-esteem than their European American peers (Gray-Little & Hafdahl, 2000; Herman, 2004; Twenge & Crocker, 2002). Scientists don't fully understand why these changes take place, but one hypothesis involves ethnic identity. The idea is that, beginning in early adolescence, many African American and Hispanic American teens take pride in belonging to a distinct social and cultural group and this raises their sense of self-worth (Gray-Little & Hafdahl, 2000; Umaña-Taylor, Diversi, & Fine, 2002).

Thus, between the late preschool years and adolescence, self-esteem becomes more complex as older children and adolescents identify distinct domains of self-worth. This growing complexity is not surprising; it reflects the older child's and adolescent's greater cognitive skill and the more extensive social world of older children and adolescents.

Influences on Adolescents' Self-Esteem

What factors contribute to adolescents' self-esteem? Research indicates two important sources. One is based on children's actual competence in domains that are important to them: Children's self-worth is greater when they are skilled in areas they value. In other words, children's interests, abilities, and self-concept are coupled. Children tend to like domains in which they do well, and their self-concepts reflect this (Denissen, Zarret, & Eccles, 2007). Mark, who likes math and gets good grades in math, has a positive math self-concept: "I'm good at math and do well when I have to learn something new in math. And I'd probably like a job that involved math."

Children's and adolescents' self-worth is also affected by how others view them, particularly other people who are important to them. Parents matter, of course—even to adolescents. Children are more likely to view themselves positively when their parents are affectionate toward them and involved with them (Lord, Eccles, & McCarthy, 1994; Ojanen & Perry, 2007). Around the world, children have higher self-esteem when families live in harmony and parents nurture their children (Scott, Scott, & McCabe, 1991). A father who routinely hugs his daughter and gladly takes her to piano lessons is saying to her, "You are important to me." When children hear this regularly from parents, they evidently internalize the message and come to see themselves positively.

Parents' discipline also is related to self-esteem. Children with high self-esteem generally have parents who have reasonable expectations for their children and are also willing to discuss rules and discipline with their children (Laible & Carlo, 2004). Parents who fail to set rules are, in effect, telling their children that they don't care—they don't value them enough to go to the trouble of creating rules and enforcing them. In much the same way, parents who refuse to discuss discipline with their children are saying, "Your opinions don't matter to me." Not surprisingly, when children internalize these messages, the result is lower overall self-worth.

Peers' views are important too. Children's and particularly adolescents' self-worth is greater when they believe that their peers think highly of them (Harter, 2005).

Maddy's self-worth increases, for example, when she hears that Pedro, Matt, and Michael think she's the hottest girl in the eighth grade.

Thus children's and adolescents' self-worth depends on their being competent at something they value and in being valued by people who are important to them. By encouraging children to find their special talents and by being genuinely interested in their progress, parents and teachers can enhance the self-esteem of all students.

The Myth of Storm and Stress

According to novelists and filmmakers, the search for identity that we've just described is inherently a struggle, a time of storm and stress for adolescents. Although this view may make for best-selling novels and hit movies, in reality the rebellious teen is vastly overstated. Adolescents generally enjoy happy and satisfying relationships with their parents (Steinberg, 2001). Most teens love their parents and feel loved by them. And they embrace many of their parents' values and look to them for advice.

Cross-cultural research provides further evidence that adolescence is not necessarily a time of turmoil and conflict. Offer and his colleagues (1988) interviewed adolescents from 10 countries: the United States, Australia, Germany, Italy, Israel, Hungary, Turkey, Japan, Taiwan, and Bangladesh. These investigators found most adolescents moving confidently and happily toward adulthood. As ■ Figure 9.1 shows, most adolescents around the world reported that they were usually happy, and few avoided their homes.

The work by Offer and his colleagues (1988) is nearly a quarter-century old but newer research paints much the same picture. In one study of Arab adolescents living in Israel (Azaiza, 2005), 82% of adolescents said they felt wanted by their family and 89% reported that they appreciated their family. In another study (Güngör & Bornstein, 2010), adolescents in Turkey and Belgium rated their mothers as being very supportive, endorsing items such as "My mother supports me in dealing with problems" and "My mother talks to me in a comforting way." These findings are consistent with the older studies and undercut the myth of adolescence as necessarily being a time when adolescent storms rain on parent–child relationships.

Of course, parent–child relations *do* change during adolescence. As teens become more independent, their relationships with their parents become more egalitarian. Parents must adjust to their children's growing sense of autonomy by treating them more like equals (Laursen & Collins, 1994). This growing independence means that teens spend less time with their parents, are less affectionate toward them, and argue more often with them about matters of style, taste, and freedom (Shanahan et al., 2007). And adolescents do have more disagreements with parents but these disputes are usually relatively mild—bickering, not all-out shouting matches—and usually concern personal choices (e.g., hairstyle, clothing). These changes are natural by-products of an evolving parent–child relationship in which the "child" is nearly a fully independent young adult (Steinberg & Silk, 2002).

Before you think that this portrait of parent–child relationships in adolescence is too good to be true, we want to add two cautionary notes. First, conflicts between parents and their adolescent children are often very distressing for *parents*, who may read far more into these conflicts than their teenagers do (Steinberg, 2001). Parents sometimes fear that arguments over attire or household chores may reflect much more fundamental disagreements about values: A mother may interpret her son's refusal to clean his room as a rejection of values concerning the need for order and cleanliness, when the son simply doesn't want to waste time cleaning a room that he knows will become a mess again in a matter of days. Second, for a minority of families (roughly 25%), parent–child conflicts in adolescence are more serious and are associated with behavior problems in adolescents (van Doorn, Branje, & Meeus, 2008). These more harmful conflicts are more common among adolescents who don't regulate their emotions well (Eisenberg et al., 2008) and they often predate adolescence—as children, these adolescents were prone to conflict with their parents (Steinberg, 2001).

Although the myth is that adolescence is inherently a period of storm and stress, in reality most adolescents worldwide claim to be happy and do not avoid their homes. Data from Offer et al. (1988).

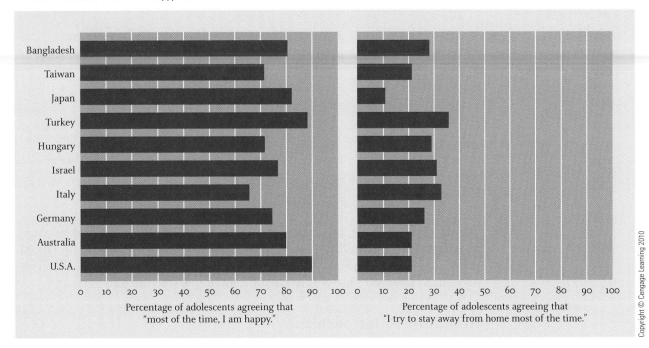

Percentage of adolescents agreeing that "most of the time, I am happy."

Percentage of adolescents agreeing that "I try to stay away from home most of the time."

Response to question on page 319 about Dea's ethnic identity. Dea, the Dutch Asian American college student, doesn't know how to integrate the Korean heritage of her biological parents with the Dutch American culture in which she was reared. This would put her in the second phase of acquiring an ethnic identity. On the one hand, she is examining her ethnic roots, which means she's progressed beyond the initial stages. On the other hand, she has not yet integrated her Asian and European roots and so has not reached the third and final phase.

Test Yourself

RECALL

1. According to Erikson, adolescents face a crisis between identity and _____.

2. The _____ status would describe an adolescent who has attained an identity based almost entirely on her parents' advice and urging.

3. A person who has simply put off searching for an identity because it seems too confusing and too overwhelming is in the _____ status.

4. _____ refers to the fact that adolescents sometimes believe that their lives are a performance with their peers watching them constantly.

5. Adolescents are most likely to achieve an identity when parents encourage them _____.

6. In the second phase of achieving an ethnic identity, adolescents _____.

7. When individuals have a strong ethnic identity, their identification with mainstream culture _____.

8. Self-esteem often drops when students enter middle school or junior high school because young adolescents _____.

INTERPRET

How do parent–child relationships change in adolescence? Do these changes indicate a period of "storm and stress"?

APPLY

The Tran family has just immigrated to the United States from Vietnam. The mother and father want their two children to grow up appreciating their Vietnamese heritage but worry that a strong ethnic identity may not be good for their kids. What advice would you give Mr. and Mrs. Tran about the impact of ethnic identity on children's development?

Recall answers: (1) role confusion, (2) foreclosure, (3) diffusion, (4) Imaginary audience, (5) to explore alternative identities but do not pressure them or provide direction, (6) start to explore the personal impact of their ethnic roots, (7) is sometimes strong and sometimes weak, depending on specific circumstances, (8) no longer know where they stand among their peers, so they must establish a new "pecking order."

For 6 months, 15-year-old Gretchen has been dating Jeff, a 17-year-old. She thinks she is truly in love for the first time, and she often imagines being married to Jeff. They have had sex a few times, each time without contraception. It sometimes crosses Gretchen's mind that if she gets pregnant she could move into her own apartment and begin a family.

THE FIRES OF ROMANTIC RELATIONSHIPS HAVE LONG WARMED THE HEARTS OF AMERICAN ADOLESCENTS. Often, as with Jeff and Gretchen, romance leads to sex. In this section, we'll explore adolescent dating and sexual behavior. As we do, you'll better understand Gretchen's reasons for having unprotected sex with Jeff.

Romantic Relationships

The social landscape adds a distinctive landmark in adolescence—romantic relationships. These are uncommon during elementary school, but by high school roughly two-thirds of U.S. adolescents have had a romantic relationship within the previous 1½ years and most have been involved in a romance lasting nearly a year (Carver, Joyner, & Udry, 2003). However, cultural factors influence the timing of romantic relationships for teenagers in America. Traditional Hispanic American and Asian American parents emphasize family ties and loyalty to parents. Because romantic relationships are a sign of independence and usually result in less time spent with family, it's not surprising that Hispanic American and Asian American adolescents often begin to date at an older age and date less frequently (Collins, Welsh, & Furman, 2009).

Romantic relationships build on friendships. Like friends, romantic partners tend to be similar in popularity and physical attractiveness. And a best friendship serves both as a prototype for and a source of support during ups-and-downs of close relationships (Collins et al., 2009). What's more, romantic relationships change over time in ways that resemble changes in friendship: for younger adolescents, romantic relationships offer companionship (like that provided by a best friend) and an outlet for sexual exploration. For older adolescents like those in the photo, intimacy, trust, and support become important features of romantic relationships (Shulman & Kipnis, 2001).

Adolescent romantic relationships build on friendships and offer companionship as well as an outlet for sexual exploration.

It's tempting to dismiss teen romances as nothing more than "puppy love," but they are often developmentally significant (Collins et al., 2009). On the one hand, adolescents involved in a romantic relationship are often more self-confident and have higher self-esteem. And high-quality adolescent romances are associated with positive relationships during adulthood. On the other hand, adolescents in romantic relationships report more emotional upheaval and conflict (Joyner & Udry, 2000). In addition, early dating with many different partners is associated with a host of problems in adolescence (e.g., drug use, lower grades) and is associated with less satisfying romantic relationships in adulthood (Collins, 2003).

Features of Sexually Transmitted Diseases (STDs)

Disease	U.S. Frequency	Symptoms	Complications
Caused by bacteria			
Chlamydia	3.3% of adolescent females and 0.7% of adolescent males	75% of women and 50% of men have no symptoms; sometimes abnormal discharge of pus from the vagina or penis or pain while urinating	Infections of the cervix and Fallopian tubes that can lead to infertility; rare in men
Gonorrhea	0.6% of adolescent females and 0.3% of adolescent males	Often no symptoms at all; pus discharged from the penis or vagina, pain associated with urination; for women, pain during intercourse; for men, swollen testicles	Pelvic inflammatory disease, a serious infection of the female reproductive tract that can lead to infertility; in men, epididymitis, an infection of the testicles that can lead to infertility
Syphilis	About 4,000 cases annually among 15- to 24-year-olds	A sore, called a chancre, at the site of the infection—usually the penis, vulva, or vagina	Left untreated, can damage internal organs such as the brain, nerves, eyes, heart, bones, and joints
Caused by virus			
Genital herpes	At least 45 million of age 12 and older (roughly 1 in 5 adolescents and adults)	Itching, burning, or pain in the genital or anal area; sores on the mouth, penis, or vagina	Recurrent sores; pregnant women can pass the virus (which can be fatal to the newborn) to the baby during birth
Genital human papilloma virus (HPV)	20 million	Usually no symptoms; sometimes genital warts or discharge from the penis or vagina	Usually goes away; in rare cases leads to cervical cancer
Hepatitis B	About 75,000 annually	Jaundice, fatigue, loss of appetite, abdominal pain	Death from chronic liver disease
HIV	About 40,000 diagnosed annually	Initially a flulike illness; later, enlarged lymph nodes, lack of energy, weight loss, frequent fevers	Loss of immune cells (AIDS), cancer, death

SOURCE: Centers for Disease Control and Prevention, 2007, 2010.

Sexual Behavior

We've already seen that sexual exploration is an important feature of romantic relationships for younger adolescents. In fact, by the end of high school, about two-thirds of American adolescents will have had intercourse at least once (Eaton et al., 2008). No single factor predicts adolescent sexual behavior. Instead, adolescents are more likely to be sexually active when they acquire (from parents and peers) permissive attitudes toward sex, when their parents don't monitor their behavior, when their peers approve and when they believe their peers are also having sex, when they are more physically mature, and when they drink alcohol regularly (Belsky et al., 2010; Hipwell et al., 2010; Zimmer-Gembeck & Helfand, 2008).

Although a majority of boys and girls have sex at some point during adolescence, sexual activity has very different meanings for boys and girls (Brooks-Gunn & Paikoff, 1993). Girls tend to describe their first sexual partner as "someone they love," but boys describe their first partner as a "casual date." Girls report stronger feelings of love for their first sexual partner than for a later partner, but boys don't. Girls have mixed feelings after their first sexual experience—fear and guilt mixed with happiness and excitement—whereas boys' feelings are more uniformly positive. Finally, when describing their sexual experiences to peers, girls' peers typically express some disapproval but boys' peers typically do not. In short: for boys, sexual behavior is viewed as recreational and self-oriented; for girls, sexual behavior is viewed as romantic and is interpreted through their capacity to form intimate interpersonal relationships (Steinberg, 1999).

THINK ABOUT IT

According to the "storm and stress" view of adolescence, sexual behavior would be one way for adolescents to rebel against their parents. Does research on adolescent sexuality support this prediction?

Sexually Transmitted Diseases

Adolescent sexual activity is cause for concern because a number of diseases are transmitted from one person to another through sexual intercourse. ● Table 9.3 lists several of the most common types of sexually transmitted diseases (STDs). Some STDs, such as chlamydia and syphilis, are caused by bacteria; others, such as herpes and hepatitis B, are caused by a virus.

Several STDs can have serious complications if left untreated. Most are cured readily with antibiotics. In contrast, the prognosis is bleak for individuals who contract the human immunodeficiency virus (HIV), which typically leads to acquired immunodeficiency syndrome (AIDS). In persons with AIDS, the immune system is no longer able to protect the body from infections, and they often die from one of these infections.

Adolescents and young adults—those age 24 and younger—account for roughly half of all new cases of AIDS in the United States (Centers for Disease Control and Prevention, 2007c). Most of these people contracted the disease during adolescence. Many factors make adolescents especially susceptible to AIDS. Teenagers and young adults are more likely than older adults to engage in unprotected sex and to use intravenous drugs, which are common pathways for the transmission of AIDS.

Teenage Pregnancy and Contraception

Adolescents' sexual behavior is also troubling because, among American adolescent girls who have ever had intercourse, approximately 1 in 6 becomes pregnant. The result is that nearly a half million babies are born to American teenagers annually. African American and Hispanic American adolescents are the most likely to become pregnant (Ventura et al., 2008).

Teenage mothers and their children usually face bleak futures. If this is the case, why do so many teens become pregnant? The answer is simple: Only about half of teenagers use contraception when they first have intercourse, and about 10% of teens who are sexually active do not use contraception. Those who do often use ineffective methods, such as withdrawal, or practice contraception inconsistently (Besharov & Gardiner, 1997; Kirby, 2001).

Why don't some sexually active teens use birth control consistently or correctly? Several factors contribute (Gordon, 1996). First, many adolescents are ignorant of basic facts of conception and many believe that they are invulnerable—that only others become pregnant. Second, some teenagers do not know how to use or where to obtain contraceptives, and others are embarrassed to buy them (Ralph & Brindis, 2010). Third, for some adolescent girls, like Gretchen from the section-opening vignette, becoming pregnant is appealing (Phipps et al., 2008). They think having a child is a way to break away from parents, gain status as an independent-living adult, and have "someone to love them."

The best way to reduce adolescent sexual behavior and teen pregnancy is with comprehensive sex education programs (Kirby & Laris, 2009). These programs teach the biological aspects of sex and emphasize responsible sexual behavior or abstaining from premarital sex altogether. They also include discussions of the pressures to become involved sexually and ways to respond to this pressure. A key element is that in role-playing sessions, students practice strategies for refusing to have sex. Youth who participate in programs like these are less likely to have intercourse; when they do have intercourse, they are more likely to use contraceptives. In contrast, there is little evidence that programs focusing solely on abstinence are effective in reducing sexual activity or encouraging contraceptive use.

THINK ABOUT IT

Suppose you had to convince a group of 15-year-olds about the hazards of adolescent sex and teenage pregnancy. What would you say?

Sexual Orientation

For most adolescents, dating and romance involve members of the opposite sex. However, in early and mid-adolescence, roughly 15% of teens experience a period of sexual questioning during which they sometimes report emotional and sexual attractions to members of their own sex (Carver, Egan, & Perry, 2004). For most adolescents, these experiences are simply a part of the larger process of role experimentation common to adolescence. However, about 5% of teenage boys and girls identify their sexual orientation as gay or lesbian (Rotherman-Borus & Langabeer, 2001).

The roots of sexual orientation are poorly understood. Scientists have, however, discredited several theories of sexual orientation. For example, it was once thought that sons become gay when raised by a domineering mother and that girls become lesbians when their father is their primary role model. However, we now know that these ideas are false (Golombok & Tasker, 1996; Patterson, 1992).

Modern accounts suggest that attraction to same-sex individuals comes about differently in males and females (Diamond, 2007). For males, genes and hormones may lead some boys to feel "different" during early adolescence; these feelings lead to an interest in gender-atypical activities and, later, attraction to other males. For females, the path to same-sex attraction is less predictable. Attraction to other females usually does not emerge until mid- or late adolescence and, in some cases, not until middle or old age. What's more, for many lesbian women, same-sex attraction grows out of deep feelings for a particular woman that, over time, extends to other females.

Although the origins of same-sex attraction are not yet well understood, it is clear that gay and lesbian individuals face many challenges. Their family and peer relationships are often disrupted, and they endure verbal and physical attacks (Heatherington & Lavner, 2008). Given these problems, it's not surprising that gay and lesbian youth often experience mental health problems (Toomey et al., 2010) and are at risk for substance abuse (Marshal et al., 2008). In recent years, social changes have helped gay and lesbian youth respond more effectively to these challenges, including more (and more visible) role models and more centers for gay and lesbian youth. These resources are making it easier for gay and lesbian youth to understand their sexual orientation and to cope with the many other demands of adolescence.

About 5% of adolescents identify themselves as gay or lesbian.

Dating Violence

As adolescents begin to explore romantic relationships and sex, many teens experience violence in dating, which can include physical violence (e.g., being hit or kicked), emotional violence (e.g., threats or bullying designed to harm self-worth), or sexual violence (being forced to engage in sexual activity against one's will). Roughly 25% of adolescents report these experiences and, as you can imagine, these youth often don't do well in school and suffer from mental health and behavioral problems (Centers for Disease Control, 2008).

Of the various kinds of dating violence, scientists know the most about factors that place adolescents at risk for sexual violence. One of the most important factors is drug and alcohol use: Heavy drinking usually impairs a female's ability to send a clear message regarding her intentions and makes males less able and less inclined to interpret such messages (Maurer & Robinson, 2008; Reyes et al., 2011). Females are also more at risk when they adhere to more traditional gender stereotypes, apparently because their view of the female gender role includes being relatively submissive to a male's desires (Foshee et al., 2004).

What factors make teenage boys likely to commit acts of violence? One contributing factor is a boy's home life: Boys are more at risk when they were abused as children or witnessed domestic violence, apparently because this leads them to believe that violence is a normal part of romantic relationships (National Center for Injury Prevention and Control, 2005).

In the Spotlight on Research feature, you'll learn about some other factors that make boys more likely to be violent while dating.

Spotlight on research | **Why Are Some Boys More Likely to Perpetrate Dating Violence?**

Who were the investigators, and what was the aim of the study? Is any teenage boy likely to perpetrate violence while dating? Or are there factors that make some boys a greater risk when it comes to dating violence? If so, what are those factors? Vangie Foshee and her colleagues (2001) designed a study to answer these questions.

How did the investigators measure the topic of interest? The investigators measured dating violence by asking teenage boys whether they had ever committed any of a list of 18 violent acts while on a date. The list included hitting, choking, slapping, and kicking a partner as well as whether they had forced a partner to have sex. In addition, several other questionnaires were created that

[continued]

measured factors that might make a boy more at risk for perpetrating dating violence. Some of these factors included peers, other problem behaviors (e.g., drinking), and personal competencies (e.g., self-esteem, communication skills).

Who were the children in the study? The study included 576 boys in eighth and ninth grades who reported that they had begun dating. About 75% of the sample was European American. (The study also included girls, but for simplicity we'll concentrate here on the results for boys.)

What was the design of the study? This study was correlational because Foshee and her colleagues were interested in the relations that existed naturally between boys' perpetration of violence and other variables that might be related to perpetration of violence. The study was longitudinal because adolescents were tested first in eighth or ninth grade and then a second time about 18 months later. Again, for simplicity we describe only the results from the first testing.

Were there ethical concerns with the study? Yes. Obviously, violence is a sensitive topic, and the investigators were careful to be sure that they obtained consent from parents and adolescents and that the adolescents' responses were confidential.

What were the results? Most boys said that they had never perpetrated dating violence. However, 10% said that they had used one of the milder forms (e.g., slapped, pushed), and 4% said that they had used one of the more severe forms (e.g., choked, burned, assaulted with a gun or knife). Personal competence was not related to dating violence, but two factors were linked. One was alcohol use: Boys were 1.31 times more likely to perpetrate violence if they reported frequent use of alcohol. The second was having a friend who had perpetrated dating violence: Boys were 3.57 times more likely to perpetrate violence if they had a friend who had perpetrated violence.

What did the investigators conclude? Some boys definitely represent a greater risk for dating violence. Boys are more likely to perpetrate violence when they drink, which parallels the finding that girls are more likely to be victims when they drink. Drinking and dating are clearly an extremely dangerous mix. Second, boys more often perpetrate violence when they believe that their friends are doing the same. As we saw in Chapter 7, friends can be powerful forces for good or for bad.

What converging evidence would strengthen these conclusions? The main limitation of the study concerns the source of the data—questionnaires completed by the boys themselves. The results hinge on the assumption that adolescent boys' reports are accurate, and there's good reason to doubt the accuracy of these reports. Clearly, some boys may be reluctant to admit that they've been violent on a date. These findings would be more compelling if there were converging information from another source about frequency of violence during dating. For example, boys and girls who are actively dating could each complete questionnaires, and researchers could compare the boy's responses to questions about perpetration of violence with the girl's responses to questions about being a victim.

 Go to Psychology CourseMate at www.cengagebrain.com to enhance your understanding of this research.

■ **Figure 9.2**
Posters like this one are designed to reduce dating violence by encouraging men to listen to and respect a woman's intentions.

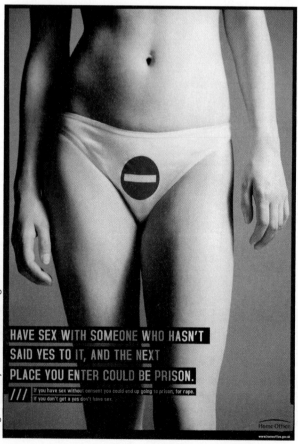

According to the Foshee et al. (2001) study, the level of dating violence is surprisingly high in boys who have just begun to date. At the start of high school, about one boy in seven admits to having perpetrated violence, which underscores the importance of effective prevention programs. One effective program for reducing sexual violence is "Safe Dates" (Foshee & Langwick, 2004). Targeted for middle- and high-school students, the program features a brief play, nine hour-long interactive sessions devoted to topics such as overcoming gender stereotypes and how to prevent sexual assault, and a poster contest. Teens who participate in Safe Date are less likely to be victims of sexual violence and are less likely to perpetrate it (Foshee et al., 2004).

Most colleges and universities offer workshops on date rape. These workshops often emphasize the importance of communication. The ad shown in ■ Figure 9.2 is part of one approach to encourage males and females to communicate about sex. Here are some useful guidelines that are often presented at such workshops (Allgeier & Allgeier, 2000).

1. Know your own sexual policies; decide when sexual intimacy is acceptable for you.
2. Communicate these policies openly and clearly.
3. Avoid being alone with a person until you have communicated these policies and believe that you can trust the person.
4. Avoid using alcohol or other drugs when you are with a person with whom you do not wish to become sexually intimate.
5. If someone tries to force you to have sex, make your objections known: Talk first, but struggle and scream if necessary.

Test Yourself

RECALL

1. For younger adolescents, romantic relationships offer companionship and _____.

2. When parents do not approve of sex, their adolescent children are _____.

3. Adolescents and young adults are at particular risk for contracting AIDS when they _____ and use intravenous drugs.

4. Adolescents often fail to use contraception because they are ignorant of the facts of conception, are attracted to becoming pregnant, and _____.

5. For some boys, the first step toward a gay sexual orientation occurs in early adolescence, when they feel different and are interested in gender-atypical activities; in contrast, for girls, the first step toward a lesbian sexual orientation often grows out of _____.

6. A girl is more likely to be a victim of sexual violence if she has been drinking and if she _____.

INTERPRET

Some sexually active teenagers do not use contraceptives. How do the reasons for this failure show connections between cognitive, social, and emotional development?

APPLY

Prepare a brief fact sheet for incoming college freshmen that summarizes the factors that put women at risk for dating violence.

Recall answers: (1) an outlet for sexual exploration, (2) less likely to be active sexually, (3) engage in unprotected sex, (4) don't know where to get contraceptives (or how to use them), (5) strong attraction to one particular female, (6) holds traditional views of gender roles.

9.3 The World of Work

LEARNING OBJECTIVES

- How do adolescents select an occupation?

- What is the impact of part-time employment on adolescents?

When 15-year-old Aaron announced that he wanted an after-school job at the local supermarket, his mother was delighted, believing that he would learn much from the experience. Five months later, she has her doubts. Aaron has lost interest in school, and they argue constantly about how he spends his money.

"WHAT DO YOU WANT TO BE WHEN YOU GROW UP?" Children are often asked this question in fun. Beginning in adolescence, however, it takes on special significance because work is such an important element of the adult life that is looming on the horizon. A job—be it as a bricklayer, reporter, or child-care worker—helps define who we are. In this section, we'll see how adolescents begin to think about possible occupations. We'll also look at adolescents' first exposure to the world of work, which usually occurs in the form of part-time jobs after school or on weekends. As we do, we'll see if Aaron's changed behavior is typical of teens who work part-time.

Career Development

In most developed nations, adolescence is a time when youth face the challenge of selecting a career. According to a theory proposed by Donald Super (1976, 1980), identity is a primary force in an adolescent's choice of a career. *At about age 13 or 14, adolescents use their emerging identity as a source of ideas about careers, a process called* **crystallization**. Teenagers use their ideas about their own talents and interests to limit potential career prospects. A teenager who is extroverted and sociable may decide that working with people would be the career for him. Decisions are provisional, and adolescents experiment with hypothetical careers, trying to envision what each might be like. Discussions with parents help adolescents refine

crystallization
first phase in Super's theory of career development, in which adolescents use their emerging identities to form ideas about careers

In the specification stage of career development, adolescents try to learn more about different careers, sometimes by serving an apprenticeship.

their emerging ideas; schools also help by providing job fairs and assessing students' job-related interests (Diemer, 2007).

At about age 18, adolescents extend the activities associated with crystallization and enter a new phase. *During* **specification**, *individuals further limit their career possibilities by learning more about specific lines of work and by starting to obtain the training required for a specific job.* The extroverted teenager who wants to work with people may decide that a career in sales would be a good match for his abilities and interests. The teen who likes math may have learned more about careers and decided she'd like to be an accountant. Some teens may begin an apprenticeship as a way to learn a trade.

The end of the teenage years or the early 20s marks the beginning of the third phase. *During* **implementation**, *individuals enter the workforce and learn firsthand about jobs.* This is a time of learning about responsibility and productivity, of learning to get along with co-workers, and of altering one's lifestyle to accommodate work. This period is often unstable; individuals may change jobs frequently as they adjust to the reality of life in the workplace.

In the Real People feature, you can see these three phases in one young woman's career development.

specification
second phase in Super's theory of career development, in which adolescents learn more about specific lines of work and begin training

implementation
third phase in Super's theory of career development, in which individuals actually enter the workforce

Real People
Applying Human Development
"The Life of Lynne": A Drama in Three Acts

Act 1: Crystallization. Throughout high school, Lynne was active in a number of organizations. She often served as the treasurer and found it very satisfying to keep the financial records in order. By the end of her junior year, Lynne decided that she wanted to study business in college, a decision that fit with her good grades in English and math.

Act 2: Specification. Lynne was accepted into the business school of a large state university.

She decided that accounting fit her skills and temperament, so this became her major. During the summers, she worked as a cashier at Target. This helped to pay for college and gave her experience in the world of retail sales.

Act 3: Implementation. A few months after graduation, Lynne was offered a junior accounting position with Wal-Mart. Her job required that she work Tuesday through Friday, auditing Wal-Mart stores in several nearby cities. Lynne

liked the pay, the company car, the pay, the feeling of independence, and the pay. However, having to hit the road every morning by 7:30 was a jolt to someone used to rising casually at 10. Also, Lynne often found it awkward to deal with store managers, many of whom were twice her age and rather intimidating. She was coming to the conclusion that there was much more to a successful career as an accountant than simply having the numbers add up correctly.

Personality Types in Holland's Theory

Personality Type	Description	Careers
Realistic	Individuals enjoy physical labor and working with their hands; they like to solve concrete problems.	Mechanic, truck driver, construction worker
Investigative	Individuals are task-oriented and enjoy thinking about abstract relations.	Scientist, technical writer
Social	Individuals are skilled verbally and interpersonally; they enjoy solving problems using these skills.	Teacher, counselor, social worker
Conventional	Individuals have verbal and quantitative skills that they like to apply to structured, well-defined tasks assigned to them by others.	Bank teller, payroll clerk, traffic manager
Enterprising	Individuals enjoy using their verbal skills in positions of power, status, and leadership.	Business executive, television producer, real estate agent
Artistic	Individuals enjoy expressing themselves through unstructured tasks.	Poet, musician, actor

"The Life of Lynne" illustrates the progressive refinement that takes place in a person's career development. An initial interest in math and finance led to a degree in business, which led to a job as an accountant. However, one other aspect of Lynne's life sheds more light on Super's theory. After 18 months on the job, Lynne's accounting group was merged with another; this would have required Lynne to move to another state, so she quit. After 6 months looking for another accounting job, Lynne gave up and began to study to become a real estate agent. The moral? Economic conditions and opportunities also shape career development. Changing times can force individuals to take new, often unexpected career paths.

Personality-Type Theory

Super's (1976, 1980) work helps to explain how self-concept and career aspirations develop hand in hand, but it does not explain why particular individuals are attracted to one line of work rather than another. Explaining the match between people and occupations has been the aim of a theory devised by John Holland (1985, 1987, 1996). *According to Holland's* **personality-type theory**, *people find work fulfilling when the important features of a job or profession fit the worker's personality.* Holland identified six prototypic personalities that are relevant to the world of work. Each one is best suited to a specific set of occupations, as indicated in the right-hand column of ● Table 9.4. Remember, these are merely prototypes. Most people do not match any one personality type exactly. Instead, their work-related personalities are a blend of the six.

This model is useful in describing the career preferences of African, Asian, European, Native, and Latino American adolescents (Gupta, Tracey, & Gore, 2008). When people have jobs that match their personality type, in the short run they are more productive employees and in the long run they have more stable career paths (Holland, 1996). For example, an enterprising youth is likely to be successful in business because he will enjoy positions of power in which he can use his verbal skills.

Of course, there's more to job satisfaction than the match between a personality type and important features of a job. Even when people are well matched to a job, some will find the work more satisfying than others because of a host of factors, including pay, stress in the workplace, and the frequency of conflicts between work and family obligations (Hammer et al., 2005). Nevertheless, the person–job match is a good place to start thinking about a vocation.

Combining Holland's work-related personality types with Super's theory of career development gives us a comprehensive picture of vocational growth. While Super's theory explains the developmental progression by which individuals translate general interests into a specific career, Holland's theory explains what makes a good match between specific interests and specific careers.

personality-type theory
view proposed by Holland that people find their work fulfilling when the important features of a job or profession fit the worker's personality

According to Holland's personality-type theory, people are satisfied with a job when it matches their personality; for example, adolescents with an enterprising personality type enjoy working in business because this allows them to use verbal skills in positions of leadership.

Of course, trying to match interests to occupations can be difficult. Fortunately, several tests can be used to describe a person's work-related personality and the jobs for which he or she is best suited. In the Strong Interest Inventory (SII), for example, people express their liking of different occupations, school subjects, activities, and types of people (e.g., the elderly or people who live dangerously). These answers are compared to the responses obtained from a representative sample of individuals from different occupations.

If you are still undecided about a career, we encourage you to visit your college's counseling center and arrange to take a test like the SII. The results will help you to focus on careers that would match your interests and help you to choose a college major that would lead to those careers.

Even if you are fairly certain of your vocational plans, you might take one of these tests anyway. As we saw with Lynne, career development does not end with the first job. People continuously refine their career aspirations over the life span, and these test results might be useful later in your life.

THINK ABOUT IT

How do the different personality types in Holland's theory relate to the different types of intelligence proposed by Howard Gardner, described in Chapter 6?

Part-Time Employment

Today, about 25% of high-school freshmen have a part-time job, and about 75% of high-school seniors do (Bachman et al., 2011). About two-thirds of these youth work in retail, and half of those working in retail are employed in the food and beverage industry (U.S. Department of Labor, 2000).

Most adults praise teens for working, believing that early exposure to the workplace teaches adolescents self-discipline, self-confidence, and important job skills (Snedeker, 1982). For most adolescents, however, the reality is quite different. Part-time work can actually be harmful, for several reasons.

1. *School performance suffers.* When students work more than approximately 20 hours per week, they become less engaged in school and are less likely to be successful in college (Bachman et al., 2011; Monahan, Lee, & Steinberg, 2011). Many high-school students apparently do not have the foresight and discipline necessary to consistently meet the combined demands of work and school.

Many American adolescents hold part-time jobs; these can be beneficial but not when adolescents work more than 15 hours weekly.

Copyright © Mark Richards / PhotoEdit

2. *Mental health and behavioral problems.* Adolescents who work long hours—more than 15 or 20 hours a week—are more likely to experience anxiety and depression, and their self-esteem often suffers. Many adolescents find themselves in jobs that are repetitive and boring but stressful, and such conditions undermine self-esteem and breed anxiety. Extensive part-time work frequently leads to substance abuse and frequent problem behavior, such as theft and cheating in school (Monahan et al., 2011).

Why employment is associated with all of these problems is not clear. Perhaps employed adolescents turn to drugs to help them cope with the anxiety and depression brought on by work. Arguments with parents may become more common because anxious, depressed adolescents are more prone to argue or because wage-earning adolescents may believe that their freedom should match their income. Whatever the exact mechanism, extensive part-time work is clearly detrimental to the mental health of most adolescents.

3. *Misleading affluence.* Adults sometimes argue that work is good for teenagers because it teaches them "the value of a dollar," but in reality the typical teenage pattern is to "earn and spend." Working adolescents spend most of their earnings on themselves: to buy clothing, snack food, or cosmetics and to pay for entertainment. Few working teens set aside much of their income for future goals, such as a college education, or use it to contribute to their family's expenses (Shanahan et al., 1996b). Because parents customarily pay for many of the essential expenses associated with truly independent living—rent, utilities, and groceries, for example—working adolescents often have a much higher percentage of their in-

come available for discretionary spending than do working adults. Thus, for many teens the part-time work experience provides unrealistic expectations about how income can be allocated (Darling et al., 2006; Zhang, Cartmill, & Ferrence, 2008).

The message that emerges repeatedly from research on part-time employment is hardly encouraging. Like Aaron, the teenage boy in the vignette, adolescents who work long hours at part-time jobs do not benefit from the experience. To the contrary, they do worse in school, are more likely to have behavioral problems, and learn how to spend money rather than how to manage it. These effects are similar for adolescents from different ethnic groups (Steinberg & Dornbusch, 1991) and are comparable for boys and girls (Bachman & Schulenberg, 1993). Ironically, though, there is a long-term benefit: Young adults who had a stressful part-time job as an adolescent are better able to cope with stressful adult jobs (Mortimer & Staff, 2004).

Does this mean that teenagers who are still in school should never work part-time? Not necessarily. Part-time employment can be a good experience, depending on the circumstances. One key is the number of hours of work. Although the exact number of hours varies from one student to the next, most students could easily work 5 hours weekly without harm, and many could work 10 hours weekly. Another key is the type of job. When adolescents have jobs that allow them to use their skills (e.g., bookkeeping, computing, or typing) and acquire new ones, self-esteem is enhanced and they learn from their work experience (Staff & Schulenberg, 2010; Vazsonyi & Snider, 2008). Yet another factor is how teens spend their earnings. When they save their money or use it to pay for clothes and school expenses, their parent–child relationships often improve (Shanahan et al., 1996a).

By these criteria, who is likely to show the harmful effects of part-time work? A teen who spends 30 hours a week bagging groceries and spends most of it on CDs or videos. Who is likely to benefit from part-time work? A teen who likes to tinker with cars and spends Saturdays working in a repair shop, setting aside some of his earnings for college.

Finally, summer jobs typically do not involve conflict between work and school. Consequently, many of the harmful effects associated with part-time employment during the school year do not hold for summer employment. In fact, such employment sometimes enhances adolescents' self-esteem, especially when they save part of their income for future plans (Marsh, 1991).

© Digital Vision / Jupiterimages Corporation

When adolescents work long hours in a part-time job, they often have trouble juggling the demands of work, school, and sleep!

THINK ABOUT IT

Think back to your own high-school years and those of your friends. Can you think of students (including yourself!) who showed harmful effects from part-time work? Can you think of people who benefited from part-time work?

Test Yourself

RECALL

1. During the _____ phase of vocational choice, adolescents learn more about specific lines of work and begin training.

2. Individuals with a(n) _____ personality type are best suited for a career as a teacher or counselor.

3. Adolescents who work extensively at part-time jobs during the school year often get lower grades, have behavior problems, and _____.

4. Part-time employment during the school year can be beneficial if adolescents limit the number of hours they work and _____.

INTERPRET

Based on the description of Lynne's career, how would you describe continuity of vocational development during adolescence and young adulthood?

APPLY

Suppose that you are a high-school guidance counselor and have been asked to prepare a set of guidelines for students who want to work part-time. What would you recommend?

Recall answers: (1) specification, (2) social, (3) experience misleading affluence, (4) hold jobs that allow them to use and develop skills

LEARNING OBJECTIVES

- Why do teenagers drink and use drugs?

- What leads some adolescents to become depressed? How can depression be treated?

- What are the causes of juvenile delinquency?

Rod was an excellent student and a starter on his high-school basketball team. He was looking forward to going to the senior prom with Peggy, his long-time girlfriend, and then going to the state college with her in the fall. Then, without a hint that anything was wrong in their relationship, Peggy dropped Rod and moved in with the drummer of a local rock band. Rod was stunned and miserable. Without Peggy, life meant so little. Basketball and college seemed pointless. Some days Rod wondered if he should just kill himself to make the pain go away.

SOME YOUNG PEOPLE DO NOT ADAPT WELL TO THE NEW DEMANDS AND RESPONSIBILITIES OF ADOLESCENCE AND RESPOND IN WAYS THAT ARE UNHEALTHY. In this last section of Chapter 9, we look at three problems, often interrelated, that create the "three D's" of adolescent development: drugs, depression, and delinquency. As we look at these problems, you'll understand why Rod feels so miserable without Peggy.

Drug Use

Teenage Drinking

Teen use of illicit drugs like cocaine and methamphetamine often makes headlines, but in reality most adolescents avoid drugs, with one glaring exception—alcohol. About two-thirds of U.S. high-school seniors have drunk alcohol within the past year and nearly half report that they've been drunk (Johnston et al., 2011).

What determines whether an adolescent joins the majority who drink? At least three factors are important.

Adolescents often drink because peers encourage them to.

© Kuttig–People / Alamy

- *Parents.* Teens are more likely to drink (a) when drinking is an important part of parents' social lives—for example, stopping at a bar after work—and (b) when parents are relatively uninvolved in their teenager's life or set arbitrary or unreasonable standards for their teens (Reesman & Hogan, 2005).

- *Peers.* Many adolescents drink because their peers do so and exert pressure on them to join the group (Popp et al., 2008).

- *Stress.* Like adults, many adolescents drink to cope with stress. Teens who report frequent life stresses—problems with parents, with interpersonal relationships, or at school—are more likely to drink and to drink more often (Chassin et al., 2003).

Because teenage drinking has so many causes, no single approach is likely to eliminate alcohol abuse. Adolescents who drink to reduce their tension can profit from therapy designed to teach them more effective means of coping with stress. School-based programs that are interactive—featuring student-led discussion—can be effective in teaching the facts about drinking and strategies for resisting peer pressure to drink (Fitzgerald, 2005; Longshore et al., 2007). Stopping teens from drinking before it becomes habitual is essential because adolescents who drink are at risk for becoming alcohol dependent, depressed, or anxious as adults (Cable & Sacker, 2008; Trim et al., 2007).

Teenage Smoking

Approximately a third of American teens experiment with cigarette smoking at some point in their teenage years (Johnston et al., 2011). American teenagers who smoke typically begin sometime between sixth and ninth grade. As was true for teenage drinking, parents and peers are influential in determining whether youth smoke. When parents smoke, their teenage children are more likely to smoke, too. But the parent–child relationship also contributes: Teens are less likely to smoke when they experience the supportive parenting associated with authoritative parenting (Foster et al., 2007). Like parents, peer influences can be direct and indirect. Teenagers more often smoke when their friends do (Mercken et al., 2007). However, a more subtle influence of peers on teen smoking comes from informal school norms. When most students in a school think it's okay to smoke—even though many of them do not themselves smoke—teens are more likely to start smoking (Kumar et al., 2002).

The dangers of cigarette smoking for adults are well known. Many teenagers (particularly those who smoke) are convinced that cigarette smoking is harmless for healthy adolescents, but they're absolutely wrong. Smoking can interfere with the growth of the lungs, and when teens smoke, they more often have a variety of health problems such as respiratory illnesses. What's more, smoking is often the fateful first step on the path to abuse of more powerful substances, including alcohol, marijuana, and cocaine (Chen et al., 2002).

Faced with these many harmful consequences of teenage smoking, health care professionals and human development researchers have worked hard to create effective programs to discourage adolescents from smoking. In fact, just as comprehensive school-based programs can reduce teenage sex, such programs are effective in reducing teenage smoking (U.S. Department of Health and Human Services, 2000). These programs typically include many common features:

- Schools have no-smoking policies for all students, staff, and school visitors.
- The program provides information about short- and long-term health and social consequences of smoking and provides students with effective ways to respond to peer pressure to smoke.
- The program goes beyond the school to involve parents and communities.

These programs can reduce teenage smoking by more than a third, but they have been implemented in only a handful of schools. By encouraging more schools to provide comprehensive anti-smoking programs, we could continue to reduce the number of teens who start to smoke (Gallagher et al., 2005).

Depression

The challenges of adolescence can lead some youth to become depressed (Fried, 2005). *When suffering from* **depression**, *adolescents have pervasive feelings of sadness, are irritable, have low self-esteem, sleep poorly, and are unable to concentrate.* About 5 to 15% of adolescents are depressed; adolescent girls are more often affected than boys, probably because social challenges in adolescence are often greater for girls than boys (Dekker et al., 2007; Hammen & Rudolph, 2003).

Research reveals that unhappiness, anger, and irritation often dominate the lives of depressed adolescents. They believe that family members, friends, and classmates are not friendly to them (Cole & Jordan, 1995) and they are often extremely lonely (Mahon et al., 2006). Rather than being satisfying and rewarding, life is empty and joyless for depressed adolescents.

Depression is often triggered when adolescents experience a serious loss, disappointment, or failure, such as the death of a loved one or when a much-anticipated date turns out to be a fiasco (Schneiders et al., 2006). Think back to Rod, the adolescent in the vignette at the beginning of this section. His girlfriend had been the center of his life. When she left him unexpectedly, he felt helpless to control his own destiny. Similarly, an athlete may play poorly in the championship game because of illness, or

depression
disorder characterized by pervasive feelings of sadness, irritability, and low self-esteem

THINK ABOUT IT

How does depression illustrate the interaction of biological, psychological, and sociocultural forces on development?

a high-school senior may get a lower score on the SAT exam because of a family crisis the night before taking the test. In each case, the adolescent could do nothing to avoid an undesirable result.

Of course, many adolescents and adults experience negative events like these, but most don't become depressed. Why? One contributing factor is temperament: Children who are less able to regulate their emotions are, as adolescents, more prone to depression (Karevold et al., 2009). Another factor is a belief system in which adolescents see themselves in an extremely negative light. Depression-prone adolescents are, for example, more likely to blame themselves for failure (Gregory et al., 2007). Thus, after the disappointing date, a depression-prone teen is likely to think "I acted like a fool," instead of placing blame elsewhere by thinking "Gee. He was a real jerk!"

Parents and families can also put an adolescent at risk for depression. Not surprisingly, adolescents more often become depressed when their parents are emotionally distant and uninvolved, and when family life is stressful due to economic disadvantage or marital conflict (Karevold et al., 2009; Yap, Allen, & Ladouceur, 2008). And because African American and Hispanic adolescents more often live in poverty, they're more often depressed (Brown, Meadows, & Elder, 2007). Finally, when parents rely on punitive discipline—hitting and shouting—adolescents often resort to the negative attributions (e.g., blaming themselves) that can lead to depression (Lau et al., 2007).

Heredity also plays a role, putting some adolescents at greater risk for depression (Haeffel et al., 2008). Neurotransmitters may be the underlying mechanism: Some adolescents may feel depressed because lower levels of neurotransmitters make it difficult for them to experience happiness, joy, and other pleasurable emotions (Kaufman & Charney, 2003).

Adolescents sometimes become depressed when they feel as if they've lost control of their lives.

iStockphoto.com/StHelena

Treating Depression

To treat depression, some adolescents take antidepressant drugs designed to correct the imbalance in neurotransmitters. However, drug treatment has no lasting effects—it only works while people are taking the drugs—and it has been linked to increased risk of suicide (Vitiello & Swedo, 2004). Consequently, psychotherapy is a better choice for treating depressed adolescents. One common approach emphasizes cognitive and social skills—adolescents learn how to have rewarding social interactions and to interpret them appropriately. These treatments *are* effective (Weisz, McCarty, & Valeri, 2006), and depressed adolescents do need help; left untreated, depression can interfere with performance in school and social relationships, and may also lead to recurring depression in adulthood (Nevid et al., 2003; Rudolph, Ladd, & Dinella, 2007). Also effective are prevention programs: They can substantially reduce the number of depressive episodes in high-risk youth (Stice et al., 2009).

Preventing Teen Suicides

Suicide is the third most frequent cause of death (after accidents and homicide) among U.S. adolescents. Roughly 10% of adolescents report having attempted suicide at least once, but only 1 in 10,000 actually commits suicide. Suicide is rare before 15 years of age, and it is uncommon in girls throughout adolescence. Suicide is far more frequent in older adolescent boys, but the rates differ across ethnic groups. Native American teenage boys have the highest suicide rate by far; Asian Americans and African Americans have the lowest rates (Anderson & Smith, 2005).

Depression is one frequent precursor of suicide; substance abuse is another (Nrugham, Larsson, & Sund, 2008; Renaud et al., 2008).

Few suicides are truly spontaneous, and in most cases there are warning signals (Atwater, 1992). Here are some common signs:

- Threats of suicide
- Preoccupation with death
- Change in eating or sleeping habits
- Loss of interest in activities that were once important
- Marked changes in personality
- Persistent feelings of gloom and helplessness
- Giving away valued possessions

If someone you know shows these signs, *don't ignore them* in the hope that they're not for real. Instead, ask the person if he or she is planning on hurting himself or herself. Be calm and supportive and, if the person appears to have made preparations to commit suicide, don't leave him or her alone. Stay with the person until other friends or relatives can come. More important, *insist* that the adolescent seek professional help. Therapy is essential to treat the feelings of depression and hopelessness that give rise to thoughts of suicide (Capuzzi & Gross, 2004).

Delinquency

Shoplifting. Selling cocaine. Murder. *Adolescents who commit acts like these, which are illegal as well as destructive to themselves or others, are engaged in* **juvenile delinquency**. Adolescents are responsible for much of the criminal activity committed in the United States. For example, adolescents account for about 15% of all arrests for violent crimes and ethnic minority youth are more likely to be arrested for violent crimes (Federal Bureau of Investigation, 2010).

juvenile delinquency
when adolescents commit illegal acts that are destructive to themselves or others

Causes of Delinquency

Why is delinquent behavior so common among adolescents? It's important to distinguish two kinds of delinquent behavior (Moffitt, 1993; Moffitt & Caspi, 2005). The most common form is relatively mild: **Adolescent-limited antisocial behavior** *refers to relatively minor criminal acts by adolescents who aren't consistently antisocial.* These youth may become involved in petty crimes, such as shoplifting or using drugs, but may be careful to follow all school rules. As the name implies, their antisocial behavior is short-lived, usually vanishing in late adolescence or early adulthood.

adolescent-limited antisocial behavior
the behavior of youth who engage in relatively minor criminal acts but aren't consistently antisocial

A second form of delinquent behavior is far more serious and, fortunately, much less common. **Life-course persistent antisocial behavior** *refers to antisocial behavior that emerges at an early age and continues throughout life.* These individuals may start with hitting at 3 years of age and progress to shoplifting at age 12 and then to car theft at age 16 (Odgers et al., 2008). Perhaps only 5% of youth fit this pattern of antisocial behavior, but they account for most adolescent criminal activity.

life-course persistent antisocial behavior
antisocial behavior that emerges at an early age and continues throughout life

Researchers have identified several forces that contribute to this type of antisocial and delinquent behavior (Vitulano, 2005).

1. *Biological contributions. Born to be Bad* is the title of at least two movies, two CDs (one by George Thorogood and one by Joan Jett), and three books. Implicit in this popular title is the idea that, from birth, some individuals follow a developmental track that leads to destructive, violent, or criminal behavior. In other words, the claim is that biology pushes people to be aggressive long before experience can affect development.

    Is there any truth to this idea? In fact, biology and heredity *do* contribute to aggressive and violent behavior, but not in the manner suggested by the epithet "born to be bad" (van Goozen, Fairchild, & Harold, 2008). Twin studies make it clear that heredity contributes: Identical twins are usually more alike in their levels of physical aggression than are fraternal twins (Brendgen et al., 2006). But these studies do not tell us that aggression per se is inherited; in-

THINK ABOUT IT

A letter to the editor of your local paper claims that "juvenile delinquents should be thrown in jail because they're born as 'bad apples' and will always be that way." Write a reply that states the facts correctly.

stead they indicate that some children inherit factors that place them at risk for aggressive or violent behavior. Temperament seems to be one such factor: Youngsters who are temperamentally difficult, overly emotional, or inattentive are, for example, more likely to be aggressive (Joussemet et al., 2008; Xu, Farver, & Zhang, 2009). Hormones represent another factor: Boys with higher levels of the hormone testosterone are often more irritable and have greater body mass (Olweus et al., 1988; Tremblay et al., 1998). A third factor is a deficit in the neurotransmitters that inhibit aggressive behavior (Van Goozen et al., 2007).

None of these factors—temperament, testosterone, or neurotransmitters—*causes* a child to be aggressive. But they do make aggressive behavior more likely: For instance, children who are emotional and easily irritated may be disliked by their peers and be in frequent conflict with them, opening the door for aggressive responses. Thus, biological factors place children at risk for aggression; to understand which children actually become aggressive, we need to look at interactions between inherited factors and children's experiences (Moffitt, 2005).

2. *Cognitive processes.* The perceptual and cognitive skills described in Chapters 6 and 8 also play a role in antisocial behavior. Adolescent boys often respond aggressively because they are not skilled at interpreting other people's intentions. Without a clear interpretation in mind, they respond aggressively by default. That is, aggressive boys far too often think, "I don't know what you're up to, and when in doubt, attack" (Crick & Dodge, 1994; Crozier et al., 2008; Fontaine et al., 2009). Antisocial adolescents are often inclined to act impulsively, and they often are unable or unwilling to postpone pleasure (Fontaine, 2007). Seeing a fancy new CD player or a car, delinquent youth are tempted to steal it simply so that they can have it *right now*. When others inadvertently get in their way, delinquent adolescents often respond without regard to the nature of the other person's acts or intentions.

3. *Family processes.* Delinquent behavior is often related to inadequate parenting. Adolescents are much more likely to become involved in delinquent acts when their parents use harsh discipline or don't monitor effectively (Patterson, 2008; Vieno et al., 2009). Parents may also contribute to delinquent behavior if their marital relationship is marked by constant conflict. When parents constantly argue and fight, their children are much more likely to be antisocial (Cummings et al., 2006; Feldman, Masalha, & Derdikman-Eiron, 2010). Of course, children have ringside seats for many of these confrontations, and thus they can see firsthand how parents use verbal and physical aggression against each other. And, sadly, children come to believe that these patterns of interacting represent "natural" ways of solving problems (Graham-Bermann & Brescoll, 2000).

4. *Poverty.* Aggressive and antisocial behavior is more common among children living in poverty than among children who are economically advantaged (Williams, Conger, & Blozis, 2007). As we've seen, living in poverty is extremely stressful for parents and often leads to the very parental behaviors that promote aggression—harsh discipline and lax monitoring (Tolan, Gorman-Smith, & Henry, 2003). In addition, violent crime is far more common in poverty-stricken neighborhoods. Older children and adolescents exposed to such violence are, as they get older, more likely to be aggressive and violent themselves (Bingenheimer, Brennan, & Earls, 2005).

Obviously, many factors contribute to make some adolescents prone to violent behavior. As you can imagine, when risk factors mount up in children's lives, they are at ever-greater risk for aggressive behavior (Greenberg et al., 1999). What's more, many of the factors operate in a cascading fashion, such that later risk factors build on prior factors: Poverty or maternal depression can lead to harsh, ineffective parenting. In turn, this leads children to be unprepared for school (both academically and socially), which leads to school failure and conduct problems. These difficulties cause some parents to become less active and less invested in parenting, which means

Aggressive teens see the world as a hostile place and typically respond aggressively by default.

© Image Source / Age Fotostock

they monitor their adolescents less often, allowing them to associate with deviant, aggressive peers (Dodge, Greenberg, & Malone, 2008).

Thus, the developmental journey that leads to a violent, aggressive, antisocial adolescent starts in early childhood but gains momentum along the way. Consequently, efforts to prevent children from taking this path must begin early, be maintained over childhood, and target children and their parents. An example of a successful intervention program is Fast Track (Conduct Problems Prevention Research Group, 2011), which is designed to teach academic and social skills to elementary-school children plus life and vocational skills to adolescents. In addition, parents are taught skills for effective child-rearing and, later, how to stay involved with their children and to monitor their behavior. At 12th grade—2 years after the program had ended—aggressive and destructive behavior among children who were at highest risk in kindergarten was cut in half compared to similar high-risk children assigned to a control condition.

Of course, this sort of successful program comes with a very expensive price tag. But costs of prevention programs are a fraction of the costs associated with the by-products of aggressive behavior: One analysis suggests that each violent and aggressive American adolescent costs $2-5 million in payments to victims, court costs, and costs of incarceration (Cohen & Piquero, 2009). Thus, programs such as Fast Track not only improve children's lives (and the lives of people around them), but they're also cost effective.

In the What Do *You* Think? feature, we describe a very different approach to dealing with adolescent crime.

What do you think?

When Juveniles Commit Serious Crimes, Should They Be Tried as Adults?

Traditionally, when adolescents under 18 commit crimes, the case is handled in the juvenile justice system. Although procedures vary from state to state, most adolescents who are arrested do not go to court; instead, law enforcement and legal authorities have considerable discretionary power. They may, for example, release arrested adolescents into the custody of their parents. However, when adolescents commit serious or violent crimes, there will be a hearing with a judge. This hearing is closed to the press and public; no jury is involved. Instead, the judge receives reports from police, probation officers, school officials, medical authorities, and other interested parties. Adolescents judged guilty can be placed on probation at home, in foster care outside the home, or in a facility for youth offenders.

Because juveniles are committing more serious crimes, many law enforcement and legal authorities believe that juveniles should be tried as adults. Advocates of this position argue for lowering the minimum age for mandatory transfer of a case to adult courts, increasing the range of offenses that must be tried in adult court, and giving prosecutors more authority to file cases with juveniles in adult criminal court. Critics argue that treating juvenile offenders as adults ignores the fact that juveniles are less able than adults to understand the nature and consequences of committing a crime. Also, they argue, punishments appropriate for adults are inappropriate for juveniles (Steinberg et al., 2009).

What do you think? Should we lower the age at which juveniles are tried as adults? Based on the theories of development we have discussed, what guidelines would you propose in deciding when a juvenile should be tried as an adult?

Test Yourself

RECALL

1. The main factors that determine whether teenagers drink include parents, peers, and _____.

2. Peers influence teenage smoking indirectly by _____.

3. Depression can be triggered when adolescents experience negative events (e.g., a loss) and they are unable to regulate their emotions or they _____.

4. Treatments for depression include drugs that correct imbalances in neurotransmitters and therapy that emphasizes _____.

5. _____ refers to antisocial behavior that begins at an early age and continues throughout life.

6. The factors that contribute to juvenile delinquency include biology, cognitive processes, _____, and poverty.

INTERPRET

Describe potential biological and environmental contributions to delinquency.

APPLY

Prepare a fact sheet that middle schools could use with antisocial adolescents to educate them about ways to resolve conflicts and achieve goals without relying upon aggression.

Recall answers: (1) stress, (2) establishing an informal school norm in which smoking is approved, (3) see themselves in a negative light; (4) improving cognitive and social skills, (5) Life-course persistent antisocial behavior, (6) family processes

9.1   IDENTITY AND SELF-ESTEEM

How do adolescents achieve an identity?

- The task for adolescents is to find an identity, a search that typically involves four statuses: Diffusion and foreclosure are more common in early adolescence; moratorium and achievement are more common in late adolescence and young adulthood. As they seek identity, adolescents often believe that others are always watching them and that no one else has felt as they do.

- Adolescents are more likely to achieve an identity when parents encourage discussion and recognize their autonomy; they are least likely to achieve an identity when parents set rules and enforce them without explanation.

What are the stages and results of acquiring an ethnic identity?

- Adolescents from ethnic groups often progress through three phases in acquiring an ethnic identity: initial disinterest, exploration, and identity achievement. Achieving an ethnic identity usually results in higher self-esteem but is not consistently related to the strength of one's identification with mainstream culture.

How does self-esteem change in adolescence?

- Social comparisons begin anew when children move from elementary school to middle or junior high school; consequently, self-esteem usually declines somewhat during this transition. However, self-esteem begins to rise in middle and late adolescence as teenagers see themselves acquiring more adult skills and responsibilities. Self-esteem is linked to adolescents' actual competence in domains that matter to them and is also linked to how parents and peers view them.

- The parent–child relationship becomes more egalitarian during the adolescent years, reflecting adolescents' growing independence. Contrary to myth, adolescence is not usually a period of storm and stress. Most adolescents love their parents, feel loved by them, rely on them for advice, and adopt their values.

9.2   ROMANTIC RELATIONSHIPS AND SEXUALITY

Why do teenagers date?

- Romantic relationships emerge in mid-adolescence. For younger adolescents, dating is for both companionship and sexual exploration; for older adolescents, it is a source of trust and support. Adolescents in romantic relationships are more self-confident but also report more emotional upheaval.

Why are some adolescents sexually active? Why do so few use contraceptives?

- By the end of adolescence, most American boys and girls have had sexual intercourse, which boys view as recreational but girls see as romantic. Adolescents are more likely to be sexually active if they believe that their parents and peers approve of sex. Adolescents do not use birth control consistently because they do not understand conception, don't know where to obtain contraceptives, and sometimes find pregnancy appealing. Because they use contraception infrequently, they are at risk for sexually transmitted diseases and becoming pregnant.

What determines an adolescent's sexual orientation?

- A small percentage of adolescents are attracted to members of their own sex. Attraction to same-sex individuals follows a different path in boys and girls. For boys, the first step involves feeling different and becoming interested in gender-atypical activities. In contrast, for girls, the first step often involves strong feelings toward one particular female. Gay and lesbian youth face many special challenges and thus often suffer from mental health problems.

What circumstances make dating violence likely?

- Many adolescents experience dating violence and teenage girls are sometimes forced into sex against their will. Girls are more likely to be victims of sexual violence when they've been drinking and when they hold traditional views of gender. Boys are more likely to perpetrate violence when they've experienced violence at home, when they drink, and when their friends perpetrate sexual violence. Date-rape workshops strive to improve communication between males and females.

9.3   THE WORLD OF WORK

How do adolescents select an occupation?

- In his theory of vocational choice, Super proposes three phases of vocational development during adolescence and young adulthood: crystallization, in which basic interests are identified; specification, in which jobs associated with interests are identified; and implementation, which marks entry into the workforce.

- Holland proposes six different work-related personalities: realistic, investigative, social, conventional, enterprising, and artistic. Each is uniquely suited to certain jobs. People are happier when their personality fits their job and less happy when it does not.

What is the impact of part-time employment on adolescents?

- Most adolescents in the United States have part-time jobs. Adolescents who are employed more than 15 hours per week during the school year typically do poorly in school, often have lowered self-esteem and increased anxiety, and have problems interacting with others. Employed adolescents save relatively little of their income. Instead, they spend it on clothing, food, and their entertainment, which can yield misleading expectations about how to allocate income.

- Part-time employment can be beneficial if adolescents work relatively few hours, if the work allows them to use existing skills or acquire new ones, and if teens save some of their earnings. Summer employment, which does not conflict with the demands of school, can also be beneficial.

9.4 THE DARK SIDE

Why do teenagers drink and use drugs?

■ Many adolescents drink alcohol regularly. The primary factors that influence whether adolescents drink are encouragement from others (parents and peers) and stress. Similarly, teenage smoking is influenced by parents and peers.

What leads some adolescents to become depressed? How can depression be treated?

■ Depressed adolescents have little enthusiasm for life, believe that others are unfriendly, and wish to be left alone. Depression can be triggered by a negative event; adolescents most likely to be affected are those who can't control their emotions and who see themselves in a negative light. Treating depression relies on medications that correct the levels of neurotransmitters and on therapy designed to improve social skills and restructure adolescents' interpretation of life events.

What are the causes of juvenile delinquency?

■ Many young people engage in antisocial behavior briefly during adolescence. In contrast, the small percentage of adolescents who engage in life-course persistent antisocial behavior are involved in one-fourth to one-half of the serious crimes committed in the United States. Life-course persistent antisocial behavior has been linked to biology, cognitive processes, family processes, and poverty. Efforts to reduce adolescent criminal activity must address all of these variables.

KEY TERMS

adolescent egocentrism (317)
imaginary audience (318)
personal fable (318)
illusion of invulnerability (318)
ethnic identity (318)

crystallization (329)
specification (330)
implementation (330)
personality-type theory (331)
depression (335)

juvenile delinquency (337)
adolescent-limited antisocial behavior (337)
life-course persistent antisocial behavior (337)

LEARN MORE ABOUT IT

Log in to **www.cengagebrain.com** to access the resources your instructor requires. For this book, you can access:

Psychology CourseMate

■ CourseMate brings course concepts to life with interactive learning, study, and exam preparation tools that support the printed textbook. A textbook-specific website, Psychology CourseMate includes an integrated interactive eBook and other interactive learning tools, including quizzes, flashcards, videos, and more.

CENGAGENOW™

■ CengageNOW Personalized Study is a diagnostic study tool containing valuable text-specific resources—and because you focus on just what you don't know, you learn more in less time to get a better grade.

WebTUTOR™

■ More than just an interactive study guide, WebTutor is an anytime, anywhere customized learning solution with an eBook, keeping you connected to your textbook, instructor, and classmates.

Becoming an Adult

Physical, Cognitive, and Personality Development in Young Adulthood

THERE COMES A TIME in life when we turn away from childhood and aspire to being adults. In some societies, the transition to adulthood is abrupt and dramatic, marked by clear rites of passage. In Western society it is fuzzier; the only apparent marker may be a birthday ritual. We may even ask "real" adults what it's like to be one. Adulthood is marked in numerous ways, some of which we explore in the first section.

Without question, young adulthood is the peak of physical processes and health. It is also a time when people who acquired unhealthy habits earlier in life may decide to adopt a healthier lifestyle. Young adulthood also marks the peak of some cognitive abilities and the continued development of others.

On a more personal level, young adulthood is a time when we make plans and dream of what lies ahead. It is a time when we think about what life as an adult will be like. But above all, it is a time when we lay the foundation for the developmental changes that we will experience during the rest of our lives. We will consider these issues as we examine young adulthood in this chapter.

LEARNING OBJECTIVES

- What role transitions mark entry into adulthood?
- How does going to college reflect the transition to adulthood?
- What behavioral criteria mark the transition to adulthood?

- How does achieving financial independence reflect the transition to adulthood?
- When do people become adults?

Marcus woke up with the worst headache he could ever remember having. "If this is adulthood, they can keep it," he muttered to himself. Like many young adults in the United States, Marcus spent his 21st birthday celebrating at a bar. But the phone call from his mother that woke him in the first place reminds him that he isn't an adult in every way; she called to see if he needs money.

THINK FOR A MINUTE ABOUT THE FIRST TIME YOU FELT LIKE AN ADULT. When was it? What was the context? Who were you with? How did you feel?

Even though becoming an adult is one of our most important life transitions, it is difficult to pin down exactly when this occurs in Western societies. Birthday celebrations marking the achievement of a certain age are inconsistent in their message. In the United States, for example, the age needed to achieve "adult" status ranges from 12 or so for movie theater tickets to 16 for driving a car (in many states), 18 for voting and joining the military, 21 for consuming alcoholic beverages, 24 for dependent status as a full-time student according to the Internal Revenue Service, 25 for renting a car (without expensive surcharges), and 26 for medical insurance coverage on one's parents' policy under the 2010 health care law. Clearly, in the United States there is no age that signals a clean break with adolescence and full achievement of adulthood. Certainly Marcus may feel like an adult because he can purchase alcohol legally, but he may not feel that way in other respects, such as supporting himself financially.

Some human developmentalists view the period from the late teens to the mid- to late 20s as distinctive. *They refer to it as* **emerging adulthood**, *a period when individuals are not adolescents but are not yet fully adults* (Arnett, 2004, 2010a, 2010b; Furstenberg,

emerging adulthood
period between late teens and mid- to late 20s when individuals are not adolescents but are not yet fully adults

Milestone birthdays such as turning 21 are often marked with celebrations.

Image 100 / Alamy

Rumbaut, & Settersten, 2005; Grant & Potenza, 2010). Emerging adulthood is a time to explore careers, self-identity, and commitments. It is also a time when certain biological and physiological developmental trends peak, brain development continues in different ways, and the risk for certain mental health problems peaks.

In this section, we examine some of the ways societies mark the transition to adulthood, and we'll see that the criteria vary widely from culture to culture.

Role Transitions Marking Adulthood

Consider the following three women. Ganika looked older than her 20 years; her son was playing quietly on the floor. She and her husband live in a small house in a village in rural India. Sheree graduated from high school a few years ago. She works full time at the Old Navy store, and rents a small apartment nearby. Claudia recently graduated from college in Germany. But due to an economic downturn, she lives with her mother while she looks for a job.

Are these women adults? Yes and no. As we will see, it all depends on how you define adulthood and the kind of role transitions cultures create. A role transition is movement into the next stage of development marked by assumption of new responsibilities and duties.

Cross-Cultural Evidence of Role Transitions

Cultures in the developing world tend to be clear about when a person becomes an adult and to put greater emphasis on specific practices that mark the transition (Nelson, Badger, & Wu, 2004). *Rituals marking initiation into adulthood, often among the most important ones in a culture, are termed* **rites of passage**. Rites of passage may involve highly elaborate steps that take days or weeks, or they may be compressed into a few minutes. Initiates are usually dressed in apparel reserved for the ritual to denote their special position. We still have traces of these rites in Western culture; consider, for example, the ritual attire for graduations or weddings.

rites of passage
rituals marking initiation into adulthood

role transitions
movement into the next stage of development marked by assumption of new responsibilities and duties

In many cultures, rites of passage to adulthood are connected with religious rituals (Levete, 2010). For example, Christian traditions use a ritual called confirmation to mark the transition from being a child spiritually to being an adult. Judaism celebrates bar and bat mitzvahs. In many Latin American countries, a girl's 15th birthday is celebrated as the transition between childhood and young womanhood. In some countries this ceremony, called the quinceañera, is preceded by a Mass if the girl's family is Catholic.

Marriage is the most important rite of passage to adulthood in most cultures because it is a prelude to childbearing, which in turn provides clear evidence of achieving adulthood (Mensch, Singh, & Casterline, 2006). In this sense, Ganika is an adult in her culture.

Some tribal cultures mark the transition to adulthood in public ways so that the whole community witnesses it. In a few cultures, it may involve pain or mutilation, such as female circumcision (Castro, 2010), or may involve being given a specific tattoo (Irish, 2009). Because rites in such cultures change little over time, they provide continuity throughout the life span (Keith, 1990); older adults lead young people through the same rites they themselves experienced years earlier. Western counterparts are much less formalized and are hard to identify; indeed, you may be hard pressed to think of any. A father buying his son his first razor or a mother helping her daughter with her first menstrual period may be as close as we get in Western society.

This couple from India reflects that the passage to adulthood for many occurs with marriage.

Role Transitions in Western Cultures

In Western cultures, the most widely used criteria for deciding whether a person has reached adulthood are **role transitions**, *which involve assuming new responsibilities and duties*. Certain role transitions have long been recognized in Western society as key markers for attaining adulthood: voting, completing education, beginning full-time employment, leaving home and establishing an independent household, being financially independent, getting married, and becoming a parent. However, many people in their 20s in industrialized countries spread these achievements over several years (Arnett, 2004), and this trend has significantly increased since the late 1970s (Burt & Masten, 2010). The result is a period some researchers call "emerging adulthood" (Arnett, 2010a; 2010b).

The point in the life span when these marker events for role transitions happen, though, has changed a lot over the years. Such changes are examples of cohort effects, described in Chapter 1. For example, in the United States the average age for completing all of one's formal schooling rose steadily during the 20th century as the proportion of people going to college at some point in adulthood increased from roughly 10% in the early part of the century to roughly two thirds today. Likewise, since the early 1970s the age at first marriage has increased in the United States by almost 5 years (U.S. Census Bureau, 2010a), a remarkable change within one generation.

What does this mean for the transition from adolescence to adulthood? It means that Western society has no clear, age-constant rituals that clearly mark the transition to adulthood (Ivory, 2004). As a result, young people create their own, such as initiations into student organizations or drinking alcohol (a topic we explore later in this chapter). The lack of clearly defined rituals makes it difficult to use any one event as the marker for becoming an adult. Although the trend is that living independently from one's parents, financial independence, and romantic involvement are associated with increased assumption of adult roles, individual patterns are extremely diverse. So, depending on how you look at it, Sheree and Claudia may—or may not—be considered adults.

THINK ABOUT IT

Why do Western cultures lack clear-cut transitions to adulthood?

Let's consider some of the components of the role transitions in Western culture, and how they help shape role transitions. First, let's consider a common one for many people, the move from high school to college.

Going to College

For about 69% of all high-school graduates in the United States, a marker of the transition to adulthood is going straight to college—although the rates vary across European Americans, African Americans, and Latinos, as shown in ■ Figure 10.1 (National Center for Educational Statistics, 2010). Considerable research has documented how students develop while they are in college (Evans et al., 2010; Kitchener, King, & DeLuca, 2006; Perry, 1970). Students start acting and thinking like adults because of advances in intellectual development and personal and social identity. It is thought that much of this developmental change in college occurs through social interaction. Indeed there is a belief that the college social experience can either facilitate or frustrate the development of one's sense of identity, including ethnic/racial identity (Guiffrida, 2009). Later in the chapter we examine some of the cognitive changes that occur during the transition from adolescence to adulthood.

Those changes may well describe the experience of those who go directly from high school to college. But take a look around, literally, at your classmates. In addition to the increased ethnic and racial diversity you see, the age diversity you also likely experience is a reflection of the rapidly changing nature of college and university campuses.

Although students who attend college full-time tend to be the traditional age (under 25), as you can see in ■ Figure 10.2, a relatively large number of undergraduates are older, with most of them attending part-time. *Colleges usually refer to students over age 25 as* **returning adult students**, *which implies that these individuals have already reached adulthood.*

returning adult students
college students over age 25

Overall, returning adult students tend to be problem solvers, self-directed, and pragmatic, may have increased stress due to work-family-school conflict, and have relevant life experiences that they can integrate with their course work (Evans et al.,

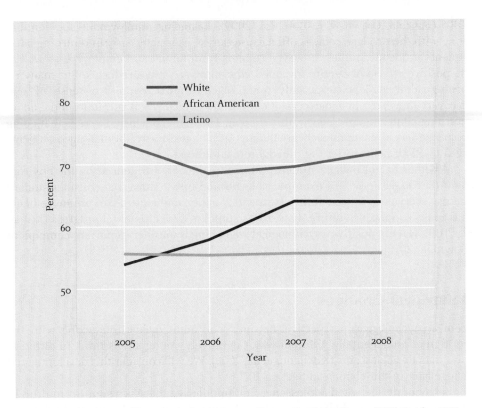

■ **Figure 10.1**
Enrollment rates of 18- to 24-year-olds in degree-granting institutions by race/ethnicity, 1975–2008.

Source: U.S. Department of Commerce, Census Bureau, Current Population Survey (CPS), October 1970–October 2008, unpublished tabulations.

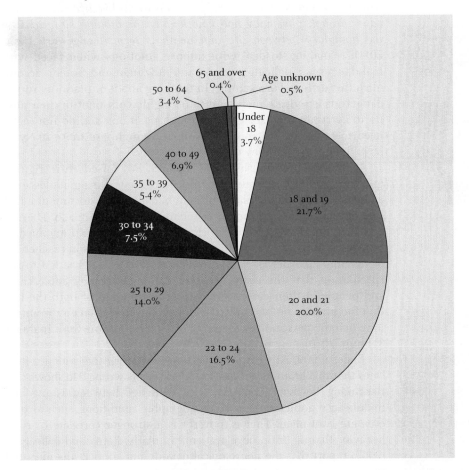

■ **Figure 10.2**
Fall 2007 distribution of students enrolled in degree-granting institutions by age. An increasing number of college students are returning adult students.

Source: U.S. Department of Education; National Center for Education Statistics, 2007; Integrated Postsecondary Education Data System (IPEDS), Spring 2007.

2010; Giancola, Grawitch, & Borchert, 2009). Balancing employment and families along with their college courses often causes stress, especially early in returning adult students' academic studies; however, support from family and employers, as well as the positive effects of continuing one's education, are stress reducers. But many returning adult students, especially middle-aged women, express a sense of self-discovery they had not experienced before (Miles, 2009). We'll encounter more about returning adult students later in this chapter. The main conclusion here is that going to college has a major impact on students of all ages; for traditional aged students (ages 18–25) it helps foster the transition to adulthood.

Whether or not college students succeed is heavily influenced by the environment they experience. For example, colleges and universities that consider student success everyone's responsibility (students, faculty, and staff alike) are more likely to create a supportive culture that fosters students' developmental progress (Kuh et al., 2010). Having positive psychological well-being is another important component (Bowman, 2010).

Behavioral Changes

edgework
the desire to live life more on the edge through physically and emotionally threatening situations on the boundary between life and death

Risky behavior such as this tends to decrease over the course of young adulthood.

Jeanne Hatch / Shutterstock.com

From a psychological perspective, becoming an adult means interacting with the world in a fundamentally different way. Cognitively, young adults think in different ways than do adolescents (Kitchener et al., 2006). We will consider this developmental change later in this chapter in Section 10.3.

Behaviorally, between adolescence and middle age there is a significant drop in the frequency of risky behaviors such as driving at high speed, having sex without contraception, engaging in extreme sports, or committing antisocial acts like vandalism (Tanner & Arnett, 2009). *The desire to live life more on the edge through physically and emotionally threatening situations on the boundary between life and death is termed* **edgework** (Lois, 2011). Managing to deal with intense emotions when faced with real danger is a delicate balance, and how men and women accomplish this differs. For example, in thinking about or planning future dangerous activities, men tend to be highly confident in their ability to extricate themselves from tough positions and do not feel a need to rehearse just in case. Women, though, are more likely to have qualms which they ease by rehearsing.

Evidence from neuroscience research has raised an intriguing question: Could continuing brain development during adolescence and early adulthood explain shifts in behavior and document a biological marker for adulthood? Some investigators think so. There is evidence that the prefrontal cortex, a part of the brain involved in high level thinking, is not fully developed until a person reaches his or her mid-20s (Berns, Moore, & Capra, 2009). That has led to the speculation that the risky behavior demonstrated by adolescents and people in their early to mid-20s is a function of both the fact that the brain is not yet fully developed and the social environment in which the person lives (Berns et al., 2009; Johnson, Sudhinaraset, & Blum, 2010).

Males (2009, 2010) argues that a closer look at the evidence indicates that this brain development hypothesis is wrong. He shows that risk-taking behavior does not vary with age between young and middle-aged adults when sociodemographic conditions, such as poverty, are controlled. That is, poverty is a stronger correlate of risky behavior than is brain development. Similarly, Berns and colleagues (2009) report that the neurological data do not support the idea that risky behavior is caused mainly by immature brains.

On the psychosocial front, young adulthood marks the transition from concern with identity (see Chapter 9) to concern with behaviors related to autonomy and intimacy, which we explore here and in Chapter 11 (Erikson, 1982). Becoming independent from one's parents entails being able to fend for oneself, but it does not imply a complete severing of the relationship. On the contrary, adult children usually establish a rewarding relationship with their parents, as we will see in Chapter 13.

Establishing Intimacy

According to Erikson, the major task for young adults is dealing with the psychosocial conflict of **intimacy versus isolation**. This is the sixth step in Erikson's theory of psychosocial development, the basic tenets of which are summarized in Chapter 1. Once a person's identity is established, Erikson (1982) believed that he or she is ready to create a shared identity with another—the key ingredient for intimacy. Without a clear sense of identity, Erikson argued, young adults would be afraid of committing to a long-term relationship with another person or might become overly dependent on the partner for his or her identity.

Some studies support this view. For example, Montgomery (2005) found that a stronger sense of identity was related to higher levels of intimacy in young adults. However, other research shows different results. For example, Berliner (2000) also found that identity formation correlated with intimacy in adults aged 35 to 45, but this relationship held even for those people who demonstrated diffusion (the lowest level of identity formation), which Erikson argued should not be the case.

A meta-analytic study (see Chapter 1) of the connection between identity and intimacy provides insight into why findings in individual studies are inconsistent. After analyzing the results of 21 studies, Årseth and colleagues (2009) concluded that resolving identity issues before intimacy issues differed between men and women.

The differences are complex. At a general level, identity is related to closeness in same-sex friendships for both men and women and also for men in cross-sex friendships; for women, however, identity is unrelated to closeness in cross-sex friendships (Johnson et al., 2007). Why is there a difference in the cross-sex friendship patterns? It turns out that most men and career-oriented women resolve identity issues before intimacy issues (Dyke & Adams, 1990). So these individuals sort through their initial career choices and complete their formal education before becoming involved in a committed relationship.

But some women resolve intimacy issues before identity issues by marrying and rearing children, and only after their children have grown and moved away do they deal with the question of their own identity. The experience of self-discovery by middle-aged women who go to college for the first time is an example of this form of identity development (Miles, 2009).

Still other women deal with both identity and intimacy issues simultaneously—for example, by entering into relationships that allow them to develop identities based on caring for others (Dyke & Adams, 1990). For example, if a woman's partner has special health or medical treatment needs, the woman could develop a strong identity as a caregiver and a strong intimate relationship with the partner simultaneously.

Thus, Erikson's idea that identity must be resolved before intimacy is most applicable in the cases of men and career-oriented women. The fact that some women show different patterns yet still resolve both issues indicate that there are likely multiple pathways to achieving identity and intimacy.

Launching One's Financial Independence

For many, the key indicator of becoming an adult is launching one's career and establishing financial independence. For the one third of high school graduates who do not go on to college and for those who do not finish high school, establishing

intimacy versus isolation
sixth stage in Erikson's theory and the major psychosocial task for young adults

THINK ABOUT IT
Why do you think it mattered to Erikson whether identity issues are resolved before intimacy issues?

financial independence may be a more pressing issue sooner than it is for college-bound students.

Some find their niche through a series of part-time jobs, full-time employment, learning a trade (such as carpentry or plumbing), or starting their own business. Others join the military, a pathway to adulthood created as a result of the all-volunteer service (Kelty, Kleykamp, & Segal, 2010). Military service paves a way to adulthood in several ways, not the least of which is through financial independence. The support services in the military can also foster a level of maturity in family and social relationships. The military also provides a stable career option, especially considering that the military will provide advanced educational opportunities that may not be available otherwise.

Eventually, though, even college-bound students face the need to establish themselves as financially independent. When this happens, though, is changing. These days, it is increasingly common for college graduates to return home to live with their parents prior to establishing financial independence (Henig, 2010). We will consider what happens when adult children return home in Chapter 13.

Regardless of when it occurs, though, reaching financial independence is a major achievement and serves as a marker of becoming an adult. And where one lives matters in this process. Research in Western cultures shows clearly that living on one's own accelerates the achievement of adulthood, whereas living with one's parents slows the process of becoming an independent adult (Kins & Beyers, 2010).

So When Do People Become Adults?

Increasingly, researchers and writers are arguing that the years between late adolescence and the late 20s to early 30s may reflect a distinct life stage researchers call emerging adulthood (Arnett, 2010a; 2010b; Grant & Potenza, 2010; Kins & Beyers, 2010). Evidence is mounting that social and demographic trends since the 1970s may be creating a new developmental period, much as similar changing circumstances created the period of adolescence in the early 20th century.

The challenges faced by adults in their 20s can be difficult. Robbins and Wilner (2001) coined the term "quarterlife crisis" to describe how life in one's 20s is far from easy as individuals struggle to find their way. Adjusting to the "real world" is increasingly difficult for college graduates who face significant debt from educational loans as well as challenges from workplace politics, networking, time management, and other daily hassles. Indeed, Byock (2010) argues that quarterlife today serves a similar purpose to midlife 50 years ago. For Byock, quarterlife represents a period of self-exploration and search for meaning in a way similar to the way Jung described the search for meaning in his original theory of the midlife crisis (a topic we will explore more in Chapter 13).

The perspectives considered in this section do not provide any definitive answers to the question of when people become adults. All we can say is that the transition depends on numerous cultural and psychological factors. In cultures without clearly defined rites of passage, defining oneself as an adult rests on one's perception of whether personally relevant key criteria have been met. In Western society, this can be very complicated—for example, when success comes at a young age, as discussed in the Real People feature.

Kristen Jaymes Stewart, born in 1990, achieved international fame for her portrayal of Isabella "Bella" Swan in the *Twilight Saga* films. Her career began in 1999 with a small, uncredited role. Her parents and family have been active in television, in supervising scripts, and in other behind-the-camera activities, so she learned about film and television early on. After acting in several films, she was thrust into major international stardom for being selected to portray "Bella," the main female character in the film version of the *Twilight* series of books by Stephenie Meyer. Kristen's educational career has been different, too. After the seventh grade, she completed the rest of her schooling through high school by correspondence.

Clearly, Kristen has not experienced many of the transitions we considered in this section that are associated with becoming an adult, such as going to college or marrying and having a child. However, there is also no question that she can be financially independent from her parents. And, she certainly lives a very public life.

Would you consider Kristen an adult? Is it enough to live in the public eye and to be extremely wealthy? What effect do these circumstances have on dealing with issues such as Erikson's intimacy versus isolation, considering that every person you may want to date will be publicly scrutinized and debated?

Kristen Jaymes Stewart

Brandon Parry / Shutterstock.com

Test Yourself

RECALL

1. The most widely used criteria for deciding whether a person has reached adulthood are _____.

2. Rituals marking initiation into adulthood are called _____.

3. Students over 25 are referred to as _____.

4. Behaviorally, a major difference between adolescence and adulthood is a significant drop in the frequency of _____.

5. Research indicates that Erikson's idea of resolving identity followed by intimacy best describes men and _____.

INTERPRET

Why are formal rites of passage important? What has Western society lost by eliminating them? What has it gained?

APPLY

When do you think people become adults? Why?

Recall answers: (1) role transitions, (2) rites of passage, (3) returning adult students, (4) reckless behavior, (5) career-oriented women

LEARNING OBJECTIVES

- In what respects are young adults at their physical peak?

- How healthy are young adults in general?

- How do smoking, drinking alcohol, and nutrition affect young adults' health?

- How does the health of young adults differ as a function of socioeconomic status, gender, and ethnicity?

Juan is a 25-year-old who started smoking cigarettes in high school to be popular. Juan wants to quit, but he knows it will be difficult. He has also heard that it doesn't really matter if he quits or not because his health will never recover. Juan wonders whether it is worthwhile to try.

JUAN IS AT THE PEAK OF HIS PHYSICAL FUNCTIONING. Most young adults are in the best physical shape of their lives. Indeed, the early 20s are the best years for strenuous work, trouble-free reproduction, and peak athletic performance. These achievements reflect a physical system at its peak. But people's physical functioning is affected by several health-related behaviors, including smoking.

Growth, Strength, and Physical Functioning

Physical functioning generally peaks during young adulthood (Schneider & Davidson, 2003). You're as tall as you will ever be (you'll likely get shorter in later life, as described in Chapter 14). Although men have more muscle mass and tend to be stronger than women, physical strength in both sexes peaks during the late 20s and early 30s, declining slowly throughout the rest of life (Whitbourne, 1996). Coordination and dexterity peak around the same time (Whitbourne, 1996). Because of these trends, few professional athletes remain at the top of their sport in their mid-30s. Indeed, individuals such as Brett Favre, a quarterback who played in the NFL into his early 40s, and Dara Torres, who set the record for the oldest swimmer to win Olympic medals by winning three in Beijing in 2008 at age 41, are famous partly because they are exceptions. Because they are at their physical peak, most professional sports stars are in their 20s.

Dara Torres, shown on the right, won three medals at the Beijing Olympics in 2008 at age 41.

Sensory acuity is also at its peak in the early 20s (Fozard & Gordon-Salant, 2001). Visual acuity remains high until middle age, when people tend to become farsighted and require glasses for reading. Hearing begins to decline somewhat by the late 20s, especially for high-pitched tones. By old age, this hearing loss may affect one's ability to understand speech. People's abilities to smell, taste, feel pain and changes in temperature, and maintain balance remain largely unchanged until late life.

Health Status

When adults aged 18–44 in the United States are asked how good their health is, 94% say that it is good, very good, or excellent (Pleis & Lethbridge-Çejku, 2007). Young adults have far fewer colds and respiratory infections than when they were children. Indeed, only about 1% of young adults are limited in their ability to function because of a health-related condition.

Because they are so healthy overall, American young adults very rarely die from disease (National Center for Health Statistics, 2010a). For example, the death rate due to cancer for people aged 15 to 24 is about 4 people per 100,000 population, compared with about 116 per 100,000 for people aged 45 to 54, and more than 1,606 per 100,000 for people over age 85. So what is the leading cause of death among young adults in the United States? Between the ages of 25 and 44, it's accidents.

There are important gender and ethnic differences in these statistics. Young adult men aged 25 to 34 are nearly 2.5 times as likely to die as women of the same age. African American and Latino young adult males are 2–2.5 times as likely to die as their European American male counterparts, but Asian and Pacific Islander young adult males are likely to die at only half the rate of their European American male counterparts (National Center for Health Statistics, 2010a).

Lifestyle Factors

To maximize the odds of being healthy, *don't* smoke, monitor your alcohol consumption, and eat a well-balanced, nutritional diet. In this section we focus on why this is good advice. We return to this theme in Chapter 13 when we examine additional aspects of health promotion, especially concerning cardiovascular disease and exercise.

Smoking

Smoking is the single biggest contributor to health problems, a fact known for decades. In the United States alone, roughly 400,000 people die each year from smoking-related illnesses, and medical treatment of smoking-related ailments costs more than $193 billion annually (Centers for Disease Control and Prevention, 2010a).

The risks of smoking are many. ■ Figure 10.3 shows the various forms of cancer and other chronic diseases that are caused by smoking. As noted in Chapter 2, nicotine in cigarettes is a potent teratogen; smoking during pregnancy can cause stillbirth, low birth weight, or perinatal death (death shortly after birth). And smoking during one's lifetime has a small but measurable negative impact on cognitive functioning in later life (Whalley et al., 2005).

Nonsmokers who breathe secondhand smoke are also at considerably higher risk for smoking-related diseases: each year, more than 3,400 adult nonsmokers die from lung cancer and 46,000 adult nonsmokers die from cardiovascular disease (American Cancer Society, 2010a). Hundreds of thousands of children suffer from lung problems annually in the United States as a result of environmental smoke. And it costs about $10 billion in health care (Centers for Disease Control and Prevention, 2010a). For

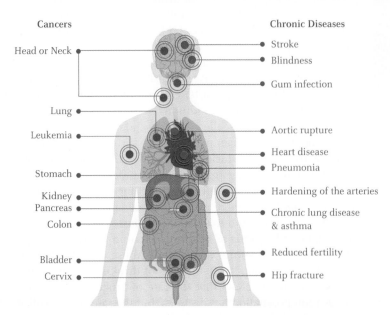

Cancers — Head or Neck, Lung, Leukemia, Stomach, Kidney, Pancreas, Colon, Bladder, Cervix

Chronic Diseases — Stroke, Blindness, Gum infection, Aortic rupture, Heart disease, Pneumonia, Hardening of the arteries, Chronic lung disease & asthma, Reduced fertility, Hip fracture

Centers for Disease Control: www.cdc.gov/VitalSigns/pdf/2010-09-vitalsigns.pdf, page 3.

■ **Figure 10.3**
Smoking can damage every part of the body

these reasons, many states and communities have passed stricter legislation banning smoking in public buildings and public outdoor spaces, and smoking is banned entirely on airline flights within the United States and on many international flights. Still, secondhand smoke remains a major problem; about 40% of nonsmokers in the United States continue to be exposed to secondhand smoke (Centers for Disease Control and Prevention, 2010b).

Juan, the young man in the vignette, is typical of people who want to stop smoking. Most people who try to stop smoking begin the process in young adulthood. Although some smokers who want to quit find formal programs helpful, more than 90% of those who stop do so on their own. But as Juan suspects, quitting is not easy; the American Cancer Society (2010b) estimates that, even with medicinal intervention, between two-thirds and three-quarters of those who try to quit smoking end up starting again within 6 months. For most people, success is attained only after a long period of stopping and relapsing.

Regardless of how it happens, quitting smoking has enormous health benefits both for smokers and for people who inhale secondhand smoke (American Cancer Society, 2010b). For example, in less than a year after quitting, the lungs regain their normal ability to move mucus out. The risks of stroke and coronary heart disease return to normal after a period of roughly 15 years. Even people who do not quit until late life show marked improvements in health (Stop Smoking, 2010). In sum, the evidence is clear: If you don't smoke, don't start. If you do, you're never too old to stop. Check out the American Cancer Society's *Guide to Quitting Smoking* (www.cancer.org/Healthy/StayAwayfromTobacco/GuidetoQuittingSmoking/index) for key information about how to quit and a quiz about whether you need help to quit (www.cancer.org/Healthy/ToolsandCalculators/Quizzes/app/smoking-habits-quiz).

binge drinking
type of drinking defined for men as consuming five or more drinks in a row and for women as consuming four or more drinks in a row within the past 2 weeks

Drinking Alcohol

If you are between the ages of 18 and 44, chances are that you sometimes drink; about two-thirds of the people in the United States drink alcohol at least occasionally (National Center for Health Statistics, 2010a). Total consumption of alcohol in industrialized countries has declined for the past few decades, and underage drinking in the United States has also declined, partly in response to tougher laws regarding underage drinking and driving while intoxicated (National Center for Health Statistics, 2010a).

For the majority of people, drinking alcohol poses no serious health problems as long as they do not drink and drive. In fact, numerous studies show that, for people who drink no more than two glasses of wine per day, alcohol consumption may be beneficial (Harvard School of Public Health, 2010). For example, moderate drinkers (one or two glasses of beer or wine per day for men, one per day for women) have a 25% to 40% reduction in risk of cardiovascular disease and stroke than either abstainers or heavy drinkers, even after controlling for hypertension, prior heart attack, and other medical conditions.

For many students, college parties and drinking alcohol are virtually synonymous. Indeed, campuses in the United States are even rated in terms of the level to which students have the reputation for "partying." Unfortunately, drinking among college students often goes beyond moderate intake to become binge drinking. **Binge drinking** *is defined for men as consuming five or more drinks in a row and for women as consuming four or more drinks in a row within the past 2 weeks.* Binge drinking has been identified as a major health problem in the United States since the 1990s (National Institute on Alcohol Abuse and Alcoholism [NIAAA], 2007; Wechsler et al., 1994). Surveys of drinking behavior among young adults consistently show that the rate of binge drinking in the general young adult population has declined, but the rate among college students has not (e.g., Grucza, Norberg, & Bierut, 2009).

Binge drinking is viewed by many as a rite of passage in college, but it is actually an especially troublesome behavior for young adults that can cause academic problems.

© David Young-Wolff / PhotoEdit

Although students between the ages of 17 and 23 are more likely than older students to binge drink, there is no relation between year in school and binge drinking rates. So which students are most likely to binge drink? You are significantly more likely to binge drink if alcohol is readily available (Truong & Sturm, 2009), if you are a member of a fraternity or sorority (McGuckin, 2007), or if you feel really positively about what you are doing (which tends to make some people behave rashly; Cyders et al., 2009). Other predictors include being male, European American, or of middle or higher socioeconomic class; having used other drugs in the past 30 days; believing that drinking will help you socially; believing there is minimal risk from drinking; believing that friends do not disapprove of binge drinking; and believing that high levels of drinking are normal (Strano, Cuomo, & Venable, 2004).

Recent research indicates that average binge drinking among college students is an international problem (Karam, Kypri, & Salamoun, 2007; Kypri et al., 2009), with rates in Europe, South America, Australia, and New Zealand roughly on par with the United States.

The notion that binge drinking is just innocent fun is simply wrong. Driving under the influence (DUI) among binge drinkers, up significantly since the late 1990s to nearly 3 million students per year in the United States, is only one aspect (NIAAA, 2007). Nearly 100,000 college students annually are victims of alcohol-related date rape and nearly 700,000 are assaulted by a student who has been drinking (NIAAA, 2007). Although no national data are routinely collected regarding alcohol poisoning, researchers estimate that about 1,700 young adult college students aged 18–24 die every year in the United States from drinking too much (Oster-Aaland et al., 2009).

No question about it—binge drinking is extremely dangerous. As you can see in ■ Figure 10.4, the rate of drinking-related problems, including missing class and engaging in unwanted sexual behavior, is much higher in binge drinkers—especially those

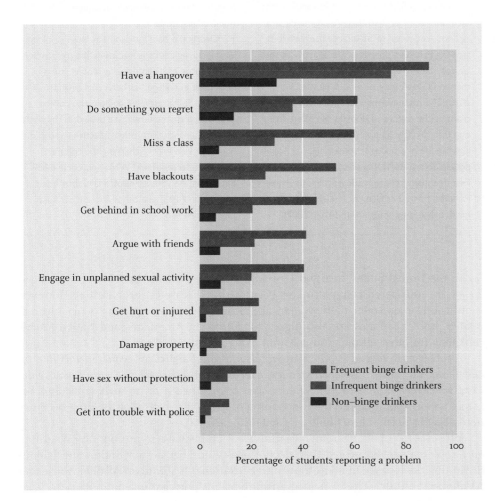

■ **Figure 10.4**
Troublesome behaviors increase with binge drinking. Note that all binge drinkers report more problems than non–binge drinkers. Adapted from a table in Wechsler et al. (2002).

who binge three or more times within a 2-week period ("frequent binge drinkers"). These problems have important long-lasting consequences that range from poorer grades and unplanned pregnancies to contracting sexually transmitted diseases.

Numerous programs aim to reduce the number of college students who binge. These efforts include establishing low tolerance levels for the antisocial behaviors associated with binge drinking; working with athletes, fraternities, and sororities; changing the expectations of incoming freshmen; and increasing the number of non-alcoholic activities available to students (Bishop, 2000; McGuckin, 2007). As discussed in the What Do *You* Think? feature, these efforts come at a time when national and international attention is directed at the problem.

What do you think? Binge Drinking on College Campuses

How can binge drinking be curtailed? The National Institute on Alcohol Abuse and Alcoholism (NIAAA; 2002, 2007) has offered several strategies. NIAAA organized campus alcohol-prevention strategies into four tiers based on their effectiveness in reducing drinking. The most effective strategy (Tier 1) provides one-on-one interventions for at-risk students and programs directly challenging students' expectations regarding alcohol use. Tier 2 strategies are also effective but are not directly implemented on campus, such as working with local officials to make alcohol more difficult to obtain illegally (e.g., by increasing ID checks at restaurants and stores). Tier 3 strategies show promise, but are not yet shown to be definitively effective; these include social norms programs, enforcement of campus alcohol policies, and designated driver programs. Tier 4 strategies are ineffective; some of the most typical strategies used by colleges are included here, such as alcohol education programs. NIAAA urges colleges to attack binge drinking in a 3-in-1 approach, focusing on the individual student, the student body as a whole, and the surrounding community.

Researchers at the University of Minnesota conducted a national study to determine how well NIAAA's recommendations were being carried out on campuses (Nelson et al., 2010). They discovered that although 22% of responding institutions were not familiar with NIAAA's recommendations, 65% had implemented at least one Tier 1 or Tier 2 strategy; however, 98% were still using strategies that NIAAA had determined were ineffective. From this research, colleges and universities have a long way to go.

Although NIAAA initially listed social norms programs as a Tier 3 strategy (a promising strategy not yet proven to be effective), evidence is growing that such programs are effective in the United States (Perkins et al., 2010) and around the world (McAlaney, Bewick, & Hughes, 2010). The social norms approach focuses on changing the culture of drinking in college from one that strongly supports binge drinking to one in which binge drinking is something that popular people do not do. This approach is based on the idea that many college students think their peers' attitudes toward drinking are much more permissive than they really are (NIAAA, 2007).

The belief that "everyone is drinking" and that drinking is entirely acceptable is a major correlate of actual binge drinking behavior. The goal of social norms intervention programs is to publicize the true rate of drinking on campus. Equally important for social norms, but often overlooked, are secondhand drinking effects: negative drinking-related consequences that are experienced by others. For example, a nondrinker may be insulted, assaulted, or have to care for an ill binge drinker, which may in turn have important academic consequences for the nondrinking student.

Clearly, the data indicate that binge drinking among young adults is not inevitable (remember that the rate of binge drinking among young adults not in college has been declining for decades). So it's something about the college experience and culture that's behind it. The social norms approach, with other effective strategies, may offer one way of changing the culture.

What do *you* think? What approaches have been tried on your campus? Have they been successful? Are there other approaches that might work too?

In the United States, nearly 8% of adults 18–24 years old and 5% of those aged 25–44 are considered heavy drinkers (Centers for Disease Control, 2010a). However, significantly more men than women are alcohol dependent or experience alcohol-related problems. Rates are also higher for European Americans and Native Americans than for other ethnic groups (Grant et al., 2006).

Alcoholism is viewed by most experts as a form of **addiction**, *which means that alcoholics demonstrate physical dependence on alcohol and experience withdrawal symptoms when they do not drink.* Dependence occurs when a drug, such as alcohol, becomes so incorporated into the functioning of the body's cells that the drug becomes necessary for normal functioning (Mayo Clinic, 2010a).

Why does alcohol affect how we think, feel, and behave so profoundly? Neuroscience research has discovered that alcohol does a number on our brain, especially in disrupting the balance in neurotransmitters (National Institute on Alcohol Abuse and Alcoholism, 2010). These neurotransmitters include gamma-aminobutyric acid

addiction
physical dependence on a substance (e.g., alcohol) such that withdrawal symptoms are experienced when deprived of that substance

(GABA), which inhibits impulsiveness; glutamate, which excites the nervous system; norepinephrine, which is released in response to stress; and dopamine, serotonin, and opioid peptides, which are responsible for pleasurable feelings. Excessive, long-term drinking can deplete or increase the levels of some of these neurotransmitters, causing the body to crave alcohol to restore good feelings or to avoid negative feelings. Additionally, other factors come into play: genetics; high stress, anxiety, or emotional pain; close friends or partners who drink excessively; and sociocultural factors that glorify alcohol.

Treating addictions such as alcoholism is difficult. Most people seeking treatment for alcohol abuse or dependence for the first time are young adults (Grant et al., 2006). The most widely known treatment option is Alcoholics Anonymous, which was founded in Akron, Ohio, in 1935 by two recovering alcoholics. Other treatment approaches include inpatient and outpatient programs at treatment centers, certain medications, and various forms of counseling (Mayo Clinic, 2010a). Typically, the goal of these approaches is abstinence. Unfortunately, we know very little about the long-term success of the various programs. What we do know is that there are multiple options that can be successful (see, for example, Dranitsaris, Selby, & Negrete, 2009), and that which approach will work with which individual is not very predictable.

Nutrition

How many times were you told to eat your vegetables? Or that "you are what you eat"? Most people have disagreements with parents about food while growing up, but as adults they later realize that those lima beans and other despised foods really are healthful. And don't taste as bad as we once thought.

Experts agree that nutrition directly affects one's mental, emotional, and physical functioning (Mayo Clinic, 2009). Diet has been linked to cancer, cardiovascular disease, diabetes, anemia, and digestive disorders. As people mature, their nutritional requirements and eating habits change. *This change is due mainly to differences in* **metabolism**, *or how much energy the body needs.* Body metabolism and the digestive process slow down with age (Whitney & Ajmera, 2010).

The U.S. Department of Agriculture (USDA) publishes guidelines to help people create healthy diets. In 2011, the USDA launched the ChooseMyPlate.gov campaign that provides a visual guide to healthy eating by showing the relative proportions of various types of food people should eat daily, shown in ■ Figure 10.5. The guidelines

Eating a heart-healthy diet is an important part of preventing cardiovascular disease.

metabolism
how much energy the body needs

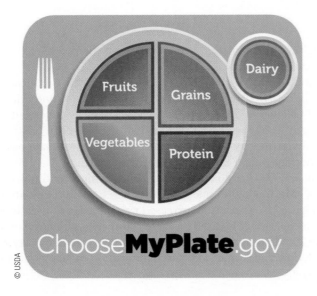

■ **Figure 10.5**
The United States Department of Agriculture depicts a healthy diet by showing the relative portions of various types of food that should be eaten.

emphasize eating less, making half of the food on one's plate fruits and vegetables, making half of the grains eaten be whole grains, switching to fat-free or low-fat (1%) milk, choosing prepared foods with lower sodium, and drinking water instead of sugary drinks (USDA, 2011). For more details on healthy eating, check out www.ChooseMyPlate.gov, especially the link to the detailed *Dietary Guidelines for Americans 2010.*

Did you ever worry as you were eating a triple-dip cone of premium ice cream that you really should be eating fat-free frozen yogurt instead? If so, you are among the people who have taken to heart (literally) the link between diet and cardiovascular disease. The American Heart Association (2010) makes it clear that foods high in saturated fat (such as our beloved ice cream) should be replaced with foods low in fat (such as fat-free frozen yogurt). (The American Heart Association provides a website at www.deliciousdecisions.org with recipes and alternatives for a heart-healthy diet.)

The main goal of these recommendations is to lower your level of cholesterol because high cholesterol is one risk factor for cardiovascular disease. There is an important difference between two types of cholesterol, which are defined by their effect on blood flow. Lipoproteins are fatty chemicals attached to proteins carried in the blood. **Low-density lipoproteins (LDLs)** *cause fatty deposits to accumulate in arteries, impeding blood flow, whereas* **high-density lipoproteins (HDLs)** *help keep arteries clear and break down LDLs*. It is not so much the overall cholesterol number but the ratio of LDLs to HDLs that matters most in cholesterol screening. High levels of LDLs are a risk factor in cardiovascular disease, and high levels of HDLs are considered a protective factor. Reducing LDL levels is effective in diminishing the risk of cardiovascular disease in adults of all ages; in healthy adults a high level of LDL (over 160 mg/dL) indicates a higher risk for cardiovascular disease (American Heart Association, 2010). In contrast, higher levels of HDL are good (in healthy adults, levels at least above 40 mg/dL for men and 50 mg/dL for women). LDL levels can be lowered and HDL levels can be raised through various interventions such as exercise and a high-fiber diet. Weight control is also an important component.

Numerous medications exist for treating cholesterol problems. The most popular of these drugs are from a family of medications called *statins* (e.g., Lipitor, Zocor). These medications lower LDL and moderately increase HDL. Because of potential side effects on liver functioning, patients taking cholesterol-lowering medications should be monitored on a regular basis.

Obesity is a growing health problem related to diet. One good way to assess your own status is to compute your body mass index. **Body mass index (BMI)** *is a ratio of body weight and height and is related to total body fat*. You can compute BMI as follows:

$$BMI = w/h^2$$

where w = weight in kilograms (or weight in pounds divided by 2.2) and h = height in meters (or inches divided by 39.37).

The National Institutes of Health and the American Heart Association (see www.heart.org/HEARTORG/GettingHealthy/WeightManagement/BodyMassIndex/Body-Mass-Index-BMI-Calculator_UCM_307849_Article.jsp for a convenient calculator) define healthy weight as having a BMI of less than 25. However, this calculation may overestimate body fat in very muscular people and underestimate body fat in those who appear of normal weight but have little muscle mass.

Body mass index is related to the risk of serious medical conditions and mortality: the higher one's BMI, the higher one's risk (Centers for Disease Control and Prevention, 2009a). ■ Figure 10.6 shows the increased risk for several diseases and mortality associated with increased BMI. Based on these estimates, you may want to lower your BMI if it's above 25. But be careful—lowering your BMI too much may not be healthy, either. Extremely low BMIs may indicate malnutrition, which is also related to increased mortality.

low-density lipoproteins (LDLs)
chemicals that cause fatty deposits to accumulate in arteries, impeding blood flow

high-density lipoproteins (HDLs)
chemicals that help keep arteries clear and break down LDLs

body mass index (BMI)
a ratio of body weight and height and is related to total body fat

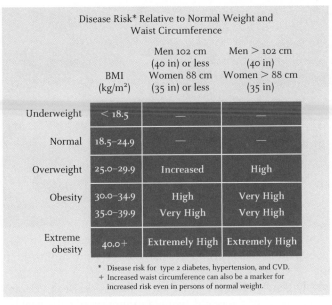

BMI (kg/m²)	Disease Risk* Relative to Normal Weight and Waist Circumference	
	Men 102 cm (40 in) or less Women 88 cm (35 in) or less	Men > 102 cm (40 in) Women > 88 cm (35 in)
Underweight < 18.5	—	—
Normal 18.5–24.9	—	—
Overweight 25.0–29.9	Increased	High
Obesity 30.0–34.9	High	Very High
35.0–39.9	Very High	Very High
Extreme obesity 40.0+	Extremely High	Extremely High

* Disease risk for type 2 diabetes, hypertension, and CVD.
\+ Increased waist circumference can also be a marker for increased risk even in persons of normal weight.

Source: www.nhlbi.nih.gov/health/public/heart/obesity/lose_wt/bmi_dis.htm.

■ **Figure 10.6**
Classification of overweight and obesity by BMI, waist circumference, and associated disease risks.

Social, Gender, and Ethnic Issues in Health

We have indicated that, although most young adults are healthy, there are important individual differences. Let's see what they are.

The two most important social influences on health are socioeconomic status (which is a strong predictor of whether a person has access to insurance and good health care) and education (a good predictor of living a healthy lifestyle and avoiding certain diseases). In the United States, regardless of ethnic group, people who live in poverty are more likely to be in poor health than people who do not.

In the United States, the worst health conditions exist in poor, inner-city neighborhoods. For example, African American men in large urban areas have a lower life expectancy than men in some developing countries (U.S. Census Bureau, 2010b). Many inner-city residents rely on overcrowded clinics. Why is this the case?

The main reason is poverty. As noted earlier, poverty is associated with inadequate health care, higher rates of chronic disease, and higher mortality throughout life in many countries (Stange, 2010; Woolf, 2009). Even when poor minorities in the United States have access to health care, they are less likely than European Americans to receive treatment for chronic disease (National Center for Health Statistics, 2010a).

College graduates are less likely to develop chronic diseases such as hypertension and cardiovascular disease than people who do not go to college. In fact, people who have less education are not only more likely to contract a chronic disease, they are also more likely to die from it. In one of the largest studies examining this issue ever conducted, results show that—in a representative sample of 5,652 working adults between the ages of 18 and 64—educational level was associated with good health even when the effects of age, gender, ethnicity, and smoking were accounted for (Pincus, Callahan, & Burkhauser, 1987). This is also true when looking at specific conditions such as rheumatoid arthritis (McCollum & Pincus, 2009).

Does education *cause* good health? Not exactly. As we discussed in Chapter 1, correlation research does not address cause and effect. In this case, higher educational level is also associated with higher income and also with more awareness of dietary and lifestyle influences on health. Thus, more highly educated people are in a better position to afford health care and to know about the kinds of foods and lifestyle that affect health.

Are men or women healthier? This question is difficult to answer, primarily because women were not routinely included in many major studies of health in the United States until the 1990s (Kolata, 1990). However, there are several clear gender differences: women live longer than men on average (for reasons described in Chap-

Many inner-city residents must rely on over-crowded clinics for their primary health care.

© Tom Carter / PhotoEdit

ter 14), and young adult males are more likely to die from homicide than are young adult females.

Internationally, gender differences matter. As the World Health Organization (2010a) points out, in some parts of the world, women cannot receive the health care they need because cultural norms prohibit women from traveling alone to a clinic, or men have much higher rates of lung cancer because smoking is considered a marker of masculinity.

Test Yourself

RECALL

1. In young adulthood, most people reach their maximum _____.

2. Sensory acuity peaks during the _____.

3. During the early 20s, death from disease is _____.

4. Young adult _____ are the most likely to die in accidents.

5. _____ is the biggest contributor to health problems.

6. Alcoholism is viewed by most experts as a form of _____.

7. The two most important social influences on health are education and _____.

8. In the United States, the poorest health conditions exist for African Americans living in _____.

INTERPRET

How could you design a health care system that provides strong incentives for healthy lifestyles during young adulthood?

APPLY

You have been asked to design a comprehensive educational campaign to promote healthy lifestyles. What information would you be certain to include?

Recall answers: (1) height, (2) 20s, (3) rare, (4) men, (5) Smoking, (6) addiction, (7) socio-economic status, (8) inner-city neighborhoods

---

10.3 | Cognitive Development

LEARNING OBJECTIVES

- What is intelligence in adulthood?
- What are primary and secondary mental abilities? How do they change?
- What are fluid and crystallized intelligence? How do they change?

- How has neuroscience research furthered our understanding of intelligence in adulthood?
- What is postformal thought? How does it differ from formal operations?
- How do emotion and logic become integrated in adulthood?

Susan, a 33-year-old woman recently laid off from her job as a secretary, slides into her seat on her first day of classes at the community college. She is clearly nervous. "I'm worried that I won't be able to compete with these younger students, that I may not be smart enough," she sighs. "Guess we'll find out soon enough, though, huh?"

MANY RETURNING ADULT STUDENTS LIKE SUSAN worry that they may not be "smart enough" to keep up with 18- or 19-year-olds. Are these fears realistic? In this section, we examine the evidence concerning intellectual performance in adulthood. We will see how the answer to this question depends on the types of intellectual skills being used.

How Should We View Intelligence in Adults?

We interrupt this section for a brief exercise. Take a sheet of paper and write down all the abilities that you think reflect intelligence in adults. When you have finished, read further to see how the items on your list match research results.

It's a safe bet that you listed more than one ability as reflecting adults' intelligence. You are not alone. *Most theories of intelligence are* **multidimensional**—*that is, they identify several types of intellectual abilities.* As discussed in Chapter 6, there is disagreement about the number and types of abilities, but nearly everyone agrees that no single generic type of intelligence is responsible for all the different kinds of mental activities we perform.

Sternberg (1985, 2003; Sternberg, Jarvin, & Grigorenko, 2009) emphasized multidimensionality in his theory of successful intelligence (discussed in Chapter 6). Based on the life-span perspective (described in Chapter 1), Baltes and colleagues (Baltes et al., 2006) introduced three other concepts as vital to intellectual development in adults: multidirectionality, interindividual variability, and plasticity. Let's look at each of these concepts in turn.

Over time, the various abilities underlying adults' intelligence show **multidirectionality**: *Some aspects of intelligence improve and other aspects decline during adulthood. Closely related to this is* **interindividual variability**: *These patterns of change also vary from one person to another.* In the next two sections, we will see evidence for both multidirectionality and interindividual variability when we examine developmental trends for specific sets of intellectual abilities. *Finally, people's abilities reflect* **plasticity**: *They are not fixed but can be modified under the right conditions at just about any point in adulthood.* Because most research on plasticity has focused on older adults, we return to this topic in Chapter 14. In general, Baltes and colleagues emphasize that intelligence has many components and that these components show varying development in different abilities and different people.

Given that intelligence in adults is a complex and multifaceted construct, how might we study it? Two common ways involve administering formal testing and assessing practical problem-solving skills. Formal testing assesses a wide range of abilities and involves tests from which we can compute overall IQ scores like those discussed in Chapter 6. Tests of practical problem-solving assess people's ability to apply intellectual skills to everyday situations. Let's see how each approach describes intellectual development.

Primary and Secondary Mental Abilities

We know that intelligence consists of many different skills and abilities. *Since the 1930s, researchers have agreed that intellectual abilities can be studied as groups of related skills (such as memory or spatial ability) organized into hypothetical constructs called* **primary mental abilities**. *In turn, related groups of primary mental abilities can be clustered into a half dozen or so broader skills, termed* **secondary mental abilities**.

Roughly 25 primary mental abilities have been identified (Horn, 1982). Because it is difficult to study all of them, researchers have focused on five representative ones:

- *Number*: the basic skills underlying our mathematical reasoning
- *Word fluency*: how easily we produce verbal descriptions of things
- *Verbal meaning*: our vocabulary ability

multidimensional
characteristic of theories of intelligence that identify several types of intellectual abilities

multidirectionality
developmental pattern in which some aspects of intelligence improve and other aspects decline during adulthood

interindividual variability
patterns of change that vary from one person to another

plasticity
concept that intellectual abilities are not fixed but can be modified under the right conditions at just about any point in adulthood

primary mental abilities
groups of related intellectual skills (such as memory or spatial ability)

secondary mental abilities
broader intellectual skills that subsume and organize the primary abilities

Testing sessions such as the one shown here are used to assess intellectual functioning.

© Frank Siteman / PhotoEdit

- *Inductive reasoning*: our ability to extrapolate from particular facts to general concepts
- *Spatial orientation*: our ability to reason in the three-dimensional world

Do these primary abilities change across adulthood? The answer is in the Spotlight on Research feature.

Spotlight on research · The Seattle Longitudinal Study

Who was the investigator and what was the aim of the study? In the 1950s, little information was available concerning longitudinal changes in adults' intellectual abilities. What there was showed a developmental pattern of relative stability or slight decline, quite different from the picture of substantial across-the-board decline obtained in cross-sectional studies. To provide a more thorough picture of intellectual change, K. Warner Schaie began the Seattle Longitudinal Study in 1956.

How did the investigator measure the topic of interest? Schaie used standardized tests of primary mental abilities to assess a wide range of abilities such as logical reasoning and spatial ability.

Who were the participants in the study? Over the course of the study, more than 5,000 individuals were tested at seven testing cycles (1956, 1963, 1970, 1977, 1984, 1991, 1998, and 2005). The participants were representative of the upper 75% of the socioeconomic spectrum and were recruited through a large health maintenance organization in Seattle. Extensions of the study include longitudinal data on second-generation family members and on the grandchildren of some of the original participants.

What was the design of the study? To provide a thorough view of intellectual change over time, Schaie invented a new type of research design—the sequential design (see Chapter 1). Participants were tested every 7 years. Like most longitudinal studies, Schaie's sequential study encountered selectivity effects—that is, people who return over the years for retesting tend to do better initially than those who fail to return (in other words, those who don't perform well initially tend to drop out of the study). However, an advantage of Schaie's sequential design is that, by bringing in new groups of participants, he was able to estimate the importance of selection effects—a major improvement over previous research.

Were there ethical concerns with the study? The most serious issue in any study in which participants are followed over time is confidentiality. Because people's names must be retained for future contact, the researchers were very careful about keeping personal information secure.

What were the results? Among the many important findings from the study are differential changes in abilities over time and cohort effects. As you can see in ■ Figure 10.7, scores on tests of primary mental abilities improve gradually until the late 30s or early 40s. Small declines begin in the 50s, increase as people age into their 60s, and become increasingly large in the 70s (Schaie & Zanjani, 2006).

Cohort differences were also found. ■ Figure 10.8 shows that, on some skills (e.g., inductive reasoning ability) but not others, more recently born younger and middle-aged cohorts performed better than cohorts born earlier. An example of the latter is that older cohorts outperformed younger ones on number skills (Schaie & Zanjani, 2006). These cohort effects probably reflect differences in educational experiences; younger groups' education emphasized figuring things out on one's own, whereas older groups' education emphasized rote learning. Additionally, older groups did not have calculators or computers, so they had to solve mathematical problems by hand.

Schaie uncovered many individual differences as well; some people showed developmental patterns closely approximating the overall trends, but others showed unusual patterns. For example, some individuals showed steady declines in most abilities beginning in their 40s and 50s, others showed declines in some abilities but not others, and some people showed little change in most abilities over a 14-year period. Such individual variation in developmental patterns means that average trends, like those depicted in the figures, must be interpreted cautiously; they

■ **Figure 10.7**
Longitudinal estimates of age changes on observed measures of five primary mental abilities.

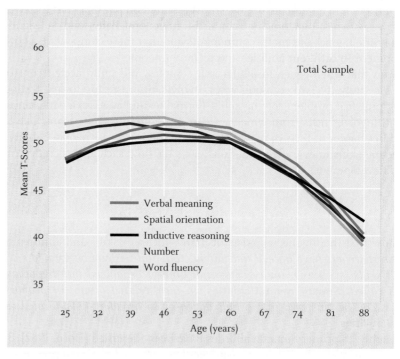

From "Intellectual Development Across Adulthood" by K. Warner Schaie and Faika A. K. Zanjani, in *Handbook of Adult Development and Learning,* ed. by C. Hoare, p. 102. Copyright © 2006 by Oxford University Press. Reprinted by permission of Oxford University Press.

reflect group averages and do not represent the patterns shown by each person in the group.

Another key finding is that people's organization of intellectual abilities does not change over time (Schaie et al., 1998). This finding is important because it means that the tests, which presuppose a particular organizational structure of intellectual abilities, can be used across different ages. Additionally, Schaie (1994) identified several variables that appear to reduce the risk of cognitive decline in old age:

• Absence of cardiovascular and other chronic diseases
• Living in favorable environmental conditions (such as good housing)
• Remaining cognitively active through reading and lifelong learning
• Having a flexible personality style in middle age

• Being married to a person with high cognitive status
• Being satisfied with one's life achievements in middle age

What did the investigator conclude? Three points are clear. First, intellectual development during adulthood is marked by a gradual levelling off of gains between young adulthood and middle age, followed by a period of relative stability, and then a time of gradual decline in most abilities. Second, these trends vary from one cohort to another. Third, individual patterns of change vary considerably from person to person.

Overall, Schaie's findings indicate that intellectual development in adulthood is influenced by a wide variety of health, environmental, personality, and relationship factors. By attending to these influences throughout adulthood, we can

at least stack the deck in favor of maintaining good intellectual functioning in late life.

What converging evidence would strengthen these conclusions? Although Schaie's study is one of the most comprehensive ever conducted, it is limited. Studying people who live in different locations around the world would provide evidence as to whether the results are limited geographically. Additional cross-cultural evidence comparing people with different economic backgrounds and differing access to health care would also provide insight into the effects of these variables on intellectual development.

 Go to Psychology CourseMate at www.cengagebrain.com to enhance your understanding of this research.

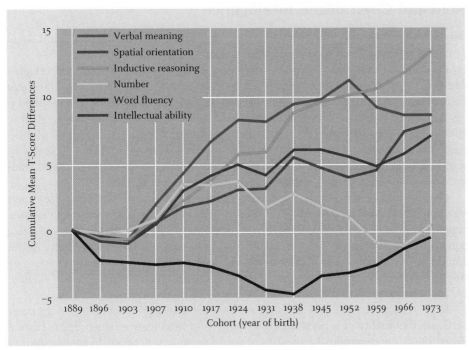

■ **Figure 10.8**
Cohort gradients showing cumulative cohort difference on five primary mental abilities for cohorts born in 1889 to 1973.

From "Intellectual Development Across Adulthood" by K. Warner Schaie and Faika A. K. Zanjani, in *Handbook of Adult Development and Learning*, ed. by C. Hoare, p. 106. Copyright © 2006 by Oxford University Press. Reprinted by permission of Oxford University Press.

Even with a relatively small number of primary mental abilities, though, it is still hard to discuss intelligence by focusing on separate abilities. As a result, theories of intelligence tend to emphasize clusters of related primary mental abilities as a framework for describing the structure of intelligence. Because they are one step removed from primary mental abilities, however, secondary mental abilities are not measured directly. This can be seen in ■ Figure 10.9 for the secondary mental abilities that we will consider next: fluid and crystallized intelligence.

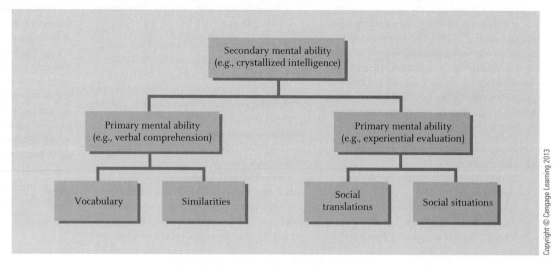

Fluid and Crystallized Intelligence

Two secondary mental abilities have received a great deal of attention in adult developmental research: fluid intelligence and crystallized intelligence (Horn, 1982).

fluid intelligence
abilities that make you a flexible and adaptive thinker, allow you to make inferences, and enable you to understand the relations among concepts

Fluid intelligence *consists of the abilities that make you a flexible and adaptive thinker, that allow you to make inferences, and that enable you to understand the relations among concepts.* It includes the abilities you need to understand and respond to any situation, but especially new ones: inductive reasoning, integration, abstract thinking, and the like (Horn, 1982). An example of a question that taps fluid abilities is the following: What letter comes next in the series *d f i m r x e*?* Other typical ways of testing fluid intelligence include mazes, puzzles, and relations among shapes. These tests are usually timed, and higher scores are associated with faster solutions.

crystallized intelligence
the knowledge you have acquired through life experience and education in a particular culture

Crystallized intelligence *is the knowledge you have acquired through life experience and education in a particular culture.* Crystallized intelligence includes your breadth of knowledge, comprehension of communication, judgment, and sophistication with information (Horn, 1982). Your ability to remember historical facts, definitions of words, knowledge of literature, and sports trivia information are some examples. Many popular television game shows (such as *Jeopardy* and *Wheel of Fortune*) are based on contestants' accumulated crystallized intelligence.

Even though crystallized intelligence involves cultural knowledge, it is based partly on the quality of a person's underlying fluid intelligence (Horn, 1982; Horn & Hofer, 1992). For example, the breadth of your vocabulary depends to some extent on how quickly you are able to make connections between new words you read and information already known, which is a component of fluid intelligence.

Developmentally, fluid and crystallized intelligence follow two very different paths, as you can see in ■ Figure 10.10. Notice that fluid intelligence declines throughout adulthood, whereas crystallized intelligence improves. Although we do not yet fully understand why fluid intelligence declines, it may be related to underlying changes in the brain from the accumulated effects of disease, injury, and aging or from lack of practice (Horn & Hofer, 1992). In contrast, the increase in crystallized intelligence (at least until late life) indicates that people continue adding knowledge every day.

What do these different developmental trends imply? First, they indicate that—although it continues through adulthood—learning becomes more difficult with age.

---

*The next letter is *m*. The rule is to increase the difference between adjacent letters in the series by one each time and use a continuous circle of the alphabet for counting. Thus, *f* is two letters from *d*, *i* is three letters from *f*, and *e* is seven letters from *x*.

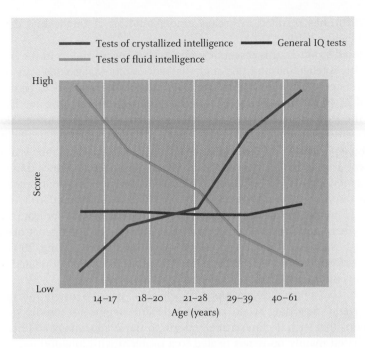

From "Organization of Data on Life-Span Development of Human Abilities" by J. L. Horn, in L. R. Goulet and P. B. Baltes (Eds.), *Life-Span Developmental Psychology: Research and Theory*, p. 463. Copyright © 1970 Academic Press. Reprinted by permission of Elsevier.

Consider what happens when Michael, age 17, and Marge, age 50, learn a second language. Although Marge's verbal skills in her native language (a component of crystallized intelligence) are probably better than Michael's, his probable superiority in the fluid abilities necessary to learn another language will usually make it easier for him to do so.

Second, these developmental trends point out once again that intellectual development varies a great deal from one set of skills to another. Beyond the differences in overall trends, differences in individuals' fluid and crystallized intelligence also vary. Whereas individual differences in fluid intelligence remain relatively uniform over time, individual differences in crystallized intelligence increase with age, largely because maintaining crystallized intelligence depends on being in situations that require its use (Horn, 1982; Horn & Hofer, 1992). For example, few adults get much practice in solving complex letter series tasks like the one on page 366. But because people can improve their vocabulary skills by reading and because people differ considerably in how much they read, differences are likely to emerge. In short, crystallized intelligence provides a rich knowledge base to draw on when material is somewhat familiar, whereas fluid intelligence provides the power to deal with learning in novel situations.

Neuroscience Research and Intelligence in Young and Middle Adulthood

Intellectual abilities have long been known to correlate with mortality in late life, but evidence is now increasing that this relation also holds in middle age. In a longitudinal study of nearly 995,000 young adult male military recruits in Sweden, Batty and colleagues (2009) showed that higher intelligence at the time these men enlisted was related to lower mortality from accidents, coronary heart disease, and suicide in middle age. Similarly, changes in specific mental abilities related to memory predicted mortality in middle age in a study of more than 10,000 workers in Britain (Sabia et al., 2010).

Recent advances in neuroscience have provided additional insights into these predictive relations. Research aimed at identifying neurobiological factors, such as continued maturation of the frontal cortex in young adulthood (Deary, Penke, &

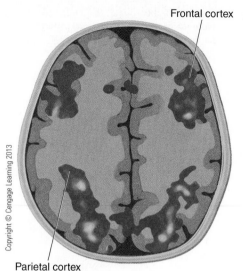

Frontal cortex

Parietal cortex

■ **Figure 10.11**

Brain imaging research indicates that active connections between the parietal and frontal lobes, shown in the colored areas here, are key to understanding intelligence.

Parieto-Frontal Integration Theory (P-FIT)

proposes that intelligence comes from a distributed and integrated network of neurons in the parietal and frontal lobes of the brain

Johnson, 2010), has resulted in greater understanding of intellectual development (Blanchard-Fields, 2010).

Considered together, this research suggests that specific areas in the brain are associated with intellectual abilities, and that developmental changes in these areas should be related to changes in performance. Indeed, this appears to be the case. On the basis of 37 studies using various brain imaging techniques, Jung and Haier (2007) proposed the Parieto-Frontal Integration Theory. **Parieto-Frontal Integration Theory (P-FIT)** *proposes that intelligence comes from a distributed and integrated network of neurons in the parietal and frontal lobes of the brain.* (The parietal lobe is at the top of the head; the frontal lobe is behind the forehead. Figure 10.11 shows these key brain areas.) In general, P-FIT accounts for individual differences in intelligence as having their origins in individual differences in brain structure and function.

The P-FIT model has been tested in several studies. Results indicate support for the theory when measures of fluid, crystallized, and spatial intelligence are related to brain structures assessed in young adults through MRI (Shih & Jung, 2009). It is also clear that performance on measures of specific abilities are likely related to specific combinations of brain structures (Haier et al., 2010).

Research from a neuroscience approach is likely to provide insight into another current controversy regarding when cognitive decline begins in adulthood. Salthouse (2010b) claims that such declines actually begin in the 20s for many abilities. His argument was based mainly on issues related to a problem with longitudinal research designs discussed in Chapter 1: the effects of repeated testing. Salthouse (2010a, 2010b) shows that once retesting effects are removed, evidence for decline in young adulthood becomes apparent. Others disagree (Nilsson et al., 2009). Schaie (2009) argues that Salthouse's claims are based on incorrect conclusions regarding research designs.

Perhaps focused research using the tools of neuroscience (e.g., brain imaging) will be able to sort through this controversy. By continuing to map changes in specific brain structures and their relation to particular changes in intellectual abilities, we will be able to determine definitively when cognitive decline actually begins.

Going Beyond Formal Operations: Thinking in Adulthood

Suppose you are faced with the following dilemma:

> You are a member of your college's or university's student judicial board and are currently hearing a case involving plagiarism. The student handbook states that plagiarism is a serious offense that results in expulsion. The student accused of plagiarizing a paper admits copying from Wikipedia but says that she has never been told that she needed to use a formal citation and quotation marks.
>
> Do you vote to expel the student?

When this and similar problems are presented to older adolescents and young adults, interesting differences emerge. Adolescents tend to point out that the student handbook is clear and the student ignored it, concluding that the student should be expelled. Adolescents thus tend to approach the problem in formal-operational terms, as discussed in Chapter 6. They reason deductively from the information given to come to a single solution grounded in their own experience. Formal-operational thinkers are certain that such solutions are right because they are based on their own experience and are logically driven.

But many adults are reluctant to draw conclusions based on the limited information in the problem, especially when the problem can be interpreted in different ways (Sinnott, 1998). They point out that there is much about the student we don't know: Has she ever been taught the proper procedure for using sources? Was the faculty member clear about what plagiarism is? For adults, the problem is much more ambiguous. Adults may eventually decide that the student is (or is not) expelled, but they do so only after considering aspects of the situation that go well beyond the information given in the problem. Such thinking shows a recognition that other people's experiences may be quite different from one's own.

Description of the Stages of Reflective Judgment

Prereflective Reasoning (Stages 1–3): Belief that "knowledge is gained through the word of an authority figure or through firsthand observation, rather than, for example, through the evaluation of evidence. [People who hold these assumptions] believe that what they know is absolutely correct, and that they know with complete certainty. People who hold these assumptions treat all problems as though they were well-structured" (King & Kitchener, 2004, p. 39). *Example statements typical of Stages 1–3:* "I know it because I see it." "If it's on Fox News it must be true."

Quasi-Reflective Reasoning (Stages 4 and 5): Recognition "that knowledge—or more accurately, knowledge claims—contain elements of uncertainty, which [people who hold these assumptions] attribute to missing information or to methods of obtaining the evidence. Although they use evidence, they do not understand how evidence entails a conclusion (especially in light of the acknowledged uncertainty), and thus tend to view judgments as highly idiosyncratic" (King & Kitchener, 2004, p. 40). *Example statements typical of stages 4 and 5:* "I would believe in climate change if I could see the proof; how can you be sure the scientists aren't just making up the data?"

Reflective Reasoning (Stages 6 and 7): People who hold these assumptions accept "that knowledge claims cannot be made with certainty, but [they] are not immobilized by it; rather, [they] make judgments that are 'most reasonable' and about which they are 'relatively certain,' based on their evaluation of available data. They believe they must actively construct their decisions, and that knowledge claims must be evaluated in relationship to the context in which they were generated to determine their validity. They also readily admit their willingness to reevaluate the adequacy of their judgments as new data or new methodologies become available" (King & Kitchener, 2004, p. 40). *Example statements typical of stages 6 and 7:* "It is difficult to be certain about things in life, but you can draw your own conclusions about them based on how well an argument is put together based on the data used to support it."

Clearly, the thought process these adults use is different from formal operations (Kitchener et al., 2006). Unlike formal-operational thinking, this approach involves considering situational constraints and circumstances, realizing that reality sometimes constrains solutions, and knowing that feelings matter.

Perry (1970) first uncovered adults' different thinking and traced its development. He found that 18-year-old first-year college students tend to rely heavily on the expertise of authority figures (e.g., experts, professors, police, parents) to determine which ways of thinking are right and which are wrong. For these students, thinking is tightly tied to logic, as Piaget had argued, and the only legitimate answers are ones that are logically derived.

Perceptions change over the next few years. Students go through a phase in which they are much less sure of which answers are right—or whether there are any right answers at all. However, by the time they are ready to graduate, students are fairly adept at examining different sides of an issue and have developed commitments to particular viewpoints. Students recognize that they are the source of their own authority, that they must take a position on an issue, and that other people may hold different positions from theirs but be equally committed. During the college years, then, individuals become able to understand many perspectives on an issue, choose one, and still acknowledge the right of others to hold differing views. Perry concluded that this kind of thinking is very different from formal operations and represents another level of cognitive development.

Based on several additional longitudinal studies and numerous cross-sectional investigations, researchers concluded that this type of thinking represents a qualitative change beyond formal operations (King & Kitchener, 2004; Kitchener et al., 2006; Sinnott, 2009). **Postformal thought** *is characterized by a recognition that truth (the correct answer) may vary from situation to situation, that solutions must be realistic to be reasonable, that ambiguity and contradiction are the rule rather than the exception, and that emotion and subjective factors usually play a role in thinking.* In general, the research evidence indicates that postformal thinking has its origins in young adulthood (Kitchener et al., 2006; Sinnott, 2009).

Several research-based descriptions of the development of thinking in adulthood have been offered. *One of the best is the description of the development of* **reflective judgment**, *a way in which adults reason through dilemmas involving current affairs, religion, science, personal relationships, and the like.* Based on decades of longitudinal and cross-sectional research, Kitchener and King (1989; Kitchener et al., 2006) refined descriptions and identified a systematic progression of reflective judgment in young adulthood, which is described in ● Table 10.1.

postformal thought
thinking characterized by recognizing that the correct answer varies from one situation to another, that solutions should be realistic, that ambiguity and contradiction are typical, and that subjective factors play a role in thinking

reflective judgment
way in which adults reason through real-life dilemmas

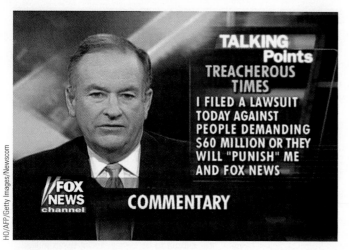

People using the initial stages of reflective judgment believe that what commentators such as Bill O'Reilly say must be true because they are perceived as authority figures.

The first three stages in the model represent prereflective thought. People in these stages typically do not acknowledge and may not even perceive that knowledge is uncertain. Consequently, they do not understand that some problems exist for which there is not a clear and absolutely correct answer. A student pressuring her instructor for the "right" theory to explain human development reflects this stage. She is also likely to hold firm positions on controversial issues and does so without acknowledging other people's ability to reach a different (but nevertheless equally logical) position.

About halfway through the developmental progression, students think very differently. In Stages 4 and 5, students are likely to say that nothing can be known for certain and to change their conclusions based on the situation and the evidence. At this point, students argue that knowledge is quite subjective. They are also less persuasive with their positions on controversial issues: "Each person is entitled to his or her own view; I cannot force my opinions on anyone else." Kitchener and King refer to thinking in these stages as "quasi-reflective" thinking.

As students continue their development into Stages 6 and 7, they begin to show true reflective judgment, understanding that people construct knowledge using evidence and argument after careful analysis of the problem or situation. They once again hold firm convictions but reach them only after careful consideration of several points of view. They also realize that they must continually reevaluate their beliefs in view of new evidence.

Efforts at using the reflective judgment model to shape classroom education are becoming common. For example, teacher training programs (Friedman & Schoen, 2009; Zeidler et al., 2009) and examination practices (Badger, 2010) have been adapted using the reflective judgment model as guidance.

Even though people are able to think at very complex levels, do they? Not usually (King & Kitchener, 2004). Why is this the case? Mostly it is because the environment does not provide the supports necessary for using one's highest-level thinking, especially for issues concerning knowledge and experience you already have. For example, people may not always purchase the product that has the least impact on the environment, such as a fully electric car, even though philosophically they are strong environmentalists, because recharging stations are currently not widely available. However, if pushed and if given the necessary supports (e.g., easily available charging stations), people demonstrate a level of thinking and performance far higher than they typically show on a daily basis. This discrepancy may explain why fewer people at each more complex level of thinking consistently employ that level of thought.

In sum, research on postformal thinking shows that many adults progress from believing in a single right way of thinking and acting to accepting the existence of multiple solutions, each potentially equally acceptable (or equally flawed). This progression is important; it allows for the integration of emotion with thought in dealing with practical, everyday problems, as we will see next.

Integrating Emotion and Logic in Life Problems

You may have noticed that a hallmark of postformal thinking is the movement from thinking "I'm right because I've experienced it" to thinking "I'm not sure who's right because your experience is different from mine." Problem situations that had seemed pretty straightforward in adolescence appear more complicated to young adults; the "right thing to do" is much tougher to figure out.

Differences in thinking styles have major implications for dealing with life problems. For example, couples who are able to understand and synthesize each other's point of view are much more likely to resolve conflicts; couples not able to do so are more likely to feel resentful, drift apart, or even break up (Kramer, 1989; Kramer et al., 1991).

In addition to an increased understanding that there is more than one "right" answer, adult thinking is characterized by the integration of emotion with logic (Jain & Labouvie-Vief, 2010; Labouvie-Vief, 2006; Labouvie-Vief, Grühn, & Studer, 2010). Beginning in young adulthood and continuing through middle age, people gradually shift from an orientation emphasizing conformity and context-free principles to one emphasizing change and context-dependent principles.

As they mature, adults tend to make decisions and analyze problems not so much on logical grounds as on pragmatic and emotional grounds. Rules and norms are viewed as relative, not absolute. Mature thinkers realize that thinking is an inherently social enterprise that demands making compromises with other people and tolerating contradiction and ambiguity. Such shifts mean that one's sense of self also undergoes a fundamental change (Jain & Labouvie-Vief, 2010; Labouvie-Vief, 2006; Labouvie-Vief et al., 2010).

A good example of this developmental shift would be the differences between how late adolescents or young adults view an emotionally charged issue—such as unethical behavior at work—compared to the views of middle-aged adults. Younger people may view such behavior as completely inexcusable, with firing of the employee an inescapable outcome. Middle-aged adults may take contextual factors into account and consider what factors may have forced the person to engage in the behavior. Some might argue that this is because the topic is too emotionally charged for adolescents to deal with intellectually whereas young adults are better able to incorporate emotion into their thinking. But is this interpretation reasonable?

It appears to be. In a now classic study, high-school students, college students, and middle-aged adults were given three dilemmas to resolve (Blanchard-Fields, 1986). One dilemma had low emotional involvement and concerned conflicting accounts of a war between two fictitious countries (North and South Livia) written by a partisan from each country. The other two dilemmas had high emotional involvement. In one, parents and their adolescent son disagreed about going to visit the grandparents (the son did not want to go). In the other, a man and a woman had to resolve an unintentional pregnancy (the man was anti-abortion, the woman was pro-choice).

The results are shown in ■ Figure 10.12. You should note two important findings. First, there were clear developmental trends in reasoning level, with the middle-aged adults best able to integrate emotion into thinking. Second, the high-school and college students were equivalent on the fictitious war dilemma, but the young adult

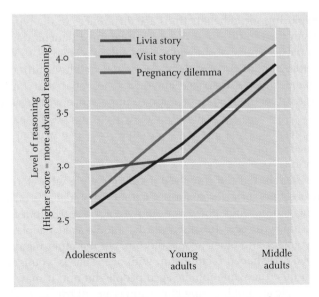

■ **Figure 10.12**
Developmental differences in reasoning level can be seen across these three story types. Middle-aged adults tend to use higher levels of thinking than adolescents or younger adults.

From "Reasoning on Social Dilemmas Varying in Emotional Saliency: An Adult Developmental Study," by F. Blanchard-Fields, 1986, *Psychology and Aging, 1,* 325–333. Copyright © 1986 by the American Psychological Association.

THINK ABOUT IT

Why are formal operations inadequate for integrating emotion and thought?

students more readily integrated emotion and thought on the visit and pregnancy dilemmas. These results support the kinds of developmental shifts suggested by Labouvie-Vief and colleagues. To continue the earlier example, dealing with an unethical co-worker may require the integration of thought and emotion, which is done better by young and best by middle-aged adults.

The mounting evidence of continued cognitive development in adulthood paints a more positive view of adulthood than that of Piaget, who focused only on logical thinking. The integration of emotion with logic that happens in adulthood provides the basis for decision making in the very personal and sometimes difficult arenas of love and work, which we examine in detail in Chapters 11 and 12, respectively. In the present context, it sets the stage for envisioning one's future life, a topic we take up later in this chapter.

Implicit Social Beliefs

The developmental integration of thought and emotion during young adulthood and middle age turns out to be influenced by life-cycle forces and cohort effects (see Chapter 1). How strongly people hold beliefs may vary as a function of how particular generations were socialized. For example, although many adults of all ages may believe that couples should not live together before marriage, older generations may be more adamant in this belief.

Social cognition researchers argue that individual differences in the strength of social representations of rules, beliefs, and attitudes are linked to specific situations (Blanchard-Fields, 2009; Blanchard-Fields & Hertzog, 2000; Labouvie-Vief et al., 2010). Such representations can be both cognitive (how we think about the situation) and emotional (how we react to the situation). When we encounter a specific situation, our cognitive belief system triggers an emotional reaction and related goals tied to the content of that situation, which demands integration of cognition and emotion. This, in turn, drives social judgments.

To see how this works, consider the belief that couples should not live together before marriage. If you were socialized from childhood to believe in this rule, then you would likely evaluate anyone violating the rule negatively. For example, suppose you were told about a man named Allen who urged Joan to live with him before they were married. They subsequently broke up. You may have a negative emotional response and blame Allen for the breakup because he pressured Joan to cohabit.

Research exploring social beliefs finds age differences in the types of social rules and evaluations evoked in different types of situations (Blanchard-Fields, 2009). For example, when participants considered a husband who chooses to work long hours instead of spending more time with his wife and children, different evaluations about the husband and the marriage emerged. The belief that "marriage is more important than a career" tended to increase in importance with age. As can be seen in ■ Figure 10.13, this was particularly evident from age 24 to age 65. The social evaluation "the marriage was already in trouble" was also evident and yielded an inverted U-shaped graph; adults between ages 30 and 55 years were most likely to give this evaluation.

When the research question involved a young couple who eloped despite their parents' objections, the social evaluations "parents should have talked to, not provoked, the young couple" and "they were too young" also exhibited an inverted U-shaped relationship with age. In other words, middle-aged people endorsed these rules, whereas younger and older people did not. On the other hand, "you can't stop true love" showed a U-shaped relationship with age. That is, younger and older age groups endorsed this rule while middle-aged people did not. It may be that between ages 30 and 45 people are not focusing mainly on issues of true love. This makes sense because they are likely at the stage of life where building a career is more important. Those in this age group also emphasized the pragmatics of age (e.g., being too young) as an important factor in marriage decisions.

There are age differences in social rules and relationships evoked in different situations. As one grows older, there is an increase in the belief that marriage is more important than achievement in one's career; note also that older and younger couples may have different explanations than middle-aged adults of why marriages fail. Adapted from Blanchard-Fields (1999).

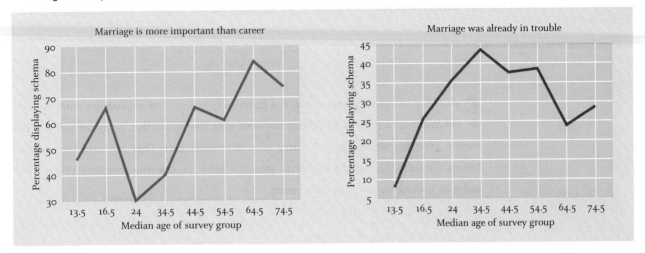

One possible explanation for these findings is that cohort effects or generational differences (as discussed in Chapter 1) influenced whether strong family social rules would be activated (that is, how cognition and emotion interact). Alternatively, the results could also reflect issues concerning different life stages of the respondents; for example, the pressures of providing for children may influence how one responds in midlife more so than in early or later adulthood. In any case, social beliefs, as expressed through social rules and evaluations, are powerful influences on how we behave in everyday life. Either way, the developmental shift in the integration of cognition and emotion is an important determinant of age differences in social problem solving.

Test Yourself

RECALL

1. Most modern theories of intelligence are _____ in that they identify many domains of intellectual abilities.

2. Number, verbal fluency, and spatial orientation are some of the _____ mental abilities.

3. _____ reflects knowledge that you have acquired through life experience and education in a particular culture.

4. Kitchener and King describe a kind of postformal thinking called _____.

5. Life problems provide a context for understanding the integration of _____ and _____ .

INTERPRET

Many young adult college students seemingly get more confused about what field they want to major in and less certain about what they know as they progress through college. From a cognitive-developmental approach, why does this happen?

APPLY

Two of LuSharon's friends explained their decision to support Senator Barack Obama during the 2008 U.S. Presidential campaign. One friend said that she had an intuition that Senator Obama's campaign statements were based on facts; the other friend said that the candidate's statements always had to be put into a specific context in order to be analyzed. Based on the reflective judgment model, what levels of thinking do LuSharon's friends demonstrate?

Recall answers: (1) multidimensional, (2) primary, (3) Crystallized intelligence, (4) reflective judgment, (5) emotion and logic

Who Do You Want to Be? Personality in Young Adulthood

- What is the life-span construct? How do adults create scenarios and life stories?

- What are possible selves? Do they show differences during adulthood?

- What are personal control beliefs?

Felicia is a 19-year-old sophomore at a community college. She expects her study of early childhood education to be difficult but rewarding. She figures that, along the way, she will meet a great guy whom she will marry soon after graduation. They will have two children before she turns 30. Felicia sees herself getting a good job teaching preschool children and someday owning her own day-care center.

IN CHAPTER **9,** we saw how children and adolescents deal with the question "what do you want to be when you grow up?" As a young adult, Felicia has arrived at the "grown up" part and is experimenting with some idealistic answers to the question. Are Felicia's answers typical of most young adults?

In this section, we examine how the search for identity gets transformed through the cognitive, social, and personal reality of adulthood. As a result, people create life scenarios and life stories, possible selves, self-concept, and personal control beliefs. Let's begin by considering how Felicia and the rest of us construct images of our adult lives.

Creating Scenarios and Life Stories

Figuring out what (and who) you want to be as an adult takes lots of thought, hard work, and time. *Based on personal experience and input from other people, young adults create a* **life-span construct** *that represents a unified sense of the past, present, and future.* Several factors influence the development of a life-span construct; identity, values, and society are only a few. Together they not only shape the creation of the life-span construct but also influence the way it is played out and whether it remains stable (Fraley & Roberts, 2005). The life-span construct represents a link between Erikson's notion of identity, which is a major focus during adolescence, and our adult view of ourselves.

The first way the life-span construct is manifested is through the **scenario**, *which consists of expectations about the future.* The scenario takes aspects of a person's identity that are particularly important now and projects them into a plan for the future. For example, you may find yourself thinking about the day you will graduate and be able to apply all of the knowledge and skills you have learned. In short, a scenario is a game plan for how your life will play out in the future.

Felicia, the sophomore human development student, has a fairly typical scenario. She plans on completing a degree in early childhood education, marrying after graduation, and having two children by age 30. *Tagging future events with a particular time or age by which they are to be completed creates a* **social clock**. This personal timetable gives people a way to track progress through adulthood, and it may use biological markers of time (such as menopause), social aspects of time (such as getting married), and historical time (such as the turn of the century) (Hagestad & Neugarten, 1985).

Felicia will use her scenario to evaluate progress toward her personal goals. With each new event, she will check where she is against where her scenario says she should be. If she is ahead of her plan, she may be proud of having made it. If she is lagging behind, she may chastise herself for being slow. But if she criticizes herself too

life-span construct
a unified sense of the past, present, and future based on personal experience and input from other people

scenario
manifestation of the life-span construct through expectations about the future

social clock
tagging future events with a particular time or age by which they are to be completed

much, she may change her scenario altogether. For example, if she does not go to college, she may decide to change her career goals entirely: Instead of owning her own day-care center, she may aim to be a manager in a department store.

McAdams's Life-Story Model

McAdams (2009; McAdams & Olson, 2010) argues that a person's sense of identity cannot be understood based on traits or personal concerns. To him, identity is not just a collection of traits, nor is it a collection of plans, strategies, or goals. Instead, it is based on the story of how each person came into being, where she has been, where she is going, and who she will become. One's **life story** *is a personal narrative that organizes past events into a coherent sequence.* Our life story becomes our autobiography as we move through adulthood.

Blogging or posting Tweets has become a modern way of writing one's autobiography.

life story
a personal narrative that organizes past events into a coherent sequence

McAdams believes that people create a life story that is an internalized narrative with a beginning, a middle, and an anticipated ending. You write your life story in young adulthood, and revise it throughout the rest of your life as you change and as the changing environment places different demands on you.

McAdams and colleagues' research indicates that people in Western society begin forming their life story in late adolescence and early adulthood, but the story has its roots in the development of one's earliest attachments in infancy. As in Erikson's theory, adolescence marks the full initiation into forming an identity and thus a coherent life story begins there. In early adulthood it is continued and refined, and from midlife and beyond the story is refashioned in the wake of major and minor life changes. *Generativity* marks the attempt to create an appealing story "ending" that will yield new beginnings for future generations (we discuss this in detail in Chapter 13).

Paramount to these life stories is the changing personal identity reflected in the emotions conveyed in the story (from tragedy to optimism or through comic and romantic descriptions). In addition, motivations change and are reflected in the person repeatedly trying to attain his or her goals over time. The two most common goal themes are agency (reflecting power, achievement, and autonomy) and communion (reflecting love, intimacy, and belongingness). Finally, stories indicate beliefs and values—or the ideology—that a person uses to set the context for his or her actions.

Every life story contains episodes that provide insight into perceived change and continuity in life. People prove to themselves and to others that they have either changed or remained the same by pointing to specific events that support the appropriate claim. The main characters in people's lives represent idealizations of the self, such as "the dutiful mother" or "the reliable worker." Integrating these various aspects of the self is a major challenge of midlife and later adulthood. Finally, all life stories need an ending through which the self is able to leave a legacy that creates new beginnings. Life stories in middle-aged and older adults have a clear quality of "giving birth to" a new generation, a notion essentially identical to generativity.

One popular method for examining the development of life stories is autobiographical memory (Rubin, Berntsen, & Hutson, 2009; Thomsen & Berntsen, 2009). When people tell their life story to others, it is a joint product of the speaker and the audience (Kihlstrom, 2009; Pasupathi, Mansour, & Brubaker, 2007; Pasupathi, McLean, & Weeks, 2009). Pasupathi and colleagues report that connections between the self and events create a stable sense of self and of how one changes over time. Additionally, the responses of the audience affect how the teller remembers his or her experiences. For example, if listeners find a particular experience interesting then the person who shared that memory will likely tell it again, but without positive listener feedback this is less likely. This is a good example of conversational remembering, much like collaborative cognition discussed in Chapter 9.

Overall, McAdams (2008; McAdams & Olson, 2010) believes identity change over time is a process of fashioning and refashioning one's life story. This process appears to be strongly influenced by culture. At times, the reformulation may be at a conscious level, such as when people make explicit decisions about changing careers. At other times, the revision process is unconscious and implicit, growing out of everyday activities. The goal is to create a life story that is coherent, credible, open to new possibilities, richly differentiated, able to reconcile opposing aspects of oneself, and integrated within the sociocultural context.

Possible Selves

possible selves
representations of what we could become, what we would like to become, and what we are afraid of becoming

When we are asked questions like, "What do you think you'll be like a few years from now?" it requires us to imagine ourselves in the future. When we speculate like this, we create a possible self (Markus & Nurius, 1986). **Possible selves** *represent what we could become, what we would like to become, and what we are afraid of becoming.* What we could or would like to become often reflects personal goals; we may see ourselves as leaders, as rich and famous, or as in shape. What we are afraid of becoming may show up in our fear of being alone, or overweight, or unsuccessful. Our possible selves are powerful motivators; indeed, how we behave is largely an effort to achieve or avoid these various possible selves and to protect the current view of self (Baumeister, 2010).

The topic of possible selves offers a way to understand how both stability and change operate in adults' personality. On one hand, possible selves tend to remain stable for at least some period of time and are measurable with psychometrically sound scales (Hooker, 1999, 2002). On the other hand, possible selves may change in response to efforts directed toward personal growth (Frazier et al., 2000, 2002), which would be expected from ego-development theory. In particular, possible selves facilitate adaptation to new roles across the life span. For example, a full-time mother who pictures herself as an executive once her child goes to school may enroll in evening courses to acquire new skills. Thus, possible selves offer a way to bridge the experience of the current self and our imagined future self.

Researchers have examined age differences in the construction of possible selves (Bardach et al., 2010; Cotter & Gonzalez, 2009; Frazier et al., 2000, 2002; Hooker, 1999). In a rare set of similar studies conducted across time and research teams by Cross and Markus (1991) and Hooker and colleagues (Frazier et al., 2000, 2002; Hooker, 1999; Hooker et al., 1996; Morfei et al., 2001), people across the adult life span were asked to describe their hoped-for and feared possible selves. The responses were grouped into categories (e.g., family, personal, material, relationships, occupation).

Several interesting age differences emerged. In terms of hoped-for selves, young adults listed family concerns—for instance, marrying the right person—as most important (Cross & Markus, 1991), and Hooker and colleagues (1996) found that getting started in an occupation was also important for this age group. In contrast, adults in their 30s listed family concerns last; their main issues involved personal concerns, such as being a more loving and caring person (Cross & Markus, 1991). By ages 40 to 59, Cross and Markus found that family issues again became most common—for example, being a parent who can "let go" of the children. Hooker and Kaus (1994) also found that reaching and maintaining satisfactory performance in one's occupational career as well as accepting and adjusting to the physiological changes of middle age were important to this age group.

As shown in ■ Figure 10.14, health becomes an increasingly important factor in defining the self as people grow older. Both sets of studies found that, for the two younger groups, being overweight and (for women) becoming wrinkled and unattractive when old were commonly mentioned as feared-for selves. For the middle-aged and older adult groups, fear of having Alzheimer's disease or being unable to care for oneself were frequent responses.

For adults over 60, researchers find that personal issues are most prominent—for example, being active and healthy for at least another decade (Cross & Markus, 1991; Smith & Freund, 2002). Similarly, Hooker and colleagues (Frazier et al., 2000, 2002;

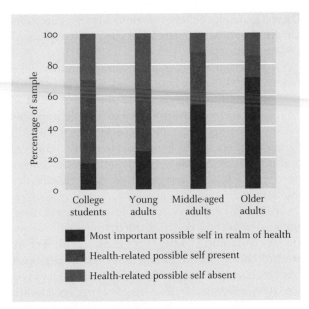

Reprinted from K. Hooker (1999), "Possible Selves in Adulthood: Incorporating Teleonomic Relevance into Studies of the Self," in T. M. Hess and F. Blanchard-Fields (Eds.), *Social Cognition and Aging*, p. 107, with permission from Elsevier.

■ **Figure 10.14**
Health plays an increasingly important role in defining the self as people age.

Morfei et al., 2001) found that continuity in possible selves was much more prevalent than change in later life, especially in regard to independence and physical and life-style areas. However, they also found that change did occur in older age. The greatest amount of change occurred in the health domain, which predominated the hoped-for and feared-for selves. The health domain is the most sensitive and central to the self in the context of aging (Frazier et al., 2000) and people's possible self related to health is quite resilient in the face of health challenges in later life (Cotter & Gonzalez, 2009).

Overall, young adults have multiple possible selves and believe that they can actually become the hoped-for self and successfully avoid the feared self. Their outlook tends to be quite positive (Remedios, Chasteen, & Packer, 2010). Life experience may dampen this outlook. By old age, both the number of possible selves and the strength of belief have decreased. Older adults are more likely to believe that neither the hoped-for nor the feared self is under their personal control. These findings may reflect differences with age in personal motivation, beliefs in personal control, and the need to explore new options.

Other researchers have examined possible selves in a different way by asking adults to describe their present, past, future, and ideal self (Busseri, Choma, & Sadava, 2009; Keyes & Ryff, 1999). Instead of examining categories of possible selves, this approach focuses on people's perceptions of change over time. The data indicate that young and middle-aged adults see themselves as improving with age and expecting to continue getting better in the future, sometimes unrealistically so. By the time people reach old age, they see themselves as having remained stable over time but foresee decline in their future. These findings suggest that the older group has internalized negative stereotypes about aging.

Personal Control Beliefs

Do you feel you have control over your life? **Personal control beliefs** *reflect the degree to which you believe your performance in a situation depends on something you do.* For example, suppose you are not offered a job when you think you should have been. Was it your fault? Or was it because the company was too shortsighted to recognize your true talent? Which option you select provides insight into a general tendency.

THINK ABOUT IT
How might the development of possible selves be related to cognitive development?

personal control beliefs
the degree to which you believe your performance in a situation depends on something you do

AP Images / Jae C. Hong

Chad Hurley, co-founder and former CEO of YouTube, is likely to have a high sense of personal control.

primary control
behavior aimed at affecting the individual's external world

secondary control
behavior or cognition aimed at affecting the individual's internal world

Do you generally believe that outcomes depend on the things you do? Or are they due to factors outside of yourself, such as luck or the power of others?

A high sense of personal control implies a belief that performance is up to you, whereas a low sense of personal control implies that your performance is under the influence of forces other than your own. Personal control is an extremely important idea in a wide variety of settings and cultures because of the way it guides behavior (Brandtstädter, 1999; Fung & Siu, 2010). Successful people like Stefani Joanne Angelina Germanotta (better known as Lady Gaga) need to exude a high sense of personal control in order to demonstrate that they are in charge.

Personal control is an important concept that can be applied broadly to several domains, including social networks, health, and careers (Antonucci, 2001; Fung & Siu, 2010). For example, personal control beliefs are important not only in personality development but also (as we will see in Chapter 14) in memory performance in late life. Research indicates that people experience four types of personal control (Tiffany & Tiffany, 1996): control from within oneself, control over oneself, control over the environment, and control from the environment.

Despite its importance, we do not have a clear picture of the developmental course of personal control beliefs. Evidence from both cross-sectional and longitudinal studies (Lang & Heckhausen, 2006) is contradictory. Some data indicate that younger adults are less likely to hold internal control beliefs (i.e., believe they are in control of outcomes) than are older adults. Other research finds the opposite.

The contradiction may derive from the complex nature of personal control beliefs in the context of the situation (Vazire & Doris, 2009). These beliefs vary depending on which domain, such as intelligence or health, is being assessed. Indeed, other research shows that perceived control over one's development declines with age whereas perceived control over marital happiness increases (Brandtstädter, 1989). Additionally, younger adults are more satisfied when attributing success in attaining a goal to their own efforts, whereas older adults are more satisfied when they attribute such success to their ability (Lang & Heckhausen, 2001, 2006). Clearly, people of all ages and cultures try to influence their environment regardless of whether they believe they will be successful.

Heckhausen, Wrosch, and Schulz (2010) pulled together the various perspectives on control beliefs and proposed a motivational theory of life-span development to describe how people optimize primary and secondary control. **Primary control** *is behavior aimed at affecting the individual's external world; working a second job to increase one's earnings is an example.* One's ability to influence the environment is heavily influenced by biological factors (e.g., stamina to work two jobs), so it changes over time—from very low influence during early childhood to high influence during middle age and then to very low again in late life. **Secondary control** *is behavior or cognition aimed at affecting the individual's internal world; an example is believing that one is capable of success even when faced with challenges.*

The developmental patterns of both types of control are shown in ■ Figure 10.15. The figure also shows that people of all ages strive to control their environment, but how they do this changes over time. Note that, for the first half of life, primary and secondary control operate in parallel. During midlife, primary control begins to decline but secondary control does not. Thus, the desire for control does not change; what differs with age is whether we can actually affect our environment or whether we need to think about things differently.

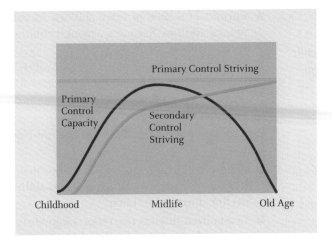

Adapted from R. Schulz and J. Heckhausen (1996), "A Lifespan Model of Successful Aging," American Psychologist, 51, 702–714. Copyright © 1996 by the American Psychological Association.

■ **Figure 10.15**
The use of primary control peaks in midlife, whereas secondary control increases across the adult life span.

Test Yourself

RECALL

1. A _____ is a unified sense of a person's past, present, and future.

2. A personal narrative that organizes past events into a coherent sequence is a _____.

3. Representations of what we could become, what we would like to be, and what we are afraid of becoming are our _____.

4. _____ reflect the degree to which a person's performance in a situation is believed to be under his or her control.

INTERPRET

How might people's scenarios, life stories, and other aspects of personality vary as a function of cognitive developmental level and self-definition as an adult?

APPLY

How could you determine whether your performance on a task or in a situation is under your control?

Recall answers: (1) life-span construct, (2) life story, (3) possible selves, (4) Personal control beliefs

---

SUMMARY

10.1 EMERGING ADULTHOOD

What role transitions mark entry into adulthood?

■ The most widely used criteria for deciding whether a person has reached adulthood are role transitions, which involve assuming new responsibilities and duties.

■ Some societies use rituals, called rites of passage, to mark this transition clearly. However, such rituals are largely absent in Western culture.

How does going to college reflect the transition to adulthood?

■ Over half of all college students are over age 25. These students tend to be more motivated and have many other positive characteristics.

■ College serves as a catalyst for cognitive development.

What behavioral criteria mark the transition to adulthood?

■ A major difference between adolescence and adulthood is a drop in the rate of participation in reckless behavior.

■ Young adults grapple with issues relating to intimacy, although whether this occurs only after the issue of identity is resolved differs across gender.

How does achieving financial independence reflect the transition to adulthood?

■ Launching oneself as financially independent is a major marker of achieving adulthood in Western society.

■ This can be achieved in many ways, such as a series of part-time jobs, joining the military, or going to college and then entering the workforce.

When do people become adults?

■ Challenges faced by many adults in their 20s have been termed the "quarterlife crisis."

■ Adjustment to adulthood is more difficult when the process of transition is less clear.

10.2 PHYSICAL DEVELOPMENT AND HEALTH

In what respects are young adults at their physical peak?

■ Young adulthood is the time when certain physical abilities peak: strength, muscle development, coordination, dexterity, and sensory acuity. Most of these abilities begin to decline in middle age.

How healthy are young adults in general?

■ Young adults are also at the peak of health. Death from disease is relatively rare, especially during the 20s. Accidents are the leading cause of death. However, homicide and violence are major factors in some groups. Poor ethnic minorities have less access to good health care, and poverty is also a major barrier to good health.

How do smoking, drinking alcohol, and nutrition affect young adults' health?

■ Smoking is the single biggest contributor to health problems. One is never too old to quit smoking.

■ Smoking is related to half of all cancers and is a primary cause of respiratory and cardiovascular disease. Although it is difficult, quitting smoking has many health benefits.

■ For most people, drinking alcohol poses few health risks. Several treatment approaches are available for alcoholics.

■ Nutritional needs change somewhat during adulthood, mostly due to changes in metabolism. Some nutrient needs, such as carbohydrates, change. The ratio of LDLs to HDLs in serum cholesterol, which can be controlled through diet or medication in most people, is an important risk factor in cardiovascular disease.

How does young adults' health differ as a function of socioeconomic status, gender, and ethnicity?

■ The two most important social factors in health are socioeconomic status and education. The poorest health conditions exist for African Americans living in poor, inner-city neighborhoods. Other ethnic groups with limited access to health care also suffer.

■ Whether women or men are healthier is difficult to answer because women have been excluded from much health research.

■ Higher education is associated with better health via better access to health care and more knowledge about proper diet and lifestyle.

10.3 COGNITIVE DEVELOPMENT

What is intelligence in adulthood?

■ Most modern theories of intelligence are multidimensional. For instance, Baltes's research shows that development in adults varies among individuals and across different categories of abilities.

What are primary and secondary mental abilities? How do they change?

■ Intellectual abilities can be studied as groups of related skills known as primary mental abilities.

■ Clusters of related primary abilities are called secondary mental abilities. Secondary mental abilities are not measured directly.

■ These abilities develop differently and change in succeeding cohorts. More recent cohorts perform better on some skills, such as inductive reasoning, but older cohorts perform better on number skills.

What are fluid and crystallized intelligence? How do they change?

■ Fluid intelligence consists of abilities that make people flexible and adaptive thinkers. Fluid abilities generally decline during adulthood.

■ Crystallized intelligence reflects knowledge that people acquire through life experience and education in a particular culture. Crystallized abilities improve until late life.

How has neuroscience research furthered our understanding of intelligence in adulthood?

■ Neuroscience research has begun mapping specific areas in the brain that relate to intelligence. One prominent theory based on this work is the parieto-frontal integration theory (P-FIT).

■ Neuroscience research may provide insight into controversies surrounding when cognitive abilities begin to show decline, which some researchers claim begins in the 20s.

What is postformal thought? How does it differ from formal operations?

■ Postformal thought is characterized by a recognition that truth may vary from one situation to another, that solutions must be realistic, that ambiguity and contradiction are the rule, and that emotion and subjectivity play a role in thinking. One example of postformal thought is reflective judgment.

How do emotion and logic become integrated in adulthood?

■ Cognition (logic) and emotion become integrated during young adulthood and middle age. This means that the way people approach and solve practical problems in life differs from adolescence through middle age.

10.4 WHO DO YOU WANT TO BE? PERSONALITY IN YOUNG ADULTHOOD

What is the life-span construct? How do adults create scenarios and life stories?

■ Young adults create a life-span construct that represents a unified sense of the past, present, and future. This is manifested in two ways: through a scenario that maps the future based on a social clock; and in the life story, which creates an autobiography.

What are possible selves? Do they show differences during adulthood?

■ People create possible selves by projecting themselves into the future and thinking about what they would like to become, what they could become, and what they are afraid of becoming.

- Age differences in these projections depend on the dimension examined. In hoped-for selves, 18- to 24-year-olds and 40- to 59-year-olds report family issues as most important, whereas 25- to 39-year-olds and older adults consider personal issues to be most important. However, all groups include physical aspects as part of their most feared selves.

What are personal control beliefs?

- Personal control is an important concept with broad applicability. However, the developmental trends are complex because personal control beliefs vary considerably from one domain to another.

KEY TERMS

emerging adulthood (346)
rites of passage (347)
role transitions (347)
returning adult students (348)
edgework (350)
intimacy versus isolation (351)
binge drinking (356)
addiction (358)
metabolism (359)
low-density lipoproteins (LDLs) (360)
high-density lipoproteins (HDLs) (360)

body mass index (BMI) (360)
multidimensional (363)
multidirectionality (363)
interindividual variability (363)
plasticity (363)
primary mental abilities (363)
secondary mental abilities (363)
fluid intelligence (366)
crystallized intelligence (366)
Parieto-Frontal Integration Theory (P-FIT) (368)

postformal thought (369)
reflective judgment (369)
life-span construct (374)
scenario (374)
social clock (374)
life story (375)
possible selves (376)
personal control beliefs (377)
primary control (378)
secondary control (378)

LEARN MORE ABOUT IT

Log in to **www.cengagebrain.com** to access the resources your instructor requires. For this book, you can access:

Psychology CourseMate

- CourseMate brings course concepts to life with interactive learning, study, and exam preparation tools that support the printed textbook. A textbook-specific website, Psychology CourseMate includes an integrated interactive eBook and other interactive learning tools, including quizzes, flashcards, videos, and more.

CENGAGENOW™

- CengageNOW Personalized Study is a diagnostic study tool containing valuable text-specific resources—and because you focus on just what you don't know, you learn more in less time to get a better grade.

WebTUTOR™

- More than just an interactive study guide, WebTutor is an anytime, anywhere customized learning solution with an eBook, keeping you connected to your textbook, instructor, and classmates.

Being With Others
Forming Relationships in Young and Middle Adulthood

IMAGINE YOURSELF YEARS FROM NOW. Your children are grown and have children and grandchildren of their own. In honor of your 80th birthday, they have all come together, along with your friends, to celebrate. Their present to you is a video created from hundreds of photographs and dozens of videos created over the decades of your life. As you watch it, you realize how lucky you've been to have so many wonderful people in your life. Your relationships have made your adult life fun and worthwhile. As you watch, you wonder what it must be like to go through life totally alone—no family, no friends (even on Facebook), no followers of your Twitter postings. You think of all the wonderful experiences you would have missed in early and middle adulthood—never knowing what friendship is all about, never being in love, never dreaming about children and becoming a parent.

That is what we'll explore in this chapter—the ways in which we share our lives with others. First, we consider what makes good friendships and love relationships. Because these relationships form the basis of our lifestyle, we examine these lifestyle influences next. In the third section, we consider what it is like to be a parent. Finally, we see what happens when marriages or partnerships end. Throughout this chapter, the emphasis is on aspects of relationships that nearly everyone experiences during young adulthood and middle age. In Chapter 12, we examine aspects of relationships specific to middle-aged adults; in Chapter 14, we do the same for relationships in later life.

LEARNING OBJECTIVES

■ What types of friendships do adults have? How do adult friendships develop?

■ What is love? How does it begin? How does it develop through adulthood?

■ What is the nature of abuse in some relationships?

Jamal and Deb, both 25, have been madly in love since they met at a party about a month ago. They spend as much time together as possible and pledge that they will stay together forever. Deb finds herself daydreaming about Jamal at work and can't wait to go over to his apartment. She wants to move in, but her co-workers tell her to slow down.

YOU KNOW WHAT JAMAL AND DEB ARE GOING THROUGH. Each of us wants to be wanted by someone else. What would your life be like if you had no one to share it with? There would be no one to go shopping with or hang out with, no one to talk to on the phone, no one to cuddle close to while watching the sunset at a mountain lake. Although there are times when being alone is desirable, for the most part we are social creatures. We need people. Without friends and lovers, life would be pretty lonely.

In the next sections, we explore both life-enhancing and life-diminishing relationships. We consider friendships, what happens when love enters the picture, and how people find mates. Unfortunately, some relationships turn violent; we'll also examine the factors underlying aggressive behaviors between partners.

Friendships

What is an adult friend? Someone who is there when you need to share? Someone not afraid to tell you the truth? Someone to have fun with? Friends, of course, are all of these and more. Researchers define friendship as a mutual relationship in which those involved influence one another's behaviors and beliefs, and they define friendship quality as the satisfaction derived from the relationship (Flynn, 2007).

The role and influence of friends for young adults is of major importance from the late teens to the mid-20s (Arnett, 2007) and continues to be a source of support throughout adulthood. Friendships are predominantly based on feelings and are grounded in reciprocity and choice. Friendships are different from love relationships in that they are less emotionally intense and involve less sexual energy or contact (Rose & Zand, 2000). Having good friendships helps boost self-esteem (Bagwell et al., 2005). They also help us become socialized into new roles throughout adulthood.

Friendship in Adulthood

From a developmental perspective, adult friendships can be viewed as having identifiable stages (Levinger, 1980, 1983): Acquaintanceship, Buildup, Continuation, Deterioration, and Ending. This ABCDE model describes not only the stages of friendships but also the processes by which they change. For example, whether a friendship will develop from Acquaintanceship to Buildup depends on several factors that include the basis of the attraction, what each person knows about the other, how good the communication is between the partners, the perceived importance of the friendship, and so on. Although many friendships reach the Deterioration stage, whether a friendship ultimately ends depends importantly on the availability of alternative relationships. If new potential friends appear, old friendships may end; if not, they may continue even though they may no longer be considered important by either person.

Longitudinal research shows how friendships change from adolescence through young adulthood, sometimes in ways that are predictable and sometimes not. For example, as you probably have experienced, life transitions (e.g., going away to college, getting married) usually result in fewer friends and less contact with the friends you keep (Flynn, 2007). People tend to have more friends and acquaintances during young adulthood than at any subsequent period (Sherman, de Vries, & Lansford, 2000). Friendships are important throughout adulthood, in part because a person's life satisfaction is strongly related to the quantity and quality of contacts with friends. College students who have strong friendship networks adjust better to stressful life events (Brissette, Scheier, & Carver, 2002) and have better self-esteem (Bagwell et al., 2005). The importance of maintaining contacts with friends cuts across ethnic lines as well. Additionally, people who have friendships that cross ethnic groups have more positive attitudes toward people with different backgrounds (Aberson, Shoemaker, & Tomolillo, 2004). Thus, regardless of one's background, friendships play a major role in determining how much we enjoy life.

Researchers have uncovered three broad themes that underlie adult friendships (de Vries, 1996):

- The most frequently mentioned dimension represents the *affective* or emotional basis of friendship. This dimension refers to self-disclosure and expressions of intimacy, appreciation, affection, and support, all of which are based on trust, loyalty, and commitment.

- A second theme reflects the *shared or communal* nature of friendship, in which friends participate in or support activities of mutual interest.

- The third dimension represents *sociability and compatibility*; our friends keep us entertained and are sources of amusement, fun, and recreation.

These three dimensions are found in friendships among adults of all ages (de Vries, 1996). They characterize both traditional (e.g., face-to-face) and new (e.g., online) forms of friendships (Ridings & Gefen, 2004).

The development of online social networks such as Facebook raised concerns among social commentators that adults' social friendship networks would decline in quality because in-depth interactions would be replaced with quick e-mails or postings. Research shows that this concern has no basis. Wang and Wellman (2010) examined friendship networks in adults aged 25 and 74. They documented that the quality of the friendship network was good overall, actually improving between 2002 and 2007. Most important, this improvement was documented whether people were non-users of the Internet or heavily virtual. In fact, they found that heavy Internet users had the most friends, both online and offline.

In the case of online friendships, trust is an important factor because visual cues may not be present to verify the information being presented. Online environments are more conducive to people who are shy, allowing opportunities to meet others in an initially more anonymous setting in which social interaction and intimacy levels can be carefully controlled (Morahan-Martin & Schumacher, 2003). This relative anonymity provides a supportive context for the subsequent development of friendships online. Online connections can facilitate strong commitment between friends; research shows that most adults who have online, committed friendships report that they can get stronger, and also report that such friendships go through the same cycles as traditional face-to-face friendships (Johnson, Becker, Craig, Gilchrist, & Haigh, 2009).

A special type of friendship exists with one's siblings. Although little research has focused on the development and maintenance of sibling friendships across adulthood, we know that sibling relationships play an important role in young adulthood (Schulte, 2006) and that the importance of these relationships varies with age. As you can see in ▪ Figure 11.1, women place more importance on sibling ties across adulthood than do men; however, for both genders the strength of such ties is

Despite concerns, social networking websites such as Facebook have not reduced the quality of friendships.

© David L. Moore - Lifestyle / Alamy

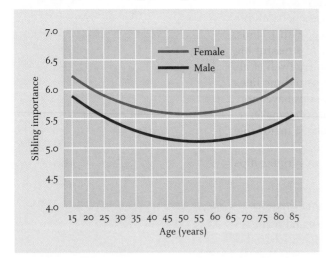

From Schmeeckle, M., Giarusso, R., and Wang, O. (November 1998). "When being a brother or sister is important to one's identity: Life stage and gender differences." Paper presented at the annual meeting of the Gerontological Society, Philadelphia.

greatest in adolescence and late life (Schmeeckle, Giarusso, & Wang, 1998). In young adulthood, siblings report that geographic distance does not affect the quality of the relationship, expressions of intimacy about the importance of the relationship really matter, and there are many kinds of relationships created between siblings that tend to change over time (Corti, 2009). We will consider sibling relationships in more detail in Chapter 15.

Men's, Women's, and Cross-Sex Friendships

Men's and women's friendships tend to differ in adulthood, reflecting continuity in the learned behaviors from childhood (Mehta & Strough, 2009). Women tend to base their friendships on more intimate and emotional sharing and use friendship as a means to confide in others. For women, getting together with friends often takes the form of getting together to discuss personal matters. Confiding in others is a basis of women's friendships. In contrast, men tend to base friendships on shared activities or interests. They are more likely to go bowling or fishing or to talk sports with their friends. For men, friendships are often, but not always, less intimate (Greif, 2009). Although men often use shared activities, rather than shared confidences, as the basis for friendships, men do tend to have a small number of friends with whom they have a close, personal relationship and with whom they share intimate information and feelings.

Women tend to have more close relationships than do men. Although you may think this puts women at an advantage, research shows that this is not always the case. Sometimes friends can get on people's nerves or make demands. When these things happen, women tend to be less happy even when they have lots of friends (Antonucci, Akiyama, & Lansford, 1998).

Why are women's friendships typically more intimate than men's? Compared to men, women have much more experience with such intimate sharing from early childhood, and they are more comfortable with vulnerability. Social pressure on men to be brave and strong may actually inhibit their ability to form close friendships (Rawlins, 1992).

What about friendships between men and women? These friendships have a beneficial effect, especially for men (Piquet, 2007). Cross-sex friendships tend to help men have lower levels of dating anxiety and higher capacity for intimacy; it is interesting, however, that such benefits are not evident for women. These patterns hold across ethnic groups, too. But cross-sex friendships can also prove troublesome as a result of misperceptions and pressures against them from third parties (e.g., spouses/partners) and organizations (e.g., companies may discourage such friendships) (Mehta & Strough, 2009). Some research shows that men tend to overperceive and women tend to underperceive their friends' sexual interest in them (Koenig, Kirkpatrick, & Ketelaar, 2007). Maintaining cross-sex friendships once individuals enter into exclusive dating relationships, marriage, or committed relationships is very difficult, and it often results in one partner feeling jealous (Williams, 2005).

Love Relationships

Love is one of those things everybody feels but nobody can define completely. (Test yourself: Can you explain fully what you mean when you look at someone special and say, "I love you"?) One way researchers have tried to understand love is to think about what components are essential. In an interesting series of studies, Sternberg (2006) found that love has three basic components: (1) *passion*, an intense physiological desire for someone; (2) *intimacy*, the feeling that one can share all one's thoughts and actions with another; and (3) *commitment*, the willingness to stay with a person through good and bad times. Ideally, a true love relationship has all three components; when couples have equivalent amounts of love and types of love, they tend to be happier. As we will see next, the balance among these components often shifts as time passes.

Love Through Adulthood

The different combinations of love can be used to understand how relationships develop (Sternberg, 2006). Research shows that the development of romantic relationships is a complex process influenced by relationships in childhood and adolescence (Collins & van Dulmen, 2006). Early in a romantic relationship, passion is usually high whereas intimacy and commitment tend to be low. This is infatuation: an intense, physically based relationship in which the two people have a high risk of misunderstanding and jealousy.

Physical attraction tends to be high early in a relationship.

But infatuation is short-lived. Whereas even the smallest touch is enough to drive each partner into wild, lustful ecstasy in the beginning, with time it takes more and more effort to get the same level of feeling. As passion fades, either a relationship acquires emotional intimacy or it is likely to end. Trust, honesty, openness, and acceptance must be a part of any strong relationship; when they are present, romantic love develops.

Although it may not be the stuff of romance novels, this pattern is a good thing. Research shows that people who select a partner for a more permanent relationship (e.g., marriage) during the height of infatuation are likely to support the idea that "love is blind"; those couples are more likely to divorce (Hansen, 2006). But if the couple gives it more time and works at their relationship, they may become committed to each other. By spending much of their time together, making decisions together, caring for each other, sharing possessions, and developing ways to settle conflicts, they increase the chances that their relationship will last. Such couples usually show outward signs of commitment, such as wearing a lover's ring, having children together, or simply sharing the mundane details of daily life, from making toast at breakfast to before-bed rituals.

Lemieux and Hale (2002) demonstrated that these developmental trends hold in romantically involved couples between 17 and 75 years of age. As the length of the relationship increases, intimacy and passion decrease but commitment increases.

Falling in Love

Everybody wants to be loved by somebody, but actually having it happen is fraught with difficulties. In his book *The Prophet*, Kahlil Gibran points out that love is two-sided: Just as it can give you great ecstasy, so can it cause you great pain. Yet most of us are willing to take the risk.

As you may have experienced, taking the risk is fun (at times) and difficult (at other times). Making a connection can be ritualized, as when people use pickup lines in a bar, or it can happen almost by accident, as when two people literally run into each other in a crowded corridor. The question that confronts us is "How do people fall in love?" Do birds of a feather flock together? Or do opposites attract?

The best explanation of the process is the theory of **assortative mating**, *which states that people find partners based on their similarity to each other.* Assortative mating occurs along many dimensions, including education, religious beliefs, physical traits, age, socioeconomic status, intelligence, and political ideology, among others (Blossfeld, 2009). Such nonrandom mating occurs most often in Western societies, which allow people to have more control over their own dating and pairing behaviors. Common activities are one basis for identifying potential mates. Except, that is, in speed dating situations. In that case, when people have very limited time to explore potentially common interests, it comes down to physical attractiveness (Luo & Zhang, 2009).

Does commonality tend to result in happier relationships? The research findings are mixed. Jenkins (2007) found that couples higher in marital satisfaction were similar

assortative mating
theory stating that people find partners based on their similarity to each other

in terms of their openness to experience but not on other aspects of personality. However, a study of 12,000 Dutch couples showed that healthy people tended to be in relationships with other healthy people and that unhealthy people tended to be in relationships with other unhealthy people; these results may be due to couples' similar levels of education, which were related to shared circumstances (Monden, 2007).

People meet people in all sorts of places. Does where people meet influence the likelihood that they will "click" on particular dimensions and will form a couple? Kalmijn and Flap (2001) found that it did. Using data from more than 1,500 couples, they found that meeting at school was most likely to result in the most forms of *homogamy*—the degree to which people are similar. Although meeting through other methods (being from the same neighborhood or through family networks) could promote homogamy, the odds are that they do not promote most forms of homogamy other than religious. Not surprisingly, the pool of available people to meet is strongly shaped by the opportunities available, which in turn constrain the type of people one is likely to meet.

Speed dating, a relatively new addition to the world of dating, provides a way to meet several people in a short period of time. Speed dating is practiced most by young adults (Whitty & Buchanan, 2009). The rules governing partner selection during a speed dating session seem quite similar to traditional dating: physically attractive people, outgoing and self-assured people, and moderately self-focused people are selected more often and their dates are rated as smoother (Eastwick, Saigal, & Finkel, 2010).

The advent of online dating makes it possible for adults who have social or dating anxiety to meet people nonetheless (Stevens & Morris, 2007; Whitty & Buchanan, 2009). But it is not only the socially anxious who are meeting this way. Surveys indicate that nearly 1 in every 5 couples in the United States meet online (compared with 1 in 10 in Australia, and 1 in 20 in Spain and the United Kingdom; Dutton et al., 2009). Not surprisingly, people who meet online tend to be young and middle-aged adults, with a slightly stronger preference among middle-aged adults for online as opposed to speed dating. Emerging research indicates that virtual dating sites offer both problems and possibilities. On the one hand, researchers note that the content of member profiles may be suspect (Small, 2004). On the other hand, many couples have met and formed committed relationships via online sites (Mazzarella, 2007).

In the online world, initial decisions whether to pursue a potential mate work similarly to the offline world. First impressions are driven mainly by the perceived attractiveness of the person's photograph, whereas more deliberative decisions are influenced by both perceived attractiveness and such self-described attributes as ambition (Sritharan et al., 2010).

We've seen that physical attractiveness matters in speed dating and online dating, as it does in traditional dating. How does physical attraction operate? Research shows that women tend to choose a more masculine looking man as a person with whom to have an exciting short-term relationship but tend to select a more feminine looking man for their husband or as the type of man their parents would want them to date (Kruger, 2006). These findings support a study of nearly 2,000 Spanish respondents, which showed that physical attractiveness is not only important in sporadic relationships but also influences the way in which people fall in love; such attractiveness is linked to feelings and thoughts associated with love (intimacy, passion, commitment) and to satisfaction with the relationship (Sangrador & Yela, 2000).

How do these couple-forming behaviors compare cross-culturally? A few studies have examined the factors that attract people to each other in different cultures. In one now classic study, Buss and a large team of researchers (1990) identified the effects of culture and gender on heterosexual mate preferences in 37 cultures worldwide. Men and women in each culture displayed unique orderings of their preferences concerning the ideal characteristics of a mate. When all of the orderings and preferences were compared, two main dimensions emerged.

In the first main dimension, the characteristics of a desirable mate changed because of cultural values—that is, whether the respondents' country has more traditional values or Western-industrial values. In traditional cultures, men place a high value on a woman's chastity, desire for home and children, and being a good cook and housekeeper; women place a high value on a man's ambition and industry, being a good financial prospect, and holding favorable social status. China, India, Iran, and Nigeria represent the traditional end of this dimension. In contrast, people in Western-industrial cultures value these qualities to a much lesser extent. The Netherlands, Great Britain, Finland, and Sweden represent this end of the dimension; people in these countries place more value on Western ideals.

The second main dimension reflects the relative importance of education, intelligence, and social refinement—as opposed to a pleasing disposition—in choosing a mate. For example, people in Spain, Colombia, and Greece highly value education, intelligence, and social refinement; in contrast, people in Indonesia place a greater emphasis on having a pleasing disposition. Note that this dimension emphasizes the same traits for both men and women.

Chastity proved to be the characteristic showing the most variability across cultures, being highly desired in some cultures but mattering little in others. It is interesting that, in their respective search for mates, men around the world value physical attractiveness in women whereas women around the world look for men capable of being good providers. But men and women around the world agree that love and mutual attraction are most important, and nearly all cultures rate dependability, emotional stability, kindness, and understanding as important factors. Attraction, it seems, has some characteristics that transcend culture.

Overall, Buss and his colleagues concluded that mate selection is a complex process no matter where you live. However, each culture has a describable set of high-priority traits that men and women look for in the perfect mate. The study also shows that socialization within a culture plays a key role in being attractive to the opposite sex; characteristics that are highly desirable in one culture may not be so desirable in another.

In the Spotlight on Research feature, Schmitt and his team of colleagues (2004) had 17,804 participants from 62 cultural regions complete the Relationship Questionnaire (RQ), a self-report measure of adult romantic attachment. They showed that secure romantic attachment was the norm in nearly 80% of cultures and that "preoccupied" romantic attachment was particularly common in East Asian cultures. In general, what these large multicultural studies show is that there are global patterns in mate selection and romantic relationships. The romantic attachment profiles of individual nations were correlated with sociocultural indicators in ways that supported evolutionary theories of romantic attachment and basic human mating strategies.

Spotlight on research

Patterns and Universals of Romantic Attachment Around the World

Who were the investigators and what was the aim of the study? One's attachment style may have a major influence on how one forms romantic relationships. In order to test this hypothesis, David Schmitt (Schmitt et al., 2004) assembled a large international team of researchers.

How did the investigators measure the topic of interest? Great care was taken to ensure equivalent translation of the survey across the 62 cultural regions included. The survey was a two-dimension four-category measure of adult romantic attachment (the Relationship Questionnaire) that measured models of self and others relative to each other: *secure* romantic attachment (high scores indicate positive models

of self and others), *dismissing* romantic attachment (high scores indicate a positive model of self and a negative model of others), *preoccupied* romantic attachment (high scores indicate a negative model of self and a positive model of others), and *fearful* romantic attachment (high scores indicate negative models of self and others). An overall score of the model of self is computed by adding together the secure and dismissing scores and then subtracting the combination of preoccupied and fearful scores. The overall model of others score is computed by adding together the secure and preoccupied scores and then subtracting the combination of dismissing and fearful scores.

Additionally, there were measures of self-esteem, personality traits, and sociocultural correlates of romantic attachment (e.g., fertility rate, national profiles of individualism versus collectivism).

Who were the participants in the study? A total of 17,804 people (7,432 men and 10,372 women) from 62 cultural regions around the world took part in the study. Such large and diverse samples are unusual in developmental research.

What was the design of the study? Data for this cross-sectional, nonexperimental study were gathered by research teams in each country. The principal researchers asked the research collabo-

[continued]

In this model of self and model of others levels across 10 world regions, note that only in East Asian cultures were model of others scores significantly higher than model of self scores. Data from Schmitt et al. (2004).

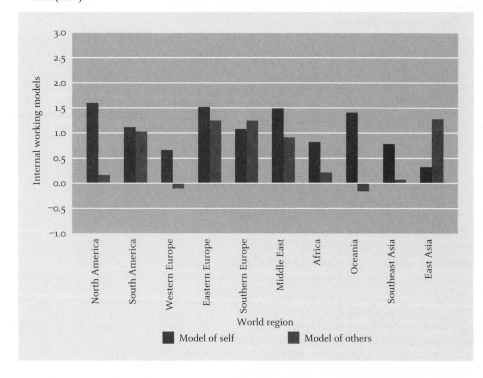

of self and model of others were valid across cultural regions, which provided general support for the independence of measures (i.e., they measure different things). Specific analyses showed that 79% of the cultural groups studied demonstrated secure romantic attachments but that North American cultures tended to be high on dismissive and East Asian cultures high on preoccupied romantic attachment. These patterns are shown in ■ Figure 11.2. Note that all the cultural regions except East Asia showed the pattern of model of self scores higher than model of others scores.

What have the investigators concluded? Overall, Schmitt and colleagues concluded that, although the same attachment pattern holds across most cultures, no one pattern holds across all of them. East Asian cultures in particular tend to fit a pattern in which people report that others do not get as emotionally close as the respondent would like, and that respondents find it difficult to trust others or to depend on them.

What converging evidence would strengthen these conclusions? Although this is one of the best designed among large cross-cultural studies, several additional lines of evidence would help bolster the conclusions. Most important, representative samples from the countries under study would provide more accurate insights into people's romantic attachment patterns.

rators to administer a nine-page survey to the participants that took 20 minutes to complete.

Were there ethical concerns with the study? Because the study involved volunteers, there were no ethical concerns. However, ensuring that all participants' rights were protected was a challenge because of the number of countries and cultures involved.

What were the results? The researchers first demonstrated that the measures used for model

 Go to Psychology CourseMate at www.cengagebrain.com to enhance your understanding of this research.

Culture is a very powerful force in shaping mate selection choices. For example, despite decades of sociopolitical change in China (the socialist transformation in the 1950s, the Cultural Revolution in the 1960s, and the economic reforms in the 1990s), research indicates that the same status hierarchy norms govern mating patterns in urban China (Xu, Ji, & Tung, 2000). Emotional investment in a romantic relationship also varies by culture (Schmitt et al., 2009). Specifically, across 48 different cultures globally, people from cultures that have good health care, education, and resources, and that permit young adults to choose their own mates tend to develop more secure romantic attachments than do people from cultures that do not have these characteristics.

Clearly, cultural norms are sometimes highly resistant to change. Arranged marriages are a major way that some cultures ensure an appropriate match on key dimensions. For example, loyalty of the individual to the family is a very important value in India, so despite many changes in mate selection about 95% of marriages in India are carefully arranged to ensure that an appropriate mate is selected (Dommaraju, 2010). Data show that this approach appears to work; among urban professionals polled in one study, 81% said their marriages had been arranged, and 94% of them rated their marriage as "very successful" (Lakshmanan, 1997). Similarly, Islamic societies use matchmaking as a way to preserve family consistency and continuity and to ensure that couples follow the prohibition on premarital relationships between men and women (Adler, 2001). Matchmaking in these societies occurs both through family connections and personal adver-

These Egyptian women, performing traditional cultural tasks, are more likely to be desired as mates.

Courtesy of John C. Cavanaugh

tisements in newspapers. To keep up with the Internet age, Muslim matchmaking has gone online too (Lo & Aziz, 2009).

THINK ABOUT IT

What are the effects of increasing interactions among cultures on mate selection?

Developmental Forces and Relationships

As you no doubt know from your own experience, finding a suitable relationship is really tough. Many things must work just right: timing, meeting the right person, luck, and effort are but a few of the factors that shape the course of a relationship. Centuries of romance stories describe this magical process and portray it as one of life's great mysteries.

From our discussion here, you know that who chooses whom, and whether the feelings will be mutual, results from the interaction of developmental forces described in the biopsychosocial model presented in Chapter 1. Recent neuroscience research is also demystifying love relationships. Let's see how.

Love is one of three discrete, interrelated emotion systems (the sex drive and attachment are the other two; Fisher, 2006). The brain circuitry involved in romantic love, maternal love, and long-term attachment overlap (Stein & Vythilingum, 2009). In terms of love, neurochemicals related to the amphetamines come into play early in the process, providing a biological explanation for the exhilaration of falling madly in love. Aron and colleagues (2005) reported that couples who were in the early stages of romantic love showed high levels of activity in the dopamine system, which is involved in all of the basic biological drives. Once the relationship settles into what some people might call long-term commitment and tranquility, the brain processes switch neurochemically to substances related to morphine, a powerful narcotic. People with a predilection to fall in love also tend to show left hemisphere chemical dominance and several changes in neurochemical processing (Kurup & Kurup, 2003). Additional research indicates that the hormone oxytocin may play an important role in attachment and women's orgasms, among other things, which has earned it the nickname of the "cuddle hormone" (Lee et al., 2009). Love really does a number on your brain!

And that's not all. It also turns out that the interactions among psychological aspects, neurological aspects, and hormonal aspects of romantic love help explain why couples tend to have exclusive relationships with each other. For women, the stronger the romantic bond with their boyfriend, the less likely they are to be able to identify the body odor of a different male friend (Lundström & Jones-Gotman, 2009). That means that women's attention is deflected from other potential male partners the more they are romantically involved with one specific male.

Psychologically, as we saw in Chapter 9, an important developmental issue is intimacy; according to Erikson, mature relationships are impossible without it. Additionally, the kinds of relationships you saw and experienced as a child (and whether they involved violence) affect how you define and act in relationships you develop as an adult. Sociocultural forces shape the characteristics you find desirable in a mate and determine whether you are likely to encounter resistance from your family when you have made your choice. Life-cycle forces matter, too; different aspects of love are more or less important depending on your stage in life. For example, romantic love tends to be most prominent in young adulthood, whereas the aspect of companionship becomes more important later in life.

In short, to understand adult relationships, we must take the forces of the biopsychosocial model into account. Relying too heavily on one or two of the forces provides an incomplete description of why people are successful (or not) in finding a partner or a friend. Unfortunately, the developmental forces do not influence only good relationships. As we will see next, sometimes relationships turn violent.

The Dark Side of Relationships: Abuse

Up to this point, we have been considering relationships that are healthy and positive. Sadly, this is not always the case. *Sometimes relationships become violent; one person becomes aggressive toward the partner, creating an* **abusive relationship**. Such relationships have received increasing attention since the early 1980s, when the U.S. criminal

abusive relationship
relationships in which one person becomes aggressive toward the partner

battered woman syndrome
situation occurring when a woman believes that she cannot leave the abusive situation and may even go so far as to kill her abuser

justice system ruled that, under some circumstances, abusive relationships can be used as an explanation for one's behavior (Walker, 1984). *For example,* **battered woman syndrome** *occurs when a woman believes that she cannot leave the abusive situation and may even go so far as to kill her abuser.*

Many college students report experiencing abuse in a dating relationship; a national representative sample of more than 4,100 respondents revealed that 40% of them had experienced victimization by young adulthood (Halpern, Spriggs, Martin, & Kupper, 2009).

Being female, Latina, African American, having an atypical family structure (something other than two biological parents), having more romantic partners, early onset of sexual activity, and being a victim of child abuse predicted victimization. Although overall national rates of sexual assault have declined more than 60% since the early 1990s, acquaintance rape or date rape is still a major problem; college women are 4 times more likely to be the victim of sexual assault than are women in other age groups (Rape, Abuse, and Incest National Network, 2010; for more information, see www.rainn.org).

What range of aggressive behaviors occur in abusive relationships? What causes such abuse? Researchers are finding answers to these and related questions. Based on considerable research on abusive partners, O'Leary (1993) argues that there is a continuum of aggressive behaviors toward a partner, which progresses as follows: verbally aggressive behaviors, physically aggressive behaviors, severe physically aggressive behaviors, and murder (see ● Table 11.1). The causes of the abuse also vary with the type of abusive behavior being expressed.

Two points about the continuum should be noted. First, there may be fundamental differences in the types of aggression independent of level of severity. Overall, each year about 4.8 million women and 2.9 million men experience partner-related physical assaults and rape in the United States (Centers for Disease Control and Prevention, 2009b); worldwide, between 10% and 69% of women report being physically assaulted or raped (World Health Organization, 2002).

The second interesting point, depicted in the table, is that the suspected underlying causes of aggressive behaviors differ as the type of aggressive behaviors change (O'Leary, 1993). Although anger and hostility in the perpetrator are associated with various forms of physical abuse, the exact nature of this relationship remains elusive (Norlander & Eckhardt, 2005).

As can be seen in the table, the number of suspected causes of aggressive behavior increases as the level of aggression increases. Thus, the causes of aggressive behavior become more complex as the level of aggression worsens. Such differences in cause imply that the most effective way to intervene with abusers is to approach each one individually and not try to apply a one-size-fits-all model (Buttell & Carney, 2007).

Men are also the victims of violence from intimate partners, though at a rate about one-third that of women (Conradi & Geffner, 2009). Studies in New Zealand and the United States revealed that both men and women showed similar patterns of holding traditional gendered beliefs, and lacking communication and anger management skills; however, intervention programs tend to focus on male perpetrators (Hines & Douglas, 2009; Robertson & Murachver, 2007). Research on violence in gay and lesbian relationships reveals similar findings. Patterns of violence among gay and lesbian couples are roughly equivalent to that shown by heterosexual couples, and reasons for abuse include dissatisfaction with the relationship and alcohol abuse (Fisher-Borne, 2007; Roberts, 2007).

Culture is also an important contextual factor in understanding partner abuse. In particular, violence against women worldwide reflects cultural traditions, beliefs, and values of patriarchal societies; this can be seen in the com-

● **TABLE 11.1**

Causes of Abuse in Relationships

Type of Abuse	Causes
Verbal abuse	Need to control Misuse of power Jealousy Marital discord
Physical abuse	Acceptance of violence as means of control Physically aggressive models Abuse as a child Aggressive personality style Alcohol abuse
Severe physical abuse	Personality disorders Emotional swings Poor self-esteem

Note: Unemployment and job stressors contribute to all types of abuse.
SOURCE: O'Leary, K. D. (1993). Through a psychological lens: Personality traits, personality disorders, and levels of violence. In Gelles & Loseke (Eds.), *Current controversies on family violence.* Newbury Park, CA: Sage.

monplace violent practices against women, which include sexual slavery, female genital cutting, intimate partner violence, and honor killing (Parrot & Cummings, 2006). For example, cultures that emphasize honor, that portray females as passive, nurturing supporters of men's activities, and that emphasize loyalty and sacrifice for the family may contribute to tolerance of abuse.

Vandello (2000) reported two studies—of Latino Americans, European Americans who live in the southern part of the United States, and European Americans who live in the northern part of the United States—that examined these ideas. Latino Americans and European Americans who lived in the southern part of the United States placed more value on honor. These groups rated a woman in an abusive relationship more positively if she stayed with the man; they also communicated less disapproval of a woman whom they witnessed being shoved and restrained if she portrayed herself as contrite and self-blaming than did European Americans who lived in the northern part of the United States, who rated the woman more positively if she left the man. Research on Mexican women who were victims of partner abuse confirm that, even when they are willing to seek help, they nevertheless still follow cultural dictates to cope (Vargas, 2007).

Chinese Americans are more likely to define domestic violence in terms of physical and sexual aggression and not include psychological forms of abuse (Yick, 2000). And South Asian immigrants to the United States report the use of social isolation (e.g., not being able to interact with family, friends, or co-workers) as a painful form of abuse that is often tied to financial dependence on the husband and traditional cultural gender roles (M. Abraham, 2000).

Additionally, international data indicate that rates of abuse are higher in cultures that emphasize female purity, male status, and family honor. For example, a common cause of women's murders in Arab countries is brothers or other male relatives killing the victim because she violated the family's honor (Kulwicki, 2002). Intimate partner violence is prevalent in China (43% lifetime risk in one study) and has strong associations with male patriarchal values and conflict resolutions (Xu et al., 2005).

Alarmed by the seriousness of abuse, many communities have established shelters for battered women and their children as well as programs that treat abusive men. However, the legal system in many localities is still not set up to deal with domestic violence; women in some locations cannot sue their husbands for assault, and restraining orders all too often offer little real protection from additional violence. Much remains to be done to protect women and their children from the fear and the reality of continued abuse.

© John Birdsall/The Image Works

Many communities have established shelters for women who have experienced abuse in relationships.

Test Yourself

RECALL

1. Friendships based on intimacy and emotional sharing are more characteristic of _____.

2. Competition is a major part of most friendships among _____.

3. Love relationships in which intimacy and passion are present but commitment is not are termed _____.

4. Chastity is an important quality that men look for in a potential female mate in _____ cultures.

5. Aggressive behavior that is based on abuse of power, jealousy, or the need to control is more likely to be displayed by _____.

INTERPRET

Why is intimacy (discussed in Chapter 9) a necessary prerequisite for adult relationships, according to Erikson? What aspects of relationships discussed here support (or refute) this view?

APPLY

Based on Schmitt and colleagues' (2004) research, what attachment pattern would Korean women likely have regarding romantic attachment?

Recall answers: (1) women, (2) men, (3) romantic love, (4) traditional, (5) men

LEARNING OBJECTIVES

- Why do some people decide not to marry, and what are these people like?
- What are the characteristics of cohabiting people?

- What are gay and lesbian relationships like?
- What is marriage like through the course of adulthood?

Kevin and Beth are on cloud nine. They got married one month ago and have recently returned from their honeymoon. Everyone who sees them can tell that they love each other a lot. They are highly compatible and have much in common, sharing most of their leisure activities. Kevin and Beth wonder what lies ahead in their marriage.

DEVELOPING RELATIONSHIPS IS ONLY PART OF THE PICTURE IN UNDERSTANDING HOW ADULTS LIVE THEIR LIVES WITH OTHER PEOPLE. Putting relationships like Kevin and Beth's in context is important for us to understand how relationships come into existence and how they change over time. In the following sections, we explore relationship lifestyles: singlehood, cohabitation, gay and lesbian couples, and marriage.

Singlehood

When Susan graduated from college with a degree in accounting, she took a job at a consulting firm. For the first several years in her job, she spent more time traveling than she did at home. During this time she had a series of love relationships, but none resulted in commitment even though she had marriage as a goal. By the time she was in her mid-30s, Susan had decided that she no longer wanted to get married. "I'm now a partner in my firm, I enjoy traveling, and I'm pretty flexible in terms of moving if something better comes along," she stated to her friend Michele. "But I do miss being with someone to share my day or to just hang around with."

Like Susan, most men and women during early adulthood are single—defined as not living with an intimate partner. Estimates are that approximately 80% of men and 70% of women between ages 20 and 24 are unmarried, with increasing numbers deciding to stay that way (U.S. Census Bureau, 2010b).

What's it like to be single in the United States? It's tougher than you might think. DePaulo (2006) points out numerous stereotypes and biases against single people. Her research found that young adults characterized married people as caring, kind, and giving about 50% of the time compared with only 2% for single people. And single people receive less compensation at work than married people do, even when age and experience are equivalent. DePaulo also found that rental agents preferred married couples 60% of the time (Morris, Sinclair, & DePaulo, 2007). Can you think of reasons why people might hold these biases against single people?

Many women and men remain single as young adults to focus on establishing their careers rather than marriage or relationships, which most do later. Others report that they simply did not meet "the right person" or prefer singlehood (Ibrahim & Hassan, 2009; Lamanna & Riedmann, 2003). However, the pressure to marry is especially strong for women; frequent questions such as "Any good prospects yet?" may leave women feeling conspicuous or left out as many of their friends marry.

Men tend to remain single longer in young adulthood because they tend to marry at a later age than women (U.S. Census Bureau, 2010a). Fewer men than women remain unmarried

Single women often face social stereotypes that make life difficult, including lower salaries at work.

© PBNJ Productions/Getty Images

throughout adulthood, though, mainly because men find partners more easily as they select from a larger age range of unmarried women.

Ethnic differences in singlehood reflect differences in age at marriage as well as social factors. For example, nearly twice as many African Americans are single during young adulthood as European Americans, and more are choosing to remain so (U.S. Census Bureau, 2010b). Professional African American women show no significant differences from partnered African American women in well-being (Williams, 2006). Singlehood is also increasing among Latinos, in part because the average age of Latinos in the United States is lower than other ethnic groups and in part because of poor economic opportunities for many Latinos (Lamanna & Riedmann, 2003). However, Latino men expect to marry (even if they do not) because it indicates achievement.

Globally, the meanings and implications of remaining single are often tied to strongly held cultural and religious beliefs. For example, Muslim women who remain single in Malaysia speak in terms of *jodoh* (the soul mate one finds through fate at a time appointed by God) as a reason; they believe that God simply has not decided to have them meet their mate at this time (Ibrahim & Hassan, 2009). But because the role of Malaysian women is to marry, they also understand their marginalized position in society through their singlehood. In Southeast Asia, the number of single adults has increased steadily as education levels have risen over the past several decades (Hull, 2009). However, family systems in these cultures have not yet adapted to these changing lifestyle patterns (Jones, 2010).

An important distinction is between adults who are temporarily single (i.e., those who are single only until they find a suitable marriage partner) and those who choose to remain single. Results from an in-depth interview study with never-married women in their 30s revealed three distinct groups: some suffer with acute distress about being single and long to be married with children; others describe experiencing the emotional continuum of desiring to be married and desiring to remain single; and others say that they are quite happy with a healthy self-image and high quality of life (Cole, 2000). For most singles, the decision to never marry is a gradual one. This transition is represented by a change in self-attributed status that occurs over time and is associated with a cultural timetable for marriage. It marks the experience of "becoming single" that occurs when an individual identifies more with singlehood than with marriage (Davies, 2003).

Still, a key question is: What prompts people to decide that they will remain single? For some, it is reaching a milestone birthday (e.g., 40) and still being single, although the particular age chosen varies a great deal (Davies, 2000). For many middle-aged single women, purchasing a house marks the decision:

> I always thought you got married, you bought a house. Well, I bought a house and I'm not married . . . I've laid down roots . . . You're sort of saying, "Okay, this is it." And it makes you feel more settled. (Davies, 2000, p. 12)

For most, though, the transition to permanent singlehood is a gradual one they drift into by circumstance rather than a lifestyle they choose, such as having to care for parents or other family members instead of attending to personal goals related to marriage, family, education, or career (Connidis, 2001). By the time they reach age 40, never-married women have defined "family" as their family of origin and friendships, and most are content with their lives (McDill, Hall, & Turell, 2006).

Cohabitation

Being unmarried does not necessarily mean living alone. *People in committed, intimate, sexual relationships but who are not married may decide that living together, or* **cohabitation,** *provides a way to share daily life.* Cohabitation is becoming an increasingly popular lifestyle choice in the United States as well as in Canada, Europe, Australia, and elsewhere, and is considered a growing hallmark of emerging adulthood. As you can see in ■ Figure 11.3, cohabitation in the United States has increased 10-fold

cohabitation
people in committed, intimate, sexual relationships who live together but are not married

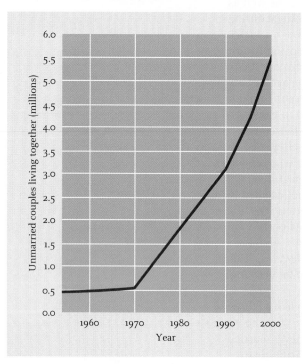

Figure 11.3
There has been a rapid growth in cohabitation in the United States since 1970.

From U.S. Census Bureau, 2003.

THINK ABOUT IT

Why might there be large differences in cohabitation rates among countries?

over the past three decades: from 523,000 in 1970 to 5.5 million in 2002, the most recent year extensive data were collected (Goodwin, Mosher, & Chandra, 2010). People with lower educational levels cohabit more, and do so in more relationships, than individuals with higher educational levels. European American, African American, and Latino men and women cohabit at about the same rates, other factors being equal.

Couples cohabit for three main reasons, most often in connection with testing their relationship in the context of potential marriage (Rhoades, Stanley, & Markman, 2009).

Some couples cohabit for reasons of convenience, such as sharing expenses and sexual accessibility. There is typically no long-term commitment for these couples, and marriage is not usually a goal. Women tend to report convenience as a factor more so than men (Rhoades et al., 2009). Most American young adult couples, though, cohabit as a step toward marriage (King & Scott, 2005); in fact, about half of cohabiting couples transition to marriage (Goodwin, Mosher, & Chandra, 2010). In this case, the couple is actually engaging in a trial marriage. If marriage does not follow, the couple usually separates. Finally, some couples use cohabitation instead of marriage.

The global picture differs by culture (Popenoe, 2009; Therborn, 2010). For example, in most European, South American, and Caribbean countries, cohabitation is a common alternative to marriage for young adults. Cohabitation is extremely common in the Netherlands, Norway, and Sweden, where this lifestyle is part of the culture; 99% of married couples in Sweden lived together before they married and nearly one in four couples are not legally married. Decisions to marry in these countries are typically made to legalize the relationship after children are born—in contrast to Americans, who marry to confirm their love and commitment to each other.

Interestingly, having cohabitated does not seem to make marriages any better; in fact, it may do more harm than good, resulting in lower quality marriages (Tach & Halpern-Meekin, 2009). These findings reflect two underlying issues: couples who have children while cohabiting, especially for European American women (as compared with African American and Latina women; Tach & Halpern-Meekin, 2009), and couples who are using cohabitation to test their relationship (Rhoades et al., 2009) are most likely to report subsequent problems.

Young adults whose parents divorced are more likely to cohabit, but this effect weakens between the late teens and early 30s (Cunningham & Thornton, 2007).

Are there differences between couples who cohabit and couples who marry right away? Longitudinal studies find few differences in couples' behavior after living together for many years regardless of whether they married without cohabiting, cohabited then married, or simply cohabited (Stafford, Kline, & Rankin, 2004). No differences are reported in relationships between parents and adult children of married versus cohabiting couples (Daatland, 2007). Additionally, many countries extend the same rights and benefits to cohabiting couples as they do to married couples, and have done so for many years. For instance, Argentina provides pension rights to cohabiting partners, Canada extends insurance benefits, and Australia has laws governing the disposition of property when cohabiting couples sever their relationship (Neft & Levine, 1997).

Gay and Lesbian Couples

Less is known about the developmental course of gay and lesbian relationships than heterosexual relationships, largely because they have historically not been the focus of research (Rothblum, 2009). To date, gay and lesbian relationships have been studied most often in comparison to married heterosexual couples. What is it like to be in a gay or lesbian relationship? One woman shares her experience in the Real People feature.

I am a 35-year-old woman who believes that each person is here with a purpose to fulfill in his or her lifetime. "Add your light to the sum of light" are words I live by in my teaching career, my personal life with friends and family, and living in general. I do not believe that our creator makes mistakes, although at times I am very discouraged by the level of hatred that is evident in the world against many groups and against homosexuals in particular.

For me, being a lesbian is the most natural state of being. I do not think of it as a mishap of genetics, a result of an unhappy or traumatic childhood, or an unnatural tendency. From the time I was a child I had a definite and strong sense of my sexual identity. However, I am aware of the homophobia that is present at all levels of my own life and in the community. That is where my sense of self and living in the world collide.

Society does not value diversity. We, as a people, do not look to people who are different and acknowledge the strength it takes to live in this society. Being gay in a homophobic, heterosexist society is a burden that manifests itself in many forms, such as through alcohol and drug abuse rates that are much higher than in the heterosexual community. The lack of acknowledgment of gay people's partners by family members, co-workers, and society at large is a stamp of nonexistence and invisibility. How can we build a life with a partner and then not share that person with society?

I consider myself a fortunate gay person in that I have a supportive family. Of the five children in my family, two of us are gay. My parents are supportive and love our partners. My siblings vary in their attitudes. One sister invited me and my partner to her wedding. Nine years later, my other sister refused to do that. Her discomfort over my sexual orientation meant that I spent a special event without my partner at my side. However, my straight brother was allowed to bring a date. It was very hurtful and hard to forgive.

In the larger community, I have been surprised by the blatant hatred I have experienced. I have demeaning comments aimed at me. The home I live in has been defaced with obscenities. But on a more positive note, I have never been more strongly certain of who I am. I am indebted to those who have supported me over the years with love and enlightenment, knowing that who I am is not a mistake. As I age, it becomes clearer to me that I am meant to share the message that our differences are to be appreciated and respected.

Like heterosexuals, gay and lesbian couples must deal with issues related to effective communication, power, and household responsibilities. For the most part, the relationships of gay and lesbian couples have many similarities to those of heterosexual couples (Kurdek, 2004). Most gay and lesbian couples are in dual-earner relationships, much like the majority of married heterosexual couples, and are likely to share household chores. However, gay and lesbian couples do differ from heterosexual couples in the degree to which both partners are similar on demographic characteristics such as race, age, and education; gay and lesbian couples tend to be more dissimilar (Schwartz & Graf, 2009). In general, though, the same factors predict long-term success of couples regardless of sexual orientation (Mackey, Diemer, & O'Brien, 2004).

Gay and lesbian couples experience stresses in relationships similar to those of heterosexual couples.

Gender differences play more of a role in determining relationship styles than do differences in sexual orientation (Huston & Schwartz, 1995). Gay men, like heterosexual men, tend to separate love and sex and have more short-term relationships (Missildine et al., 2005); both lesbian and heterosexual women are more likely to connect sex and emotional intimacy in fewer, longer-lasting relationships. Lesbians tend to make a commitment and cohabit faster than heterosexual couples (Ganiron, 2007). Men in any type of relationship tend to want more power if they earn more money. Women in any type of relationship are likely to be more egalitarian and to view money as a way to maintain independence from one's partner.

Gay and lesbian couples report receiving less support from family members than do either married or cohabiting couples (Rothblum, 2009). The more that one's family holds traditional ethnic or religious values, the less likely it is that the family will provide support. At a societal level, marriage or civil unions between same-sex couples remains highly controversial in America, with several states passing constitutional amendments or statutes defining marriage as between a man and a woman; the federal government, also, since passage of the Defense of Marriage Act (DOMA) in 1996, defines marriage as between one man and one woman. The lack of legal recognition for gay and lesbian relationships in the United States also means that certain rights and privileges, such as marriage, certain insurance benefits, and hospital visitation rights, are not always

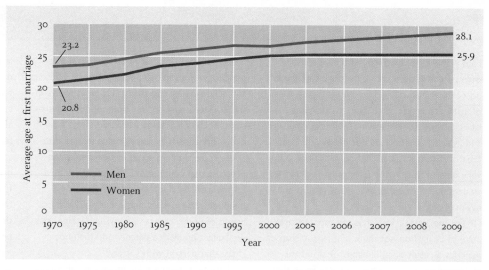

From U.S. Census Bureau, Current Population Survey, March and Annual Social and Economic Supplements, 2010 and earlier.

granted. For example, it is difficult for gay and lesbian partners to inherit property from their partners in the absence of a will, and sometimes they are denied visitation rights when their partner is hospitalized. Although the legal status of gay and lesbian couples is changing in more countries (most notably in Scandinavia), and a few states in the United States (including New York), most countries and states in the United States do not provide them with the same legal rights as married couples.

Marriage

Most adults want their love relationships to result in marriage. However, U.S. residents are in less of a hurry to achieve this goal; the median age at first marriage for adults in the United States has been rising for several decades. As you can see in ■ Figure 11.4, between 1970 and 2009, the median age for first marriage rose nearly 4 years for both men and women, from roughly 23 to 28.1 for men, and from roughly 21 to 25.9 for women (U.S. Census Bureau, 2010a). This trend has some benefits in that, for women anyway, marrying at a later age lessens the likelihood of divorce: women under age 20 at the time they are first married are 3 times more likely to end up divorced than women who first marry in their 20s, and 6 times more likely to end up divorced than first-time wives in their 30s (U.S. Census Bureau, 2010b). Let's explore age and other factors that keep marriages going strong over time.

What Is a Successful Marriage and What Predicts It?

You undoubtedly know couples who appear to have a very successful marriage. But what does that mean, really? Is success in marriage defined as subjective happiness and contentment, personal fulfillment, or simply the fact that the couple is still married? *Minnotte (2010) differentiates* **marital success**, *which is an umbrella term referring to any marital outcome (such as divorce rate),* **marital quality**, *which is a subjective evaluation of the couple's relationship on a number of different dimensions,* **marital adjustment**, *the degree to which a husband and wife accommodate to each other over a certain period of time, and* **marital satisfaction**, *which is a global assessment of one's marriage.* Each of these provides a unique insight into the workings of a marriage.

At a general level, why do some marriages succeed? Part of it has to do with what people believe about marriage and what makes a marriage succeed. Answering the questions in ● Table 11.2 will provide you with some good insights. Take time to think about your responses and why you answered the way you did. Your responses are the result of many factors, including the socialization you had about marriage. As we ex-

marital success
an umbrella term referring to any marital outcome

marital quality
a subjective evaluation of the couple's relationship on a number of different dimensions

marital adjustment
the degree to which a husband and wife accommodate to each other over a certain period of time

marital satisfaction
a global assessment of one's marriage

Beliefs About Marriage

1. A husband's marital satisfaction is usually lower if his wife is employed full time than if she is a full-time homemaker	True	False
2. Marriages that last many years almost always have a higher level of satisfaction than marriages that last only a few years	True	False
3. In most marriages, having a child improves marital satisfaction for both spouses	True	False
4. The best single predictor of marital satisfaction is the quality of the couple's sex life	True	False
5. Overall, married women are physically healthier than married men	True	False
6. African American women are happier in marriage than African American men	True	False
7. Marital satisfaction for a wife is usually lower if she is employed full time than if she is a full-time homemaker	True	False
8. "If my spouse loves me, he/she should instinctively know what I want and need to make me happy"	True	False
9. In a marriage in which the wife is employed full time, the husband usually shares equally in the housekeeping tasks	True	False
10. "No matter how I behave, my spouse should love me because he/she is my spouse"	True	False
11. European American husbands spend more time on household work than do Latino husbands	True	False
12. Husbands usually make more lifestyle adjustments in marriage than do wives	True	False
13. "I can change my spouse by pointing out his/her inadequacies and bad habits"	True	False
14. The more a spouse discloses positive and negative information to his/her partner, the greater the marital satisfaction of both partners	True	False
15. For most couples, maintaining romantic love is the key to marital happiness over the life span	True	False

All of the items are false. The more "True" responses you gave, the greater your belief in stereotypes about marriage.

SOURCE: Benokraitis, Nijole V., Marriages and Families: Changes, Choices, and Constraints, 5th Edition, © 2005, p. 262. Reprinted by permission of Pearson Education, Inc., Upper Saddle River, NJ.

plore the research data about marital success, think about these and other widely held beliefs about marriage.

Marriages, like other relationships, differ from one another, but some important predictors of future success can be identified. One key factor in enduring marriages is the relative maturity of the two partners at the time they are married. In general, the younger the partners are, the lower the odds that the marriage will last—especially when the people are in their teens or early 20s (U.S. Census Bureau, 2010b). In part, the age issue relates to Erikson's (1982) belief that intimacy cannot be achieved until after one's identity is established (see Chapter 10). Other reasons that increase or decrease the likelihood that a marriage will last include financial security and pregnancy at the time of the marriage.

A second important predictor of successful marriage is **homogamy**, *or the similarity of values and interests a couple shares.* As we saw in relation to choosing a mate, the extent that the partners share similar age, values, goals, attitudes (especially the desire for children), socioeconomic status, certain behaviors (such as drinking alcohol), and ethnic background increases the likelihood that their relationship will succeed (Kippen, Chapman, & Yu, 2009).

A third factor in predicting marital success is a feeling that the relationship is equal. *According to* **exchange theory**, *marriage is based on each partner contributing something to the relationship that the other would be hard-pressed to provide.* Satisfying and happy marriages result when both partners perceive that there is a fair exchange, or equity, in all the dimensions of the relationship. Problems achieving such equity can arise because of the competing demands of work and family, an issue we take up again in Chapter 12.

Cross-cultural research supports these factors. Couples in the United States and Iran (Asoodeh et al., 2010; Hall, 2006; McKenzie, 2003) say that trust, consulting each other, honesty, making joint decisions, and commitment make the difference between

homogamy
similarity of values and interests

exchange theory
relationship, such as marriage, based on each partner contributing something to the relationship that the other would be hard pressed to provide

Marital satisfaction is highest early on and in later life, dropping off during the child-rearing years.

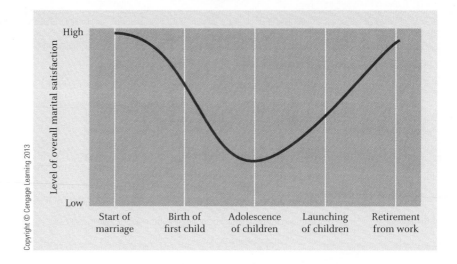

a successful marriage and an unsuccessful marriage. Couples for whom religion is important also point to commonly held faith.

Do Married Couples Stay Happy?

Few sights are happier than a couple on their wedding day. Newlyweds, like Kevin and Beth in the vignette, are at the peak of marital bliss. The beliefs people bring into a marriage (which you identified in Table 11.2 in the quiz you took) influence how satisfied they will be as the marriage develops. But as you may have experienced, feelings change over time, sometimes getting better and stronger, sometimes not.

Research shows that, for most couples, overall marital satisfaction is highest at the beginning of the marriage, falls until the children begin leaving home, and rises again in later life; this pattern holds for both married and never-married cohabiting couples with children (see ■ Figure 11.5; Hansen, Moum, & Shapiro, 2007). However, for some couples, satisfaction never rebounds and remains low; in essence, they have become emotionally divorced.

Overall, marital satisfaction ebbs and flows over time. The pattern of a particular marriage over the years is determined by the nature of the dependence of each spouse on the other. When dependence is mutual and about equal, and both people hold similar values that form the basis for their commitment to each other, the marriage is strong and close (Givertz, Segrin, & Hanzal, 2009). When the dependence of one partner is much higher than that of the other, however, the marriage is likely to be characterized by stress and conflict. Changes in individual lives over adulthood shift the balance of dependence from one partner to the other; for example, one partner may go back to school, become ill, or lose status. Learning how to deal with these changes is the secret to long and happy marriages.

The fact that marital satisfaction has a general downward trend but varies widely across couples led Karney and Bradbury (1995) to propose a vulnerability–stress–adaptation model of marriage, depicted in ■ Figure 11.6. *The* **vulnerability–stress–adaptation model** *sees marital quality as a dynamic process resulting from the couple's ability to handle stressful events in the context of their particular vulnerabilities and resources.* For example, as a couple's ability to adapt to stressful situations gets better over time, the quality of the marriage probably will improve. How well couples adapt to various stresses on the relationship determines whether the marriage continues or they get divorced. Let's see how this works over time.

Setting the Stage: The Early Years of Marriage

Marriages are most intense in their early days. Early on, husbands and wives share many activities and are open to new experiences together, so bliss results (Olson & McCubbin, 1983). But bliss doesn't come from avoiding tough issues. Discussing financial matters honestly is a key to bliss, as many newly married couples experience

vulnerability–stress–adaptation model
model that proposes that marital quality is a dynamic process resulting from the couple's ability to handle stressful events in the context of their particular vulnerabilities and resources.

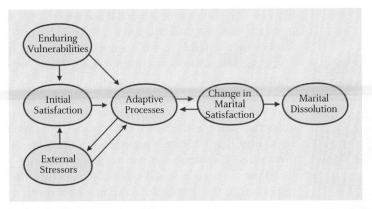

Source: Karney, B. R. (2010). *Keeping marriages healthy, and why it's so difficult.* Retrieved from http://www.apa.org/science/about/psa/2010/02/sci-brief.aspx.

Figure 11.6

The vulnerability-stress-adaptation model shows how adapting to vulnerabilities and stress can result in either adaptation or dissolution of the marriage.

their first serious marital stresses around money issues (Parkman, 2007). How tough issues early in the marriage are handled sets the stage for the years ahead. When there is marital conflict, the intensity of the early phase may create the basis for considerable unhappiness (Faulkner, Davey, & Davey, 2005).

Early in a marriage, the couple must learn to adjust to the different perceptions and expectations each person has for the other. Many wives tend to be more concerned than their husbands with keeping close ties with their friends. Research indicates that men and women both recognize and admit when problems occur in their marriage (Moynehan & Adams, 2007). The couple must also learn to handle confrontation. Indeed, learning effective strategies for resolving conflict is an essential component of a strong marriage, because these strategies provide ways for couples to discuss their problems maturely.

Early in a marriage, couples tend to have global adoration for their spouse regarding the spouse's qualities (Karney, 2010; Neff & Karney, 2005). For wives, but not for husbands, more accurate specific perceptions of what their spouses are really like were associated with more supportive behaviors, feelings of control in the marriage, and a decreased risk of divorce. Thus, for women, love grounded in accurate perceptions of a spouse's qualities appears to be stronger than love that is "blind" to a spouse's true qualities. Still, couples who are happiest in the early stage of their marriage tend to focus on the good aspects, not the annoyances; nit-picking and nagging do not bode well for long-term wedded bliss (Karney, 2010).

As time goes on, and stresses increase, marital satisfaction tends to decline (Lamanna & Riedmann, 2003). Researchers have shown that, for most couples, the primary reason for this drop is having children (Jokela et al., 2009). But it's not just a matter of having a child. The temperament of the child matters, with fussier babies creating more marital problems (Greving, 2007; Meijer & van den Wittenboer, 2007). Parenthood also means having substantially less time to devote to the marriage. Most couples are ecstatic over having their first child, a tangible product of their love for each other. But soon the reality of child care sets in, with 2 a.m. feedings, diaper changing, and the like—not to mention the long-term financial obligations that will continue at least until the child becomes an adult. Both African American and European American couples report an increase in conflict after the birth of their first child (Crohan, 1996).

However, using the birth of a child as the explanation for the drop in marital satisfaction is much too simplistic, because child-free couples also experience a decline in marital satisfaction (Hansen et al., 2007). It appears that a decline in general marital satisfaction over time is a common developmental phenomenon, even for couples who choose to remain childless (Clements & Markman, 1996). Additionally, couples who have no children as a result of infertility face the stress associated with the inability to have children, which exacerbates existing stresses in the relationship and can lower marital satisfaction (Spector, 2004). Longitudinal research indicates that disillusionment—as demonstrated by a decline in feeling in love, in demonstrations of affection, and in the feeling that one's spouse is responsive, as well as an

Sandy Huffaker/Getty Images

Young married military couples face special types of stress on their relationship.

increase in feelings of ambivalence—is a key predictor of marital dissatisfaction (Huston et al., 2001).

During the early years of their marriage, many couples may spend significant amounts of time apart. This is especially true of marriages involving individuals in the military (Fincham & Beach, 2010). Recent research has focused on the special types of stress these couples face. Spouses that serve in combat areas on active duty assignment who suffer from post-traumatic stress disorder (PTSD) are particularly vulnerable, as they are at greater risk for other spouse-directed aggression.

What the non-deployed spouse believes turns out to be very important. If the non-deployed spouse believes that the deployment will have negative effects on the marriage, then problems are much more likely. In contrast, if the non-deployed spouse believes that such challenges make the relationship stronger, then they typically can do so (Renshaw, Rodrigues, & Jones, 2008). Research indicates that the effects of deployment may be greater on wives than husbands; divorce rates for women service members who are deployed is higher than for their male counterparts (Karney & Crown, 2007).

Keeping Marriages Happy

Although no two marriages are exactly the same, couples must be flexible and adaptable. Couples who have been happily married for many years show an ability to roll with the punches and to adapt to changing circumstances in the relationship. For example, a serious problem of one spouse may not be detrimental to the relationship and may even make the bond stronger if the couple use good stress- and conflict-reduction strategies. Successful couples also find a way to keep the romance in the relationship, a very important determinant of marital satisfaction over the long run (Acevedo & Aron, 2009).

Sharing religious beliefs and spirituality with one's spouse is another good way to help insure higher quality marriages, and that's especially the case among couples in lower socioeconomic groups (Lichter & Carmalt, 2009). It appears that this effect goes beyond merely doing an activity together, as religion and spirituality may provide a framework for conflict resolution and a way to put one's marriage in a bigger, more significant context.

But when you get down to basics, it's how well couples communicate their thoughts, actions, and feelings to each other and show intimacy and support each other that largely determines the level of conflict couples experience and, by extension, how happy they are likely to be over the long term (Patrick et al., 2007). This is especially important regarding high-stress areas such as children and work. And this evidence of the importance of good communication skills forms the basis of most marriage education programs we will consider a bit later.

So what are the best ways to stack the deck in favor of a long, happy marriage? Based on research, here are the best:

THINK ABOUT IT

What types of interventions would help keep married couples happier?

- Make time for your relationship.
- Express your love to your spouse.
- Be there in times of need.
- Communicate constructively and positively about problems in the relationship.
- Be interested in your spouse's life.
- Confide in your spouse.
- Forgive minor offenses, and try to understand major ones.

Test Yourself

11.3 The Family Life Cycle

LEARNING OBJECTIVES

- What are the common forms of families?
- Why do people have children?
- What is it like to be a parent? What differences are there in different types of parenting?

Bob, 32, and Denise, 33, just had their first child, Matthew, after several years of trying. They've heard that having children while in their 30s can have advantages, but Bob and Denise wonder whether people are just saying that to be nice to them. They are also concerned about the financial obligations they are likely to face.

"WHEN ARE YOU GOING TO START A FAMILY?" is a question that young couples like Bob and Denise are asked frequently. Most couples want children because they believe they will bring great joy, which they often do. But once the child is born, adults may feel inadequate because children don't come with instructions. Young adults may be surprised when the reality of being totally responsible for another person hits them. Experienced middle-aged parents often smile knowingly to themselves.

Frightening as it might be, the birth of a child transforms a couple (or a single parent) into a family. *The most common form of family in Western societies is the* **nuclear family**, *consisting only of parent(s) and child(ren). The most common family form around the world is the* **extended family**, *in which grandparents and other relatives live with parents and children.* Because we have discussed families from the child's perspective in earlier chapters, here we focus on families from the parents' point of view.

Deciding Whether to Have Children

One of the biggest decisions couples have to make is whether to have children. This decision appears complicated. You would think that potential parents must weigh the many benefits of child rearing—such as personal satisfaction, fulfilling personal needs, continuing the family line, and companionship—with the many drawbacks,

nuclear family
most common form of family in Western societies, consisting only of parent(s) and child(ren)

extended family
most common form of family around the world; one in which grandparents and other relatives live with parents and children

including expense and lifestyle changes, especially the balance between work and family. But apparently, this is not what most people actually do.

Rijken (2009) reports that potential parents actually don't think very deliberately or deeply about when to have a child, and that those who are career oriented or like their freedom do not often deliberately postpone parenthood because of those factors. Rather, thoughts about having children seem to just not cross their minds until they are ready to start thinking about having children. Whether the pregnancy is planned or not (and over half of all U.S. pregnancies are unplanned), a couple's first pregnancy is a milestone event in a relationship, with both benefits and costs (Greving, 2007; Meijer & van den Wittenboer, 2007). Having a child raises many important matters for consideration, such as relationships with one's own parents, marital stability, career satisfaction, and financial issues. Parents largely agree that children add affection, improve family ties, and give parents a feeling of immortality and a sense of accomplishment. Most parents willingly sacrifice a great deal for their children and hope that they grow up to be happy and successful. In this way, children bring happiness to their parents (Angeles, 2010).

Nevertheless, finances are of great concern to most parents because children are expensive. How expensive? According to the U.S. Department of Agriculture (2010), a family who had a child in 2009 would spend the following estimated amounts for food, shelter, and other necessities by the time the child turned 17: in the lowest income bracket, $206,000; in the middle income bracket, $286,000; and in the highest income bracket, $476,000. College expenses would be an additional expense. These costs do not differ significantly between two-parent and single-parent households but clearly are a bigger financial burden for single parents. No wonder parents are concerned.

For many reasons that include personal choice, financial instability, and infertility, an increasing number of couples are child-free. Attitudes toward childless couples have improved since the 1970s, with women having more positive views than men (Koropeckyj-Cox & Pendell, 2007). Social attitudes in many countries (Austria, Germany, Great Britain, Ireland, Netherlands, and United States) are also improving toward child-free couples (Gubernskaya, 2010). Couples without children also have some advantages: higher marital satisfaction, more freedom, and higher standards of living. A major international study of older adult couples without children in Australia, Finland, Germany, Japan, the Netherlands, the United Kingdom, and the United States revealed highly similar patterns across all countries except Japan (Koropeckyj-Cox & Call, 2007). In Japan, the cultural norm of children caring for older parents created difficulties for childless older couples.

The Parental Role

Today, couples in the United States typically have fewer children and have their first child later than in the past. The average age at the time of the birth of a woman's first child is about 25. This average age has been increasing steadily since 1970 as a result of two major trends: Many women postpone children because they are marrying later, they want to establish careers, or they make a choice to delay childbearing. Additionally, the teen birthrate dropped dramatically between the early 1990s and 2005 (but has increased since 2006; U.S. Census Bureau, 2010b).

Being older at the birth of one's first child is advantageous. Older mothers, like Denise in the vignette, are more at ease being parents, spend more time with their babies, and are more affectionate, sensitive, and supportive to them (Berlin, Brady-Smith, and Brooks-Gunn, 2002). The age of the father also makes a difference in how he interacts with children (Palkovitz & Palm, 2009). Remember Bob, the 32-year-old first-time father in the vignette? Compared to men who become fathers in their 20s, men (like Bob) who become fathers in their 30s are generally more invested in their paternal role and spend up to 3 times as much time caring for their preschool children as younger fathers do. Father involvement has increased significantly, due in part to social attitudes that support it (Fogarty & Evans, 2010).

Parenting skills do not come naturally; they must be acquired. Having a child changes all aspects of couples' lives. As we have seen, children place a great deal of stress on a relationship. Both motherhood and fatherhood require major commitment and co-operation. Parenting is full of rewards, but it also takes a great deal of work. Caring for young children is demanding. It may create disagreements over division of labor, especially if both parents are employed outside the home (see Chapters 4 and 11). Even when mothers are employed outside the home (and more than 70% of women with children under age 18 are), they still perform most of the child-rearing tasks. Even when men take employment leave, though more likely to share tasks they still do not spend more time with children than fathers who do not take leave (Seward et al., 2006).

In general, parents manage to deal with the many challenges of child rearing reasonably well. They learn how to compromise when necessary and when to apply firm but fair discipline. Given the choice, most parents do not regret their decision to have children.

Having a child later in adulthood has many benefits.

THINK ABOUT IT

Should there be mandatory programs for parent education?

Ethnic Diversity and Parenting

Ethnic background matters a great deal in terms of family structure and the parent–child relationship. African American husbands are more likely than their European American counterparts to help with household chores, and they help regardless of their wives' employment status (Dixon, 2009). In low-income families, African American parents may buffer their children from involvement with drugs and other problems owing to their more conservative views about illegal substance abuse (Paschal, Lewis, & Sly, 2007). Overall, most African American parents provide a cohesive, loving environment that often exists within a context of strong religious beliefs (Anderson, 2007; Dixon, 2009), pride in cultural heritage, self-respect, and cooperation with the family (Brissett-Chapman & Issacs-Shockley, 1997).

As a result of several generations of oppression, many Native American parents have lost the cultural parenting skills that were traditionally part of their culture: children were valued, women were considered sacred and honored, and men cared for and provided for their families (Witko, 2006). Thus, retaining a strong sense of tribalism is an important consideration for Native American families. This helps promote strong ties to parents, siblings, and grandparents (Garrod & Larimore, 1997). Native American children are viewed as important family members, and tribal members spend great amounts of time with them imparting the cultural values—such as cooperation, sharing, personal integrity, generosity, harmony with nature, and spirituality—that differ from European American values, which emphasize competitiveness and individuality (Stauss, 1995). Many Native American parents worry that their children will lose their values if they are overexposed (e.g., during college) to European American values.

Family ties among Native Americans tend to be very strong.

Latino families make up a growing percentage of American families; 22% of all children under 18 in the United States are Latino, and most are at least second generation (Fry, 2009). Among two-parent families, Mexican American mothers and fathers both tend to adopt similar authoritative behaviors in dealing with their preschool children, but mothers use these behaviors more frequently (Gamble, Ramakumar, & Diaz, 2007). Latino families demonstrate two key values: familism and the extended family. **Familism** *refers to the idea that the well-being of the family takes precedence over the concerns of individual family members.* This value is a defining

familism
the idea that the family's well-being takes precedence over the concerns of individual family members

characteristic of Latino families; for example, Brazilian and Mexican families consider familism a cultural strength (Carlo et al., 2007; Lucero-Liu, 2007). Indeed, familism helps account for the significantly higher trend for Latino college students to live at home (Desmond & López Turley, 2009). The extended family is also very strong among Latino families and serves as the venue for a wide range of exchanges of goods and services, such as child care and financial support. Emphasis on traditional values such as familism is especially strong in Latino families from lower socioeconomic backgrounds (Almeida et al., 2009).

Like Latinos, Asian Americans value familism (Meyer, 2007) and place an even higher value on extended family. Other key values include obtaining good grades in school, maintaining discipline, being concerned about what others think, and conformity. Asian American adolescents report very high feelings of obligation to their families compared with European American adolescents (Kiang & Fuligni, 2009). In general, males enjoy higher status in traditional Asian families (Tsuno & Homma, 2009). Among recent immigrants, though, women are expanding their role by working outside the home. Research shows that Chinese American parents experience less marital stress during the transition to parenthood than European American couples, perhaps because of the clearer traditional cultural division of tasks between husbands and wives (Burns, 2005).

Raising multi-ethnic children presents challenges not experienced by parents of same-race children. For example, parents of biracial children report feeling discrimination and that they are targets of prejudicial behavior from others (Hubbard, 2010; Kilson & Ladd, 2009). These parents also worry that their children may be rejected by members of both racial communities. Perhaps that is why parents of multiracial children tend to provide more economic and cultural resources to their children than do parents of single race children (Cheng & Powell, 2007).

In multi-ethnic families, you might think that the parent from a minority group takes primary responsibility for guiding that aspect of the child's ethnic identity. However, a study of children of European mothers and Maori fathers in New Zealand showed that the mothers played a major role in establishing the child's Maori identity (Kukutai, 2007). Similarly, European American mothers of biracial children whose fathers were African American tended to raise them as African American in terms of public ethnic identity (O'Donoghue, 2005). As adults, most European American–African American biracial individuals describe themselves as African American (Khanna, 2010).

It is clear that ethnic groups vary a great deal in how they approach the issue of parenting and what values are most important. Considered together, there is no one parenting standard that applies equally to all groups.

Single Parents

Although the overall number of single-parent households in the United States has remained at about 9% since 1994, the proportion of births to unwed mothers is at an all-time high, now over 40% (Livingston & Cohn, 2010). The number of single parents, most of whom are women, continues to be high in some ethnic groups. Over 70% of births to African American mothers, over 50% of births to Latina mothers, and nearly 30% of births to European American mothers are to unmarried women (Livingston & Cohn, 2010). Among the reasons are simply wanting to have children, failure to use contraception, high divorce rates, the decision to keep children born out of wedlock, and different fertility rates across ethnic groups. Being a single parent raises important questions.

Two main questions arise concerning single parents: How are children affected when only one adult is responsible for child care? And how do single parents meet their own needs for emotional support and intimacy?

Many divorced single parents report complex feelings such as frustration, failure, guilt, and a need to be overindulgent (Lamanna & Riedmann, 2003). Loneliness can be especially difficult to deal with (Anderson et al., 2004). Separation anxiety is a common and strong feeling among military parents who are about to be deployed (Roper,

2007). Feelings of guilt may lead to attempts to make up for the child's lack of a father or mother. Some single parents make the mistake of trying to be peers to their children, using inconsistent discipline or (if they are the noncustodial parent) spoiling their children with lots of monetary or material goods.

Single parents, regardless of gender, face considerable obstacles. Financially, they are usually much less well-off than their married counterparts. Having only one source of income puts additional pressure on single parents to provide all of the necessities. Integrating the roles of work and parenthood are difficult enough for two people; for the single parent, the hardships are compounded. Financially, single mothers are hardest hit, mainly because women typically are paid less than men and because single mothers may not be able to afford enough child care to provide the work schedule flexibility needed for higher-paying jobs.

One particular concern for many divorced single parents is dating. Several common questions asked by single parents involve dating: "How do I become available again?" "How will my children react?" "How do I cope with my own sexual needs?" Single parents often feel insecure about sexuality and how they should behave around their children in terms of having partners stay overnight (Lampkin-Hunter, 2010). When single parenting happens through divorce, dating often begins fairly quickly; we will consider those situations a bit later in this chapter.

Alternative Forms of Parenting

Not all parents raise their own biological children. In fact, roughly one third of North American couples become stepparents or foster or adoptive parents some time during their lives.

To be sure, the parenting issues we have discussed so far are just as important in these situations as when people raise their own biological children. In general, there are few differences among parents who have their own biological children or who become parents in some other way, but there are some unique challenges (McKay & Ross, 2010).

A big issue for foster parents, adoptive parents, and stepparents is how strongly the child will bond with them. Although infants less than 1 year old will probably bond well, children who are old enough to have formed attachments with their biological parents may have competing loyalties. For example, some stepchildren remain strongly attached to the noncustodial parent and actively resist attempts to integrate them into the new family ("My real mother wouldn't make me do that"), or they may exhibit behavioral problems. As a result, the dynamics in blended families can best be understood as a complex system (Dupuis, 2010). Stepparents must often deal with continued visitation by the noncustodial parent, which may exacerbate any difficulties. These problems are a major reason that second marriages are at high risk for dissolution, as discussed later in this chapter. They are also a major reason why behavioral and emotional problems are more common among stepchildren (Crohn, 2006).

Still, many stepparents and stepchildren ultimately develop good relationships with each other. Stepparents must be sensitive to the relationship between the stepchild and his or her biological, noncustodial parent. Allowing stepchildren to develop a relationship with the stepparent at their own pace also helps. What style of stepparenting ultimately develops is influenced by the expectations of the stepparent, stepchild, spouse, and nonresidential parent, but there are several styles that result in positive outcomes (Crohn, 2006).

Adoptive parents also contend with attachment to birth parents, but in different ways. Even if they don't remember them, adopted children may wish to locate and meet their birth parents. Wanting to know one's origins is understandable, but such searches can strain the relationships between these children and their adoptive parents, who may interpret these actions as a form of rejection, and create difficulties for the adopted person (Curtis & Pearson, 2010).

Families with children adopted from another culture pose unique issues in terms of how to establish and maintain connection with the child's culture of origin. For mothers of transracially adopted Chinese and Korean children, becoming connected to the appropriate Asian American community is an important way to accomplish this (Johnston et al., 2007). Research in the Netherlands found that children adopted from Columbia, Sri Lanka, and Korea into Dutch homes struggled with looking different, and many expressed desires to be white (Juffer, 2006). Research in Sweden also revealed challenges in maintaining the culture from the child's country of origin (Yngvesson, 2010).

Foster parents tend to have the most tenuous relationship with their children because the bond can be broken for any of a number of reasons having nothing to do with the quality of the care being provided. For example, a court may award custody back to the birth parents, or another couple may legally adopt the child. Dealing with attachment is difficult; foster parents want to provide secure homes, but they may not have the children long enough to establish continuity. Furthermore, because many children in foster care have been unable to form attachments at all, they are less likely to form ones that will inevitably be broken. Thus, foster parents must be willing to tolerate considerable ambiguity in the relationship and to have few expectations about the future. Despite the challenges, the good news is that placement in good foster care does result in the development of attachment between foster parents and children who were placed out of institutional settings (Smyke et al., 2010).

Finally, many gay men and lesbian women also want to be parents. Some have biological children themselves, whereas others are increasingly choosing adoption or foster parenting (Braun, 2007; Goldberg, 2009). Although gay men and lesbian women make good parents, they often experience resistance to their having children (Clifford, Hertz, & Doskow, 2010); for example, some states in the United States have laws preventing gay and lesbian couples from adopting. Actually, research indicates that children reared by gay or lesbian parents do not experience any more problems than children reared by heterosexual parents and are as psychologically healthy as children of heterosexual parents (Biblarz & Savci, 2010). Substantial evidence exists that children raised by gay or lesbian parents do not develop sexual identity problems or any other problems any more than children raised by heterosexual parents (Goldberg, 2009; Macatee, 2007). Children of gay and lesbian parents were no more likely than children of heterosexual parents to identify as gay, lesbian, bisexual, transgendered, or questioning.

The evidence is clear that children raised by gay or lesbian parents suffer no adverse consequences compared with children raised by heterosexual parents. Children of lesbian couples and heterosexual couples are equally adjusted behaviorally, show equivalent cognitive development, have similar behaviors in school, and do not show any different rates of use of illegal drugs or delinquent behavior (Biblarz & Savci, 2010).

Some evidence shows that children raised by gay or lesbian parents may even have some advantages over children raised by heterosexual parents (Macatee, 2007). Children of gay or lesbian parents might be better adjusted than adult children of heterosexual parents in that the adult children of gay and lesbian parents exhibit lower levels of homophobia and lower fear of negative evaluation than do the adult children of heterosexual parents. Gay men are often especially concerned about being good and nurturing fathers, and they try hard to raise their children with nonsexist, egalitarian attitudes (Goldberg, 2009). Evidence shows that gay parents have more egalitarian sharing of child rearing than do fathers in heterosexual households (Biblarz & Savci, 2010).

These data will not eliminate the controversy, much of which is based on long-held beliefs (often religion-based) and prejudices. In the United States, the topic of lesbian and gay couples' right to be parents is likely to continue to play out in political agendas for years to come.

Research evidence indicates that children raised by gay or lesbian parents may have some advantages in terms of exposure to egalitarian attitudes.

© WorldFoto / Alamy

Test Yourself

RECALL

1. The series of relatively predictable changes that families experience is called _____.

2. Major influences on the decision to have children are marital factors, career factors, lifestyle factors, and _____.

3. A new father who is invested in his parental role, but who may also feel ambivalent about time lost to his career, is probably over age _____.

4. A major issue for foster parents, adoptive parents, and stepparents is _____.

INTERPRET

What difference do you think it would make to view children as a financial asset (i.e., a source of income) as opposed to a financial burden (i.e., mainly an expense)? Which of these attitudes do you think characterizes most Western societies? Can you think of an example of the other type?

APPLY

Would northern European cultures be likely to demonstrate familism? Why or why not?

Recall answers: (1) the family life cycle, (2) psychological factors, (3) 30, (4) how strongly the child will bond with them

11.4 Divorce and Remarriage

LEARNING OBJECTIVES

- Who gets divorced? How does divorce affect parental relationships with children?

- What are remarriages like? How are they similar to and different from first marriages?

Frank and Marilyn, both in their late 40s, thought their marriage would last forever. However, they weren't so lucky and have just been divorced. Although two of their children are married, their youngest daughter is still in college. The financial pressures Marilyn feels now that she's on her own are beginning to take their toll. She wonders whether her financial situation is similar to that of other recently divorced women.

DESPITE WHAT FRANK AND MARILYN PLEDGED ON THEIR WEDDING DAY, their marriage did not last until death parted them; they dissolved their marriage through divorce. Even though divorce is stressful and difficult, thousands of people each year also choose to try again. Most enter their second (or third or fourth) marriage with renewed expectations of success. Are these new dreams realistic? As we'll see, it depends on many things; among the most important is whether children are involved.

Divorce

Most couples enter marriage with the idea that their relationship will be permanent. Unfortunately, fewer and fewer couples experience this permanence. Rather than growing together, many couples grow apart.

Who Gets Divorced and Why?

You, or someone you know, has experienced divorce. No wonder. Divorce in the United States is very common—couples who marry in the United States today have about a 50–50 chance of divorce (National Center for Health Statistics, 2010b). The odds are even worse if you marry young: for couples between 20 and 24 at the time of marriage, the odds are 60% for divorce.

But there is good news. The divorce rate has been slowly declining in the United States since it peaked in the late 1970s and early 1980s. In part that's due to people being more serious about marriage and waiting longer to marry, and in part to a greater social acceptance of cohabitation as an alternative.

How about other countries? As you can see in ■ Figure 11.7, the divorce rate in nearly every other country is lower than that of the United States (National Center for Health Statistics, 2010b; United Nations, 2010). However, divorce rates in nearly every developed country have increased over the past several decades (United Nations, 2005).

One factor consistently related to divorce rates in the United States is ethnicity. Of those marriages ending in divorce, African American and Asian American couples tended to be married longer at the time of divorce than European American couples (U.S. Census Bureau, 2010b). Ethnically mixed marriages are at greater risk of divorce than ethnically homogenous ones (U.S. Census Bureau, 2010b).

Men and women tend to agree on the reasons for divorce (Amato & Previti, 2003). Infidelity is the most commonly reported cause, followed by incompatibility, drinking or drug use, and growing apart. People's specific reasons for divorcing vary with gender, social class, and life-course variables. Former husbands and wives are more likely to blame their ex-spouses than themselves for the problems that led to the divorce. Former husbands and wives claim, however, that women are more likely to have initiated the divorce.

Why people divorce has been the focus of much research. Much attention has been devoted to the notion that success or failure depends critically on how couples handle conflict. Although conflict management is important, it has become clear from research in couples therapy that the reasons couples split are complex (Kayser, 2010).

Gottman and Levenson (2000) proposed a bold framework for understanding divorce. They developed two models that predict divorce early (within the first 7 years of marriage) and later (when the first child reaches age 14) with 93% accuracy over the 14-year period of their study. Negative emotions displayed during conflict between the couple predict early divorce but not later divorce. Longitudinal research with European and African American couples over a 16-year period demonstrates that how couples deal with conflict changes over time (Birditt et al., 2010). In general, European American wives and African American couples use more accommodating and fewer

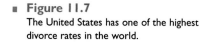
■ **Figure 11.7**
The United States has one of the highest divorce rates in the world.

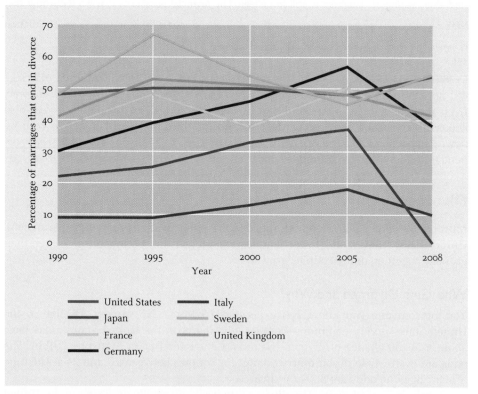

National Center for Health Statistics. (2010b). *Marriage and divorce.* Retrieved October 17, 2010 from www.cdc.gov/nchs/fastats/divorce.htm. United Nations. (2010). *Divorces and crude divorce rates by urban/rural residence: 2004 – 2008.* Retrieved October 17, 2010 from http://unstats.un.org/unsd/demographic/products/dyb/dyb2008/Table25.pdf.

destructive and quiet withdrawal behaviors over time, indicating that they are looking for ways to defuse conflict and are working through difficult issues more effectively. European American husbands tend to remain consistent in their behaviors, perhaps because they use less withdrawal early in the marriage. These findings explain why the odds for divorce are higher earlier in marriage: couples married for shorter times are less able to deal effectively with conflict.

Gottman's framework and the research it has generated is important because it clearly shows that how couples express emotion is critical to marital success. Couples who divorce earlier typically do so because of high levels of negative feelings (e.g., contempt, criticism, defensiveness, stonewalling) experienced as a result of intense marital conflict. But for many couples, such intense conflict is generally absent. Although this makes it easier to stay in a marriage longer, the absence of positive emotions eventually takes its toll and results in later divorce. For a marriage to last, people need to be told that they are loved and that what they do and feel really matters to their partner.

But we must be cautious about applying Gottman's model to all married couples. Kim, Capaldi, and Crosby (2007) reported that Gottman's variables predicting early divorce did not hold in a sample of lower-income, high-risk couples. However, Coan and Gottman (2007) point out that sample differences among the various studies means that, as noted in Chapter 1, conclusions about the predictive model must be drawn carefully.

Why people divorce is certainly complex, as shown in ■ Figure 11.8. Lots of factors enter in, such as cultural values, age, presence of children, income, and conflict. The

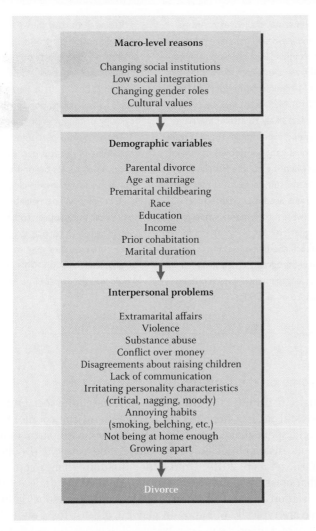

■ **Figure 11.8**
Many factors on different levels enter into the decision to divorce.

From Benokraitis, Nijole V., *Marriages and Families: Changes, Choices, and Constraints*, 4th ed., © 2002, p. 401. Reprinted with permission of Pearson Education, Inc., Upper Saddle River, NJ.

high divorce rate in the United States and the reasons typically cited for getting divorced have sparked new approaches, some of them controversial, designed to help keep couples together.

Covenant marriage *expands the marriage contract to a lifelong commitment between the partners within a supportive community.* This approach is based on the idea that if getting married and getting divorced were grounded in religious and cultural values, and divorce was made more difficult, couples would be more likely to stay together. The couple agrees to participate in mandatory premarital counseling, and that, should problems arise later, the grounds for divorce become very limited (White, 2010).

The U.S. government is also involved in efforts to strengthen marriage, most notably with the Healthy Marriage Initiative, which was created as part of the Deficit Reduction Act of 2005 and provides $150 million per year for promotion of healthy marriages and fatherhood. These in turn resulted in the National Healthy Marriage Center and the National Center for Marriage Research. Research related to these initiatives has focused on the positive aspects of marriage and on the need to do a better job with marriage education (Fincham & Beach, 2010). Will they succeed in helping couples stay married longer? That remains to be seen.

covenant marriage
expands the marriage contract to a lifelong commitment between the partners within a supportive community

marriage education
the idea that the more couples are prepared for marriage, the better the relationship will survive over the long run

What do you think? Does Marriage Education Work?

The Healthy Marriage Initiative really focused a great deal of attention on ways to lower the divorce rate (Fincham & Beach, 2010). *One approach endorsed by many groups, called* **marriage education**, *is based on the idea that the more couples are prepared for marriage, the better the relationship will survive over the long run.* More than 40 states have initiated some type of education program. Do they work?

Most education programs focus on communication between the couple; the programs provide general advice, not specific ways to deal with issues within a particular couple. Because only a minority of couples currently attend a marriage education program, there is plenty of room for improvement. Several religious denominations have their own version of marriage education programs; the Catholic's Pre-Cana program is one example.

There are numerous challenges to more extensive community-based marriage education programs. For example, in some cases the education programs were originally developed to address poverty (Administration for Children and Families, 2010). Many nonreligious couples cohabit and are less likely to attend marriage education programs, even though there is little evidence that cohabitation improves communication skills between the couple (Fincham & Beach, 2010). As a result, versions of marriage education programs are being adapted for younger adults (who, if they marry while young, have a much higher risk for divorce) and for single adults (to teach them about communication skills). Additionally, programs timed at key transition points (e.g., engagement) have also been developed (Halford et al., 2008).

Research to date shows that these skills-based education programs have modest but consistently positive effects on marital quality and communication (Cowan, Cowan, & Knox, 2010; Fincham & Beach, 2010). Perhaps not surprisingly, couples who report more problems at the beginning of the program appear to benefit most.

These positive outcomes are resulting in a broadening of the approaches used by marriage educators to topics beyond communication. How these programs develop, and whether more couples will participate, remains to be seen. What does appear to be the case is that if couples agree to participate in a marriage education program, they may well lower their risk for problems later on. What do *you* think? Would you participate in a marriage education program?

Effects of Divorce on the Couple

Although changes in attitudes toward divorce have eased the social trauma associated with it, divorce still takes a high toll on the psyche of the couple. Research in the United States and Spain shows great similarity in how both partners in a failed marriage feel: deeply disappointed, misunderstood, and rejected (Doohan, Carrère, & Riggs, 2010; Yárnoz-Yaben, 2010). Unlike the situation when a spouse dies, divorce often means that one's ex-spouse is present to provide a reminder of the unpleasant aspects of the relationship and, in some cases, feelings of personal failure. As a result, divorced people are typically unhappy in general, at least for a while (Doohan et al., 2010). The effects of a divorce can even be traced to generations not yet born because of the long-term negative consequences on education and parent–child relations in future generations (Amato & Cheadle, 2005).

Divorced people sometimes find the transition difficult; researchers refer to these problems as "divorce hangover" (Walther, 1991). Divorce hangover reflects divorced partners' inability to let go, develop new friendships, or reorient themselves as single parents. Indeed, ex-spouses who are preoccupied with thoughts of—and who have high feelings of hostility toward—their former partner have significantly poorer emotional well-being than ex-spouses who are not so preoccupied or who have feelings of friendship toward the former partner (Masheter, 1997). Forgiving the ex-spouse is also important for eventual adjustment after divorce (Rye et al., 2004). Both low preoccupation and forgiveness may be indicators that ex-spouses are able to move on with their lives.

Divorce in middle age or late life has some special characteristics. If women initiate the divorce, they report self-focused growth and optimism; if they did not initiate the divorce, they tend to ruminate and feel vulnerable (Sakraida, 2005). However, in both cases they report changes in their social networks. Middle-aged and elderly women are at a significant disadvantage for remarriage—an especially traumatic situation for women who obtained much of their identity from their roles as wife and mother. Support groups help people adjust; for men this works best in large groups and for women it works best when the group provides emotional support (Oygard & Hardeng, 2001).

We must not overlook the financial problems faced by many divorced women (Malone et al., 2010). These problems are especially keen for the middle-aged divorcee who may have spent years as a homemaker and has few marketable job skills. For her, divorce presents an especially difficult financial hardship, which is intensified if she has children in college and if the father provides little support (Lamanna & Riedmann, 2003).

collaborative divorce
a voluntary, contractually based alternative dispute resolution process for couples who want to negotiate a resolution of their situation rather than having a ruling imposed upon them by a court or arbitrator

Relationships With Young Children

When it involves children, divorce becomes a very complicated matter (Clarke-Stewart & Brentano, 2006). In most countries, mothers tend to obtain custody but often do not obtain sufficient financial resources to support the children. This puts an extreme financial burden on divorced mothers, whose standard of living is typically reduced.

Single mothers often face difficult challenges financially as well as the stress of raising their children.

In contrast, divorced fathers pay a psychological price. Although many would like to remain active in their children's lives, few actually do. Child-support laws in some states also may limit fathers' contact with their children (Wadlington, 2005).

One hopeful direction that addresses the usually difficult custody situations following divorce is the Collaborative Divorce Project, based on collaborative law (Mosten, 2009; Pruett, Insabella, & Gustafson, 2005). **Collaborative divorce** *is a voluntary, contractually based alternative dispute resolution process for couples who want to negotiate a resolution of their situation rather than having a ruling imposed upon them by a court or arbitrator* (DiFonzo, 2010; Mosten, 2009). Collaborative divorce is an intervention designed to assist the parents of children 6 years old or younger as they begin the separation/divorce process.

Early results from this approach are positive (DeLucia-Waack, 2010). In addition to positive evaluations from both parents, couples benefited in terms of less conflict, greater father involvement, and better outcomes for children than the control group. Attorneys and court records indicate that intervention families were more cooperative and were less likely to need custody evaluations and other costly services. The Collaborative Divorce Project is evidence that programs can be designed and implemented that benefit all members of the family.

Divorce and Relationships With Adult Children

We saw in Chapter 5 that young children can be seriously affected by their parents' divorce. But what happens when the parents of adult children divorce? Are adult children affected too? It certainly looks that way. Young adults whose parents divorce experience a great deal of emotional vulnerability and stress (Cooney & Uhlenberg, 1990). One young man put it this way:

> the difficult thing was that it was a time where, you know [you're] making the transition from high school to college . . . your high school friends are dispersed . . . they're all over the place . . . It's normally a very difficult transition [college], new atmosphere, new work load, meeting new people. You've got to start deciding what you want to do, you've got to sort of start getting more independent, and so forth. And then, at the same time you find out about a divorce. You know, it's just that much more adjustment you have to make. (Cooney et al., 1986)

THINK ABOUT IT

Given the serious impact of divorce, what changes in mate selection might lower the divorce rate?

The effects of experiencing the divorce of one's parents while growing up can be quite long-lasting. College-age students report poorer relations with their parents if their parents are divorced (Yu et al., 2010). Parental divorce also affects young adults' views on intimate relationships and marriage, often having negative effects on them (Ottaway, 2010). Wallerstein and Lewis (2004) report the findings from a 25-year follow-up study of individuals whose parents divorced when they were between 3 and 18 years old. Results show an unexpected gulf between growing up in intact versus divorced families as well as the difficulties that children of divorce encounter in achieving love, sexual intimacy, and commitment to marriage and parenthood. These themes are echoed in a study of college students whose parents divorced while they were in college (Bulduc, Caron, & Logue, 2007). These students said that experiencing their parents' divorce negatively affected their own intimate relationships and their relationships with their fathers but that it brought them closer to their mothers. Clearly, experiencing divorce at any age alters lives.

Remarriage

The trauma of divorce does not deter people from beginning new relationships, which often lead to another marriage. In the United States, about 25% of the adult population has been married more than once (Elliott & Lewis, 2010). Typically, men and women both wait about 3½ years before they remarry (U.S. Census Bureau, 2010b). However, remarriage rates vary somewhat across ethnic groups and education levels: European Americans are more likely to be married two or more times compared to other ethnic groups, as do people with lower educational levels (Elliott & Lewis, 2010). Military veterans are also more likely to be married more than once compared to non-veterans.

An interesting emerging trend reflects an important cohort difference. Compared to older generations, younger generations tend to remarry at a lower rate (Elliott & Lewis, 2010). This may be a reflection of greater social acceptability of cohabitation in the United States, as well as a reluctance in younger people to jump back into a marriage.

In contrast to first marriages, remarriage has few norms or guidelines for couples, especially in how to deal with stepchildren and extended families (Elliott & Lewis, 2010). The lack of clear role definitions may be a major reason why the divorce rate for remarriages is significantly higher (about 25%; even higher if stepchildren are involved) than for first marriages.

Although women are more likely to initiate a divorce, they are less likely to remarry (Birditt et al., 2010; Fincham & Beach, 2010;

Although remarriage is common, adjusting to it can be difficult.

© Masterfile

Kayser, 2010) unless they are poor (Elliott & Lewis, 2010). However, women in general tend to benefit more from remarriage than do men, particularly if they have children (Ozawa & Yoon, 2002). Although many people believe that divorced individuals should wait before remarrying to avoid the so-called "rebound effect," there is no evidence that those who remarry sooner have less success in remarriage than those who wait longer (Wolfinger, 2007).

Adapting to new relationships in remarriage is stressful. For example, partners may have unresolved issues from the previous marriage that may interfere with satisfaction with the new marriage (Faber, 2004). The effects of remarriage on children is positive, at least for young adult children who report a positive effect on their own intimate relationships as an effect of their parent(s) remarrying happily (Yu & Adler-Baeder, 2007).

Test Yourself

RECALL

1. Following divorce, most women suffer disproportionately in the _____ domain compared with most men.

2. On average, within 2 years after a divorce, _____ fathers remain central in their children's lives.

3. Even many years later, divorced _____ may not experience positive relationships with their adult children.

4. For African American couples, divorce rates for remarried couples are _____ than for first marriages.

INTERPRET

Despite greatly increased divorce rates over the past few decades, the rate of marriage has not changed very much. Why do you think this is?

APPLY

Ricardo and Maria are engaged to be married. Ricardo works long hours as the manager of a local coffee shop, while Maria works regular hours as an administrative assistant at a large communications company. Based on what you know about why couples get divorced, what factors may increase the likelihood that Ricardo and Maria's marriage will fail?

Recall answers: (1) financial, (2) few, (3) fathers, (4) lower

SUMMARY

11.1 RELATIONSHIPS

What types of friendships do adults have? How do adult friendships develop?

■ People tend to have more friendships during young adulthood than during any other period. Friendships are especially important for maintaining life satisfaction throughout adulthood.

■ Men tend to have fewer close friendships and to base them on shared activities, such as sports. Women tend to have more close friendships and to base them on intimate and emotional sharing. Gender differences in same-gender friendship patterns may explain the difficulties men and women have in forming cross-gender friendships.

What is love? How does it begin? How does it develop through adulthood?

■ Passion, intimacy, and commitment are the key components of love.

■ Although styles of love change with age, the priorities within relationships do not. Men tend to be more romantic earlier in relationships than women, who tend to be cautious pragmatists. As the length of the relationship increases, intimacy and passion decrease but commitment increases.

■ Selecting a mate works best when there are shared values, goals, and interests. There are cross-cultural differences with regard to the specific aspects of these that are considered most important.

What is the nature of abuse in some relationships?

■ Levels of aggressive behavior range from verbal aggression to physical aggression to actually killing one's partner. The causes of aggressive behaviors become more complex as the level of aggression increases. People remain in abusive relationships for many reasons, including low self-esteem and the belief that they cannot leave.

11.2 LIFESTYLES

Why do some people decide not to marry, and what are these people like?

■ Most adults decide by age 30 whether they plan on getting married. Never-married adults often develop a strong network of close friends. Dealing with other people's expectations that they should marry is often difficult for single people.

What are the characteristics of cohabiting people?

■ Young adults usually cohabit as a step toward marriage, and adults of all ages may also cohabit for financial reasons. Cohabitation is only rarely seen as an alternative to marriage. Overall, more similarities than differences exist between cohabiting and married couples.

What are gay and lesbian relationships like?

■ Gay and lesbian relationships are similar to heterosexual marriages in terms of relationship issues. Some countries and some states in the United States now permit same-sex marriages. Lesbian couples tend to be more egalitarian and are more likely to remain together than gay couples.

What is marriage like through the course of adulthood?

■ The most important factors in creating marriages that endure are a stable sense of identity as a foundation for intimacy, similarity of values and interests, effective communication, and the contribution of unique skills by each partner.

■ For couples with children, marital satisfaction tends to decline until the children leave home, although individual differences are apparent, especially in long-term marriages. Most long-term marriages are happy.

11.3 THE FAMILY LIFE CYCLE

What are the common forms of families?

■ Although the nuclear family is the most common form of family in Western societies, the most common form around the world is the extended family. Families experience a series of relatively predictable changes called the *family life cycle*. This cycle provides a framework for understanding the changes families go through as children mature.

Why do people have children?

■ Although having children is stressful and very expensive, most people do it anyway because of the many emotional rewards they bring. However, the number of child-free couples is increasing.

What is it like to be a parent? What differences are there in different types of parenting?

■ The timing of parenthood is important in how involved parents are in their families as opposed to their careers.

■ Single parents are faced with many problems, especially if they are women and are divorced. The main problem is significantly reduced financial resources.

■ A major issue for adoptive parents, foster parents, and stepparents is how strongly the child will bond with them. Each of these relationships has some special characteristics.

■ Gay and lesbian parents also face numerous obstacles, but they usually prove to be good parents.

11.4 DIVORCE AND REMARRIAGE

Who gets divorced? How does divorce affect parental relationships with children?

■ Currently, odds are about 50–50 that a new marriage will end in divorce. Conflict styles can predict who divorces. Recovery from divorce is different for men and women. Men tend to have a tougher time in the short run, but women clearly have a harder time in the long run, often for financial reasons.

■ Difficulties between divorced partners usually involve visitation and child support. Disruptions also occur in divorced parents' relationships with their children, whether the children are young or are adults themselves.

What are remarriages like? How are they similar to and different from first marriages?

■ Most divorced couples remarry. Second marriages are especially vulnerable to stress if spouses must adjust to having stepchildren. Remarriage in middle age and beyond tends to be happy.

KEY TERMS

assortative mating (387)

abusive relationship (391)

battered woman syndrome (392)

cohabitation (395)

marital success (398)

marital quality (398)

marital adjustment (398)

marital satisfaction (398)

homogamy (399)

exchange theory (399)

vulnerability–stress–adaptation model (400)

nuclear family (403)

extended family (403)

familism (405)

covenant marriage (412)

marriage education (412)

collaborative divorce (413)

Log in to **www.cengagebrain.com** to access the resources your instructor requires. For this book, you can access:

■ CourseMate brings course concepts to life with interactive learning, study, and exam preparation tools that support the printed textbook. A textbook-specific website, Psychology CourseMate includes an integrated interactive eBook and other interactive learning tools including quizzes, flashcards, videos, and more.

CENGAGENOW™

■ CengageNOW Personalized Study is a diagnostic study tool containing valuable text-specific resources—and because you focus on just what you don't know, you learn more in less time to get a better grade.

WebTUTOR™

■ More than just an interactive study guide, WebTutor is an anytime, anywhere customized learning solution with an eBook, keeping you connected to your textbook, instructor, and classmates.

Work and Leisure

Occupational and Lifestyle Issues in Young and Middle Adulthood

12

iStockphoto.com/track5

FROM THE TIME WE ARE SMALL CHILDREN, we think and plan what we will "be" when we grow up. Of course, what we will "be" means what occupation we will have. When we are "grown up," we simply change that question to ". . . and what do you do?" We are socialized, throughout the life span, that work is a central aspect of life. Remember those chores we did as children? Or how we learned to juggle school and part-time jobs? Or how so many people end up putting in 12-hour days at the office? We are taught that working is a natural part of life. For some, work *is* life; for all, it is at least a source of identity.

In this chapter, we explore the world of work first by considering how people choose occupations and develop in them. After that, we examine how women and minorities contend with barriers to their occupational selection and development. Dealing with occupational transitions is considered in the third section. How to balance work and family obligations is a difficult issue for many people; this is discussed in the fourth section. Finally, we will see how people spend their time away from work in leisure activities.

As in Chapter 11, our focus in this chapter is on issues faced by both young and middle-aged adults. No longer is it only young adults who must deal with occupational selection issues: It is increasingly common for middle-aged people to confront the issues of occupational selection all over again as their industry changes or their company downsizes. The "last hired, first fired" approach taken in many businesses can affect people of all ages who recently started jobs. Similarly, the other topics we consider apply to both younger and middle-aged adults.

LEARNING OBJECTIVES

■ How do people view work? How do occupational priorities vary with age?

■ How do people choose their occupations?

■ What factors influence occupational development?

■ What is the relationship between job satisfaction and age?

Monique, a 28-year-old senior communications major, wonders about careers. Should she enter the broadcast field as a behind-the-scenes producer, or would she be better suited as a public relations spokesperson? She thinks her outgoing personality is a factor that she should consider in making this decision.

CHOOSING ONE'S WORK IS SERIOUS BUSINESS. Like Monique, we try to select a field in which we are trained and that is also appealing. Work colors much of what we do in life. You may be taking this course as part of your preparation for work. Work is a source for friends. People arrange personal activities around work schedules. Parents often choose child-care centers on the basis of proximity to where they work.

In this section, we explore what work means to adults. We also revisit issues pertaining to occupational selection, first introduced in Chapter 9, and examine occupational development. Finally, we will see how satisfaction with one's job changes during adulthood.

Hassling with commuting makes us think about why we work.

Thinkstock/Getty Images

The Meaning of Work

Why do so many of us fight rush hour traffic and the commuting crowds of people to get to work? Studs Terkel, author of the fascinating, classic book *Working* (1974), writes that work is "a search for daily meaning as well as daily bread, for recognition as well as cash, for astonishment rather than torpor; in short, for a sort of life rather than a Monday through Friday sort of dying" (p. xiii). Kahlil Gibran (1923), in his mystical book *The Prophet*, put it this way: "Work is love made visible."

For some of us, work is a source of prestige, social recognition, and a sense of worth. For others, the excitement, the creativity, and the opportunity to give something of themselves make work meaningful. But for most, the main purpose of work is to earn a living. This is not to imply, of course, that money is the only reward in a job; friendships, the chance to exercise power, and feeling useful are also important. The meaning most of us derive from working includes both the money that can be exchanged for life's necessities (and perhaps a few luxuries too) and the possibility of personal growth (Rosso, Dekas, & Wrzesniewski, in press).

What specific occupation a person holds appears to have no effect on his or her need to derive meaning from work. Even when their occupation consists of highly repetitive work (Isaksen, 2000) or is in a declining industry experiencing a high number of layoffs (Dorton, 2001), people find great personal meaning in what they do. The specific meanings people derive from their work vary with the type of occupation and are influenced by socialization (Chetro-Szivos, 2001). Finding meaning in one's work can mean the difference between feeling that work is the source of one's life problems or a source of fulfillment and contentment (Grawitch, Barber, & Justice, 2010).

What meanings do people derive from their work? Lips-Wiersma (2003) sought answers to this question by interviewing people in depth about the meanings they derive from work and whether and how these meanings determine work behavior. Despite widely diverse backgrounds in the participants in her study, Lips-Wiersma found four common meanings: developing self, union with others, expressing self, and serving others. To the extent that these meanings can all be achieved, people experience the workplace as an area of personal fulfillment. This viewpoint also provides a framework for understanding occupational transitions as a means to find opportunities for better balance among the four main meanings.

Contemporary business theory also supports the idea that meaning matters. *Using a concept called* **meaning-mission fit**, *French (2007) showed that corporate executives with a better alignment between their personal intentions and their firm's mission cared more about their employees' happiness, job satisfaction, and emotional well-being.*

Given the various meanings that people derive from work, occupation is clearly a key element of a person's sense of identity and self-efficacy (Lang & Lee, 2005). This can be readily observed when adults introduce themselves socially. When asked to tell something about themselves, you've probably noticed that people usually provide information about what they do for a living. Occupation affects your life in a host of ways and often influences where you live, what friends you make, and even what clothes you wear. In short, the impact of work cuts across all aspects of life. Work, then, is a major social role and influence on adult life. Occupation is an important anchor that complements the other major role of adulthood—love relationships.

As we will see, occupation is part of human development. Young children, in their pretend play, are in the midst of the social preparation for work. Adults are always asking them, "What do you want to be when you grow up?" School curricula, especially in high school and college, are geared toward preparing people for particular occupations. Young adult college students as well as older returning students have formulated perspectives on the meanings they believe they will get from work. Hance (2000) organized these beliefs into three main categories: working to achieve social influence; working to achieve personal fulfillment; and working because of economic reality. These categories reflect fairly well the actual meanings that working adults report.

Because work plays such a key role in providing meaning for people, an important question is how people select an occupation. Let's turn our attention to two theories explaining how and why people choose the occupations they do.

meaning-mission fit
alignment between an executive's personal intentions and their firm's mission

Occupational Choice Revisited

In Chapter 9, we saw that early decisions about what people want to do in the world of work are related to their personalities. For adults, two primary theories have influenced research. First, Holland's (1997) personality-type theory proposes that people choose occupations to optimize the fit between their individual traits (such as personality, intelligence, skills, and abilities) and their occupational interests. *Second,* **social cognitive career theory** *(SCCT) proposes that career choice is a result of the application of Bandura's social cognitive theory, especially the concept of self-efficacy (see Chapter 1).*

Recall from ●Table 9.4 (page 331) that Holland categorizes occupations in two ways: by the interpersonal settings in which people must function and by their associated lifestyles. From this perspective, he identifies six personality types that combine these factors: investigative, social, realistic, artistic, conventional, and enterprising, which he believes are optimally related to occupations.

How does Holland's theory help us understand the continued development of occupational interests in adulthood? Monique's situation helps illustrate this point. Monique, the college senior in the vignette, found a good match between her outgoing nature and her major, communications. Indeed, college students of all ages tend to like best the courses and majors that fit well with their own personalities. Thus, early occupational choices in adolescence continue to be modified and fine-tuned in adulthood.

social cognitive career theory (SCCT)
proposes that career choice is a result of the application of Bandura's social cognitive theory, especially the concept of self-efficacy

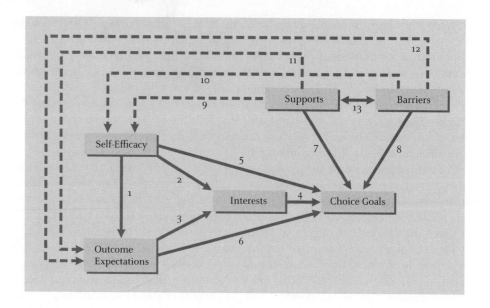

Social cognitive career theory complements this approach by emphasizing that how people choose careers is also influenced by what they think they can do and how well they can do it, as well as how motivated they are to pursue a career (Sheu et al., 2010). In short, it proposes that people's career choices are heavily influenced by their interests. As depicted in ■ Figure 12.1, SCCT has two versions. The simplest includes four main factors: Self-Efficacy (your belief in your ability), Outcome Expectations (what you think will happen in a specific situation), Interests (what you like), and Choice Goals (what you want to achieve). The more complex version also includes Supports (environmental things that help you) and Barriers (environmental things that block or frustrate you).

How well do these theories work in actual practice? Certainly, the relations among occupation, personality, and demographic variables are complex (Clark, 2007). First, if we consider the gender and ethnic/racial distributions of Holland's personality types, adult men and women are distributed differently although there are minimal differences across ethnic/racial groups (Fouad & Mohler, 2004). Regardless of age, women are more likely than men to have a social, artistic, or conventional personality type. In part, gender differences reflect different experiences in growing up (e.g., hearing that girls grow up to be nurses whereas boys grow up to be firefighters), differences in personality (e.g., gender-role identity), and differences in socialization (e.g., women are expected to be more outgoing and people-oriented than men). However, if we look within a specific occupational type, women and men are quite similar and their interests correspond closely to those that Holland describes (Betz, Harmon, & Borgen, 1996).

When these types are viewed through the lens of SCCT, meta-analyses (see Chapter 1) of several studies show support for the six-variable version of the model (Sheu et al., 2010). This means that although people may have underlying tendencies that relate to certain types of occupations, unless they believe they could be successful in those occupations and careers they are unlikely to choose them. These beliefs can be influenced by external factors. For example, occupational prestige and gender-related factors need to be taken into account (Deng, Armstrong, & Rounds, 2007).

Occupational selection is a complex developmental process involving interactions among personal beliefs, ethnic, gender, and economic factors. As research continues to document how these factors interrelate, we will continue to understand better how people choose what they would like to do for a living. Of course, life does not always play out according to what we would like to do. We will see how this happens as we consider occupational development and occupational changes in this chapter.

THINK ABOUT IT

How does one's level of cognitive development relate to one's choice of occupation?

Occupational Development

For most of us, getting a job is not enough; we would also like to move up the ladder. Promotion is a measure of how well one is doing in one's career. How quickly occupational advancement does (or does not) occur may lead to such labels as "fast-tracker" or "dead-ender." Barack Obama, who was elected president of the United States at age 47, is an example of a fast-tracker. People who want to advance learn quickly how long they should stay at one level and how to seize opportunities when they arise, but others may experience the frustration of remaining in the same job with no chance for promotion.

How a person advances in a career seems to depend on professional socialization, which is the socialization that occurs when people learn the unwritten rules of an organization. These rules include several factors *other* than those that are important in choosing an occupation, including expectations, support from co-workers, priorities, and job satisfaction. Before we consider these particular aspects, let's look at a general scheme of occupational development.

Super's Theory

Over four decades, Donald Super (1957, 1980; Super, Savickas, & Super, 1996) developed a theory of occupational development based on self-concept, first introduced in Chapter 9. He proposed a progression through five distinct stages during adulthood as a result of changes in individuals' self-concept and adaptation to an occupational role: implementation, establishment, maintenance, deceleration, and retirement (see ■ Figure 12.2). *People are located along a continuum of* **vocational maturity** *through their working years; the more congruent their occupational behaviors are with what is expected of them at different ages, the more vocationally mature they are.*

Barack Obama is an example of a "fast-tracker."

vocational maturity
degree of congruence between people's occupational behavior and what is expected of them at different ages

Super proposed five developmental tasks, the first two of which (crystallization and specification) occur primarily in adolescence. The remaining three (implementation, stabilization, and consolidation) occur over adulthood. Each of the tasks in adulthood has distinctive characteristics, as follows.

- The *implementation* task begins in the early 20s, when people take a series of temporary jobs to learn firsthand about work roles and to try out some possible career choices. Summer internships that many students use to gain experience are one example.

- The *stabilization* task begins in the mid-20s with selecting a specific occupation during young adulthood. It continues until the mid-30s as the person confirms the occupational choice that was made.

- The *consolidation* task begins in the mid-30s and continues throughout the rest of the person's working life as people advance up the career ladder. Taking a position in a law firm and working one's way up to partner or beginning as a sales clerk in a store in a mall and moving up to store manager are two examples.

These adult tasks overlap a sequence of developmental stages, beginning at birth, that continues during adulthood: exploratory (age 15–24), establishment (age 24–44), maintenance (age 45–64), and decline (age 65 and beyond). These stages reflect the overall occupational cycle from choosing what one wants to do through achieving the maximum possible in a career to the reduction in work in late adulthood.

■ **Figure 12.2**
Super's occupational stages during adulthood.

Implementation → Establishment → Maintenance → Deceleration → Retirement

Copyright © Cengage Learning 2010

Super's theory can be placed into context with Holland's theory and social cognitive career theory to form a more comprehensive view of occupational selection and career development (Walker, 2010). This more comprehensive view emphasizes that people's occupations evolve in response to changes in their self-concept and self-efficacy (Walker, 2010) and reflect an individual's personal view of occupation and career related to each task and stage (Hurley-Hanson, 2006). So, for example, a high-level executive could, either voluntarily or as the result of losing her job, decide to take a lower-ranking position if her sense of self changed from a high need to be in charge to one with a lower need for that responsibility. Consequently, how a person's occupation and career unfolds is part of a developmental process that reflects and explains important life changes.

Super's theory applies to people who enter and stay in a particular career their entire adult lives and also to those who change occupations. Because it is now typical for Americans to have a series of careers, Super's notion is that we cycle and recycle through the tasks and stages as we adapt to changes in ourselves and the workplace (Super et al., 1996).

Super may be right. A longitudinal study of 7,649 individuals born in the United Kingdom showed that occupational aspirations at age 16 in science and health fields predicted actual occupational attainments in science, health professions, or engineering at age 33 (Schoon, 2001). Adult occupational attainment was also related to belief in one's own ability, mathematical test performance, several personality characteristics, sociocultural background, and gender. These results point to the importance of viewing occupational development as a true developmental process, as Super claimed, as well as the importance of personal characteristics, as proposed in occupational selection theories.

Occupational Expectations

As we saw in Chapter 9, adolescents form opinions about what work in a particular occupation will be like based on what they learn in school and from their parents, peers, other adults, and the media. These expectations influence what they want to become and when they hope to get there.

As they enter and traverse adulthood, people continue to refine and update their occupational expectations. This usually involves trying to achieve their occupational goal, monitoring progress toward it, and changing or even abandoning it as necessary. Modifying the goal happens for many reasons, such as realizing that interests have changed, the occupation was not a good fit for them, they never got the chance to pursue the level of education necessary to achieve the goal, or because they lack certain essential skills and cannot acquire them. (The lack of outstanding athletic skill was a major reason why neither of your textbook authors could have achieved a goal of playing professional sports.) Still other people modify their goals because of age, race, or sex discrimination, a point we will consider later in this chapter.

Some goal modification is expected and common; think about how you or other people you know have changed their major in college. Still, it often surprises us to realize that we could have been wrong about what seemed to be a logical choice in the past. As Marie, a 38-year-old advertising manager, put it, "I really thought I wanted to be an airline pilot; the travel sounded highly interesting. But it just wasn't what I expected."

Research continues to support these personal experiences as typical of career development (Fouad, 2007). People do change their goals based on experience. Research shows, though, that people who know they have both the talent and the opportunity to achieve their occupational and career goals often do attain those goals. When high school students who were identified as academically talented were asked about their career expectations and outcomes, it turned out that 10 and even 20 years later they had been surprisingly accurate (Perrone et al., 2010).

Understanding the connection between adolescents' expectations and young adults' actual experiences helps explain the transition from school to work. What is

THINK ABOUT IT

What biological, psychological, sociocultural, and life-cycle forces influence the progression of one's career?

clear from research since the 1970s is that the biggest change has been in women's occupational and career expectations (Jacob & Wilder, 2010). In general, this research shows that young adults modify their expectations at least once, usually on the basis of new information, especially about their academic ability. The connection between adolescent expectations and adult reality reinforces the developmental aspects of occupations and careers.

Many writers believe that occupational expectations vary not only as a function of age but also of generation. Nowhere has this belief been stronger than in the supposed differences between the baby boom generation (born between 1946 and 1964) and the current millennial generation (born since 1983). As many commentators have noted, what people in these generations, on average, expect in occupations appears to be very different (Hershatter & Epstein, 2010). Millennials are more likely to change jobs more often than the older generations did, and are likely to view traditional organizations with more distrust and cynicism. But contrary to most stereotypes, millennials are no more egotistical, and are just as happy and satisfied as were young adults in every generation since the 1970s (Trzesniewski & Donnellan, 2010). So in many ways, as tech-savvy as they may be, the current generation of young adults is, psychologically speaking, quite similar to their parents. From an occupational development perspective, then, we would expect the same forces to shape their occupational expectations and ultimately their occupational development.

The importance of occupational expectations can be seen clearly in the transition from school to the workplace (Moen & Roehling, 2005). The 21st-century workplace is not one in which hard work and long hours necessarily lead to a stable career. *It can also be a place in which you experience* **reality shock**, *a situation in which what you learn in the classroom does not always transfer directly into the "real world" and does not represent all that you need to know.* When reality shock sets in, things never seem to happen the way we expect. Reality shock befalls everyone; for example, you can imagine how a new teacher feels when her long hours preparing a lesson result in students who act bored and unappreciative of her efforts.

Many professions, such as nursing and teaching, have gone to great lengths to alleviate reality shock (Alhija & Fresko, 2010; Hinton & Chirgwin, 2010). This global problem is one that is best addressed through internship and practicum experiences for students under the careful guidance of experienced people in the field. These experienced people serve critical roles as mentors and coaches, a topic we turn to next.

Reality shock typically hits younger workers soon after they begin an occupation.

reality shock
situation in which what you learn in the classroom does not always transfer directly into the "real world" and does not represent all that you need to know

The Role of Mentors and Coaches

Imagine how hard it would be to figure out everything you needed to know in a new job with no support from the people around you. Talk about reality shock! Entering an occupation involves more than the relatively short formal training a person receives. Indeed, much of the most critical information is not taught in training seminars. Instead, most people are shown the ropes by co-workers. In many cases, an older, more experienced person makes a specific effort to do this, taking on the role of a *mentor* or *coach*. Although mentors and coaches are not the only source of guidance in the workplace, they have been studied fairly closely.

A **mentor** *or* **developmental coach** *is part teacher, part sponsor, part model, and part counselor who facilitates on-the-job learning to help the new hire do the work required in his or her present role and to prepare for future roles* (Hunt & Weintraub, 2006). The mentor helps a young worker avoid trouble ("Be careful what you say around Harry") and also provides invaluable information about the unwritten rules that govern day-to-day activities in the workplace (not working too fast on the assembly line, wearing the right clothes, and so on), with mentors being sensitive to the employment situation (such as guarding against playing favorites in a union environment or if the protégé is an immediate subordinate or of the opposite gender) (Smith,

mentor or developmental coach
person who is part teacher, part sponsor, part model, and part counselor who facilitates on-the-job learning to help the new hire do the work required in his or her present role and to prepare for future roles

Howard, & Harrington, 2005). As part of the relationship, a mentor makes sure that the protégé is noticed and receives credit from supervisors for good work. Thus, occupational success often depends on the quality of the mentor–protégé relationship and the protégé's perceptions of its importance (Eddleston, Baldridge, & Veiga, 2004). In times of economic downturns, mentors can also provide invaluable advice on finding another job (Froman, 2010).

Kram (1985) theorized that the mentor relationship develops through four phases: initiation (mentors and protégés begin the relationship), cultivation (mentors work with protégés), separation (protégés and mentors spend less time together), and redefinition (the mentor–protégé relationship either ends or is transformed into a different type of relationship). Research supports this developmental process as well as the benefits of having a mentor (Day & Allen, 2004). Having a mentor and a high level of career motivation are especially important to career development (Day & Allen, 2004). Clearly, protégés gain tangible benefits from having a mentor, as described in Zachary and Fischler's (2009) practical guide to mentoring from the protégé's perspective (a companion to Zachary's [2000] guide to mentoring).

What do mentors get from the relationship? In Chapter 1, we saw that the ideas in Erikson's theory (1982) included important aspects of adulthood related to work. Helping a younger employee learn the job is one way to fulfill aspects of Erikson's phase of generativity. As we will see in more detail in Chapter 13, generativity reflects middle-aged adults' need to ensure the continuity of society through activities such as socialization or having children. In work settings, generativity is most often expressed through mentoring. In particular, mentors ensure that there is some continuity in the corporation or profession by passing on the knowledge and experience they have gained over the years. Additionally, leaders may need to serve as mentors to activate transformational leadership (leadership that changes the direction of an organization) and promote positive work attitudes and career expectations of followers, enabling the mentor to rise to a higher level in his or her own career (Scandura & Williams, 2004).

Women and minorities have an especially important need for mentors (Pratt, 2010). When paired with mentors, women benefit by having higher expectations; mentored women also have better perceived career development (Enslin, 2007). For example, Latina nurses in the U.S. Army benefitted from mentors in terms of staying in the military and getting better assignments (Aponte, 2007). Female lawyers with mentors earn more, are promoted more often, are treated more fairly, and are integrated better in the firm than women without mentors (Wallace, 2001). It is also critical to adopt a culturally conscious model of mentoring in order to enhance the advantages of mentoring for minority mentees (Campinha-Bacote, 2010). Culturally conscious mentoring involves understanding how an organization's culture, for example, affects employees and building those assumptions and behaviors into the mentoring situation. It can also involve addressing the cultural background of an employee and incorporating that into the mentoring relationship.

The demonstrable benefits to new employees from having mentors has led some business organizations and colleges and universities to strongly encourage senior professionals to serve as mentors (Enslin, 2007; McLaughlin, 2010). These advantages are especially important for women and minorities, so enlisting women and minorities to serve as mentors is especially important.

Despite the evidence that having a mentor can have many positive effects on one's occupational development, there is an important caveat. Having a poor mentor is worse than having no mentor at all (Ragins, Cotton, & Miller, 2000). A mentoring program is only as good as the mentor. Consequently, prospective protégés must be carefully matched with a mentor, and mentorship programs need to select motivated and skilled individuals who are provided with extensive training. How can prospective mentors and protégés meet

Women employees typically prefer and may achieve more from a female mentor.

Bloom Productions / Getty Images

more effectively? Some organizations have taken a page from dating and created speed mentoring as a way to help create better matches (Berk, 2010; Cook, Bahn, & Menaker, 2010). However it's done, care must be taken to ensure that the match between mentor and protégé is the best one possible for both.

Job Satisfaction

What does it mean to be satisfied with one's job or occupation? **Job satisfaction** *is the positive feeling that results from an appraisal of one's work.* Research indicates that job satisfaction is a multifaceted concept but that certain characteristics—including hope, resilience, optimism, and self-efficacy—predict both job performance and job satisfaction (Luthans et al., 2007).

Satisfaction with some aspects of one's job tends to increase gradually with age, and successful aging includes a workplace component (Robson et al., 2006). Why is this the case?

The factors that predict job satisfaction differ somewhat across cultures (Klassen, Usher, & Bong, 2010). This is explored in more detail in the Spotlight on Research feature. To the extent that this is the case, developmental age differences in job satisfaction may simply reflect the fact that what makes people happy in their jobs varies depending on the cultural values the person holds.

job satisfaction

the positive feeling that results from an appraisal of one's work

Spotlight on research | Cross-Cultural Aspects of Teachers' Job Satisfaction

Who were the investigators and what was the aim of the study? Robert Klassen, Ellen Usher, and Mimi Bong wondered about the similarities and differences in teachers' job satisfaction, self-efficacy, and job stress. To find out, they studied teachers in the United States, Korea, and Canada. Their main question was whether teachers' cultural values, self-efficacy, and job stress would predict job satisfaction across the three countries.

How did the investigators measure the topic of interest? The researchers measured self-efficacy with the Collective Teacher Efficacy Belief Scale, which assesses teachers' individual perceptions about their school's collective capabilities to influence student achievement; the scale is based on teachers' analysis of the teaching staff's capabilities to effectively teach all students. Job satisfaction was measured through four questions: (1) "I am satisfied with my job," (2) "I am happy with the way my colleagues and superiors treat me," (3) "I am satisfied with what I achieve at work," and (4) "I feel good at work." Job stress was measured using a single item ("I find teaching to be very stressful"). Collectivism, a cultural value, was measured with a 6-item scale in which the first part of the question was, "In your opinion, how important is it that you and your family . . . ," with the conclusion of the items including the following: (1) "take responsibility for caring for older family members?" (2) "turn to each other in times of trouble?" (3) "raise each other's children whenever there is a need?" (4) "do everything you can to help each other move ahead in life?" (5) "take responsibility for caring

for older family members?" and (6) "call, write, or see each other often?". The Korean version of the scales was created using a translation–back-translation process to ensure that the meaning of the items was preserved.

Who were the participants in the study? A total of 500 elementary and middle school teachers from the United States ($n = 137$), Canada ($n = 210$), and Korea ($n = 153$) participated. The sample from the United States was included to connect this study to other research on teachers' job satisfaction. Canadian teachers were included to determine the degree to which findings from the United States could be generalized to a country holding similar (but not identical) cultural values. The Korean teachers represented a group with a different geographic and demographic profile (East Asian, Confucian, collectivist). Careful analyses showed no significant differences in age, teaching experience, job satisfaction, collective efficacy, job stress, or cultural values across the three countries.

What was the design of the study? The study used a cross-sectional design.

Were there ethical concerns with the study? Because the study involved voluntary completion of a survey, there were no ethical concerns.

What were the results? The analyses revealed that North American teachers scored higher on all the variables than the Korean teachers. However, there were no differences across countries regarding the efficacy of the teachers and either the strength or direction of its relation to job satisfaction. In contrast, analyses also revealed that

job stress had a bigger impact for North American teachers whereas the cultural value of collectivism was more important for Korean teachers.

What have the investigators concluded? The most important finding from the study is the similarity across countries in the connection between the efficacy teachers believe they have and their job satisfaction—the less efficacy, the lower a teacher's satisfaction is likely to be. Second, the higher importance of the cultural value of collectivism for Korean teachers probably reflects a cultural norm of avoiding conflict and working for the betterment of the group. Finally, the finding that job stress was a negative predictor of job satisfaction for North American teachers (the higher the stress, the lower the satisfaction) but a positive predictor for Korean teachers (higher job stress predicted higher satisfaction) indicates that job stress may have different components as a function of culture. For Korean teachers, feeling stressed by the presence of more competent teachers may create an urge to improve, rather than a feeling of defeat. In sum, some predictors of job satisfaction transcend countries; others do not.

What converging evidence would strengthen these conclusions? Klassen and colleagues' study examined only a few countries and cultures, so their study needs to be repeated with others. Also, additional types of teachers (high school, college/university) need to be included.

 Go to Psychology CourseMate at www.cengagebrain.com to enhance your understanding of this research.

So how does job satisfaction evolve over young and middle adulthood? As you probably suspect, it's complicated. You may be pleased to learn that research shows that, given sufficient time, most people eventually find a job with which they are reasonably happy (Hom & Kinicki, 2001). Optimistically, this indicates that there is a job out there, somewhere, in which you will be happy. That's good, because research grounded in positive psychology theory indicates that happiness fuels success (Achor, 2010). Consider also that middle-aged workers have had more time to find a job that they like or may have resigned themselves to the fact that things are unlikely to improve, resulting in a better congruence between worker desires and job attributes (Glickman, 2001).

It's also true that job satisfaction does not increase in all areas and job types with age. For example, middle-aged workers are more satisfied with the intrinsic personal aspects of their jobs, such as perceived control and self-efficacy, than they are with the extrinsic aspects, such as pay (Glickman, 2001; Mirabella, 2001). White-collar professionals show an increase in job satisfaction with age, whereas those in blue-collar positions generally do not, and these findings hold with both men and women (Aasland, Rosta, & Nylenna, 2010). This is also true across cultures. A study of Filipino and Taiwanese workers in the long-term health care industry in Taiwan showed that workers with 4 or 5 years' experience had lower job satisfaction than workers with less experience, but job satisfaction among older physicians in Norway increases over time (Aasland et al., 2010; Tu, 2007).

As you might suspect, the type of job one has and the kinds of family responsibilities one has at different career stages—as well as the flexibility of work options such as telecommuting and family leave benefits to accommodate those responsibilities—influence the relationship between age and job satisfaction (Marsh & Musson, 2008). This suggests that the accumulation of experience, changing context, and the stage of one's career development may contribute to the increase in job satisfaction over time.

Finally, job satisfaction may be cyclical. With the current pattern of multiple occupations and careers, job satisfaction may wax and wane. The idea is that job satisfaction increases over time because people change jobs or responsibilities on a regular basis, thereby keeping their occupation interesting and challenging (Moen & Roehling, 2005).

Alienation and Burnout

As you've most likely heard or experienced, all jobs create a certain level of stress. There are deadlines, times during which you are extremely busy, performance standards that have little room for deviation. For most workers, such negatives are merely annoyances. But for others, extremely stressful situations on the job may result in alienation and burnout.

When workers feel that what they are doing is meaningless and that their efforts are devalued, or when they do not see the connection between what they do and the final product, a sense of **alienation** *is likely to result.* The alienated workers interviewed in Studs Terkel's (1974) classic study all expressed the feeling that they were nameless, faceless, unappreciated cogs in a large machine. He reported that employees are most likely to feel alienated when they perform routine, repetitive actions such as those on an assembly line, jobs that are most likely to be replaced by technology. But other workers can become alienated, too. Given the large number of companies that have collectively downsized by millions of workers over the past few decades, even white-collar managers and executives do not have the same level of job security that they once had. The financial crisis that began in 2008 and resulted in record levels of job loss is only the most recent example of employees feeling abandoned by their employers.

As employees have experienced and employers have been reminded over the past several years, it is essential for companies to provide positive work environments to ensure that the workforce remains stable and committed (Griffin et al., 2010). Em-

alienation
feeling that results in workers when their work seems meaningless and their efforts devalued or when they see no connection between their own work and the final product

ployee alienation happens most often when workers become cynical and dissatisfied (R. Abraham, 2000). How can employers avoid alienating workers and improve organizational commitment? Research indicates that trust is key (Chen, Aryee, & Lee, 2005). Also key is a perception among employees that the employer deals with people fairly and impartially (Howard & Cordes, 2010). It is also helpful to involve employees in the decision-making process, create flexible work schedules, and institute employee development and enhancement programs. Indeed, many organizations have instituted new practices such as total quality management and related programs partly as a way of addressing worker alienation. These approaches make a concerted effort to involve employees in the operation and administration of their plant or office. Contributing positively to the work environment reduces alienation and improves employee commitment and satisfaction (Freund, 2005).

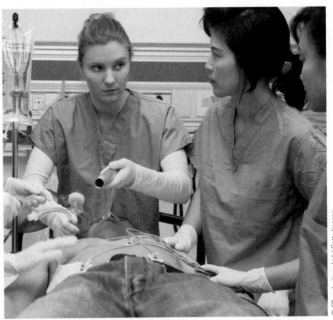

High-stress jobs such as intensive care nursing often result in burnout.

Sometimes the pace and pressure of the occupation becomes more than a person can bear, resulting in **burnout**, *a depletion of a person's energy and motivation, the loss of occupational idealism, and the feeling that one is being exploited.* Burnout is a state of physical, emotional, and mental exhaustion as a result of job stress (Malach-Pines, 2005). Burnout is most common among people in the helping professions—such as teaching, social work, health care (Bozikas et al., 2000), and occupational therapy (Bird, 2001)—and for those in the military (Harrington et al., 2001). For example, nurses have high levels of burnout from stress and are thus more likely to verbally abuse other nurses (Rowe & Sherlock, 2005).

People in these professions must constantly deal with other people's complex problems, usually under difficult time constraints. Dealing with these pressures every day, along with bureaucratic paperwork, may become too much for the worker to bear. Ideals are abandoned, frustration builds, and disillusionment and exhaustion set in. In short, the worker is burned out. And burnout can negatively affect the people who are supposed to receive services from the burned-out employee (Rowe & Sherlock, 2005).

But we know that burnout does not affect everyone is a particular profession. For example, some nurses feel burned out whereas others do not. Why? Vallerand (2008) proposes that the difference relates to people feeling different types of passion (obsessive and harmonious) toward their jobs. *A* **passion** *is a strong inclination toward an activity that individuals like (or even love), that they value (and thus find important), and in which they invest time and energy* (Vallerand et al., 2010). Vallerand's (2008) Passion Model proposes that people develop a passion toward enjoyable activities that are incorporated into identity.

Vallerand's model differentiates between two kinds of passion: obsessive and harmonious. Obsessive passion happens when people experience an uncontrollable urge to engage in the activity, which actually interferes with positive feelings and satisfaction, perhaps even creating negative feelings when they do it. A critical aspect of obsessive passion is that the internal urge to engage in the passionate activity makes it very difficult for the person to fully disengage from thoughts about the activity, leading to conflict with other activities in the person's life (Vallerand et al., 2010).

In contrast, harmonious passion results when individuals have freely accepted the activity as important for them without any contingencies attached to it. Individuals do not feel compelled to engage in the enjoyable activity; rather, they freely choose to do so. With this type of passion, the activity occupies a significant but not overpowering space in the person's identity and is in harmony with other aspects of the person's life (Vallerand et al., 2010).

Research in France and Canada indicate that the Passion Model accurately predicts employees' feelings of burnout (Vallerand, 2008; Vallerand et al., 2010). As

burnout
a depletion of a person's energy and motivation, the loss of occupational idealism, and the feeling that one is being exploited

passion
strong inclination toward an activity that individuals like (or even love), that they value (and thus find important), and in which they invest time and energy

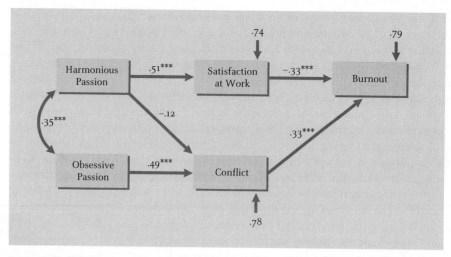

Source: Vallerand, R. J., Paquet, Y., Philippe, F. L., & Charest, J. (2010)., On the role of passion for work in burnout: A process model. *Journal of Personality, 78,* 289-312. Figure 1(p. 300).

shown in ■ Figure 12.3, obsessive passion predicts higher levels of conflict, which in turn predicts higher levels of burnout. In contrast, harmonious passion predicts higher levels of satisfaction at work, which in turn predicts lower levels of burnout.

The best ways to lower burnout are intervention programs that focus on both the organization and the employee (Awa, Plaumann, & Walter, 2010). At the organizational level, job restructuring and employee-provided programs are important. Research on what employees themselves can do shows that practicing stress-reduction techniques, lowering other people's expectations, cognitive restructuring of the work situation, and finding alternative ways to enhance personal growth and identity are most effective (van Dierendonck, Garssen, & Visser, 2005).

In short, alienation and burnout can be reduced by making workers feel that they are important to the organization, involving them in decisions, keeping expectations realistic, ensuring good communication, and promoting teamwork. As organizations adopt different management styles, perhaps these goals can be achieved.

Test Yourself

RECALL

1. For most people, the main reason to work is _____.

2. Holland's theory deals with the relationship between occupation and _____.

3. Super believes that through their working years, people are located along a continuum of _____

4. The role of a mentor is part teacher, part sponsor, part model, and part _____.

5. Recent research has shown that job satisfaction does not increase consistently as a person ages; rather, satisfaction may be _____.

6. Two salient aspects of job dissatisfaction are alienation and _____.

INTERPRET

What is the relation between Holland's theory, occupational development, and job satisfaction? Would these relations be different in the case of a person with a good match between personality and occupation versus one with a poor match?

APPLY

If you were the director of the campus career services office, what would you do to provide students with realistic and accurate information about potential careers?

Recall answers: (1) to earn a living, (2) personality, (3) vocational maturity, (4) counselor, (5) cyclical, (6) burnout

LEARNING OBJECTIVES

- How do women's and men's occupational expectations differ? How are people viewed when they enter occupations that are not traditional for their gender?

- What factors are related to women's occupational development?

- What factors affect ethnic minority workers' occupational experiences and occupational development?

- What types of bias and discrimination hinder the occupational development of women and ethnic minority workers?

Janice, a 35-year-old African American manager at a business consulting firm, is concerned because her career is not progressing as rapidly as she had hoped. Janice works hard and has received excellent performance ratings every year. But she has noticed that there are very few women in upper management positions in her company. Janice wonders whether she will ever be promoted.

OCCUPATIONAL CHOICE AND DEVELOPMENT ARE NOT EQUALLY AVAILABLE TO ALL, as Janice is experiencing. Gender, ethnicity, and age may create barriers to achieving one's occupational goals. Men and women in similar occupations may nonetheless have different life experiences and probably received different socialization as children and adolescents that made it easier or harder for them to set their sights on a career. Bias and discrimination also create barriers to occupational success. In this section, we'll get a better appreciation of the personal and structural barriers that exist for many people.

Gender Differences in Occupational Selection

About 60% of all women over age 16 in the United States are working, and they represent roughly 47% of the total workforce (Bureau of Labor Statistics, 2010a). Across ethnic groups, African American women participate the most (about 62%) and Latina women the least (about 56%). However, major structural barriers to women's occupational selection remain (Maume, 2004; Probert, 2005). Let's take a look at the situation in both traditional and nontraditional occupations for women.

Structural Barriers for Women: Traditional and Nontraditional Occupations

In the past, women employed outside the home tended to enter traditional, female-dominated occupations such as secretarial, teaching, and social work jobs. This was due mainly to their socialization into these occupational tracks. However, as more women enter the workforce and as new opportunities are opened for women, a growing number work in occupations that have been traditionally male-dominated, such as construction and engineering. The U.S. Department of Labor (2010a) categorizes women's nontraditional occupations as those in which women constitute 25% or less of the total number of people employed; the skilled trades (electricians, plumbers, carpenters) still have among the lowest participation rates of women.

Why some women end up in nontraditional occupations appears to be related to personal feelings and experiences as well as to self-efficacy and expectations about the occupation (Aguayo, 2005). Despite the efforts to counteract gender stereotyping of occupations, women who choose nontraditional occupations and

Although more women are entering nontraditional occupations, they still have to contend with negative stereotypes.

© Jeff Greenberg / Alamy

THINK ABOUT IT

What changes in children's school and other socialization experiences will enable girls to acquire different occupational skills?

are successful in them tend to be viewed negatively; they are described as being cold and as having strongly undesirable interpersonal characteristics (e.g., in one study, they were described as bitter, quarrelsome, selfish, deceitful, and devious) as compared with similarly successful male managers (Heilman et al., 2004).

In a study conducted in India, both women and men gave higher "respectability" ratings to males than to females in the same occupation (Kanekar, Kolsawalla, & Nazareth, 1989). In the United States, research shows that men still prefer to date women who are in traditional careers (Kapoor et al., 2010). Worst of all, people are less likely to perceive incidents of sexual coercion as harassing when a woman is in a nontraditional occupation (Burgess & Borgida, 1997). Additionally, compared to women who work in traditional occupations, women who work in nontraditional occupations are less likely to believe that they are being sexually harassed when confronted with the same behavior (Bouldin & Grayson, 2010; Maeder, Wiener, & Winter, 2007).

Taken together, these studies show that we still have a long way to go before people can choose any occupation they want without having to contend with gender-related stereotypes. Although differences in opportunities for women in traditional and nontraditional occupations are narrowing, key differences remain.

Women and Occupational Development

If you were to guess what a young woman who has just graduated from college will be doing occupationally 10 years from now, what would you say? Would you guess that she will be strongly committed to her occupation? Will she have abandoned it for other things?

Women who graduate from college now have more opportunities in the workplace than their grandmothers did.

Occupational development for women has undergone major changes over the past several decades. The characteristics and aspirations of women who entered the workforce in the 1950s and those from the Baby Boomers (born between 1946 and 1964), Generation X (born between 1965 and 1982), and the Millennials (born since 1983) are significantly different (Howe & Strauss, 1992, Piscione, 2004; Strauss & Howe, 2007). Women in previous generations had fewer opportunities for employment choice and had to overcome more barriers.

These differences play out in women's occupational development. In the 21st century, women entrepreneurs are starting small businesses at a faster rate than men and are finding that a home-based business can solve many of the challenges they face in balancing employment and a home life. For those seeking work outside the home, Gen-X women negotiate beyond the first offer of a job, salary, and benefits package to make deals with prospective employers on a work environment that is best for their career interests and the needs of their family. In this scenario, the employer wins as well; happy employees can provide their employer with cost savings, increased retention, reduced absenteeism, and greater productivity. These beneficial opportunities for working women include flexible work options, increased personal and vacation time, child-care assistance, and benefits that fit individual needs.

Still, some women who work in traditional work environments cite challenges that include pressure to work longer hours, increased commute time, rising child-care costs, and limited health care options. The working women in Piscione's (2004) study also report increased financial and emotional stress during their children's summer breaks or during the after-school hours when their children are home alone. Equally difficult to contend with is the perception in some workplaces of a working mother not being a team player or not being able to do the "tough" work because she may be pulled away by child-care needs at a critical time.

As the millennial generation heads into the workforce, it will be interesting to see whether their high degree of technological sophistication will provide still more occupational and career options. Technologically mediated workplaces may provide solutions to many traditional issues, such as work–family conflict. But traditional gender roles have a powerful influence; a survey of millennial generation college women showed that, even when they explored nontraditional occupational options, they tended to commit more often to traditional occupations (Garbrecht, 2005). When millennial generation women choose nontraditional occupations, their attitudes toward them are more similar to than different from previous generations' attitudes (Real, Mitnick, & Maloney, 2010).

In the corporate world, unsupportive or insensitive work environments, organizational politics, and the lack of occupational development opportunities are most important for women working full-time (Silverstein, 2001; Yamini-Benjamin, 2007). Female professionals leave their jobs for two main reasons. First, the organizations in which women work are felt to hold contrary or nonsupportive values. Organizations that women perceive to be unsupportive are those that idealize and reward masculine values of working—individuality, self-sufficiency, and individual contributions—while emphasizing tangible outputs, competitiveness, and rationality. Most women prefer organizations that more highly value relationships, interdependence, and collaboration.

Second, women may feel disconnected from the workplace—disconnected from their colleagues, clients, and co-workers, deriving less meaning from work and feeling alienated from themselves. By midcareer, women may conclude that they must leave these unsupportive organizations in order to achieve satisfaction, growth, and development at work and to be rewarded for the relational skills they consider essential for success. Clearly, women are focusing on issues that create barriers to their occupational development and personal satisfaction and are looking for ways around these barriers.

Such barriers are a major reason why women's workforce participation is discontinuous. Because they cannot find affordable and dependable child care (or because they choose to take on this responsibility), many women stay home while their children are young. Discontinuous participation makes it difficult to maintain an upward trajectory in one's career through promotion and in terms of maintaining skills. Some women make this choice willingly; however, many find themselves forced into it.

Ethnicity and Occupational Development

What factors are related to occupational selection and development for people from ethnic minorities? Unfortunately, little research has been conducted from a developmental perspective. Rather, most researchers have focused on the limited opportunities that ethnic minorities have and on the structural barriers, such as discrimination, that they face. Most of the developmental research to date focuses on occupational selection issues and variables that foster occupational development. Three topics have received the most focus: nontraditional occupations, vocational identity, and issues pertaining to occupational aspirations.

Women do not differ significantly in terms of participation in nontraditional occupations across ethnic groups (Bureau of Labor Statistics, 2010a). However, African American women who choose nontraditional occupations tend to plan for more formal education than necessary to achieve their goal. This may actually make them overqualified for the jobs they get; for example, a woman with a college degree may be working in a job that does not require that level of education. In the area of leadership aspiration—the degree to which one aspires to a leadership position in one's company—African American workers in general report lower levels than do European American workers (Clarke-Anderson, 2005).

Latino Americans differ from European Americans in some ways and are similar to them in others. For Mexican American

Ethnic-minority workers face more significant barriers to career development.

© Jeff Greenberg / Alamy

college students, perceived career barriers and adherence to career myths predicted their tendency to unnecessarily limit career choices (Leal-Muniz & Constantine, 2005). However, Latino Americans are similar to European Americans in occupational development and work values.

Research on occupational development of ethnic minority workers is clear on one point: Whether an organization is responsive to the needs of ethnic minorities makes a big difference for employees. Ethnic minority employees of a diverse organization in the Netherlands reported more positive feelings about their workplace when they perceived their organizations as responsive and communicative in supportive ways (Dinsbach, Fiej, & de Vries, 2007). But much still remains to be accomplished. For years, African American managers have reported less job choice, less acceptance, more career dissatisfaction, lower performance evaluations and promotability ratings, and more reaching of plateaus in their careers than European American managers (Greenhaus, Parasuraman, & Wormley, 1990). There is a shortage of African American mentors and career coaches, meaning that most African American protégés have European American mentors. This is problematic because it has been known for many years that same-ethnicity mentors provide more psychosocial support than cross-ethnicity mentors (Thomas, 1990). Nevertheless, having any good mentor is more beneficial than having none (Bridges, 1996).

Bias and Discrimination

Since the 1960s, organizations in the United States have been sensitized to the issues of bias and discrimination in the workplace. Hiring, promotion, and termination procedures have come under close scrutiny in numerous court cases, resulting in judicial rulings governing these processes.

Gender Bias and the Glass Ceiling

By the end of the first decade of the 21st century, women accounted for more than half of all people employed in management, professional, and related occupations (Bureau of Labor Statistics, 2010a). However, women are still underrepresented at the very top; for example, in 2010 only 15 women were the CEOs of Fortune 500 companies, a 50% increase over the number in 2006 but still very few. Janice's observation in the vignette, that few women serve in the highest ranks of major corporations, is accurate. Moreover, progress is also slow in other sectors; for example, in the 2010 U.S. elections the number of women in Congress actually declined, and women tend to be underrepresented at the highest faculty ranks in most colleges and universities.

gender discrimination
denying a job to someone solely on the basis of whether the person is a man or a woman

Why are there so few women in such positions? *The most important reason is* **gender discrimination**: *denying a job to someone solely on the basis of whether the person is a man or a woman.* Gender discrimination is still pervasive in too many aspects of the workplace (Purcell, MacArthur, & Samblanet, 2010): Women are being kept out of high-status jobs by the men at the top (Barnes, 2005; Reid, Miller, & Kerr, 2004).

Research in the United States and Britain also confirms that women are forced to work harder than men (Gorman & Kmec, 2007). Neither differences in job characteristics nor family obligations account for this difference; the results clearly point to stricter job performance standards being applied to women.

glass ceiling
the level to which women may rise in an organization but beyond which they may not go

Women themselves refer to a **glass ceiling**, *the level to which they may rise in an organization but beyond which they may not go.* The glass ceiling is a major barrier for women (Maume, 2004; Purcell et al., 2010), and the greatest barrier facing them is at the boundary between lower-tier and upper-tier grades. Men are largely blind to the existence of the glass ceiling (Heppner, 2007). Women like Janice tend to move to the top of the lower tier and remain there, whereas men are more readily promoted to the upper tier even when other factors (e.g., personal attributes, qualifications, job performance) are controlled (Lovoy, 2001).

The glass ceiling is pervasive across higher management and professional workplace settings (Heppner, 2007). The glass ceiling has also been used to account for why

African Americans and Asian Americans do not advance as much in their careers as do European American men (Hwang, 2007; Johnson, 2000; Phelps & Constantine, 2001). It also provides a framework for understanding limitations to women's careers in many countries around the world (Mugadza, 2005; Zafarullah, 2000).

Interestingly, a different trend emerges if one examines who is appointed to critical positions in organizations in times of crisis. Research shows that at such times, it is women who are more likely to be put into leadership positions. *Consequently, women often confront a* **glass cliff** *in which their leadership position is precarious.* For example, evidence shows that companies are more likely to appoint a woman to their board of directors if their financial performance had been poor in the recent past, and women are more likely to be political candidates if the seat is a highly contested one (Ryan, Haslam, & Kulich, 2010). This evidence indicates that women's experience of and opportunities for high level leadership in organizations are different than men's, and tend to be much riskier.

What can be done to begin eliminating the glass ceiling and minimizing the glass cliff? Kolb, Williams, and Frohlinger (2010) argue that women can and must be assertive in getting their rightful place at the table by focusing on five key things: drilling deep into the organization so you can make informed decisions, getting critical support, getting the necessary resources, getting buy-in, and making a difference. On the organizational side, Mitchell (2000) suggests that companies must begin to value the competencies women develop, such as being more democratic and interpersonally oriented than men, and to assist men in feeling more comfortable with their female colleagues. Mentoring is also an important aspect. Lovoy (2001) adds that companies must be more proactive in promoting diversity, provide better and more detailed feedback about performance and where employees stand regarding promotion, and establish ombuds offices (company offices where employees can complain about working conditions or their supervisor without retribution) that help women deal with difficulties on the job.

Women like Hillary Clinton who make it to the highest ranks of their profession have had to contend with both the glass ceiling and the glass cliff.

glass cliff
situation that women confront in which their leadership position is precarious

Equal Pay for Equal Work

In addition to discrimination in hiring and promotion, women are also subject to pay discrimination. According to the Bureau of Labor Statistics (2010b), women are paid about 80% of what men are paid on an average annual basis. As you can see in ■ Figure 12.4, the wage gap depends on age and has been narrowing since the late 1970s.

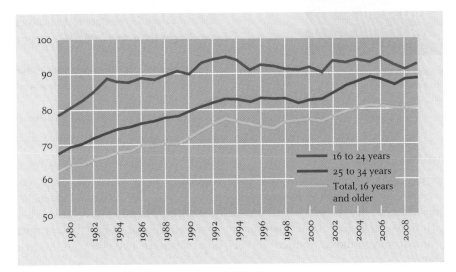

■ **Figure 12.4**
Women's earnings as a percentage of men's. Median usual weekly earnings of full-time wage and salary workers, by age, 1979–2009.

Legend:
— 16 to 24 years
— 25 to 34 years
— Total, 16 years and older

Source: U.S. Bureau of Labor Statistics. (2010b). *Women's to men's earnings ratio by age, 2009.* Retrieved from www.bls.gov/opub/ted/2010/ted_20100708.htm.

When the first equal pay legislation passed in the United States in 1963, women were paid only 59 cents for every dollar men were paid.

Given that equal pay law has been in effect for roughly 50 years, what can be done to eliminate the problem permanently? In their comprehensive look at pay inequity, Dey and Hill (2007) suggested several actions to address the problem: encouraging women to negotiate salary more effectively, rethinking the use of hours worked as the primary measure of productivity, creating more work options for working mothers, and ending gender discrimination in the workplace. The What Do *You* Think? feature explores this issue in more detail.

What do you think? Unequal Pay for Equal Work

Despite the progress that women have made in American society over many decades, there remains one area where gender differences are so entrenched they almost seem normal: pay equity. In the United States, the first law regarding pay equity was passed by Congress in 1963. Forty-six years later in 2009, President Obama signed the Lilly Ledbetter Fair Pay Act, showing clearly that the problem of pay inequity still exists. In their comprehensive and insightful analysis of the continuing gap between men's and women's paychecks for the same work, Dey and Hill (2007) make a clear case that much needs to be done, and now.

Why? Consider this: Only one year out of college, a woman earns on average about $0.80 for every $1.00 a male college graduate earns. A mere 10 years later, she's down to about $0.69. This is even after controlling for such important variables as occupation, hours worked, parenthood, and other factors associated with pay.

What if women choose a college major that is associated with high-paying jobs, such as those in science, technology, engineering, and mathematics? Will that help reduce the pay differential? No. Choosing a traditionally "male-dominated" major will not solve the problem alone. For example, women in mathematics occupations earn only about $0.76 for every $1.00 a male mathematics graduate earns.

A woman is also significantly disadvantaged when it comes to the division of labor at home if she is married or living with a man. Despite decades of effort in getting men to do more of the housework and child-care tasks, little has changed in terms of the amount of time men actually spend on these tasks. In effect, this means that women have two careers, one in the workplace and the other at home. And if a college-educated woman stays at home to care for a child or parent, then her return to the workforce will be at a lower salary than it would have been otherwise.

So what can be done about pay inequity? First, we must recognize that it exists; only then can other steps be taken. Women and men deserve to be treated fairly and paid the same for the same work. Workplaces need to make accommodations for people who take care of a child or parent, and carefully analyze salary policies. Making this a reality will take a cooperative effort by everyone.

What do *you* think? Do you think it's fair that women make less than men? How would you eliminate the differences?

SAUL LOEB/AFP/Getty Images

The Fair Pay Act of 2009 was named for Lilly Ledbetter (standing behind President Obama), who refused to accept being paid $500–$1,000 less a month than men doing the same job.

Sexual Harassment

Suppose you have been working very hard on a paper for a course and think you've done a good job. When indeed you receive an "A" for the paper, you are elated. When you discuss your paper (and your excitement) with your instructor, you receive a big hug. How do you feel? What if this situation involved a major project at work and the hug came from your boss? Your co-worker? What if it were a kiss on your lips instead of a hug?

Whether such behavior is acceptable, or whether it constitutes sexual harassment, depends on many situational factors, including the setting and the people involved and the relationship between them. Interest among U.S. researchers increased dramatically after the 1991 Senate hearings involving Supreme Court nominee Clarence Thomas and Anita Hill (who accused him of sexual harassment), charges against President Bill Clinton in the late 1990s, and scandals involving major corporate and political leaders. Sexual harassment is not merely a U.S. phenomenon; sad to say, it occurs around the world (Zippel, 2006).

Generally, sexual harassment research focuses on situations in which there is a power differential between two people, most often involving men with more power over women (Bornstein & Adya, 2007). This research has influenced and been influenced by court decisions and the emerging definitions of sexual harassment.

How many people have been sexually harassed? That's a very hard question to answer for two reasons: harassment involves a subjective assessment, and good research on how often it occurs is scarce. Available evidence suggests that about 58% of women report they have experienced potentially harassing behaviors, which typically involve comments, jokes, brushes of hands on shoulders, and so on (Ilies et al., 2003). Reliable statistics on more serious forms of harassment, such as touching and sexual activity, are more difficult to obtain; this is due in part to the unwillingness of many victims to report harassment and in part to differences in reporting procedures. Even given these difficulties, evidence from meta-analytic research (see Chapter 1 for a discussion of meta-analyses) indicates that at least 28% of women report having been sexually harassed in the workplace (Ilies et al., 2003). Victims are most often single or divorced young adult women (Zippel, 2006), but about 16% of workplace cases that result in formal legal charges involve male victims (Equal Employment Opportunity Commission, 2010).

What are the effects of being sexually harassed? As you might expect, research evidence clearly shows negative job-related, psychological, and physical health outcomes (Lim & Cortina, 2005). Establishing the degree of the problem is difficult, though, because many women try to minimize or hide their reactions or feelings (Tang & McCollum, 1996). It is becoming evident, however, that one need not experience the worst kinds of sexual harassment before being affected. Even low-level but frequent encounters with sexual harassment can have significant negative consequences for women (Schneider, Swan, & Fitzgerald, 1997).

Cultural differences in labeling behaviors as sexually harassing are also important. For example, one study found that U.S. women judged specific interactions as more harassing than U.S. men, although women and men from Australia, Brazil, and Germany did not differ in their judgments (Pryor et al., 1997). Research comparing countries in the European Union reveals differences across these countries in terms of definitions and corrective action (Zippel, 2006). Unfortunately, little research has been done to identify what aspects of organizations foster harassment or to determine the impact of educational programs aimed at addressing the problem.

In 1998, the U.S. Supreme Court (in *Oncale v. Sundowner Offshore Services*) ruled that sexual harassment is not limited to female victims and that the relevant laws also protect men. This case is significant for establishing that same-sex harassment is barred by the same laws that ban heterosexual harassment (Knapp & Kustis, 2000). Thus, the standard by which sexual harassment is judged could now be said to be a "reasonable person" standard.

What can be done to provide people with safe work and learning environments, free from sexual harassment? Training in gender awareness is a common approach that often works (Tang & McCollum, 1996), especially given that gender differences exist in perceptions of behavior (Lindgren, 2007). Clearly differentiating between workplace romance and sexual harassment is another essential element.

THINK ABOUT IT

What are the key biological, psychological, sociocultural, and life-cycle factors that should inform training programs concerning sexual harassment?

Age Discrimination

age discrimination
denying a job or promotion to someone solely on the basis of age

Another structural barrier to occupational development is **age discrimination**, *which involves denying a job or promotion to someone solely on the basis of age.* The U.S. Age Discrimination in Employment Act of 1986 protects workers over age 40. A similar law, the Employment Equality (Age) Regulations, went into effect in the United Kingdom in 2006, and more European countries are protecting middle-aged and older workers (Lahey, 2010). These laws stipulate that people must be hired based on their ability, not their age. Under these laws, employers are banned from refusing to hire (and from discharging) workers solely on the basis of age. Furthermore, employers cannot segregate or classify workers or otherwise denote their status on the basis of age.

Employment prospects for middle-aged people around the world are lower than for their younger counterparts (Lahey, 2010). For example, age discrimination toward those over age 45 is common in Germany (Frerichs & Naegele, 1997) and Hong Kong (Cheung, Kam, & Ngan, 2011), resulting in longer periods of unemployment, early retirement, or negative attitudes. Such practices may save companies money in the short run, but the loss of expertise and knowledge comes at a high price. Indeed, global corporations are retraining and integrating middle-aged workers as a necessary part of remaining competitive (Frerichs & Naegele, 1997).

Age discrimination may occur in several ways but is not typically demonstrated by professional human resources staff (Kager, 2000). Age discrimination usually happens prior to or after interaction with human resources staff by other employees making the hiring decisions, and it can be covert (Lahey, 2010; Pillay, Kelly, & Tones, 2006). For example, employers can make certain types of physical or mental performance a job requirement and argue that older workers cannot meet the standard prior to an interview. Or they can attempt to get rid of older workers by using retirement incentives. Supervisors' stereotyped beliefs sometimes factor in performance evaluations for raises or promotions or in decisions about which employees are eligible for additional training (Chiu et al., 2001).

Dennis Wise / Getty Images

Employers cannot make a decision not to hire this woman solely on the basis of her age.

Test Yourself

RECALL

1. Women who choose nontraditional occupations are viewed _____ by their peers.

2. Among the reasons women in well-paid occupations leave, _____ are most important for part-time workers.

3. Ethnic minority workers are more satisfied with and committed to organizations that are responsive and provide _____.

4. Three barriers to women's occupational development are sex discrimination, the glass ceiling, and _____.

INTERPRET

What steps need to be taken to eliminate gender, ethnic, and age bias in the workplace?

APPLY

Suppose that you are the CEO of a large organization and that you need to make personnel reductions through layoffs. Many of your most expensive employees are over age 40. How could you accomplish this without being accused of age discrimination?

Recall answers: (1) negatively, (2) family obligations, (3) positive work environments, (4) pay discrimination

12.3 Occupational Transitions

LEARNING OBJECTIVES

- Why do people change occupations?
- Is worrying about potential job loss a major source of stress?
- How does job loss affect the amount of stress experienced?

Fred has 32 years of service for an automobile manufacturer. Over the years, more and more assembly-line jobs have been eliminated by new technology (including robots) and by the export of manufacturing jobs to other countries. Although Fred has been assured that his job is safe, he isn't so sure. He worries that he could be laid off at any time.

IN THE PAST, people like Fred commonly chose an occupation during young adulthood and stayed in it throughout their working years. Today, however, not many people take a job with the expectation that it will last a lifetime. Corporations have restructured globally so often that employees now assume occupational changes to be part of the career process. Such corporate actions mean that people's conceptions of work and career are in flux and that losing one's job no longer has only negative meanings (Haworth & Lewis, 2005).

Several factors have been identified as important in determining who will remain in an occupation and who will change. Some factors—such as whether the person likes the occupation—lead to self-initiated occupation changes. For example, people who really like their occupation may seek additional training or accept overtime assignments in hopes of acquiring new skills that will enable them to get better jobs. Others will use the training to become more marketable.

However, other factors—such as obsolete skills and economic trends—may cause forced occupational changes. For example, continued improvement of robots has caused some auto industry workers to lose their jobs; corporations send jobs overseas to increase profits; and economic recessions usually result in large-scale layoffs and high levels of unemployment. But even forced occupational changes can have benefits in some cases. As we saw in Chapter 10, for instance, many adults go to college after a job loss. Some are able to take advantage of educational benefits offered as part of a severance package at the time they are laid off in order to gain new and more marketable skills.

In this section, we explore the positive and negative aspects of occupational transitions. First we examine the retraining of midcareer and older workers. The increased use of technology, corporate downsizing, and an aging workforce have focused attention on the need to keep older workers' skills current. Later, we will examine occupational insecurity and the effects of job loss.

Retraining Workers

When you are hired into a specific job, you are selected because your employer believes you offer the best fit between the abilities you already have and those needed to perform the job. As most people can attest, though, the skills needed to perform a job usually change over time. Such changes may be due to the introduction of new technology, additional responsibilities, or promotion.

Unless a person's skills are kept up-to-date, the outcome is likely to be either job loss or a career plateau (McCleese & Eby, 2006; Rose & Gordon, 2010). **Career plateauing** *occurs when there is a lack of challenge in one's job or promotional opportunity in the organization or when a person decides not to seek advancement.* Research in Canada (Lemire, Saba, & Gagnon, 1999), Asia (Lee, 2003), and Australia (Rose & Gordon, 2010) shows that feeling one's career has plateaued usually results

career plateauing
when promotional advancement is either not possible or not desired by the worker

in less organizational commitment, lower job satisfaction, and a greater tendency to leave. But attitudes can remain positive if it is only the lack of challenge and not a lack of promotion opportunity that is responsible for the plateauing (McCleese & Eby, 2006).

In cases of job loss or a career plateau, retraining may be an appropriate response. Around the world, large numbers of employees participate each year in programs and courses that are offered by their employer or by a college or university and are aimed at improving existing skills or adding new job skills. For midcareer employees, retraining might focus on how to advance in one's occupation or on how to find new career opportunities—for example, through résumé preparation and career counseling. Increasingly, such programs are offered online in order to make them easier and more convenient for people to access (Githens & Sauer, 2010).

Many corporations, as well as community and technical colleges, offer retraining programs in a variety of fields. Organizations that promote employee development typically promote in-house courses to improve employee skills. They may also offer tuition reimbursement programs for individuals who successfully complete courses at colleges or universities.

Seminars such as this are taken by thousands of workers around the world each year as part of worker training and retraining programs.

The retraining of midcareer and older workers highlights the need for lifelong learning (Armstrong-Stassen & Templer, 2005; Sinnott, 1994). If corporations are to meet the challenges of a global economy, it is imperative that they include retraining in their employee development programs. Such programs will help improve people's chances of advancement in their chosen occupations and can also assist people in making successful transitions from one occupation to another.

Occupational Insecurity

Over the past few decades, changing U.S. economic conditions (e.g., the move toward a global economy), changing demographics, and a global recession have forced many people out of their jobs. Heavy manufacturing and support businesses (such as the steel, oil, and automotive industries) and farming were the hardest-hit sectors during the 1970s and 1980s. But no one is immune. Indeed, the corporate takeover frenzy of the 1980s and the recessions of the early 1990s and 2000s put many middle- and upper-level corporate executives out of work worldwide.

As a result of these trends, many people feel insecure about their jobs. Economic downturns create significant levels of stress, especially when such downturns create massive job loss (Sinclair et al., 2010). Like Fred, the autoworker in the vignette, many worried workers have numerous years of dedicated service to a company. Unfortunately, people who worry about their jobs tend to have poorer physical and psychological well-being (McKee-Ryan et al., 2005). For example, anxiety about one's job may result in negative attitudes about one's employer or even about work in general, which in turn may result in diminished desire to be successful. Whether there is any actual basis for people's feelings of job insecurity may not matter; sometimes what people *think* is true about their work situation is more important than what is actually the case. Just the possibility of losing one's job can negatively affect physical and psychological health.

So how does the possibility of losing one's job affect employees? Mantler and colleagues (2005) examined coping strategies for comparable samples of laid-off and employed high-technology workers. They found that although unemployed participants reported higher levels of stress compared with employed participants, employment uncertainty mediated the association between employment status and perceived stress. That is, people who believe that their job is in jeopardy—even if it is not—show levels of stress similar to unemployed participants. This result is due to differences in coping strategies. There are several different ways that people deal with stress, and two of the more common are emotion-focused coping and problem-focused coping. Some people focus on how the stressful situation makes them feel, so they cope by making themselves feel better about it. Others focus on the problem itself and do something to solve it. People who used emotional avoidance as a strategy reported higher levels of stress, particularly when they were fairly certain of the outcome. Thus, even people whose jobs aren't really in jeopardy can report high levels of stress if they tend to use emotion-focused coping strategies.

Coping With Unemployment

What does it feel like to lose one's job after many years of dedicated service? Read the Real People feature to get a sense of what it's like.

Real People
Applying Human Development | Experiencing Layoff

When Jo Ann graduated from college, she thought she had hit the jackpot by getting a great job with a major technology company. She was a rising star for her first 16 years with the company, being named to all-company lists of the best systems marketing employees, exceeding her sales goals every year, being promoted quickly up the ranks. So when the rumors of layoffs began circulating through the company, she thought she had little to worry about.

She was wrong.

To Jo Ann's great shock, she was laid off. The entire division of the company was eliminated and outsourced. She was totally devastated. The severance package the company provided was no consolation. Unemployment benefits did not pay the mortgage. Her savings would soon be gone. She felt as if someone had punched her, very hard, right in the gut. She had lost much of her identity. She felt ashamed. She felt at fault. She became seriously depressed. She realized it was going to be a difficult process finding a new job with so many others like herself out of work.

As Jo Ann's case so poignantly reflects, losing one's job can have enormous personal impact that can last a very long time (Lin & Leung, 2010; McKee-Ryan et al., 2005). When unemployment rates hit 10.6% in January 2010, millions of people could relate to these feelings. When unemployment lasts, unemployed people commonly experience a wide variety of negative effects (Lin & Leung, 2010) that range from a decline in immune system functioning (Cohen et al., 2007) to decreases in well-being (McKee-Ryan et al., 2005).

In a comprehensive meta-analysis (see Chapter 1) of the research on the effects of unemployment, McKee-Ryan and colleagues (2005) found several specific results from losing one's job. Unemployed workers had significantly lower mental health, life satisfaction, marital or family satisfaction, and subjective physical health (how they perceive their health to be) than their employed counterparts. With reemployment, these negative effects disappear. ■ Figure 12.5 shows that physical and psychological health following job displacement is influenced by several factors (McKee-Ryan et al., 2005).

The effects of job loss vary with age and gender. In the United States, middle-aged men are more vulnerable to negative effects than older or younger men—largely because they have greater financial responsibilities than the other two groups—but women report more negative effects over time (Bambra, 2010). Research in Spain indicates that gender differences in responding to job loss are complexly related to family responsibilities and social class (Artazcoz et al., 2004). Specifically, to the extent that work is viewed as your expected contribution to the family, losing one's job has a more substantial negative effect. Because this tends to be more the case for men than for women, it helps explain the gender differences.

As we noted in Chapter 11, single mothers are particularly vulnerable to financial pressures. Losing a job for them can be simply devastating.

Because unemployment rates for many ethnic minority groups are substantially higher than for European Americans (Bureau of Labor Statistics, 2010c), the effects of unemployment are experienced by a greater proportion of people in these groups. As far as is known, however, the nature of the distress resulting from job loss is the same

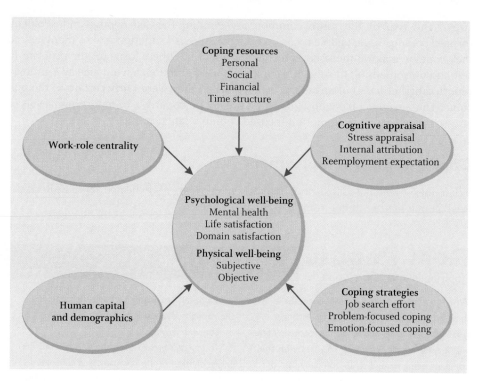

From McKee-Ryan et al., 2005, p. 56.

regardless of ethnicity. Compared to European Americans, however, it usually takes minority workers longer to find another job.

Research also offers some advice for adults who are trying to manage occupational transitions (Ebberwein, 2001):

- Approach job loss with a healthy sense of urgency.
- Consider your next career move and what you must do to achieve it, even if there are no prospects for it in sight.
- Acknowledge and react to change as soon as it is evident.
- Be cautious of stopgap employment.
- Identify a realistic goal and then list the steps you must take to achieve it.

Additionally, the U.S. Department of Labor (www.dol.gov/dol/audience/aud-unemployed .htm) offers tips for job seekers, as do online services such as LinkedIn (www.linkedin .com), which also provides networking groups. These steps may not guarantee that you will find a new job quickly, but they will help create a better sense that you are in control.

THINK ABOUT IT

What are some of the broader effects of unemployment on an individual's personal and family life?

Test Yourself

RECALL

1. One response to the pressures of a global economy and an aging workforce is to provide _____.

2. Two factors that could cause involuntary occupational change are economic trends and _____.

3. Fear of job loss is often a more important determinant of stress than _____.

4. The age group that is most at risk for negative effects of job loss is _____.

INTERPRET

It is likely that the trend toward multiple careers will continue and become the norm. What implications will this have for theories of career development?

APPLY

You have been asked to design a program to help employees cope with losing their job. What key components would you include in this program?

Recall answers: (1) worker retraining, (2) obsolete skills, (3) actual likelihood of job loss, (4) middle-aged adults.

LEARNING OBJECTIVES

■ What are the issues faced by employed people who care for dependents?

■ How do partners view the division of household chores? What is work–family conflict, and how does it affect couples' lives?

Jennifer, a 38-year-old sales clerk at a department store, feels that her husband, Bill, doesn't do his share of the housework or child care. Bill says that real men don't do housework and that he's really tired when he comes home from work. Jennifer thinks that this isn't fair, especially because she works as many hours as her husband.

ONE OF THE MOST DIFFICULT CHALLENGES FACING ADULTS LIKE JENNIFER IS TRYING TO BALANCE THE DEMANDS OF OCCUPATION WITH THE DEMANDS OF FAMILY. Over the past few decades, the rapid increase in the number of families in which both parents are employed has fundamentally changed how we view the relationship between work and family. This can even mean taking a young child to work as a way to deal with the pushes and pulls of being an employed parent. In roughly 60% of two-parent households today, both adults work outside the home, a rate slightly lower than previous years due to the economic recession (Bureau of Labor Statistics, 2010d). The main reason? Families need the dual income to pay their bills and maintain a moderate standard of living.

We will see that dual-earner couples with children experience both benefits and disadvantages. The stresses of living in this arrangement are substantial, and gender differences are clear—especially in the division of household chores.

The Dependent Care Dilemma

Many employed adults must also provide care for dependent children or parents. As we will see, the issues they face are complex.

Employed Caregivers Revisited

Many mothers have no option but to return to work after the birth of a child. In fact, about 55% of married and unmarried mothers with children under the age of 3 years work for pay, with another 9% officially considered unemployed but looking for employment (U.S. Department of Labor, 2010b).

Some women, though, grapple with the decision of whether they want to return to work. Surveys of mothers with preschool children reveal that the motivation for returning to work tends to be related to financial need and how attached mothers are to their work. The amount of leave time a woman has matters; the passage of the Family and Medical Leave Act in 1993 entitled workers to take unpaid time off to care for their dependents with the right to return to their jobs. This Act resulted in an increase in the number of women who returned to work at least part-time (Schott, 2010). Although working part-time work may seem appealing, what matters more is whether mothers are working hours that are close to what they consider ideal and are accommodating to their family's needs (Kim, 2000). Perceptions of ideal working hours differ as a function of gender and life-cycle stage regarding children, as shown in ■ Figure 12.6.

A concern for many women is whether stepping out of their occupations following childbirth will negatively affect their career paths. Indeed, evidence clearly indicates that it does (Aisenbrey, Evertsson, & Grunow, 2009). Women in the United States are punished, even for short leaves. But even in women-friendly countries such as Sweden, long leaves typically result in a negative effect on upward career movement.

The ideal number of hours that men want to work stays about the same regardless of whether they have children; women's ideal number of hours depends on whether they have children and how old the children are.

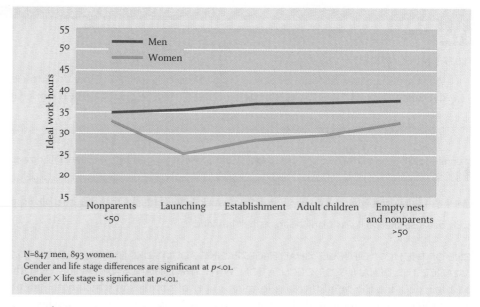

N=847 men, 893 women.
Gender and life stage differences are significant at $p<.01$.
Gender × life stage is significant at $p<.01$.

Source: The Career Mystique: Cracks in the American Dream, 2005, Phyllis Moen & Patricia Roehling. Boulder: Rowman & Littlefield.

Often overlooked is the increasing number of workers who must also care for a parent or partner. Of women caring for parents or parents-in-law, more than 80% provide an average of 23 hours per week of care and 70% contribute money (Pierret, 2006). Because most of these women are middle-aged, we will consider their situation in more detail in Chapter 13.

Whether assistance is needed for one's children or one's parents, key factors in selecting an appropriate care site are quality of care, price, and hours of availability (Helpguide.org, 2005; Mitchell & Messner, 2003–2004). Depending on one's economic situation, it may not be possible to find affordable and quality care that is available when needed. In such cases, there may be no option but to drop out of the workforce or enlist the help of friends and family.

Dependent Care and Effects on Workers

Workers who care for dependents face tough choices. Especially when both partners are employed, dependent care becomes the primary determiner of how the couple organizes their time.

Being responsible for dependent care has significant negative effects on caregivers. For example, whether responsible for the care of an older parent or a child, women and men report negative effects on their work, higher levels of stress, and problems with coping (Neal & Hammer, 2006). Because women serve as caregivers more than men, the related problems are greater for women; for example, women experience far more negative consequences on their career advancement if they are both employed full-time and serve as a primary caregiver (Roxburgh, 2002).

How can these negative effects be lessened? When women's partners provide good support and women have average or high control over their jobs, employed mothers are significantly less distressed than employed nonmothers (Roxburgh, 2002) or mothers without support (Rwampororo, 2001). Research focusing on single working mothers also shows that those who have support from their families manage to figure out a balance between work and family obligations (Son & Bauer, 2010). However, when support and job control are lacking, employed mothers (single or married) are significantly more distressed than employed nonmothers. Clearly, the most important factors are partner or family support and a job that allows one to control such things as one's sched-

Balancing work and family obligations is especially difficult for women.

Keith Brofsky / Getty Images

ule. This latter point demonstrates that what employers provide in terms of a supportive workplace is also important, as we see next.

Dependent Care and Employer Responses

Employed parents with small children or dependent parents are confronted with the difficult prospect of leaving them in the care of others. This is especially problematic when the usual care arrangement is unavailable. *A growing need in the workplace is for* **backup care**, *which provides emergency care for dependent children or adults so that the employee does not need to lose a day of work.* Does providing a workplace care center or backup care make a difference in terms of an employee's feelings about work, absenteeism, and productivity?

The answer is that there is no simple answer. For example, just making a child-care center available to employees does not necessarily reduce parents' work–family conflict or their absenteeism, particularly among younger employees (Connelly, Degraff, & Willis, 2004). A "family-friendly" company must also pay attention to the attitudes of their employees and make sure that the company provides broad-based support (Allen, 2001; Grandey, 2001). The keys are how supervisors act and the number and type of benefits the company provides. When the organization adopts a "justice" approach—in which supervisors are sympathetic and supportive regarding family issues and child care—and provides benefits that workers consider important, employees report less work–family conflict, have lower absenteeism, and report higher job satisfaction. The most important single thing a company can do is allow the employee to leave work without penalty to tend to family needs (Lawton & Tulkin, 2010).

Research also indicates that there may not be differences for either mothers or their infants between work-based and nonwork-based child-care centers in terms of the mothers' ease in transitioning back to work or the infants' ability to settle into day care (Skouteris, McNaught, & Dissanayake, 2007).

Specific working conditions and benefits that help caregivers perform optimally on the job also matter. Caregivers fare better to the extent that employers provide job sharing, nursery facilities, better job security, autonomy, lower productivity demands, supervisor support, and flexible schedules (Aryee & Luk, 1996; Brough, O'Driscoll, & Kalliath, 2005). Job applicants tend to perceive an organization more positively if it provides flexible work schedules and assistance with dependent care and nonwork life (Casper, 2000).

It will be interesting to watch how these issues—especially flexible schedules—play out in the United States, where such practices are not yet common. A global study of parental leave, such as that granted under the U.S. Family and Medical Leave Act, showed that the more generous parental leave policies are, the lower the infant mortality rates, clearly indicating that parental leave policies are a good thing (Ferrarini & Norström, 2010).

Juggling Multiple Roles

When both members of a heterosexual couple with dependents are employed, who cleans the house, cooks the meals, and takes care of the children when they are ill? This question goes to the heart of the core dilemma of modern, dual-earner couples: How are household chores divided? How are work and family role conflicts handled?

Dividing Household Chores

Despite much media attention and claims of increased sharing in the duties, women still perform the lion's share of housework, regardless of employment status. As shown in ▪ Figure 12.7, this is true globally (Ruppanner, 2010). This unequal division

THINK ABOUT IT

How do the effects of dependent care on mothers relate to the controversy concerning whether children should be placed in day care?

backup care
emergency care for dependent children or adults so that the employee does not need to lose a day of work

Employers who provide day-care centers on-site have more satisfied employees.

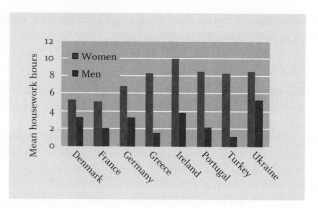

Source: Data from Ruppanner, L. E. (2010). Cross-national reports of housework: An investigation of the gender empowerment measure. Social Science Research, 19, 963-975. Table 1 p. 968.

of labor causes the most arguments and the most unhappiness for dual-earner couples. This is the case with Jennifer and Bill, the couple in the vignette; Jennifer does most of the housework.

Although women still do most of the household chores, things are getting a bit better. A great deal of evidence indicates that, since the 1970s, women have reduced the amount of time they spend on housework (especially when they are employed) and that men have increased the amount of time they spend on such tasks (Saginak & Saginak, 2005). The increased participation of men in these tasks is not all that it seems, however. Most of the increase is on weekends, involves specific tasks that they agree to perform, and is largely unrelated to women's employment status. In short, the increase in men's participation has not done much to lower women's burdens around the house.

The best good news for women is that, the longer they spend in the workforce, the more household tasks their partners perform (Cunningham, 2007). Over the course of a marriage (in Cunningham's study, a 31-year period), women who had longer employment histories had husbands who helped out more around the house.

Men and women view the division of labor differently. Men are often most satisfied with an equitable division of labor based on the number of hours spent, especially if the amount of time needed to perform household tasks is relatively small. Women are often most satisfied when men are willing to perform women's traditional chores (Saginak & Saginak, 2005). When ethnic minorities are studied, much the same is true concerning satisfaction. For example, in African American dual-earner couples, women were twice as likely as men to feel overburdened with housework and to be dissatisfied with their family life (Dillaway & Broman, 2001). Among Navajos not living on a reservation, women do more of the cooking and child care, but both men and women spend about the same amount of time on home maintenance (Hossain, 2001).

Ethnic differences in the division of household labor are also apparent. In Mexican American families with husbands born in Mexico, men help more when family income is lower and their wives contribute a proportionately higher share of the household income (Pinto & Coltrane, 2009). Comparisons of Latino, African American, and European American men consistently show that European American men help with the chores less than Latino or African American men (Omori & Smith, 2009).

In sum, the available evidence from heterosexual couples indicates that women still perform more household tasks than men but that the difference varies with ethnic groups. The discrepancy is greatest when the male endorses traditional masculine gender roles and is least when the male endorses more feminine or androgynous gender roles (Saginak & Saginak, 2005).

Work–Family Conflict

When people have both occupations and children, they must figure out how to balance the demands of each. Parents agonize over how to watch their daughter's ball game when they must also attend an important business meeting. *These competing demands cause* **work–family conflict,** *which is the feeling of being pulled in multiple directions by incompatible demands from one's job and one's family.*

work–family conflict
the feeling of being pulled in multiple directions by incompatible demands from one's job and one's family

Dual-earner couples must find a balance between their occupational and family roles. Because nearly 60% of married couples with children consist of dual-earner households (Bureau of Labor Statistics, 2010b), how to divide the household chores and how to care for the children have become increasingly important questions.

Many people believe that, in such cases, work and family roles influence each other: When things go badly at work, the family suffers, and when there are troubles at home, work suffers. That's true, but the influence is not the same in each direction (Andreassi, 2007). Whether work influences family or vice versa is a complex function of support resources, type of job, and a host of other issues (Saginak & Saginak, 2005). One key but often overlooked factor is whether the work schedules of both partners allow them to coordinate activities such as child care (Jacobs & Gerson, 2001).

Of course, it is important that the partners negotiate agreeable arrangements of household and child-care tasks, but we've noted that truly equitable divisions of labor are clearly the exception. Most U.S. households with heterosexual dual-worker couples still operate under a gender-segregated system: There are traditional chores for men and for women. These important tasks must be performed to keep homes safe, clean, and sanitary, and these tasks also take time. The important point for women is not how much time is spent performing household chores so much as which tasks are performed. What bothers wives the most is when their husbands are unwilling to do "women's work." Men may mow the lawn, wash the car, and even cook, but they are much less likely to vacuum, scrub the toilet, or change the baby's diaper.

Dual-earner couples must learn how to grapple with work-family conflict in balancing job and family demands.

This division of labor apparently occurs because that's how people saw their parents do it and so that's what they are comfortable with. John Cavanaugh had an interesting experience in this regard. While doing some volunteer maintenance work at a battered women's shelter in Appalachia, he had to use the vacuum cleaner. He soon became aware that several women were following him around, pointing at him and talking excitedly. A little later, he asked why they did that. He was told that it was the first time in these adult women's lives that they had ever seen a man use a vacuum cleaner.

So how and when will things change? An important step would be to talk about these issues with your partner. Keep communication lines open all the time, and let your partner know if something is bothering you. Teaching your children that men and women are equally responsible for household chores will also help end the problem. Only by creating true gender equality—without differentiating among household tasks—will this unfair division of labor be ended.

Understanding work–family conflict requires taking a life-stage approach (see Chapter 1) to the issue (Blanchard-Fields, Baldi, & Constantin, 2004). For example, several studies have found that the highest conflict between the competing demands of work and family occurs during the peak parenting years, when there are at least two preschool children in the home. Inter-role conflict diminishes in later life stages, especially when the quality of the marriage is high.

Dual-earner couples often have difficulty finding time for each other, especially if both work long hours. The amount of time together is not necessarily the most important issue; as long as the time is spent in shared activities such as eating, playing, and conversing, couples tend to be happy (Jacobs & Gerson, 2001). Especially when both partners are employed, getting all of the schedules to work together smoothly can be a major challenge. Unfortunately, many couples find themselves in the same position

as Hi and Lois; by the time they have an opportunity to be alone together, they are too tired to make the most of it.

Some effects are global: for example, burnout from the dual demands of work and parenting is more likely to affect women (Spector et al., 2005). Japanese career women's job satisfaction declines, and turnover becomes more likely, to the extent they have high work–family conflict (Honda-Howard & Homma, 2001). Research comparing sources of work–family conflict in the United States and China reveals that when work demands do not differ, that work pressure is a significant source of work–family conflict in both countries (Yang et al., 2000).

So exactly what effects do family matters have on work performance and vice versa? Evidence suggests that work–family conflict is a major source of stress in couples' lives. In general, women feel the work-to-family spillover to a greater extent than men, but both men and women feel the pressure (Saginak & Saginak, 2005).

Test Yourself

RECALL

1. Parents report lower work–family conflict and have lower absenteeism when supervisors are sympathetic and supportive regarding _____.

2. Men are satisfied with an equitable division of labor based on _____, whereas women are satisfied _____.

INTERPRET

What can organizations do to help ease work–family conflict?

APPLY

Suppose you are the vice president for human resources and you are thinking about creating a way to help employees who face dependent care issues. What factors will you need to consider before implanting a plan?

Recall answers: (1) family issues and child care, (2) the number of hours spent; when men perform traditionally female chores

LEARNING OBJECTIVES

■ What activities are leisure activities? How do people choose among them?

■ What changes in leisure activities occur with age?

■ What do people derive from leisure activities?

Claude is a 55-year-old electrician who has enjoyed outdoor activities his whole life. From the time he was a boy, he has fished and water-skied in the calm inlets of coastal Florida. Although he doesn't compete in slalom races any more, Claude still skis regularly, and he participates in fishing competitions every chance he gets.

ADULTS DO NOT WORK EVERY WAKING MOMENT OF THEIR LIVES. As each of us knows, we need to relax sometimes and engage in leisure activities. Intuitively, leisure consists of activities not associated with work. **Leisure** *is discretionary activity that includes simple relaxation, activities for enjoyment, and creative pursuits.* As you might expect, men and women differ in their views of leisure, as do people in different ethnic and age groups (van der Pas & Koopman-Boyden, 2010).

A major issue with leisure is simply finding the time. Young and middle-aged adults must fit leisure into an already busy schedule, so leisure becomes another component in our overall time management problem (Corbett & Hilty, 2006).

leisure
discretionary activity that includes simple relaxation, activities for enjoyment, and creative pursuits

Types of Leisure Activities

Leisure can include virtually any activity. To organize the options, researchers have classified leisure activities into several categories. Jopp and Hertzog (2010) developed an empirically based set of categories that includes a wide variety of activities: physical (e.g., lifting weights, backpacking, jogging), crafts (e.g., woodworking, household repairs), games (e.g., board/online games, puzzles, card games), watching TV, social-private (e.g., going out with a friend, visiting relatives, going out to dinner), social-public (e.g., attending a club meeting, volunteering), religious (e.g., attending a religious service, praying), travel (e.g., travel abroad, travel out of town), experiential (e.g., collect stamps, read for leisure, gardening, knitting), developmental (e.g., read as part of a job, study a foreign language, attend public lecture), and technology use (e.g., photography, use computer software, play an instrument).

More complete measures of leisure activities not only provide better understanding of how adults spend their time, but can help in clinical settings. For example, declines in the frequency of leisure activities is associated with depression (Schwerdtfeger & Friedrich-Mei, 2009) and with a later diagnosis of dementia (Hertzog et al., 2009). Monitoring changes in leisure activity levels during and after intervention programs can provide better outcomes assessments of these interventions.

Given the wide range of options, how do people pick their leisure activities? Apparently, each of us has a leisure repertoire, a personal library of intrinsically motivated activities that we do regularly and that we take with us into retirement (Nimrod, 2007a, b). The activities in our repertoire are determined by two things: perceived competence (how good we think we are at the activity compared to other people our age) and psychological comfort (how well we meet our personal goals for performance).

A study of French adults revealed that, as for occupations, personality factors are related to one's choice of leisure activi-

Adults engage in many different types of leisure activities, including backpacking.

© Bernard van Dierendonck / Getty Images

ties (Gaudron & Vautier, 2007). Other factors are important as well: income, interest, health, abilities, transportation, education, and social characteristics. For example, some leisure activities, such as downhill skiing, are relatively expensive and require transportation and reasonably good health and physical coordination for maximum enjoyment. In contrast, reading requires minimal finances (if one uses a public library) and is far less physically demanding. Women in all ethnic groups tend to participate less in leisure activities that involve physical activity (Eyler et al., 2002).

The use of computer technology in leisure activities has increased dramatically (Bryce, 2001). Most usage involves e-mail, Facebook, Twitter, or other social networking tools for such activities as keeping in touch with family and friends, pursuing hobbies, and lifelong learning. Computer gaming on the Internet has also increased among adult players.

Developmental Changes in Leisure

Cross-sectional studies report age differences in leisure activities. Young adults participate in a greater range of activities than middle-aged adults. Furthermore, young adults tend to prefer intense leisure activities, such as scuba diving and hang gliding. In contrast, middle-aged adults focus more on home- and family-oriented activities. In later middle age, they spend less of their leisure time in strenuous physical activities and more in sedentary activities such as reading and watching television (van der Pas & Koopman-Boyden, 2010).

Longitudinal studies of changes in individuals' leisure activities over time show considerable stability over reasonably long periods, and that level of activity in young adulthood predicts activity level later in life (Hillsdon et al., 2005; Patel et al., 2006). Claude, the 55-year-old in the vignette who likes to fish and ski, is a good example of this overall trend. As Claude demonstrates, frequent participation in particular leisure activities during childhood tends to continue into adulthood. Similar findings hold for the pre- and postretirement years. Apparently, one's preferences for certain types of leisure activities are established early in life; they tend to change over the life span primarily in terms of how physically intense they are.

Participating in leisure activities improves one's well-being.

Consequences of Leisure Activities

What do people gain from participating in leisure activities? Researchers have long known that involvement in leisure activities is related to well-being (Warr, Butcher, & Robertson, 2004). Research shows that participating in leisure activities helps promote better mental health in women (Ponde & Santana, 2000) and buffers the effects of stress and negative life events. It even helps lower the risk of mortality (Talbot et al., 2007).

Studies show that leisure activities provide an excellent forum for the interaction of biological, psychological, and sociocultural forces (Kleiber, Hutchinson, & Williams, 2002). Leisure activities are a good way to deal with stress, which—as we have seen—has significant biological effects. This is especially true for unforeseen negative events (Janoff-Bulman, & Berger, 2000). Psychologically, leisure activities have been well documented as one of the primary coping mechanisms that people use (Patry, Blanchard, & Mask, 2007). How people cope by using leisure varies across cultures depending on the various types of activities that are permissible and available. Likewise, leisure activities vary across social class; basketball is one activity that cuts across social class because it is inexpensive, whereas downhill skiing is more associated with people who can afford to travel to ski resorts and pay the fees.

How do leisure activities provide protection against stress? Kleiber and colleagues (2002) offer four ways that leisure activities serve as a buffer against negative life events:

- Leisure activities distract us from negative life events.
- Leisure activities generate optimism about the future because they are pleasant.
- Leisure activities connect us to our personal past by allowing us to participate in the same activities over much of our lives.
- Leisure activities can be used as vehicles for personal transformation.

Whether the negative life events we experience are personal, such as the loss of a loved one, or societal, such as a terrorist attack, leisure activities are a common and effective way to deal with them. They truly represent the confluence of biopsychosocial forces and are effective at any point in the life cycle.

Participating with others in leisure activities may also strengthen feelings of attachment to one's partner, friends, and family (Carnelley & Ruscher, 2000). Adults use leisure as a way to explore interpersonal relationships or to seek social approval. In fact, some research suggests that marital satisfaction is helped more when couples spend some leisure time with others than if they spend it just as a couple (Shebilske, 2000). But there's no doubt that couples who play together are happier (Johnson, Zabriskie, & Hill, 2006).

But what if leisure activities are pursued very seriously? In some cases, people create leisure–family conflict by engaging in leisure activities to extremes (Goff, Fick, & Opplinger, 1997). Only when there is support from others for such extreme involvement are problems avoided (Goff et al., 1997). As in most things, moderation in leisure activities is probably best.

You have probably heard the saying that "no vacation goes unpunished." It appears to be true. Workers report that high postvacation workloads eliminate most of the positive effects of a vacation (Strauss-Blasche, Ekmekcioglu, & Marktl, 2002). Restful vacations do not prevent declines in mood or in sleep due to one's postvacation workload.

One frequently overlooked outcome of leisure activity is social acceptance. For persons with disabilities, this is a particularly important consideration (Devine & Lashua, 2002). There is a positive connection between frequency of leisure activities and social acceptance, friendship development, and acceptance of differences. These findings highlight the importance of designing inclusive leisure activity programs.

Test Yourself

RECALL

1. Activities in which people engage for relaxation, enjoyment, or as creative pursuits can all be considered _____ activities.

2. Compared to younger adults, middle-aged adults prefer leisure activities that are more family- and home-centered and _____.

3. Being involved in leisure activities is related to _____.

INTERPRET

How are choices of leisure activities related to physical, cognitive, and social development?

APPLY

Workers in the United States tend to take fewer vacation days than workers in European countries. What might the consequences of this be for U.S. workers?

Recall answers: (1) leisure, (2) less physically intense, (3) well-being

12.1 OCCUPATIONAL SELECTION AND DEVELOPMENT

How do people view work?

- Although most people work for money, other reasons are highly variable.

How do people choose their occupations?

- Holland's theory is based on the idea that people choose occupations to optimize the fit between their individual traits and their occupational interests. Six personality types, representing different combinations of these, have been identified. Support for these types has been found in several studies.

- Social cognitive career theory emphasizes that how people choose careers is also influenced by what they think they can do and how well they can do it, as well as how motivated they are to pursue a career.

What factors influence occupational development?

- Super's developmental view of occupations is based on self-concept and adaptation to an occupational role. Super describes five stages in adulthood: implementation, establishment, maintenance, deceleration, and retirement.

- Reality shock is the realization that one's expectations about an occupation are different from what one actually experiences. Reality shock is common among young workers.

- Few differences exist across generations in terms of their occupational expectations.

- A mentor or developmental coach is a co-worker who teaches a new employee the unwritten rules and fosters occupational development. Mentor–protégé relationships, like other relationships, develop through stages over time.

What is the relationship between job satisfaction and age?

- Older workers report higher job satisfaction than younger workers, but this may be partly due to self-selection; unhappy workers may quit. Other reasons include intrinsic satisfaction, good fit, lower importance of work, finding nonwork diversions, and life-cycle factors.

- Alienation and burnout are important considerations in understanding job satisfaction. Both involve significant stress for workers.

- Vallerand's Passion Model proposes that people develop a passion toward enjoyable activities that are incorporated into identity. Obsessive passion happens when people experience an uncontrollable urge to engage in the activity; harmonious passion results when individuals have freely accepted the activity as important for them without any contingencies attached to it.

12.2 GENDER, ETHNICITY, AND DISCRIMINATION ISSUES

How do women's and men's occupational expectations differ? How are people viewed when they enter occupations that are not traditional for their gender?

- Boys and girls are socialized differently for work, and their occupational choices are affected as a result. Women choose nontraditional occupations for many reasons, including expectations and personal feelings. Women in such occupations are still viewed more negatively than men in the same occupations.

What factors are related to women's occupational development?

- Women leave well-paid occupations for many reasons, including family obligations and workplace environment. Women who continue to work full-time have adequate child care and look for ways to further their occupational development.

- The glass ceiling, which limits women's occupational attainment, and the glass cliff, which puts women leaders in a precarious position, affect how often women achieve top executive positions and how successful women leaders are.

What factors affect ethnic minority workers' occupational experiences and occupational development?

- Vocational identity and vocational goals vary in different ethnic groups. Whether an organization is sensitive to ethnicity issues is a strong predictor of satisfaction among ethnic minority employees.

What types of bias and discrimination hinder the occupational development of women and ethnic minority workers?

- Gender bias remains the chief barrier to women's occupational development. In many cases, this operates as a glass ceiling. Pay inequity is also a problem; women are often paid less than what men earn in similar jobs.

- Sexual harassment is a problem in the workplace. Current criteria for judging harassment are based on the "reasonable person" standard. Denying employment to anyone over 40 because of age is age discrimination.

12.3 OCCUPATIONAL TRANSITIONS

Why do people change occupations?

- Important reasons people change occupations include personality, obsolescence, and economic trends.

- To adapt to the effects of a global economy and an aging workforce, many corporations are providing retraining opportunities for workers. Retraining is especially important in cases of outdated skills and career plateauing.

Is worrying about potential job loss a major source of stress?

- Occupational insecurity is a growing problem. Fear that one may lose one's job is a better predictor of anxiety than the actual likelihood of job loss.

How does job loss affect the amount of stress experienced?

■ Job loss is a traumatic event that can affect every aspect of a person's life. Degree of financial distress and the extent of attachment to the job are the best predictors of distress.

12.4 WORK AND FAMILY

What are the issues faced by employed people who care for dependents?

■ Caring for children or aging parents creates dilemmas for workers. Whether a woman returns to work after having a child depends largely on how attached she is to her work. Simply providing child care on-site does not always result in higher job satisfaction. A more important factor is the degree to which supervisors are sympathetic.

How do partners view the division of household chores? What is work–family conflict? How does it affect couples' lives?

■ Although women have reduced the amount of time they spend on household tasks over the past two decades, they still do most of the work. European American men are less likely than either African American or Latino American men to help with traditionally female household tasks.

■ Flexible work schedules and the number of children are important factors in role conflict. Recent evidence shows that work stress has a much greater impact on family life than family stress has on work performance. Some women pay a high personal price for having careers.

12.5 TIME TO RELAX: LEISURE ACTIVITIES

What activities are leisure activities? How do people choose among them?

■ Leisure activities can be simple relaxation, activities for enjoyment, or creative pursuits. Views of leisure activities vary by gender, ethnicity, and age.

What changes in leisure activities occur with age?

■ As people grow older, they tend to engage in leisure activities that are less strenuous and more family-oriented. Leisure preferences in adulthood reflect those earlier in life.

What do people derive from leisure activities?

■ Leisure activities enhance well-being and can benefit all aspects of people's lives.

KEY TERMS

meaning-mission fit (421)
social cognitive career theory (SCCT) (421)
vocational maturity (423)
reality shock (425)
mentor or developmental coach (425)

job satisfaction (427)
alienation (428)
burnout (429)
passion (429)
gender discrimination (434)
glass ceiling (434)

glass cliff (434)
age discrimination (438)
career plateauing (439)
backup care (445)
work–family conflict (447)
leisure (449)

LEARN MORE ABOUT IT

Log in to **www.cengagebrain.com** to access the resources your instructor requires. For this book, you can access:

■ CourseMate brings course concepts to life with interactive learning, study, and exam preparation tools that support the printed textbook. A textbook-specific website, Psychology CourseMate includes an integrated interactive eBook and other interactive learning tools including quizzes, flashcards, videos, and more.

CENGAGENOW™

■ CengageNOW Personalized Study is a diagnostic study tool containing valuable text-specific resources—and because you focus on just what you don't know, you learn more in less time to get a better grade.

WebTUTOR™

■ More than just an interactive study guide, WebTutor is an anytime, anywhere customized learning solution with an eBook, keeping you connected to your textbook, instructor, and classmates.

Making It in Midlife
The Biopsychosocial Challenges of Middle Adulthood

THERE'S AN OLD SAYING that life begins at 40. That's good news for middle-aged adults. As we will see, they face many stressful events, but they also leave many of the pressures of young adulthood behind. In some respects, middle age is the prime of life: People's health is generally good, and their earnings are at their peak.

Of course, during middle age people typically get wrinkles, gray hair, and a bulging waistline. But middle-aged adults also achieve new heights in cognitive development, reevaluate their personal goals and change their behavior if they choose, develop adult relationships with their children, and ease into grandparenthood. Along the way, they must deal with stress, changes in the way they learn, and the challenges of helping their aging parents.

Some of these issues are based more on social stereotypes than on hard evidence. Which is which? You will know by the end of this chapter.

LEARNING OBJECTIVES

- How does appearance change in middle age?
- What changes occur in bones and joints?
- What reproductive changes occur in men and women in middle age?

- What is stress? How does it affect physical and psychological health?
- What benefits are there to exercise?

By all accounts, Dean is extremely successful. Among other things, he became the head of a moderate-sized manufacturing firm by the time he was 43. Dean has always considered himself to be a rising young star in the company. Then one day he found more than the usual number of hairs in his brush. "Oh no!" he exclaimed. "I can't be going bald! What will people say?" What does Dean think about these changes?

ONE MORNING, WHEN YOU LEAST EXPECT IT, YOU'LL BE STARING INTO THE BATHROOM MIRROR WHEN THE REALITY OF MIDDLE AGE STRIKES. Standing there, peering through half-awake eyes, you see *it*. One solitary gray hair or one tiny wrinkle at the corner of your eye—or, like Dean, some excess hairs falling out—and you worry that your youth is gone, your life is over, and you will soon be acting the way your parents did when they totally embarrassed you in your younger days. Middle-aged people become concerned that they are over the hill, sometimes going to great lengths to prove that they are still vibrant.

Crossing the boundary to middle age in the United States is typically associated with turning 40 (or "the big four-oh," as many people term it). This event is frequently marked with a special party, and the party often has an "over the hill" motif. Such events are society's attempt to create a rite of passage between youth and maturity.

As people move into middle age, they begin experiencing some of the physical changes associated with aging. In this section, we focus on the changes most obvious in middle-aged adults: appearance, reproductive capacity, and stress and coping. In Chapter 14, we will consider changes that may begin in middle age but are usually not apparent until later in life, such as slower reaction time and sensory changes. A critical factor in setting the stage for healthy aging is living a healthy lifestyle in young adulthood and middle age. Eating a healthy diet and exercising regularly across adulthood can help reduce the chances of chronic disease later in life (Leventhal et al., 2002).

Copyright © David Young-Wolff / Photo Edit

Turning age 40 usually marks the beginning of middle age.

Changes in Appearance

On that fateful day when the hard truth stares at you in the bathroom mirror, it probably doesn't matter to you that getting wrinkles and gray hair is universal and inevitable. Wrinkles are caused by changes in the structure of the skin and its connective and supporting tissues as well as by the cumulative effects of damage from exposure to sunlight and smoking cigarettes (Aldwin & Gilmer, 2004). It may not make you feel better to know that gray hair is perfectly natural and caused by a normal cessation of pigment production in hair follicles. Male pattern baldness, a genetic trait in which hair is lost progressively beginning with the top of the head, often begins to appear in middle age. No, the scientific evidence that these changes occur to many people isn't what matters most. What matters is that these changes are affecting *you*.

To make matters worse, you may also have noticed that your clothes aren't fitting properly even though you carefully watch what you eat. You remember a time not very long ago when you could eat whatever you wanted; now it seems that as soon as you look at food you put on weight. Your perceptions are correct; most people gain weight between their early 30s and mid-50s, producing the infamous "middle-aged bulge" as metabolism slows down (Aldwin & Gilmer, 2004).

People's reactions to these changes in appearance vary. Dean wonders how people will react to him now that he's balding. Some people rush out to purchase hair coloring and wrinkle cream. Others just take it as another stage in life. You've probably experienced several different reactions yourself. There is a wide range of individual differences, especially those between men and women and across cultures. As the cartoon depicts, certain changes on men in Western society are viewed as positive, but the same changes in women are not.

Changes in Bones and Joints

The bones and the joints change with age, sometimes in potentially preventable ways and sometimes because of genetic predisposition or disease. Let's take a closer look.

Osteoporosis

One physical change that can be potentially serious is loss of bone mass. **Skeletal maturity,** *the point at which bone mass is greatest and the skeleton is at peak development, occurs at around 19 for women and 20 in men* (National Institute of Arthritis and Musculoskeletal and Skin Diseases, 2010a). Bone mass stays about the same until women experience menopause and men reach late life. For women, there is a rapid loss of bone mass in the first few years after menopause, which greatly increases the risk of problems with disease and broken bones (National Institute of Arthritis and Musculoskeletal and Skin Diseases, 2009a).

Loss of bone mass makes bones weaker and more brittle, thereby making them easier to break. Because there is less bone mass, bones also take longer to heal in middle-aged and older adults. *Severe loss of bone mass results in* **osteoporosis,** *a disease in which bones become porous and extremely easy to break* (see ■ Figure 13.1). In severe cases, osteoporosis can cause spinal vertebrae to collapse, causing the person to stoop and to become shorter (National Institutes of Health, 2000a; see ■ Figure 13.2). About 40 million Americans either have osteoporosis or are at high risk due to low bone density. Non-Latina white women and Asian women are at highest risk, and osteoporosis is the leading cause of broken bones in older women (National Institute of Arthritis and Musculoskeletal and Skin Diseases, 2010b). Although the severe effects of osteoporosis typically are not observed until later life, this disease can occur in people in their 50s.

Osteoporosis is more common in women than men, for several reasons: women have less bone mass in general, some girls and women do not consume enough calcium

skeletal maturity
the point at which bone mass is greatest and the skeleton is at peak development, around 19 for women and 20 in men

osteoporosis
a disease in which bones become porous and extremely easy to break from severe loss of bone mass

The difference between normal bone (on the left) and osteoporosis (on the right) is easy to see in these sections of bone tissue from the hip.

Solid bone matrix

Weakened bone matrix

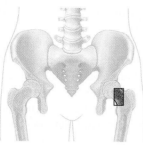

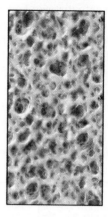

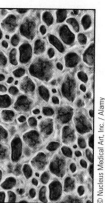

Bone section through hip

© Nucleus Medical Art, Inc. / Alamy

■ **Figure 13.2**
Notice how osteoporosis eventually causes a person to stoop and to lose height owing to compression of the vertebrae.

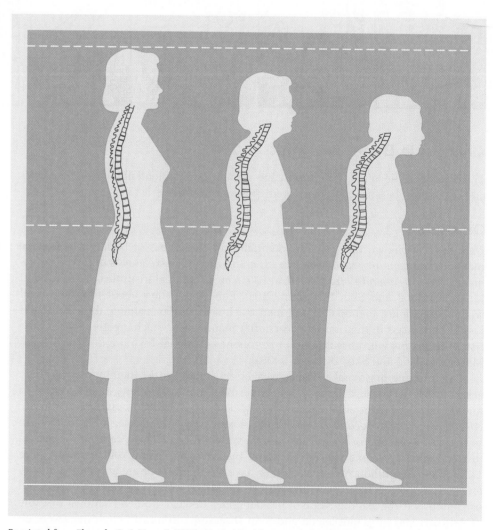

Reprinted from Ebersole, P., & Hess, P. (1998). *Toward healthy aging* (5th ed., p. 395), with permission from Elsevier Science.

Recommended Calcium and Vitamin D Intakes

Age	Calcium (milligrams)	Vitamin D (International Units)
Infants		
Birth–6 months	200	400
6 months–1 year	260	400
Children/Young Adults		
1–3 years	700	600
4–8 years	1,000	600
9–13 years	1,300	600
14–18 years	1,300	600
Adult Women and Men		
19–30 years	1,000	600
31–50 years	1,000	600
51- to 70-year-old males	1,000	600
51- to 70-year-old females	1,200	600
Over 70 years	1,200	800
Pregnant or Lactating Women		
18 years or younger	1,300	600
19–50 years	1,000	600

SOURCE: Food and Nutrition Board, Institute of Medicine, National Academy of Sciences, 2010
(www.niams.nih.gov/Health_Info/Bone/Osteoporosis/osteoporosis_hoh.asp#calcium)

to build strong bones (i.e., build bone mass) from childhood to young adulthood, and the decrease in estrogen following menopause greatly accelerates bone loss (National Institute of Arthritis and Musculoskeletal and Skin Diseases, 2010b).

Osteoporosis is caused in part by having low bone mass at skeletal maturity (the point at which your bones reach peak development), deficiencies of calcium and vitamin D, estrogen depletion, and lack of weight-bearing exercise that builds up bone mass. Other risk factors include smoking, high-protein diets, and excessive intake of alcohol, caffeine, and sodium. Women who are being treated for asthma, cancer, rheumatoid arthritis, thyroid problems, and epilepsy are also at increased risk because the medications used can lead to loss of bone mass (National Institute of Arthritis and Musculoskeletal and Skin Diseases, 2010b).

The National Institute of Arthritis and Musculoskeletal and Skin Diseases (2010b) recommends getting enough dietary calcium and vitamin D as ways to prevent osteoporosis. There is evidence that calcium and vitamin D supplements after menopause may slow the rate of bone loss and delay the onset of osteoporosis (National Institute of Arthritis and Musculoskeletal and Skin Diseases, 2010b). The best prevention occurs in youth, before skeletal maturity is reached, but research shows few children get the minimum daily requirements of calcium and vitamin D. People should consume foods (such as milk and broccoli) that are high in calcium and vitamin D. Recommended calcium and vitamin D intakes for men and women of various ages as determined by the Institute of Medicine (2010) are shown in ● Table 13.1. Data clearly show that metabolizing vitamin D directly affects rates of osteoporosis; however, whether supplementary dietary vitamin D retards bone loss is less certain (National Institute of Arthritis and Musculoskeletal and Skin Diseases, 2010b).

Women over age 65 are especially encouraged to have their bone mineral density (BMD) tested by having a **dual-energy X-ray absorptiometry (DXA) test**, *which measures bone density at the hip and spine* (National Institute of Arthritis and Musculoskeletal and Skin Diseases, 2009b). The DXA test results are usually compared to the

dual-energy X-ray absorptiometry (DXA) test
test of bone mineral density (BMD) at the hip and spine

ideal or peak bone mineral density of a healthy 30-year-old adult, and you are given a T-score. A score of 0 means your bone mineral density is equal to the norm for a healthy young adult. Differences between your bone mineral density and that of the healthy young adult norm are measured in units called standard deviations (SDs). The more standard deviations below 0, indicated as negative numbers, the lower your bone mineral density and the higher your risk of fracture. A T-score between +1 and −1 is considered normal or healthy. A T-score between −1 and −2.5 indicates low bone mass, although not low enough to be diagnosed with osteoporosis. A T-score of −2.5 or lower indicates that you have osteoporosis. The greater the negative number, the more severe the osteoporosis.

In terms of medication interventions, biophosphonates are the most commonly used and are highly effective (Kennel, 2007). Fosamax, Actonel, and Boniva are three common examples of this family of medications. Biophosphonates slow the bone breakdown process by helping to maintain bone density during menopause. Research indicates that using biophosphonates for up to 5 years appears relatively safe if followed by stopping the medication (called a "drug holiday"); there is evidence for protective effects lasting up to 5 years more.

Raloxifene (e.g., Evista) is also approved for the treatment and prevention of osteoporosis. It is one of a relatively new group of drugs known as selective estrogen receptor modulators. **Selective estrogen receptor modulators (SERMs)** *are not estrogens, but they are compounds that have estrogen-like effects on some tissues and estrogen-blocking effects on other tissues.* Raloxifene mimics the effects of estrogen on bones, but does not have estrogen's potentially harmful effects on breast tissue or the uterus. Raloxifene has been shown to prevent bone loss, have beneficial effects on bone mass, and reduce the risk of spine fractures (National Institute of Arthritis and Musculoskeletal and Skin Diseases, 2010b).

To reduce the risk of osteoporosis, the National Institutes of Health recommends dietary, medication, and activity approaches to prevent osteoporosis. Some evidence also supports the view that oral ingestion of magnesium, zinc, vitamin K, and special forms of fluoride may be effective. Estrogen replacement is effective in preventing women's bone loss after menopause but is controversial because of potential side effects (as discussed later). There is also some evidence that regular exercise is beneficial, but results vary depending on the type and intensity of the regimen. The best results come from a regular regimen of moderate weight- or load-bearing exercise, such as weight lifting, jogging, or other exercises that require you to work against gravity.

Arthritis

Many middle-aged adults complain of aching joints. They have good reason. Beginning in one's 20s, the protective cartilage in joints shows signs of deterioration, such as thinning and becoming cracked and frayed. *Over time the bones underneath the cartilage become damaged, which can result in* **osteoarthritis**, *a disease marked by gradual onset of bone damage with progression of pain and disability together with minor signs of inflammation.*

Osteoarthritis is an example of a **wear-and-tear disease**, *a degenerative disease caused by injury or overuse* (National Arthritis Foundation, 2010). This most common form of arthritis usually becomes noticeable in late middle age or early old age, progresses slowly, and is especially common in people whose joints are subjected to routine overuse and injury, such as athletes and manual laborers. Pain typically is worse when the joint is used, but skin redness, heat, and swelling are minimal or absent. Osteoarthritis usually affects the hands, spine, hips, and knees, sparing the wrists, elbows, shoulders, and ankles. Effective management approaches consist mainly of certain steroids and anti-inflammatory drugs, rest and nonstressful exercises that focus on range of motion, dietary modifications, and a variety of homeopathic remedies.

A second form of arthritis is **rheumatoid arthritis**, *a more destructive disease of the joints that also develops slowly; it typically affects different joints and causes different*

selective estrogen receptor modulators (SERMs)
compounds that are not estrogens but that have estrogen-like effects on some tissues and estrogen-blocking effects on other tissues

osteoarthritis
most common form of arthritis, a disease marked by gradual onset of bone damage with progression of pain and disability together with minor signs of inflammation from wear-and-tear

wear-and-tear disease
a degenerative disease caused by injury or overuse

rheumatoid arthritis
a more destructive disease of the joints that also develops slowly; it typically affects different joints and causes different types of pain than osteoarthritis

Similarities and Differences Among Osteoporosis, Osteoarthritis, and Rheumatoid Arthritis

	Osteoporosis	Osteoarthritis	Rheumatoid Arthritis
Risk Factors	X	X	
Age-related	X	X	
Menopause	X		
Family history	X	X	X
Use of certain medications such as glucocorticoids or seizure medications	X		
Calcium deficiency or inadequate vitamin D	X		
Inactivity	X		
Overuse of joints		X	
Smoking	X		
Excessive alcohol	X		
Anorexia nervosa	X		
Excessive weight		X	
Physical Effects			
Affects entire skeleton	X		
Affects joints		X	X
Is an autoimmune disease			X
Bony spurs		X	X
Enlarged or malformed joints	X	X	
Height loss	X		

SOURCES: National Institute of Arthritis and Musculoskeletal and Skin Diseases (2010), www.niams.nih.gov/Health_Info/Bone/Osteoporosis/Conditions_Behaviors/osteoporosis_arthritis.asp

types of pain than osteoarthritis. Most often, a pattern of morning stiffness and aching develops in the fingers, wrists, and ankles on both sides of the body. Joints appear swollen. There is no cure, but there are several treatment approaches (Matsumoto, Bathon, & Bingham, 2010; National Institute of Arthritis and Musculoskeletal and Skin Diseases, 2010c).

There are three general classes of medications commonly used in the treatment of rheumatoid arthritis: nonsteroidal anti-inflammatory agents (NSAIDs, such as Advil or Aleve), corticosteroids (such as prednisone), and disease modifying anti-rheumatic drugs (DMARDs, such as methotrexate) (Matsumoto et al., 2010). Because cartilage and bone damage frequently occur within the first 2 years of disease, physicians now move more aggressively to a DMARD agent early in the course of disease, usually as soon as a diagnosis is confirmed.

Rest and passive range-of-motion exercises are also helpful. Contrary to popular belief, rheumatoid arthritis is neither contagious nor self-induced by any known diet, habit, job, or exposure. Interestingly, the symptoms often come and go in repeating patterns (National Institute of Arthritis and Musculoskeletal and Skin Diseases, 2010c). Although it is not directly inherited, family history plays a role because researchers believe you can inherit a predisposition for the disease (Mayo Clinic, 2007).

Telling the differences among osteoporosis, osteoarthritis, and rheumatoid arthritis can be tricky. A comparison of risk factors and effects of each are shown in ●Table 13.2.

Reproductive Changes

Besides changes in the way we look, middle age brings transitions in our reproductive systems. These changes differ dramatically for women and men. Yet even in the context of these changes, middle-aged adults continue to have active sex lives. Belsky

(2007) reports that couples can and often do have sexual relationships that are very much alive and may be based on a newfound or re-found respect and love for each other. A major national survey by AARP (Fisher, 2010) found that middle-aged adults not only tend to continue to enjoy active sex lives but also enjoy romantic weekends, and about 6 of every 10 middle-aged men and women report that a satisfying sex life is important for their quality of life. However, both the frequency of sexual intercourse and satisfaction with sexual activity declined about 10 points between 2004 and 2009, perhaps due to an increase in overall stress levels in people's lives. These changes strongly emphasize the importance of environmental factors in sexual activity and satisfaction.

The Climacteric and Menopause

As women enter midlife, they experience a major biological process, called the **climacteric**, *during which they pass from their reproductive to nonreproductive years.* **Menopause** *is the point at which menstruation stops.* Men do not endure such sweeping biological changes but experience several gradual changes. These changes have important psychological implications because midlife is thought by many to be a key time for redefining ourselves, an issue we will examine later in this chapter. For example, some women view climacteric as the loss of the ability to have children, whereas others view it as liberating because they no longer need to worry about pregnancy.

The major reproductive change in women during adulthood is the loss of the ability to bear children. This change begins in the 40s as menstrual cycles become irregular, and by age 50 to 55 it is usually complete (Robertson, 2006). *This time of transition from regular menstruation to menopause is called* **perimenopause**, *and how long it lasts varies considerably* (Mayo Clinic, 2010b). The gradual loss and eventual end of monthly periods is accompanied by decreases in estrogen and progesterone levels, changes in the reproductive organs, and changes in sexual functioning (Aldwin & Gilmer, 2004).

A variety of physical and psychological symptoms may accompany perimenopause and menopause as a result of decreases in hormonal levels (Mayo Clinic, 2010b; Robertson, 2006): hot flashes, night sweats, headaches, mood changes, difficulty concentrating, vaginal dryness, changing cholesterol levels, and a variety of aches and pains. Negative effects on sexuality, such as low libido, are common (Myskow, 2002). Chinese women reported increased sleep disturbances and fatigue (Chang et al., 2010). Many women report no symptoms at all but most women experience at least some, and there are large ethnic and cultural group differences in how they are expressed (Banger, 2003). For example, studies of European American women reveal a decrease in reported physical symptoms after climacteric. In contrast, African American women reported more physical symptoms after climacteric than before. Although these differences could be a function of the different age groups included in the various studies, they also draw attention to the experiences of women from different ethnic and racial backgrounds, where much of the focus has been on Asian women (Shea, 2006).

Cultural differences are exemplified in Lock's (1991) classic study of Japanese women. Fewer than 13% of Japanese women whose menstrual periods were becoming irregular reported having hot flashes during the previous 2 weeks, compared with nearly half of Western women. In fact, fewer than 20% of Japanese women in the study had ever had a hot flash, compared with nearly 65% of Western women. However, Japanese women reported more headaches, shoulder stiffness, ringing in the ears, and dizziness than Western women. Why? The answer seems to be the power of sociocultural forces. In Japan, the government considered "menopausal syndrome" to be a modern affliction of women with too much time on their hands. With this official attitude, it is hard to know whether Japanese women actually experience menopause differently or may simply be reluctant to describe their true experience.

Similar findings were reported by Fu, Anderson, and Courtney (2003), who compared Taiwanese and Australian women. Significant differences were found in their attitudes toward menopause, menopausal symptoms, and physical vitality. These re-

---

climacteric
biological process during which women pass from their reproductive to nonreproductive years

menopause
the point at which menstruation stops

perimenopause
the individually varying time of transition from regular menstruation to menopause

sults clearly indicate that sociocultural factors are critical in understanding women's experience during menopause. But Shea (2006) found that Chinese women reported symptoms at a level more similar to American women, a result that emphasizes the cultural differences within Asia. Understanding cultural differences in reporting of various symptoms is a complex issue that may reflect cultural norms about self-reporting medical symptoms as well as accurate reporting of what is experienced.

Women's genital organs undergo progressive change after menopause (Aldwin & Gilmer, 2004). The vaginal walls shrink and become thinner, the size of the vagina decreases, vaginal lubrication is reduced and delayed, and some shrinkage of the external genitalia occurs. These changes have important effects on sexual activity, such as an increased possibility of painful intercourse and a longer time and more stimulation needed to reach orgasm. Failure to achieve orgasm is more common than in a woman's younger years. However, maintaining an active sex life throughout adulthood lowers the degree to which problems are encountered.

Despite the physical changes, there is no physiological reason most women cannot continue sexual activity and enjoy it well into old age. Whether this happens depends more on the availability of a suitable partner than on a woman's desire for sexual relations. This is especially true for older women. All three AARP surveys of sexual activity among middle-aged and older adults (AARP, 1999; Fisher, 2010; Jacoby, 2005) found that older married women were far more likely to have an active sex life than unmarried women. The primary reason for the decline in women's sexual activity with age is the lack of a willing or appropriate partner, not a lack of physical ability or desire (AARP, 1999; Fisher, 2010; Jacoby, 2005).

Reproductive technology such as fertility drugs and in vitro fertilization (see Chapter 2) has made it possible for postmenopausal women to have children. Indeed, in 2008 Rajo Devi Lohan of India gave birth at age 70 through in vitro fertilization, making her the world's oldest woman to become pregnant and deliver. Scientists have thus fundamentally changed the rules of reproduction, although the risks of doing so are great to the older mother and child. Still, even though a woman has gone through the climacteric, she can still have children. Technology can make her pregnant if she so chooses and if she has access to the proper medical centers.

What does this do to our understanding of human reproduction? It changes the whole notion of menopause as an absolute end to childbearing. Some of the women who have given birth after menopause have done so because their daughters were unable to have children; they consider this act another way to show their parental love. Others view it as a way to equalize reproductive potential in middle age between men and women, because men remain fertile throughout adulthood.

Clearly, these are complicated issues that currently affect a very small number of women. But as reproductive technology continues to advance faster than our ability to think through the issues, we will be confronted with increasingly complex ethical questions (Lindlaw, 1997). Should children be born to older parents? Might not there be some advantage, considering the life experience such parents would have, compared to young parents? Are such births merely selfish acts? Are they a viable alternative way for younger adults to have a family? What dangers are there to older pregnant women? How do you feel about it?

Despite physical changes associated with middle age, women and men continue to enjoy sexual activity.

Treating Symptoms of Menopause

The decline in estrogen that women experience after menopause is related to increased risk of osteoporosis, cardiovascular disease, stress urinary incontinence (involuntary loss of urine during physical stress, as when exercising, sneezing, or laughing), weight gain, and memory loss (Dumas et al., 2010; Mayo Clinic, 2010c). In the case of cardiovascular disease, at age 50 (prior to menopause) women have 3 times less risk of heart attacks than men on average. Ten years after menopause, when women are about 60, their risk equals that of men.

In response to these increased risks and to the estrogen-related symptoms that women experience, one approach is the use of **menopausal hormone therapy (MHT)**:

menopausal hormone therapy (MHT)

medication therapy in which women take low doses of estrogen, which is often combined with progestin (synthetic form of progesterone) to counter symptoms associated with menopause

women take low doses of estrogen, which is often combined with progestin (synthetic form of progesterone). Hormone therapy is controversial and has been the focus of many research studies with conflicting results (Bach, 2010; Mayo Clinic, 2010d; Shapiro, 2007). There appear to be both benefits and risks with MHT, as discussed in the What Do *You* Think? Feature.

What do you think? Menopausal Hormone Therapy

For many years, women have had the choice of taking medications to replace the female hormones that are not produced naturally by the body after menopause. Hormone therapy may involve taking estrogen alone or in combination with progesterone (or progestin in its synthetic form). Until about 2003, it was thought that menopausal hormone therapy (MHT) was beneficial for most women, and results from several studies were positive. But results from the Women's Health Initiative research in the United States and from the Million Women Study in the United Kingdom indicated that, for some types of MHT, there were several potentially serious side effects. As a result, physicians are now far more cautious in recommending MHT.

The Women's Health Initiative (WHI), begun in the United States in 1991, was a very large study (National Heart, Lung, and Blood Institute, 2003). The estrogen plus progestin trial used 0.625 milligram of estrogens taken daily plus 2.5 milligrams of medroxyprogesterone acetate (Prempro) taken daily. This combination was chosen because it is the mostly commonly prescribed form of the combined hormone therapy in the United States and, in several observational studies, had appeared to benefit women's health. The women in the WHI estrogen plus progestin study were aged 50 to 79 when they enrolled in the study between 1993 and 1998. The health of study participants was carefully monitored by an independent panel called the Data and Safety Monitoring Board (DSMB). The study was stopped in July 2002 because investigators discovered a significant increased risk for breast cancer and that overall the risks outnumbered the benefits. However, in addition to the increased risk of breast cancer, heart attack, stroke, and blood clots, MHT resulted in fewer hip fractures and lower rates of colorectal cancer.

The Million Women Study began in 1996 and includes 1 in 4 women over age 50 in the United Kingdom, the largest study of its kind ever conducted. Like the Women's Health Initiative, the study examined how MHT (both estrogen/progestin combinations and estrogen alone) affects breast cancer, cardiovascular disease, and other aspects of women's health. Results from this study confirmed the Women's Health Initiative outcome of increased risk for breast cancer associated with MHT.

The combined results from the WHI and the Million Women Study led physicians to recommend that women over age 60 should not begin MHT to relieve menopausal symptoms or protect their health. In fact, women over age 60 who begin MHT are at increased risk for certain cancers.

In sum, women face difficult choices when deciding whether to use MHT as a means of combatting certain menopausal symptoms and protecting themselves against other diseases. For example, MHT can help reduce hot flashes and night sweats, help reduce vaginal dryness and discomfort during sexual intercourse, slow bone loss, and perhaps ease mood swings. On the other hand, MHT can increase a woman's risk of blood clots, heart attack, stroke, breast cancer, and gallbladder disease.

The best course of action is to consult closely with one's physician to weigh the benefits and risks. It's also a good idea to keep in mind several key points (Womenshealth.gov, 2010):

- Once a woman reaches menopause, MHT is recommended only as a short-term treatment.
- Doctors very rarely recommend MHT to prevent certain chronic diseases like osteoporosis.
- Women who have gone through menopause should not take MHT to prevent heart disease.
- MHT should not be used to prevent memory loss, dementia, or Alzheimer's disease.

What do *you* think is the best course of action? How would you decide whether to use MHT?

There are lifestyle and alternative medicine approaches to addressing both estrogen-related and somatic symptoms of menopause (Mayo Clinic, 2010d, 2010e). Herbal remedies, especially those rich in phytoestrogens (such as soybeans, chickpeas, and other legumes), are used effectively in Asian cultures and may be one reason why Asian American women report the fewest symptoms (Mayo Clinic, 2010e). The use of a nonpetroleum-based lubrication (such as K-Y Jelly) usually solves the problem of vaginal dryness, which often makes intercourse painful. Additionally, exercise (especially Kegel pelvic exercises), yoga, and good sleep habits also reduce symptoms (Mayo Clinic, 2010d).

Reproductive Changes in Men

Unlike women, men do not have a clear physiological (or cultural) event to mark reproductive changes, although there is a gradual decline in testosterone levels (Bribiescas, 2010). Most men never experience a complete loss of the ability to father children, but men do experience a normative decline in the quantity of sperm (Knowles, 2006). However, even at age 80 a man is still half as fertile as he was at age 25 and is quite capable of fathering a child.

With increasing age the prostate gland enlarges, becomes stiffer, and may obstruct the urinary tract. Prostate cancer becomes a real threat during middle age, and its diagnosis and treatment are controversial (Wolf et al., 2010). Following an extensive review of research and clinical evidence in 2009, the American Cancer Society currently recommends that men with average risk of developing prostate cancer be informed at age 50 of the benefits and risks of the available diagnostic tests (e.g., digital rectal exam, prostate-specific antigen [PSA] blood test). The PSA test, though potentially useful, has a high error rate in that even though many men with elevated PSA levels actually have prostate cancer a majority of men with elevated PSA levels do not. Men over age 50 should work with their physician to determine the best course of action for diagnosis.

Testosterone levels decline about 1% per year after age 40, and about 20% of men over age 60 have levels below the lower limit of the normal range (Gupta & Agarwal, 2010). A few men who experience an abnormally rapid decline in testosterone production during midlife or early old age report symptoms similar to those experienced by some menopausal women, such as hot flashes, chills, rapid heart rate, and nervousness (Pines, in press).

Men experience some physiological changes in sexual performance. By old age, men report less perceived demand to ejaculate, a need for longer time and more stimulation to achieve erection and orgasm, and a much longer resolution phase during which erection is impossible (Saxon & Etten, 1994). Older men also report more frequent failures to achieve orgasm and loss of erection during intercourse (Fisher, 2010). However, the advent of Viagra and other medications to treat erectile dysfunction has provided easy-to-use medical treatments and the possibility of an active sex life well into later life.

As with women, as long as men enjoy sex and have a willing partner, sexual activity is a lifelong option. As for women, the most important ingredient of sexual intimacy for men is a strong relationship with a partner (Fisher, 2010). For example, married men in early middle age tend to have intercourse 4–8 times per month. The loss of an available partner is a significant reason that frequency of intercourse drops on average by 2 and 3 times per month in men over age 50 and 60, respectively (Araujo, Mohr, & McKinlay, 2004).

THINK ABOUT IT

Why does sexual desire remain largely unchanged despite the biological changes that are occurring?

Work-related stress is a major problem around the world and can have serious negative effects on physical and psychological health.

Stress and Health

There's no doubt about it—life is full of stress. Think for a moment about all the things that bother you, such as exams, jobs, relationships, and finances. For most people, this list lengthens quickly. But, you may wonder, isn't this true for people of all ages? Is stress more important in middle age?

Although stress affects people of all ages, it is during middle age that the effects of both short- and long-term stress become most apparent. In part, this is because it takes time for stress disorders to manifest themselves, and in part it is due to the gradual loss of physical capacity as the normal changes accompanying aging begin to take their toll. As we will see, psychological factors also play a major role.

You may think that stress affects health mainly in people who hold certain types of jobs, such as air traffic controllers and high-level business executives. In fact, business executives actually have *fewer* stress-related health problems than waitresses, construction workers, secretaries, laboratory technicians, machine operators, farm workers, and painters. Why? Even though business executives are often under great stress and tend to be isolated and lonely (Cooper & Quick, 2003), they have better outlets for their stress, such as the ability to delegate problems, and they are in control. What do all of these truly high-stress jobs have in common? These workers have little direct control over their jobs.

Although we understand some important workplace factors related to stress, our knowledge is largely based on research examining middle-aged men. Unfortunately, the relation of stress to age, gender, and ethnic status remains to be researched. Women tend to rate their stressful experiences as more negative and uncontrollable than do men, and they report stress most often in family and health areas compared to men's reports of financial and work-related stressors (Matud, 2004). Middle-aged people report the highest levels of stress, whereas people over age 65 report the lowest. Why? As we will see in the next section, part of the reason may be the number of pressures that middle-aged people feel: Children may be in college, the job has high demands and fears of losing one's job are especially scary, the mortgage payment and other bills always need paying, the marriage needs some attention, parents and parents-in-law are experiencing health issues and need assistance, and on it goes.

What Is Stress?

Think about the last time you felt stressed. What was it about the situation that made you feel stressed? How did you feel? *The answers to questions like these provide a way to understand the dominant framework used to study stress, the* **stress and coping paradigm**, *which emphasizes the transactions between a person and the environment.* Because the stress and coping paradigm emphasizes these transactions, it fits well with the biopsychosocial framework. An example of a transactional model of stress is shown in ■ Figure 13.3.

Physiologically, stress refers to a number of specific changes in the body, including increased heart rate, sweaty palms, and hormone secretion (Lyness, 2007). In the short run, stress can be beneficial and may even allow you to perform at your peak. In the long run, though, a high physical and psychological toll and even death may result (Lyness, 2007).

One of the most widely supported theories of stress over the past several decades is based on how we think about events. In this view, whether you report feeling stressed depends on how you interpret a situation or event (Folkman, 2008; Lazarus & Folkman, 1984). What the situation or event is, or what you do to deal with it, does not matter. Stress results from an appraisal of a situation or event as taxing or exceeding your personal, social, or other resources and endangering your well-being. It is the

stress and coping paradigm
the dominant framework used to study stress, which emphasizes the transactions between a person and the environment

■ **Figure 13.3**
The physical markers of stress are the result of complex and dynamic psychological processes.

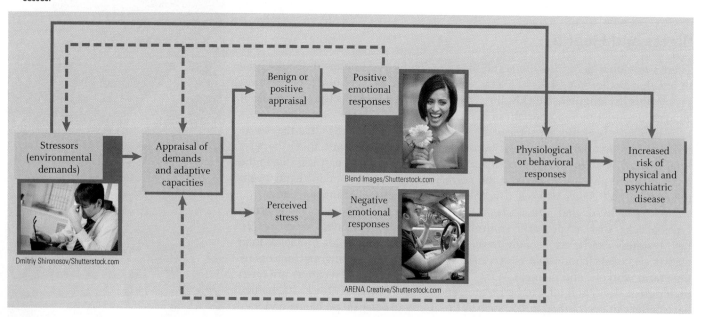

Based on *Measuring stress: A guide for health and social scientists,* edited by Sheldon Cohen, Ronald C. Kessler, and Lynn Underwood Gordon, copyright © 1995 by Oxford University Press, Inc.

day-to-day hassles, or the things that upset and annoy us, that prove to be particularly stressful.

Interestingly, culture plays an important role in how people perceive stress (Laungani, 2001). These differences are grounded in the values people hold. For example, what constitutes stressors varies a great deal between Eastern societies such as India and Western societies such as England. Indians tend to believe that much of life is determined, whereas the British tend to emphasize personal choice and free will. Consequently, frustrations that Britons may feel when free will is thwarted may not be perceived as stressful by Indians. These differences point out the importance of understanding a culture when studying a concept such as stress.

Likewise, understanding ethnic issues and how they relate to stress is important. Fernander, Schumacher, and Nasim (2008) found that cultural distrust and race-related stress predicted the use of tobacco as a stress-reducing behavior.

Coping *is any attempt to deal with stress.* People cope in several different ways (Kinney et al., 2003). Sometimes people cope by trying to solve the problem at hand; for example, you may cope with a messy roommate by moving out. At other times, people focus on how they feel about the situation and deal with things on an emotional level; feeling sad after breaking up with your partner would be one way of coping with the stress of being alone. Sometimes people cope by simply redefining the event as not stressful—an example of this approach would be saying that it was no big deal you failed to get the job you wanted. Still others focus on religious or spiritual approaches, often through prayer, perhaps asking God for help.

People appraise different types of situations or events as stressful at different times during adulthood. For example, the pressures from work and raising a family are typically greater for younger and middle-aged adults than for older adults. However, stressors due to chronic disease are often more important to older adults than to their younger counterparts. Similarly, the same kind of event may be appraised differently at different ages. For example, uncertainty about one's job security may be less stressful in young adulthood, when one might get another job more easily, than in middle age, when alternative job prospects might diminish. From a biopsychosocial perspective, such life-cycle factors must be taken into account when considering what kinds of stress adults of different ages are experiencing.

coping
any attempt to deal with stress

THINK ABOUT IT

How might life experience and cognitive developmental level influence the appraisal of and coping with stress?

How Are Stress and Coping Related to Physical Health?

A great deal of research has been conducted over the years examining links between stress and physical health. Being under chronic stress suppresses the immune system (resulting in increased susceptibility to viral infections), increases risk of atherosclerosis (buildup of plaque along the walls of arteries so that the arteries become stiffer and restrict blood flow) and hypertension (high blood pressure), and leads to impaired memory and cognition (McEwen & Gianaros, 2010; Yancura & Aldwin, 2010). However, these effects depend on the kind of event, as well as one's socioeconomic status (Kemeny, 2003; McEwen & Gianaros, 2010). Experiencing negative events tends to lower immune function, whereas experiencing positive events tends to improve immune functioning.

Many specific diseases and conditions are caused or exacerbated by stress (McEwen & Gianaros, 2010; Yancura & Aldwin, 2010). Stress serves as a major trigger for angina (pain caused by interrupted blood flow to the heart); causes arrhythmias (irregular pulse) and blood to become stickier (making it more likely to cause a clot in an artery); raises cholesterol; reduces estrogen in women; increases production of certain proteins that damage cells; causes sudden increases in blood pressure; increases the risk of irritable bowel syndrome; causes weight fluctuations; is associated with the development of insulin resistance (a primary factor in diabetes); causes tension headaches, sexual dysfunction, and infertility; and results in poorer memory and cognitive performance. Clearly, chronic stress is harmful to one's health!

Surprisingly, little research has been conducted testing whether successful coping strategies reverse these health effects of stress. At best, we can only surmise that if stress causes these health problems, effective coping strategies may prevent them.

How Are Stress and Coping Related to Behavior and Psychological Health?

Probably the most well-known connection between stress and behavior involves the link with cardiovascular disease. Due mostly to the pioneering work of Friedman and Rosenman (1974), we know that two behavior patterns differ dramatically in terms of risk of cardiovascular disease. *People who demonstrate a* **Type A** *behavior pattern tend to be intensely competitive, angry, hostile, restless, aggressive, and impatient. In contrast, people who show a* **Type B** *behavior pattern tend to be just the opposite.* Type A individuals are at least twice as likely as Type B people to develop cardiovascular disease, even when other risk factors such as smoking and hypertension are taken into account. In fact, Type A behavior is a more important predictor of cardiovascular disease than body weight, alcohol intake, or activity level (Zmuda et al., 1997).

How do these behavior types relate to *recovery* from a heart attack? Although it is relatively rare, Type B people sometimes do have heart attacks. Who recovers better, Type A people or Type B people?

The answer—based on a set of classic studies—may surprise you. Ragland and Brand (1988) conducted a 22-year longitudinal follow-up of the original Friedman and Rosenman study and discovered that Type A people recover from a heart attack better than Type B people. Why? Some of the characteristics of being Type A may help motivate people to stick to diet and exercise regimens after heart attacks and to have a more positive attitude toward recovery (Ivancevich & Matteson, 1988). Indeed, although the anger and hostility components of Type A behavior increase the risk for cardiovascular disease, the other components appear to aid the recovery process (Ivancevich & Matteson, 1988). In contrast, the laid-back approach to life of Type B people may actually work against them during recovery.

Experiencing stress can trigger psychological processes and reactions. Although stress does not directly cause psychopathology, it does influence how people react and behave. For example, the stress many people experienced after the terrorist attacks of September 11, 2001, resulted in higher levels of anxiety experienced through nightmares, flashbacks, insomnia, traumatic grief, emotional numbing, and avoidance (LeDoux & Gorman, 2001). *The National Institute of Mental Health (2010a) defines* **post-traumatic stress disorder (PTSD)** *as an anxiety disorder that can develop after exposure to a terrifying event or ordeal in which grave physical harm occurred or was threatened.* The kinds of traumatic events that may trigger PTSD include violent personal assaults, natural or human-caused disasters, accidents, or military combat.

Data examining ethnic group differences highlight the importance of self-esteem in how a person deals with stress. For example, a national study of Latina American professionals showed that higher self-esteem predicted lower levels of stress, marital stress, family–cultural conflict, and occupational–economic stress (Arellano, 2001). Additionally, results indicated that emotion-focused coping (focusing on controlling emotional reactions to a problem) reflects Western concepts of coping; other traditional approaches to coping do not capture the dynamic process of coping that these women showed. That is, Latinas use more complex and not-so-neatly categorized coping styles that don't fit traditional labels. Mexican immigrant farmworkers who reported high levels of stress from cultural pressures also reported lower levels of self-esteem and higher levels of symptoms of depression (Hovey & Magana, 2000). Cross-cultural research in Hong Kong indicates that, with increased age, the effects of stress on one's well-being are reduced (Siu et al., 2001). This could be a result of people learning how to cope better as they gain experience in dealing with stress. Research on job stress in Taiwan shows that culture needs to be taken into account when designing stress-reduction interventions and that results from U.S. research may not generalize to other cultural contexts (Chang & Lu, 2007).

Another way to lessen the effects of stress is to disclose and discuss one's health problems. For example, women who disclose the fact that they have breast cancer, which is a source of considerable stress in their lives, had more optimism and lower reported levels of stress than women who did not disclose their disease (Henderson et al., 2002). How people disclose such information matters. Pennebaker and Graybeal (2001) showed

---

Type A behavior pattern
a behavior pattern in which people tend to be intensely competitive, angry, hostile, restless, aggressive, and impatient

Type B behavior pattern
a behavior pattern that is the opposite of Type A

post-traumatic stress disorder (PTSD)
an anxiety disorder that can develop after exposure to a terrifying event or ordeal in which grave physical harm occurred or was threatened

that particular patterns of word use can be analyzed by a computer to predict health and personality style. Such analyses may prove useful to physicians and clinicians in providing guidance to individuals who need help in discussing stressful situations.

On a larger scale, the Health and Safety Commission in the United Kingdom developed an extensive program to lower work-related stress (Cox & Griffiths, 2010; Health and Safety Executive, 2010). The management standards that were developed address six key areas, each with a goal and specific behaviors that organizations must address: demands, control, support, relationships, role, and organizational change. This independent agency monitors numerous health aspects of the workplace, and has had a direct hand in developing many health-related policies in the United Kingdom.

Exercise

Ever since the time of Hippocrates, physicians and researchers have known that exercise significantly slows the aging process. Indeed, evidence suggests that a program of regular exercise, in conjunction with the healthy lifestyles discussed in Chapter 10, can slow the physiological aging process (Rogers, 2010). Being sedentary is hazardous to your health.

Adults benefit from **aerobic exercise**, *which places moderate stress on the heart by maintaining a pulse rate between 60 and 90% of the person's maximum heart rate.* You can calculate your maximum heart rate by subtracting your age from 220. Thus, if you are 40 years old, your target range would be 108–162 beats per minute. Examples of aerobic exercise include jogging, step aerobics, swimming, and cross-country skiing.

How much exercise is ideal? The U.S. Department of Health and Human Services (2008) established the first-ever guidelines for physical activity in 2008. Adults should average 150 minutes per week of moderate-intensity aerobic exercise, 75 minutes of vigorous-intensity aerobic activity, or a combination of the two. Strengthening exercises are recommended at least twice per week.

What happens when a person follows these guidelines and exercises aerobically (besides becoming tired and sweaty)? Physiologically, adults of all ages show improved cardiovascular functioning and maximum oxygen consumption; lower blood pressure; and better strength, endurance, flexibility, and coordination (Mayo Clinic, 2010f). Psychologically, people who exercise aerobically report lower levels of stress, better moods, and better cognitive functioning (Mayo Clinic, 2010f).

aerobic exercise
exercise that places moderate stress on the heart by maintaining a pulse rate between 60 and 90% of the person's maximum heart rate

Engaging in an aerobic exercise program throughout middle age is a great way to stay fit and stay healthy.

The best way to gain the benefits of aerobic exercise is to maintain physical fitness through the life span, beginning at least in middle age. The Mayo Clinic's Fitness website (www.mayoclinic.com/health/fitness/MY00396/TAB=indepth) provides an excellent place to start. In planning an exercise program, three points should be remembered. First, check with a physician before beginning an aerobic exercise program. Second, bear in mind that moderation is important. Third, just because you intend to exercise doesn't mean you will; you must take the necessary steps to turn your intention into action (Schwarzer, 2008).

Test Yourself

RECALL

1. Severe bone loss may result in the disease _____.

2. The cessation of menstruation is termed _____.

3. Reduction of fertility in men usually occurs _____.

4. The stress and _____ paradigm defines stress on the basis of the person's appraisal of a situation as taxing his or her well-being.

5. Research indicates that Type _____ individuals have a better chance of recovering from a heart attack than Type _____ individuals.

INTERPRET

The media are full of advertisements for anti-aging creams, diets, and exercise plans. Based on what you have read in this section, how would you evaluate these ads?

APPLY

What would be an ideal stress reduction exercise program for middle-aged adults?

Recall answers: (1) osteoporosis, (2) menopause (3) gradually, (4) coping, (5) A; B

13.2 Cognitive Development

LEARNING OBJECTIVES

- How does practical intelligence develop in adulthood?
- How does a person become an expert?

- What is meant by lifelong learning? What differences are there between adults and young people in how they learn?

Kesha, a 54-year-old social worker, is widely regarded as the resident expert when it comes to working the system of human services. Her co-workers admire her ability to get several agencies to cooperate, which they do not do normally, and to keep clients coming in for routine matters and follow-up visits. Kesha claims there is nothing magical about it—it's just her experience that makes the difference.

COMPARED TO THE RAPID COGNITIVE GROWTH OF CHILDHOOD OR THE CONTROVERSIES ABOUT POSTFORMAL COGNITION IN YOUNG ADULTHOOD, cognitive development in middle age is relatively quiet. For the most part, the trends in intellectual development discussed in Chapter 10 are continued and solidified. The hallmark of cognitive development in middle age involves developing higher levels of expertise like Kesha shows and flexibility in solving practical problems, such as dealing with complex forms like the one shown in ■ Figure 13.4. We will also see how important it is to continue learning throughout adulthood.

Practical Intelligence

Take a moment to think about the following problems (Denney, 1989, 1990; Denney, Pearce, & Palmer, 1982):

- A middle-aged woman is frying chicken in her home when, all of a sudden, a grease fire breaks out on top of the stove. Flames begin to shoot up. What should she do?

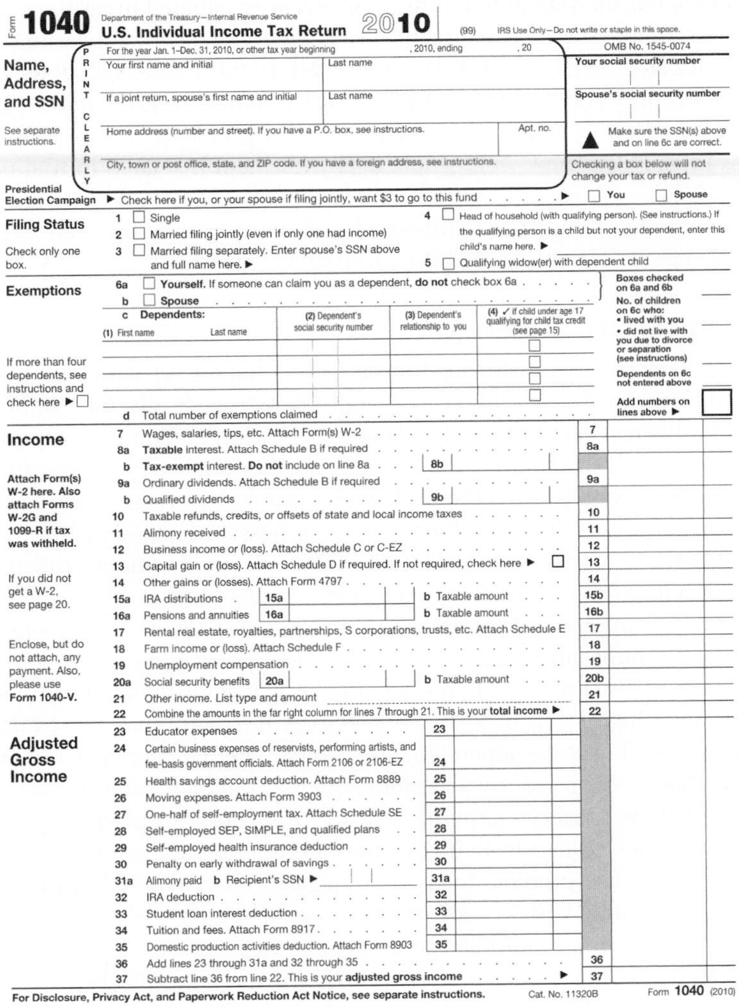

Form **1040** Department of the Treasury—Internal Revenue Service
U.S. Individual Income Tax Return **2010** (99) IRS Use Only—Do not write or staple in this space.

For the year Jan. 1–Dec. 31, 2010, or other tax year beginning , 2010, ending , 20 OMB No. 1545-0074

Name, Address, and SSN — PRINT CLEARLY
See separate instructions.

- Your first name and initial | Last name | **Your social security number**
- If a joint return, spouse's first name and initial | Last name | **Spouse's social security number**
- Home address (number and street). If you have a P.O. box, see instructions. | Apt. no. | ▲ Make sure the SSN(s) above and on line 6c are correct.
- City, town or post office, state, and ZIP code. If you have a foreign address, see instructions. | Checking a box below will not change your tax or refund.

Presidential Election Campaign ▶ Check here if you, or your spouse if filing jointly, want $3 to go to this fund ▶ ☐ You ☐ Spouse

Filing Status
Check only one box.

1 ☐ Single
2 ☐ Married filing jointly (even if only one had income)
3 ☐ Married filing separately. Enter spouse's SSN above and full name here. ▶
4 ☐ Head of household (with qualifying person). (See instructions.) If the qualifying person is a child but not your dependent, enter this child's name here. ▶
5 ☐ Qualifying widow(er) with dependent child

Exemptions

6a ☐ **Yourself.** If someone can claim you as a dependent, **do not** check box 6a
b ☐ **Spouse**
c **Dependents:**

(1) First name Last name	(2) Dependent's social security number	(3) Dependent's relationship to you	(4) ✓ if child under age 17 qualifying for child tax credit (see page 15)
			☐
			☐
			☐
			☐

If more than four dependents, see instructions and check here ▶ ☐

d Total number of exemptions claimed

Boxes checked on 6a and 6b
No. of children on 6c who:
• lived with you
• did not live with you due to divorce or separation (see instructions)
Dependents on 6c not entered above
Add numbers on lines above ▶

Income

Attach Form(s) W-2 here. Also attach Forms W-2G and 1099-R if tax was withheld.

If you did not get a W-2, see page 20.

Enclose, but do not attach, any payment. Also, please use **Form 1040-V.**

Line	Description	Line box
7	Wages, salaries, tips, etc. Attach Form(s) W-2	7
8a	**Taxable** interest. Attach Schedule B if required	8a
b	**Tax-exempt** interest. **Do not** include on line 8a — 8b	
9a	Ordinary dividends. Attach Schedule B if required	9a
b	Qualified dividends — 9b	
10	Taxable refunds, credits, or offsets of state and local income taxes	10
11	Alimony received	11
12	Business income or (loss). Attach Schedule C or C-EZ	12
13	Capital gain or (loss). Attach Schedule D if required. If not required, check here ▶ ☐	13
14	Other gains or (losses). Attach Form 4797	14
15a	IRA distributions — 15a — b Taxable amount	15b
16a	Pensions and annuities — 16a — b Taxable amount	16b
17	Rental real estate, royalties, partnerships, S corporations, trusts, etc. Attach Schedule E	17
18	Farm income or (loss). Attach Schedule F	18
19	Unemployment compensation	19
20a	Social security benefits — 20a — b Taxable amount	20b
21	Other income. List type and amount	21
22	Combine the amounts in the far right column for lines 7 through 21. This is your **total income** ▶	22

Adjusted Gross Income

Line	Description	Line box
23	Educator expenses	23
24	Certain business expenses of reservists, performing artists, and fee-basis government officials. Attach Form 2106 or 2106-EZ	24
25	Health savings account deduction. Attach Form 8889	25
26	Moving expenses. Attach Form 3903	26
27	One-half of self-employment tax. Attach Schedule SE	27
28	Self-employed SEP, SIMPLE, and qualified plans	28
29	Self-employed health insurance deduction	29
30	Penalty on early withdrawal of savings	30
31a	Alimony paid b Recipient's SSN ▶	31a
32	IRA deduction	32
33	Student loan interest deduction	33
34	Tuition and fees. Attach Form 8917	34
35	Domestic production activities deduction. Attach Form 8903	35
36	Add lines 23 through 31a and 32 through 35	36
37	Subtract line 36 from line 22. This is your **adjusted gross income** ▶	37

For Disclosure, Privacy Act, and Paperwork Reduction Act Notice, see separate instructions. Cat. No. 11320B Form **1040** (2010)

SOURCE: http://www.irs.gov/formspubs/index.html

■ **Figure 13.4**
The tax returns that people complete are an example of everyday problem-solving tasks.

- A man finds that the heater in his apartment is not working. He asks his landlord to send someone out to fix it, and the landlord agrees. But after a week of cold weather and several calls to the landlord, the heater is still not fixed. What should the man do?

These practical problems are different from the examples of measures of fluid and crystallized intelligence in Chapter 10. They are more realistic; they reflect real-world situations that people routinely face. Many researchers argue that using such problems to assess cognition provides a better assessment of the kinds of skills adults actually use in everyday life (Diehl et al., 2005; Marsiske & Margrett, 2006; Margrett et al., 2010). Most people spend more time at tasks such as managing their personal finances, dealing with uncooperative people, and juggling busy schedules than they do solving esoteric mazes.

The shortcomings of traditional tests of adults' intelligence led to different ways of viewing intelligence that differentiate academic (or traditional) intelligence from other skills (Diehl et al., 2005; Marsiske & Margrett, 2006; Margrett et al., 2010; Sternberg & Grigorenko, 2000). *The broad range of skills related to how individuals shape, select, or adapt to their physical and social environments is termed* **practical intelligence**. The examples at the beginning of this section illustrate how practical intelligence is measured. Such real-life problems differ in three main ways from traditional tests (Diehl et al., 2005): People are more motivated to solve them; personal experience is more relevant; and they have more than one correct answer. Research evidence supports the view that practical intelligence is distinct from general cognitive ability (Margrett et al., 2010; Taub et al., 2001).

practical intelligence
the broad range of skills related to how individuals shape, select, or adapt to their physical and social environments

When people's answers to practical problems are evaluated in terms of how likely their answers are to be effective, practical intelligence does not appear to decline appreciably until late life (Heidrich & Denney, 1994). Diehl (1998; Diehl et al., 2005) and Allaire and Marsiske (1999) showed that practical intelligence is related to psychometric intelligence. That is, to the extent that everyday problems reflect well-structured challenges in daily life, how well people deal with them is related to traditional psychometric abilities (fluid and crystallized intelligence; Marsiske & Margrett, 2006). However, when problems are ill-structured and more open-ended and vague, these relations are not as straightforward (Margrett et al., 2010).

Applications of Practical Intelligence

Practical intelligence and postformal thinking (see Chapter 10) across adulthood have been linked (Blanchard-Fields, Janke, & Camp, 1995). Specifically, the extent to which a practical problem evokes an emotional reaction, in conjunction with experience and one's preferred mode of thinking, determines whether one will use a cognitive analysis (thinking one's way through the problem), a problem-focused action (tackling the problem head-on by doing something about it), passive-dependent behavior (withdrawing from the situation), or avoidant thinking and denial (rationalizing to redefine the problem and so minimize its seriousness). Adults tend to blend emotion with cognition in their approach to practical problems, whereas adolescents tend not to because they get hung up in the logic. Summarizing over a decade of her research, Blanchard-Fields (2007) notes that, for late middle-aged adults, highly emotional problems (issues with high levels of feelings, such as dealing with unexpected deaths) are associated most with passive-dependent and avoidant-denial approaches. It is interesting, though, that problems concerned more with instrumental issues (issues related to daily living such as grocery shopping, getting from place to place, etc.) and home management (issues related to living in one's household) are dealt with differently. Middle-aged adults use problem-focused strategies more frequently in dealing with instrumental problems than do adolescents or young adults. Clearly, we cannot characterize problem solving in middle age in any one way.

How different cultural groups show practical intelligence has been examined in many different parts of the world. For example, links between practical intelligence and behaviors have been documented with Russians' ability to deal with rapid change (Grigorenko & Sternberg, 2001), Alaskan Yup'ik community members' competence in life tasks (Grigorenko et al., 2004), and American leaders' ability to convince people that their vision is worth pursuing (Sternberg, 2002). Overall, people higher in practical intelligence are able to deal with a more rapid pace of change (as found in the Russian example), come up with new and more effective ways of solving daily life problems (the Alaskan example), and persuade people to change the way they do things (the American leadership example). Finally, one study comparing European American, African American, and Caribbean American adults showed no differences in practical intelligence (Castro, 2000).

Mechanics and Pragmatics of Intelligence

When we combine the research on practical intelligence with the research on the components or mechanics of intelligence discussed in Chapter 10, we have a more complete description of cognition in adulthood. The two-component model of life-span intelligence (Baltes, Lindenberger, & Staudinger, 2006) is grounded in the dynamic interplay among the biopsychosocial forces (see Chapter 1). *The* **mechanics of intelligence** *reflects those aspects of intelligence comprising fluid intelligence* (see Chapter 10). *The* **pragmatics of intelligence** *refers to those aspects of intelligence reflecting crystallized intelligence* (see Chapter 10). However, as Baltes and colleagues point out, the biopsychosocial forces differentially influence the mechanics and pragmatics of intelligence. Whereas the mechanics of intelligence is more directly an expression of the neurophysiological architecture of the mind, the pragmatics of intelligence is associated more with the bodies of knowledge that are available from and mediated through one's culture (Baltes et al., 2006).

mechanics of intelligence
those aspects of intelligence comprising fluid intelligence

pragmatics of intelligence
those aspects of intelligence reflecting crystallized intelligence

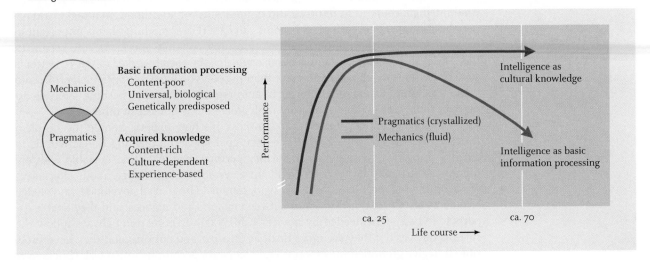

Figure 13.5
The pragmatics of intelligence remains optimal across adulthood whereas the mechanics of fluid intelligence declines.

SOURCE: *Annual Review of Psychology*, volume 50, 1999 by Spence, Janet T. Copyright 1999. Reproduced with permission of ANNUAL REVIEWS, INC. in the format Textbook via Copyright Clearance

These concepts are illustrated in the left portion of ■ Figure 13.5. The mechanics of intelligence in later life is more associated with the fundamental organization of the central nervous system (i.e., biological forces). Thus, it is more closely linked with a gradual loss of brain efficiency with age (Horn & Hofer, 1992), a finding supported by brain imaging studies (Blanchard-Fields, 2010).

On the other hand, the pragmatics of intelligence is more closely associated with psychological and sociocultural forces. At the psychological level, knowledge structures change as a function of the accumulated acquisition of knowledge over time. For example, the more you learn about the American Revolution, the more differentiated your knowledge system becomes, going perhaps from knowing just that the Americans declared their independence from Great Britain and won the war to a much more elaborated knowledge about the various battles, how the Americans nearly lost the war, and that many people in America sided with the British. At the sociocultural level, knowledge structures are also influenced by how we are socialized given the particular historical period in which we are raised. For example, people who grew up in the Cold War era were taught that the people in the former Soviet Union were our enemies; those who grew up after the breakup of the former Soviet Union in 1991 were taught that the Russians could be our allies. Such differences reflect the sociocultural and historical contexts of particular points in time. Overall, these knowledge structures influence the way we implement our professional skills, solve everyday problems, and conduct the business of life (Baltes et al., 2006).

Finally, as the right portion of ■ Figure 13.5 suggests, different weightings of the forces of intelligence lead to specific predictions about the developmental pathway they take across the course of adult life. Given that biological and genetic forces govern the mechanics more, there is a downward trajectory with age. However, given that the pragmatics of intelligence is governed more by environmental and cultural factors, there is an upward trajectory that is maintained across the adult life span.

Becoming an Expert

One day John Cavanaugh was driving along when his car suddenly began coughing and sputtering. As deftly as possible, he pulled over to the side of the road, turned off the engine, opened the hood, and proceeded to look inside. It was hopeless; to him, it looked like a jumble of unknown parts. After the car was towed to a garage, a middle-aged mechanic set about fixing it. Within a few minutes, the car was running like new. How?

We saw in Chapter 10 that aspects of intelligence grounded in experience (crystallized intelligence) tend to improve throughout most of adulthood. In a real-world experiential perspective, each of us becomes an expert at something that is important to us, such as our work, interpersonal relationships, cooking, sports, or auto repair. In this sense, an expert (like the mechanic or Kesha, the social worker in the vignette) is someone who is much better at a task than people who have not put much effort into it (such as John Cavanaugh, in terms of auto repair). We tend to become selective experts in some areas while remaining rank amateurs or novices at others.

What makes experts better than novices? It's how experts handle the problem (Ericsson & Towne, 2010). For novices, the goal for accomplishing the activity is to reach as rapidly as possible a satisfactory performance level that is stable and "autonomous." In contrast, experts build up a wealth of knowledge about alternative ways of solving problems or making decisions. These well-developed knowledge structures are the major difference between experts and novices, and they enable experts to bypass steps needed by novices (Chi, 2006). Experts don't always follow the rules as novices do; they are more flexible, creative, and curious; and they have superior strategies grounded on superior knowledge for accomplishing a task (Ericsson & Towne, 2010). Even though experts may be slower in terms of raw speed because they spend more time planning, their ability to skip steps puts them at a decided advantage. In a way, this represents "the triumph of knowledge over reasoning" (Charness & Bosman, 1990).

Research evidence indicates that expert performance tends to peak by middle age and drops off slightly after that (Masunaga & Horn, 2001). However, the declines in expert performance are not nearly as great as they are for the abilities of information processing, memory, and fluid intelligence that underlie expertise, and expertise may sometimes compensate for declines in underlying cognitive abilities (Masunaga & Horn, 2001; Taylor et al., 2005). Thus, it appears that knowledge based on experience is an important component of expertise.

But why are expertise and information processing, memory, and fluid intelligence not strongly related? After all, we saw in Chapter 10 that the latter abilities underlie good cognitive performance. Rybash, Hoyer, and Roodin (1986) proposed a process called *encapsulation* as the answer. **Encapsulation** *occurs when the processes of thinking (information processing, memory, fluid intelligence) become connected or encapsulated to the products of thinking (expertise).* This process of encapsulation allows expertise to compensate for declines in underlying abilities, perhaps by making thinking more efficient, and has proven to be a useful approach to understanding expertise (Hoyer & Rybash, 1994). It also mirrors neurological evidence about how the way the brain is structured into different areas that "specialize" in different types of processing (Karmiloff-Smith, 2010).

Let's consider how encapsulation might work with auto mechanics. As a rule, people who become auto mechanics are taught to think as if they were playing a game of Twenty Questions, in which the optimal strategy is to ask a question such that the answer eliminates half of the remaining possibilities. In the beginning, the mechanic learns the thinking strategy and the content knowledge about automobiles separately. But as the person's experience with repairing automobiles increases, the thinking strategy and content knowledge merge; instead of having to go through a Twenty Questions approach, the expert mechanic just "knows" how to proceed. This cognitive-developmental pattern in adults is very different from the one that occurs in children (Hoyer & Rybash, 1994). In the adult's case, development is directed toward mastery and adaptive competency in specific domains, whereas children (or, more generally, novices) tend to show less specialization and more uniform performance across content domains (Ericsson & Towne, 2010).

One of the outcomes of encapsulation appears to be a decrease in the ability to explain how one arrives at a particular answer (Hoyer & Rybash, 1994). It seems that the increased efficiency that comes through merging the process with the product of

encapsulation

occurs when the processes of thinking (information processing, memory, fluid intelligence) become connected or encapsulated to the products of thinking (expertise)

thinking comes at the cost of being able to explain to others what one is doing. This could be why some instructors have a difficult time explaining the various steps involved in solving a problem to novice students but an easier time explaining it to graduate students who have more background and experience. Because these instructors may skip steps, it's harder for those with less elaborated knowledge to fill in the missing steps.

We will return to the topic of expertise in Chapter 14 when we discuss wisdom, which some believe to be the outcome of becoming an expert in living.

THINK ABOUT IT

Can expertise be taught? Why or why not?

Lifelong Learning

Many people work in occupations in which information and technology change rapidly. To keep up with these changes, many organizations and professions now emphasize the importance of learning how to learn, rather than learning specific content that may become outdated in a couple of years. For most people, a college education will probably not be the last educational experience they have in their careers. Workers in many professions—such as medicine, nursing, social work, psychology, auto mechanics, and teaching—are now required to obtain continuing education credits to stay current in their fields. Online learning has made lifelong learning more accessible to professionals and interested adults alike (Fretz, 2001; Ranwez, Leidig, & Crampes, 2000), but open access to computers for these programs needs to be in supportive, quiet environments (Eaton & Salari, 2005).

College campuses are an obvious site for lifelong learning; you probably have seen returning adult students on your campus. Lifelong learning also takes place in settings other than college campuses. Many organizations offer workshops for their employees on a wide range of topics, from specific job-related tasks to leisure-time activities. Additionally, many channels on cable television offer primarily educational programming, and online courses, computer networks, and bulletin boards are available for educational exchanges. Only a few generations ago, a high-school education was the ticket to a lifetime of secure employment. Today, lifelong learning is rapidly becoming the norm.

Lifelong learning is gaining acceptance as the best way to approach the need for continuing education and for retraining displaced workers. But should lifelong learning be approached as merely an extension of earlier educational experiences? Knowles, Swanson, and Holton (2005) argue that teaching aimed at children and youth differs from teaching aimed at adults. Adult learners differ from their younger counterparts in several ways:

- Adults have a higher need to know why they should learn something before undertaking it.
- Adults enter a learning situation with more and different experience on which to build.

Most occupations require the acquisition of new information over time through lifelong learning in order to do the job well and to stay up to date on the latest information.

- Adults are most willing to learn those things they believe are necessary to deal with real-world problems rather than abstract, hypothetical situations.
- Most adults are more motivated to learn by internal factors (such as self-esteem or personal satisfaction) than by external factors (such as a job promotion or pay raise).

Lifelong learning is becoming increasingly important, but educators need to keep in mind that learning styles change as people age. Effective lifelong learning requires smart decisions about how to keep knowledge updated and which approach will work best among the many different learning options available (Janssen et al., 2007).

Test Yourself

RECALL

1. The skills and knowledge necessary for people to function in everyday life make up _____.
2. Even though they may be slower in terms of raw speed, experts are at a distinct advantage over novices because they _____.
3. The way in which the process of thinking becomes connected to the products of thinking is termed _____.
4. Due to rapidly changing technology and information, many educators now support the concept of _____.

INTERPRET

Based on the cognitive-developmental changes described in this section, what types of jobs would be done best by middle-aged adults?

APPLY

If you were asked to design a cognitive training program for middle-aged adults, what strategies would be included?

Recall answers: (1) practical intelligence, (2) can skip steps, (3) encapsulation, (4) lifelong learning

---

13.3 Personality

LEARNING OBJECTIVES

- What is the five-factor model? What evidence is there for stability in personality traits?
- What changes occur in people's priorities and personal concerns? How does a person achieve generativity? How is midlife best described?

Jim showed all the signs. He divorced his wife of nearly 20 years to enter into a relationship with a woman 15 years younger, sold his ordinary-looking midsize sedan and bought a red sports car, and began working out regularly at the health club after years of being a couch potato. Jim claims he hasn't felt this good in years; he is happy to be making this change in middle age. All of Jim's friends agree: This is a clear case of midlife crisis. Or is it?

THE TOPIC OF PERSONALITY DEVELOPMENT IN MIDDLE AGE IMMERSES US IN ONE OF THE HOTTEST DEBATES IN THEORY AND RESEARCH ON ADULT DEVELOPMENT AND AGING. Take Jim's case. Many people believe strongly that middle age brings with it a normative crisis called the midlife crisis. There would appear to be lots of evidence to support this view based on case studies like Jim's. But is everything as it seems? We'll find out in this section.

Unlike most of the other topics we have covered in this chapter, research on personality in middle-aged adults is grounded in several competing theories, one being the psychoanalytic approach we encountered in Chapter 1. Another difference is that much of the research we will consider is longitudinal research, also discussed in Chapter 1.

First, we examine the evidence that personality traits remain fairly stable in adulthood. This position makes the claim that what you are like in young adulthood predicts pretty well what you will be like for the rest of your life. Second, we consider the evidence that people's priorities and personal concerns change throughout adulthood, requiring adults to reassess themselves from time to time. This alternative position claims that change is the rule during adulthood.

At no other point in the life span is the debate about stability versus change as heated as it is concerning personality in middle age. In this section, we consider the evidence for both positions.

Stability Is the Rule: The Five-Factor Model

In the past few decades, one of the most important advances in research on adult development and aging has been the emergence of a personality theory aimed specifically at describing adults. Due mostly to the efforts of Paul Costa Jr. and Robert McCrae (1997; McCrae, 2002), we are now able to describe adults' personality traits using five dimensions: neuroticism, extraversion, openness to experience, agreeableness, and conscientiousness. These dimensions (the so-called Big Five traits) are strongly grounded in cross-sectional, longitudinal, and sequential research. First, though, let's take a closer look at each dimension.

- *People who are high on the* **neuroticism** *dimension tend to be anxious, hostile, self-conscious, depressed, impulsive, and vulnerable.* They may show violent or negative emotions that interfere with their ability to get along with others or to handle problems in everyday life. People who are low on this dimension tend to be calm, even-tempered, self-content, comfortable, unemotional, and hardy.

- *Individuals who are high on the* **extraversion** *dimension thrive on social interaction, like to talk, take charge easily, readily express their opinions and feelings, like to keep busy, have boundless energy, and prefer stimulating and challenging environments.* Such people tend to enjoy people-oriented jobs, such as social work and sales, and they often have humanitarian goals. People who are low on this dimension tend to be reserved, quiet, passive, serious, and emotionally unreactive.

- *Being high on the* **openness to experience** *dimension tends to mean a vivid imagination and dream life, appreciation of art, and a strong desire to try anything once.* These individuals tend to be naturally curious about things and to make decisions based on situational factors rather than absolute rules. People who are readily open to new experiences place a relatively low emphasis on personal economic gain. They tend to choose jobs such as the ministry or counseling, which offer diversity of experience rather than high pay. People who are low on this dimension tend to be down-to-earth, uncreative, conventional, uncurious, and conservative.

- *Scoring high on the* **agreeableness** *dimension is associated with being accepting, willing to work with others, and caring.* People who score low on this dimension (i.e., demonstrate high levels of antagonism) show many of the characteristics of the Type A behavior pattern discussed earlier in this chapter. They tend to be ruthless, suspicious, stingy, antagonistic, critical, and irritable.

- *People who show high levels of* **conscientiousness** *tend to be hard working, ambitious, energetic, scrupulous, and persevering.* Such people have a strong desire to make something of themselves. People at the opposite end of this scale tend to be negligent, lazy, disorganized, late, aimless, and nonpersistent.

The five-factor model has been examined cross-culturally. Research evidence generally shows that the same five factors appear across at least 50 cultures, including rarely studied Arabic and Black African groups (McCrae & Terracciano, 2005). Heine and Buchtel (2009) point out, though, that much of this research has been conducted by Westerners, so it remains to be seen whether similar studies conducted by local researchers will have the same outcomes.

neuroticism
a personality trait reflected as the tendency to be anxious, hostile, self-conscious, depressed, impulsive, and vulnerable

extraversion
a personality trait dimension associated with the tendency to thrive on social interaction, to like to talk, to take charge easily, to readily express opinions and feelings, to like to keep busy, to have boundless energy, and to prefer stimulating and challenging environments

openness to experience
a personality dimension that reflects a tendency to have a vivid imagination and dream life, an appreciation of art, and a strong desire to try anything once

agreeableness
a dimension of personality associated with being accepting, willing to work with others, and caring

conscientiousness
a dimension of personality in which people tend to be hard working, ambitious, energetic, scrupulous, and persevering

What's the Evidence for Trait Stability?

Costa and McCrae have investigated whether the general traits that make up their model remain stable across adulthood (e.g., Costa & McCrae, 1988, 1997; McCrae & Costa, 1994). In fact, they suggest that personality traits stop changing by age 30 and appear to be "set in plaster" (McCrae & Costa, 1994, p. 21). The data from the Costa and McCrae studies came from the Baltimore Longitudinal Study of Aging for the 114 men who took the Guilford-Zimmerman Temperament Survey (GZTS) on three occasions, with each of the two follow-up testings about 6 years apart. What Costa and McCrae found was surprising. Even over a 12-year period, the 10 traits measured by the GZTS remained highly stable; the correlations ranged from .68 to .85. In much of personality research we might expect to find this degree of stability over a week or two, but to see it over 12 years is noteworthy.

We would normally be skeptical of such consistency over a long period, but similar findings were obtained in other studies. In a study of 684 adults aged 17–76, Terracciano, McCrae, and Costa (2010) found that the stability of personality traits plateaus in adulthood. A longitudinal study of 60-, 80-, and 100-year-old men and women by Martin, Long, and Poon (2003) found no significant changes across age groups in overall personality patterns. However, some interesting changes did occur in the very old. There was an increase in suspiciousness and sensitivity. This could be explained by increased wariness of victimization in older adulthood. Stability was also observed in past longitudinal data collected over an 8-year span by Siegler, George, and Okun (1979) at Duke University and over a 30-year span by Leon and colleagues (1979) in Minnesota, as well as in other longitudinal studies (Schaie & Willis, 1995; Schmitz-Scherzer & Thomae, 1983). Even more amazing was the finding that personality ratings by spouses of each other showed no systematic changes over a 6-year period (Costa & McCrae, 1988). Thus, it appears that individuals change very little in personality traits (either self-reported or rated by spouses) over periods of up to 30 years and over the age range of 20 to 90.

This is an important conclusion. Clearly, lots of things change in people's lives over 30 years. They marry, divorce, have children, change jobs, face stressful situations, move, and maybe even retire. Social networks and friendships come and go. Society changes, and economic ups and downs have important effects. Personal changes in appearance and health occur. People read volumes, see dozens of movies, and watch thousands of hours of television. But their underlying personality dispositions hardly change at all. Or do they?

Perhaps. There is growing evidence for personality change based on more careful analysis of the issues. First, there are data indicating that certain personality traits (self-confidence, cognitive commitment, outgoingness, and dependability) show some change over a 30- to 40-year period (Jones & Meredith, 1996). Second, there are a growing number of studies suggesting that neuroticism may increase and extraversion may decrease as we grow older (Maiden et al., 2003; Small et al., 2003). Third, Srivastava and colleagues (2003) conducted a large Internet study—more than 130,000 people ranging in age from 21 to 60—of the Big Five traits. This study, described in detail in the Spotlight on Research feature, is a testament to how changes in our technology allow more in-depth analyses of larger samples of individuals. They found that none of the Big Five personality traits remained stable after age 30.

Spotlight on research — Is Personality in Young and Middle Adulthood Set in Plaster?

Who were the investigators and what was the aim of the study? Srivastava and colleagues (2003) wanted to test the notion that the Big Five personality traits are "set in plaster" in adulthood against the contextualist view that they should change over time.

How did the investigators measure the topic of interest? All participants completed the Big Five inventory that was available through the study websites.

Who were the participants in the study? Srivastava and colleagues had 132,515 people

aged 21 to 60 (54% female; 86% European descended) complete a Big Five personality measure on the Internet. This is one of the largest samples ever collected. Participants were all residents of either the United States (90.8%) or Canada (9.2%).

Age differences in Big Five personality traits across young adulthood and middle age.

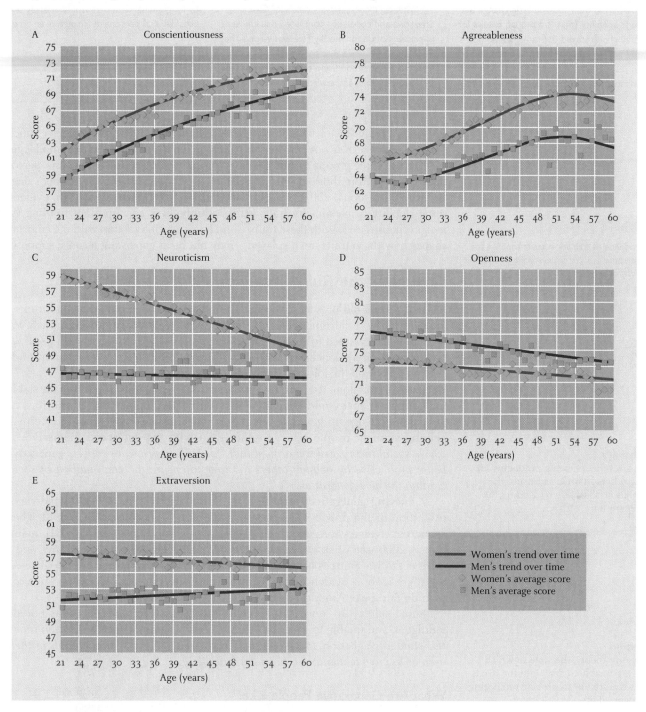

From Srivastava et al., 2003, p. 1047.

What was the design of the study? The study used a cross-sectional design. To attract a broad and diverse sample, they used two types of web pages. One was a guide called "all about you," which informed individuals that they would take a test on what psychologists considered to be the fundamental dimensions of personality. The second was a "Find your *Star*

Wars Twin," which included feedback about the characters from *Star Wars* with whom the participant was most similar based on the Big Five personality test.

Were there ethical concerns with the study? Because the study used volunteers who completed surveys containing no questions about sensitive topics, there were no ethical concerns.

What were the results? The developmental patterns for each of the Big Five dimensions are shown in ■ Figure 13.6. They found that none of the Big Five personality traits remained completely stable after age 30. For example, conscientiousness showed the most differences across early adulthood, a time when adults are advancing in the workforce and forming intimate rela-

[continued]

tionships. Agreeableness increased most during the 30s. Neuroticism differed considerably for women, whereas men showed little change but increased variability (that is, although results for the overall group stayed the same, some men increased and some decreased). Despite differences in respondents across the two websites (more women responded to the "all about you"

site and more men responded to the "*Star Wars*" site), the developmental patterns were the same.

What did the investigators conclude? Overall, Srivastava and colleagues concluded that the set-in-plaster notion for the Big Five was wrong. They found much evidence that personality traits differed across adulthood, and they argue that this is the result of traits and environments interacting.

What converging evidence would strengthen their conclusions? The findings would be strengthened by longitudinal data that would actually track the possibility of personality change over time.

 Go to Psychology CourseMate at www.cengagebrain.com to enhance your understanding of this research.

THINK ABOUT IT

Does evidence of stability in traits support the idea that some aspects of personality are genetic? Why or why not?

Although the five-factor model enjoys great popularity and appears to have much supporting evidence, it is not perfect. For example, the degree to which the model generalizes across ethnic groups is questionable; in one study, Hmong Americans' personality traits differed from those of European Americans depending on the former's degree of acculturation (Moua, 2007). Even acknowledging the problems, though, evidence for some stability in personality traits across adulthood is an important finding. What a person chooses to do with these traits—and how their interaction with the environment shapes how those traits are displayed—may not be as consistent as once supposed.

Change Is the Rule: Changing Priorities in Midlife

Joyce, a 52-year-old preschool teacher, thought carefully about what she thinks is important in life. "I definitely feel differently about what I want to accomplish. When I was younger, I wanted to advance and be a great teacher. Now, although I still want to be good, I'm more concerned with providing help to the new teachers around here. I've got lots of on-the-job experience that I can pass along."

Joyce is not alone. Despite the evidence that personality traits remain stable during adulthood, many middle-aged people report that their personal priorities change during middle age. In general, they report that they are increasingly concerned with helping younger people achieve rather than with getting ahead themselves. *In his psychosocial theory, Erikson argued that this shift in priorities reflects* **generativity**, *or being productive by helping others in order to ensure the continuation of society by guiding the next generation.*

generativity
in Erikson's theory, being productive by helping others in order to ensure the continuation of society by guiding the next generation

Achieving generativity can be most enriching. It is grounded in the successful resolution of the previous six phases of Erikson's theory (see Chapter 1). There are numerous avenues for generativity, such as parenting (Pratt et al., 2001), mentoring (Lucas, 2000; see Chapter 12), volunteering, foster grandparent programs, and many other activities. Sources of generativity do not vary across ethnic groups (Bates, 2009), but there is some evidence that African Americans express more generative concern than do European Americans (Hart et al., 2001).

Some adults do not achieve generativity. Instead, they become bored, self-indulgent, and unable to contribute to the continuation of society. *Erikson referred to this state as* **stagnation**, *in which people are unable to deal with the needs of their children or to provide mentoring to younger adults.*

stagnation
in Erikson's theory, the state in which people are unable to deal with the needs of their children or to provide mentoring to younger adults

What Are Generative People Like?

Several researchers have constructed various descriptions of generativity so that we can recognize it more easily (Washko, 2001). Research shows that generativity is different from traits; for example, generativity is more related to societal engagement than are traits (Cox et al., 2010).

One of the best approaches to generativity is McAdams's model (McAdams, 2001a, 2008; McAdams & Olson, 2010), shown in ■ Figure 13.7. This multidimensional model shows how generativity results from the complex interconnections among societal and inner forces, which create a concern for the next generation and a belief in the goodness of the human enterprise; this leads to generative commitment, which produces generative actions. *A person derives personal meaning from being generative by constructing a life story or* **narrative**, *which helps create the person's identity* (see Chapter 10).

narrative
a way in which a person derives personal meaning from being generative and by constructing a life story, which helps create the person's identity

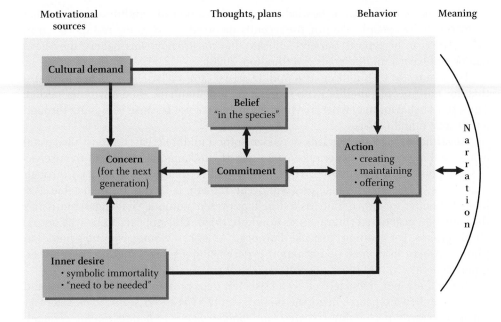

Motivational sources   Thoughts, plans   Behavior   Meaning

■ **Figure 13.7**
McAdams's model of generativity. Note that how one shows generativity (action) is influenced by several factors.

The components of McAdams's model constitute an approach that differs from personality traits (McAdams, 2008; McAdams & Olson, 2010). In McAdams's model, generative *concern* (a trait) relates to life satisfaction and overall happiness, whereas generative *action* does not (de St. Aubin & McAdams, 1995). For instance, new grandparents may derive much satisfaction from their grandchildren and are greatly concerned with their well-being, but they have little desire to engage in the daily hassles of caring for them on a regular basis. Women who exhibit high generativity tend to have prosocial personality traits, to have positive personality characteristics, and to be satisfied with marriage and motherhood (Peterson & Duncan, 2007); they are personally invested in being a parent, express generative attitudes at work, and exhibit caring behaviors toward others outside their immediate families (Peterson & Klohnen, 1995); and they report high well-being in their role as a spouse (MacDermid, De Haan, & Heilbrun, 1996). These results have led to the creation of positive and negative generativity indices that reliably identify differences between generative and nongenerative individuals (Himsel et al., 1997).

How well do these ideas generalize across ethnic groups and cultures? A study of second-generation Chinese American women found similar trends in generativity with European American women (Grant, 2007). In one of the few studies to examine generativity across cultures, Hofer and colleagues (2008) examined it in Cameroon, Costa Rica, and Germany. They found that McAdams's model could be successfully applied across the three cultures.

The growing evidence on generativity indicates that the personal concerns and priorities of middle-aged adults are different from those of younger adults. But is this view consistent with other aspects of personality? Let's consider the evidence.

Life Transition Theories and the Midlife Crisis

We have seen that theorists such as Erikson believe that adults face several important challenges and that, by struggling with these issues, people develop new aspects of themselves. Erikson's notion that people experience fundamental changes in their priorities and personal concerns was grounded in the possibility that middle adulthood includes other important changes. Carl Jung, one of the founders of psychoanalytic theory, believed that adults may experience a midlife crisis. This belief led to the development of several theories suggesting that adulthood consists of alternating periods of stability and transition that people experience in a fixed sequence.

These theoretical approaches led to a popularization of "midlife crisis," so that many considered people like Jim, the recently divorced guy with the red sports car in the vignette, to be typical. Surveys indicate that most Americans believe they have had or will have a midlife crisis (Wethington, 2000).

Is there really such a thing as a midlife crisis? Not really. The evidence indicates that, for most people, midlife is no more or no less traumatic than any other period. Thus, Jim's behavior may have an explanation, but it's not because he's going through a universal midlife crisis.

Nevertheless, despite this lack of evidence for a midlife crisis, there is substantial evidence that people do experience some sort of fundamental change in themselves at some point during adulthood. Thus, it may well be that most adults pass through transitions; when those transitions will occur, though, is largely unpredictable. Perhaps it is better to view midlife as a time that presents unique challenges and issues that must be negotiated (Bumpass & Aquilino, 1995). This appears to be true across ethnic groups. For example, just like European American women, second-generation Chinese American women do not report experiencing a general midlife crisis but do report generativity and other positive aspects of personality (Grant, 2007).

ego resilience
a powerful personality resource that enables people to handle midlife changes

If midlife is not characterized by a crisis but does present unique challenges and issues, then how do people negotiate it successfully? *The secret seems to be* **ego resilience**, *a powerful personality resource that enables people to handle midlife changes.* Longitudinal data from two samples indicate that people who enter middle age with high ego resilience are more likely to experience it as an opportunity for change and growth, whereas people with low ego resilience are more likely to experience it as a time of stagnation or decline (Klohnen, Vandewater, & Young, 1996). There are significant individual differences in the timing of such experiences and how people deal with midlife, which probably accounts for the failure to find a universal midlife crisis (Klohnen et al., 1996). Ego resilience may also be the resource that could account for the two outcomes (generativity and stagnation) of Erikson's view of midlife, and it is an important aspect of development (Luthar, 2006).

In sum, perhaps the best way to view the life transitions associated with middle age is through the words of a 52-year-old woman (Klohnen et al., 1996):

> Middle age . . . The time when you realize you've moved to the caretaker, senior responsibility role . . . A time of discomfort because you watch the generation before you, whom you have loved and respected and counted on for emotional back-up, for advice . . . become more dependent on you and then die. Your children grow up, move out, try their wings . . .; indeed, they attempt to teach you the "truths" they've discovered about life . . . It's time to make some new choices—groups, friends, activities need not be so child related anymore.

Test Yourself

RECALL

1. The dimensions in the five-factor theory of personality include neuroticism, extraversion, openness to experience, agreeableness, and _____.

2. According to Erikson, an increasing concern with helping younger people achieve is termed _____.

3. According to McAdams, the meaning one derives from being generative happens through the process of _____.

4. Research indicates that _____ is a key personality factor in predicting who will negotiate midlife successfully.

INTERPRET

How can you reconcile the data from trait research, which indicates little change, with the data from other research, which shows substantial change in personality during adulthood?

APPLY

If psychotherapy assumes that a person can change behavior over time, what is the relation between personality and behavior from this perspective?

Recall answers: (1) conscientiousness, (2) generativity, (3) narrative, (4) ego resilience

LEARNING OBJECTIVES

- Who are the kinkeepers in families?

- How does the relationship between middle-aged parents and their young adult children change?

- How do middle-aged adults deal with their aging parents?

- What styles of grandparenthood do middle-aged adults experience? How do grandchildren and grandparents interact?

Esther is facing a major milestone: Her youngest child, Megan, is about to head off to college. But instead of feeling depressed, as she thought she would, Esther feels almost elated at the prospect. She and Bill are finally free of the day-to-day parenting duties of the past 30 years. Esther is looking forward to getting to know her husband again. She wonders whether there is something wrong with her for being excited that her daughter is moving away.

PEOPLE LIKE ESTHER CONNECT GENERATIONS. Family ties across the generations provide the context for socialization and for continuity in the family's identity. At the center agewise are members of the middle-aged generation, like Esther, who serve as the links between their aging parents and their own maturing children (Hareven, 2001). *Middle-aged mothers (more than fathers) tend to take on this role of **kinkeeper**, the person who gathers family members together for celebrations and keeps them in touch with each other.*

Think about the major issues confronting a typical middle-aged couple: maintaining a good marriage, parenting responsibilities, dealing with children who are becoming adults themselves, handling job pressures, and worrying about aging parents, just to name a few. Middle-aged adults truly have quite a lot to deal with every day in balancing their responsibilities to their children and their aging parents (Riley & Bowen, 2005). *Indeed, middle-aged adults are sometimes referred to as the **sandwich generation** because they are caught between the competing demands of two generations: their parents and their children.* Being in the sandwich generation means different things for women and men. When middle-aged women assess how well they are dealing with the challenges of midlife, their most pressing issues relate more to their adolescent children than to their aging parents; for middle-aged men, it is the other way around (Riley & Bowen, 2005).

In this section, we first examine the dynamics of middle-aged parents and their maturing children and discover whether Esther's feelings are typical. Next, we consider the issues facing middle-aged adults and their aging parents. Later, we consider what happens when people become grandparents.

kinkeeper
the person who gathers family members together for celebrations and keeps them in touch with each other, usually a middle-aged mother

sandwich generation
middle-aged adults who are caught between the competing demands of two generations: their parents and their children

Adult children's relationships with their parents often include a friendship dimension.

Letting Go: Middle-Aged Adults and Their Children

Being a parent has a rather strange side when you think about it. After creating children out of love, parents spend considerable time, effort, and money preparing them to become independent and leave. For most parents, the leaving (and sometimes returning) occurs during midlife. Let's take a closer look.

Becoming Friends and the Empty Nest

Sometime during middle age, most parents experience two positive developments with regard to their children. Suddenly their children see them in a new light, and the children leave home.

After the strain of raising adolescents, parents generally appreciate the transformation that occurs when their children head into young adulthood. In general, parent–child relationships improve when children become young adults; the relationships become more symmetrical in that both the adult child and the parent now relate on a mature level (Buhl, 2008). The difference can be dramatic, as in the case of Deb, a middle-aged mother. "When Sacha was 15, she acted as if I was the dumbest person on the planet. But now that she's 21, she acts as if I got smart all of a sudden. I like being around her. She's a great kid, and we're really becoming friends."

A key factor in making this transition as smoothly as possible is the extent to which parents foster and approve of their children's attempts at being independent. Most parents are like Esther, the mother in the vignette, and manage the transition to an empty nest successfully; many mothers, in particular, use this as a time for growth (Owen, 2005). That's not to say that parents are heartless. As depicted in the cartoon, when children leave home, emotional bonds are disrupted. Parents feel the change, although differently; women who define themselves more in their role as a mother tend to report more distress and negative mood (Hobdy, 2000). But most parents see the launching of children as a positive event; for example, mothers in all ethnic groups report feeling sad at the time that children leave, but have many more positive feelings about the potential for growth in their relationships with their children (Feldman, 2010).

Still, parents provide considerable emotional support (by staying in touch) and financial help (such as paying college tuition, providing a free place to live until the child finds employment) when possible (Mitchell, 2006; Warner, Henderson-Wilson, & Andrew, 2010). Most help in other ways, ranging from the mundane (such as making the washer and dryer available to their college-age children) to the extraordinary (providing the down payment for their child's house). Adult children and their parents generally believe that they have strong, positive relationships and that they can count on each other for help when necessary (Connidis, 2001).

It appears that a positive experience with launching children is strongly influenced by the extent to which parents perceive a job well done and that their children have turned out well (Mitchell, 2010). Children are regarded as successes when they meet parents' culturally based developmental expectations, and they are seen as "good kids" when there is agreement between parents and children in basic values. These feelings are part of the reason adult children increasingly come back home, as we see next.

When Children Come Back

Parents' satisfaction with the empty nest is sometimes short-lived. Roughly half of young adults in the United States return to their parents' home at least once after moving out (Osgood et al., 2005). There is evidence that these young adults, called "boomerang kids" (Mitchell, 2006), reflect a less permanent, more mobile contemporary society.

Why do children move back? Those that do typically arrive back home about the time they enter the workplace, and a major impetus is the increased costs of living on one's own when saddled with college debt, especially if the societal economic situation is bad and jobs are not available. Several demographic and psychological factors influence the decision. Men are more likely to move back than women, as are children who had low college GPAs, a low sense of autonomy, or an expectation that their parents would provide a large portion of their income following graduation (Mitchell, 2006; Osgood et al., 2005). Adult children whose parents were verbally or physically abusive are not likely to move back, and neither are those who have married.

The U.S. trend for young adults to move back home differs from the trend in some southern European countries (e.g., Italy) for young adults to simply stay at home until they marry or obtain a full-time job (L'Abate, 2006). In contrast, the trend in other countries has resulted in terms such as *Nesthocker* in Germany and KIPPERS (Kids in Parents' Pockets Eroding Retirement Savings) in the United Kingdom (Blatterer, 2005).

This trend reflects the changing definition of adulthood we considered in Chapter 10. As the ages at which young people take on the roles of adulthood increase, we are likely to see more children living at home longer or returning to their parents' home after graduating from college.

Giving Back: Middle-Aged Adults and Their Aging Parents

No matter how old you may be, being someone's child is a role that people still play well into adulthood and, sometimes, into their 60s and 70s. How do middle-aged adults relate to their parents? What happens when their parents become frail? How do middle-aged adults deal with the need to care for their parents?

Caring for Aging Parents

Most middle-aged adults have parents who are in reasonably good health. For a growing number of people, however, being a middle-aged child of aging parents involves providing some level of care. The job of caring for older parents usually falls to a daughter or a daughter-in-law (Stephens & Franks, 1999), and daughters also tend to coordinate care provided by multiple siblings (Friedman & Seltzer, 2010). Even after ruling out all other demographic characteristics of adult child caregivers and their care recipients, daughters are more than 3 times as likely to provide care as sons (Stephens et al., 2001). The cartoon depicts a common situation: a daughter worrying about her aging parents and wondering whether she should be doing more. This gender difference is also found in other cultures. In Japan, even though the oldest son is responsible for parental care, it is his wife who actually does the day-to-day caregiving for her own parents and her in-laws (Lee, 2010).

In some situations, older parents must move in with one of their children. Such moves usually occur after decades of both generations living independently. This his-

tory of independent living sets the stage for adjustment difficulties following the move; both lifestyles must be accommodated. Most of the time, an adult child provides care for her mother, who may have provided care for her own husband before he died. (Spousal caregiving is discussed in Chapter 14.) In other situations, adult daughters must try to manage care from a distance. As we will see later, women are under considerable stress from the pressures of caregiving irrespective of the location of care.

As described in the Real People feature, caring for one's parent presents a dilemma, especially for women (Baek, 2005; Lai, 2010; Lee, 2010; Stephens et al., 2001). *Most adult children feel a sense of responsibility, termed* **filial obligation**, *to care for their parents if necessary.* For example, adult child caregivers sometimes express the feeling that they "owe it to Mom or Dad" to care for them; after all, their parents provided for them for many years, and now the shoe is on the other foot (Gans, 2007). Adult children often provide the majority of care when needed to their parents in all Western and non-Western cultures studied, but especially in Asian cultures (Hareven & Adams, 1996; Lai, 2010). Worldwide, caregiving situations tend to be better when the economic impact on the caregiving family is minimal; in rural China, for example, when middle-aged children care for aging parents it is the financial impact on the caregivers that matters most (Zhan, 2006). Filial obligation also knows no borders; in an Australian study, middle-aged adults were found to be caring for parents in numerous other countries in Europe, the Middle East, Asia, and New Zealand (Baldassar, Baldock, & Wilding, 2007).

filial obligation
a sense of obligation to care for one's parents if necessary

Real People
Applying Human Development
Taking Care of Mom

Everything seemed to be going well for Joan. Her career was really taking off, her youngest daughter Kelly had just entered high school, and her marriage to Bill was better than ever. So when her phone rang one June afternoon, she was really taken by surprise.

The voice on the other end was matter-of-fact. Joan's mother had suffered a major stroke and would need someone to care for her. Because her mother did not have sufficient medical and long-term care insurance to afford a nursing home, Joan made the only decision she could—her mom would move in with her, Bill,

and Kelly. Joan firmly believed that, because her mom had provided for her, Joan owed it to her mom to do the same now that she was in need.

What Joan didn't count on was that taking care of her mom was both the most difficult yet most rewarding thing she had ever done. Joan quickly realized that her days of lengthy business trips and seminars were over, as was her quick rise up the company leadership ladder. Other employees were now the ones who brought back the great new ideas and could respond to out-of-town crises quickly. Hard as it was, Joan knew that her career trajectory had taken a dif-

ferent turn. And she and Bill had more disagreements than she could ever remember, usually about the decreased amount of time they had to spend with each other. Kelly's demands to be driven here and there also added to Joan's stress.

But Joan and her mom were able to develop the kind of relationship that they could not have otherwise and to talk about issues that they had long suppressed. Although caring for a physically disabled mother was extremely taxing, Joan and her mother's ability to connect on a different level made it worthwhile.

Caring for a parent places significant demands on the adult child. Joan's experience embodies the notion of the "sandwich generation" noted earlier. Joan's need to balance caring for her daughter and for her mother can create conflict, both within herself and between the individuals involved (Neal & Hammer, 2006). Being pulled in different directions can put considerable stress on the caregiver, a topic we consider in the next section. But the rewards of caregiving are also great, and relationships can be strengthened as a result.

Roughly 50 million Americans provide unpaid care for older parents, in-laws, grandparents, and other older loved ones (National Alliance for Caregiving & AARP, 2010). The typical caregiver is a 48-year-old woman who is employed outside the home and who provides more than 20 hours per week of unpaid caregiving. These family caregivers spend as much as $7,000 per year, on average, in support of their loved one (National Endowment for Financial Education, 2010).

Caring for an older parent is often not easy. It usually doesn't happen by choice; each party would just as soon live independently. The potential for conflict over daily routines and lifestyles can be high. Indeed, one major source of conflict between middle-aged daughters and their older mothers is differences in perceived need for care, with middle-aged daughters believing that their mothers needed care more than the mothers believed they did (Fingerman, 1996). The balance between independence and connection can be a difficult one (McGraw & Walker, 2004); among Japanese immigrants, one source of conflict is between caregiving daughters and older mothers who give unsolicited advice (Usita & Du Bois, 2005).

Caregiving Stresses and Rewards

Caregiving is a major source of both stresses and rewards. On the stress side, adult children and other family caregivers are especially vulnerable from two main sources (Pearlin et al., 1990):

- Adult children may have trouble coping with declines in their parents' functioning, especially those involving cognitive abilities and problematic behavior, and with work overload, burnout, and loss of the previous relationship with a parent.

- If the caregiving situation is perceived as confining or seriously infringes on the adult child's other responsibilities (spouse, parent, employee, etc.), then the situation is likely to be perceived negatively, which may lead to family or job conflicts, economic problems, loss of self-identity, and decreased competence.

When caring for an aging parent, even the most devoted adult child caregiver will at times feel depressed, resentful, angry, or guilty (Cavanaugh, 1999; Stephens et al., 2001). Many middle-aged caregivers are hard pressed financially: they may still be paying child care or college tuition expenses, perhaps trying to save adequately for their own retirement, and having to work more than one job to do it. Financial pressures are especially serious for those caring for parents with chronic conditions, such as Alzheimer's disease, that require services, such as adult day care, not adequately covered by medical insurance even if the older parent has supplemental coverage. In some cases, adult children may even need to quit their jobs to provide care if adequate alternatives, such as adult day care, are unavailable or unaffordable.

The stresses of caring for one's parent are especially difficult for women. In terms of its timing in the life course, caring for a parent is typically something that coincides with women's peak employment years (ages 35–64). Longitudinal research clearly shows that employment status has no effect on women's decisions to become caregivers (many have little choice), but it has long been the case that becoming a caregiver makes it likely that a woman will reduce employment hours or stop working (Pavalko & Artis, 1997). When you consider that most women caring for parents are also mothers, wives, and employees, it should come as no surprise that stress from these other roles exacerbates the effects of stress due to caregiving (Baek, 2005; National Alliance for Caregiving & AARP, 2010). The stresses of caring for a parent mean that the caregiver needs to carefully monitor his or her own health. Indeed, many professionals point out that "caregiving for the caregiver" is an important consideration to avoid caregiver burnout (Tamayo et al., 2010).

On the plus side, caring for an aging parent also has rewards. Caring for aging parents can bring parents and their adult children closer together and can provide a way for adult children to feel that they are giving back to their parents (Miller et al., 2008). Cross-cultural research examining Taiwanese (Lee, 2007) and Chinese (Zhan, 2006) participants confirms that adults caring for aging parents can find the experience rewarding.

Cultural values enter into the caregiving relationship in an indirect way (Knight & Sayegh, 2010). Caregivers in all cultures studied to date show a common set of outcomes: caregiver stressors are appraised as burdensome, which creates negative health consequences for the caregiver. However, cultural values influence the kinds of social support that are available to the caregiver.

Caring for an older parent creates both stresses and rewards.

Jed Share/Getty Images

Ethnic differences among U.S. groups in adult children's experiences of caregiving reflect the impact of cultural values. A focused study of Mexican American caregivers showed that, given time and a supportive context from service providers, they come to accept services from social service agencies within their cultural norm of providing family care themselves (Crist, Garcia-Smith, & Phillips, 2006). African Americans prefer family caregiving to other options (e.g., placement in a long-term care facility) more strongly than European Americans do; European Americans are more open to having nonfamily members (such as professional paid caregivers) provide care (Foley, Tung, & Mutran, 2002). Such differences show that the relation between caregiving and stress is mediated by beliefs in family cohesiveness versus individual independence and also by one's socialization. And caregivers also report experiencing rewards (Stephens & Franks, 1999).

Things aren't always rosy from the parents' perspective, either. Independence and autonomy are important traditional values in some ethnic groups, and their loss is not taken lightly. Older adults in these groups are more likely to express the desire to pay a professional for assistance rather than ask a family member for help; they may find it demeaning to live with their children and express very strong feelings about "not wanting to burden them" (Cahill et al., 2009). Most move in only as a last resort. As many as two thirds of older adults who receive help with daily activities feel negatively about the help they receive (Newsom, 1999).

Determining whether older parents are satisfied with the help their children provide is a complex issue (Cahill et al., 2009; Newsom, 1999). Based on a critical review of the research, Newsom (1999) proposes a model of how certain aspects of care can produce negative perceptions of care directly or by affecting the interactions between caregiver and care recipient (see ■ Figure 13.8). The important thing to conclude from the model is that, even under the best circumstances, there is no guarantee that the help adult children provide their parents will be well received. Misunderstandings can occur, and the frustration that caregivers feel may be translated directly into negative interactions.

In sum, taking care of one's aging parents is a difficult task. Despite the numerous challenges and risks of negative psychological and financial outcomes, many caregivers nevertheless experience positive outcomes.

■ **Figure 13.8**

Whether a care recipient perceives care to be good depends on interactions with the caregiver and whether those interactions are perceived negatively.

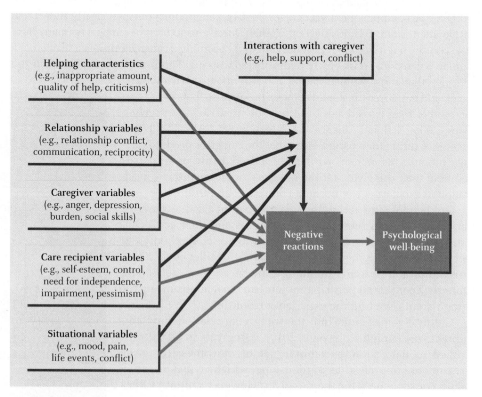

From Newsom, J. T., 1999, "Another side to caregiving: Negative reactions to being helped," *Current Directions in Psychological Science*, 8, 185. Reprinted by permission of Sage Publications. Blackwell Publishing Ltd.

Grandparenthood

Becoming a grandparent takes some help. Being a parent yourself, of course, is a prerequisite. But it is your children's decisions and actions that determine whether you will experience the transition to grandparenthood, making this role different from most others we experience throughout life. Most people become grandparents in their 40s and 50s, though some are older or perhaps as young as their late 20s or early 30s. In most cases, grandparents are quite likely to still be employed and to have living parents themselves. Thus, although being a grandparent may be an exciting time, it is often only one part of their busy lives.

How Do Grandparents Interact With Grandchildren?

Keisha, an 8-year-old girl, smiled brightly when asked to describe her grandparents. "Nana Mary gives me chocolate ice cream, and that's my favorite! Poppy Bill sometimes takes care of me when Momma and Daddy go out, and plays ball with me." Kyle, a 14-year-old, had a different view. "My grandparents generally tell me stories of what life was like back when they were young."

Grandparents can forge strong relationships with their grandchildren and get great enjoyment from them.

As Keisha's and Kyle's experiences show, grandparents have many different ways of interacting with their grandchildren. Categorizing these styles has been attempted over many decades (e.g., Neugarten & Weinstein, 1964), but none of these attempts has been particularly successful because grandparents use different styles with different grandchildren and styles change as grandparents and grandchildren age (Stephens & Clark, 1996).

An alternative approach involves considering the many functions grandparents serve, and the changing nature of families (Hills, 2010). The social dimension includes societal needs and expectations of what grandparents are to do, such as passing on family history to grandchildren. The personal dimension includes the personal satisfaction and individual needs that are fulfilled by being a grandparent. Many grandparents pass on skills—as well as religious, social, and vocational values (social dimension)—through storytelling and advice, and they may feel great pride and satisfaction (personal dimension) from working with grandchildren on joint projects.

Grandchildren give grandparents a great deal in return. For example, grandchildren keep grandparents in touch with youth and the latest trends. Sharing the excitement of surfing the web in school may be one way in which grandchildren keep grandparents on the technological forefront.

Being a Grandparent Is Meaningful

Does being a grandparent matter to people? You bet it does, at least to the vast majority of grandparents. In her groundbreaking research, Kivnick (1982, 1985) identified five dimensions of meaning that grandparents often assign to their roles. However, additional research shows that grandparents can derive multiple meanings, and that they are linked with generativity (Hayslip, Henderson, & Shore, 2003; Thiele & Whelan, 2010). For some, grandparenting is the most important thing in their lives. For others, meaning comes from being seen as wise, from spoiling grandchildren, from recalling the relationship they had with their own grandparents, or from taking pride in the fact that they will be followed by not one but two generations.

Most grandparents derive several different meanings, regardless of the style of their relationship with the grandchildren (Alley, 2004). Similar findings are reported when overall satisfaction with being a grandparent is examined; no matter what their style is, grandparents find their role meaningful (Hayslip et al., 2003; Thiele & Whelan, 2010). These findings have resulted in viewing grandparenthood as an aspect of generativity from which most grandparents derive a great deal of satisfaction (Thiele & Whelan, 2010).

Grandchildren also highly value their relationships with grandparents, even when they are young adults (Alley, 2004). Grandparents are valued as role models as well as for their personalities, the activities they share, and the attention they show to grandchildren. Grandchildren also note that, when their grandparents are frail, helping their grandparents is a way for them to act on their altruistic beliefs (Kennedy, 1991). Young adult grandchildren (ages 21–29) derive both stress and rewards from caring for grandparents, much the same way that middle-aged adults do when they care for their aging parents (Fruhauf, 2007).

Ethnic Differences

How grandparents and grandchildren interact varies in different ethnic groups. Intergenerational relationships are especially important and have historically been a source of strength in African American families (Waites, 2009) and Latino families (Gladding, 2002). African American grandparents play an important role in many aspects of their grandchildren's lives, such as religious education (King et al., 2006). African American grandfathers, in particular, tend to perceive grandparenthood as a central role to a greater degree than do European American grandfathers (Kivett, 1991). And Latino American grandparents are more likely to participate in child rearing owing to a cultural core value of family (Burnette, 1999).

Native American grandparents appear to have some interactive styles that differ from those of other groups (Weibel-Orlando, 1990). *Fictive grandparenting* is a style that allows adults to fill in for missing or deceased biological grandparents, functionally creating the role of surrogate grandparent. These adults provide a connection to the older generation that would otherwise be absent for these children. In the *cultural conservator* style, grandparents request that their grandchildren be allowed to live with them to ensure that the grandchildren learn the native ways. These grandparents provide grandchildren with a way to connect with their cultural heritage, and they are also likely to provide a great deal of care for their grandchildren (Mutchler, Baker, & Lee, 2007). In general, Native American grandmothers take a more active role in these styles than do grandfathers, and are more likely to pass on traditional rituals (Woodbridge, 2008).

Asian American grandparents, particularly if they are immigrants, serve as a primary source of traditional culture for their grandchildren (Yoon, 2005). When these grandparents become heavily involved in caring for their grandchildren, they especially want and need services that are culturally and linguistically appropriate. Grandparents caring for grandchildren is a topic to which we now turn.

How grandparents and grandchildren interact varies across ethnic groups.

When Grandparents Care for Grandchildren

Grandparenthood today is tougher than it used to be. Families are more mobile, which means that grandparents are more often separated from their grandchildren by geographical distance. Grandparents are more likely to have independent lives apart from their children and grandchildren. What being a grandparent entails in the 21st century is more ambiguous than it once was (Fuller-Thompson, Hayslip, & Patrick, 2005).

Perhaps the biggest change worldwide for grandparents is the increasing number who serve as custodial parents or primary caregivers for their grandchildren (Moorman & Greenfield, 2010). Estimates are that about 6.4 million U.S. grandparents have grandchildren living with them, and 2.6 million of these grandparents provide basic needs (food, shelter, clothing) for one or more of their grandchildren (U.S. Census Bureau, 2010c). These situations result most often when both parents are employed outside the home (Uhlenberg & Cheuk, 2010); when the parents are deceased, addicted, incarcerated, or unable to raise their children for some other reason (Backhouse, 2006; Moorman & Greenfield, 2010); or when discipline or behavior problems have been exhibited by the grandchild (Giarusso et al., 2000). Lack of legal recognition stemming from the grandparents' lack of legal guardianship also poses problems and challenges—for example, in dealing with schools and obtaining records. Typically, social service workers must assist grandparents in navigating the many unresponsive policies and systems they encounter when trying to provide the best possible assistance to their grandchildren (Cox, 2007). Clearly, public policy changes are needed to address these issues, especially regarding grandparents' rights regarding schools and health care for their grandchildren (Ellis, 2010).

Raising grandchildren is not easy. Financial stress, cramped living space, and social isolation are only some of the issues facing custodial grandmothers (Bullock, 2004). Rates of problem behavior, hyperactivity, and learning problems in grandchildren are high and may negatively affect the grandparent–grandchild relationship (Hayslip et al., 1998). The grandchildren's routines, activities, and school-related issues also cause stress (Musil & Standing, 2005). Native American custodial grandparents report more symptoms of depression than European American custodial grandparents, especially those who had lower household income and spent less time with their grandchild (Litiecq, Bailey, & Kurtz, 2008). All of these stresses are also reported cross-culturally; for example, full-time custodial grandmothers in Kenya reported higher levels of stress than part-time caregivers (Oburu & Palmérus, 2005).

Even custodial grandparents raising grandchildren without these problems report more stress and role disruption than noncustodial grandparents (Emick & Hayslip, 1999). Custodial grandmothers who are employed report that they arrive late, miss work, must leave work suddenly, or leave early to tend to the grandchild's needs (Pruchno, 1999). But most custodial grandparents consider their situation better for their grandchild than any other alternative and report surprisingly few negative effects on their marriages.

Test Yourself

RECALL

1. The term _____ refers to middle-aged adults who have both living parents and children of their own.

2. The people who gather the family together for celebrations and keep family members in touch are called

   _____ .

3. Most caregiving for aging parents is provided by

   _____ .

4. The sense of personal responsibility to care for one's parents is called _____ .

5. _____ grandparents have a style called cultural conservator.

INTERPRET

If you were to create a guide to families for middle-aged adults, what would your most important pieces of advice be? Why did you select these?

APPLY

What are the connections between the first part of this chapter, on health, and this section, on caregiving?

Recall answers: (1) sandwich generation, (2) kinkeepers, (3) daughters and daughters-in-law, (4) filial obligation, (5) Native American

13.1 PHYSICAL CHANGES AND HEALTH

How does appearance change in middle adulthood?

- Some of the signs of aging appearing in middle age include wrinkles, gray hair, and weight gain.

What changes occur in bones and joints?

- An important change—especially in women—is loss of bone mass, which in severe form may result in the disease osteoporosis.

- Osteoarthritis generally becomes noticeable in late middle or early old age. Rheumatoid arthritis is a more common form affecting fingers, wrists, and ankles.

What reproductive changes occur in men and women in middle age?

- The climacteric (loss of the ability to bear children by natural means) and menopause (cessation of menstruation) occur in the 40s and 50s and constitute a major change in reproductive ability in women.

- Most women do not have severe physical symptoms associated with the hormonal changes. Menopausal hormone therapy is a controversial approach to treatment of menopausal symptoms.

- Reproductive changes in men are much less dramatic; even older men are usually still fertile. Physical changes do affect sexual response.

What is stress? How does it affect physical and psychological health?

- In the stress and coping paradigm, stress results from a person's appraisal of an event as taxing his or her resources. Daily hassles are viewed as the primary source of stress.

- The types of situations people appraise as stressful change through adulthood. Family and career issues are more important for young and middle-aged adults; health issues are more important for older adults.

- Type A behavior pattern is characterized by intense competitiveness, anger, hostility, restlessness, aggression, and impatience. It is linked with a person's first heart attack and with cardiovascular disease. Type B behavior pattern is the opposite of Type A; it is associated with lower risk of first heart attack, but the prognosis after an attack is poorer. Following an initial heart attack, Type A behavior pattern individuals have a higher recovery rate.

- Although stress is unrelated to serious psychopathology, it is related to social isolation and distrust.

What benefits are there to exercise?

- Aerobic exercise has numerous benefits, especially to cardiovascular health and fitness. The best results are obtained with a moderate exercise program maintained throughout adulthood.

13.2 COGNITIVE DEVELOPMENT

How does practical intelligence develop in adulthood?

- Research on practical intelligence reveals differences between optimally exercised ability and unexercised ability. This gap closes during middle adulthood. Practical intelligence appears not to decline appreciably until late life.

How does a person become an expert?

- People tend to become experts in some areas and not in others. Experts tend to think in more flexible ways than novices and are able to skip steps in solving problems. Expert performance tends to peak in middle age.

What is meant by lifelong learning? What differences are there between adults and young people in how they learn?

- Adults learn differently than children and youth. Older students need practical connections and a rationale for learning, and they are more motivated by internal factors.

13.3 PERSONALITY

What is the five-factor model? What evidence is there for stability in personality traits?

- The five-factor model postulates five dimensions of personality: neuroticism, extraversion, openness to experience, agreeableness, and conscientiousness. Several longitudinal studies indicate that personality traits show long-term stability, but increasing evidence shows that traits change across adulthood.

What changes occur in people's priorities and personal concerns? How does a person achieve generativity? How is midlife best described?

- Erikson believed that middle-aged adults become more concerned with doing for others and passing social values and skills to the next generation—a set of behaviors and beliefs he labeled *generativity*. Those who do not achieve generativity are thought to experience stagnation.

- For the most part, there is little support for theories based on the premise that all adults go through predictable life stages at specific points in time. Individuals may face similar stresses, but transitions can occur at any time in adulthood. Research indicates that not everyone experiences a crisis at midlife.

13.4 FAMILY DYNAMICS AND MIDDLE AGE

Who are the kinkeepers in families?

- Middle-aged mothers tend to adopt the role of kinkeepers in order to keep family traditions alive and as a way of linking generations.

- Middle age is sometimes referred to as the sandwich generation.

How does the relationship between middle-aged parents and their young adult children change?

■ Parent–child relations improve dramatically when children grow out of adolescence. Most parents look forward to having an empty nest. Difficulties emerge to the extent that raising children has been a primary source of personal identity for parents. However, once children have left home, parents still provide considerable support.

■ Children move back home primarily for financial or child-rearing reasons. Neither parents nor children generally prefer this arrangement.

How do middle-aged adults deal with their aging parents?

■ Caring for aging parents usually falls to a daughter or daughter-in-law. Caregiving creates a stressful situation due to conflicting feelings and roles. The potential for conflict is high, as is financial pressure.

■ Caregiving stress is usually greater in women, who must deal with multiple roles. Older parents are often dissatisfied with the situation as well.

What styles of grandparenthood do middle-aged adults experience? How do grandchildren and grandparents interact?

■ Becoming a grandparent means assuming new roles. Styles of interaction vary across grandchildren and with the age of the grandchild. Also relevant are the social and personal dimensions of grandparenting.

■ Grandparents derive several different types of meaning regardless of style. Most children and young adults report positive relationships with grandparents, and young adults feel a responsibility to care for them if necessary.

■ Ethnic differences are found in the extent to which grandparents take an active role in their grandchildren's lives.

■ In an increasingly mobile society, grandparents are more frequently assuming a distant relationship with their grandchildren. An increasing number of grandparents serve as the custodial parent. These arrangements are typically stressful.

KEY TERMS

skeletal maturity (457)

osteoporosis (457)

dual-energy X-ray absorptiometry (DXA) test (459)

selective estrogen receptor modulators (SERMs) (460)

osteoarthritis (460)

wear-and-tear disease (460)

rheumatoid arthritis (460)

climacteric (462)

menopause (462)

perimenopause (462)

menopausal hormone therapy (MHT) (463)

stress and coping paradigm (466)

coping (467)

Type A behavior pattern (468)

Type B behavior pattern (468)

post-traumatic stress disorder (PTSD) (468)

aerobic exercise (469)

practical intelligence (472)

mechanics of intelligence (472)

pragmatics of intelligence (471)

encapsulation (474)

neuroticism (477)

extraversion (477)

openness to experience (477)

agreeableness (477)

conscientiousness (477)

generativity (480)

stagnation (480)

narrative (480)

ego resilience (482)

kinkeeper (483)

sandwich generation (483)

filial obligation (486)

Log in to **www.cengagebrain.com** to access the resources your instructor requires. For this book, you can access:

- CourseMate brings course concepts to life with interactive learning, study, and exam preparation tools that support the printed textbook. A textbook-specific website, Psychology CourseMate includes an integrated interactive eBook and other interactive learning tools including quizzes, flashcards, videos, and more.

CENGAGENOW™

- CengageNOW Personalized Study is a diagnostic study tool containing valuable text-specific resources—and because you focus on just what you don't know, you learn more in less time to get a better grade.

WebTUTOR™

- More than just an interactive study guide, WebTutor is an anytime, anywhere customized learning solution with an eBook, keeping you connected to your textbook, instructor, and classmates.

PART FOUR | Late Adulthood

CHAPTER 14
The Personal Context of Later Life
Physical, Cognitive, and Mental Health Issues

CHAPTER 15
Social Aspects of Later Life
Psychosocial, Retirement, Relationship, and Societal Issues

CHAPTER 16
The Final Passage
Dying and Bereavement

The Personal Context of Later Life

Physical, Cognitive, and Mental Health Issues

STOP! Before you read this chapter, do the following exercise. Take out a piece of paper and write down all the adjectives you can think of that describe aging and older adults, as well as all of the "facts" about aging that you know.

Now that you have your list, look it over carefully. Are most of your descriptors positive or negative? Do you have lots of "facts" written down, or just a few? Most people's lists contain at least some words and phrases that reflect images of older adults as portrayed by the media. Many of the media's images are stereotypes of aging that are only loosely based on reality. For example, people over age 60 are almost never pictured in ads for perfume, but they are shown in ads for wrinkle removers.

In this chapter, our journey through old age begins. Our emphasis will be on physical and cognitive changes. To begin, we consider the key physical changes and health issues confronting older adults. Changes in cognitive abilities, as well as interventions to help remediate the changes, are discussed next. Finally, some well-known mental health issues are considered, including depression, anxiety disorders, and Alzheimer's disease.

LEARNING OBJECTIVES

■ What are the characteristics of older adults in the population?

■ How long will most people live? What factors influence this?

■ What is the distinction between the third and fourth age?

Sarah is an 87-year-old African American woman who comes from a family of long-lived individuals. She has never been to a physician in her entire life, and she has never really been seriously ill. Sarah figures it's just as well that she has never needed a physician, because for most of her life she had no health insurance. Because she feels healthy and has more living that she wants to do, Sarah believes that she'll live for several more years.

WHAT IS IT LIKE TO BE OLD? Do you want your own late life to be described by the words and phrases you wrote at the beginning of the chapter? Do you look forward to becoming old, or are you afraid of what may lie ahead? Most of us probably want to be like Sarah and enjoy a long healthy life. Growing old is not something we think about very much until we have to. Most of us experience the coming of old age as a surprise. It's as if we go to bed one night middle-aged and wake up the next day feeling old. But we can take comfort in knowing that, when the day comes, we will have plenty of company.

The Demographics of Aging

Did you ever stop to think about how many older adults you see in your daily life? Did you ever wonder whether your great-grandparents had the same experience? Actually, you are privileged—there have never been as many older adults alive as there are now, so you see many more older people than your great-grandparents (or even your parents) did. The proportion of older adults in the population of industrialized countries has increased tremendously in this century, which is due mainly to better health care and to lowering women's mortality rate during childbirth.

demographers
people who study population trends

population pyramid
graphic technique for illustrating population trends

People who study population trends, called **demographers**, *use a graphic technique called a* **population pyramid** *to illustrate these changes.* ■ Figure 14.1 shows population pyramids for both developed and developing countries combined around the world. Let's consider developed countries first (they're designated by the darker color in the figure). Notice the shape of the population pyramid in 1950, shown in the top panel of the figure. In the middle of the 20th century there were fewer people over age 60 than under age 60, so the figure tapers toward the top. By 2011, the shape had begun to change as more older adults were still alive. Compare this to projections for 2050, and you can see that a dramatic change will occur in the number of people over 65.

These changes also occur in developing countries, shown in the lighter color. Notice that the figures for both 1950 and 2011 look more like pyramids because there are substantially fewer older adults than younger people. But by 2050 the number of older adults in developing countries will also have increased dramatically, changing the shape of the figure.

The rapid increase in the number of older adults (individuals over age 60) will bring profound changes to everyone's lives. Older adults are already a major marketing target and wield considerable political and economic power. In the United States the sheer number of older adults will place enormous pressure on pension systems (especially Social Security), health care (especially Medicare, Medicaid, and long-term care), and other human services. The costs will be borne by smaller cohorts of taxpaying workers behind them.

Note the changing shapes of the distributions in terms of the proportion of the population that is young versus old over time and as a function of whether countries are considered developed or developing.

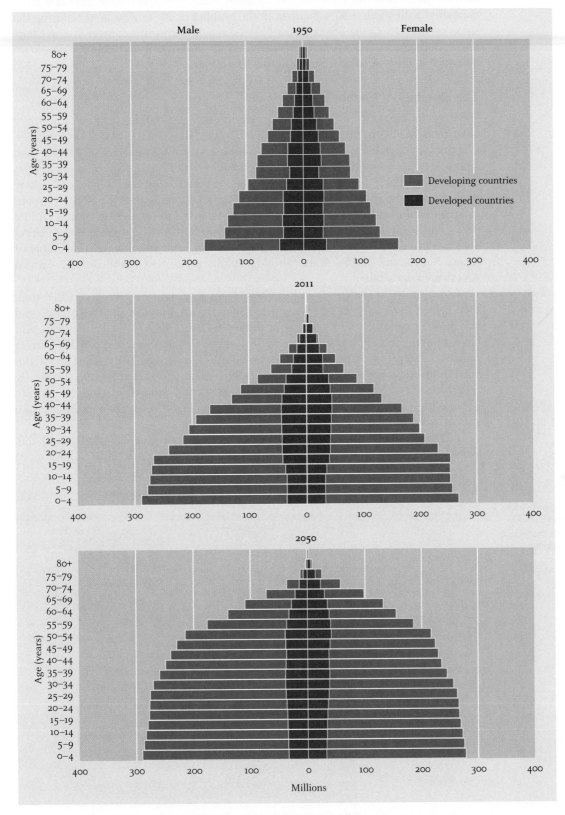

From U.S. Census Bureau, 2010. Retrieved from www.census.gov/ipc/www/idb

THINK ABOUT IT

How will the demographic changes in the first 30 years of the 21st century affect social policy?

The growing strain on social service systems will intensify because the most rapidly growing segment of the U.S. population is the group of people over age 85. In fact, the number of such people will increase nearly 500% between 2000 and 2050, compared with about a 50% increase in the number of 20- to 29-year-olds during the same period (U.S. Census Bureau, 2010b). As we will see in this chapter and in Chapter 15, individuals over age 85 generally need more assistance with daily living than do people under 85, further straining the health care system.

The Diversity of Older Adults

Older adults are no more alike than people at other ages. Older women outnumber older men in all ethnic groups in the United States, for reasons we will explore later. The number of older adults among ethnic minority groups is increasing faster than among European Americans. For example, the number of Native American elderly has increased by nearly two thirds in recent decades; Asian and Pacific Islander elderly have quadrupled; older adults are the fastest-growing segment of the African American population; and the number of Latino American elderly is also increasing rapidly (U.S. Census Bureau, 2010b). Projections for the future diversity of the U.S. population are shown in ■ Figure 14.2. You should note the very large increases in the number of Asian, Native, and Latino American older adults relative to European and African American older adults.

Older adults in the future will be better educated, too. At present, a little more than half of those over age 65 have only a high-school diploma or some college, and about 18% have a bachelor's degree or higher. By 2030 it is estimated that 85% will have a high-school diploma and about 75% will have a college degree (U.S. Census Bureau, 2010b). These dramatic changes will be due mainly to better educational opportunities for more students and the greater need for formal schooling (especially college) to find a good job. Also, better-educated people tend to live longer—mostly because they have higher incomes, which give them better access to good health care and a chance to follow healthier lifestyles.

Internationally, the number of older adults is also growing rapidly, and this growth is affecting nearly all countries globally (United Nations, 2010). These rapid

■ **Figure 14.2**

Projected changes in the U.S. minority population of older adults. Note that the number of older Latinos will increase the fastest. Data from U.S. Census Bureau (2008).

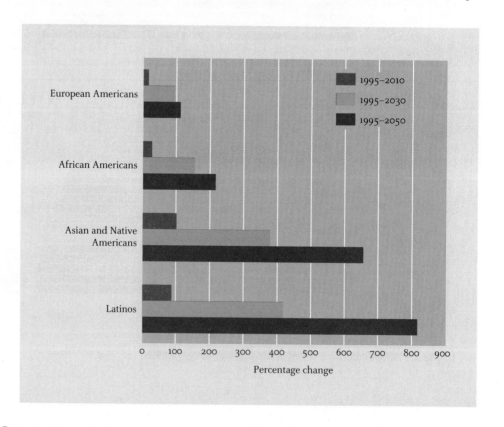

increases are due mostly to improved health care in developing countries. Such increases will literally change the face of the population as more people live to old age.

Economically powerful countries around the world, such as Japan and China, are trying to cope with increased numbers of older adults that strain the country's resources. Indeed, the rate of growth of older adults in Japan is the highest in the industrialized world; because of a declining birth rate, by 2025 there will be twice as many adults over age 65 as there will be children (Ministry of Internal Affairs and Communication, 2010). China is grappling with increased needs for health care for their older adult population. By 2040, China expects to have more than 300 million people over age 60. So they are already addressing issues related to providing services for many more older adults, especially regarding care for older adults with dementia, such as Alzheimer's disease (Barboza, 2011).

Japan, China, and the United States are not alone in facing increased numbers of older adults. As you can see in ■ Figure 14.3, many countries will have substantially more older adults in the population over the next few decades. All of them will need to deal with increased needs for services to older adults and, in some cases, competing demands with children and younger and middle-aged adults for limited resources.

Even though the financial implications of an aging population are predictable, the United States and most other countries have done surprisingly little to prepare. For example, little research has been done on the characteristics of older workers (even though mandatory retirement has been virtually eliminated for some time), on differences between the young-old (ages 65–80) and the oldest-old (over age 80), or on specific health care needs of older adults with regard to chronic illness—despite a call for such work in the early 1990s (American Psychological Society, 1993). As of 2011, the U.S. Congress had yet to adopt long-term plans for funding Social Security and Medicare, even though the first Baby Boomers became eligible for reduced Social Security benefits in 2008 and for Medicare in 2011. Unfortunately, even though the issues have been clear for decades, no legislative action has resulted.

Longevity

The number of years a person can expect to live, termed **longevity,** *is jointly determined by genetic and environmental factors.* Researchers distinguish between three types of longevity: average life expectancy, useful life expectancy, and maximum life expectancy. **Average life expectancy** *(or median life expectancy) is the age at which half of the people born in a particular year will have died.* As you can see in ■ Figure 14.4, average life expectancy at birth for people in the United States increased steadily dur-

longevity
number of years a person can expect to live

average life expectancy
the age at which half of the people born in a particular year will have died

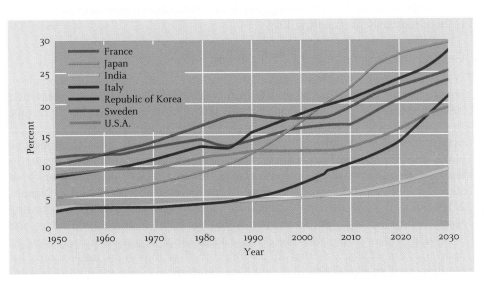

From United Nations, Statistics Bureau, MPHPT, Ministry of Health, Labour and Welfare.

■ **Figure 14.3**
The proportion of older adults (aged 65 years and over) is increasing in many countries and will continue to do so in the coming decades.

Life expectancy at birth and at 65 years of age in the United States, 1900–2007. Data from Centers for Disease Control and Prevention, National Center for Health Statistics, National Vital Statistics System (2010). Retrieved from www.cdc.gov/nchs/data/hus/hus2009tables/Table024.pdf

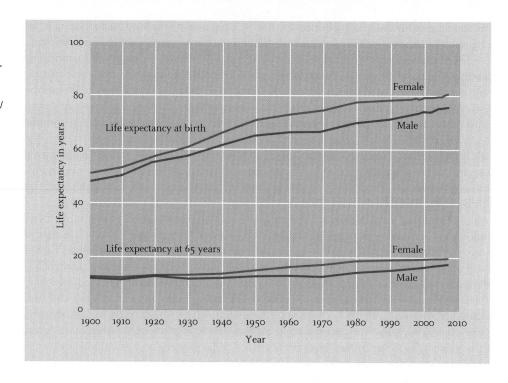

useful life expectancy
the number of years that a person is free from debilitating chronic disease and impairment

maximum life expectancy
the oldest age to which any person lives

Being able to continue doing things one likes is one hallmark of useful life expectancy.

ing the 20th century. This increase was due mainly to significant declines in infant mortality and in the number of women dying during childbirth, the elimination of major diseases such as smallpox and polio, and improvements in medical technology that prolong the lives of people with chronic disease. Currently, average life expectancy at birth for American women is 80.4 years; for men it's 75.4 (National Center for Health Statistics, 2010c).

Useful life expectancy *is the number of years that a person is free from debilitating chronic disease and impairment.* Ideally, useful life expectancy would match the actual length of a person's life. However, medical technology sometimes enables people to live on for years even though they may no longer be able to perform routine daily tasks. Accordingly, when making medical treatment decisions people are now placing greater emphasis on useful life expectancy and less emphasis on the sheer number of years they may live.

Maximum life expectancy *is the oldest age to which any person lives.* Currently, scientists estimate that the maximum limit for humans is around 120 years, mostly because the heart and other key organ systems are limited in how long they can last without replacement (Hayflick, 1998).

Genetic and Environmental Factors in Longevity

We have known for some time that a good way to increase one's chances of a long life is to come from a family with a history of long-lived individuals (Hayflick, 1998). Researchers have suspected that this is due in large part to genetic factors. One exciting line of contemporary research, the Human Genome Project, has mapped the basic human genetic code. This research—and its spinoffs in microbiology, behavior genetics, and aging—has produced some astounding results in terms of genetic linkages to both physical health and disease, with the promise of allowing even more people to reach old age in relatively good health (Vaupel, 2010). Some attempts are even being made to treat genetic diseases by implanting "corrected" genes in people in the hopes that the good genes will reproduce and eventually wipe out the defective genes. The fact that we can identify certain genetic markers of diseases that affect longevity has important implications for families, such as whether they should have children and even whether they want to know their genetic makeup (Miller & Martin, 2008).

Payoffs from such research are already helping us understand how increasing numbers of people are living to 100 or older. For example, genomic studies of human aging are advancing our understanding of why longevity tends to run in certain families (Slagboom et al., 2011). Perls and Terry (2003) showed that genetic factors play a major role in determining how well centenarians (people over age 100) cope with disease. The oldest-old are hardy because they have a high threshold for disease and show slower rates of disease progression than their peers, who develop chronic diseases at younger ages and die earlier.

Although heredity is a major determinant of longevity, environmental factors also affect the life span (Perls & Terry, 2003; Slagboom et al., 2011). Some environmental factors are more obvious; diseases, toxins, lifestyle, and social class are among the most important. Diseases, such as cardiovascular disease and Alzheimer's disease, and life-style issues, such as smoking and exercise, receive a great deal of attention from researchers. Environmental toxins, encountered mainly as air and water pollution, are a continuing problem. For example, toxins in fish, bacteria and cancer-causing chemicals in drinking water, and airborne pollutants are major agents in shortening longevity.

The impact of social class on longevity results from the reduced access to goods and services, especially medical care, that characterizes the poor (regardless of ethnic group) and many older adults (National Center for Health Statistics, 2007c). Most poor people (and even many with full-time jobs whose employers do not provide health benefits) have little or no health insurance, and many cannot afford the cost of a more healthful lifestyle. For example, living in large urban areas exposes people to serious problems such as lead poisoning from old water pipes, air pollution, and poor drinking water, but many people simply cannot afford to move.

How environmental factors influence average life expectancy changes over time. For example, acquired immunodeficiency syndrome (AIDS) became a new factor in longevity during the 1980s and continues to kill millions of people around the world. In contrast, the effect of cardiovascular diseases on life expectancy is lessening somewhat in developed countries as the rates of those diseases decline owing to overall healthier lifestyles.

The sad part about most environmental factors is that people are responsible for them. Denying adequate health care to everyone, continuing to pollute our environment, failing to address the underlying causes of poverty, and not living a healthy life style have undeniable consequences: They needlessly shorten lives and dramatically increase the cost of health care.

Ethnic and Gender Differences in Longevity

Ethnic differences in average life expectancy are clear (National Center for Health Statistics, 2010c). For example, African Americans' average life expectancy at birth is roughly 6 years lower for men and 4 years lower for women than that of European Americans. By age 65, though, the gap narrows somewhat—the average life expectancy for African Americans is only about 2 years less for men and about 1 year less for women than it is for European Americans, and by age 85 African Americans tend to live longer. Perhaps because they do not typically have access to the same quality of care that European Americans usually do and thus are at greater risk for disease and accidents, African Americans who survive to age 85 tend to be healthier than their European American counterparts. Like Sarah, the 87-year-old woman in the vignette, they may well have needed little medical care throughout their lives. But this is not the whole story. For example, Latino Americans' average life expectancy exceeds European Americans' at all ages despite having, on average, less access to health care (National Center for Health Statistics, 2010d).

A visit to a senior center or to a nursing home can easily lead to the question "Where are all the very old men?" The answer is that women tend to live longer. Women's average longevity is about 5 years more than men's at birth, narrowing to roughly 1 year by age 85 (National Center for Health Statistics, 2010d). These differences are fairly typical of most developed countries but not of developing countries.

THINK ABOUT IT

How do ethnic and gender differences in life expectancy relate to biological, psychological, sociocultural, and life-cycle factors?

Indeed, the female advantage in average longevity in the United States became apparent only in the early 20th century (Hayflick, 1996). Why? Until then, so many women died in childbirth that their average longevity as a group was reduced to that of men. Death in childbirth still partially explains the lack of a female advantage in developing countries today; however, part of the difference in some countries also results from infanticide of baby girls. In developed countries, socioeconomic factors such as access to health care, work and educational opportunities, and athletics also helped account for the emergence of the female advantage (Hayflick, 1998).

What accounts for women's longevity advantage, especially in developed countries? Overall, men's rates of dying from 12 of the top 15 causes of death are significantly higher than women's at nearly every age, and men are also more susceptible to infectious diseases (Pinkhasov et al., 2010). Explanations based on such ideas as women's second X-chromosome, more active immune functioning, and the protective effects of estrogen have not explained the difference. These results have led some to speculate that perhaps it is not simply a gender-related biological difference at work in longevity but rather a more complex interaction of lifestyle, much greater susceptibility in men of contracting certain fatal diseases, and genetics (Pinkhasov, 2010).

Despite their longer average longevity, women do not have all the advantages. Interestingly, older men who survive beyond age 90 are the hardiest segment of their birth cohort in terms of performance on cognitive tests (Perls & Terry, 2003). Between ages 65 and 89, women score higher on cognitive tests; beyond age 90, men do much better. At this point, we don't know why.

International Differences in Longevity

Countries around the world differ dramatically in how long their populations live on average. As you can see in ■ Figure 14.5, the current range extends from 38 years in Sierra Leone, Africa, to over 82 years in Japan. Such a wide divergence in life expectancy reflects vast discrepancies in genetics, sociocultural and economic conditions, health care, disease, and the like across developed and developing nations.

Latina women live longer on average than other ethnic groups, and are the fastest growing group of older adults in the United States.

The Third–Fourth Age Distinction

The development of the science of gerontology, the study of older adults, in the latter part of the 20th century led to cultural, medical, and economic advances for older adults (e.g., longer average longevity, increased quality of life) that in turn resulted in fundamental, positive changes in how older people are viewed in society. Gerontologists and policy makers became optimistic that old age was a time of potential growth rather than of decline. This combination of factors is termed the Third Age (Baltes & Smith, 2003). As we will see in this chapter and in Chapter 15, much research has documented that the young-old (ages 60–80) do indeed have much to look forward to.

However, recent research shows conclusively that the oldest-old (over age 80) typically have a much different experience, which is referred to as the Fourth Age (Baltes & Smith, 2003; Gilleard & Higgs, 2010). The oldest-old are at the limits of their functional capacity, and few interventions to reverse the effects of aging have been successful to date. We will see that the rates of diseases such as cancer and dementia increase dramatically in the oldest-old and that other aspects of psychological functioning (e.g., memory) also undergo significant and fairly rapid decline.

Baltes and Smith (2003) view the differences between the Third Age and the Fourth Age as important for research and social policy. They characterize the Third Age as the "good news" about aging and the Fourth Age as the "bad news."

International data on average life expectancy at birth. Note the differences between developed and developing countries.

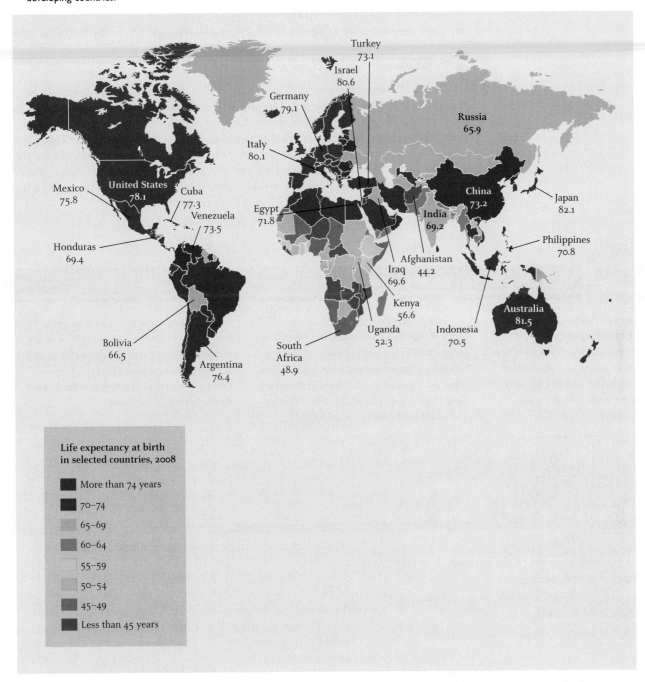

Life expectancy at birth in selected countries, 2008

- More than 74 years
- 70–74
- 65–69
- 60–64
- 55–59
- 50–54
- 45–49
- Less than 45 years

From www.census.gov/ipc/www/idb/tables.html

The "Good News": The Third Age (Young-Old)

- Increased life expectancy, with more older people living longer and aging successfully
- Substantial potential for physical and mental fitness, with improvement in each generation
- Evidence of cognitive and emotional reserves in the aging mind
- High levels of emotional and personal well-being
- Effective strategies to master the gains and losses of later life

The "Bad News": The Fourth Age (Oldest-Old)

- Sizeable losses in cognitive potential and ability to learn
- Increases in the negative effects of chronic stress
- High prevalence of dementia (50% in people over age 90), frailty, and multiple chronic conditions
- Problems with quality of life and dying with dignity

The Third and Fourth Ages approach is grounded in the "selective optimization with compensation" model described in Chapter 1. The description of gains and losses in the Third and Fourth Ages flows naturally from this life-span perspective. As you proceed through this and the next chapter, keep the distinction between the Third and the Fourth Ages in mind. Note the different developmental patterns shown by the young-old and oldest-old. In Chapter 15, we will consider some of the social policy implications of this distinction. In the meantime, think about the issues raised in the What Do *You* Think? Feature—how long do *you* want to live?

What do you think?

How Long Do You Want to Live?

We have considered evidence that average longevity has increased significantly over the past century. This means that there have never been as many older adults alive at one time than there are right now, and this will only increase during the next several decades at least.

Getting older brings with it many positives (in terms of experience, well-being, and other things people enjoy in life) and negatives (especially biological and physical changes, as we will see). This raises an important question for current generations: How long do you want to live? What, for you, would be the optimal length of life?

For some people, the optimal length of life is the number of years they can continue to live independently and well. For others, it's as long as their life has meaning. What is it for you? What are the things that help you define your answer? What do *you* think?

Test Yourself

RECALL

1. The fastest growing segment of the population in the United States is people over age _____.
2. The age at which half of the people born in a particular year will have died is called _____.
3. The Fourth Age refers to the _____-old.

INTERPRET

Think back to the lifestyle influences on health discussed in Chapter 12. If most people actually exhibited a healthy life-style, what do you think would happen to average life expectancy?

APPLY

If you were to design an intervention program to maximize the odds that people will live to maximum longevity, what would you emphasize?

Recall answers: (1) 85, (2) average life expectancy, (3) oldest

LEARNING OBJECTIVES

- What are the major biological theories of aging?
- What physiological changes normally occur in later life?
- What are the principal health issues for older adults?

Frank is an 80-year-old man who has been physically active his whole life. He still enjoys sailing, long-distance biking, and cross-country skiing. Although he considers himself to be in excellent shape, he has noticed that his endurance has decreased and that his hearing isn't quite as sharp as it used to be. Frank wonders: Can he do something to stop these declines, or are they an inevitable part of growing older?

IF YOUR FAMILY HAS KEPT PHOTOGRAPH ALBUMS OVER MANY YEARS, you are able to see how your grandparents or great-grandparents changed over the years. Some of the more visible differences are changes in the color and amount of hair and the addition of wrinkles, but many other physical changes are harder to see. In this section, we consider some of these changes, as well as a few things that adults can do to improve their health. As noted in Chapter 13, many aging changes begin during middle age. However, most of these changes typically do not affect people's daily lives until later in life, as Frank is discovering. But first, we will ask a basic question: Why do people grow old in the first place?

Biological Theories of Aging

Why does everyone who lives long enough grow old and eventually die? Despite major advances in molecular biology and genetics, there is no adequate theory or explanation (Kunlin, 2010). Instead, there are two major groups of biological theories of aging that provide partial explanations: programmed theories and damage or error theories.

Programmed theories *suggest that aging is due to a biological or genetic program.* This possibility seems more likely as the explosion of knowledge about human genetics continues to unlock the secrets of our genetic code. Genetic programming may control cell death, the endocrine system, and the immune system. Researchers now believe that the decline in functioning in these areas may be part of a master genetic program, a kind of biological clock (Kunlin, 2010; Pankow & Solotoroff, 2007; Slagboom et al., 2011). For example, programmed aging appears to be a function of physiological processes, the innate ability of cells to self-destruct, the clocklike changes in some hormones, the decline in effectiveness of the immune system, and the ability of dying cells to trigger key processes in other cells. At present, we do not know how this self-destruct program is activated, nor do we understand how it works. However, understanding how programmed cell death occurs throughout the body may be the key to understanding how genes and physiological processes interact with psychological and sociocultural forces to produce aging (Kunlin, 2010; Pankow & Solotorff, 2007).

Damage or error theory includes several components (Kunlin, 2010). **Wear-and-tear theory** *suggests that the body, much like any machine, gradually deteriorates and finally wears out.* This theory explains some diseases, such as osteoarthritis, rather well. Years of use of the joints causes the protective cartilage lining to deteriorate, resulting in pain and stiffness. However, wear-and-tear theory fails to explain most other aspects of aging.

Another damage or error theory stresses the destructive effects that certain substances have on cellular functioning. *For example, some researchers believe that* **free radicals**—*chemicals produced randomly during normal cell metabolism and that bond easily to other substances inside cells—cause cellular damage that impairs functioning.*

programmed theories
theory that aging is biologically or genetically programmed

wear-and-tear theory
suggests that the body, much like any machine, gradually deteriorates and finally wears out

free radicals
chemicals produced randomly during normal cell metabolism and that bond easily to other substances inside cells

THINK ABOUT IT

What would be the psychological and sociocultural effects of discovering a single, comprehensive biological theory of aging?

Eating a healthy diet can delay the appearance of age-related diseases.

According to this theory, aging is caused by the cumulative effects of free radicals over the life span. Free radicals may play a role in some diseases, such as atherosclerosis and cancer. The formation of free radicals can be prevented by substances called antioxidants. Although there is growing evidence that taking antioxidants, such as vitamins A, C, and E, in appropriate levels, postpones the appearance of some age-related diseases, there is little evidence that taking antioxidants or other "anti-aging medicine" increases average longevity (Olshansky, Hayflick, & Perls, 2004a, 2004b). Nonetheless, including antioxidants in one's diet (e.g., by consuming berries, red beans, and apples) may prove beneficial as part of a good nutrition plan (Jacka & Berk, 2007).

Another damage or error theory focuses on **cross-linking**, *in which some proteins interact randomly with certain body tissues, such as muscles and arteries.* The result of cross-linking is that normal, elastic tissue becomes stiffer, so that muscles and arteries are less flexible over time. The results in some cases can be serious; for example, stiffening in the heart muscle forces the heart to work harder, which may increase the risk of heart attacks. Although we know that these protein substances accumulate, there is little evidence that cross-linking causes all aspects of aging (Timiras, 2002).

Finally, **cellular theories** *explain aging by focusing on processes that occur within individual cells, which may lead to the buildup of harmful substances or the deterioration of cells over a lifetime.* One notion focuses on the number of times cells can divide, which presumably places limits on the life span of a complex organism. Cells grown in laboratory culture dishes undergo only a fixed number of divisions before dying, with the number of possible divisions dropping depending on the age of the donor organism; this phenomenon is called the *Hayflick limit* after its discoverer, Leonard Hayflick (Hayflick, 1996). For example, cells from human fetal tissue are capable of 40 to 60 divisions; cells from a human adult are capable of only about 20. What causes cells to limit their number of divisions? *Evidence suggests that the tips of the chromosomes, called* **telomeres**, *play a major role* (Sahin & DePinho, 2010). An enzyme called *telomerase* is needed in DNA replication to fully replicate the telomeres. But telomerase normally is not present in cells, so with each replication the telomeres become shorter. Eventually, the chromosomes become unstable and cannot replicate because the telomeres become too short with age (Lung et al., 2005), and they are also susceptible to prolonged stress (Epel, Burke, & Wolkowitz, 2007). Some researchers believe that cancer cells proliferate so quickly in some cases because they can activate telomerase. Current thinking is that cancer cells may thus become func-

cross-linking
random interaction of some proteins with certain body tissues, such as muscles and arteries

cellular theories
explanation of aging that focuses on processes that occur within individual cells that may lead to the buildup of harmful substances or the deterioration of cells over a lifetime

telomeres
tips of the chromosomes that shorten and break with increasing age

tionally immortal and that effective cancer therapy may involve targeting telomerase (Artandi & DePinho, 2010; Harley, 2008). Some good news comes from research indicating that aerobic exercise may help maintain telomere length, which may help slow the aging process itself (LaRocca, Seals, & Pierce, 2010).

Much remains to be learned about human aging before we can create a unified theory of why it happens. Given recent research outcomes, and the great complexity of the processes involved, the discovery of that unified theory that answers the fundamental question of why we age may be many years away.

Physiological Changes

Growing older brings with it several inevitable physiological changes. Like Frank, whom we met in the vignette, older adults find that their endurance has declined, relative to what it was 20 or 30 years earlier, and that their hearing has declined. In this section, we consider some of the most important physiological changes that occur in neurons, the cardiovascular and respiratory systems, the motor system, and the sensory systems. We also consider general health issues such as sleep, nutrition, and cancer. Throughout this discussion you should keep in mind that, although the changes we will consider happen to everyone, the rate and the amount of change varies a great deal among individuals.

Changes in the Neurons

Neuroscience research indicates that the most important normative changes with age involve structural changes in the neurons, the basic cells in the brain, and in how they communicate (Bishop, Lu, & Yankner, 2010; Deary, Penke, & Johnson, 2010; Reuter-Lorenz & Park, 2010). Recall the basic structures of the neuron we encountered in Chapter 3, shown again here in ▪ Figure 14.6. Two structures in neurons are most important in understanding aging: the dendrites, which pick up information from other neurons, and the axon, which transmits information inside a neuron from the dendrites to the terminal branches. Each of the changes we consider in this section impairs the neurons' ability to transmit information, which ultimately affects how well the person functions (Bishop & Yankner, 2010). Three structural changes are most important in normal aging: neurofibrillary tangles, dendritic changes, and neuritic plaques.

For reasons that are not understood, fibers that compose the axon sometimes become twisted together to form spiral-shaped masses called **neurofibrillary tangles**. These tangles interfere with the neuron's ability to transmit information down the axon. Some degree of tangling occurs normally with age, but large numbers of neurofibrillary tangles are associated with Alzheimer's disease and other forms of dementia (Ribe et al., 2011; Risacher & Saykin, 2011).

neurofibrillary tangles
spiral-shaped masses formed when fibers that compose the axon become twisted together

▪ **Figure 14.6**
Basic structure of the neuron.

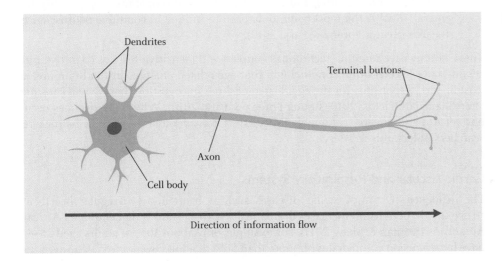

Dendrites

Terminal buttons

Axon

Cell body

Direction of information flow

Changes in the dendrites are more complicated. Some dendrites shrivel up and die, making it more difficult for neurons to communicate with each other and transmit information (von Bohlen und Halbach, 2010). However, research indicates that dendrites continue to grow in some areas of the brain, and embryonic stem cell research indicates that inducing growth may be a future way to treat brain disease and injury (M. West, 2010). This may help explain why older adults continue to improve in some areas, as we will discover later in this chapter. Why some dendrites degenerate and others do not is poorly understood; it may reflect the existence of two different families of neurons.

Damaged and dying neurons sometimes collect around a core of protein and produce **neuritic plaques**. It is likely that plaques interfere with normal functioning of healthy neurons. Although large numbers of plaques are considered a defining criteria of dementia (e.g., Alzheimer's disease), researchers have not established an "allowable number" of plaques that indicate a healthy aging brain (Takata, Kitamura, & Taniguchi, 2011). As we will see later in the chapter, the current focus of many researchers is on understanding how plaques can be prevented or eliminated as ways to treat dementia.

Because neurons do not physically touch each other, they must communicate via chemicals called **neurotransmitters**. With age, the levels of these neurotransmitters decline (Björklund & Dunnett, 2007). These declines are believed to be responsible for numerous age-related behavioral changes, including those in memory and sleep, and perhaps for afflictions such as Parkinson's disease (Björklund & Dunnett, 2007).

These changes in neurons are a normal part of aging. However, when these changes occur at a much greater rate, they cause considerable problems and are associated with Alzheimer's or related diseases, conditions we discuss in more detail on pages 532–536. This point is important because it means that serious behavioral changes (e.g., severe memory impairment) are not a result of normative age changes in the brain; rather, they are indicators of disease.

We are learning a great deal about the relations between changes in the brain and changes in behavior through technological advances in noninvasive imaging and in assessing psychological functioning (Blanchard-Fields, 2010; Reuter-Lorenz & Park, 2010). Neuroimaging is an important tool for understanding both normal and abnormal cognitive aging. Two neuroimaging techniques are used:

- *Structural neuroimaging* provides highly detailed images of anatomical features in the brain. The most commonly used are X-rays, computerized tomography (CT) scans, and magnetic resonance imaging (MRI).

- *Functional neuroimaging* provides an indication of brain activity but not high anatomical detail. The most commonly used are single photon emission computerized tomography (SPECT), positron emission tomography (PET), functional magnetic resonance imaging (fMRI), magnetoencephalograpy (or multichannel encephalograpy), and near infrared spectroscopic imaging (NIRSI). In general, fMRI is the most commonly used technique in cognitive neuroscience research (Reuter-Lorenz & Park, 2010).

These noninvasive imaging techniques coupled with sensitive tests of cognitive processing have shown quite convincingly that age-related changes in the brain are, at least in part, responsible for the age-related declines in cognition that we will consider later (Blanchard-Fields, 2010; Reuter-Lorenz & Park, 2010). Why these declines occur has yet to be discovered, although fMRI offers considerable promise in helping researchers unlock this mystery.

Cardiovascular and Respiratory Systems

The incidence of cardiovascular diseases such as heart attack, irregular heartbeat, stroke, and hypertension increases dramatically with age and is higher among African Americans (Keenan & Shaw, 2011). For example, only about 10% of adults aged 25–44 have hypertension compared with more than 50% of adults over age 65. However, the

neuritic plaques
structural change in the brain produced when damaged and dying neurons collect around a core of protein

neurotransmitters
chemicals released by neurons in order for them to communicate with each other

overall death rates from these diseases have been declining over recent decades, mainly because fewer adults smoke cigarettes and many people have reduced the amount of fat in their diets. However, the death rate for some ethnic groups, such as African Americans, remains much higher because of poorer preventive health care and less healthy lifestyles due to lack of financial resources (Keenan & Shaw, 2011).

Normative changes in the cardiovascular system that contribute to disease begin by young adulthood. Fat deposits are found in and around the heart and in the arteries (National Institute on Aging, 2010b). Eventually, the amount of blood that the heart can pump per minute will decline roughly 30%, on average, by the late 70s to 80s. The amount of muscle tissue in the heart also declines as it is replaced by connective tissue. There is also a general stiffening of the arteries due to calcification. These changes appear irrespective of lifestyle, but they occur more slowly in people who exercise, eat low-fat diets, and manage to lower stress effectively (see Chapter 13).

As people grow older, their chances of having a stroke increase. **Strokes**, *or* **cerebral vascular accidents (CVAs)**, *are caused by interruptions in the blood flow in the brain due to blockage or a hemorrhage in a cerebral artery*. Blockages of arteries may be caused by clots or by deposits of fatty substances due to the disease atherosclerosis. Hemorrhages are caused by ruptures of the artery. CVAs are the leading cause of disability (and third leading cause of death) in the United States. Research indicates that quick intervention and aggressive rehabilitation result in much better recovery (DeAngelis, 2010).

Older adults often experience **transient ischemic attacks (TIAs)**, *which involve an interruption of blood flow to the brain and are often early warning signs of stroke*. A single, large cerebral vascular accident may produce serious cognitive impairment, such as the loss of the ability to speak, or physical problems, such as the inability to move an arm. The nature and severity of the impairment in functioning a person experiences are usually determined by which specific area of the brain is affected. Recovery from a single stroke depends on many factors, including the extent and type of the loss, the ability of other areas in the brain to assume the functions that were lost, and personal motivation.

Numerous small cerebral vascular accidents can result in a disease termed **vascular dementia**. Unlike Alzheimer's disease, another form of dementia discussed later in this chapter, vascular dementia can have a sudden onset and may progress slowly (Oh et al., 2011). Typical symptoms include hypertension, specific and extensive alterations on an MRI, and differential impairment on neuropsychological tests. The differential impairment refers to a pattern of scores showing some functions intact and others significantly below average. The tests assess the ability to establish or maintain a mental set (i.e., remain focused on a particular task or situation) and visual imagery, and what clinicians look for are relatively higher scores on tests of delayed recognition memory (Gustafson et al., 2011; Way, 2011). Individuals' specific symptom patterns may vary a great deal, depending on which specific areas of the brain are damaged. In some cases, vascular dementia has a much faster course than Alzheimer's disease, resulting in death an average of 2 to 3 years after onset; in other cases, the disease may progress much more slowly with idiosyncratic symptom patterns.

Single cerebral vascular accidents and vascular dementia are diagnosed similarly. Evidence of damage may be obtained from diagnostic structural imaging (e.g., CT scan or MRI), which provides pictures like the one shown in ■ Figure 14.7, that is then confirmed by neuropsychological tests. Known risk factors for both conditions include hypertension and a family history of the disorders.

Although the size of the lungs does not change with age, the maximum amount of air in one breath drops 40% from age 25 to age 85, due mostly to stiffening of the rib cage and air passages with age and to destruction of the air sacs in the lungs by pollution and smoking (Pride, 2005). This decline is the main cause of shortness of breath after physical exertion in later life. Because of the cumulative effects of breathing polluted air over a lifetime, it is hard to say how much of these changes is strictly age-related. *The most common form of inca-*

stroke, or cerebral vascular accident (CVA)
an interruption of the blood flow in the brain due to blockage or a hemorrhage in a cerebral artery

transient ischemic attack (TIA)
an interruption of blood flow to the brain; often an early warning sign of stroke

vascular dementia
disease caused by numerous small cerebral vascular accidents

■ **Figure 14.7**
Neuroimaging is especially helpful in diagnosing a cerebrovascular accident (shown in red).

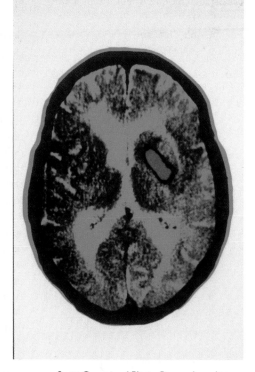

Scott Camazine / Photo Researchers, Inc.

AP Images/The Canadian Press

Actor Michael J. Fox is one of millions of people who have Parkinson's disease.

chronic obstructive pulmonary disease (COPD)

the most common form of incapacitating respiratory disease among older adults

Parkinson's disease

brain disease known primarily for its characteristic motor symptoms: very slow walking, difficulty getting into and out of chairs, and a slow hand tremor

pacitating respiratory disease among older adults is **chronic obstructive pulmonary disease (COPD)**. COPD can be an extremely debilitating condition and may result in depression, anxiety, and the need to be continually connected to oxygen (Borson, 2011; Vestbo, 2011). Emphysema is the most common form of COPD; although most cases of emphysema are due to smoking, some forms are genetic. Asthma is another common type of COPD.

Parkinson's Disease

Parkinson's disease *is known primarily for its characteristic motor symptoms: very slow walking, difficulty getting into and out of chairs, and a slow hand tremor.* These problems are caused by a deterioration of neurons in the midbrain that produce the neurotransmitter dopamine. Former boxing champion Muhammad Ali and actor Michael J. Fox are some of the more famous individuals who have Parkinson's disease. Over 1 million people in the United States (over 4 million globally) have Parkinson's disease (National Parkinson's Foundation, 2011a).

Symptoms of Parkinson's disease are treated effectively with two primary approaches: medication and surgery (Bhidayasiri & Brenden, 2011; National Parkinson's Foundation, 2011b). The most common medications are levodopa, which raises the functional level of dopamine in the brain; Sinemet (a combination of levodopa and carbidopa), which gets more levodopa to the brain; and Stalevo (a combination of Sinemet and entacapone), which extends the effective dosage time of Sinemet.

Surgical intervention involves using a device called a *neurostimulator,* which acts like a brain pacemaker by regulating brain activity when implanted deep inside the brain. Research indicates that a neurostimulator may prove effective in significantly reducing tremors, shaking, rigidity, stiffness, and walking problems when medications fail (National Institute of Neurological Disorders and Stroke, 2010). For reasons we do not yet understand, roughly 30 to 50% of the time Parkinson's disease also involves severe cognitive impairment and eventually dementia (Schapira & Olanow, 2004).

Sensory Changes

Growing older brings with it several normative changes in sensory abilities. These changes can affect people's ability to enjoy life but in most cases can be adequately compensated for through various types of interventions.

VISION. Two major kinds of age-related structural changes occur in the eye as we reach old age. One is a decrease in the amount of light that passes through the eye, resulting in the need for more light to do tasks such as reading. As you might suspect, this change is one reason older adults do not see as well in the dark, which may account in part for their reluctance to go places at night. One possible logical response to the need for more light would be to increase illumination levels in general. However, this solution does not work in all situations because we also become increasingly sensitive to glare (Lighthouse International, 2011). The second major change is that our ability to adjust to changes in illumination, called *adaptation*, declines. Going from outside into a darkened movie theater involves dark adaptation; going back outside involves light adaptation. Research indicates that the time it takes for both types of adaptation increases with age (Fozard & Gordon-Salant, 2001). These changes are especially important for older drivers, who have more difficulty seeing after being confronted with the headlights of an oncoming car.

The other key structural changes involve the lens. As we grow older, the lens becomes more yellow, causing poorer color discrimination in the green–blue–violet end of the spectrum, and the ability of the lens to adjust and focus declines as the muscles around it stiffen (Fozard & Gordon-Salant, 2001). *This is what causes* **presbyopia**, *difficulty in seeing close objects clearly, necessitating either longer arms or corrective lenses.* To complicate matters further, the time our eyes need to change focus from near to far (and vice versa) increases (Fozard & Gordon-Salant, 2001). This also poses a major problem in driving. Because drivers are constantly changing their focus from the instrument panel to other autos and signs on the highway, older drivers may miss important information because of their slower refocusing time.

presbyopia
difficulty in seeing close objects clearly

Besides these normative structural changes, some people experience diseases caused by abnormal structural changes. First, opaque spots called cataracts may develop on the lens, which limits the amount of light transmitted. Cataracts often are treated by surgical removal and use of corrective lenses. Second, the fluid in the eye may not drain properly, causing very high pressure; this condition, called glaucoma, can cause internal damage and loss of vision. Glaucoma is a fairly common disease in middle and late adulthood and is usually treated with eye drops.

The second major family of changes in vision result from changes in the retina. The retina lines approximately two thirds of the interior of the eye. The specialized receptor cells for vision, the rods and the cones, are contained in the retina. They are most densely packed toward the rear and especially at the focal point of vision, a region called the *macula*. At the center of the macula is the *fovea*, where incoming light is focused for maximum acuity, as when one is reading. With increasing age the probability of degeneration of the macula increases (Fozard & Gordon-Salant, 2001). Macular degeneration involves the progressive and irreversible destruction of receptors from any of a number of causes. This disease results in the loss of the ability to see details; for example, reading becomes extremely difficult and television is often reduced to a blur. Roughly 1 in 5 people over age 75, especially smokers and European American women, have macular degeneration, making it the leading cause of functional blindness in older adults.

A second age-related retinal disease is a by-product of diabetes. Diabetes is accompanied by accelerated aging of the arteries, with blindness being one of the more serious side effects. Diabetic retinopathy, as this condition is called, can involve fluid retention in the macula, detachment of the retina, hemorrhage, and aneurysms (Fozard & Gordon-Salant, 2001). Because it takes many years to develop, diabetic retinopathy is more common among people who developed diabetes early in life.

The combined effects of these structural changes in the eye create two other types of changes. First, the ability to see detail and to discriminate different visual patterns, called *acuity*, declines steadily between ages 20 and 60, with a more rapid decline thereafter. Loss of acuity is especially noticeable at low light levels (Fozard & Gordon-Salant, 2001).

HEARING. The age-related changes in vision we have considered can significantly affect people's ability to function in their environment. Similarly, age-related changes in hearing can also have this effect and interfere with people's ability to communicate with others. Hearing loss, especially for high-pitched tones, is one of the well-known normative changes associated with aging (National Institute on Aging, 2010a). A visit to any housing complex for older adults will easily verify this point; you will quickly notice that television sets and radios are turned up fairly loud in most of the apartments. But you don't have to be old to experience significant hearing problems.

When it became difficult to hear what was being said to him, former President Bill Clinton obtained two hearing aids. He was 51 years old at the time, and he attributed his hearing loss to too many high-school bands and rock concerts when he was young. His situation is far from unique. Loud noise is the enemy of hearing at any age. You probably have seen people who work in noisy environments (such as factories and airports) wearing protective gear on their ears so that they are not exposed to loud noise over extended periods of time.

However, you can do serious damage to your hearing with short exposure, too; in 1984, San Francisco punk rock bassist Kathy Peck—performing with her all-female punk band "The Contractions" at the Oakland Coliseum—played so loud that she had ringing in her ears for 3 days and suffered permanent hearing loss. As a result, she

Exercising while wearing headphones or ear-buds and listening to loud music when you are young can result in serious hearing loss in later life.

presbycusis
reduced sensitivity to high-pitched tones

founded Hearing Education and Awareness for Rockers (HEAR; www.hearnet.com) shortly thereafter to educate musicians about the need to protect their ears (Noonan, 2005). You don't need to be at a concert to damage your hearing, either. Using headphones or earbuds, especially at high volume, can cause the same serious damage and should be avoided. It is especially easy to cause hearing loss with headphones or earbuds if you wear them while exercising; the increased blood flow to the ear during exercise makes hearing receptors more vulnerable to damage. Because so many young people wear them with iPods and other devices, serious hearing problems may increase in frequency at much younger ages in the future.

The cumulative effects of noise and normative age-related changes create the most common age-related hearing problem: reduced sensitivity to high-pitched tones or **presbycusis**, *which occurs earlier and more severely than the loss of sensitivity to low-pitched tones.* Research indicates that, by their late 70s, roughly half of older adults have presbycusis. Men typically have greater loss than women, but this may be due to differential exposure to noisy environments. Hearing loss usually is gradual at first but accelerates during the 40s, a pattern seen clearly in ■ Figure 14.8.

Presbycusis results from four types of changes in the inner ear (Fozard & Gordon-Salant, 2001): sensory, consisting of atrophy and degeneration of receptor cells; neural, consisting of a loss of neurons in the auditory pathway in the brain; metabolic, consisting of a diminished supply of nutrients to the cells in the receptor area; and mechanical, consisting of atrophy and stiffening of the vibrating structures in the receptor area. Knowing the cause of a person's presbycusis is important because the different causes have different implications for other aspects of hearing. Sensory presbycusis has little effect on other hearing abilities. Neural presbycusis seriously affects the ability to understand speech. Metabolic presbycusis produces severe loss of sensitivity to all pitches. Finally, mechanical presbycusis also produces loss across all pitches, but the loss is greatest for high pitches.

Because hearing plays a major role in social communication, its progressive loss can have an equally important effect on social adjustment. Loss of hearing in later life may cause numerous adverse emotional reactions, such as loss of independence, social isolation, irritation, paranoia, and depression. Much research indicates that hearing loss per se does not cause social maladjustment or emotional disturbance. However, friends and relatives of an older person with hearing loss often attribute emotional changes to hearing loss, which strains the quality of interpersonal relationships, and the older person's emotional well-being can be negatively affected (Sprinzl & Riechelmann, 2010). In fact, such problems often start with family and friends becoming impatient at having to re-

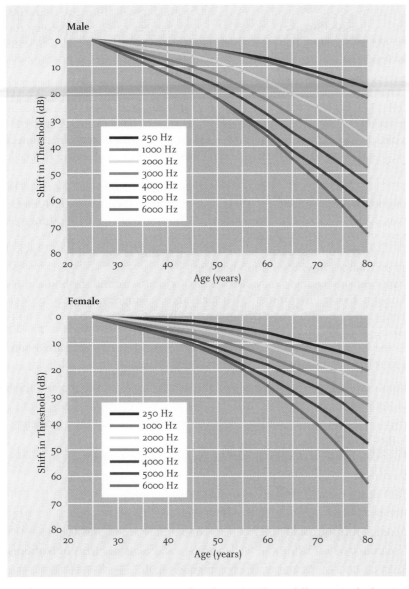

■ **Figure 14.8**
Hearing loss occurs in all adults but is greatest for high-pitched tones and greater for men than for women. As a reference, the highest note on a piano is 4,186 Hz; normal human hearing ranges from 27 Hz to 20,000 Hz.

Based on Ordy, J. M., Brizzee, K. R., Beavers, T., and Medart, P. (1979). Age differences in the functional and structural organization of the auditory system in man. In J. M. Ordy and K. R. Brizzee (Eds.), *Sensory systems and communication in the elderly.*

peat everything to the person with hearing loss. Thus, while hearing loss may not directly affect older adults' self-concept or emotions, it may negatively affect how they feel about interpersonal communication. By understanding hearing-loss problems and ways to overcome them, those without hearing loss can play a large part in minimizing the effects of hearing loss on the older people in their lives.

Fortunately, many people with hearing loss can be helped through two types of amplification systems and cochlear implants. Analog hearing aids are the most common and least expensive, but they provide the lowest-quality sound. Digital hearing aids include microchips that can be programmed for different hearing situations. Cochlear implants do not amplify sound; rather, a microphone transmits sound to a receiver, which stimulates auditory nerve fibers directly. Although technology continues to improve, none of these devices can duplicate our original equipment, so be kind to your ears.

TASTE AND SMELL. The sense of taste remains largely intact in older adults, as do sensitivity to touch, temperature, and pain (Smith & Gove, 2005). Some changes in taste have also been associated with mild cognitive impairment and with Alzheimer's disease (Steinbach et al., 2010). However, substantial age declines in smell occur after age 70 in many people (Ashendorf et al., 2005), and large declines are characteristic of Alzheimer's disease (Suzuki et al., 2004). These changes can be dangerous; for example, very old adults often have difficulty detecting the substance added to natural gas to make leaks noticeable, which can prove fatal.

FALLS. Changes in eyesight, hearing, muscle tone, reflexes, and balance make older people increasingly likely to fall. Indeed, the fear of falling and becoming injured is a real concern for many older adults and can affect their willingness to engage in certain types of activities (Li et al., 2005). Staying active and fit, along with taking precautions such as ensuring that there is sufficient light and no loose carpets, can also help reduce falls.

THINK ABOUT IT

How might fear of falling and osteoporosis (see Chapter 13) be linked?

EFFECTS ON EVERYDAY LIFE. The sensory changes that people experience as they age have important implications for their everyday lives (Schneider, Pichora-Fuller, & Daneman, 2010; Whitbourne, 1996). Some, such as difficulty reading things close up, are minor annoyances that are easily corrected (by wearing reading glasses). Others are more serious and less easily addressed. For example, the ability to drive a car is affected by changes in vision and in hearing.

Because sensory changes may also lead to accidents around the home, it is important to design a safer environment that takes these changes into account. Many accidents can be prevented by maintaining health through prevention and conditioning. But making some relatively simple environmental changes also helps. For example, falls are the most common cause of accidental serious injury and death among older adults. Here are some steps that can help reduce the potential for falls:

- Illuminate stairways and provide light switches at both the top and the bottom of the stairs.
- Avoid high-gloss floor finishes because of their glare and their tendency to be slippery when wet.
- Provide nightlights or bedside remote-control light switches.
- Be sure that both sides of stairways have sturdy handrails.
- Tack down carpeting on stairs or use nonskid treads.
- Remove throw rugs or area rugs that tend to slide on the floor.
- Arrange furniture and other objects so that they are not obstacles.
- Use grab bars on bathroom walls and nonskid mats or strips in bathtubs.
- Keep outdoor steps and walkways in good repair.

Health Issues

In Chapter 13, we examined how lifestyle factors can lower the risk of many chronic diseases. The importance of health promotion does not diminish with increasing age. As we will see, lifestyle factors influence sleep, nutrition, and cancer. In addition, whether an older adult is an immigrant can, at least in the United States, make a significant difference in health status.

Yet the biggest issue facing older adults is what the American Psychological Association (2007) calls a "broken healthcare system for older adults." The *Healthy People 2020* (HealthyPeople.gov, 2011) initiative also documents the great need for better health care. Older adults are especially disadvantaged because of the system's insensitivity to multiple health conditions, fragmented care, and stereotyping/ageism on the part of providers. Only when the health care system becomes better integrated will overall care improve.

Sleep

Older adults have more trouble sleeping than do younger adults, which is probably related to a decreased "ability" to sleep (Ancoli-Israel & Alessi, 2005). Compared to younger adults, older adults report that it takes roughly twice as long to fall asleep, that they get less sleep on an average night, and that they feel more negative effects following a night with little sleep. Some of these problems are due to mental health problems such as depression; physical diseases such as heart disease, arthritis, diabetes, lung diseases, stroke, and osteoporosis; and other conditions such as obesity (Foley et al., 2004). *Sleep problems can disrupt a person's* **circadian rhythm**, *or sleep–wake cycle.* Circadian rhythm disruptions can cause problems with attention and memory. Research shows that interventions, such as properly timed exposure to bright light, are effective in correcting circadian rhythm sleep disorders (Terman, 1994).

circadian rhythm
the sleep–wake cycle

Nutrition

Most older adults do not require vitamin or mineral supplements as long as they are eating a well-balanced diet (Ahluwalia, 2004). Even though body metabolism declines with age, older adults need to consume the same amounts of proteins and carbohydrates as young adults because of changes in how readily the body extracts the nutrients from these substances. Because they are typically in poor health, residents of nursing homes (Pauly, Stehle, & Volkert, 2007) and frail older adults (Keller, 2004) are especially prone to malnutrition or deficiencies of such nutrients as vitamin B12 and folic acid unless their diets are closely monitored.

A good nutritional guide for older adults is the Modified MyPyramid for Older Adults developed by Tufts University (Lichtenstein et al., 2008). Based on the U.S. Department of Agriculture's guidelines, the Modified MyPyramid takes into account the changes that occur with age. You can find the Modified MyPyramid at nutrition.tufts.edu/1197972031385/Nutrition-Page-nl2w_1198058402614.html.

Cancer

One of the most important health promotion steps people can take is cancer screening. In many cases, screening procedures involve little more than tests performed in a physician's office (e.g., visual screening for signs of skin cancer), at home (e.g., breast self-exams), blood tests (e.g., screening for prostate cancer), or X-rays (e.g., mammograms).

Why is cancer screening so important? As shown in ■ Figure 14.9, the risk of getting cancer increases markedly with age (American Cancer Society, 2010c). Why this happens is not fully understood. Unhealthy lifestyles (smoking and poor diet), genetics, and exposure to cancer-causing chemicals certainly are important, but they do not fully explain the age-related increase in risk (American Cancer Society, 2010c). Early detection of cancer—even in older adults—is essential to maximize the odds of surviving, and survival rates for most cancers are improving (American Cancer Society, 2010c). As our knowledge of the genetic underpinnings of some forms of cancer continues to increase, early detection and lifestyle changes will be increasingly important.

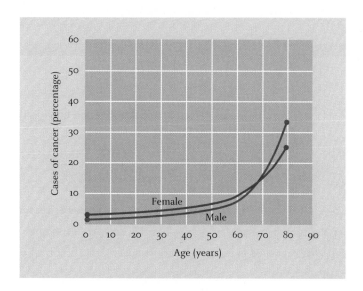

■ **Figure 14.9**
The risk of cancer increases greatly with age. Adapted from data from ACS (2010): www.cancer.org/Cancer/CancerBasics/lifetime-probability-of-developing-or-dying-from-cancer.

Immigrant Status

Whether an older adult was born in the United States or emigrated from another country affects health. Overall, immigrants in the U.S. face numerous barriers in obtaining adequate health care (Glick, 2010). However, these barriers largely do not exist in Canada (Prus, Tfaily, & Lin, 2010).

Whether or not there are structural barriers, immigrants still face challenges. Immigrants may have difficulty communicating problems with family members or professionals if they do not speak English fluently (Usita & Blieszner, 2002). Similarly, language and cultural differences need to be considered in performing examinations with immigrants; cultures vary regarding how comfortable people are about allowing strangers (i.e., physicians) to examine them (Bylsma, Ostendorf, & Hofer, 2002), and cultural differences can result in mislabeling or misdiagnosing problems (McConatha, Stoller, & Oboudiat, 2001).

Research comparing the health status of immigrants and U.S.-born older adults shows that even when socioeconomic status is controlled, immigrants show poorer health than U.S.-born people with the same ethnic background (e.g., Angel, Buckley, & Sakamoto, 2001; Berdes & Zych, 2000). Higher rates of depression are reported among older immigrant Mexican Americans who are the least acculturated, which may indicate that language and other barriers affect not only physical health but mental health as well (González, Haan, & Hinton, 2001). One mitigating factor is when immigrants have an excellent relationship with their child; when that is the case, older immigrants have fewer chronic illnesses (Ajrouch, 2007).

Test Yourself

RECALL

1. Two major groups of biological theories of aging are _____ theories and damage or error theories.

2. Damaged and dying neurons that collect around a core of protein produce _____.

3. The risk of getting cancer _____ markedly with age.

INTERPRET

In this section we have concentrated on the biological forces in development. Think about the other forces (psychological, social, and life-cycle), and list some reasons why scientists have yet to propose a purely biological theory that accounts for all aspects of aging.

APPLY

How do the sensory changes that occur with age affect an older adult's everyday life?

14.3 Cognitive Processes

LEARNING OBJECTIVES

- What changes occur in information processing as people age? How do these changes relate to everyday life?

- What changes occur in memory with age? What can be done to remediate these changes?

- What are creativity and wisdom, and how do they relate to age?

Rocio is a 75-year-old widow who feels that she does not remember recent events—such as whether she took her medicine—as well as she used to, but she has no trouble remembering things that happened in her 20s. Rocio wonders if this is normal or whether she should be worried.

ROCIO, LIKE MANY OLDER PEOPLE, TAKES MEDICATIONS FOR ARTHRITIS, ALLERGIES, AND HIGH BLOOD PRESSURE. However, each drug has its own pattern; some are taken only with meals, others are taken every 8 hours, and still others are taken twice daily. Keeping these regimens straight is important to avoid potentially dangerous interactions and side effects, and older people face the problem of remembering to take each medication at the proper time.

Such situations place a heavy demand on cognitive resources such as attention and memory. In this section, we examine age-related changes in these and other cognitive processes, including reaction time, intelligence, and wisdom.

Information Processing

In Chapter 1 we saw that one theoretical framework for studying cognition is information-processing theory. This framework provides a way to identify and study the basic mechanisms by which people take in, store, and remember information. Innovations and discoveries in neuroscience have resulted in major advances in our understanding of how people process information across the life span. Neuroscience has guided investigators as they examine age-related differences in basic processes such as attention and reaction time, particularly through the use of neuroimaging (Blanchard-Fields, 2010; Reuter-Lorenz & Park, 2010).

Cognitive aging research using neuroimaging is based on two techniques. The first and older technique involves taking a resting image of brain activity through magnetic resonance imaging (MRI) or positron emission tomography (PET) and correlating that with a behavioral measure, such as memory for a list of words. The second and main approach involves using a task-related neuroimaging technique, such as functional MRI (fMRI), to measure brain activity while the person is performing the cognitive task.

Psychomotor Speed

You are driving home from a friend's house when all of a sudden a car pulls out of a driveway right into your path. You must hit the brakes as fast as possible, or you will have an accident. How quickly can you move your foot from the accelerator to the brake?

This real-life situation is an example of **psychomotor speed**, *the speed with which a person can make a specific response.* Psychomotor speed (also called reaction time) is one of the most studied phenomena of aging, and hundreds of studies all point to the same conclusion: People slow down as they get older. In fact, the slowing-with-age finding is so well documented that many researchers accept it as the only universal behavioral change in aging discovered so far (Salthouse, 2000, 2006, 2010a, 2010b). As the cartoon shows, even Garfield feels the effects but blames faster spiders. For people, though, it's not that the world really speeds up; we definitely slow down in general. However, data suggest that the rate at which cognitive processes slow down from young adulthood to late life varies a great deal depending on the task (Salthouse, 2010a, b).

The most important reason reaction times slow down is that older adults take longer to decide that they need to respond, especially when the situation involves

psychomotor speed
the speed with which a person can make a specific response

SPIDERS SEEM TO HAVE GOTTEN FASTER AS I'VE GROWN OLDER

ambiguous information (Salthouse, 2010a, b). Even when the information presented indicates that a response will definitely be needed, there is an orderly slowing of responding with age. As the uncertainty of whether a response is needed increases, older adults become differentially slower; the difference in reaction time between older adults and middle-aged adults increases as the uncertainty level increases.

Although response slowing is inevitable, the amount of the decline can be reduced if older adults are allowed to practice making quick responses or if they are experienced in the task. In a classic study, Salthouse (1984) showed that, although older secretaries' reaction times (measured by how fast they could tap their finger) were slower than those of younger secretaries, their computed typing speed was no slower than that of their younger counterparts. Why? Typing speed is calculated on the basis of words typed including those corrected for errors; because older typists are more accurate, their final speeds were just as good as those of younger secretaries, whose work tended to include more errors. Also, older secretaries are better at anticipating what letters come next (Kail & Salthouse, 1994).

Because psychomotor slowing is a universal phenomenon, researchers have argued that it may explain a great deal of the age differences in cognition (e.g., Salthouse, 2010a, b). Indeed, psychomotor slowing is a very good predictor of cognitive performance, but there's a catch. The prediction is best when the task requires little effort (Park et al., 1996). If the task requires more effort and is more difficult, then working memory (which we consider later) is a better predictor of performance (Park et al., 1996). Also, exercise can mediate the effects of normative aging on cognitive slowing (Spirduso, Poon, & Chodzko-Zajko, 2008).

Neuroscience has shed light on the brain changes that may underlie the decline in psychomotor speed. Kennedy and Raz (2009) summarize research using diffusion tensor imaging (DTI) that documents declines in brain white matter (which aids in neural transmission), which in turn has been associated with measures of processing speed, among other things.

Psychomotor slowing with age has also sparked considerable controversy concerning whether older adults should be allowed to drive. News headlines fuel the debate. For example, on July 16, 2003, an 86-year-old driver killed 10 people when he lost control of his car while driving through a farmers' market in Santa Monica, California. He reportedly confused his brake and gas pedals as he tried to stop (Bowles, 2003). Also, a 70-year-old man injured 13 middle-school students in Belmont, California, in 2007 (Manekin, 2008).

Especially in societies that promote individual independence and that also do not provide extensive public transportation systems, driving a car is necessary to accomplish many daily tasks, such as purchasing food. However, as we have seen, age-related changes in vision, hearing, attention, and reaction time affect people's competence as drivers. Moreover, the number of older adults is rapidly increasing.

As you can see in ■ Figure 14.10, statistics compiled by the Insurance Institute for Highway Safety (Cheung & McCartt, 2010) show that, although the fatality rate for older drivers over age 75 has declined, it is higher than that for middle-aged adults. Research shows that the higher fatality rate is due to older drivers' age-related decline in key sensory, attentional, and psychomotor abilities (Horswill et al., 2010; Krishnasamy & Unsworth, 2011).

Experts agree that decisions about whether "at risk" drivers should be allowed to continue driving must be based on performance measures rather than age or medical diagnosis alone. Since the mid-1980s, researchers have been working to develop these diagnostic measures.

Karlene Ball and her colleagues developed the **useful field of view (UFOV)** *measure, an area from which one can extract visual information in a single glance without turning one's head or moving one's eyes* (Ball & Owsley, 1993), which can easily be assessed via a personal desktop computer (Edwards et al., 2005). A prototype assessment that can be used during normal driving has also been developed (Danno et al., 2010). The size of the UFOV is important; it may mean the difference between "seeing" a car running a stop sign, or a child running out between two parked cars, and "not seeing" such information.

useful field of view (UFOV)
an area from which one can extract visual information in a single glance without turning one's head or moving one's eyes

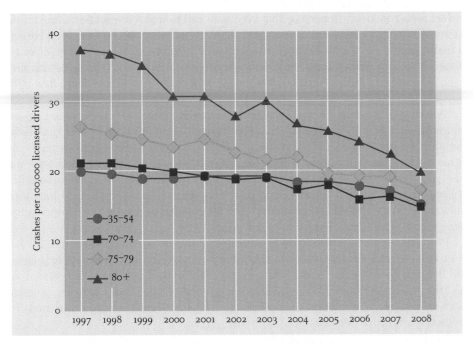

From Declines in Fatal Crashes of Older Drivers: Changes in Crash Risk and Survivability, by Ivan Sheung, Anne T. McCartt, June 2010. Insurance Institute for Highway Safety. www.iihs.org.

Clearly, "seeing" and "not seeing" may mean the difference between having an accident and avoiding one (Ball et al., 1993). The UFOV test simulates driving in that it demands quick processing of information, simultaneous monitoring of central and peripheral stimuli, and the extraction of relevant target stimuli from irrelevant background information while performing a task. Performance on the UFOV predicts driving performance (e.g., Danno et al., 2010; Hoffman et al., 2005). In fact, driving performance improves after training in how to expand one's UFOV; for example, people reduce the number of dangerous maneuvers made while driving (Ball, 1997).

Other researchers have focused on alternative diagnostic methods. For example, Freund and colleagues (2005) showed that a new method of scoring the "clock drawing test" (in which people reproduce various configurations of clock faces from memory) was a highly reliable predictor of how well older adults preformed on a driving simulator. McKenna and colleagues (2004) reported that a neuropsychological test battery was 85% accurate in predicting which of the over-70-year-old participants would fail an on-road driving test.

Researchers and clinicians alike agree that better testing of older drivers is needed (Carr & Ott, 2010). To assist states in adopting more uniform standards, the American Automobile Association (2005) created the *AAA Roadwise Review Online: A Tool to Help Seniors Drive Safely Longer*. The *Roadwise Review* is a screening tool developed by AAA and transportation safety researchers and validated in research. Designed to be administered online, the *Roadwise Review* assesses eight key functional areas: leg strength and general mobility, head and neck flexibility, high-contrast visual acuity, low visual acuity, working memory, visualization of missing information, visual search, and visual information processing speed. Drivers with a significant loss in the functional capabilities tested by *Roadwise Review* are 2 to 5 times more likely to cause a motor vehicle crash than drivers without losses in these key safe driving abilities (American Automobile Association, 2005). You can find out more about *Roadwise Review* at the website www. seniordrivers.org/driving/driving.cfm?button=roadwiseonline.

Working Memory

One evening while you are watching television, you suddenly remember that your significant other's birthday is a week from tomorrow. You decide that a nice romantic dinner would be a great way to celebrate, so you go online, find a restaurant that is perfect for a romantic dinner, see that you must call to make reservations, look at the phone number, pick up your phone, and call. Remembering the number long enough to dial it successfully requires good working memory. **Working memory** *involves the processes and structures involved in holding information in mind and simultaneously using it to solve a problem, make a decision, perform some function, or learn new information.*

Working memory is an umbrella term for many similar short-term holding and computational processes relating to a wide range of cognitive skills and knowledge domains (Braver & West, 2008). Working memory has a relatively small capacity. Because working memory deals with information that is being used right at the moment, it acts as a kind of mental scratchpad or blackboard. Unless we take some action to keep the information active (perhaps by rehearsal) or pass it along to long-term storage, the "page" we are using will be filled up quickly; to handle more information, some of the old information must be discarded.

Working memory generally declines with age, and several researchers use this fact to explain age-related differences in cognitive performance on tasks that are difficult and demand considerable effort and resources. These findings can be integrated under "the goal maintenance account" (Braver & West, 2008). Taken together, working memory and psychomotor speed provide a powerful set of explanatory constructs for predicting cognitive performance (Salthouse, 2010a, b).

Neuroimaging studies reveal why these age differences occur. It turns out that both younger and older adults activate the prefrontal area of their brains (an area behind the forehead) during working memory tasks. But older adults activate more of it on easier tasks, and they also exhaust their resources sooner (Cappell, Gmeindl, & Reuter-Lorenz, 2010; Reuter-Lorenz & Park, 2010). In a sense, older adults have to devote more "brain power" to working memory than younger adults, so older adults run out of resources sooner, resulting in poorer performance.

Memory

"Memory is power" (Johnson-Laird, 1988, p. 41). Indeed it is, when you think of the importance of remembering tasks, faces, lists, instructions, and our personal past and identity. Perhaps that is why people put such a premium on maintaining a good memory in old age—many older adults use it to judge whether their mind is intact. Poor memory is often viewed as an inevitable part of aging. But is it? And if it is, what aspects of memory change happen to everyone and which ones don't? Many people like Rocio, the woman in the vignette, believe that forgetting a loaf of bread at the store when one is 25 is not a big deal, but forgetting it when one is 65 is cause for alarm—a sign of Alzheimer's disease or some other malady. Is this true? In this section, we sort out the myth and the reality of memory changes with age.

What Changes?

The study of memory aging generally focuses on two types of memory: **explicit memory**, *the deliberate and conscious remembering of information that is learned and remembered at a specific time, and* **implicit memory**, *the unconscious remembering of information learned at some earlier time. Explicit memory is further divided into* **episodic memory**, *the general class of memory having to do with the conscious recollection of information from a specific time or event, and* **semantic memory**, *the general class of memory concerning the remembering of meanings of words or concepts not tied to a specific time or event.*

The results from hundreds of studies point to several conclusions (Berry et al., 2010; Craik & Salthouse, 2008; Negash et al., 2011). Older adults tend to perform worse than younger adults on tests of episodic memory recall in that they omit more information, include more intrusions, and repeat more previously recalled items. These age differences have been well documented and are large; for example, more than 80% of a sample of adults in their 20s will do better than adults in their 70s

Older adults perform as well as younger adults at semantic memory tasks, like the TV show *Jeopardy* shown here, that involve remembering facts or words.

Amanda Edwards/Getty Images

(Verhaeghen & Salthouse, 1997). These differences are not reliably lowered by a slower presentation or by giving cues or reminders during recall. On recognition tests, age differences are smaller but still present (Zacks, Hasher, & Li, 2000). Older adults also tend to be less efficient at spontaneously using memory strategies to help themselves remember (Hertzog & Dunlosky, 2004), but can learn to use such strategies effectively (Berry et al., 2010).

In contrast, age differences on semantic memory tasks are typically absent in normative aging but are found in persons with dementia, making this difference one way to diagnose probable cases of abnormal cognitive aging (Paraita, Díaz, & Anllo-Vento, 2008). Similarly, age differences are typically absent on tests of implicit memory, such as with picture priming. Picture priming tasks are tasks in which people are shown pictures of objects, not asked to remember them, distracted, and then asked to name a series of pictures, some of which they had seen before. Priming occurs when people name the pictures they had seen before more quickly (Ballesteros, Reales, & Mayas, 2007). However, one area in which older adults have difficulty is in word finding, such as delays in coming up with the right word based on a definition and having more tip-of-the-tongue experiences. Neuroimaging research has discovered that word finding problems are related to how much white matter is intact in the brain (Stamatakis et al., 2011).

A final area of memory research concerns **autobiographical memory**, *memory for events that occur during one's life.* An interesting phenomenon arises when the distribution of highly memorable autobiographical events across the life span is examined. ■ Figure 14.11 shows that, for both younger and older adults, vivid memories experienced between ages 10 and 30 are reported more often than those occurring after age 30 (Fitzgerald, 1999). This same pattern holds when people are asked to name Academy Award winners, news stories, and teams that played in the World Series (Rubin, Rahhal, & Poon, 1998). It is possible that this earlier period of life has greater importance in defining oneself and thus helps people organize their memories (Fitzgerald, 1999). Older adults also tend to report fewer details from previous events than do younger adults (Addis, Wong, & Schacter, 2008). Interestingly, neuroimaging research indicates that similar brain processes underlie autobiographical, episodic, and semantic memory, all of which have to do with retrieving "facts" (Burianova, McIntosh, & Grady, 2010).

In general, neuroscience research provides evidence that structural changes in the brain underlie changes in memory. Neuroimaging results show that the areas of the brain involved in encoding memories (e.g., the hippocampus and medial temporal

autobiographical memory
memory for events that occur during one's life

Both younger and older adults remember
more life events from their teens and 20s
than from any other period of life.

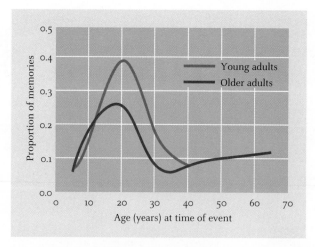

Reprinted from Fitzgerald, J. (1999). Autobiographical memory and social cognition. In T. M. Hess & F. Blanchard-Fields (Eds.), *Social cognition and aging* (p. 161) with permission from Elsevier.

lobe) shrink in older adults, making it harder for them to get information stored properly (Mitchell & Johnson, 2009). Older adults tend to activate their prefrontal cortex as a way to try to compensate, but their resources there are limited as we have seen (Reuter-Lorenz & Park, 2010).

In sum, contrary to social stereotypes of a broad-based decline in memory ability with age, research shows that the facts are more complex, and related to underlying changes in the brain. Whether memory declines with age depends on the type of memory.

When Is Memory Change Abnormal?

The older man in the *For Better or For Worse* cartoon voices a concern that many older adults have: that their forgetfulness is indicative of something much worse. Because people are concerned that memory failures may reflect disease, identifying true cases of memory-impairing disease is extremely important. Differentiating normal and abnormal memory changes is usually accomplished through a wide array of tests that are grounded in the research findings that document the various developmental patterns discussed previously (American Psychological Association, 2004). Such testing focuses on measuring performance and identifying declines in aspects of memory that typically do not change, such as tertiary memory (which is essentially long-term memory) (Stoner, O'Riley, & Edelstein, 2010).

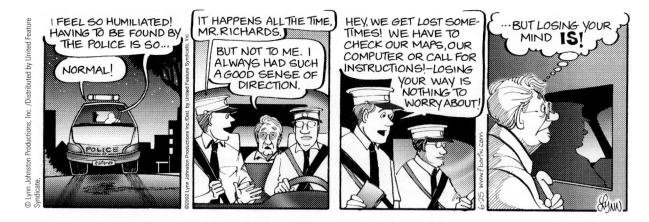

Even if a decline is identified in an aspect of memory that is cause for concern, it does not automatically follow that there is a serious problem. A first step is to find out whether the memory problem is interfering with everyday functioning. When the memory problem does interfere with functioning, such as not remembering your spouse's name or how to get home, it is appropriate to suspect a serious, abnormal underlying reason.

Once a serious problem is suspected, the next step is to obtain a thorough examination (Stoner et al., 2010). This should include a complete physical and neurological examination and a complete battery of neuropsychological tests. These may help identify the nature and extent of the underlying problem and provide information about what steps, if any, can be taken to alleviate the difficulties. Neuroimaging can help sort out the specific type of problem or disease the individual may be experiencing.

The most important point to keep in mind is that there is no magic number of times that a person must forget something before it becomes a matter for concern. Indeed, many memory-impairing diseases progress slowly, and poor memory performance may only be noticed gradually over an extended period of time. The best course is to have the person examined; only with complete and thorough testing can these concerns be checked appropriately.

Remediating Memory Problems

Remember Rocio, the person in the vignette who had to remember when to take several different medications? In the face of normal age-related declines, how can her problem be solved?

Support programs can be designed for people to help them remember. Sometimes, people like Rocio who are experiencing normal age-related memory changes need extra help because of the high memory demands they face. At other times, people need help because the memory changes they are experiencing are greater than normal.

Camp and colleagues (1993; Camp, 2001; Malone & Camp, 2007) developed the E-I-E-I-O framework to handle both situations. The E-I-E-I-O framework combines two types of memory: explicit and implicit. The framework also includes two types of memory aids. **External aids** *are memory aids that rely on environmental resources, such as notebooks or calendars.* **Internal aids** *are memory aids that rely on mental processes, such as imagery.* The "aha" experience that comes with suddenly remembering something (as in, "Oh, now I remember!") is the O that follows these Es and Is. As you can see in ∎ Figure 14.12, the E-I-E-I-O framework allows different types of memory to be combined with different types of memory aids to provide a broad range of intervention options to help people remember.

You are probably most familiar with the explicit-external and explicit-internal types of memory aids. Explicit-internal aids such as rehearsal help people remember phone numbers. Explicit-external aids are used when information needs to be better organized and remembered, such as using a smartphone to remember appointments. Implicit-internal aids represent nearly effortless learning, such as the association between the color of the particular wing of the apartment building one lives in and the fact that one's residence is there. Implicit-external aids such as icons representing time of day and the number of pills to take help older adults remember their medication (Murray et al., 2004).

external aids
memory aids that rely on environmental resources, such as notebooks or calendars

internal aids
memory aids that rely on mental processes, such as imagery

Type of memory	Type of memory aid	
	External	Internal
Explicit	Appointment book	Mental imagery
	Grocery list	Rote rehearsal
Implicit	Color-coded maps	Spaced retrieval
	Sandpaper letters	Conditioning

∎ **Figure 14.12**
The E-I-E-I-O model of memory helps categorize different types of memory aids.

In general, explicit-external interventions are the most frequently used to remediate the kinds of memory problems that older adults face, probably because such methods are easy to use and widely available (Berry et al., 2010). For example, virtually everyone owns either a smartphone or an address book in which they store addresses and phone numbers.

Explicit-external interventions have other important applications, too. Ensuring that older adults take the proper medication at the proper time is a problem best solved by an explicit-external intervention: a pillbox that is divided into compartments corresponding to days of the week and different times of the day, which research shows to be the easiest to load and results in the fewest medication errors (Park, Morrell, & Shifren, 1999). Electronic pillboxes that vibrate when it is time to take medication work well for people with mild to moderate memory decline (Kaldy, 2010). This type of device is especially helpful for older adults with limited literacy skills (Wolf et al., 2007). Memory interventions like this can help older adults maintain their independence. Nursing homes also use explicit-external interventions, such as bulletin boards with the date and weather conditions or activities charts, to help residents keep in touch with current events.

The E-I-E-I-O framework can be used to design remediation strategies for any kind of memory problem, including those due to disease or abnormal patterns of aging. Neuroscience research, supported by neuroimaging studies, has conclusively shown that older adults use work-arounds in their brains as an attempt to compensate for structural changes that impair certain aspects of memory (Reuter-Lorenz & Park, 2010). This research supports the notion that memory aids work, and that matching the type of memory intervention with the specific memory problem will maximize the likelihood of success over the long run. Later, we will see how the E-I-E-I-O framework provides insight into how people with Alzheimer's disease can be helped to improve their memory. In the meantime, see how many different categories of memory interventions you can discover.

Creativity and Wisdom

Two aspects of cognition that have been examined for age-related differences are creativity and wisdom. Each has been the focus of stereotypes: creativity is assumed to be a function of young people, whereas wisdom is assumed to be the province of older adults. Let's see whether these views are accurate.

Creativity

What makes a person creative? Is it exceptional productivity? Quincy Jones has written hundreds of musical pieces and produced numerous albums (including Michael Jackson's classic *Thriller*), Diego Rivera painted hundreds of pictures, and Thomas Edison had 1,093 patents (still the record for one person). But Gregor Mendel had only seven scientific papers, yet he endures as a major figure in the history of genetics. Lao Tzu is remembered mostly for one work—the enduring *Tao Te Ching*. Does creativity mean having a career marked by precocity and longevity? Wolfgang Goethe wrote poetry as a teenager, a best-selling novel in his 20s, popular plays in his 30s and 40s, Part I of *Faust* at 59, and Part II at 83. But others are "early bloomers" and decline thereafter, whereas still others are relatively unproductive early and are "late bloomers."

Researchers define creativity in adults as the ability to produce work that is novel, high in demand, and task appropriate (Sternberg & Lubart, 2001). Creative output, in terms of the number of creative ideas a person has or the major contributions a person makes, varies across the adult life span and across disciplines (Jones, 2010; Kozbelt & Durmysheva, 2007; Simonton, 1997, 2007). When considered as a function of age, the overall number of creative contributions a person has in Western culture tends to increase through one's 30s, peak in the early 40s, and decline thereafter; see ■ Figure 14.13. Analyses focusing on non-Western topics, such as Japanese

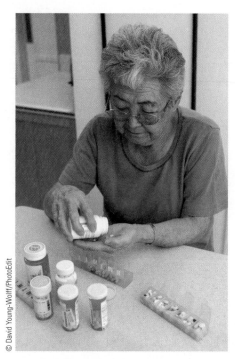

External memory aids such as pill organizers help people remember when certain medications need to be taken.

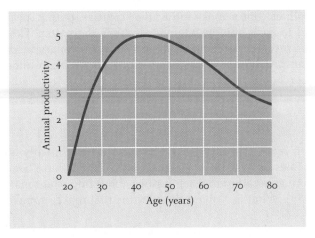

Data from Simonton (1997).

Ukiyo-e printmakers, find similarly shaped developmental functions. The age-related decline does *not* mean that people stop being creative altogether; rather, it means that older people produce fewer creative ideas than when they were younger (Dixon & Hultsch, 1999). In fact, the age at which people made major creative contributions, such as research that resulted in winning the Nobel Prize, increased throughout the 20th century (Jones, 2010). The fact that creativity never stops is illustrated in the Real People feature.

Real People
Applying Human Development
Creativity Never Stops

Susan Perlstein never believed that creativity stopped at a certain age. She believes that no matter how old you are, you need to be striving to reach your potential. For her, being creative is one of those potentials.

To help older adults reach their creative potential, Susan founded the National Center for Creative Aging (www.creativeaging.org). The NCCA showcases the creative work of older adults through its blog, which includes works by older artists and photographers, among others,

and personal stories of ways in which older adults are creative.

Examples of people who found their creative expression through Susan's efforts abound. Among them are numerous older military veterans who started painting in later life. The NCCA also showcases active, creative people, such as women who returned to complete their education. In one case, a woman returned at age 52 as a young widow to complete her degree, and worked to educate the military about

human sexuality through the 1990s. She was reelected to her city council at age 78 (for a fifth term), works out with her Wii, and enjoys living with her daughter, son-in-law, and grandchildren.

Susan Perlstein's efforts to promote creative aging remind us that although the quantity of creative output may decrease, the fact of creative output never ends. So no matter how old you are, you can, and should, be creative in some way.

Exciting new neuroimaging research is supporting previous research that one's most innovative contribution tends to happen most often during the 30s or 40s, as well as showing that creative people's brains are different. This new research shows that white matter brain structures that connect distant brain regions, and coordinate the cognitive control of information among them, are related to creativity and are more apparent in creative people (Jung et al., 2010; Takeuchi et al., 2010). This research supports the belief that creativity involves connecting disparate ideas in new ways, as different areas of the brain are responsible for processing different kinds of information. Because white matter tends to change with age, this finding also suggests that there are underlying brain maturation reasons why innovative thinking tends to occur most often during late young adulthood and early middle age.

Taken together, these analyses provide a powerful model for explaining individual differences in creative output across adulthood. The trend is clear: Across a variety of disciplines, one's most innovative creative output peaks during late young adulthood to early middle age and declines thereafter. This may help explain why senior researchers include many younger scholars in their work. The senior scholar can provide the overall context of the problem under study, while the younger researchers may provide a continuous flow of innovative ideas (Dixon & Hultsch, 1999).

Wisdom

For thousands of years, cultures around the world have greatly admired people who were wise. Tales of wise people, usually older adults, have been passed down from generation to generation to teach lessons about important matters of life and love (Chinen, 1989). What is it about these truths that makes someone who knows them wise?

From a psychological perspective, wisdom has been viewed from three main aspects (Sternberg & Lubart, 2001): the orchestration of mind and virtue, involving the ability to solve difficult real-world problems (see Chapter 13); postformal thinking (see Chapter 10); and action-oriented knowledge—acquired without direct help from others—that allows people to achieve goals they value. A growing body of research has been examining these aspects.

Based on years of research using in-depth think-aloud interviews with young, middle-aged, and older adults about normal and unusual problems that people face, Baltes and colleagues (Ardelt, 2010; Baltes & Staudinger, 2000; Scheibe, Kunzmann, & Baltes, 2007) describe four characteristics of wisdom:

- Wisdom deals with important or difficult matters of life and the human condition.
- Wisdom is truly "superior" knowledge, judgment, and advice.
- Wisdom is knowledge with extraordinary scope, depth, and balance that is applicable to specific situations.
- Wisdom, when used, is well intended and combines mind and virtue (character).

Researchers have used this framework to discover that people who are wise are experts in the basic issues in life (Ardelt, 2010; Baltes & Staudinger, 2000). Wise people know a great deal about how to conduct life, how to interpret life events, and what life means. Kunz (2007) refers to this as the strengths, knowledge, and understanding learned only by living through the earlier stages of life.

Research studies indicate that, contrary to what many people expect, there is no association between age and wisdom (Ardelt, 2010; Baltes & Staudinger, 2000; De Andrade, 2000; Hartman, 2001). As depicted in the model of Baltes and colleagues, whether a person is wise depends on whether he or she has extensive life experience with the type of problem given and has the requisite cognitive abilities and personality. Thus, wisdom could be related to crystallized intelligence, knowledge that builds over time and through experience (Ardelt, 2010).

So what specific factors help one become wise? Baltes (1993) identified three factors: (1) *general personal conditions*, such as mental ability; (2) *specific expertise conditions*, such as mentoring or practice; and (3) *facilitative life contexts*, such as education or leadership experience. Other researchers point to additional criteria. For example, Kramer (1990) argues that the *integration of affect and cognition* that occurs during adulthood results in the ability to act wisely. Personal growth during adulthood, reflecting Erikson's concepts of generativity and integrity, also helps foster the process, as do facing and dealing with life crises (Ardelt, 2010). All of these factors take time. Thus, although growing old is no guarantee of wisdom, it does provide the time that, if used well, creates a supportive context for developing wisdom.

Test Yourself

RECALL

1. One universal factor regarding information processing and aging is the slowing of _____.

2. The two types of explicit memory are _____ and _____.

3. Three factors that help a person become wise are general personal conditions, special expertise conditions, and _____.

INTERPRET

How would the view that wisdom involves life experience fit into the discussion of expertise in Chapter 13?

APPLY

If you were to design a training program for older drivers, what elements would you include?

Recall answers: (1) psychomotor speed, (2) episodic and semantic, (3) facilitative life contexts

14.4 Mental Health and Intervention

LEARNING OBJECTIVES

- How does depression in older adults differ from depression in younger adults? How is it diagnosed and treated?

- How are anxiety disorders treated in older adults?

- What is Alzheimer's disease? How is it diagnosed and managed? What causes it?

Mary lived by herself for 30 years after her husband died. For all but the last 5 years or so, she managed very well. Little by little, family members and friends began noticing that Mary wasn't behaving quite right. For example, her memory slipped, she sounded confused sometimes, and her moods changed without warning. Her appearance deteriorated. Some of her friends attribute these changes to the fact that Mary is in her 80s. But others wonder whether this is something more than normal aging.

SUPPOSE MARY IS A RELATIVE OF YOURS. How would you deal with the situation? How would you decide whether her behavior is normal? What would you do to try to improve Mary's life? If Mary has a serious disease like Alzheimer's disease, we'll consider interventions that might help later in this section.

Every day, families turn to mental health professionals for help in dealing with the psychological problems of their aging relatives. Unfortunately, myths interfere with appropriate mental health diagnoses and interventions for older adults. For example, many people mistakenly believe that nearly all older adults are either depressed, demented, or both. When they observe older adults behaving in these ways, they take no action because they believe that nothing can be done.

In this section, we will see that such beliefs are wrong. Only a minority of older adults have mental health problems, and most such problems respond to therapy. Sometimes these problems manifest themselves differently in younger and older adults, so we need to know what to look for. Accurate diagnosis is essential. Let's examine some of the most commonly occurring and widely known disorders: depression, anxiety disorders, and Alzheimer's disease.

Depression

Most people feel down or sad from time to time, perhaps in reaction to a problem at work or in one's relationships. But does this mean that most people are depressed? How is depression diagnosed? Are there age-related differences in the symptoms examined in diagnosis? How is depression treated?

Although rates of serious depression decline with age, it remains a significant problem for many older adults.

dysphoria
feeling sad or down

First of all, let's dispense with a myth. Contrary to the popular belief that most older adults are depressed, for healthy people the rate of severe depression *declines* from young adulthood to old age; the average age of onset is one's early 30s (NIMH, 2010b), a fact that also holds cross-culturally (Chou & Chi, 2005). Rates for depression tend to be equivalent for Latino and for European American older adults, while rates for African American and Asian older adults are lower (Jimenez et al., 2010). Immigrant Latino older adults, though, have rates higher than their native-born counterparts (Jimenez et al., 2010). In the United States, only about 4.5% of older adults living in the community show signs of depression in a given year (compared with nearly 9% of young adults), but this figure rises to over 13% among those who require home health care (NIMH, 2010b). For those people who do experience depression, let's examine its diagnosis and treatment.

How Is Depression Diagnosed in Older Adults?

Depression in later life is usually diagnosed on the basis of two clusters of symptoms that must be present for at least 2 weeks: feelings and physical changes. *As with younger people, the most prominent symptom of depression in older adults is feeling sad or down, termed* **dysphoria**. But whereas younger people are likely to label these feelings directly as "feeling depressed," older adults may refer to them as "feeling helpless" or in terms of physical health such as "feeling tired" (Stoner et al., 2010). Older adults are also more likely than younger people to appear apathetic and expressionless, to confine themselves to bed, to neglect themselves, and to make derogatory statements about themselves.

The second cluster of symptoms includes physical changes such as loss of appetite, insomnia, and trouble breathing (Stoner et al., 2010). In young people, these symptoms usually indicate an underlying psychological problem, but in older adults they may simply reflect normal, age-related changes. Thus, older adults' physical symptoms of depression must be evaluated very carefully (Whitbourne & Spiro, 2010). Memory problems are also a common long-term feature of depression in older adults (González, Bowen, & Fisher, 2008).

An important step in diagnosis is ruling out other possible causes of the symptoms. For example, other physical health problems, neurological disorders, medication side effects, metabolic conditions, and substance abuse can all cause behaviors that resemble depression (Stoner et al., 2010; Whitbourne & Spiro, 2010). For many minorities, immigration status and degree of acculturation and assimilation are key factors to consider (Jimenez et al., 2010), as is the stress from multiple roles. For example, Native American custodial grandparents show more symptoms of depression than their European American counterparts (Letiecq, Bailey, & Kurtz, 2008). An important criterion to be established is that the symptoms interfere with daily life; clinical depression involves significant impairment of daily living (Stoner et al., 2010).

What Causes Depression?

There are two main schools of thought about the causes of depression. One focuses on biological and physiological processes, particularly on imbalances of specific neurotransmitters, in ways that parallel other diseases. Because most neurotransmitter levels decline with age, some researchers believe that depression in later life is likely to be a biochemical problem (Ciraulo et al., 2011; Way, 2011). The general view that depression has a biochemical basis underlies current approaches to drug therapies, discussed a bit later.

The second view focuses on psychosocial factors, such as loss and internal belief systems. Although several types of loss have been associated with depression—including loss of a spouse, a job, or good health—it is how a person interprets a loss, rather than the event itself, that causes depression (Gaylord & Zung, 1987). *In this approach,* **internal belief systems,** *or what one tells oneself about why certain things are happening, are emphasized as the cause of depression.* For example, experiencing an unpredictable and uncontrollable event such as the death of a spouse may cause depression if you believe it happened because you are a bad person (Beck, 1967). People who are depressed tend to believe that they are personally responsible for all the bad things that happen to them, that things are unlikely to get better, and that their whole life is a shambles.

Gatz (2000) takes a comprehensive view that depression depends on the balance between biological dispositions, stress, and protective factors. Developmentally, biological factors become more important with age while stress factors diminish in importance. But given the changes in neurotransmitters with age and increases in losses, one would predict an age-related *increase* in depression. That's not what happens. So it is protective factors, such as coping skills learned through experience, that may help explain why the rate of depression decreases across the adult life span.

How Is Depression Treated in Older Adults?

Regardless of how severe depression is, people benefit from treatment, often through a combination of medication and psychotherapy (Ciraulo et al., 2011; Qualls & Layton, 2010). Medications work by altering the balance of specific neurotransmitters in the brain. *For very severe cases of depression, medications such as* **heterocyclic antidepressants (HCAs), monoamine oxidase (MAO) inhibitors,** *or* **selective serotonin reuptake inhibitors (SSRIs)** *can be administered.* SSRIs are the medication of first choice because they have the lowest overall side effects of any antidepressant. SSRIs work by boosting the level of serotonin, which is a neurotransmitter involved in regulating moods. One of the SSRIs, Prozac, has been the subject of controversy because it has been linked in some cases to the serious side effect of high levels of agitation. Other SSRIs, such as Zoloft, appear to have fewer adverse reactions on average (typically sexual dysfunction, headache, insomnia, and agitation), but Serzone has been associated with serious liver side effects in some patients (NIMH, 2009).

If SSRIs are not effective, the next family of medications are the HCAs. However, HCAs cannot be used if the person is also taking medications to control hypertension or has certain metabolic conditions. As a last resort, MAO inhibitors may be used. But MAO inhibitors cause dangerous, potentially fatal interactions with foods—such as cheddar cheese, wine, and chicken liver—containing tyramine or dopamine.

Either as an alternative to medication or in conjunction with it, psychotherapy is also a popular approach to treating depression. Two forms of psychotherapy have been shown to be effective with older adults. *The basic idea in* **behavior therapy** *is that depressed people experience too few rewards or reinforcements from their environment.* Thus, the goal of behavior therapy is to increase the good things that happen and minimize the negative things (Lewinsohn, 1975). This is often accomplished by having people increase their activities; simply by doing more, the likelihood that something nice will happen is increased. In addition, behavior therapy seeks to get people to reduce the negative things that happen by learning how to avoid them. The net increase in positive events and net decrease in negative events comes about through practice and homework assignments during the course of therapy, such as going out more or joining a club to meet new people.

A second effective approach is **cognitive therapy,** *which is based on the idea that maladaptive beliefs or cognitions about oneself are responsible for depression.* From this perspective, those who are depressed view themselves as unworthy and inadequate, the world as insensitive and ungratifying, and the future as bleak and unpromising (Beck et al., 1979). In a cognitive therapy session, a person is taught how to recognize these thoughts and to reevaluate the self, the world, and the future more positively, resulting in a change in the underlying beliefs. Cognitive therapy is the psychotherapy approach of choice for older adults (Laidlaw, 2007).

The most important fact to keep in mind about depression is that it *is* treatable. Thus, if an older person behaves in ways that indicate depression, it is a good idea to have him or her examined by a mental health professional. Even if the malady turns out not to be depression, another underlying and possibly treatable condition may be uncovered. A major health care problem in the United States is that less than 40% of adults of all ages receive minimally adequate treatment for depression (NIMH, 2010b).

Anxiety Disorders

Imagine you are about to give a speech to an audience of several hundred people. During the last few minutes before you begin, you start to feel nervous, your heart begins to pound, your mouth gets very dry, and your palms get sweaty. These feelings, common even to veteran speakers, are similar to those experienced more frequently and intensely by people with anxiety disorders.

anxiety disorders
problems such as feelings of severe anxiety, phobias, and obsessive-compulsive behaviors

Anxiety disorders *involve excessive, irrational dread in everyday situations, and include problems such as feelings of severe anxiety for no apparent reason, phobias with regard to specific things or places, and obsessions or compulsions in which thoughts or actions are performed repeatedly* (NIMH, 2010c). Although anxiety disorders occur in adults of all ages, they are particularly common in older adults owing to loss of health, relocation stress, isolation, fear of losing independence, and many other reasons. Anxiety disorders are diagnosed in about 17% of older men and 21% of older women (Fitzwater, 2008). The reasons for this gender difference are unknown.

Anxiety disorders can be treated with medication and psychotherapy (Qualls & Layton, 2010). The most commonly used medications are benzodiazepine (e.g., Valium and Librium), SSRIs (Paxil, among others), buspirone, and beta-blockers. Though moderately effective, these drugs must be monitored carefully in older adults because the amount needed to treat the disorder is very low and the potential for harmful side effects is great. For older adults, the clear treatment of choice is psychotherapy, especially relaxation therapy (Beck & Averill, 2004). Relaxation therapy is highly effective, is easily learned, and presents a technique that is useful in many situations (e.g., falling asleep at night).

To this point, we have focused on psychopathologies that can be treated effectively. In the next section we consider Alzheimer's disease, which at present cannot be treated effectively over the long run and progressively worsens until the person dies.

Dementia: Alzheimer's Disease

dementia
family of diseases involving serious impairment of behavioral and cognitive functioning

Arguably the most serious age-related condition is **dementia**, *a family of diseases involving serious impairment of behavioral and cognitive functioning.* Of these disorders, Alzheimer's disease is the most common.

Alzheimer's disease causes people to change from thinking, communicative human beings to confused, bedridden victims unable to recognize their family members and close friends. Because these symptoms can be so life-changing, the *fear* of Alzheimer's disease among healthy older adults—especially those who are married to or related to a person with Alzheimer's disease—is often a significant concern (Kaiser & Panegyres, 2007).

Millions of people are afflicted with Alzheimer's disease, including such notable individuals as former U.S. President Ronald Reagan, who died from it in 2004. About 5.3 million Americans have Alzheimer's disease, which cuts across ethnic, racial, and socioeconomic groups (Alzheimer's Association, 2010). The prevalence increases with age, rising from extremely low rates in the 50s to about half of all people aged 85 and older. As the number of older adults increases rapidly over the next several decades, the number of cases is expected to roughly triple.

What Are the Symptoms of Alzheimer's Disease?

Alzheimer's disease
a disease marked by gradual declines in memory, attention, and judgment; confusion as to time and place; difficulties in communicating; decline in self-care skills; inappropriate behavior; and personality changes

The key symptoms of **Alzheimer's disease** *are gradual declines in memory, learning, attention, and judgment; confusion as to time and place; difficulties in communicating and finding the right words; decline in personal hygiene and self-care skills; inappropri-*

ate social behavior; and changes in personality. These classic symptoms may be vague and may occur only occasionally in the beginning with little behavioral impact, but as the disease progresses the symptoms become much more pronounced and are exhibited much more regularly (Roberson, 2011). Wandering away from home and not being able to remember how to return increases. Delusions, hallucinations, and other related behaviors develop and get worse over time. Spouses become strangers. Patients may not even recognize themselves in a mirror; they wonder who is looking back at them. *In its advanced stages, Alzheimer's disease often causes* **incontinence,** *the loss of control of bladder or bowels.* It may also result in a total loss of mobility. Victims eventually become completely dependent on others for care. At this point many caregivers seek facilities such as adult day-care centers and other sources of help, such as family and friends, in order to provide a safe environment for the Alzheimer's patient while the primary caregiver is at work or needs to run basic errands.

The rate of deterioration in Alzheimer's disease varies widely from one patient to another, but averages around 12 years from onset of symptoms, although progression usually is faster when onset occurs earlier in adulthood (Roberson, 2011). It is very difficult to predict how long a specific patient will survive, which only adds to the stress experienced by the caregiver (Cavanaugh & Nocera, 1994).

incontinence
loss of bladder or bowel control

THINK ABOUT IT
How do the memory problems in Alzheimer's disease differ from those in normal aging?

How Is Alzheimer's Disease Diagnosed?

Given that the behavioral symptoms of Alzheimer's disease eventually become quite obvious, one would assume that diagnosis would be straightforward. Quite the contrary. In fact, despite intensive research to find specific indicators, absolute certainty that a person has Alzheimer's disease cannot even be achieved while the individual is still alive (Roberson, 2011). Definitive diagnosis must be based on an autopsy of the brain after death, because the defining criteria for diagnosing Alzheimer's disease involve documenting large numbers of amyloid plaques and neurofibrillary tangles, structural changes in neurons that occur normally with age but in very large numbers and much earlier in Alzheimer's disease.

Of course, one is still left with the issue of figuring out whether a person probably has Alzheimer's disease while he or she is still alive. Although not definitive, the number and severity of behavioral changes lead clinicians to make fairly accurate diagnoses of *probable* Alzheimer's disease (Roberson, 2011). Several brief screening measures have been developed, with some, such as the 7-Minute Screen (Ijuin et al., 2008), showing about 90% accuracy. Greater accuracy depends on a broad-based and thorough series of medical and psychological tests, including complete blood tests, metabolic and neurological tests, and neuropsychological tests (Roberson, 2011). A great deal of diagnostic work goes into ruling out virtually all other possible causes of the observed symptoms. This effort is essential. Because Alzheimer's disease is an incurable, fatal disease, every treatable cause of the symptoms must be explored first. In essence, Alzheimer's disease is diagnosed by excluding all other possible explanations. A model plan for making sure the diagnosis is correct is shown in ■ Figure 14.14.

In an attempt to be as thorough as possible, clinicians usually interview family members about their perceptions of the observed behavioral symptoms. Most clinicians view this information as critical to understanding the history of the difficulties the person is experiencing. However, research indicates that spouses are often inaccurate in their assessments of the level of their partner's impairment (McGuire & Cavanaugh, 1992). In part, this inaccuracy is due to lack of knowledge about the disease; if people do not understand or know what to look for, they will be less accurate in reporting changes in their spouse's behavior. Also, spouses may wish to portray

Alzheimer's disease involves memory loss to an extent that may include forgetting the names of family members.

■ **Figure 14.14**
Diagnosing Alzheimer's disease requires a
thorough process of ruling out other possibilities.

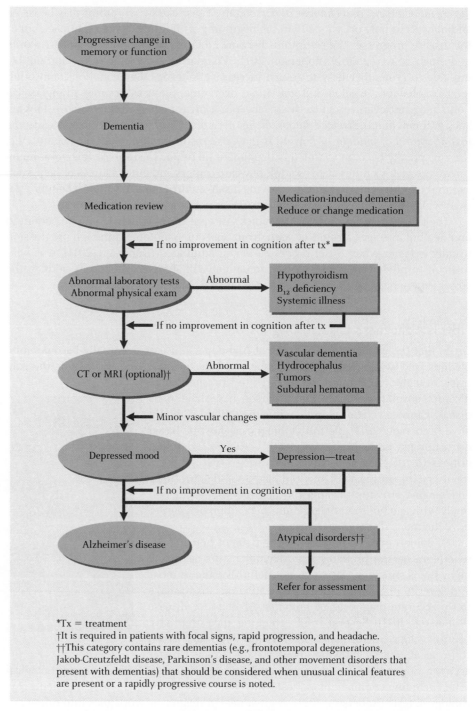

*Tx = treatment
†It is required in patients with focal signs, rapid progression, and headache.
††This category contains rare dementias (e.g., frontotemporal degenerations, Jakob-Creutzfeldt disease, Parkinson's disease, and other movement disorders that present with dementias) that should be considered when unusual clinical features are present or a rapidly progressive course is noted.

Alzheimer's Association online document, developed and endorsed by the TriAD Advisory Board. Copyright 1996 Pfizer Inc. and Esai Inc. with special thanks to J. L. Cummings. Algorithm reprinted from TriAD, *Three for the Management of Alzheimer's Disease.*

themselves as being in control, either by denying that the symptoms are actually severe or by exaggerating the severity in order to give the appearance that they are coping well in a very difficult situation. Some spouses describe their partner's symptoms accurately, but family reports should not be the only source of information about the person's ability to function.

A great deal of attention has been given to the development of more definitive tests for Alzheimer's disease while the person is still alive. *Much of this work has focused on* **amyloid**, *a protein that is produced in abnormally high levels in Alzheimer's patients, perhaps causing the neurofibrillary tangles and neuritic plaques described*

amyloid
protein that is produced in abnormally
high levels in Alzheimer's patients

earlier. Research is progressing toward developing a way to measure amyloid concentrations in cerebrospinal fluid and blood, but there is no definitive test as yet, especially in predicting later-onset cases (Okonkwo et al., 2011; Roe et al., 2011).

What Causes Alzheimer's Disease?

We do not know for sure what causes Alzheimer's disease (Roberson, 2011). Currently, most research concentrates on identifying genetic links (Bekris et al., 2011). To understand the evidence better, we need to think about two general types of Alzheimer's disease: early onset (before age 60) and later onset (after age 60).

The early onset version tends to run in families. *It has an* **autosomal dominant inheritance** *in that the presence of certain genes means that there is a 100% chance of the person eventually getting the disease.* Familial Alzheimer's disease is linked to three causative genes: *APP, PSEN1,* and *PSEN2*. If you have one of these genes, symptoms always appear before age 60, and sometimes as early as the 30s or 40s. These genes are the ones most often included in studies regarding cerebrospinal fluid indicators of Alzheimer's disease.

Later onset Alzheimer's disease may be linked to **risk genes**, *that is, genes that increase one's risk of getting the disease.* The most common later onset risk gene is the *APOE-e4* gene, which appears to be related to the formation of amyloid plaques (Kester, 2011). *APOE-e4* is one of three common forms of the *APOE* gene; the others are *APOE-e2* and *APOE-e3.* Everyone inherits a copy of some form of *APOE* from each parent. If you inherit *APOE-e4* from one parent, you have an increased risk of Alzheimer's disease. If you inherit *APOE-e4* from both parents, you have an even higher risk, but still it is not a certainty that you will get the disease. No one knows for sure yet how *APOE-e4* actually works.

Neuroimaging studies of persons with Alzheimer's disease are providing supportive evidence of the structural changes caused by the genes identified so far (Risacher & Saykin, 2011). An important opportunity in this regard is the Alzheimer's Disease Neuroimaging Initiative (ADNI; www.adni-info.org) that is following 800 people (200 with early onset Alzheimer's disease, 400 with mild cognitive impairment, and 200 healthy individuals as controls). All of the data discovered in the ADNI are made public.

Although the mechanisms of specific genes in causing Alzheimer's disease are being studied extensively, exactly how they work is still unclear, suggesting that other associated genes or environmental triggers remain to be identified. Perhaps additional advances in the area of genetics will give us insights into what we can do to prevent this devastating disease.

What Can Be Done for Victims of Alzheimer's Disease?

Currently there is no effective treatment for Alzheimer's disease and no way to prevent it. The best we can do today is alleviate some of the symptoms. Most of the research is focused on drugs aimed at improving cognitive functioning. However, most medications approved by the Food and Drug Administration to date provide little relief over the long run, and few medications in development show promising results.

However, numerous behavioral and educational interventions have also been developed. *One behavioral intervention, grounded in the E-I-E-I-O model discussed earlier, involves using the implicit-internal memory intervention called* **spaced retrieval**. Adapted by Camp and colleagues (Camp, 2001; Camp & McKitrick, 1991), spaced retrieval involves teaching persons with Alzheimer's disease to remember new information by gradually increasing the time between retrieval attempts. This easy, almost magical technique has been used to teach names of staff members and other information, and it holds considerable potential for broad application. Research shows that spaced retrieval is superior to other techniques (Haslam, Hodder, & Yates, 2011), and combining spaced retrieval with additional memory encoding aids helps even more (Kinsella et al., 2007).

In designing interventions for those with Alzheimer's disease, the guiding principle should be optimizing the person's functioning. Regardless of the level of impair-

THINK ABOUT IT

If an accurate diagnostic test for Alzheimer's disease *is* developed, and there is no treatment for the disease, should the test be made available?

autosomal dominant inheritance
the presence of certain genes that means there is a 100% chance of the person eventually getting the disease

risk genes
genes that increase one's risk of getting the disease

spaced retrieval
memory intervention based on the E-I-E-I-O approach to memory intervention

ment, attempts should be made to help the person cope as well as possible with the symptoms. The key is helping all individuals maintain their dignity as human beings. This can be achieved in some very creative ways, such as adapting the principles of Montessori methods of education to bring older adults with Alzheimer's disease together with preschool children so that they can perform tasks together (Camp et al., 1997; Malone & Camp, 2007). One example of this approach is discussed in the Spotlight on Research feature.

Spotlight on research

Training Persons With Dementia to Be Group Activity Leaders

Who were the investigators, and what was the aim of the study? Dementia is marked by progressive and severe cognitive decline. But despite these losses, can people with dementia be trained to be group leaders? Most people might think the answer is "no," but Cameron Camp and Michael Skrajner (2005) decided to find out by using a training technique based on the Montessori method.

How did the investigators measure the topic of interest? The Montessori method is based on self-paced learning and developmentally appropriate activities. As Camp and Skrajner point out, many techniques used in rehabilitation (e.g., task breakdown, guided repetition, moving from simple to complex and concrete to abstract) and in intervention programs for people with dementia (e.g., use of external cues and implicit memory) are consistent with the Montessori method.

For this study, a program was developed to train group leaders for memory bingo (see Camp, 1999a and 1999b, for details about this game). Group leaders had to learn which cards to pick for the game, where the answers were located on the card, where to "discard" the used (but not the winning) cards, and where to put the winning cards. Success in the program was measured by research staff raters, who made

ratings of the type and quality of engagement in the task shown by the group leader.

Who were the participants in the study? Camp and Skrajner tested four people who had been diagnosed as probably having dementia who were also residents of a special care unit of a nursing home.

What was the design of the study? The study used a longitudinal design so that Camp and Skrajner could track participants' performance over several weeks.

Were there ethical concerns with the study? Having persons with dementia as research participants raises important issues regarding informed consent. Because of their serious cognitive impairments, these individuals may not fully understand the procedures. Thus, family members such as a spouse or adult child caregiver are also asked to give informed consent. Additionally, researchers must pay careful attention to participants' emotions; if participants become agitated or frustrated, the training or testing session must be stopped. Camp and Skrajner took all these precautions.

What were the results? Results showed that at least partial adherence to the established game protocols was achieved at a very high rate. Indeed, staff assistance was not required at all

for most of the game sessions for any leader. All of the leaders said that they enjoyed their role, and one recruited another resident to become a leader in the next phase of the project.

What did the investigators conclude? It appears that persons with dementia can be taught to be group activity leaders through a procedure based on the Montessori method. This is important because it provides a way for such individuals to become more engaged in an activity and to be more productive.

Although more work is needed to continue refining the technique, applications of the Montessori method offer a promising intervention approach for people with cognitive impairments.

What converging evidence would strengthen these conclusions? Camp and Skrajner studied only four residents; more evidence that the approach works with different types of people would bolster their conclusions. Although the Montessori method is effective for training persons with dementia, the approach has not yet been demonstrated to be effective with other diseases that cause serious memory loss.

 Go to Psychology CourseMate at www.cengagebrain.com to enhance your understanding of this research.

One of the best ways to find out about the latest medical and behavioral research, and also about the educational and support programs available in your area, is to contact your local chapter of the Alzheimer's Association (www.alz.org). The chapter in your area will be happy to supply a range of educational material and information about local programs.

Test Yourself

RECALL

1. Compared to younger adults, older adults are less likely to label their feelings of sadness as _____.

2. A form of psychotherapy that focuses on people's beliefs about the self, the world, and the future is called _____.

3. Relaxation techniques are an effective therapy for _____.

4. The only way to definitively diagnose Alzheimer's disease is through a _____.

5. Twisted fibers called _____ occur in the axon of neurons in persons with Alzheimer's disease.

INTERPRET

After reading about the symptoms of Alzheimer's disease, what do you think would be the most stressful aspects of caring for a parent who has the disease? (You may want to refer to the section on caring for aging parents in Chapter 13.)

APPLY

If a friend asked you the difference between dementia and the normative increases in forgetting that occur with age, what would you tell him or her?

Recall answers: (1) depression, (2) cognitive therapy, (3) anxiety disorders, (4) brain autopsy, (5) neurofibrillary tangles

SUMMARY

14.1 WHAT ARE OLDER ADULTS LIKE?

What are the characteristics of older adults in the population?

■ The number of older adults is growing rapidly, especially the number of people over age 85. In the future, older adults will be more ethnically diverse and better educated than they are now.

How long will most people live? What factors influence this?

■ Average life expectancy has increased dramatically in this century, and this is due mainly to improvements in health care. Useful life expectancy refers to the number of years that a person is free from debilitating disease. Maximum life expectancy is the longest time any human can live.

■ Genetic factors that can influence longevity include familial longevity and a family history of certain diseases. Environmental factors include acquired diseases, toxins, pollutants, and lifestyle.

■ Women have a longer average life expectancy at birth than men. Ethnic group differences are complex; depending on how old people are, the patterns of differences change.

What is the distinction between the Third Age and the Fourth Age?

■ The Third Age refers to changes in research that led to cultural, medical, and economic advances for older adults (e.g., longer average longevity, increased quality of life). In contrast, the Fourth Age reflects that the oldest-old are at the limits of their functional capacity, the rates of diseases such as cancer and dementia increase dramatically, and other aspects of psychological functioning (e.g., memory) also undergo significant and fairly rapid decline.

14.2 PHYSICAL CHANGES AND HEALTH

What are the major biological theories of aging?

■ There are two main groups of theories of biological aging. First, programmed theories argue that aging is the result of a biological or genetic program. The damage or error theory approach includes wear-and-tear theories

that postulate that aging is caused by body systems simply wearing out, and cellular theories that focus on reactions within cells that involve telomeres, free radicals, and cross-linking. No single theory is sufficient to explain aging.

What physiological changes normally occur in later life?

■ Three important structural changes in the neurons are neurofibrillary tangles, dendritic changes, and neuritic plaques. These have important consequences for functioning because they reduce the effectiveness with which neurons transmit information.

■ The risk of cardiovascular disease increases with age. Normal changes in the cardiovascular system include buildup of fat deposits in the heart and arteries, a decrease in the amount of blood the heart can pump, a decline in heart muscle tissue, and stiffening of the arteries. Most of these changes are affected by lifestyle. Stroke and vascular dementia cause significant cognitive impairment, depending on the location of the brain damage.

■ Strictly age-related changes in the respiratory system are hard to identify because of the lifetime effects of pollution. However, older adults may suffer shortness of breath and face an increased risk of chronic obstructive pulmonary disorder.

■ Parkinson's disease is caused by insufficient levels of dopamine but can be effectively managed with levodopa. In a minority of cases, dementia develops.

■ Age-related declines in vision and hearing are well documented. The main changes in vision concern the structure of the eye and the retina. Changes in hearing mainly involve presbycusis, reduced sensitivity to high-pitched tones. However, similar changes in taste, smell, touch, pain, and temperature are not as clear.

What are the principal health issues for older adults?

■ Older adults have more sleep disturbances than younger adults. Nutritionally, most older adults do not need vitamin or mineral supplements. Cancer risk increases sharply with age. The poorer health status of aging immigrants is largely due to communication problems and barriers to care.

14.3 COGNITIVE PROCESSES

What changes occur in information processing as people age? How do these changes relate to everyday life?

- Older adults' psychomotor speed is slower than younger adults'. However, the amount of slowing is lessened if older adults have practice or expertise in the task.

- Sensory and information-processing changes create problems for older drivers. Working memory is another powerful explanatory concept for changes in information processing with age.

What changes occur in memory with age? What can be done to remediate these changes?

- Older adults typically do worse on tests of episodic recall; age differences are less on recognition tasks. Semantic and implicit memory are both largely unaffected by aging. People tend to remember best those events that occurred between ages 10 and 30.

- Distinguishing memory changes associated with aging from memory changes due to disease should be accomplished through comprehensive evaluations.

- Memory training can be achieved in many ways. A useful framework is to combine explicit-implicit memory distinctions with external-internal types of memory aids.

What are creativity and wisdom, and how do they relate to age?

- Research indicates that creative output peaks in late young adulthood or early middle age and declines thereafter, but the point of peak activity varies across disciplines and occupations.

- Wisdom has more to do with being an expert in living than with age per se. Three factors that help people become wise are personal attributes, specific expertise, and facilitative life contexts.

14.4 MENTAL HEALTH AND INTERVENTION

How does depression in older adults differ from depression in younger adults? How is it diagnosed and treated?

- The key symptom of depression is persistent sadness. Other psychological and physical symptoms also occur, but the importance of these depends on the age of the person reporting them.

- Major causes of depression include imbalances in neurotransmitters and psychosocial forces such as loss and internal belief systems.

- Depression can be treated with medications (e.g., heterocyclic antidepressants, MAO inhibitors, selective serotonin reuptake inhibitors) and through psychotherapy, such as behavioral or cognitive therapy.

How are anxiety disorders treated in older adults?

- Many older adults are afflicted with a variety of anxiety disorders. All of them can be effectively treated with either medications or psychotherapy.

What is Alzheimer's disease? How is it diagnosed and managed? What causes it?

- Dementia is a family of diseases that cause severe cognitive impairment. Alzheimer's disease is the most common form of irreversible dementia.

- Symptoms of Alzheimer's disease include memory impairment, personality changes, and behavioral changes. These symptoms usually worsen gradually, with rates varying considerably among individuals.

- Definitive diagnosis of Alzheimer's disease can only be made following a brain autopsy. Diagnosis of probable Alzheimer's disease in a living person involves a thorough process by which other potential causes are eliminated.

- Most researchers are focusing on a probable genetic cause of Alzheimer's disease.

- Although Alzheimer's disease is incurable, various therapeutic interventions may improve the quality of the patient's life.

KEY TERMS

demographers (498)
population pyramid (498)
longevity (501)
average life expectancy (501)
useful life expectancy (502)
maximum life expectancy (502)
programmed theories (507)
wear-and-tear theory (507)
free radicals (507)
cross-linking (508)
cellular theories (508)
telomeres (508)

neurofibrillary tangles (509)
neuritic plaques (510)
neurotransmitters (510)
stroke, or cerebral vascular accident (CVA) (511)
transient ischemic attack (TIA) (511)
vascular dementia (511)
chronic obstructive pulmonary disease (COPD) (512)
Parkinson's disease (512)
presbyopia (513)
presbycusis (514)

circadian rhythm (517)
psychomotor speed (519)
useful field of view (UFOV) (520)
working memory (522)
explicit memory (522)
implicit memory (522)
episodic memory (522)
semantic memory (522)
autobiographical memory (523)
external aids (525)
internal aids (525)
dysphoria (530)

internal belief systems (531)

heterocyclic antidepressants (HCAs), monoamine oxidase (MAO) inhibitors, and selective serotonin reuptake inhibitors (SSRIs) (531)

behavior therapy (531)

cognitive therapy (531)

anxiety disorders (532)

dementia (532)

Alzheimer's disease (532)

incontinence (533)

amyloid (534)

autosomal dominant inheritance (535)

risk gene (535)

spaced retrieval (535)

LEARN MORE ABOUT IT

Log in to **www.cengagebrain.com** to access the resources your instructor requires. For this book, you can access:

Psychology CourseMate

■ CourseMate brings course concepts to life with interactive learning, study, and exam preparation tools that support the printed textbook. A textbook-specific website, Psychology CourseMate includes an integrated interactive eBook and other interactive learning tools including quizzes, flashcards, videos, and more.

CENGAGENOW™

■ CengageNOW Personalized Study is a diagnostic study tool containing valuable text-specific resources—and because you focus on just what you don't know, you learn more in less time to get a better grade.

WebTUTOR™

■ More than just an interactive study guide, WebTutor is an anytime, anywhere customized learning solution with an eBook, keeping you connected to your textbook, instructor, and classmates.

Social Aspects of Later Life

Psychosocial, Retirement, Relationship, and Societal Issues

WHAT'S IT REALLY LIKE to be an older adult? As we saw in Chapter 14, aging brings with it both physical limits (such as declines in vision and hearing) and psychological gains (such as increased expertise). Old age also brings social challenges. Older adults are sometimes stereotyped as being marginal and powerless in society, much like children. Psychosocial issues confront older adults as well. How do people think about their lives and bring meaning and closure to them as they approach death? What constitutes well-being for older people? How do they use their time once they are no longer working full-time? Do they like being retired? What roles do relationships with friends and family play in their lives? How do older people cope if their partner is ill and requires care? What if their partner should die? When older people need assistance, where do they live?

These are a few of the issues we will examine in this chapter. As in Chapter 14, our main focus will be on the majority of older adults who are healthy and live in the community. The distinction made in Chapter 1 between young-old (60- to 80-year-olds) and old-old (80-year-olds and up) adults is important. We know the most about young-old people, even though the old-old reflect the majority of frail elderly and those who live in nursing homes.

Just as at other times in life, getting along in the environment is a complicated issue. We begin by considering a few ideas about how to optimize our fit with the environment. Next, we examine how we bring the story of our lives to a culmination. After that, we consider how interpersonal relationships and retirement provide contexts for life satisfaction. We conclude with an examination of the social contexts of aging.

LEARNING OBJECTIVES

■ What is continuity theory?

■ What is the competence and environmental press model, and how do docility and proactivity relate to the model?

Since Sandy retired from her job as secretary at the local African Methodist Episcopal Church, she has hardly slowed down. She sings in the gospel choir, is involved in the Black Women's Community Action Committee, and volunteers one day a week at a local Head Start school. Sandy's friends say that she has to stay involved, because that's the only way she's ever known. They claim you'd never know that Sandy is 71 years old.

UNDERSTANDING HOW PEOPLE GROW OLD IS NOT AS SIMPLE AS ASKING SOMEONE HOW OLD HE OR SHE IS, as Sandy shows. As we saw in Chapter 14, aging is an individual process involving many variations in physical changes, cognitive functioning, and mental health. As Dennis the Menace notes, older adults are often marginalized in society. Psychosocial approaches to aging recognize these issues. Sandy's life reflects several key points. Her level of activity has remained constant across her adult life. This consistency fits well in continuity theory, the first framework considered in this section. Her ability to maintain this level of commitment indicates that the match between her abilities and her environment is just about right, as discussed in competence–environmental press theory a bit later in this section.

Continuity Theory

continuity theory
theory based on idea that people tend to cope with daily life in later adulthood by applying familiar strategies based on past experience to maintain and preserve both internal and external structures

People tend to keep doing whatever works for them (Atchley, 1989). *According to continuity theory, people tend to cope with daily life in later adulthood by applying familiar strategies based on past experience to maintain and preserve both internal and external structures.* By building on and linking to one's past life, change becomes part of continuity. Thus Sandy's new activities represent both change (because they are new) and continuity (because she has always been engaged in her community). In this sense, continuity represents an evolution, not a complete break with the past (Atchley, 1989).

The degree of continuity in life falls into one of three general categories: too little, too much, and optimal (Atchley, 1989). Too little continuity results in feeling that life is too unpredictable. Too much continuity can create utter boredom or a rut of predictability; there is simply not enough change to make life interesting. Optimal continuity provides just enough change to be challenging and provide interest, but not so much as to overly tax one's resources.

Continuity can be either internal or external (Atchley, 1989). Internal continuity refers to a remembered inner past, such as temperament, experiences, emotions, and skills; in brief, it is one's personal identity. Internal continuity enables you to see that how you are now is connected with your past, even if your current behavior looks different. Internal continuity provides feelings of competence, mastery, ego integrity (discussed later in the chapter), and self-esteem. In contrast, internal discontinuity, if severe enough, can seriously affect mental health. Indeed, one of the most destructive aspects of Alzheimer's disease is that it destroys internal continuity as it strips away one's identity.

External continuity concerns remembered physical and social environments, role relationships, and activities. A person feels external continuity from being in familiar environments or with familiar people. For example, continuity theory provides a framework for understanding how social participation and volunteer activity in recently widowed older adults helps them

"WE HAVE A LOT IN COMMON, DON'T WE? I'M TOO YOUNG TO DO MOST EVERYTHING AND YOU'RE TOO OLD TO DO MOST EVERYTHING."

maintain connections with people, sometimes over many years (Donnelly & Hinterlong, 2010). Similarly, phasing from full-time employment to retirement offers some people (e.g., university faculty) a way to maintain connections with their professional lives and facilitates the adjustment to retirement (Kim & Feldman, 2000). And some people continue a relationship with deceased loved ones (Filanosky, 2004). In contrast, external discontinuity can have serious consequences for adaptation. For example, if your physical environment becomes much more difficult to negotiate, the resulting problems can eat away at your identity as well.

Research indicates that there is considerable evidence that people in late life typically continue to engage in similar activities as they did earlier in adulthood (Agahi, Ahacic, & Parker, 2006; Donnelly & Hinterlong, 2010). Additional evidence also indicates that, within broad continuity of activities, people also strike out in some new directions; this is due in part to increased flexibility in their time and in part to emerging personal interests (Agahi et al., 2006; Nimrod & Kleiber, 2007). This aspect of exploration has been called an "innovation theory" of successful aging (Nimrod & Kleiber, 2007). It turns out that innovation in such things as leisure activities helps preserve a sense of continuity (Nimrod & Hutchinson, 2010).

Clearly, continuing to engage in areas of personal interest—while expanding one's opportunities—characterizes many older adults. So, both internal and external continuity matter. But what happens if people experience major challenges? How continuity affects adaptation in later life is the focus of the competence–environmental press framework, to which we now turn.

Competence and Environmental Press

Understanding psychosocial aging requires attention to individuals' needs rather than treating all older adults alike. One way of doing this is to focus on the relation between the person and the environment (Wahl & Oswald, 2010). As discussed in Chapter 1, the competence–environmental press approach is a good example of a theory that incorporates elements of the biopsychosocial model into the person–environment relation (Lawton & Nahemow, 1973; Nahemow, 2000; Wahl & Oswald, 2010).

Competence *is defined as the upper limit of a person's ability to function in five domains: physical health, sensory-perceptual skills, motor skills, cognitive skills, and ego strength.* We discussed age-related changes for most of these domains in Chapter 14; ego strength, which is related to Erikson's concept of integrity, is discussed later in this chapter. These domains are viewed as underlying all other abilities and reflect biological and psychological forces. **Environmental press** *refers to the physical, interpersonal, or social demands that environments put on people.* Physical demands might include having to walk up three flights of stairs to your apartment. Interpersonal demands include having to adjust your behavior patterns to different types of people. Social demands include dealing with laws or customs that place certain expectations on people. These aspects of the theory reflect biological, psychological, and social forces. Both competence and environmental press change as people move through the life span; what you are capable of doing as a 5-year-old differs from what you are capable of doing as a 25-, 45-, 65-, or 85-year-old. Similarly, the demands put on you by the environment change as you age. Thus, the competence–environmental press framework reflects life-cycle factors as well.

The competence and environmental press model, depicted in ■ Figure 15.1, shows how the two are related. Low to high competence is represented on the vertical axis, and weak to strong environmental press is represented on the horizontal axis. Points in the figure represent various combinations of the two. Most important, the shaded areas show that adaptive behavior and positive affect (emotion) can result from many different combinations of competence and environmental press levels. **Adaptation level** *is the area where press level is average for a particular level of competence; this is where behavior and affect are normal. Slight increases in press tend to improve performance; this area on the figure is labeled the* **zone of maximum performance potential**. *Slight decreases in press create the* **zone of maximum comfort**, *in which people are able*

THINK ABOUT IT
How do the five-factor theory of personality and the life-story approach to personality fit with continuity theory?

competence
the upper limit of a person's ability to function in five domains: physical health, sensory-perceptual skills, motor skills, cognitive skills, and ego strength

environmental press
the physical, interpersonal, or social demands that environments put on people

adaptation level
when press level is average for a particular level of competence

zone of maximum performance potential
when press level is slightly higher, tending to improve performance

zone of maximum comfort
when press level is slightly lower, facilitating a high quality of life

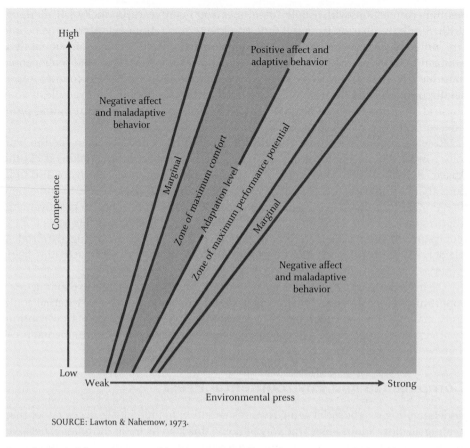

SOURCE: Lawton & Nahemow, 1973.

From "Ecology of the Aging Process," by M. P. Lawton and L. Nahemow. In C. Eisdorfer and M. P. Lawton (Eds.), *The Psychology of Adult Development and Aging*, pp. 619–674. Copyright © 1973 American Psychological Association.

to live happily without worrying about environmental demands. Combinations of competence and environmental press that fall within either of these two zones result in adaptive behavior and positive affect, which translate into a high quality of life.

As one moves away from these areas, behavior becomes increasingly maladaptive and affect becomes negative. Notice that these outcomes, too, can result from several different combinations and for different reasons. For example, too many environmental demands on a person with low competence and too few demands on a person with high competence both result in maladaptive behaviors and negative affect.

What does this mean with regard to late life? Is aging merely an equation relating certain variables? The important thing to realize about the competence–environmental press model is that each person has the potential of being well adapted to some but not all living situations. Whether people are functioning well depends on whether their abilities fit the demands of their environment. When their abilities match these demands, people adapt; when there is a mismatch, they don't. In this view, aging is more than an equation, as the best fit must be determined on an individual basis.

How do people deal with changes in their particular combinations of environmental press (such as adjusting to a new living situation) and competence (perhaps due to illness)? People respond in two basic ways (Lawton, 1989; Nahemow, 2000). *When people choose new behaviors to meet new desires or needs, they exhibit **proactivity** and exert control over their lives. In contrast, when people allow the situation to dictate their options, they demonstrate **docility** and have little control.* Lawton (1989) argues that proactivity is more likely to occur in people with relatively high competence and docility in people with relatively low competence.

This model has considerable research support. For example, it explains why people choose the activities they do (Lawton, 1982), how well people adhere to medication

proactivity
when people choose new behaviors to meet new desires or needs and exert control over their lives

docility
when people allow their situation to dictate the options they have

regimens (LeRoux & Fisher, 2006; Morrow & Wilson, 2010), and how people adapt to changing housing needs over time (Nygren et al., 2007; Pynoos, Caraviello, & Cicero, 2010) and the need to exert some degree of control over their lives (Langer & Rodin, 1976). It also helps us understand how well people adapt to various care situations, such as adult day care (Moore, 2005). In short, there is considerable merit to the view that aging is a complex interaction, mediated by choice, between a person's competence level and environmental press. This model can be applied in many different settings.

Understanding how people age usually entails taking a broader perspective than any single theory can offer. The Real People feature about Nelson Mandela, a Nobel Peace Prize winner and world leader from South Africa, shows that both continuity theory and competence–environmental press theory are important.

THINK ABOUT IT

How does the competence–environmental press approach help explain which coping strategies might work best in a particular situation?

Real People
Applying Human Development
Nelson Mandela: A Heroic Leader

Few people have had the impact on their native country that Nelson Mandela has. Born on July 18, 1918, Mandela was the first member of his family to attend a school, eventually earning his bachelor's degree at the University of South Africa. In 1948, he began his political career by opposing the Afrikaner-dominated National Party, which supported the apartheid policy of racial segregation. It was a decision that changed his life.

Mandela was initially dedicated to nonviolent opposition, and was influenced by Mahatma Gandhi, who had begun his efforts at social activism in South Africa years earlier. However, after Mandela's arrest for treason in 1956, and his subsequent 5-year trial (he was acquitted), he changed his view about nonviolent opposition. The Sharpeville Massacre in 1960, in which 69 peaceful protesters were killed by South African police, convinced him that armed struggle was now necessary to overthrow the apartheid government. So in 1961 he formed the armed wing of the African National Congress and began a guerrilla campaign of sabotage against military and government targets.

Mandela was arrested in 1962, convicted of sabotage and treason, and sentenced to life imprisonment. He remained in jail until February 11, 1990, when he was released by President F. W. de Klerk. During a speech right after his release, Mandela said his main focus was to bring peace to the black majority and give them the right to vote in both national and local elections. Between 1990 and 1994, he negotiated the first multiracial elections in South Africa's history.

Mandela was elected president, and served from 1994 to 1999. He helped the country move from white minority apartheid rule to a multiracial model of government built on reconciliation. His support of the Springboks rugby team that won the 1996 world title was especially important, and was the subject of the 2009 film *Invictus*.

Since his retirement in 1999, Mandela has remained politically active. He became an advocate for human rights organizations, and has been active in the fight against AIDS. He has founded three organizations: the Nelson Mandela Foundation, the Nelson Mandela Children's Fund, and the Mandela Rhodes Foundation.

Nelson Mandela is a true world leader, a person who reshaped the history of his country. He shows continuity in his life through his political activity, as well as showing the match between competence and environmental press through his changing approach to situations across his life.

© Gallo Images / Alamy

Nelson Mandela

Test Yourself

RECALL

1. A central premise of _____ theory is that people make adaptive choices to maintain and preserve existing internal and external structures.

2. A person's ability to function in several key domains is termed _____, whereas demands put on a person from external sources are termed _____.

INTERPRET

How does continuity theory incorporate aspects of the biopsychosocial model?

APPLY

How would a new state law requiring older adults to pass a vision test before renewing their driver's license be an example of changes in environmental press?

Recall answers: (1) continuity, (2) competence, environmental press

15.2 Personality, Social Cognition, and Spirituality

LEARNING OBJECTIVES

- What is integrity in late life? How do people achieve it?
- How is well-being defined in adulthood? How do people view themselves differently as they age?
- What role does spirituality play in late life?

Olive is a spry 88-year-old who spends more time thinking and reflecting about her past than she used to. She also tends to be much less critical now of decisions made years ago than she was at the time. Olive remembers her visions of the woman she wanted to become and concludes that she's come pretty close. Olive wonders if this process of reflection is something that most older adults go through.

THINK FOR A MINUTE ABOUT THE OLDER ADULTS YOU KNOW. What are they really like? How do they see themselves today? How do they visualize their lives a few years from now? Do they see themselves as the same or different from the way they were in the past? These questions have intrigued authors since the early days of psychology. In the late 19th century, William James (1890), one of the early pioneers in psychology, wrote that a person's personality traits are set by young adulthood. Some researchers agree; as we saw in Chapter 13, some aspects of personality remain relatively stable throughout adulthood. But people also change in important ways, as Carl Jung (1960/1933) argued, by integrating such opposite tendencies as masculine and feminine traits. As we have seen, Erik Erikson (1982) was convinced that personality development takes a lifetime, unfolding over a series of eight stages from infancy through late life.

In this section, we explore how people like Olive assemble the final pieces in the personality puzzle and see how important aspects of personality continue to evolve in later life. We begin with Erikson's issue of integrity, the process by which people try to make sense of their lives. Next, we see how well-being is achieved and how personal aspirations play out. Finally, we examine how spirituality is an important aspect of many older adults' lives.

Integrity Versus Despair

integrity versus despair
according to Erikson, the process in late life by which people try to make sense of their lives

As people enter late life, they begin the struggle of **integrity versus despair**, *which involves the process by which people try to make sense of their lives.* According to Erikson (1982), this struggle comes about as older adults like Olive try to understand their lives in terms of the future of their family and community. Thoughts of a person's own death are balanced by the realization that they will live on through children, grandchildren, great-grandchildren, and the community as a whole. This realization produces what Erikson calls a "life-affirming involvement" in the present.

The struggle of integrity versus despair requires people to engage in a **life review,** *the process by which people reflect on the events and experiences of their lifetimes.* To achieve integrity, a person must come to terms with the choices and events that have made his or her life unique. There must also be an acceptance of the fact that one's life is drawing to a close. Looking back on one's life may resolve some of the second-guessing of decisions made earlier in adulthood (Erikson, Erikson, & Kivnick, 1986). People who were unsure whether they made the right choices concerning their children, for example, now feel satisfied that things eventually worked out well. In contrast, others feel bitter about their choices, blame themselves or others for their misfortunes, see their lives as meaningless, and greatly fear death. These people end up in despair rather than integrity.

Research shows a connection between engaging in a life review and achieving integrity, so life review forms the basis for effective mental health interventions (Westerhof, Bohlmeijer, & Webster, 2010). In one study, life-review activities done in a group improved the quality of life of older adults, an effect that lasts at least 3 months following the sessions (Hanaoka & Okamura, 2004). A study in Australia showed a negative correlation between "accepting the past" and symptoms of depression (Rylands & Rickwood, 2001): Older women who accepted the past were less likely to show symptoms of depression than older women who did not. A therapeutic technique called "structured life review" (Haight & Haight, 2007) has been shown to be effective in helping people deal with stressful life events.

Who reaches integrity? Erikson (1982) emphasizes that people who demonstrate integrity come from various backgrounds and cultures and arrive there having taken different paths. Such people have made many different choices and follow many different lifestyles; the point is that everyone has this opportunity to achieve integrity if they strive for it. Those who reach integrity become self-affirming and self-accepting; they judge their lives to have been worthwhile and good. They are glad to have lived the lives they did.

life review
the process by which people reflect on the events and experiences of their lifetimes

Well-Being and Emotion

How is your life going? Are you reasonably content, or do you think you could be doing better? Answers to these questions provide insight into your **subjective well-being,** *an evaluation of one's life that is associated with positive feelings.* In life-span developmental psychology, subjective well-being is usually assessed by measures of life satisfaction, happiness, and self-esteem (Oswald & Wu, 2010).

subjective well-being
an evaluation of one's life that is associated with positive feelings

Overall, young-older adults are characterized by improved subjective well-being compared to earlier in adulthood, although the extent of the difference depends on several factors, such as hardiness, chronic illness, marital status, the quality of one's social network, and stress (Charles & Carstensen, 2010). The differences in people's typical level of happiness across adulthood are illustrated in results from the United Kingdom, shown in ■ Figure 15.2. These happiness-related factors hold across cultures as well; for example, a study of Taiwanese and Tanzanian older adults showed similar predictors of successful aging (Hsu, 2005; Mwanyangala et al., 2010). Although gender differences in subjective well-being have been found to increase with age, they are most likely due to the fact that older women are particularly disadvantaged compared with older men with regard to chronic illness and its effect on ability to care for oneself, everyday competence, quality of social network, socioeconomic status, and widowhood (Charles & Carstensen, 2010; Hsu, 2005; Pinquart & Sörensen, 2001). Such gender differences are smaller in more recent cohorts, indicating that societal changes over the past few decades have led to improvements in the way that older women view themselves.

Given the findings about improved well-being in older adults, researchers began wondering how well-being was related to emotions. People's feelings are clearly important, as they get expressed in daily moods and underlie mental health problems

■ **Figure 15.2**

The pattern of a typical person's happiness through life.

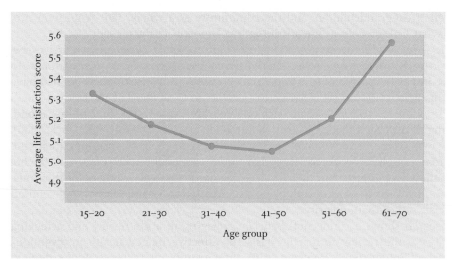

From A. Oswald "Happiness, Health, and Economics," Warwick University, http://imechanica.org/files/andrew_oswald_presentation_071129.pdf

such as depression (Cacioppo et al., 2011). So how emotions are regulated in later life may provide insights into people's subjective well-being.

Emotion-focused research in neuroscience is providing answers to the question of why subjective well-being tends to increase with age (Cacioppo et al., 2011). A brain structure called the amygdala, an almond-shaped set of nuclei deep in the brain, helps regulate emotion. Evidence is growing that age-related changes in how the amygdala functions may play a key role in understanding emotional regulation in older adults. Here's how. In young adults, arousal of the amygdala is associated with negative emotional arousal. When negative emotional arousal occurs, for example, memory for events associated with the emotion are stronger. But the situation is different for older adults—both amygdala activation and emotional arousal are lower. That may be one reason why older adults experience less negative emotion, lower rates of depression, and better well-being (Cacioppo et al., 2011; Winecoff et al., 2011). But that's not the whole story. In Chapter 13, we discovered that brain activity in the prefrontal cortex, which is associated with cognition, changes with age. Neuroimaging research shows that changes in cognitive processing in the prefrontal cortex is also associated with changes in emotional regulation in older adults, as described in the Spotlight on Research feature.

Spotlight on research

The Aging Emotional Brain

Who were the investigators, and what was the aim of the study? Although much research has examined the behavioral side of emotions, very little has examined the specific underlying neural mechanisms in the brain. Winecoff and her colleagues (2011) decided to examine these mechanisms and discover whether they differed with age.

How did the investigators measure the topic of interest? Winecoff and colleagues used a battery of tests to measure cognitive performance and emotional behavior. They tested participants' immediate recall, delayed recall, and rec-

ognition for 16 target words as measures of memory. They also administered a response-time test to measure psychomotor speed (see Chapter 14), and a digit-span test to measure working memory (see Chapter 14). The researchers also had participants complete three questionnaires to measure various types of emotions.

After these measures were obtained, participants were given the cognitive reappraisal task depicted in ■ Figure 15.3. In brief, participants learned a reappraisal strategy that involved thinking of themselves as an emotion-

ally detached and objective third party. During the training session they told the experimenter how they were thinking about the image to ensure task compliance, but they were instructed not to speak during the scanning session. During the functional magnetic resonance imaging (fMRI) session, participants completed 60 positive image trials (30 "Experience" and 30 "Reappraise"), 60 negative image trials (30 "Experience" and 30 "Reappraise") trials, and 30 neutral image trials (all "Experience"). Within each condition, half of the images contained people, and the other half did

Cognitive reappraisal task. Participants were trained in the use of a reappraisal strategy for emotion regulation. (A) On "Experience" trials, participants viewed an image and then received an instruction to experience naturally the emotions evoked by that image. The image then disappeared, but participants continued to experience their emotions throughout a 6-second delay period. At the end of the trial, the participants rated the perceived emotional valence of that image using an 8-item rating scale. (B) "Reappraise" trials had similar timing, save that the cue instructed participants to decrease their emotional response to the image by reappraising the image (e.g., distancing oneself from the scene). Shown are examples of the negative (A) and positive (B) images used in the study.

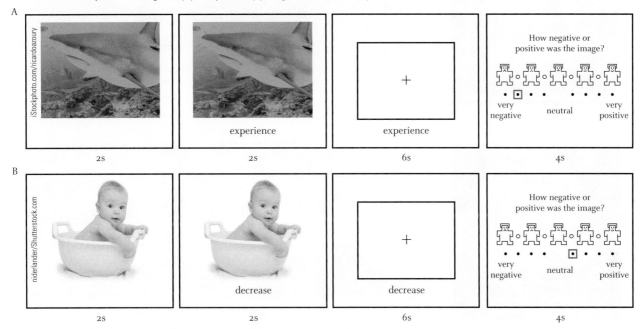

Winecoff, A. et al. (2011). Cognitive and neural contributions to emotion regulation in aging. *Social Cognitive and Affective Neuroscience, 6,* 165-176.

not. The fMRI session provided images of ongoing brain activity.

Who were the participants in the study? The sample consisted of 22 younger adults (average age = 23, range = 19–33 years) and 20 older adults (average age = 69; range = 59–73 years). Participants were matched on demographic variables including education. Participants received the cognitive/memory/emotion tests on one day and the reappraisal task in the fMRI session on a second day. Participants were paid $55.

What was the design of the study? The study used a cross-sectional design, with testing of two age groups over two sessions.

Were there ethical concerns with the study? All participants provided written consent under a protocol approved by the Institutional Review Board of Duke University Medical Center.

What were the results? Younger and older adults performed the reappraisal tasks similarly; that is, in the reappraisal condition, positive images were reported as less positive and negative images reported as less negative. However,

older adults' reports of negative emotion were higher than younger adults' in the negative reappraisal situation.

Examination of the fMRI results showed that reappraisals involved significant activation of specific areas in the prefrontal cortex for both positive and negative emotions. For both age groups, activity in the prefrontal area increased and activity in the amygdala decreased during the reappraisal phase. These patterns are shown in ■ Figure 15.4. As you can see in the top figure, certain areas in the prefrontal cortex showed a pattern of activation that followed participants' self-reports of emotion regulation. Shown here are voxels activated in the contrast between "Reappraise-Negative" and "Experience-Negative" conditions. The top graphs show that for both positive and negative stimuli, and for both younger and older adults, prefrontal activation increased in "Reappraise" trials compared to "Experience" trials. In contrast, the lower graphs show that in the amygdala (Amy) there was a systematic decrease in activation during emotion regulation between

"Experience-Negative" and "Reappraise-Negative" conditions.

Additional analyses of the fMRI data showed that emotion regulation modulates the functional interaction between the prefrontal cortex and the amygdala. Younger adults showed more activity in the prefrontal cortex during "Reappraise" trials for negative pictures than older adults did. No age difference in brain activation for positive pictures was found. Cognitive abilities were related to the degree of decrease in amygdala activation, independent of age.

What did the investigators conclude? Winecoff and her colleagues concluded that the prefrontal cortex plays a major role in emotional regulation, especially for older adults. In essence, the prefrontal cortext may help suppress (regulate) emotions in the same way as that area of the brain is involved in inhibiting other behaviors. Importantly, the degree of emotional regulation was predicted by cognitive ability, with higher cognitive ability associated with higher emotional regulation. This may mean that as cognitive abilities decline, people may be less able to regulate their emotions, a pattern typical in such

[continued]

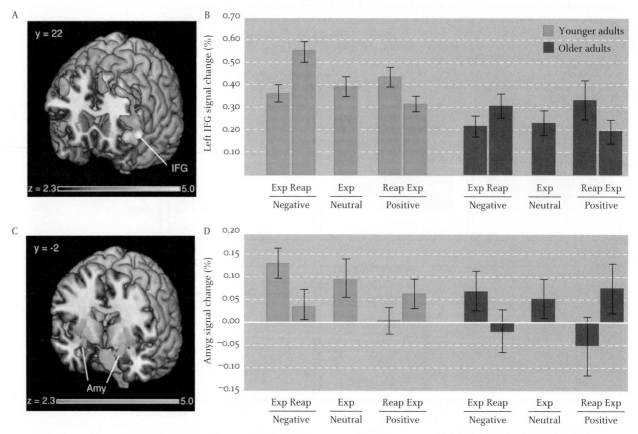

Modulation of prefrontal and amygdalar activation by emotion regulation.

Winecoff, A. et al. (2011). Cognitive and neural contributions to emotion regulation in aging. *Social Cognitive and Affective Neuroscience.*

diseases as dementia. Thus, not only is there evidence of underlying brain structures playing critical roles in emotion regulation, but there may be a neurological explanation for the kinds of emotional outbursts that occur in dementia and related disorders.

What converging evidence would strengthen these conclusions? Winecoff and her colleagues studied only two age groups of healthy adults, and did not include either old-old participants or adults with demonstrable cognitive impairment. It will be important to study these groups

to map brain function changes and behavior more completely.

 Go to Psychology CourseMate at www.cengagebrain.com to enhance your understanding of this research.

Spirituality in Later Life

When faced with the daily problems of living, how do many older adults cope? According to research, older adults in many countries use their religious faith and spirituality, often more than they use family or friends (Ai, Wink, & Ardelt, 2010; Fischer et al., 2007; Hank & Schaan, 2008). For some older adults, especially African Americans, a strong attachment to God is what they believe helps them deal with the challenges of life (Cicirelli, 2004; Dilworth-Anderson, Boswell, & Cohen, 2007).

There is considerable evidence linking spirituality and health (Krause, 2006; Park, 2007). In general, older adults who are more involved with and committed to their faith have better physical and mental health than older adults who are not religious (Ai et al., 2010). For example, older Mexican Americans who pray to the saints and the Virgin Mary on a regular basis tend to have greater optimism and better health (Krause & Bastida, 2011). Spirituality also helps improve psychological well-being, particularly among frail older adults (Kirby, Coleman, & Daley, 2004; Reinhoudt, 2005) and in patients following cardiac surgery (Ai et al., 2010). Upchurch and Mueller

(2005) found that older African Americans were more likely to be able to perform key activities of daily living if they had higher levels of spirituality, and Troutman, Nies, and Mavellia (2011) found that older African Americans report that spirituality is an important part of their concept of successful aging.

When asked to describe ways of dealing with problems in life that affect physical and mental health, many people list coping strategies associated with spirituality (Ai et al., 2010; White, Peters, & Shim, 2011). Of these, the most frequently used were placing trust in God, praying, and getting strength and help from God. These strategies can also be used to augment other ways of coping. Caregivers for people with Alzheimer's disease also report using religion and spiritual practices as primary coping mechanisms (Kinney et al., 2003; Stuckey, 2001).

Researchers have increasingly focused on **spiritual support**—*which includes seeking pastoral care, participation in organized and nonorganized religious activities, and expressing faith in a God who cares for people—as a key factor in understanding how older adults cope.* Even when under high levels of stress, such as during critical illness or other major life trauma, people who rely on spiritual support report greater personal well-being (Ai et al., 2010; White et al., 2011). Krause (2006) reports that feelings of self-worth are lowest in older adults who have very little religious commitment, a finding supported by cross-cultural research with Muslims, Hindus, and Sikhs (Mehta, 1997). However, Pargament (1997) also notes the importance of individual differences in the effectiveness of spiritual support: some people are helped more than others, some problems are more amenable to religious coping, and certain types of religious coping may be more effective than others.

When people rely on spirituality to cope, how do they do it? Krause and colleagues (2000) were among the first to ask older adults what they meant when they said that they were "turning it all over to God" and "letting God have it." The older adults in this study reported that turning problems over to God really was a three-step process: (1) differentiating between things that can and cannot be changed; (2) focusing one's own efforts on the parts of the problem that can be changed; and (3) emotionally disconnecting from those aspects of the problem that cannot be changed by focusing on the belief that God will provide the best outcome possible for those. These findings show that reliance on spiritual beliefs acts to help people focus their attention on parts of the problem that may be under their control.

Reliance on religion in times of stress appears to be especially important for many African Americans, who as a group are more intensely involved in religious activities (Taylor, Chatters, & Levin, 2004; Troutman et al., 2011). African Americans tend to identify with their race and religion much more strongly than do European Americans, and they are more committed to their religion (Fife, 2005). They also are more likely to rely on God for support than are European Americans (Lee & Sharpe, 2007). Churches have historically offered considerable social support for the African American community and have also served an important function in advocating social justice, and ministers play a major role in providing support in times of personal need (Chatters et al., 2011). For example, the civil rights movement in the 1950s and 1960s was led by Dr. Martin Luther King Jr., a Baptist minister, and contemporary congregations often champion equal rights. The role of the church in the majority of African Americans' lives is central; indeed, one of the key predictors of life satisfaction among African Americans is regular church attendance and commitment to their religion (Fife, 2005).

Many older persons of Mexican heritage adopt a different approach. Research indicates that they use *la fé de la gente* ("the faith of the people") as a coping strategy (Villa & Jaime, 1993). The notion of *fé* incorporates varying degrees of faith, spiritual-

spiritual support
type of coping strategy that includes seeking pastoral care, participation in organized and nonorganized religious activities, and expressing faith in a God who cares for people

© Kim-Jae-Hwan / AFP / Getty Images

These Buddhist monks' spirituality can serve as an important coping strategy.

THINK ABOUT IT

What psychological and sociocultural factors make religion and spiritual support important for minority groups?

ity, hope, cultural values, and beliefs. *Fé* does not necessarily imply that people identify with a specific religious community; rather, they identify with a cultural value or ideology. Research indicates that spirituality is a key component of well-being for Mexican Americans (Rivera, 2007). A study of Mexican American women experiencing chronic pain showed that they rely on spirituality as a means by which to cope (Flores, 2008). What's more, spiritual healers play an important part in Mexican culture and share many common aspects with cultures such as the !Kung of South Africa and Native American tribes in North America—other cultures in which traditional healing methods continue to play an important role (Finkler, 2004).

Among many Native Americans, the spiritual elders are the wisdom-keepers, the repositories of the sacred ways and philosophies that extend back indefinitely in time (Wall & Arden, 1990). The wisdom-keepers also share dreams and visions, perform healing ceremonies, and may make apocalyptic prophecies. The place of the wisdom-keepers in the tribe is much more central than that of religious leaders in Western society. Incorporating Native American spiritual traditions into intervention programs is important for maximizing positive outcomes (Holkup et al., 2007).

Similar effects of spirituality are observed in Asian and Asian American groups. For example, the risk of dying in a given year among the old-old in China was found to be 21% lower among frequent religious participants compared to nonparticipants, after initial health condition was equated (Zeng, Gu, & George, 2011). Asian caregivers of dementia patients who are more religious report being able to handle the stresses and burden of caregiving better than nonreligious caregivers (Chan, 2010).

And neuroscience research has shown that there is a connection between certain practices and brain activity. For example, there is evidence that people who have practiced meditation show more organized attention systems and less activity in areas of the brain that focus on the self (Davidson, 2010; Lutz et al., 2009). Thus, neurological evidence indicates there may be changes in brain activity associated with spiritual practices that help people cope.

Health care and social service providers would be well advised to keep in mind the self-reported importance of spirituality in the lives of many older adults when designing interventions to help them adapt to life stressors. For example, older adults may be more willing to talk with their minister or rabbi about a personal problem than they would be to talk with a psychotherapist. However, when working with people of Mexican heritage, providers should realize that a major source of distress for this group is lack of familial interaction and support. Overall, many churches offer a wide range of programs to assist poor or homebound older adults in the community. Such programs may be more palatable to the people served than programs based in social service agencies. To be successful, service providers should try to view life as their clients see it.

Test Yourself

RECALL

1. The Eriksonian struggle that older adults face is termed _____.

2. An evaluation of one's life that is associated with positive feelings is termed _____.

3. The most commonly reported method for coping with life stress among older adults is _____.

INTERPRET

How might different spiritual traditions influence personal well-being?

APPLY

Given that the prefrontal cortex is activated for both positive and negative emotions for both younger and older adults, what might account for older adults' increased reaction to negative emotions?

Recall answers: (1) integrity versus despair, (2) subjective well-being, (3) religion or spiritual support

LEARNING OBJECTIVES

■ What does being retired mean?

■ Why do people retire?

■ How satisfied are retired people?

■ How do retirees keep busy?

Marcus is a 77-year-old retired construction worker who labored hard all of his life. He managed to save a little money, but he and his wife live primarily off of his monthly Social Security checks. Though not rich, they have enough to pay the bills. Marcus is largely happy with retirement, and he stays in touch with his friends. He thinks maybe he's a little strange, though, since he has heard that retirees are supposed to be isolated and lonely.

YOU PROBABLY TAKE IT FOR GRANTED THAT SOMEDAY, after working for many productive years, you will retire. But did you know that until 1934, when a railroad union sponsored a bill promoting mandatory retirement, and 1935, when Social Security was inaugurated, retirement was not even considered a possibility by most Americans like Marcus (McClinton, 2010)? Only since World War II has there been a substantial number of retired people in the United States (McClinton, 2010). Today, the notion that people work a specified time and then retire is built into our most fundamental expectations about work. However, the series of economic downturns that began in the early 1990s had a major disruptive effect on people's retirement decisions and plans—after declining for decades, the number of people over age 65 still in the workforce has increased significantly from a low of about 15% in the early 1990s to more than 20% in 2010 (Sterns & Chang, 2010).

As more people retire and take advantage of longer lives, a significant social challenge is created regarding how to fund retiree benefits and how to view older adults who are still very active (Bengtsson & Scott, 2011; McClinton, 2010; Tsao, 2004). In this section, we explore what retirement is like for older adults. We consider people like Marcus as we examine how retirement is defined, why people retire, how people adjust to being retired, and how retirement affects interpersonal relationships.

Retirement provides many people the opportunity to do things they want to do rather than things they must do.

What Does Being Retired Mean?

Defining retirement is difficult (Beehr & Bennett, 2007; McClinton, 2010). Retirement means different things to people in different ethnic groups (Luborsky & LeBlanc, 2003). One traditional way is to equate retirement with complete withdrawal from the workforce. But this definition is inadequate; many "retired" people continue to work full- or part-time, often because they have no choice in order to make ends meet. Still others use retirement to explore new outlets, such as spending more time on hobbies that may turn into more serious pursuits.

Part of the reason it is difficult to define retirement precisely is that the decision to retire involves the loss of occupational identity (see Chapter 12) and not what people may add to their lives. What people do for a living is a major part of their identity; we introduce ourselves as postal workers, teachers, builders, or nurses as a way to tell people something about ourselves. Not doing those jobs any more means that we either put that aspect of our lives in the past tense—"I used to work as a manager at the Hilton"—or say nothing at all. Loss of this aspect of ourselves can be difficult to face, so some look for a label other than "retired" to describe themselves.

A useful way to view retirement is as another one of many transitions people experience in life (Schlossberg, 2004; Sterns & Chang, 2010). This view makes retirement a complex process by which people withdraw from full-time participation in an occupation (Beehr & Bennett, 2007; Henretta, 2001), recognizing that there are many pathways to this end (Everingham, Warner-Smith, & Byles, 2007). This withdrawal process can be described as either "crisp" (making a clean break from employment by stopping work entirely) or "blurred" (repeatedly leaving and returning to work, with some unemployment periods) (Sterns & Chang, 2010). Bob is a good example of a "crisp" retirement. He retired from the Ford Motor Company at age 65; now in his late 80s, he has done nothing work-related in the interim.

Whereas many people think of retirement as a crisp transition, the evidence shows that fewer than half of older men who retire fit this pattern (McClinton, 2010; Sterns & Chang, 2010). Most men adopt a more gradual or "blurred" process involving part-time work in an effort to maintain economic status. Jack is one of these men. When he retired from DuPont at age 62, he and a friend began a small consulting company. For about 5 years, Jack worked when he wanted, gradually cutting back over time.

The current lack of widespread crisp retirement creates another complicating factor—the idea of a "normal" retirement age such as age 65 is no longer appropriate because few jobs have mandatory retirement ages (Sargeant, 2004). Indeed, in the absence of a mandatory retirement age, the concept of "early" retirement has no meaning. Instead, the notion of a typical retirement age spans a range of ages and depends on one's acceptance of transition, further blurring the meaning of retirement (Beehr & Bennett, 2007; Schlossberg, 2004).

To reflect these changes, researchers describe a transition phase from career job, the career one has throughout most of adulthood, to **bridge job**, *the job one holds between one's exit from the career job and final retirement.* Research indicates that many workers hold bridge jobs as a way to blur the shift from full-time employment to retirement (McClinton, 2010; Ulrich & Brott, 2005). For some workers, bridge jobs are a continuation of a work history characterized by short-term employment. For others, they reflect a desire to continue working even if it is not financially necessary. In this latter case, generativity may also be a factor in deciding to bridge from full-time employment to retirement (Dendinger, Adams, & Jacobson, 2005). Bridge jobs can be incorporated into retirement plans, and have been shown to be related both to retirement satisfaction and to overall life satisfaction (Ekerdt, 2010; McClinton, 2010).

bridge job
transitional job held between one's exit from a career job and final retirement

This former full-time cook uses a part-time cafeteria job as a bridge job to transition to retirement.

Why Do People Retire?

Provided that they have good health, more workers retire by choice than for any other reason (Ederdt, 2010; McClinton, 2010; Sterns & Chang, 2010). Individuals usually retire when they feel financially secure after considering projected income from Social Security, pensions and other structured retirement programs, and personal savings. Of course, some people are forced to retire because of health problems or because they lose their jobs. As corporations downsize during economic downturns or after corporate mergers, some older workers accept buyout packages involving supplemental payments if they retire. Others are permanently furloughed, laid off, or dismissed.

The decision to retire is complex and is influenced by one's occupational history and goal expectations (Brougham & Walsh, 2005; Ekerdt, 2010; McClinton, 2010). Whether people perceive that they will achieve their personal goals through work or retirement influences the decision to retire and its connection with health and disability.

THINK ABOUT IT

In the absence of mandatory retirement, what does the term "early retirement" really mean?

Yellow Dog Productions / Getty Images

Gender Differences

Most of the existing models of the retirement process are based on research focusing on men (Everingham et al., 2007), yet women's experience can be quite different (Everingham et al., 2007; Frye, 2008). For example, women may enter the workforce after they have stayed home and raised children and in general have more discontinuous work histories; also, having fewer financial resources may affect women's decisions to retire. Women also tend to spend less time planning their retirement (Jacobs-Lawson, Hershey, & Neukam, 2004). These different patterns mean that, in order to understand the process of women's retirement, it is necessary to understand their specific employment history (Everingham et al., 2007). Note, however, that neither women nor men have much influence on a partner's decision to retire (van Solinge & Henkens, 2005).

For women who were never employed outside the home, the process of retirement is especially unclear (Gardiner et al., 2007). Because they most likely were not paid for all of their work raising children and caring for the home, it is rare for them to have their own pensions or other sources of income in retirement. Additionally, the work they have always done in caring for the home continues, often nearly uninterrupted.

Ethnic Differences

There has not been much research examining the process of retirement as a function of ethnicity. African American older adults are likely to continue working beyond age 65 (Troutman et al., 2011). Both African American and European American women, already disadvantaged in terms of lower pay (compared to men) during their employment years, are even more disadvantaged financially during retirement because their retirement plans (e.g., pensions) are based on their wages (Hogan & Perrucci, 2007). As a result, these women are much more likely than their male counterparts to continue working at least part-time. However, there are no ethnic-based differences in health outcomes between African American women and men following retirement (Curl, 2007).

Adjustment to Retirement

You might imagine that people may experience a loss of identity when transitioning to retirement after being employed full-time, especially in a job that they really enjoyed and found meaningful. Similarly, you may also imagine that people who are no longer working 40 hours a week would find it satisfying to have much more time for doing the things they enjoy. And you may even know people who go from being extremely busy in their employment life to being even busier in "retirement." So how do people who go through the process of retirement adjust to it?

Researchers agree on one point: Retirement is a major and personally important life transition. As is true of any major life transition, new patterns of personal involvement must be developed in the context of changing roles and lifestyles in retirement (McClinton, 2010; Schlossberg, 2004). Researchers support the idea that people's adjustment to retirement evolves over time as a result of complex interrelations involving physical health, financial status, the degree to which their retirement was voluntary, and feelings of personal control (Ekerdt, 2010; McClinton, 2010).

How do most people fare? As long as people have financial security, health, a supportive network of relatives and friends, and an internally driven sense of motivation, they report feeling very good about being retired (Ekerdt, 2010; Stephan, Fouquereau, & Fernandez, 2007).

But what about couples? Just because one or the other partner is satisfied doesn't necessarily mean that the couple as a whole is. That's exactly what Smith and Moen (2004) found. The couples most likely to report being satisfied with retirement, individually and jointly, are retired husbands and wives who reported that their husbands did not influence their retirement decision. Barnes and Parry (2004) report that both gender roles and finances are the most important factors in predicting satisfaction with retirement in their sample of U.K. respondents, with traditional gender roles creating more difficulties for older retired men. Clearly, we need to view satisfaction in retirement as an outcome that depends on one's gender and that is experienced at both the individual and couple levels.

One widely held view of retirement is that being retired has negative effects on health. Research findings show that the relation between health and retirement is complex. On the one hand, there is no evidence that voluntary retirement has any immediate negative effects on health (Weymouth, 2005). In contrast, there is ample evidence that being forced to retire likely leads to significant declines in physical and mental health (Donahue, 2007). Individuals in this circumstance experience poorer physical functioning, loss of mobility, and poorer mental health—even when health status at the time of retirement is taken into account. Health issues are also a major predictor of when a person retires, as a longitudinal study in England showed (Rice et al., 2010).

For some, retirement can provide a new lease on life. In a longitudinal study of socially disadvantaged men from adolescence until age 75, many men who had reported risk factors during adolescence through midlife (e.g., low IQ, dropping out of school, poor mental health, and being part of a multiproblem family) but in later life had some positive resources (e.g., a good marriage, enjoyment of vacations, a capacity for play, and a low level of neuroticism) experienced high levels of satisfaction in retirement (Vaillant, DiRago, & Mukamal, 2006). Clearly, experiencing many challenges earlier in life does not preclude the possibility of satisfying retirement years.

Keeping Busy in Retirement

Retirement is an important life transition, one that is best understood through a life-course perspective (see Chapter 1; McClinton, 2010; Schlossberg, 2004). This life change means that retirees must look for ways to maintain social integration and to be active in various ways.

For some people, being active means being employed, either part- or full-time. Employment may be a financial necessity for people without sufficient means to make ends meet, especially for those whose entire income would consist only of Social Security benefits. For others, the need to stay employed at least part-time represents a way to stay connected with their former lives and careers.

The past few decades have witnessed a rapid growth of organizations devoted to offering such opportunities to retirees. National groups such as AARP provide the chance to learn, through magazines and pamphlets, about other retirees' activities and about services such as insurance and discounts. Groups at the local community level, including senior centers and clubs, promote the notion of life-long learning and help keep older adults cognitively active. Many also offer travel opportunities specifically designed for active older adults.

Some retired adults take up hobbies to develop their creative side.

Leisure activities provide retirees with many sources of satisfaction (Nimrod, 2007a). A study of Israeli retirees showed that a wide array of leisure activities (e.g., hobbies, cultural and spiritual pursuits) were related to higher life satisfaction (Nimrod, 2007a). Similarly, participation in leisure activities was associated with higher satisfaction for retired Turkish men (Sener, Terzioglu, & Karabulut, 2007).

Healthy, active retired adults also maintain community ties by volunteering (Moen et al., 2000a, 2000b). Older adults report they volunteer for many reasons that benefit their well-being (Greenfield & Marks, 2005): to provide service to others, to maintain social interactions and improve their communities, and to keep active. There are many opportunities for retirees to help

Some retired adults do volunteer work as a way to stay active.

others, both at the local and national levels. One federal agency, the Corporation for National and Community Service, administers the Senior Corps, which in turn oversees three programs that have hundreds of local chapters: Foster Grandparents, Senior Companion Program, and the Retired Senior Volunteer Program (RSVP; a "one-stop shop" for all older adult volunteers). The AmeriCorp program, another of the Corporation's divisions, also has many older adult participants. Why do so many people volunteer?

Several factors are responsible (Tang, Morrow-Howell, & Choi, 2010): developing a new aspect of the self, finding a personal sense of purpose, desire to share one's skills and expertise, a redefinition of the nature and merits of volunteer work, a more highly educated and healthy population of older adults, and greatly expanded opportunities for people to become involved in volunteer work that they enjoy. Given the demographic trends of increased numbers and educational levels of older adults (discussed in Chapter 14), still higher rates of volunteerism are expected during the next few decades. Brown and colleagues (2011) argue that volunteerism offers a way for society to tap into the vast resources that older adults offer.

THINK ABOUT IT

What might the opportunity for more older adults to volunteer for organizations mean politically? Check your answer with the research data cited later in the chapter.

Test Yourself

RECALL

1. One useful way to view retirement is as a _____.
2. The most common reason people retire is _____.
3. Overall, most retirees are _____ with retirement.
4. Many retirees keep contacts in their communities by _____.

INTERPRET

Why does forced retirement have a negative effect on health?

APPLY

Using the information from Chapter 12 on occupational development, create a developmental description of occupations that incorporates retirement.

Recall answers: (1) complex process by which people gradually withdraw from employment, (2) by choice, (3) satisfied, (4) volunteering

LEARNING OBJECTIVES

- What role do friends and family play in late life?
- What are older adults' marriages and same-sex partnerships like?
- What is it like to provide basic care for one's partner?
- How do people cope with widowhood? How do men and women differ?
- What special issues are involved in being a great-grandparent?

Alma was married to Charles for 46 years. Even though he died 20 years ago, Alma still speaks about him as if he had only recently passed away. Alma still gets sad on special dates—their anniversary, Charles's birthday, and the date on which he died. Alma tells everyone that she and Chuck, as she called him, had a wonderful marriage and that she still misses him terribly even after all these years.

TO OLDER ADULTS LIKE ALMA, the most important thing in life is relationships. In this section, we consider many of the relationships older adults have. Whether it is friendship or family ties, having relationships with others is what keeps us connected. Thus, when one's partner is in need of care, it is not surprising to find wives and husbands devoting themselves to caregiving. Widows like Alma also feel close to departed partners. For a growing number of older adults, becoming a great-grandparent is an exciting time.

We have seen throughout this text how our lives are shaped and shared by the company of others. *The term* **social convoy** *is used to suggest how a group of people journeys with us throughout our lives, providing support in good and bad times.* People form the convoy, and under ideal conditions that convoy provides a protective, secure cushion that permits the person to explore and learn about the world (Antonucci, 2001; Luong, Charles, & Fingerman, in press). Especially for older adults, the social convoy also provides a source of affirmation of who they are and what they mean to others, which leads to better mental health and well-being.

Several studies have shown that the size of one's social convoy and the amount of support it provides do not differ across generations. This finding strongly supports the conclusion that friends and family are essential aspects of all adults' lives. Social support is especially important in the African American community, as these networks provide all sorts of informal assistance for health-related and other issues (Warren-Findlow & Issel, 2010). Social networks also play a critical role in helping immigrants; for example, older Mexican American immigrants are assisted by family members in their network when settling in their new country (Miller-Martinez & Wallace, 2007).

social convoy
a group of people that journeys with us throughout our lives, providing support in good times and bad

Friends and Siblings

By late life, some members of a person's social network have been friends for several decades. Research consistently finds that older adults have the same need for friends as do people in younger generations; it also shows that their life satisfaction is poorly correlated with the number or quality of relationships with younger family members yet is strongly correlated with the number and quality of their friendships (Rawlins, 2004). Why? As will become clear, friends serve as confidants and sources of support in ways that relatives (e.g., children or nieces and nephews) typically do not.

Friendships

The quality of late-life friendships is particularly important (Moorman & Greenfield, 2010; Rawlins, 2004). Having at least one close friend or confidant provides a buffer against the losses of roles and status that accompany old age, such as retirement or the death of a loved one, and can increase people's happiness and self-esteem (Moorman & Greenfield, 2010; Rawlins, 2004). Patterns of friendship among older adults tend to mirror those in young adulthood described in Chapter 11 (Rawlins, 2004). For example, older women have more numerous and more intimate friendships than older men do. As noted previously, these differences help explain why women are in a better position to deal with the stresses of life. Men's friendships, like women's, evolve over time and become important sources of support in late life (Adams & Ueno, 2006). One major difference is that cross-sex friendships in late life are more important, especially for men whose male friendship network may have been depleted through death (Adams & Ueno, 2006; Moorman & Greenfield, 2010).

In general, older adults have fewer relationships and develop fewer new relationships than younger or middle-aged adults, and do not replace friends lost through death or other reasons with people in younger generations (Moorman & Greenfield, 2010). This decline in numbers does not simply reflect the loss of relationships to death or other causes; instead, the changes reflect a more complicated process (Carstensen, 1993, 1995). *This process, termed* **socioemotional selectivity**, *implies that social contact is motivated by many goals, including information seeking, self-concept, and emotional regulation.* Each of these goals is differentially relevant at different times and results in different social behaviors. For example, information seeking tends to lead to meeting more people, whereas emotional regulation results in being particular about one's choice of social partners, with a strong preference for people who are familiar.

With time, older adults begin to lose members of their friendship network, usually through death. It is loneliness that matters a great deal in explaining the relation between social network and life satisfaction for older adults (Gow et al., 2007). There is evidence that newer cohorts of older adults have both more friends and more long-term friends on average than previous cohorts (Stevens & Van Tilburg, in press).

Online friendship opportunities have been embraced by older adults. For example, Ledbetter and Kuznekoff (in press) found that some online gamers were young-old adults (about 25% of online gamers are over age 50). Regardless of age, heavy Internet users tend to have more friends online and offline than Internet light users or nonusers (Wang & Wellman, 2010). And as we'll see a bit later, older adults use online services for finding friends that could lead to dating relationships (McIntosh et al., 2011).

socioemotional selectivity
process by which social contact is motivated by many goals, including information seeking, self-concept, and emotional regulation

Sibling Relationships

For many older adults, the preference for long-term friendships may explain their desire to keep in touch with siblings. As we saw in Chapter 11, maintaining connections with a sibling is important for most adults. Indeed, siblings constitute the longest-lasting relationships in most people's lives (Van Volkom, 2006); on average, sibling relationships last 70 or more years (Moorman & Greenfield, 2010). The importance of sibling relationships varies a lot within and across families, but are typically more important in late life than earlier in adulthood (Moorman & Greenfield, 2010).

As social emotional selectivity theory would predict, sibling relationships provide a way for older adults to have close, emotionally-based relationships because siblings have a long, shared past. Older adult siblings often provide assistance or care. But it turns out that if one wants a friendship with a sibling, it takes more effort than it does with one's other friends, especially regarding emotional support (Voorpostel & van der Lippe, 2007). That's because you can choose your friends but you can't choose your siblings.

Even though there are many grounds for good relationships among older siblings, that's not always the case. Sometimes siblings become rivals, even to the point of having hostile relationships.

Siblings play an important role in the lives of older adults.

© Brownie Harris / Corbis

Older adult siblings whose lives took very different paths are least likely to be close (Moorman & Greenfield, 2010).

Sibling closeness depends on many factors (Moorman & Greenfield, 2010). Among the key predictors are whether one's siblings live close by, genetic relatedness (twins report closer relationships than non-twins), health, and the presence of other relationships. Regarding this last point, sibling relationships are particularly important to unmarried older adults.

Marriage and Same-Sex Partnerships

"It's great to be 72 and still married," said Lucia. "Yeah, it's great to have Juan around to share old times with and have him know how I feel even before I tell him." Lucia and Juan are typical of most older married couples. Marital satisfaction improves once the children leave home and remains fairly high in older couples (see Chapter 13). However, recent research shows conflicting results regarding couples' satisfaction with marriage in late life (Moorman & Greenfield, 2010). There is some evidence that couples who had children report a rebound in satisfaction, whereas couples without children report a decrease. Other studies show that older couples are more likely to report positive behaviors in their spouse than middle-aged couples (Henry et al., 2007).

Older married couples show several specific characteristics (O'Rourke & Cappeliez, 2005). Many older couples exhibit selective memory regarding the occurrence of negative events and perceptions of their partner. Like the older couple in the "For Better or Worse" cartoon, older couples have a reduced potential for marital conflict and greater potential for pleasure, are more likely to be similar in terms of mental and physical health, and show fewer gender differences in sources of pleasure. In short, most older married couples have developed adaptive ways to avoid conflict.

Being married in late life has several benefits. A study of 9,333 European Americans, African Americans, and Latino Americans showed that marriage helps people deal better with chronic illness, functional problems, and disabilities (Pienta, Hayward, & Jenkins, 2000). The division of household chores becomes more egalitarian after the husband retires than it was when the husband was employed, irrespective of whether the wife was working outside the home (Kulik, 2001a, 2001b).

Very little research has been conducted on long-term gay and lesbian partnerships (Moorman & Greenfield, 2010). Based on the available data, it appears that such relationships do not differ in quality from long-term heterosexual marriages (Connidis, 2001; O'Brien & Goldberg, 2000). As is true for heterosexual married couples, relationship satisfaction is better when partners communicate well and are basically happy themselves. Research on lesbians indicates they are flexible and adapt to the challenges they face, including social marginalization and discrimination (Averett & Jenkins, in press.)

A major issue for older gay and lesbian couples is their legal rights (Stone, 2011). The social debate over civil unions and gay marriage has focused attention on the fact that gay and lesbian couples may not be able to visit their partner in a hospital, be eligible for survivor's benefits, or provide the same level of support to them as married couples are permitted to do. This inequality is a major social issue that continues to be addressed in U.S. courts and legislatures.

Caring for a Partner

When couples pledge their love to each other "in sickness and in health," most envision the sickness part to be no worse than an illness lasting a few weeks. That may be the case for many couples, but for some the illness they experience severely tests their pledge.

Francine and Ron are one such couple. After 42 years of mainly good times together, Ron was diagnosed as having Alzheimer's disease. When first contacted by researchers, Francine had been caring for Ron for 6 years. "At times it's very hard, especially when he looks at me and doesn't have any idea who I am. Imagine, after all these years, not to recognize me. But I love him, and I know that he would do the same for me. But, to be perfectly honest, we're not the same couple we once were. We're just not as close; I guess we really can't be."

Francine and Ron are typical of couples in which one partner cares for the other. Caring for a chronically ill partner presents different challenges than caring for a chronically ill parent (see Chapter 13). The partner caregiver is assuming a new role after decades of shared responsibilities. Often without warning, the division of labor that had worked for years must be readjusted. Such change inevitably puts stress on the relationship (Cavanaugh & Kinney, 1994). This is especially true in cases involving Alzheimer's disease or other dementias because of the cognitive and behavioral consequences of the disease (see Chapter 14), but it is also the case with diseases such as AIDS. Caregiving challenges are felt by partner caregivers in any type of long-term, committed relationship.

Studies of spousal caregivers of persons with Alzheimer's disease show that marital satisfaction is much lower than for healthy couples (Bouldin & Andresen, 2010; Cavanaugh & Kinney, 1994; Dow & Meyer, 2010). Spousal caregivers report a loss of companionship and intimacy over the course of caregiving but also more rewards than adult child caregivers (Raschick & Ingersoll-Dayton, 2004). Marital satisfaction is also an important predictor of spousal caregivers' reports of depressive symptoms: the better the perceived quality of the marriage, the fewer symptoms caregivers report (Dow & Meyer, 2010), a finding that holds across European and African American spousal caregivers (Parker, 2008). Caring for a partner can also change one's retirement plans and dramatically affect one's financial status (Dow & Meyer, 2010).

Caring for a spouse can be both extremely stressful and highly rewarding.

Most partner caregivers are forced to respond to an environmental challenge that they did not choose—their partner's illness. They adopt the caregiver role out of necessity. Once they adopt the role, caregivers assess their ability to carry out the duties required. Although evidence about the mediating role of caregivers' appraisal of stressors is unclear, interventions that help improve the functional level of the ill partner generally improve the caregiving partner's situation (Van Den Wijngaart, Vernooij-Dassen, & Felling, 2007). However, spousal caregivers do not always remember their major hassles accurately over time; in one study, caregivers remembered only about two thirds of their major hassles after a 1-month delay (Cavanaugh & Kinney, 1998). This finding points out that health care professionals should not rely exclusively on partner caregivers' reports about the caregiving situation when making diagnostic judgments.

The importance of feeling competent as a partner caregiver fits with the docility component of the competence–environmental press model presented earlier in this chapter. Caregivers attempt to balance their perceived competence with the environmental demands of caregiving. Perceived competence allows them to be proactive rather than merely reactive (and docile), which gives them a better chance to optimize their situation.

Providing full-time care for a partner is both stressful and rewarding in terms of the marital relationship (Baek, 2005; Carbonneau, Caron, & Desrosiers, 2010). For example, coping with a wife who may not remember her husband's name, who may act strangely, and who has a chronic and fatal disease presents serious challenges even to the happiest of couples, as depicted in the Doonesbury cartoon. Yet even in that situation, the caregiving husband may experience no change in marital happiness despite the changes in his wife due to the disease.

Widowhood

Alma, the woman in our vignette, still feels the loss of her husband, Chuck. "There are lots of times when I feel him around. We were together for so long that you take it for granted that your husband is just there. And there are times when I just don't want to go on without him. But I suppose I'll get through it."

Traditional marriage vows proclaim that the union will last "'til death do us part." Like Alma and Chuck, virtually all older married couples will see their marriages end because one partner dies. For most people, the death of a partner is one of the most traumatic events they will ever experience, causing an increased risk of death among older European Americans (but not African Americans), an effect that lasts several years (Moorman & Greenfield, 2010). For example, an extensive study of widowed adults in Scotland showed that the increased likelihood of dying lasted for at least

10 years (Boyle, Feng, & Raab, 2011). Despite the stress of losing one's partner, though, most widowed older adults manage to cope reasonably well (Moorman & Greenfield, 2010).

Women are much more likely to be widowed than are men. More than half of all women over age 65 are widows, but only 15% of men the same age are widowers. The reasons for this discrepancy are related to biological and social forces: As we saw in Chapter 14, women have longer life expectancies. Also, women typically marry men older than themselves, as discussed in Chapter 11. Consequently, the average married woman can expect to live at least 10 years as a widow.

The impact of widowhood goes well beyond the ending of a long-term partnership (Boyle et al., 2011; Guiaux, 2010). Loneliness is a major problem. Widowed people may be left alone by family and friends who do not know how to deal with a bereaved person (see Chapter 16). As a result, widows and widowers may lose not only a partner but also those friends and family who feel uncomfortable with including a single person rather than a couple in social functions (Guiaux, 2010). Feelings of loss do not dissipate quickly, as the case of Alma shows clearly. As we will see in Chapter 16, feeling sad on important dates is a common experience, even many years after a loved one has died.

Men and women react differently to widowhood. In general, those who were most dependent on their partners during the marriage report the highest increase in self-esteem in widowhood because they have learned to do the tasks formerly done by their partners (Carr, 2004). Widowers may recover more slowly unless they have strong social support systems (Bennett, 2010). Some people believe that the loss of a wife presents a more serious problem for a man than the loss of a husband for a woman. Older men are often ill equipped to handle such routine and necessary tasks as cooking, shopping, and keeping house, and they may become emotionally isolated from family members.

Widowers are less likely than widows to form new friendships, continuing a trend throughout adulthood that men have fewer close friendships than women have.

Although both widows and widowers suffer financial loss, widows often suffer more because survivor's benefits are usually only half of their husband's pensions, so most widows suffer a drop in standard of living (Weaver, 2010). For many women, widowhood results in difficult financial circumstances, particularly regarding medical expenses (McGarry & Schoeni, 2005).

An important factor to keep in mind about gender differences in widowhood is that men are usually older than women when they are widowed. To some extent, the difficulties reported by widowers may be partly due to this age difference. Regardless of age, men are perceived to have a clear advantage over women regarding opportunities to form new heterosexual relationships, because there are fewer social restrictions on relationships between older men and younger women (Moorman & Greenfield, 2010). Actually, older widowers are less likely to form new, close friendships than are widows. Perhaps this is simply a continuation of men's lifelong tendency to have few close friendships (see Chapter 11).

For many reasons, including the need for companionship and financial security, some widowed people cohabit or remarry. A newer variation on re-partnering is "living alone together," an arrangement in which two older adults form a romantic relationship but maintain separate living arrangements (Moorman & Greenfield, 2010). Re-partnering in widowhood can be difficult due to family objections (e.g., resistance from children), objective limitations (decreased mobility, poorer health, poorer finances), absence of incentives common to younger ages (desire for children), and social pressures to protect one's estate (Moorman & Greenfield, 2010).

Great-Grandparenthood

As discussed in Chapter 13, grandparenting is an important and enjoyable role for many adults. With increasing numbers of people—especially women—living to very old age, more people are experiencing great-grandparenthood. Age at first marriage and age at parenthood also play a critical role; people who reach these milestones at relatively younger ages are more likely to become great-grandparents. Most current great-grandparents are women who married relatively young and had children and grandchildren who also married and had children relatively early in adulthood.

Although surprisingly little research has been conducted on great-grandparents, their investment in their roles as parents, grandparents, and great-grandparents forms a single family identity (Drew & Silverstein, 2005; Moorman & Greenfield, 2010). That is, great-grandparents see a true continuity of the family through the passing on of the genes. However, their sources of satisfaction and meaning apparently differ from those of grandparents (Doka & Mertz, 1988; Wentkowski, 1985). Compared to grandparents, great-grandparents are much more similar as a group in what they derive from the role, largely because they are less involved with the children than are grandparents. Three aspects of great-grandparenthood appear to be most important (Doka & Mertz, 1988).

First, being a great-grandparent provides a sense of personal and family renewal—important components for achieving integrity. Their grandchildren have produced new life, renewing their own excitement for life and reaffirming the continuance of their lineage. Seeing their families stretch across four generations may also provide psychological support, through feelings of symbolic immortality, to help them face death. They take pride and comfort in knowing that their families will live many years beyond their own lifetime.

Becoming a great-grandparent is a meaningful role and a way to ensure the continuity of one's lineage.

Second, great-grandchildren provide new diversions in great-grandparents' lives. There are now new people with whom they can share their experiences. Young children can learn from a person they perceive as "really old" (Mietkiewicz & Venditti, 2004).

Third, becoming a great-grandparent is a major milestone, a mark of longevity that most people never achieve. The sense that one has lived long enough to see the fourth generation is perceived very positively.

As you might expect, people with at least one living grandparent and great-grandparent interact more with their grandparent, who is also perceived as more influential (Roberto & Skoglund, 1996). Unfortunately, some great-grandparents must assume the role of primary caregiver to their great-grandchildren, a role for which few great-grandparents are prepared (Bengtson, Mills, & Parrott, 1995; Burton, 1992). As more people live longer, it will be interesting to see whether the role of great-grandparents changes and becomes more prominent.

Test Yourself

RECALL

1. The longest relationship most people have is with their _____.

2. In general, older couples have reduced likelihood of _____.

3. Compared to non-caregiver spouses, marital satisfaction in spousal caregivers is _____.

4. _____ are at special risk of experiencing a drop in living standard following the death of their spouse.

5. Three aspects of being a great-grandparent that are especially important are personal and family renewal, diversion, and _____.

INTERPRET

Why would widows and widowers want to establish a new cohabitation relationship?

APPLY

How do the descriptions of marital satisfaction and spousal caregiving presented here fit with the descriptions of marital satisfaction in Chapter 11 and caring for aging parents in Chapter 13? What similarities and differences are there?

Recall answers: (1) siblings, (2) marital conflict, (3) lower, (4) Widows, (5) the fact that it is a milestone

LEARNING OBJECTIVES

- Who are frail older adults? How common is frailty?

- What housing options are there for older adults?

- How do you know whether an older adult is abused or neglected? Which people are most likely to be abused and to be abusers?

- What are the key social policy issues affecting older adults?

Rosa is an 82-year-old woman who still lives in the same neighborhood where she grew up. She has been in relatively good health for most of her life, but in the last year she has needed help with tasks, such as preparing meals and shopping for personal items. Rosa wants very much to continue living in her own home. She dreads being placed in a nursing home, but her family wonders whether that might be the best option.

OUR CONSIDERATION OF LATE LIFE THUS FAR HAS FOCUSED ON THE EXPERIENCES OF THE TYPICAL OLDER ADULT. In this final section, we consider people like Rosa, who represent a substantial number but still a minority of all older adults. Like Rosa, some older adults experience problems completing such common tasks as taking care of themselves. We consider the prevalence and kinds of problems such people face. Although most older adults live in the community, some reside in other settings; we consider the kinds of housing options older adults have. Unfortunately, some older adults are the victims of abuse or neglect; we will examine some of the key issues relating to how elder abuse happens. Finally, we conclude with an overview of the most important emerging social policy issues.

All of these issues are critical when viewed from Baltes and Smith's (2003) Fourth Age perspective, as described in Chapter 14. We will see that it is the oldest-old who make up most of the frail and who live in nursing homes. With the rapid increase in the number of oldest-old on the horizon (the baby boom generation), finding ways to deal with these issues is essential.

frail older adults
older adults who have physical disabilities, are very ill, and may have cognitive or psychological disorders

Older adults over age 85 are much more likely to be frail and to need help with basic daily tasks.

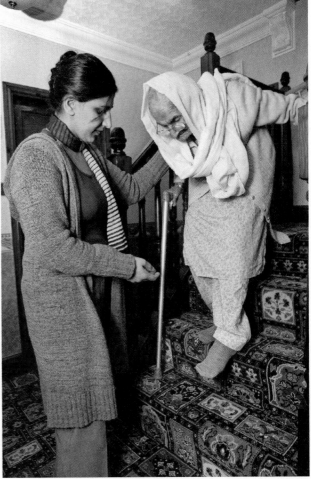

© John Birdsall / The Image Works

Frail Older Adults

In our discussion about aging, to this point we have focused on the majority of older adults who are healthy, cognitively competent, financially secure, and have secure family relationships. Some older adults are not as fortunate. *They are the* **frail older adults** *who have physical disabilities, are very ill, and may have cognitive or psychological disorders*. These frail older adults constitute a minority of the population over age 65, but it is a proportion that increases with age.

Frail older adults are people whose competence (in terms of the competence–environmental press model presented earlier) is declining. They do not have one specific problem that differentiates them from their active, healthy counterparts; instead, they tend to have multiple problems (Crews, 2011). Given the rise in the population of older adults, the number of people who could be considered frail is increasing.

Assessing everyday competence consists of examining how well people can complete activities of daily living and instrumental

activities of daily living (ADLs)
basic self-care tasks such as eating, bathing, toileting, walking, and dressing

instrumental activities of daily living (IADLs)
actions that require some intellectual competence and planning

activities of daily living (Crews, 2011). **Activities of daily living (ADLs)** *are basic self-care tasks such as eating, bathing, toileting, walking, or dressing.* A person could be considered frail if he or she needs help with one of these tasks. Other tasks are also deemed important for living independently. *These* **instrumental activities of daily living (IADLs)** *are actions that require some intellectual competence and planning.* Which actions constitute IADLs vary considerably from one culture to another and factor into cross-cultural differences in conceptions of competence (Sternberg & Grigorenko, 2004). For example, for most older adults in Western cultures, IADLs would include shopping for personal items, paying bills, making telephone calls, taking medications appropriately, and keeping appointments. In other cultures, IADLs might include caring for animal herds, making bread, threshing grain, and tending crops.

Prevalence of Frailty

How common are people like Rosa, the 82-year-old woman in the vignette who still lives in the same neighborhood in which she grew up? As you can see in ■ Figure 15.5, about 40% of people over age 65 report a functional limitation of some kind (Federal Interagency Forum on Aging-Related Statistics, 2010). As you can see in ■ Figure 15.6, the percentage of people needing assistance increases with age, from 7.6% of people aged 65–69 to 29.2% of those over age 80 (Brault, 2008). The percentage of people reporting a disability varies somewhat across ethnic groups, with African Americans over age 65 having the highest rate (20.5%), European Americans (19.7%) and Latino Americans (13.1%) reporting lower rates, and Asian Americans (12.4%) reporting the lowest rate (Brault, 2008). Rates for Native Americans were not reported. Disability rates for females (20.1%) were higher than for males (17.3); this difference was found in all ethnic groups studied (Brault, 2008).

In addition to basic assistance with ADLs and IADLs, frail older adults have other needs. Research shows that these individuals are also prone to higher rates of anxiety disorders and depression, especially if they are living in a long-term care facility (Qualls & Layton, 2010).

Although frailty becomes more likely with increasing age, especially during the last year of life, there are many ways to provide a supportive environment for frail older adults. We have already seen how many family members provide care. Exercise can also help improve the quality of life of some frail older adults, and "exercise prescriptions" can be created for them (Liu & Fielding, 2011). Next, we will consider the role that nursing homes play. The key to providing a supportive context for frail older adults is to create an optimal match between the person's competence and the environmental demands.

Housing Options

Where one lives carries meaning well beyond having a roof over one's head and a place to eat and sleep. Having a stable home offers a place where things are familiar, where services are nearby, and a way to add to one's sense of identity (Pynoos et al., 2010). Living independently, like most older adults do, provides a measure of one's ability to provide self-care, rooted in their attachment to their home, and a way to fulfill their desire to age in place.

Communities must take the rising number of older adults into account when planning housing. As we just saw, the number of older adults with disabilities is rising, so the design and location of housing and support services is now a major issue. Such planning should be conducted within the framework of the competence–environmental press model considered earlier in this chapter, and the selective optimization with compensation model discussed in Chapter 1. Specifically, there should be a match between the person's competence and the level of supports provided by the environment. Let's take a closer look at the various types of places where older adults live.

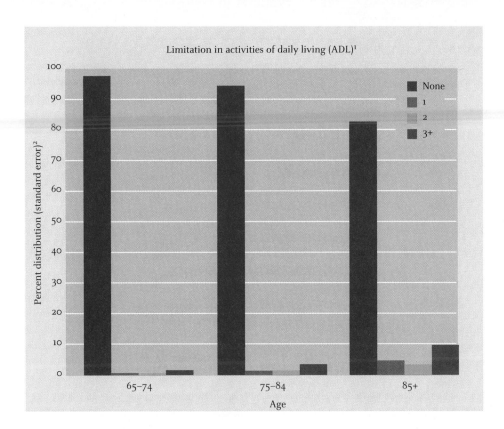

Limitation in activities of daily living (ADL)[1]

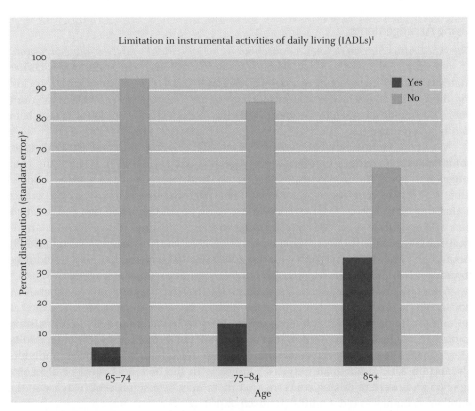

Limitation in instrumental activities of daily living (IADLs)[1]

From www.cdc.gov/nchs/health_policy/ADL_tables.htm

■ **Figure 15.5**
Limitations in activities of daily living and instrumental activities of daily living, 2003–2007.

■ **Figure 15.6**
Prevalence of disability and the need for
assistance by age, 2005 (percentage).

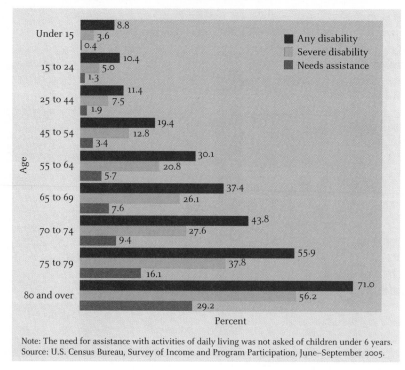

Note: The need for assistance with activities of daily living was not asked of children under 6 years.
Source: U.S. Census Bureau, Survey of Income and Program Participation, June–September 2005.

From U.S. Census Bureau, Survey of Income and Program Participation, June–September 2005.

Living Arrangements

household
an individual who lives alone or a group
of individuals who live together

The U.S. Census Bureau defines a **household** *as an individual who lives alone or a group
of individuals who live together.* About one in every five households in the United
States is headed by someone at least 65 years old (U.S. Census Bureau, 2011). Including those aged 60–64 brings that proportion to one in every three. By 2015, the number of households headed by people 60–69 years old is expected to double as the baby
boom generation ages.

Most older adults do whatever they can to adapt their homes and activities to accommodate the changes that occur with age (Pynoos et al., 2010). Despite the challenges that arise due to physical changes, the death of a spouse or partner, sudden
illness, or other event, nearly all older adults report that their very strong preference
is that they age in place. Doing so provides a sense of self-determination and independence, even if they need assistance to achieve these outcomes. These feelings are reflected in the results from a nationally representative sample of adults (Keenan, 2009).
About half of adults over age 55 who thought it would be likely that they would need
to move in with family or friends said they would not like this arrangement.

A few older adults are considered homeless in that they are not members of a
household (and consequently are unaccounted for in national databases). The mortality rate for homeless older adults is very high; most homeless people do not survive
to old age, which helps account in part for the low numbers of homeless older adults.
Mental health and substance abuse problems are much higher in homeless individuals, and the lack of adequate services coupled with the increased number of people
aged 65 and over are raising concerns that there may be a significant increase in the
number of homeless older adults (Deutsch, 2010).

In general, older adults live in their own home or apartment, in an assisted living
situation (a formal assisted living facility or a shared single-family home with family
or friend), or in long-term care facilities. Which of these arrangements provides the
optimal setting for a particular person depends on the person's functional health.

functional health
the ability to perform the activities of
daily living (ADLs) and instrumental
activities of daily living (IADLs)

Functional health *refers to the ability to perform the activities of daily living (ADLs) and
instrumental activities of daily living (IADLs) discussed earlier in this section.* As a per-

son's functional health diminishes, the level of supports needed from the environment increases, and the optimal housing situation changes. Let's take a closer look at each of the major types of housing to see how this works.

Independent Living Situations

Like nearly all older adults, Annie wanted to live independently. She's lived in the same neighborhood of a large city her whole life, so it's become a part of her. Besides, everything is familiar to her. An important reason why Annie feels this way is that where one lives usually takes on special meaning. *A* **sense of place** *refers to the cognitive and emotional attachments that a person puts on their place of residence, by which a "house" is made into a "home."* Scheidt and Schwarz (2010) point out that a sense of place comprises an important part of people's identity. As a result, aging in place in one's home carries enormous psychological meaning for older adults.

Because aging in place is so important, there are many approaches to ensuring that this is at least a possibility. One common way of achieving this is through home modification, one approach to changing the environment to maintain the optimal match with the person's competence level. Home modifications can range from minor changes, such as replacing knobs on cabinets with pull handles that are more easily grasped, to extensive renovation such as widening doorways and bathrooms to provide access for wheelchairs.

Little research has been done to evaluate the benefits of home modification. Wahl and colleagues (2009) concluded that home modification tends to help people address ADL and IADL challenges. These outcomes argue for using home modification as a way to help older adults age in place successfully.

One way older adults can age in place is to renovate their home to accommodate wheelchairs.

sense of place
the cognitive and emotional attachments that a person puts on their place of residence, by which a "house" is made into a "home"

assisted living facilities
a supportive living arrangement for people who need assistance with ADLs or IADLs but who are not so impaired physically or cognitively that they need 24-hour care

Assisted Living

Bessie lived in the same home for 57 years. Her familiar surroundings enabled her to manage the challenges of failing eyesight and worsening arthritis that made it hard to walk. Finally, her children convinced her that she needed to relocate to an assisted living facility that would provide enough support for her to still enable her to have her own space.

Bessie is pretty typical of people who move to assisted living facilities. **Assisted living facilities** *provide a supportive living arrangement for people who need assistance with ADLs or IADLs but who are not so impaired physically or cognitively that they need 24-hour care.* Estimates are that nearly two-thirds of residents of assisted living facilities over age 65 have an ADL or IADL limitation (Federal Interagency Forum on Aging-Related Statistics, 2010), and about half have some degree of memory impairment (Pynoos et al., 2010). Assisted living facilities usually provide support for activities of daily living (e.g., assistance with bathing) as well as meals and other services. There are health care personnel to assist with medications and certain other procedures.

Assisted living facilities provide a range of services and care, but are not designed to provide intensive, around-the-clock medical care. As residents become frailer, their needs may go beyond those the facility can provide. Policies governing discharge for this reason, usually to a long-term care facility, may result in competing interests between providing for the residents increasing medical needs and providing a familiar environment.

Nursing Homes

The last place Sadie thought she would ever end up was in a bed in nursing home. "That's a place where old people go to die," she would tell her friends. "It's not gonna be for me." But here she is. Sadie fell a few weeks ago and broke her hip.

Percentage of Medicare enrollees age 65 and over residing in selected residential settings by age group, 2007. Data from Federal Interagency Forum on Age-Related Statistics, July 2010.

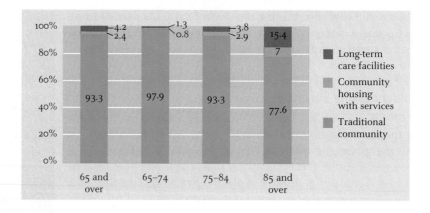

Because she lives alone, she needs to stay in the facility until she recovers. She detests the food; "tasteless," she calls it. A few doors down lives Doris, who is 87 and has dementia.

Sadie and Doris are representative of the people who live in nursing homes, some temporarily, some permanently. If given the choice, the vast majority of older adults do not want to live there; they and their families would prefer to age in place. Sometimes, though, placement in a nursing home is necessary because of the older person's needs or the family's circumstances.

Nursing homes provide medical care 24-hours a day, 7 days a week by a team of health care professionals that includes physicians (who must be on call at all times), nurses, therapists (e.g., physical, occupational), and others. Misconceptions about nursing homes are common. Contrary to what some people believe, only about 5% of older adults in the United States live in nursing homes. As you can see in ■ Figure 15.7, the percentage of older adults who live in a long-term care facility at any given point in time increases from 1% in those aged 65–74 to about 15% of adults over age 85 (Centers for Disease Control, 2010c); however, over their lifetime, adults over age 65 have about a 50% chance of spending at least some time in a nursing home (Centers for Disease Control, 2010c).

Who is the typical resident of a nursing home? She is very old, European American, financially disadvantaged (and eligible for Medicaid), probably widowed or divorced, possibly without living children, and has lived in the nursing home for more than a year (Centers for Disease Control and Prevention, 2010c).

The decision to place a family member in a nursing home is a very difficult one (Caron, Ducharme, & Griffith, 2006), and often is made quickly in reaction to a crisis, such as a person's impending discharge from a hospital or other health emergency. The decision tends to be made by partners or adult children, a finding that generalizes across ethnic groups—especially when there is evidence of cognitive impairment (Almendarez, 2008; Caron et al., 2006).

Selecting a nursing home should be done carefully. The Centers for Medicare and Medicaid Services of the U.S. Department of Health and Human Services provides a detailed *Nursing Home Quality Initiatives* website (www.cms.gov/NursingHomeQualityInits) that is a guide for choosing a nursing home based on several key quality factors. Among the most important things to consider are quality of life for residents (e.g., whether residents are well groomed, the food is tasty, and rooms contain comfortable furniture); quality of care (whether staff respond quickly to calls, whether staff and family are involved in care decisions); safety (whether there are enough staff, whether hallways are free of clutter); and other issues (whether there are outdoor areas for residents to use). These aspects of nursing homes reflect those dimensions considered by states in their inspections and licensing process.

The Eden Alternative, Green House Project, and Cohousing Initiatives

In response to the need to provide support for older adults who need assistance with ADLs and IADLs, and their desire to age in place, new approaches to housing options have emerged that provide both. These movements include programs that infuse a different culture into nursing homes as well as those that create small-scale living (usually 6–10 residents) in a community-based setting with an emphasis on living well rather than on receiving care (Pynoos et al., 2010).

The Eden Alternative (www.edenalt.org) seeks to eliminate loneliness, helplessness, and boredom from the lives of those living in long-term care facilities and to create a community in which life is worth living. This can be achieved by rethinking how care is provided in the older person's own home or in long-term care facilities through training.

The Green House Project (www.thegreenhouseproject.org) creates small neighborhood-integrated homes for 6–10 residents in which older adults receive a high level of personal and professional care. The Green House Project takes the principles of the Eden Alternative and creates a different culture of care in the community.

Various cohousing options provide another alternative approach. Cohousing is a planned community that is modest in size, and that is built around an open, walkable space designed to foster social interaction among neighbors (Pynoos et al., 2010). Neighbors provide care for each other when it is needed. Personal autonomy is a core value for the people who create cohousing developments (Nusbaum, 2010).

These alternatives to traditional housing options for older adults indicate that the choices for how one spends late life and ensures that appropriate support systems are in place are becoming more varied. Such alternatives will be important as the baby boom generation enters the years in which support services will be needed more. Researchers need to focus their attention on documenting the types of advantages these alternatives have and exploring their relative effectiveness.

Elder Abuse and Neglect

Arletta, an 82-year-old woman in relatively poor health, has been living with her 60-year-old daughter, Sally, for the past 2 years. Recently, neighbors became concerned because they had not seen Arletta very much for several months. When they did, she looked rather worn and extremely thin, and as if she had not bathed in weeks. Finally, the neighbors decided that they should do something, so they called the local office of the department of human services. Upon hearing the details of the situation, a caseworker immediately investigated. The caseworker found that Arletta was severely malnourished, had not bathed in weeks, and appeared disoriented. Based on these findings, the agency concluded that Arletta was a victim of neglect. She was moved to a county nursing home temporarily.

Unfortunately, some older adults who need quality caregiving by family members or in nursing homes do not receive it. In some cases, older adults like Arletta are treated inappropriately. Arletta's case is representative of this sad but increasing problem: elder abuse and neglect. In this section, we consider what elder abuse and neglect are, how often they happen, and what victims and abusers are like.

Defining Elder Abuse and Neglect

Like child abuse (see Chapter 7) and partner abuse (see Chapter 11), elder abuse has been extremely difficult to define precisely (Nerenberg, 2010). In general, researchers and public policy advocates describe seven different categories of elder abuse (National Center on Elder Abuse, 2010a; Nerenberg, 2010).

- *Physical abuse*: the use of physical force that may result in bodily injury, physical pain, or impairment.
- *Sexual abuse*: nonconsensual sexual contact of any kind.
- *Emotional or psychological abuse*: infliction of anguish, pain, or distress.

- *Financial or material exploitation*: illegal or improper use of an older adult's funds, property, or assets.
- *Abandonment*: desertion of an older adult by an individual who had physical custody or otherwise had assumed responsibility for providing care for the older adult.
- *Neglect*: refusal or failure to fulfill any part of a person's obligation or duties to an older adult.
- *Self-neglect*: behaviors of an older person that threaten his or her own health or safety, excluding those conscious and voluntary decisions by a mentally competent and healthy adult.

Part of the problem in agreeing on definitions of elder abuse and neglect is that perceptions of what constitutes elder abuse differ among ethnic groups. For example, different models are needed to describe how African American, Latino American, Asian American, Native American, and European American adults decide whether elder abuse has occurred and, if so, how serious a problem it is (Parra-Cardona et al., 2007; Tauriac & Scruggs, 2006). Cultural values, such as multiple families sharing an older adult's public benefits or pension, a group's history of oppression, or an unwillingness to report problems to strangers also create difficulties in understanding and preventing elder abuse across ethnic groups (Horsford et al., 2011; Nerenberg, 2010).

Prevalence

As many as 5 million older adults in the United States may be victims of elder abuse (National Center on Elder Abuse, 2010b), but only about 1 in every 6 cases come to the attention of authorities. The most common forms are neglect (roughly 60%), physical abuse (16%), and financial or material exploitation (12%). Thus, Arletta's case of neglect would be one of the most common types.

Risk Factors

Who is likely to be at risk for elder abuse? It's hard to say precisely; studies of risk factors in elder abuse have produced conflicting results. For example, there is evidence both that people living with others and that people living alone are at greater risk depending on the nature of the living situation (National Center on Elder Abuse, 2010c; Nerenberg, 2010).

In roughly two thirds of elder abuse and neglect incidents, the abuser is a family member—typically the victim's partner or an adult child (National Center on Elder Abuse, 2010c). People who provide care (e.g., nursing home employees) or are in positions of trust (e.g., bankers, accountants, attorneys, and clergy) are also in a position to take advantage of an older adult (National Center on Elder Abuse, 2010c; Nerenberg, 2010). Telemarketing and Internet fraud against older adults, including fraudulent investment schemes and sweepstakes, is a growing problem (Barnard, 2010; Grimes et al., 2010). As a Canadian study (Cohen, 2006) showed, perpetrators of telemarketing fraud prey on the fears and needs of their targets by focusing on older adults' loneliness, desire to please, need to help, and other weaknesses.

Why elder abuse occurs is a matter of debate and often conflicting data (National Center on Elder Abuse, 2010c). Nerenberg (2010) summarizes decades of research by pointing to several possible causes, all of which have some research support: external stresses (e.g., poverty, family problems), social isolation, dependency on the victim (usually financial dependency), retribution for spousal abuse, caregiver stress, potential financial gain, and feeling a sense of power or control.

Clearly, elder abuse is an important social problem that has received insufficient attention from researchers and policy makers. Increased educational efforts, better reporting and investigation, more options for placement of victims, and better physical and mental health intervention options for victims are all needed for the problem to be adequately addressed (Nerenberg, 2010).

THINK ABOUT IT

What kinds of interventions might reduce the risk of elder abuse?

Social Security and Medicare

Without a doubt, the 20th century saw a dramatic improvement in the everyday lives of older adults in industrialized countries. Perhaps the most important gain was the creation of government-funded retirement programs such as Social Security (Polivka, 2010). Such programs have helped reduce the number of older adults below the poverty line; for example, in the 1950s, roughly 35% of older adults were below the federal poverty line compared to only about 10% in 2008 (Johnson & Wilson, 2010).

It remains to be seen whether this downward trend in poverty rate continues. Economists argue that older adults should be evaluated by different standards than other adults in terms of financial health (Johnson & Wilson, 2010). They argue that older adults need 200% of the federal poverty limit to make ends meet, especially with respect to health care costs. By this measure, women and minorities are especially at risk financially. To capture these issues, a new index, the Elder Economic Security Standard index, is targeted for completion in all U.S. states and counties by 2012. It will provide the most geographically specific measure of income needed for an older adult to live in his or her community based on a market basket of goods and services.

The aging of the Baby Boomers presents difficult and expensive problems (Office of Management and Budget, 2010). In fiscal year 2011, federal spending on Social Security and Medicare alone was expected to top $1.2 trillion. As you can see in ■ Figure 15.8, if spending patterns do not change, then by 2039 (when most Baby Boomers will have reached old age) the expenditures for Social Security and Medicare alone are projected to consume roughly 12% of the U.S. gross domestic product (GDP), a 50% increase over rates in 2010. Without major reforms in these programs, such growth will force extremely difficult choices in how to pay for them.

Clearly, the political and social issues concerning benefits to older adults are quite complex. As the first of the Baby Boomers become eligible for Medicare benefits in 2011 and full Social Security benefits in 2012, the need for action to ensure that these programs remain solvent becomes increasingly urgent. Let's look more closely at the Social Security and Medicare issues.

Social Security

Social Security began in 1935 as an initiative by President Franklin D. Roosevelt to "frame a law which will give some measure of protection to the average citizen and to his family against the loss of a job and against poverty-ridden old age" (Roosevelt, 1935). Thus, Social Security was originally intended to provide a *supplement* to savings and other means of financial support.

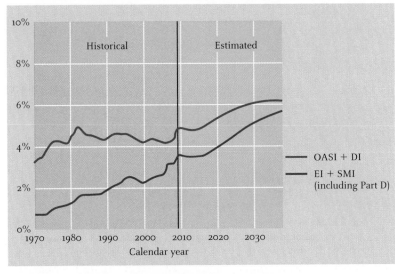

■ **Figure 15.8**
Cost of Social Security (blue line) and Medicare (red line) as percentage of U.S. gross domestic product.

From Federal Interagency Forum on Aging-Related Statistics (2010).

Over the years, revisions to the original law have changed Social Security to the point that now it represents the primary source of financial support after retirement for most U.S. citizens and the only source for many (Polivka, 2010). Since the 1970s, however, increasing numbers of workers have been included in employer-sponsored pension plans such as 401(k), 403(b), and 457 plans as well as mutual funds and various types of Individual Retirement Accounts (IRAs) (McGill et al., 2010; Polivka, 2010). These various retirement plans, especially savings options, plus financial pressure to reform Social Security, may force future retirees to use Social Security as the supplemental financial source for which it was intended, thereby shifting responsibility for retirement financial planning to the individual (Polivka, 2010).

The primary challenge facing Social Security is the aging of the Baby Boomers and the much smaller generation that follows. Because Social Security is not a savings account but is funded by current workers' payroll taxes, the amount of money each worker must pay depends on the ratio of the number of people paying Social Security taxes to the number of people collecting benefits. By 2030 this ratio will have been reduced by nearly half of what it was in 2010; that is, by the time most Baby Boomers have retired, there will be nearly twice as many people collecting Social Security per worker paying into the system (Social Security Administration, 2011). Despite numerous plans having been proposed since the 1970s to address this issue, Congress has not yet taken the actions necessary to ensure the long-term financial stability of the system (Social Security Administration, 2011). As discussed in the What Do *You* Think? feature, the suggestions for doing this present difficult choices for politicians.

What do you think? Saving Social Security

Few political issues have been around as long and are as politically sensitive as those that concern making Social Security fiscally sound for the long term. The basic issues have been well known for decades: the present method for raising and distributing revenues in Social Security is not sustainable (Office of Management and Budget, 2010; Social Security Administration, 2011).

Because Social Security is based on current workers paying a tax to support current retirees, the looming funding problems depend critically on the worker-to-retiree ratio. As ■ Figure 15.9 shows, this ratio has declined precipitously since Social Security began and will continue to decline until about 2030 when there will only be about two workers paying to support every beneficiary. This declining ratio places an increasing financial burden on workers to provide the level of benefits to retirees that people have come to expect. Because of this declining ratio, unless major structural changes are made, the Social Security system will be totally bankrupt around 2040; major reductions in benefits would be required (Social Security Administration, 2011). So it's no wonder that young adults have little faith that Social Security will be there for them.

What steps can be taken to keep Social Security sound in the long term? For decades, U.S. presidents, Congress, and others have proposed numerous strategies (Polivka, 2010).

Among the changes proposed over the years are the following.

- *Privatization:* Various proposals have been made for allowing or requiring workers

■ **Figure 15.9**

Ratio of covered workers to Social Security beneficiaries, 2009 data.

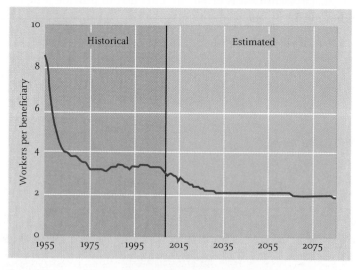

From *2010 Annual Report of the Board of Trustees of the Federal Old-Age and Survivors Insurance and Disability Insurance Trust Funds.* Retrieved from www.ssa.gov/policy/docs/chartbooks/fast_facts/2010/fast_facts10.html

to invest at least part of their money in personal retirement accounts managed by either the federal government or private investment companies. A variation would take trust funds and invest them in private-sector equity markets. Another option would be to allow individuals to create personal accounts with a portion of the funds paid in payroll taxes.

- *Means-test benefits:* This proposal would reduce or eliminate benefits to people with high incomes.
- *Increase the number of years used to compute the benefit:* Currently, benefits are based on one's history of contributions over a 35-year period. This proposal would increase that period to 38 or 40 years.
- *Increase the retirement age:* The age of eligibility for full Social Security benefits is increasing slowly from age 65 in 2000 to age 67 in 2027. Various proposals have been made to speed up the increase, to increase the age to 70, or to connect age at which a person becomes fully eligible to average longevity statistics.
- *Adjust cost-of-living increases downward:* Some proposals have been made to lower those increases given to beneficiaries that result from increases in the cost of living.
- *Increase the payroll tax rate:* One direct way to address the coming funding shortfall is to increase revenues through a higher tax rate.
- *Increase the earnings cap for payroll tax purposes:* This proposal would either raise or remove the cap on income subject to the Social Security payroll tax (the maximum taxable earnings for Social Security was $106,800 in 2011).
- *Make across-the-board reductions in Social Security pension benefits:* A reduction in benefits of 3 to 5% would resolve most of the funding problem.

None of these proposals has universal support. Many proposed solutions would significantly disadvantage certain people—especially minorities and older widows—who depend almost entirely on Social Security for their retirement income (Polivka, 2010). Given the political difficulties inherent in tackling the issue and the lack of perfect solutions, it is likely that Social Security will remain a major controversy.

Solving the funding problems facing Social Security will become increasingly important in the next few years. What do *you* think should be done to stabilize the system?

Medicare

Roughly 40 million U.S. citizens depend on Medicare for their medical insurance (Kaiser Family Foundation, 2010). To be eligible, a person must meet one of the following criteria: be over age 65, be disabled, or have permanent kidney failure. Medicare consists of four parts (Medicare.gov, 2011): Part A, which covers inpatient hospital services, skilled nursing facilities, home health services, and hospice care; Part B, which covers the cost of physician services, outpatient hospital services, medical equipment and supplies, and other health services and supplies; and Part D, which provides some coverage for prescription medications. Part C, also called "Medicare Advantage," is offered by private companies approved by Medicare, and includes all of the benefits of Parts A and B, plus additional coverage (e.g., vision, dental), and usually Part D.

Expenses relating to most long-term care needs are funded by Medicaid, another major health care program funded by the U.S. government and targeted to people who are poor. Out-of-pocket expenses associated with co-payments and other charges are often paid by supplemental insurance policies, sometimes referred to as "Medigap" policies (Medicare.gov, 2011).

Like Social Security, Medicare is funded by a payroll tax. Hence the funding problems facing Medicare are similar to those facing Social Security and are exacerbated by the aging of the Baby Boomers. In addition, Medicare costs have increased dramatically as a result of the rapidly increasing costs of health care.

Cost containment of health care remains a major political concern, and the issue has been caught in the overall debate about health care reform in the United States. Also, because Medicare is a government-run health care program, it continues to be controversial among those who oppose such approaches to health care. But unlike Social Security, Medicare has already been subjected to significant cuts in expenditures, typically through reduced payouts to health care providers. Whether this practice will continue is unclear, especially when Baby Boomers find out that their coverage could be significantly reduced or limited to certain health conditions.

Taken together, the challenges facing society concerning older adults' financial security and health will continue to be major political issues throughout the first few decades of the 21st century. There are no easy answers, but open discussion of the various arguments will be essential for creating the optimal solution.

Test Yourself

RECALL

1. Activities of daily living (ADLs) include functioning in the areas of bathing, toileting, walking, dressing, and _____.

2. Most people who live in nursing homes are _____.

3. The people who most often abuse older adults are _____.

4. The two most important public policy issues in the United States that are being affected by the aging baby boom generation are Social Security and _____.

INTERPRET

How might the large generation now graduating from high school affect Social Security in the future?

APPLY

How would the competence–environmental press framework, presented earlier in this chapter, apply specifically to the various types of housing and nursing homes discussed in this section?

Recall answers: (1) eating, (2) older European American women who are very ill, (3) partners or adult children, (4) Medicare

---

SUMMARY

15.1 THEORIES OF PSYCHOSOCIAL AGING

What is continuity theory?

- Continuity theory is based on the view that people tend to cope with daily life in later adulthood by applying familiar strategies based on past experience in order to maintain and preserve both internal and external structures.

What is the competence and environmental press model, and how do docility and proactivity relate to the model?

- According to competence–environmental press theory, people's optimal adaptation occurs when there is a balance between their ability to cope and the level of environmental demands placed on them. When balance is not achieved, behavior becomes maladaptive. Several studies indicate that competence–environmental press theory can be applied to a variety of real-world situations.

15.2 PERSONALITY, SOCIAL COGNITION, AND SPIRITUALITY

What is integrity in late life? How can people achieve it?

- Older adults face the Eriksonian struggle of integrity versus despair primarily through a life review. Integrity involves accepting one's life for what it is; despair involves bitterness about one's past. People who reach integrity become self-affirming and self-accepting, and they judge their lives to have been worthwhile and good.

How is well-being defined in adulthood? How do people view themselves differently as they age?

- Subjective well-being is an evaluation of one's life that is associated with positive feelings. In life-span developmental psychology, subjective well-being is usually assessed by measures of life satisfaction, happiness, and self-esteem.

- Neuroscience research shows a developmental connection with brain activity in the prefrontal cortex and the amygdala.

What role does spirituality play in late life?

- Older adults use religion and spiritual support more often than any other strategy to help them cope with problems of life. This is especially true for African Americans, but all ethnic groups use spiritual-based coping to some degree.

15.3 I USED TO WORK AT . . . : LIVING IN RETIREMENT

What does being retired mean?

- Retirement is a complex process by which people withdraw from full-time employment. No single definition is adequate for all ethnic groups; self-definition involves several factors, including eligibility for certain social programs.

Why do people retire?

- People generally retire because they choose to, although some people are forced to retire or do so because of serious health problems, such as cardiovascular disease or cancer. However, there are important gender and ethnic differences in why people retire and how they label themselves after retirement. Most of the research is based on European American men from traditional marriages.

How satisfied are retired people?

■ Retirement is an important life transition. Most people are satisfied with retirement. Most retired people maintain their health, friendship networks, and activity levels—at least in the years immediately following retirement. For men, personal life priorities are all- important; little is known about women's retirement satisfaction.

How do retirees keep busy?

■ Most retired people stay busy in activities such as volunteer work and helping others. From a life-course perspective, it is important to maintain social integration in retirement. Participation in community organizations and volunteering are primary ways of achieving this.

15.4 FRIENDS AND FAMILY IN LATE LIFE

What role do friends and family play in late life?

■ A person's social convoy is an important source of satisfaction in late life. Patterns of friendships among older adults are similar to those among young adults, but older adults are more selective. Sibling relationships are especially important in old age. Because people cannot choose their siblings like they do their friends, these relationships often take more work than friendships.

What are older adults' marriages and same-sex partnerships like?

■ Long-term marriages tend to be happy until one partner develops serious health problems. Older married couples show a lower potential for marital conflict and greater potential for pleasure. Long-term gay and lesbian relationships tend to be similar in characteristics to long-term heterosexual marriages; issues pertaining to legal rights continue to present challenges.

What is it like to provide basic care for one's partner?

■ Caring for a partner puts considerable strain on the relationship. The degree of marital satisfaction strongly affects how spousal caregivers perceive stress. Although caught off guard initially, most spousal caregivers are able to provide adequate care. Perceptions of competence among spousal caregivers at the outset of caregiving may be especially important.

How do people cope with widowhood? How do men and women differ?

■ Widowhood is a difficult transition for most people. Feelings of loneliness are hard to cope with, especially during the first few months following bereavement. Men generally have problems in social relationships and in household tasks; women tend to have more severe financial problems. Some widowed people remarry, partly to solve loneliness and financial problems.

What special issues are involved in being a great-grandparent?

■ Becoming a great-grandparent is an important source of personal satisfaction for many older adults. Great-grandparents as a group are more similar to each other than grandparents are. Three aspects of great-grandparenthood are most important: sense of personal and family renewal, new diversions in life, and a major life milestone.

15.5 SOCIAL ISSUES AND AGING

Who are frail older adults? How common is frailty?

■ The number of frail older adults is growing. Frailty is defined in terms of impairment in activities of daily living (basic self-care skills) and instrumental activities of daily living (actions that require intellectual competence or planning). As many as half of those over age 85 may need assistance with ADLs or IADLs. Supportive environments are useful in optimizing the balance between competence and environmental press.

What housing options are there for older adults?

■ Most older adults prefer to stay in place in their community; home modification offers one option to achieve that. Assisted living facilities offer support for ADLs and IADLs while providing a significant degree of independence. Nursing homes provide 24×7 medical care for those who need continual assistance. Newer alternatives include the Eden Alternative, the Green House Project, and cohousing options that focus on providing greater support for people to remain in the community.

How do you know whether an older adult is abused or neglected? Which people are most likely to be abused and to be abusers?

■ Abuse and neglect of older adults is an increasing problem. However, abuse and neglect are difficult to define precisely. Several categories are used, including physical abuse, sexual abuse, emotional or psychological abuse, financial or material exploitation, abandonment, neglect, and self-neglect. Most perpetrators are family members, usually partners or adult children of the victims. Research indicates that abuse results from a complex interaction of characteristics of the caregiver and care recipient.

What are the key social policy issues affecting older adults?

■ Although initially designed as an income supplement, Social Security has become the primary source of retirement income for most U.S. citizens. The aging of the Baby Boom generation will place considerable stress on the system's financing.

■ Medicare is the principal health insurance program for adults in the United States over age 65. Cost containment is a major concern, and there are no easy answers for ensuring the system's future viability.

continuity theory (542)

competence (543)

environmental press (543)

adaptation level (543)

zone of maximum performance potential (543)

zone of maximum comfort (543)

proactivity (544)

docility (544)

integrity versus despair (546)

life review (547)

subjective well-being (547)

spiritual support (551)

bridge job (554)

social convoy (558)

socioemotional selectivity (559)

frail older adults (565)

activities of daily living (ADLs) (566)

instrumental activities of daily living (IADLs) (566)

household (568)

functional health (568)

sense of place (569)

assisted living facilities (569)

LEARN MORE ABOUT IT

Log in to **www.cengagebrain.com** to access the resources your instructor requires. For this book, you can access:

Psychology CourseMate

■ CourseMate brings course concepts to life with interactive learning, study, and exam preparation tools that support the printed textbook. A textbook-specific website, Psychology CourseMate includes an integrated interactive eBook and other interactive learning tools including quizzes, flashcards, videos, and more.

CENGAGENOW™

■ CengageNOW Personalized Study is a diagnostic study tool containing valuable text-specific resources—and because you focus on just what you don't know, you learn more in less time to get a better grade.

WebTUTOR™

■ More than just an interactive study guide, WebTutor is an anytime, anywhere customized learning solution with an eBook, keeping you connected to your textbook, instructor, and classmates.

The Final Passage

Dying and Bereavement

16

WE HAVE A PARADOXICAL RELATIONSHIP with death. Sometimes we are fascinated by it. As tourists, we visit places where famous people died or are buried. We watch as television newscasts show scenes of devastation in natural disasters and war. But when it comes to pondering our own death or that of people close to us, we have many problems. As French writer and reformer La Rochefoucauld wrote more than 300 years ago, looking into the sun is easier than contemplating our death. When death is personal, we become uneasy. Looking at the sun is hard indeed.

In this chapter we delve into thanatology. **Thanatology** *is the study of death, dying, grief, bereavement, and social attitudes toward these issues.* We will first consider definitional and ethical issues surrounding death. Next, we look specifically at the process of dying. Dealing with grief is important for survivors, so we consider this topic in the third section. Finally, we examine how people view death at different points in the life span.

LEARNING OBJECTIVES

- How is death defined?
- What legal and medical criteria are used to determine when death occurs?

- What are the ethical dilemmas surrounding euthanasia?

Greta, a college sophomore, was very upset when she learned that her roommate's mother had died suddenly. Her roommate is Jewish, and Greta had no idea what customs would be followed during the funeral. When Greta arrived at her roommate's house, she was surprised to find all of the mirrors in the house covered. Greta realized for the first time that death rituals vary in different religious traditions.

thanatology

the study of death, dying, grief, bereavement, and social attitudes toward these issues

WHEN ONE FIRST THINKS ABOUT IT, death seems a very simple concept to define: It is the point at which a person is no longer alive. Similarly, dying is simply the process of making the transition from being alive to being dead. It all seems clear enough, doesn't it? But death and dying are actually far more complicated concepts.

As we will see, Greta's experience reflects the many cultural and religious differences in the definition of death and the customs surrounding it. The meaning of death depends on the observer's perspective as well as on the specific medical and biological criteria one uses.

The symbols we use when people die, such as these caskets from Ghana, provide insights into how cultures think about death.

Max Milligan / John Warburton-Lee Photography / Photolibrary

Sociocultural Definitions of Death

What comes to mind when you hear the word *death*? A driver killed in a traffic accident? A transition to an eternal reward? Flags at half-staff? A cemetery? A car battery that doesn't work anymore? Each of these possibilities represents a way in which death can be considered in Western culture, which has its own set of specific rituals (Bustos, 2007; Penson, 2004). All cultures have their own views. Among Melanesians, the term *mate* includes the very sick, the very old, and the dead; the term *toa* refers to all other living people (Counts & Counts, 1985). Other South Pacific cultures believe that the life force leaves the body during sleep or illness; sleep, illness, and death are considered together. Thus people "die" many times before experiencing "final death" (Counts & Counts, 1985). The Kwanga of Papua New Guinea believe that most deaths are caused by sorcery (Brison, 1995).

In Ghana people are said to have a "peaceful" or "good" death if the dying person finished all business and made peace with others before death, which implies being at peace with his or her own death (van der Geest, 2004). A good and peaceful death comes "naturally" after a long and well-spent life. Such a death preferably takes place at home, which is the epitome of peacefulness, surrounded by children and grandchildren. Finally, a good death is a death that is accepted by the relatives.

Mourning rituals and states of bereavement also vary in different cultures (Lee, 2010; Rosenblatt, 2001). There is great variability across cultures in the meaning of death and whether there are rituals or other behaviors to express grief. Some cultures have formalized periods of time during which certain prayers or rituals are performed. For example, after the death of a close relative, Orthodox Jews recite ritual prayers and cover all the mirrors in the house. The men slash their ties as a symbol of loss. (These are the customs that Greta, the college stu-

The large international public displays of grief at the death of Pope John Paul II show that death can bring together people from around the world.

dent in the vignette, experienced.) In Papua New Guinea, there are accepted time periods for phases of grief (Herner, 2010). In the United States, the Muscogee Creek tribe's rituals include digging the grave by hand and giving a "farewell handshake" by throwing a handful of dirt into the grave before covering it (Walker & Balk, 2007). Ancestor worship, a deep respectful feeling toward individuals from whom a family is descended or who are important to them, is an important part of customs of death in many Asian cultures (Roszko, 2010). We must keep in mind that the experiences of our culture or particular group may not generalize to other cultures or groups.

Death can be a truly cross-cultural experience. The international outpouring of grief over the death of world leaders such Pope John Paul II in 2005, the thousands killed in the terrorist attacks against the United States in September 2001, and the hundreds of thousands killed in such natural disasters as the earthquake in Haiti in 2010 drew much attention to the ways in which the deaths of people we do not know personally can still affect us. It is at these times we realize that death happens to us all and that death can simultaneously be personal and public.

Altogether, death can be viewed in at least 10 different ways (Kalish, 1987; Kastenbaum, 1985). Look at the list that follows and think about the examples given for these definitions. Then take another moment to think up additional examples of your own.

DEATH AS AN IMAGE OR OBJECT

A flag at half-staff

Sympathy cards

Tombstone

Black crepe paper

Monument or memorial

DEATH AS A STATISTIC

Mortality rates

Number of AIDS patients who die

Murder and suicide rates

Life expectancy tables

DEATH AS AN EVENT

Funeral

Family gathering

Memorial service

Viewing or wake

DEATH AS A STATE OF BEING

Time of waiting

Nothingness

Being happy with God all the time

State of being; pure energy

DEATH AS AN ANALOGY

Dead as a doornail

Dead-letter box

Dead-end street

You're dead meat.

In the dead of winter

DEATH AS A MYSTERY

What is it like to die?

Will we meet family?

What happens after death?

Will I learn everything when I die?

DEATH AS A BOUNDARY

How many years do I have left?

What happens to my family?

What do I do now?

You can't come back.

DEATH AS A THIEF OF MEANING

I feel so cheated.

Why should I go on living?

Life doesn't mean much anymore.

I have much left to do.

DEATH AS FEAR AND ANXIETY

Will dying be painful?

I worry about my family.

I'm afraid to die.

Who will care for the kids?

DEATH AS REWARD OR PUNISHMENT

Live long and prosper.

The wicked go to hell.

Heaven awaits the just.

Purgatory prepares you for heaven.

The many ways of viewing death can be seen in various customs involving funerals. You may have experienced a range of different types of funeral customs, from very small, private services to elaborate rituals. Variations in the customs surrounding death are reflected in some of the most iconic structures on earth, such as the pyramids in Egypt, and some of the most beautiful, such as the Taj Mahal in India.

Legal and Medical Definitions

Sociocultural approaches help us understand the different ways in which people conceptualize and understand death. But they do not address a very fundamental question: How do we determine that someone has died? The medical and legal communities have grappled with this question for centuries and continue to do so today. Let's see what the current answers are.

Determining when death occurs has always been subjective. *For hundreds of years, people accepted and applied the criteria that now define* **clinical death**: *lack of heartbeat and respiration. Today, however, the most widely accepted criteria are those that characterize* **whole-brain death**. In 1981, the President's Commission for the Ethical Study of Problems in Medicine and Biomedical and Behavioral Research established several criteria still used today that must be met for the determination of whole-brain death:

- No spontaneous movement in response to any stimuli
- No spontaneous respirations for at least 1 hour
- Total lack of responsiveness to even the most painful stimuli
- No eye movements, blinking, or pupil responses
- No postural activity, swallowing, yawning, or vocalizing
- No motor reflexes
- A flat electroencephalogram (EEG) for at least 10 minutes
- No change in any of these criteria when they are tested again 24 hours later

For a person to be declared dead, all eight criteria must be met. Moreover, other conditions that might mimic death—such as deep coma, hypothermia, or drug overdose—must be ruled out. Finally, according to most hospitals, the lack of brain activity must occur both in the brainstem, which involves vegetative functions such as heartbeat and respiration, and in the cortex, which involves higher processes such as thinking. In the United States, all 50 states and the District of Columbia use the whole-brain standard to define death.

clinical death
lack of heartbeat and respiration

whole-brain death
declared only when the deceased meets eight criteria established in 1981

A major problem facing the medical profession is how brain death is to be diagnosed in practice (Sung & Greer, 2011). In part this is due to variable intervals taken to make the second assessment required as the final item in the President's Commission's criteria listed above (Lustbader et al., 2011). Because patients declared brain dead on first examination do not spontaneously recover brainstem function, and because long delays in second assessments lower the rate at which patients' families agree to organ donation, some medical professionals are calling for a single assessment or at least a simpler, more direct process (Sung & Greer, 2011).

Brain death is also controversial from some religious perspectives. For example, some Islamic scholars argue that brain death is not complete death; complete death must include the cessation of respiration (Bedir & Aksoy, 2011). Roman Catholics focus on what they term "natural death" (Verheijde, 2010).

It is possible for a person's cortical functioning to cease while brainstem activity continues; this is a **persistent vegetative state,** *from which the person does not recover.* This condition can occur following disruption of the blood flow to the brain, a severe head injury, or a drug overdose. Persistent vegetative state allows for spontaneous heartbeat and respiration but not for consciousness. The whole-brain standard does not permit a declaration of death for someone who is in a persistent vegetative state. Because of conditions like persistent vegetative state, family members sometimes face difficult ethical decisions concerning care for the individual. These issues are the focus of the next section.

> **persistent vegetative state**
> situation in which a person's cortical functioning ceases while brainstem activity continues

Ethical Issues

An ambulance screeches to a halt, and emergency personnel rush a woman into the emergency room. As a result of an accident at a swimming pool, she has no pulse and no respiration. Working rapidly, the trauma team reestablishes a heartbeat through electric shock. A respirator is connected. An EEG and other tests reveal extensive and irreversible brain damage—she is in a persistent vegetative state. What should be done?

This is an example of the kinds of problems faced in the field of **bioethics,** *the study of the interface between human values and technological advances in health and life sciences.* Bioethics grew from two bases: respect for individual freedom and the impossibility of establishing any single version of morality by rational argument or common sense. Both of these factors are increasingly based on empirical evidence and cultural contexts (Holloway, 2011; Sherwin, 2011). In practice, bioethics emphasizes the importance of individual choice and the minimization of harm over the maximization of good. That is, bioethics requires people to weigh how much the patient will benefit from a treatment relative to the amount of suffering he or she will endure as a result of the treatment. Examples of the tough choices required are those facing cancer patients about aggressive treatment for cancer that is quite likely to be fatal in any case and those facing family members about whether to turn off a life-support machine that is attached to their loved one.

> **bioethics**
> study of the interface between human values and technological advances in health and life sciences

In the arena of death and dying, the most important bioethical issue is **euthanasia**—*the practice of ending life for reasons of mercy.* The moral dilemma posed by euthanasia becomes apparent when we try to decide the circumstances under which a person's life should be ended, which implicitly forces one to place a value on the life of another (Bedir & Aksoy, in press; Elliott & Oliver, 2008; Verheijde, 2010). It also makes us think about the difference between "killing" and "letting die" at the end of life (Dickens, Boyle, & Ganzini, 2008). In our society, this dilemma occurs most often when a person is being kept alive by machines or when someone is suffering from a terminal illness.

> **euthanasia**
> the practice of ending life for reasons of mercy

Euthanasia

Euthanasia can be carried out in two different ways: actively and passively (Moeller, Lewis, & Werth, 2010). **Active euthanasia** *involves the deliberate ending of someone's life, which may be based on a clear statement of the person's wishes or be a decision made by someone else who has the legal authority to do so.* Usually, this involves situ-

> **active euthanasia**
> the deliberate ending of someone's life

passive euthanasia
allowing a person to die by withholding
available treatment

THINK ABOUT IT

How do sociocultural forces shape attitudes
about euthanasia?

ations in which people are in a persistent vegetative state or suffer from the end
stages of a terminal disease. Examples of active euthanasia would be administering a
drug overdose or ending a person's life through so-called mercy killing.

A second form of euthanasia, **passive euthanasia**, involves allowing a person to die
by withholding available treatment. For example, a ventilator might be disconnected,
chemotherapy might be withheld from a patient with terminal cancer, a surgical procedure might not be performed, or food could be withdrawn.

Some ethicists and medical professionals do not differentiate active and passive
euthanasia. For example, the European Association of Palliative Care (EAPC, 2011)
established an ethics task force opposing euthanasia, and the group claims that the
expression "passive euthanasia" is a contradiction in terms because any ending of a
life is by definition active. Despite these concerns, Garrard and Wilkinson (2005) conclude that there is really no reason to abandon the category provided that it is properly and narrowly understood and provided that "euthanasia reasons" for withdrawing or withholding life-prolonging treatment are carefully distinguished from other
reasons, such as family members not wanting to wait to divide the patient's estate.
Still, whether there is a difference between active and passive euthanasia remains
controversial (Busch & Rodogno, 2011).

Most Americans favor such actions as disconnecting life support in situations involving patients in a persistent vegetative state, withholding treatment if the person
agrees or is already in the later stages of a terminal illness, and even the concept of
assisted death. But feelings also run strongly against such actions for religious or
other reasons (Bedir & Aksoy, in press; Dickens et al., 2008; Verheijde, 2010). Even
political debates can incorporate the issue, as demonstrated in the summer of 2009 in
the United States when opponents of President Obama's health care reform falsely
claimed that "death panels" would make decisions about terminating life support if
the reform measure passed.

Globally, opinions about euthanasia vary (Bosshard & Materstvedt, 2011). A systematic survey of laypersons and health care professionals in the Netherlands and
Belgium found that most said that they would support euthanasia under certain specific conditions (Teisseyre, Mullet, & Sorum, 2005). Respondents assigned most importance to patients' specific requests for euthanasia and supported these requests,
but they did not view patients' willingness to donate organs—without another compelling reason—as an acceptable reason to request euthanasia. Other analyses show
that opinions are often related to religious or political beliefs (Swinton & Payne,
2009). For example, Western Europeans tend to view active euthanasia more positively due to a lesser influence of religion and more social welfare services than residents of Eastern European and Islamic countries, who tend to be more influenced by
religious beliefs that argue against such practices (Baumann et al., 2011; Góra & Mach,
2010; Nayernouri, 2011).

Disconnecting a life support system is one thing; withholding nourishment from
a terminally ill person is quite another for many people. Indeed, such cases often end
up in court. The first high-profile legal case involving passive euthanasia in the United
States was brought to the courts in 1990; the U.S. Supreme Court took up the case of
Nancy Cruzan, whose family wanted to end her forced feeding. The court ruled that,
unless clear and incontrovertible evidence is presented that an individual desires to
have nourishment stopped, such as through a durable health care power of attorney
or living will, a third party (such as a parent or partner) cannot decide to end it.

The most widely publicized and politicized case of euthanasia in the United States
involved Terri Schiavo, who died in Florida in 2005. This extremely controversial case
involving the withdrawal of forced feeding had its origins in a disagreement between
Terri's husband Michael, who said that Terri would have wanted to die with dignity
and therefore the feeding tube should be removed, and her parents, who argued the
opposite. The debate resulted in the involvement of government officials, state and
federal legislators, and the courts. As discussed in the What Do *You* Think? feature,
such cases reveal the difficult legal, medical, and ethical issues as well as the high
degree of emotion surrounding the topic of euthanasia and death with dignity.

On February 25, 1990, 26-year-old Terri Schiavo collapsed in her home from a possible potassium imbalance caused by an eating disorder, temporarily stopping her heart and cutting off oxygen to her brain. On March 31, 2005, Terri Schiavo died after her feeding tube had been removed 13 days earlier. On these two points everyone connected with Terri's case agreed. But on all other essential aspects of it, Terri's husband Michael and Terri's parents deeply disagreed.

The central point of disagreement was Terri's medical condition. Terri's husband and numerous physicians argued that she was in a persistent vegetative state. Based on this diagnosis, Michael Schiavo requested that Terri's feeding tube be withdrawn and that she be allowed to die with dignity in the way he asserted she would have wanted to.

Terri's parents and some other physicians said she was not in a persistent vegetative state and that she was capable of recognizing them and others. Based on this diagnosis, their belief that Terri would not want the intervention stopped, and their contention that passive euthanasia is morally wrong, they fought Michael's attempts to remove the feeding tube.

What made this case especially difficult was that Terri had left no written instructions that clearly stated her thoughts and intentions on the issue. So the ensuing legal and political debates became based on what various people thought Terri would have wanted and reflected various aspects of people's positions on personal rights regarding life and death.

The legal and political battles began in 1993, when Terri's parents tried unsuccessfully to have Michael removed as Terri's guardian. But the most heated aspects of the case began in 2000, when a circuit court judge ruled that Terri's feeding tube could be removed based on his belief that she had told Michael that she would not have wanted it. In April 2001, the feeding tube was removed after state courts and the U.S. Supreme Court refused to hear the case. However, the tube was reinserted 2 days later upon another judge's order. In November 2002, the original circuit court judge ruled that Terri had no hope of recovery and again ordered the tube removed, an order eventually carried out in October 2003. Within a week, however, Florida Governor Jeb Bush signed a bill passed by the Florida legislature requiring that the tube be reinserted. This law was ruled unconstitutional by the Florida Supreme Court in September 2004. In February 2005, the original circuit court judge again ordered the tube removed. Between March 16 and March 27, the Florida House introduced and passed a bill that would have required the tube be reinserted, but the Florida Senate defeated a somewhat different version of the bill. From March 19 to 21, bills that would have allowed a federal court to review the case passed in the U.S. House of Representatives and the U.S. Senate, but the two versions could not be reconciled. Over the next 10 days, the Florida Supreme Court, the U.S. district court, and a U.S. circuit court refused to hear the case, as did the U.S. Supreme Court. The original circuit court judge rejected a final attempt by Terri's parents to have the feeding tube reinserted.

The public debate on the case was as long and complex as the legal and political arguments. The debate had several positive outcomes. The legal and political complexities dramatically illustrated the need for people to reflect on end-of-life issues and to make their wishes known to family members and others (e.g., health care providers) in writing. The case also brought to light the high cost of long-term care, the difficulties in actually determining whether someone is in a persistent vegetative state (and, in turn, what that implies about life), the tough moral and ethical issues surrounding the withdrawal of nutrition, and the individual's personal feelings about death. The legal and medical communities have proposed reforms concerning how these types of cases are heard in the courts and the processes used to resolve them (Dickinson, in press; Moran, 2008).

What do *you* think? Should Terri Schiavo's feeding tube have been removed?

The legal and political debate over the removal of Terri Schiavo's feeding tube raised people's awareness of the need to make one's wishes about end-of-life issues known in writing.

Physician-Assisted Suicide

Taking one's own life through suicide has never been popular in the United States because of religious and other prohibitions. In other cultures, such as Japan, suicide is viewed as an honorable way to die under certain circumstances (Joiner, 2010). American Indian and Alaskan Native males, closely followed by non-Latino white males, have the highest suicide rates in the United States (Centers for Disease Control and Prevention, 2009).

But attitudes regarding suicide in certain situations are changing. *Much of this change concerns the topic of* **physician-assisted suicide**, *in which physicians provide dying patients with a fatal dose of medication that the patient self-administers.* A

physician-assisted suicide
process in which physicians provide dying patients with a fatal dose of medication that the patient self-administers

Harris Poll released in 2011 indicated that 70% of all adult respondents (and 62% of those over age 65) agreed that people who are terminally ill, in great pain, and have no chance of recovery should have the right to choose to end their lives. Only 17% of the respondents disagreed. By a margin of 58% to 20%, respondents supported physician-assisted suicide for such patients (Harris Interactive, 2011).

Several countries—including Switzerland, Belgium, and Colombia—tolerate physician-assisted suicide. In 1984, the Dutch Supreme Court eliminated prosecution of physicians who assist in suicide if five criteria are met:

1. The patient's condition is intolerable with no hope for improvement.
2. No relief is available.
3. The patient is competent.
4. The patient makes a request repeatedly over time.
5. Two physicians have reviewed the case and agree with the patient's request.

The Dutch Parliament approved the policy in April 2001, making the Netherlands the first country to have an official policy legalizing physician-assisted suicide (Deutsch, 2001).

Voters in Oregon passed the Death With Dignity Act in 1994, the first physician-assisted suicide law in the United States (I-1000 passed in Washington state in 2008, modeled after the Oregon law). These laws make it legal for people to request a lethal dose of medication if they have a terminal disease and make the request voluntarily. Although the U.S. Supreme Court ruled in two cases in 1997 (*Vacco v. Quill* and *Washington v. Glucksberg*) that there is no right to assisted suicide, the Court decided in 1998 not to overturn the Oregon law.

The Oregon and Washington laws are more restrictive than the law in the Netherlands (Deutsch, 2001). Both laws provide for people to obtain and use prescriptions for self-administered lethal doses of medication. The law requires that a physician inform the person that he or she is terminally ill and to describe alternative options (e.g., hospice care, pain control), and the person must be mentally competent and make two oral requests and a written one, with at least 15 days between each oral request. Such provisions are included to ensure that people making the request fully understand the issues and that the request is not made hastily.

Several studies have examined the impact of the Oregon law. The numbers of patients who received prescriptions and who died between 1998 and early 2011 are shown in ■ Figure 16.1. Over the period, a total of 525 patients died under the terms

The Oregon Death With Dignity Act provides the option of obtaining a prescription for a lethal dose of medication if the patient meets certain strict criteria.

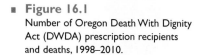

■ **Figure 16.1**
Number of Oregon Death With Dignity Act (DWDA) prescription recipients and deaths, 1998–2010.

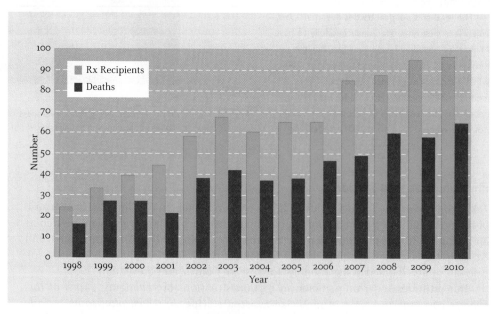

From Oregon Public Health Division, DWDA, at Oregon.gov.

of the law (Oregon Department of Human Services, 2011). Comprehensive reviews of the implementation of the Oregon law soon after its passage concluded that all safeguards worked and that such things as depression, coercion, and misunderstanding of the law were carefully screened (Orentlicher, 2000). Available data also indicate that Oregon's law has psychological benefits for patients, who are comforted by knowing they have this option (Cerminara & Perez, 2000).

There is no question that the debate over physician-assisted suicide will continue. As the technology to keep people alive continues to improve, the ethical issues about active euthanasia in general and physician-assisted suicide in particular will continue to become more complex, and will likely focus increasingly on quality of life.

Making Your End-of-Life Intentions Known

As has been clearly shown, euthanasia raises complex legal, political, and ethical issues. In most jurisdictions, euthanasia is legal only when a person has made known his or her wishes concerning medical intervention. Unfortunately, many people fail to take this step, perhaps because it is difficult to think about such situations or because they do not know the options available to them. But without clear directions, medical personnel may be unable to take a patient's preferences into account.

There are two ways to make one's intentions known. *In a* **living will**, *a person simply states his or her wishes about life support and other treatments. In a* **durable power of attorney for health care**, *an individual appoints someone to act as his or her agent for health care decisions* (see ■ Figure 16.2). A major purpose of both is to make one's wishes known about the use of life support interventions in the event that the person is unconscious or otherwise incapable of expressing them, along with other related end-of-life issues such as organ transplantation and other health care options (Baumann et al., 2011; Castillo et al., 2011). A durable power of attorney for health care has an additional advantage: It names an individual who has the legal authority to speak for the person if necessary.

Although there is considerable support for both mechanisms, there are several problems as well (Castillo et al., 2011). States vary in their laws relating to advance directives. Many people fail to inform their relatives and physicians about their health care decisions. Others do not tell the person named in a durable power of attorney where the document is kept. Obviously, this puts relatives at a serious disadvantage if decisions concerning the use of life-support systems need to be made.

A living will or a durable power of attorney for health care can be the basis for a "Do Not Resuscitate" medical order. *A* **Do Not Resuscitate (DNR) order** *means that cardiopulmonary resuscitation (CPR) is not started should one's heart and breathing*

living will
a document in which a person states his or her wishes about life support and other treatments

durable power of attorney for health care
a document in which an individual appoints someone to act as his or her agent for health care decisions

Do Not Resuscitate (DNR) order
a medical order that means that cardiopulmonary resuscitation (CPR) is not started should one's heart and breathing stop

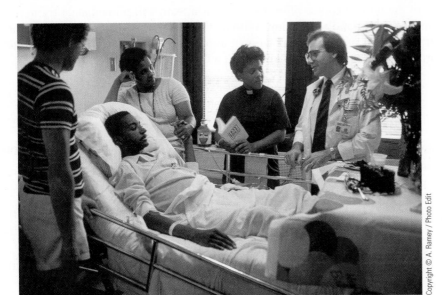

Frank discussions of end-of-life issues with patients and their families by health care workers provides a better context for handling these issues.

California Medical Association
DURABLE POWER OF ATTORNEY FOR HEALTH CARE DECISIONS
(California Probate Code Sections 4600-4753)

WARNING TO PERSON EXECUTING THIS DOCUMENT

This is an important legal document. Before executing this document, you should know these important facts:

This document gives the person you designate as your agent (the attorney-in-fact) the power to make health care decisions for you. Your agent must act consistently with your desires as stated in this document or otherwise made known.

Except as you otherwise specify in this document, this document gives your agent power to consent to your doctor not giving treatment or stopping treatment necessary to keep you alive.

Notwithstanding this document, you have the right to make medical and other health care decisions for yourself so long as you can give informed consent with respect to the particular decision. In addition, no treatment may be given to you over your objection, and health care necessary to keep you alive may not be stopped or withheld if you object at the time.

This document gives your agent authority to consent, to refuse to consent, or to withdraw consent to any care, treatment, service, or procedure to maintain, diagnose, or treat a physical or mental condition. This power is subject to any statement of your desires and any limitations that you include in this document. You may

state in this document any types of treatment that you do not desire. In addition, a court can take away the power of your agent to make health care decisions for you if your agent (1) authorizes anything that is illegal, (2) acts contrary to your known desires or (3) where your desires are not known, does anything that is clearly contrary to your best interests.

This power will exist for an indefinite period of time unless you limit its duration in this document.

You have the right to revoke the authority of your agent by notifying your agent or your treating doctor, hospital, or other health care provider orally or in writing of the revocation.

Your agent has the right to examine your medical records and to consent to their disclosure unless you limit this right in this document.

Unless you otherwise specify in this document, this document gives your agent the power after you die to (1) authorize an autopsy, (2) donate your body or parts thereof for transplant or therapeutic or educational or scientific purposes, and (3) direct the disposition of your remains.

If there is anything in this document that you do not understand, you should ask a lawyer to explain it to you.

1. CREATION OF DURABLE POWER OF ATTORNEY FOR HEALTH CARE

By this document I intend to create a durable power of attorney by appointing the person designated below to make health care decisions for me as allowed by Sections 4600 to 4753, inclusive, of the California Probate Code. This power of attorney shall not be affected by my subsequent incapacity. I hereby revoke any prior durable power of attorney for health care. I am a California resident who is at least 18 years old, of sound mind, and acting of my own free will.

2. APPOINTMENT OF HEALTH CARE AGENT

(Fill in below the name, address and telephone number of the person you wish to make health care decisions for you if you become incapacitated. You should make sure that this person agrees to accept this responsibility. The following may not serve as your agent: (1) your treating health care provider; (2) an operator of a community care facility or residential care facility for the elderly; or (3) an employee of your treating health care provider, a community care facility, or a residential care facility for the elderly, unless that employee is related to you by blood, marriage or adoption, or unless you are also an employee of the same treating provider or facility. If you are a conservatee under the Lanterman-Petris-Short Act (the law governing involuntary commitment to a mental health facility) and you wish to appoint your conservator as your agent, you must consult a lawyer, who must sign and attach a special declaration for this document to be valid.)

I, _____ , hereby appoint:
(insert your name)

Name _____

Address _____

Work Telephone (_____) _____ Home Telephone (_____) _____

as my agent (attorney-in-fact) to make health care decisions for me as authorized in this document. I understand that this power of attorney will be effective for an indefinite period of time unless I revoke it or limit its duration below.

(Optional) This power of attorney shall expire on the following date: _____ .

© California Medical Association 1996 (revised)

stop. In the normal course of events, a medical team will immediately try to restore normal heartbeat and respiration. With a DNR order, this treatment is not done. As with living wills and durable powers of attorney, it is extremely important to let all appropriate medical personnel know that a DNR order is in effect.

Test Yourself

RECALL

1. The phrase "dead as a doornail" is an example of the sociocultural definition of death as _____.

2. The difference between brain death and a persistent vegetative state is _____.

3. Withholding an antibiotic from a person who dies as a result is an example of _____.

INTERPRET

What is the difference between the Oregon Death With Dignity Act and active euthanasia?

APPLY

Describe how people at each level of Kohlberg's theory of moral reasoning (described in Chapter 8) would deal with the issue of euthanasia.

Recall answers: (1) an analogy, (2) the brainstem still functions in a persistent vegetative state, (3) passive euthanasia

LEARNING OBJECTIVES

■ How do feelings about death change over adulthood?

■ How do people deal with their own death?

■ What is death anxiety, and how do people show it?

■ How do people deal with end-of-life issues and create a final scenario?

■ What is hospice?

Jean is a 72-year-old woman who was recently diagnosed with advanced colon cancer. She has vivid memories of her father dying a long, protracted death in great pain. Jean is afraid that she will suffer the same fate. She has heard that the hospice in town emphasizes pain management and provides a lot of support for families. Jean wonders whether this is something she should explore in the time she has left.

LIKE JEAN, MOST PEOPLE ARE UNCOMFORTABLE THINKING ABOUT THEIR OWN DEATH, ESPECIALLY IF THEY THINK IT WILL BE UNPLEASANT. As one research participant put it, "You are nuts if you aren't afraid of death" (Kalish & Reynolds, 1976). Still, death is a paradox, as we noted at the beginning of the chapter. That is, we are afraid of or anxious about death but we are drawn to it, sometimes in very public ways. We examine this paradox at the personal level in this section. Specifically, we focus on two questions: (1) How do people's feelings about death differ with age? (2) What is it about death that we fear or that makes us anxious?

Before proceeding, however, take a few minutes to complete the following exercise.

A SELF-REFLECTIVE EXERCISE ON DEATH

■ In 200 words or less, write your own obituary. Be sure to include your age and cause of death. List your lifetime accomplishments. Don't forget to list your survivors.

■ Think about all the things you will have done that are not listed in your obituary. List some of them.

■ Think of all the friends you will have made and how you will have affected them.

■ Would you make any changes in your obituary now?

A Life-Course Approach to Dying

How do you feel about dying? Do you think people of different ages feel the same way? It probably doesn't surprise you to learn that feelings about dying vary across adulthood. For example, adults of various ages who live with a person who has a life-threatening illness each comes to terms with death in an individual and a family-based way (Carlander, 2011).

Although not specifically addressed in research, the shift from formal operational thinking to postformal thinking (see Chapter 10) could be important in young adults' contemplation of death. Presumably, this shift in cognitive development is accompanied by a lessening of the feeling of immortality as young adults begin to integrate personal feelings and emotions with their thinking.

Midlife is the time when most people in developed countries confront the death of their parents. Up until that point, people tend not to think much about their own death; the fact that their parents are still alive buffers them from reality. After all, in the normal course of events, our parents are supposed to die before we do.

Once their parents have died, people realize that they are now the oldest generation of their family—the next in line to die. Reading the obituary pages, they are re-

Dealing with the death of friends is often especially difficult for young adults.

minded of this, as the ages of many of the people who have died get closer and closer to their own.

Probably as a result of this growing realization of their own mortality, middle-aged adults' sense of time undergoes a subtle yet profound change. It changes from an emphasis on how long they have already lived to how long they have left to live, a shift that increases into late life (Cicirelli, 2006; Neugarten, 1969). This may lead to occupational change or other redirection such as improving relationships that had deteriorated over the years.

In general, older adults are less anxious about death and more accepting of it than any other age group. Still, because the discrepancy between desired and expected number of years left to live is greater for young-old than for mid-old adults, anxiety is higher for young-old adults (Cicirelli, 2006). In part, the greater overall acceptance of death results from the achievement of ego integrity, as described in Chapter 15. For many older adults, the joy of living is diminishing (Kalish, 1987). More than any other group, they have experienced loss of family and friends and have come to terms with their own mortality. Older adults have more chronic diseases, which are not likely to go away. They may feel that their most important life tasks have been completed (Kastenbaum, 1999).

Understanding how adults deal with death and their consequent feelings of grief is best approached from the perspective of attachment theory (Mercer, 2011; Stroebe, Schut, & Stroebe, 2005). In this view, a person's reactions are a natural consequence of forming attachments and then losing them. We consider adult grief a bit later in the chapter.

Dealing With One's Own Death

Thinking about death from an observer's perspective is one thing. Thinking about one's own death, like Jean is doing, is quite another. The reactions people have to their own impending death, long thought to be the purview of religion and philosophy, were not researched until well into the 20th century.

Many authors have tried to describe the dying process, often using the metaphor of a trajectory that captures the duration of time between the onset of dying (e.g., from the diagnosis of a fatal disease) as well as death and the course of the dying process (Field & Cassel, 2010; Wilkinson & Lynn, 2001). These dying trajectories vary a great deal across diseases, as illustrated in ■ Figure 16.3. Some diseases, such as some cancers (represented by the green line), have a clear and rapid period of decline. Other diseases, such as a series of cerebral vascular accidents (represented by the red line), have some periods of rapid decline, followed by some recovery, but gradual overall

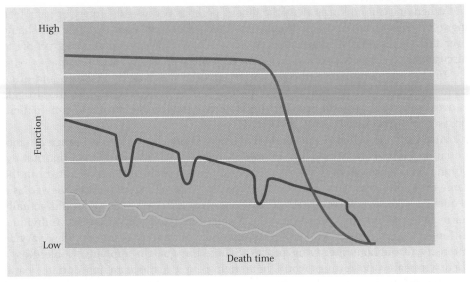

■ **Figure 16.3**
The three main trajectories of decline at the end of life.

High

Function

Low

Death time

Reproduced from Murray, S.A., and Sheikh, A. (2008). Care for all at the end of life. Figure 1, the three main trajectories of decline at the end of life. *British Medical Journal, 336,* 958–959. Reprinted with permission from BMJ Publishing Group Ltd.

decline over time. A third situation, represented by the gold line, involve a long-term chronic condition that ends in death.

Kübler-Ross's Theory

Elisabeth Kübler-Ross changed the way we approach dying. She became interested in the experience of dying when she was an instructor in psychiatry at the University of Chicago in the early 1960s. When she began her investigations into the dying process, such research was controversial; her physician colleagues initially were outraged, and some even denied that their patients were terminally ill. Still, she persisted. More than 200 interviews with terminally ill people convinced her that most people experienced several emotional reactions. Using her experiences, she described five reactions that represented the ways in which people dealt with death: denial, anger, bargaining, depression, and acceptance (Kübler-Ross, 1969). Although they were first presented as a sequence, it was subsequently realized that the emotions can overlap and can be experienced in different order.

When people are told that they have a terminal illness, their first reaction is likely to be shock and disbelief. Denial is a normal part of getting ready to die. Some want to shop around for a more favorable diagnosis, and most feel that a mistake has been made. Others try to find assurance in religion. Eventually, though, reality sets in for most people.

At some point, people express anger as hostility, resentment, and envy toward health care workers, family, and friends. People ask, "Why me?" and express a great deal of frustration. The fact that they are going to die when so many others will live seems so unfair. With time and work, most people confront their anger and resolve it.

In the bargaining phase, people look for a way out. Maybe a deal can be struck with someone, perhaps God, that would allow survival. For example, a woman might promise to be a better mother if only she could live. Or a person sets a timetable: "Just let me live until my daughter graduates from college." Eventually, the person becomes aware that these deals will not work.

Dr. Elisabeth Kübler-Ross revolutionized the study of death and dying.

©Ken Ross

When one can no longer deny the illness, perhaps because of surgery or pain, feelings of depression are common. People report feeling deep loss, sorrow, guilt, and shame over their illness and its consequences. Kübler-Ross believes that allowing people to discuss their feelings with others helps move them to an acceptance of death.

In the acceptance stage, the person accepts the inevitability of death and often seems detached from the world and at peace. "It is as if the pain is gone, the struggle is over, and there comes a time for the 'final rest before the journey' as one patient phrased it" (Kübler-Ross, 1969, p. 100).

Although she believed that these five stages represent the typical range of emotional development in the dying process, Kübler-Ross (1974) cautioned that not everyone experiences all of them or progresses through them at the same rate or in the same order. Research supports the view that her "stages" should not be viewed as a sequence (Charlton & Verghese, 2010; Neimeyer, 1997). In fact, we could actually harm dying people by considering these stages as fixed and universal. Individual differences are great, as Kübler-Ross points out. Emotional responses may vary in intensity throughout the dying process. Thus, the goal in applying Kübler-Ross's theory to real-world settings would be to help people achieve an appropriate death: one that meets the needs of the dying person, allowing him or her to work out each problem as it comes.

A Contextual Theory of Dying

One of the difficulties with most theories of dying is a general lack of research evaluating them in a wide variety of contexts (Kastenbaum & Thuell, 1995). By their very nature, stages or sequences imply a particular directionality. Stage theories, in particular, emphasize qualitative differences between the various stages. However, the duration of a particular stage, or a specific phase, varies widely from person to person. Such theories assume some sort of underlying process for moving through the stages or phases but do not clearly state what causes a person to move from one to another.

One reason for these problems is the realization that there is no one right way to die, although there may be better or worse ways of coping (Corr, 1991–1992, 2010a; Corr, Corr, & Nabe, 2008). A perspective that recognizes this realization would approach the issue from the mind-set of the dying person and the issues or tasks that he or she must face. Corr identified four dimensions of such tasks: bodily needs, psychological security, interpersonal attachments, and spiritual energy and hope. This holistic approach acknowledges individual differences and rejects broad generalizations. Corr's task work approach also recognizes the importance of the coping efforts of family members, friends, and caregivers as well as those of the dying person.

Kastenbaum and Thuell (1995) argue that what is needed is an even broader contextual approach that takes a more inclusive view of the dying process. They point out that theories must be able to handle people who have a wide variety of terminal illnesses and be sensitive to dying people's own perspectives and values related to death. The socio-environmental context within which dying occurs, which often changes over time, must be recognized. For example, a person may begin the dying process living independently but end up in a long-term care facility. Such moves may have profound implications for how the person copes with dying. A contextual approach would provide guidance for health care professionals and families for discussing how to protect the quality of life, provide better care, and prepare caregivers for dealing with the end of life. Such an approach would also provide research questions. For example, how does one's acceptance of dying change across various stages?

We do not yet have such a comprehensive theory of dying. But as Kastenbaum and Thuell point out, we can move in that direction by rejecting a reductionistic approach that focuses on set stages in favor of a truly holistic one. One way to accomplish this is to examine people's experiences as a narrative that can be written from many points of view (e.g., the patient, family members, caregivers). What would emerge would be a rich description of a dynamically changing process.

Death Anxiety

We have seen that how people view death varies with age. In the process, we encountered the notion of feeling anxious about death. **Death anxiety** *refers to people's anxiety or even fear of death and dying.* Death anxiety is tough to pin down; indeed, it is the ethereal nature of death, rather than something about it in particular, that usually makes us feel so uncomfortable. We cannot put our finger on something specific about death that is causing us to feel uneasy. Because of this, we must look for indirect behavioral evidence to document death anxiety. Research findings suggest that death anxiety is a complex, multidimensional construct.

For nearly three decades, researchers have applied terror management theory as a framework to study death anxiety (Burke, Martens, & Faucher, 2010). **Terror management theory** *addresses the issue of why people engage in certain behaviors to achieve particular psychological states based on their deeply rooted concerns about mortality* (Arndt & Vess, 2008). The theory proposes that ensuring the continuation of one's life is the primary motive underlying behavior and that all other motives can be traced to this basic one. Additionally, some suggest that older adults present an existential threat for the younger and middle-aged adults because they remind us all that death is inescapable, the body is fallible, and the bases by which we may secure self-esteem (and manage death anxiety) are transitory (Martens, Goldenberg, & Greenberg, 2005). Thus, death anxiety is a reflection of one's concern over dying, an outcome that would violate the prime motive.

Neuroimaging research shows that terror management theory provides a useful framework for studying brain activity related to death anxiety. Quirin and colleagues (in press) found that brain activity in the right amygdala, left rostral anterior cingulate cortex, and right caudate nucleus was greater when male participants were answering questions about fear of death and dying than when they were answering questions about dental pain.

On the basis of several diverse studies using many different measures, researchers conclude that death anxiety consists of several components. Each of these components is most easily described with terms that resemble examples of great concern (anxiety) but cannot be tied to any one specific focus. Some research on U.S. and Canadian adults indicates that components of death anxiety included pain, body malfunction, humiliation, rejection, nonbeing, punishment, interruption of goals, being destroyed, and negative impact on survivors (Fortner & Neimeyer, 1999; Power & Smith, 2008). To complicate matters further, each of these components can be assessed at any of three levels: public, private, and nonconscious. That is, what we admit feeling about death in public may differ greatly from what we feel when we are alone with our own thoughts. In short, the measurement of death anxiety is complex, and researchers need to specify which aspects they are assessing.

Much research has been conducted to learn what demographic and personality variables are related to death anxiety. Although the results often are ambiguous, some patterns have emerged. For example, older adults tend to have lower death anxiety than younger adults, perhaps because of their tendency to engage in life review, to have a different perspective about time, and to have a higher level of religious motivation (Henrie, 2010). Men show greater fear of the unknown than women, but women report more specific fear of the dying process (Cicirelli, 2001). In Taiwan, higher death anxiety among patients with cancer is associated with not having a purpose in life and fearing disease relapse (Tang et al., in press).

Strange as it may seem, death anxiety may have a beneficial side. For one thing, being afraid to die means that we often go to great lengths to make sure we stay alive, as argued by terror management theory (Burke et al., 2010; Pyszczynski et al., 1997, 1999). Because staying alive helps to ensure the continuation and socialization of the species, fear of death serves as a motivation to have children and raise them properly.

death anxiety
people's anxiety or even fear of death and dying

terror management theory
addresses the issue of why people engage in certain behaviors to achieve particular psychological states based on their deeply rooted concerns about mortality

THINK ABOUT IT
Why does death anxiety have so many components?

Learning to Deal With Death Anxiety

Although some degree of death anxiety may be appropriate, we must guard against letting it become powerful enough to interfere with our normal daily routines. Several ways exist to help us in this endeavor. Perhaps the one most often used is to live life to the fullest. Kalish (1984, 1987) argues that people who do this enjoy what they have; although they may still fear death and feel cheated, they have few regrets. Adolescents are particularly likely to do this; research shows that teenagers, especially males, engage in risky behavior that is correlated with low death anxiety (Cotter, 2001).

Koestenbaum (1976) proposes several exercises and questions to increase one's death awareness. Some of these are to write your own obituary (like you did earlier in this chapter) and to plan your own death and funeral services. You can also ask yourself: "What circumstances would help make my death acceptable?" "Is death the sort of thing that could happen to me right now?"

These questions serve as a basis for an increasingly popular way to reduce anxiety: death education. Most death education programs combine factual information about death with issues aimed at reducing anxiety and fear. These programs vary widely in orientation; they can include such topics as philosophy, ethics, psychology, drama, religion, medicine, art, and many others. Additionally, they can focus on death, the process of dying, grief and bereavement, or any combination of them. In general, death education programs help primarily by increasing our awareness of the complex emotions felt and expressed by dying people and their families. It is important to make education programs reflect the diverse backgrounds of the participants (Fowler, 2008). Research shows that participating in experiential workshops about death significantly lowers death anxiety in younger, middle-aged, and older adults and raises awareness about the importance of advance directives (Abengozar, Bueno, & Vega, 1999; Moeller et al., 2010).

Facing death on a regular basis often forces people to confront their death anxiety.

BOB STRONG/Reuters /Landov

Creating a Final Scenario

When given the chance, many adults would like to discuss a variety of issues, collectively called **end-of-life issues***: management of the final phase of life, after-death disposition of their body and memorial services, and distribution of assets* (Green, 2008; Kleespies, 2004). How these issues are confronted represents a significant generational shift (Green, 2008). Parents and grandparents of the baby boom generation spoke respectfully about those who had "passed away." Baby Boomers are far more likely to plan and be more matter-of-fact. People want to manage the final part of their lives by thinking through the choices between traditional care (e.g., that provided by hospitals and nursing homes) and alternatives (such as hospices, which we discuss in the next section), completing advance directives (e.g., durable power of attorney, living will), resolving key personal relationships, and perhaps choosing the alternative of ending one's life prematurely through euthanasia.

What happens to one's body and how one is memorialized is very important to most people. Is a traditional burial preferred over cremation? A traditional funeral over a memorial service? Such choices often are based in people's religious beliefs and their desire for privacy for their families after they have died.

Making sure that one's estate and personal effects are passed on appropriately often is overlooked. Making a will is especially important in ensuring that one's wishes are carried out. Providing for the informal distribution of personal effects also helps prevent disputes between family members.

end-of-life issues
issues pertaining to the management of the final phase of life, after-death disposition of the body and memorial services, and distribution of assets

Deciding whether to have a traditional funeral is part of the creation of one's final scenario.

Whether people choose to address these issues formally or informally, it is important that they be given the opportunity to do so. In many cases, family members are reluctant to discuss these matters with the dying relative because of their own anxiety about death. *Making such choices known about how they do and do not want their lives to end constitutes a* **final scenario**.

One of the most difficult and important parts of a final scenario for most people is the process of separation from family and friends (Corr et al., 2008; Wanzer & Glenmullen, 2007). The final days, weeks, and months of life provide opportunities to affirm love, resolve conflicts, and provide peace to dying people. The failure to complete this process often leaves survivors feeling that they did not achieve closure in the relationship, which can result in bitterness toward the deceased.

Health care workers realize the importance of giving dying patients the chance to create a final scenario and recognize the uniqueness of each person's final passage. A key part of their role is to ease this process (Wanzer & Glenmullen, 2007). Any given final scenario reflects the individual's personal past, which is the unique combination of the development forces that person experienced. Primary attention is paid to how people's total life experiences have prepared them to face end-of-life issues (Moeller et al., 2010).

One's final scenario helps family and friends interpret one's death, especially when the scenario is constructed jointly, such as between spouses, and when communication is open and honest (Byock, 1997; Green, 2008). The different perspectives of everyone involved are unlikely to converge without clear communication and discussion. Respecting each person's perspective is key and greatly helps in creating a good final scenario.

Encouraging people to decide for themselves how the end of their lives should be handled has helped people take control of their dying (Wass, 2001). Taking personal control over one's dying process is a trend that is occurring even in cultures like Japan that traditionally defer to physician's opinions (Hayashi et al., 2000). The emergence of final scenarios as an important consideration fits well with the emphasis on addressing pain through palliative care, an approach underlying hospice.

final scenario
making one's choices known about how they do and do not want their lives to end

The Hospice Option

As we have seen, most people would like to die at home among family and friends. An important barrier to this choice is the availability of support systems when the person has a terminal disease. Most people believe that they have no choice but to go

hospice
an approach to assisting dying people that emphasizes pain management, or palliative care, and death with dignity

palliative care
care that is focused on providing relief from pain and other symptoms of disease at any point during the disease process

to a hospital or nursing home. However, another alternative exists. **Hospice** *is an approach to assisting dying people that emphasizes pain management, or palliative care, and death with dignity* (Knee, 2010; Russo, 2008). The emphasis in a hospice is on the dying person's quality of life. This approach grows out of an important distinction between the prolongation of life and the prolongation of death, a distinction that is important to Jean, the woman we met in the vignette. In a hospice the concern is to make the person as peaceful and comfortable as possible, not to delay an inevitable death. Although medical care is available at a hospice, it is aimed primarily at controlling pain and restoring normal functioning. *The approach to care in hospice is called* **palliative care** *and is focused on providing relief from pain and other symptoms of disease at any point during the disease process* (Reville, 2011).

Modern hospices are modeled after St. Christopher's Hospice in England, founded in 1967 by Dr. Cicely Saunders. Hospice services are requested only after the person or physician believes that no treatment or cure is possible, making the hospice program markedly different from hospital or home care. The differences are evident in the principles that underlie hospice care: Clients and their families are viewed as a unit, clients should be kept free of pain, emotional and social impoverishment must be minimal, clients must be encouraged to maintain competencies, conflict resolution and fulfillment of realistic desires must be assisted, clients must be free to begin or end relationships, an interdisciplinary team approach is used, and staff members must seek to alleviate pain and fear (Knee, 2010).

Two types of hospices exist: inpatient and outpatient. Inpatient hospices provide all care for clients; outpatient hospices provide services to clients who remain in their own homes. The outpatient variation, in which a hospice nurse visits clients in their home, is becoming increasingly popular, largely because more clients can be served at a lower cost. Having hospice services available to people at home is a viable option for many more people, especially in helping home-based caregivers cope with loss (Grande et al., 2004).

Hospices do not follow a hospital model of care. The role of the staff in a hospice is not so much to treat the client as it is just to be with the client. A client's dignity is always maintained; often more attention is paid to appearance and personal grooming than to medical tests. Hospice staff members also provide a great deal of support to the client's family. The Real People feature provides some insight into one family's experience with a hospice.

Hospice outpatient health care workers provide help for people with terminal diseases who choose to die at home.

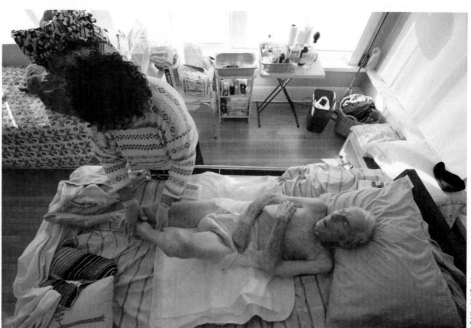

The news from Roseanne's oncologist had been expected, but it still came as a shock to her and her husband, Harry, and two daughters: her cancer had spread and she had only about 3 months to live. At first, the family didn't know what to do. But her oncologist made a suggestion that turned out to be extremely helpful: Roseanne and her family should contact the local hospice for help.

The hospice in Roseanne's city was typical. The nurse who visited the house, along with an entire backup team, were the most caring people the family had ever met. Far from being in a rush to complete the workup, the nurse spent a great deal of time asking Roseanne about her pain and how she wanted to manage it, her wishes and desires about her pro-

cess of dying, and many other personal topics. This approach made Roseanne and her family feel much more at ease. They knew that their feelings mattered and that Roseanne would be well cared for.

As Roseanne's condition deteriorated, the home hospice nurse made sure that her pain medication was adequate to provide physical comfort. Counselors worked with the family to help them discuss their feelings about Roseanne's pending death, and with Roseanne to help her prepare to die. They also explained to the family that they should call the hospice first if Roseanne died when no one from the hospice was present. This would ensure that Roseanne's wishes concerning life support and resuscitation would be honored.

Two and a half months after contacting the hospice, Roseanne died at home—surrounded by her family and the hospice nurse, just as she wanted. Because her pain was well managed, she was comfortable even at the end. Harry's and Roseanne's daughters' grief was made easier through the constant support they received and the counseling they needed.

The hospice staff who worked with Roseanne and her family have the client's physical comfort as their primary goal, followed by supporting the family. The way they helped Roseanne die, and Harry and their daughters to grieve with support, made a very difficult process a bit easier.

Hospice and hospital patients differ in important ways (Knee, 2010). Hospice clients are more mobile, less anxious, and less depressed; spouses visit hospice clients more often and participate more in their care; and hospice staff members are perceived as more accessible. Significant improvements in clients' quality of life have been documented after hospice placement (Cohen et al., 2001).

Although the hospice is a valuable alternative for many people, it may not be appropriate for everyone. Those who trust their physician regarding medical care options are more likely to select hospice than those who do not trust their physician, especially among African Americans (Ludke & Smucker, 2007). Most people who select hospice are suffering from cancer, AIDS, cardiovascular disease, pulmonary disease, or a progressive neurological condition; two-thirds are over age 65; and most are in the last 6 months of life (Hospice Foundation of America, 2011a).

Needs expressed by hospice staff, family, and clients differ (Hiatt et al., 2007). Staff and family members tend to emphasize pain management, whereas many clients want more attention paid to personal issues, such as spirituality and the process of dying. This difference means that the staff and family members may need to ask clients more often what they need instead of making assumptions about what they need.

How do people decide to explore the hospice option? Families need to consider several things (Hospice Foundation of America, 2011b; Kastenbaum, 1999):

- *Is the person completely informed about the nature and prognosis of his or her condition?* Full knowledge and the ability to communicate with health care personnel are essential to understanding what hospice has to offer.

- *What options are available at this point in the progress of the person's disease?* Knowing about all available treatment options is critical. Exploring treatment options also requires health care professionals to be aware of the latest approaches and be willing to disclose them.

- *What are the person's expectations, fears, and hopes?* Some older adults, like Jean, remember or have heard stories about people who suffered greatly at the end of their lives. This can produce anxiety about one's own death. Similarly, fears of becoming dependent play an important role in a person's decision making. Discovering and discussing these anxieties helps clarify options.

- How well do people in the person's social network communicate with each other? Talking about death is taboo in many families (Book, 1996). In others, intergenerational communication is difficult or impossible. Even in families with good communication, the pending death of a loved relative is difficult. As a result, the dying person may have difficulty expressing his or her wishes. The decision to explore the hospice option is best made when it is discussed openly.

- Are family members available to participate actively in terminal care? Hospice relies on family members to provide much of the care, which is supplemented by professionals and volunteers. We saw in Chapter 13 that being a primary caregiver can be highly stressful. Having a family member who is willing to accept this responsibility is essential for the hospice option to work.

- Is a high-quality hospice care program available? Hospice programs are not uniformly good. As with any health care provider, patients and family members must investigate the quality of local hospice programs before making a choice. The Hospice Foundation of America (www.hospicefoundation.org) provides excellent material for evaluating a hospice.

- Is hospice covered by insurance? Hospice services are reimbursable under Medicare in most cases, but any additional expenses may or may not be covered under other forms of insurance (Knee, 2010).

THINK ABOUT IT

How might the availability of hospices relate to physician-assisted suicide?

Hospice provides an important end-of-life option for many terminally ill people and their families. Moreover, the supportive follow-up services they provide are often used by surviving family and friends. Most important, the success of the hospice option has had important influences on traditional health care. For example, the American Academy of Pain Medicine (2009) published an official position paper advocating the use of medical and behavioral interventions to provide pain management.

Despite the importance of the hospice option for end-of-life decisions, terminally ill persons face the barriers of family reluctance to face the reality of terminal illness and participate in the decision-making process and health care providers hindering access to hospice care (Knee, 2010; Melhado & Byers, 2011; Reville, 2011).

As the end of life approaches, the most important thing to keep in mind is that the dying person has the right to state-of-the-art approaches to treatment and pain management. Irrespective of the choice of traditional health care or hospice, the wishes of the dying person should be honored, and family members must participate.

Test Yourself

RECALL

1. _____ are most likely to face the death of their parents.

2. A _____ approach to dying acknowledges individual differences and rejects broad generalizations.

3. The primary framework for studying death anxiety is _____.

4. Making choices known about how people do and do not want their lives to end constitutes a _____.

5. _____ is an approach to assisting dying people that emphasizes pain management, or palliative care, and death with dignity.

INTERPRET

Why is there a difference in treatment approach between hospitals and hospices?

APPLY

Using Erikson's theory as a framework, explain how death anxiety changes from adolescence to late life.

Recall answers: (1) Middle-aged adults, (2) holistic, (3) terror management theory, (4) final scenario, (5) Hospice

LEARNING OBJECTIVES

- How do people experience the grief process?
- What feelings do grieving people have?
- What is the difference between normal and complicated or prolonged grief disorder?

After 67 years of marriage, Bertha recently lost her husband. At 90, Bertha knew that neither she nor her husband was likely to live much longer, but the death was a shock just the same. Bertha thinks about him much of the time and often finds herself making decisions on the basis of "what John would have done" in the same situation.

EACH OF US SUFFERS MANY LOSSES OVER A LIFETIME. Whenever we lose someone close to us through death or other separation, like Bertha we experience bereavement, grief, and mourning. **Bereavement** *is the state or condition caused by loss through death.* **Grief** *is the sorrow, hurt, anger, guilt, confusion, and other feelings that arise after suffering a loss.* **Mourning** *concerns the ways in which we express our grief.* For example, you can tell that people in some cultures are bereaved and in mourning because of the clothing they wear. Mourning is highly influenced by culture. For some, mourning may involve wearing black, attending funerals, and observing an official period of grief; for others, it means drinking, wearing white, and marrying the deceased spouse's sibling. Grief corresponds to the emotional reactions following loss, whereas mourning is the culturally approved behavioral manifestations of those feelings. Even though mourning rituals may be fairly standard within a culture, how people grieve varies, as we see next. We will also see how Bertha's reactions are fairly typical of most people.

bereavement
the state or condition caused by loss through death

grief
the sorrow, hurt, anger, guilt, confusion, and other feelings that arise after suffering a loss

mourning
the ways in which we express our grief

The Grief Process

How do people grieve? What do they experience? Perhaps you already have a good idea about the answers to these questions from your own experience. If so, you already know that the process of grieving is a complicated and personal one. Just as there is no right way to die, there is no right way to grieve. Recognizing that there are plenty of individual differences, we consider these patterns in this section.

The grieving process is often described as reflecting many themes and issues that people confront (Kübler-Ross & Kessler, 2005). Like the process of dying, grieving does not have clearly demarcated stages through which we pass in a neat sequence, although there are certain issues people must face that are similar to those faced by dying people. When someone close to us dies, we must reorganize our lives, establish new patterns of behavior, and redefine relationships with family and friends. Indeed, Attig (1996) provided one of the best descriptions of grief when he wrote that grief is the process by which we relearn the world.

Unlike bereavement, over which we have no control, grief is a process that involves choices in coping, from confronting the reality and emotions to using religion to ease one's pain (Ivancovich & Wong, 2008). From this perspective, grief is an active process in which a person must do several things (Worden, 1991):

- *Acknowledge the reality of the loss.* We must overcome the temptation to deny the reality of our loss; we must fully and openly acknowledge it and realize that it affects every aspect of our life.
- *Work through the emotional turmoil.* We must find effective ways to confront and express the complete range of emotions we feel after the loss and must not avoid or repress them.

Going through the personal effects of a deceased loved one can be a difficult process for survivors.

- *Adjust to the environment where the deceased is absent.* We must define new patterns of living that adjust appropriately and meaningfully to the fact that the deceased is not present.
- *Loosen ties to the deceased.* We must free ourselves from the bonds of the deceased in order to reengage with our social network. This means finding effective ways to say good-bye.

The notion that grief is an active coping process emphasizes that survivors must come to terms with the physical world of things, places, and events as well as our spiritual place in the world; the interpersonal world of interactions with family and friends, the dead, and, in some cases, God; and aspects of our inner selves and our personal experiences (Ivancovich & Wong, 2008; Papa & Litz, 2011). Bertha, the woman in the vignette, is in the middle of this process. Even the matter of deciding what to do with the deceased's personal effects can be part of this active coping process (Attig, 1996).

In considering the grief process, we must avoid making several mistakes. First, grieving is a highly individual experience (Mallon, 2008; Papa & Litz, 2011). A process that works well for one person may not be the best for someone else. Second, we must not underestimate the amount of time people need to deal with the various issues. To a casual observer, it may appear that a survivor is "back to normal" after a few weeks. Actually, it takes much longer to resolve the complex emotional issues that are faced during bereavement (Mallon, 2008; Papa & Litz, 2011). Researchers and therapists alike agree that a person needs at least a year following the loss to begin recovery, and 2 years is not uncommon. Finally, "recovery" may be a misleading term. It is probably more accurate to say that we learn to live with our loss rather than that we recover from it (Attig, 1996). The impact of the loss of a loved one lasts a very long time, perhaps for the rest of one's life. Still, most people reach a point of moving on with their lives in a reasonable time frame (Bonanno, 2009; Mancini & Bonnano, 2010).

Recognizing these aspects of grief makes it easier to know what to say and do for bereaved people. Among the most useful things are to simply let the person know that you are sorry for his or her loss, that you are there for support, and mean what you say.

Risk Factors in Grief

Bereavement is a life experience that most people have many times, and most people eventually handle it, often better than we might suspect (Bonanno, 2009). However, there are some risk factors that may make bereavement much more difficult. Several of the more important are the mode of death, personal factors (e.g., personality, religiosity, age, gender), income, and interpersonal context (social support, kinship relationship) (Kersting et al., in press; W. Stroebe & Schut, 2001).

Most people believe that the circumstances or mode of death affects the grief process. A person whose family member was killed in an automobile accident has a different situation to deal with than a person whose family member died after a long period of suffering with Alzheimer's disease. *It is believed that when death is anticipated, people go through a period of* **anticipatory grief** *before the death that supposedly serves to buffer the impact of the loss when it does come and to facilitate recovery* (Lane, 2007). However, the research evidence for this is mixed. Anticipating the loss of a loved one from cancer or another terminal disease can provide a framework for understanding family members' reactions (Coombs, 2010). Not all family members actually experience it, though. However, people who do experience anticipatory grief tend to disengage from the dying person (Lane, 2007).

The strength of attachment to the deceased person does make a difference in dealing with a sudden as opposed to an unexpected death. Attachment theory provides a framework for understanding different reactions (Stroebe, Schut, & Boerner, 2010). When the deceased person was one with whom the survivor had a strong and close attachment and the loss was sudden, greater grief is experienced (Wayment & Vierthaler, 2002). However, such secure attachment styles tend to result in less depression after the loss due to less guilt over unresolved issues (because there are fewer of them), things not provided (because more were likely provided), and so on.

anticipatory grief
grief that is experienced during the period before an expected death occurs that supposedly serves to buffer the impact of the loss when it does come and to facilitate recovery

Few studies of personal risk factors have been done, and few firm conclusions can be drawn. To date there are no consistent findings regarding personality traits that either help buffer people from the effects of bereavement or exacerbate them (Lane, 2007; Stroebe et al., 2010). There is some evidence to suggest that church attendance or spirituality in general helps people deal with bereavement and subsequent grief through the post-grief period (Bratkovich, 2010; Ivancovich & Wong, 2008). There are, however, consistent findings regarding gender. Men have higher mortality rates following bereavement than women, who have higher rates of depression and complicated grief (discussed later in this section) than men, but the reasons for these differences are unclear (Kersting et al., in press). Research also consistently shows that older adults suffer the fewest health consequences following bereavement, with the impact perhaps being strongest for middle-aged adults, but strong social support networks lessen these effects to varying degrees (Papa & Litz, 2011).

Two interpersonal risk factors have been examined: lack of social support and kinship. Studies indicate that social support and mastery help buffer the effects of bereavement more for older adults than for middle-aged adults (Onrust et al., 2007; Papa & Litz, 2011). The type of kinship relationship involved in the loss matters a great deal. Research for decades consistently shows that the loss of a child is the most difficult, followed by loss of a spouse or partner and of a parent (Leahy, 1993; Nolen-Hoeksema & Larson, 1999).

THINK ABOUT IT

How are risk factors in grief influenced by sociocultural factors?

Normal Grief Reactions

The feelings experienced during grieving are intense, which not only makes it difficult to cope but can also make a person question her or his own reactions. The feelings involved usually include sadness, denial, anger, loneliness, and guilt. A summary of these feelings is presented in the following list (Vickio, Cavanaugh, & Attig, 1990). Take a minute to read through them to see whether they agree with what you expected.

Disbelief	Denial	Shock
Sadness	Anger	Hatred
Guilt	Fear	Anxiety
Confusion	Helplessness	Emptiness
Loneliness	Acceptance	Relief
Happiness	Lack of enthusiasm	Absence of emotion

Many authors refer to the psychological side of coming to terms with bereavement as **grief work**. Whether the loss is ambiguous and lacking closure (e.g., waiting to learn the fate of a missing loved one) or certain (e.g., verification of death through a dead body), people need space and time in which to grieve (Boss, 2006; Papa & Litz, 2011). Even without personal experience of the death of close family members, people recognize the need to give survivors time to deal with their many feelings. One study asked college students to describe the feelings they thought were typically experienced by a person who had lost particular loved ones (such as a parent, child, sibling, or friend). The students were well aware of the need for grief work, recognized the need for at least a year to do it, and were sensitive to the range of emotions and behaviors demonstrated by the bereaved (Vickio et al., 1990).

Muller and Thompson (2003) examined people's experience of grief in a detailed interview study and found five themes. *Coping* concerns what people do to deal with their loss in terms of what helps them. *Affect* refers to people's emotional reactions to the death of their loved one; for example, most people have certain topics that serve as emotional triggers for memories of their loved one. *Change* involves the ways in which survivors' lives change as a result of the loss; personal growth (e.g., "I didn't think I could deal with something that painful, but I did") is a common experience. *Narrative* relates to the stories survivors tell about their

grief work
the psychological side of coming to terms with bereavement

How openly grief is expressed varies considerably across cultures.

anniversary reaction
changes in behavior related to feelings of sadness on the anniversary date of a loss

deceased loved one, which sometimes includes details about the process of the death. Finally, *relationship* reflects who the deceased person was and the nature of the ties between that person and the survivor. Collectively, these themes indicate that the experience of grief is complex and involves dealing with one's feelings as a survivor as well as memories of the deceased person.

How people show their feelings of grief varies across ethnic groups (Papa & Litz, 2011). For example, Latino American men show more of their grief behaviorally than do European American men (Sera, 2001). Such differences also are found across cultures. For example, families in KwaZulu-Natal, South Africa, have a strong desire for closure and need for dealing with the "loneliness of grief" (Brysiewicz, 2008). In many cultures the bereaved construct a relationship with the person who died, but how this happens differs widely, from "ghosts" to appearances in dreams to connection through prayer (Rosenblatt, 2001).

In addition to psychological grief reactions, there are also physiological ones. Widows report sleep disturbances as well as neurological and circulatory problems (Kowalski & Bondmass, 2008). Physical health may decline, illness may result, and use of health care services may increase (Stroebe, Schut, & Stroebe, 2007). In some cases it is necessary to treat severe depression following bereavement; research indicates that using SSRIs (see Chapter 14) is one effective medical intervention (Simon et al., 2007).

In the time following the death of a loved one, dates that have personal significance may reintroduce feelings of grief. For example, holidays such as Thanksgiving or birthdays that were spent with the deceased person may be difficult times. The actual anniversary of the death can be especially troublesome. *The term* **anniversary reaction** *refers to changes in behavior related to feelings of sadness on this date.* Personal experience and research show that recurring feelings of sadness or other examples of the anniversary reaction are common in normal grief (Holland & Neimeyer, 2010). Such feelings also accompany remembrances of major catastrophes across cultures, such as Thais remembering the victims of a major flood (Assanangkornchai et al., 2007).

Most research on how people react to the death of a loved one is cross-sectional. This work shows that grief tends to peak within the first 6 months following the death of a loved one (Maciejewski et al., 2007). However, some work has been done to examine how people continue grieving many years after the loss. Some widows show no sign of lessening of grief after 5 years (Kowalski & Bondmass, 2008). Rosenblatt (1996) reported that people still felt the effects of the deaths of family members 50 years after the event. The depth of the emotions over the loss of loved ones never totally went away, as people still cried and felt sad when discussing the loss despite the length of time that had passed. In general, though, people move on with their lives within a relatively short period of time and deal with their feelings reasonably well (Bonanno, 2009).

Coping With Grief

Thus far, we have considered the behaviors people show when they are dealing with grief. We have also seen that these behaviors change over time. How does this happen? How can we explain the grieving process?

Numerous theories have been proposed to account for the grieving process, such as general life-event theories, psychodynamic theories, attachment theories, and cognitive process theories (Stroebe et al., 2010). All of these approaches to grief are based on more general theories, which results in none of them providing an adequate explanation of the grieving process. Two integrative approaches have been proposed that are specific to the grief process: the four-component model and the dual-process model of coping with bereavement.

The Four-Component Model

The **four-component model** *proposes that understanding grief is based on four things*: (1) *the context of the loss*, referring to the risk factors such as whether the death was expected; (2) *continuation of subjective meaning associated with loss*, ranging from evaluations of everyday concerns to major questions about the meaning of life; (3) *changing representations of the lost relationship over time*; and (4) *the role of coping and emotion regulation processes* that cover all coping strategies used to deal with grief (Bonanno, 2009; Bonanno & Kaltman, 1999). The four-component model relies heavily on emotion theory, has much in common with the transactional model of stress, and has empirical support. According to the four-component model, dealing with grief is a complicated process that can only be understood as a complex outcome that unfolds over time.

There are several important implications of this integrative approach. One of the most important is that helping a grieving person involves helping them make meaning from the loss (Bratkovich, 2010; Wong, 2008). Second, this model implies that encouraging people to express their grief may actually not be helpful. *An alternative view, called the* **grief work as rumination hypothesis**, *not only rejects the necessity of grief processing for recovery from loss but views extensive grief processing as a form of rumination that may actually increase distress* (Bonanno, Papa, & O'Neill, 2001). Although it may seem that people who think obsessively about their loss or who ruminate about it are confronting the loss, rumination is actually considered a form of avoidance because the person is not dealing with his or her real feelings and moving on (Stroebe et al., 2007).

One prospective study has shown, for instance, that bereaved individuals who were not depressed prior to their spouse's death but then evidenced chronically elevated depression through the first year and a half of bereavement (i.e., a chronic grief pattern) had also tended to report more frequently thinking about and talking about their recent loss at the 6-month point in bereavement (Bonanno, Wortman, & Neese, 2004). Thus, some bereaved individuals engage in minimal grief processing, whereas others are predisposed toward more extensive grief processing. Furthermore, the individuals who engage in minimal grief processing will show a relatively favorable grief outcome, whereas those who are predisposed toward more extensive grief processing will tend toward ruminative preoccupation and, consequently, toward a more prolonged grief course (Bonanno, 2009; Bonanno et al., 2001; Nolen-Hoeksema, 2001).

As noted earlier, the grief work as rumination hypothesis also views grief avoidance as an independent but maladaptive form of coping with loss (Stroebe et al., 2007). In contrast to the traditional perspective, which equates the absence of grief processing with grief avoidance, the grief work as rumination framework assumes that resilient individuals are able to minimize processing of a loss through relatively automated processes, such as distraction or shifting attention toward more positive emotional experiences (Bonanno, 2009). The grief work as rumination framework argues that the deliberate avoidance or suppression of grief represents a less effective form of coping (Wegner & Gold, 1995) that tends to exacerbate rather than minimize the experience of grief (Bonanno, 2009).

The Spotlight on Research feature explores grief work regarding the loss of a spouse and the loss of a child in two cultures, the United States and China. As you read it, pay special attention to the question of whether encouraging people to express and deal with their grief is necessarily a good idea.

four-component model
model for understanding grief that is based on (1) the context of the loss; (2) continuation of subjective meaning associated with loss; (3) changing representations of the lost relationship over time; and (4) the role of coping and emotion-regulation processes

grief work as rumination hypothesis
an approach that not only rejects the necessity of grief processing for recovery from loss but views extensive grief processing as a form of rumination that may actually increase distress

Spotlight on research — Grief Processing and Avoidance in the United States and China

Who were the investigators and what was the aim of the study? Bonanno and colleagues (2005) noted that grief following the loss of a loved one often tends to be denied. However, research evidence related to positive benefits of resolving grief is largely lacking. Thus, whether unresolved grief is "bad" remains an open issue. Likewise, cross-cultural evidence is also lacking.

How did the investigators measure the topic of interest? Collaborative meetings between U.S. and Chinese researchers resulted in a 13-item grief processing scale and a 7-item grief avoidance scale, with both English and Mandarin Chinese versions. Self-reported psychological symptoms and physical health were also collected.

[continued]

Figure 16.4

Grief processing and deliberate grief avoidance across time in the People's Republic of China and the United States.

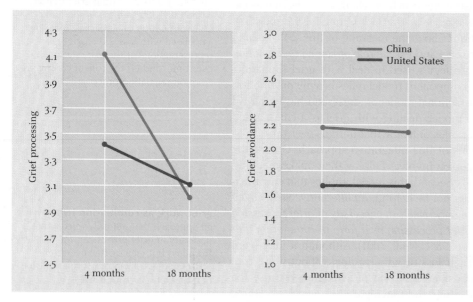

Bonanno, G. A., Papa, A., Lalande, K., Zhang, N., & Noll, J. G. (2005). Grief processing and deliberate grief avoidance: A prospective comparison of bereaved spouses and parents in the United States and the People's Republic of China. *Journal of Consulting and Clinical Psychology, 73*, 86-98.

Who were the participants in the study? Adults under age 66 who had experienced the loss of either a spouse or child approximately 4 months prior to the start of data collection were asked to participate through solicitation letters. Participants were from either the metropolitan areas of Washington, D.C., or Nanjing, Jiangsu province in China.

What was the design of the study? Two sets of measures were collected at approximately 4 months and 18 months after the loss.

Were there ethical concerns in the study? Because participation was voluntary, there were no ethical concerns.

What were the results? Consistent with the grief work as rumination view, scores on the two grief measures were uncorrelated. Overall, women tended to show more grief processing than men, and grief processing decreased over time. As you can see in ■ Figure 16.4, Chinese participants reported more grief processing and grief avoidance than U.S. participants at the first time of measurement, but differences disappeared by the second measurement for grief processing.

What did the investigators conclude? Based on converging results from the United States and China, the researchers concluded that the data supported the grief work as rumination view. The results support the notion that excessive processing of grief may actually increase a bereaved person's stress and feelings of discomfort rather than being helpful. These findings contradict the idea that people should be encouraged to work through their grief and that doing so will always be helpful.

What converging evidence would strengthen these conclusions? Although the data were collected in two cities in two countries, additional areas (e.g., rural and urban) and more cross-cultural data would be helpful. Also, the sample was limited to people under age 66 and to those who had recently experienced the loss of either a spouse or child. Older adults and people experiencing different types of loss (parent, partner, sibling, or friend) would provide a richer data set.

Go to Psychology CourseMate at www.cengagebrain.com to enhance your understanding of this research.

The Dual Process Model

dual process model (DPM)
view of coping with bereavement that integrates loss-oriented stressors and restoration-oriented stressors

The **dual process model (DPM)** *of coping with bereavement integrates existing ideas regarding stressors* (M. Stroebe & Schut, 2001; Stroebe et al., 2010). As shown in ■ Figure 16.5, the DPM defines two broad types of stressors. *Loss-oriented stressors* concern the loss itself, such as the grief work that needs to be done. *Restoration-oriented stressors* are those that involve adapting to the survivor's new life situation, such as building new relationships and finding new activities. The DPM proposes that dealing with these stressors is a dynamic process, as indicated by the lines connecting them in the figure. This is a distinguishing feature of DPM. It shows how bereaved people cycle back and forth between dealing mostly with grief and trying to move on with life. At times the emphasis will be on grief; at other times on moving forward.

The DPM captures well the process that bereaved people themselves report—at times they are nearly overcome with grief, while at other times they handle life well. The DPM also helps us understand how, over time, people come to a balance between the long-term effects of bereavement and the need to live life. Understanding how people handle grief requires understanding of the various contexts in which people live and interact with others (Sandler, Wolchik, & Ayers, 2008).

Complicated or Prolonged Grief Disorder

Not everyone is able to cope with grief well and begin rebuilding a life. Sometimes the feelings of hurt, loneliness, and guilt are so overwhelming that they become the focus of the survivor's life to such an extent that there is never any closure and the

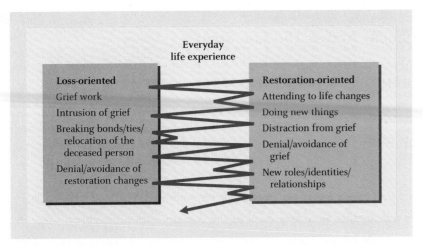

Stroebe, M. S., & Schut, H. (2001). Models of coping with bereavement: A review. In M. S. Stroebe, R. O., Hansson, W. Stroeve, H. Schut (Eds.), *Handbook of bereavement research: Consequences, coping, and care* (pp. 375-403). Washington, DC: American Psychological Assocation.

Figure 16.5
The dual process model of coping with bereavement shows the relation between dealing with the stresses of the loss itself (loss-oriented) and moving on with one's life (restoration-oriented).

grief continues to interfere indefinitely with one's ability to function. *When this occurs, individuals are viewed as having* **complicated or prolonged grief disorder**, *which is distinguished from depression and from normal grief in terms of separation distress and traumatic distress* (Boelen & Prigerson, 2007; Kersting et al., 2011; Shear et al., 2011). *Symptoms of* **separation distress** *include preoccupation with the deceased to the point that it interferes with everyday functioning, upsetting memories of the deceased, longing and searching for the deceased, and isolation following the loss. Symptoms of* **traumatic distress** *include disbelief about the death; mistrust, anger, and detachment from others as a result of the death; feeling shocked by the death; and the experience of physical presence of the deceased.*

Complicated grief forms a separate set of symptoms from depression (Bonanno, 2009; Kersting et al., 2011; Papa & Litz, 2011; Shear et al., 2011). Individuals experiencing complicated grief report high levels of separation distress (such as yearning, pining, or longing for the deceased person), along with specific cognitive, emotional, or behavioral indicators (such as avoiding reminders of the deceased, diminished sense of self, difficulty in accepting the loss, feeling bitter or angry), as well as increased morbidity, increased smoking and substance abuse, and difficulties with family and other social relationships. Similar distinctions have been made between complicated or prolonged grief disorder and anxiety disorders.

complicated or prolonged grief disorder
expression of grief which is distinguished from depression and from normal grief in terms of separation distress and traumatic distress

separation distress
expression of complicated or prolonged grief disorder that includes preoccupation with the deceased to the point that it interferes with everyday functioning, upsetting memories of the deceased, longing and searching for the deceased, and isolation following the loss

traumatic distress
expression of complicated or prolonged grief disorder that includes disbelief about the death; mistrust, anger, and detachment from others as a result of the death; feeling shocked by the death; and the experience of physical presence of the deceased

Test Yourself

RECALL

1. Feeling sad on the date when your grandmother died the previous year is an example of an _____.

2. Compared to other age groups, _____ show the most negative effects following bereavement.

3. Separation distress and _____ are two characteristics of complicated or prolonged grief disorder.

INTERPRET

What connections might there be between bereavement and stress?

APPLY

If you were to create a brochure listing the five most important things to do and not to do in reacting to someone who just lost a close family member or friend through death, what would you include? Why?

Recall answers: (1) anniversary reaction, (2) middle-aged adults, (3) traumatic distress

- What do children understand about death? How should adults help them deal with it?

- How do adolescents deal with death?

- How do adults deal with death? What are the special issues they face concerning the death of a child or parent?

- How do older adults face the loss of a child, grandchild, or partner?

Donna and Carl have a 6-year-old daughter, Jennie, whose grandmother just died. Jennie and her grandmother were very close, as the two saw each other almost every day. Other adults have told her parents not to take Jennie to the funeral. Donna and Carl aren't sure what to do. They wonder whether Jennie will understand what happened to her grandmother, and they worry about how she will react.

COMING TO GRIPS WITH THE REALITY OF DEATH IS PROBABLY ONE OF THE HARDEST THINGS WE HAVE TO DO IN LIFE. American society does not help much either, as it tends to distance itself from death through euphemisms, such as "passed away" or "dearly departed," and by eliminating many rituals from the home (for example, viewings no longer take place there, no more official mourning visits to the bereaved's house, and so on).

These trends make it difficult for people like Donna, Carl, and Jennie to learn about death in its natural context. Dying itself has been moved from the home to hospitals and other institutions such as nursing homes. The closest most people get to death is a quick glance inside a nicely lined casket at a corpse that has been made to look as if the person were still alive.

What do people, especially children like Jennie, understand about death? How do Donna and Carl feel? How do the friends of Jennie's grandmother feel? In this section, we consider how our understanding of death changes throughout the life span.

Childhood

Parents often take their children to funerals of relatives and close friends. But many adults, like Donna and Carl in the vignette, wonder whether young children really know what death means. Children's understanding of death changes with their development (Webb, 2010a). Preschoolers tend to believe that death is temporary and magical, something dramatic that comes to get you in the middle of the night like a burglar or a ghost. Not until children are 5 to 7 years of age do they realize that death is permanent, that it eventually happens to everyone, and that dead people no longer have any biological functions.

Why does this shift occur? There are three major areas of developmental change in children that affect their understanding of death and grief (Oltjenbruns, 2001; Webb, 2010a): cognitive-language ability, psychosocial development, and coping skills. In terms of cognitive-language ability, think back to Chapters 4 and 6, especially to the discussion of Piaget's theory of cognitive development. Take Jennie, the 6-year-old daughter of Donna and Carl in the vignette. Where would she be in Piaget's terms? In this perspective, the ages 5 to 7 include the transition from preoperational to concrete-operational thinking. Concrete-operational thinking permits children to know that death is final and permanent. Therefore, Jennie is likely to understand what happened to her grandmother. With this more mature understanding of death comes a lower fear of death, too (Slaughter & Griffiths, 2007).

Children's expressions of grief at the loss of a loved one vary with age, too (Webb, 2010b). Several common manifestations of grief among children are shown in ■ Fig-

Children show their grief in many ways, including physiological (somatic), emotional (intrapsychic), and behavioral ones.

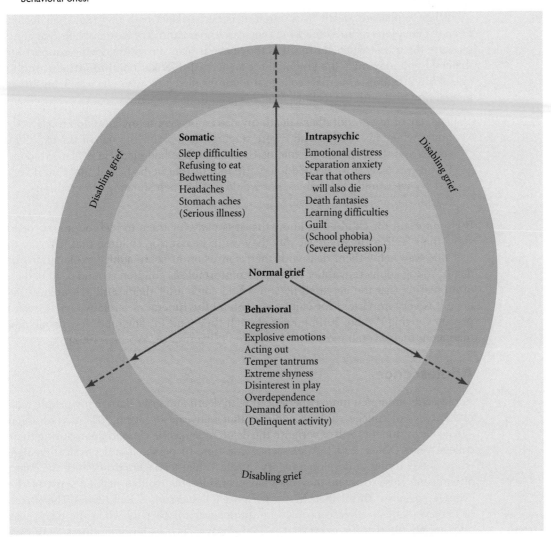

Somatic
Sleep difficulties
Refusing to eat
Bedwetting
Headaches
Stomach aches
(Serious illness)

Intrapsychic
Emotional distress
Separation anxiety
Fear that others
   will also die
Death fantasies
Learning difficulties
Guilt
(School phobia)
(Severe depression)

Normal grief

Behavioral
Regression
Explosive emotions
Acting out
Temper tantrums
Extreme shyness
Disinterest in play
Overdependence
Demand for attention
(Delinquent activity)

Disabling grief

From Oltjenbruns, K. A. (2001). "Developmental context of childhood: Grief and regrief phenomena." *Handbook of Bereavement Research: Consequences, Coping, and Care* edited by M. S. Stroebe, R. O. Hansson, W. Stroebe, & H. Schut, Fig. 8-1, p. 177. Copyright © 2001 by the American Psychological Association.

ure 16.6. Typical reactions in early childhood include regression, guilt for causing the death, denial, displacement, repression, and wishful thinking that the deceased will return. In later childhood, common behaviors include problems at school, anger, and physical ailments. As children mature, they acquire more coping skills that permit a shift to problem-focused coping, which provides a better sense of personal control. Children will often flip between grief and normal activity, a pattern they may learn from adults (Stroebe & Schut, 1999). Sensitivity to these feelings and how they get expressed is essential so the child can understand what happened and that he or she did not cause the death.

Research shows that bereavement per se during childhood typically does not have long-lasting effects, such as depression (Oltjenbruns & Balk, 2007; Webb, 2010b). Problems are more likely to occur if the child does not receive adequate care and attention following the death.

Understanding death can be particularly difficult for children when adults are not open and honest with them, especially about the meaning of death (Buchsbaum, 1996). The use of euphemisms, such as "Grandma has gone away" or "Mommy is only sleeping," is unwise. Young children do not understand the deeper level of meaning in such statements and are likely to take them literally (Attig, 1996; Silverman & Nickman, 1996).

When explaining death to children, it is best to deal with them on their terms. Keep explanations simple, at a level they can understand. Try to allay their fears and reassure them that whatever reaction they have is okay. Providing loving support for the child will maximize the potential for a successful (albeit painful) introduction to one of life's realities. One male college student recalled how, when he was 9, his father helped him deal with his feelings after his grandfather's death:

> The day of my grandfather's death my dad came over to my aunt and uncle's house where my brother and I were staying. He took us into one of the bedrooms and sat us down. He told us Grandaddy Doc had died. He explained to us that it was okay if we needed to cry. He told us that he had cried, and that if we did cry we wouldn't be babies, but would just be men showing our emotions. (Dickinson, 1992, pp. 175–176)

It is important for children to know that it is okay for them to feel sad, to cry, or to show their feelings in whatever way they want. Reassuring children that it's okay to feel this way helps them deal with their confusion at some adults' explanations of death. Young adults remember feeling uncomfortable as children around dead bodies, often fearing that the deceased person would come after them. Still, researchers believe it is very important for children to attend the funeral of a relative or to have a private viewing (Webb, 2010b). Even though they tend to remember few details immediately, their overall recovery is enhanced (Silverman & Worden, 1992).

Adolescence

Adolescents are much more experienced with death and grief than many people realize. Surveys of college students indicate that between 40% and 70% of traditional aged college students will experience the death of someone close to them during their college years (Knox, 2007). Adolescence is a time of personal and physical change, when one is trying to develop a theory of self. When teenagers experience the death of someone close to them, they may have considerable trouble making sense of the event, especially if this is their first experience (Oltjenbruns & Balk, 2007). The effects of bereavement in adolescence can be quite severe, especially when the death was unexpected, and can be expressed in many ways, such as chronic illness, enduring guilt, low self-esteem, poorer performance in school and on the job, substance abuse, problems in interpersonal relationships, and suicidal thinking (Malone, 2010; Morgan & Roberts, 2010).

In reaction to the loss of a sibling, younger adolescents are particularly reluctant to discuss their grief, mainly because they do not want to appear different from their peers (Fleming & Balmer, 1996). This reluctance leaves them particularly vulnerable to psychosomatic symptoms such as headaches and stomach pains that signal underlying problems.

Adolescents often do not demonstrate a clear end point to their grief over the loss of a sibling or parent (Hogan & DeSantis, 1996; Knox, 2007). Bereaved adolescent siblings continue to miss and to love their dead loved ones and try to find ways to keep them in their lives. However, few nonbereaved peers were willing to talk with the bereaved students about their experience or even felt comfortable being with them (Balk & Corr, 2001).

However, grief does not interfere with normative developmental processes. Bereaved adolescent siblings experience continued personal growth following the death of a loved one in much the same way as adolescents who did not experience such a loss (Morgan & Roberts, 2010).

Adolescents may not always openly express their feelings after the loss of a loved one; sometimes their feelings are manifested in problems at school.

Adulthood

Because young adults are just beginning to pursue the family, career, and personal goals they have set, they tend to be more intense in their feelings toward death. When asked how they feel about death, young adults report a strong sense that those who die at this point in their lives would be cheated out of their future (Attig, 1996).

Wrenn (1999) relates that one of the challenges faced by bereaved college students is learning "how to respond to people who ignore their grief, or who tell them that they need to get on with life, that it's not good for them to continue to grieve" (p. 134). College students have a need to express their grief like other bereaved people do, so providing them the opportunity to do so is crucial (Servaty-Seib & Taub, 2010).

Experiencing the loss of one's partner in young adulthood can be traumatic, not only because of the loss itself but also because such loss is unexpected. As Trish Straine, a 32-year-old widow whose husband was killed in the World Trade Center attack, put it: "I suddenly thought, 'I'm a widow.' Then I said to myself, 'A widow? That's an older woman, who's dressed in black. It's certainly not a 32-year-old like me'" (Lieber, 2001). One of the most difficult aspects for young widows and widowers is that they must deal with both their own and their young children's grief and provide the support their children need. But that can be extremely difficult. "Every time I look at my children, I'm reminded of Mark," said Stacey, a 35-year-old widow whose husband died of bone cancer. "And people don't want to hear you say that you don't feel like moving on, even though there is great pressure from them to do that." Stacey is a good example of what research shows: Young adult widows report that their level of grief does not typically diminish significantly until 5 to 10 years after the loss, and they maintain strong attachments to their deceased husbands for at least that long (Derman, 2000). Young Canadian widows also report intense feelings and a desire to stay connected through memories (Lowe & McClement (2010–2011).

Losing one's spouse in midlife often results in the survivor challenging basic assumptions about self, relationships, and life options, especially if one is a parent of a child still at home (Glazer, Clark, Thomas, & Haxton, 2010). By the first-year anniversary of the loss, the surviving spouse has usually begun transforming his or her perspectives on these issues. The important part of this process is to make meaning of the death and to continue working on and revising it over time (Gillies & Neimeyer, 2006).

Becoming a widow as a young adult can be especially traumatic.

Death of One's Child in Young and Middle Adulthood

Many people believe that the death of one's child is the worst type of loss (Woodgate, 2006). Because children are not supposed to die before their parents, it is as if the natural order of things has been violated, shaking parents to their core (Rubin & Malkinson, 2001). Mourning is always intense, and some parents never recover or reconcile themselves to the death of their child and may terminate their relationship with each other (Rosenbaum, Smith, & Zollfrank, 2011). The intensity of feelings is due to the strong parent–child bond that begins before birth and that lasts a lifetime (Maple et al., 2010; Rosenbaum et al., 2011).

Young parents who lose a child due to Sudden Infant Death Syndrome (SIDS) report high anxiety, a more negative view of the world, and much guilt, which results in a devastating experience (Seyda & Fitzsimons, 2010). The most overlooked losses of a child are those that happen through stillbirth, miscarriage, abortion, or neonatal death (Rosenbaum et al., 2011). Attachment to the child begins before birth, especially for mothers, so the loss hurts deeply. For this reason, ritual is extremely important to acknowledge the death and validate parents' feelings of grief (Kobler, Limbo, & Kavanaugh, 2007). Yet parents who experience this type of loss are expected to recover quickly. The lived experience of parents tells a very different story (Seyda &

THINK ABOUT IT

How might young adults' thoughts about death change with each different level of reflective judgment?

Fitzsimons, 2010). These parents talk about a life-changing event, and report a deep sense of loss and hurt, especially when others do not understand their feelings. Worst of all, if societal expectations for quick recovery are not met, the parents may be subjected to unfeeling comments. As one mother notes, parents often just wish somebody would acknowledge the loss (Okonski, 1996).

The loss of a young adult child for a middle-aged parent is experienced differently but is equally devastating (Maple et al., 2010; Rubin & Malkinson, 2001). For example, parents who lost sons in wars (Rubin, 1996) and in traffic accidents (Shalev, 1999) still report strong feelings of anxiety, problems in functioning, and difficulties in relationships with both surviving siblings and the deceased as long as 13 years after the loss.

Death of One's Parent

Most parents die after their children are grown. But whenever parental death occurs, it hurts. Losing a parent in adulthood is a rite of passage as one is transformed from being a "son" or "daughter" to being "without parents" (Edwards, 2006). We, the children, are now next in line. Indeed, the death of a parent often leads the surviving children to redefine the meaning of their relationships with their siblings, children, and other family members (Moss, Moss, & Hansson, 2001).

Regardless of one's age, the loss of a parent is typically a difficult experience.

The death of a parent deprives people of many important things: a source of guidance and advice, a source of love, and a model for their own parenting style (Buchsbaum, 1996). It also cuts off the opportunity to improve aspects of their relationship with the parent. Expressing feelings toward a parent before he or she dies is important. In some cases, the death has a negative effect on the adult child's own marital relationship (Henry, 2006).

The loss of a parent is perceived as a significant one; no matter how old we are, society allows us to grieve for a reasonable length of time. For young adult women transitioning to motherhood, losing their own mother during adolescence raises many feelings, such as deep loss at not being able to share their pregnancies with their mothers and fear of dying young themselves (Franceschi, 2005). Middle-aged women who lose a parent report feeling a complex set of emotions (Westbrook, 2002): they have intense emotional feelings of both loss and freedom, they remember both positive and negative aspects of their parent, and they experience shifts in their own sense of self.

The feelings accompanying the loss of an older parent reflect a sense of letting go, loss of a buffer against death, better acceptance of one's own eventual death, and a sense of relief that the parent's suffering is over (Moss et al., 2001). Yet, if the parent died from a cause, such as Alzheimer's disease, that involves the loss of the parent–child relationship along the way, then bodily death can feel like the second time the parent died (Shaw, 2007). Whether the adult child now tries to separate from the deceased parent's expectations or finds comfort in the memories, the impact of the loss is great.

Late Adulthood

In general, older adults are less anxious about death and more accepting of it than any other age group (Kastenbaum, 1999). They may feel that their most important life tasks have been completed (Kastenbaum, 1999).

Death of One's Child or Grandchild in Late Life

The loss of a child can happen at any point over the adult life span. Older bereaved parents tend to reevaluate their grief as experienced shortly after the loss and years and decades later. Even more than 30 years after the death of a child, older adults still feel a keen sense of loss and have continued difficulty coming to terms with it (Malkinson & Bar-Tur, 2004–2005). The long-lasting effects of the loss of a child are often accompanied by a sense of guilt that the pain affected the parents' relationships with the surviving children.

The loss of a grandchild results in similar feelings: intense emotional upset, survivor guilt, regrets about the relationship with the deceased grandchild, and a need to restructure relationships with the surviving family. However, bereaved grandparents tend to control and hide their grief behavior in an attempt to shield their child (the bereaved parent) from the level of pain being felt. In cases in which older adults were the primary caregivers for grandchildren, feelings can be especially difficult. For example, custodial grandparents in South Africa whose grandchildren in their care died from AIDS go through emotionally difficult times due both to the loss and to the social stigma regarding the disease (Boon et al., 2010).

Death of One's Partner

Experiencing the loss of one's partner is the type of loss in late life we know most about. The death of a partner differs from other losses. It clearly represents a deep personal loss, especially when the couple has had a long and close relationship (Moss et al., 2001). In a very real way, when our partner dies, a part of ourself dies, too.

There is pressure from society to mourn the loss of one's partner for a period of time and then to "move on" (Jenkins, 2003). Typically, this pressure is manifested if the survivor begins to show interest in finding another partner before an "acceptable" period of mourning has passed. Although Americans no longer specify the length of the period, many feel that about a year is appropriate. The fact that such pressure and negative commentary usually do not accompany other losses is another indication of the seriousness with which most people take the death of a partner.

Older bereaved spouses may grieve a great deal for a long time (Hansson & Stroebe, 2007); results from one study showed that such grief can last for at least 30 months (Thompson et al., 1991). Given that, you might wonder whether having a supportive social network might help people cope. Research findings on this topic are mixed, however. Some studies find that social support plays a significant role in the outcome of the grieving process. For example, during the first 2 years after the death of a partner, some data show that the quality of the support system—rather than simply the number of friends—is especially important for the grieving partner. Survivors who have confidants are better off than survivors who have many acquaintances (Hansson & Stroebe, 2007). In contrast, other studies find that having a supportive social network plays little role in helping people cope. For example, Miller, Smerglia, and Bouchet (2004) reported that the type of social support available to a widow had no relation to her adjustment to widowhood. It may be that there is a complex relationship involving the bereaved person, whether he or she wants to have contact with others, who in the social network is willing to provide support, and whether that support is of high quality.

When one's partner dies, how he or she felt about the relationship could play a role in coping with bereavement. One study of spousal bereavement measured how the surviving spouse rated the marriage. Bereaved older widows/widowers rated their relationships at 2, 12, and 30 months after the death of their spouses; nonbereaved older adults served as a comparison group. The results are summarized in ■ Figure 16.7. Bereaved widows and widowers gave their marriages more positive ratings than nonbereaved older adults. A marriage lost through death left a positive bias in memory. However, bereaved spouses' ratings were related to depression in an interesting way. The more depressed the bereaved spouse, the more positive the marriage's rating. In contrast, depressed nonbereaved spouses gave their marriages negative ratings. This result suggests that depression following bereavement signifies positive aspects of a relationship, whereas depression not connected with bereavement indicates a troubled relationship (Futterman et al., 1990).

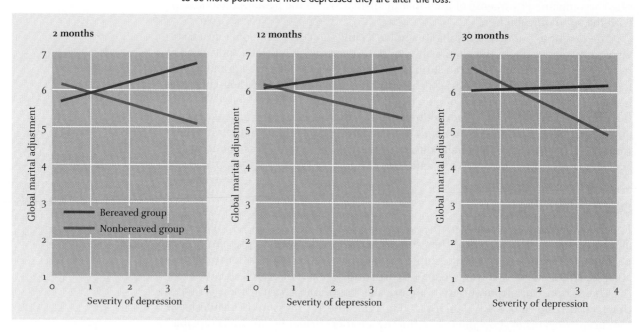

■ **Figure 16.7**

In general, bereaved spouses rate their marriages more positively than nonbereaved spouses, and they tend to be more positive the more depressed they are after the loss.

From "Retrospective Assessment of Marital Adjustment and Depression During the First Two Years of Spousal Bereavement," by A. Futterman, D. Gallagher, L. W. Thompson, S. Lovett, and M. Gilewski, 1990, *Psychology and Aging, 5,* 277–283. Copyright © 1990 American Psychological Association.

Several studies of widows document a tendency for some older women to "sanctify" their husbands (Lopata, 1996). Sanctification involves describing deceased husbands in idealized terms, and it serves several functions: validating that the widow had a strong marriage, is a good and worthy person, and is capable of rebuilding her life. European American women who view being a wife as above all other roles a woman can perform are somewhat more likely to sanctify their husbands (Lopata, 1996). In fact, the higher the quality of the relationship, the more bereaved spouses yearn for their lost spouse (Stroebe, Abakoumkin, & Stroebe, 2010).

Older bereaved spouses who can talk about their feelings concerning their loss exhibit reduced feelings of hopelessness, fewer intrusive thoughts, and fewer obsessive-compulsive behaviors (Segal et al., 1999). Cognitive-behavioral therapy is one especially effective intervention to help bereaved people make sense of the loss and deal with their other feelings and thoughts (Fleming & Robinson, 2001). A key to this process is helping people make meaning from the death (Neimeyer & Wogrin, 2008).

Gay and lesbian couples may experience other feelings and reactions in addition to typical feelings of grief (Clarke, Ellis, Peel & Riggs, 2010). For example, family members of the deceased may not make the partner feel welcome at the funeral, making it hard for the partner to bring closure to the relationship. For gay partners who were also caregivers, the loss affects one's sense of identity in much the same way as the death of a spouse, and making sense of the death becomes the primary issue (Cadell & Marshall, 2007). Lesbian widows report similar feelings (Bent & Magilvy, 2006).

Conclusion

Death is not as pleasant a topic as children's play or occupational development. It's not something we can go to college to master. What it represents to many people is the end of their existence, and that is a scary prospect. But because we all share in this fear at some level, each of us is equipped to provide support and comfort for grieving survivors.

Death is the last life-cycle force we encounter, the ultimate triumph of biological forces that limit the length of life. Yet the same psychological and social forces that are so influential throughout life help us deal with death, either our own or someone else's. As we come to the end of our life journey, we understand death through an interaction of psychological forces—such as coping skills and intellectual and emotional understanding of death—and the sociocultural forces expressed in a particular society's traditions and rituals.

Learning about and dealing with death is clearly a developmental process across the life span that fits well in the biopsychosocial framework. Most apparent is that biological forces are essential to understanding death. The very definition of death is based on whether certain biological functions are present; these same definitions create numerous ethical dilemmas that must be dealt with psychologically and socioculturally. Life-cycle forces also play a key role. We have seen that, depending on a person's age, the concept of death has varied meanings beyond the mere cessation of life.

How a person's understanding of death develops is also the result of psychological forces. As the ability to think and reflect undergoes fundamental change, the view of death changes from a mostly magical approach to one that can be transcendent and transforming. As we have seen, people who are facing their own imminent death experience certain feelings. Having gained experience through the deaths of friends and relatives, a person's level of comfort with his or her own death may increase. Such personal experience may also come about by sharing the rituals that are defined through sociocultural forces. People observe how others deal with death and how the culture sets the tone and prescribes behavior for survivors. The combined action of forces also determines how they cope with the grief that accompanies the loss of someone close. Psychologically, confronting grief depends on many things, including the quality of the support system we have.

Thus, just as the beginning of life represents a complex interaction of biological, psychological, sociocultural, and life-cycle factors, so does death. What people believe about what happens after death is also an interaction of these factors. So, as we bring our study of human development to a close, we end where we began: What we experience in our lives cannot be understood from only a single perspective.

Test Yourself

RECALL

1. In general, adults should be _____ when discussing death with children.
2. Adolescents are usually _____ to talk about their grief experiences.
3. The most devastating type of loss for an adult is the loss of a _____.
4. In general, a marriage ended by death is rated _____ than a marriage ended in some other way.

INTERPRET

What similarities and differences would you expect to find between the survivors in heterosexual marriages and in gay or lesbian relationships when the spouse or partner dies?

APPLY

How do the different ways that adults view death relate to the stages of Erikson's theory discussed in Chapters 10, 13, and 15?

Recall answers: (1) honest, (2) reluctant, (3) child, (4) more positively

---

SUMMARY

16.1 DEFINITIONS AND ETHICAL ISSUES

How is death defined?

■ Death is a difficult concept to define precisely. Different cultures have different meanings for death. Among the meanings in Western culture are images, statistics, events, state of being, analogy, mystery, boundary, thief of meaning, basis for anxiety, and reward or punishment.

What legal and medical criteria are used to determine when death occurs?

■ For many centuries, a clinical definition of death was used: the absence of a heartbeat and respiration. Currently, whole-brain death is the most widely used definition. It is based on several highly specific criteria, including brain activity and responses to specific stimuli.

What are the ethical dilemmas surrounding euthanasia?

■ Two types of euthanasia are distinguished. Active euthanasia consists of deliberately ending someone's life, such as turning off a life-support system. Physician-assisted suicide is a controversial issue and a form of active euthanasia. Passive euthanasia is ending someone's life by withholding some type of intervention or treatment (e.g., by stopping nutrition). It is essential that people make their wishes known through either a durable power of attorney or a living will.

16.2 THINKING ABOUT DEATH: PERSONAL ASPECTS

How do feelings about death change over adulthood?

■ Young adults report a sense of being cheated by death. Cognitive developmental level is important for understanding how young adults view death.

■ Middle-aged adults begin to confront their own mortality and undergo a change in their sense of time lived and time until death.

■ Older adults are more accepting of death.

How do people deal with their own death?

■ Kübler-Ross's theory includes five stages: denial, anger, bargaining, depression, and acceptance. Some people do not progress through all these stages, and some people move through them at different rates. People may be in more than one stage at a time and do not necessarily go through them in order.

■ A contextual theory of dying emphasizes the tasks a dying person must face. Four dimensions of these tasks have been identified: bodily needs, psychological security, interpersonal attachments, and spiritual energy and hope. A contextual theory incorporates differences in reasons people die and the places people die.

What is death anxiety, and how do people show it?

■ Most people exhibit some degree of anxiety about death, even though it is difficult to define and measure. Individual difference variables include gender, religiosity, age, ethnicity, and occupation. Death anxiety may have some benefits.

■ The main ways death anxiety is shown are by avoiding death (e.g., refusing to go to funerals) and deliberately challenging it (e.g., engaging in dangerous sports).

■ Several ways to deal with anxiety exist: living life to the fullest, personal reflection, and education. Death education has been shown to be extremely effective.

How do people deal with end-of-life issues and create a final scenario?

■ Managing the final aspects of life, after-death disposition of the body and memorial services, and distribution of assets are important end-of-life issues. Making choices about what people do and do not want done constitutes making a final scenario.

What is hospice?

■ The goal of a hospice is to maintain the quality of life and to manage the pain of terminally ill patients. Hospice clients typically have cancer, AIDS, or a progressive neurological disorder. Family members tend to stay involved in the care of hospice clients.

16.3 SURVIVING THE LOSS: THE GRIEVING PROCESS

How do people experience the grief process?

■ Grief is an active process of coping with loss. Four aspects of grieving must be confronted: the reality of the loss, the emotional turmoil, adjusting to the environment, and loosening the ties with the deceased. When death is expected, survivors go through anticipatory grief; unexpected death is usually more difficult for people to handle.

What feelings do grieving people have?

■ Dealing with grief, called *grief work*, usually takes at least 1 to 2 years. Grief is equally intense for both expected and unexpected death, but it may begin before the actual death when the patient has a terminal illness. Normal grief reactions include sorrow, sadness, denial, disbelief, guilt, and anniversary reactions.

■ In terms of dealing with normal grief, middle-aged adults have the most difficult time. Poor copers tend to have low self-esteem before losing a loved one.

What is the difference between normal and prolonged grief?

■ The four-component model proposes that the grief process is described by context of the loss, continuation of subjective meaning associated with the loss, changing representations of the lost relationship over time, and the role of coping and emotion-regulation processes.

■ The dual-process model of coping with bereavement focuses on loss-oriented and restoration-oriented stressors.

■ Prolonged grief involves symptoms of separation distress and symptoms of traumatic distress. Excessive guilt and self-blame are common manifestations of traumatic grief.

16.4 DYING AND BEREAVEMENT EXPERIENCES ACROSS THE LIFE SPAN

What do children understand about death? How should adults help them deal with it?

■ The cognitive and psychosocial developmental levels of children determine their understanding of and ability to cope with death. This is especially evident in the behaviors children use to display their grief.

■ Research indicates that there are few long-lasting effects of bereavement in childhood.

How do adolescents deal with death?

■ Adolescents may have difficulty making sense of death and are often severely affected by bereavement. Adolescents may be reluctant to discuss their feelings of loss, and peers often provide little support.

How do adults deal with death? What are the special issues they face concerning the death of a child or parent?

- Young and middle-aged adults usually have intense feelings about death. Attachment theory provides a useful framework for understanding these feelings.
- Midlife is a time when people usually deal with the death of their parents and confront their own mortality.
- The death of one's child is especially difficult to cope with.
- The death of one's parent deprives an adult of many important things, and the feelings accompanying it are often complex.

How do older adults face the loss of a child, grandchild, or partner?

- Older adults are usually less anxious about death and deal with it better than any other age group.
- The death of a grandchild can be very traumatic for older adults, and the feelings of loss may never go away.
- The death of one's partner represents a deep personal loss, especially when the couple had a long and close relationship. Older widowers often have a difficult time coping, whereas older widows often have a difficult time financially.

KEY TERMS

thanatology (582)
clinical death (584)
whole-brain death (584)
persistent vegetative state (585)
bioethics (585)
euthanasia (585)
active euthanasia (585)
passive euthanasia (586)
physician-assisted suicide (587)
living will (589)
durable power of attorney for health care (589)

Do Not Resuscitate (DNR) order (589)
death anxiety (595)
terror management theory (595)
end-of-life issues (596)
final scenario (597)
hospice (598)
palliative care (598)
bereavement (601)
grief (601)
mourning (601)
anticipatory grief (602)

grief work (603)
anniversary reaction (604)
four-component model (605)
grief work as rumination hypothesis (605)
dual process model (DPM) (606)
complicated or prolonged grief disorder (607)
separation distress (607)
traumatic distress (607)

LEARN MORE ABOUT IT

Log in to **www.cengagebrain.com** to access the resources your instructor requires. For this book, you can access:

 Psychology CourseMate

- CourseMate brings course concepts to life with interactive learning, study, and exam preparation tools that support the printed textbook. A textbook-specific website, Psychology CourseMate includes an integrated interactive eBook and other interactive learning tools including quizzes, flashcards, videos, and more.

CENGAGENOW™

- CengageNOW Personalized Study is a diagnostic study tool containing valuable text-specific resources—and because you focus on just what you don't know, you learn more in less time to get a better grade.

WebTUTOR™

- More than just an interactive study guide, WebTutor is an anytime, anywhere customized learning solution with an eBook, keeping you connected to your textbook, instructor, and classmates.

Glossary

abusive relationship relationships in which one person becomes aggressive toward the partner

accommodation according to Piaget, changing existing knowledge based on new knowledge

active euthanasia the deliberate ending of someone's life

activities of daily living (ADLs) basic self-care tasks such as eating, bathing, toileting, walking, and dressing

adaptation level when press level is average for a particular level of competence

addiction physical dependence on a substance (e.g., alcohol) such that withdrawal symptoms are experienced when deprived of that substance

adolescent egocentrism self-absorption that is characteristic of teenagers as they search for identity

adolescent-limited antisocial behavior the behavior of youth who engage in relatively minor criminal acts but aren't consistently antisocial

aerobic exercise exercise that places moderate stress on the heart by maintaining a pulse rate between 60 and 90% of the person's maximum heart rate

age discrimination denying a job or promotion to someone solely on the basis of age

age of viability age at which a fetus can survive because most of its bodily systems function adequately; typically at 7 months after conception

agreeableness a dimension of personality associated with being accepting, willing to work with others, and caring

alert inactivity state in which a baby is calm with eyes open and attentive; the baby seems to be deliberately inspecting the environment

alienation feeling that results in workers when their work seems meaningless and their efforts devalued or when they see no connection between their own work and the final product

alleles variations of genes

altruism prosocial behavior such as helping and sharing in which the individual does not benefit directly from his or her behavior

Alzheimer's disease a disease marked by gradual declines in memory, attention, and judgment; confusion as to time and place; difficulties in communicating; decline in self-care skills; inappropriate behavior; and personality changes

amniocentesis prenatal diagnostic technique that uses a syringe to withdraw a sample of amniotic fluid through the mother's abdomen

amnion inner sac in which the developing child rests

amniotic fluid fluid that surrounds the fetus

amyloid protein that is produced in abnormally high levels in Alzheimer's patients

analytic ability in Sternberg's theory of intelligence, the ability to analyze problems and generate different solutions

animism crediting inanimate objects with life and lifelike properties such as feelings

anniversary reaction changes in behavior related to feelings of sadness on the anniversary date of a loss

anorexia nervosa persistent refusal to eat accompanied by an irrational fear of being overweight

anticipatory grief grief that is experienced during the period before an expected death occurs that supposedly serves to buffer the impact of the loss when it does come and to facilitate recovery

anxiety disorders problems such as feelings of severe anxiety, phobias, and obsessive-compulsive behaviors

assimilation according to Piaget, taking in information that is compatible with what one already knows

assisted living facilities a supportive living arrangement for people who need assistance with ADLs or IADLs but who are not so impaired physically or cognitively that they need 24-hour care

assortative mating theory stating that people find partners based on their similarity to each other

attachment enduring socioemotional relationship between infants and their caregivers

attention processes that determine which information will be processed further by an individual

authoritarian parenting parents who show high levels of control and low levels of warmth toward their children

authoritative parenting parents who use a moderate amount of control and are warm and responsive to their children

autobiographical memory memories of the significant events and experiences of one's own life

autosomal dominant inheritance the presence of certain genes that means there is a 100% chance of the person eventually getting the disease

autosomes first 22 pairs of chromosomes

average children as applied to children's popularity, children who are liked and disliked by different classmates, but with relatively little intensity

average life expectancy the age at which half of the people born in a particular year will have died

avoidant attachment relationship in which infants turn away from their mothers when they are reunited following a brief separation

axon tubelike structure that emerges from the cell body and transmits information to other neurons

babbling speechlike sounds that consist of vowel–consonant combinations; common at about 6 months

backup care emergency care for dependent children or adults so that the employee does not need to lose a day of work

basal metabolic rate the speed at which the body consumes calories

basic cry cry that starts softly and gradually becomes more intense; often heard when babies are hungry or tired

basic emotions emotions experienced by humankind and that consist of three elements: a subjective feeling, a physiological change, and an overt behavior

battered woman syndrome situation occurring when a woman believes that she cannot leave the abusive situation and may even go so far as to kill her abuser

behavior therapy type of therapy based on the notion that depressed people experience too few rewards or reinforcements from their environment

bereavement the state or condition caused by loss through death

binge drinking type of drinking defined for men as consuming five or more drinks in a row and for women as consuming four or more drinks in a row within the past 2 weeks

bioethics study of the interface between human values and technological advances in health and life sciences

biopsychosocial framework a useful way to organize the biological, psychological, and sociocultural forces on human development

blended family family consisting of a biological parent, a stepparent, and children

body mass index (BMI) an adjusted ratio of weight to height; used to define "overweight"

bridge job transitional job held between one's exit from a career job and final retirement

bulimia nervosa disease in which people alternate between binge eating—periods when they eat uncontrollably—and purging with laxatives or self-induced vomiting

burnout a depletion of a person's energy and motivation, the loss of occupational idealism, and the feeling that one is being exploited

cardinality principle counting principle that the last number name denotes the number of objects being counted

career plateauing when promotional advancement is either not possible or not desired by the worker

cell body center of the neuron that keeps the neuron alive

cellular theories explanation of aging that focuses on processes that occur within individual cells that may lead to the buildup of harmful substances or the deterioration of cells over a lifetime

centration according to Piaget, narrowly focused type of thought characteristic of preoperational children

cephalocaudal principle a principle of physical growth that states that structures nearest the head develop first

cerebral cortex wrinkled surface of the brain that regulates many functions that are distinctly human

cesarean section (C-section) surgical removal of infant from the uterus through an incision made in the mother's abdomen

chorionic villus sampling (CVS) prenatal diagnostic technique that involves taking a sample of tissue from the chorion

chromosomes threadlike structures in the nuclei of cells that contain genetic material

chronic obstructive pulmonary disease (COPD) the most common form of incapacitating respiratory disease among older adults

circadian rhythm the sleep–wake cycle

classical conditioning a form of learning that involves pairing a neutral stimulus and a response originally produced by another stimulus

climacteric biological process during which women pass from their reproductive to nonreproductive years

clinical death lack of heartbeat and respiration

clique small group of friends who are similar in age, sex, and race

cognitive self-regulation skill at identifying goals, selecting effective strategies, and accurate monitoring; a characteristic of successful students

cognitive therapy type of therapy based on the idea that maladaptive beliefs or cognitions about oneself are responsible for depression

cohabitation people in committed, intimate, sexual relationships who live together but are not married

cohort effects problem with cross-sectional designs in which differences between age groups (cohorts) may result as easily from environmental events as from developmental processes

collaborative divorce a voluntary, contractually based alternative dispute resolution process for couples who want to negotiate a resolution of their situation rather than having a ruling imposed upon them by a court or arbitrator

competence the upper limit of a person's ability to function in five domains: physical health, sensory-perceptual skills, motor skills, cognitive skills, and ego strength

complicated or prolonged grief disorder expression of grief which is distinguished from depression and from normal grief in terms of separation distress and traumatic distress

comprehension the process of extracting meaning from a sequence of words

cones specialized neurons in the back of the eye that sense color

conscientiousness a dimension of personality in which people tend to be hard working, ambitious, energetic, scrupulous, and persevering

constricting actions interaction in which one partner tries to emerge as the victor by threatening or contradicting the other

continuity theory theory based on idea that people tend to cope with daily life in later adulthood by applying familiar strategies based on past experience to maintain and preserve both internal and external structures

continuity–discontinuity issue whether a particular developmental phenomenon represents a smooth progression throughout the life span (continuity) or a series of abrupt shifts (discontinuity)

controversial children as applied to children's popularity, children who are intensely liked or disliked by classmates

conventional level second level of reasoning in Kohlberg's theory, where moral reasoning is based on society's norms

convergent thinking using information to arrive at one standard and correct answer

cooing early vowel-like sounds that babies produce

cooperative play play that is organized around a theme, with each child taking on a different role; begins at about 2 years of age

coping any attempt to deal with stress

core knowledge hypothesis infants are born with rudimentary knowledge of the world, which is elaborated based on experiences

corpus callosum thick bundle of neurons that connects the two hemispheres

correlation coefficient an expression of the strength and direction of a relation between two variables

correlational study investigation looking at relations between variables as they exist naturally in the world

co-rumination conversations about one's personal problems, common among adolescent girls

counterimitation learning what should not be done by observing the behavior

covenant marriage expands the marriage contract to a lifelong commitment between the partners within a supportive community

creative ability in Sternberg's theory of intelligence, the ability to deal adaptively with novel situations and problems

cross-linking random interaction of some proteins with certain body tissues, such as muscles and arteries

cross-sectional study study in which developmental differences are identified by testing people of different ages

crowd large group including many cliques that have similar attitudes and values

crowning appearance of the top of the baby's head during labor

crying state in which a baby cries vigorously, usually accompanied by agitated but uncoordinated movement

crystallization first phase in Super's theory of career development, in which adolescents use their emerging identities to form ideas about careers

crystallized intelligence the knowledge you have acquired through life experience and education in a particular culture

culture-fair intelligence tests intelligence tests devised using items common to many cultures

death anxiety people's anxiety or even fear of death and dying

deductive reasoning drawing conclusions from facts; characteristic of formal-operational thought

dementia family of diseases involving serious impairment of behavioral and cognitive functioning

demographers people who study population trends

dendrite end of the neuron that receives information; it looks like a tree with many branches

deoxyribonucleic acid (DNA) molecule composed of four nucleotide bases that is the biochemical basis of heredity

dependent variable the behavior being observed

depression disorder characterized by pervasive feelings of sadness, irritability, and low self-esteem

differentiation distinguishing and mastering individual motions

direct instruction telling a child what to do, when, and why

disorganized (disoriented) attachment relationship in which infants don't seem to understand what's happening when they are separated and later reunited with their mothers

divergent thinking thinking in novel and unusual directions

dizygotic twins the result of two separate eggs fertilized by two sperm; also called fraternal twins

Do Not Resuscitate (DNR) order a medical order that means that cardiopulmonary resuscitation (CPR) is not started should one's heart and breathing stop

docility when people allow their situation to dictate the options they have

dominance hierarchy ordering of individuals within a group in which group members with lower status defer to those with greater status

dominant form of an allele whose chemical instructions are followed

dual process model (DPM) view of coping with bereavement that integrates loss-oriented stressors and restoration-oriented stressors

dual-energy X-ray absorptiometry (DXA) test test of bone mineral density (BMD) at the hip and spine

durable power of attorney for health care a document in which an individual appoints someone to act as his or her agent for health care decisions

dynamic systems theory theory that views motor development as involving many distinct skills that are organized and reorganized over time to meet specific needs

dysphoria feeling sad or down

ecological theory theory based on idea that human development is inseparable from the environmental contexts in which a person develops

ectoderm outer layer of the embryo, which will become the hair, the outer layer of skin, and the nervous system

edgework the desire to live life more on the edge through physically and emotionally threatening situations on the boundary between life and death

ego resilience a person's ability to respond adaptively and resourcefully to new situations

egocentrism difficulty in seeing the world from another's point of view; typical of children in the preoperational period

elaboration memory strategy in which information is embellished to make it more memorable

electroencephalography the study of brain waves recorded from electrodes that are placed on the scalp

embryo term given to the zygote once it is completely embedded in the uterine wall

emerging adulthood period between late teens and mid- to late 20s when individuals are not adolescents but are not yet fully adults

emotional intelligence ability to use one's own and others' emotions effectively for solving problems and living happily

empathy experiencing another person's feelings

enabling actions individuals' actions and remarks that tend to support others and sustain the interaction

encapsulation occurs when the processes of thinking (information processing, memory, fluid intelligence) become connected or encapsulated to the products of thinking (expertise)

endoderm inner layer of the embryo, which becomes the lungs and the digestive system

end-of-life issues issues pertaining to the management of the final phase of life, after-death disposition of the body and memorial services, and distribution of assets

environmental press the physical, interpersonal, or social demands that environments put on people

epigenetic principle in Erikson's theory, the idea that each psychosocial strength has its own special period of particular importance

episodic memory the general class of memory having to do with the conscious recollection of information from a specific time or event

equilibration according to Piaget, a process by which children reorganize their schemes to return to a state of equilibrium when disequilibrium occurs

essentialism children's belief that all living things have an essence that can't be seen but gives a living thing its identity

ethnic identity feeling of belonging to a specific ethnic group

eugenics effort to improve the human species by letting only people whose characteristics are valued by a society mate and pass along their genes

euthanasia the practice of ending life for reasons of mercy

evolutionary psychology theoretical view that many human behaviors represent successful adaptations to the environment

exchange theory relationship, such as marriage, based on each partner contributing something to the relationship that the other would be hard pressed to provide

exosystem social settings that a person may not experience firsthand but that still influence development

experience-expectant growth process by which the wiring of the brain is organized by experiences that are common to most humans

experiment a systematic way of manipulating the key factor(s) that the investigator thinks causes a particular behavior

explicit memory the deliberate and conscious remembering of information that is learned and remembered at a specific time

expressive style language-learning style of children whose vocabularies include many social phrases that are used like one word

extended family most common form of family around the world; one in which grandparents and other relatives live with parents and children

external aids memory aids that rely on environmental resources, such as notebooks or calendars

extraversion a personality trait dimension associated with the tendency to thrive on social interaction, to like to talk, to take charge easily, to readily express opinions and feelings, to like to keep busy, to have boundless energy, and to prefer stimulating and challenging environments

extremely low birth weight newborns who weigh less than 1,000 grams (2 pounds)

familism the idea that the family's well-being takes precedence over the concerns of individual family members

fast mapping a child's connections between words and referents that are made so quickly that he or she cannot consider all possible meanings of the word

fetal alcohol spectrum disorder (FASD) disorder affecting babies whose mothers consumed large amounts of alcohol while they were pregnant

fetal medicine field of medicine concerned with treating prenatal problems before birth

filial obligation a sense of obligation to care for one's parents if necessary

final scenario making one's choices known about how they do and do not want their lives to end

fine motor skills motor skills associated with grasping, holding, and manipulating objects

fluid intelligence abilities that make you a flexible and adaptive thinker, allow you to make inferences, and enable you to understand the relations among concepts

four-component model model for understanding grief that is based on (1) the context of the loss; (2) continuation of subjective meaning associated with loss; (3) changing representations of the lost relationship over time; and (4) the role of coping and emotion-regulation processes

frail older adults older adults who have physical disabilities, are very ill, and may have cognitive or psychological disorders

free radicals chemicals produced randomly during normal cell metabolism and that bond easily to other substances inside cells

friendship voluntary relationship between two people involving mutual liking

frontal cortex brain region that regulates personality and goal-directed behavior

functional health the ability to perform the activities of daily living (ADLs) and instrumental activities of daily living (IADLs)

functional magnetic resonance imaging (fMRI) method of studying brain activity by using magnetic fields to track blood flow in the brain

gender constancy understanding that maleness and femaleness do not change over situations or personal wishes

gender discrimination denying a job to someone solely on the basis of whether the person is a man or a woman

gender identity sense of oneself as male or female

gender labeling young children's understanding that they are either boys or girls and naming themselves accordingly

gender stability understanding in preschool children that boys become men and girls become women

gender stereotypes beliefs and images about males and females that are not necessarily true

gender-schema theory theory that states that children want to learn more about an activity only after first deciding whether it is masculine or feminine

gene group of nucleotide bases that provides a specific set of biochemical instructions

generativity in Erikson's theory, being productive by helping others in order to ensure the continuation of society by guiding the next generation

genotype person's hereditary makeup

germ disc small cluster of cells near the center of the zygote that will eventually develop into a baby

glass ceiling the level to which women may rise in an organization but beyond which they may not go

glass cliff situation that women confront in which their leadership position is precarious

grammatical morphemes words or endings of words that make a sentence grammatical

grief the sorrow, hurt, anger, guilt, confusion, and other feelings that arise after suffering a loss

grief work the psychological side of coming to terms with bereavement

grief work as rumination hypothesis an approach that not only rejects the necessity of grief processing for recovery from loss but views extensive grief processing as a form of rumination that may actually increase distress

guided participation children's involvement in structured activities with others who are more skilled, typically producing cognitive growth

habituation becoming unresponsive to a stimulus that is presented repeatedly

hemispheres right and left halves of the cortex

heritability coefficient a measure (derived from a correlation coefficient) of the extent to which a trait or characteristic is inherited

heterocyclic antidepressants (HCAs), monoamine oxidase (MAO) inhibitors, and selective serotonin reuptake inhibitors (SSRIs) medications for severe depression

heterozygous when the alleles in a pair of chromosomes differ from each other

high-density lipoproteins (HDLs) chemicals that help keep arteries clear and break down LDLs

homogamy similarity of values and interests

homozygous when the alleles in a pair of chromosomes are the same

hope according to Erikson, an openness to new experience tempered by wariness that occurs when trust and mistrust are in balance

hospice an approach to assisting dying people that emphasizes pain management, or palliative care, and death with dignity

hostile aggression unprovoked aggression that seems to have the sole goal of intimidating, harassing, or humiliating another child

household an individual who lives alone or a group of individuals who live together

human development the multidisciplinary study of how people change and how they remain the same over time

Huntington's disease progressive and fatal type of dementia caused by dominant alleles

hypoxia a birth complication in which umbilical blood flow is disrupted and the infant does not receive adequate oxygen

illusion of invulnerability adolescents' belief that misfortunes cannot happen to them

imaginary audience adolescents' feeling that their behavior is constantly being watched by their peers

imitation or observational learning learning that occurs by simply watching how others behave

implantation step in which the zygote burrows into the uterine wall and establishes connections with a woman's blood vessels

implementation third phase in Super's theory of career development, in which individuals actually enter the workforce

implicit memory the unconscious remembering of information learned at some earlier time

in vitro fertilization process by which sperm and an egg are mixed in a petri dish to create a zygote, which is then placed in a woman's uterus

incomplete dominance situation in which one allele does not dominate another completely

incontinence loss of bladder or bowel control

independent variable the factor being manipulated

infant mortality the number of infants out of 1,000 births who die before their first birthday

infant-directed speech speech that adults use with infants that is slow and has exaggerated changes in pitch and volume; it is thought to aid language acquisition

information-processing theory theory proposing that human cognition consists of mental hardware and mental software

instrumental activities of daily living (IADLs) actions that require some intellectual competence and planning

instrumental aggression aggression used to achieve an explicit goal

instrumental orientation characteristic of Kohlberg's Stage 2, in which moral reasoning is based on the aim of looking out for one's own needs

integration linking individual motions into a coherent, coordinated whole

integrity versus despair according to Erikson, the process in late life by which people try to make sense of their lives

intellectual disability substantially below-average intelligence and problems adapting to an environment that emerge before the age of 18

intelligence quotient (IQ) mathematical representation of how a person scores on an intelligence test in relation to how other people of the same age score

interindividual variability patterns of change that vary from one person to another

internal aids memory aids that rely on mental processes, such as imagery

internal belief systems what one tells oneself about why certain things are happening

internal working model infant's understanding of how responsive and dependable the mother is; thought to influence close relationships throughout the child's life

interpersonal norms characteristic of Kohlberg's Stage 3, in which moral reasoning is based on winning the approval of others

intersensory redundancy infants' sensory systems are attuned to information presented simultaneously to different sensory modes

intersubjectivity mutual, shared understanding among participants in an activity

intimacy versus isolation sixth stage in Erikson's theory and the major psychosocial task for young adults

irregular or rapid-eye-movement (REM) sleep irregular sleep in which an infant's eyes dart rapidly beneath the eyelids while the body is quite active

job satisfaction the positive feeling that results from an appraisal of one's work

joint custody when both parents retain legal custody of their children following divorce

juvenile delinquency when adolescents commit illegal acts that are destructive to themselves or others

kinetic cues cues to depth perception in which motion is used to estimate depth

kinkeeper the person who gathers family members together for celebrations and keeps them in touch with each other, usually a middle-aged mother

knowledge-telling strategy writing down information as it is retrieved from memory, a common practice for young writers

knowledge-transforming strategy deciding what information to include and how best to organize it to convey a point

learning disability when a child with normal intelligence has difficulty mastering at least one academic subject

leisure discretionary activity that includes simple relaxation, activities for enjoyment, and creative pursuits

life review the process by which people reflect on the events and experiences of their lifetimes

life story a personal narrative that organizes past events into a coherent sequence

life-course persistent antisocial behavior antisocial behavior that emerges at an early age and continues throughout life

life-course perspective description of how various generations experience the biological, psychological, and sociocultural forces of development in their respective historical contexts

life-span construct a unified sense of the past, present, and future based on personal experience and input from other people

life-span perspective view that human development is multiply determined and cannot be understood within the scope of a single framework

linear perspective a cue to depth perception based on the fact that parallel lines come together at a single point in the distance

living will a document in which a person states his or her wishes about life support and other treatments

locomote ability to move around in the world

longevity number of years a person can expect to live

long-term memory permanent storehouse for memories that has unlimited capacity

longitudinal study longitudinal study research design in which the same individuals are observed or tested repeatedly at different points in their lives

low birth weight newborns who weigh less than 2,500 grams (5 pounds)

low-density lipoproteins (LDLs) chemicals that cause fatty deposits to accumulate in arteries, impeding blood flow

macrosystem the cultures and subcultures in which the microsystem, mesosystem, and exosystem are embedded

mad cry more intense version of a basic cry

malnourished being small for one's age because of inadequate nutrition

marital adjustment the degree to which a husband and wife accommodate to each other over a certain period of time

marital quality a subjective evaluation of the couple's relationship on a number of different dimensions

marital satisfaction a global assessment of one's marriage

marital success an umbrella term referring to any marital outcome

marriage education the idea that the more couples are prepared for marriage, the better the relationship will survive over the long run

maximum life expectancy the oldest age to which any person lives

meaning-mission fit alignment between an executive's personal intentions and their firm's mission

mechanics of intelligence those aspects of intelligence comprising fluid intelligence

menarche onset of menstruation

menopausal hormone therapy (MHT) medication therapy in which women take low doses of estrogen, which is often combined with progestin (synthetic form of progesterone) to counter symptoms associated with menopause

menopause the point at which menstruation stops

mental age (MA) in intelligence testing, a measure of children's performance corresponding to the chronological age of those whose performance equals the child's

mental hardware mental and neural structures that are built-in and that allow the mind to operate

mental operations cognitive actions that can be performed on objects or ideas

mental software mental "programs" that are the basis for performing particular tasks

mentor or developmental coach person who is part teacher, part sponsor, part model, and part counselor who facilitates on-the-job learning to help the new hire do the work required in his or her present role and to prepare for future roles

mesoderm middle layer of the embryo, which becomes the muscles, bones, and circulatory system

mesosystem provides connections across microsystems

meta-analysis a tool that enables researchers to synthesize the results of many studies to estimate relations between variables

metabolism how much energy the body needs

metacognitive knowledge a person's knowledge and awareness of cognitive processes

metamemory person's informal understanding of memory; includes the ability to diagnose memory problems accurately and to monitor the effectiveness of memory strategies

microsystem the people and objects in an individual's immediate environment

monozygotic twins the result of a single fertilized egg splitting to form two new individuals; also called identical twins

motion parallax kinetic cue to depth perception based on the fact that nearby moving objects move across our visual field faster than do distant objects

motor skills coordinated movements of the muscles and limbs

mourning the ways in which we express our grief

multidimensional characteristic of theories of intelligence that identify several types of intellectual abilities

multidirectionality developmental pattern in which some aspects of intelligence improve and other aspects decline during adulthood

myelin fatty sheath that wraps around neurons and enables them to transmit information more rapidly

narrative a way in which a person derives personal meaning from being generative and by constructing a life story which helps create the person's identity

naturalistic observation technique in which people are observed as they behave spontaneously in some real-life situation

nature–nurture issue the degree to which genetic or hereditary influences (nature) and experiential or environmental influences (nurture) determine the kind of person you are

negative reinforcement trap unwittingly reinforcing a behavior you want to discourage

neglected children as applied to children's popularity, children who are ignored—neither liked nor disliked—by their classmates

neural plate flat group of cells present in prenatal development that becomes the brain and spinal cord

neuritic plaques structural change in the brain produced when damaged and dying neurons collect around a core of protein

neurofibrillary tangles spiral-shaped masses formed when fibers that compose the axon become twisted together

neuron basic cellular unit of the brain and nervous system that specializes in receiving and transmitting information

neuroscience the study of the brain and nervous system, especially in terms of brain-behavior relationships

neuroticism a personality trait reflected as the tendency to be anxious, hostile, self-conscious, depressed, impulsive, and vulnerable

neurotransmitters chemicals released by neurons in order for them to communicate with each other

neurotransmitters chemicals released by the terminal buttons that allow neurons to communicate with each other

niche-picking process of deliberately seeking environments that are compatible with one's genetic makeup

nonshared environmental influences forces within a family that make siblings different from one another

nuclear family most common form of family in Western societies, consisting only of parent(s) and child(ren)

obedience orientation characteristic of Kohlberg's Stage 1, in which moral reasoning is based on the belief that adults know what is right and wrong

object permanence understanding, acquired in infancy, that objects exist independently of oneself

one-to-one principle counting principle that states that there must be one and only one number name for each object counted

openness to experience a personality dimension that reflects a tendency to have a vivid imagination and dream life, an appreciation of art, and a strong desire to try anything once

operant conditioning learning paradigm in which the consequences of a behavior determine whether a behavior is repeated in the future

organization as applied to children's memory, a strategy in which information to be remembered is structured so that related information is placed together

orienting response an individual views a strong or unfamiliar stimulus, and changes in heart rate and brain-wave activity occur

osteoarthritis a disease marked by gradual onset of bone damage with progression of pain and disability together with minor signs of inflammation from wear-and-tear

osteoporosis a disease in which bones become porous and extremely easy to break from severe loss of bone mass

overextension when children define words more broadly than adults do

overregularization grammatical usage that results from applying rules to words that are exceptions to the rule

pain cry cry that begins with a sudden long burst, followed by a long pause and gasping

palliative care care that is focused on providing relief from pain and other symptoms of disease at any point during the disease process

parallel play when children play alone but are aware of and interested in what another child is doing

Parieto-Frontal Integration Theory (P-FIT) proposes that intelligence comes from a distributed and integrated network of neurons in the parietal and frontal lobes of the brain

Parkinson's disease brain disease known primarily for its characteristic motor symptoms: very slow walking, difficulty getting into and out of chairs, and a slow hand tremor

passion strong inclination toward an activity that individuals like (or even love), that they value (and thus find important), and in which they invest time and energy

passive euthanasia allowing a person to die by withholding available treatment

perception processes by which the brain receives, selects, modifies, and organizes incoming nerve impulses that are the result of physical stimulation

perimenopause the individually varying time of transition from regular menstruation to menopause

period of the fetus longest period of prenatal development, extending from the 9th until the 38th week after conception

permissive parenting style of parenting that offers warmth and caring but little parental control over children

persistent vegetative state situation in which a person's cortical functioning ceases while brainstem activity continues

personal control beliefs the degree to which you believe your performance in a situation depends on something you do

personal fable attitude of many adolescents that their feelings and experiences are unique and have never been experienced by anyone else before

personality-type theory view proposed by Holland that people find their work fulfilling when the important features of a job or profession fit the worker's personality

phenotype physical, behavioral, and psychological features that result from the interaction between one's genes and the environment

phenylketonuria (PKU) inherited disorder in which the infant lacks a liver enzyme

phonemes unique sounds used to create words; the basic building blocks of language

phonological awareness the ability to hear the distinctive sounds of letters

phonological memory ability to remember speech sounds briefly; an important skill in acquiring vocabulary

physician-assisted suicide process in which physicians provide dying patients with a fatal dose of medication that the patient self-administers

pictorial cues cues to depth perception that are used to convey depth in drawings and paintings

placenta structure through which nutrients and wastes are exchanged between the mother and the developing child

plasticity concept that intellectual abilities are not fixed but can be modified under the right conditions at just about any point in adulthood

polygenic inheritance when phenotypes are the result of the combined activity of many separate genes

popular children children who are liked by many classmates

population pyramid graphic technique for illustrating population trends

populations broad groups of people that are of interest to researchers

possible selves representations of what we could become, what we would like to become, and what we are afraid of becoming

postconventional level third level of reasoning in Kohlberg's theory, in which morality is based on a personal moral code

postformal thought thinking characterized by recognizing that the correct answer varies from one situation to another, that solutions should be realistic, that ambiguity and contradiction are typical, and that subjective factors play a role in thinking

post-traumatic stress disorder (PTSD) an anxiety disorder that can develop after exposure to a terrifying event or ordeal in which grave physical harm occurred or was threatened

practical ability in Sternberg's theory of intelligence, the ability to know which problem solutions are likely to work

practical intelligence the broad range of skills related to how individuals shape, select, or adapt to their physical and social environments

pragmatics of intelligence those aspects of intelligence reflecting crystallized intelligence

preconventional level first level of reasoning in Kohlberg's theory, where moral reasoning is based on external forces

prejudice a view of other people, usually negative, that is based on their membership in a specific group

prenatal development the many changes that turn a fertilized egg into a newborn human

presbycusis reduced sensitivity to high-pitched tones

presbyopia difficulty in seeing close objects clearly

preterm (premature) babies born before the 36th week after conception

primary control behavior aimed at affecting the individual's external world

primary mental abilities groups of related intellectual skills (such as memory or spatial ability)

primary sex characteristics physical signs of maturity that are directly linked to the reproductive organs

private speech a child's comments that are not intended for others but are designed instead to help regulate the child's own behavior

proactivity when people choose new behaviors to meet new desires or needs and exert control over their lives

programmed theories theory that aging is biologically or genetically programmed

prosocial behavior any behavior that benefits another person

proximodistal principle principle of physical growth that states that structures nearest the center of the body develop first

psychodynamic theories theories proposing that development is largely determined by how well people resolve conflicts they face at different ages

psychometricians psychologists who specialize in measuring psychological traits such as intelligence and personality

psychomotor speed the speed with which a person can make a specific response

psychosocial theory Erikson's proposal that personality development is determined by the interaction of an internal maturational plan and external societal demands

puberty collection of physical changes that marks the onset of adolescence, including a growth spurt and the growth of breasts or testes

punishment a consequence that decreases the future likelihood of the behavior that it follows

purpose according to Erikson, balance between individual initiative and the willingness to cooperate with others

qualitative research method that involves gaining in-depth understanding of human behavior and what governs it

reality shock situation in which what you learn in the classroom does not always transfer directly into the "real world" and does not represent all that you need to know

recessive allele whose instructions are ignored in the presence of a dominant allele

recursive thinking thoughts that focus on what another person is thinking

referential style language-learning style of children whose vocabularies are dominated by names of objects, persons, or actions

reflective judgment way in which adults reason through real-life dilemmas

reflexes unlearned responses triggered by specific stimulation

regular (nonREM) sleep sleep in which heart rate, breathing, and brain activity are steady

reinforcement a consequence that increases the likelihood that a behavior will be repeated in the future

rejected children as applied to children's popularity, children who are disliked by many classmates

relational aggression aggression used to hurt others by undermining their social relationships

reliability extent to which a measure provides a consistent index of a characteristic

resistant attachment relationship in which, after a brief separation, infants want to be held but are difficult to console

retinal disparity way of inferring depth based on differences in the retinal images in the left and right eyes

returning adult students college students over age 25

rheumatoid arthritis most common form of arthritis, a more destructive disease of the joints that also develops slowly; it typically affects different joints and causes different types of pain than osteoarthritis

risk genes genes that increase one's risk of getting the disease

rites of passage rituals marking initiation into adulthood

role transitions movement into the next stage of development marked by assumption of new responsibilities and duties

sample a subset of the population

sandwich generation middle-aged adults who are caught between the competing demands of two generations: their parents and their children

scaffolding a style in which teachers gauge the amount of assistance they offer to match the learner's needs

scenario manifestation of the life-span construct through expectations about the future

scheme according to Piaget, a mental structure that organizes information and regulates behavior

secondary control behavior or cognition aimed at affecting the individual's internal world

secondary mental abilities broader intellectual skills that subsume and organize the primary abilities

secondary sex characteristics physical signs of maturity that are not directly linked to reproductive organs

secure attachment relationship in which infants have come to trust and depend on their mothers

selective estrogen receptor modulators (SERMs) compounds that are not estrogens but that have estrogen-like effects on some tissues and estrogen-blocking effects on other tissues

selective optimization with compensation (SOC) model model in which three processes (selection, optimization, and compensation) form a system of behavioral action that generates and regulates development and aging

self-efficacy people's beliefs about their own abilities and talents

self-reports people's answers to questions about the topic of interest

semantic memory the general class of memory concerning the remembering of meanings of words or concepts not tied to a specific time or event

sense of place the cognitive and emotional attachments that a person puts on their place of residence, by which a "house" is made into a "home"

sensorimotor period first of Piaget's four stages of cognitive development, which lasts from birth to approximately 2 years

separation distress expression of complicated or prolonged grief disorder that includes preoccupation with the deceased to the point that it interferes with everyday functioning, upsetting memories of the deceased, longing and searching for the deceased, and isolation following the loss

sequential design developmental research design based on cross-sectional and longitudinal designs

sex chromosomes 23rd pair of chromosomes; these determine the sex of the child

sickle-cell trait disorder in which individuals show signs of mild anemia only when they are seriously deprived of oxygen; occurs in individuals who have one dominant allele for normal blood cells and one recessive sickle-cell allele

simple social play play that begins at about 15 to 18 months; toddlers engage in similar activities as well as talk and smile at each other

skeletal maturity the point at which bone mass is greatest and the skeleton is at peak development, around 19 for women and 20 in men

sleeping state in which a baby alternates from being still and breathing regularly to moving gently and breathing irregularly; the eyes are closed throughout

social clock tagging future events with a particular time or age by which they are to be completed

social cognitive career theory (SCCT) proposes that career choice is a result of the application of Bandura's social cognitive theory, especially the concept of self-efficacy

social contract characteristic of Kohlberg's Stage 5, in which moral reasoning is based on the belief that laws are for the good of all members of society

social convoy a group of people that journeys with us throughout our lives, providing support in good times and bad

social referencing behavior in which infants in unfamiliar or ambiguous environments look at an adult for cues to help them interpret the situation

social role set of cultural guidelines about how one should behave, especially with other people

social smiles smile that infants produce when they see a human face

social system morality characteristic of Kohlberg's Stage 4, in which moral reasoning is based on maintenance of order in society

socialization teaching children the values, roles, and behaviors of their culture

socioemotional selectivity process by which social contact is motivated by many goals, including information seeking, self-concept, and emotional regulation

spaced retrieval memory intervention based on the E-I-E-I-O approach to memory intervention

specification second phase in Super's theory of career development, in which adolescents learn more about specific lines of work and begin training

spermarche first spontaneous ejaculation of sperm

spina bifida disorder in which the embryo's neural tube does not close properly

spiritual support type of coping strategy that includes seeking pastoral care, participation in organized and nonorganized religious activities, and expressing faith in a God who cares for people

stable-order principle counting principle that states that number names must always be counted in the same order

stagnation in Erikson's theory, the state in which people are unable to deal with the needs of their children or to provide mentoring to younger adults

stereotype threat an evoked fear of being judged in accordance with a negative stereotype about a group to which you belong

stranger wariness first distinct signs of fear that emerge around 6 months of age when infants become wary in the presence of unfamiliar adults

stress physical and psychological responses to threatening or challenging conditions

stress and coping paradigm the dominant framework used to study stress, which emphasizes the transactions between a person and the environment

stroke, or cerebral vascular accident (CVA) an interruption of the blood flow in the brain due to blockage or a hemorrhage in a cerebral artery

structured observations technique in which a researcher creates a setting that is likely to elicit the behavior of interest

subjective well-being an evaluation of one's life that is associated with positive feelings

sudden infant death syndrome (SIDS) when a healthy baby dies suddenly for no apparent reason

synaptic pruning gradual reduction in the number of synapses, beginning in infancy and continuing until early adolescence

systematic observation watching people and carefully recording what they do or say

telegraphic speech speech used by young children that contains only the words necessary to convey a message

teleological explanations children's belief that living things and parts of living things exist for a purpose

telomeres tips of the chromosomes that shorten and break with increasing age

temperament consistent style or pattern of behavior

teratogen an agent that causes abnormal prenatal development

terminal buttons small knobs at the end of the axon that release neurotransmitters

terror management theory addresses the issue of why people engage in certain behaviors to achieve particular psychological states based on their deeply rooted concerns about mortality

texture gradient perceptual cue to depth based on the fact that the texture of objects changes from coarse and distinct for nearby objects to finer and less distinct for distant objects

thanatology the study of death, dying, grief, bereavement, and social attitudes toward these issues

theory an organized set of ideas that is designed to explain development

theory of mind ideas about connections between thoughts, beliefs, intentions, and behavior that create an intuitive understanding of the link between mind and behaviour

time-out punishment that involves removing children who are misbehaving from a situation to a quiet, unstimulating environment

toddlers young children who have just learned to walk

toddling early, unsteady form of walking done by infants

transient ischemic attack (TIA) an interruption of blood flow to the brain; often an early warning sign of stroke

traumatic distress expression of complicated or prolonged grief disorder that includes disbelief about the death; mistrust, anger, and detachment from others as a result of the death; feeling shocked by the death; and the experience of physical presence of the deceased

Type A behavior pattern a behavior pattern in which people tend to be intensely competitive, angry, hostile, restless, aggressive, and impatient

Type B behavior pattern a behavior pattern that is the opposite of Type A

ultrasound prenatal diagnostic technique that uses sound waves to generate an image of the fetus

umbilical cord structure containing veins and arteries that connects the developing child to the placenta

underextension when children define words more narrowly than adults do

uninvolved parenting style of parenting that provides neither warmth nor control and that minimizes the amount of time parents spend with children

universal ethical principles characteristic of Kohlberg's Stage 6, in which moral reasoning is based on moral principles that apply to all

universal versus context-specific development issue whether there is just one path of development or several paths

useful field of view (UFOV) an area from which one can extract visual information in a single glance without turning one's head or moving one's eyes

useful life expectancy the number of years that a person is free from debilitating chronic disease and impairment

validity extent to which a measure actually assesses what researchers think it does

vascular dementia disease caused by numerous small cerebral vascular accidents

vernix substance that protects the fetus's skin during development

very low birth weight newborns who weigh less than 1,500 grams (3 pounds)

visual acuity smallest pattern that one can distinguish reliably

visual cliff glass-covered platform that appears to have a "shallow" and a "deep" side; used to study infants' depth perception

visual expansion kinetic cue to depth perception that is based on the fact that an object fills an ever-greater proportion of the retina as it moves closer

vocational maturity degree of congruence between people's occupational behavior and what is expected of them at different ages

vulnerability–stress–adaptation model model that proposes that marriage is based on each partner contributing something to the relationship that the other would be hard-pressed to provide

waking activity state in which a baby's eyes are open but seem unfocused while the arms or legs move in bursts of uncoordinated motion

wear-and-tear disease a degenerative disease caused by injury or overuse

wear-and-tear theory suggests that the body, much like any machine, gradually deteriorates and finally wears out

whole-brain death declared only when the deceased meets eight criteria established in 1981

will according to Erikson, a young child's understanding that he or she can act on the world intentionally; this occurs when autonomy, shame, and doubt are in balance

word recognition the process of identifying a unique pattern of letters

work–family conflict the feeling of being pulled in multiple directions by incompatible demands from one's job and one's family

working memory type of memory in which a small number of items can be stored briefly

zone of maximum comfort when press level is slightly lower, facilitating a high quality of life

zone of maximum performance potential when press level is slightly higher, tending to improve performance

zone of proximal development difference between what children can do with assistance and what they can do alone

zygote fertilized egg

References

AAIDD Ad Hoc Committee on Terminology and Classification. (2010). *Intellectual disability* (11th ed.). Washington, DC: American Association on Intellectual and Developmental Disabilities.

AARP. (1999). *AARP/Modern Maturity sexuality survey: Summary of findings.* Retrieved from http://assets.aarp.org/rgcenter/health/mmsexsurvey.pdf

Aasland, O. G., Rosta, J., & Nylenna, M. (2010). Healthcare reforms and job satisfaction among doctors in Norway. *Scandinavian Journal of Public Health, 38,* 253–258.

Abengozar, M. C., Bueno, B., & Vega, J. L. (1999). Intervention on attitudes toward death along the life span. *Educational Gerontology, 25,* 435–447.

Aberson, C. L., Shoemaker, C., & Tomolillo, C. (2004). The role of interethnic friendships. *Journal of Social Psychology, 144,* 335–347.

Aboud, F. E. (1993). The developmental psychology of racial prejudice. *Transcultural Psychiatric Research Review, 30,* 229–242.

Aboud, F. E. (2003). The formation of in-group favoritism and out-group prejudice in young children: Are they distinct attitudes? *Developmental Psychology, 39,* 48–60.

Abraham, M. (2000). Isolation as a form of marital violence: The South Asian immigrant experience. *Journal of Social Distress and the Homeless, 9,* 221–236.

Abraham, R. (2000). Organizational cynicism: Bases and consequences. *Genetic, Social, and General Psychology Monographs, 126,* 269–292.

Acevedo, B. P., & Aron, A. (2009). Does a long-term relationship kill romantic love? *Review of General Psychology, 13,* 59–65.

Achor, S. (2010). *The happiness advantage: The seven principles of positive psychology that fuel success and performance at work.* New York: Random House.

Ackerman, B. P. (1993). Children's understanding of the speaker's meaning in referential communication. *Journal of Experimental Child Psychology, 55,* 56–86.

Adams, J. (1999). On neurodevelopmental disorders: Perspectives from neurobehavioral teratology. In H. Tager-Flusberg (Ed.), *Neurodevelopmental disorders* (pp. 451–468). Cambridge, MA: MIT Press.

Adams, M. J., Treiman, R., & Pressley, M. (1998). Reading, writing, and literacy. In W. Damon (Ed.), *Handbook of child psychology* (Vol. 4) (pp. 275–356). New York: Wiley.

Adams, R. G., & Ueno, K. (2006). Middle-aged and older adult men's friendships. In V. H. Bedford & B. Formaniak Turner (Eds.), *Men in relationships: A new look from a life course perspective* (pp. 103–124). New York: Springer.

Adams, R. J., & Courage, M. L. (1995). Development of chromatic discrimination in early infancy. *Behavioural Brain Research, 67,* 99–101.

Addis, D. R., Wong, A. T., & Schacter, D. L. (2008). Age-related changes in the episodic simulation of future events. *Psychological Science, 19,* 33–41.

Adler, L. L. (2001). Women and gender roles. In L. L. Adler & U. P. Gielen (Eds.), *Cross-cultural topics on psychology* (2nd ed., pp. 103–114). Westport, CT: Praeger/Greenwood.

Administration for Children and Families. (2010). *Administration for Children and Families healthy marriage initiative, 2002-2009.* Retrieved from http://www.healthymarriageinfo.org/docs/ACFGuideto09.pdf

Administration for Children and Families. (2010). *Head Start facts.* Author.

Adolph, K. E. (1997). Learning in the development of infant locomotion. *Monographs of the Society for Research in Child Development, 62,* 1–140.

Adolph, K. E. (2000). Specificity of learning: Why infants fall over a veritable cliff. *Psychological Science, 11,* 290–295.

Adolph, K. E. (2003). Learning to keep balance. In R. V. Kail (Ed.), *Advances in child development and behavior* (Vol. 30). Orlando, FL: Academic Press.

Adzick, N. S. (2010). Fetal myelomeningocele: Natural history, pathophysiology, and in-utero intervention. *Seminars in Fetal and Neonatal Medicine, 15,* 9–14.

Agahi, N., Ahacic, K., & Parker, M. G. (2006). Continuity of leisure participation from middle age to old age. *Journals of Gerontology: Psychological Sciences and Social Sciences, 61B,* S340–S346.

Aguayo, G. M. (2005). The effect of occupational environment and gender traditionality on self-efficacy for a nontraditional occupation in community college women. *Dissertation Abstracts International. Section B. Sciences and Engineering, 65(7-B),* 3696.

Ahluwalia, N. (2004). Aging, nutrition and immune function. *Journal of Nutrition, Health & Aging, 8,* 2–6.

Ahnert, L., Pinquart, M., & Lamb, M. E. (2006). Security of children's relationships with nonparental care providers: A meta-analysis. *Child Development, 77,* 664–679.

Ai, A. L., Wink, P., & Ardelt, M. (2010). Spirituality and aging: A journey for meaning through deep interconnection in humanity. In J. C. Cavanaugh & C. K. Cavanaugh (Eds.), *Aging in America: Vol. 3: Societal issues* (pp. 222–246). Santa Barbara, CA: Praeger Perspectives.

Ainsworth, M. S. (1978). The development of infant–mother attachment. In B. M. Caldwell & H. N. Ricciuti (Eds.), *Review of child development research* (Vol. 3, pp. 1–94). Chicago, IL: University of Chicago Press.

Ainsworth, M. S. (1993). Attachment as related to mother–infant interaction. *Advances in Infancy Research, 8,* 1–50.

Aisenbrey, S., Evertsson, M., & Grunow, D. (2009). Is there a career penalty for mothers' time out? A comparison of Germany, Sweden, and the United States. *Social Forces, 88,* 573–605.

Ajrouch, K. J. (2007). Health disparities and Arab-American elders: Does intergenerational support buffer the inequality-health link? *Journal of Social Issues, 63,* 745–758.

Akhtar, N., Jipson, J., & Callanan, M. (2001). Learning words through overhearing. *Child Development, 72,* 416–430.

Alberts, A. E. (2005). Neonatal behavioral assessment scale. In C. B. Fisher & R. M. Lerner (Eds.), *Encyclopedia of applied developmental science* (Vol. 1, pp. 111–115). Thousand Oaks, CA: Sage.

Aldridge, V., Dovey, T. M., & Halford, J. C. G. (2009). The role of familiarity in dietary development. *Developmental Review, 29,* 32–44.

Aldwin, C. M., & Gilmer, D. F. (2004). *Health, illness, and optimal aging: Biological and psychosocial perspectives.* Thousand Oaks, CA: Sage.

Alhija, F. N.-A., & Fresko, B. (2010). Socialization of new teachers: Does induction matter? *Teaching and Teacher Education, 26,* 1592–1597.

Allaire J. C., & Marsiske, M. (1999). Everyday cognition: Age and intellectual ability correlates. *Psychology and Aging, 14,* 627–644.

Allen, T. D. (2001). Family-supportive work environments: The role of organizational perceptions. *Journal of Vocational Behavior, 58,* 414–435.

Alley, J. L. (2004). The potential meaning of the grandparent-grandchild relationship as perceived by young adults: An exploratory study. *Dissertation Abstracts International. Section B. Sciences and Engineering, 65(3-B),* 1536.

Allgeier, E. R., & Allgeier, A. R. (2000). *Sexual interactions* (5th ed.). Belmont, CA: Wadsworth.

Almeida, J., Molnar, B. E., Kawachi, I., & Subramanian, S. V. (2009). Ethnicity and nativity status as determinants of perceived social support: Testing the concept of familism. *Social Science and Medicine, 68,* 1852–1858.

Almendarez, B. L. (2008). Mexican American elders and nursing home transition. *Dissertation Abstracts International. Section B. Sciences and Engineering, 68(7-B),* 4384.

Altschul, I., Oyserman, D., & Bybee, D. (2006). Racial-ethnic identity in midadolescence: Content and change as predictors of academic achievement. *Child Development, 77,* 1155–1169.

Alzheimer's Association. (2010). *2010 Alzheimer's disease facts and figures.* Retrieved from http://www.alz.org/documents_custom/report_alzfactsfigures2010.pdf

Amato, P. R. (2001). Children of divorce in the 1990s: An update of the Amato and Keith (1991) meta-analysis. *Journal of Family Psychology, 15,* 355–370.

Amato, P. R., & Cheadle, J. (2005). The long reach of divorce: Divorce and child well-being across three generations. *Journal of Marriage and Family, 67,* 191–206.

Amato, P. R., & Fowler, F. (2002). Parenting practices, child adjustment, and family diversity. *Journal of Marriage and Family, 64,* 703–716.

Amato, P. R., & Keith, B. (1991). Parental divorce and the well-being of children: A meta-analysis. *Psychological Bulletin, 110,* 26–46.

Amato, P. R., & Previti, D. (2003). People's reasons for divorcing: Gender, social class, the life course, and adjustment. *Journal of Family Issues, 24,* 602–626.

American Academy of Pain Management. (2009). Pain medicine position paper. *Pain Medicine, 10,* 972-1000. Retrieved March 5, 2011 from http://onlinelibrary.wiley.com/doi/10.1111/j.1526-4637.2009.00696.x/pdf.

American Academy of Pediatrics. (2005). Policy statement: Use of performance-enhancing substances. *Pediatrics, 115,* 1103–1106.

American Academy of Pediatrics. (2008). *Feeding kids right isn't always easy: Tips for preventing food hassles.* Elk Grove Village, IL: Author.

American Automobile Association. (2005). *AAA roadwise review—A tool to help seniors drive safely longer: Overview.* Retrieved from http://www.seniordrivers.org/driving/driving.cfm?button=roadwiseonline

American Cancer Society. (2010a). *Secondhand smoke.* Retrieved from http://www.cancer.org/Cancer/CancerCauses/TobaccoCancer/secondhand-smoke

American Cancer Society. (2010b). *Guide to quitting smoking.* Retrieved from http://www.cancer.org/Healthy/StayAwayfromTobacco/GuidetoQuitting Smoking/index

American Cancer Society. (2010c). *Cancer facts and figures 2010.* Retrieved from http://www.cancer.org/acs/groups/content/@epidemiology surveilance/documents/document/acspc-026238.pdf

American Heart Association. (2010). *Cholesterol.* Retrieved from http://www.heart.org/HEARTORG/Conditions/Cholesterol/CholestrolATH_UCM_001089_SubHomePage.jsp

American Psychiatric Association. (1994). *Diagnostic and statistical manual of mental disorders* (4th ed.). Washington, DC: Author.

American Psychological Association. (2004). Guidelines for psychological practice with older adults. *American Psychologist, 59,* 236–260.

American Psychological Association. (2007). *Blueprint for change: Achieving integrated health care for an aging population.* Retrieved from http://www.apa.org/pi/aging/programs/integrated/integrated-healthcare-report.pdf

American Psychological Society. (1993, December). Vitality for life. *APS Observer,* pp. 1–24.

Amso, D., & Johnson, S. P. (2006). Learning by selection: Visual search and object perception in young infants. *Developmental Psychology, 42,* 1236–1245.

Ancoli-Israel, S., & Alessi, C. (2005). Sleep and aging. *American Journal of Geriatric Psychiatry, 13,* 341–343.

Andersen, A. M. N., Wohlfahrt, J., Christens, P., Olsen, J., & Melbye, M. (2000). Maternal age and fetal loss: Population based register linkage study. *British Medical Journal, 320,* 1708–1712.

Anderson, C. A., Shibuya, A., Ihori, N., Swing, E. L., Bushman, B. J., Sakamoto, A., et al. (2010). Violent video game effects on aggression, empathy, and prosocial behavior in Eastern and Western countries: A meta-analytic review. *Psychological Bulletin, 136,* 151–173.

Anderson, D. R., Huston, A. C., Schmitt, K. L., Linebarger, D. L., & Wright, J. C. (2001). Early childhood television viewing and adolescent behavior. *Monographs of the Society for Research in Child Development, 66*(Serial No. 264).

Anderson, E. R., Greene, S. M., Walker, L., Malerba, C., Forgatch, M. S., & DeGarmo, D. S. (2004). Ready to take a chance again: Transitions into dating among divorced parents. *Journal of Divorce & Remarriage, 40,* 61–75.

Anderson, K. G., Tomlinson, K., Robinson, J. M., & Brown, S. A. (2011). Friends or foes: Social anxiety, peer affiliation, and drinking in middle school. *Journal of Studies on Alcohol and Drugs, 72,* 61–69.

Anderson, R. N., & Smith, B. L. (2005). Deaths: Leading causes for 2002. *National Vital Statistics Reports* (Vol. 53, No. 17). Hyattsville, MD: National Center for Health Statistics.

Anderson, S. W., Damasio, H., Tranel, D., & Damasio, A. R. (2001). Long-term sequelae of prefrontal cortex damage acquired in early childhood. *Developmental Neuropsychology, 18,* 281–296.

Anderson, V. D. (2007). Religiosity as it shapes parenting processes in preadolescence: A contextualized process model. *Dissertation Abstracts International. Section B. Sciences and Engineering, 67(9-B),* 5439.

Andreassi, J. K. (2007). The role of personality and coping in work-family conflict: New directions. *Dissertation Abstracts International. Section A. Humanities and Social Sciences, 67(8-A),* 3053.

Angel, J. L., Buckley, C. J., & Sakamoto, A. (2001). Duration or disadvantage? Exploring nativity, ethnicity, and health in midlife. *Journal of Gerontology: Social Sciences, 56B,* S275–S284.

Angeles, L. (2010). Children and life satisfaction. *Journal of Happiness Studies, 11,* 523–538.

Anisfeld, M. (1991). Neonatal imitation. *Developmental Review, 11,* 60–97.

Anisfeld, M. (1996). Only tongue protrusion modeling is matched by neonates. *Developmental Review, 16,* 149–161.

Annett, M. (2008). Test of the right shift genetic model for two new samples of family handedness and for the data of McKeever (2000). *Laterality, 13,* 105–123.

Antonucci, T. (2001). Social relations: An examination of social networks, social support, and sense of control. In J. E. Birren & K. W. Schaie (Eds.), *Handbook of the psychology of aging* (5th ed., pp. 427–453). San Diego, CA: Academic Press.

Antonucci, T. C., Akiyama, H., & Lansford, J. E. (1998). Negative effects of close social relations. *Family Relations, 47,* 379–384.

Apfelbaum, E. P., Pauker, K., Ambady, N., Sommers, S. R., & Norton, M. I. (2008). Learning (not) to talk about race: When older children underperform in social categorization. *Developmental Psychology, 44,* 1513–1518.

Apgar, V. (1953). A proposal for a new method of evaluation of the newborn infant. *Current Researches in Anesthesia and Analgesia, 32,* 260–267.

Aponte, M. (2007). Mentoring: Career advancement of Hispanic army nurses. *Dissertation Abstracts International. Section A. Humanities and Social Sciences, 68(4-A),* 1609.

Appleyard, K., Yang, C. M., & Runyan, D. K. (2010). Delineating the maladaptive pathways of child maltreatment: A mediated moderation analysis of the roles of self-perception and social support. *Development and Psychopathology, 22,* 337–352.

Araujo, A. B., Mohr, B. A., & McKinlay, J. B. (2004). Changes in sexual function in middle-aged and older men: Longitudinal data from the Massachusetts Aging Study. *Journal of the American Geriatrics Society, 52,* 1502–1509.

Archer, N., & Bryant, P. (2001). Investigating the role of context in learning to read: A direct test of Goodman's model. *British Journal of Psychology, 92,* 579–591.

Ardelt, M. (2010). Age, experience, and the beginning of wisdom. In D. Dannefer & C. Phillipson (Eds.), *The SAGE handbook of social gerontology* (pp. 306–316). Thousand Oaks, CA: Sage Publications.

Arellano, L. M. (2001). The psychological experiences of Latina professionals. *Dissertation Abstracts International. Section B. Sciences and Engineering, 62(1-B),* 534.

Armstrong-Stassen, M., & Templer, A. (2005). Adapting training for older employees: The Canadian response to an aging workforce. *Journal of Management Development, 24,* 57–67.

Arndt, J., & Vess, M. (2008). Tales from existential oceans: Terror management theory and how the awareness of our mortality affects us all. *Social and Personality Psychology Compass, 2,* 909–928.

Arnett, J. (2010b). Emerging adulthood(s): The cultural psychology of a new life stage. In L. A. Jensen (Ed.), *Bridging cultural and developmental approaches to psychology: New syntheses in theory, research, and policy* (pp. 255–275). New York: Oxford University Press.

Arnett, J. J. (2004). *Emerging adulthood: The winding road from the late teens through the twenties.* New York: Oxford University Press.

Arnett, J. J. (2007). Socialization in emerging adulthood: From the family to the wider world, from socialization to self-socialization. In J. E. Grusec & P. D. Hastings (Eds.), *Handbook of socialization: Theory and research* (pp. 208–231). New York: Guilford Press.

Arnett, J. J. (2010a). Oh, grow up! Generational grumbling and the new life stage of emerging adulthood—Commentary on Trzesniewski & Donnellan (2010). *Perspectives on Psychological Science, 5,* 89–92.

Aron, A., Fisher, H., Mashek, D. J., Strong, G., Li, H., & Brown, L. L. (2005). Reward, motivation, and emotion systems associated with early-stage intense romantic love. *Journal of Neurophysiology, 94,* 327–337.

Arseneault, L., Tremblay, R. E., Boulerice, B., & Saucier, J. F. (2002). Obstetrical complications and violent delinquency: Testing two developmental pathways. *Child Development, 73,* 496–508.

Årseth, A. K., Kroger, J., Martinussen, M., & Marcia, J. E. (2009). Meta-analytic studies of identity status and the relational issues of attachment and intimacy. *Identity, 9,* 1–32.

Artandi, S. E., & DePinho, R. A. (2010). Telomeres and telomerase in cancer. *Carcinogenesis, 31,* 9–18.

Artazcoz, L., Benach, J., Borrell, C., & Cortès, I. (2004). Unemployment and mental health: Understanding the interactions among gender, family roles, and social class. *American Journal of Public Health, 94,* 82–88.

Aryee, S., & Luk, V. (1996). Work and nonwork influences on the career satisfaction of dual-earner couples. *Journal of Vocational Behavior, 49,* 38–52.

Asbury, K., Dunn, J. F., Pike, A., & Plomin, R. (2003). Nonshared environmental influences on individual differences in early behavioral development: A monozygotic twin differences study. *Child Development, 74,* 933–943.

Asendorpf, J. B., Denissen, J. J. A., & van Aken, M. A. G. (2008). Inhibited and aggressive preschool children at 23 years of age: Personality and social transitions into adulthood. *Developmental Psychology, 44,* 997–1011.

Ashcraft, M. H. (1982). The development of mental arithmetic: A chronometric approach. *Developmental Review, 2,* 212–236.

Ashendorf, L., Constantinou, M., Duff, K., & McCaffrey, R. J. (2005). Performance of community-dwelling adults ages 55 to 75 on the University of Pennsylvania Smell Identification Test: An item analysis. *Applied Neuropsychology, 12,* 24–29.

Asher, S. R., & Paquette, J. A. (2003). Loneliness and peer relations in childhood. *Current Directions in Psychological Science, 12,* 75–78.

Aslin, R. N., Jusczyk, P. W., & Pisoni, D. B. (1998). Speech and auditory processing during infancy: Constraints on and precursors to language. In W. Damon (Ed.), *Handbook of child psychology* (Vol. 2). New York: Wiley.

Aslin, R. N., Saffran, J. R., & Newport, W. L. (1998). Computation of conditional probability statistics by 8-month-old infants. *Psychological Science, 9,* 321–324.

Asoodeh, M. H., Khalili, S., Daneshpour, N., & Lavasani, M. Gh. (2010). Factors of successful marriage: Accounts from self described happy couples. *Procedia Social and Behavioral Sciences, 5,* 2042–2046.

Aspenlieder, L., Buchanan, C. M., McDougall, P., & Sippola, L. K. (2009). Gender nonconformity and peer victimization in pre- and early adolescence. *European Journal of Developmental Science, 3,* 3–16.

Assanangkornchai, S., Tangboonngam, S., Samangsri, N., & Edwards, J. G. (2007). A Thai community's anniversary reaction to a major catastrophe. *Stress and Health: Journal of the International Society for the Investigation of Stress, 23,* 43–50.

Associated Press. (2007). *SUV hits several outside California middle school.* Retrieved from http://www.foxnews.com/story/0,2933,269717,00.html

Atchley, R. C. (1989). A continuity theory of normal aging. *The Gerontologist, 29,* 183–190.

Attig, T. (1996). *How we grieve: Relearning the world.* New York: Oxford University Press.

Atwater, E. (1992). *Adolescence.* Englewood Cliffs, NJ: Prentice Hall.

Au, T. K., & Glusman, M. (1990). The principle of mutual exclusivity in word learning: To honor or not to honor? *Child Development, 61,* 1474–1490.

Aunola, K., Stattin, H., & Nurmi, J.-E. (2000). Parenting styles and adolescents' achievement strategies. *Journal of Adolescence, 23,* 205–222.

Auyeung, B., Baron-Cohen, S., Ashwin, E., Knickmeyer, R., Taylor, K., Hackett, G., et al. (2009). Fetal testosterone predicts sexually differentiated childhood behavior in girls and in boys. *Psychological Science, 20,* 144–148.

Averett, P., & Jenkins, C. (in press). Review of the literature on older lesbians: Implications for education, practice, and research. *Journal of Applied Gerontology.*

Awa, W. L., Plaumann, M., & Walter, U. (2010). Burnout prevention: A review of intervention programs. *Patient Education and Counseling, 78,* 184–190.

Azaiza, F. (2005). Parent-child relationships as perceived by Arab adolescents living in Israel. *International Journal of Social Welfare, 14,* 297–304.

Bach, P. B. (2010). Postmenopausal hormone therapy and breast cancer: An uncertain trade-off. *JAMA, 304,* 1719–1720.

Bachman, J. G., & Schulenberg, J. (1993). How part-time work intensity relates to drug use, problem behavior, time use, and satisfaction among high school seniors: Are these consequences or merely correlates? *Developmental Psychology, 29,* 229–230.

Bachman, J. G., Staff, J. G., O'Malley, P. M., Schulenberg, J. E., & Freedman-Doan, P. (2011). Twelfth-grade student work intensity linked to later educational attainment and substance use: New longitudinal evidence. *Developmental Psychology, 47,* 344–363.

Backhouse, J. (2006). Grandparents-as-parents: Social change and its impact on grandparents who are raising their grandchildren. In: *Social Change in the 21st Century Conference,* Carseldine QUT, Brisbane, Australia. Retrieved from http://eprints.qut.edu.au/6072/1/6072.pdf

Backscheider, A. G., Shatz, M., & Gelman, S. A. (1993). Preschoolers' ability to distinguish living kinds as a function of regrowth. *Child Development, 64,* 1242–1257.

Badger, J. (2010). Assessing reflective thinking: Pre-service teachers' and professors' perceptions of an oral examination. *Assessment in Education: Principles, Policy, and Practice, 17,* 77–89.

Baek, J. (2005). Individual variations in family caregiving over the caregiving career. *Dissertation Abstracts International. Section B. Sciences and Engineering, 65(B),* 3769.

Baer, D. M., & Wolf, M. M. (1968). The reinforcement contingency in preschool and remedial education. In R. D. Hess & R. M Baer (Eds.), *Early education.* Chicago, IL: Aldine.

Bagwell, C. L. (2004). Friendships, peer networks and antisocial behavior. In J. B. Kupersmidt & K. A. Dodge (Eds.), *Children's peer relations* (pp. 37–57). Washington, DC: American Psychological Association.

Bagwell, C. L., Bender, S. E., Andreassi, C. L., Kinoshita, T. L., Montarello, S. A., & Muller, J. G. (2005). Friendship quality and perceived relationship changes predict psychosocial adjustment in early adulthood. *Journal of Social & Personal Relationships, 22,* 235–254.

Bagwell, C. L., Newcomb, A. F., & Bukowski, W. M. (1998). Preadolescent friendship and peer rejection as predictors of adult adjustment. *Child Development, 69,* 140–153.

Bahrick, L. E., & Lickliter, R. (2002). Intersensory redundancy guides early perceptual and cognitive development. In R. V. Kail (Ed.), *Advances in child development and behavior* (Vol. 30). Orlando, FL: Academic Press.

Bahrick, L. E., Lickliter, R., & Flom, R. (2004). Intersensory redundancy guides the development of selective attention, perception, and cognition in infancy. *Current Directions in Psychological Science, 13,* 99–102.

Bailey, D. A., & Rasmussen, R. L. (1996). Sport and the child: Physiological and skeletal issues. In F. L. Smoll & R. E. Smith (Eds.), *Children and youth in sport: A biopsychological perspective* (pp. 187–199). Dubuque, IA: Brown & Benchmark.

Bailey, J. A., Hill, K. G., Oesterle, S., & Hawkins, J. D. (2009). Parenting practices and problem behavior across three generations: Monitoring, harsh discipline, and drug use in the intergenerational transmission of externalizing behavior. *Developmental Psychology, 45,* 1214–1226.

Baillargeon, R. (1987). Object permanence in 3½- and 4½-month-old infants. *Developmental Psychology, 23,* 655–664.

Baillargeon, R. (1994). How do infants learn about the physical world? *Current Directions in Psychological Science, 3,* 133–140.

Baillargeon, R. (2004). Infants' reasoning about hidden objects: Evidence for event-general and event-specific expectations. *Developmental Science, 7,* 391–424.

Baillargeon, R. H., Zoccolillo, M., Keenan, K., Côté, S., Pérusse, D., Wu, H., et al., (2007). Gender differences in physical aggression: A prospective population-based survey of children before and after 2 years of age. *Developmental Psychology, 43,* 13–26.

Baker, L. (1994). Fostering metacognitive development. In H. W. Reese (Ed.), *Advances in child development and behavior* (Vol. 25) (pp. 201–239). San Diego, CA: Academic Press.

Baker, L., & Brown, A. L. (1984). Metacognitive skills and reading. In P. D. Pearson (Ed.), *Handbook of reading research* (Pt. 2) (pp. 353–394). New York: Longman.

Bakermans-Kranenburg, M., van IJzendoorn, M. H., & Juffer, F. (2003). Less is more: Meta-analyses of sensitivity and attachment interventions in early childhood. *Psychological Bulletin, 129,* 195–215.

Baldassar, L., Baldock, C. V., & Wilding, R. (2007). *Families caring across borders: Migration, ageing and transnational caregiving.* New York: Palgrave Macmillan.

Balk, D. E. & Corr, C. A. (2001). Bereavement during adolescence: A review of research. In M. S. Stroebe, R. O. Hansson, W. Stroebe, & H. Schut (Eds.), *Handbook of bereavement research: Consequences, coping, and care* (pp. 169–197). Washington, DC: American Psychological Association.

Ball, K. (1997). Enhancing mobility in the elderly: Attentional interventions for driving. In S. M. C. Dollinger & L. F. Dilalla (Eds.), *Assessment and intervention issues across the lifespan* (pp. 267–292). Mahwah, NJ: Erlbaum.

Ball, K., & Owsley, C. (1993). The Useful Field of View Test: A new technique for evaluating age-related declines in visual function. *Journal of the American Optometric Association, 64,* 71–79.

Ball, K., Owsley, C., Sloane, M. E., Roenker, D. L., & Bruni, J. R. (1993). Visual attention problems as a predictor of vehicle accidents among older drivers. *Investigative Ophthalmology and Visual Science, 34*(11), 3110–3123.

Ballesteros, S., Reales, J. M., & Mayas, J. (2007). Picture priming in normal aging and Alzheimer's disease. *Psicothema, 19,* 239–244.

Baltes, B. B., & Heydens-Gahir, H. A. (2003). Reduction of work-family conflict through the use of selection, optimization, and compensation behaviors. *Journal of Applied Psychology, 88,* 1005–1018.

Baltes, M. M., & Carstensen, L. L. (1999). Social-psychological theories and their applications to aging: From individual to collective. In V. L. Bengtson & K. W. Schaie (Eds.), *Handbook of theories of aging* (pp. 209–226). New York: Springer.

Baltes, P. B. (1987). Theoretical propositions of life-span developmental psychology: On the dynamics between growth and decline. *Developmental Psychology, 23,* 611–626.

Baltes, P. B. (1993). The aging mind: Potential and limits. *The Gerontologist, 33,* 580–594.

Baltes, P. B. (1997). On the incomplete architecture of human ontogeny: Selection, optimization, and compensation as foundation of developmental theory. *American Psychologist, 52,* 366–380.

Baltes, P. B., & Smith, J. (2003). New frontiers in the future of aging: From successful aging of the young old to the dilemmas of the fourth age. *Gerontology, 49,* 123–135.

Baltes, P. B., & Staudinger, U. M. (2000). Wisdom: A metaheuristic (pragmatic) to orchestrate mind and virtue toward excellence. *American Psychologist, 55,* 122–136.

Baltes, P. B., Lindenberger, U., & Staudinger, U. M. (2006). Life span theory in developmental psychology. In R. M. Lerner & W. Damon (Eds.), *Handbook of child psychology: Vol. 1. Theoretical models of human development* (6th ed., pp. 569–664). Hoboken, NJ: Wiley.

Baltes, P. B., Staudinger, U. M., & Lindenberger, U. (1999). Lifespan psychology: Theory and application to intellectual functioning. *Annual Review of Psychology, 50,* 471–507.

Bambra, C. (2010). Yesterday once more? Unemployment and health in the 21st century. *Journal of Epidemiology and Community Health, 64,* 213–215.

Bandura, A. (1977). *Social learning theory.* Englewood Cliffs, NJ: Prentice Hall.

Bandura, A. (1986). *Social foundations of thought and action: A social-cognitive theory.* Englewood Cliffs, NJ: Prentice Hall.

Bandura, A., & Bussey, K. (2004). On broadening the cognitive, motivational, and sociostructural scope of theorizing about gender development and functioning: Comment on Martin, Ruble, and Szkrybalo (2002). *Psychological Bulletin, 130,* 691–701.

Banger, M. (2003). Affective syndrome during perimenopause. *Maturitas, 41*(Suppl. I), S13–S18.

Bannard, C., & Mathews, D. (2008). Stored word sequences in language learning: The effect of familiarity on children's repetition of four-word combinations. *Psychological Science, 19,* 241–248.

Banse, R., Gawronski, B., Rebetez, C., Gutt, H., & Morton, J. B. (2010). The development of spontaneous gender stereotyping in childhood: Relations to stereotype knowledge and stereotype flexibility. *Developmental Science, 13,* 298–306.

Barber, B. K., & Olsen, J. A. (1997). Socialization in context: Connection, regulation, and autonomy in the family, school, and neighborhood, and with peers. *Journal of Adolescent Research, 12,* 287–315.

Barboza, D. (2011, January 12). *China, in a shift, takes on its Alzheimer's problem.* Retrieved from http://www.nytimes.com/2011/01/13/world/asia/13shanghai.html?_r=1&scp=1&sq=china%20alzheimer%27s&st=cse

Barca, L., Ellis, A. W., & Burani, C. (2007). Context-sensitive rules and word naming in Italian children. *Reading and Writing, 20,* 495–509.

Bardach, S. H., Gayer, C. C., Clinkinbeard, T., Zanjani, F., & Watkins, J. F. (2010). The malleability of possible selves and expectations regarding aging. *Educational Gerontology, 36,* 407–424.

Barenboim, C. (1981). The development of person perception in childhood and adolescence: From behavioral comparisons to psychological constructs to psychological comparisons. *Child Development, 52,* 129–144.

Barkley, R. A. (1996). Attention-deficit hyperactivity disorder. In E. J. Mash & R. A. Barkley (Eds.), *Child psychopathology* (pp. 63–112). New York: Guilford Press.

Barkley, R. A. (2003). Attention-deficit/hyperactivity disorder. In E. J. Mash & R. A. Barkley (Eds.), *Child psychopathology* (2nd ed., pp. 63–112). New York: Guilford Press.

Barkley, R. A. (2004). Adolescents with attention deficit/hyperactivity disorder: An overview of empirically based treatments. *Journal of Psychiatric Review, 10,* 39–56.

Barling, J., Zacharatos, A., & Hepburn, C. G. (1999). Parents' job insecurity affects children's academic performance through cognitive difficulties. *Journal of Applied Psychology, 84,* 437–444.

Barnard, J. W. (2010). Deception, decisions, and investor education. *The Elder Law Journal, 17,* 201.

Barnes, A. (Ed.). (2005). *The handbook of women, psychology, and the law.* New York: Wiley.

Barnes, H., & Parry, J. (2004). Renegotiating identity and relationships: Men and women's adjustments to retirement. *Ageing and Society, 24,* 213–233.

Baron-Cohen, S, (1995). *Mindblindness: An essay on autism and theory of mind.* Cambridge, MA: MIT Press/Bradford Books.

Barr, R., & Hayne, H. (1999). Developmental changes in imitation from television during infancy. *Child Development, 70,* 1067–1081.

Barton, M. E., & Tomasello, M. (1991). Joint attention and conversation in mother-infant-sibling triads. *Child Development, 62,* 517–529.

Bartsch, K., & Wellman, H. M. (1995). *Children talk about the mind.* New York: Oxford University Press.

Baschat, A. A. (2007). From amniocentesis to selective laser coagulation. *Ob.Gyn. News, 42* (17), 17–18.

Bascoe, S. M., Davies, P. T., Sturge-Apple, M. L., & Cummings, E. M. (2009). Children's representations of family relationships, peer information processing, and school adjustment. *Developmental Psychology, 45,* 1740–1751.

Baskett, L. M. (1985). Sibling status effects: Adult expectations. *Developmental Psychology, 21,* 441–445.

Basso, K. H. (1970). *The Cibecue Apache.* New York: Holt, Rinehart, & Winston.

Bates, E., Bretherton, I., & Snyder, L. (1988). *From first words to grammar: Individual differences and dissociable mechanisms.* New York: Cambridge University Press.

Bates, J. E., Pettit, G. S., Dodge, K. A., & Ridge, B. (1998). Interaction of temperamental resistance to control and restrictive parenting in the development of externalizing behavior. *Developmental Psychology, 34,* 982–995.

Bates, J. S. (2009). *Generative grandfathering, commitment, and contact: How grandfathers nurture relationships with grandchildren and the relational and mental health benefits for aging men.* Dissertation submitted in partial fulfillment of the doctor of philosophy degree at Syracuse University.

Batty, G. D., Wennerstad, K. M., Smith, G. D., Gunnell, D., Deary, I. J., Tynelius, P., & Rasmussen, F. (2009). IQ in early adulthood and morality in middle age: Cohort study of 1 million Swedish men. *Epidemiology, 20,* 100–109.

Bauer, P. J. (2006). Event memory. In W. Damon & R. M. Lerner (Eds.), *Handbook of child psychology* (6th ed., Vol. 2). New York: Wiley.

Bauer, P. J. (2007). *Remembering the times of our lives: Memory in infancy and beyond.* Mahwah, NJ: Erlbaum.

Bauer, P. J., & Lukowski, A. F. (2010). The memory is in the details: Relations between memory for the specific features of events and long-term recall during infancy. *Journal of Experimental Child Psychology, 107,* 1–14.

Baumann, A., Claudot, F., Audibert, G., Mertes, P.-M., & Puybasset, M. (2011). The ethical and legal aspects of palliative sedation in severely brain-injured patients: a French perspective. *Philosophy, Ethics, and Humanities in Medicine, 6.* Retrieved February 27, 2011 from http://preview.peh-med.com/content/pdf/1747-5341-6-4.pdf.

Baumeister, R.F. (2010). The self. In R. F. Baumeister & E. J. Finkel (Eds.), *Advanced social psychology: The state of the science* (pp. 139–175). New York: Oxford University Press.

Baumrind, D. (1975). *Early socialization and the discipline controversy.* Morristown, NJ: General Learning Press.

Baumrind, D. (1991). Parenting styles and adolescent development. In R. M. Lerner, A. C. Petersen, & J. Brooks-Gunn (Eds.), *Encyclopedia of adolescence* (pp. 746–758). New York: Garland.

Bauserman, R. (2002). Child adjustment in joint-custody versus sole-custody arrangements: A meta-analytic review. *Journal of Family Psychology, 16,* 91–102.

Bava, S., & Tapert, S. F. (2010). Adolescent brain development and the risk for alcohol and other drug problems. *Neuropsychology Review, 20,* 398–413.

Beal, C. R. (1996). The role of comprehension monitoring in children's revision. *Educational Psychology Review, 8,* 219–238.

Beal, C. R., & Belgrad, S. L. (1990). The development of message evaluation skills in young children. *Child Development, 61,* 705–712.

Beck, A. T. (1967). *Depression: Clinical, experimental, and theoretical aspects.* New York: Harper & Row.

Beck, A. T., Rush, J., Shaw, B., & Emery, G. (1979). *Cognitive therapy of depression.* New York: Guilford Press.

Beck, E., Burnet, K. L., & Vosper, J. (2006). Birth-order effects on facets of extraversion. *Personality and Individual Differences, 40,* 953–959.

Beck, J. G., & Averill, P. M. (2004). Older adults. In R. G. Heimberg, C. L. Turk, & D. S. Mennin (Eds.), *Generalized anxiety disorder: Advances in research and practice* (pp. 409–433). New York: Guilford Press.

Becker, B. J. (1986). Influence again: An examination of reviews and studies of gender differences in social influence. In J. S. Hyde & M. C. Linn (Eds.), *The psychology of gender differences: Advances through meta-analysis* (pp. 178–209). Baltimore, MD: Johns Hopkins University Press.

Becker, C. (2010). Another selective estrogen-receptor modulator for osteoporosis. *New England Journal of Medicine, 362,* 752–754.

Bedir, A., & Aksoy, Ş. (in press). Brain death revisited: It is not 'complete death' according to Islamic sources. *Journal of Medical Ethic.*

Beehr, T. A., & Bennett, M. M. (2007). Examining retirement from a multilevel perspective. In K. S. Shultz & G. A. Adams (Eds.), *Aging and work in the 21st century* (pp. 277–302). Mahwah, NJ: Erlbaum.

Behnke, M., & Eyler, F. D. (1993). The consequences of prenatal substance use for the developing fetus, newborn, and young child. *International Journal of the Addictions, 28,* 1341–1391.

Bekris, L. M., Yu, C.-E., Bird, T. D., & Tsuang, D. (2011). The genetics of Alzheimer's disease and Parkinson's disease. In J. P. Blass (Ed.), *Neurochemical mechanisms in disease* (pp. 695–756). New York: Springer.

Belsky, G. (2007). *Over the hill and between the sheets: Sex, love and lust in middle age.* New York: Springboard Press.

Belsky, J., Bakermans-Kranenburg, M. J., & van IJzendoorn, M. H. (2007). For better and for worse: Differential susceptibility to environmental influences. *Current Directions in Psychological Science, 16,* 300–304.

Belsky, J., Houts, R. M., & Fearon, R. M. P. (2010). Infant attachment security and the timing of puberty: Testing an evolutionary hypothesis. *Psychological Science, 21,* 1195–1201.

Belsky, J., Steinberg, L. D., Houts, R. M., Friedman, S. L., DeHart, G., Cauffman, E., et al. (2007). Family rearing antecedents of pubertal timing. *Child Development, 78,* 1302–1321.

Belsky, J., Steinberg, L., & Draper, P. (1991). Childhood experience, interpersonal development, and reproductive strategy: An evolutionary theory of socialization. *Child Development, 62,* 647–670.

Belsky, J., Steinberg, L., Houts, R. M., Halpern-Felsher, B., and the NICHD Early Child Care Research Network. (2010). The development of reproductive strategy in females: Early maternal harshness → earlier menarche → increased sexual risk taking. *Developmental Psychology, 46,* 120–128.

Ben Bashat, D., Ben Sira, L., Graif, M., Pianka, P., Hendler, T., Cohen, Y., et al. (2005). Normal white matter development from infancy to adulthood: Comparing diffusion tensor and high b value diffusion weighted MR images. *Journal of Magnetic Resonance Imaging, 21,* 503–511.

Benenson, J. F., & Christakos, A. (2003). The greater fragility of females' versus males' closest same-sex friendships. *Child Development, 74,* 1123–1129.

Bengtson, V. L., Mills, T. L., & Parrott, T. M. (1995). Ageing in the United States at the end of the century. *Korea Journal of Population and Development, 24,* 215–244.

Bengtsson, T., & Scott, K. (2011). Population aging and the future of the welfare state: The example of Sweden. *Population and Review, 37 (Supplement S1),* 158–170.

Bennett, D. S., Bendersky, M., & Lewis, M. (2008). Children's cognitive ability from 4 to 9 years old as a function of prenatal cocaine exposure, environmental risk, and maternal verbal intelligence. *Developmental Psychology, 44,* 919–928.

Bennett, K. M. (2010). How to achieve resilience as an older widower: Turning points or gradual change? *Ageing and Society, 30,* 369–382.

Benokraitis, N. V. (1999). *Marriages and families: Changes, choices, and constraints.* Upper Saddle River, NJ: Prentice Hall.

Bent, K. N., & Magilvy, J. K. (2006). When a partner dies: Lesbian widows. *Issues in Mental Health Nursing, 27,* 447–459.

Benton, D. (2010). The influence of dietary status on the cognitive performance of children. *Molecular Nutrition and Food Research, 54,* 457–470.

Benton, S. L., Corkill, A. J., Sharp, J. M., Downey, R. G., et al. (1995). Knowledge, interest, and narrative writing. *Journal of Educational Psychology, 87,* 66–79.

Berch, D. B. (2005). Making sense of number sense: Implications for children with mathematical disabilities. *Journal of Learning Disabilities, 38,* 333–339.

Berdes, C., & Zych, A. A. (2000). Subjective quality of life of Polish, Polish-immigrant, and Polish-American elderly. *International Journal of Aging & Human Development, 50,* 385–395.

Bereiter, C., & Scardamalia, M. (1987). *The psychology of written composition.* Hillsdale, NJ: Erlbaum.

Berenbaum, S. A., & Snyder, E. (1995). Early hormonal influences on childhood sex-typed activity and playmate preferences: Implications for the development of sexual orientation. *Developmental Psychology, 31,* 31–42.

Berenbaum, S. A., Duck, S. C., & Bryk, K. (2000). Behavioral effects of prenatal vs. postnatal androgen excess in children with 21-hydroxylase-deficient congenital adrenal hyperplasia. *Journal of Clinical Endocrinology and Metabolism, 85,* 727–733.

Bergen, D., & Mauer, D. (2000). Symbolic play, phonological awareness, and literacy skills at three age levels. In K. A. Roskos & J. F. Christie (Eds.), *Play and literacy in early childhood: Research from multiple perspectives* (pp. 45–62). Mahwah, NJ: Erlbaum.

Berger, S. E., Adolph, K. E., & Lobo, S. A. (2005). Out of the toolbox: Toddlers differentiate wobbly and wooden handrails. *Child Development, 76,* . 1294–1307.

Berk, L. E. (2003). Vygotsky, Lev. In L. Nadel (Ed.), *Encyclopedia of cognitive science* (Vol. 6). London: Macmillan.

Berk, R. A. (2010). Where's the chemistry in mentor-mentee academic relationships? Try spend mentoring. *The International Journal of Mentoring and Coaching, 8,* 85–92. Retrieved from http://www.ronberk.com/articles/2010_mentor.pdf

Berko, J. (1958). The child's learning of English morphology. *Word, 14,* 150–177.

Berkowitz, M. W., Sherblom, S., Bier, M., & Battistich, V. (2006). Educating for positive youth development. In M. Killen & J. G. Smetana (Eds.), *Handbook of moral development* (pp. 683–702). Mahwah, NJ: Erlbaum.

Berlin, L. J., Appleyard, K., & Dodge, K. A. (2011). Intergenerational continuity in child maltreatment: Mediating mechanisms and implications for prevention. *Child Development, 82,* 162–176.

Berlin, L. J., Brady-Smith, C., & Brooks-Gunn, J. (2002). Links between childbearing age and observed maternal behaviors with 14-month-olds in the Early Head Start Research and Evaluation Project. *Infant Mental Health Journal, 23,* 104–129.

Berlin, L. J., Ispa, J. M., Fine, M. A., Malone, P. S., Brooks-Gunn, J., Brady-Smith, C., et al. (2009). Correlates and consequences of spanking and verbal punishment for low-income White, African Americans and Mexican American toddlers. *Child Development, 80,* 1403–1420.

Berliner, A. J. (2000). Re-visiting Erikson's developmental model: The impact of identity crisis resolution on intimacy motive, generativity formation, and psychological adaptation in never-married, middle-aged adults. *Dissertation Abstracts International. Section B. Sciences and Engineering, 61,* 560.

Bernal, S., Dehaene-Lambertz, G., Millotte, S., & Christophe, A. (2010). Two-year-olds compute syntactic structure on-line. *Developmental Science, 13,* 69–76.

Berndt, T. J., & Keefe, K. (1995). Friends' influence on adolescents' adjustment to school. *Child Development, 66,* 1312–1329.

Berndt, T. J., & Murphy, L. M. (2002). Influences of friends and friendships: Myths, truths, and research recommendations. *Advances in Child Development and Behavior, 30,* 275–310.

Berndt, T. J., & Perry, T. B. (1990). Distinctive features and effects of adolescent friendships. In R. Montemayer, G. R. Adams, & T. P. Gullotta (Eds.), *From childhood to adolescence: A transition period?* London: Sage.

Berns, G. S., Moore, S., & Capra, C. M. (2009). *Adolescent engagement in dangerous behaviors is associated with increased white matter maturity of frontal cortex.* Retrieved from http://www.ncbi.nlm.nih.gov/pmc/articles/PMC2728774/

Berry, J. M., Hastings, E., West, R. L., Lee, C., & Cavanaugh, J. C. (2010). Memory aging: Deficits, beliefs, and interventions. In J. C. Cavanaugh & C. K. Cavanaugh (Eds.), *Aging in America: Vol. 1: Psychological aspects* (pp. 255–299). Santa Barbara, CA: Praeger Perspectives.

Bertenthal, B. H., & Clifton, R. K. (1998). Perception and action. In W. Damon (Ed.), *Handbook of child psychology* (Vol. 2). New York: Wiley.

Berthier, N. E. (1996). Learning to reach: A mathematical model. *Developmental Psychology, 32,* 811–823.

Besharov, D. J., & Gardiner, K. N. (1997). Trends in teen sexual behavior. *Children and Youth Services Review, 19,* 341–367.

Best, C. T. (1995). Learning to perceive the sound pattern of English. In C. Rovee-Collier (Ed.), *Advances in infancy research.* Norwood, NJ: Ablex.

Best, J. R. (in press). Effects of physical activity on children's executive function: Contributions of experimental research on aerobic exercise. *Developmental Review.*

Betz, N. E., Harmon, L. W., & Borgen, F. H. (1996). The relationships of self-efficacy for the Holland themes to gender, occupational group membership, and vocational interests. *Journal of Counseling Psychology, 43,* 90–98.

Bhatt, R. S., Bertin, E., Hayden, A., & Reed, A. (2005). Face processing in infancy: Developmental changes in the use of different kinds of relational information. *Child Development, 76,* 169–181.

Bhidayasiri, R., & Brenden, N. (2011). 10 commonly asked questions about Parkinson's disease. *The Neurologist, 17,* 57–62.

Bialystok, E. (1988). Levels of bilingualism and levels of linguistic awareness. *Developmental Psychology, 24,* 560–567.

Bialystok, E. (2010). Global-local and trail-making tasks by monolingual and bilingual children: Beyond inhibition. *Developmental Psychology, 46,* 93–105.

Biblarz, T. J., & Savci, E. (2010). Lesbian, gay, bisexual, and transgender families. *Journal of Marriage and Family, 72,* 480–497.

Biederman, J., Monuteaux, M. C., Mick, E., Spencer, T., Wilens, T. E., Silva, J. M., et al. (2006). Young adult outcome of attention deficit hyperactivity disorder: A controlled 10-year follow-up study. *Psychological Medicine, 362,* 167–179.

Biederman, J., Petty, C. R., Evans, M., Small, J., & Faraone, S. V. (2010). How persistent is ADHD? A controlled 10-year follow-up study of boys with ADHD. *Psychiatry Research, 177,* 299–304.

Bigler, R. S., & Liben, L. S. (2007). Developmental intergroup theory: Explaining and reducing children's social stereotyping and prejudice. *Current Directions in Psychological Science, 16,* 162–166.

Bigler, R. S., Jones, L. C., & Lobliner, D. B. (1997). Social categorization and the formation of intergroup attitudes in children. *Child Development, 68,* 530–543.

Bingenheimer, J. B., Brennan, R. T., & Earls, F. J. (2005). Firearm violence exposure and serious violent behavior. *Science, 308,* 1323–1326.

Birch, S. A., Akmal, N., & Frampton, K. (2010). Two-year-olds are vigilant of others non-verbal cues to credibility. *Developmental Science, 13,* 363–369.

Bird, D. J. (2001). The influences and impact of burnout on occupational therapists. *Dissertation Abstracts International. Section B. Sciences and Engineering, 62(1-B),* 204.

Birditt, K. S., Brown, E., Orbuch, T. L., & McIlvane, J. M. (2010). Marital conflict behaviors and implications for divorce over 16 years. *Journal of Marriage and Family, 72,* 1188–1204.

Biro, S., & Leslie, A. M. (2007). Infants' perception of goal-directed actions: Development through cue-based bootstrapping. *Developmental Science, 8,* 36–43.

Bishop, J. B. (2000). An environmental approach to combat binge drinking on college campuses. *Journal of College Student Psychotherapy, 15,* 15–30.

Bishop, N. A., Lu, T., & Yankner, B. A. (2010). Neural mechanisms of ageing and cognitive decline. *Nature, 464,* 529–535.

Bjerkedal, T., Kristensen, P., Skjeret, G. A., & Brevik, J. I. (2007). Intelligence test scores and birth order among young Norwegian men (conscripts) analyzed within and between families. *Intelligence, 35,* 503–514.

Björklund, A., & Dunnett, S. B. (2007). Dopamine neuron systems in the brain: An update. *Trends in Neurosciences, 30,* 194–202.

Bjorklund, D. F. (2005). *Children's thinking: Cognitive development and individual differences* (4th ed.). Belmont, CA: Wadsworth.

Bjorklund, D. F., & Pellegrini, A. D. (2000). Child development and evolutionary psychology. *Child Development, 71,* 1687–1708.

Bjorklund, D. F., Yunger, J. L., & Pellegrini, A. D. (2002). The evolution of parenting and evolutionary approaches to childrearing. In M. H. Bornstein (Ed.), *Handbook of parenting: Vol. 2. Biology and ecology of parenting* (pp. 3–30). Mahwah, NJ: Erlbaum.

Black, J. E. (2003). Environment and development of the nervous system. In I. B. Weiner, M. Gallagher, & R. J. Nelson (Eds.), *Handbook of psychology: Vol. 3. Biological psychology* (pp. 655–668). Hoboken, NJ: Wiley.

Black-Gutman, D., & Hickson, F. (1996). The relationship between racial attitudes and social-cognitive development in children: An Australian study. *Developmental Psychology, 32,* 448–456.

Blanchard-Fields, F. (1986). Reasoning on social dilemmas varying in emotional saliency: An adult developmental study. *Psychology and Aging, 1,* 325–333.

Blanchard-Fields, F. (1999). Social schematicity and causal attributions. In T. M. Hess & F. Blanchard-Fields (Eds.), *Social cognition and aging* (pp. 219–236). San Diego, CA: Academic Press.

Blanchard-Fields, F. (2007). Everyday problem solving and emotion: An adult developmental perspective. *Current Directions in Psychological Science, 16,* 26–31.

Blanchard-Fields, F. (2009). Flexible and adaptive socio-emotional problem solving in adult development and aging. *Restorative Neurology and Neuroscience, 27,* 539–550.

Blanchard-Fields, F. (2010). Neuroscience and aging. In J. C. Cavanaugh & C. K. Cavanaugh (Eds.), *Aging in America: Vol. 1: Psychological aspects* (pp. 1–25). Santa Barbara, CA: Praeger Perspectives.

Blanchard-Fields, F., & Hertzog, C. (2000). Age differences in social schematicity. In U. von Hecker, S. Dutke, & G. Sedek (Eds.), *Processes of generative mental representation and psychological adaptation* (pp. 175–198). Dordrecht, Netherlands: Kluwer.

Blanchard-Fields, F., Baldi, R. A., & Constantin, L. P. (2004). *Interrole conflict across the adult lifespan: The role of parenting stage, career stages and quality of experiences.* Unpublished manuscript, School of Psychology, Georgia Institute of Technology.

Blanchard-Fields, F., Janke, H. C., & Camp, C. J. (1995). Age differences in problem-solving style: The role of emotional salience. *Psychology and Aging, 10,* 173–180.

Blatterer, H. (2005). *New adulthood: Personal or social transition?* Paper delivered at the Social Change in the 21st Century Conference, Queensland University. Retrieved from http://eprints.qut.edu.au/3485/1/3485.pdf

Bloom, L. (1998). Language acquisition in its developmental context. In D. Kuhn & R. S. Siegler (Eds.), *Handbook of child psychology: Vol. 2. Cognition, perception, and language* (5th ed., pp. 309–370). New York: Wiley.

Bloom, L., & Tinker, E. (2001). The intentionality model and language acquisition. *Monographs of the Society for Research in Child Development, 66*(Serial No. 267).

Bloom, L., Margulis, C., Tinker, E., & Fujita, N. (1996). Early conversations and word learning: Contributions from child and adult. *Child Development, 67,* 3154–3175.

Blossfeld, H.-P. (2009). Educational assortative marriage in comparative perspective. *Annual Review of Sociology, 35,* 513–530.

Boelen, P. A., & Prigerson, H. G. (2007). The influence of symptoms of prolonged grief disorder, depression, and anxiety on quality of life among bereaved adults: A prospective study. *European Archives of Psychiatry and Clinical Neuroscience, 257,* 444–452.

Boivin, M., Vitaro, F., & Gagnon, C. (1992). A reassessment of the self-perception profile for children: Factor structure, reliability, and convergent validity of a French version among second through sixth grade children. *International Journal of Behavioral Development, 15,* 275–290.

Bonanno, G. A. (2009). *The other side of sadness: What the new science of bereavement tells us about life after loss.* New York, NY: Basic Books.

Bonanno, G. A., Papa, A., & O'Neill, K. (2001). Loss and human resilience. *Applied and Preventive Psychology, 10,* 193–206.

Bonanno, G. A., Papa, A., Lalande, K., Zhang, N., & Noll, J. G. (2005). Grief processing and deliberate grief avoidance: A prospective comparison of bereaved spouses and parents in the United States and the People's Republic of China. *Journal of Consulting and Clinical Psychology, 73,* 86–98.

Bonanno, G. A., Wortman, C. B., & Neese, R. M. (2004). Prospective patterns of resilience and maladjustment during widowhood. *Psychology and Aging, 19,* 260–271.

Bonanno, G., & Kaltman, S. (1999). Toward an integrative perspective on bereavement. *Psychological Bulletin, 125,* 760–776.

Book, P. L. (1996). How does the family narrative influence the individual's ability to communicate about death? *Omega: The Journal of Death and Dying, 33,* 323–342.

Boon, H., Ruiter, R. U. C., James, S., van den Borne, B., Williams, E., & Reddy, P. (2010). Correlates of grief among older adults caring for children and grandchildren as a consequence of HIV and AIDS in South Africa. *Journal of Aging and Health, 22,* 48–67.

Bornstein, B. H., & Adya, M. (2007). What can researchers tell the courts, and what can the courts tell researchers about sexual harassment? In R. L. Wiener, B. H. Bornstein, R. Schopp & S. L. Willborn (Eds.), *Social consciousness in legal decision making: Psychological perspectives* (pp. 197–206). New York: Springer.

Bornstein, M. H., & Arterberry, M. E. (2003). Recognition, discrimination, and categorization of smiling by 5-month-old infants. *Developmental Science, 6,* 585–599.

Bornstein, M. H., Putnick, D. L., Suwalsky, J. T., & Gini, M. (2006). Maternal chronological age, prenatal and perinatal history, social support, and parenting of infants. *Child Development, 77,* 875–892.

Borson, S. (2011). Depression and anxiety in COPD: Diagnosis and management issues. In N. A. Hanania & A. Sharafkhaneh (Eds.), *COPD: A guide to diagnosis and clinical management* (pp. 271–281). New York: Springer.

Bosco, F. M., Friedman, O., & Leslie, A. M. (2006). Recognition of pretend and real actions in play by 1- and 2-year-olds: Early success and why they fail. *Cognitive Development, 21,* 3–10.

Boseovski, J. J. (2010). Evidence for "rose-colored glasses": An examination of the positivity bias in young children's personality judgments. *Child Development Perspectives, 4,* 212–218.

Bosma, H. A., & Kunnen, E. S. (2001). Determinants and mechanisms in ego identity development: A review and synthesis. *Developmental Review, 21,* 39–66.

Boss, P. (2006). *Loss, trauma, and resilience: Therapeutic work with ambiguous loss.* New York: Norton.

Bosshard, G., & Materstvedt, L. J. (2011). Medical and societal issues in euthanasia and assisted suicide. In R. Chadwick, H. ten Have, & E. M. Meslin (Eds.), *The SAGE handbook of health care ethics* (pp.). thousand Oaks, CA: Sage Publications.

Bot, S. M., Engels, R. C. M. E., Knibbe, R. A., & Meeus, W. H. J. (2005). Friends' drinking behavior and adolescent alcohol consumption: The moderating role of friendship characteristics. *Addictive Behaviors, 30,* 929–947.

Bouchard, T. J. (2004). Genetic influence on human psychological traits. *Current Directions in Psychological Science, 13,* 148–151.

Bouchard, T. J. (2009). Genetic influence on human intelligence (Spearman's g): How much? *Annals of Human Biology, 36,* 527–544.

Bouldin, E. D., & Andresen, E. (2010). Caregiving and health. Personal relationships in later life. In J. C. Cavanaugh & C. K. Cavanaugh (Eds.), *Aging in America: Vol. 2: Physical and mental health* (pp. 81–99). Santa Barbara, CA: Praeger Perspectives.

Bouldin, P.L., & Grayson, A. M. (2010). *Perceptions of sexual harassment and sexual assault: A study of gender differences among U.S. Navy officers.* Master's thesis completed at the Naval Postgraduate School, Monterey, CA. Retrieved from http://edocs.nps.edu/npspubs/scholarly/theses/2010/Mar/10Mar_Bouldin.pdf

Bowker, A. (2006). The relationship between sports participation and self-esteem during early adolescence. *Canadian Journal of Behavioral Science, 38,* 214–229.

Bowlby, J. (1969). *Attachment and loss* (Vol. 1). New York: Basic Books.

Bowlby, J. (1991) Ethological light on psychoanalytical problems. In P. Bateson (Ed.), *The development and integration of behaviour: Essays in honour of Robert Hinde* (pp. 301–313). New York: Cambridge University Press.

Bowles, S. (2003, July 18–20). More older drivers in car accidents. *USA Today,* p. 1.

Bowman, N. A. (2010). The development of psychological well-being among first-year college students. *Journal of College Student Development, 51,* 180–200.

Boyle, P. J., Feng, Z., & Raab, G. M. (2011). Does widowhood increase mortality risk? Testing for selection effects by comparing causes of spousal death. *Epidemiology, 22,* 1–5.

Bozikas, V., Kioseoglou, V., Palialia, M., Nimatoudis, I., Iakovides, A., Karavatos, A., & Kaprinis, G. (2000). Burnout among hospital workers and community-based mental health staff. *Psychiatriki, 11,* 204–211.

Bradley, R. H., Corwyn, R. F., Burchinal, M., McAdoo, H. P., & Coll, C. G. (2001). The home environment of children in the United States Part II: Relations with behavioral development through age thirteen. *Child Development, 72,* 1868–1886.

Braine, M. D. S. (1976). Children's first word combinations. *Monographs of the Society for Research in Child Development, 41*(Serial No. 164).

Braine, M. D. S. (1992). What sort of innate structure is needed to "bootstrap" into syntax? *Cognition, 45,* 77–100.

Brainerd, C. J. (1996). Piaget: A centennial celebration. *Psychological Science, 7,* 191–203.

Brandone, A. C., &Wellman, H. M. (2009). You can't always get what you want: Infants understand failed goal-directed actions. *Psychological Science, 20,* 85-91.

Brandtstädter, J. (1989). Personal self-regulation of development: Cross-sequential analyses of development-related control beliefs and emotions. *Developmental Psychology, 25,* 96–108.

Brandtstädter, J. (1999). Sources of resilience in the aging self. In T. M. Hess & F. Blanchard-Fields (Eds.), *Social cognition and aging* (pp. 123–141). San Diego, CA: Academic Press.

Bratkovich, K. L. (2010). *The relationship of attachment and spirituality with posttraumatic growth following a death for college students.* Doctoral dissertation submitted to the Department of Psychology at Oklahoma State University.

Brault, M. W. (2008). *Americans with disabilities: 2005.* Retrieved from http://www.census.gov/prod/2008pubs/p70-117.pdf

Braun, S. D. (2007). Gay fathers with children adopted from foster care: Understanding their experiences and predicting adoption outcomes. *Dissertation Abstracts International. Section B. Sciences and Engineering, 68(2-B)*, 1296.

Braungart, J. M., Plomin, R., DeFries, J. C., & Fulker, D. W. (1992). Genetic influence on tester-rated infant temperament as assessed by Bayley's Infant Behavior Record: Nonadoptive and adoptive siblings and twins. *Developmental Psychology, 28*, 40–47.

Braungart-Rieker, J. M., Hill-Soderlund, A. L., & Karrass, J. (2010). Fear and anger reactivity trajectories from 4 to 16 months: The roles of temperament, regulation, and maternal sensitivity. *Developmental Psychology, 46*, 791–804.

Braver, T. S., & West, R. F. (2008). Working memory, executive control, and aging. In F. I. M. Craik & T. A. Salthouse (Eds.) *The handbook of aging and cognition* (pp. 311–372). New York: Psychology Press.

Brazelton, T. B., & Nugent, J. K. (1995). *Neonatal behavioral assessment scale* (3rd ed.). London: Mac Keith.

Brazelton, T. B., Nugent, J. K., & Lester, B. M. (1987). Neonatal behavioral assessment scale. In J. D. Osofsky (Ed.), *Handbook of infant development* (2nd ed.). New York: Wiley.

Brechwald, W. A., & Prinstein, M. J. (2011). Beyond homophily: A decade of advances in understanding peer influence processes. *Journal of Research on Adolescence, 21*, 166–179.

Brendgen, M., Lamarche, V., Wanner, B., & Vitaro, F. (2010). Links between friendship relations and early adolescents' trajectories of depressed mood. *Developmental Psychology, 46*, 491–501.

Brendgen, M., Vitaro, F., Boivin, M., Dionea, G., & Perusse, D. (2006). Examining genetic and environmental effects on reactive versus proactive aggression. *Developmental Psychology, 42*, 1299–1312.

Brennan, P. A., Grekin, E. R., Mortensen, E. L., & Mednick, S. A. (2002). Relationship of maternal smoking during pregnancy with criminal arrest and hospitalization for substance abuse in male and female adult offspring. *American Journal of Psychiatry, 159*, 48–54.

Bridges, C. R. (1996). The characteristics of career achievement perceived by African American college administrators. *Journal of Black Studies, 26*, 748–767.

Briebiescas, R. G. (2010). An evolutionary and life history perspective on human male reproductive senescence. *Annals of the New York Academy of Sciences, 1204*, 54–64.

Brison, K. J. (1995). You will never forget: Narrative, bereavement, and worldview among Kwanga women. *Ethos, 23*, 474–488.

Brissett-Chapman, S., & Isaacs-Shockley, M. (1997). *Children in social peril: A community vision for preserving family care of African American children and youth.* Washington, DC: Child Welfare League of America.

Brissette, I., Scheier, M. F., & Carver, C. S. (2002). The role of optimism in social network development, coping, and psychological adjustment during a life transition. *Journal of Personality and Social Psychology, 82*, 102–111.

Brockington, I. (1996). *Motherhood and mental health.* Oxford, England: Oxford University Press.

Brody, G. H. (1998). Sibling relationship quality: Its causes and consequences. *Annual Review of Psychology, 49*, 1–24.

Brody, G. H., & Ge, X. (2001). Linking parenting processes and self-regulation to psychological functioning and alcohol use during early adolescence. *Journal of Family Psychology, 15*, 82–94.

Brody, G. H., Kim, S., Murry, V. M., & Brown, A. C. (2003). Longitudinal direct and indirect pathways linking older sibling competence to the development of younger sibling competence. *Developmental Psychology, 39*, 618–628.

Brody, G. H., Stoneman, A., & McCoy, J. K. (1994). Forecasting sibling relationships in early adolescence from child temperament and family processes in middle childhood. *Child Development, 65*, 771–784.

Brody, N. (1992). *Intelligence* (2nd ed.). San Diego, CA: Academic Press.

Brodzinsky, D. M., & Pinderhughes, E. (2002). Parenting and child development in adoptive families. In M. H. Bornstein (Ed), *Handbook of parenting: Vol. 1: Children and parenting* (2nd ed., pp. 279–311). Mahwah, NJ: Lawrence Erlbaum.

Bronfenbrenner, U. (1979). Contexts of child rearing: Problems and prospects. *American Psychologist, 34*, 844–850.

Bronfenbrenner, U. (1989). Ecological systems theory. In R. Vasta (Ed.), *Annals of child development: Vol. 6. Theories of child development: Revised formulations and current issues.* Greenwich, CT: JAI Press.

Bronfenbrenner, U. (1995). Developmental ecology through space and time: A future perspective. In P. Moen, G. H. Elder, Jr., & K. Luscher (Eds.), *Examining lives in context: Perspectives on the ecology of human development* (pp. 619–647). Washington, DC: American Psychological Association.

Bronfenbrenner, U., & Morris, P. (2006). The ecology of developmental processes. In W. Damon & R. M. Lerner (Eds.), *Handbook of child psychology* (6th ed., Vol. 1 pp. 793–828). New York: Wiley.

Brooks-Gunn, J., & Paikoff, R. (1993). "Sex is a gamble, kissing is a game": Adolescent sexuality, contraception, and sexuality. In S. P. Millstein, A. C. Petersen, & E. O. Nightingale (Eds.), *Promoting the health behavior of adolescents* (pp. 180–208). New York: Oxford University Press.

Brooks-Gunn, J., & Ruble, D. N. (1982). The development of menstrual-related beliefs and behaviors during early adolescence. *Child Development, 53*, 1567–1577.

Brough, P., O'Driscoll, M. P., & Kalliath, T. J. (2005). The ability of "family friendly" organizational resources to predict work-family conflict and job and family satisfaction. *Stress and Health: Journal of the International Society for the Investigation of Stress, 21*, 223–234.

Brougham, R. R., & Walsh, D. A. (2005). Goal expectations as predictors of retirement intentions. *International Journal of Aging & Human Development, 61*, 141–160.

Brown, B. B., & Klute, C. (2003). Friendships, cliques, and crowds. In G. R. Adams & M. D. Berzonsky (Eds.), *Blackwell handbook of adolescence* (pp. 330–348). Malden, MA: Blackwell Publishing.

Brown, B. B., Lohr, M. J., & McClenahan, E. L. (1986). Early adolescents' perceptions of peer pressure. *Journal of Early Adolescence, 6*, 139–154.

Brown, B. B., Mounts, N., Lamborn, S. D., & Steinberg, L. (1993). Parenting practices and peer group affiliation in adolescence. *Developmental Psychology, 64*, 467–482.

Brown, J. R., & Dunn, J. (1996). Continuities in emotion understanding from three to six years. *Child Development, 67*, 789–802.

Brown, J. V., Bakeman, R., Coles, C. D., Platzman, K. A., & Lynch, M. E. (2004). Prenatal cocaine exposure: A comparison of 2-year-old children in parental and nonparental care. *Child Development, 75*, 1282–1295.

Brown, J. W., Chen, S-l., Mefford, L., Brown, A., Callen, B., & McArthur, P. (2011). Becoming an older volunteer: A grounded theory study. *Nursing Research and Practice, 2011.* doi: 10.1155/2011/361250. Retrieved from http://www.hindawi.com/journals/nrp/2011/361250.html

Brown, J., & Dunn, J. (1992). Talk with your mother or your sibling? Developmental changes in early family conversations about feelings. *Child Development, 63*, 336–349.

Brown, J., Meadows, S. O., & Elder, G. H., Jr. (2007). Race-ethnic inequality and psychological distress: Depressive symptoms from adolescence to young adulthood. *Developmental Psychology, 43*, 1295–1311.

Brown, R., Pressley, M., Van Meter, P., & Schuder, T. (1996). A quasi-experimental validation of transactional strategies instruction with low-achieving second-grade readers. *Journal of Educational Psychology, 88*, 18–37.

Browne, K. D., & Hamilton-Giachritsis, C. (2005). The influence of violent media on children and adolescents: A public-health approach. *Lancet, 365*, 702–710.

Bruno, J. L., Manis, F. R., Keating, P., Sperling, A. J., Nakamoto, J., & Seidenberg, M. S. (2007). Auditory word identification in dyslexic and normally achieving readers. *Journal of Experimental Child Psychology, 97*, 183–204.

Bryce, J. (2001). The technological transformation of leisure. *Social Science Computer Review, 19*, 7–16.

Brysiewicz, P. (2008). The lived experience of losing a loved one to a sudden death in KwaZulu-Natal, South Africa. *Journal of Clinical Nursing, 17*, 224–231.

Buchanan, C. M., & Heiges, K. L. (2001). When conflict continues after the marriage ends: Effects of postdivorce conflict on children. In J. Grych & F. D. Fincham (Eds.), *Interparental conflict and child development* (pp. 337–362). New York: Cambridge University Press.

Buchanan, C. M., Eccles, J. S., & Becker, J. B. (1992). Are adolescents the victims of raging hormones? Evidence for activational effects of hormones on moods and behavior at adolescence. *Psychological Bulletin, 111*, 62–107.

Buchsbaum, B. C. (1996). Remembering a parent who has died: A developmental perspective. In D. Klass, P. R. Silverman, & S. L. Nickman (Eds.), *Continuing bonds: New understandings of grief* (pp. 113–124). Washington, DC: Taylor & Francis.

Bugental, D. B., & Happaney, K. (2004). Predicting infant maltreatment in low-income families: The interactive effects of maternal attributions and child status at birth. *Developmental Psychology, 40,* 234–243.

Bugental, D. B., & Schwartz, A. (2009). A cognitive approach to child maltreatment prevention among medically at-risk infants. *Developmental Psychology, 45,* 284–288.

Buhl, H. (2008). Development of a model describing individuated adult child-parent relationships. *International Journal of Behavioral Development, 32,* 381–389.

Buhrmester, D., & Furman, W. (1990). Perceptions of sibling relationships during middle childhood and adolescence. *Child Development, 61,* 1387–1398.

Bukowski, W. M., Sippola, L. K., & Hoza, B. (1999). Same and other: Interdependency between participation in same- and other-sex friendships. *Journal of Youth and Adolescence, 28,* 439–459.

Bulduc, J. L., Caron, S. L., & Logue, M. E. (2007). The effects of parental divorce on college students. *Journal of Divorce & Remarriage, 46,* 83–104.

Bullard, L., Wachlarowicz, M., DeLeeuw, J., Snyder, J., Low, S., Forgatch, M., et al. (2010). Effects of the Oregon model of Parent Management Training (PMTO) on marital adjustment in new stepfamilies: A randomized trial. *Journal of Family Psychology, 24,* 485–496.

Bullock, K. (2004). Family social support. In P. J. Bomar (Ed.), *Promoting health in families: Applying family research and theory to nursing practice* (3rd ed., pp. 142–161). Philadelphia: W. B. Saunders.

Bullock, M., & Lütkenhaus, P. (1990). Who am I? The development of self-understanding in toddlers. *Merrill-Palmer Quarterly, 36,* 217–238.

Bumpass, L. L., & Aquilino, W. S. (1995). *A social map of midlife: Family and work over the middle years.* Madison, WI: University of Wisconsin, Center for Demography and Ecology.

Burack, J. A., Flanagan, T., Peled, T., Sutton, H. M., Zygmuntowicz, C., & Manly, J. T. (2006). Social perspective-taking skills in maltreated children and adolescents. *Developmental Psychology, 42,* 207–217.

Burchinal, M. R., Roberts, J. E., Riggins, R., Zeisel, S. A., Neebe, E., & Bryant, D. (2000). Relating quality of center-based child care to early cognitive and language development longitudinally. *Child Development, 71,* 338–357.

Bureau of Labor Statistics. (2010a). *Women in the labor force: A databook (2009 edition).* Retrieved from http://www.bls.gov/cps/wlf-intro-2009.htm

Bureau of Labor Statistics. (2010b). *Employment Characteristics of Families survey.* Retrieved from http://www.bls.gov/news.release/famee.nr0.htm

Bureau of Labor Statistics. (2010c). *Labor force statistics from the Current Population Survey.* Retrieved from http://www.bls.gov/cps

Bureau of Labor Statistics. (2010d). *Employment Characteristics of Families survey.* Retrieved from http://www.bls.gov/news.release/famee.nr0.htm

Bureau of Labor Statistics. (2010e). *Employment characteristics of families summary.* Retrieved from http://www.bls.gov/news.release/famee.nr0.htm

Burgess, D., & Borgida, E. (1997). Sexual harassment: An experimental test of sex-role spillover theory. *Personality and Social Psychology Bulletin, 23,* 63–75.

Burianova, H., McIntosh, A. R., & Grady, C. L. (2010). A common functional brain network for autobiographical, episodic, and semantic memory retrieval. *Neuroimage, 49,* 865–874.

Burk, W. J., & Laursen, B. (2005). Adolescent perceptions of friendship and their associations with individual adjustment. *International Journal of Behavioral Development, 29,* 156–164.

Burke, B. L., Martens, A., & Faucher, E. H. (2010). Two decades of terror management theory: a meta-analysis of mortality salience research. *Personality and Social Psychology Review, 14,* 155–195.

Burnette, D. (1999). Social relationships of Latino grandparent caregivers: A role theory perspective. *The Gerontologist, 39,* 49–58.

Burns, S. T. (2005). The transition to parenthood: The effects of gender role attitudes and level of culture change in Chinese-American couples. *Dissertation Abstracts International. Section B. Sciences and Engineering, 66(6-B),* 3461.

Burt, K. B., & Masten, A. S. (2010). Development in the transition to adulthood: Vulnerabilities and opportunities. In J. E. Grant & M. N. Potenza (Eds.), *Young adult mental health* (pp. 5–18). Oxford, England: Oxford University Press.

Burton, L. M. (1992). Black grandparents rearing children of drug-addicted parents: Stressors, outcomes, and social service needs. *The Gerontologist, 32,* 744–751.

Busch, J., & Rodogno, R. (2011). Life support and euthanasia, a perspective on Shaw's new perspective. *Journal of Medical Ethics, 37,* 81–83.

Buss, D. M., Abbott, M., Angeleitner, A., Asherian, A., Biaggio, A., Blanco Villasenor, A., et al. (1990). International preferences in selecting mates: A study of 37 cultures. *Journal of Cross-Cultural Psychology, 21,* 5–47.

Buss, K. A., & Goldsmith, H. H. (1998). Fear and anger regulation in infancy: Effects on the temporal dynamics of affective expression. *Child Development, 69,* 359–374.

Buss, K. A., & Kiel, E. J. (2004). Comparison of sadness, anger, and fear facial expressions when toddlers look at their mothers. *Child Development, 75,* 1761–1773.

Busseri, M. A., Choma, B. L., & Sadava, S. W. (2009). Function or fantasy? Examining the implications of subjective temporal perspective "trajectories" for life satisfaction. *Personality and Social Psychology Bulletin, 35,* 295–308.

Bustos, M. L. C. (2007). La muerte en la cultura occidental: Antropología de la muerte [Death in Western culture: Anthropology of death]. *Revista Colombiana de Psiquiatría, 36,* 332–339.

Buttell, F., & Carney, M. M. (2007). Emerging trends in batterer intervention programming: Equipping forensic workers for effective practice. In D. W. Springer & A. R. Roberts (Eds.), *Handbook of forensic mental health with victims and offenders: Assessment, treatment, and research* (pp. 151–169). New York: Springer.

Bylsma, F. W., Ostendorf, C. A., & Hofer, P. J. (2002). Challenges in providing neuropsychological and psychological services in Guam and the Commonwealth of the Northern Marianas Islands (CNMI). In F. R. Ferraro (Ed.), *Minority and cross-cultural aspects of neuropsychological assessment: Studies on neuropsychology, development, and cognition* (pp. 145–157). Bristol, PA: Swets & Zeitlinger.

Byock, I. (1997). *Dying well.* New York: Riverhead.

Byock, S. D. (2010, August). *The quarterlife crisis and the path to individuation in the first half of life.* Master's thesis completed at the Pacifica Graduate Institute, Carpinteria, CA.

Byrne, B. M., & Gavin, D. A. W. (1996). The Shavelson model revisited: Testing for the structure of academic self-concept across pre-, early, and late adolescents. *Journal of Educational Psychology, 88,* 215–228.

Cable, N., & Sacker, A. (2008). Typologies of alcohol consumption in adolescence: Predictors and adult outcomes. *Alcohol and Alcoholism, 43,* 81–90.

Cacioppo, J. T., Berntson, G. G., Bechara, A., Tranel, D., & Hawkley, H. C. (2011). Could an aging brain contribute to subjective well-being? The value added by a social neuroscience perspective. In A. Todorov, S. Fiske, & D. Prentice (Eds.), *Social neuroscience: toward understanding the underpinnings of the social mind* (pp. 249–262). New York: Oxford University Press.

Cade, R. (2010). Covenant marriage. *The Family Journal, 18,* 230–233.

Cadell, S., & Marshall, S. (2007). The (re)construction of self after the death of a partner to HIV/AIDS. *Death Studies, 31,* 537–548.

Cahill, E., Lewis, L. M., Barg, F. K., & Bogner, H. R. (2009). "You don't want to burden them": Older adults' views on family involvement in care. *Journal of Family Nursing, 27,* 295–317.

Cain, K. (1999). Ways of reading: How knowledge and use of strategies are related to reading comprehension. *British Journal of Developmental Psychology, 17,* 293–312.

Callaghan, R., Rochat, P., Lillard, A., Claux, M. L., Odden, H., Itakura, S., et al. (2005). Synchrony in the onset of mental-state reasoning: Evidence from five cultures. *Psychological Science, 16,* 378–384.

Callahan, C. M. (2000). Intelligence and giftedness. In R. J. Sternberg (Ed.), *Handbook of intelligence* (pp. 159–175). Cambridge, England: Cambridge University Press.

Callanan, M. A., & Sabbagh, M. A. (2004). Multiple labels for objects in conversations with young children: Parents' language and children's developing expectations about word meanings. *Developmental Psychology, 40,* 746–763.

Cameron, L., Rutland, A., Brown, R., & Douch, R. (2006). Changing children's intergroup attitudes toward refugees: Testing different models of extended contact. *Child Development, 77,* 1208–1219.

Camp, C. J. (1999a). Memory interventions for normal and pathological older adults. In R. Schulz, M. P. Lawton, & G. Maddox (Eds.), *Annual review of gerontology and geriatrics* (Vol. 18, pp. 155–189). New York: Springer.

Camp, C. J. (2001). From efficacy to effectiveness to diffusion: Making the transitions in dementia intervention research. *Neuropsychological Rehabilitation, 11,* 495–517.

Camp, C. J. (Ed.). (1999b). *Montessori-based activities for persons with dementia: Volume 1.* Beachwood, OH: Menorah Park Press.

Camp, C. J., & McKitrick, L. A. (1991). Memory interventions in Alzheimer-type dementia populations: Methodological and theoretical issues. In R. L. West & J. D. Sinnott (Eds.), *Everyday memory and aging: Current research and methodology* (pp. 155–172). New York: Springer.

Camp, C. J., & Skrajner, M. J. (2005). Resident-assisted Montessori programming (RAMP): Training persons with dementia to serve as group activity leaders. *The Gerontologist, 44,* 426–431.

Camp, C. J., Foss, J. W., Stevens, A. B., Reichard, C. C., McKitrick, L. A., & O'Hanlon, A. M. (1993). Memory training in normal and demented elderly populations: The E-I-E-I-O model. *Experimental Aging Research, 19,* 277–290.

Camp, C. J., Judge, K. S., Bye, C. A., Fox, K. M., Bowden, J., Bell, M., et al. (1997). An intergenerational program for persons with dementia using Montessori methods. *The Gerontologist, 37,* 688–692.

Campbell, F. A., Pungello, E. P., Miller-Johnson, S., Burchinal, M., & Ramey, C. T. (2001). The development of cognitive and academic abilities: Growth curves from an early childhood educational experiment. *Developmental Psychology, 37,* 231–242.

Campbell, R., & Sais, E. (1995). Accelerated metalinguistic (phonological) awareness in bilingual children. *British Journal of Developmental Psychology, 13,* 61–68.

Campinha-Bacote, J. (2010). A culturally conscious model of mentoring. *Nurse Educator, 35,* 130–135.

Campos, J. J., Anderson, D. I., Barbu-Roth, M. A., Hubbard, E. M., Hertenstein, M. J., & Witherington, D. (2000). Travel broadens the mind. *Infancy, 1,* 149–219.

Campos, J. J., Hiatt, S., Ramsay, D., Henderson, C., & Svejda, M. (1978). The emergence of fear on the visual cliff. In M. Lewis & L. Rosenblum (Eds.), *The origins of affect.* New York: Plenum.

Camras, L. A., Chen, Y., Bakeman, R., Norris, K., & Cain, R. T. (2006). Culture, ethnicity, and children's facial expressions: A study of European American, Mainland Chinese, Chinese American, and Adopted Chinese girls. *Emotion, 6,* 103–114.

Camras, L. A., Ester, H., Campos, J., Campos, R., Ujiie, T., Miyake, K., et al. (1998). Production of emotional facial expressions in European, American, Japanese, and Chinese infants. *Developmental Psychology, 34,* 616–628.

Candel, I., Hayne, H., Strange, D., & Prevoo, E. (2009). The effect of suggestion on children's recognition memory for seen and unseen details. *Psychology, Crime, and Law, 15,* 29–39.

Cannon, T. D., Rosso, I. M., Hollister, J. M., Bearden, C. E., Sanchez, L. E., & Hadley, T. (2000). A prospective cohort study of genetic and perinatal influences in the etiology of schizophrenia. *Schizophrenia Bulletin, 26,* 351–366.

Cappell, K. A., Gmeindl, L., & Reuter-Lorenz, P.A. (2010). Age differences in prefrontal recruitment during verbal working memory maintenance depend on memory load. *Cortex, 46,* 462–473.

Capuzzi, D., & Gross, D. R. (2004). Counseling suicidal adolescents. In D. Capuzzi (Ed.), *Suicide across the lifespan: Implications for counselors* (pp. 235–270). Alexandria, VA: American Counseling Association.

Carbonneau, H., Caron, C., & Desrosiers, J. (2010). Development of a conceptual framework of positive aspects of caregiving in dementia. *Dementia, 9,* 327–353.

Card, N. A., Stucky, B. D., Sawalani, G. M., & Little, T. D. (2008). Direct and indirect aggression during childhood and adolescence: A meta-analytic review of gender differences, intercorrelations, and relations to maladjustment. *Child Development, 79,* 1185–1229.

Carlander, I., Ternestedt, B.-M., Sahlberg-Blom, E., Hellström, I., & Sandberg, J. (in press). Being me and being us in a family living close to death at home. *Qualitative Health Research, 5,* 683–695.

Carlo, G., Koller, S., Raffaelli, M., & de Guzman, M. R. T. (2007). Culture-related strengths among Latin American families: A case study of Brazil. *Marriage & Family Review, 41,* 335–360.

Carlson Jones, D. (2004). Body image among adolescent girls and boys: A longitudinal study. *Developmental Psychology, 40,* 823–835.

Carlson, S. M., & Meltzoff A. N. (2008). Bilingual experience and executive functioning in young children. *Developmental Science, 11,* 282–298.

Carnelley, K., & Ruscher, J. B. (2000). Adult attachment and exploratory behavior in leisure. *Journal of Social Behavior and Personality, 15,* 153–165.

Caron, C. D., Ducharme, F., & Griffith, J. (2006). Deciding on institutionalization for a relative with dementia: The most difficult decision for caregivers. *Canadian Journal on Aging, 25,* 193–205.

Carpenter, P. A., & Daneman, M. (1981). Lexical retrieval and error recovery in reading: A model based on eye fixations. *Journal of Verbal Learning and Verbal Behavior, 20,* 137–160.

Carr, D. (2004). Gender, preloss marital dependence, and older adults' adjustment to widowhood. *Journal of Marriage and Family, 66,* 220–235.

Carr, D. B., & Ott, B. R. (2010). The older adult driver with cognitive impairment: "It's a very frustrating life." *JAMA, 303,* 1632–1641.

Carrere, S., & Gottman, J. M. (1999). Predicting the future of marriages. In E. M. Hetherington (Ed.), *Coping with divorce, single parenting, and remarriage: A risk and resiliency perspective.* (pp. 3–22). Mahwah, NJ: Erlbaum.

Carroll, J. B. (1993). *Human cognitive abilities: A survey of factor-analytic studies.* New York: Cambridge University Press.

Carroll, J. B. (1996). A three-stratum theory of intelligence: Spearman's contribution. In I. Dennis & P. Tapsfield (Eds.), *Human abilities: Their nature and measurement* (pp. 1–18). Mahwah, NJ: Erlbaum.

Carroll, J. L., & Loughlin, G. M. (1994). Sudden infant death syndrome. In F. A. Oski, C. D. DeAngelis, R. D. Feigin, J. A. McMillan, & J. B. Warshaw (Eds.), *Principles and practice of pediatrics.* Philadelphia: Lippincott.

Carstensen, L. L. (1993). Motivation for social contact across the life span: A theory of socioemotional selectivity. In J. E. Jacobs (Ed.), *Nebraska symposium on motivation: Vol. 40. Developmental perspectives on motivation* (pp. 209–254). Lincoln: University of Nebraska Press.

Carstensen, L. L. (1995). Evidence for a life-span theory of socioemotional selectivity. *Current Directions in Psychological Science, 4,* 151–156.

Carver, K., Joyner, K., & Udry, J. R. (2003). National estimates of adolescent romantic relationships. In P. Florsheim (Ed.), *Adolescent romantic relations and sexual behavior: Theory, research, and practical implications* (pp. 23–56). Mahwah, NJ: Erlbaum.

Carver, P. R., Egan, S. K., & Perry, D. G. (2004). Children who question their heterosexuality. *Developmental Psychology, 40,* 43–53.

Casaer, P. (1993). Old and new facts about perinatal brain development. *Journal of Child Psychology and Psychiatry, 34,* 101–109.

Casey, B. J., Jones, R. M., & Hare, T. A. (2008). The adolescent brain. *Annals of the New York Academy of Sciences, 1124,* 111–126.

Casey, B. J., Tottenham, N., Liston, C., & Durston, S. (2005). Imaging the developing brain: What have we learned about cognitive development? *Trends in Cognitive Sciences, 9,* 104–110.

Casey, M. B. (1996). Understanding individual differences in spatial ability within females: A nature/nurture interactionist framework. *Developmental Review, 16,* 241–260.

Casper, W. J. (2000). The effects of work-life benefits and perceived organizational support on organizational attractiveness and employment desirability. *Dissertation Abstract International Section B. Sciences and Engineering, 61(5-B),* 2803.

Caspi, A., Roberts, B. W., & Shiner, R. L. (2005). Personality development: Stability and change. *Annual Review of Psychology, 56,* 453–484.

Castelli, L., Zogmaister, C., & Tomelleri, S. (2009). The transmission of racial attitudes within the family. *Developmental Psychology, 45,* 586–591.

Castillo, L. S., Williams, B. A., Hooper, S. M., Sabatino, C. P., Weithorn, L. A., & Sudore, R. L. (2011). Lost in translation: the unintended consequences of advance directive law on clinical care. *Annals of Internal Medicine, 154,* 121–128.

Castro, A. S. (2010). *The rite to womanhood: An interdisciplinary study of female circumcision among the Gikuyu of Kenya.* Bachelor's degree thesis completed at Wesleyan University. Retrieved from http://wesscholar.wesleyan.edu/cgi/viewcontent.cgi?article=1516&context=etd_hon_theses

Castro, W. (2000). The assessment of practical intelligence in a multicultural context. *Dissertation Abstracts International. Section A. Humanities and Social Sciences, 60(10-A)*, 3638.

Cattell, R. B. (1965). *The scientific analysis of personality.* Baltimore, MD: Penguin.

Cavanagh, S. E. (2004). The sexual debut of girls in early adolescence: The intersection of race, pubertal timing, and friendship group characteristics. *Journal of Research on Adolescence, 14*, 285–312.

Cavanaugh, J. C. (1999). Caregiving to adults: A life event challenge. In I. H. Nordhus, G. R. VandenBos, S. Berg, & P. Fromholt (Eds.), *Clinical geropsychology* (pp. 131–135). Washington, DC: American Psychological Association.

Cavanaugh, J. C., & Kinney, J. M. (1994, July). *Marital satisfaction as an important contextual factor in spousal caregiving.* Paper presented at the 7th International Conference on Personal Relationships, Groningen, The Netherlands.

Cavanaugh, J. C., & Kinney, J. M. (1998). Accuracy of caregivers' recollections of caregiving hassles. *Journal of Gerontology: Psychological Sciences, 53B*, P40–P42.

Cavanaugh, J. C., & Nocera, R. (1994). Cognitive aspects and interventions in Alzheimer's disease. In J. D. Sinnott (Ed.), *Interdisciplinary handbook of adult lifespan learning* (pp. 389–407). New York: Greenwood Press.

Ceci, S. J., & Bruck, M. (1998). Children's testimony: Applied and basic issues. In W. Damon (Ed.), *Handbook of child psychology* (Vol. 4). New York: Wiley.

Centers for Disease Control and Prevention. (2007). *Sexually transmitted disease surveillance, 2006.* Atlanta, GA: Author.

Centers for Disease Control and Prevention. (2007d). *2005 assisted reproductive technology success rates.* Atlanta, GA: Author.

Centers for Disease Control and Prevention. (2008). *Understanding teen dating violence fact sheet.* Atlanta, GA: Author.

Centers for Disease Control and Prevention. (2009a). *Body mass index.* Retrieved from http://www.cdc.gov/healthyweight/assessing/bmi

Centers for Disease Control and Prevention. (2009b). *Understanding intimate partner violence.* Retrieved from http://www.cdc.gov/violenceprevention/pdf/IPV_factsheet-a.pdf

Centers for Disease Control and Prevention. (2010). *Sexually transmitted disease surveillance 2009.* Atlanta, GA: Author.

Centers for Disease Control and Prevention. (2010a). *Smoking and tobacco use: Fast facts.* Retrieved from http://www.cdc.gov/tobacco/data_statistics/fact_sheets/fast_facts/index.htm

Centers for Disease Control and Prevention. (2010b). *Vital signs: Tobacco use.* Retrieved from http://www.cdc.gov/VitalSigns/pdf/2010-09-vitalsigns.pdf

Centers for Disease Control and Prevention. (2010c). *Nursing home care.* Retrieved from http://www.cdc.gov/nchs/fastats/nursingh.htm

Cerminara, K. L., & Perez, A. (2000). Therapeutic death: A look at Oregon's law. *Psychology, Public Policy, and Law, 6*, 503–525.

Cervantes, C. A., & Callanan, M. A. (1998). Labels and explanations in mother-child emotion talk: Age and gender differentiation. *Developmental Psychology, 34*, 88–98.

Champion, T. B. (2003). A "matter of vocabulary": Performance of low-income African-American Head Start children on the Peabody Picture Vocabulary Test. *Communication Disorders Quarterly, 24*, 121–127.

Chan, S. W-C. (2010). Family caregiving in dementia: The Asian perspective of a global problem. *Dementia and Geriatric Cognitive Disorders, 30*, 469–478.

Chandler, M. J., & Carpendale, J. I. M. (1998). Inching toward a mature theory of mind. In M. D. Ferrari & R. J. Sternberg (Eds.), *Self-awareness: Its nature and development* (pp. 148–190). New York: Guilford Press.

Chang, K., & Lu, L. (2007). Characteristics of organizational culture, stressors and wellbeing: The case of Taiwanese organizations. *Journal of Managerial Psychology, 22*, 549–598.

Chang, Y.-C., Jou, H.-J., Hsiao, H.-C., & Tsao, L.-I. (2010). Sleep quality, fatigue, and related factors among perimenopausal women in Taipei city. *Journal of Nursing Research, 18*, 275–282.

Chanquoy, L. (2001). How to make it easier for children to revise their writing: A study of text revision from 3rd to 5th grades. *British Journal of Educational Psychology, 71*, 15–41.

Chao, R. K. (2001). Extending research on the consequences of parenting style for Chinese Americans and European Americans. *Child Development, 72*, 1832–1843.

Chapman, P. D. (1988). *Schools as sorters: Lewis M. Terman, applied psychology, and the intelligence testing movement, 1890–1930.* New York: NYU Press.

Charles, S. T., & Carstensen, L. L. (2010). Social and emotional aging. *Annual Review of Psychology, 61*, 383–409.

Charlton, B., & Verghese, A. (2010). Caring for Ivan Ilyich. *Journal of General Internal Medicine, 25*, 93–95.

Charness, N., & Bosman, E. A. (1990). Expertise and aging: Life in the lab. In T. M. Hess (Ed.), *Aging and cognition: Knowledge organization and utilization* (pp. 343–385). Amsterdam, Netherlands: North-Holland.

Chassin, L., Ritter, J., Trim, R. S., & King, K. M. (2003). Adolescent substance use disorders. In E. J. Mash & R. A. Barkely (Eds.), *Child psychopathology* (2nd ed.) (pp. 199–230). New York: Guilford Press.

Chatters, L. M., Mattis, J. S., Woodward, A. T., Taylor, R. J., Neighbors, H. J., & Grayman, N. A. (2011). Use of ministers for a serious personal problem among African Americans: Findings from the National Survey of American Life. *American Journal of Orthopsychiatry, 81*, 118–127.

Chavous, T. M., Bernat, D. H., Schmeelk-Cone, K., Caldwell, C. H., Kohn-Wood, L., & Zimmerman, M. A. (2003). Racial identity and academic attainment among African American adolescents. *Child Development, 74*, 1076–1090.

Chen, J., & Gardner, H. (2005). Assessment based on multiple intelligences theory. In D. P. Flanagan & P. L. Harrison (Eds.), *Contemporary intellectual assessment: Theories, tests, and issues* (pp. 77–102). New York: Guilford Press.

Chen, X., Chen, H., Li, D., & Wang, L. (2009). Early childhood behavioral inhibition and social and school adjustment in Chinese children: A 5-year longitudinal study. *Child Development, 80*, 1692–1704.

Chen, X., Unger, J. B., Palmer, P., Weiner, M. D., Johnson, C. A., Wong, M. M., & Austin, G. (2002). Prior cigarette smoking initiation predicting current alcohol use: Evidence for a gateway drug effect among California adolescents from eleven ethnic groups. *Addictive Behaviors, 27*, 799–817.

Chen, Z. X., Aryee, S., & Lee, C. (2005). Test of a mediation model of perceived organizational support. *Journal of Vocational Behavior, 66*, 457–470.

Cheng, S., & Powell, B. (2007). Under and beyond constraints: Resource allocation to young children from biracial families. *American Journal of Sociology, 112*, 1044–1094.

Chetro-Szivos, J. (2001). Exploring the meaning of work: A CMM analysis of the grammar of working Acadian-Americans. *Dissertation Abstracts International. Section A. Humanities and Social Sciences, 62(1-A)*, 14.

Cheung, C-k., Kam, P. K., & Ngan, R. M-h. (2011). Age discrimination in the labour market from the perspectives of employers and older workers. *International Social Work, 54, 118–136.*

Cheung, I., & McCartt, A. T. (2010). *Declines in fatal crashes of older drivers: Changes in crash risk and survivability.* Insurance Institute for Highway Safety. Retrieved from http://www.iihs.org/research/topics/pdf/r1140.pdf

Chi, M. T. H. (2006). Laboratory methods for assessing experts' and novices' knowledge. In K. A. Ericsson, N. Charness, P. J. Feltovich, & R. R. Hoffman (Eds), *The Cambridge handbook of expertise and expert performance* (pp. 167–184). New York: Cambridge University Press.

Chilman, C. S. (1983). *Adolescent sexuality in a changing American society* (2nd ed.). New York: Wiley.

Chinen, A. B. (1989). *In the ever after.* Willmette, IL: Chiron.

Chiu, W. C. K., Chan, A. W., Snape, E., & Redman, T. (2001). Age stereotypes and discriminatory attitudes towards older workers: An East-West comparison. *Human Relations, 54*, 629–661.

Chomitz, V. R., Cheung, L. W. Y., & Lieberman, E. (1995). The role of lifestyle in preventing low birth weight. *The Future of Children, 5*, 121–138.

Chomsky, N. (1957). *Syntactic structures.* The Hague: Mouton.

Chomsky, N. (1995). *The minimalist program.* Cambridge, MA: MIT Press.

Chorpita, B. F., & Barlow, D. H. (1998). The development of anxiety: The role of control in the early environment. *Psychological Bulletin, 124*, 3–21.

Chou, K-L., & Chi, I. (2005). Prevalence and correlates of depression in Chinese oldest-old. *International Journal of Geriatric Psychiatry, 20,* 41–50.

Chow, B. W., McBride-Chang, C., Cheung, H., & Chow, C. S. (2008). Dialogic reading and morphology training in Chinese children: Effects on language and literacy. *Developmental Psychology, 44,* 233–244.

Christensen, A., & Heavey, C. L. (1999). Intervention for couples. *Annual Review of Psychology, 50,* 165–190.

Cicchetti, D., & Toth, S. L. (2006). Developmental psychopathology and preventive intervention. In W. Damon & R. M. Lerner (Eds.), *Handbook of child psychology* (Vol. 4) (pp. 497–547). New York: Wiley.

Cicirelli, V. G. (2001). Personal meaning of death in older adults and young adults in relation to their fears of death. *Death Studies, 25,* 663–683.

Cicirelli, V. G. (2004). God as the ultimate attachment figure for older adults. *Attachment & Human Development, 6,* 371–388.

Cicirelli, V. G. (2006). Fear of death in mid-old age. *Journals of Gerontology: Psychological Sciences, 61B,* P75–P81.

Cillessen, A. H. N., & Rose, A. (2005). Understanding popularity in the peer system. *Current Directions in Psychological Science, 14,* 102–105.

Ciraulo, D. A., Evans, J. A., Qiu, W. Q., Shader, R. I., & Salzman, C. (2011). Antidepressant treatment of geriatric depression. In D. A. Ciraulo & R. I. Shader (Eds.), *Pharmacotherapy of depression* (2nd ed., pp. 125 www.iihs.org 183). New York: Humana Press.

Clark, L. A. (2007). Relationships between the big five personality dimensions and attitudes toward telecommuting. *Dissertation Abstracts International. Section A. Humanities and Social Sciences, 68(5-A),* 2041.

Clarke, P. J., Snowling, M. J., Truelove, E., & Hulme, C. (2010). Ameliorating children's reading-comprehension difficulties: A randomized controlled trial. *Psychological Science, 21,* 1106–1116.

Clarke, V., Ellis, S. J., Peel, E., & Riggs, D. W. (2010). *Lesbian, gay, bisexual, trans and queer psychology: An introduction.* Cambridge, England: Cambridge University Press.

Clarke-Anderson, P. (2005). Minorities' attainability to leadership positions in business settings: A study of self-efficacy and leadership aspirations. *Dissertation Abstracts International. Section A. Humanities and Social Sciences, 65(9-A),* 3456.

Clarke-Stewart, A., & Brentano, C. (2006). *Divorce: Causes and consequences.* New Haven, CT: Yale University Press.

Clarke-Stewart, K. A., & Brentano, C. (2005). *Till divorce do us part.* New Haven, CT: Yale University Press.

Clements, D. H. (1995). Teaching creativity with computers. *Educational Psychology Review, 7,* 141–161.

Clements, M., & Markman, H. J. (1996). The transition to parenthood: Is having children hazardous to marriage? In N. Vanzetti & S. Duck (Eds.), *A lifetime of relationships* (pp. 290–310). Pacific Grove, CA: Brooks/Cole.

Clifford, D., Hertz, F., & Doskow, E. (2010). *A legal guide for lesbian and gay couples* (15th ed.). Berkeley, CA: Nolo.

Cnattingius, S. (2004). The epidemiology of smoking during pregnancy: Smoking prevalence, maternal characteristics, and pregnancy outcomes. *Nicotine & Tobacco Research, 6,* S125–S140.

Coan, J. A., & Gottman, J. M. (2007). Sampling, experimental control, and generalizability in the study of marital process models. *Journal of Marriage and Family, 69,* 73–80.

Coatsworth, J. D., & Conroy, D. E. (2009). The effects of autonomy-supporting coaching, need satisfaction, and self-perceptions on initiative and identity in youth swimmers. *Developmental Psychology, 45,* 320–328.

Coelho, J. S., Jansen, A., Roefs, A., & Nederkoom, C. (2009). Eating behavior in response to food-cue exposure: Examining the cue-reactivity and counteractive-control models. *Psychology of Addictive Behaviors, 23,* 131–139.

Cohen, C. A. (2006). Consumer fraud and the elderly: A review of Canadian challenges and initiatives. *Journal of Gerontological Social Work, 46,* 137–144.

Cohen, F., Kemeny, M. E., Zegans, L., Johnson, P., Kearney, K. A., & Sites, D. P. (2007). Immune function declines with unemployment and recovers after stressor termination. *Psychosomatic Medicine, 69,* 225–234.

Cohen, G. L., Garcia, J., Apfel, N., & Master, A. (2006). Reducing the racial achievement gap: A social-psychological intervention. *Science, 313,* 1307–1310.

Cohen, M., & Piquero, A. R. (2009). New evidence on the monetary value of saving a high risk youth. *Journal of Quantitative Criminology, 25,* 25–49.

Cohen, R. R., Boston, P., Mount, B. M., & Porterfield, P. (2001). Changes in quality of life following admission to palliative care units. *Palliative Medicine, 15,* 363–371.

Cohen, R. W., Martinez, M. E., & Ward, B. W. (2010). *Health insurance coverage: Early release of estimates from the National Health Interview Survey, 2009.* Hyattsville, MD: National Center for Health Statistics.

Cohen, S., & Williamson, G. M. (1991). Stress and infectious disease in humans. *Psychological Bulletin, 109,* 5–24.

Cohen, S., Kessler, R. C., & Gordon, L. U. (Eds.). (1995). *Measuring stress: A guide for health and social scientists.* Oxford, England: Oxford University Press.

Coie, J. D., Dodge, K. A., Terry, R., & Wright, V. (1991). The role of aggression in peer relations: An analysis of aggression episodes in boys' play groups. *Child Development, 62,* 812–826.

Colby, A., Kohlberg, L., Gibbs, J. C., & Lieberman, M. (1983). A longitudinal study of moral development. *Monographs of the Society for Research in Child Development, 48* (Serial No. 200).

Cole, D. A., & Jordan, A. E. (1995). Competence and memory: Integrating psychosocial and cognitive correlates of child depression. *Child Development, 66,* 459–473.

Cole, M. (2006). Culture and cognitive development in phylogenetic, historical and ontogenetic perspective. In W. Damon & R. M. Lerner (Eds.), *Handbook of child psychology* (6th ed., Vol. 2, pp. 636–683). New York: Wiley.

Cole, M. L. (2000). The experience of never-married women in their thirties who desire marriage and children. *Dissertation Abstracts International. Section A. Humanities and Social Sciences, 60(9-A),* 3526.

Cole, P. M., Tamang, B. L., & Shrestha, S. (2006). Cultural variations in the socialization of young children's anger and shame. *Child Development, 77,* 1237–1251.

Collins, W. A. (2003). More than myth: The developmental significance of romantic relationships during adolescence. *Journal of Research on Adolescence, 13,* 1–24.

Collins, W. A., & van Dulmen, M. (2006). "The course of true love(s) . . .": Origins and pathways in the development of romantic relationships. In A. C. Crouter & A. Booth (Eds.), *Romance and sex in adolescence and emerging adulthood: Risks and opportunities* (pp. 63–86). Mahwah, NJ: Erlbaum.

Collins, W. A., Welsh, D. P., & Furman, W. (2009). Adolescent romantic relationships. *Annual Review of Psychology, 60,* 631–652.

Conduct Problems Prevention Research Group. (2011). The effects of the Fast Track preventive intervention on the development of conduct disorder across childhood. *Child Development, 82,* 331–345.

Connelly, R., Degraff, D. S., & Willis, R. A. (2004). The value of employer-sponsored child care to employees. *Industrial Relations: A Journal of Economy & Society, 43,* 759–792.

Connidis, I. A. (2001). *Family ties and aging.* Thousand Oaks, CA: Sage.

Connor, C. M., Morrison, F. J., Fishman, B. J., Schatschneider, C., & Underwood, P. (2007). Algorithm-guided individualized reading instruction. *Science, 315,* 464–465.

Conradi, L., & Geffner, R. (2009). Introduction to Part I of the special issue on female offenders of intimate partner violence. *Journal of Aggression, Maltreatment and Trauma, 18,* 547–551.

Cook, D. A., Bahn, R. S., & Menaker, R. (2010). Speed mentoring: An innovative method to facilitate mentoring relationships. *Medical Teacher, 32,* 692–694.

Coombs, M. A. (2010). The mourning that comes before: Can anticipatory grief theory inform family care in adult intensive care? *International Journal of Palliative Nursing, 16,* 580–584.

Cooney, T. M., & Uhlenberg, P. (1990). The role of divorce in men's relations with their adult children after mid-life. *Journal of Marriage and the Family, 52,* 677–688.

Cooney, T. M., Smyer, M. A., Hagestad, G. O., & Klock, R. (1986). Parental divorce in young adulthood: Some preliminary findings. *American Journal of Orthopsychiatry, 56,* 470–477.

Cooper, C. L., & Quick, J. C. (2003). The stress and loneliness of success. *Counselling Psychology Quarterly, 16,* 1–7.

Coplan, R. J., & Armer, M. (2007). A "multitude" of solitude: A closer look at social withdrawal and nonsocial play in early childhood. *Child Development Perspectives, 1,* 26–32.

Coplan, R. J., Gavinski-Molina, M. H., Lagace-Seguin, D. G., & Wichman, C. (2001). When girls versus boys play alone: Nonsocial play and adjustment in kindergarten. *Developmental Psychology, 37,* 464–474.

Copper, R. L., Goldenberg, R. L., Das, A., Elder, N., Swain, M., Norman, G., et al. (1996). The preterm prediction study: Maternal stress is associated with spontaneous preterm birth at less than thirty-five weeks' gestation. *American Journal of Obstetrics & Gynecology, 175,* 1286–1292.

Corbett, B. A., & Hilty, D. M. (2006). Managing your time. In L. W. Roberts & D. M. Hilty (Eds.), *Handbook of career development in academic psychiatry and behavioral sciences* (pp. 83–91). Washington, DC: American Psychiatric Publishing.

Cordes, S., & Brannon, E. M. (2009). The relative salience of discrete and continuous quantity in young infants. *Developmental Science, 12,* 453–463.

Cornelius, M., Taylor, P., Geva, D., & Day, N. (1995). Prenatal tobacco exposure and marijuana use among adolescents: Effects on offspring gestational age, growth, and morphology. *Pediatrics, 95,* 738–743.

Corr, C. A. (1991–1992). A task-based approach to coping with dying. *Omega: The Journal of Death and Dying, 24,* 81–94.

Corr, C. A., Corr, D. M., & Nabe, C. M. (2008). *Death and dying: Life and living.* Belmont, CA: Wadsworth.

Cortesi, F., Giannotti, F., Sebastiani, T., & Vagnoni, C. (2004). Cosleeping and sleep behavior in Italian school-aged children. *Journal of Developmental & Behavioral Pediatrics, 25,* 28–33.

Corti, J. K. (2009). *Sibling relationships during the young adult years: An analysis of closeness, relational satisfaction, everyday talk, and turning points.* Retrieved from https://adr.coalliance.org/codu/fez/eserv/codu:55639/Corti_denver_0061D_10200.pdf

Cosmides, L., & Tooby, J. (2000). Evolutionary psychology and the emotions. In M. Lewis & J. Haviland-Jones (Eds.), *Handbook of emotions* (2nd ed., pp. 91–115). New York: Guilford.

Costa, P. T., Jr., & McCrae, R. R. (1988). Personality in adulthood: A six-year longitudinal study of self-reports and spouse ratings on the NEO Personality Inventory. *Journal of Personality and Social Psychology, 54,* 853–863.

Costa, P. T., Jr., & McCrae, R. R. (1997). Longitudinal stability of adult personality. In R. Hogan, J. Johnson, & S. Briggs (Eds.), *Handbook of personality psychology* (pp. 269–292). San Diego, CA: Academic Press.

Costigan, C. L., & Dokis, D. P. (2006). Relations between parent-child acculturation differences and adjustment within immigrant Chinese families. *Child Development, 77,* 1252–1267.

Costin, S. E., & Jones, D. C. (1992). Friendship as a facilitator of emotional responsiveness and prosocial interventions among young children. *Developmental Psychology, 28,* 941–947.

Cotter, R. P. (2001). High-risk behaviors in adolescence and their relationship to death anxiety and death personification. *Dissertation Abstracts International. Section B. Sciences and Engineering, 61(8-B),* 4446.

Cotter, V. T., & Gonzalez, E. W. (2009). Self-concept in older adults: An integrative review of empirical literature. *Holistic Nursing Practice, 23,* 335–348.

Coulton, C. J., Crampton, D. S., Irwin, M., Spilsbury, J. C., & Korbin, J. E. (2007). How neighborhoods influence child maltreatment: A review of the literature and alternative pathways. *Child Abuse and Neglect, 31,* 1117–1142.

Counts, D., & Counts, D. (Eds.). (1985). *Aging and its transformations: Moving toward death in Pacific societies.* Lanham, MD: University Press of America.

Coutelle, C. et al. (2005). Gene therapy progress and prospects: Fetal gene therapy—first proofs of concept—some adverse effects. *Gene Therapy, 12,* 1601–1607.

Cowan, P. A., Cowan, C. P., & Knox, V. (2010). Marriage and fatherhood programs. *Fragile Families, 20.* Retrieved from http://futureofchildren.org/publications/journals/article/index.xml?journalid=73&articleid=537&sectionid=3702

Cox, C. B. (2007). Grandparent-headed families: Needs and implications for social work interventions and advocacy. *Families in Society, 88,* 561–566.

Cox, K. S., Wilt, J., Olson, B. D., & McAdams, D. P. (2010). Generativity, the Big Five, and psychosocial adaptation in midlife adults. *Journal of Personality, 78,* 1185–1208.

Cox, M. J., Paley, B., & Harter, K. (2001). Interparental conflict and parent–child relationships. In J. H. Grych & F. D. Fincham (Eds.), *Interparental conflict and child development* (pp. 249–272). New York: Cambridge University Press.

Cox, T., & Griffiths, A. (2010). Work-related stress: A theoretical perspective. In S. Leka & J. Houdmont (Eds.), *Occupational health psychology* (pp. 31–56). Chichester, UK: Blackwell.

Craig, K. D., Whitfield, M. F., Grunau, R. V. E., Linton, J., & Hadjistavropoulos, H. D. (1993). Pain in the preterm neonate: Behavioural and physiological indices. *Pain, 52,* 238–299.

Craik, F. I. M., & Salthouse, T. A. (Eds.). (2008). *The handbook of aging and cognition* (3rd ed.). New York: Psychology Press.

Crews, J. (2011). Aging, disability, and public health. In D. L. Lollar & E. M. Andresen (Eds.), *Public health perspectives on disability: Epidemiology to ethics and beyond* (pp. 163–183). New York: Springer.

Crick, N. R., & Dodge, K. A. (1994). A review and reformulation of social-information processing mechanisms in children's social adjustment. *Psychological Bulletin, 115,* 74–101.

Crick, N. R., Ostrov, J. M., Appleyard, K., Jansen, E. A., & Casas, J. F. (2004). Relational aggression in early childhood: "You can't come to my birthday party unless." In M. Puttalaz & K. L. Bierman (Eds.), *Aggression, antisocial behavior, and violence among girls* (pp. 71–89). New York: Guilford.

Crist, J. D., Garcia-Smith, D., & Phillips, L. R. (2006). Accommodating the stranger en casa: How Mexican American elders and caregivers decide to use formal care. *Research and Theory for Nursing Practice: An International Journal, 20,* 109–126.

Cristia, A. (2010). Phonetic enhancement of sibilants in infant-directed speech. *Journal of the Acoustical Society of America, 128,* 424–434.

Crocetti, E., Rubini, M., Luyckx, K., & Meeus, W. (2008). Identity formation in early and middle adolescents from various ethnic groups: From three dimensions to five statuses. *Journal of Youth and Adolescence, 37,* 983–996.

Crohan, S. E. (1996). Marital quality and conflict across the transition to parenthood in African American and white couples. *Journal of Marriage and Family, 58,* 933–944.

Crohn, H. M. (2006). Five styles of positive stepmothering from the perspective of young adult stepdaughters. *Journal of Divorce & Remarriage, 46,* 119–134.

Cross, S., & Markus, H. (1991). Possible selves across the lifespan. *Human Development, 34,* 230–255.

Crouter, A. C., & Bumpus, M. F. (2001). Linking parents' work stress to children's and adolescents' psychological adjustment. *Current Directions in Psychological Science, 10,* 156–159.

Crouter, A. C., Whiteman, S. D., McHale, S. M., & Osgood, D. (2007). Development of gender attitude traditionality across middle childhood and adolescence. *Child Development, 78,* 911–926.

Crozier, J. C., Dodge, K. A., Fontaine, R. G., Lansford, J. E., Bates, J. E., Pettit, G. S., et al. (2008). Social information processing and cardiac predictors of adolescent antisocial behavior. *Journal of Abnormal Psychology, 117,* 253–267.

Csikszentmihalyi, M., & Larson, R. (1984). *Being adolescent: Conflict and growth in the teenage years.* New York: Basic Books.

Cuellar, I., Nyberg, B., Maldonado, R. E., & Roberts, R. E. (1997). Ethnic identity and acculturation in a young adult Mexican-origin population. *Journal of Community Psychology, 25,* 535–549.

Cullerton-Sen, C., Cassidy, A. R., Murray-Close, D., Cicchetti, D., Crick, N. R., & Rogosch, F. A. (2008). Child maltreatment and the development of relational and physical aggression: The importance of a gender-informed approach. *Child Development, 79,* 1736–1751.

Cummings, E., Schermerhorn, A. C., Davies, P. T., Goeke-Morey, M. C., & Cummings, J. S. (2006). Interparental discord and child adjustment: Prospective investigations of emotional security as an explanatory mechanism. *Child Development, 77,* 132–152.

Cunningham, A. E., Perry, K. E., Stanovich, K. E., & Share, D. L. (2002). Orthographic learning during reading: Examining the role of self-teaching. *Journal of Experimental Child Psychology, 82,* 185–199.

Cunningham, M. (2007). Influences of women's employment on the gendered division of household labor over the life course: Evidence from a 31-year panel study. *Journal of Family Issues, 28,* 422–444.

Cunningham, M., & Thornton, A. (2007). Direct and indirect influences of parents' marital instability on children's attitudes toward cohabitation in young adulthood. *Journal of Divorce & Remarriage, 46,* 125–143.

Curl, A. L. (2007). The impact of retirement on trajectories of physical health of married couples. *Dissertation Abstracts International. Section A. Humanities and Social Sciences, 68(4-A),* 1606.

Curtis, R., & Pearson, F. (2010). Contact with birth parents: Differential psychological adjustment for adults adopted as infants. *Journal of Social Work, 10,* 347–367.

Cyders, M. A., Flory, K., Rainer, S., & Smith, G. T. (2009). The role of personality dispositions to risky behavior in predicting first-year college drinking. *Addiction, 104,* 193v202.

D'Onofrio, B. M., Goodnight, J. A., Van Hulle, C. A., Rodgers, J. L., Rathouz, P. J., Waldman, I. D., & Lahey, B. B. (2009). Maternal age at childbirth and offspring disruptive behaviors: testing the causal hypothesis. *Journal of Child Psychology and Psychiatry, 50,* 1018–1028.

D'Onofrio, B. M., Singh, A. L., Iliadou, A., Lambe, M., Hultman, C. M., Neiderhiser, J. M., Långström, N., & Lichtenstein, P. (2010). A quasi-experimental study of maternal smoking during pregnancy and offspring academic achievement. *Child Development, 81,* 80–100.

Daatland, S. O. (2007). Marital history and intergenerational solidarity: The impact of divorce and unmarried cohabitation. *Journal of Social Issues, 63,* 809–825.

Daley, T. C., Whaley, S. E., Sigman, M. D., Espinosa, M. P., & Neumann, C. (2003). IQ on the rise: The Flynn effect in rural Kenyan children. *Psychological Science, 14,* 215–219.

Daly, M., & Wilson, M. (1996). Violence against stepchildren. *Current Directions in Psychological Science, 5,* 77–81.

Dannefer, D., & Miklowski, C. (2006). Developments in the life course. In J. A. Vincent, C. R. Phillipson, & M. Downs (Eds.), *The futures of old age* (pp. 30–39). Thousand Oaks, CA: Sage.

Dannemiller, J. L. (1998). Color constancy and color vision during infancy: Methodological and empirical issues. In V. Walsh & J. Kulikowski (Eds.), *Perceptual constancy: Why things look as they do.* New York: Cambridge University Press.

Danno, M., Miyaji, M., Kortelainen, J. M., & Kutila, M. (2010, September). *Detection of a driver's visual attention using the online UFOV method.* Paper presented at the 13th International Conference on Intelligent Transportation Systems, Funchal. Portugal.

Darling, H., Reeder, A. I., McGee, T., & Williams, S. (2006). Brief report: Disposable income, and spending on fast food, alcohol, cigarettes, and gambling by New Zealand secondary school students. *Journal of Adolescence, 29,* 837–843.

Darling, N., Cumsille, P., & Martínez, M. L. (2008). Individual differences in adolescents' beliefs about the legitimacy of parental authority and their own obligation to obey: A longitudinal investigation. *Child Development, 79,* 1103–1118.

David, A., & Rodeck, C. H. (2009). Fetal gene therapy. In C. H. Rodeck & M. J. Whittle (Eds.), (2009). *Fetal medicine: Basic science and clinical practice.* London: Churchill Livingstone.

Davidson, F. H., & Davidson, M. M. (1994). *Changing childhood prejudice: The caring work of the schools.* Westport, CT: Bergin & Garvey/Greenwood.

Davidson, R. J. (2010). Empirical explorations of mindfulness: Conceptual and methodological conundrums. *Emotion, 10,* 8–11.

Davies, L. (2000). *Transitions and singlehood: The forgotten life course.* Unpublished manuscript.

Davies, L. (2003). Singlehood: Transitions within a gendered world. *Canadian Journal on Aging, 22,* 343–352.

Davies, P. T., & Cummings, E. M. (1998). Exploring children's emotional security as a mediator of the link between marital relations and child adjustment. *Child Development, 69,* 124–139.

Davis, E. P., & Sandman, C. A. (2010). The timing of prenatal exposure to maternal cortisol and psychosocial stress is associated with human cognitive development. *Child Development, 81,* 131–148.

Day, R., & Allen, T. D. (2004). The relationship between career motivation and self-efficacy with protégé career success. *Journal of Vocational Behavior, 64,* 72–91.

De Andrade, C. E. (2000). Becoming the wise woman: A study of women's journeys through midlife transformation. *Dissertation Abstracts International. Section B. Sciences and Engineering, 61(2-B),* 1109.

De Beni, R., & Palladino, P. (2000). Intrusion errors in working memory tasks: Are they related to reading comprehension ability? *Learning and Individual Differences, 12,* 131–143.

de Haan, M., Wyatt, J. S., Roth, S., Vargha-Khadem, F., Gadian, D., & Mishkin, M. (2006). Brain and cognitive-behavioral development after asphyxia at term birth. *Developmental Science, 9,* 441–442.

De Neys W., & Everaerts, D. (2008). Developmental trends in everyday conditional reasoning: The retrieval and inhibition interplay. *Journal of Experimental Child Psychology, 100,* 252–263.

de St. Aubin, E., & McAdams, D. P. (1995). The relations of generative concern and generative action to personality traits, satisfaction/happiness with life, and ego development. *Journal of Adult Development, 2,* 99–112.

de Vries, B. (1996). The understanding of friendship: An adult life course perspective. In C. Magai & S. H. McFadden (Eds.), *Handbook of emotion, adult development, and aging.* San Diego, CA: Academic Press.

De Wolff, M. S., & van IJzendoorn, M. H. (1997). Sensitivity and attachment: A meta-analysis on parental antecedents of infant attachment. *Child Development, 68,* 571–591.

DeAngelis, T. (2010). New help for stroke survivors. *Monitor on Psychology, 41(3),* 52–55.

Deary, I. J., Batty, G. D., & Gale, C. R. (2008). Bright children become enlightened adults. *Psychological Science, 19,* 1–6.

Deary, I. J., Penke, L., & Johnson, W. (2010). The neuroscience of human intelligence differences. *Nature Reviews Neuroscience, 11,* 201–211.

DeCasper, A. J., & Spence M. J. (1986). Prenatal maternal speech influences newborn's perception of speech sounds. *Infant Behavior and Development, 9,* 133–150.

Dekker, M. C. Ferdinand, R. F., van Lang, D. J., Bongers, I. L. van der Ende, J., & Verhulst, F. C. (2007). Developmental trajectories of depressive symptoms from early childhood to late adolescence: gender differences and adult outcome. *Journal of Child Psychology and Psychiatry, 48,* 657–666.

Dekovic, M., & Janssens, J. M. (1992). Parents' child-rearing style and child's sociometric status. *Developmental Psychology, 28,* 925–932.

Del Giudice, M., & Colle, L. (2007). Differences between children and adults in the recognition of enjoyment smiles. *Developmental Psychology, 43,* 796–803.

del Valle, J. F., Bravo, A., & Lopez, M. (2010). Parents and peers as providers of support in adolescents' social network: A developmental perspective. *Journal of Community Psychology, 38,* 16–27.

Delaney, C. (2000). Making babies in a Turkish village. In J. S. DeLoache & A. Gottlieb (Eds.), *A world of babies: Imagined childcare guides for seven societies.* New York: Cambridge University Press.

DeLoache, J. S. (1995). Early understanding and use of models: The model model. *Current Directions in Psychological Science, 4,* 109–113.

DeLoache, J. S., et al. (2010). Do babies learn from baby media? *Psychological Science, 21,* 1570–1574.

DeLucia-Waack, J. L. (2010). Children of divorce groups. In G. L. Greif & P. H. Ephross (Eds.), *Group work with populations at risk* (3rd ed., pp. 93–114). New York: Oxford University Press.

Demir, A., Levine, S. C., & Goldin-Meadow, S. (2010). Narrative skill in children with unilateral brain injury: A possible limit to functional plasticity. *Developmental Science, 13,* 636–647.

Dendinger, V. M., Adams, G. A., & Jacobson, J. D. (2005). Reasons for working and their relationship to retirement attitudes, job satisfaction and occupational self-efficacy of bridge employees. *International Journal of Aging & Human Development, 61,* 21–35.

Deng, C.-P., Armstrong, P. I., & Rounds, J. (2007). The fit of Holland's RIASEC model to US occupations. *Journal of Vocational Behavior, 71,* 1–22.

Denissen, J. J., Zarrett, N. R., & Eccles, J. S. (2007). I like to do it, I'm able, and I know I am: Longitudinal couplings between domain-specific achievement, self-concept, and interest. *Child Development, 78,* 430–447.

Denney, N. W. (1989). Everyday problem solving: Methodological issues, research findings, and a model. In L. W. Poon, D. C. Rubin, & B. A. Wilson (Eds.), *Everyday cognition in adulthood and late life* (pp. 330–351). Cambridge, England: Cambridge University Press.

Denney, N. W. (1990). Adult age differences in traditional and practical problem solving. In E. A. Lovelace (Ed.), *Aging and cognition: Mental processes, self-awareness, and interventions* (pp. 329–349). Amsterdam, Netherlands: North-Holland.

Denney, N. W., Pearce, K. A., & Palmer, A. M. (1982). A developmental study of adults' performance on traditional and practical problem-solving tasks. *Experimental Aging Research, 8,* 115–118.

Dennis, T., Bendersky, M., Ramsay, D., & Lewis, M. (2006). Reactivity and regulation in children prenatally exposed to cocaine. *Developmental Psychology, 42,* 84–97.

DePaulo, B. M. (2006). *Singled out: How singles are stereotyped, stigmatized, and ignored, and still live happily ever after.* New York: St. Martin's Press.

Derman, D. S. (2000). Grief and attachment in young widowhood. *Dissertation Abstracts International. Section A. Humanities and Social Sciences, 60(7-A),* 2383.

Desmond, N., & López Turley, R. N. (2009). The role of familism in explaining the Hispanic-white college application gap. *Social Problems, 56,* 311–334.

Deutsch, A. (2001, April 11). Dutch parliament OKs strict euthanasia bill. *Wilmington (NC) Morning Star,* p. 2A.

Deutsch, D. (2010). *Increased mental health services for the homeless.* Master's thesis submitted to California State University, Humboldt. Retrieved from http://gradworks.umi.com/1486303.pdf

Devine, M. A., & Lashua, B. (2002). Constructing social acceptance in inclusive leisure contexts: The role of individuals with disabilities. *Therapeutic Recreation Journal, 36,* 65–83.

Dewey, C., Fleming, P., Goldin, J., & the ALSPAC Study Team. (1998). Does the supine sleeping position have any adverse effects on the child? II. Development in the first 18 months. *Pediatrics, 101,* E5.

Dewey, K. G. (2001). Nutrition, growth, and complementary feeding of the breastfed infant. *Pediatric Clinics of North America, 48,* 87–104.

Dey, J. G., & Hill, C. (2007). *Behind the pay gap.* Washington, DC: American Association of University Women Educational Foundation.

Diamond, A. (2007). Interrelated and interdependent. *Developmental Science, 10,* 152–158.

Diamond, A., Prevor, M. B., Callender, G., & Druin, D. P. (1997). Prefontal cortex deficits in children treated early and continuously for PKU. *Monographs of the Society for Research in Child Development, 62*(4, Serial No. 252).

Dick, D. M., & Rose, R. J. (2002). Behavior genetics: What's new? What's next? *Current Directions in Psychological Science, 11,* 70–74.

Dickens, B. M., Boyle, J. M., Jr., & Ganzini, L. (2008). Euthanasia and assisted suicide. In P. A. Singer & A. M. Viens (Eds.), *The Cambridge textbook of bioethics* (pp. 72–77). New York: Cambridge University Press.

Dickens, W. T., & Flynn, J. R. (2001). Heritability estimates versus large environmental effects: The IQ paradox resolved. *Psychological Review, 108,* 346–369.

Dickinson, G. E. (1992). First childhood death experiences. *Omega: The Journal of Death and Dying, 25,* 169–182.

Dickinson, G. E. (in press). Thirty-five years of end-of-life issues in US medical schools. *American Journal of Hospice and Palliative Medicine.*

Dick-Read, G. (1959). *Childbirth without fear.* New York: Harper.

Diehl, M. (1998). Everyday competence in later life: Current status and future directions. *The Gerontologist, 4,* 422–433.

Diehl, M., Marsiske, M., Horgas, A. L., Rosenberg, A., Saczynski, J. S., & Willis, S. L. (2005). The Revised Observed Tasks of Daily Living: A performance-based assessment of everyday problem solving in older adults. *Journal of Applied Gerontology, 24,* 211–230.

Diemer, M. A. (2007). Parental and school influences upon the career development of poor youth of color. *Journal of Vocational Behavior, 70,* 502–524.

DiFonzo, J. H. (2010). A vision for collaborative practice: The final report of the Hofstra Collaborative Law Conference. *Hofstra Law Review, 39,* 101.

Dillaway, H., & Broman, C. (2001). Race, class and gender differences in marital satisfaction and divisions of household labor among dual-earner couples. *Journal of Family Issues, 22,* 309–327.

Dilworth-Anderson, P., Boswell, G., & Cohen, M. D. (2007). Spiritual and religious coping values and beliefs among African American caregivers: A qualitative study. *Journal of Applied Gerontology, 26,* 355–369.

Dinsbach, A. A., Fiej, J. A., & de Vries, R. E. (2007). The role of communication content in an ethnically diverse organization. *International Journal of Intercultural Relations, 31,* 725–745.

Dionne, G., Dale, P. S., Boivin, M., & Plomin, R. (2003). Genetic evidence for bidirectional effects of early lexical and grammatical development. *Child Development, 74,* 394–412.

DiPietro, J. A., Bornstein, M. H., Hahn, C. S., Costigan, K., & Achy-Brou, A. (2007). Fetal heart rate and variability: Stability and prediction to developmental outcomes in early childhood. *Child Development, 78,* 1788–1798.

DiPietro, J. A., Caulfield, L., Costigan, K. A., Merialdi, M., Nguyen, R. H. N., Zavaleta, N., & Gurewitsch, E. D. (2004). Fetal neurobehavioral development: A tale of two cities. *Developmental Psychology, 40,* 445–456.

DiPietro, J. A., Hodgson, D. M., Costigan, K. A., & Johnson, T. R. B. (1996). Fetal antecedents of infant temperament. *Child Development, 67,* 2568–2583.

DiPietro, J. A., Novak, M. F., Costigan, K. A., Atella, L. D., & Reusing, S. P. (2006). Maternal psychological distress during pregnancy in relation to child development at age two. *Child Development, 77,* 573–587.

Dirix, C. E. H., Nijhuis, J. G., Jongsma, H. W., & Hornstra, G. (2009). Aspects of fetal learning and memory. *Child Development, 80,* 1251–1258.

Dishion, T. J., Poulin, F., & Burraston, B. (2001). Peer group dynamics associated with iatrogenic effects in group interventions with high-risk young adolescents. In D. W. Nangle & C. A. Erdley (Eds.), *The role of friendship in psychological adjustment* (pp. 79–92). San Francisco, CA: Jossey-Bass.

Divan, H. A., Kheifets, L., Obel, C., & Olsen, J. (2008). Prenatal and postnatal exposure to cell phone use and behavioral problems in children. *Epidemiology, 19,* 523–529.

Dixon, J. A., & Marchman, V. A. (2007). Grammar and the lexicon: Developmental ordering in language acquisition. *Child Development, 78,* 190–212.

Dixon, P. (2009). Marriage among African Americans: What does the research reveal? *Journal of African American Studies, 13,* 29–46.

Dixon, R. A., & Hultsch, D. F. (1999). Intelligence and cognitive potential in late life. In J. C. Cavanaugh & S. K. Whitbourne (Eds.), *Gerontology: An interdisciplinary perspective.* New York: Oxford University Press.

Dodge, K. A., Coie, J. D., & Lynam, D. (2006). Aggression and antisocial behavior in youth. In N. Eisenberg (Ed.), *Handbook of child psychology: Vol. 3. Social, emotional, and personality development* (6th ed., pp. 719–788). New York: Wiley.

Dodge, K. A., Greenberg, M. T., & Malone, P. S. (2008). Testing an idealized dynamic cascade model of the development of serious violence in adolescence. *Child Development, 79,* 1907–1927.

Dohnt, H., & Tiggemann, M. (2006). The contribution of peer and media influences to the development of body satisfaction and self-esteem in young girls: A prospective study. *Developmental Psychology, 42,* 929–936.

Doka, K. J., & Mertz, M. E. (1988). The meaning and significance of great-grandparenthood. *The Gerontologist, 28,* 192–197.

Dommaraju, P. (2010). *The changing demography of marriage in India.* Podcast retrieved from http://www.ari.nus.edu.sg/publication_details.asp?pubtypeid=AU&pubid=1630

Donahue, P. J. D. (2007). Retirement reconceptualized: Forced retirement and its relationship to health. *Dissertation Abstracts International. Section A. Humanities and Social Sciences, 67(7-A),* 2750.

Donaldson, S. J., & Ronan, K. R. (2006). The effects of sports participation on young adolescents' emotional well-being. *Adolescence, 41,* 369–389.

Donnelly, E. A., & Hinterlong, J. E. (2010). Changes in social participation and volunteer activity among recently widowed older adults. *The Gerontologist, 50,* 158–169.

Doohan, E.-A. M., Carrère, S., & Riggs, M. L. (2010). Using relational stories to predict the trajectory toward marital dissolution: The oral history interview and spousal feelings of flooding, loneliness, and depression. *Journal of Family Communication, 10,* 57–77.

Dorton, H. E., Jr. (2001). Job exit and the meaning of work in declining industries: Comparing retirement, displacement by retirement, and displacement by layoff at Weirton Steel Corporation. *Dissertation Abstracts International. Section A. Humanities and Social Sciences, 61(12-A)*, 4952.

Dow, B., & Meyer, C. (2010). Caring and retirement: Crossroads and consequences. *International Journal of Health Services, 40*, 645–665.

Doyle, J. M., & Kao, G. (2007). Are racial identifies of multiracials stable? Change self-identification among single and multiple race individuals. *Social Psychology Quarterly, 70*, 405–423.

Draghi-Lorenz, R., Reddy, V., & Costall, A. (2001). Rethinking the development of "nonbasic" emotions: A critical review of existing theories. *Developmental Review, 21*, 263–304.

Dranitsaris, G., Selby, P., & Negrete, J. C. (2009). Meta-analyses of placebo-controlled trials of acamprosate for the treatment of alcohol dependence: Impact of the combined pharmacotherapies and behavior interventions study. *Journal of Addiction Medicine, 3*, 74–82.

Drew, L. M., & Silverstein, M. (2005). Inter-generational role investments of great-grandparents: Consequences for psychological well-being. *Ageing and Society, 24*, 95–111.

Driscoll, A. K., Russell, S. T., & Crockett, L. J. (2008). Parenting styles and youth well-being across immigrant generations. *Journal of Family Issues, 29*, 185–209.

Duckworth, A. L., & Seligman, M. E. (2005). Self-discipline outdoes IQ in predicting academic performance of adolescents. *Psychological Science, 16*, 939–944.

Duff, L. J., Ericsson, K. A., & Baluch, B. (2007). In search of the loci for sex differences in throwing: The effects of physical size and differential recruitment rates on high levels of dart performance. *Research Quarterly for Exercise and Sport, 78*, 71–78.

Dumas, J. A., McDonald, B. C., Saykin, A. J., McAllister, T. W., Hynes, M. L., West, J. D., et al. (2010). Cholinergic modulation of hippocampal activity during episodic memory encoding in postmenopausal women: A pilot study. *Menopause, 17*, 852–859.

Duncan, G. J., & Brooks-Gunn, J. (2000). Family poverty, welfare reform, and child development. *Child Development, 71*, 188–196.

Dunham, P. J., Dunham, F., & Curwin, A. (1993). Joint attentional states and lexical acquisition at 18 months. *Developmental Psychology, 29*, 827–831.

Dunn, J., & Brophy, M. (2005). Communication, relationships, and individual differences in children's understanding of mind. In J. W. Astington & J. A. Baird (Eds.), *Why language matters for theory of mind* (pp. 50–69). New York: Oxford University Press.

Dunn, J., & Davies, L. (2001). Sibling relationships and interpersonal conflict. In J. Grych & F. D. Fincham (Eds.), *Interparental conflict and child development* (pp. 273–290). New York: Cambridge University Press.

Dunn, J., & Kendrick, C. (1981). Social behavior of young siblings in the family context: Differences between same-sex and different-sex dyads. *Child Development, 52*, 1265–1273.

Dunn, J., O'Connor, T. G., & Cheng, H. (2005). Children's responses to conflict between their different parents: Mothers, stepfathers, nonresident fathers, and nonresident stepmothers. *Journal of Clinical Child and Adolescent Psychology, 34*, 223–234.

Dunn, J., Slomkowski, C., & Beardsall, L. (1994). Sibling relationships from the preschool period through middle childhood and early adolescence. *Developmental Psychology, 30*, 315–324.

Dunn, M., & White, V. (2011). The epidemiology of anabolic-androgenic steroid use among Australian secondary school students. *Journal of Science and Medicine in Sport, 14*, 10–14.

Dunson, D. B., Colombo, B., & Baird, D. D. (2002). Changes in age in the level and duration of fertility in the menstrual cycle. *Human Reproduction, 17*, 1399–1403.

Dupuis, S. (2010). Examining the blended family: The application of systems theory toward an understanding of the blended family system. *Journal of Couple and Relationship Therapy, 9*, 239–251.

Durik, A., Hyde, J. S., & Clark, R. (2000). Sequelae of cesarean and vaginal deliveries: Psychosocial outcomes for mothers and infants. *Developmental Psychology, 36*, 251–260.

Durston, S., Davidson, M. C., Tottenham, N., Galvan, A., Spicer, J., Fossella, J. A., et al. (2006). A shift from diffuse to focal cortical activity with development. *Developmental Science, 9*, 1–8.

Dutton, W. H., Helsper, E. J., Whitty, M. T., Li, N., Buckwalter, J. G., & Lee, E. (2009). The role of the Internet in reconfiguring marriages: A cross-national study. *Interpersona: An International Journal on Personal Relationships, 3(Suppl. 2)*. Retrieved from http://www.interpersona.org/pdf/interpersona3(suppl.2).pdf#page=4

Dwyer, T., & Ponsonby, A. L. (2009). Sudden infant death syndrome and prone sleeping position. *Annals of Epidemiology, 19*, 245–249.

Dyke, P. H., & Adams, G. R. (1990). Identity and intimacy: An initial investigation of three theoretical models using cross-lag panel correlations. *Journal of Youth and Adolescence, 19*, 91–110.

Eagly, A. H., Karau, S. J., & Makhijani, M. G. (1995). Gender and the effectiveness of leaders: A meta-analysis. *Psychological Bulletin, 117*, 125–145.

Eastwick, P. W., Saigal, S. D, & Finkel, E. J. (2010). *Smooth operating: A structural analysis of social behavior (SASB) perspective on initial romantic encounters*. Retrieved from http://faculty.wcas.northwestern.edu/eli-finkel/documents/55_EastwickSaigalFinkelInPress_SPPS.pdf

Eaton, D. K., Kann, L., Kinchen, S., Shanklin, S., Ross, J., Hawkins, J., et al. (2008). Youth risk behavior surveillance—United States, 2007. *Morbidity and Mortality Weekly Report, 57*, 1–131.

Eaton, J., & Salari, S. (2005). Environments for lifelong learning in senior centers. *Educational Gerontology, 31*, 461–480.

Ebberwein, C. A. (2001). Adaptability and the characteristics necessary for managing adult career transition: A qualitative investigation. *Dissertation Abstracts International. Section B. Sciences and Engineering, 62(1-B)*, 545.

Ebersole, P., & Hess, P. (1998). *Toward healthy aging* (5th ed.). Amsterdam, Netherlands: Elsevier.

Eddleston, K. A., Baldridge, D. C., & Veiga, J. F. (2004). Toward modeling the predictors of managerial career success: Does gender matter? *Journal of Managerial Psychology, 19*, 360–385.

Edwards, C. A. (1994). Leadership in groups of school-age girls. *Developmental Psychology, 30*, 920–927.

Edwards, J. D., Vance, D. E., Wadley, V. G., Cissell, G. M., Roenker, D. L., & Ball, K. K. (2005). Reliability and validity of useful field of view test scores as administered by personal computer. *Journal of Clinical & Experimental Neuropsychology, 27*, 529–543.

Edwards, M. B. (2006). The relationship between the internal working model of attachment and patterns of grief experienced by college students after the death of a parent. *Dissertation Abstracts International. Section A. Humanities and Social Sciences, 66(11-A)*, 4197.

Egan, S. K., Monson, T. C., & Perry, D. G. (1998). Social-cognitive influences on change in aggression over time. *Developmental Psychology, 34*, 996–1006.

8newsnow.com. (2010). *Fatal accident raises questions about senior drivers*. Retrieved from http://www.8newsnow.com/story/12299621/fatal-accident-raises-questions-about-senior-drivers

Einolf, C. J. (in press). The timing of generative concern: Evidence from a longitudinal survey. *The Journal of Adult Development*.

Eisenberg, N. (2000). Emotion, regulation, and moral development. *Annual Review of Psychology, 51*, 665–697.

Eisenberg, N., & Fabes, R. A. (1998). Prosocial development. In W. Damon (Ed.), *Handbook of child psychology* (Vol. 3, pp. 701–778). New York: Wiley.

Eisenberg, N., & Morris, A. S. (2002). Children's emotion-related regulation. *Advances in Child Development and Behavior, 30*, 189–229.

Eisenberg, N., & Shell, R. (1986). Prosocial moral judgment and behavior in children: The mediating role of cost. *Personality and Social Psychology Bulletin, 12*, 426–433.

Eisenberg, N., Cumberland, A., Spinrad, T. L., Fabes, R. A., Shepard, S. A., Reiser, M., et al. (2001). The relations of regulation and emotionality to children's externalizing and internalizing problem behavior. *Child Development, 72*, 1112–1134.

Eisenberg, N., Fabes, R. A., & Spinrad, T. (2006). Prosocial development. In W. Damon & R. M. Lerner (Eds.), *Handbook of child psychology, Vol. 3* (6th ed.) (pp. 646–718). New York: Wiley.

Eisenberg, N., Hofer, C., Spinrad, T., Gershoff, E., Valiente, C., Losoya, S. L., et al. (2008). Understanding parent-adolescent conflict discussions: Concurrent and across-time prediction from youths' dispositions and parenting. *Monographs of the Society for Research in Child Development, 73,* Serial No. 290.

Eisenberg, N., Michalik, N., Spinrad, T. L., Hofer, C., Kupfer, A., Valiente, C., et al. (2007). The relations of effortful control and impulsivity to children's symptoms: A longitudinal study. *Cognitive Development, 22,* 544–567.

Eisenberg, N., Sadovsky, A., Spinrad, T. L., Fabes, R. A., Losoya, S. H., Valienta, C., et al. (2005). The relations of problem behavior status to children's negative emotionality, effortful control, and impulsivity: Concurrent relations and predictions of change. *Developmental Psychology, 41,* 193–211.

Eizenman, D. R., & Bertenthal, B. I. (1998). Infants' perception of object unity in translating and rotating displays. *Developmental Psychology, 34,* 426–434.

Ekerdt, D. J. (2010). Frontiers of research on work and retirement. *Journal of Gerontology: Social Sciences, 65(B),* S69–S80.

El Nokali, N. E., Bachman, H. J., & Votruba-Drzal, E. (2010). Parent involvement and children's academic and social development in elementary school. *Child Development, 81,* 988–1005.

Elbert, T., Pantev, C., Weinbruch, C., Rockstroh, B., & Taub, E. (1995). Increased cortical representation of the fingers of the left hand in string players. *Science, 270,* 305–307.

Elkind, D. (1978). *The child's reality: Three developmental themes.* Hillsdale, NJ: Erlbaum.

Elkind, D., & Bowen, R. (1979). Imaginary audience behavior in children and adolescents. *Developmental Psychology, 15,* 38–44.

Elliott, D. B., & Lewis, J. M. (2010). *Embracing the institution of marriage: The characteristics of remarried Americans.* Retrieved from http://www.census.gov/population/www/socdemo/marr-div/Remarriage.pdf

Elliott, J., & Oliver, I. (2008). Choosing between life and death: Patient and family perceptions of the decision not to resuscitate the terminally ill cancer patient. *Bioethics, 22,* 179–189.

Ellis, B. J. (2004). Timing of pubertal maturation in girls: An integrated life history approach. *Psychological Bulletin, 130,* 920–958.

Ellis, B. J., & Essex, M. J. (2007). Family environments, adrenarche, and sexual maturation: A longitudinal test of a life history model. *Child Development, 78,* 1799–1817.

Ellis, B. J., & Garber, J. (2000). Psychosocial antecedents of variation in girls' pubertal timing: Maternal depression, stepfather presence, and marital family stress. *Child Development, 71,* 485–501.

Ellis, B. J., Bates, J. E., Dodge, K. A., Fergusson, D. M., Horwood, L. J., Pettit, G. S., et al. (2003). Does father absence place daughters at special risk for early sexual activity and teenage pregnancy? *Child Development, 74,* 801–821.

Ellis, J. W. (2010). Comment: Yours, mine, ours? Why the Texas legislature should simplify caretaker consent capabilities for minor children and the implications of the addition of Chapter 34 to the Texas Family Code. *Texas Tech Law Review, 42,* 987.

Ellis, W. K., & Rusch, F. R. (1991). Supported employment: Current practices and future directions. In J. L. Matson & J. A. Mulick (Eds.), *Handbook of mental retardation* (2nd ed.). (pp. 479–488). New York: Pergamon.

Else-Quest, N. M., Hyde, J. S., & Linn, M. C. (2010). Cross-national patterns of gender differences in mathematics: A meta-analysis. *Psychological Bulletin, 136,* 103–127.

Else-Quest, N. M., Hyde, J. S., Goldsmith, H., & Van Hulle, C. A. (2006). Gender differences in temperament: A meta-analysis. *Psychological Bulletin, 132,* 33–72.

Emick, M. A., & Hayslip, B., Jr. (1999). Custodial grandparenting: Stresses, coping skills, and relationships with grandchildren. *International Journal of Aging & Human Development, 48,* 35–61.

Engel, S. M., Berkowitz, G. S., Wolff, M. S., & Yehuda, R. (2005). Psychological trauma associated with the World Trade Center attacks and its effect on pregnancy outcome. *Paediatric and Perinatal Epidemiology, 19,* 334–341.

Engel, S. M., et al. (2009). Prenatal phthalate exposure and performance on the Neonatal Behavioral Assessment Scale in a multiethnic birth cohort. *Neurotoxicology, 30,* 522–528.

Engle, P., & Huffman, S. L. (2010). Growing children's bodies and minds: Maximizing child nutrition and development. *Food and Nutrition Bulletin, 31,* S186–S197.

Enslin, C. (2007). Women in organizations: A phenomenological study of female executives mentoring junior women in organizations. *Dissertation Abstracts International. Section A. Humanities and Social Sciences, 68(4-A),* 1692.

Epel, E. S., Burke, H. M., & Wolkowitz, O. M. (2007). The psychoneuroendocrinology of aging: Anabolic and catabolic hormones. In C. M. Aldwin, C. L. Park, & A. Spiro III (Eds.), *Handbook of health psychology and aging* (pp. 119–141). New York: Guilford Press.

Epstein, L. H., & Cluss, P. A. (1986). Behavioral genetics of childhood obesity. *Behavior Therapy, 17,* 324–334.

Epstein, L. H., Leddy, J. J., Temple, J. L., & Faith, M. S. (2007). Food reinforcement and eating: A multilevel analysis. *Psychological Bulletin, 133,* 884–906.

Equal Employment Opportunity Commission. (2010). *Sexual harassment charges: EEOC and FEPAs combined: FY1997-FY2009.* Retrieved from http://www.eeoc.gov/eeoc/statistics/enforcement/sexual_harassment.cfm

Erel, O., & Burman, B. (1995). Interrelatedness of marital relations and parent–child relations: A meta-analytic review. *Psychological Bulletin, 118,* 108–132.

Erel, O., Margolin, G., & John, R. S. (1998). Observed sibling interaction: Links with the marital and the mother–child relationship. *Developmental Psychology, 34,* 288–298.

Ericsson, K. A., & Towne, T. J. (2010). Expertise. *Wiley Interdisciplinary Reviews: Cognitive Science, 1,* 404–416.

Erikson, E. H. (1968). *Identity: Youth and crisis.* New York: Norton.

Erikson, E. H. (1982). *The life cycle completed: Review.* New York: Norton.

Erikson, E. H., Erikson, J. M., & Kivnick, H. Q. (1986). *Vital involvement in old age.* New York: Norton.

Ernst, M., Moolchan, E. T., & Robinson, M. L. (2001). Behavioral and neural consequences of prenatal exposure to nicotine. *Journal of the American Academy of Child & Adolescent Psychiatry, 40,* 630–641.

Eskritt, M., & McLeod, K. (2008). Children's note taking as a mnemonic tool. *Journal of Experimental Child Psychology, 101,* 52–74.

European Association for Palliative Care. (2011). *The EAPC ethics task force on palliative care and euthanasia.* Retrieved February 27, 2011 from http://www.eapcnet.eu/Themes/Ethics/PCeuthanasiataskforce/tabid/232/Default.aspx.

Evans, M., Platt, L., & De La Cruz, F. (Eds.). (2001). *Fetal therapy.* New York: Parthenon.

Evans, N. J., Forney, D. S., Guido, F. M., Patton, L. D., & Renn, K. A. (2010). *Student development in college: Theory, research, and practice* (2nd ed.). San Francisco, CA: Jossey-Bass.

Everingham, C., Warner-Smith, P., & Byles, J. (2007). Transforming retirement: Re-thinking models of retirement to accommodate the experiences of women. *Women's Studies International Forum, 30,* 512–522.

Eyler, A. E., Wilcox, S., Matson-Koffman, D., Evenson, K. R., Sanderson, B., Thompson, J., et al. (2002). Correlates of physical activity among women from diverse racial/ethnic groups. *Journal of Women's Health and Gender Based Medicine, 11,* 239–253.

Faber, A. J. (2004). Examining remarried couples through a Bowenian family systems lens. *Journal of Divorce & Remarriage, 40,* 121–133.

Fabes, R. A., Eisenberg, N., Jones, S., Smith, M., Guthrie, I., Poulin, R., et al. (1999). Regulation, emotionality, and preschoolers' socially competent peer interactions. *Child Development, 70,* 432–442.

Fabricius, W. V., & Luecken, L. J. (2007). Postdivorce living arrangements, parent conflict, and long-term physical health correlates for children of divorce. *Journal of Family Psychology, 21,* 195–205.

Fagot, B. I. (1985). Changes in thinking about early sex role development. *Developmental Review, 5,* 83–98.

Faith, M. S., Fontaine, K. R., Baskin, M. L., & Allison, D. B. (2007). Toward the reduction of population obesity: Macrolevel environmental approaches to the problems of food, eating, and obesity. *Psychological Bulletin, 133,* 205–226.

Falbo, T., & Polit, E. F. (1986). Quantitative review of the only child literature: Research evidence and theory development. *Psychological Bulletin, 100,* 176–186.

Farver, J. M., & Shin, Y. L. (1997). Social pretend play in Korean- and Anglo-American preschoolers. *Child Development, 68,* 544–556.

Fasig, L. G. (2000). Toddlers' understanding of ownership: Implications for self-concept development. *Social Development, 9,* 370–382.

Faulkner, R. A., Davey, M., & Davey, A. (2005). Gender-related predictors of change in marital satisfaction and marital conflict. *American Journal of Family Therapy, 33,* 61–83.

Fearon, R. P., Bakermans-Kranenburg, M. J., van IJzendoorn, M. H., Lapsley, A., & Roisman, G. I. (2010). The significance of insecure attachment and disorganization in the development of children's externalizing behavior: A meta-analytic study. *Child Development, 81,* 435–456.

Feddes, A. R., Noack, P. & Rutland, A. (2009). Direct and extended friendship effects on minority and majority children's interethnic attitudes: A longitudinal study. *Child Development, 80,* 377–390.

Federal Bureau of Investigation. (2010). *Crime in the United States, 2009.* Retrieved from http://www.fbi.gov/ucr/09cius.htm

Federal Interagency Forum on Aging-Related Statistics. (2010). *Older Americans 2010: Key indicators of well-being.* Retrieved from http://www.agingstats.gov/agingstatsdotnet/Main_Site/Data/2010_Documents/Docs/OA_2010.pdf

Federal Interagency Forum on Child and Family Statistics. (2005). *America's children: Key national indicators of well-being, 2005.* Washington, DC: U.S. Government Printing Office.

Feigenson, L., Carey, S., & Hauser, M. (2002). The representations underlying infants' choice of more: Object files versus analog magnitudes. *Psychological Science, 13,* 150–156.

Feinberg, M. E., McHale, S. M., Crouter, A. C., & Cumsille, P. (2003). Sibling differentiation: Sibling and parental relationships trajectories in adolescence. *Child Development, 74,* 1261–1274.

Feldhusen, J. F. (1996). Motivating academically able youth with enriched and accelerated learning experiences. In C. P. Benbow & D. J. Lubinski (Eds.), *Intellectual talent: Psychometric and social issues.* Baltimore, MD: Johns Hopkins University Press.

Feldman, K. (2010). *Post parenthood redefined: Race, class, and family structure differences in the experience of launching children.* Doctoral dissertation completed at Case Western Reserve University. Retrieved from http://etd.ohiolink.edu/send-pdf.cgi/Feldman%20Karie%20Ellen.pdf?case1267730564

Feldman, R., Masalha, S., & Derdikman-Eiron, R. (2010). Conflict resolution in the parent-child, marital, and peer context and children's aggression in the peer group: A process-oriented cultural perspective. *Developmental Psychology, 46,* 310–325.

Fennell, C. T., Byers-Heinlein, K., & Werker, J. F. (2007). Using speech sounds to guide word learning: The case of bilingual infants. *Child Development, 78,* 1510–1525.

Fenson, L., Dale, P. S., Reznick, J. S., Bates, E., Thal, D. J., & Pethick, S. J. (1994). Variability in early communicative development. *Monographs of the Society for Research in Child Development, 59*(5, Serial No. 242).

Fergusson, D. M., & Woodward, L. J. (2000). Teenage pregnancy and female educational underachievement: A prospective study of a New Zealand birth cohort. *Journal of Marriage and Family, 62,* 147–161.

Fernander, A., Schumacher, M., & Nasim, A. (2008). Sociocultural stress, smoking risk, and cessation among African American women. *Journal of Black Psychology, 34,* 49–69.

Fernyhough, C. (2010). Inner speech. In H. Pashler (Ed.), *Encyclopedia of the mind.* Thousand Oaks, CA: Sage.

Ferrarini, T., & Norström, T. (2010). Family policy, economic development, and infant mortality: a longitudinal comparative analysis. *Internaitonal Journal of Social Welfare, 19,* S89–S102.

Ferreol-Barbey, M., Piolat, A., & Roussey, J. (2000). Text recomposition by eleven-year-old children: Effects of text length, level of reading comprehension, and mastery of prototypical schema. *Archives de Psychologie, 68,* 213–232.

Field, T. (2010). Postpartum depression effects on early interactions, parenting, and safety practices: A review. *Infant Behavior and Development, 33,* 1–6.

Field, T. M. (1990). *Infancy.* Cambridge, MA: Harvard University Press.

Field, T. M., & Widmayer, S. M. (1982). Motherhood. In B.-J. Wolman (Ed.), *Handbook of developmental psychology* (pp. 681–701). Englewood Cliffs, NJ: Prentice Hall.

Field, M. J., & Cassel, C. K. (2010). Approaching death: Improving care at the end of life. In D. Meier, S. L. Isaacs, & R. G. Hughes (Eds.), *Palliative care: Transforming the care of serious illness* (pp. 79–91). San Francisco: Jossey-Bass.

Field, T., & Diego, M. (2010). Preterm infant massage therapy research: A review. *Infant Behavior and Development, 33,* 115–124.

Fife, J. E. (2005). The relationship between religious commitment, self identification, and life satisfaction among African Americans and European Americans. *Dissertation Abstracts International. Section B. Sciences and Engineering, 65(7-B),* 3704.

Filanosky, C. A., Jr. (2004). The nature of continuing bonds with the deceased and their effect on bereavement outcome. *Dissertation Abstracts International. Section B. Sciences and Engineering, 65(2-B),* 1027.

Fincham, F. D., & Beach, S. R. H. (2010). Marriage in the new millennium: A decade in review. *Journal of Marriage and Family, 72,* 630–649.

Fingerman, K. L. (1996). Sources of tension in the aging mother and adult daughter relationship. *Psychology and Aging, 11,* 591–606.

Finkler, K. (2004). Traditional healers in Mexico: The effectiveness of spiritual practices. In U. P. Gielen, J. M. Fish, & J. G. Draguns (Eds.), *Handbook of culture, therapy, and healing* (pp. 161–174). Mahwah, NJ: Erlbaum.

Fischer, R. S., Nygren, B., Lundman, B., & Norberg, A. (2007). Living amidst consolation in the presence of God perceptions of consolation among the oldest old: The Umeå 85+ study. *Journal of Religion, Spirituality & Aging, 19,* 3–20.

Fisher, C. (1996). Structural limits on verb mapping: The role of analogy in children's interpretations of sentences. *Cognitive Psychology, 31,* 41–81.

Fisher, H. (2006). The drive to love: The neural mechanism for mate selection. In R. J. Sternberg & K. Weis (Eds.), *The new psychology of love* (pp. 87–115). New Haven, CT: Yale University Press.

Fisher, L. L. (2010). Sex, romance, and relationships: AARP survey of midlife and older adults. Retrieved from http://assets.aarp.org/rgcenter/general/srr_09.pdf

Fisher-Borne, M. (2007). Making the link: Domestic violence in the GLBT community. In L. Messinger & D. F. Morrow (Eds.), *Case studies on sexual orientation and gender expression in social work practice* (pp. 96–98). New York: Columbia University Press.

FitzGerald, D. P., & White, K. J. (2003). Linking children's social worlds: Perspective-taking in parent-child and peer contexts. *Social Behavior and Personality, 31,* 509–522.

Fitzgerald, H. E. (2005). Alcoholism prevention programs for children. In C. B. Fisher & R. M. Lerner (Eds.), *Encyclopedia of applied developmental science* (Vol. 1, pp. 73–76). Thousand Oaks, CA: Sage.

Fitzgerald, H. E., & Brackbill, Y. (1976). Classical conditioning in infancy: Development and constraints. *Psychological Bulletin, 83,* 353–375.

Fitzgerald, J. (1987). Research on revision in writing. *Review of Educational Research, 57,* 481–506.

Fitzgerald, J. M. (1999). Autobiographical memory and social cognition: Development of the remembered self in adulthood. In T. M. Hess & F. Blanchard-Fields (Eds.), *Social cognition and aging* (pp. 143–171). San Diego, CA: Academic Press.

Fitzwater, E. L. (2008). *Older adults and mental health: Pt. 2. Anxiety disorder.* Retrieved from http://www .netwellness.org/healthtopics/aging/anxietydisorder.cfm

Fivush, R., Reese, E., & Haden, C. A. (2006). Elaborating on elaborations: Role of maternal reminiscing style in cognitive and socioemotional development. *Child Development, 77,* 1568–1588.

Flavell, J. H. (1985). *Cognitive development* (2nd ed.). Englewood Cliffs, NJ: Prentice Hall.

Flavell, J. H. (1996). Piaget's legacy. *Psychological Science, 7,* 200–203.

Flavell, J. H. (2000). Development of children's knowledge about the mental world. *International Journal of Behavioral Development, 24,* 15–23.

Fleming, S. J., & Balmer, L. E. (1996). Bereavement in adolescence. In C. A. Coor & D. E. Balk (Eds.), *Handbook of adolescent death and bereavement* (pp. 139–154). New York: Springer.

Fleming, S., & Robinson, P. (2001). Grief and cognitive-behavioral therapy: The reconstruction of meaning. In M. S. Stroebe, R. O. Hansson, W. Stroebe, & H. Schut (Eds.), *Handbook of bereavement research: Consequences, coping, and care* (pp. 647–669). Washington, DC: American Psychological Association.

Flom, R., & Bahrick, L. E. (2007). The development of infant discrimination of affect in multimodal and unimodal stimulation: The role of intersensory redundancy. *Developmental Psychology, 43,* 238–252.

Flores, C. M. (2008). Beliefs and experiences of Mexican-American women with chronic pain. *Dissertation Abstracts International. Section B. Sciences and Engineering, 68(9-B),* 6303.

Flores, E., Cicchetti, D., & Rogosch, F. A. (2005). Predictors of resilience in maltreated and nonmaltreated Latino children. *Developmental Psychology, 41,* 338–351.

Flynn, H. K. (2007). Friendship: A longitudinal study of friendship characteristics, life transitions, and social factors that influence friendship quality. *Dissertation Abstracts International. Section A. Humanities and Social Sciences, 67(9-A),* 3608.

Flynn, J. R. (1999). Searching for justice: The discovery of IQ gains over time. *American Psychologist, 54,* 5–20.

Flynn, J. R., & Weiss, L. G. (2007). American IQ gains from 1932 to 2002: The WISC subtests and educational progress. *International Journal of Testing, 7,* 209–224.

Fogarty, K., & Evans, G. D. (2010). Being an involved father: What does it mean? Retrieved from http://edis.ifas.ufl.edu/he141

Foley, D., Ancoli-Israel, S., Britz, P., & Walsh, J. (2004) Sleep disturbances and chronic disease in older adults: Results of the 2003 National Sleep Foundation Sleep in America Survey. *Journal of Psychosomatic Research, 56,* 497–502.

Foley, K. L., Tung, M.-J., & Mutran, E. J. (2002). Self-gain and self-loss among African American and White caregivers. *Journals of Gerontology: Social Sciences, 57B,* S14–S22.

Folkman, S., (2008). The case for positive emotions in the stress process. *Anxiety, Stress & Coping, 21,* 3–14.

Fontaine, R. G. (2007). On-line social decision making and antisocial behavior: Some essential but neglected issues. *Clinical Psychology Review, 28,* 17–35.

Fontaine, R. G., Yang, C., Dodge, K. A., Pettit, G. S., & Bates, J. E. (2009). Development of response evaluation and decision (RED) and antisocial behavior in childhood and adolescence. *Developmental Psychology, 45,* 447–459.

Fonzi, A., Schneider, B. H., Tani, F., & Tomada, G. (1997). Predicting children's friendship status from their dynamic interaction in structured situations of potential conflict. *Child Development, 68,* 496–506.

Foreyt, J. P., & Goodrick, G. K. (1995). Obesity. In R. T. Ammerman & M. Hersen (Eds.), *Handbook of child behavior therapy in the psychiatric setting.* (pp. 409–426). New York: Wiley.

Fortner, B. V., & Neimeyer, R. A. (1999). Death anxiety in older adults: A quantitative review. *Death Studies, 23,* 387–411.

Foshee, V. A., & Langwick, S. (2004). *Safe dates: An adolescent dating abuse prevention curriculum.* Center City, MN: Hazelden Publishing and Educational Services.

Foshee, V. A., Benefield, T. S., Ennett, S. T., Bauman, K. E., & Suchindran, S. (2004). Longitudinal predictors of serious physical and sexual dating violence victimization during adolescence. *Preventive Medicine, 39,* 1007–1016.

Foshee, V. A., Linder, F., MacDougall, J. E., & Bangdiwala, S. (2001). Gender differences in the longitudinal predictors of adolescent dating violence. *Preventive Medicine, 32,* 128–141.

Foster, E. M., & Watkins, S. (2010). The value of reanalysis: TV viewing and attention problems. *Child Development, 81,* 368–375.

Foster, S. E., Jones, D. J., Olson, A. L., Forehand, R., Gaffney, C. A., Zens, M. S., et al. (2007). Family socialization of adolescents' self-reported cigarette use: The role of parents' history of regular smoking and parenting style. *Journal of Pediatric Psychology, 32,* 481–493.

Foster, S. H. (1986). Learning discourse topic management in the preschool years. *Journal of Child Language, 13,* 231–250.

Fouad, N. A. (2007). Work and vocational psychology: Theory, research, and applications. *Annual Review of Psychology, 58,* 543–564.

Fouad, N. A., & Mohler, C. J. (2004). Cultural validity of Holland's theory and the Strong Interest Inventory for five racial/ethnic groups. *Journal of Career Assessment, 12,* 423–439.

Fowler, K. L. (2008). "The wholeness of things": Infusing diversity and social justice into death education. *Omega: The Journal of Death and Dying, 57,* 53–91.

Fox, S. E., Levitt, P., & Nelson, C. A. (2010). How the timing and quality of early experiences influence the development of brain architecture. *Child Development, 81,* 28–40.

Fozard, J. L. & Gordon-Salant, S. (2001). Changes in vision and hearing with aging. In J. E. Birren & K. W. Schaie (Eds.), *Handbook of the psychology of aging* (5th ed., pp. 241–266). San Diego, CA: Academic Press.

Fraley, R. C., & Roberts, B. W. (2005). Patterns of continuity: A dynamic model for conceptualizing the stability of individual differences in psychological constructs across the life course. *Psychological Review, 112,* 60–74.

Franceschi, K. A. (2005). The experience of the transition to motherhood in women who have suffered maternal loss in adolescence. *Dissertation Abstracts International. Section B. Sciences and Engineering, 65(8-B),* 4282.

Frank, D. A., Augustyn, M., Knight, W. G., Pell, T., & Zuckerman, B. (2001). Growth, development, and behavior in early childhood following prenatal cocaine exposure: A systematic review. *Journal of the American Medical Association, 285,* 1613–1625.

Franklin, A., Pilling, M., & Davies, I. (2005). The nature of infant color categorization: Evidence from eye movements on a target detection task. *Journal of Experimental Child Psychology, 91,* 227–248.

Frazier, L. D., Hooker, K., Johnson, P. M., & Kaus, C. R. (2000). Continuity and change in possible selves in later life: A 5-year longitudinal study. *Basic and Applied Social Psychology, 22,* 237–243.

Frazier, L. D., Johnson, P. M., Gonzalez, G. K., & Kafka, C. L. (2002) Psychosocial influences on possible selves: A comparison of three cohorts of older adults. *International Journal of Behavioral Development, 26,* 308–317.

Fredricks, J. A., & Eccles, J. S. (2005). Family socialization, gender, and sport motivation and involvement. *Journal of Sport & Exercise Psychology, 27,* 3–31.

French, M. L. (2007). The alignment between personal meaning and organizational mission among music executives: A study of happiness, job satisfaction, and responsibility toward employees. *Dissertation Abstracts International. Section A. Humanities and Social Sciences, 67(11-A),* 4247.

French, S. E., Seidman, E., Allen, L., & Aber, J. (2006). The development of ethnic identity during adolescence. *Developmental Psychology, 42,* 1–10.

Frerichs, F., & Naegele, G. (1997). Discrimination of older workers in Germany: Obstacles and options for the integration into employment. *Journal of Aging and Social Policy, 9,* 89–101.

Fretz, B. R. (2001). Coping with licensing, credentialing, and lifelong learning. In S. Walfish & A. K. Hess (Eds.), *Succeeding in graduate school: The career guide for psychology students* (pp. 353–367). Mahwah, NJ: Erlbaum.

Freund, A. (2005). Commitment and job satisfaction as predictors of turnover intentions among welfare workers. *Administration in Social Work, 29,* 5–21.

Freund, B., Gravenstein, S., Ferris, R., Burke, B. L., & Shaheen, E. (2005). Drawing clocks and driving cars: Use of brief tests of cognition to screen driving competency in older adults. *Journal of General Internal Medicine, 20,* 240–244.

Fried, A. (2005). Depression in adolescence. In C. B. Fisher & R. M. Lerner (Eds.), *Encyclopedia of applied developmental science* (Vol. 1, pp. 332–334). Thousand Oaks, CA: Sage.

Fried, P. A., O'Connell, C. M., & Watkinson, B. (1992). 60- and 72-month follow-up of children prenatally exposed to marijuana, cigarettes, and alcohol: Cognitive and language assessment. *Journal of Developmental & Behavioral Pediatrics, 13,* 383–391.

Friedman, A., & Schoen, L. (2009). Reflective practice interventions: Raising levels of reflective judgment. *Action in Teacher Education, 31,* 61–73.

Friedman, E. M., & Seltzer, J. A. (2010). *Providing for older parents: Is it a family affair?* California Center for Population Research paper #PWP-CCPR-2010-12. Retrieved from http://www.n4a.org/pdf/PWP-CCPR-2010-012.pdf

Friedman, J. M., & Polifka, J. E. (1996). *The effects of drugs on the fetus and nursing infant: A handbook for health care professionals.* Baltimore: Johns Hopkins University Press.

Friedman, M., & Rosenman, R. H. (1974). *Type A behavior and your heart.* New York: Random House.

Friend, A., DeFries, J., & Olson, R. (2008). Parental education moderates genetic and environmental influences on reading disability. *Psychological Science, 19*, 1124–1130.

Froman, L. (2010). Positive psychology in the workplace. *Journal of Adult Development, 17*, 59–69.

Fruhauf, C. A. (2007). Grandchildren's perceptions of caring for grandparents. *Dissertation Abstracts International. Section A. Humanities and Social Sciences, 68(3-A)*, 1120.

Fry, R. (2009). *Latino children: A majority are U.S.-born offspring of immigrants.* Retrieved from http://pewhispanic.org/reports/report.php?ReportID=110

Frye, D. (1993). Causes and precursors of children's theories of mind. In D. F. Hay & A. Angold (Eds.), *Precursors and causes in development and psychopathology.* Chichester, England: Wiley.

Frye, K. L. (2008). Perceptions of retirement and aging as experienced by self-identified lesbians ages 51 through 60. *Dissertation Abstracts International. Section B. Sciences and Engineering, 68(7-B)*, 4886.

Fu, S.-Y., Anderson, D., & Courtney, M. (2003). Cross-cultural menopausal experience: Comparison of Australian and Taiwanese women. *Nursing & Health Sciences, 5*, 77–84.

Fujita, F., & Diener, E. (2005). Life satisfaction set point: Stability and change. *Journal of Personality and Social Psychology, 88*, 158–164.

Fuligni, A. J., Kiang, L., Witkow, M. R., & Baldelomar, O. (2008). Stability and change in ethnic labeling among adolescents from Asian and Latin American immigrant families. *Child Development, 79*, 944–956.

Fuller-Thompson, E., Hayslip, B., Jr., & Patrick, J. H. (2005). Introduction to the special issue: Diversity among grandparent caregivers. *International Journal of Aging & Human Development, 60*, 269–272.

Fung, H. H., & Siu, T. M. Y. (2010). Time, culture, and life-cycle changes of social goals. In T. W. Miller (Ed.), *Handbook of stressful transitions across the lifespan* (pp. 441–464). New York: Springer.

Furstenberg, F. F., Jr., Rumbaut, R., & Settersten, R. A., Jr. (2005). On the frontier of adulthood: Emerging themes and new directions. In R. A. Settersten, Jr., F. F. Furstenberg, Jr., & R. Rumbaut (Eds.), *On the frontier of adulthood: Theory, research, and public policy* (pp. 3–25). Chicago: University of Chicago Press.

Futterman, A., Gallagher, D., Thompson, L. W., Lovett, S., & Gilewski, M. (1990). Retrospective assessment of marital adjustment and depression during the first two years of spousal bereavement. *Psychology and Aging, 5*, 277–283.

Gaddis, A., & Brooks-Gunn, J. (1985). The male experience of pubertal change. *Journal of Youth and Adolescence, 14*, 61–69.

Gagliardi, A. (2005). Postpartum depression. In C. B. Fisher & R. M. Lerner (Eds.), *Encyclopedia of applied developmental science* (Vol. 2, pp. 867–870). Thousand Oaks, CA: Sage.

Gallagher, R., Bruzzese, J.-M., & McCann-Doyle, S. (2005). Cigarette smoking in adolescents. In C. B. Fisher & R. M. Lerner (Eds.), *Encyclopedia of applied developmental science* (Vol. 1, pp. 254–256). Thousand Oaks, CA: Sage.

Gamble, W. C., Ramakumar, S., & Diaz, A. (2007). Maternal and paternal similarities and differences in parenting: An examination of Mexican-American parents of young children. *Early Childhood Research Quarterly, 22*, 72–88.

Ganiron, E. E. (2007). Mutuality and relationship satisfaction in the formation of lesbian relationships across the life cycle: Unpacking the U-Haul myth. *Dissertation Abstracts International. Section B. Sciences and Engineering, 68(3-B)*, 1924.

Gans, D. (2007). Normative obligations and parental care in social context. *Dissertation Abstracts International. Section A. Humanities and Social Sciences, 68(5-A)*, 2115.

Garbrecht, L. S. (2005). Gender roles, identity development, and the gender traditionalism of career aspirations. *Dissertation Abstracts International. Section A. Humanities and Social Sciences, 65(11-A)*, 4102.

Garciaguirre, J. S., Adolph, K. E., & Shrout, P. E. (2007). Baby carriage: Infants walking with loads. *Child Development, 78*, 664–680.

Gardiner, J., Stuart, M., Forde, C., Greenwood, I., MacKenzie, R., & Perrett, R.

(2007). Work-life balance and older workers: Employees' perspectives on retirement transitions following redundancy. *International Journal of Human Resource Management, 18*, 476–489.

Gardner, H. (1983). *Frames of mind: The theory of multiple intelligences.* New York: Basic Books.

Gardner, H. (1993). *Multiple intelligences: The theory in practice.* New York: Basic Books.

Gardner, H. (1995). Reflections on multiple intelligences: Myths and messages. *Phi Delta Kappan, 77*, 200–203, 206–209.

Gardner, H. (1999). *Intelligence reframed: Multiple intelligences for the 21st century.* New York: Basic Books.

Gardner, H. (2002). *MI millennium: Multiple intelligences for the new millennium* [Video recording]. Los Angeles, CA: Into the Classroom Media.

Gardner, H. (2006). *Multiple intelligences: New horizons.* New York: Basic Books.

Gardner, M., Roth, J., & Brooks-Gunn, J. (2009). Sports participation and juvenile delinquency: The role of the peer context among adolescent boys and girls with varied histories of problem behavior. *Developmental Psychology, 45*, 341–353.

Garrard, E., & Wilkinson, S. (2005). Passive euthanasia. *Journal of Medical Ethics, 31*, 64–68.

Garrod, A., & Larimore, C. (Eds.). (1997). *First person, first peoples: Native American college graduates tell their life stories.* Ithaca, NY: Cornell University Press.

Gartstein, M. A., et al. (2010). A latent growth examination of fear development in infancy: Contributions of maternal depression and the risk for toddler anxiety. *Developmental Psychology, 46*, 651–668.

Gartstein, M. A., Knyazev, G. G., & Slobodskaya, H. R. (2005). Cross-cultural differences in the structure of infant temperament: United States of America and Russia. *Infant Behavior and Development, 28*, 54–61.

Gartstein, M. A., Slobodskaya, H. R., & Kinsht, I. A. (2003). Cross-cultural differences in temperament in the first year of life: United States of America (U.S.) and Russia. *International Journal of Behavioral Development, 27*, 316–328.

Garvey, C., & Berninger, G. (1981). Timing and turn taking in children's conversations. *Discourse Processes, 4*, 27–59.

Gass, K., Jenkins, J., & Dunn, J. (2007). Are sibling relationships protective? A longitudinal study. *Journal of Child Psychology and Psychiatry, 48*, 167–175.

Gathercole, S. E., Willis, C. S., Emslie, H., & Baddeley, A. D. (1992). Phonological memory and vocabulary development during the early school years: A longitudinal study. *Developmental Psychology, 28*, 887–898.

Gatz, M. (2000). Variations on depression in later life. In S. H. Qualls & N. Abeles (Eds.), *Psychology and the aging revolution* (pp. 239–254). Washington, DC: American Psychological Association.

Gaudron, J.-P., & Vautier, S. (2007). Analyzing individual differences in vocational, leisure, and family interests: A multitrait-multimethod approach. *Journal of Vocational Behavior, 70*, 561–573.

Gaulin, S. J. C., & McBurney, D. H. (2001). *Psychology: An evolutionary approach.* Upper Saddle River, NJ: Prentice-Hall.

Gavin, L. A., & Furman, W. (1996). Adolescent girls' relationships with mothers and best friends. *Child Development, 67*, 375–386.

Gaylord, S. A., & Zung, W. W. K. (1987). Affective disorders among the aging. In L. L. Carstensen & B. A. Edelstein (Eds.), *Handbook of clinical gerontology* (pp. 76–95). New York: Pergamon Press.

Ge, X., Brody, G. H., Conger, R. D., Simons, R. L., & Murphy, V. M. (2002). Contextual amplification of pubertal transition effects on deviant peer affiliation and externalizing behavior among African American children. *Developmental Psychology, 38*, 45–54.

Ge, X., Conger, R. D., & Elder, G. H. (2001). Pubertal transition, stressful life events, and the emergence of gender differences in adolescent depressive symptoms. *Developmental Psychology, 37*, 404–417.

Ge, X., Kim, I. J., Brody, G. H., Conger, R. D., Simons, R. L., Gibbons, F. X., & Cutrona, C. E. (2003). It's about timing and change: Pubertal transition effects on symptoms of major depression among African American youths. *Developmental Psychology, 39*, 430–439.

Geary, D. C. (2002). Sexual selection and human life history. In R. V. Kail (Ed.), *Advances in child development and behavior* (Vol. 30, pp. 41–102). San Diego, CA: Academic Press.

Geary, D. C. (2005). *The origin of mind: Evolution of brain, cognition, and general intelligence.* Washington, DC: American Psychological Association.

Geary, D. C., Byrd-Craven, J., Hoard, M. K., Vigil, J., & Numtee, C. (2003). Evolution and development of boys' social behavior. *Developmental Review, 23,* 444–470.

Geary, D. C., Hoard, M. K., Byrd-Craven, J., Nugent, L., & Numtee, C. (2007). Cognitive mechanisms underlying achievement deficits in children with mathematical learning disability. *Child Development, 78,* 1343–1359.

Gelman, R., & Meck, E. (1986). The notion of principle: The case of counting. In J. Hiebert (Ed.), *Conceptual and procedural knowledge: The case of mathematics.* (pp. 29–57). Hillside, NJ: Erlbaum.

Gelman, S. A. (2003). *The essential child.* New York: Oxford University Press.

Gelman, S. A., & Gottfried, G. M. (1996). Children's casual explanations of animate and inanimate motion. *Child Development, 67,* 1970–1987.

Gelman, S. A., & Wellman, H. M. (1991). Insides and essences: Early understandings of the non-obvious. *Cognition, 38,* 213–244.

Gelman, S. A., Coley, J. D., Rosengren, K. S., Hartman, E., & Pappas, A. (1998). Beyond labeling: The role of maternal input in the acquisition of richly structured categories. *Monographs of the Society for Research in Child Development, 63*(Serial No. 253).

Gelman, S. A., Taylor, M. G., & Nguyen, S. P. (2004). Mother-child conversations about gender. *Monographs of the Society for Research in Child Development, 69*(Serial No. 275).

Gentile, D. (2009). Pathological video-game use among youth ages 8 to 18: A national study. *Psychological Science, 20,* 594–602.

George, J. B. F., & Franko, D. L. (2010). Cultural issues in eating pathology and body image among children and adolescents. *Journal of Pediatric Psychology, 35,* 231–242.

Gerry, D. W., Faux, A. L., & Trainor, L. J. (2010). Effects of Kindermusik training on infants rhymic enculturation. *Developmental Science, 13,* 545–551.

Gershkoff-Stowe, L., & Smith, L. B. (2004). Shape and the first hundred nouns. *Child Development, 75,* 1098–1114.

Gershoff, E. T., & Bitensky, S. H. (2007). The case against corporal punishment of children: Converging evidence from social science research and international human rights law and implications for U. S. public policy. *Psychology, Public Policy, and the Law, 13,* 231–272.

Gershoff, E. T., Grogan-Kaylor, A., Lansford, J. E., Chang, L., Zelli, A. Deater-Deckard, K., et al. (2010). Parent discipline practices in an international sample: Associations with child behavior and moderation by perceived normativeness. *Child Development, 81,* 487–502.

Ghetti, S. (2008). Rejection of false events in childhood: A metamemory account. *Current Directions in Psychological Science, 17,* 16–20.

Giancola, J., Grawitch, M. J., & Borchert, D. (2009). Dealing with the stress of college: A model for adult students. *Adult Education Quarterly, 59,* 246–263.

Giarusso, R., Feng, D., Silverstein, M., & Marenco, A. (2000). Primary and secondary stressors of grandparents raising grandchildren: Evidence from a national survey. *Journal of Mental Health and Aging, 6,* 291–310.

Gibbs, D. A., Martin, S. L., Kupper, L. L., & Johnson, R. E. (2007). Child maltreatment in enlisted soldiers' families during combat-related deployments. *Journal of the American Medical Association, 298,* 528–535.

Gibbs, J. C., Clark, P. M., Joseph, J. A., Green, J. L., Goodrick, T. S., & Makowski, D. (1986). Relations between moral judgment, moral courage, and field independence. *Child Development, 57,* 185–193.

Gibran, K. (1923). *The prophet.* New York: Knopf.

Gibson, E. J., & Walk, R. D. (1960). The visual cliff. *Scientific American, 202,* 64–71.

Giles, J. W., & Heyman, G. D. (2005). Reconceptualizing children's suggestibility: Bidirectional and temporal properties. *Child Development, 76,* 40–53.

Gill, S. V., Adolph, K. E., & Vereijken, B. (2009). Change in action: How infants learn to walk down slopes. *Developmental Science, 12,* 888–902.

Gilleard, C., & Higgs, P. (2010). Aging without agency: Theorizing the fourth age. *Aging and Mental Health, 14,* 121–128.

Gillies, J., & Neimeyer, R. A. (2006). Loss, grief, and the search for significance: Toward a model of meaning reconstruction in bereavement. *Journal of Constructivist Psychology, 19,* 31–65.

Gilligan, C. (1982). *In a different voice: Psychological theory and women's development.* Cambridge, MA: Harvard University Press.

Gilligan, C., & Attanucci, J. (1988). Two moral orientations: Gender differences and similarities. *Merrill-Palmer Quarterly, 34,* 223–237.

Giordano, G. (2005). *How testing came to dominate American schools: The history of educational assessment.* New York: Peter Lang.

Githens, R., & Sauer, T. (2010). Going green online: Distance learning prepares students for success in green-collar job markets. *Community College Journal, 80,* 32–35.

Givertz, M., Segrin, C., & Hansal, A. (2009). The association between satisfaction and commitment differs across marital couple types. *Communication Research, 36,* 561–584.

Gladding, S. T. (2002). *Family therapy: History, theory, and practice* (3rd ed.). Upper Saddle River, NJ: Merrill Prentice Hall.

Glazer, H. R., Clark, M. D, Thomas, R., & Haxton, H. (2010). Parenting after the death of a spouse. *Journal of Hospice and Palliative Care, 27,* 532–536.

Gleason, T. R., & Hohmann, L. M. (2006). Concepts of real and imaginary friendships in early childhood. *Social Development, 15,* 128–144.

Glick, J. E. (2010). Connecting complex processes: A decade of research on immigrant families. *Journal of Marriage and Family, 72,* 498–515.

Glickman, H. M. (2001). The relationship between person-organization value congruence and global job satisfaction. *Dissertation Abstracts International. Section B. Sciences and Engineering, 61(12-B),* 6745.

Global Initiative to End All Corporal Punishment of Children. (2011). States with full prohibition. Retrieved from http://www.endcorporalpunishment.org

Goeke-Morey, M. C., Cummings, E. M., Harold, G. T., & Shelton, K. H. (2003). Categories and continua of destructive and constructive conflict tactics from the perspective of U.S. and Welsh children. *Journal of Family Psychology, 17,* 327–338.

Goff, S. J., Fick, D. S., & Opplinger, R. A. (1997). The moderating effect of spouse support on the relation between serious leisure and spouses' perceived leisure-family conflict. *Journal of Leisure Research, 29,* 47–60.

Goldberg, A. E. (2009). *Lesbian and gay parents and their children: Research on the family life cycle.* Washington, DC: American Psychological Association.

Goldenberg, R. L., & Klerman, L. V. (1995). Adolescent pregnancy—Another look. *New England Journal of Medicine, 332,* 1161–1162.

Goldfield, B. A., & Reznick, J. S. (1990). Early lexical acquisition: Rate, content, and the vocabulary spurt. *Journal of Child Language, 17,* 171–184.

Goldin-Meadow, S., Mylander, C., & Franklin, A. (2007). How children make language out of gesture: Morphological structure in gesture systems developed by American and Chinese deaf children. *Cognitive Psychology, 55,* 87–135.

Goldman, L. S., Genel, M., Bezman, R. J., & Slanetz, P. J. (1998). Diagnosis and treatment of attention-deficit/hyperactivity disorder in children and adolescents. *Journal of the American Medical Association, 279,* 1100–1107.

Goldsmith, H. H., & Harman, C. (1994). Temperament and attachment: Individuals and relationships. *Current Directions in Psychological Science, 3,* 53–57.

Goldsmith, H. H., Buss, K. A., & Lemery, K. S. (1997). Toddler and childhood temperament: Expanded content, stronger genetic evidence, new evidence for the importance of environment. *Developmental Psychology, 33,* 891–905.

Goldsmith, H. H., Pollak, S. D., & Davidson, R. J. (2008). Developmental neuroscience perspectives on emotion regulation. *Child Development Perspectives, 2,* 132–140.

Goldstein, M. H., & Schwade, J. A. (2008). Social feedback to infants' babbling facilitates rapid phonological learning. *Psychological Science, 19,* 515–523.

Goleman, D. (1995). *Emotional intelligence: Why it can matter more than IQ.* New York: Bantam Books.

Golinkoff, R. M. (1993). When is communication a "meeting of minds"? *Journal of Child Language, 20,* 199–207.

Golombok, S., & Tasker, F. (1996). Do parents influence the sexual orientation of their children? Findings from a longitudinal study of lesbian families. *Developmental Psychology, 32,* 3–11.

Golombok, S., Murray, C., Jadva, V., MacCallum, F., & Lycett, E. (2004). Families created through surrogacy arrangements: Parent-child relationships in the first year of life. *Developmental Psychology, 40,* 400–411.

Gonzales, P., Williams, T., Jocelyn, L., Roey, S., Kastberg, D., & Brenwald, S. (2008). *Highlights from TIMSS 2007: Mathematics and science achievement of U.S. fourth- and eighth-grade students in an international context.* Washington, DC: U.S. Department of Education.

González, H. M., Bowen, M. E., & Fisher, G. G. (2008). Memory decline and depressive symptoms in a nationally representative sample of older adults: The Health and Retirement Study (1998–2004). *Dementia and Geriatric Cognitive Disorders, 25,* 266–271.

González, H. M., Haan, M. N., & Hinton, L. (2001). Acculturation and the prevalence of depression in older Mexican Americans: Baseline results of the Sacramento Area Latino Study on Aging. *Journal of the American Geriatrics Society, 49,* 948–953.

Good, T. L., & Brophy, J. E. (1994). *Looking in classrooms* (6th ed.). New York: HarperCollins.

Good, T. L., & Brophy, J. E. (2008). *Looking in classrooms.* Boston, MA: Pearson/Allyn & Bacon.

Goodnow, J. J. (1992). *Parental belief systems: The psychological consequences for children.* Hillsdale, NJ: Erlbaum.

Goodwin, P.Y., Mosher, W. D., & Chandra, A. (2010). *Marriage and cohabitation in the United States: A statistical portrait based on Cycle 6 (2002) of the National Survey of Family Growth.* Retrieved from http://www.cdc.gov/nchs/data/series/sr_23/sr23_028.pdf

Goodwyn, S. W., & Acredolo, L. P. (1993). Symbolic gesture versus word: Is there a modality advantage for onset of symbol use? *Child Development, 64,* 688–701.

Goossens, L. (2001). Global versus domain-specific statuses in identity research: A comparison of two self-report measures. *Journal of Adolescence, 24,* 681–699.

Góra, M., & Mach, Z. (2010). Between old fears and new challenges: The Polish debate on Europe. In J. Lacroix & K. Nicholaïdis. *European stories: Intellectual debates on Europe in national contexts.* (pp. 221–240).

Gordon, B. N., Baker-Ward, L., & Ornstein, P. A. (2001). Children's testimony: A review of research on memory for past experiences. *Clinical Child & Family Psychology Review, 4,* 157–181.

Gordon, C. P. (1996). Adolescent decision making: A broadly based theory and its application to the prevention of early pregnancy. *Adolescence, 31,* 561–584.

Gordon, R. A., Chase-Lansale, P. L., & Brooks-Gunn, J. (2004). Extended households and the life course of young mothers: Understanding the associations using a sample of mothers with premature, low birth weight babies. *Child Development, 75,* 1013–1038.

Gorman, E. H., & Kmec, J. A. (2007). We (have to) try harder: Gender and required work effort in Britain and the United States. *Gender & Society, 21,* 828–856.

Gosso, Y., Morais, M. L. S., & Otta, E. (2007). Pretend play of Brazilian children: A window into different cultural worlds. *Journal of Cross-Cultural Psychology, 38,* 539–558.

Gottlieb, L. N., & Mendelson, M. J. (1990). Parental support and firstborn girls' adaptation to the birth of a sibling. *Journal of Applied Developmental Psychology, 11,* 29–48.

Gottman, J. M. (1986). The world of coordinated play: Same- and cross-sex friendships in children. In J. M. Gottman & J. G. Parker (Eds.), *Conversations of friends.* (pp. 131–191). New York: Cambridge University Press.

Gottman, J. M., & Levenson, R. W. (2000). The timing of divorce: predicting when a couple will divorce over a 14-year period. *Journal of Marriage and the Family, 62,* 737–745.

Gottman, J. M., Katz, L. F., & Hooven, C. (1996). Parental meta-emotion philosophy and the emotional life of families: Theoretical models and preliminary data. *Journal of Family Psychology, 10,* 243–268.

Goubet, N., Clifton, R. K., & Shah, B. (2001). Learning about pain in preterm newborns. *Journal of Developmental & Behavioral Pediatrics, 22,* 418–424.

Gough, P. B., & Tunmer, W. E. (1986). Decoding, reading and reading disability. *Remedial and Special Education, 7,* 6–10.

Gow, A. J., Pattie, A., Whiteman, M. C., Whalley, L. J., & Deary, I. J. (2007). Social support and successful aging: Investigating the relationships between lifetime cognitive change and life satisfaction. *Journal of Individual Differences, 28,* 103–115.

Graham, S., & Perin, D. (2007). A meta-analysis of writing instruction for adolescent students. *Journal of Educational Psychology, 99,* 445–476.

Graham, S., Berninger, V. W., Abbott, R. D., Abbott, S. P., & Whitaker, D. (1997). Role of mechanics in composing of elementary school students: A new methodological approach. *Journal of Educational Psychology, 89,* 170–182.

Graham, S., Harris, K. R., & Fink, B. (2000). Is handwriting causally related to learning to write? Treatment of handwriting problems in beginning writers. *Journal of Educational Psychology, 92,* 620–633.

Graham-Bermann, S. A., & Brescoll, V. (2000). Gender, power, and violence: Assessing the family stereotypes of the children of batterers. *Journal of Family Psychology, 14,* 600–612.

Grande, G. E., Farquhar, M. C., Barclay, S. I. G., & Todd, C. J. (2004). Caregiver bereavement outcome: Relationship with hospice at home, satisfaction with care, and home death. *Journal of Palliative Care, 20,* 69–77.

Grandey, A. A. (2001). Family friendly policies: Organizational justice perceptions of need-based allocations. In R. Cropanzano (Ed.), *Justice in the workplace: From theory to practice* (pp. 145–173). Mahwah, NJ: Erlbaum.

Grant, B. F., Dawson, D. A., Stinson, F. S., Chou, S. P., Dufour, M. C., & Pickering, R. P. (2006). The 12-month prevalence and trends in DSM–IV alcohol abuse and dependence: United States, 1991–1992 and 2001–2002. *Alcohol Research and Health, 29,* 79–91.

Grant, J. E. & Potenza, M. N. (Eds.). (2010). *Young adult mental health.* Oxford, England: Oxford University Press.

Grant, L. M. (2007). Second generation Chinese American women and the midlife transition: A qualitative analysis. *Dissertation Abstracts International. Section B. Sciences and Engineering, 67(10-B),* 6095.

Graves, R., & Landis, T. (1990). Asymmetry in mouth opening during different speech tasks. *International Journal of Psychology, 25,* 179–189.

Grawitch, M. J., Barber, L. K., & Justice, L. (2010). Rethinking the work-life interface: It's not about balance, it's about resource allocation. *Applied Psychology: Health and Well-Being, 2,* 127–159.

Gray-Little, B., & Hafdahl, A. R. (2000). Factors influencing racial comparisons of self-esteem: A quantitative review. *Psychological Bulletin, 126,* 26–54.

Graziano, P. A., Keane, S. P., & Calkins, S. D. (2007). Cardiac vagal regulation and early peer status. *Child Development, 78,* 264–278.

Green, J. S. (2008). *Beyond the good death: An anthropology of modern dying.* Baltimore, MD: University of Pennsylvania Press.

Greenberg, M. T., & Crnic, K. A. (1988). Longitudinal predictors of developmental status and social interaction in premature and full-term infants at age two. *Child Development, 59,* 554–570.

Greenberg, M. T., Lengua, L. J., Coie, J. D., Pinderhughes, E. E., & the Conduct Problems Prevention Research Group. (1999). Predicting developmental outcomes at school entry using a multiple-risk model: Four American communities. *Developmental Psychology, 35,* 403–417.

Greenberger, E., O'Neil, R., & Nagel, S. K. (1994). Linking workplace and homeplace: Relations between the nature of adults' work and their parenting behaviors. *Developmental Psychology, 30,* 990–1002.

Greenfield, E. A., & Marks, N. F. (2005). Formal volunteering as a protective factor for older adults' psychological well-being. *Journals of Gerontology: Social Sciences, 59,* S258–S264.

Greenfield, P. M. (1998). The cultural evolution of IQ. In U. Neisser (Ed.), *The rising curve: Long-term gains in IQ and related measures* (pp. 81–123). Washington, DC: American Psychological Association.

Greenhaus, J. H., Parasuraman, S., & Wormley, W. M. (1990). Effects of race on organizational experiences, job performance evaluations, and career outcomes. *Academy of Management Journal, 33,* 64–86.

Greenough, W. T., & Black, J. E. (1992). Induction of brain structure by experience: Substrates for cognitive development. In M. Gunnar & C. Nelson (Eds.), *Minnesota symposia on child psychology: Vol. 24. Developmental behavioral neuroscience* (pp. 155–200). Hillsdale, NJ: Erlbaum.

Gregory, A. M., Light-Häusermann, J. H., Rijsdijk, F., & Eley, T. C. (2009). Behavioral genetic analyses of prosocial behavior in adolescents. *Developmental Science, 12,* 165–174.

Gregory, A. M., Rijsdijk, F., Lau, J. Y. F., Napolitano, M., McGuffin, P., & Eley, T. C. (2007). Genetic and environmental influences on interpersonal cognitions and associations with depressive symptoms in 8-year-old twins. *Journal of Abnormal Psychology, 116,* 762–775.

Greif, G. L. (2009). *Buddy system: Understanding male friendships*. New York: Oxford University Press.

Greving, K. A. (2007). Examining parents' marital satisfaction trajectories: Relations with children's temperament and family demographics. *Dissertation Abstracts International. Section A. Humanities and Social Sciences, 68(4-A)*, 1676.

Grice, H. P. (1975). Logic and conversation. In P. Cole & J. Morgan (Eds.), *Speech acts: Syntax and semantics*. (Vol. 3, pp. 41–58). New York: Academic Press.

Griffin, M. L., Hogan, N. L., Lambert, E. G., Tucker-Gail, K. A., & Baker, D. N. (2010). Job involvement, job stress, job satisfaction, and organizational commitment and the burnout of correctional staff. *Criminal Justice and Behavior, 37*, 239–255.

Grigorenko, E. L., & Sternberg, R. J. (2001). Analytical, creative, and practical intelligence as predictors of self-reported adaptive functioning: A case study in Russia. *Intelligence, 29*, 57–73.

Grigorenko, E. L., Jarvin, L., & Sternberg, R. J. (2002). School-based tests of the triarchic theory of intelligence: Three settings, three samples, three syllabi. *Contemporary Educational Psychology, 27*, 167–208.

Grigorenko, E. L., Meier, E., Lipka, J., Mohatt, G., Yanez, E., & Sternberg, R. J. (2004). Academic and practical intelligence: A case study of the Yup'ik in Alaska. *Learning & Individual Differences, 14*, 183–207.

Grimes, G., Hough, M., Mazur, E., and Signorella, M. (2010). Older adults' knowledge of Internet hazards. *Educational Gerontology, 36*(3), 173–192.

Grucza, R. A., Norberg, K. E., & Bierut, L. J. (2009). Binge drinking among youths and young adults in the United States: 1979–2006. *Child and Adolescent Psychiatry, 48*, 692–702.

Grusec, J. E., Goodnow, J. J., & Cohen, L. (1996). Household work and the development of concern for others. *Developmental Psychology, 32*, 999–1007.

Gubernskaya, Z. (2010). Changing attitudes toward marriage and children in six countries. *Sociological Perspectives, 53*, 179–200.

Guerra, N. G., Williams, K. R., & Sadek, S. (2011). Understanding bullying and victimization during childhood and adolescence: A mixed methods study. *Child Development, 82*, 295–310.

Guiaux, M. (2010). *Social adjustment to widowhood: Changes in personal relationships and loneliness before and after partner loss*. Doctoral dissertation completed at Vrije Universiteit Amsterdam. Retrieved from http://dspace.ubvu.vu.nl/bitstream/1871/17427/2/2010%20PhD%20Dissertation%20Guiaux.pdf

Guiffrida, D. A. (2009). Theories of human development that enhance an understanding of the college transition process. *Teachers College Record, 111*, 2419–2443.

Guillemin, J. (1993). Cesarean birth: Social and political aspects. In B. K. Rothman (Ed.), *Encyclopedia of childbearing*. Phoenix, AZ: Oryx Press.

Güngör, D., & Bornstein, M. H. (2010). Culture-general and -specific associations of attachment avoidance and anxiety with perceived parental warmth and psychological control among Turk and Belgian adolescents. *Journal of Adolescence, 33*, 593–602.

Gunnar, M. R., Bruce, J., & Grotevant, H. D. (2000). International adoption of institutionally reared children: Research and policy. *Development and Psychopathology, 12*, 677–693.

Gupta, N., & Agarwal, J. L. (2010). Male menopause: Myth or reality. *Medico-Legal Update: An International Journal, 10(2)*. Retrieved from http://www.indianjournals.com/ijor.aspx?target=ijor:mlu&volume=10&issue=2&article=018

Gupta, S., Tracey, T. J. G., & Gore, P. A. (2008). Structural examination of RIASEC scales in high school students: Variation across ethnicity and method. *Journal of Vocational Behavior, 72*, 1–13.

Gurucharri, C., & Selman, R. L. (1982). The development of interpersonal understanding during childhood, preadolescence, and adolescence: A longitudinal follow-up study. *Child Development, 53*, 924–927.

Gustafson, L., Erikson, C., Warkentin, S., Brun, A., Englund, E., & Passant, U. (2011). A factor analytic approach to symptom patterns in dementia. *International Journal of Alzheimer's Disease*.

Haeffel, G. J., Gibb, B. E., Metalsky, G. I., Alloy, L. B., Abramson, L. Y., Hankin, B. L., et al. (2008). Measuring cognitive vulnerability to depression: Development and validation of the cognitive style questionnaire. *Clinical Psychology Review, 28*, 824–836.

Hagestad, G. O. & Dannefer, D. (2001). Concepts and theories of aging: Beyond microfication in social science approaches. In R. H. Binstock & L. K. George (Eds.), *Handbook of aging and the social sciences* (5th ed., pp. 3–21). San Diego, CA: Academic Press.

Hagestad, G. O., & Neugarten, B. L. (1985). Age and the life course. In R. H. Binstock & E. Shanas (Eds.), *Handbook of aging and the social sciences* (2nd ed., pp. 35–61). New York: Van Nostrand Reinhold.

Haier, R. J., Schroeder, D. H., Tang, C., Head, K., & Colom, R. (2010). Gray matter correlates of cognitive ability tests used for vocational guidance. *BMC Research Notes, 3*. Retrieved from http://www.biomedcentral.com/content/pdf/1756-0500-3-206.pdf

Haight, B. K., & Haight, B. S. (2007). *The handbook of structured life review*. Baltimore, MD: Health Professions Press.

Halford, W. K., Markman, H. J., & Stanley, S. (2008). Strengthening couples' relationships with education: Social policy and public health perspectives. *Journal of Family Psychology, 22*, 497–505.

Halgunseth, L. C., Ispa, J. M., & Rudy, D. (2006). Parental control in Latino families: An integrated review of the literature. *Child Development, 77*, 1282–1297.

Hall, D. G., Lee, S. C., & Belanger, J. (2001). Young children's use of syntactic cues to learn proper names and count nouns. *Developmental Psychology, 37*, 298–307.

Hall, J. A., & Halberstadt, A. G. (1981). Sex roles and nonverbal communication skills. *Sex Roles, 7*, 273–287.

Hall, S. S. (2006). Marital meaning: exploring young adult's belief systems about marriage. *Journal of Family Issues, 27*, 1437–1458.

Halpern, C. T., Spriggs, A. L., Martin, S. L., & Kupper, L. L. (2009). Patterns of intimate partner violence victimization from adolescence to young adulthood in a nationally representative sample. *Journal of Adolescent Health, 45*, 508–516.

Halpern, D. F., Benbow, C. P., Geary, D. C., Gur, R. C., Hyde, J. S., & Gernsbacher, M. A. (2007). The science of sex differences in science and mathematics. *Psychological Science in the Public Interest, 8*, 1–51.

Halpern L. F., MacLean, W. E., & Baumeister, A. A. (1995). Infant sleep-wake characteristics: Relation to neurological status and the prediction of developmental outcome. *Developmental Review, 15*, 255–291.

Halpern-Felsher, B. L., & Cauffman, E. (2001). Costs and benefits of a decision: Decision-making competence in adolescents and adults. *Journal of Applied Developmental Psychology, 22*, 257–273.

Halpern-Meekin, S., & Tach, L. (2008). Heterogeneity in two-parent families and adolescent well-being. *Journal of Marriage and Family, 70*, 435–451.

Hamilton, B. E., Martin, J. A., & Ventura, S. J. (2010). Births: Preliminary data for 2008. *National Vital Statistics Reports*, Vol. 58. Hyattsville, MD: National Center for Health Statistics.

Hamilton, C. E. (2000). Continuity and discontinuity of attachment from infancy through adolescence. *Child Development, 71*, 690–694.

Hamm, J. V. (2000). Do birds of a feather flock together? The variable bases for African American, Asian American, and European American adolescents' selection of similar friends. *Developmental Psychology, 36*, 209–219.

Hammen, C., & Rudolph, K. D. (2003). Childhood mood disorders. In E. J. Mash & R. A. Barkley (Eds.), *Child psychopathology* (2nd ed., pp. 233–278). New York: Guilford.

Hammer, L. B., Neal, M. B., Newsom, J. T., Brockwood, K.-J., & Colton, C. L. (2005). A longitudinal study of the effects of dual-earner couples' utilization of family-friendly workplace supports on work and family outcomes. *Journal of Applied Psychology, 90*, 799–810.

Hanaoka, H., & Okamura, H. (2004). Study on effects of life review activities on the quality of life of the elderly: A randomized controlled trial. *Psychotherapy and Psychosomatics, 73*, 302–311.

Hance, V. M. (2000). An existential perspective describing undergraduate students' ideas about meaning in work: A q-method study. *Dissertation Abstracts International. Section A. Humanities and Social Sciences, 61(3-A)*, 878.

Hancox, R. J., & Poulton, R. (2006). Watching television is associated with childhood obesity—but is it clinically important? *International Journal of Obesity, 30*, 171–175.

Hane, A. A., & Fox, N. W. (2006). Ordinary variations in maternal caregiving influence human infants' stress reactivity. *Psychological Science, 17,* 550–556.

Hank, K., & Schaan, B. (2008). Cross-national variations in the correlation between frequency of prayer and health among older Europeans. *Research on Aging, 30,* 36–54.

Hannon, E. E., & Trehub, S. E. (2005). Metrical categories in infancy and adulthood. *Psychological Science, 16,* 48–55.

Hansen, S. R. (2006). Courtship duration as a correlate of marital satisfaction and stability. *Dissertation Abstracts International. Section B. Sciences and Engineering, 67(4-B),* 2279.

Hansen, T., Moum, T., & Shapiro, A. (2007). Relational and individual well-being among cohabitors and married individuals in midlife: Recent trends from Norway. *Journal of Family Issues, 28,* 910–933.

Hansson, R. O., & Stroebe, M. S. (2007). *Bereavement in late life: Coping, adaptation, and developmental influences.* Washington, DC: American Psychological Association.

Hare, T. A., Tottenham, N., Galvan, A., Voss, H. U., Glover, G. H., & Casey, B. J. (2008). Biological substrates of emotional reactivity and regulation in adolescence during an emotional go-nogo task. *Biological Psychiatry, 63,* 927–934.

Hareven, T. K. (1995). Introduction: Aging and generational relations over the life course. In T. K. Hareven (Ed.), *Aging and generational relations over the life course: A historical and cross-cultural perspective* (pp. 1–12). Berlin, Germany: de Gruyter.

Hareven, T. K. (2001). Historical perspectives on aging and family relations. In R. H. Binstock & L. K. George (Eds.), *Handbook of aging and the social sciences* (5th ed., pp. 141–159). San Diego, CA: Academic Press.

Hareven, T. K., & Adams, K. (1996). The generation in the middle: Cohort comparisons in assistance to aging parents in an American community. In T. K. Hareven (Ed.), *Aging and generational relations: Life course and cross-cultural perspectives* (pp. 3–29). New York: Aldine de Gruyter.

Harkness, K. L., Lumley, M. N., & Truss, A. E. (2008). Stress generation in adolescent depression: The moderating role of child abuse and neglect. *Journal of Abnormal Child Psychology, 36,* 421–432.

Harley, C. B. (2008). Telomerase and cancer therapeutics. *Nature Reviews: Cancer, 8,* 167–179.

Harre, N. (2007). Community service or activism as an identity project for youth. *Journal of Community Psychology, 35,* 711–724.

Harrington, D., Bean, N., Pintello, D., & Mathews, D. (2001). Job satisfaction and burnout: Predictors of intentions to leave a job in a military setting. *Administration in Social Work, 25,* 1–16.

Harris, B., Lovett, L., Newcombe, R. G., Read, G. F., Walker, R., & Riad-Fahmy, D. (1994). Maternity blues and major endocrine changes: Cardiff puerperal mood and hormone study II. *British Medical Journal, 308,* 949–953.

Harris, K. R., Graham, S., Mason, L., & Friedlander, B. (2008). *Powerful writing strategies for all students.* Baltimore, MD: Brookes.

Harris, P. L., Brown, E., Marriot, C., Whithall, S., & Harmer, S. (1991). Monsters, ghosts, and witches: Testing the limits of the fantasy-reality distinction in young children. *British Journal of Developmental Psychology, 9,* 105–123.

Harris Interactive. (2011). *Large majorities support doctor assisted suicide for terminally ill patients in great pain.* Retrieved February 27, 2011 from http://www.harrisinteractive.com/NewsRoom/HarrisPolls/tabid/447/mid/1508/articleId/677/ctl/ReadCustom%20Default/Default.aspx.

Harrist, A. W., Zaia, A. F., Bates, J. E., Dodge, K. A., & Pettit, G. S. (1997). Subtypes of social withdrawal in early childhood: Sociometric status and social-cognitive differences across four years. *Child Development, 68,* 278–294.

Hart, H. M., McAdams, D. P., Hirsch, B. J., & Bauer, J. J. (2001). Generativity and social involvement among African Americans and White adults. *Journal of Research in Personality, 35,* 208–230.

Harter, S. (1994). Developmental changes in self-understanding across the 5 to 7 shift. In A. Sameroff & M. M. Haith (Eds.), *Reason and responsibility: The passage through childhood.* Chicago: University of Chicago Press.

Harter, S. (2005). Self-concepts and self-esteem, children and adolescents. In C. B. Fisher & R. M. Lerner (Eds.), *Encyclopedia of applied developmental science* (Vol. 2., pp. 972–977). Thousand Oaks, CA: Sage.

Harter, S. (2006). The self. In W. Damon & R. M. Lerner (Eds.), *Handbook of child psychology* (6th ed., Vol. 3). New York: Wiley.

Harter, S., Waters, P., & Whitesell, N. R. (1998). Relational self-worth: Differences in perceived worth as a person across interpersonal contexts among adolescents. *Child Development, 69,* 756–766.

Harter, S., Whitesell, N. R., & Kowalski, P. S. (1992). Individual differences in the effects of educational transitions on young adolescents' perceptions of competence and motivational orientation. *American Educational Research Journal, 29,* 777–807.

Hartman, P. S. (2001). Women developing wisdom: Antecedents and correlates in a longitudinal sample. *Dissertation Abstracts International. Section B. Sciences and Engineering, 62(1-B),* 591.

Hartup, W. W. (1992). Friendships and their developmental significance. In H. McGurk (Ed.), *Contemporary issues in childhood social development.* (pp. 175–205). London: Routledge.

Hartup, W. W., & Stevens, N. (1999). Friendships and adaptation across the life span. *Current Directions in Psychological Science, 8,* 76–79.

Harvard School of Public Health. (2010). *The nutrition source: Alcohol: Balancing risks and benefits.* Retrieved from http://www.hsph.harvard.edu/nutritionsource/what-should-you-eat/alcohol-full-story

Haslam, C., & Lawrence, W. (2004). Health-related behavior and beliefs of pregnant smokers. *Health Psychology, 23,* 486–491.

Haslam, C., Hodder, K. I., & Yates, P. J. (2011). Errorless learning and spaced retrieval: How do these methods fare in healthy and clinical populations? *Journal of Clinical and Experimental Neuropsychology, 33,* 1–16.

Hastings, P. D., & Rubin, K. H. (1999). Predicting mothers' beliefs about preschool-aged children's social behavior: Evidence for maternal attitudes moderating child effects. *Child Development, 70,* 722–741.

Hastings, P. D., Zahn-Waxler, C., & McShane, K. (2006). We are, by nature, moral creatures: Biological bases of concern for others. In M. Killen & J. G. Smetana (2006). *Handbook of moral development* (pp. 483–516). Mahwah, NJ: Erlbaum.

Hatania, R., & Smith, L. B. (2010). Selective attention and attention switching: Toward a unified developmental approach. *Developmental Science, 13,* 622–635.

Haviland, J. M., & Lelwica, M. (1987). The induced affect response: 10-week-old infants' responses to three emotion expressions. *Developmental Psychology, 23,* 97–104.

Hawley, P. H. (1999). The ontogenesis of social dominance: A strategy-based evolutionary perspective. *Developmental Review, 19,* 7–132.

Haworth, J., & Lewis, S. (2005). Work, leisure and well-being. *British Journal of Guidance & Counselling, 33,* 67–78.

Hayashi, M., Hasui, C., Kitamura, F., Murakami, M., Takeuchi, M., Katoh, H., & Kitamura, T. (2000). Respecting autonomy in difficult medical settings: A questionnaire study in Japan. *Ethics and Behavior, 10,* 51–63.

Hayflick, L. (1996). *How and why we age* (2nd ed.). New York: Ballantine.

Hayflick, L. (1998). How and why we age. *Experimental Gerontology, 33,* 639–653.

Hayslip, B., Jr., Henderson, C. E., & Shore, R. J. (2003). The structure of grandparental role meaning. *Journal of Adult Development, 10,* 1–11.

Hayslip, B., Jr., Shore, R. J., Hendereson, C. E., & Lambert, P. L. (1998). Custodial grandparenting and the impact of grandchildren with problems on role satisfaction and role meaning. *Journal of Gerontology: Social Sciences, 53B,* S164–S173.

Hazell, P. L. (2009). 8-year follow-up of the MTA sample. *Journal of the American Academy of Child and Adolescent Psychiatry, 48,* 461–462.

Health and Safety Executive. (2010). *About us.* Retrieved from http://www.hse.gov.uk/aboutus/index.htm

HealthyPeople.gov (2011). *Healthy People 2020.* Retrieved from http://www.healthypeople.gov/2020/default.aspx

Heatherington, L., & Lavner, J. A. (2008). Coming to terms with coming out: Review and recommendations for family systems-focused research. *Journal of Family Psychology, 22,* 329–343.

Heckhausen, J., Wrosch, C., & Schulz, R. (2010). A motivational theory of lifespan development. *Psychological Review, 117,* 32–60.

Heidrich, S. M., & Denney, N. W. (1994). Does social problem solving differ from other types of problem solving during the adult years? *Experimental Aging Research, 20,* 105–126.

Heilman, M. E., Wallen, A. S., Fuchs, D., & Tamkins, M. M. (2004). Penalties for success: Reactions to women who succeed at male gender-typed tasks. *Journal of Applied Psychology, 89,* 416–427.

Heine, S. J., & Buchtel, E. E. (2009). Personality: the universal and the culturally specific. *Annual Review of Psychology, 60,* 369–394.

Hellerstedt, W. L., Madsen, N. J., Gunnar, M. R., Grotevant, H. D., Lee, R. M., & Johnson, D. E. (2008). The International Adoption Project: Population-based surveillance of Minnesota parents who adopted children internationally. *Maternal & Child Health Journal, 12,* 162–171.

Helpguide.org. (2005). *Adult day care centers: A guide to options and selecting the best center for your needs.* Retrieved from http://www.helpguide.org/elder/adult_day_care_centers.htm

Hemer S. R. (2010). Grief as social experience: Death and bereavement in Lihir, Papua New Guinea. *The Australian Journal of Anthropology, 21,* 281–297.

Henderson, B. N., Davison, K. P., Pennebaker, J. W., Gatchel, R. J., & Baum, A. (2002). Disease disclosure patterns among breast cancer patients. *Psychology and Health, 17,* 51–62.

Henderson, H. A., & Wachs, T. D. (2007). Temperament theory and the study of cognition-emotion interactions across development. *Developmental Review, 27,* 396–427.

Henderson, N. D. (2010). Predicting long-term firefighter performance from cognitive and physical ability measures. *Personnel Psychology, 63,* 999–1039.

Henig, R. M. (2010). *What is it about 20-somethings?* Retrieved from http://www.nytimes.com/2010/08/22/magazine/22Adulthood-t.html?_r=1&ref=families_and_family_life

Henretta, J. C. (2001). Work and retirement. In R. H. Binstock & L. K. George (Eds.), *Handbook of aging and the social sciences* (pp. 255–271). San Diego, CA: Academic Press.

Henrie, J. A. (2010). *Religiousness, future time perspective, and death anxiety among adults.* Dissertation submitted to West Virginia University.

Henry, D. B., Schoeny, M. E., Deptula, D. P., & Slavick, J. T. (2007). Peer selection and socialization effects on adolescent intercourse without a condom and attitudes about the costs of sex. *Child Development, 78,* 825–838.

Henry, N. J. M., Berg, C. A., Smith, T. W., & Florsheim, P. (2007). Positive and negative characteristics of marital interaction and their association with marital satisfaction in middle-aged and older couples. *Psychology and Aging, 22,* 428–441.

Henry, R. G. (2006). Parental death and its impact on the marital relationship of the surviving adult child. *Dissertation Abstracts International. Section B. Sciences and Engineering, 67(2-B),* 1181.

Heppner, R. S. (2007). A paradox of diversity: Billions invested, but women still leave. *Dissertation Abstracts International. Section A. Humanities and Social Sciences, 68(4-A),* 1527.

Herman, M. (2004). Forced to choose: Some determinants of racial identification in multiracial adolescents. *Child Development, 75,* 730–748.

Herrnstein, R. J., & Murray, C. (1994). *The bell curve: Intelligence and class structure in American life.* New York: Free Press.

Hershatter, A., & Epstein, M. (2010). Millennials and the world of work: An organization and management perspective. *Journal of Business and Psychology, 25,* 211–223.

Hertenstein, M. J., & Campos, J. J. (2004). The retention effects of an adult's emotional displays on infant behavior. *Child Development, 75,* 595–613.

Hertzog, C., & Dunlosky, J. (2004). Aging, metacognition, and cognitive control. In B. H. Ross (Ed.), *The psychology of learning and motivation: Advances in research and theory* (Vol. 45, pp. 215–251). San Diego, CA: Elsevier.

Hertzog, C., Kramer, A. F., Wilson, R. S., & Lindenberger, U. (2009). Enrichment effects on adult cognitive development: Can the functional capacity of older adults be preserved and enhanced? *Psychological Science in the Public Interest* (Vol. 9, No. 1). Washington, DC: Association for Psychological Science.

Hespos, S. J., Ferry, A. L., & Rips, L. J. (2009). Five-month-old infants have different expectations for solids and liquids. *Psychological Science, 20,* 603–611.

Hess, U., & Kirouac, G. (2000). Emotion expression in groups. In M. Lewis &

J. Haviland-Jones (Eds.), *Handbook of emotions* (2nd ed., pp. 368–381). New York: Guilford Press.

Hetherington, E. M., & Kelly, J. (2002). *For better or for worse: Divorce reconsidered.* New York: W. W. Norton.

Heyman, G. D. (2009). Children's reasoning about traits. In P. Bauer (Ed.), *Advances in child development and behavior,* Vol. 37 (pp. 105–143). London: Elsevier.

Heyman, G. D., & Legare, C. H. (2004). Children's beliefs about gender differences in the academic and social domains. *Sex Roles, 50,* 227–239.

Hiatt, K., Stelle, C., Mulsow, M., & Scott, J. P. (2007). The importance of perspective: Evaluation of hospice care from multiple stakeholders. *American Journal of Hospice & Palliative Medicine, 24,* 376–382.

Hill, J. L., Brooks-Gunn, J., & Waldfogel, J. (2003). Sustained effects of high participation in an early intervention for low-birth-weight premature infants. *Developmental Psychology, 39,* 730–744.

Hill, N. E., & Taylor, L. E. (2004). Parental school involvement and children's academic achievement: Pragmatics and issues. *Current Directions in Psychological Science, 13,* 161–164.

Hillman, C. H., Buck, S. M., Themanson, J. R. Pontifex, M. B., & Castelli, D. M. (2009). Aerobic fitness and cognitive development: Event-related brain potential and task performance indices of executive control in preadolescent children. *Developmental Psychology, 45,* 114–129.

Hills, W. E. (2010). Grandparenting roles in the evolving American family. In D. Wiseman (Ed.), *The American family: Understanding its changing dynamics and place in society* (pp. 65–78.) Springfield, IL: Charles C. Thomas.

Hillsdon, M., Brunner, E., Guralnik, J., & Marmot, M. (2005). Prospective study of physical activity and physical function in early old age. *American Journal of Preventive Medicine, 28*(3), 245–250.

Himsel, A. J., Hart, H., Diamond, A., & McAdams, D. P. (1997). Personality characteristics of highly generative adults as assessed in Q-sort ratings of life stories. *Journal of Adult Development, 4,* 149–161.

Hines, D. A., & Douglas, E. M. (2009). Women's use of intimate partner violence against men: Prevalence, implications, and consequences. *Journal of Aggression, Maltreatment and Trauma, 18,* 572–586.

Hinton, A., & Chirgwin, S. (2010). Nursing education: Reducing reality shock for graduate indigenous nurses—It's all about time. *Australian Journal of Advanced Nursing, 28,* 60–66.

Hipwell, A. E., Keenan, K., Loeber, R., & Battista, D. (2010). Early predictors of sexually intimate behaviors in an urban sample of young girls. *Developmental Psychology, 46,* 366–378.

Hirsh-Pasek, K., & Golinkoff, R. M. (2008). King Solomon's take on word learning: An integrative account from the radical middle. In R. V. Kail (Ed.), *Advances in child development and behavior* (Vol. 36, pp. 1–29). San Diego, CA: Elsevier.

Hitlin, S., Brown, J., & Elder, G. H., Jr. (2006). Racial self-categorization in adolescence: Multiracial development and social pathways. *Child Development, 77,* 1298–1308.

Hobdy, J. (2000). The role of individuation processes in the launching of children into adulthood. *Dissertation Abstracts International. Section B. Sciences and Engineering, 60(9-B),* 4929.

Hofer, J., Busch, H., Chasiotis, A., Kärtner, J., & Campos, D. (2008). Concern for generativity and its relation to implicit pro-social power motivation, generative goals, and satisfaction with life: A cross-cultural investigation. *Journal of Personality, 76,* 1–30.

Hofer, M. A. (2006). Psychobiological roots of early attachment. *Current Directions in Psychological Science, 15,* 84–88.

Hoff, E. (2003). The specificity of environmental influence: Socioeconomic status affects early vocabulary developmental via maternal speech. *Child Development, 74,* 1368–1378.

Hoff, E. (2009). *Language development* (4th ed.). Belmont, CA: Wadsworth Cengage Learning.

Hoff, E., & Naigles, L. (2002). How children use input to acquire a lexicon. *Child Development, 73,* 418–433.

Hoff-Ginsberg, E. (1997). *Language development.* Pacific Grove, CA: Brooks/Cole.

Hoff-Ginsberg, E., & Tardif, T. (1995). Socioeconomic status and parenting. In M. H. Bornstein (Ed.), *Handbook of parenting* (Vol. 2, pp. 161–188). Mahwah, NJ: Erlbaum.

Hoffman, L., McDowd, J. M., Atchley, P., & Dubinsky, R. (2005). The role of visual attention in predicting driving impairment in older adults. *Psychology and Aging, 20,* 610–622.

Hoffman, M. L. (1988). Moral development. In M. H. Bornstein and M. E. Lamb (Eds.), *Developmental psychology: An advanced textbook* (2nd ed., pp. 497–538). Hillsdale, NJ: Erlbaum.

Hoffman, M. L. (1994). Discipline and internalization. *Developmental Psychology, 30,* 26–28.

Hogan, A. M., de Haan, M., Datta, A., & Kirkham, F. J. (2006). Hypoxia: An acute, intermittent and chronic challenge to cognitive development. *Developmental Science, 9,* 335–337.

Hogan, N., & DeSantis, L. (1996). Basic constructs of a theory of adolescent sibling bereavement. In D. Klass, P. R. Silverman, & S. L. Nickman (Eds.), *Continuing bonds: New understandings of grief* (pp. 235–254). Washington, DC: Taylor & Francis.

Hogan, R., & Perrucci, C. C. (2007). Black women: Truly disadvantaged in the transition from employment to retirement income. *Social Science Research, 36,* 1184–1199.

Hogge, W. A. (1990). Teratology. In I. R. Merkatz & J. E. Thompson (Eds.), *New perspectives on prenatal care.* New York: Elsevier.

Holden, G. W., & Miller, P. C. (1999). Enduring and different: A meta-analysis of the similarity in parents' child rearing. *Psychological Bulletin, 125,* 223–254.

Holkup, P. A., Salois, E. M., Tripp-Reimer, T., & Weinert, C. (2007). Drawing on wisdom from the past: An elder abuse intervention with tribal communities. *The Gerontologist, 47,* 248–254.

Holland, J. L. (1985). *Making vocational choices: A theory of vocational personalities and work environments* (2nd ed.). Englewood Cliffs, NJ: Prentice Hall.

Holland, J. L. (1987). Current status of Holland's theory of careers: Another perspective. *Career Development Quarterly, 36,* 24–30.

Holland, J. L. (1996). Exploring careers with a typology: What we have learned and some new directions. *American Psychologist, 51,* 397–406.

Holland, J. L. (1997). *Making vocational choices: A theory of vocational personalities and work environments* (3rd ed.). Baltimore, MD: Johns Hopkins University Press.

Holland, J. M., & Neimeyer, R. A. (2010). An examination of stage theory of grief among individuals bereaved by natural and violent causes: A meaning-oriented contribution. *OMEGA: Journal of Death and Dying, 61,* 103–120.

Hollich, G. J., Golinkoff, R. M., & Hirsh-Pasek, K. (2007). Young children associate novel words with complex objects rather than salient parts. *Developmental Psychology, 43,* 1051–1061.

Hollich, G. J., Hirsh-Pasek, K., & Golinkoff, R. M. (2000). Breaking the language barrier: An emergentist coalition model for the origins of word learning. *Monographs of the Society for Research in Child Development, 65*(Serial No. 262).

Holloway, K. F. C. (2011). *Private bodies, public texts: Race, gender, and a cultural bioethics.* Durham, NC: Duke University Press.

Holowka, S., & Petitto, L. A. (2002). Left hemisphere cerebral specialization for babies while babbling. *Science, 297,* 1515.

Hom, P. W., & Kinicki, A. J. (2001). Toward a greater understanding of how dissatisfaction drives employee turnover. *Academy of Management Journal, 44,* 975–987.

Honda-Howard, M., & Homma, M. (2001). Job satisfaction of Japanese career women and its influence on turnover intention. *Asian Journal of Social Psychology, 4,* 23–38.

Hood, B., Carey, S., & Prasada, S. (2000). Predicting the outcomes of physical events: Two-year-olds fail to reveal knowledge of solidity and support. *Child Development, 71,* 1540–1554.

Hooker, K. (1999). Possible selves in adulthood. In T. M. Hess & F. Blanchard-Fields (Eds.), *Social cognition and aging* (pp. 97–122). San Diego, CA: Academic Press.

Hooker, K. (2002). New directions for research in personality and aging: A comprehensive model for linking levels, structures, and processes. *Journal of Research in Personality, 36,* 318–334.

Hooker, K., & Kaus, C. R. (1994). Health-related possible selves in young and mid-adulthood. *Psychology and Aging, 9,* 126–133.

Hooker, K., Fiese, B. H., Jenkins, L., Morfei, M. Z., & Schwagler, J. (1996). Possible selves among parents of infants and preschoolers. *Developmental Psychology, 32,* 542–550.

Hopkins, B., & Westra, T. (1988). Maternal handling and motor development: An intercultural study. *Genetic, Social, and General Psychology Monographs, 14,* 377–420.

Horn, J. L. (1982). The aging of human abilities. In B. B. Wolman (Ed.), *Handbook of developmental psychology* (pp. 847–870). Englewood Cliffs, NJ: Prentice Hall.

Horn, J. L., & Hofer, S. M. (1992). Major abilities and development in the adult period. In R. J. Sternberg & C. A. Berg (Eds.), *Intellectual development* (pp. 44–99). Cambridge, England: Cambridge University Press.

Horsford, S. R., Parra-Cardona, J. R., Post, L. A., & Schiamberg, L. (2011). Elder abuse and neglect in African American families: Informing best practice based on ecological and cultural frameworks. *Journal of elder Abuse and Neglect, 23,* 75–88.

Horswill, M. S., Kemala, C. N., Wetton, M., Scialfa, C. T., & Pachana, N. A. (2010). Improving older drivers' hazard perception ability. *Psychology and Aging, 25,* 464–469.

Hospice Foundation of America. (2011a). *Hospice patients and staff.* Retrieved March 5, 2011 from http://www.hospicefoundation.org/pages/page.asp?page_id=53123.

Hospice Foundation of America. (2011b). *Choosing hospice: questions to ask.* Retrieved March 5, 2011 from http://www.hospicefoundation.org/pages/page.asp?page_id=71942.

Hossain, Z. (2001). Division of household labor and family functioning in off-reservation Navajo Indian families. *Family Relations, 50,* 255–261.

Houston, D. M., & Jusczyk, P. W. (2003). Infants' long-term memory for the sound patterns of words and voices. *Journal of Experimental Psychology: Human Perception and Performance, 29,* 1143–1154.

Hovey, J. D., & Magana, C. (2000). Acculturative stress, anxiety, and depression among Mexican farmworkers in the Midwest United States. *Journal of Immigrant Health, 2,* 119–131.

Howard, L. W., & Cordes, C. L. (2010). Flight from unfairness: Effects of perceived injustice on emotional exhaustion and employee withdrawal. *Journal of Business and Psychology, 25,* 409–428.

Howe, N., & Ross, H. S. (1990). Socialization, perspective taking and the sibling relationship. *Developmental Psychology, 26,* 160–165.

Howe, N., & Strauss, W. (1992). *Generations: The history of America's future, 1584–2069.* New York: Harper Perennial.

Howell, K. K., Lynch, M. E., Platzman, K. A., Smith, G. H., & Coles, C. D. (2006). Prenatal alcohol exposure and ability, academic achievement, and school functioning in adolescence: A longitudinal follow-up. *Journal of Pediatric Psychology, 31,* 116–126.

Howes, C., & Matheson, C. C. (1992). Sequences in the development of competent play with peers: Social and social pretend play. *Developmental Psychology, 28,* 961–974.

Howes, C., Unger, O., & Seidner, L. B. (1990). Social pretend play in toddlers: Parallels with social play and with solitary pretend. *Child Development, 60,* 77–84.

Hoyer, W. J., & Rybash, J. M. (1994). Characterizing adult cognitive development. *Journal of Adult Development, 1,* 7–12.

Hsu, H.-C. (2005). Gender disparity of successful aging in Taiwan. *Women & Health, 42,* 1–21.

Hubbard, F. O. A., & van IJzendoorn, M. H. (1991). Maternal unresponsiveness and infant crying across the first 9 months: A naturalistic longitudinal study. *Infant Behavior and Development, 14,* 299–312.

Hubbard, R. R. (2010). *Afro-German biracial identity development.* Retrieved from http://digarchive.library.vcu.edu/bitstream/10156/2804/1/Afro-German%20HEMBAG1%20%28F3%29.pdf

Huesmann, L. R. (2007). The impact of electronic media violence: Scientific theory and research. *Journal of Adolescent Health Care, 41,* S6–S13.

Hughes, J. M., Bigler, R. S., & Levy, S. R. (2007). Consequences of learning about historical racism among European American and African American children. *Child Development, 78,* 1689–1705.

Huizink, A., Robles de Medina, P., Mulder, E., Visser, G., & Buitelaar, J. (2002). Psychological measures of prenatal stress as predictors of infant temperament. *Journal of the American Academy of Child & Adolescent Psychiatry, 41,* 1078–1085.

Hull, T. H. (2009). *Fertility prospects in south-eastern Asia: Report of the United Nations Expert Group Meeting on Recent and Future Trends in Fertility.* Retrieved from http://www.un.org/esa/population/meetings/EGM-Fertility2009/P14_Hull.pdf

Hulme, C., & Snowling, M. J. (2009). *Developmental disorders of language learning and cognition.* Chichester, England: Wiley-Blackwell.

Human Genome Project. (2003). *Genomics and its impact on science and society: A 2003 primer.* Washington, DC: U.S. Department of Energy.

Hunt, E., & Carlson, J. (2007). Considerations relating to the study of group differences in intelligence. *Perspectives on Psychological Science, 2,* 194–213.

Hunt, J. M., & Weintraub, J. R. (2006). *The coaching organization: A strategy for developing leaders.* Thousand Oaks, CA: Sage.

Hurley-Hanson, A. (2006). Applying image norms across Super's career development stages. *Career Development Quarterly.* Retrieved from http://www.allbusiness.com/human-resources/careers/1184672-1.html

Hurtado, N., Marchman, V. A., & Fernald, A. (2008). Does input influence uptake? Links between maternal talk, processing speed, and vocabulary size in Spanish-learning children. *Developmental Science, 11,* F31–F39.

Huston, A. C., & Wright, J. C. (1998). Mass media and children's development. In W. Damon (Ed.), *Handbook of child psychology* (Vol. 4; pp. 998–1058). New York: Wiley.

Huston, M., & Schwartz, P. (1995). The relationships of lesbians and of gay men. In J. T. Wood & S. Duck (Eds.), *Understudied relationships: Off the beaten track* (pp. 89–121). Thousand Oaks, CA: Sage.

Huston, T. L., Caughlin, J. P., Houts, R. M., Smith, S. E., & George, L. J. (2001). The connubial crucible: Newlywed years as a predictor of marital delight, distress, and divorce. *Journal of Personality and Social Psychology, 80,* 237–252.

Hutchinson, D. M., Rapee, R. M., & Taylor, A. (2010). Body dissatisfaction and eating disturbances in early adolescence: A structural modeling investigation examining negative affect and peer factors. *Journal of Early Adolescence, 30,* 489–517.

Huth-Bocks, A. C., Levendosky, A. A., Bogat, G. A., & von Eye, A. (2004). The impact of maternal characteristics and contextual variables on infant-mother attachment. *Child Development, 75,* 480–496.

Huttenlocher, J., Haight, W., Bryk, A., Seltzer, M., & Lyons, T. (1991). Early vocabulary growth: Relation to language input and gender. *Developmental Psychology, 27,* 236–248.

Huttenlocher, J., Waterfall, H., Vasilyeva, M., Vevea, J., & Hedges, L. V. (2010). Sources of variability in children's language growth. *Cognitive Psychology, 61,* 343–365.

Hwang, M. J. (2007). Asian social workers' perceptions of glass ceiling, organizational fairness and career prospects. *Journal of Social Service Research, 33,* 13–24.

Hyde, J. S. (2007). New directions in the study of gender similarities and differences. *Current Directions in Psychological Science, 16,* 259–263.

Hymel, S., Vaillancourt, T., McDougall, P., & Renshaw, P. D. (2004). Peer acceptance and rejection in childhood. In P. K. Smith & C. H. Hart (Eds.), *Blackwell handbook of childhood social development* (pp. 265–284). Malden, MA: Blackwell.

Ibrahim, R., & Hassan, Z. (2009). Understanding singlehood from the experiences of never-married malay muslim women in Malaysia: Some preliminary findings. *European Journal of Social Sciences, 8,* 395–405.

Iervolino, A. C., Hines, M., Golombok, S. E., Rust, J., & Plomin, R. (2005). Genetic and environmental influences on sex-typed behavior during the preschool years. *Child Development, 76,* 826–840.

Ijuin, M., Homma, A., Mimura, M., Kitamura, S., Kawai, Y., Imai, Y., & Gondo, Y. (2008). Validation of the 7-minute screen for the detection of early-stage Alzheimer's disease. *Dementia and Geriatric Cognitive Disorders, 25,* 248–255.

Ilies, R., Hauserman, N., Schwochau, S., & Stibal, J. (2003). Reported incidence rates of work-related sexual harassment in the United States: Using meta-analysis to explain reported rate disparities. *Personnel Psychology, 56,* 607–631.

Inhelder, B., & Piaget, J. (1958). *The growth of logical thinking from childhood to adolescence.* New York: Basic Books.

Institute of Medicine. (1990). *Nutrition during pregnancy.* Washington, DC: National Academy Press.

Institute of Medicine. (2010). *DRIs for calcium and vitamin D.* Retrieved from http://iom.edu/Reports/2010/Dietary-Reference-Intakes-for-Calcium-and-Vitamin-D/~/media/Files/Report%20Files/2010/Dietary-Reference-Intakes-for-Calcium-and-Vitamin-D/calciumvitd_lg.jpg

Irish, L. S. (2009). *Modern tribal tattoo designs.* East Petersburg, PA: Fox Chapel Publishing.

Isaksen, J. (2000). Constructing meaning despite drudgery of repetitive work. *Journal of Humanistic Psychology, 40,* 84–107.

Israel, A. C., Guile, C. A., Baker, J. E., & Silverman, W. K. (1994). An evaluation of enhanced self-regulation training in the treatment of childhood obesity. *Journal of Pediatric Psychology, 19,* 737–749.

Ivancevich, J. M., & Matteson, M. T. (1988). Type A behavior and the healthy individual. *British Journal of Medical Psychology, 61,* 37–56.

Ivancovich, D. A., & Wong, T. P. (2008). The role of existential and spiritual coping in anticipatory grief. In A. Tomer, G. T. Eliason, & P. T. P. Wong (Eds.), *Existential and spiritual issues in death attitudes* (pp. 209–233). Mahwah, NJ: Lawrence Erlbaum.

Iverson, J. M., & Goldin-Meadow, S. (2005). Gesture paves the way for language development. *Psychological Science, 16,* 367–371.

Ivory, B. T. (2004). A phenomenological inquiry into the spiritual qualities and transformational themes associated with a self-styled rite of passage into adulthood. *Dissertation Abstracts International. Section A. Humanities & Social Sciences, 65(2-A),* 429.

Izard, C. E. (2007). Basic emotions, natural kinds, emotion schemas, and a new paradigm. *Perspectives on Psychological Science, 2,* 260–280.

Izard, C. E., & Ackerman, B. P. (2000). Motivational, organizational, and regulatory functions of discrete emotions. In M. Lewis & J. Haviland-Jones (Eds.), *Handbook of emotions* (2nd ed., pp. 253–264). New York: Guilford.

Jack, F., MacDonald, S., Reese, E., & Hayne, H. (2009). Maternal reminiscing style during early childhood predicts the age of adolescents' earliest memories. *Child Development, 80,* 496–505.

Jacka, F., & Berk, M. (2007). Food for thought. *Acta Neuropsychiatrica, 19,* 321–323.

Jacob, B., & Wilder, T. (2010). *Educational expectations and attainment.* National Bureau of Economic Research Working Paper No. w156983. Retrieved from http://papers.ssrn.com/sol3/papers.cfm?abstract_id=1540987

Jacobi, C., Hayward, C., de Zwaan, M., Kraemer, H. C., & Agras, W. S. (2004). Coming to terms with risk factors for eating disorders: Application of risk terminology and suggestions for a general taxonomy. *Psychological Bulletin, 130,* 19–65.

Jacobs, J. A., & Gerson, K. (2001). Overworked individuals or overworked families? Explaining trends in work, leisure, and family time. *Work and Occupations, 28,* 40–63.

Jacobs, J. E., & Eccles, J. S. (1992). The impact of mothers' gender-role stereotypic beliefs on mothers' and children's ability perceptions. *Journal of Personality and Social Psychology, 63,* 932–944.

Jacobs-Lawson, J. M., Hershey, D. A., & Neukam, K. A. (2004). Gender differences in factors that influence time spent planning for retirement. *Journal of Women & Aging, 16,* 55–69.

Jacobson, J. L., & Jacobson, S. W. (1996). Intellectual impairment in children exposed to polychlorinated biphenyls in utero. *New England Journal of Medicine, 335,* 783–789.

Jacobson, J. L., Jacobson, S. W., & Humphrey, H. E. B. (1990). Effects of in utero exposure to polychlorinated biphenyls and related contaminants on cognitive functioning in young children. *Journal of Pediatrics, 116,* 38–45.

Jacobson, S. W., & Jacobson, J. L. (2000). Teratogenic insult and neurobehavioral function in infancy and childhood. In C. A. Nelson (Ed.), *The Minnesota symposium on child psychology: Vol. 31. The effects of early adversity on neurobehavioral development* (pp. 61–112). Mahwah, NJ: Erlbaum.

Jacoby, S. (2005). *Sex in America.* AARP report. Retrieved from http://assets.aarp.org/rgcenter/general/2004_sexuality.pdf

Jaffee, S., & Hyde, J. S. (2000). Gender differences in moral orientation: A meta-analysis. *Psychological Bulletin, 126,* 703–726.

Jain, E., & Labouvie-Vief, G. (2010). Compensatory effects of emotional avoidance in adult development. *Biological Psychology, 84,* 497–513.

James, W. (1890). *The principles of psychology.* New York: Holt.

Janoff-Bulman, R., & Berger, A. R. (2000). The other side of trauma: Toward a psychology of appreciation. In J. H. Harvey & E. D. Miller (Eds.), *Loss and trauma: General and close relationship perspectives* (pp. 29–44). Philadelphia, PA: Brunner-Routledge.

Jansen, J., de Weerth, C., & Riksen-Walraven, J. M. (2008). Breastfeeding and the mother-infant relationship—A review. *Developmental Review, 28,* 503–521.

Janson, H., & Mathiesen, K. S. (2008). Temperament profiles from infancy to middle childhood: Development and associations with behavioral problems. *Developmental Psychology, 44,* 1314–1328.

Janssen, J., Tattersall, C., Waterink, W., van den Berg, B., van Es, R., Bolman, C., & Koper, R. (2007). Self-organising navigational support in lifelong learning: How predecessors can lead the way. *Computers & Education, 49,* 781–793.

Jenkins, C. L. (Ed.). (2003). *Widows and divorcees in later life: On their own again.* Binghamton, NY: Haworth Press.

Jenkins, J. (2007). An investigation of marital satisfaction: Assortative mating and personality similarity. *Dissertation Abstracts International. Section B. Sciences and Engineering, 67(12-B),* 7376.

Jiao, S., Ji, G., & Jing, Q. (1996). Cognitive development of Chinese urban only children and children with siblings. *Child Development, 67,* 387–395.

Jiao, Z. (1999, April). *Which students keep old friends and which become new friends across a school transition?* Paper presented at the 1999 meeting of the Society for Research in Child Development, Albuquerque, New Mexico.

Jimenez, D. E., Alegria, M., Chen, C-n., Chan, D., & Laderman, N. (2010). Prevalence of psychiatric illnesses in older ethnic minority adults. *Journal of the American Geriatrics Society, 58,* 256–264.

Jipson, J. L., & Gelman, S. A. (2007). Robots and rodents: Children's inferences about living and nonliving kinds. *Child Development, 78,* 1675–1688.

Johanson, R. B., Rice, C., Coyle, M., Arthur, J., Anyanwu, L., Ibrahim, J., et al. (1993). A randomized prospective study comparing the new vacuum extractor policy with forceps delivery. *British Journal of Obstetrics and Gynecology, 100,* 524–530.

John, O. P., & Gross, J. J. (2007). Individual differences in emotion regulation. In J. J. Gross (Ed.), *Handbook of emotion regulation* (pp. 351–372). New York: Guilford.

Johnson, A. J., Becker, J. A. H., Craig, E. A., Gilchrist, E. S., & Haigh, M. M. (2009). Changes in friendship commitment: Comparing geographically close and long-distance young-adult friendships. *Communication Quarterly, 57,* 395–415.

Johnson, D. L. (2000). The Black corporate experience: Perceptions of the impact of skin color and gender on Black professionals' success. *Dissertation Abstracts International. Section B. Sciences and Engineering, 60(8-B),* 4282.

Johnson, H. A., Zabriskie, R. B., & Hill, B. (2006). The contribution of couple leisure involvement, leisure time, and leisure satisfaction to marital satisfaction. *Marriage & Family Review, 40,* 69–91.

Johnson, H. D., Brady, E., McNair, R., Congdon, D., Niznik, J., & Anderson, S. (2007). Identity as a moderator of gender differences in the emotional closeness of emerging adults' same- and cross-sex friendships. *Adolescence, 42,* 1–23.

Johnson, K., & Wilson, K. (2010). *Current economic status of older adults in the United States: demographic analysis.* Report prepared for the National Council on Aging. Retrieved from http://www.ncoa.org/assets/files/pdf/Economic-Security-Trends-for-Older-Adults-65-and-Older_March-2010.pdf

Johnson, M. H., Grossman, T., & Cohen Kadosh, K. (2009). Mapping functional brain development: Building a social brain through interactive specialization. *Developmental Psychology, 45,* 151–159.

Johnson, M. J., Grossman, T., & Farroni, T. (2008). The social cognitive neuroscience of infancy: Illuminating the early development of social brain functions. In R. V. Kail (Ed.), *Advances in child development and behavior* (Vol. 36, pp. 331–372). San Diego, CA: Academic Press.

Johnson, R. M., Miller, M., Vriniotis, M., Azrael, D., & Hemenway, D. (2006). Are household firearms stored less safely in home with adolescents? Analysis of a national random sample of parents. *Archives of Pediatrics and Adolescent Medicine, 160,* 788–792.

Johnson, S. B., Sudhinaraset, R., & Blum, R. W. (2010). Neuromaturation and adolescent risk taking: Why development is not determinism. *Journal of Adolescent Research, 25,* 4–23.

Johnson, S. C., Dweck, C. S., & Chen, F. S. (2007). Evidence for infants' internal working models of attachment. *Psychological Science, 18,* 501–502.

Johnson, S. P. (2001). Visual development in human infants: Binding features, surfaces, and objects. *Visual Cognition, 8,* 565–578.

Johnson-Laird, P. N. (1988). *The computer and the mind: An introduction to cognitive science.* London: Fontana.

Johnston, K. E., Swim, J. K., Saltsman, B. M., Deater-Deckard, K., & Petrill, S. A. (2007). Mothers' racial, ethnic, and cultural socialization of transracially adopted Asian children. *Family Relations, 56,* 390–402.

Johnston, L. D., Delva, J., & O'Malley, P. M. (2007). Sports participation and physical education in American secondary schools: Current levels and racial/ethnic and socioeconomic disparities. *American Journal of Preventive Medicine, 33,* S195–S208.

Johnston, L. D., O'Malley, P. M., Bachman, J. G., & Schulenberg, J. E. (2011). *Monitoring the Future national results on adolescent drug use: Overview of key findings, 2010.* Ann Arbor, MI: Institute for Social Research, The University of Michigan.

Joiner, T. (2010). *Myths about suicide.* Cambridge, MA: Harvard University Press.

Jokela, M. (2010). Characteristics of the first child predict the parents' probability of having another child. *Developmental Psychology, 46,* 915–926.

Jokela, M., Kivimäki, M., Elovainio, M., & Keltikangas-Järvinen, L. (2009). Personality and having children: A two-way relationship. *Journal of Personality and Social Psychology, 96,* 218–230.

Jones, B. F. (2010). Age and great invention. *Review of Economics and Statistics, 92,* 1–14. Retrieved from http://www.mitpressjournals.org/doi/pdfplus/10.1162/rest.2009.11724

Jones, C. J., & Meredith, W. (1996). Patterns of personality change across the life span. *Psychology and Aging, 11,* 57–65.

Jones, G. W. (2010). *Changing marriage patterns in Asia.* (Asia Research Institute Working Paper Series No. 131). Retrieved from http://www.ari.nus.edu.sg/docs/wps/wps10_131.pdf

Jones, H. E. (2006). Drug addiction during pregnancy: Advances in maternal treatment and understanding child outcomes. *Current Directions in Psychological Science, 15,* 126–130.

Jopp, D. S., & Hertzog, C. (2010). Assessing adult leisure activities: An extension of a self-report activity questionnaire. *Psychological Assessment, 22,* 108–120.

Jordan, N. C. (2007). The need for number sense. *Educational Leadership, 65,* 63–66.

Jordan, N. C., Kaplan, D., Ramineni, C., & Locuniak, M. N. (2008). Development of number combination skill in the early school years: When do fingers help? *Developmental Science, 11,* 662–668.

Joseph, D. L., & Newman, D. A. (2010). Emotional intelligence: An integrative meta-analysis and cascading model. *Journal of Applied Psychology, 95,* 54–78.

Joseph, R. (2000). Fetal brain behavior and cognitive development. *Developmental Review, 20,* 81–98.

Joseph, R. M., Keehn, B., Connolly, C., Wolfe, J. M., & Horowitz, T. S. (2009). Why is visual search superior in autism spectrum disorder? *Developmental Science, 12,* 1083–1096.

Joussemet, M., Vitaro, F., Barker, E. D., Côté, S., Zoccolillo, M., Nagin, D. S., et al. (2008). Controlling parenting and physical aggression during elementary school. *Child Development, 79,* 411–425.

Joyner, K., & Udry, J. R. (2000). You don't bring me anything but down: Adolescent romance and depression. *Journal of Health and Social Behavior, 41,* 369–391.

Juffer, F. (2006). Children's awareness of adoption and their problem behavior in families with 7-year-old internationally adopted children. *Adoption Quarterly, 9,* 1–22.

Juffer, F., & van IJzendoorn, M. H. (2007). Adoptees do not lack self-esteem: A meta-analysis of studies on self-esteem of transracial, international, and domestic adoptees. *Psychological Bulletin, 133,* 1067–1083.

Jung, C. G. (1960/1933). The stages of life. In G. Adler, M. Fordham, & H. Read (Eds.), *The collected works of C. J. Jung: Vol. 8. The structure and dynamics of the psyche.* London: Routledge & Kegan Paul.

Jung, R. E., & Haier, R. J. (2007). The parieto-frontal integration theory (P-FIT) of intelligence: Converging neuroimaging evidence. *Behavioral and Brain Sciences, 30,* 135–154.

Jung, R. E., Segall, J. M., Bockholt, H. J., Flores, R. A., Smith, S. M., Chavez, R. S., et al. (2010). Neuroanatomy of creativity. *Human Brain Mapping, 31,* 398–409.

Jusczyk, P. W. (1995). Language acquisition: Speech sounds and phonological development. In J. L. Miller & P. D. Eimas (Eds.), *Handbook of perception and cognition: Vol. 11. Speech, language, and communication.* (pp. 263–301). Orlando, FL: Academic Press.

Jusczyk, P. W. (2002). How infants adapt speech-processing capacities to native-language structure. *Current Directions in Psychological Science, 11,* 15–18.

Justice, L. M., Pullen, P. C., & Pence, K. (2008). Influence of verbal and nonverbal references to print on preschoolers' visual attention to print during storybook reading. *Developmental Psychology, 44,* 855–866.

Kagan, J., Arcus, D., Snidman, N., Feng, W. Y., Hendler, J., & Greene, S. (1994). Reactivity in infants: A cross-national comparison. *Developmental Psychology, 30,* 342–345.

Kager, M. B. (2000). Factors that affect hiring: A study of age discrimination and hiring. *Dissertation Abstracts International. Section A. Humanities and Social Sciences, 60(11-A),* 4201.

Kaijura, H., Cowart B. J., & Beauchamp, G. K. (1992). Early developmental change in bitter taste responses in human infants. *Developmental Psychobiology, 25,* 375–386.

Kail, R. (2004). Cognitive development includes global and domain-specific processes. *Merrill-Palmer Quarterly, 50,* 445–455.

Kail, R. V., & Salthouse, T. A. (1994). Processing speed as a mental capacity. *Acta Psychologica, 86,* 199–225.

Kail, R., & Bisanz, J. (1992). The information-processing perspective on cognitive development in childhood and adolescence. In R. J. Sternberg & C. A. Berg (Eds.), *Intellectual development.* (pp. 229–260). New York: Cambridge University Press.

Kaiser Family Foundation. (2010). *Medicare chartbook, fourth edition, 2010.* Retrieved from http://facts.kff.org/chart.aspx?cb=58&sctn=162&p=1

Kaiser, S., & Panegyres, P. K. (2007). The psychosocial impact of young onset dementia on spouses. *American Journal of Alzheimer's Disease and Other Dementias, 21,* 398–402.

Kaldy, J. (2010). Making memories matter: The pharmacist's role in memory care. *The Consultant Pharmacist, 25,* 534–543.

Kalil, A., & Ziol-Guest, K. M. (2005). Single mothers' employment dynamics and adolescent well-being. *Child Development, 76,* 196–211.

Kalish, R. A. (1984). *Death, grief, and caring relationships* (2nd ed.). Pacific Grove, CA: Brooks/Cole.

Kalish, R. A. (1987). Death and dying. In P. Silverman (Ed.), *The elderly as modern pioneers* (pp. 320–334). Bloomington, IN: Indiana University Press.

Kalish, R. A., & Reynolds, D. (1976). *Death and ethnicity: A psychocultural study.* Los Angeles, CA: University of Southern California Press.

Kalmijn, M., & Flap, H. (2001). Assortative meeting and mating: Unintended consequences of organized settings for partner choices. *Social Forces, 79,* 1289–1312.

Kanayama, G., Hudson, J. I., & Pope, H. G. (2008). Long-term psychiatric and medical consequences of anabolic-androgenic steroid use: A looming public health concern? *Drug and Alcohol Dependence, 98,* 1–12.

Kanekar, S., Kolsawalla, M. B., & Nazareth, T. (1989). Occupational prestige as a function of occupant's gender. *Journal of Applied Social Psychology, 19,* 681–688.

Kann L., Collins, J. L., Pateman, B. C., Small, M. L., Ross, J.-G., & Kolbe L. J. (1995). The School Health Policies and Programs Study (SHPPS): Rationale for a nationwide status report on school health programs. *Journal of School Health, 65,* 291–294.

Kaplan, H., & Dove, H. (1987). Infant development among the Ache of eastern Paraguay. *Developmental Psychology, 23,* 190–198.

Kapoor, U., Pfost, K. S., House, A. E., & Pierson, E. (2010). Relation of success and nontraditional career choice to selection for dating and friendship. *Psychological Reports, 107,* 177–184.

Karam, E., Kypri, K., & Salamoun, M. (2007). Alcohol use among college students: An international perspective. *Current Opinion in Psychiatry, 20,* 213–221.

Karevold, E., Røysamb, E., Ystrom, E., & Mathiesen, K. S. (2009). Predictors and pathways from infancy to symptoms of anxiety and depression in early adolescence. *Developmental Psychology, 45,* 1051–1060.

Karmiloff-Smith, A. (2010). A developmental perspective on modularity. In B. Glatzeder, V. Goel, & A. Müller (Eds.), *Towards a theory of thinking: Building blocks for a conceptual framework* (Part 3, pp. 179–187). New York: Springer.

Kärnä, A., Voeten, M., Little, T. D., Poskiparta, E., Kaljonen, A., & Salmivalli, C. (2011). A large-scale evaluation of the KiVa antibullying program: Grades 4–6. *Child Development, 82,* 311–330.

Karney, B. R. (2010). *Keeping marriages healthy, and why it's so difficult.* Retrieved from http://www.apa.org/science/about/psa/2010/02/sci-brief.aspx

Karney, B. R., & Bradbury, T. N. (1995). The longitudinal course of marital quality and stability: A review of theory, method, and research. *Psychological Bulletin, 118,* 3–34.

Karney, B. R., & Crown, J. S. (2007). *Families under stress: An assessment of data, theory, and research on marriage and divorce in the military* (MG-599-OSD). Santa Monica, CA: RAND Corporation.

Karniol, R. (1989). The role of manual manipulative states in the infant's acquisition of perceived control over objects. *Developmental Review, 9,* 205–233.

Karraker, K. H., Vogel, D. A., & Lake, M. A. (1995). Parents' gender-stereotyped perceptions of newborns: The eye of the beholder revisited. *Sex Roles, 33,* 687–701.

Kastenbaum, R. (1985). Dying and death: A life-span approach. In J. E. Birren & K. W. Schaie (Eds.), *Handbook of the psychology of aging* (2nd ed., pp. 619–643). New York: Van Nostrand Reinhold.

Kastenbaum, R. (1999). Dying and bereavement. In J. C. Cavanaugh & S. K. Whitbourne (Eds.), *Gerontology: An interdisciplinary perspective.* New York: Oxford University Press.

Kastenbaum, R., & Thuell, S. (1995). Cookies baking, coffee brewing: Toward a contextual theory of dying. *Omega: The Journal of Death and Dying, 31,* 175–187.

Katz, L. F., & Woodin, E. M. (2002). Hostility, hostile detachment, and conflict engagement in marriages: Effects on child and family functioning. *Child Development, 73,* 636–652.

Katz-Wise, S. L., Priess, H. A., & Hyde, J. S. (2010). Gender-role attitudes and behavior across the transition to parenthood. *Developmental Psychology, 46,* 18–28.

Kaufman, J., & Charney, D. (2003). The neurobiology of child and adolescent depression: Current knowledge and future directions. In D. Cicchetti & E. Walker (Eds.), *Neurodevelopmental mechanisms in psychopathology* (pp. 461–490). New York: Cambridge University Press.

Kavsek, M., & Bornstein, M. H. (2010). Visual habituation and dishabituation in preterm infants: A review and meta-analysis. *Research in Developmental Disabilities, 31,* 951–975.

Kayser, K. (2010). Couples therapy. In J. R. Brandell (Ed.), *Theory and practice in clinical social work* (2nd ed., pp. 259–288). Thousand Oaks, CA: Sage Publications.

Keane, S. P., Brown, K. P., & Crenshaw, T. M. (1990). Children's intention-cue detection as a function of maternal social behavior: Pathways to social rejection. *Developmental Psychology, 26,* 1004–1009.

Keenan, N. L., & Shaw, K. M. (2011). *Coronary heart disease and stroke deaths—United States 2006.* Retrieved from http://www.cdc.gov/mmwr/preview/mmwrhtml/su6001a13.htm?s_cid=su6001a13_w

Keenan, T. (2009). *Multi-generational housing patterns*. Washington, DC: AARP.

Keith, J. (1990). Age in social and cultural context: Anthropological perspectives. In R. H. Binstock & L. K. George (Eds.), *Handbook of aging and the social sciences* (3rd ed., pp. 91–111). San Diego, CA: Academic Press.

Kelch-Oliver, K. (2010). The experiences of African American grandmothers in grandparent-headed families. *The Family Journal, 19*, 73–82.

Kelemen, D. (2003). British and American children's preferences for teleo-functional explanations of the natural world. *Cognition, 88*, 201–221.

Kelemen, D., & DiYanni, C. (2005). Intuitions about origins: Purpose and intelligent design in children's reasoning about nature. *Journal of Cognition and Development, 6*, 3–31.

Keller, H. H. (2004). Nutrition and health-related quality of life in frail older adults. *Journal of Nutrition, Health & Aging, 8*, 245–252.

Kellman, P. J., & Arterberry, M. E. (2006). Infant visual perception. In W. Damon & R. M. Lerner (Eds.), *Handbook of child psychology: Vol. 2. Cognition, perception, and language* (6th ed., pp. 109–160). Hoboken, NJ: Wiley.

Kelly, D. J., et al. (2009). Development of the other-race effect during infancy: Evidence toward universality? *Journal of Experimental Child Psychology, 104*, 105–114.

Kelly, P. (2011). Corporal punishment and child maltreatment in New Zealand. *Acta Paediatrica, 100*, 14–20.

Kelty, R., Kleykamp, M., & Segal, D. R. (2010). The military and the transition to adulthood. *The Future of Children, 20*, 181–207.

Kemeny, M. E. (2003). The psychobiology of stress. *Current Directions in Psychological Science, 12*, 124–129.

Kendall, J., & Hatton, D. (2002). Racism as a source of health disparity in families with children with attention deficit hyperactivity disorder. *Advances in Nursing Science, 25*, 22–39.

Kennedy, G. E. (1991). Grandchildren's reasons for closeness with grandparents. *Journal of Social Behavior and Personality, 6*, 697–712.

Kennedy, K. M., & Raz, N. (2009). Aging white matter and cognition: Differential effects of regional variations in diffusion properties on memory, executive functions, and speed. *Neuropsychologia, 47*, 916–927.

Kennel, K. (2007). *Osteoporosis treatment puts brakes on bone loss*. Retrieved from http://www.mayoclinic.com/print/osteoporosis-treatment/WO00127/METHOD=print

Kersting, A., Brähler, E., Glaesmer, H., & Wagner, B. (in press). Prevalence of complicated grief in a representative population-based sample. *Journal of Affective Disorders*.

Kester, M. I. (2011). *Biomarkers for Alzheimer's pathology: Monitoring, predicting, and understanding the disease*. Retrieved from http://dare.ubvu.vu.nl/bitstream/1871/18384/2/abstract_english.pdf

Keyes, C. M., & Ryff, C. D. (1999). Psychological well-being in midlife. In S. Willis & J. D. Reid (Eds.), *Life in the middle* (pp. 161–181). San Diego, CA: Academic Press.

Khanna, N. (2010). "If you're half black, you're just black": Reflected appraisals and the persistence of the one-drop rule. *Sociological Quarterly, 51*, 96–121.

Kiang, L., & Fuligni, A. J. (2009). Ethnic identity and family processes among adolescents from Latin American, Asian, and European backgrounds. *Journal of Youth and Adolescence, 38*, 228–241.

Kiang, L., Yip, T., Gonzales-Backen, M., Witkow, M., & Fuligni, A. J. (2006). Ethnic identity and the daily psychological well-being of adolescents from Mexican and Chinese backgrounds. *Child Development, 77*, 1338–1350.

Kidd, E., & Holler, J. (2009). Children's use of gesture to resolve lexical ambiguity. *Developmental Science, 12*, 903–913.

Kihlstrom, J. F. (2009). 'So that we might have roses in December': The functions of autobiographical memory. *Applied Cognitive Psychology, 23*, 1179–1192.

Kilgore, K., Snyder, J., & Lentz, C. (2000). The contribution of parental discipline, parental monitoring, and school risk to early-onset conduct problems in African American boys and girls. *Developmental Psychology, 36*, 835–845.

Killen, M., & McGlothlin, H. (2005). Prejudice in children. In C. B. Fisher & R. M. Lerner (Eds.), *Encyclopedia of applied developmental science* (Vol. 2, pp. 870–872). Thousand Oaks, CA: Sage.

Kilson, M., & Ladd, F. (2009). *Is that your child? Mothers talking about rearing biracial children*. Lanham, MD: Lexington Books.

Kim, H. K., Capaldi, D. M., & Crosby, L. (2007). Generalizability of Gottman and colleagues' affective process models of couples' relationship outcomes. *Journal of Marriage and Family, 69*, 55–72.

Kim, J.-Y., McHale, S. M., Crouter, A. C., & Osgood, D. (2007). Longitudinal linkages between sibling relationships and adjustment from middle childhood through adolescence. *Developmental Psychology, 43*, 960–973.

Kim, J.-Y., McHale, S. M., Osgood, D., & Crouter, A. C. (2006). Longitudinal course and family correlates of sibling relationships from childhood through adolescence. *Child Development, 77*, 1746–1761.

Kim, S. S. (2000). Gradual return to work: The antecedents and consequences of switching to part-time work after first childbirth. *Dissertation Abstract International. Section A. Humanities and Social Sciences, 61(3-A)*, 1182.

Kim, S., & Feldman, D. C. (2000). Working in retirement: The antecedents of bridge employment and its consequences for quality of life in retirement. *Academy of Management Journal, 43*, 1195–1210.

Kim, Y. H., & Goetz, E. T. (1994). Context effects on word recognition and reading comprehension of good and poor readers: A test of the interactive compensatory hypothesis. *Reading Research Quarterly, 29*, 178–188.

Kim-Cohen, J., Moffitt, T. E., Taylor, A., Pawlby, S. J., & Caspi, A. (2005). Maternal depression and child antisocial behavior: Nature and nurture effects. *Archives of General Psychiatry, 62*, 173–181.

Kindermann, T. A. (2007). Effects of naturally existing peer groups on changes in academic engagement in a cohort of sixth graders. *Child Development, 78*, 1186–1203.

King, P. E., & Furrow, J. L. (2004). Religion as a resource for positive youth development: Religion, social capital, and moral outcomes. *Developmental Psychology, 40*, 703–713.

King, P. M., & Kitchener, K. S. (2004). Reflective judgment: Theory and research on the development of epistemic assumptions through adulthood. *Educational Psychologist, 39*, 5–18.

King, S. V., Burgess, E. O., Akinyela, M., Counts-Spriggs, M., & Parker, N. (2006). The religious dimensions of the grandparent role in three-generation African American households. *Journal of Religion, Spirituality & Aging, 19*, 75–96.

King, V., & Scott, M. E. (2005). A comparison of cohabiting relationships among older and younger adults. *Journal of Marriage and Family, 67*, 271–285.

King, V., Elder, G. H., & Whitbeck, L. B. (1997). Religious involvement among rural youth: An ecological and life-course perspective. *Journal of Research on Adolescence, 7*, 431–456.

Kinney, J. M., Ishler, K. J., Pargament, K. I., & Cavanaugh, J. C. (2003). Coping with the uncontrollable: The use of general and religious coping by caregivers to spouses with dementia. *Journal of Religious Gerontology, 14*, 171–188.

Kins, E., & Beyers, W. (2010). Failure to launch, failure to achieve criteria for adulthood? *Journal of Adolescent Research, 25*, 743–777.

Kinsella, G. J., Ong, B., Storey, E., Wallace, J., & Hester, R. (2007). Elaborated spaced-retrieval and prospective memory in mild Alzheimer's disease. *Neuropsychological Rehabilitation, 17*, 688–706.

Kippen, R., Chapman, B., & Yu, P. (2009). *What's love got to do with it? Homogamy and dyadic approaches to understanding marital instability*. Retrieved from http://www.datingsitesreviews.com/images/other/Kippen-Rebecca-paper.pdf.

Kirby, D. (2001). *Emerging answers: Research findings on programs to reduce teen pregnancy* [Summary]. Washington, DC: National Campaign to Prevent Teen Pregnancy.

Kirby, D., & Laris, B. A. (2009). Effective curriculum-based sex and STD/HIV education programs for adolescents. *Child Development Perspectives, 3*, 21–29.

Kirby, S. E., Coleman, P. G., & Daley, D. (2004). Spirituality and well-being in frail and nonfrail older adults. *Journal of Gerontology: Psychological Sciences, 59*, P123–P129.

Kitchener, K. S., & King, P. M. (1989). The reflective judgment model: Ten years of research. In M. L. Commons, C. Armon, L. Kohlberg,

F. A. Richards, T. A. Grotzer, & J. D. Sinnott (Eds.), *Adult development: Vol. 2. Models and methods in the study of adolescent and adult thought* (pp. 63–78). New York: Praeger.

Kitchener, K. S., King, P. M., & DeLuca, S. (2006). Development of reflective judgment in adulthood. In C. Hoare (Ed.), *Handbook of adult development and learning* (pp. 73–98). New York: Oxford University Press.

Kivett, V. R. (1991). Centrality of the grandfather role among older rural black and white men. *Journal of Gerontology: Social Sciences, 46,* S250–S258.

Kivnick, H. Q. (1982). *The meaning of grandparenthood.* Ann Arbor, MI: UMI Research.

Kivnick, H. Q. (1985). Grandparenthood and mental health: Meaning, behavior, and satisfaction. In V. L. Bengtson & J. F. Robertson (Eds.), *Grandparenthood* (pp. 151–158). Beverly Hills, CA: Sage.

Klaczynski, P. A. (2004). A dual-process model of adolescent development: Implications for decision making, reasoning, and identity. In R. Kail (Ed.), *Advances in child development and behavior* (Vol. 32, pp. 73–123). San Diego, CA: Elsevier.

Klaczynski, P. A., & Lavallee, K. L. (2005). Domain-specific identity, epistemic regulation, and intellectual ability as predictors of belief-biased reasoning: A dual-process perspective. *Journal of Experimental Child Psychology, 92,* 1–24.

Klaczynski, P. A., & Narasimham, G. (1998). Development of scientific reasoning biases: Cognitive versus ego-protective explanations. *Developmental Psychology, 34,* 175–187.

Klassen, R. M., Usher, E. L., & Bong, M. (2010). Teachers' collective efficacy, job satisfaction, and job stress in cross-cultural context. *Journal of Experimental Education, 78,* 464–486.

Kleespies, P. M. (2004). *Life and death decisions: Psychological and ethical considerations in end-of-life care.* Washington, DC: American Psychological Association.

Kleiber, D. A., Hutchinson, S. L., & Williams, R. (2002). Leisure as a resource in transcending negative life events: Self-protection, self-restoration, and personal transformation. *Leisure Sciences, 24,* 219–235.

Klohnen, E. C., Vandewater, E. A., & Young, A. (1996). Negotiating the middle years: Ego-resiliency and successful midlife adjustment in women. *Psychology and Aging, 11,* 431–442.

Klump, K. L., & Culbert, K. M. (2007). Molecular genetic studies of eating disorders: Current status and future directions. *Current Directions in Psychological Science, 16,* 37–41.

Knapp, D. E., & Kustis, G. A. (2000). Same-sex sexual harassment: A legal assessment with implications for organizational policy. *Employee Responsibilities and Rights Journal, 12,* 105–119.

Knee, D. O. (2010). Hospice care for the aging population in the United States. In J. C. Cavanaugh & C. K. Cavanaugh (Eds.), *Aging in America: Vol. 3: societal issues* (pp. 203–221). Santa Barbara, CA: Praeger Perspectives.

Knight, B. G., & Sayegh, P. (2010). Cultural values and caregiving: The updated sociocultural stress and coping model. *Journal of Gerontology: Psychological Sciences and Social Sciences, 65B,* P5-P13.

Knowles, D. R. (2006). *Aging changes in the male reproductive system.* Retrieved from http://www.nlm.nih.gov/medlineplus/ency/article/004017.htm

Knowles, M. S., Swanson, R. A., & Holton, E. F. (2005). *The adult learner: The definitive classic in adult education and human resource development.* New York: Elsevier.

Knox, D. (2007). Counseling students who are grieving: Finding meaning in loss. In J. A. Lippincott & R. B. Lippincott (Eds.), *Special populations in college counseling: A handbook for mental health professionals* (pp. 187–199). Alexandria, VA: American Counseling Association.

Kobler, K., Limbo, R., & Kavanaugh, K. (2007). Meaningful moments: The use of ritual in perinatal and pediatric death. *MCN: The American Journal of Maternal/Child Nursing, 32,* 288–297.

Kochanska, G., Aksan, N., & Joy, M. E. (2007). Children's fearfulness as a moderator of parenting in early socialization: Two longitudinal studies. *Developmental Psychology, 43,* 222–237.

Kochanska, G., Gross, J. N., Lin, M., & Nichols, K. E. (2002). Guilt in young children: Development, determinants, and relations with a broader system of standards. *Child Development, 73,* 461–482.

Kochenderfer-Ladd, B., & Waldrop, J. L. (2001). Chronicity and instability of children's peer victimization experiences as predictors of loneliness and social satisfaction trajectories. *Child Development, 72,* 134–151.

Koenig, B. L., Kirkpatrick, L. A., & Ketelaar, T. (2007). Misperception of sexual and romantic interests in opposite-sex friendships: Four hypotheses. *Personal Relationships, 14,* 411–429.

Koenig, M. A., & Woodward, A. L. (2010). Sensitivity of 24-month-olds to the prior inaccuracy of the source: Possible mechanisms. *Developmental Psychology, 46,* 815–826.

Koestenbaum, P. (1976). *Is there an answer to death?* Englewood Cliffs, NJ: Prentice Hall.

Kogan, N. (1983). Stylistic variation in childhood and adolescence: Creativity, metaphor, and cognitive style. In P. H. Mussen (Ed.), *Handbook of child psychology* (Vol. 3, pp. 630–706). New York: Wiley.

Kohlberg, L. (1966). A cognitive-developmental analysis of children's sex-role concepts and attitudes. In E. E. Maccoby (Ed.), *The development of sex differences.* (pp. 87–123). Stanford, CA: Stanford University Press.

Kohlberg, L. (1969). Stage and sequence: The cognitive-developmental approach to socialization. In D. Goslin (Ed.), *Handbook of socialization theory and research* (pp. 347–480). Chicago: Rand McNally.

Kohlberg, L., & Ullian, D. Z. (1974). Stages in the development of psychosexual concepts and attitudes. In R. C. Friedman, R. M. Richart, & R. L. Van Wiele (Eds.), *Sex differences in behavior.* (pp. 209–222). New York: Wiley.

Kokis, J. V., Macpherson, R., Toplak, M. E., West, R. F., & Stanovich, K. E. (2002). Heuristic and analytic processing: Age trends and associations with cognitive ability and cognitive styles. *Journal of Experimental Child Psychology, 83,* 26–52.

Kolata, G. (1990, June 19). *NIH neglects women, study says.* Retrieved from http://www.nytimes.com/1990/06/19/science/nih-neglects-women-study-says.html

Kolata, G. (1990, February 6). Rush is on to capitalize on test for gene causing cystic fibrosis. *New York Times,* p. C3.

Kolb, B. (1989). Brain development, plasticity, and behavior. *American Psychologist, 44,* 1203–1212.

Kolb, D. M., Williams, J., & Frohlinger, C. (2010). *Her place at the table: A woman's guide to negotiating five key challenges to leadership success.* San Francisco, CA: Jossey-Bass.

Kolberg, K. J. S. (1999). Environmental influences on prenatal development and health. In T. L. Whitman & T. V. Merluzzi (Eds.), *Life-span perspectives on health and illness* (pp. 87–103). Mahwah, NJ: Erlbaum.

Konijn, E. A., Nije Bijvank, M., & Bushman, B. J. (2007). I wish I were a warrior: The role of wishful identification in the effects of violent video games on aggression in adolescent boys. *Developmental Psychology, 43,* 1038–1044.

Kornhaber, M., Fierros, E., & Veenema, S. (2004). *Multiple intelligences: Best ideas from research and practice.* Boston, MA: Allyn & Bacon.

Koropeckyj-Cox T., & Call, V. R. A. (2007). Characteristics of older childless persons and parents: Cross-national comparisons. *Journal of Family Issues, 28,* 1362–1414.

Koropeckyj-Cox T., & Pendell, G. (2007). The gender gap in attitudes about childlessness in the United States. *Journal of Marriage and Family, 69,* 899–915.

Kowal, A., & Kramer, L. (1997). Children's understanding of parental differential treatment. *Child Development, 68,* 113–126.

Kowalski, S. D., & Bondmass, M. D. (2008). Physiological and psychological symptoms of grief in widows. *Research in Nursing & Health, 31,* 23–30.

Kozbelt, A., & Durmysheva, Y. (2007). Lifespan creativity in a non-Western artistic tradition: A study of Japanese Ukiyo-e printmakers. *International Journal of Aging & Human Development, 65,* 23–51.

Kram, K. E. (1985). *Mentoring at work: Developmental relationships in organizational life.* Glenview, IL: Scott, Foresman.

Kramer, D. A. (1989). A developmental framework for understanding conflict resolution processes. In J. D. Sinnott (Ed.), *Everyday problem solving: Theory and applications* (pp. 138–152). New York: Praeger.

Kramer, D. A. (1990). Conceptualizing wisdom: The primacy of affect-cognition relations. In R. J. Sternberg (Ed.), *Wisdom: Its nature, origins, and development* (pp. 279–313). Cambridge, England: Cambridge University Press.

Kramer, D. A., Angiuld, N., Crisafi, L., & Levine, C. (1991, August). *Cognitive processes in real-life conflict resolution.* Paper presented at the annual meeting of the American Psychological Association, San Francisco.

Kramer, L. (2010). The essential ingredients of successful sibling relationships: An emerging framework for advancing theory and practice. *Child Development Perspectives, 4,* 80–86.

Krause, N. (2006). Religion and health in late life. In J. E. Birren & K. W. Schaie (Eds.), *Handbook of the psychology of aging* (6th ed., pp. 499–518). Amsterdam: Elsevier.

Krause, N., & Bastida, E. (2011). Prayer to the saints or the Virgin and health among older Mexican Americans. *Hispanic Journal of Behavioral Sciences, 33,* 71–87.

Krause, N., Morgan, D., Chatters, L., & Meltzer, T. (2000). Using focus groups to explore the nature of prayer in late life. *Journal of Aging Studies, 14,* 191–212.

Krebs, D. L., & Denton, K. (2005). Toward a more pragmatic approach to morality: A critical evaluation of Kohlberg's model. *Psychological Review, 113,* 672–675.

Krebs, D., & Gillmore, J. (1982). The relationships among the first stages of cognitive development, role-taking abilities, and moral development. *Child Development, 53,* 877–886.

Krishnasamy, C., & Unsworth, C. A. (2011). Normative data, preliminary inter-rater reliability and predictive validity of the Drive Home Maze Test. *Clinical Rehabilitation, 25,* 88–95.

Krispin, O., Sternberg, K. J., & Lamb, M. E. (1992). The dimensions of peer evaluation in Israel: A cross-cultural perspective. *International Journal of Behavioral Development, 15,* 299–314.

Kroger, J., & Greene, K. E. (1996). Events associated with identity status change. *Journal of Adolescence, 19,* 477–490.

Kruger, D. J. (2006). Male facial masculinity influences attributions of personality and reproductive strategy. *Personal Relationships, 13,* 451–463.

Kübler-Ross, E. (1969). *On death and dying.* New York: Macmillan.

Kübler-Ross, E. (1974). *Questions and answers on death and dying.* New York: Macmillan.

Kübler-Ross, E., & Kessler, D. (2005). *On grief and grieving: On finding meaning of grief through the five stages of loss.* New York: Scribner.

Kuh, G. D., Kinzie, J., Schuh, J. H., & Whitt, K. J. (2010). *Student success in college: Creating conditions that matter.* San Francisco, CA: Jossey-Bass.

Kuhl, P. K., Andruski, J. E., Chistovich, I. A., Chistovich, L.-A., Kozhevnikova, E. V., Ryskina, V. L., et al. (1997). Cross-language analysis of phonetic units in language addressed to infants. *Science, 277,* 684–686.

Kuhl, P. K., Stevens, E., Hayashi, A., Deguchi, T., Kiritani, S., & Iverson, P. (2006). Infants show a facilitation effect for native language phonetic perception between 6 and 12 months. *Developmental Science, 9,* F13–F21.

Kukutai, T. H. (2007). White mothers, brown children: Ethnic identification of Maori-European children in New Zealand. *Journal of Marriage and Family, 69,* 1150–1161.

Kulik, L. (2001a) The impact of men's and women's retirement on marital relations: A comparative analysis. *Journal of Women & Aging, 13,* 21–37.

Kulik, L. (2001b). Marital relationships in late adulthood: Synchronous versus asynchronous couples. *International Journal of Aging & Human Development, 52,* 323–339.

Kulkofsky, S., Wang, Q., & Koh, J. B. K. (2009). Functions of memory sharing and mother-child reminiscing behaviors: Individual and cultural variations. *Journal of Cognition and Development, 10,* 92–114.

Kulwicki, A. D. (2002). The practice of honor crimes: A glimpse of domestic violence in the Arab world. *Issues in Mental Health Nursing, 23,* 77–87.

Kumar, R., O'Malley, P. M., Johnston, L. D., Schulenberg, J.-E., & Bachman, J. G. (2002). Effects of school-level norms on student substance use. *Prevention Science, 3,* 105–124.

Kunlin, J. (2010). Modern biological theories of aging. *Aging and Disease, 1,* 72-74.

Kunz, J. A. (2007). Older adult development. In J. A. Kunz & F. G. Soltys (Eds.), *Transformational reminiscence: Life story work* (pp. 19–39). New York: Springer.

Kurdek, L. A. (2004). Are gay and lesbian cohabiting couples really different from heterosexual married couples? *Journal of Marriage and Family, 66,* 880–900.

Kurup, R. K., & Kurup, P. A. (2003). Hypothalamic digoxin, hemispheric dominance, and neurobiology of love and affection. *International Journal of Neuroscience, 113,* 721–729.

Kypri, K., Paschall, M. J., Langley, J., Baxter, J., Cashell-Smith, M., & Bourdeau, B. (2009). Drinking and alcohol-related harm among New Zealand university students: findings from a national web-based survey. *Alcoholism: Clinical and Experimental Research, 33,* 307–314.

L'Abate, L. (2006). You can go home again, but should you stay? *PsycCRITIQUES, 51,* 25.

Labouvie-Vief, G. (2006). Emerging structures of adult thought. In J. J. Arnett & J. L. Tanner (Eds.), *Emerging adults in America: Coming of age in the 21st century* (pp. 59–84). Washington, DC: American Psychological Association.

Labouvie-Vief, G., Grühn, D., & Studer, J. (2010). Dynamic integration of emotion and cognition: Equilibrium regulation in development and aging. In M. E. Lamb & A. M. Freund (Eds.), *The handbook of life-span development: Vol. 2. Social and emotional development.* Hoboken, NJ: Wiley.

Labouvie-Vief. G. (2006). Emerging structures of adult thought. In J. Arnett (Ed.), *Psychological development during emerging adulthood* (pp. 60–84). Washington, DC: American Psychological Association.

Ladd, G. W. (1998). Peer relationships and social competence during early and middle childhood. *Annual Review of Psychology, 50,* 333–359.

Ladd, G. W. (2003). Probing the adaptive significance of children's behavior and relationships in the school context: A child by environment perspective. In R. V. Kail (Ed.), *Advances in child development and behavior* (Vol. 31, pp. 44–104). San Diego, CA: Academic Press.

Ladd, G. W. (2006). Peer rejection, aggressive or withdrawn behavior, and psychological maladjustment from ages 5 to 12: An examination of four predictive models. *Child Development, 77,* 822–846.

Ladd, G. W., & Ladd, B. K. (1998). Parenting behaviors and parent–child relationships: Correlates of peer victimization in kindergarten? *Developmental Psychology, 34,* 1450–1458.

Ladd, G. W., & Pettit, G. S. (2002). Parents and children's peer relationships. In M. Bornstein (Ed.), *Handbook of parenting: Vol. 4* (2nd ed., pp. 377–409). Hillsdale, NJ: Erlbaum.

Lagattuta, K. H., & Wellman, H. M. (2002). Differences in early parent–child conversations about negative versus positive emotions: Implications for the development of psychological understanding. *Developmental Psychology, 38,* 564–580.

LaGreca, A. M. (1993). Social skills training with children: Where do we go from here? *Journal of Clinical Child Psychology, 22,* 288–298.

Lahey, J. N. (2010). International comparison of age discrimination laws. *Research on Aging, 32,* 679–697.

Lai, D. W. L. (2010). Filial piety, caregiving appraisal, and caregiving burden. *Research on Aging, 32,* 200–223.

Laible, D. J., & Carlo, G. (2004). The differential relations of maternal and paternal support and control to adolescent social competence, self-worth, and sympathy. *Journal of Adolescent Research, 19,* 759–782.

Laidlaw, K. (2007). Cognitive behavior therapy with older adults. In S. H. Qualls & B. G. Knight (Eds.), *Psychotherapy for depression in older adults* (pp. 83–109). Hoboken, NJ: Wiley.

Lako, M., Trounson, A., & Daher, S. (2010). Law, ethics, and clinical translation in the 21st century—A discussion with Stephen Bellamy. *Stem Cells, 28,* 177–180.

Lakshmanan, I. A. R. (1997, September 22). Marriage? Think logic, not love. *Baltimore Sun,* p. A2.

Lamanna, M. A., & Riedmann, A. (2003). *Marriages and families: Making choices in a diverse society* (8th ed.). Belmont, CA: Wadsworth.

Lamaze, F. (1958). *Painless childbirth.* London: Burke.

Lamb, M. E. (1999). Nonparental child care. In M. E. Lamb (Ed.), *Parenting and child development in "nontraditional" families.* (pp. 39–55). Mahwah, NJ: Erlbaum.

Lamb, M. E., Orbach, Y., Hershkowitz, I., Esplin, P. W., & Horowitz, D. (2007). A structured forensic interview protocol improves the quality and informativeness of investigative interviews with children: A review of research using the NICHD Investigative Interview Protocol. *Child Abuse and Neglect, 31,* 1201–1231.

Lamb, M. E., Thompson, R. A., Gardner, W., Charnov, E. L., & Connell, J. P. (1985). *Infant-mother attachment: The origins and developmental significance of individual differences in Strange Situation behavior.* Hillsdale, NJ: Erlbaum.

Lampkin-Hunter, T. (2010). *Single parenting.* Bloomington, IN: Xlibris.

Landerl, K., Fussenegger, B., Moll, K., & Willburger, E. (2009). Dyslexia and dyscalculia: Two learning disorders with different cognitive profiles. *Journal of Experimental Child Psychology, 103,* 309–324.

Lane, B. N. (2007). Understanding anticipatory grief: Relationship to coping style, attachment style, caregiver strain, gender role identification, and spirituality. *Dissertation Abstracts International. Section B. Sciences and Engineering, 67(8-B),* 4714.

Lang, F. R., & Heckhausen, J. (2001). Perceived control over development and subjective well-being: Differential benefits across adulthood. *Journal of Personality and Social Psychology, 81,* 509–523.

Lang, F. R., & Heckhausen, J. (2006). Motivation and interpersonal regulation across adulthood: Managing the challenges and constraints of social contexts. In C. Hoare (Ed.), *Handbook of adult development and learning* (pp. 149–166). New York: Oxford University Press.

Lang, J. C., & Lee, C. H. (2005). Identity accumulation, others' acceptance, job-search self-efficacy, and stress. *Journal of Organizational Behavior, 26,* 293–312.

Langer, E. J., & Rodin, J. (1976). The effects of choice and enhanced personal responsibility for the aged: A field experiment in an institutional setting. *Journal of Personality and Social Psychology, 34,* 191–198.

Lansford, J. E. (2009). Parental divorce and children's adjustment. *Perspectives on Psychological Science, 4,* 140–152.

Laplante, D. P., Barr, R. G., Brunet, A., Du Fort, G. G., Meaney, M. L., Saucier, J., et al. (2004). Stress during pregnancy affects general intellectual and language functioning in human toddlers. *Pediatric Research, 56,* 400–410.

LaRocca, T. J., Seals, D. R., & Pierce, G. L. (2010). Leukocyte telomere length is preserved with aging in endurance exercise-trained adults and related to maximal aerobic capacity. *Mechanisms of Ageing and Development, 131,* 165–167.

Lau, J. Y., Rijsdijk, F., Gregory, A. M., McGuffin, P., & Eley, T. C. (2007). Pathways to childhood depressive symptoms: The role of social, cognitive, and genetic risk factors. *Developmental Psychology, 43,* 1402–1414.

Laungani, P. (2001). The influence of culture on stress: India and England. In L. L. Adler & U. P. Gielen (Eds.), *Cross-cultural topics in psychology* (2nd ed., pp. 149–169). Westport, CT: Praeger.

Laursen, B., & Collins, W. A. (1994). Interpersonal conflict during adolescence. *Psychological Bulletin, 115,* 197–209.

Laursen, B., Bukowski, W. M., Nurmi, J. E., Marion, D., Salmela-Aro, K., & Kiuru, N. (2010). Opposites detract: Middle school peer group antipathies. *Journal of Experimental Child Psychology, 106,* 240–256.

Lawton, L. E., & Tulkin, D. O. (2010). Work-family balance, family structure and family-friendly employer programs. Paper presented at the annual meeting of the Population Association of America, Dallas. Retrieved from http://paa2010.princeton.edu/download.aspx?submissionId=100573

Lawton, M. P. (1982). Competence, environmental press, and the adaptation of old people. In M. P. Lawton, P. G. Windley, & T. O. Byerts (Eds.), *Aging and the environment: Theoretical approaches* (pp. 33–59). New York: Springer.

Lawton, M. P. (1989). Environmental proactivity in older people. In V. L. Bengtson & K. W. Schaie (Eds.), *The course of later life: Research and reflections* (pp. 15–23). New York: Springer.

Lawton, M. P., & Nahemow, L. (1973). Ecology of the aging process. In C. Eisdorfer & M. P. Lawton (Eds.), *The psychology of adult development and aging* (pp. 619–674). Washington, DC: American Psychological Association.

Lazarus, R. S., & Folkman, S. (1984). *Stress, appraisal, and coping.* New York: Springer.

Leahy, J. M. (1993). A comparison of depression in women bereaved of a spouse, a child, or a parent. *Omega: The Journal of Death and Dying, 26,* 207–217.

Leal-Muniz, V., & Constantine, M. G. (2005). Predictors of the career commitment process in Mexican American college students. *Journal of Career Assessment, 13,* 204–215.

Leaper, C., & Smith, T. E. (2004). A meta-analytic review of gender variations in children's language use: Talkativeness, affiliative speech, and assertive speech. *Developmental Psychology, 40,* 993–1027.

Lecanuet, J. P., Granier-Deferre, C., & Busnel, M. C. (1995). Human fetal auditory perception. In J. P. Lecanuet, W. P. Fifer, N. A. Krasnegor, & W. P. Smotherman (Eds.), *Fetal development: A psychobiological perspective.* Hillsdale, NJ: Erlbaum.

Leclercq, A-L., & Majerus, S. (2010). Serial-order short-term memory predicts vocabulary development: Evidence from a longitudinal study. *Developmental Psychology, 46,* 417–427.

Ledbetter, A. M., & Kuznekoff, J. H. (in press). More than a game: friendship relational maintenance and attitudes toward Xbox LIVE communication. *Communication Research, 38.*

Ledebt, A. (2000). Changes in arm posture during the early acquisition of walking. *Infant Behavior and Development, 23,* 79–89.

Ledebt, A., van Wieringen, P. C. W., & Saveslsbergh, G. J. P. (2004). Functional significance of foot rotation in early walking. *Infant Behavior and Development, 27,* 163–172.

LeDoux, J. E., & Gorman, J. M. (2001). A call to action: Overcoming anxiety through active coping. *American Journal of Psychiatry, 158,* 1953–1955.

Lee, E.-K. O., & Sharpe, T. (2007). Understanding religious/spiritual coping and support resources among African American older adults: A mixed-method approach. *Journal of Religion, Spirituality & Aging, 19,* 55–75.

Lee, H. J., Macbeth, A. H., Pagani, J. H., & Young, W. S. (2009). Oxytocin: The great facilitator of life. *Progress in Neurobiology, 88,* 127–151.

Lee, K. S. (2010). Gender, care work, and the complexity of family membership in Japan. *Gender & Society, 24,* 647–671.

Lee, K. T., Mattson, S. N., & Riley, E. P. (2004). Classifying children with heavy prenatal alcohol exposure using measures of attention. *Journal of the International Neuropsychological Society, 10,* 271–277.

Lee, M.-D. (2007). Correlates of consequences of intergenerational caregiving in Taiwan. *Journal of Advanced Nursing, 59,* 47–56.

Lee, P. C. B. (2003). Going beyond career plateau: Using professional plateau to account for work outcomes. *Journal of Management Development, 22,* 538–551.

Lee, R. M., Seol, K. O., Miyoung, M., Miller, M. J., & the Minnesota International Adoption Project Team. (2010). The behavioral development of Korean children in institutional care and international adoptive families. *Developmental Psychology, 46,* 468–478.

Leerkes, E. M., Blankson, A. M., & O'Brien, M. (2009). Differential effects of maternal sensitivity to infant distress and nondistress on social-emotional functioning. *Child Development, 80,* 762–775.

Legare, C. H., Wellman, H. M., & Gelman, S. A. (2009). Evidence for an explanation advantage in naïve biological reasoning. *Cognitive Psychology, 58,* 177–194.

LeMare, L. J., & Rubin, K. H. (1987). Perspective taking and peer interaction: Structural and developmental analyses. *Child Development, 58,* 306–315.

Lemery, K. S., Goldsmith, H. H., Klinnert, M. D., & Mrazek, D. A. (1999). Developmental models of infant and childhood temperament. *Developmental Psychology, 35,* 189–204.

Lemieux, R., & Hale, J. L. (2002). Cross-sectional analysis of intimacy, passion, and commitment: Testing the assumptions of the triangular theory of love. *Psychological Reports, 90,* 1009–1014.

Lemire, L., Saba, T., & Gagnon, Y. C. (1999). Managing career plateauing in the Quebec public sector. *Public Personnel Management, 28,* 375–391.

Lengua, L. J., Sandler, I. N., West, S. G., Wolchik, S. A., & Curran, P. J. (1999). Emotionality and self-regulation, threat appraisal, and coping in children of divorce. *Development & Psychopathology, 11,* 15–37.

Leon, G. R., Gillum, B., Gillum, R., & Gouze, M. (1979). Personality stability and change over a 30-year period: Middle to old age. *Journal of Consulting and Clinical Psychology, 47,* 517–524.

Leon, K. (2003). Risk and protective factors in young children's adjustment to parental divorce: A review of the research. *Family Relations, 52,* 258–270.

Leppanen, J. M., Moulson, M. C., Vogel-Farley, V. K., & Nelson, C. A. (2007). An ERP study of emotional face processing in the adult and infant brain. *Child Development, 78,* 232–245.

LeRoux, H., & Fisher, J. E. (2006). Strategies for enhancing medication adherence in the elderly. In W. T. O'Donohue & E. R. Levensky (Eds.), *Promoting treatment adherence: A practical handbook for health care providers* (pp. 353–362). Thousand Oaks, CA: Sage.

Lervåg, A., Bråten, I., & Hulme, C. (2009). The cognitive and linguistic foundations of early reading development: A Norwegian latent variable longitudinal study. *Developmental Psychology, 45,* 764–781.

Letiecq, B. L., Bailey, S. J., & Kurtz, M. A. (2008). Depression among rural Native American and European American grandparents rearing their grandchildren. *Journal of Family Issues, 29,* 334–356.

Leventhal, H., Rabin, C., Leventhal, E. A., & Burns, E. (2002). Health risk behaviors and aging. In J. E. Birren & K. W. Schaie (Eds.), *Handbook of the psychology of aging* (5th ed., pp. 186–214). San Diego, CA: Academic Press.

Levete, S. (2010). *Coming of age.* New York: Wayland/The Rosen Publishing Group.

Levine, L. E. (1983). *Mine:* Self-definition in 2-year-old boys. *Developmental Psychology, 19,* 544–549.

Levine, S., Suiyakham, L., Huttenlocher, J., Rowe, M., & Gunderson, E. (2008). What counts in toddlers' development of cardinality knowledge? Unpublished manuscript.

Levinger, G. (1980). Toward the analysis of close relationships. *Journal of Experimental Social Psychology, 16,* 510–544.

Levinger, G. (1983). Development and change. In H. H. Kelley, E. Berscheid, A. Christensen, J. H. Harvey, T. L. Hutson, G. Levinger, et al. (Eds.), *Close relationships* (pp. 315–359). New York: Freeman.

Levitt, A. G., & Utman, J. A. (1992). From babbling towards the sound systems of English and French: A longitudinal two-case study. *Journal of Child Language, 19,* 19–49.

Levitt, M. J., Guacci-Franco, N., & Levitt, J. L. (1993). Convoys of social support in childhood and early adolescence: Structure and function. *Developmental Psychology, 29,* 811–818.

Levy, B. A., Gong, Z., Hessels, S., Evans, M. A., & Jared, D. (2006). Understanding print: Early reading development and the contributions of home literacy experiences. *Journal of Experimental Child Psychology, 93,* 63–93.

Levy, G. D., Taylor, M. G., & Gelman, S. A. (1995). Traditional and evaluative aspects of flexibility in gender roles, social conventions, moral rules, and physical laws. *Child Development, 66,* 515–531.

Levy, J. (1976). A review of evidence for a genetic component in the determination of handedness. *Behavior Genetics, 6,* 429–453.

Lewinsohn, P. M. (1975). The behavioral study and treatment of depression. In M. Hersen, R. M. Eisler, & P. M. Miller (Eds.), *Progress in behavior modification* (Vol. 1, pp. 19–64). New York: Academic Press.

Lewis, M. (1997). The self in self-conscious emotions. In J. G. Snodgrass & R. L. Thompson (Eds.), *The self across psychology: Self-awareness, self-recognition, and the self-concept* (pp. 119–142). New York: New York Academy of Science.

Lewis, M. (2000). The emergence of human emotions. In M. Lewis & J. Haviland-Jones (Eds.), *Handbook of emotions* (2nd ed., pp. 265–280). New York: Guilford Press.

Lewis, M., & Brooks-Gunn, J. (1979). *Social cognition and the acquisition of self.* New York: Plenum.

Lewis, M., & Ramsay, D. (2004). Development of self-recognition, personal pronoun use, and pretend play during the second year. *Child Development, 75,* 1821–1831.

Lewis, M., Takai-Kawakami, K., Kawakami, K., & Sullivan, M. W. (2010). Cultural differences in emotional responses to success and failure. *International Journal of Behavioral Development, 34,* 53–61.

Lewkowicz, D. J. (2000). The development of intersensory perception: An epigenetic systems/limitations view. *Psychological Bulletin, 126,* 281–308.

Lewontin, R. (1976). Race and intelligence. In N. J. Block & G. Dworkin (Eds.), *The IQ controversy* (pp. 78–92). New York: Pantheon.

Li, F., Fisher, K. J., Harmer, P., & McAuley, E. (2005). Falls self-efficacy as a mediator of fear of falling in an exercise intervention for older adults. *Journals of Gerontology: Psychological Sciences & Social Sciences, 60B,* P34–P40.

Li, Y., Anderson, R. C., Nguyen-Jahiel, K., Dong, T., Archodidou, A., Kim, I-H, Clark, A. M., et al. (2007). Emergent leadership in children's discussion groups. *Cognition and Instruction, 25,* 75–111.

Liang, H., & Eley, T. C. (2005). A monozygotic twin differences study of non-shared environmental influence on adolescent depressive symptoms. *Child Development, 76,* 1247–1260.

Liben, L. S., & Bigler, R. S. (2002). The developmental course of gender differentiation. *Monographs of the Society for Research in Child Development, 67*(Serial No. 269).

Lichtenstein, A. H., Rasmussen, H., Yu, W.W., Epstein, S.R., & Russell, R.M. (2008). Modified MyPyramid for older adults. *Journal of Nutrition, 138,* 78–82.

Lichter, D. T., & Carmalt, J. H. (2009). Religion and marital quality among low-income couples. *Social Science Research, 38,* 168–187.

Liebal, K., Behne, T., Carpenter, M., & Tomasello, M. (2009). Infants use shared experience to interpret pointing gestures. *Developmental Science, 12,* 264–271.

Lieber, J. (2001, October 10). Widows of tower disaster cope, but with quiet fury. *USA Today,* pp. A1–A2.

Lighthouse International. (2011). *Illumination.* Retrieved from http://www.lighthouse.org/for-professionals/practice-management/professional-products/illumination

Lim, S., & Cortina, L. M. (2005). Interpersonal mistreatment in the workplace: The interface and impact of general incivility and sexual harassment. *Journal of Applied Psychology, 90,* 483–496.

Lin, C. C., & Fu, V. R. (1990). A comparison of childrearing practices among Chinese, immigrant Chinese, and Caucasian-American parents. *Child Development, 61,* 429–433.

Lin, X., & Leung, K. (2010). Differing effects of coping strategies on mental health during prolonged unemployment: A longitudinal analysis. *Human Relations, 63,* 637–665.

Lindberg, S. M., Hyde, J. S., Petersen, J. L., & Linn, M. C. (2010). New trends in gender and mathematics performance: A meta-analysis. *Psychological Bulletin, 136,* 1123–1135.

Lindgren, K. P. (2007). Sexual intent perceptions: Review and integration of findings, investigation of automatic processes, and development and implementation of a dynamic assessment methodology. *Dissertation Abstracts International. Section B. Sciences and Engineering, 67(9-B),* 5469.

Lindlaw, S. (1997, April 25). Ethical issues surround oldest new mom. *News Journal* (Wilmington, DE), p. A13.

Lindsey, E. W., & Colwell, M. J. (2003). Preschoolers' emotional competence: Links to pretend and physical play. *Child Study Journal, 33,* 39–52.

Lindsey, E. W., & Mize, J. (2000). Parent-child physical and pretense play: Links to children's social competence. *Merrill-Palmer Quarterly, 46,* 565–591.

Linebarger, D. L., & Vaala, S. E. (2010). Screen media and language development in infants and toddlers: An ecological perspective. *Developmental Review, 30,* 176–202.

Linver, M. R., Roth, J. L., & Brooks-Gunn, J. (2009). Patterns of adolescents' participation in organized activities: Are sports best when combined with other activities? *Developmental Psychology, 45,* 354–367.

Lipsitt, L. P. (1990). Learning and memory in infants. *Merrill-Palmer Quarterly, 36,* 53–66.

Lipsitt, L. P. (2003). Crib death: A biobehavioral phenomenon. *Psychological Science, 12,* 164–170.

Lips-Wiersma, M. S. (2003). Making conscious choices in doing research on workplace spirituality: Utilizing the "holistic development model" to articulate values, assumptions and dogmas of the knower. *Journal of Organizational Change Management, 16,* 2003, 406–425.

Litiecq, B. L., Bailey, S. J., & Kurtz, M. A. (2008). Depression among rural Native American and European American grandparents rearing their grandchildren. *Journal of Family Issues, 29,* 334–356.

Liu, C. K., & Fielding, R. A. (2011). Exercise as an intervention for frailty. *Clinics in Geriatric Medicine, 27,* 101–110.

Liu, D., Gelman, S. A., & Wellman, H. M. (2007). Components of young children's trait understanding: Behavior-to-trait and trait-to-behavior predictions. *Child Development, 78,* 1543–1558.

Liu, D., Wellman, H. M., Tardif, T., & Sabbagh, M. A. (2008). Theory of mind development in Chinese children: A meta-analysis of false-belief understanding across cultures and languages. *Developmental Psychology, 44,* 523–531.

Liu, H.-M., Kuhl, P. K., & Tsao, F.-M. (2003). An association between mothers' speech clarity and infants' speech discrimination skills. *Developmental Science, 6*, F1–F10.

Liu, H.-M., Tsao, F.-M., & Kuhl, P. K. (2007). Acoustic analysis of lexical tone in Mandarin infant-directed speech. *Developmental Psychology, 43*, 912–917.

Liu, R X., Lin, W., & Chen, Z. Y. (2010). School performance, peer association, psychological and behavioral adjustments: A comparison between Chinese adolescents with and without siblings. *Journal of Adolescence, 33*, 411–417.

Livesley, W. J., & Bromley, D. B. (1973). *Person perception in childhood and adolescence.* New York: Wiley.

Livingston, G., & Cohn, D'V. (2010). *The new demography of American motherhood.* Retrieved from http://pewresearch.org/pubs/1586/changing-demographic-characteristics-american-mothers?src=prc-latest&proj=peoplepress

Lo, M., & Aziz, T. (2009). Muslim marriage goes online: The use of Internet matchmaking by American Muslims. *Journal of Religion and Popular Culture, 21*(3). Retrieved from http://www.usask.ca/relst/jrpc/art21%283%29-MuslimMarriage.html.

Lobelo, F., Dowda, M., Pfeiffer, K. A., & Pate, R. R. (2009). Electronic media exposure and its association with activity-related outcomes in female adolescents: Cross-sectional and longitudinal analyses. *Journal of Physical Activity and Health, 6*, 137–143.

LoBue, V., & DeLoache, J. S. (2010). Superior detection of threat-relevant stimuli in infancy. *Developmental Science, 13*, 221–228.

Lock, M. (1991). Contested meanings of the menopause. *Lancet, 337*, 1270–1272.

Lockl, K., & Schneider, W. (2007). Knowledge about the mind: Links between theory of mind and later metamemory. *Child Development, 78*, 148–167.

Lois, J. (2011). Gender and emotion management in the stages of edgework. In Spade, J. Z., & Valentine, C. G. (Eds.), *The kaleidoscope of gender: Prisms, patterns, and possibilities* (3rd ed., pp. 333–343). Thousand Oaks, CA: Pine Forge Press.

Longshore, D., Ellickson, P. L., McCaffrey, D. F., & St. Clair, P. A. (2007). School-based drug prevention among at-risk adolescents: Effects of ALERT Plus. *Health Education and Behavior, 34*, 651–668.

Lopata, H. Z. (1996). Widowhood and husband sanctification. In D. Klass, P. R. Silverman, & S. L. Nickman (Eds.), *Continuing bonds: New understandings of grief* (pp. 149–162). Washington, DC: Taylor & Francis.

Lord, S. E., Eccles, J. S., & McCarthy, K. A. (1994). Surviving the junior high transition: Family processes and self-perception as protective and risk factors. *Journal of Early Adolescence, 14*, 162–199.

Lovoy, L. (2001). A historical survey of the glass ceiling and the double bind faced by women in the workplace: Options for avoidance. *Law and Psychology Review, 25*, 179–203.

Low, J. (2010). Preschoolers' implicit and explicit false-belief understanding: Relations with complex syntactical mastery. *Child Development, 81*, 597–615.

Lowe, M. E., & McClement, S. E. (2010-2011). Spousal bereavement: The lived experience of young Canadian widows. *OMEGA: The Journal of Death and Dying, 62*, 127–148.

Lowry, R., Wechsler, H., Kann, L., & Collins, J. L. (2001). Recent trends in participation in physical education among U.S. high school students. *Journal of School Health, 71*, 145–152.

Lozoff, B., Wolf, A. W., & Davis, N. S. (1985). Sleep problems seen in pediatric practice. *Pediatrics, 75*, 477–483.

Lubinski, D., Benbow, C. P, Webb, R. M., & Bleske-Rechek, A. (2006). Tracking exceptional human capital over two decades. *Psychological Science, 17*, 194–199.

Luborsky, M. R., & LeBlanc, I. M. (2003). Cross-cultural perspectives on the concept of retirement: An analytic redefinition. *Journal of Cross-Cultural Gerontology, 18*, 251–271.

Lucas, J. L. (2000). Mentoring as a manifestation of generativity among university faculty. *Dissertation Abstracts International. Section A. Humanities and Social Sciences, 61(3-A),* 881.

Lucero-Liu, A. A. (2007). Exploring intersections in the intimate lives of Mexican origin women. *Dissertation Abstracts International. Section A. Humanities and Social Sciences, 68(3-A),* 1175.

Ludke, R. L., & Smucker, D. R. (2007). Racial differences in the willingness to use hospice services. *Journal of Palliative Medicine, 10,* 1329–1337.

Ludwig, J., & Phillips, D. (2007). The benefits and costs of Head Start. *SRCD Social Policy Report, 21,* 3–11, 16–18.

Luecke-Aleksa, D., Anderson, D. R., Collins, P. A., & Schmitt, K. L. (1995). Gender constancy and television viewing. *Developmental Psychology, 31,* 773–780.

Lundström, J. N., & Jones-Gotman, M. (2009). Romantic love modulates women's identification of men's body odors. *Hormones and Behavior, 55,* 280–284.

Lung, F.-W., Fan, P.-L., Chen, N. C., & Shu, B.-C. (2005). Telomeric length varies with age and polymorphisms of the MAOA gene promoter in peripheral blood cells obtained from a community in Taiwan. *Psychiatric Genetics, 15,* 31–35.

Luo, S., & Zhang, G. (2009). What leads to romantic attraction: similarity, reciprocity, security, or beauty? Evidence from a speed-dating study. *Journal of Personality, 77,* 933–964.

Luo, Y., Kaufman, L., & Baillargeon, R. (2009). Young infants' reasoning about physical events involving inert and self-propelled objects. *Cognitive Psychology, 58,* 441–486.

Luong, G., Charles, S. T., & Fingerman, K. L. (in press). Better with age: social relationships across adulthood. *Journal of Social and Personal Relationships.*

Lustbader, D., O'Hara, M. S., Wijdicks, E. F. M., MacLean, L., Tajik, W., Ying, A., et al. (2011). Second brain death examination may negatively affect organ donation. *Neurology, 76,* 119–124.

Luthans, F., Avolio, B. J., Avey, J. B., & Norman, S. M. (2007). Positive psychological capital: Measurement and relationship with performance and satisfaction. *Personnel Psychology, 60,* 541–572.

Luthar, S. S. (2006). Resilience in development: A synthesis of research across five decades. In D. Cicchetti & D. J. Cohen (Eds.), *Developmental psychopathology: Vol. 3. Risk, disorder, and adaptation* (2nd ed., pp. 739–795). Hoboken, NJ: Wiley.

Luthar, S. S., Zigler, E., & Goldstein, D. (1992). Psychosocial adjustment among intellectually gifted adolescents: The role of cognitive-developmental and experiential factors. *Journal of Child Psychology and Psychiatry and Allied Disciplines, 33,* 361–373.

Lutz, A., Slagter, H. A., Rawlings, N. B., Francis, A. D., Greischar, L. L., & Davidson, R. J. (2009). Mental training enhances attentional stability: Neural and behavioral evidence. *The Journal of Neuroscience, 29,* 13418–13427.

Luyckx, K., Soenens, B., Vansteeenkiste, M., Goossens, L., & Berzonsky, M. D. (2007). Parental psychological control and dimensions of identity formation in emerging adulthood. *Journal of Family Psychology, 21,* 546–550.

Lyness, D. (2007). *Stress.* Retrieved from http://kidshealth.org/teen/your_mind/emotions/stress.html

Lynne-Landsman, S. D., Graber, J. A., & Andrews, J. A. (2010). Do trajectories of household risk in childhood moderate pubertal timing effects on substance initiation in middle school? *Developmental Psychology, 46,* 853–868.

Lytton, H. (2000). Toward a model of family-environmental and child-biological influences on development. *Developmental Review, 20,* 150–179.

Lytton, H., & Romney, D. M. (1991). Parents' differential socialization of boys and girls: A meta-analysis. *Psychological Bulletin, 109,* 267–296.

Mabbott, D. J., Noseworthy, M., Bouffet, E., Laughlin S., & Rockel, C. (2006). White matter growth as a mechanism of cognitive development in children. *Neuroimage, 33,* 936–946.

Macatee, T. C. (2007). Psychological adjustment of adult children raised by a gay or lesbian parent. *Dissertation Abstracts International. Section B. Sciences and Engineering, 68(3-B),* 1983.

MacCallum, F., Golombok, S., & Brinsden, P. (2007). Parenting and child development in families with a child conceived through embryo donation. *Journal of Family Psychology, 21,* 278–287.

Maccoby, E. E. (1990). Gender and relationships: A developmental account. *American Psychologist, 45,* 513–520.

Maccoby, E. E. (1998). *The two sexes: Growing up apart, coming together.* Cambridge, MA: Belknap Press.

Maccoby, E. E., & Jacklin, C. N. (1974). *The psychology of sex differences.* Stanford, CA: Stanford University Press.

MacDermid, S. M., De Haan, L. G., & Heilbrun, G. (1996). Generativity in multiple roles. *Journal of Adult Development, 3,* 145–158.

Machatkova, M., Brouckova, M., Matejckova, M., Krebsova, A., Sperling, K., Vorsanova, S., et al. (2005). QF-PCR examination of parental and meiotic origin of trisomy 21 in central and eastern Europe. *Journal of Histochemistry and Cytochemistry, 53,* 371–373.

Maciejewski, P. K., Zhang, B, Block, S. D., & Prigerson, H. G. (2007). An empirical examination of the stage theory of grief. *Journal of the American Medical Association, 297,* 716–723.

Mackey, A. P., Hill, S. S., Stone, S. I., & Bunge, S. A. (2011). Differential effects of reasoning and speed training in children. *Developmental Science, 14,* 582–590.

Mackey, R. A., Diemer, M. A., & O'Brien, B. A. (2004). Relational factors in understanding satisfaction in the lasting relationships of same-sex and heterosexual couples. *Journal of Homosexuality, 47,* 111–136.

MacWhinney, B. (1998). Models of the emergence of language. *Annual Review of Psychology, 49,* 199–227.

Maeder, E. M., Wiener R. L., & Winter, R. (2007). Does a truck driver see what a nurse sees? The effects of occupation type on perceptions of sexual harassment. *Sex Roles, 56,* 801–810.

Maggie, S., Ostry, A., Tansey, J., Dunn, J., Hershler, R., Chen, L., & Hertzman, C. (2008). Paternal psychosocial work conditions and mental health outcomes: A case-control study. *BMC Public Health, 8,* 104.

Magnuson, K., & Duncan, G. (2006). The role of family socioeconomic resources in black and white test score gaps among young children. *Developmental Review, 26,* 365–399.

Maguire, A. M., et al. (2009). Age-dependent effects of RPE65 gene therapy for Leber's congenital amaurosis: A phase 1 dose-escalation trial. *Lancet, 374,* 1597–1605.

Maguire, E. A., Woollett, K., & Spiers, H. J. (2006). London taxi drivers and bus drivers: A structural MRI and neuropsychological analysis. *Hippocampus, 16,* 1091–1101.

Mahon, N. E., Yarcheski, A., Yarcheski, T., Cannella, B. L., & Hanks, M. M. (2006). A meta-analytic study of predictors for loneliness during adolescence. *Nursing Research, 55,* 308–315.

Maiden, R. J., Peterson, S. A., Caya, M., & Hayslip, B. (2003). Personality changes in the old-old: A longitudinal study. *Journal of Adult Development, 10,* 31–39.

Maimburg, R. D., Vaeth, M., Durr, J., Hvidman, L., & Olsen, J. (2010). Randomised trial of structured antenatal training sessions to improve the birth process. *BJOG—An International Journal of Obstetrics and Gynaecology, 117,* 921–927.

Malach-Pines, A. (2005). The Burnout Measure, Short Version. *International Journal of Stress Management, 12,* 78–88.

Males, M. (2009). Does the adolescent brain make risk taking inevitable?: A skeptical appraisal. *Journal of Adolescent Research, 24,* 3–20.

Males, M. (2010). Is jumping off the roof always a bad idea? A rejoinder on risk taking and the adolescent brain. *Journal of Adolescent Research, 25,* 48–63.

Malinosky-Rummell, R., & Hansen, D. J. (1993). Long-term consequences of childhood physical abuse. *Psychological Bulletin, 114,* 68–79.

Malkinson, R., & Bar-Tur, L. (2004–2005). Long term bereavement processes of older parents: The three phases of grief. *Omega: The Journal of Death and Dying, 50,* 103–129.

Mallon, B. (2008). *Dying, death, and grief: Working with adult bereavement.* Thousand Oaks, CA: Sage.

Malone, K., Stewart, S. D., Wilson, J., & Korsching, P. F. (2010). Perceptions of financial well-being among American women in diverse families. *Journal of Family and Economic Issues, 31,* 63–81.

Malone, M. L., & Camp, C. J. (2007). Montessori-Based Dementia Programming: Providing tools for engagement. *Dementia: The International Journal of Social Research and Practice, 6,* 150–157.

Malone, P. A. (2010). *The impact of peer death on adolescent girls: an efficacy study of the Adolescent Grief and Loss group.* Doctoral dissertation submitted to The University of Texas at Austin.

Malti, T., Gummerum, M., Keller, M., & Buchmann, M. (2009). Children's moral motivation, sympathy, and prosocial behavior. *Child Development, 80,* 442–460.

Mancini, A. D., & Bonanno, G. A. (2010). Resilience to potential trauma: Toward a lifespan approach. In J. W. Reich, A. Zautra, & J. S. Hall (Eds.), *Handbook of adult resilience* (pp. 258–280). New York: Guilford Press.

Mandara, J., Gaylord-Harden, N. K., Richard, M H., & Ragsdale, B. L. (2009). The effects of changes in racial identity and self-esteem on changes in African American adolescents' mental health. *Child Development, 80,* 1660–1675.

Mandel, D. R., Jusczyk, P. W., & Pisoni, D. B. (1995). Infants' recognition of the sound patterns of their own names. *Psychological Science, 6,* 314–317.

Manekin, M. (2008). *Two victims in car crash sue elderly driver.* Retrieved from http://sanmateodailynews.com/article/smdn/2008-4-19-suit

Mange, A. P., & Mange, E. J. (1990). *Genetics: Human aspects* (2nd ed.). Sunderland, MA: Sinhauer Associates.

Mangelsdorf, S. C. (1992). Developmental changes in infant-stranger interaction. *Infant Behavior and Development, 15,* 191–208.

Mangelsdorf, S. C., Shapiro, J. R., & Marzolf, D. (1995). Developmental and temperamental differences in emotional regulation in infancy. *Child Development, 66,* 1817–1828.

Mangelsdorf, S., Gunnar, M., Kestenbaum, R., Lang, S., & Andreas, D. (1990). Infant proneness-to-distress temperament, maternal personality, and mother–infant attachment: Associations and goodness of fit. *Child Development, 61,* 820–831.

Mantler, J., Matejicek, A., Matheson, K., & Anisman, H. (2005). Coping with employment uncertainty: A comparison of employed and unemployed workers. *Journal of Occupational Health Psychology, 10,* 200–209.

Maple, M., Edwards, H., Plummer, D., & Minichiello, V. (2010). Silenced voices: Hearing the stories of parents bereaved through the suicide death of a young adult child. *Health and Social Care in the Community, 18,* 241–248.

Maratsos, M. (1998). The acquisition of grammar. In W. Damon (Ed.), *Handbook of child psychology* (Vol. 2, pp. 421–455). New York: Wiley.

Marcia, J. E. (1980). Identity in adolescence. In J. Adelson (Ed.), *Handbook of adolescent psychology.* (pp. 159–187). New York: Wiley.

Marcus, G. F., Pinker, S., Ullman, M., Hollander, M., Rosen, T. J., & Xu, F. (1992). Overregularization in language acquisition. *Monographs of the Society for Research in Child Development, 58*(4, Serial No. 228).

Margrett, J. A., Allaire, J. C., Johnson, T. L., Daugherty, K. E., & Weatherbee, S. R. (2010). Everyday problem solving. In J. C. Cavanaugh & C. K. Cavanaugh (Eds.), *Aging in America: Vol. 1: Psychological aspects* (pp. 79–101). Santa Barbara, CA: Praeger Perspectives.

Markovits, H., Benenson, J., & Dolenszky, E. (2001). Evidence that children and adolescents have internal models of peer interactions that are gender differentiated. *Child Development, 72,* 879–886.

Marks, A. K., Patton, F., & García Coll, C. (2011). Being bicultural: A mixed-methods study of adolescents' implicitly and explicitly measured multiethnic identities. *Developmental Psychology, 47,* 270–288.

Markus, H., & Nurius, P. (1986). Possible selves. *American Psychologist, 41,* 954–969.

Marschik, P. B., Einspieler, C., Strohmeier, A., Plienegger, J., Garzarolli, B., & Prechtl, H. F. R. (2008). From the reaching behavior at 5 months of age to hand preference at preschool age. *Developmental Psychobiology, 50,* 511–518.

Marsh, H. W. (1991). Employment during high school: Character building or a subversion of academic goals? *Sociology of Education, 64,* 172–189.

Marsh, H. W., & Yeung, A. S. (1997). Causal effects of academic self-concept on academic achievement: Structural equation models of longitudinal data. *Journal of Educational Psychology, 89,* 41–54.

Marsh, K., & Musson, G. (2008). Men at work and at home: Managing emotion in telework. *Gender, Work & Organization, 15,* 31–48.

Marshal, M. P., Friedman, M. S., Stall, R., King, K. M., Miles, J., Gold, M. A., et al. (2008). Sexual orientation and adolescent substance use: A meta-analysis and methodological review. *British Journal of Addiction, 103,* 546–556.

Marsiske, M., & Margrett, J. A. (2006). Everyday problem solving with decision making. In J. E. Birren & K. W. Schaie (Eds.), *Handbook of the psychology of aging* (6th ed., pp. 315–342). Boston, MA: Academic Press.

Martens, A., Goldenberg, J. L., & Greenberg, J. (2005). A terror management perspective on ageism. *Journal of Social Issues, 61,* 223–239.

Martin, C. L., & Fabes, R. A. (2001). The stability and consequences of young children's same-sex peer interactions. *Developmental Psychology, 37,* 431–446.

Martin, C. L., & Halverson, C. F. (1987). The roles of cognition in sex role acquisition. In D. B. Carter (Ed.), *Current conceptions of sex roles and sex typing: Theory and research* (pp. 123–137). New York: Praeger Publishers.

Martin, C. L., & Ruble, D. (2004). Children's search for gender cues: Cognitive perspectives on gender development. *Current Directions in Psychological Science, 13,* 67–70.

Martin, C. L., Fabes, R. A., Evans, S. M., & Wyman, H. (1999). Social cognition on the playground: Children's beliefs about playing with girls versus boys and their relationships to sex-segregated play. *Journal of Social and Personal Relationships, 16,* 751–772.

Martin, C. L., Ruble, D. N., & Szkrybalo, J. (2002). Cognitive theories of early gender development. *Psychological Bulletin, 128,* 903–933.

Martin, J. L., & Ross, H. S. (2005). Sibling aggression: Sex differences and parents' reactions. *International Journal of Behavioral Development, 29,* 129–138.

Martin, M., Long, M. V., & Poon, L. W. (2003). Age changes and differences in personality traits and states of the old and very old. *Journal of Gerontology: Psychological Sciences, 57B,* 144–152.

Mascolo, M. F., Fischer, K. W., & Li, J. (2003). Dynamic development of component systems of emotions: Pride, shame, and guilt in China and the United States. In R. J. Davidson, K. R. Scherer, and H. H. Goldsmith (Eds.), *Handbook of affective sciences* (pp. 375–408). Oxford, England: Oxford University Press.

Mash, E. J., & Wolfe, D. A. (2010). *Abnormal child psychology* (4th ed). Belmont, CA: Cengage.

Masheter, C. (1997). Healthy and unhealthy friendship and hostility between ex-spouses. *Journal of Marriage and Family, 59,* 463–475.

Masunaga, H., & Horn, J. (2001). Expertise and age-related changes in components of intelligence. *Psychology and Aging, 16,* 293–311.

Masur, E. F. (1995). Infants' early verbal imitation and their later lexical development. *Merrill-Palmer Quarterly, 41,* 286–306.

Matsumoto, A. K., Bathon, J., & Bingham III, C. O. (2010). *Rheumatoid arthritis treatment.* Retrieved from http://www.hopkins-arthritis.org/arthritis-info/rheumatoid-arthritis/rheum_treat.html

Mattys, S. L., & Jusczyk, P. W. (2001). Phonotactic cues for segmentation of fluent speech by infants. *Cognition, 78,* 91–121.

Mattys, S. L., Jusczyk, P. W., Luce, P. A., & Morgan, J. L. (1999). Phonotactic and prosodic effects on word segmentation in infants. *Cognitive Psychology, 38,* 465–494.

Matud, M. P. (2004). Gender differences in stress and coping styles. *Personality and Individual Differences, 37,* 1401–1415.

Maughan, A., & Cicchetti, D. (2002). Impact of child maltreatment and interadult violence on children's emotion regulation abilities and socioemotional adjustment. *Child Development, 73,* 1525–1542.

Maume, D. J., Jr. (2004). Is the glass ceiling a unique form of inequality? Evidence from a random-effects model of managerial attainment. *Work and Occupations, 31,* 250–274.

Maurer, T. W., & Robinson, D. W. (2008). Effects of attire, alcohol, and gender on perceptions of date rape. *Sex Roles, 58,* 423–434.

Maye, J., Weiss, D. J., & Aslin, R. N. (2008). Statistical phonetic learning in infants: Facilitation and feature generalization. *Developmental Science, 11,* 122–134.

Mayer, H. U. (2009). New directions in life course research. *Annual Review of Sociology, 35,* 413–433.

Mayer, J. D., Roberts, R. D., Barsade, S. G. (2008). Human abilities: Emotional intelligence. *Annual Review of Psychology, 59,* 507–536.

Mayer, J. D., Salovey, P., & Caruso, D. R. (2008). Emotional intelligence: New ability or eclectic traits. *American Psychologist, 63,* 503–517.

Maynard, A. E. (2002). Cultural teaching: The development of teaching skills in Maya sibling interactions. *Child Development, 73,* 969–982.

Mayo Clinic. (2007). *Rheumatoid arthritis.* Retrieved from http://www.mayoclinic.com/health/rheumatoid-arthritis/DS00020/DSECTION=causes

Mayo Clinic. (2009). *Nutrition and healthy eating.* Retrieved from http://www.mayoclinic.com/health/nutrition-and-healthy-eating/MY00431

Mayo Clinic. (2010a). *Alcoholism.* Retrieved from http://www.mayoclinic.com/health/alcoholism/DS00340

Mayo Clinic. (2010b). *Perimenopause.* Retrieved from http://www.mayoclinic.com/health/perimenopause/DS00554

Mayo Clinic. (2010c). *Hormone therapy: Is it right for you?* Retrieved from http://www.mayoclinic.com/health/hormone-therapy/WO00046

Mayo Clinic. (2010d). *Lifestyle and home remedies.* Retrieved from http://www.mayoclinic.com/health/menopause/DS00119/DSECTION=lifestyle-and-home-remedies

Mayo Clinic. (2010e). *Alternative medicine.* Retrieved from http://www.mayoclinic.com/health/menopause/DS00119/DSECTION=alternative-medicine

Mayo Clinic. (2010f). *Aerobic exercise: Top 10 reasons to get physical.* Retrieved from http://www.mayoclinic.com/health/aerobic-exercise/EP00002/NSECTIONGROUP=2

Mayo Clinic. (2010g). *Osteoporosis treatment puts brakes on bone loss.* Retrieved from http://www.mayoclinic.com/health/osteoporosis-treatment/WO00127/METHOD=print

Mazur, E., Wolchik, S. A., Virdin, L., Sandler, I. N., & West, S. G. (1999). Cognitive moderators of children's adjustment to stressful divorce events: The role of negative cognitive errors and positive illusions. *Child Development, 70,* 231–245.

Mazzarella, S. R. (2007). Cyberdating success stories and the mythic narrative of living "Happily-Ever-After with the One." In M.-L. Galician & D. L. Merskin (Eds.), *Critical thinking about sex, love, and romance in the mass media* (pp. 23–37). Mahwah, NJ: Erlbaum.

McAdams, D. P. (2001a). Generativity at midlife. In M. E. Lachman (Ed.), *Handbook of midlife development* (pp. 395–443). New York: Wiley.

McAdams, D. P. (2008). Personal narratives and the life story. In O. P. John, R. W. Robins, & L. A. Pervin. (Eds.) *Handbook of personality: Theory and research.* (3rd ed., pp. 241–61). New York: Guilford.

McAdams, D. P. (2009). The problem of meaning in personality psychology from the standpoints of dispositional traits, characteristic adaptations, and life stories. *Japanese Journal of Psychology, 18,* 173–186.

McAdams, D. P., & Olson, B. D. (2010). Personality development: Continuity and change over the life course. *Annual Review of Psychology, 61,* 517–542.

McAlaney, J., Bewick, B., & Hughes, C. (2011). The international development of the 'social norms' approach to drug education and prevention. *Drugs: Education, Prevention, and Policy, 18,* 81-89.

McCall, R. B. (1979). *Infants.* Cambridge, MA: Harvard University Press.

McCarty, M. E., & Ashmead, D. H. (1999). Visual control of reaching and grasping in infants. *Developmental Psychology, 35,* 620–631.

McCarty, M. E., Clifton, R. K., Ashmead, D. H., Lee, P., & Goubet, N. (2001). How infants use vision for grasping objects. *Child Development, 72,* 973–987.

McCleese, C. S., & Eby, L. T. (2006). Reactions to job content plateaus: Examining role ambiguity and hierarchical plateaus as moderators. *Career Development Quarterly, 55,* 64–76.

McClinton, B. E. (2010). *Preparing for the third Age: A retirement planning course outline for lifelong learning programs.* Master's thesis from California State University, Long Beach. Retrieved from http://gradworks.umi.com/1486345.pdf

McClure, E. B. (2000). A meta-analytic review of sex differences in facial expression processing and their development in infants, children, and adolescents. *Psychological Bulletin, 126,* 424–453.

McCollum, L., & Pincus, T. (2009). A biopsychosocial model to complement a biomedical model: Patient questionnaire data and socioeconomic status usually are more significant than laboratory tests and imaging studies in prognosis of rheumatoid arthritis. *Rheumatoid Disease Clinics of North America, 35,* 699–712.

McConatha, J. T., Stoller, P., & Oboudiat, F. (2001). Reflections of older Iranian women: Adapting to life in the United States. *Journal of Aging Studies, 15,* 369–381.

McCormick, C. B. (2003). Metacognition and learning. In I. B. Weiner (Editor-in-Chief) and W. M. Reynolds & G. E. Miller (Vol. Eds.), *Handbook of psychology: Vol. 7. Educational Psychology* (pp. 79–102). New York: Wiley.

McCrae, R. R. (2002). The maturation of personality psychology: Adult personality development and psychological well-being. *Journal of Research in Personality, 36,* 307–317.

McCrae, R. R., & Costa, P. T. (1994). The stability of personality: Observation and evaluations. *Current Directions in Psychological Sciences, 3,* 173–175.

McCrae, R. R., & Terracciano, A. (2005). Universal features of personality traits from the observer's perspective: Data from 50 cultures. *Journal of Personality & Social Psychology, 88,* 547–561.

McCrink, K., & Wynn, K. (2007). Ratio abstraction by 6-month-old infants. *Psychological Science, 18,* 740–745.

McCutchen, D., Covill, A., Hoyne, S. H, & Mildes, K. (1994). Individual differences in writing: Implications of translating fluency. *Journal of Educational Psychology, 86,* 256–266.

McCutchen, D., Francis, M., & Kerr, S. (1997). Revising for meaning: Effects of knowledge and strategy. *Journal of Educational Psychology, 89,* 667–676.

McDill, T., Hall, S. K., & Turell, S. C. (2006). Aging and creating families: Never-married heterosexual women over forty. *Journal of Women & Aging, 18,* 37–50.

McDonald, K. L., Bowker, J. C., Rubin, K. H., Laursen, B., & Duchene, M. S. (2010). Interactions between rejection sensitivity and supportive relationships in the prediction of adolescents' internalizing difficulties. *Journal of Youth and Adolescence, 39,* 563–574.

McElwain, N. L., Booth-LaForce, C., Lansford, J. E., Wu, X., & Justin Dyer, W. (2008). A process model of attachment-friend linkages: Hostile attribution biases, language ability and mother-child affective mutuality as intervening mechanisms. *Child Development, 79,* 1891–1906.

McEwen, B. S., & Gianaros, P. J. (2010). Central role of the brain in stress and adaptation: Links to socioeconomic status, health, and disease. *Annals of the New York Academy of Sciences, 1186,* 190–222.

McGarry, K., & Schoeni, R. F. (2005). Widow(er) poverty and out-of-pocket medical expenditures near the end of life. *Journal of Gerontology: Social Sciences, 60,* S160–S168.

McGill, D., Brown, K., Haley, J., Schieber, S., & Warshawsky, M. (2010). *Fundamentals of private pensions* (9th ed.). New York: Oxford University Press.

McGraw, L. A., & Walker, A. J. (2004). Negotiating care: Ties between aging mothers and their caregiving daughters. *Journals of Gerontology: Social Sciences, 59B,* S324–S332.

McGraw, M. B. (1935). *Growth: A study of Johnny and Jimmy.* East Norwalk, CT: Appleton-Century-Crofts.

McGuckin, T. L. (2007). *Examining the patterns of alcohol use on campus and the perceptions of faculty related to student alcohol use.* Unpublished doctoral dissertation, University of West Florida.

McGuire, L. C., & Cavanaugh, J. C. (1992, April). *Objective measures versus spouses' perceptions of cognitive status in dementia patients.* Paper presented at the biennial Cognitive Aging Conference, Atlanta, GA.

McGuire, S., & Shanahan, L. (2010). Sibling experiences in diverse family contexts. *Child Development Perspectives, 4,* 72–79.

McHale, J. P., Laurette, A., Talbot, J., & Pourquette, C. (2002). Retrospect and prospect in the psychological study of coparenting and family group process. In J. P. McHale & W. Grolnick (Eds.), *Retrospect and prospect in the psychological study of families* (pp. 127–165). Mahwah, NJ: Erlbaum.

McHale, S. M., Kim, J. Y., Whiteman, S. D., & Crouter, A. C. (2004). Links between sex-typed activities in middle childhood and gender development in early adolescence. *Developmental Psychology, 40,* 868–881.

McHale, S. M., Whiteman, S. D., Kim, J.-Y., & Crouter, A. C. (2007). Characteristics and correlates of sibling relationships in two-parent African American families. *Journal of Family Psychology, 21,* 227–235.

McIntosh, W. D., Locker, L., Briley, K., Ryan, R., & Scott, A. (2011). What do older adults seek in their potential romantic partners? Evidence from online personal ads. *The International Journal of Aging and Human Development, 72,* 67–82.

McKay, K., & Ross, L. E. (2010). The transition to adoptive parenthood: A pilot study of parents adopting in Ontario, Canada. *Children and Youth Services Review, 32,* 604-610.

McKee-Ryan, F., Song, Z., Wanberg, C. R., & Kinicki, A. J. (2005). Psychological and physical well-being during unemployment: A meta-analytic study. *Journal of Applied Psychology, 90,* 53–76.

McKenna, P., Jefferies, L., Dobson, A., & Frude, N. (2004). The use of a cognitive battery to predict who will fail an on-road driving test. *British Journal of Clinical Psychology, 43,* 325–336.

Mckenzie, P. T. (2003). *Factors of Successful Marriage: Accounts from Self-Described Happy Couples.* In partial fulfillment of the requirements for the degree of doctor of philosophy, Howard University.

McKusick, V. A. (1995). *Mendelian inheritance in man: Catalogs of autosomal dominant, automosal recessive, and X-linked phenotypes* (10th ed.). Baltimore: Johns Hopkins University Press.

McLanahan, S. (1999). Father absence and the welfare of children. In E. M. Hetherington (Ed.), *Coping with divorce, single parenting, and remarriage: A risk and resiliency perspective* (pp. 117–145). Mahwah, NJ: Erlbaum.

McLaughlin, C. (2010). Mentoring: What is it? How do we do it and how do we get more of it? *Health Services Research, 45,* 871–884.

McLellan, J. A., & Youniss, J. (2003). Two systems of youth service: Determinants of voluntary and required youth community service. *Journal of Youth and Adolescence, 32,* 47–58.

McMurray, B. (2007) Defusing the childhood vocabulary explosion. *Science, 317,* 631.

Meaney, M. J. (2010) Epigenetics and the biological definition of gene x environment interactions. *Child Development, 81,* 41–79.

Medicare.gov. (2011). *Medicare basics.* Retrieved from http://www.medicare.gov/navigation/medicare-basics/medicare-basics-overview.aspx

Meeus, W., Oosterwegel, A., & Vollebergh, W. (2002). Parental and peer attachment and identity development in adolescence. *Journal of Adolescence, 25,* 93–106.

Meeus, W., van de Schoot, R., Keijsers, L., Schwartz, S. J., & Branje, S. (2010). On the progression and stability of adolescent identity formation: A five-wave longitudinal study in early-to-middle and middle-to-late adolescence. *Child Development, 81,* 1565–1581.

Mehta, C. M., & Strough, J. (2009). Sex segregation in friendships and normative contexts across the life span. *Developmental Review, 29,* 201–220.

Mehta, K. K. (1997). The impact of religious beliefs and practices on aging: A cross-cultural comparison. *Journal of Aging Studies, 11,* 101–114.

Meijer A. M., & van den Wittenboer, G. L. H. (2007). Contribution of infants' sleep and crying to marital relationship of first-time parent couples in the 1st year after childbirth. *Journal of Family Psychology, 21,* 49–57.

Melby, J. N., Conger, R. D., Fang, S., Wickrama, K. A. S., & Conger, K. J. (2008). Adolescent family experiences and educational attainment during early adulthood. *Developmental Psychology, 44,* 1519–1536.

Melhado, L. W., & Byers, J. F. (2011). Patients' and surrogates' decision-making characteristics: withdrawing, withholding, and continuing life-sustaining treatments. *Journal of Hospice and Palliative Nursing, 13,* 16–28.

Meltzoff, A. N., & Moore, M. K. (1989). Imitation in newborn infants: Exploring the range of gestures imitated and the underlying mechanisms. *Developmental Psychology, 25,* 954–962.

Meltzoff, A. N., & Moore, M. K. (1994). Imitation, memory, and the representation of persons. *Infant Behavior and Development, 17,* 83–99.

Mendle, J., Turkheimer, E., & Emery, R. E. (2007). Detrimental psychological outcomes associated with early pubertal timing in adolescent girls. *Developmental Review, 27,* 151–171.

Mennella, J. A., & Beauchamp, G. K. (1996). The human infant's response to vanilla flavors in mother's milk and formula. *Infant Behavior and Development, 19,* 13–19.

Mennella, J. A., Jagnow, C. P., & Beauchamp, G. K. (2001). Prenatal and postnatal flavor learning by human infants. *Pediatrics, 107,* E88.

Mennella, J., & Beauchamp, G. K. (1997). The ontogeny of human flavor perception. In G. K. Beauchamp & L. Bartoshuk (Eds.), *Tasting and smelling. Handbook of perception and cognition.* San Diego, CA: Academic Press.

Mensch, B. S., Singh, S., & Casterline, J. B. (2006). Trends in the timing of first marriage among men and women in the developing world. In C. B. Lloyd, J. R. Behrman, N. P. Stromquist, & B. Cohen (Eds.), *The changing transitions to adulthood in developing countries: Selected studies* (pp. 1180–171). Washington, DC: National Research Council.

Mercer, J. (2011). Attachment theory and its vicissitudes: Toward an updated theory. *Theory and Psychology, 21,* 25–45.

Mercken, L., Candel, M., Willems, P., & de Vries, H. (2007). Disentangling social selection and social influence effects on adolescent smoking: The importance of reciprocity in friendships. *British Journal of Addiction, 102,* 1483–1492.

Mervis, C. B., & Johnson, K. E. (1991). Acquisition of the plural morpheme: A case study. *Developmental Psychology, 27,* 222–235.

Metcalfe, J. S., McDowell, K., Chang, T.-Y., Chen, L.-C., Jeka, J. J., & Clark, J. E. (2005). Development of somatosensory-motor integration: An event-related analysis of infant posture in the first year of independent walking. *Developmental Psychobiology, 46,* 19–35.

Meyer, J. F. (2007). Confucian "familism" in America. In D. S. Browning & D. A. Clairmont (Eds.), *American religions and the family: How faith traditions cope with modernization and democracy* (pp. 168–184). New York: Columbia University Press.

Midgette, E., Haria, P., & MacArthur, C. (2008). The effects of content and audience awareness goals for revision on the persuasive essays of fifth- and eighth-grade students. *Reading and Writing, 21,* 131–151.

Mietkiewicz, M.-C., & Venditti, L. (2004). Les arrière-grands-pères le point de vue de leurs arrière-petits-enfants [Great-grandfathers from their great-grandchildren's point of view]. *Psychologie & NeuroPsychiatrie du Vieillissement, 2,* 275–283.

Milberger, S., Biederman, J., Faraone, S. V., Guite, J., & Tsuang, M. T. (1997). Pregnancy, delivery and infancy complication, and attention deficit hyperactivity disorder: Issues of gene-environment interaction. *Biological Psychiatry, 41,* 65–75.

Miles, J. (2009). *Autobiographical reflection and perspective transformation in adult learners returning to study: research in progress.* Retrieved from http://www.avetra.org.au/papers-2009/papers/37.00.pdf

Miller, B. C., Fan, X., Christensen, M., Grotevant, H. D., & van Dulmen, M. (2000). Comparisons of adopted and nonadopted adolescents in a large, nationally representative sample. *Child Development, 71,* 1458–1473.

Miller, G. E., & Chen, E. (2010). Harsh family climate in early life presages the emergence of proinflammatory phenotype in adolescence. *Psychological Science, 21,* 848–856.

Miller, J. G., & Bersoff, D. M. (1992). Culture and moral judgment: How are conflicts between justice and interpersonal responsibilities resolved? *Journal of Personality & Social Psychology, 62,* 541–554.

Miller, K. I., Shoemaker, M. M., Willyard, J., & Addison, P. (2008). Providing care for elderly parents: A structurational approach to family caregiver identity. *Journal of Family Communication, 8,* 19–43.

Miller, N. B., Smerglia, V. L., & Bouchet, N. (2004). Women's adjustment to widowhood: Does social support matter? *Journal of Women & Aging, 16,* 149–167.

Miller, P. M., Danaher, D. L., & Forbes, D. (1986). Sex-related strategies of coping with interpersonal conflict in children aged five to seven. *Developmental Psychology, 22,* 543–548.

Miller, S. A. (2009). Children's understanding of second-order mental states. *Psychological Bulletin, 135,* 749–773.

Miller, T. W., Nigg, J. T., & Miller, R. L. (2009). Attention deficit hyperactivity disorder in African American children: What can be learned from the past ten years? *Clinical Psychology Review, 29,* 77–86.

Miller, V. L., & Martin, A. M. (2008). The Human Genome Project: Implications for families. *Health and Social Work, 33,* 73–76.

Miller-Martinez, D., & Wallace, S. P. (2007). Structural contexts and life-course processes in the social networks of older Mexican immigrants in the United States. In S. Carmel, C. Morse, & F. Torres-Gil (Eds.), *Lessons on aging from three nations: Volume I. The art of aging well* (pp. 141–154). Amityville, NY: Baywood.

Millstein, S. G., & Halpern-Felsher, B. L. (2002). Judgments about risk and perceived invulnerability in adolescents and young adults. *Journal of Research on Adolescence, 12,* 399–422.

Ministry of Internal Affairs and Communications. (2010). *Statistical handbook of Japan 2010.* Retrieved from http://www.stat.go.jp/english/data/handbook/index.htm

Minnotte, K. L. (2010). *Methodologies of assessing marital success.* Retrieved from http://wfnetwork.bc.edu/encyclopedia_entry.php?id=16779&area=All

Mirabella, R. L. (2001). Determinants of job satisfaction in psychologists. *Dissertation Abstracts International. Section B. Sciences and Engineering, 61(12-B),* 6714.

Mischel, W. (1970). Sex-typing and socialization. In P. H. Mussen (Ed.), *Carmichael's manual of child psychology* (Vol. 2, pp. 3–72). New York: Wiley.

Missildine, W., Feldstein, G., Punzalan, J. C., & Parsons, J. T. (2005). S/he loves me, s/he loves me not: Questioning heterosexist assumptions of gender differences for romantic and sexually motivated behaviors. *Sexual Addiction & Compulsivity, 12,* 65–74.

Mitchell, B. A. (2006). *The boomerang age: Transitions to adulthood in families.* New Brunswick, NJ: AldineTransaction.

Mitchell, B. A. (2010). Happiness in midlife parental roles: A contextual mixed methods analysis. *Family Relations, 59,* 326–339.

Mitchell, C. V. (2000). Managing gender expectations: A competency model for women in leadership. *Dissertation Abstracts International. Section B. Sciences and Engineering, 61(3-B),* 1682.

Mitchell, K. J., & Johnson, M. K. (2009). Source monitoring 15 years later: What have we learned from fMRI about the neural mechanisms of source memory? *Psychological Bulletin, 135,* 638–677.

Mitchell, L. M., & Messner, L. (2003–2004). Relative child care: Supporting the providers. *Journal of Research in Childhood Education, 18,* 105–113.

Mix, K. S., Huttenlocher, J., & Levine, S. C. (2002). Multiple cues for quantification in infancy: Is number one of them? *Psychological Bulletin, 128,* 278–294.

Mize, J., & Ladd, G. W. (1990). A cognitive social-learning approach to social skill training with low-status preschool children. *Developmental Psychology, 26,* 388–397.

Mize, J., & Pettit, G. S. (1997). Mothers' social coaching, mother-child relationship style, and children's peer competence: Is the medium the message? *Child Development, 68,* 312–332.

Mize, J., Pettit, G. S., & Brown, E. G. (1995). Mothers' supervision of their children's peer play: Relations with beliefs, perceptions, and knowledge. *Developmental Psychology, 31,* 311–321.

Mizes, J., & Palermo, T. M. (1997). Eating disorders. In M. Hersen & R. T. Ammerman (Eds.), *Handbook of prevention and treatment with children and adolescents: Intervention in the real world context.* (pp 572–603). New York: Wiley.

Moeller, J. R., Lewis, M. M., & Werth, J. L., Jr. (2010). End of life issues. In J. C. Cavanaugh & C. K. Cavanaugh (Eds.), *Aging in America: Vol. 1: Psychological aspects* (pp. 202–231). Santa Barbara, CA: Praeger Perspectives.

Moen, P. (1999). *The Cornell couples and careers study.* Ithaca, NY: Cornell University.

Moen, P., & Roehling, P. (2005). *The career mystique: Cracks in the American dream.* Lanham, MD: Rowman & Littlefield.

Moen, P., Fields, V., Meador, R., & Rosenblatt, H. (2000a). Fostering integration: A case study of the Cornell Retirees Volunteering in Service (CRVIS) program. In K. Pillemer & P. Moen (Eds.), *Social integration in the second half of life* (pp. 247–264). Baltimore: Johns Hopkins University Press.

Moen, P., Fields, V., Quick, H. E., & Hofmeister, H. (2000b). A life course approach to retirement and social integration. In K. Pillemer & P. Moen (Eds.), *Social integration in the second half of life* (pp. 75–107). Baltimore, MD: Johns Hopkins University Press.

Moerk, E. L. (2000). *The guided acquisition of first language skills.* Westport, CT: Ablex.

Moffitt, T. E. (1993). Adolescence-limited and life-course-persistent antisocial behavior: A developmental taxonomy. *Psychological Review, 100,* 674–701.

Moffitt, T. E. (2005). The new look of behavioral genetics in developmental psychopathology: Gene-environment interplay in antisocial behaviors. *Psychological Bulletin, 131,* 533–554.

Moffitt, T. E., & Caspi, A. (2005). Life-course persistent and adolescence-limited antisocial males: Longitudinal follow-up to adulthood. In D. M. Stoff & E. J. Susman (Eds.), *Developmental psychobiology of aggression* (pp. 161–186). New York: Cambridge University Press.

Moffitt, T. E., Caspi, A., Belsky, J., & Silva, P. A. (1992). Childhood experience and the onset of menarche: A test of a sociobiological model. *Child Development, 63,* 47–58.

Molfese, D. L., & Burger-Judisch, L. M. (1991). Dynamic temporal-spatial allocation of resources in the human brain: An alternative to the static view of hemisphere differences. In F. L. Ketterle (Ed.), *Cerebral laterality: Theory and research. The Toledo symposium.* Hillsdale, NJ: Erlbaum.

Molina, B. S. G., et al. (2009). The MTA at 8 years: Prospective follow-up of children treated for combined-type ADHD in a multisite study. *Journal of the American Academy of Child and Adolescent Psychiatry, 48,* 484–500.

Molloy, L. E., Gest, S. D., & Rulison, K. L. (2011). Peer influences on academic motivation: Exploring multiple methods of assessing youths' most "influential" peer relationships. *Journal of Early Adolescence, 31,* 13–40.

Monahan, K. C., Lee, J. M., & Steinberg, L. (2011). Revisiting the impact of part-time work on adolescent adjustment: Distinguishing between selection and socialization using propensity score matching. *Child Development, 82,* 96–112.

Monden, C. (2007). Partners in health? Exploring resemblance in health between partners in married and cohabiting couples. *Sociology of Health & Illness, 29,* 391–411.

Monk, C., Fifer, W. P., Myers, M. M., Sloan, R. P., Trien, L., & Hurtado, A. (2000). Maternal stress responses and anxiety during pregnancy: Effects on fetal heart rate. *Developmental Psychobiology, 36,* 67–77.

Montague, D. P., & Walker-Andrews, A. S. (2001). Peekaboo: A new look at infants' perception of emotion expressions. *Developmental Psychology, 37,* 826–838.

Montgomery, M. J. (2005). Psychosocial intimacy and identity: From early adolescence to emerging adulthood. *Journal of Adolescent Research, 20,* 346–374.

Moore, C. (2007). Understanding self and others in the second year. In C. A. Brownell & C. B. Kopp (Eds.), *Socioemotional development in the toddler years.* New York: Guilford.

Moore, K. D. (2005). Using place rules and affect to understand environmental fit: A theoretical exploration. *Environment and Behavior, 37,* 330–363.

Moore, K. L., & Persaud, T. V. N. (1993). *Before we are born* (4th ed.). Philadelphia: Saunders.

Moore, M. R., & Brooks-Gunn, J. (2002). Adolescent parenthood. In M. H. Bornstein (Ed.), *Handbook of parenting: Vol. 3. Being and becoming a parent* (2nd ed., pp. 173–214). Mahwah, NJ: Erlbaum.

Moorman, S. M., & Greenfield, E. A. (2010). Personal relationships in later life. In J. C. Cavanaugh & C. K. Cavanaugh (Eds.), *Aging in America: Vol. 3: Societal issues* (pp. 20–52). Santa Barbara, CA: Praeger Perspectives.

Morahan-Martin, J., & Schumacher, P. (2003). Loneliness and social uses of the Internet. *Computers in Human Behavior, 19,* 659–671.

Moran, J. D. (2008). Families, courts, and the end of life: Schiavo and its implications for the family justice system. *Family Court Review, 46,* 297–330.

Morfei, M. Z., Hooker, K., Fiese, B. H., & Cordeiro, A. M. (2001). Continuity and change in parenting possible selves: A longitudinal follow-up. *Basic and Applied Social Psychology, 23,* 217–223.

Morgan, B., & Gibson, K. R. (1991). Nutritional and environmental interactions in brain development. In K. R. Gibson & A. C. Peterson (Eds.), *Brain maturation and cognitive development: Comparative and crosscultural perspectives.* New York: de Gruyter.

Morgan, J. P., & Roberts, J. E. (2010). Helping bereaved children and adolescents: Strategies and implications for counselors. *Journal of Mental Health Counseling, 32,* 206–217.

Morgane, P. J., Austin-Lafrance, R., Bronzino, J. D., Tonkiss, J., Diaz-Cintra, S., Cintra, L., et al. (1993). Prenatal malnutrition and development of the brain. *Neuroscience & Biobehavioral Reviews, 17,* 91–128.

Morris, S. C., Taplin, J. E., & Gelman, S. A. (2000). Vitalism in naïve biological thinking. *Developmental Psychology, 36,* 582–595.

Morris, W. L., Sinclair, S., & DePaulo, B. M. (2007). No shelter for singles: The perceived legitimacy of marital status discrimination. *Group Processes & Intergroup Relation, 10,* 457–470.

Morrow, D. G., & Wilson, E. A. H. (2010). Medication adherence among older adults: A systems perspective. In J. C. Cavanaugh & C. K. Cavanaugh (Eds.), *Aging in America: Vol. 2: Physical and mental health* (pp. 211–239). Santa Monica, CA: Praeger Perspectives.

Morrow, J. R. (2005). Are American children and youth fit? It's time we learned. *Research Quarterly for Exercise and Sport, 76,* 377–388.

Mortimer, J. T., & Staff, J. (2004). Early work as a source of developmental discontinuity during the transition to adulthood. *Development and Psychopathology, 16,* 1047–1070.

Morton, J., & Johnson, M. H. (1991). CONSPEC and CONLERN: A two-process theory of infant face recognition. *Psychological Review, 98,* 164–181.

Moses, L. J., Baldwin, D. A., Rosicky, J. G., & Tidball, G. (2001). Evidence for referential understanding in the emotions domain at twelve and eighteen months. *Child Development, 72,* 718–735.

Moss, E., Cyr, C., Bureau, J.-F., Tarabulsy, G. M., & Dubois-Comtois, K. (2005). Stability of attachment during the preschool period. *Developmental Psychology, 41,* 773–783.

Moss, E., Smolla, N., Guerra, I., Mazzarello, T., Chayer, D., & Berthiaume, C. (2006). Attachement et problèmes de comportements intériorisés et extériorisés auto-rapportés á la période scolaire. *Canadian Journal of Behavioural Science, 38,* 142–157.

Moss, M. S., Moss, S. Z., & Hansson, R. O. (2001). Bereavement and old age. In M. S. Stroebe, R. O. Hansson, W. Stroebe, & H. Schut (Eds.), *Handbook of bereavement research: Consequences, coping, and care* (pp. 241–260). Washington, DC: American Psychological Association.

Mosten, F. S. (2009). *Collaborative divorce handbook: Helping families without going to court.* San Francisco, CA: Jossey-Bass.

Moua, G. K. (2007). Trait structure and levels in Hmong Americans: A test of the five factor model of personality. *Dissertation Abstracts International. Section B. Sciences and Engineering, 67(8-B),* 4750.

Moynehan, J., & Adams, J. (2007). What's the problem? A look at men in marital therapy. *American Journal of Family Therapy, 35,* 41–51.

Mugadza, T. (2005). Discrimination against women in the world of human rights: The case of women in southern Africa. In A. Barnes (Ed.), *The handbook of women, psychology, and the law* (pp. 354–365). New York: Wiley.

Muller, E. D., & Thompson, C. L. (2003). The experience of grief after bereavement: A phenomenological study with implications for mental health counseling. *Journal of Mental Health Counseling, 25,* 183–203.

Mumme, D. L., Fernald, A., & Herrera, C. (1996). Infants' responses to facial and vocal emotional signals in a social referencing paradigm. *Child Development, 67,* 3219–3237.

Murphy, K. R., Barkley, R. A., & Bush, T. (2002). Young adults with attention deficit hyperactivity disorder: Subtype differences in comorbidity, educational, and clinical history. *Journal of Nervous and Mental Disease, 190,* 147–157.

Murphy, N., & Messer, D. (2000). Differential benefits from scaffolding and children working alone. *Educational Psychology, 20,* 17–31.

Murray, M. D., Morrow, D. G., Weiner, M., Clark, D. O., Tu, W., Deer, M. M., et al. (2004). A conceptual framework to study medication adherence in older adults. *American Journal of Geriatric Pharmacotherapy, 2,* 36–43.

Murray, S. A., and Sheikh, A. (2008). Care for all at the end of life. *British Medical Journal, 336,* 958–959.

Murray-Close, D., et al. (2010). Developmental processes in peer problems of children with attention-deficit/hyperactivity disorder in the Multimodal Treatment Study of Children with ADHD: Developmental cascades and vicious cycles. *Development and Psychopathology, 22,* 785–802.

Musil, C. M., & Standing, T. (2005). Grandmothers' diaries: A glimpse at daily lives. *International Journal of Aging & Human Development, 60,* 317–329.

Mustanski, B. S., Viken, R. J., Kaprio, J., Pulkkinen, L., & Rose, R. J. (2004). Genetic and environmental influences on pubertal development: Longitudinal data from Finnish twins at ages 11 and 14. *Health Psychology, 26,* 610–617.

Mutchler, J. E., Baker, L. A., & Lee, S. A. (2007). Grandparents responsible for grandchildren in Native-American families. *Social Science Quarterly, 88,* 990–1009.

Muter V., Hulme, C., Snowling, M. J., & Stevenson, J. (2004). Phonemes, rimes, vocabulary, and grammatical skills as foundations of early reading development: Evidence from a longitudinal study. *Developmental Psychology, 40,* 663–681.

Mwanyangala, M. A., Mayombana, C., Urassa, H., Charles, J., Mahutanga, C., Abdullah, S., et al. (2010). Health status and quality of life among older adults in rural Tanzania. *Global Health Aciton, 3.* Retrieved from http://journals.sfu.ca/coaction/index.php/gha/article/viewArticle/2142/6055

Myskow, L. (2002). Perimenopausal issues in sexuality. *Sexual & Relationship Therapy, 17,* 253–260.

Nadig, A. S., & Sedivy, J. C. (2002). Evidence of perspective-taking constraints in children's on-line reference resolution. *Psychological Science, 13,* 329–336.

Nahemow, L. (2000). The ecological theory of aging: Powell Lawton's legacy. In R. L. Rubinstein & M. Moss (Eds.), *The many dimensions of aging* (pp. 22–40). New York: Springer.

Naigles, L. G., & Gelman, S. A. (1995). Overextensions in comprehension and production revisited: Preferential-looking in a study of dog, cat, and cow. *Journal of Child Language, 22,* 19–46.

Nánez, J., Sr., & Yonas, A. (1994). Effects of luminance and texture motion on infant defensive reactions to optical collision. *Infant Behavior and Development, 17,* 165–174.

Nation, K., Adams, J. W., Bowyer-Crane, C. A., & Snowling, M. J. (1999). Working memory deficits in poor comprehenders reflect underlying language impairments. *Journal of Experimental Child Psychology, 73,* 139–158.

Nation, K., Clark, P., Marshall, C., & Durand, M. (2004). Hidden language impairments in children: Parallels between poor reading comprehension and specific language impairment? *Journal of Speech, Language and Hearing Research, 47,* 199–211.

National Alliance for Caregiving and AARP. (2010). *Caregiving in the U.S.: Executive Summary.* Retrieved from http://assets.aarp.org/rgcenter/il/caregiving_09_es.pdf

National Arthritis Foundation. (2010). *Osteoarthritis fact sheet.* Retrieved from http://www.arthritis.org/media/newsroom/media-kits/Osteoarthritis_fact_sheet.pdf

National Association for Sport and Physical Education. (2004). *Appropriate practices for high school physical education.* Reston, VA: Author.

National Cancer Institute. (2006). *DES: Questions and Answers.* Retrieved from http://www.cancer.gov/cancertopics/factsheet/Risk/DES

National Cancer Institute. (2006). *Understanding cancer series: Estrogen receptors/SERMs.* Retrieved from http://www.cancer.gov/cancertopics/understandingcancer/estrogenreceptors-archive

National Center for Educational Statistics. (2010). *The condition of education 2010.* Retrieved from http://nces.ed.gov/pubsearch/pubsinfo.asp?pubid=2010028

National Center for Health Statistics. (2007c). *The state of aging and health in America.* Retrieved from http://www.cdc.gov/aging/pdf/saha_2007.pdf

National Center for Health Statistics. (2010a). *Health, United States, 2009: With special feature on medical technology.* Retrieved from http://www.cdc.gov/nchs/data/hus/hus09.pdf

National Center for Health Statistics. (2010b). *Marriage and divorce.* Retrieved from http://www.cdc.gov/nchs/fastats/divorce.htm

National Center for Health Statistics. (2010c). *Life expectancy.* Retrieved from http://www.cdc.gov/nchs/data/hus/hus2009tables/Table024.pdf

National Center for Health Statistics. (2010d). *United States life tables by Hispanic origin.* Retrieved from http://www.cdc.gov/nchs/data/series/sr_02/sr02_152.pdf

National Center for Injury Prevention and Control. (2005). *Sexual violence: Fact sheet.* Retrieved from http://www.cdc.gov/ncipc/factsheets/svfacts.htm

National Center on Elder Abuse. (2010a). *Major types of elder abuse.* Retrieved from http://www.ncea.aoa.gov/NCEAroot/Main_Site/FAQ/Basics/Types_Of_Abuse.aspx

National Center on Elder Abuse. (2010b). *Frequently asked questions.* Retrieved from http://www.ncea.aoa.gov/NCEAroot/Main_Site/FAQ/Questions.aspx

National Center on Elder Abuse. (2010c). *Who are the abusers?* Retrieved from http://www.ncea.aoa.gov/NCEAroot/Main_Site/FAQ/Basics/Abusers.aspx

National Endowment for Financial Education. (2010). *Costs of careing for an aging parent.* Retrieved from http://www.smartaboutmoney.org/LifeEventsFinancialDecisions/HealthandFamilySupport/AgingParents/CostsofCaregiving/tabid/365/Default.aspx

National Federation of State High School Associations. (2008). *NFHS participation survey, 2007–2008.* Indianapolis, IN: Author.

National Federation of State High School Associations. (2011). Participation data. Retrieved from www.nfhs.org/Participation

National Heart, Lung, and Blood Institute. (2003). *Facts about postmenopausal hormone therapy.* Retrieved from http://www.nhlbi.nih.gov/health/women/pht_facts.htm

National High Blood Pressure Education Program Working Group on Hypertension Control in Children and Adolescents. (1996). Update on the 1987 task force report on high blood pressure in children and adolescents: A working group report from the National High Blood Pressure Education Program. *Pediatrics, 98,* 649–658.

National Institute of Arthritis and Musculoskeletal and Skin Diseases. (2009a). *Osteoporosis: Peak bone mass in women.* Retrieved from http://www.niams.nih.gov/Health_Info/Bone/Osteoporosis/bone_mass.asp

National Institute of Arthritis and Musculoskeletal and Skin Diseases. (2009b). *Bone mass measurement: What the numbers mean.* Retrieved from http://www.niams.nih.gov/Health_Info/Bone/Bone_Health/bone_mass_measure.asp

National Institute of Arthritis and Musculoskeletal and Skin Diseases. (2010a). *Bone health for life.* Retrieved from http://www.niams.nih.gov/Health_Info/Bone/Bone_Health/bone_health_for_life.asp

National Institute of Arthritis and Musculoskeletal and Skin Diseases. (2010b). *Osteoporosis handout on health.* Retrieved from http://www.niams.nih.gov/Health_Info/Bone/Osteoporosis/osteoporosis_hoh.asp#5

National Institute of Arthritis and Musculoskeletal and Skin Diseases. (2010c). *Osteoarthritis.* Retrieved from http://www.niams.nih.gov/Health_Info/Osteoarthritis

National Institute of Mental Health. (2009). *How is depression detected and treated?* Retrieved from http://www.nimh.nih.gov/health/publications/depression/how-is-depression-detected-and-treated.shtml

National Institute of Mental Health. (2010a). *Post-traumatic stress disorder (PTSD).* Retrieved from http://www.nimh.nih.gov/health/publications/post-traumatic-stress-disorder-ptsd/complete-index.shtml

National Institute of Mental Health. (2010b). *Major depressive disorder among adults.* Retrieved from http://www.nimh.nih.gov/statistics/1MDD_ADULT.shtml

National Institute of Mental Health. (2010c). *Anxiety disorders.* Retrieved from http://www.nimh.nih.gov/health/topics/anxiety-disorders/index.shtml

National Institute of Neurological Disorders and Stroke. (2010). *NINDS deep brain stimulation for Parkinson's disease information page.* Retrieved from http://www.ninds.nih.gov/disorders/deep_brain_stimulation/deep_brain_stimulation.htm

National Institute on Aging. (2010a). *Hearing loss.* Retrieved from http://www.nia.nih.gov/HealthInformation/Publications/hearing.htm

National Institute on Aging. (2010b). *Aging hearts and arteries: A scientific quest.* Retrieved from http://www.nia.nih.gov/HealthInformation/Publications/AgingHeartsandArteries

National Institute on Alcohol Abuse and Alcoholism (NIAAA). (2002). *A call to action: Changing the culture of drinking at U.S. colleges. Final report of the Task Force on College Drinking.* NIH Pub. No. 02–5010. Rockville, MD: Author.

National Institute on Alcohol Abuse and Alcoholism (NIAAA). (2007). *What colleges need to know now: An update on college drinking research.* Washington, DC: National Institutes of Health.

National Institute on Alcohol Abuse and Alcoholism. (2010). *Neuroscience: pathways to alcohol dependence.* Retrieved from http://pubs.niaaa.nih.gov/publications/AA77/AA77.htm

National Institutes of Health. (2000a). *Osteoporosis prevention, diagnosis, and therapy: Consensus statement.* Retrieved from http://consensus.nih.gov/2000/2000Osteoporosis111html.htm

National Parkinson's Foundation. (2011a). *Parkinson's disease overview.* Retrieved from http://www.parkinson.org/parkinson-s-disease.aspx

National Parkinson's Foundation. (2011b). *How is PD treated?* Retrieved from http://www.parkinson.org/Parkinson-s-Disease/Treatment

National Radiological Protection Board. (2004). Review of the scientific evidence for limiting exposure to electromagnetic fields (0-300 GHz). *Documents of the NRPB, 15,* No. 3.

Natsuaki, M. N., Biehl, M. C., & Ge, X. (2009). Trajectories of depressed mood from early adolescence to young adulthood: The effects of pubertal timing and adolescent dating. *Journal of Research on Adolescence, 19,* 47–74.

Nayernouri, T. (2011). Euthanasia, terminal illness and quality of life. *Archives of Iranian Medicine, 14,* 54-55. Retrieved February 27, 2011 from http://www.ams.ac.ir/aim/011141/0011.pdf.

Neal, M. B., & Hammer, L. B. (2006). *Working couples caring for children and aging parents: Effects on work and well-being.* Mahwah, NJ: Erlbaum.

Neff, L. A., & Karney, B. R. (2005). To know you is to love you: The implications of global adoration and specific accuracy for marital relationships. *Journal of Personality and Social Psychology, 88,* 480–497.

Neft, N., & Levine, A. D. (1997). *Where women stand: An international report on the status of women in over 140 countries, 1997–1998.* New York: Random House.

Negash, S., Bennett, D. A., Wilson, R. S., Schneider, J. A., & Arnold, S. E. (2011). Cognition and neuropathology in aging: Multidimensional perspectives from the Rush Religious Orders Study and Rush Memory and Aging Project. *Current Alzheimer's Research, 8,* 336–340.

Neimeyer, R. (1997). Knowledge at the margins. *The Forum Newsletter 23,* 2, 10. (Available from the Association for Death Education and Counseling, 342 N. Main St., West Hartford, CT 06117-2507.)

Neimeyer, R. A., & Wogrin, C. (2008). Psychotherapy for complicated bereavement: A meaning-oriented approach. *Illness, Crisis, & Loss, 16,* 1–20.

Neisser, U., Boodoo, G., Bouchard, T. J., Boykin, A. W., Brody, N., Ceci, S. J., et al. (1996). Intelligence: Knowns and unknowns. *American Psychologist, 51,* 77–101.

Nell, V. (2002). Why young men drive dangerously: Implications for injury prevention. *Current Directions in Psychological Science, 11,* 75–79.

Nelson, E. A. S., Schiefenhoevel, W., & Haimerl, F. (2000). Child care practices in nonindustrialized societies. *Pediatrics, 105,* E75.

Nelson, K. (1973). Structure and strategy in learning to talk. *Monographs of the Society for Research in Child Development, 38*(Serial No. 149).

Nelson, K. (2001). Language and the self: From the "Experiencing I" to the "Continuing Me." In C. Moore & K. Lemmon (Eds.), *The self in time: Developmental perspectives* (pp. 15–33). Mahwah, NJ: Erlbaum.

Nelson, K., & Fivush, R. (2004). The emergence of autobiographical memory: A social cultural developmental theory. *Psychological Review, 111,* 486–511.

Nelson, L. J., Badger, S., & Wu, B. (2004). The influence of culture in emerging adulthood: Perspectives of Chinese college students. *International Journal of Behavioral Development, 28,* 26–36.

Nelson, M. A. (1996). Protective equipment. In O. Bar-Or (Ed.), *The child and adolescent athlete.* (pp 214–223). Oxford, England: Blackwell.

Nelson, T. L., Toomey, T. L., Lenk, K. M., Erickson, D. J., & Winters, K. C. (2010). Implementation of NIAAA college drinking task force recommendations: How are colleges doing 6 years later? *Alcoholism: Clinical and Experimental Research, 34,* 1687–1693.

Nerenberg, L. (2010). Elder abuse prevention: A review of the field. In J. C. Cavanaugh & C. K. Cavanaugh (Eds.), *Aging in America: Vol.3: Societal Issues* (pp. 53–80). Santa Barbara, CA: Praeger Perspectives.

Nesdale, D., Maass, A., Durkin, K., & Griffiths, J. (2005). Groups norms, threat, and children's racial prejudice. *Child Development, 76,* 652–663.

Neugarten, B. L. (1969). Continuities and discontinuities of psychological issues into adult life. *Human Development, 12,* 121–130.

Neugarten, B. L., & Weinstein, K. K. (1964). The changing American grandparent. *Journal of Marriage and Family, 26,* 299–304.

Nevid, J. S., Rathus, S. A., & Greene, B. (2003). *Abnormal psychology in a changing world* (5th ed.). Upper Saddle River, NJ: Prentice Hall.

Newcomb, A. F., & Bagwell, C. L. (1995). Children's friendship relations: A meta-analytic review. *Psychological Bulletin, 117,* 306–347.

Newman, B., & Newman, P. (2007). *Theories of human development.* London: Psychology Press.

Newport, E. L. (1991). Contrasting conceptions of the critical period for language. In S. Carey & R. Gelman (Eds.), *The epigenesis of mind: Essays on biology and cognition* (pp. 111–130). Hillsdale, NJ: Erlbaum.

Newsom, J. T. (1999). Another side to caregiving: Negative reactions to being helped. *Current Directions in Psychological Science, 8,* 183–187.

NICHD Early Child Care Research Network. (1997). The effects of infant child care of infant–mother attachment security: Results of the NICHD Study of Early Child Care. *Child Development, 68,* 860–879.

NICHD Early Child Care Research Network. (2001). Child-care and family predictors of preschool attachment and stability from infancy. *Developmental Psychology, 37,* 847–862.

NICHD. (2004). *The NICHD community connection.* Washington, DC: Author.

Nilsson, L.-G., Sternäng, O., Rönnlund, M., & Nyberg, L. (2009). Challenging the notion of an early-onset of cognitive decline. *Neurobiology of Aging, 30,* 521–524.

Nimrod, G. (2007a). Retirees' leisure: Activities, benefits, and their contribution to life satisfaction. *Leisure Studies, 26,* 65–80.

Nimrod, G. (2007b). Expanding, reducing, concentrating and diffusing: Post retirement leisure behavior and life satisfaction. *Leisure Sciences, 29,* 91–111.

Nimrod, G., & Hutchinson, S. (2010). Innovation among older adults with chronic health conditions. *Journal of Leisure Research, 42,* 1–23.

Nimrod, G., & Kleiber, D. A. (2007). Reconsidering change and continuity in later life: Toward an innovation theory of successful aging. *International Journal of Aging & Human Development, 65,* 1–22.

NINDS. (2009). *Autism fact sheet.* NIH Publication No. 09-1877. Bethesda, MD. Author.

Nishida, T. K., & Lillard, A. S. (2007). The informative value of emotional expressions: "Social referencing" in mother-child pretense. *Developmental Science, 10,* 205–212.

Nishina, A., & Juvonen, J. (2005). Daily reports of witnessing and experiencing peer harassment in middle school. *Child Development, 76,* 435–450.

Nistor, G. I., Totoiu, M. O., Haque, N., Carpenter, M. K., & Keirstead, H. S. (2005). Human embryonic stem cells differentiate into oligodendrocytes in high purity and myelinate after spinal cord transplantation. *Glia, 49,* 385–396.

Nolen-Hoeksema, S. (2001). Ruminative coping and adjustment to bereavement. In M. S. Stroebe, R. O. Hansson, W. Stroebe, & H. Schut (Eds.), *Handbook of bereavement research* (pp. 545–562). Washington, DC: American Psychological Association.

Nolen-Hoeksema, S., & Larson, J. (1999). *Coping with loss.* Mahwah, NJ: Erlbaum.

Noonan, D. (2005, June 6). *A little bit louder, please.* Retrieved from http://www.newsweek.com/2005/06/05/a-little-bit-louder-please.html

Nord, M., Andrews, M., & Carlson, S. (2007). *Household food security in the United States, 2006.* Washington, DC: U.S. Department of Agriculture.

Norlander, B., & Eckhardt, C. (2005). Anger, hostility, and male perpetrators of intimate partner violence: A meta-analytic review. *Clinical Psychology Review, 25,* 119–152.

Nrugham, L., Larsson, B., & Sund, A. M. (2008). Predictors of suicidal acts across adolescent: Influences of familial, peer and individual factors. *Journal of Affective Disorders, 109,* 35–45.

Nurmsoo, E., & Bloom, P. (2008). Preschoolers' perspective taking in word learning: Do they blindly follow eye gaze? *Psychological Science, 19,* 211–215.

Nusbaum, L. E. (2010). *How the elder co-housing model of living affects residents' experience of autonomy: A self-determination theory perspective.* Dissertation submitted to the Wright Institute. Retrieved from http://proquest.umi.com/pqdlink?Ver=1&Exp=02-05-2016&FMT=7&DID=2189312001&RQT=309&attempt=1&cfc=1

Nygren, C., Oswald, F., Iwarsson, S., Fänge, A., Sixsmith, J., Schilling, O., et al. (2007). Relationships between objective and perceived housing in very old age. *The Gerontologist, 47,* 85–95.

O'Brien, C.-A., & Goldberg, A. (2000). Lesbians and gay men inside and outside families. In N. Mandell & A. Duffy (Eds.), *Canadian families: Diversity, conflict, and change* (2nd ed., pp. 115–145). Toronto: Harcourt Brace.

O'Connor, T., Heron, J., Golding, J., Beveridge, M., & Glover, V. (2002). Maternal antenatal anxiety and children's behavioural/emotional problems at 4 years. *British Journal of Psychiatry, 180,* 502–508.

O'Donoghue, M. (2005). White mothers negotiating race and ethnicity in the mothering of biracial, Black-White adolescents. *Journal of Ethnic & Cultural Diversity in Social Work, 14,* 125–156.

O'Hara, M. W. (2009). Postpartum depression: What we know. *Journal of Clinical Psychology, 65,* 1258–1269.

O'Leary, K. D. (1993). Through a psychological lens: Personality traits, personality disorders, and levels of violence. In R. J. Gelles & D. R. Loseke (Eds.), *Current controversies on family violence* (pp. 7–30). Newbury Park, CA: Sage.

O'Neill, D. K. (1996). Two-year-old children's sensitivity to a parent's knowledge state when making requests. *Child Development, 67,* 659–677.

O'Rourke, N., & Cappeliez, P. (2005). Marital satisfaction and self-deception: Reconstruction of relationship histories among older adults. *Social Behavior and Personality, 33,* 273–282.

Oakhill, J. V., & Cain, K. E. (2004). The development of comprehension skills. In T. Nunes & P. E. Bryant (Eds.), *Handbook of children's literacy* (pp. 155–180). Dordrecht, The Netherlands: Kluwer.

Oaten M., Stevenson, R. J., & Case, T. I. (2009). Disgust as a disease-avoidance mechanism. *Psychological Bulletin, 135,* 303–321.

Oburu, P. O., & Palmérus, K. (2005). Stress related factors among primary and part-time caregiving grandmothers of Kenyan grandchildren. *International Journal of Aging & Human Development, 60,* 273–282.

Odgers, C. L., Moffitt, T. E., Broadbent, J. M., Dickson, N., Hancox, R. J., Harrington, H., et al. (2008). Female and male antisocial trajectories: From childhood origins to adult outcomes. *Development and Psychopathology, 20,* 673–716.

OECD. (2006). *Starting strong II: Early childhood education and care.* Paris: OECD Publishing.

Offer, D., Ostrov, E., Howard, K. I., & Atkinson, R. (1988). *The teenage world: Adolescents' self-image in ten countries.* New York: Plenum.

Office of Management and Budget. (2010). *Historical tables: Budget of the U. S. Government, Fiscal Year 2011.* Retrieved from http://www.whitehouse.gov/sites/default/files/omb/budget/fy2011/assets/hist.pdf

Oh, E. S., Lee, J. H., Jeong, S.-H., Sohn, E. H., & Lee, A. Y. (2011). Comparisons of cognitive deterioration rates by dementia subtype. *Archives of Gerontology and Geriatrics, 53(3):*320–322.

Ojanen, T., & Perry, D. G. (2007). Relational schemas and the developing self: Perceptions of mother and of self as joint predictors of early adolescents' self-esteem. *Developmental Psychology, 43,* 1474–1483.

Okagaki, L., & Sternberg, R. J. (1993). Parental beliefs and children's school performance. *Child Development, 64,* 36–56.

Okami, P., Weisner, T., & Olmstead, R. (2002). Outcome correlates of parent-child bedsharing: An eighteen-year longitudinal study. *Developmental and Behavioral Pediatrics, 23,* 244–253.

Okonkwo, O. C., Mielke, M. M., Griffith, H. R., Moghekar, A. R., O'Brien, R. J., Shaw L. M., et al. (2011). Cerebrospinal fluid profiles and prospective course and outcome in patients with amnestic mild cognitive impairment. *Archives of Neurology, 68,* 113–119.

Okonski, B. (1996, May 6). Just say something. *Newsweek,* p. 14.

Oliner, S. P., & Oliner, P. M. (1988). *The altruistic personality: Rescuers of Jews in Nazi Europe.* New York: Free Press.

Olinghouse, N. G. (2008). Student- and instruction-level predictors of narrative writing in third-grade students. *Reading and Writing, 21,* 3–26.

Olshansky, S. J., Hayflick, L., & Perls, T. (2004a). Anti-aging medicine: The hype and the reality—Part I. *Journal of Gerontology: Biological Sciences, 59A,* 513–514.

Olshansky, S. J., Hayflick, L., & Perls, T. (2004b). Anti-aging medicine: The hype and the reality—Part II. *Journal of Gerontology: Biological Sciences, 59A,* 649–651.

Olson, D. H., & McCubbin, H. (1983). *Families: What makes them work.* Newbury Park, CA: Sage.

Olson, S. L., Sameroff, A. J., Kerr, D. C. R., Lopez, N. L., & Wellman, H. M. (2005). Developmental foundations of externalizing problems in young children: The role of effortful control. *Development and Psychopathology, 17,* 25–45.

Oltjenbruns, K. A. (2001). Developmental context of childhood: Grief and regrief phenomena. In M. S. Stroebe, R. O. Hansson, W. Stroebe, & H. Schut (Eds.), *Handbook of bereavement research: Consequences, coping, and care* (pp. 169–197). Washington, DC: American Psychological Association.

Oltjenbruns, K. A., & Balk, D. E. (2007). Life span issues and loss, grief, and mourning: Part 1. The importance of a developmental context: Childhood and adolescence as an example; Part 2. Adulthood. In D. Balk, C. Wogrin, G. Thornton, & D. Meagher (Eds.), *Handbook of thanatology: The essential body of knowledge for the study of death, dying, and bereavement* (pp. 143–163). New York: Routledge/Taylor & Francis.

Olweus, D., Mattson, A., Schalling, D., & Low, H. (1988). Circulating testosterone levels and aggression in adolescent males: A causal analysis. *Psychosomatic Medicine, 50,* 261–272.

Omori, M., & Smith, D. T. (2009). The impact of occupational status on household chore hours among dual earner couples. *Sociation Today, 7.* Retrieved from http://www.ncsociology.org/sociationtoday/v71/chore.htm

Onrust, S., Cuijpers, P., Smit, F., & Bohlmeijer, E. (2007). Predictors of psychological adjustment after bereavement. *International Psychogeriatrics, 19,* 921–934.

Opfer, J. E., & Siegler, R. S. (2004). Revisiting preschoolers *living things* concept: A microgenetic analysis of conceptual change in basic biology. *Cognitive Psychology, 49,* 301–332.

Oregon Department of Human Services. (2011). *Oregon's Death with Dignity Act—2010.* Retrieved February 27, 2011 from http://www.oregon.gov/DHS/ph/pas/docs/year13.pdf.

Orentlicher, D. (2000). The implementation of Oregon's Death with Dignity Act: Reassuring, but more data are needed. *Psychology, Public Policy, and Law, 6,* 489–502.

Osgood, D. W., Ruth, G., Eccles, J. S., Jacobs, J. E., & Barber, B. L. (2005). Six paths to adulthood: Fast starters, parents without careers, educated partners, educated singles, working singles, and slow starters. In R. A. Settersten, Jr., F. F. Furstenberg, Jr., & R. G. Rumbaut (Eds.), *On the frontier of adulthood: Theory, research, and public policy* (pp. 320–355). Chicago, IL: University of Chicago Press.

Oster-Aaland, L., Lewis, M. A., Neighbors, C., Vangsness, J., & Larimer, M. E. (2009). Alcohol poisoning among colleges students turning 21: Do they recognize the symptoms and how do they help? *Journal of Studies on Alcohol and Drugs, Suppl. 16,* 122–130.

Ostrov, J. M., & Godleski, S. A. (2010). Toward an integrated gender-linked model of aggression subtypes in early and middle childhood. *Psychological Review, 117,* 233–242.

Oswald, A. J., & Wu, S. (2010). Objective confirmation of subjective measures of human well-being: Evidence from the USA. *Science, 327,* 576–579.

Ottaway, A. J. (2010). The impact of parental divorce on the intimate relationships of adult offspring: A review of the literature. *Graduate Journal of Counseling Psychology, 2(1),* Article 5. Retrieved from http://epublications.marquette.edu/cgi/viewcontent.cgi?article=1037&context=gjcp

Over, H., & Carpenter, M. (2009). Eighteen-month-old infants show increased helping following priming with affiliation. *Psychological Science, 20,* 1189–1193.

Owen, C. J. (2005). The empty nest transition: The relationship between attachment style and women's use of this period as a time for growth and change. *Dissertation Abstracts International. Section B. Sciences and Engineering, 65(7-B),* 3747.

Oygard, L., & Hardeng, S. (2001). Divorce support groups: How do group characteristics influence adjustment to divorce? *Social Work with Groups, 24,* 69–87.

Ozawa, M. N., & Yoon, H. S. (2002). The economic benefit of remarriage: Gender and class income. *Journal of Divorce & Remarriage, 36,* 21–39.

Paarlberg, K. M., Vingerhoets, A. J. J. M., Passchier, J., Dekker, G. A., et al. (1995). Psychosocial factors and pregnancy outcome: A review with emphasis on methodological issues. *Journal of Psychosomatic Research, 39,* 563–595.

Palkovitz, R., & Palm, G. (2009). Transitions within fathering. *Fathering, 7,* 3–22.

Pankow, L. J., & Solotoroff, J. M. (2007). Biological aspects and theories of aging. In J. A. Blackburn & C. N. Dulmus (Eds.), *Handbook of gerontology: Evidence-based approaches to theory, practice, and policy* (pp. 19–56). Hoboken, NJ: Wiley.

Papa, A., & Litz, B. (2011). Grief. In W. T O'Donohue & C. Draper (Eds.), *Stepped care and e-health: Practical applications to behavioral disorders* (pp. 223–245). New York: Springer.

Paquette, D. (2004). Theorizing the father-child relationship: Mechanisms and developmental outcomes. *Human Development, 47,* 193–219.

Paraita, H., Díaz, C., & Anllo-Vento, L. (2008). Processing of semantic relations in normal aging and Alzheimer's disease. *Archives of Clinical Neuropsychology, 23,* 33–46.

Parault, S. J., & Schwanenflugel, P. J. (2000). The development of conceptual categories of attention during the elementary school years. *Journal of Experimental Child Psychology, 75,* 245–262.

Parcel, G. S., Simons-Morton, B. G., O'Hara, N. M., Baranowksi, T., Kilbe, L. J., & Bee, D. E. (1989). School promotion of healthful diet and exercise behavior: An integration of organizational change and social learning theory interventions. *Journal of School Health, 57,* 150–156.

Pargament, K. I. (1997). *The psychology of religion and coping: Theory, research, and practice.* New York: Guilford.

Park, C. L. (2007). Religiousness/spirituality and health: A meaning systems perspective. *Journal of Behavioral Medicine, 30,* 319–328.

Park, D. C., Morrell, R. W., & Shifren, K. (Eds.). (1999). *Processing of medical information in aging patients: Cognitive and human factors perspectives.* Mahwah, NJ: Erlbaum.

Park, D. C., Smith, A. D., Lautenschlager, G., Earles, J. L., Frieski, D., Zwahr, M., & Gaines, C. L. (1996). Mediators of long-term memory performance across the life span. *Psychology and Aging, 11,* 621–637.

Park, G., Lubinski, D., & Benbow, C. P. (2008). Ability differences among people who have commensurate degrees matter for scientific creativity. *Psychological Science, 19,* 957–961.

Parke, R. D., & Buriel, R. (1998). Socialization in the family: Ethnic and ecological perspectives. In W. Damon (Ed.), *Handbook of child psychology* (Vol. 3, pp. 463–552). New York: Wiley.

Parke, R. D., & O'Neil, R. (2000). The influence of significant others on learning about relationships: From family to friends. In R. S. L. Mills & S. Duck (Eds.), *The developmental psychology of personal relationships* (pp. 15–47). New York: Wiley.

Parke, R. D., Coltrane, S., Duffy, S., Buriel, R., Dennis, J., Powers, J., et al. (2004). Economic stress, parenting, and child adjustment in Mexican American and European American families. *Child Development, 75,* 1632–1656.

Parker, J. G., & Seal, J. (1996). Forming, losing, renewing, and replacing friendships: Applying temporal parameters to the assessment of children's friendship experiences. *Child Development, 67,* 2248–2268.

Parker, L. D. (2008). A study about older African American spousal caregivers of persons with Alzheimer's disease. *Dissertation Abstracts International. Section B. Sciences and Engineering, 68(10-B),* 6589.

Parkman, A. M. (2007). *Smart marriage: Using your (business) head as well as your heart to find wedded bliss.* Westport, CT: Praeger.

Parra-Cardona, J. R., Meyer, E., Schiamberg, L., & Post, L. (2007). Elder abuse and neglect in Latino families: An ecological and culturally relevant theoretical framework for clinical practice. *Family Process, 46,* 451–470.

Parritz, R. H. (1996). A descriptive analysis of toddler coping in challenging circumstances. *Infant Behavior and Development, 19,* 171–180.

Parrot, A., & Cummings, N. (2006). *Forsaken females: The global brutalization of women.* Lanham, MD: Rowman & Littlefield.

Parten, M. (1932). Social participation among preschool children. *Journal of Abnormal and Social Psychology, 27,* 243–269.

Pascalis, O., de Hann, M., & Nelson, C. A. (2002). Is face processing species-specific during the first year of life? *Science, 296,* 1321–1323.

Paschal, A. M., Lewis, R. K., & Sly, J. (2007). African American parents' behaviors and attitudes about substance use and abuse. *Journal of Ethnicity in Substance Abuse, 6,* 67–79.

Pascual, B., Aguardo, G., Sotillo, M., & Masdeu, J. C. (2008). Acquisition of mental state language in Spanish children: A longitudinal study of the relationship between the production of mental verbs and linguistic development. *Developmental Science, 11,* 454–466.

Pascual, C. (2000, October 3). Asians have highest elderly suicide rate. *Wilmington (NC) Morning Star,* p. 5D.

Pasterski, V. L., Geffner, M. E., Brain, C., Hindmarsh, P., Brook, C., & Hines, M. (2005). Prenatal hormones and postnatal socialization by parents as determinants of male-typical toy play in girls with congenital adrenal hyperplasia. *Child Development, 76,* 264–278.

Pasupathi, M., Mansour, E., & Brubaker, J. R. (2007). Developing a life story: Constructing relations between self and experience in autobiographical narratives. *Human Development, 50,* 85–110.

Pasupathi, M., McLean, K. C., & Weeks, T. (2009). To tell or not to tell: Disclosure and the narrative self. *Journal of Personality, 77,* 89–124.

Patel, K., Coppin, A., Manini, T., Lauretani, F., Bandinelli, S., Ferrucci, L., & Guralnik, J. (2006). Midlife physical activity and mobility in older age: The CHIANTI study, *American Journal of Preventive Medicine, 31* (3), 217–224.

Patrick, S., Sells, J. N., Giordano, F. G., & Tollerud, T. R. (2007). Intimacy, differentiation, and personality variables as predictors of marital satisfaction. *The Family Journal, 15,* 359–367.

Patry, D. A., Blanchard, C. M., & Mask, L. (2007). Measuring university students' regulatory leisure coping styles: Planned breathers or avoidance? *Leisure Sciences, 29,* 247–265.

Patterson, C. J. (1992). Children of lesbian and gay parents. *Child Development, 63,* 1025–1042.

Patterson, G. R. (1980). Mothers: The unacknowledged victims. *Monographs of the Society for Research in Child Development, 45*(5, Serial No. 186).

Patterson, G. R. (2008). A comparison of models for interstate wars and for individual violence. *Perspectives on Psychological Science, 3,* 203–223.

Patterson, M. M., & Bigler, R. S. (2006). Preschool children's attention to environmental messages about groups: Social categorization and the origins of intergroup bias. *Child Development, 77,* 847–860.

Pauker, K., Ambady, N., & Apfelbaum, E. P. (2010). Race salience and essentialist thinking in racial stereotype development. *Child Development, 81,* 1799–1813.

Paulussen-Hoggeboom, M. C., Stams, G. J. J., Hermanns, J. M., & Peetsma, T. T. (2007). Child negative emotionality and parenting from infancy to preschool: A meta-analytic review. *Developmental Psychology, 43,* 438–453.

Pauly, L., Stehle, P., & Volkert, D. (2007). Ernährungssituation älterer Heimbewohner. [Nutritional situation of elderly nursing home residents]. *Zeitschrift für Gerontologie und Geriatrie, 40,* 3–12.

Pavalko, E. K., & Artis, J. E. (1997). Women's caregiving and paid work: Causal relationships in late midlife. *Journals of Gerontology: Social Sciences, 52B,* S170–S179.

Paxton, S. J., Eisenberg, M. E., & Neumark-Sztainer, D. (2006). Prospective predictors of body dissatisfaction in adolescent girls and boys: A five-year longitudinal study. *Developmental Psychology, 42,* 888–899.

Pearlin, L. I., Mullan, J. T., Semple, S. J., & Skaff, M. M. (1990). Caregiving and the stress process: An overview of concepts and their measures. *The Gerontologist, 30,* 583–594.

Pellicano, E. (2010). Individual differences in executive function and central coherence predict developmental changes in theory of mind in autism. *Developmental Psychology, 46,* 530–544.

Pelphrey, K. A., Reznick, J. S., Davis Goldman, B., Sasson, N., Morrow, J., Donahoe, A., & Hodgson, K. (2004). Development of visuospatial memory in the second half of the first year. *Developmental Psychology, 40,* 836–851.

Peltola, M. J., Leppänen, J. M., Palokangas, T., & Hietanen, J. K. (2008). Fearful faces modulate looking duration and attention disengagement in 7-month-old infants. *Developmental Science, 11,* 60–68.

Pelucci, B., Hay, J. F., & Saffran, J. R. (2009). Statistical learning in a natural language by 8-month-old infants. *Child Development, 80,* 674–685.

Pennebaker, J. W., & Graybeal, A. (2001). Patterns of natural language use: Disclosure, personality, and social integration. *Current Directions in Psychological Science, 10,* 90–93.

Pennington, B. F., Willcutt, E., & Rhee, S. H. (2005). Analyzing comorbidity. In R. V. Kail (Ed.), *Advances in child development and behavior* (Vol. 33, pp. 263–304). San Diego, CA: Elsevier.

Pennisi, E. (2005). Why do humans have so few genes? *Science, 309,* 80.

Penson, R. T. (2004). Bereavement across cultures. In R. J. Moore & D. Spiegel (Eds.), *Cancer, culture, and communication* (pp. 241–279). New York: Kluwer/Plenum.

Perkins, D. F., Jacobs, J. E., Barber, B. L., & Eccles, J. S. (2004). Childhood and adolescent sports participation as predictors of participation in sports and physical fitness activities during young adulthood. *Youth & Society, 35,* 495–520.

Perkins, H. W., Linkenbach, J. W., Lewis, M. A., & Neighbors, C. (2010). Effectiveness of social norms media marketing in reducing drinking and driving: A statewide campaign. *Addictive Behaviors, 35,* 866–874.

Perls, T., & Terry, D. (2003). Genetics of exceptional longevity. *Experimental Gerontology, 38,* 725–730.

Perrone, K. M., Tschopp, M. K., Snyder, E. R., Boo, J. N., & Hyatt, C. (2010). Expectations and outcomes of academically talented students 10 and 20 years post-high school graduation. *Journal of Career Development, 36,* 291–309.

Perry, M. (2000). Explanations of mathematical concepts in Japanese, Chinese, and U.S. first- and fifth-grade classrooms. *Cognition & Instruction, 18,* 181–207.

Perry, W. I. (1970). *Forms of intellectual and ethical development in the college years.* New York: Holt, Rinehart & Winston.

Peters, A. M. (1995). Strategies in the acquisition of syntax. In P. Fletcher & B. MacWhinney (Eds.), *The handbook of child language* (pp. 462–483). Oxford, England: Blackwell.

Peterson, B. E., & Duncan, L. E. (2007). Midlife women's generativity and authoritarianism: Marriage, motherhood, and 10 years of aging. *Psychology and Aging, 22,* 411–419.

Peterson, B. E., & Klohnen, E. C. (1995). Realization of generativity in two samples of women at midlife. *Psychology and Aging, 10,* 20–29.

Peterson, C., & Slaughter, V. (2003). Opening windows into the mind: Mothers' preferences for mental state explanations and children's theory of mind. *Cognitive Development, 18,* 399–429.

Peterson, C., Wang, Q., & Hou, Y. (2009). When I was little: Childhood recollections in Chinese and European Canadian school children. *Child Development, 80,* 506–518.

Peterson, L. (1983). Role of donor competence, donor age, and peer presence on helping in an emergency. *Developmental Psychology, 19,* 873–880.

Pettit, G. S., Bates, J. E., & Dodge, K. A. (1997). Supportive parenting, ecological context, and children's adjustment: A seven-year longitudinal study. *Child Development, 68,* 908–923.

Pettito, L. A., Katerelos, M., Levy, B. G., Gauna, K., Tetreault, K., et al. (2001). Bilingual signed and spoken language acquisition from birth: Implications for the mechanisms underlying early bilingual language acquisition. *Journal of Child Language, 28,* 453–496.

Phelps, J. A., Davis, J. O., & Schartz, K. M. (1997). Nature, nurture, and twin research strategies. *Current Directions in Psychological Science, 6,* 117–121.

Phelps, R. E., & Constantine, M. G. (2001). Hitting the roof: The impact of the glass-ceiling effect on the career development of African Americans. In W. B. Walsh, R. P. Bingham, et al. (Eds.), *Career counseling for African Americans* (pp. 161–175). Mahwah, NJ: Erlbaum.

Phinney, J. (1989). Stage of ethnic identity in minority group adolescents. *Journal of Early Adolescence, 9,* 34–49.

Phinney, J. S. (1990). Ethnic identity in adolescents and adults: Review of research, *Psychological Bulletin 108,* 499–514.

Phinney, J. S. (2005). Ethnic identity development in minority adolescents. In C. B. Fisher & R. M. Lerner (Eds.), *Encyclopedia of applied developmental science* (Vol. 1, pp. 420–423). Thousand Oaks, CA: Sage.

Phipps, M. G., Rosengard, C., Weitzen, S., Meers, A., & Billinkoff, Z. (2008). Age group differences among pregnant adolescents: Sexual behavior, health habits, and contraceptive use. *Journal of Pediatric and Adolescent Gynecology, 21,* 9–15.

Piaget, J. (1929). *The child's conception of the world.* New York: Harcourt Brace.

Piaget, J. (1951). *Plays, dreams, and imitation in childhood.* New York: Norton.

Piaget, J. (1952). *The origins of intelligence in children.* New York: International Universities Press.

Piaget, J. (1954). *The construction of reality in the child.* New York: Basic Books.

Piaget, J., & Inhelder, B. (1956). *The child's conception of space.* Boston: Routledge & Kegan Paul.

Piehler, T. F., & Dishion, T. J. (2007). Interpersonal dynamics within adolescent friendships: Dyadic mutuality, deviant talk, and patterns of antisocial behavior. *Child Development, 78,* 1611–1624.

Pienta, A. M., Hayward, M. D., & Jenkins, K. R. (2000). Health consequences of marriage for the retirement years. *Journal of Family Issues, 21,* 559–586.

Pierret, C. R. (2006). The "sandwich generation": Women caring for parents and children. *Monthly Labor Review, September.* Retrieved from http://www.bls.gov/opub/mlr/2006/09/art1full.pdf

Pillay, H., Kelly, K., & Tones, M. (2006). Career aspirations of older workers: An Australian study. *International Journal of Training and Development, 10,* 298–305.

Pincus, T., Callahan, L. F., & Burkhauser, R. V. (1987). Most chronic diseases are reported more frequently by individuals with fewer than 12 years of formal education in the age 18–64 United States population. *Journal of Chronic Diseases, 40,* 865–874.

Pines, A. (in press). Male menopause: Is it a real clinical syndrome? *Climacteric, 14(1):*15–17.

Pinkhasov, R. M., Shteynshlyuger, A., Hakimian, P., Lindsay, G. K., Samadi, D. B., & Shabsigh, R. (2010). Are men shortchanged on health? Perspective on life expectancy, morbidity, and mortality in men and women in the United States, *International Journal of Clinical Practice, 64,* 465–474.

Pinquart, M., & Sörensen, S. (2001). Gender differences in self-concept and psychological well-being in old age: A meta-analysis. *Journal of Gerontology: Psychological Sciences, 56B,* P195–P213.

Pinto, K. M., & Coltrane, S. (2009). Division of labor in Mexican origin and Anglo families: Structure and culture. *Sex Roles, 60,* 482–495.

Piquet, B. J. (2007). That's what friends are for. *Dissertation Abstracts International. Section B. Sciences and Engineering, 67(7-B),* 4114.

Piscione, D. P. (2004). *The many faces of 21st century working women: A report to the Women's Bureau of the U.S. Department of Labor.* McLean, VA: Education Consortium.

Pleck, J. H., & Masciadrelli, B. P. (2004). Paternal involvement by U.S. residential fathers: Levels, sources, and consequences. In M. E. Lamb (Ed.), *The role of the father in child development* (4th ed; pp. 222–271). Hoboken NJ: Wiley.

Pleis, J. R., & Lethbridge-Çejku, M. (2007). Summary health statistics for U.S. adults: National Health Interview Survey, 2006. *Vital Health Statistics, 10(235).* Washington, DC: National Center for Health Statistics.

Plomin, R. (1990). *Nature and nurture.* Pacific Grove, CA: Brooks/Cole.

Plomin, R., & Crabbe, J. (2000). DNA. *Psychological Bulletin, 126,* 806–828.

Plomin, R., & Petrill, S. A. (1997). Genetics and intelligence: What's new? *Intelligence, 24,* 53–77.

Plomin, R., & Spinath, F. (2004). Intelligence: Genes, genetics, and genomics. *Journal of Personality and Social Psychology, 86,* 112–129.

Plomin, R., Fulker, D. W., Corley, R., & DeFries, J. C. (1997). Nature, nurture, and cognitive development from 1 to 16 years: A parent-offspring adoption study. *Psychological Science, 8,* 442–447.

Plunkett, K. (1996). *Connectionism and development: Neural networks and the study of change.* New York: Oxford University Press.

Polivka, L. (2010). Neoliberalism and the new politics of aging and retirement security. In J. C. Cavanaugh & C. K. Cavanaugh (Eds.), *Aging in America: Vol.3: Societal Issues* (pp. 160-202). Santa Barbara, CA: Praeger Perspectives.

Pollitt, E. (1995). Does breakfast make a difference in school? *Journal of the American Dietetic Association, 95,* 1134–1139.

Ponde, M. P., & Santana, V. S. (2000). Participation in leisure activities: Is it a protective factor for women's mental health? *Journal of Leisure Research, 32,* 457–472.

Poole, D. A., & Lindsay, D. S. (1995). Interviewing preschoolers: Effects of non-suggestive techniques, parental coaching, and leading questions on reports of nonexperienced events. *Journal of Experimental Child Psychology, 60*, 129–154.

Popenoe, D. (2009). Cohabitation, marriage, and child wellbeing: A cross-national perspective. *Society: Social Science and Public Policy, 46*, 429–436.

Popp, D., Laursen, B., Kerr, M., Stattin, H., & Burk, W. K. (2008). Modeling homophily over time with an actor-partner interdependence model. *Developmental Psychology, 44*, 1028–1039.

Porter, R. H., & Winberg, J. (1999). Unique salience of maternal breast odors for newborn infants. *Neuroscience & Biobehavioral Reviews, 23*, 439–449.

Poulin, F., & Chan, A. (2010). Friendship stability and change in childhood and adolescence. *Developmental Review, 30*, 257–272.

Poulson, C. L., Kymissis, E., Reeve, K. F., Andreatos, M., & Reeve, L. (1991). Generalized vocal imitation in infants. *Journal of Experimental Child Psychology, 51*, 267–279.

Power, T. L., & Smith, S. M. (2008). Predictors of fear of death and self-mortality: An Atlantic Canadian perspective. *Death Studies, 32*, 252–272.

Pratt, H. D. (2010). Perspectives from a non-traditional mentor. In C. A. Rayburn, F. L. Denmark, M. E. Reuder, & A.M. Austria, (Eds.), *A handbook for women mentors: Transcending barriers of stereotype, race, and ethnicity* (pp. 223–232). Santa Barbara, CA: ABC-CLIO.

Pratt, M. W., Danso, H. A., Arnold, M. L., Norris, J. E., & Filyer, R. (2001). Adult generativity and the socialization of adolescents: Relations to mothers' and fathers' parenting beliefs, styles, and practices. *Journal of Personality, 69*, 89–120.

President's Council on Physical Fitness and Sports. (2004). Physical activity for children: Current patterns and guidelines. *Research Digest* (Series 5, No. 2).

Pressley, M., & Hilden, K. (2006). Cognitive strategies. In D. Kuhn & R. S. Siegler (Eds.), *Handbook of child psychology* (6th ed., Vol. 2, pp. 511–556). Hoboken, NJ: Wiley.

Pride, N. B. (2005). Ageing and changes in lung mechanics. *European Respiratory Journal, 26*, 563–565.

Prinstein, M. J., & Cillessen, A. H. N. (2003). Forms and functions of adolescent peer aggression associated with high levels of peer status. *Merrill Palmer Quarterly, 49*, 310–342.

Prinstein, M. J., & La Greca, A. M. (2004). Childhood peer rejection and aggression as predictors of adolescent girls' externalizing and health risk behaviors: A 6-year longitudinal study. *Journal of Consulting and Clinical Psychology, 72*, 103–112.

Probert, B. (2005). "I just couldn't fit it in": Gender and unequal outcomes in academic careers. *Gender, Work & Organization, 12*, 50–72.

Pruchno, R. (1999). Raising grandchildren: The experiences of black and white grandmothers. *The Gerontologist, 39*, 209–221.

Pruett, M. K., Insabella, G. M., & Gustafson, K. (2005). The collaborative divorce project: A court-based intervention for separating parents with young children. *Family Court Review, 43*, 38–51.

Prus, S. G., Tfaily, R., & Lin, Z. (2010). Comparing racial and immigrant health status and health care access in late life in Canada and the United States. *Canadian Journal on Aging, 29*, 383–395.

Pryor, J. B., Desouza, E. R., Fitness, J., & Hutz, C. (1997). Gender differences in the interpretation of social-sexual behavior: A cross-cultural perspective on sexual harassment. *Journal of Cross-Cultural Psychology, 28*, 509–534.

Puhl, R. M., & Brownell, K. D. (2005). Bulimia nervosa. In C. B. Fisher & R. M. Lerner (Eds.), *Encyclopedia of applied developmental science* (Vol. 1, pp. 192–195). Thousand Oaks, CA: Sage.

Puhl, R. M., & Latner, J. D. (2007). Stigma, obesity, and the health of the nation's children. *Psychological Bulletin, 133*, 557–580.

Pungello, E. P., et al. (2010). Early educational intervention, early cumulative risk, and the early home environment as predictors of young adult outcomes within a high-risk sample. *Child Development, 81*, 410–426.

Purcell, D., MacArthur, K. R. and Samblanet, S. (2010), Gender and the glass ceiling at work. *Sociology Compass, 4*, 705–717.

Pynoos, J. Caraviello, R., & Cicero, C. (2010). Housing in an aging America. In J. C. Cavanaugh & C. K. Cavanaugh (Eds.), *Aging in America: Vol. 3: Societal issues* (pp. 129–159). Santa Barbara, CA: Praeger Perspectives.

Pyszczynski, T., Greenberg, J., & Solomon, S. (1997). Why do we need what we need? A terror management perspective on the roots of human social motivation. *Psychological Inquiry, 8*, 1–20.

Pyszczynski, T., Greenberg, J., & Solomon, S. (1999). A dual-process model of defense against conscious and unconscious death-related thoughts: An extension of terror management theory. *Psychological Review, 106*, 835–845.

Qin, L., Pomerantz, E. M., & Wang, Q. (2009). Are gains in decision-making autonomy during early adolescence beneficial for emotional functioning? The case of the United States and China. *Child Development, 80*, 1705–1721.

Qualls, S. H., & Layton, H. (2010). Mental health and adjustment. In J. C. Cavanaugh & C. K. Cavanaugh (Eds.), *Aging in America: Vol. 2: Physical and mental health* (pp. 171–187). Santa Barbara, CA: Praeger Perspectives.

Quillian, L., & Campbell, M. E. (2003). Beyond black and white: The present and future of multiracial friendship segregation. *American Sociological Review, 68*, 540–566.

Quinn, P. C., & Liben, L. S. (2008). A sex difference in mental rotation in young infants. *Psychological Science, 19*, 1067–1070.

Quirin, M., Loktyushin, A., Arndt, J., Küstermann, E., Lo, Y.-Y., Kuhl, J., et al. (in press). Existential neuroscience: a functional magnetic resonance imaging investigation of neural responses to reminders of one's mortality. *Social Cognitive and Affective Neuroscience.*

Radford, A. (1995). Phrase structure and functional categories. In P. Fletcher & B. MacWhinney (Eds.), *The handbook of child language* (pp. 483–507). Oxford, England: Blackwell.

Raedeke, T. D., & Smith, A. L. (2004). Coping resources and athlete burnout: An examination of stress mediated and moderation hypotheses. *Journal of Sport & Exercise Psychology, 26*, 525–541.

Ragins, B. R., Cotton, J. L., & Miller, J. S. (2000). Marginal mentoring: The effects of type of mentor, quality of relationship, and program design on work and career attitudes. *Academy of Management Journal, 43*, 1177–1194.

Ragland, O. R., & Brand, R. J. (1988). Type A behavior and mortality from coronary heart disease. *New England Journal of Medicine, 318*, 65–69.

Raikes, H., Luze, G., Brooks-Gunn, J., Raikes, H., Pan, B. A., Tamis-LeMonda, C. S., et al. (2006). Mother-child bookreading in low-income families: Correlates and outcomes during the first three years of life. *Child Development, 77*, 924–953.

Raine, A., Moffitt, T. E., Caspi, A., Loeber, R., Stouthamer-Loeber, M., & Lynam, D. (2005). Neurocognitive impairments in boys on the life-course persistent antisocial path. *Journal of Abnormal Psychology, 114*, 38–49.

Rakic, P. (1995). Corticogenesis in human and nonhuman primates. In M. S. Gazzaniga (Ed.), *The cognitive neurosciences.* Cambridge, MA: MIT Press.

Rakison, D. H., & Hahn, E. R. (2004). The mechanisms of early categorization and induction: Smart or dumb infants? *Advances in Child Development and Behavior, 32*, 281–322.

Rakoczy, H. (2008). Taking fiction seriously: Young children understand the normative structure of joint pretence games. *Developmental Psychology, 44*, 1195–1201.

Ralph, L. J., & Brindis, C. D. (2010). Access to reproductive healthcare for adolescents: establishing healthy behaviors at a critical juncture in the life-course. *Current Opinion in Obstetrics and Gynecology, 22*, 369–374.

Raman, L., & Gelman, S. A. (2005). Children's understanding of the transmission of genetic disorders and contagious illnesses. *Developmental Psychology, 41*, 171–182.

Ransjoe-Arvidson, A. B., Matthiesen, A. S., Lilja, G., Nissen, E., Widstroem, A. M., & Uvnaes-Moberg, K. (2001). Maternal analgesia during labor disturbs newborn behavior: Effects on breastfeeding, temperature, and crying. *Birth: Issues in Perinatal Care, 28*, 5–12.

Ranwez, S., Leidig, T., & Crampes, M. (2000). Formalization to improve life-long learning. *Journal of Interactive Learning Research, 11,* 389–409.

Rapaport, J. L., & Ismond, D. R. (1990). *DSM-III-R training guide for diagnosis of childhood disorders.* New York: Brunner/Mazel.

Rape, Abuse, and Incest National Network. (2010). *Statistics.* Retrieved from http://www.rainn.org/statistics

Rapport, M. D. (1995). Attention-deficit hyperactivity disorder. In M. Hersen & R. T. Ammerman (Eds.), *Advanced abnormal child psychology.* Hillsdale, NJ: Erlbaum.

Raschick, M., & Ingersoll-Dayton, B. (2004). The costs and rewards of caregiving among aging spouses and adult children. *Family Relations: Interdisciplinary Journal of Applied Family Studies, 53,* 317–325.

Raskauskas, J., & Stoltz, A. D. (2007). Involvement in traditional and electronic bullying among adolescents. *Developmental Psychology, 43,* 564–575.

Rathunde, K. R., & Csikszentmihalyi, M. (1993). Undivided interest and the growth of talent: A longitudinal study of adolescents. *Journal of Youth and Adolescence, 22,* 385–405.

Rawlins, W. K. (1992). *Friendship matters.* Hawthorne, NY: de Gruyter.

Rawlins, W. K. (2004). Friendships in later life. In J. F. Nussbaum & J. Coupland (Eds.), *Handbook of communication and aging research* (2nd ed., pp. 273–299). Mahwah, NJ: Erlbaum.

Rayner, K., Foorman, B. R., Perfetti, C. A. Pesetsky, D., & Seidenberg, M. S. (2001). How psychological science informs the teaching of reading. *Psychological Science in the Public Interest, 2,* 31–75.

Real, K., Mitnick, A. D., & Maloney, W. F. (2010). More similar than different: Millennials in the U.S. building trades. *Journal of Business and Psychology, 25,* 303–313.

Reese, E., & Cox, A. (1999). Quality of adult book reading affects children's emergent literacy. *Developmental Psychology, 35,* 20–28.

Reesman, M. C., & Hogan, J. D. (2005). Substance use and abuse across the life span. In C. B. Fisher & R. M. Lerner (Eds.), *Encyclopedia of applied developmental science* (pp. 1069–1072). Thousand Oaks, CA: Sage.

Reich, P. A. (1986). *Language development.* Englewood Cliffs, NJ: Prentice Hall.

Reid, D. H., Wilson, P. G., & Faw, G. D. (1991). Teaching self-help skills. In J. L. Matson & J. A. Mulick (Eds.), *Handbook of mental retardation* (2nd ed., pp. 429–442). New York: Pergamon Press.

Reid, M., Miller, W., & Kerr, B. (2004). Sex-based glass ceilings in U.S. state-level bureaucracies, 1987–1997. *Administration & Society, 36,* 377–405.

Reimer, M. S. (1996). "Sinking into the ground": The development and consequences of shame in adolescence. *Developmental Review, 16,* 321–363.

Reinhoudt, C. J. (2005). Factors related to aging well: The influence of optimism, hardiness and spiritual well-being on the physical health functioning of older adults. *Dissertation Abstracts International. Section B. Sciences and Engineering, 65(7-B),* 3762.

Remedios, J. D., Chasteen, A. L., & Packer, D. J. (2010). Sunny side up: The reliance on positive age stereotypes in descriptions of future older selves. *Self and Identity, 9,* 257–275.

Renaud, J., Berlim, M. T., McGirr, A., Tousignant, M., & Turecki, G. (2008). Current psychiatry morbidity, aggression/impulsivity, and personality dimensions in child and adolescent suicide: A case-control study. *Journal of Affective Disorders, 105,* 221–228.

Renshaw, K. D., Rodrigues, C., & Jones, D. H. (2008). Psychological symptoms and marital satisfaction in spouses of operation Iraqi freedom veterans: Relationships with spouses' perceptions of veterans' experiences and symptoms. *Journal of Family Psychology, 22,* 586–594.

Repacholi, B. M. (1998). Infants' use of attentional cues to identify the referent of another person's emotional expression. *Developmental Psychology, 34,* 1017–1025.

Repacholi, B. M., & Meltzoff, A. N. (2007). Emotional eavesdropping: Infants selectively respond to indirect emotional signals. *Child Development, 78,* 503–521.

Repacholi, B. M., Meltzoff, A. N., & Olsen, B. (2008). Infants' understanding of the link between visual perception and emotion: "If she can't see me doing it, she won't get angry." *Developmental Psychology, 44,* 561–574.

Reuter-Lorenz, P. A., & Park, D. C. (2010). Human neuroscience and the aging mind: A new look at old problems. *Journal of Gerontology: Psychological Sciences, 65B,* P405–P415.

Reville, B. (2011). Utilization of palliative care: providers still hinder access. *Health Policy Newsletter, 24(1).* Retrieved March 5, 2011 from http://jdc.jefferson.edu/cgi/viewcontent.cgi?article=1714&context=hpn.

Reyes, H. L. M., Foshee, V. A., Bauer, D. J., & Ennett, S. T. (2011). The role of heavy alcohol use in the developmental process of desisting aggression during adolescence. *Journal of Abnormal Child Psychology, 39,* 239–250.

Reyna, V. F., & Farley, F. (2006). Risk and rationality in adolescent decision making. *Psychological Science in the Public Interest, 7,* 1–44.

Reynolds, A. J., & Robertson, D. L. (2003). School-based early intervention and later child maltreatment in the Chicago Longitudinal Study. *Child Development, 74,* 3–26.

Rhoades, G. K., Stanley, S. M., & Markman, H. J. (2009). Couples' reasons for cohabitation: Associations with individual well-being and relationship quality. *Journal of Family Issues, 30,* 233–258.

Rhoades, K. A. (2008). Children's responses to interparental conflict: A meta-analysis of their associations with child adjustment. *Child Development, 79,* 1942–1956.

Ribe, E. M., Heidt, L., Beaubier, N., & Troy, C. M. (2011). Molecular mechanisms of neuronal death. *Neurochemical Mechanisms in Disease, 1,* 17–47.

Ricciardelli, L. A., & McCabe, M. P. (2004). A biopsychosocial model of disordered eating and the pursuit of muscularity in adolescent boys. *Psychological Bulletin, 130,* 179–205.

Ricciuti, H. N. (1993). Nutrition and mental development. *Current Directions in Psychological Science, 2,* 43–46.

Rice, N. E., Lang, I. A., Henley, W., & Melzer, D. (2010). Common health predictors of early retirement: Findings from the English Longitudinal Study of Ageing. *Age and Ageing, 40,* 54–61.

Rice, S. G. (1993). Injury rates among high school athletes 1979–1992. Unpublished raw data.

Richmond, J., & Nelson, C. A. (2007). Accounting for change in declarative memory: A cognitive neuroscience perspective. *Developmental Review, 27,* 349–373.

Richters, J. E., Arnold, L. E., Jensen, P. S., Abikoff, H., Conners, C. K., & Greenhill, L. L., et al. (1995). NIMH collaborative multisite multimodal treatment study of children with ADHD: I. Background and rationale. *Journal of the American Academy of Child & Adolescent Psychiatry, 34,* 987–1000.

Ridings, C., & Gefen, D. (2004). Virtual community attraction: Why people hang out online. *Journal of Computer-Mediated Communication, 10.* Retrieved from http://jcmc.indiana.edu/vol10/issue1/ ridings_gefen.html

Rijken, A. J. (2009). *Happy families, high fertility?: Childbearing choices in the context of family and partner relationships.* Dissertation submitted in partial fulfillment of the doctor of philosophy degree, University of Utrecht.

Riley, L. D., & Bowen, C. (2005). The sandwich generation: Challenges and coping strategies of multigenerational family. *Counseling & Therapy for Couples & Families, 13,* 52–58.

Riley, M. W. (1979). Introduction. In M. W. Riley (Ed.), *Aging from birth to death: Interdisciplinary perspectives* (pp. 3–14). Boulder, CO: Westview Press.

Risacher, S. L., & Saykin, A. J. (2011). Neuroimaging of Alzheimer's disease, mild cognitive impairment, and other dementias. In R. A. Cohen & H. L. Sweet (Eds.), *Brain imaging in behavioral medicine and clinical neuroscience* (pp. 309–339). New York: Springer.

Ritchie, K. L. (1999). Maternal behaviors and cognitions during discipline episodes: A comparison of power bouts and single acts of noncompliance. *Developmental Psychology, 35,* 580–589.

Rivera, S. (2007). Acculturation and ethnic identity as they relate to the psychological well-being of adult and elderly Mexican Americans. *Dissertation Abstracts International. Section B. Sciences and Engineering, 68(6-B),* 4141.

Robbins, A., & Wilner, A. (2001). *Quarterlife crisis: The unique challenges of life in your twenties.* New York: Putnam.

Roberson, E. D. (2011). Contemporary approaches to Alzheimer's disease and frontotemporal dementia. In E. D. Roberson (Ed.), *Alzheimer's disease and frontotemporal dementia: Methods and protocols* (pp. 1–9). New York: Humana Press.

Roberto, K. A., & Skoglund, R. R. (1996). Interactions with grandparents and great-grandparents: A comparison of activities, influences, and relationships. *International Journal of Aging & Human Development, 43,* 107–117.

Roberts, J. A. (2007). Developing an explanatory model of the chronicity of psychologically aggressive behavior among coupled gay and bisexual men. *Dissertation Abstracts International. Section A. Humanities and Social Sciences, 68(2-A),* 739.

Roberts, J. E., Burchinal, M., & Durham, M. (1999). Parents' report of vocabulary and grammatical development of African American preschoolers: Child and environmental associations. *Child Development, 70,* 92–106.

Roberts, R. E., Phinney, J. S., Masse, L. C., Chen, Y. R., Roberts, C. R., & Romero, A. (1999). The structure of ethnic identity of young adolescents from diverse ethnocultural groups. *Journal of Early Adolescence, 19,* 301–322.

Robertson, A. (2006). *Aging changes in the female reproductive system.* Retrieved from http://www.nlm.nih.gov/medlineplus/ency/article/004016.htm

Robertson, K., & Murachver, T. (2007). It takes two to tangle: Gender symmetry in intimate partner violence. *Basic and Applied Social Psychology, 29,* 109–118.

Robinson, A., & Clinkenbeard, P. R. (1998). Giftedness: An exceptionality examined. *Annual Review of Psychology, 49,* 117–139.

Robson, S. M., Hansson, R. O., Abalos, A., & Booth, M. (2006). Successful aging: Criteria for aging well in the workplace. *Journal of Career Development, 33,* 156–177.

Roby, A. C., & Kidd, E. (2008). The referential communication skills of children with imaginary companions. *Developmental Science, 11,* 531–540.

Rodeck, C. H., & Whittle, M. J. (Eds.), (2009). *Fetal medicine: Basic science and clinical practice.* London: Churchill Livingstone.

Roe, C. M., Fagan, A. M., Williams, M. M., Ghoshal, N., Aeschleman, M., Grant, E. A., et al. (2011). Improving CSF biomarker accuracy in predicting prevalent and incident Alzheimer disease. *Neurology, 76,* 501–510.

Roffwarg, H. P., Muzio, J. N., & Dement, W. C. (1966). Ontogenetic development of the human sleep-dream cycle. *Science, 152,* 604–619.

Rogers, L. A., & Graham, S. (2008). A meta-analysis of single subject design writing intervention research. *Journal of Educational Psychology, 100,* 879–906.

Rogers, S. D. (2010). Health promotion and chronic disease management. In J. C. Cavanaugh & C. K. Cavanaugh (Eds.), *Aging in America: Vol. 2: Physical and mental health* (pp. 57–80). Santa Barbara, CA: Praeger Perspectives.

Rogoff, B. (2003). *The cultural nature of human development.* New York: Oxford University Press.

Rogoff, B., Mistry, J., Goncu, A., & Mosier, C. (1993). Guided participation in cultural activity by toddlers and caregivers. *Monographs of the Society for Research in Child Development, 58* (Serial No. 236).

Rogosch, F. A., Cicchetti, D., Shields, A., & Toth, S. L. (1995). Parenting dysfunction in child maltreatment. In M. H. Bornstein (Ed.), *Handbook of parenting* (Vol. 4, pp. 127–159). Mahwah, NJ: Erlbaum.

Rönnqvist, L., & Domellöff, E. (2006). Quantitative assessment of right and left reaching movements in infants: A longitudinal study from 6 to 36 months. *Developmental Psychobiology, 48,* 444–459.

Roosevelt, F. D. (1935). *Statement on signing the Social Security Act.* Retrieved July 1, 2011 from http://docs.fdrlibrary.marist.edu/odssast.html.

Roper, L. L. (2007). Air force single parent mothers and maternal separation anxiety. *Dissertation Abstracts International. Section A. Humanities and Social Sciences, 67(11-A),* 4349.

Rose, A. J., & Asher, S. R. (1999). Children's goals and strategies in response to conflicts within a friendship. *Developmental Psychology, 35,* 69–79.

Rose, A. J., & Rudolph, K. D. (2006). A review of sex differences in peer relationship processes: Potential trade-offs for the emotional and behavioral development of girls and boys. *Psychological Bulletin, 132,* 98–131.

Rose, A. J., Carlson, W., & Waller, E. M. (2007). Prospective associations of co-rumination with friendship and emotional adjustment: Considering the socioemotional trade-offs of co-rumination. *Developmental Psychology, 43,* 1019–1031.

Rose, D. M., & Gordon, R. (2010). Retention practices for engineering and technical professionals in an Australian public agency. *Australian Journal of Public Administration, 69,* 314–325.

Rose, S. A., Feldman, J. F., & Jankowski, J. J. (2009). Information processing in toddlers: Continuity from infancy and persistence of preterm deficits. *Intelligence, 37,* 311–320.

Rose, S., & Zand, D. (2000). Lesbian dating and courtship from young adulthood to midlife. *Journal of Gay and Lesbian Social Services, 11,* 77–104.

Rosenbaum, J. L., Smith, J. R., & Zollfrank, B. C. C. (2011). Neonatal end-of-life spiritual support care. *Journal of Perinatal and Neonatal Nursing, 25,* 61–69.

Rosenblatt, P. C. (1996). Grief that does not end. In D. Klass, P. R. Silverman, & S. L. Nickman (Eds.), *Continuing bonds: New understandings of grief* (pp. 45–58). Washington, DC: Taylor & Francis.

Rosenblatt, P. C. (2001). A social constructivist perspective on cultural differences in grief. In M. S. Stroebe, R. O. Hansson, W. Stroebe, & H. Schut (Eds.), *Handbook of bereavement research: Consequences, coping, and care* (pp. 285–300). Washington, DC: American Psychological Association.

Rosengren, K. S., Gelman, S. A., Kalish, C., & McCormick, M. (1991) As time goes by: Children's early understanding of growth in animals. *Child Development, 62,* 1302–1320.

Rosso, B. D., Dekas, K. H., & Wrzesniewski, A. (in press). On the meaning of work: A theoretical integration and review. *Research in organizational Behavior.*

Rostenstein, D., & Oster, H. (1997). Differential facial responses to four basic tastes in newborns. In P. Ekman & E. L. Rosenberg (Eds), *What the face reveals: Basic and applied studies of spontaneous expression using the Facial Action Coding System (FACS). Series in affective science.* New York: Oxford University Press.

Roszko, E. (2010). Commemoration and the state: Memory and legitimacy in Vietnam. *Sojourn: Journal of Social Issues in Southeast Asia, 25,* 1–28.

Rotenberg, K. J., & Mayer, E. V. (1990). Delay gratification in native and white children: A cross-cultural comparison. *International Journal of Behavioral Development, 13,* 23–30.

Rothbart, M. K. (2007). Temperament, development, and personality. *Current Directions in Psychological Science, 16,* 207–212.

Rothbart, M. K., & Rueda, M. R. (2005). The development of effortful control. In U. Mayr, E. Awh, & S. W. Keele (Eds.), *Developing individuality in the human brain: A tribute to Michael I. Posner* (pp. 167–188). Washington, DC: American Psychological Association.

Rothbart, M. K., & Sheese, B. E. (2007). Temperament and emotion regulation. In J. J. Gross (Ed.), *Handbook of emotion regulation* (pp. 331–350). New York: Guilford.

Rothbaum, F., Weisz, J., Pott, M., Miyake, K., & Morelli, G. (2000). Attachment and culture: Security in the United States and Japan. *American Psychologist, 55,* 1093–1104.

Rothblum, E. D. (2009). An overview of same-sex couples in relationships: A research area still at sea. *Nebraska Symposium on Motivation: Contemporary Perspectives on Lesbian, Gay, and Bisexual Identities, 54,* 113–139.

Rotherman-Borus, M. J., & Langabeer, K. A. (2001). Developmental trajectories of gay, lesbian, and bisexual youths. In A. R. D'Augelli & C. Patterson (Eds.), *Lesbian, gay, and bisexual identities among youth: Psychological perspectives* (pp. 97–128). New York: Oxford University Press.

Rovee-Collier, C. (1987). Learning and memory in infancy. In J. D. Osofsky (Ed.), *Handbook of infant development* (2nd ed., pp. 98–148). New York: Wiley.

Rovee-Collier, C. (1997). Dissociations in infant memory: Rethinking the development of implicit and explicit memory. *Psychological Review, 104,* 467–498.

Rovee-Collier, C. (1999). The development of infant memory. *Current Directions in Psychological Science, 8,* 80–85.

Rowe, M. L., & Goldin-Meadow, S. (2009). Early gesture selectively predicts later language learning. *Developmental Science, 12,* 182–187.

Rowe, M. M., & Sherlock, H. (2005). Stress and verbal abuse in nursing: Do burned out nurses eat their young? *Journal of Nursing Management, 13,* 242–248.

Roxburgh, S. (2002). Racing through life. The distribution of time pressures by roles and roles resources among full-time workers. *Journal of Family and Economic Issues, 23,* 121–145.

Rubin, D. C., Berntsen, D., & Hutson, M. (2009). The normative and the personal life: Individual differences in life scripts and the life story events among USA and Danish undergraduates. *Memory, 17,* 54–68.

Rubin, D. C., Rahhal, T., & Poon, L. W. (1998). Things learned in early adulthood are remembered best: Effects of a major transition on memory. *Memory and Cognition, 26,* 3–19.

Rubin, K., Bukowski, W., & Parker, J. (2006). Peer interaction and social competence. In W. Damon & R. M. Lerner (Eds.), *Handbook of child psychology* (6th ed., Vol. 3, pp. 571–645). New York: Wiley.

Rubin, K. H., Bukowski, W., & Parker, J. G. (1998). Peer interactions, relationships, and groups. In W. Damon (Ed.), *Handbook of child psychology* (Vol. 3, pp. 619–700). New York: Wiley.

Rubin, K. H., Coplan, R. J., & Bowker, J. C. (2009). Social withdrawal in childhood. *Annual Review of Psychology, 60,* 141–171.

Rubin, K. H., Stewart, S., & Chen, X. (1995). Parents of aggressive and withdrawn children. In M. Bornstein (Ed.), *Handbook of parenting* (Vol. 1, pp. 255–284). Hillsdale, NJ: Erlbaum.

Rubin, S. S. (1996). The wounded family: Bereaved parents and the impact of adult child loss. In D. Klass, P. R. Silverman, & S. L. Nickman (Eds.), *Continuing bonds: New understandings of grief* (pp. 217–232). Philadelphia: Taylor & Francis.

Rubin, S. S., & Malkinson, R. (2001). Parental response to child loss across the life cycle: Clinical and research perspectives. In M. S. Stroebe, R. O. Hansson, W. Stroebe, & H. Schut (Eds.), *Handbook of bereavement research: Consequences, coping, and care* (pp. 169–197). Washington, DC: American Psychological Association.

Ruble, D. N., Martin, C. L., & Berenbaum, S. A. (2006). Gender development. In N. Eisenberg, W. Damon, & R. M. Lerner (Eds.), *Handbook of child psychology: Vol. 3, Social, emotional, and personality development* (6th ed., pp. 858–932). Hoboken, NJ: John Wiley & Sons.

Ruble, D. N., Taylor, L. J., Cyphers, L., Greulich, F. K., Lurye, L. E., & Shrout, P. E. (2007). The role of gender constancy in early gender development. *Child Development, 78,* 1121–1136.

Rudolph, K. D., & Troop-Gordon, W. (2010). Personal-accentuation and contextual-amplification models of pubertal timing: Predicting youth depression. *Development and Psychopathology, 22,* 433–451.

Rudolph, K. D., Ladd, G., & Dinella, L. (2007). Gender differences in the interpersonal consequences of early-onset depressive symptoms. *Merrill-Palmer Quarterly, 53,* 461–488.

Rudolph, K. D., Troop-Gordon, W., & Flynn, M. (2009). Relational victimization predicts children's social-cognitive and self-regulatory responses in a challenging peer context. *Developmental Psychology, 45,* 1444–1454.

Ruppanner, L. E. (2010). Cross-national reports of housework: An investigation of the gender empowerment measure. *Social Science Research, 19,* 963–975.

Rushton, J. P., & Bonds, T. A. (2005). Mate choice and friendship in twins. *Psychological Science, 16,* 555–559.

Russell, A., & Finnie, V. (1990). Preschool children's social status and maternal instructions to assist group entry. *Developmental Psychology, 26,* 603–611.

Russo, R. (Ed.). (2008). *A healing touch: True stories of life, death, and hospice.* Camden, ME: Down East Books.

Rutland, A., Killen, M., & Abrams, D. (2010). A new social-cognitive developmental perspective on prejudice: The interplay between morality and group identity. *Perspectives on Psychological Science, 5,* 279–291.

Rutter, M. (2007). Gene-environment interdependence. *Developmental Science, 10,* 12–18.

Rwampororo, R. K. (2001). Social support: Its mediation of gendered patterns in work-family stress and health for dual-earner couples. *Dissertation Abstract International. Section A. Humanities and Social Sciences, 61(9-A),* 3792.

Ryan, M. K., Haslam, S. A., & Kulich, C. (2010). Politics and the glass cliff: Evidence that women are preferentially selected to contest hard-to-win seats. *Psychology of Women Quarterly, 34,* 56–64.

Rybash, J. M., Hoyer, W. J., & Roodin, P. A. (1986). *Adult cognition and aging.* New York: Pergamon Press.

Rye, M. S., Folck, C. D., Heim, T. A., Olszewski, B. T., & Traina, E. (2004). Forgiveness of an ex-spouse: How does it relate to mental health following a divorce? *Journal of Divorce & Remarriage, 41,* 31–51.

Rylands, K., & Rickwood, D. J. (2001). Ego integrity versus despair: The effect of "accepting the past" on depression in older women. *International Journal of Aging & Human Development, 53,* 75–89.

Rymer, R. (1993). *Genie.* New York: HarperCollins.

Saarni, C., Campos, J. J., Camras, L. A., & Witherington, D. (2006). Emotional development: Action, communication, and understanding. In N. Eisenberg (Ed.), *Handbook of child psychology: Vol. 3. Social, emotional, and personality development.* (pp. 226–299). Hoboken, NJ: Wiley.

Sabia, S., Guéguen, A., Marmot, M. G., Shipley, M. J., Ankri, J., & Singh-Manoux, A. (2010). Does cognition predict mortality in midlife? Results from the Whitehall II cohort study. *Neurobiology of Aging, 31,* 688–695.

Saffran, J. R., Aslin, R. N., & Newport, E. L. (1996). Statistical learning by 8-month-old infants. *Science, 274,* 1926–1928.

Saffran, J. R., Werker, J. F., & Werner, L. A. (2006). The infant's auditory world: Hearing, speech, and the beginnings of language. In W. Damon & R. M. Lerner (Eds.), *Handbook of child psychology: Vol. 2. Cognition, perception, and language* (6th ed., pp. 58–108). Hoboken, NJ: Wiley.

Sagi, A., Koren-Karie, N., Gini, M., Ziv, Y., & Joels, T. (2002). Shedding further light on the effects of various types and quality of early child care on infant–mother attachment relationship: The Haifa study of early child care. *Child Development, 73,* 1166–1186.

Sagi, A., van IJzendoorn, M. H., Aviezer, O., Donnell, F., & Mayseless, O. (1994). Sleeping out of home in a kibbutz communal arrangement: It makes a difference for infant-mother attachment. *Child Development, 65,* 992–1004.

Saginak, K. A., & Saginak, M. A. (2005). Balancing work and family: Equity, gender, and marital satisfaction. *Family Journal: Counseling & Therapy for Couples & Families, 13,* 162–166.

Sahin, E., & DePinho, R. A. (2010). Linking functional decline of telomeres, mitochondria and stem cells during ageing. *Nature, 464,* 520–528.

Sahni, R., Fifer, W. P., & Myers, M. M. (2007). Identifying infants at risk for sudden infant death syndrome. *Current Opinion in Pediatrics, 19,* 145–149.

Sakraida, T. J. (2005). Divorce transition differences of midlife women. *Issues in Mental Health Nursing, 26,* 225–249.

Salmivalli, C., & Isaacs, J. (2005). Prospective relations among victimization, rejection, friendlessness, and children's self- and peer-perceptions. *Child Development, 76,* 1161–1171.

Salovey, P., & Grewal, D. (2005). The science of emotional intelligence. *Current Directions in Psychological Science, 14,* 281–285.

Salthouse, T. A. (1984). Effects of age and skill in typing. *Journal of Experimental Psychology: General, 113,* 345–371.

Salthouse, T. A. (2000). Steps toward the explanation of adult age differences in cognition. In T. Perfect & E. Maylor (Eds.), *Theoretical debate in cognitive aging* (pp. 19–49). Oxford, England: Oxford University Press.

Salthouse, T. A. (2006). Mental exercise and mental aging. *Perspectives on Psychological Science, 1,* 68–87.

Salthouse, T. A. (2010a). Influence of age on practice effects in longitudinal neurocognitive change. *Neuropsychology, 24,* 563–572.

Salthouse, T. A. (2010b). *Major issues in cognitive aging.* New York: Oxford University Press.

Sandler, I. N., Tein, J.-Y., Mehta, P., Wolchik, S., & Ayers, T. (2000). Coping efficacy and psychological problems of children of divorce. *Child Development, 71,* 1099–1118.

Sandler, I. N., Wolchik, S. A., & Ayers, T. S. (2008). Resilience rather than recovery: A contextual framework on adaptation following bereavement. *Death Studies, 32,* 59–73.

Sangrador, J. L., & Yela, C. (2000). "What is beautiful is loved": Physical attractiveness in love relationships in a representative sample. *Social Behavior and Personality, 28,* 207–218.

Sangrigoli, S., Pallier, C., Argenti, A.-M., Ventureyra, V. A. G., & de Schonen, S. (2005). Reversibility of the other-race effect in face recognition during childhood. *Psychological Science, 16,* 440–444.

Sann, C., & Streri, A. (2007). Perception of object shape and texture in human newborns: Evidence from cross-modal transfer tasks. *Developmental Science, 10,* 399–410.

Sanson, A., Prior, M., Smart, D., & Oberklaid, F. (1993). Gender differences in aggression in childhood: Implications for a peaceful world. *Australian Psychologist, 28,* 86–92.

Sargeant, M. (2004). Mandatory retirement age and age discrimination. *Employee Relations, 26,* 151–166.

Sarnecka, B. W., Kamenskaya, V. G., Yamana, Y., Ogura, T., & Yudovina, Y. (2007). From grammatical number to exact numbers: Early meanings of "one," "two," and "three" in English, Russian, and Japanese. *Cognitive Psychology, 55,* 136–168.

Saudino, K. J. (2009). Do different measures tap the same genetic influences? A multi-method study of activity level in young twins. *Developmental Science, 12,* 626–633.

Saudino, K. J., & Plomin, R. (2007). Why are hyperactivity and academic achievement related? *Child Development, 78,* 972–986.

Savage-Rumbaugh, E. S. (2001). *Apes, language, and the human mind.* New York, NY: Oxford University Press.

Saxe, G. B. (1988). The mathematics of child street vendors. *Child Development, 59,* 1415–1425.

Saxon, S. V., & Etten, M. J. (1994). *Physical changes and aging* (3rd ed.). New York: Tiresias.

Scandura, T. A., & Williams, E. A. (2004). Mentoring and transformational leadership: The role of supervisory career mentoring. *Journal of Vocational Behavior, 65,* 448–468.

Scarr, S. (1992). Developmental theories for the 1990s: Development and individual differences. *Child Development, 63,* 1–19.

Scarr, S., & McCartney, K. (1983). How people make their own environments: A theory of genotype environment effects. *Child Development, 54,* 424–435.

Schaal, B., Marlier, L., & Soussignan, R. (1998). Olfactory function in the human fetus: Evidence from selective neonatal responsiveness to the odor of amniotic fluid. *Behavioral Neuroscience, 112,* 1438–1449.

Schaie, K. W. (1994). The course of adult intellectual development. *American Psychologist, 49,* 304–313.

Schaie, K. W. (2009). "When does age-related cognitive decline begin?": Salthouse again reifies the "cross-sectional fallacy." *Neurobiology of Aging, 30,* 528–533.

Schaie, K. W., & Willis, S. L. (1995). Perceived family environment across generations. In V. L. Bengston & K. W. Schaie (Eds.), *Adult intergenerational relations: Effects of societal change* (pp. 174–226). New York: Springer.

Schaie, K. W., & Zanjani, F. (2006). Intellectual development across adulthood. In C. Hoare (Ed.), *Oxford handbook of adult development and learning* (pp. 99–122). New York: Oxford University Press.

Schaie, K. W., Maitland, S. B., Willis, S. L., & Intrieri, R. L. (1998). Longitudinal invariance of adult psychometric ability factor structures across seven years. *Psychology and Aging, 13,* 8–20.

Schapira, A. H. V., & Olanow, C. W. (2004). Neuroprotection in Parkinson disease: Mysteries, myths, and misconceptions. *Journal of the American Medical Association, 291,* 358–364.

Scheibe, S., Kunzmann, U., & Baltes, P. B. (2007). Wisdom, life longings, and optimal development. In J. A. Blackburn & C. N. Dulmus (Eds.), *Handbook of gerontology: Evidence-based approaches to theory, practice, and policy* (pp. 117–142). Hoboken, NJ: Wiley.

Scheidt, R. J., & Schwarz, B. (2010). Environmental gerontology: A sampler of issues and applications. In J. C. Cavanaugh & C. K. Cavanaugh (Eds.), *Aging in America: Vol. 1: Psychological Aspects* (pp. 156–176). Santa Barbara, CA: Praeger Perspectives.

Scherf, K. S., Behrmann, M., Humphreys, K., & Luna, B. (2007). Visual category-selectivity for faces, places and objects emerges along different developmental trajectories. *Developmental Science, 10,* F15–F30.

Schermerhorn, A. C., & Cummings, E. M. (2008). Transactional family dynamics: A new framework for conceptualizing family influence processes. In R. V. Kail (Ed.), *Advances in child development and behavior* (Vol. 36, pp. 187–250). Amsterdam, Netherlands: Academic Press.

Schermerhorn, A. C., Chow, S., & Cummings, E. M. (2010). Developmental family processes and interparental conflict: Patterns of microlevel influences. *Developmental Psychology, 46,* 869–885.

Schlossberg, N. K. (2004). *Retire smart, retire happy: Finding your true path in life.* Washington, DC: American Psychological Association.

Schmale, R., & Seidl, A. (2009). Accommodating variability in voice and foreign accent: Flexibility of early word representations. *Developmental Science, 12,* 583–601.

Schmeeckle, M., Giarusso, R., & Wang, Q. (1998, November). *When being a brother or sister is important to one's identity: Life stage and gender differences.* Paper presented at the annual meeting of the Gerontological Society, Philadelphia.

Schmidt, F. L., & Hunter, J. E. (1998). The validity and utility of selection methods in personnel psychology: Practical and theoretical implications of 85 years of research findings. *Psychological Bulletin, 124,* 262–274.

Schmidt, F. L., & Hunter, J. E. (2004). General mental ability in the world of work: Occupational attainment and job performance. *Journal of Personality and Social Psychology, 86,* 162–173.

Schmithorst, V. J., & Yuan, W. (2010). White matter development during adolescence as shown by diffusion MRI. *Brain & Cognition, 72,* 16–25.

Schmitt, D. P., Alcalay, L., Allensworth, M., Allik, J., Ault, L., Austers, I., et al. (2004). Patterns and universals of adult romantic attachment across 62 cultural regions: Are models of self and of other pancultural constructs? *Journal of Cross-Cultural Psychology, 35,* 367–402.

Schmitt, D. P., Youn, G., Bond, B., Brooks, S., Frye, H., Johnson, S., et al. (2009). When will I feel love? The effects of culture, personality, and gender on the psychological tendency to love. *Journal of Research in Personality, 43,* 830–846.

Schmitz-Scherzer, R., & Thomae, H. (1983). Constancy and change of behavior in old age: Findings from the Bonn Longitudinal Study on Aging. In K. W. Schaie (Ed.), *Longitudinal studies of adult psychological development* (pp. 191–221). New York: Guilford.

Schneider, B. A., Pichora-Fuller, K., & Daneman, M. (2010). Effects of senescent changes in audition and cognition on spoken language comprehension. In S. Gordon-Salant, R. D. Frisina, A. N. Popper, & R. R. Fay (Eds.), *The aging auditory system* (pp. 167–210). New York: Springer.

Schneider, E. L. and L. Davidson (2003). Physical health and adult well-being. In The Center for Child Well-Being (Ed.), *Well-being: Positive development across the life course* (pp. 407–423). Mahwah, NJ: Erlbaum.

Schneider, K. T., Swan, S., & Fitzgerald, L. F. (1997). Job-related and psychological effects of sexual harassment in the workplace: Empirical evidence from two organizations. *Journal of Applied Psychology, 82,* 401–415.

Schneider, M. L., Roughton, E. C., Koehler, A. J., & Lubach, G. R. (1999). Growth and development following prenatal stress exposure in primates: An examination of ontogenetic vulnerability. *Child Development, 70,* 253–274.

Schneider, W., & Bjorklund, D. F. (1998). Memory. In W. Damon (Ed.), *Handbook of child psychology* (Vol. 2, pp. 467–521). New York: Wiley.

Schneider, W., & Pressley, M. (1997). *Memory development between 2 and 20* (2nd ed.). Mahwah, NJ: Erlbaum.

Schneiders, J., Nicolson, N. A., Berkhof, J., Feron, F. J., van Os., J., & deVries, M. W. (2006). Mood reactivity to daily negative events in early adolescence: Relationship to risk for psychopathology. *Developmental Psychology, 42,* 543–554.

Schofield, T. J., Parke, R. D., Kim, Y., & Coltrane, S. (2008). Bridging the acculturation gap: Parent-child relationship quality as a moderator in Mexican American families. *Developmental Psychology, 44,* 1190–1194.

Schoon, I. (2001). Teenage job aspirations and career attainment in adulthood: A 17-year follow-up study of teenagers who aspired to become scientists, health professionals, or engineers. *International Journal of Behavioral Development, 25,* 124–132.

Schott, W. (2010). *Going back part-time: Federal leave legislation and women's return to work.* Retrieved from http://www.sas.upenn.edu/~wschott/Schott-Part-time-return-052010.pdf

Schouten, A. P., Valkenburg, P. M., & Peter, J. (2007). Precursors and underlying processes of adolescents' online self-disclosure: Developing and testing an "Internet-attribute-perception" model. *Media Psychology, 10,* 292–314.

Schulte, H. A. (2006). Family of origin and sibling influence on the experience of social support in adult friendships. *Dissertation Abstracts International. Section B. Sciences and Engineering, 67(6-B),* 3510.

Schwartz, C. E., Wright, C. I., Shin, L. M., Kagan, J., & Rauch, S. L. (2003). Inhibited and uninhibited infants "grow up": Adult amygdalar response to novelty. *Science, 300,* 1952–1953.

Schwartz, C. R., & Graf, N. L. (2009). Assortative matching among same-sex and different-sex couples in the United States. *Demographic Research, 21,* 843–878.

Schwartz, D., Dodge, K. A., Pettit, G. S., Bates, J. E., & The Conduct Problems Prevention Research Group. (2000). Friendship as a moderating factor in the pathway between early harsh home environment and later victimization in the peer group. *Developmental Psychology, 36,* 646–662.

Schwarzer, G., Zauner, N., & Jovanovic, B. (2007). Evidence of a shift from featural to configural face processing in infancy. *Developmental Science, 10,* 452–463.

Schwarzer, R. (2008). Modeling health behavior change: How to predict and modify the adoption and maintenance of health behaviors. *Applied Psychology: An International Review, 57,* 1–29.

Schwerdtfeger, A., & Friedrich-Mai, P. (2009). Social interaction moderate the relationship between depressive mood and heart rate variability: Evidence from an ambulatory monitoring study. *Health Psychology, 28,* 501–509.

Scott, W. A., Scott, R., & McCabe, M. (1991). Family relationships and children's personality: A cross-cultural, cross-source comparison. *British Journal of Social Psychology, 30,* 1–20.

Segal, D. L., Bogaards, J. A., Becker, L. A., & Chatman, C. (1999). Effects of emotional expression on adjustment to spousal loss among older adults. *Journal of Mental Health and Aging, 5,* 297–310.

Segrin, C., Taylor, M. E., & Altman, J. (2005). Social cognitive mediators and relational outcomes associated with parental divorce. *Journal of Social and Personal Relationships, 22,* 361–377.

Seidl, A., & Johnson, E. L. (2006). Infants' word segmentation revisited: Edge alignment facilitates target extraction. *Developmental Science, 9,* 565–573.

Seifer, R., Schiller, M., Sameroff, A. J., Resnick, S., & Riordan, K. (1996). Attachment, maternal sensitivity, and infant temperament during the first year of life. *Developmental Psychology, 32,* 12–25.

Selman, R. L. (1980). *The growth of interpersonal understanding: Development and clinical analyses.* New York: Academic Press.

Selman, R. L. (1981). The child as a friendship philosopher: A case study in the growth of interpersonal understanding. In S. R. Asher & J. M. Gottman (Eds.), *The development of children's friendships.* Cambridge, England: Cambridge University Press.

Sénéchal, M., & LeFevre, J. (2002). Parental involvement in the development of children's reading skill: A five-year longitudinal study. *Child Development, 73,* 445–460.

Sénéchal, M., Thomas, E., & Monker, J. (1995). Individual differences in 4-year-old children's acquisition of vocabulary during storybook reading. *Journal of Educational Psychology, 87,* 218–229.

Sener, A., Terzioglu, R. G., & Karabulut, E. (2007). Life satisfaction and leisure activities during men's retirement: A Turkish sample. *Aging & Mental Health, 11,* 30–36.

Sera, E. J. (2001). Men and spousal bereavement: A cross-cultural study of majority-culture and Hispanic men and the role of religiosity and acculturation on grief. *Dissertation Abstracts International. Section B. Sciences and Engineering, 61(11-B),* 6149.

Serbin, L., & Karp, J. (2003). Intergenerational studies of parenting and the transfer of risk from parent to child. *Current Directions in Psychological Science, 12,* 138–142.

Serbin, L. A., Poulin-Dubois, D., Colburne, K. A., Sen, M. G., & Eichstedt, J. A. (2001). Gender stereotyping in infancy: Visual preferences for and knowledge of gender-stereotyped toys in the second year. *International Journal of Behavioral Development, 25,* 7–15.

Serbin, L. A., Powlishta, K. K., & Gulko, J. (1993). The development of sex typing in middle childhood. *Monographs of the Society for Research in Child Development, 58*(Serial No. 232).

Servaty-Seib, H. L., & Taub, D. J. (2010). Bereavement and college students: the role of counseling psychology. *The Counseling Psychologist, 38,* 947–975.

Servin, A., Nordenstroem, A., Larsson, A., & Bohlin, G. (2003). Prenatal androgens and gender-typed behavior: A study of girls with mild and severe forms of congenital-adrenal hyperplasia. *Developmental Psychology, 39,* 440–450.

Seward, R. R., Yeatts, D. E., Amin, I., & DeWitt, A. (2006). Employment leave and fathers' involvement with children: According to mothers and fathers. *Men and Masculinities, 8,* 405–427.

Seyda, B. A., & Fitzsimons, A. M. (2010). Infant deaths. In C. A. Corr & D. A. Balk (Eds.), *Children's encounters with death, bereavement, and coping* (pp. 83–107). New York: Springer.

Seyfarth, R., & Cheney, D. (1996). Inside the mind of a monkey. In M. Bekoff & D. Jamieson (Eds.), *Readings in animal cognition.* (pp. 337–343). Cambridge, MA: MIT Press.

Shalev, R. (1999). *Comparison of war-bereaved and motor vehicle accident-bereaved parents.* Unpublished master's thesis, University of Haifa.

Shanahan, L., McHale, S. M., Crouter, A. C., & Osgood, D. (2007). Warmth with mothers and father from middle childhood to late adolescence: Within- and between-families comparisons. *Developmental Psychology, 43,* 551–563.

Shanahan, M. J., Elder, G. H., Burchinal, M., & Conger, R.-D. (1996a). Adolescent earnings and relationships with parents: The work-family nexus in urban and rural ecologies. In J. T. Mortimer & M. D. Finch (Eds.), *Adolescents, work, and family: An intergenerational developmental analysis.* Thousand Oaks, CA: Sage.

Shanahan, M. J., Elder, G. H., Burchinal, M., & Conger, R.-D. (1996b). Adolescent paid labor and relationships with parents: Early work-family linkages. *Child Development, 67,* 2183–2200.

Shapiro, S. (2007). Recent epidemiological evidence relevant to the clinical management of the menopause. *Climacteric, 10*(Suppl. 2), 2–15.

Share, D. L. (2008). Orthographic learning, phonological recoding, and self-teaching. In R. V. Kail (Ed.), *Advances in child development and behavior* (Vol. 36, pp. 31–84). San Diego, CA: Elsevier.

Shaw, D. S., Winslow, E. B., & Flanagan, C. (1999). A prospective study of the effects of marital status and family relations on young children's adjustment among African American and European American families. *Child Development, 70,* 742–755.

Shaw, G. M., Schaffer, D., Velie, E. M., Morland, K., & Harris, J. A. (1995). Periconceptional vitamin use, dietary folate, and the occurrence of neural tube defects. *Epidemiology, 6,* 219–226.

Shaw, S. S. (2007). Losing a parent twice. *American Journal of Alzheimer's Disease and Other Dementias, 21,* 389–390.

Shea, J. L. (2006). Cross-cultural comparison of women's midlife symptom-reporting: A China study. *Culture, Medicine and Psychiatry, 30,* 331–362.

Shear, M. K., Simon, N., Wall, M., Zisook, S., Neimeyer, R., Duan, N., et al., (2011). Complicated grief and related bereavement issues for DSM-5. *Depression and Anxiety, 28,* 103–117.

Shebilske, L. J. (2000). Affective quality, leisure time, and marital satisfaction: A 13-year longitudinal study. *Dissertation Abstracts International. Section A. Humanities and Social Sciences, 60(9-A),* 3545.

Shelly, W., Draper, M. W., Krishnan, V., Wong, M., & Jaffe., R. B. (2008). Selective estrogen receptor modulators: An update on recent clinical findings. *Obstetrical and Gynecological Survey, 63,* 163–181.

Sherman, A. M., de Vries, B., & Lansford, J. E. (2000). Friendship in childhood and adulthood: Lessons across the life span. *International Journal of Aging & Human Development, 51,* 31–51.

Sherwin, S. (2011). Looking backwards, looking forward: Hope for *Bioethics'* next twenty-five years. *Bioethics, 25,* 75–82.

Sheu, H.-B., Lent, R. W., Brown, S. D., Miller, M. J., Hennessy, K. D., & Duffy, R. D. (2010). Testing the choice model of social cognitive career theory across Holland themes: A meta-analytic path analysis. *Journal of Vocational Behavior, 76,* 252–264.

Shi, R., & Lepage, M. (2008). The effect of functional morphemes on word segmentation in preverbal infants. *Developmental Science, 11,* 407–413.

Shi, R., & Werker, J. F. (2001). Six-month old infants' preference for lexical words. *Psychological Science, 12,* 70–75.

Shih, P. C., & Jung, R. E. (2009). Gray matter correlates of fluid, crystallized, and spatial intelligence: Testing the P-FIT model. *Intelligence, 37,* 124–135.

Shiwach, R. (1994). Psychopathology in Huntington's disease patients. *Acta Psychiatrica Scandinavica, 90,* 241–246.

Shoemaker, L. B., & Furman, W. (2009). Interpersonal influences on late adolescent girls' and boys' disordered eating. *Eating Behaviors, 10,* 97–106.

Shulman, S., & Kipnis, O. (2001). Adolescent romantic relationships: A look from the future. *Journal of Adolescence, 24,* 337–351.

Shutts, K., Banaji, M. R., & Spelke, E. S. (2010). Social categories guide young children's preferences for novel objects. *Developmental Science, 13,* 599–610.

Sicotte, N. L., Woods, R. P., & Mazziotta, J. C. (1999). Handedness in twins: A meta-analysis. *Laterality: Asymmetries of Body, Brain, and Cognition, 4,* 265–286.

Siddiqui, A. (1995). Object size as a determinant of grasping in infancy. *Journal of Genetic Psychology, 156,* 345–358.

Sidebotham, P., Heron, J., & the ALSPAC Study Team. (2003). Child maltreatment in the "children of the nineties": The role of the child. *Child Abuse and Neglect, 27,* 337–352.

Siegler, I. C., George, L. K., & Okun, M. A. (1979). A cross-sequential analysis of adult personality. *Developmental Psychology, 15,* 350–351.

Siegler, R. S. (1981). Developmental sequences within and between concepts. *Monographs of the Society for Research in Child Development, 46* (Serial No. 189).

Siegler, R. S. (1986). Unities in strategy choices across domains. In M. Perlmutter (Ed.), *Minnesota symposia on child development* (Vol. 19, pp. 1–48). Hillsdale, NJ: Erlbaum.

Siegler, R. S. (1988). Strategy choice procedures and the development of multiplication skill. *Journal of Experimental Psychology: General, 117,* 258–278.

Siegler, R. S., & Alibali, M. W. (2004). *Children's thinking* (4th ed.). Upper Saddle River, NJ: Prentice Hall.

Siegler, R. S., & Jenkins, E. (1989). *How children discover new strategies.* Hillsdale, NJ: Erlbaum.

Siegler, R. S., & Robinson, M. (1982). The development of numerical understandings. In H. W. Reese & L. P. Lipsitt (Eds.), *Advances in child development and behavior* (Vol. 16, pp. 242–312). New York: Academic Press.

Siegler, R. S., & Shrager, J. (1984). Strategy choices in addition and subtraction: How do children know what to do? In C. Sophian (Ed.), *Origins of cognitive skills.* Hillsdale, NJ: Erlbaum.

Silk, J. S., Morris, A. S., Kanaya, T., & Steinberg, L. D. (2003). Psychological control and autonomy granting: Opposite ends of a continuum or distinct constructs? *Journal of Research on Adolescence, 13,* 113–128.

Silverman, P. R., & Nickman, S. L. (1996). Children's construction of their dead parents. In D. Klass, P. R. Silverman, & S. L. Nickman (Eds.), *Continuing bonds: New understandings of grief* (pp. 73–86). Washington, DC: Taylor & Francis.

Silverman, P. R., & Worden, J. W. (1992). Children's understanding of funeral ritual. *Omega, 25,* 319–331.

Silverman, W. K., La Greca, A. M., & Wasserstein, S. (1995). What do children worry about? Worries and their relations to anxiety. *Child Development, 66,* 671–686.

Silverstein, J. S. (2001). Connections and disconnections: Towards an understanding of reasons for mid-career professional women leave large corporations. *Dissertation Abstracts International. Section B. Sciences and Engineering, 62(1-B),* 581.

Simmons, R., & Blyth, D. (1987). *Moving into adolescence.* New York: Aldine de Gruyter.

Simon, N. M., Thompson, E. H., Pollack, M. H., & Shear, M. K. (2007). Complicated grief: A case series using escitalopram. *American Journal of Psychiatry, 164,* 1760–1761.

Simons, D. J., & Keil, F. C. (1995). An abstract to concrete shift in the development of biological thought: The insides story. *Cognition, 56,* 129–163.

Simonton, D. K. (1997). Creative productivity: A predictive and explanatory model of career trajectories and landmarks. *Psychological Review, 104,* 66–89.

Simonton, D. K. (2007). Creativity: Specialised expertise or general cognitive processes? In M. J. Roberts (Ed). *Integrating the mind: Domain general vs. domain specific processes in higher cognition* (pp. 351–367). New York: Psychology Press.

Simonton, D. K., & Song, A. V. (2009). Eminence, IQ, physical and mental health, and achievement domain: Cox's 282 geniuses revisited. *Psychological Science, 20,* 429–434.

Simpson, E. L. (1974). Moral development research: A case study of scientific cultural bias. *Human Development, 17,* 81–106.

Simpson, J. M. (2001). Infant stress and sleep deprivation as an aetiological basis for the sudden infant death syndrome. *Early Human Development, 61,* 1–43.

Sinclair, R. R., Sears, L. E., Zajack, M., & Probst, T. (2010). A multilevel model of economic stress and employee well-being. In J. Houdmont & S. Leka (Eds.), *Contemporary occupational health psychology: Global perspectives on research and practice* (Vol. 1, pp. 1–20). Malden, MA: Wiley-Blackwell.

Singer, J. D., Fuller, B., Keiley, M. K., & Wolf, A. (1998). Early child-care selection: Variation by geographic location, maternal characteristics, and family structure. *Developmental Psychology, 34,* 1129–1144.

Sinnott, J. (2009). Complex thought and construction of the self in the face of aging and death. *Journal of Adult Development, 16,* 155–165.

Sinnott, J. D. (1994). New science models for teaching adults: Teaching as a dialogue with reality. In J. D. Sinnott (Ed.), *Interdisciplinary handbook of adult lifespan learning* (pp. 90–104). Westport, CT: Greenwood Press.

Sinnott, J. D. (1998). *The development of logic in adulthood: Postformal thought and its applications.* New York: Plenum.

Siu, O.-L., Spector, P. E., Cooper, C. L., & Donald, I. (2001). Age differences in coping and locus of control: A study of managerial stress in Hong Kong. *Psychology and Aging, 16,* 707–710.

Skinner, B. F. (1957). *Verbal behavior.* New York: Appleton-Century-Crofts.

Skinner, E. A. (1985). Determinants of mother-sensitive and contingent-responsive behavior: The role of childbearing beliefs and socioeconomic status. In I. E. Sigel (Ed.), *Parental belief systems: The psychological consequences for children* (pp. 51–82). Hillsdale, NJ: Erlbaum.

Skouteris, H., McNaught, S., & Dissanayake, C. (2007). Mothers' transition back to work and infants' transition to child care: Does work-based child care make a difference? *Child Care in Practice, 13,* 33–47.

Slagboom, P. E., Beekman, M., Passtoors, W. M., Deelen, J., Vaarhorst, A. A. M., Boer, J. M., et al. (2011). Genomics of human longevity. *Philosophical Transactions of the Royal Society, 366,* 35–42.

Slaughter, V., & Griffiths, M. (2007). Death understanding and fear of death in young children. *Clinical Child Psychology and Psychiatry, 12,* 525–535.

Slobin, D. I. (1985). Cross-linguistic evidence for the language-making capacity. In D. I. Slobin (Ed.), *The cross-linguistic study of language acquisition: Vol. 2. Theoretical issues.* (pp. 1157–1256). Hillsdale, NJ: Erlbaum.

Small, B. (2004). Online personals and narratives of the self: Australia's RSVP. *Convergence: The Journal of Research into New Media Technologies, 10,* 93–107.

Small, B. J., Hertzog, C., Hultsch, D. F., & Dixon, R. A. (2003). Stability and change in adult personality over 6 years: Findings from the Victoria longitudinal study. *Journal of Gerontology: Psychological Sciences, 58B,* P166–P176.

Smith, D. B., & Moen, P. (2004). Retirement satisfaction for retirees and their spouses: Do gender and the retirement decision-making process matter? *Journal of Family Issues, 25,* 262–285.

Smith, E. R., & Mackie, D. M. (2000). *Social psychology* (2nd ed.). Philadelphia, PA: Psychology Press.

Smith, J., & Freund, A. M. (2002). The dynamics of possible selves in old age. *Journal of Gerontology: Psychological Sciences, 57B,* P492–P500.

Smith, L. B. (2000). How to learn words: An associative crane. In R. Golinkoff & K. Hirsch-Pasek (Eds.), *Breaking the word learning barrier* (pp. 51–80). Oxford, England: Oxford University Press.

Smith, L. B. (2009). From fragments to geometric shape: Changes in visual object recognition between 18 and 24 months. *Current Directions in Psychological Science, 18,* 290–294.

Smith, R. E., & Smoll, F. L. (1996). The coach as the focus of research and intervention in youth sports. In F. L. Smoll & R. E. Smith (Eds.), *Children and youth in sport: A biopsychological perspective* (pp. 125–141). Dubuque, IA: Brown & Benchmark.

Smith, S., & Gove, J. E. (2005). *Physical changes in aging.* Retrieved from http://edis.ifas.ufl.edu/he019

Smith, W. J., Howard, J. T., & Harrington, K. V. (2005). Essential formal mentor characteristics and functions in governmental and non-governmental organizations from the program administrator's and the mentor's perspective. *Public Personnel Management, 34,* 31–58.

Smits, I., Soenens, B., Vansteenkiste, M., Luyckx, K., & Goossens, L. (2010). Why do adolescents gather information or stick to parental norms? Examining autonomous and controlled motives behind adolescents' identity style. *Journal of Youth & Adolescence, 39,* 1343–1356.

Smoll, F. L., & Schutz, R. W. (1990). Quantifying gender differences in physical performance: A developmental perspective. *Developmental Psychology, 26,* 360–369.

Smyke, A. T., Zeanah, C. H. Fox, N. A., Nelson, C. A., & Guthrie, D. (2010). Placement in foster care enhances quality of attachment among young institutionalized children. *Child Development, 81,* 212–223.

Snedeker, B. (1982). *Hard knocks: Preparing youth for work.* Baltimore, MD: Johns Hopkins University Press.

Snow, C. W. (1998). *Infant development* (2nd ed.). Upper Saddle River, NJ: Prentice Hall.

Snow, D. (2006). Regression and reorganization of intonation between 6 and 23 months. *Child Development, 77,* 281–296.

Snow, M. E., Jacklin, C. N., & Maccoby, E. E. (1983). Sex-of-child differences in father–child interaction at one year of age. *Child Development, 54,* 227–232.

Social Security Administration. (2011). *Annual statistical supplement, 2010.* Retrieved from http://www.ssa.gov/policy/docs/statcomps/supplement/

Sokol, R. J., Delaney-Black, V., & Nordstrom, B. (2003). Fetal alcohol spectrum disorder. *JAMA, 290,* 2996–2999.

Somerville, L. H., & Casey, B. J. (2010). Developmental neurobiology of cognitive control and motivational systems. *Current Opinion in Neurobiology, 20,* 236–241.

Son, S., & Bauer, J. W. (2010). Employed rural, low income, single mothers' family and work over time. *Journal of Family and Economic Issues, 31,* 107–120.

Sousa, P., Altran, S., & Medin, D. (2002). Essentialism and folkbiology: Further evidence from Brazil. *Journal of Cognition and Culture, 2,* 195–223.

Spearman, C. (1904). "General intelligence" objectively determined and measured. *American Journal of Psychology, 15,* 201–293.

Spector, A. R. (2004). Psychological issues and interventions with infertile patients. *Women & Therapy, 27,* 91–105.

Spector, F., & Maurer, D. (2009). Synesthesia: A new approach to understanding the development of perception. *Developmental Psychology, 45,* 175–189.

Spector, P. E., Allen, T. D., Poelmans, S., Cooper, C. L., Bernin, P., Hart, P., et al. (2005). An international comparative study of work-family stress and occupational strain. In S. A. Y. Poelmans (Ed.), *Work and family: An international research perspective* (pp. 71–84). Mahwah, NJ: Erlbaum.

Spelke, E. S., & Kinzler, K. D. (2007). Core knowledge. *Developmental Science, 10,* 89–96.

Spirduso, W. W., Poon, L. W., & Chodzko-Zajko, W. (2008). Using resources and reserves in an exercise-cognition model. In W. W. Spirduso, L. W. Poon, & W. Chodzko-Zajko (Eds.), *Exercise and its mediating effects on cognition* (pp. 3–11). Champaign, IL: Human Kinetics.

Springer, K., & Keil, F. C. (1991). Early differentiation of causal mechanisms appropriate to biological and nonbiological kinds. *Child Development, 62,* 767–781.

Sprinzl, G. H., & Riechelmann, H. (2010). Current trends in treating hearing loss in elderly people: A review of the technology and treatment options. *Gerontology, 56,* 351–358.

Sritharan, R., Heilpern, K., Wilbur, C. J., & Gawronski, B. (2010). I think I like you: Spontaneous and deliberate evaluations of potential romantic partners in an online dating context. *European Journal of Social Psychology, 40,* 1062–1077.

Srivastava, S., John, O. P., Gosling, S. D., & Potter, J. (2003). Development of personality in early and middle adulthood: Set like plaster or persistent change? *Journal of Personality and Social Psychology, 84,* 1041–1053.

Sroufe, L. A., & Waters, E. (1976). The ontogenesis of smiling and laughter: A perspective on the organization of development in infancy. *Psychological Review, 83,* 173–189.

St. George, I. M., Williams, S., & Silva, P. A. (1994). Body size and the menarche: The Dunedin study. *Journal of Adolescent Health, 15,* 573–576.

St. James-Roberts, I., & Plewis, I. (1996). Individual differences, daily fluctuations, and developmental changes in amounts of infant waking, fussing, crying, feeding, and sleeping. *Child Development, 67,* 2527–2540.

Staff, J., & Schulenberg, J. E. (2010). Millenials and the world of work: Experiences in paid work during adolescence. *Journal of Business and Psychology, 25,* 247–255.

Stafford, L., Kline, S. L., & Rankin, C. T. (2004). Married individuals, cohabiters, and cohabiters who marry: A longitudinal study of relational and individual well-being. *Journal of Social & Personal Relationships, 21,* 231–248.

Stamatakis, E. A., Shafto, M. A., Williams, G., Tam, P., & Tyler, L. K. (2011). White matter changes and word finding failures with increasing age. *PLoS ONE, 6,* e14496. doi:10.1371/journal.pone.0014496. Retrieved from http://www.plosone.org/article/info%3Adoi%2F10.1371%2Fjournal.pone.0014496

Stange, K. C. (2010). Power to advocate for health. *Annals of Family Medicine, 8,* 100–107.

Stanovich, K. E., Toplak, M. E., & West, R. F. (2008). The development of rational thought: A taxonomy of heuristics and biases. In R. V. Kail (Ed.), *Advances in child development and behavior* (Vol. 36, pp. 251–285). San Diego, CA: Elsevier.

Starko, A. J. (1988). Effects of the Revolving Door Identification Model on creative productivity and self-efficacy. *Gifted Child Quarterly, 32,* 291–297.

Stauss, J. H. (1995). Reframing and refocusing American Indian family strengths. In C. K. Jacobson (Ed.), *American families: Issues in race and ethnicity* (pp. 105–118). New York: Garland.

Steele, C. M. (1997). A threat in the air: How stereotypes shape intellectual identity and performance. *American Psychologist, 52,* 613–629.

Steele, C. M., & Aronson, J. (1995). Stereotype threat and the intellectual test performance of African Americans. *Journal of Personality and Social Psychology, 69,* 797–811.

Stein, D. J., & Vythilingum, B. (2009). Love and attachment: The psychobiology of social bonding. *CNS Spectrums, 14,* 239–242.

Stein, J. H., & Reiser, L. W. (1994). A study of White middle-class adolescent boys' responses to "semenarche" (the first ejaculation). *Journal of Youth and Adolescence, 23,* 373–384.

Steinbach, S., Hundt, W., Vaitl, A., Heinrich, P., Förster, S., Bürger, K., et al., (2010). Taste in mild cognitive impairment and Alzheimer's disease. *Journal of Neurology, 257,* 238–246.

Steinberg, L. (2001). We know some things: Parent-adolescent relationships in retrospect and prospect. *Journal of Research on Adolescence, 11,* 1–19.

Steinberg, L. D. (1999). *Adolescence* (5th ed.). Boston, MA: McGraw-Hill.

Steinberg, L., & Dornbusch, S. M. (1991). Negative correlates of part-time employment during adolescence: Replication and elaboration. *Developmental Psychology, 27,* 304–313.

Steinberg, L., & Monahan, K. C. (2007). Age differences in resistance to peer influence. *Developmental Psychology, 43,* 1531–1543.

Steinberg, L., & Silk, J. (2002). Parenting adolescents. In M. Bornstein (Ed.), *Handbook of parenting: Vol. 1* (2nd ed., pp. 103–133). Hillsdale, NJ: Erlbaum.

Steinberg, L., Cauffman, E., Woolard, J., Graham. S., & Banich, M. (2009). Are adolescents less mature than adults? *American Psychologist, 64,* 583–594.

Steiner, J. E., Glaser, D., Hawilo, M. E., & Berridge, K. C. (2001). Comparative expression of hedonic impact: Affective reactions to taste by human infants and other primates. *Neuroscience & Biobehavioral Reviews, 25,* 53–74.

Stephan, Y., Fouquereau, E., & Fernandez, A. (2007). The relation between self-determination and retirement satisfaction among active retired individuals. *International Journal of Aging & Human Development, 66,* 329–345.

Stephens, M. A. P., & Clark, S. L. (1996). Interpersonal relationships in multi-generational families. In N. Vanzetti & S. Duck (Eds.), *A lifetime of relationships* (pp. 431–454). Pacific Grove, CA: Brooks/Cole.

Stephens, M. A. P., & Franks, M. M. (1999). Intergenerational relationships in later-life families: Adult daughters and sons as caregivers to aging parents. In J. C. Cavanaugh & S. K. Whitbourne (Eds.), *Gerontology: An interdisciplinary perspective* (pp. 329–354). New York: Oxford University Press.

Stephens, M. A. P., Townsend, A. L., Martire, L. M., & Druley, J. A. (2001). Balancing parent care with other roles: Interrole conflict of adult daughter caregivers. *Journal of Gerontology: Psychological Sciences, 56B*, P24–P34.

Sternberg, R. J. (1985). *Beyond IQ: A triarchic theory of human intelligence.* Cambridge, England: Cambridge University Press.

Sternberg, R. J. (1999). The theory of successful intelligence. *Review of General Psychology, 3*, 292–316.

Sternberg, R. J. (2002). Successful intelligence: A new approach to leadership. In R. E. Riggio & S. E Murphy (Eds.), *Multiple intelligences and leadership* (pp. 9–28). Mahwah, NJ: Erlbaum.

Sternberg, R. J. (2003). Issues in the theory and measurement of successful intelligence: A reply to Brody. *Intelligence, 31*, 331–337.

Sternberg, R. J. (2006). A duplex theory of love. In R. J. Sternberg & K. Weis (Eds.), *The new psychology of love* (pp. 184–199). New Haven, CT: Yale University Press.

Sternberg, R. J. (2008). The triarchic theory of successful intelligence. In N. Salkind (Ed.), *Encyclopedia of educational psychology* (pp. 988–994). Thousand Oaks, CA: Sage.

Sternberg, R. J., & Grigorenko, E. L. (2000). Practical intelligence and its development. In R. Bar-On & D. A. Parker (Eds.), *The handbook of emotional intelligence: Theory, development, assessment, and application at home, school, and in the workplace* (pp. 215–243). San Francisco, CA: Jossey-Bass.

Sternberg, R. J., & Grigorenko, E. L. (Eds.). (2004). *Culture and competence: Contexts of life success.* Washington, DC: American Psychological Association.

Sternberg, R. J., & Kaufman, J. C. (1998). Human abilities. *Annual Review of Psychology, 49*, 479–502.

Sternberg, R. J., & Lubart, T. I. (2001). Wisdom and creativity. In J. E. Birren & K. W. Schaie (Eds.), *Handbook of the psychology of aging* (5th ed., pp. 500–522). San Diego, CA: Academic Press.

Sternberg, R. J., Grigorenko, E. L, & Kidd, K. K. (2005). Intelligence, race, and genetics. *American Psychologist, 60*, 46–59.

Sternberg, R. J., Jarvin, L., & Grigorenko, E. L. (2009). *Teaching for wisdom, intelligence, creativity, and success.* Thousand Oaks, CA: Corwin.

Sterns, H. L., & Chang, B. (2010). Workforce issues and retirement. In J. C. Cavanaugh & C. K. Cavanaugh (Eds.), *Aging in America: Vol. 3: Societal issues* (pp. 81–105). Santa Barbara, CA: Praeger Perspectives.

Stevens, J., & Ward-Estes, J. (2006). Attention-deficit/hyperactivity disorder. In M. Hersen & J. C. Thomas (Series Eds.) and R. T. Ammerman (Vol. Ed.), *Comprehensive handbook of personality and psychopathology: Vol. 3. Child psychopathology* (pp. 316–329). Hoboken, NJ: Wiley.

Stevens, J., Harman, J. S., & Kelleher, K. J. (2005). Race/ethnicity and insurance status as factors associated with ADHD treatment patterns. *Journal of Child and Adolescent Psychopharmacology, 15*, 88–96.

Stevens, N. L., & Van Tilburg, T. G. (in press). Cohort differences in having and retaining friends in personal networks in later life. *Journal of Social and Personal Relationships.*

Stevens, S. B., & Morris, T. L. (2007). College dating and social anxiety: Using the Internet as a means of connecting to others. *CyberPsychology & Behavior, 10*, 680–688.

Stevenson, H. W., & Lee, S. (1990). Contexts of achievement. *Monographs of the Society for Research in Child Development, 55*(Serial No. 221).

Stevenson, H. W., & Stigler, J. W. (1992). *The learning gap.* New York: Summit Books.

Stevenson, R. J., Oaten, M. J., Case, T. I., Repacholi, B. M., & Wagland, P. (2010). Children's response to adult disgust elicitors: Development and acquisition. *Developmental Psychology, 46*, 165–177.

Stewart, L., & Pascual-Leone, J. (1992). Mental capacity constraints and the development of moral reasoning. *Journal of Experimental Child Psychology, 54*, 251–287.

Stewart, R. B., Mobley, L. A., Van Tuyl, S. S., & Salvador, W. A. (1987). The firstborn's adjustment to the birth of a sibling: A longitudinal assessment. *Child Development, 58*, 341–355.

Stice, E., & Shaw, H. (2004). Eating disorder prevention programs: A meta-analytic review. *Psychological Bulletin, 130*, 206–227.

Stice, E., Shaw, H., & Marti, C. (2006). A meta-analytic review of obesity prevention programs for children and adolescents: The skinny on interventions that work. *Psychological Bulletin, 132*, 667–691.

Stice, E., Shaw, H., Bohon, C., Martin, C. N., & Rohde, P. (2009). A meta-analytic review of depression prevention programs for children and adolescents: Factors that predict magnitude of intervention effects. *Journal of Consulting and Clinical Psychology, 77*, 486–503.

Stifter, C. A., & Fox, N. A. (1990). Infant reactivity: Physiological correlates of newborn and 5-month temperament. *Developmental Psychology, 26*, 582–588.

Stigler, J. W., Gallimore, R., & Hiebert, J. (2000). Using video surveys to compare classrooms and teaching across cultures: Examples and lessons from the TIMSS video studies. *Educational Psychologists, 35*, 87–100.

Stiles, J., Reilly, J., Paul, B., & Moses, P. (2005). Cognitive development following early brain injury: Evidence for neural adaptation. *Trends in Cognitive Sciences, 9*, 136–143.

Stjernqvist, K. (2009). Predicting development for extremely low birthweight infants: Sweden. In K. Nugent, B. J. Petrauskas, & T. B. Brazelton (Eds.), *The newborn as a person: Enabling healthy infant development worldwide.* Hoboken, NJ: Wiley.

Stone, R. (2011). Gender equity in marriage and civil unions. *American Journal of Public Health, 101*, 201.

Stoner, S., O'Riley, A., & Edelstein, B. (2010). Assessment of mental health. In J. C. Cavanaugh & C. K. Cavanaugh (Eds.), *Aging in America: Vol. 2: Physical and mental health* (pp. 141–170). Santa Barbara, CA: Praeger Perspectives.

Stop Smoking. (2010). *Quit smoking even in old age: It's going to benefit.* Retrieved from http://www.stop-smoking-updates.com/quitsmoking/quitting-benefits/psychological-benefits/quit-smoking-even-in-old-age-its-going-to-benefit.htm

Strano, D. A., Cuomo, M. J., & Venable, R. H. (2004). Predictors of undergraduate student binge drinking. *Journal of College Counseling, 7*, 50–63.

Strauss, W., & Howe, N. (2007). *Millennials go to college: Strategies for a new generation on campus* (2nd ed.). Ithaca, NY: Paramount.

Strauss-Blasche, G., Ekmekcioglu, C., & Marktl, W. (2002). Moderating effects of vacation on reactions to work and domestic stress. *Leisure Sciences, 24*, 237–249.

Strayer, D. L., Drews, F. A., & Crouch, D. J. (2006). A comparison of the cell phone driver and the drunk driver. *Human Factors, 48*, 381–391.

Strayer, J., & Roberts, W. (2004). Children's anger, emotional expressiveness, and empathy: Relations with parents' empathy, emotional expressiveness, and parenting practices. *Social Development, 13*, 229–254.

Strenze, T. (2007). Intelligence and socioeconomic success: A meta-analytic review of longitudinal research. *Intelligence, 35*, 401–426.

Striano, T., Tomasello, M., & Rochat, P. (2001). Social and object support for early symbolic play. *Developmental Science, 4*, 442–455.

Stroebe, M. S., & Schut, H. (1999). The dual process model of bereavement: Rationale and description. *Death Studies, 23*, 197–224.

Stroebe, M. S., & Schut, H. (2001). Models of coping with bereavement: A review. In M. S. Stroebe, R. O. Hansson, W. Stroebe, & H. Schut (Eds.), *Handbook of bereavement research: Consequences, coping, and care* (pp. 375–403). Washington, DC: American Psychological Association.

Stroebe, M., Schut, H., & Boerner, K. (2010). Continuing bonds in adaptation to bereavement: Toward theoretical integration. *Clinical Psychology Review, 30*, 259–268.

Stroebe, M., Boelen, P. A., van den Hout, M., Stroebe, W., Salemink, E., & van den Bout, J. (2007). Ruminative coping as avoidance: A reinterpretation of its function in adjustment to bereavement. *European Archives of Psychiatry and Clinical Neuroscience, 257*, 462–472.

Stroebe, M., Schut, H., & Stroebe, W. (2005). Attachment in coping with bereavement: A theoretical integration. *Review of General Psychology, 9*, 48–66.

Stroebe, M., Schut, H., & Stroebe, W. (2007). Health outcomes of bereavement. *Lancet, 370,* 1960–1973.

Stroebe, W., Abakoumkin, G., & Stroebe, M. (2010). Beyond depression: Yearning for the loss of a loved one. *OMEGA: Journal of Death and Dying, 61,* 85–101.

Strough, J., & Berg, C. A. (2000). Goals as a mediator of gender differences in high-affiliation dyadic conversations. *Developmental Psychology, 36,* 117–125.

Stuckey, J. C. (2001). Blessed assurance: The role of religion and spirituality in Alzheimer's disease caregiving and other significant life events. *Journal of Aging Studies, 15,* 69–84.

Studelska, J. V. (2006, Spring). At home in birth. *Midwifery Today,* pp. 32–33.

Stunkard, A. J., Sorensen, T. I. A., Hanis, C., Teasdale, T. W., Chakraborty, R., Schull, W. J., & Schulsinger, F. (1986). An adoption study of human obesity. *New England Journal of Medicine, 314,* 193–198.

Sturge-Apple, M. L., Davies, P. T., & Cummings, E. M. (2010). Typologies of family functioning and children's adjustment during the early school years. *Child Development, 81,* 1320–1335.

Sturge-Apple, M. L., Davies, P. T., Winter, M. A., Cummings, E. M., & Schermerhorn, A. (2008). Interparental conflict and children's school adjustment: The explanatory role of children's internal representations of interparental and parent-child relationships. *Developmental Psychology, 44,* 1678–1690.

Subrahmanyam, K., Greenfield, P., Kraut, R., & Gross, E. (2001). The impact of computer use on children's and adolescents' development. *Journal of Applied Developmental Psychology, 22,* 7–30.

Sullivan, L. W. (1987). The risks of the sickle-cell trait: Caution and common sense. *New England Journal of Medicine, 317,* 830–831.

Sullivan, M. W., & Lewis, M. (2003). Contextual determinants of anger and other negative expressions in young infants. *Developmental Psychology, 39,* 693–705.

Sung, G., & Greer, D. (2011). The case for simplifying brain death criteria. *Neurology, 76,* 113-114.

Super, C. M. (1981). Cross-cultural research on infancy. In H. C. Triandis & A. Heron (Eds.), *Handbook of cross-cultural psychology: Vol. 4. Developmental psychology.* Boston: Allyn & Bacon.

Super, C. M., Herrera, M. G., & Mora, J. O. (1990). Long-term effects of food supplementation and psychosocial intervention on the physical growth of Colombian infants at risk of malnutrition. *Child Development, 61,* 29–49.

Super, D. E. (1957). *The psychology of careers.* New York: Harper & Row.

Super, D. E. (1976). *Career education and the meanings of work.* Washington, DC: U.S. Offices of Education.

Super, D. E. (1980). A life-span, life space approach to career development. *Journal of Vocational Behavior, 16,* 282–298.

Super, D., Savickas, M., & Super, C. (1996). The life-span, life-space approach to careers. In D. Brown, L. Brooks, & Associates (Eds.), *Career choice & development* (3rd ed., pp. 121–178). San Francisco, CA: Jossey-Bass.

Sussman, S., Pokhrel, P., Ashmore, R. D., & Brown, B. B. (2007). Adolescent peer group identification and characteristics: a review of the literature. *Addictive Behaviors, 32,* 1602–1627.

Suzuki, L., & Aronson, J. (2005). The cultural malleability of intelligence and its impact on the racial/ethnic hierarchy. *Psychology, Public Policy, and Law, 11,* 320–327.

Suzuki, Y., Yamamoto, S., Umegaki, H., Onishi, J., Mogi, N., Fujishiro, H., & Iguchi, A. (2004). Smell identification test as an indicator for cognitive impairment in Alzheimer's disease. *International Journal of Geriatric Psychiatry, 19,* 727–733.

Sweeney, M. M. (2010). Remarriage and stepfamilies: Strategic sites for family scholarship in the 21st century. *Journal of Marriage and Family, 72,* 667–684.

Swinton, J., & Payne, R. (Eds.). (2009). *Living well and dying faithfully: Christian practices for end-of-life care.* Grand Rapids, MI: Eerdmans.

Tach, L., & Halpern-Meekin, S. (2009). How does premarital cohabitation affect trajectories of marital quality? *Journal of Marriage and the Family, 71,* 298–317.

Tager-Flusberg, H. (2007). Evaluating the theory-of-mind hypothesis of autism. *Current Directions in Psychological Science, 16,* 311–315.

Takata, K., Kitamura, Y., & Taniguchi, T. (2011). Pathological changes induced by amyloid-β in Alzheimer's disease. *Yakugaku Zasshi, 131,* 3–11.

Takeuchi, H., Taki, Y., Sassa, Y., Hashizume, H., Sekiguchi, A., Fukushima, A., et al. (2010). White matter structures associated with creativity: Evidence from diffusion tensor imaging. *Neuroimage, 51,* 11–18.

Talbot, L. A., Morrell, C. H., Fleg, J. L., & Metter, E. J. (2007). Changes in leisure time physical activity and risk of all-cause mortality in men and women: The Baltimore longitudinal study of aging. *Preventive Medicine, 45,* 169–176.

Tamayo, G. J., Broxson, A., Munsell, M., & Cohen, M Z. (2010). Caring for the caregiver. *Oncology Nursing Forum, 37,* E50–E57.

Tamis-LeMonda, C. S., & Bornstein, M. H. (1996). Variation in children's exploratory, nonsymbolic, and symbolic play: An explanatory multidimensional framework. In C. Rovee-Collier & L. P. Lipsitt (Eds.), *Advances in infancy research* (Vol. 10, pp. 37–78). Norwood, NJ: Ablex.

Tamis-Lemonda, C. S., Adolph, K. E., Lobo, S. A., Karasik, L. B., Ishak, S., & Dimitropoulou, K. A. (2008). When infants take mothers' advice: 18-month-olds integrate perceptual and social information to guide motor action. *Developmental Psychology, 44,* 734–746.

Tamis-LeMonda, C. S., Shannon, J. D., Cabrera, N., & Lamb, M. E. (2004). Father and mothers at play with their 2- and 3-year-olds: Contributions to language and cognitive development. *Child Development, 75,* 1806–1820.

Tan, K. L. (2009). Bed sharing among mother-infant pairs in Kiang District, Peninsular Malaysia and its relationship to breast-feeding. *Journal of Developmental and Behavioral Pediatrics, 30,* 420–425.

Tang, F., Morrow-Howell, N., & Choi, E. (2010). Why do older adult volunteers stop volunteering? *Ageing and Society, 30,* 859–878.

Tang, P.-L., Chiou, C.-P., Lin, H.-S., Wang, C., & Liand, S.-L. (in press). Correlates of death anxiety among Taiwanese cancer patients. *Cancer Nursing.*

Tang, T. L. P., & McCollum, S. L. (1996). Sexual harassment in the workplace. *Public Personnel Management, 25,* 53–58.

Tanner, J. L., & J. J. Arnett (2009). The emergence of emerging adulthood: the new life stage between adolescence and young adulthood. In A. Furlong (Ed.), *Handbook of youth and young adulthood: New perspectives and agendas.* London: Routledge.

Tanner, J. M. (1970). Physical growth. In P. H. Mussen (Ed.), *Carmichael's manual of child psychology* (3rd ed., pp. 77–155). New York: Wiley.

Tanner, J. M. (1990). *Fetus into man: Physical growth from conception to maturity* (2nd ed.). Cambridge, MA: Harvard University Press.

Tardif, T., Fletcher, P., Liang, W., Zhang, Z., Kaciroti, N., & Marchman, V. A. (2008). Baby's first 10 words. *Developmental Psychology, 44,* 929–938.

Taub, G. E., Hayes, B. G., Cunningham, W. R., & Sivo, S. A. (2001). Relative roles of cognitive ability and practical intelligence in the prediction of success. *Psychological Reports, 88,* 931–942.

Taumoepeau, M., & Ruffman, T. (2008). Stepping stones to others' minds: Maternal talk relates to child mental state language and emotion understanding at 15, 24, and 33 months. *Child Development, 79,* 284–302.

Tauriac, J. J., & Scruggs, N. (2006). Elder abuse among African Americans. *Educational Gerontology, 32,* 37–48.

Taylor, J. L., O'Hara, R., Mumenthaler, M. S., Rosen, A. C., & Yesavage, J. A. (2005). Cognitive ability, expertise, and age differences in following air-traffic control instructions. *Psychology and Aging, 20,* 117–133.

Taylor, M., Carlson, S. M., Maring, B. L., Gerow, L., & Charley, C. M. (2004). The characteristics of fantasy in school-age children: Imaginary companions, impersonation, and social understanding. *Developmental Psychology, 40,* 1173–1187.

Taylor, M., Hulette, A. C., & Dishion, T. J. (2010). Longitudinal outcomes of young high-risk adolescents with imaginary companions. *Developmental Psychology, 46,* 632–1636.

Taylor, R. J., Chatters, L. M., & Levin, J. (2004). *Religion in the lives of African Americans.* Thousand Oaks, CA: Sage.

Teachman, J. (2008). Complex life course patterns and the risk of divorce in second marriages. *Journal of Marriage and Family, 70,* 294–305.

Teichman, Y. (2001). The development of Israeli children's images of Jews and Arabs and their expression in human figure drawings. *Developmental Psychology, 37,* 749–761.

Teilmann, G., Pedersen, C. B., Skakkeback, N. E., & Jensen, T. K. (2006). Increased risk of precious puberty in internationally adopted children in Denmark. *Pediatrics, 118,* 391–399.

Teisseyre, N., Mullet, E., & Sorum, P. C. (2005). Under what conditions is euthanasia acceptable to lay people and health professionals? *Social Science and Medicine, 60,* 357–368.

Tenenbaum, H. R., & Leaper, C. (2002). Are parents' gender schemas related to their children's gender-related cognitions? A meta-analysis. *Developmental Psychology, 38,* 615–630.

Terkel, S. (1974). *Working.* New York: Pantheon Books.

Terman, M. (1994). Light therapy. In M. H. Kryger, T. Roth, & W. C. Dement (Eds.), *Principles and practice of sleep medicine* (2nd ed., pp. 1012–1029). Philadelphia: Saunders.

Terracciano, A., McRae, R. R., & Costa, P. T., Jr. (2010). Intra-individual change in personality stability and age. *Journal of Research in Personality, 44,* 31–37.

The MTA Cooperative Group. (1999). Moderates and mediators of treatment response for children with attention-deficit/hyperactivity disorder. *Archives of General Psychiatry, 56,* 1088–1096.

Thelen, E., & Smith, L. B. (1998). Dynamic systems theories. In W. Damon (Ed.), *Handbook of child psychology* (Vol. 1). New York: Wiley.

Thelen, E., & Ulrich, B. D. (1991). Hidden skills. *Monographs of the Society for Research in Child Development, 56*(Serial No. 223).

Thelen, E., Ulrich, B. D., & Jensen, J. L. (1989). The developmental origins of locomotion. In M. H. Woollacott and A. Shumway-Cook (Eds.), *Development of posture and gait across the lifespan.* Columbia, SC: University of South Carolina Press.

Therborn, G. (2010). Families in global perspective. In A. Giddens & P. W. Sutton (Eds.), *Sociology: Introductory readings* (3rd ed., pp. 119–123). Malden, MA: Polity Press.

Thiele, D. M., & Whelan, T. A. (2010). The relationship between grandparent satisfaction, meaning, and generativity. *International Journal of Aging and Human Development, 66,* 21–48.

Thiessen, E. D., & Saffran, J. R. (2003). When cues collide: Use of stress and statistical cues to word boundaries by 7- to 9-month-old infants. *Developmental Psychology, 39,* 706–716.

Thiessen, E. D., Hill, E., & Saffran, J. R. (2005). Infant-directed speech facilitates word segmentation. *Infancy, 7,* 53–71.

Thomas, A., & Chess, S. (1977). *Temperament and development.* New York: Brunner/Mazel.

Thomas, A., Chess, S., & Birch, H. G. (1968). *Temperament and behavior disorders in children.* New York: NYU Press.

Thomas, D. A. (1990). The impact of race on managers' experiences of developmental relationships (mentoring and sponsorship): An intra-organizational study. *Journal of Organizational Behavior, 11,* 479–492.

Thomas, J. W., Bol, L., Warkentin, R. W., Wilson, M., Strage, A., & Rohwer, W. D. (1993). Interrelationships among students' study activities, self-concept of academic ability, and achievement as a function of characteristics of high-school biology courses. *Applied Cognitive Psychology, 7,* 499–532.

Thomas, L. A., De Bellis, M. D., Graham, R., & LaBar, K. S. (2007). Development of emotional facial recognition in late childhood and adolescents. *Developmental Science, 10,* 547–558.

Thomas, N. G., & Berk, L. E. (1981). Effects of school environments on the development of young children's creativity. *Child Development, 52,* 1152–1162.

Thomas, R., & Zimmer-Gembeck, M. J. (2011). Accumulating evidence for parent-child interaction therapy in the prevention of child maltreatment. *Child Development, 82,* 177–192.

Thompson, L. W., Gallagher-Thompson, D., Futterman, A., Gilewski, M. J., & Peterson, J. (1991). The effects of late-life spousal bereavement over a 30-month interval. *Psychology and Aging, 6,* 434–441.

Thompson, R. A. (2000). The legacy of early attachments. *Child Development, 71,* 145–152.

Thompson, R. A. (2006). The development of the person: Social understanding, relationships, conscience, self. In N. Eisenberg (Ed.), *Handbook of child psychology: Vol. 3. Social, emotional, and personality development* (6th ed, pp. 24–98). Hoboken, NJ: Wiley.

Thompson, R. A., & Limber, S. (1990). "Social anxiety" in infancy: Stranger wariness and separation distress. In H. Leitenberg (Ed.), *Handbook of social and evaluation anxiety* (pp. 85–137). New York: Plenum.

Thompson, R. A., Laible, D. J., & Ontai, L. L. (2003). Early understandings of emotion, morality, and self: Developing a working model. *Advances in Child Development and Behavior, 31,* 137–172.

Thompson, R. A., Lewis, M. D., & Calkins, S. D. (2008). Reassessing emotion regulation. *Child Development Perspectives, 2,* 124–131.

Thomsen, D. K., & Berntsen, D. (2009). The long-term impact of emotionally stressful events on memory characteristics and life story. *Applied Cognitive Psychology, 23,* 579–598.

Thurstone, L. L., & Thurstone, T. G. (1941). *Factorial studies of intelligence.* Chicago, IL: University of Chicago Press.

Tiffany, D. W., & Tiffany, P. G. (1996). Control across the life span: A model for understanding self-direction. *Journal of Adult Development, 3,* 93–108.

Timiras, P. (2002). *Physiological bases of aging and geriatrics* (3rd ed.). Boca Raton, FL: CRC Press.

Tincoff, R., & Jusczyk, P. W. (1999). Some beginnings of word comprehension in 6-month-olds. *Psychological Science, 10,* 172–175.

Tither, J. M., & Ellis, B. J. (2008). Impact of fathers on daughters' age at menarche: A genetically and environmentally controlled sibling study. *Developmental Psychology, 44,* 1409–1420.

Toga, A. W., Thompson, P. M., & Sowell, E. R. (2006). Mapping brain maturation. *Trends in Neuroscience, 29,* 148–159.

Tolan, P. H., Gorman-Smith, D., & Henry, D. B. (2003). The developmental ecology of urban males' youth violence. *Developmental Psychology, 39,* 274–291.

Tomasello, M., Carpenter, M., & Liszkowski, U. (2007). A new look at infant pointing. *Child Development, 78,* 705–722.

Tomlinson, M., Cooper, P., & Murray, L. (2005). The mother–infant relationship and infant attachment in a South African peri-urban settlement. *Child Development, 76,* 1044–1054.

Toomey, R. B., Ryan, C., Diaz, R. M., Card, N. A., & Russell, S. T. (2010). Gender-noncomforming lesbian, gay, bisexual, and transgender youth: School victimization and young adult psychosocial adjustment. *Developmental Psychology, 46,* 1589–1589.

Torgesen, J. K. (2004). Learning disabilities: An historical and conceptual overview. In B. Y. L. Wong (Ed.), *Learning about learning disabilities* (3rd ed., pp. 3–40). San Diego, CA: Elsevier.

Tracy, B., Reid, R., & Graham, S. (2009). Teaching young students strategies for planning and drafting stories: The impact of self-regulated strategy development. *Journal of Educational Research, 102,* 323–331.

Trainor, L. J., & Heinmiller, B. M. (1998). The development of evaluative responses to music: Infants prefer to listen to consonance over dissonance. *Infant Behavior and Development, 21,* 77–88.

Trainor, L. J., Austin, C. M., & Desjardins, R. N. (2000). Is infant-directed speech prosody a result of the vocal expression of emotion? *Psychological Science, 11,* 188–195.

Trainor, L. J., Wu, L., & Tsang, C. D. (2004). Long-term memory for music: Infants remember tempo and timbre. *Developmental Science, 7,* 289–296.

Treiman, R., & Kessler, B. (2003). The role of letter names in the acquisition of literacy. *Advances in Child Development and Behavior, 31,* 105–135.

Tremblay, R. E., Schall, B., Boulerice, B., Arsonault, L., Soussignan, R. G., & Paquette, D. (1998). Testosterone, physical aggression, and dominance and physical development in adolescence. *International Journal of Behavioral Development, 22,* 753–777.

Trim, R. S., Meehan, B. T., King, K. M., & Chassin, L. (2007). The relation between adolescent substance use and young adult internalizing symptoms: Findings from a high-risk longitudinal sample. *Psychology of Addictive Behaviors, 21,* 97–107.

Troseth, G. L., Pierroutsakos, S. L., & DeLoache, J. S. (2004). From the innocent to the intelligent eye: The early development of pictorial competence. In R. V. Kail (Ed.), *Advances in child development and behavior* (Vol. 32, pp. 1–35). New York: Academic Press.

Troutman, M., Nies, M. A., & Mavellia, H. (2011). Perceptions of successful aging in black older adults. *Journal of Psychosocial Nursing and Mental Health Services, 49(1),* 28–34.

Truong, K. D., & Sturm, R. (2009). Alcohol environments and disparities in exposure associated with adolescent drinking in California. *American Journal of Public Health, 99,* 264–270.

Trzesniewski, K., H., & Donnellan, M. B. (2010). Rethinking "Generation Me": A study of cohort effects from 1976–2006. *Perspective on Psychological Science, 5,* 58–75.

Tsao, T.-C. (2004). New models for future retirement: A study of college/university-linked retirement communities. *Dissertation Abstracts International. Section A. Humanities and Social Sciences, 64(10-A),* 3511.

Tsuno, N., & Homma, A. (2009). Aging in Asia—The Japan experience. *Ageing International, 34,* 1–14.

Tu, M. C.-H. (2007). Culture and job satisfaction: A comparative analysis between Taiwanese and Filipino caregivers working in Taiwan's long-term care industry (China). *Dissertation Abstracts International. Section A. Humanities and Social Sciences, 67(9-A),* 3488.

Turiel, E. (2006). The development of morality. In W. Damon & R. M. Lerner (Eds.), *Handbook of child psychology* (6th ed., Vol. 3, pp. 789–857). Hoboken, NJ: Wiley.

Turley, R. N. L. (2003). Are children of young mothers disadvantaged because of their mother's age or family background? *Child Development, 74,* 465–474.

Twenge, J. M., & Campbell, W. K. (2001). Age and birth cohort differences in self-esteem: A cross-temporal meta-analysis. *Personality and Social Psychology Review, 5,* 321–344.

Twenge, J., & Crocker, J. (2002). Race and self-esteem: Meta-analysis comparing Whites, Blacks, Hispanics, and American Indians and comment on Gray-Little and Hafdahl (2000). *Psychological Bulletin, 128,* 371–408.

Tynes, B. M. (2007). Role taking in online "classrooms": What adolescents are learning about race and ethnicity. *Developmental Psychology, 43,* 1312–1320.

U.S. Bureau of Labor Statistics. (2010b). *Women's to men's earnings ratio by age, 2009.* Retrieved from http://www.bls.gov/opub/ted/2010/ted_20100708.htm

U.S. Census Bureau. (2003). *Married-couple and unmarried-partner households: 2000.* Retrieved from http://www.census.gov/prod/2003pubs/censr-5.pdf

U.S. Census Bureau. (2008). *The 2008 statistical abstract: 2008 edition.* Retrieved from http://www.census.gov/compendia/statab/2008edition.html

U.S. Census Bureau. (2010a). *Estimated median age at first marriage by sex: 1890 to the present.* Retrieved from http://search.census.gov/search?q=cache:aJo0LSIH0h0J:www.census.gov/population/socdemo/hh-fam/ms2.xls+median+age+first+marriage&output=xml_no_dtd&ie=UTF-8&client=default_frontend&proxystylesheet=default_frontend&site=census&access=p&oe=ISO-8859-1

U.S. Census Bureau. (2010b). *The 2010 statistical abstract: PDF version.* Retrieved from http://www.census.gov/compendia/statab/2010edition.html

U.S. Census Bureau. (2010c). *Facts for features: Grandparents Day: Sept. 12.* Retrieved from http://www.census.gov/newsroom/releases/archives/facts_for_features_special_editions/cb10-ff16.html

U. S. Census Bureau. (2011). *The 2011 statistical abstract.* Retrieved July 1, 2011 from http://www.census.gov/compendia/statab/.

U.S. Department of Agriculture. (2010). *Expenditures on children by families, 2009.* Retrieved from http://www.cnpp.usda.gov/Publications/CRC/crc2009.pdf

U.S. Department of Health and Human Services. (1997). *Vital statistics of the United States, 1994: Vol. 2. Mortality* (Pt. A). Hyattsville, MD: U.S. Public Health Service.

U.S. Department of Health and Human Services. (2000). *Reducing tobacco use: A report of the Surgeon General—Executive summary.* Atlanta, GA: U.S. Department of Health and Human Services, Centers for Disease Control and Prevention, National Center for Chronic Disease Prevention and Health Promotion, Office on Smoking and Health.

U.S. Department of Health and Human Services. (2001). *The Surgeon General's call to action to prevent and decrease overweight and obesity.* Rockville, MD: Author.

U.S. Department of Health and Human Services. (2008). *Physical activity guidelines for Americans.* Retrieved from http://www.health.gov/paguidelines/committeereport.aspx

U.S. Department of Health and Human Services. (2010). *Child Maltreatment 2009.* Retrieved from http://www.acf.hhs.gov/programs/cb/stats_research/index.htm#can

U.S. Department of Health and Human Services. (2010). *The Surgeon General's vision for a healthy and fit nation.* Rockville, MD: Author.

U.S. Department of Labor. (2000). *Report on the youth labor force.* Washington, DC: Author.

U.S. Department of Labor. (2010a). *Nontraditional occupations for women in 2009.* Retrieved from http://www.dol.gov/wb/factsheets/nontra2009.htm

U.S. Department of Labor. (2010b) *Women in the labor force: A Databook (2010 edition).* Retrieved from http://www.bls.gov/cps/wlftable6-2010.htm

Uhlenberg, P., & Cheuk, M. (2010). The significance of grandparents to grandchildren: An international perspective. In D. Dannefer & C. Phillipson (Eds.), *The SAGE handbook of social gerontology* (pp. 447–458). Thousand Oaks, CA: Sage Publications.

Ulrich, L. B., & Brott, P. E. (2005). Older workers and bridge employment: Redefining retirement. *Journal of Employment Counseling, 42,* 159–170.

Umaña-Taylor, A., Diversi, M., & Fine, M. (2002). Ethnic identity and self-esteem among Latino adolescents: Distinctions among Latino populations. *Journal of Adolescent Research, 17,* 303–327.

Umbel, V. M., Pearson, B. Z., Fernandez, M. C., & Oller, D.-K. (1992). Measuring bilingual children's receptive vocabularies. *Child Development, 63,* 1012–1020.

UNICEF. (2006). *Progress for children: A report card on Nutrition, 2000–2006.* New York: Author.

UNICEF. (2007). *The state of the world's children, 2008.* New York: Author.

UNICEF. (2008). *The state of the world's children, 2009: Maternal and newborn health.* Author.

United Nations. (2005). *Demographic yearbook 2005.* Retrieved from http://unstats.un.org/unsd/demographic/products/dyb/dyb2005/Table25.pdf

United Nations. (2010). *Divorces and crude divorce rates by urban/rural residence: 2004–2008.* Retrieved from http://unstats.un.org/unsd/demographic/products/dyb/dyb2008/Table25.pdf

United Nations. (2010). *World population ageing 2009.* Retrieved from http://www.un.org/esa/population/publications/WPA2009/WPA2009-report.pdf

Upchurch, S., & Mueller, W. H. (2005). Spiritual influences on ability to engage in self-care activities among older African Americans. *International Journal of Aging & Human Development, 60,* 77–94.

Updegraff, K. A., Thayer, S. M., Whiteman, S. D., Denning, D. J., & McHale, S. M. (2005). Aggression in adolescents' sibling relationships: Links to sibling and parent–adolescents relationship quality. *Family Relations: Interdisciplinary Journal of Applied Family Studies, 54,* 373–385.

Updegraff, K. A., McHale, S. M., Whiteman, S. D., Thayer, S. M., & Delgado, M. Y. (2005). Adolescent sibling relationships in Mexican American families: Exploring the role of familism. *Journal of Family Psychology, 19,* 512–522.

Usher, E. L., & Pajares, F. (2009). Sources of self-efficacy in mathematics: A validation study. *Contemporary Educational Psychology, 34,* 89–101.

Usita, P. M., & Blieszner, R. (2002). Immigrant family strengths: Meeting communication challenges. *Journal of Family Issues, 23,* 266–286.

Usita, P. M., & Du Bois, B. C. (2005). Conflict sources and responses in mother-daughter relationships: Perspectives of adult daughters of aging immigrant women. *Journal of Women & Aging, 17,* 151–165.

Vaillant, G. E., DiRago, A. C., & Mukamal, K. (2006). Natural history of male psychological health, XV: Retirement satisfaction. *American Journal of Psychiatry, 163,* 682–688.

Vaish, A., Carpenter, M., & Tomasello, M. (2009). Sympathy through affective perspective taking and its relation to prosocial behavior in toddlers. *Developmental Psychology, 45,* 534–543.

Vaish, A., Woodward, A., & Grossmann, T. (2008). Not all emotions are created equal: The negativity bias in social-emotional development. *Psychological Bulletin, 134,* 383–403.

Valentino, K., Ciccetti, D., Rogosch, F. A., & Toth, S. L. (2008). Memory, maternal representations, and internalizing symptomatology among abused, neglected, and nonmaltreated children. *Child Development, 79*, 705–719.

Valkenburg, P. M., & Jochen, P. (2009). Social consequences of the Internet for adolescents: A decade of research. *Current Directions in Psychological Science, 18*, 1–5.

Valkenburg, P. M., & van der Voort, T. H. A. (1994). Influence of TV on daydreaming and creative imagination: A review of research. *Psychological Bulletin, 116*, 316–339.

Valkenburg, P. M., & van der Voort, T. H. A. (1995). The influence of television on children's daydreaming styles: A 1-year-panel study. *Communication Research, 22*, 267–287.

Vallerand, R. J. (2008). On the psychology of passion: In search of what makes people's lives most worth living. *Canadian Psychology, 49*, 1–13.

Vallerand, R. J., Paquet, Y., Philippe, F. L., & Charest, J. (2010). On the role of passion for work in burnout: A process model. *Journal of Personality, 78*, 289–312.

van den Boom, D. C. (1994). The influence of temperament and mothering on attachment and exploration: An experimental manipulation of sensitive responsiveness among lower-class mothers with irritable infants. *Child Development, 65*, 1457–1477.

van den Boom, D. C. (1995). Do first-year intervention effects endure? Follow-up during toddlerhood of a sample of Dutch irritable infants. *Child Development, 66*, 1798–1816.

Van Den Wijngaart, M. A. G., Vernooij-Dassen, M. J. F. J., & Felling, A. J. A. (2007). The influence of stressors, appraisal and personal conditions on the burden of spousal caregivers of persons with dementia. *Aging & Mental Health, 11*, 626–636.

van der Geest, S. (2004). Dying peacefully: Considering good death and bad death in Kwahu-Tafo, Ghana. *Social Science and Medicine, 58*, 899–911.

van der Mark, I. L., van IJzendoorn, M. H., & Bakermans-Kranenburg, M. J. (2002). Development of empathy in girls during the second year of life: Associations with parenting, attachment, and temperament. *Social Development, 11*, 451–468.

van der Pas, S., & Koopman-Boyden, P. (2010). Leisure and recreation activities and wellbeing among midlife New Zealanders. In C. Waldegrave & P. Koopman-Boyden (Eds.), *Midlife New Zealanders aged 40–64 in 2008: Enhancing well-being in an aging society* (pp. 111–128). Hamilton, New Zealand: Family Centre Social Policy Research Unit, Lower Hutt, Wellington and the Population Studies Centre, University of Waikato. Retrieved from http://www.ewas.net.nz/Publications/filesEWAS/EWAS_M2.pdf#page=126

van Dierendonck, D., Garssen, B., & Visser, A. (2005). Burnout prevention through personal growth. *International Journal of Stress Management, 12*, 62–77.

Van Doorn, M. D., Branje, S. J. T., & Meeus, W. H. J. (2008). Conflict resolution in parent-adolescent relationships and adolescent delinquency. *Journal of Early Adolescence, 28*, 503–527.

van Goozen, S. H. M., Fairchild G., & Harold, G. T. (2008). The role of neurobiological deficits in childhood antisocial behavior. *Current Directions in Psychological Science, 17*, 224–228.

van Goozen, S. H., Fairchild, G., Snoek, H., & Harold, G. T. (2007). The evidence for a neurobiological model of childhood antisocial behavior. *Psychological Bulletin, 133*, 149–182.

Van Hof, P., van der Kamp, J., & Savelsbergh, G. J. P. (2002). The relation of unimanual and bimanual reaching to crossing the midline. *Child Development, 73*, 1352–1362.

Van IJzendoorn, M. H., Vereijken, C. M. J. L., Bakermans-Kranenburg, M. J., & Riksen-Walraven, J. (2004). Assessing attachment security with the Attachment Q Sort: Meta-analytic evidence for the validity of the observer AQS. *Child Development, 75*, 1188–1213.

van Solinge, H., & Henkens, K. (2005). Couples' adjustment to retirement: A multi-actor panel study. *Journal of Gerontology: Social Sciences, 60*, S11–S20.

Van Volkom, M. (2006). Sibling relationships in middle and older adulthood: A review of the literature. *Marriage & Family Review, 40*, 151–170.

Van Zalk, M., Herman, W., Kerr, M., Branje, S. J. T., Stattin, H., & Meeus, W. H. J. (2010). It takes three: Selection, influence, and de-selection processes of depression in adolescent friendship networks. *Developmental Psychology, 46*, 927–938.

Vandello, J. A. (2000). Domestic violence in cultural context: Male honor, female fidelity, and loyalty. *Dissertation Abstracts International. Section B. Sciences and Engineering, 61(5-B)*, 2821.

vandenBerg, P., Neumark-Sztainer, D., & Wall, M. (2007). Steroid use among adolescents: Longitudinal findings from Project EAT. *Pediatrics, 119*, 476–486.

Vander Wal, J. S., & Thelen, M. H. (2000). Eating and body image concerns among obese and average-weight children. *Addictive Behaviors, 25*, 775–778.

Vargas, M. G. (2007). An existential approach to Mexican women's ways of coping with domestic violence: An exploratory study. *Dissertation Abstracts International. Section A. Humanities and Social Sciences, 67(7-A)*, 2782.

Vaupel, J. W. (2010). Biodemography of human ageing. *Nature, 464*, 536–542.

Vazire, S., & Doris, J. M. (2009). Personality and personal control. *Journal of Research in Personality, 43*, 274–275.

Vazsonyi, A. T., & Snider, J. B. (2008). Mentoring, competencies, and adjustment in adolescents: American part-time employment and European apprenticeships. *International Journal of Behavioral Development, 32*, 46–55.

Veenstra, R., Lindberg, S., Oldenhinkel, A. J., De Winter, A. F., Verhulst, F. C., & Ormel, J. (2005). Bullying and victimization in elementary schools: A comparison of bullies, victims, bully/victims, and uninvolved preadolescents. *Developmental Psychology, 41*, 672–682.

Veenstra, R., Lindenberg, S., Munniksma, A., Dijkstra, J. K. (2010). The complex relation between bullying, victimization, acceptance, and rejection: Giving special attention to status, affection, and sex differences. *Child Development, 81*, 480–486.

Vélez, C. E., Wolchik, S. A., Tein, J., & Sandler, I. (2011). Protecting children from the consequences of divorce: A longitudinal study of the effects of parenting on children's coping processes. *Child Development, 82*, 244–257.

Ventura, S. J., Abma, J. C., Mosher, W. D., & Henshaw, S. K. (2008). Estimated pregnancy rates by outcome for the United States, 1990–2004. *National Vital Statistics Reports, 56*, 1–26.

Verhaeghen, P., & Salthouse, T. A. (1997). Meta-analysis of age–cognition relations in adulthood: Establishment of linear and non-linear age effects and structural models. *Psychological Bulletin, 122*, 231–249.

Verheijde, J. L. (2010). Commentary on the concept of brain death within the Catholic bioethical framework. *Christian Bioethics, 16*, 246–256.

Véronneau, M.-H., Vitaro, F., Brendgen, M., Dishion, T. J., & Tremblay, R. E. (2010). Transactional analysis of the reciprocal linkages between peer relationships and academic achievement from middle childhood to early adolescence. *Developmental Psychology, 46*, 773–790.

Verschaeve, L. (2009). Genetic damage in subjects exposed to radiofrequency radiation. *Mutation Research—Reviews in Mutation Research, 681*, 259–270.

Vestbo, J. (2011). Clinical diagnosis of COPD. In N. A. Hanania & A. Sharafkhaneh (Eds.), *COPD: A guide to diagnosis and clinical management* (pp. 21–31). New York: Springer.

Vickio, C. J., Cavanaugh, J. C., & Attig, T. (1990). Perceptions of grief among university students. *Death Studies, 14*, 231–240.

Vieno, A., Nation, M., Pastore, M., & Santinello, M. (2009). Parenting and antisocial behavior: A model of the relationship between adolescent self-disclosure, parental closeness, parental control, and adolescent antisocial behavior. *Developmental Psychology, 45*, 1509–1519.

Villa, R. F., & Jaime, A. (1993). *La fé de la gente*. In M. Sotomayor & A. Garcia (Eds.), *Elderly Latinos: Issues and solutions for the 21st century*. Washington, DC: National Hispanic Council on Aging.

Visher, E. G., Visher, J. S., & Pasley, K. (2003). Remarriage families and stepparenting. In F. Walsh (Ed.), *Normal family processes* (pp. 153–175). New York: Guilford.

Vitiello, B., & Swedo, S. (2004). Antidepressant medications in children. *New England Journal of Medicine, 350*, 1489–1491.

Vitulano, L. A. (2005). Delinquency. In C. B. Fisher & R. M. Lerner (Eds.), *Encyclopedia of applied developmental science* (Vol. 1, pp. 327–328). Thousand Oaks, CA: Sage.

Volling, B. L., & Belsky, J. (1992). The contribution of mother–child and father–child relationships to the quality of sibling interaction: A longitudinal study. *Child Development, 63,* 1209–1222.

von Bohlen und Halbach, O. (2010) Involvement of BDNF in age-dependent alterations in the hippocampus. *Frontiers in Aging Neuroscience, 2.* Retrieved from http://www.frontiersin.org/aging_neuroscience/10.3389/fnagi.2010.00036/full

Voorpostel, M., & van der Lippe, T. (2007). Support between siblings and between friends: Two worlds apart? *Journal of Marriage and Family, 69,* 1271–1282.

Vorhees, C. V., & Mollnow, E. (1987). Behavior teratogenesis: Long-term influences on behavior. In J. D. Osofsky (Ed.), *Handbook of infant development* (2nd ed.). New York: Wiley.

Vouloumanos, A., Hauser, M. D., Werker, J. F., & Martin, A. (2010). The tuning of human neonates preference for speech. *Child Development, 81,* 517–527.

Vrijheid, M., et al. (2010). Prenatal exposure to cell phone use and neurodevelopment at 14 months. *Epidemiology, 21,* 259–262.

Vygotsky, L. S. (1986). *Thought and language* (A. Kozulin, Trans.). Cambridge, MA: MIT Press. (Original work published in 1934)

Wachs, T. D., & Bates, J. E. (2001). Temperament. In G. Bremner & A. Fogel (Eds.), *Blackwell handbook of infant development* (pp. 465–501). Malden, MA: Blackwell.

Wachs, T. D., Black, M. M., & Engle, P. L. (2009). Maternal depression: A global threat to children's health, development, and behavior and to human rights. *Child Development Perspectives, 3,* 51–59.

Wadlington, W. (2005). Family law in America. *Family Court Review, 43,* 178–179.

Wagoner, B. (2009). The experimental methodology of constructive microgenesis. In J. Valsiner, P. C. M. Molenaar, M. C. D. P. Lyra, & N. Chaudhary (Eds.), *Dynamic process methodology in the social and developmental sciences* (pp. 99–121). New York: Springer.

Wahl, H. W., Fänge, A., Oswald, F., Gitlin, L., & Iwarsson, S. (2009). The home environment and disability-related outcomes in aging individuals: What is the empirical evidence? *The Gerontologist, 49,* 355–367.

Wahl, H.-W., & Oswald, F. (2010). Environmental perspectives on ageing. In D. Dannefer & C. Phillipson (Eds.), *The SAGE handbook of social gerontology* (pp. 111–124). Thousand Oaks, CA: Sage Publications.

Waites, C. (2009). Building on strengths: Intergenerational practice with African-American families. *Social Work, 54,* 278–287.

Wakschlag, L. S., Leventhal, B. L., Pine, D. S., Pickett, K. E., & Carter, A. S. (2006). Elucidating early mechanisms of developmental psychopathology: The case of prenatal smoking and disruptive behavior. *Child Development, 77,* 893–906.

Walberg, H. J. (1995). General practices. In G. Cawelti (Ed.), *Handbook of research on improving student achievement.* Arlington, VA: Educational Research Service.

Walden, T., Kim, G., McCoy, C., & Karrass, J. (2007). Do you believe in magic? Infants' social looking during violation of expectations. *Developmental Science, 10,* 654–663.

Walker, A. C., & Balk, D. E. (2007). Bereavement rituals in the Muscogee Creek tribe. *Death Studies, 31,* 633–652.

Walker, L. E. A. (1984). *The battered woman syndrome.* New York: Springer.

Walker, L. J. (1980). Cognitive and perspective-taking prerequisites for moral development. *Child Development, 51,* 131–139.

Walker, L. J., & Taylor, J. H. (1991). Family interactions and the development of moral reasoning. *Child Development, 62,* 264–283.

Walker, L. J., Hennig, K. H., & Krettenauer, T. (2000). Parent and peer contexts for children's moral reasoning development. *Child Development, 71,* 1033–1048.

Walker, P., et al. (2010). Preverbal infants' sensitivity to synaesthetic cross-modality correspondences. *Psychological Science, 21,* 21–25.

Walker, Q. D. (2010). An investigation of the relationship between career maturity, career decision self-efficacy, and self-advocacy of college students with and without disabilities. *University of Iowa Theses and Dissertations.* Paper 617. Retrieved from http://ir.uiowa.edu/etd/617.

Wall, S., & Arden, H. (1990). *Wisdomkeepers: Meetings with Native American spiritual elders.* Hillsboro, OR: Beyond Words.

Wallace, J. E. (2001). The benefits of mentoring for female lawyers. *Journal of Vocational Behavior, 58,* 366–391.

Wallerstein, J. S., & Lewis, J. M. (2004). The unexpected legacy of divorce: Report of a 25-year study. *Psychoanalytic Psychology, 21,* 353–370.

Walther, A. N. (1991). *Divorce hangover.* New York: Pocket Books.

Wang, H., & Wellman, B. (2010). Social connectivity in America: changes in adult friendship network size from 2002 to 2007. *American Behavioral Scientist, 53,* 1148–1169.

Wang, Q. (2006). Culture and the development of self-knowledge. *Current Directions in Psychological Science, 15,* 182–187.

Wang, Q. (2007). "Remember when you got the big, big bulldozer?" Mother-child reminiscing over time and across cultures. *Social Cognition, 25,* 455–471.

Wang, Q., Pomerantz, E. M., & Chen, H. (2007). The role of parents' control in early adolescents' psychological functioning: A longitudinal investigation in the United States and China. *Child Development, 78,* 1592–1610.

Wang, S. S., & Brownell, K. D. (2005). Anorexia nervosa. In C. B. Fisher & R. M. Lerner (Eds.), *Encyclopedia of applied developmental science* (Vol. 1, pp. 83–85). Thousand Oaks, CA: Sage.

Wang, S.-H., & Baillargeon, R. (2005). Inducing infants to detect a physical violation in a single trial. *Psychological Science, 16,* 542–549.

Wansink, B., & Sobal, J. (2007). Mindless eating: The 200 daily food decisions we overlook. *Environment and Behavior, 39,* 106–123.

Wanzer, S. H., & Glenmullen, J. (2007). *To die well: Your right to comfort, calm, and choice in the last days of life.* Cambridge, MA: Da Capo Press.

Warneken, F., & Tomasello, M. (2006). Altruistic helping in human infants and young chimpanzees. *Science, 311,* 1301–1303.

Warner, B., Altimier, L., & Crombleholme, T. M. (2007). Fetal surgery. *Newborn and Infant Nursing Reviews, 7,* 181–188.

Warner, E., Henderson-Wilson, C., & Andrew, F. (2010). Flying the coop: why is the move out of the home proving unsustainable? In B. Randolph, T. Burke, K. Hulse, & V. Milligan, (Eds.). Refereed papers presented at the 4th Australasian Housing Researchers Conference, Sydney, University of New South Wales. Retrieved from http://www.fbe.unsw.edu.au/cf/apnhr/papers/Attachments/Warner.pdf

Warnock, F., & Sandrin, D. (2004). Comprehensive description of newborn distress behavior in response to acute pain (newborn male circumcision). *Pain, 107,* 242–255.

Warr, P., Butcher, V., & Robertson, I. (2004). Activity and psychological well-being in older people. *Aging & Mental Health, 8,* 172–183.

Warren-Findlow, J., & Issel, I. M. (2010). Stress and coping in African American women with chronic heart disease: A cultural cognitive coping model. *Journal of Transcultural Nursing, 21,* 45–54.

Washko, M. (2001). *An examination of generativity: Past, present, and future research directions.* Unpublished master's thesis, University of Delaware.

Wass, H. (2001). Past, present, and future of dying. *Illness, Crisis, and Loss, 9,* 90–110.

Waterhouse, L. (2006). Multiple intelligences, the Mozart effect, and emotional intelligence: A critical review. *Educational Psychologist, 41,* 207–225.

Waters, E., & Cummings, E. M. (2000). A secure base from which to explore close relationships. *Child Development, 71,* 164–172.

Waters, E., Merrick, S., Treboux, D., Crowell, J., & Albersheim, L. (2000). Attachment security in infancy and early adulthood. *Child Development, 71,* 684–689.

Waters, H. S. (1980). "Class news": A single-subject longitudinal study of prose production and schema formation during childhood. *Journal of Verbal Learning and Verbal Behavior, 19,* 152–167.

Watson, J. B. (1925). *Behaviorism.* New York: Norton.

Watts, F. (2007). Emotion regulation and religion. In J. J. Gross (Ed.), *Handbook of emotion regulation* (pp. 504–522). New York: Guilford.

Wax, J. R., Pinette, M. G., & Cartin, A. (2010). Home versus hospital birth—process and outcome. *Obstetrical & Gynecological Survey, 65,* 132–140.

Way, D. (2011). Dementia, delirium, and depression. In P. A. Fenstemacher & P. Winn (Eds.), *Long-term care medicine: A pocket guide* (pp. 225–245). New York: Springer.

Wayment, H. A., & Vierthaler, J. (2002). Attachment style and bereavement reactions. *Journal of Loss & Trauma, 7,* 129–149.

Weaver, D. A. (2010). Widows and Social Security. *Social Security Bulletin, 70,* 89–109.

Webb, N. B. (2010a). The child and death. In N. B. Webb (Ed.), *Helping bereaved children: A handbook for practitioners* (3rd ed., pp. 3–21). New York: Guilford Press.

Webb, N. B. (2010b). Assessment of the bereaved child. In N. B. Webb (Ed.), *Helping bereaved children: A handbook for practitioners* (3rd ed., pp. 22–47). New York: Guilford Press.

Webb, S. J., Monk, C. S., & Nelson, C. A. (2001). Mechanisms of postnatal neurobiological development: Implications for human development. *Developmental Neuropsychology, 19,* 147–171.

Webber, L. S., Wattigney, W. A., Srinivasan, S. R., & Berenson, G. S. (1995). Obesity studies in Bogalusa. *American Journal of Medical Science, 310,* S53–S61.

Wechsler, H., Davenport, A., Dowdall, G., Moeykens, B., & Castillo, S. (1994). Health and behavioral consequences of binge drinking in college. *Journal of the American Medical Association, 272,* 1672–1677.

Wechsler, H., Lee, J. E., Kuo, M., Seibrung, M., Nelson, T. F., & Lee, H. (2002). Trends in college binge drinking during a period of increased prevention efforts. *Journal of American College Health, 2002,* 203–217.

Wegman, M. E. (1994). Annual summary of vital statistics—1993. *Pediatrics, 95,* 792–803.

Wegner, D. M., & Gold, D. G. (1995). Fanning old flames: Emotional and cognitive effects of suppressing thoughts of a past relationship. *Journal of Personality and Social Psychology, 68,* 782–792.

Weibel-Orlando, J. (1990). Grandparenting styles: Native American perspectives. In J. Sokolovsky (Ed.), *The cultural context of aging* (pp. 109–125). New York: Bergin & Garvey.

Weichold, K., & Silbereisen, R. K. (2005). Puberty. In C. B. Fisher & R. M. Lerner (Eds.), *Encyclopedia of applied developmental science* (Vol. 2, pp. 893–898). Thousand Oaks, CA: Sage.

Weinberg, M. K., Tronick, E. Z., Cohn, J. F., & Olson, K. L. (1999). Gender differences in emotional expressivity and self-regulation during early infancy. *Developmental Psychology, 35,* 175–188.

Weinstock, M. P., Neuman, Y., & Glassner, A. (2006). Identification of informal reasoning fallacies as a function of epistemological level, grade level, and cognitive ability. *Journal of Educational Psychology, 89,* 327–341.

Weis, R., & Cerankosky, B. C. (2010). Effects of video-game ownership on young boys' academic and behavioral functioning: A randomized, controlled study. *Psychological Science, 21,* 463–470.

Weisberg, D. S., & Bloom, P. (2009). Young children separate multiple pretend worlds. *Developmental Science, 12,* 699–705

Weisgram, E. S., Bigler, R. S., & Liben, L. S. (2010). Gender, values, and occupational interests among children, adolescents, and adults. *Child Development, 81,* 778–796.

Weisner, T. S., & Wilson-Mitchell, J. E. (1990). Nonconventional family lifestyles and sex typing in six-year-olds. *Child Development, 61,* 1915–1933.

Weissman, M. D., & Kalish, C. W. (1999). The inheritance of desired characteristics: Children's view of the role of intention in parent-offspring resemblance. *Journal of Experimental Child Psychology, 73,* 245–265.

Weisz, J. R., McCarty, C. A., & Valeri, S. M. (2006). Effects of psychotherapy for depression in children and adolescents: A meta-analysis. *Psychological Bulletin, 132,* 132–149.

Wellman, H. M. (1993). Early understanding of mind: The normal case. In S. Baron-Cohen, H. Tager-Flusberg, & D. J. Cohen (Eds.), *Understanding other minds: Perspectives from autism.* Oxford, England: Oxford University Press.

Wellman, H. M. (2002). Understanding the psychological world: Developing a theory of mind. In U. Goswami (Ed.), *Blackwell handbook of childhood cognitive development* (pp. 167–187). Malden, MA: Blackwell.

Wellman, H. M., & Gelman, S. A. (1998). Knowledge acquisition in foundational domains. In W. Damon (Ed.), *Handbook of child psychology* (Vol. 2, 523–573. New York: Wiley.

Wellman, H. M., Cross, D., & Watson, J. (2001). Meta-analysis of theory-of-mind development: The truth about false belief. *Child Development, 72,* 655–684.

Wentkowski, G. (1985). Older women's perceptions of greatgrandparenthood: A research note. *The Gerontologist, 25,* 593–596.

Wentworth, N., Benson, J. B., & Haith, M. M. (2000). The development of infants' reaches for stationary and moving targets. *Child Development, 71,* 576–601.

Wentzel, K. R., Filisetti, L., & Looney, L. (2007). Adolescent prosocial behavior: The role of self-processes and contextual cues. *Child Development, 78,* 895–910.

Werner, E. (1994). Overcoming the odds. *Journal of Developmental & Behavioral Pediatrics, 15,* 131–136.

Werner, E. E. (1989). Children of Garden Island. *Scientific American, 260,* 106–111.

Werner, E. E. (1995). Resilience in development. *Current Directions in Psychological Science, 4,* 81–85.

Werner, E. E., & Smith, R. S. (1992). *Overcoming the odds: High risk children from birth to adulthood.* Ithaca, NY: Cornell University Press.

Werner, H. (1948). *Comparative psychology of mental development.* Chicago: Follet.

Wertsch, J. V., & Tulviste, P. (1992). L. S. Vygotsky and contemporary developmental psychology. *Developmental Psychology, 28,* 548–557.

West, M. D. (2010). Embryonic stem cells: Prospects of regenerative medicine for the treatment of human aging. In G. M. Fahy, M. D. West, L. S. Coles, & S. B. Harris (Eds.), *The future of aging* (pp. 451–487). New York: Springer.

Westbrook, L. A. (2002). The experience of mid-life women in the years after the deaths of their parents. *Dissertation Abstracts International. Section A. Humanities and Social Sciences, 62(8A),* 2884.

Westerhof, G. J., Bohlmeijer, E., Webster, J. D. (2010). Reminiscence and mental health: a review of recent progress in theory, research, and interventions. *Ageing and Society, 30,* 697–721.

Wethington, E. (2000). Expecting stress: Americans and the "midlife crisis." *Motivation and Emotion, 24,* 85–103.

Weymouth, P. L. (2005). A longitudinal look at the predictors of four types of retirement. *Dissertation Abstracts International. Section B. Sciences and Engineering, 65(7-B),* 3760.

Whalley, L. J., Fox, H. C., Deary, I. J., & Starr, J. M. (2005). Childhood IQ, smoking, and cognitive change from age 11 to 64 years. *Addictive Behaviors, 30,* 77–88.

Whitbourne, S. B., & Spiro, A., III. (2010). The intersection of physical and mental health in aging: Minding the gap? In J. C. Cavanaugh & C. K. Cavanaugh (Eds.), *Aging in America: Vol. 2: Physical and mental health* (pp. 119–140). Santa Barbara, CA: Praeger Perspectives.

Whitbourne, S. K. (1996). *The aging individual.* New York: Springer.

White, K. D. (2010). Note: Covenant marriage: An unnecessary second attempt at fault-based divorce. *Alabama Law Review, 61,* 869.

White, L., & Gilbreth, J. G. (2001). When children have two fathers: Effects of relationships with stepfathers and noncustodial fathers on adolescent outcomes. *Journal of Marriage and the Family, 63,* 155–167.

White, M. L., Peters, R., Shim, S. M. (2011). Spirituality and spiritual self-care: Expanding self-care deficit nursing theory. *Nursing Science quarterly, 24,* 48–56.

Whitehurst, G. J., & Vasta, R. (1975). Is language acquired through imitation? *Journal of Psycholinguistic Research, 4,* 37–59.

Whiting, B. B., & Edwards, P. E. (1988). *Children of different worlds.* Cambridge, MA: Harvard University Press.

Whiting, J. W. M., & Child, I. L. (1953). *Child training and personality: A cross-cultural study.* New Haven, CT: Yale University Press.

Whitney, E. N., & Hamilton, E. M. N. (1987). *Understanding nutrition* (4th ed.). St. Paul, MN: West.

Whitney, L. D., & Ajmera, S. (2010). Nutrition and aging. In R. H. Robnett & W. C. Chop (Eds.), *Gerontology for the health care professional* (2nd ed., pp. 199–234). Sudbury, MA: Jones & Bartlett.

Whitty, M. T., & Buchanan, T. (2009). Looking for love in so many places: Characteristics of online daters and speed daters. *Interpersona: An International Journal on Personal Relationships, 3(Suppl. 2).* Retrieved from http://www.interpersona.org/pdf/interpersona3%28suppl.2%29.pdf#page=64

Wicks-Nelson, R., & Israel, A. C. (2006). *Behavior disorders of childhood* (6th ed.). Upper Saddle River, NJ: Pearson Education.

Wilkinson, A. M., & Lynn, J. (2001). The end of life. In R. H. Binstock & L. K. George (Eds.), *Handbook of aging and the social sciences* (5th ed., pp. 444–461). San Diego, CA: Academic Press.

Williams, N. R. (2006). The influence of intimate relationship on the psychological well-being of African-American professional women. *Dissertation Abstracts International. Section B. Sciences and Engineering, 66(8-B),* 4539.

Williams, S. A. (2005). Jealousy in the cross-sex friendship. *Journal of Loss & Trauma, 10,* 471–485.

Williams, S. T., Conger, K. J., & Blozis, S. A. (2007). The development of interpersonal aggression during adolescence: The importance of parents, siblings, and family economics. *Child Development, 78,* 1526–1542.

Willinger, M. (1995). Sleep position and sudden infant death syndrome. *Journal of the American Medical Association, 273,* 818–819.

Wilson, B. J. (2008). Media and children's aggression, fear, and altruism. *Future of Children, 18,* 87–118.

Wilson, D. B. (2010). Meta-analysis. In A. R. Piquero & D. Wesiburd (Eds.), *Handbook of quantitative criminology* (Part 2, pp. 181–208). New York: Springer.

Wilson, G. T., Heffernan, K., & Black, C. M. D. (1996). Eating disorders. In E. J. Marsh & R. A. Barkley (Eds.), *Child psychopathology* (pp. 541–571). New York: Guilford.

Wilson, R. D. (2000). Amniocentesis and chorionic villus sampling. *Current Opinion in Obstetrics & Gynecology, 12,* 81–86.

Winecoff, A., LaBar, K. S. Madden, D. J., Cabeza, R., & Huettel, S. A. (2011). Cognitive and neural contributions to emotion regulation in aging. *Social Cognitive and Affective Neuroscience, 6,* 165–176.

Winner, E. (2000). Giftedness: Current theory and research. *Current Directions in Psychological Science, 9,* 153–156.

Witko, T. M. (2006). A framework for working with American Indian parents. In T. M. Witko (Ed.), *Mental health care for urban Indians: Clinical insights from Native practitioners* (pp. 155–171). Washington, DC: American Psychological Association.

Wolf, A. M. D., Wender, R. C., Etzioni, R. B., Thompson, I. M., D'Amico, A. V., Volk, R. J., et al. (2010). American Cancer Society guideline for the early detection of prostate cancer: 2010 update. *CA: A Cancer Journal for Clinicians, 60,* 70–98.

Wolf, M. S., Davis, T. C., Shrank, W., Rapp, D. N., Bass, P. F., Connor, U. M., et al. (2007). To err is human: Patient misinterpretations of prescription drug label instructions. *Patient Education and Counseling, 67,* 293–300.

Wolfe, D. A. (1985). Child-abusive parents: An empirical review and analysis. *Psychological Bulletin, 97,* 462–482.

Wolff, P. H. (1987). *The development of behavioral states and the expression of emotions in early infancy.* Chicago: University of Chicago Press.

Wolfinger, N. H. (2007). Does the rebound effect exist? Time to remarriage and subsequent union stability. *Journal of Divorce & Remarriage, 46,* 9–20.

Wolraich, M. L., Lindgren, S. D., Stumbo, P. J., Stegink, L. D., Appelbaum, M. I., & Kiritsy, M. C. (1994). Effects of diets high in sucrose or aspartame on the behavior and cognitive performance of children. *New England Journal of Medicine, 330,* 301–307.

Womenshealth.gov. (2010). *Menopausal hormone therapy (MHT).* Retrieved from http://www.womenshealth.gov/menopause/treatment/hormone-therapy.cfm

Wong, P. T. P. (2008). Transformation of grief through meaning: Meaning-centered counseling for bereavement. In A. Tomer, G. T. Eliason, & P. T. P. Wong (Eds.), *Existential and spiritual issues in death attitudes* (pp. 375–396). Mahwah, NJ: Erlbaum.

Wood, J. J., Emmerson, N. A., & Cowan, P. A. (2004). Is early attachment security carried forward into relationships with preschool peers? *British Journal of Developmental Psychology, 22,* 245–253.

Woodbridge, S. (2008). Sustaining families in the 21st century: The role of grandparents. *International Journal of Environmental, Cultural, Economic and Social Sustainability.* Retrieved from http://www98.griffith.edu.au/dspace/bitstream/10072/27417/1/50932_1.pdf

Woodgate, R. L. (2006). Living in a world without closure: Reality for parents who have experienced the death of a child. *Journal of Palliative Care, 22,* 75–82.

Woodward, A. L., & Markman, E. M. (1998). Early word learning. In W. Damon (Ed.), *Handbook of child psychology* (Vol. 2, pp. 371–420). New York: Wiley.

Woolf, S. H. (2009). Social policy as health policy. *JAMA, 301,* 1166–1169.

Woollacott, M. H., Shumway-Cook, A., & Williams, H. (1989). The development of balance and locomotion in children. In M. H. Woollacott & A. Shumway-Cook (Eds.), *Development of posture and gait across the lifespan.* Columbia, SC: University of South Carolina Press.

Worden, W. (1991). Grief counseling and grief therapy: A handbook for the mental health practitioner (2nd ed.). New York: Springer.

World Health Organization. (2002). *Worldwide report on violence and health: Chapter 4: Violence by intimate partners.* Retrieved from http://www.who.int/violence_injury_prevention/violence/global_campaign/en/chap4.pdf

World Health Organization. (2005). *The world health report: 2005: Make every mother and child count.* Geneva: Author.

World Health Organization. (2010a). *Why gender and health?* Retrieved from http://www.who.int/gender/genderandhealth/en/index.html

World Health Organization. (2010b). *Preventing intimate partner and sexual violence against women: Taking action and generating evidence.* Retrieved from http://whqlibdoc.who.int/publications/2010/9789241564007_eng.pdf

Worobey, J. (2005). Effects of malnutrition. In C. B. Fisher & R. M. Lerner (Eds.), *Encyclopedia of applied developmental science* (Vol. 2, pp. 673–676). Thousand Oaks, CA: Sage.

Worthman, C. M., & Brown, R. A. (2007). Companionable sleep: Social regulation of sleep and cosleeping in Egyptian families. *Journal of Family Psychology, 21,* 124–135.

Wozniak, J. R., & Lim, K. O. (2006). Advances in white matter imaging: A review of in vivo magnetic resonance methodologies and their applicability to the study of development and aging. *Neuroscience & Biobehavioral Reviews, 30,* 762–774.

Wray-Lake, L., Crouter, A. C., & McHale, S. M. (2010). Developmental patterns in decision-making autonomy across middle childhood and adolescence: European American parents perspectives. *Child Development, 81,* 636–651.

Wrenn, R. L. (1999). The grieving college student. In J. D. Davidson & K. J. Doka (Eds.), *Living with grief: At work, at school, at worship* (pp. 131–141). Levittown, PA: Brunner/Mazel.

Wright, J. C., et al. (2001). The relations of early television viewing to school readiness and vocabulary of children from low-income families: The Early Window Project. *Child Development, 72,* 1347–1366.

Wynn, K. (1992). Addition and subtraction by human infants. *Nature, 358,* 749–750.

Wynn, K. (1996). Infants' individuation and enumeration of actions. *Psychological Science, 7,* 164–169.

Xie, H., Li, Y., Boucher, S. M., Hutchins, B. C., & Cairns, B. D. (2006). What makes a girl (or a boy) popular (or unpopular)? African American children's perceptions and developmental differences. *Developmental Psychology, 42,* 599–612.

Xu, X., Ji, J., & Tung, Y. Y. (2000). Social and political assortative mating in urban China. *Journal of Family Issues, 21,* 47–77.

Xu, X., Zhu, F., O'Campo, P., Koenig, M. A., Mock, V., & Campbell, J. (2005). Prevalence of and risk factors for intimate partner violence in China. *American Journal of Public Health, 95,* 78–85.

Xu, Y., Carver, J. A., & Zhang, Z. (2009). Temperament, harsh and indulgent parenting, and Chinese children's proactive and reactive aggression. *Child Development, 80,* 244–258.

Yamini-Benjamin, Y. I. (2007). Moving toward a better understanding of Black women's work adjustment: The role of perceived discrimination and self-efficacy in predicting job satisfaction and psychological distress in Black women. *Dissertation Abstracts International. Section B. Sciences and Engineering, 67(10-B),* 6085.

Yancura, & Aldwin, C. (2010). Does psychological stress accelerate the aging process? In J. C. Cavanaugh & C. K. Cavanaugh (Eds.), *Aging in America: Vol. 2: Physical and mental health* (pp. 100–118). Santa Barbara, CA: Praeger Perspectives.

Yang, N., Chen, C. C., Choi, J., & Zou, Y. (2000). Sources of work-family conflict: A Sino-U.S. comparison of the effects of work and family. *Academy of Management Journal, 43,* 113–123.

Yang, Q., Rasmussen, S. A., & Friedman, J. M. (2002). Mortality associated with Down's syndrome in the USA from 1983 to 1997: A population-based study. *Lancet, 359,* 1019–1025.

Yap, M. B. H., Allen, N. B., & Ladouceur, C. D. (2008). Maternal socialization of positive affect: The impact of invalidation on adolescent emotion regulation and depressive symptomatology. *Child Development, 79,* 1415–1431.

Yárnoz-Yaben, S. (2010). Attachment style and adjustment to divorce. *Spanish Journal of Psychology, 13,* 210–219.

Yick, A. G. (2000). Domestic violence beliefs and attitudes in the Chinese American community. *Journal of Social Service Research, 27,* 29–51.

Yngvesson, B. (2010). *Belonging in an adopted world: Race, identity, and transnational adoption.* Chicago: University of Chicago Press.

Yoon, S. M. (2005). The characteristics and needs of Asian-American grandparent caregivers: A study of Chinese-American and Korean-American grandparents in New York City. *Journal of Gerontological Social Work, 44,* 75–94.

Young, S. K., Fox, N. A., & Zahn-Waxler, C. (1999). The relations between temperament and empathy in 2-year-olds. *Developmental Psychology, 35,* 1189–1197.

Youniss, J., McLellan, J. A., & Yates, M. (1999). Religion, community service, and identity in American youth. *Journal of Adolescence, 22,* 243–253.

Yu, T., Pettit, G. S., Lansford, J. E., Dodge, K. A., & Bates, J. E. (2010). The interactive effects of marital conflict and divorce on parent-adult children's relationships. *Journal of Marriage and Family, 72,* 282–292.

Yu, T., & Adler-Baeder, F. (2007). The intergenerational transmission of relationship quality: The effects of parental remarriage quality on young adults' relationships. *Journal of Divorce & Remarriage, 47,* 87–102.

Yuan, S., & Fisher, C. (2009). "Really? She blicked the baby?": Two-year-olds learn combinatorial facts about verbs by listening. *Psychological Science, 20,* 619–626.

Yumoto, C., Jacobson, S. W., & Jacobson, J. L. (2008). Fetal substance exposure and cumulative environmental risk in an African American cohort. *Child Development, 79,* 1761–1776.

Zachary, L. J. (2000). *The mentor's guide: Facilitating effective learning relationships.* San Francisco, CA: Jossey-Bass.

Zachary, L. J., & Fischler, L. A. (2009). *The mentee's guide: Making mentoring work for you.* San Francisco, CA: Jossey-Bass.

Zacks, R. T., Hasher, L., & Li, K. Z. H. (2000). Human memory. In F. I. M. Craik & T. A. Salthouse (Eds.), *Handbook of aging and cognition* (2nd ed., pp. 293–357). Mahwah, NJ: Erlbaum.

Zafarullah, H. (2000). Through the brick wall and the glass ceiling: Women in the civil service in Bangladesh. *Gender, Work & Organization, 7,* 197–209.

Zahn-Waxler, C., Radke-Yarrow, M., Wagner, E., & Chapman, M. (1992). Development of concern for others. *Developmental Psychology, 28,* 126–136.

Zarrett, N., Fay, K., Li, Y., Carrano, J., Phelps, E., & Lerner, R. M. (2009). More than child's play: Variable- and pattern-centered approaches for examining effects of sports participation on youth development. *Developmental Psychology, 45,* 368–382.

Zaslow, M. J., & Hayes, C. D. (1986). Sex differences in children's responses to psychosocial stress: Toward a cross-context analysis. In M. E. Lamb, A. L. Brown, & B. Rogoff (Eds.), *Advances in developmental psychology* (Vol. 4, pp. 285–337). Hillsdale, NJ: Erlbaum.

Zeanah, C. H., Smyke, A. T., Koga, S. F., & Carlson, E. (2005). Attachment in institutionalized and community children in Romania. *Child Development, 76,* 1015–1028.

Zeidler, D. L., Sadler, T. D., Applebaum, S., & Callahan, B. E. (2009). Advancing reflective judgment through socioscientific issues. *Journal of Research in Science Teaching, 46,* 74–101.

Zelazo, P. D., & Cunningham, W. A. (2007). Executive function: Mechanisms underlying emotion regulation. In J. J. Gross (Ed.), *Handbook of emotion regulation* (pp. 135–158). New York: Guilford.

Zelazo, P. R. (1993). The development of walking: New findings and old assumptions. *Journal of Motor Behavior, 15,* 99–137.

Zeng, Y., Gu, D., & George, L. K. (2011). Association of religious participation with mortality among Chinese old adults. *Research on Aging, 33,* 51–83.

Zhan, H. J. (2006). Joy and sorrow: Explaining Chinese caregivers' reward and stress. *Journal of Aging Studies, 20,* 27–38.

Zhang, B., Cartmill, C., & Ferrence, R. (2008). The role of spending money and drinking alcohol in adolescent smoking. *Addiction, 103,* 310–319.

Zhou, Q., Eisenberg, N., Losoya, S. H., Fabes, R. A., Reiser, M., Guthrie, I. K., et al. (2002). The relations of parental warmth and positive expressiveness to children's empathy-related responding and social functioning: A longitudinal study. *Child Development, 73,* 893–915.

Zhou, Q., Wang, Y., Eisenberg, N., Wolchik, S., Tein, J-W., & Deng, X. (2008). Relations of parenting and temperament to Chinese children's experience of negative life events, coping efficacy, and externalizing problems. *Child Development, 79,* 493–513.

Ziegler, J. C., Pech-Georgel, C., Dufau, S., & Grainger, J. (2010). Rapid processing of letters, digits, and symbols: What purely visual-attentional deficit in developmental dyslexia? *Developmental Science, 13,* F8–F14.

Zigler, E., & Finn-Stevenson, M. (1992). Applied developmental psychology. In M. H. Bornstein & M. E. Lamb (Eds.), *Developmental psychology: An advanced textbook* (pp. 677–729). Hillsdale, NJ: Erlbaum.

Zimiles, H., & Lee, V. E. (1991). Adolescent family structure and educational progress. *Developmental Psychology, 27,* 314–320.

Zimmer-Gembeck, M. J., & Helfand, M. (2008). Ten years of longitudinal research on U.S. adolescent sexual behavior: Developmental correlates of sexual intercourse, and the importance of age, gender and ethnic background. *Developmental Review, 28,* 153–224.

Zimmerman, B. J. (2001). Theories of self-regulated learning and academic achievement: An overview and analysis. In B. J. Zimmerman & D. H. Schunk (Eds.), *Self-regulated learning and academic achievement: Theoretical perspectives* (2nd ed., pp. 1–37). Mahwah, NJ: Erlbaum.

Zimmerman, F. J., Christakis, D. A., & Meltzoff, A. N. (2007). Associations between media viewing and language development in children under age 2 years. *Journal of Pediatrics, 151,* 364–368.

Zinar, S. (2000). The relative contributions of word identification skill and comprehension-monitoring behavior to reading comprehension ability. *Contemporary Educational Psychology, 25,* 363–377.

Zippel, K. S. (2006). *The politics of sexual harassment: A comparative study of the United States, the European Union, and Germany.* New York: Cambridge University Press.

Zmuda, J. M., Cauley, J. A., Kriska, A., Glynn, N. W., Gutai, J. P., & Kuller, L. H. (1997). Longitudinal relation between endogenous testosterone and cardiovascular disease risk factors in middle-aged men: A 13-year follow-up of former Multiple Risk Factors Intervention Trial participants. *American Journal of Epidemiology, 146,* 609–617.

Zosuls, K. M., Ruble, D. N., Tamis-LeMonda, C. S., Shrout, P. E., Bornstein, M. H., & Greulich, F. K. (2009). The acquisition of gender labels in infancy: Implications for gender-typed play. *Developmental Psychology, 45,* 688–701.

Zukow-Goldring, P. (2002). Sibling caregiving. In M. H. Bornstein (Ed.), *Handbook of parenting: Vol. 3. Status and social conditions of parenting* (2nd ed., pp. 253–286). Mahwah, NJ: Erlbaum.

Name Index

Page numbers in *italic* type indicate photographs.

Subject Index

Note: Page numbers in *italic* type indicate figures, illustrations, or tables.

coordinating skills, 101–102
COPD (chronic obstructive pulmonary disease), 512
coping, 603
   with grief, 604–606
   with leisure activities, 450–451
   with stress, 467, 468
core knowledge hypothesis, 132
corpus callosum, 94
correlation coefficient, 26, *27*
correlational study, 26–27, *27*
co-rumination, 268
co-sleeping, 85
counterimitation, 251
counting, 142–143, 234
covenant marriage, 412
creative ability, 214
creativity, 222–223, *223*, 526–528, *527*
cross-linking, 508
cross-sectional study, 30
crowd, 268
crowning, 70
crying, in newborns, 84–85
crystallization, 329, 330
crystallized intelligence, 212, 366–367, *367*
culture, 8–9
   abusive relationships and, 392–393
   adoptive families and, 407–408
   caregiving, 487–488
   child abuse, 262
   couple-forming behaviors, 388–391
   generativity, 481
   job satisfaction, 427
   menopause, 462–463
   moral reasoning, 309
   mourning and bereavement, 582–583
   parenting styles, 250
   popular children, 271
   role transitions, 347
   sexual harassment, 437
   singlehood, 395
   sociocultural forces, 8–9, 391
   stress and, 467, 468
culture-fair intelligence tests, 219–220, *220*
custody, 259, 261, 491
CVAs (cerebral vascular accidents), 511

date rape, 392
dating
   abusive relationships, 392
   adolescent, 324
   by divorced single parents, 407
   online, 388
   speed, 388
   violence, 327–328
dead-ender, 423
death
   adolescence, *299*, 299–300, 610
   anxiety, 595–596
   of child, 611–613
   childhood and, 608–610, *609*
   clinical, 584
   contextual theory of dying, 594
   end-of-life intentions, 589–590, *590*
   ethical issues, 585–590
   euthanasia, 585–586
   final scenarios, 596–597
   grieving process, 601–607
   hospice and, 597–600
   Kübler-Ross theory, 593–594
   late adulthood, 612–614
   legal and medical definitions, 584–585
   life-course approach to, 591–592
   middle adulthood, 611–612
   one's own, dealing with, 592–594
   of parent, 612
   of partner, 613–614, *614*

physician-assisted suicide, 587–589, *588*
   sociocultural definitions, 582–584
   thinking about, 591–600
   whole-brain, 584
   young adults, 354–355, 611–612
death anxiety, 595–596
Death With Dignity Act, *588*, 588–589
deductive reasoning, 207
Defense of Marriage Act (DOMA), 397
delinquency
   causes of, 337–339
   defined, 337
dementia
   Alzheimer's disease, 532–536
   defined, 532
   vascular, 511
demographers, 498
demographics of aging, 498–501, *499*
dendrites, 93, *93*, 95, 510
deoxyribonucleic acid (DNA), 43, *43*
dependent care, 443–445
dependent variable, 27
depression
   in adolescence, 335–337
   causes, 530–531
   defined, 335
   diagnosis, 530
   late adulthood, 529–532
   postpartum, 72–73
   suicide prevention, 336–337
   treatment, 336, 531–532
depth, perception of, 107–109
describing others, 276–277, *277*
developmental coach, 425
developmental dyslexia, 224, 225
developmental research. *See* research
developmental theories, 11–22
   cognitive-developmental theory, 14–16
   ecological and systems approach, 16–18
   learning theory, 13–14
   life-course perspective, 20–21
   life-span perspective, 18–19
   psychodynamic theory, 12–13
   selective optimization with compensation (SOC) model, 19–20
   summary, *21*
diabetes, 513
Dietary Guidelines for Americans 2010, 360
diethylstilbestrol (DES), 66
dieting, 296, 297
differentiation, 101
direct instruction, 251
disability, *568*
   intellectual, 223–224
   learning, 224–226, *225*
discipline
   adolescent self-esteem and, 321
   prosocial behavior, effect on, 189
discontinuity, 6
discrimination
   age, 438
   gender, 434–435
diseases, teratogenic, 63
disgust, 177, 178
disorganized (disoriented) attachment, 172
distress, 177
divergent thinking, 222–223
diversity, of older adults, 500–501
divorce, 258–260, 409–414
   adult children and, 414
   blended families, 261
   children affected by, 258–260
   collaborative, 413
   effects on couple, 412–413
   reasons for, 410–412, *411*
   remarriage, 414–415

single parenting, 406–407
   statistics on, 409–410, *410*
   young children and, 413
divorce hangover, 413
dizygotic twins, 49
DNA (deoxyribonucleic acid), 43, *43*
docility, 544
dominance hierarchy, 269
dominant allele, 44
Do Not Resuscitate (DNR) order, 589–590
dopamine, 391
doubt, 168–169
Down syndrome, 46–47
dressing, fine motor skill development and, 103
drinking
   binge, 356–358, *357*
   teenage, 334
   in young adulthood, 356–359
driving under the influence (DUI), 357
drugs
   adolescence, 334–335
   sports and, 298–299
   teratogenic, 62, *62*
dual-earner couples, 445–448
dual-energy X-ray absorptiometry (DXA) test, 459–460
dual process model (DPM) of coping with bereavement, 606, *607*
durable power of attorney for health care, 589, *590*
dynamic systems theory, 100, 101
dyslexia, 224, 225
dysphoria, 530

early childhood. *See also* infancy
   cognitive development in, 123–162
   socioemotional development in, 167–199
eating disorders, 296–297
ecological and systems approach, 16–18, *21*
   Bronfenbrenner's theory, *17*, 17–18
   competence-environmental press theory, 18
ecological theory, 16
ectoderm, 55
Eden Alternative, 571
edgework, 350
education
   academic skills, 229–238
   aptitudes for school in middle childhood, 211–221
   college, 348–350, *349*
   comparing U.S. students with students in other countries, 234–236, *235*
   effective schools/teachers, 237–238
   IQ as predictor of school success, 216
   lifelong learning, 475–476
   theory of multiple intelligences, implications of, 214
   theory of successful intelligence, implications of, 215
ego resilience, 264, 482
egocentrism, 128, *128*, *131*, 278, 317–318, *319*
E-I-E-I-O model, 526, 535
elaboration, 208
elder abuse and neglect, 571–572
elective selection, 19
electroencephalography, 96
electronic media, 273–275
   computers, 275
   television, 273–274
embryo, 55–56
emerging adulthood, 346–353
   behavioral changes, 350–351
   college, 348–350, *349*
   defined, 346
   financial independence, launching, 351–352
   role transitions, 347–348
emotion(s), 176–182
   basic, 177–178, *179*
   cultural differences in expression, 179–180

emergence of complex, 178–179
function of, 177
integrating with logic in life problems, 370–373
neural mechanisms, 548–550
others' emotions, reorganizing and using, 180–181, *181*
regulating, 181–182
well-being and, 547–548
emotional intelligence, 213–214
emotional sensitivity, gender differences in, 193
empathy, 188
employment. *See* work
empty nest, 483–484
enabling actions, 185
encapsulation, 474
endoderm, 55
end-of-life intentions, 589–590, *590*
end-of-life issues, 596–597
environment
adapting and exploring, 126
intelligence, impact on, 217–219
temperament, 87–88
environmental factors in longevity, 503
environmental press, 18, 543–545
environmental teratogens, 63–64, *64*
epigenetic principle, 12, 50
episodic memory, 522
equal pay for equal work, 435–436
equilibration, 125–126
Erikson's theory, *12*, 12–13, *169*, 169–169, 351
essentialism, 135
estrogen, 289, 460, 463
ethics
bioethics, 585
death, 585–590
research, 32–33
universal ethical principles, 307
ethnic identity, 318–320
ethnicity
caregiving, 488
children's growth in middle childhood, 240
death rate in young adulthood, 355
divorce, 410
grandparenthood, 490
grief expression, 604
health, 361–362
household chores, division of, 446
impact of, 8–9
intelligence and, 219–221
longevity, 503
occupational development, 433–434
older adults, 500
parenting, 405–406
retirement, 555
self-esteem, 321
sibling relationships, 256–257
singlehood, 395
spirituality, 550–552
terminology, 9
unemployment, 441–442
eugenics, 54, 55
European American
abusive relationships, 393
adoption by, 257
attention-deficit hyperactivity disorder, 228
caregiving, 488
children's growth in middle childhood, 240
climacteric, 462
household chores, division of, 446
life expectancy, 503
parenting styles, 250
self-esteem, 321
single parents, 406
terminology, 9
European Association of Palliative Care (EAPC), 586

euthanasia, 585–586
evolutionary psychology, 169
exchange theory, 399
exercise
aerobic, 469–470
in middle adulthood, 469–470
exosystem, 17
expectations, occupational, 424–425
experience-dependent growth, 97
experience-expectant growth, 97
experiment, 27
experimental studies, 27–28
expertise, 473–475
explicit memory, 522, 525–526
expressive style, 155
extended family, 403
external aids for memory, 525
extraversion, 477
extremely low birth weight, 74

fable, personal, 318, *319*
Facebook, 275, 385
faces, perception of, 110–111, 112
Fair Pay Act, 436
fallacies, identification of, 303–304, *304*
falls, 516
false-belief task, 116, *117*, 279
families
adoptive, 407–408
blended, 261
children, decision to have, 403–404
children's influence on, 254–255
divorce, 409–414
ethnicity and parenting, 405–406
extended, 403
foster, 408
gay and lesbian, 408
late adulthood, 559–560
life cycle of, 403–408
middle adulthood, 483–491
middle childhood, 248–265
nuclear, 403
parents role in, 248–255
sibling friendships, 385–386, *386*
siblings, 255–258, 559–560
single parents, 406–407
step, 407
as systems, 248–249, *249*
work and, 443–448
young and middle adulthood, 403–408
familism, 405–406
Family and Medical Leave Act, 443
fast mapping, 151–154
fast-tracker, 423
fathers. *See also* parents
age, 404
father-infant relationships, 170
stepfathers, 261
fear
adaptive value of, 177
age differences in, 179
regulating the emotion, 182
stranger wariness, 178
feedback, 252–253
feeding, fine motor skill development and, 103
females
abusive relationships, 392–393
caring for aging parents, 487
climacteric and menopause, 462–464
friendships, 267, 385–386, *386*
longevity, 503–504
occupational development, 432–433
osteoporosis, 457–460, *458*
physical development, 239–240, *240, 286*
play, 185–186
puberty, 286–294

retirement, 555
work, 431–433, 434–436
fertilization, 54
fetal alcohol spectrum disorder, 62
fetal medicine, 68–69
fetus, 56–58, *57, 58*
filial obligation, 486
final scenario, 596–597
financial independence, establishing, 351–352
fine motor skills, 102–104, 240–241
defined, 99
dressing skills, 103
feeding skills, 103
handedness, 104
reaching and grasping, 102–104
firearms, deaths from, 299, *299*
fitness. *See* physical fitness
five-factor model, 477–480, *479*
fluid intelligence, 366–367, *367*
formal-operational period, 206–208, 316
formula, 91–92
foster parents, 408
four-component model of grieving, 605
Fourth Age, 504, 506
frail older adults, 565–566, *567*
free radicals, 507–508
friendships
ABCDE model, 384
adulthood, 384–386
cross-sex, 386
defined, 266
late adulthood, 558–559
middle childhood, 266–268
online, 385
parents and children, 483–484
parties in, 267
quality and consequences of, 267–268
sibling, 385–386, *386*
themes of, 385
frontal cortex, 94, *94*
functional health, 568–569
functional magnetic resonance imaging (fMRI), 96
functional neuroimaging, 510

Gardner's theory of multiple intelligences, 212–214, *213*
gay adolescent sexual orientation, 326–327
gay and lesbian parents, 408
gay couples, 396–398, 561, 614
gender, 191–199
death rate in young adulthood, 355
differences, 192–194, *193*
friendships, 267, 385–386, *386*
health, 361–362
identity, 196–198
longevity, 503–504
moral reasoning, 310
motor skills, 241
occupational selection, 431–432
physical development, 239–240, *240*
play, 185–186
retirement, 555
roles, 194–196, 198–199
sports participation, 298, *298*
stereotypes, 191–192, *195*
stress and, 466
work, 431–433, 434–437, *435*
gender constancy, 196
gender discrimination, 434–435
gender identity, 196–198
gender labeling, 196
gender-schema theory, 197, *197*
gender stability, 196
gender stereotypes, 191–192, *195*
gender typing, 194–196
generativity, 375, 480–481, *481*